Dictionary of Plants Containing
Secondary Metabolites

Dictionary of Plants Containing
Secondary Metabolites

J.S. Glasby

Taylor & Francis
London • New York • Philadelphia

UK   Taylor & Francis Ltd, 4 John St., London WC1N 2ET

USA   Taylor & Francis Inc., 1900 Frost Road, Suite 101, Bristol, PA 19007

British Library Cataloguing in Publication Data

A catalogue record for this book is available from the British Library.

ISBN 0-85066-423-3

Library of Congress Cataloging in Publication Data
is available

Printed in Great Britain by Burgess Science Press, Basingstoke on paper which has a specified pH value on final paper manufacture of not less than 7.5 and is therefore 'acid free'.

Preface

The plant kingdom has proved to be the richest source of organic compounds, many of which have been used in medicine and perfumery since ancient times.

A voluminous literature has grown up regarding the discovery of new compounds, their occurrence, structures, biosynthesis and physiological properties. Unfortunately, this information is to be found in a plethora of scientific journals and it is far from easy for the research worker to determine which plant has been examined for its chemical content and, equally important, which classes of compound have been isolated from individual species.

This dictionary, which covers the literature up to the end of 1987, is intended to provide this information in a readily accessible form. Each plant is listed alphabetically according to genus and species. Wherever possible, the original papers have been consulted and references to these are given for each class of compound.

In a field as large as this, it is inevitable there will be some omissions and for these the author accepts full responsibility. However, it is hoped that the volume will prove useful to chemists, biochemists and botanists working in the field of organic chemistry.

J.S. Glasby

ABELIA

A. grandiflora.
IRIDOIDS: abelioside A, abelioside B.
Murai *et al., Phytochemistry* **24**, 2329 (1985)

ABELOMOSCHUS

A. moschatus.
MONOTERPENOID: pineol.
SESQUITERPENOID: farnesol.
Kerschbaun, *Ber. Dtsch. Chem. Ges.*, **46**, 1732 (1913)

ABIES (Pinaceae)

A. alba Mill.
DITERPENOIDS: abietic acid, abietospiran.
ACID: salicylic acid.
MONOTERPENOIDS: bornyl acetate, camphene, limonene, phellandrene, pinene.
Steglich *et al., Angew. Chem.*, **91**, 751 (1979)

A. balsamea (L.) Mill.
DITERPENOIDS: abienol, abieslactone.
MONOTERPENOIDS: camphene, 4,4-dimethylcyclohept-2-en-1-ol, limonene, (-)-α-phellandrene, pinene.
SESQUITERPENOIDS: dehydrojuvabione, juvabione.
Uyeo *et al., J. Pharm. Soc., Japan,* **84**, 453 (1964)
Balogh *et al., Phytochemistry* **11**, 1481 (1972)

A. concolor var. Lowiana.
MONOTERPENOID: camphene.
Zavarin *et al., Chem. Abstr.*, 87 180724s

A. excelsa.
MONOTERPENOIDS: bornyl acetate, camphene, phellandrene, pinene.
Wallach, Guttman, *Justus Liebigs Ann. Chem.*, **357**, 79 (1907)

A. grandis (Dougl.) Lindl.
MONOTERPENOIDS: bornyl acetate, camphene, 3-carene, limonene, myrcene, β-phellandrene, α-pinene, β-pinene.
SESQUITERPENOID: tricyclene.
TRITERPENOIDS: cyclograndisolide, epi-cyclograndisolide.
Allen *et al., Tetrahedron Lett.*, 282 (1971)
Zavarin *et al., Chem. Abstr.*, 87 180724s

A. lasiocarpa.
SESQUITERPENOIDS: 1'(E)-dehydrojuvabione, 1'(Z)-dehydrojuvabione, 2'-dehydrojuvani-1'-ol, juvabiol, juvabione, lasiocarpenone.
Bowers *et al., Science,* **154**, 1020 (1966)
Manville *et al., Can. J. Chem.*, **55**, 2547 (1977)

A. magnifica Murray.
MONOTERPENOIDS: phellandrene, pinene.
SESQUITERPENOID: cyclosativene.
Smedman, Zavarin, *Tetrahedron Lett.*, 3833 (1968)

A. mariesii.
DITERPENOID: abieslactone.
TRITERPENOID: mariesiic acid.
Uyeo *et al., J. Pharm. Soc., Japan,* **84**, 453 (1964)
Hasegawa *et al., Chem. Lett.*, 1589 (1985)

A. recurvata var. ernestii (Rehd.) C.T.Kuan.
MONOTERPENOIDS: bornyl acetate, camphene, β-caryophyllene, limonene, myrcene, α-pinene, β-pinene.
Huang *et al., Chem. Abstr.*, 104 3422r

A. sachalinensis.
SESQUITERPENOID: todomatuic acid.
Manville *et al., Can. J. Chem.*, **55**, 2547 (1977)

A. sibirica.
DITERPENOIDS: abienol, abieslactone, 14,15-bisnor-8-hydroxylabd-11-en-13-acetate, isoabienol, neoabienol.
MONOTERPENOIDS: 3-carene, β-phellandrene, α-pinene, β-pinene.
Chirkova *et al., Khim. Prir. Soedin.*, 86 (1967)
Chirkova, Pentegova, *ibid.*, 247 (1969)

A. webbiana.
BIFLAVONE: abiesin.
Chatterjee *et al., Phytochemistry* **23**, 704 (1984)

ABROMA

A. angusta.
UNCHARACTERIZED ALKALOID: abromine.
Srivastava, Basu, *Indian J. Pharm.*, **18**, 472 (1956)

ABRUS (Leguminosae)

A. cantoniensis.
TRITERPENOID: cantoniensistriol.
Chiang *et al., Planta Med.*, **46**, 52 (1982)

A. precatorius L.
INDOLE ALKALOID: (+)-abrine.
PYRIDINE ALKALOID: precatorine.
UNCHARACTERIZED ALKALOIDS: abrasine, precasine.
ANTHOCYANINS: (p-coumaroylgalloyl)-glucodelphinidin, delphinidin.
STEROIDS: abricin, abrisin.
TRITERPENOIDS: abrusgenic acid, abrusgenic acid methyl ester, abruslactone.
(A) Khaleque *et al., Sci. Res. (Dacca),* **3**, 203 (1966)
(An) Karawya *et al., Fitoterapia,* **52**, 175 (1982)
(S) Mandava *et al., Steroids,* **23**, 357 (1974)
(T) Chiang *et al., Planta Med.*, **49**, 165 (1983)

ABUTA (Menispermaceae)

A. bullata.
ALKALOID: saulatine.
Hocquemiller, Fournet, *J. Nat. Prod.*, **47**, 539 (1984)

A. grandifolia (Martius) Sandwith.
ISOQUINOLINE ALKALOID: grandirubrine.
Menachery, Cava, *Heterocycles,* **14**, 943 (1980)

A. grisabachii Triana et Planchon.
BISBENZYLISOQUINOLINE ALKALOIDS: grisabine, grisabutine, macolidine, macoline.
Ahmad, Cava, *J. Org. Chem.*, **42**, 2271 (1977)
Galeffi *et al., Farmaco Ed. Sci.*, **32**, 853 (1977)

A. imene.
OXODIBENZOQUINOLINE ALKALOIDS: imelutine, imenine, norrufescine.
Cava *et al., Tetrahedron,* **31**, 1667 (1975)

A. panurensis.
BISBENZYLISOQUINOLINE ALKALOIDS: norpanurensine, panurensine.
Cava *et al., J. Org. Chem.*, **40**, 2607 (1975)

A. rufescens.
APORPHINE ALKALOID: splendidine.
OXOBENZOQUINOLINE ALKALOIDS: imenine, imerubrine, imulutine, norrufescine, rufescine.
Cava et al., Tetrahedron, **31**, 1667 (1975)
Skiles et al., Can. J. Chem., **57**, 1642 (1979)

A. velutina.
STEROID: abutasterone.
Pinheiro et al., Phytochemistry **22**, 2320 (1983)

ABUTILON

A. pakistanicum.
ALIPHATICS: ceryl alcohol, hentriacontane.
AMINO ACIDS: aspartic acid, glutaric acid, histidine, leucine, serine, threonine.
PHENOLICS: caffeic acid, p-coumaric acid, p-hydroxybenzoic acid, vanillic acid.
STEROID: β-sitosterol.
TRITERPENOIDS: α-amyrin, β-amyrin.
Bi et al., Journal of Pharmaceuticals of University of Karachi, **4**, 85 (1986); Chem. Abstr., **105** 222782g

ACACIA (Leguminosae)

A. angustissima.
AMINE ALKALOID: N-methyl-β-phenethylamine.
Camp, Norvell, Econ. Bot., **20**, 274 (1966)

A. argentea.
HISTAMINE ALKALOID: cinnamoylhistamine.
Johns, Lamberton, Aust. J. Chem., **20**, 555 (1967)

A. aroma.
FLAVONOIDS: apigenin, 7-hydroxyflavone, isorhamnetin, isorhamnetin 3-O-glucoside, luteolin, luteolin-7-O-glucoside, myricetin, myricetin 3-O-glucoside, myricetin-3-O-rhamnoside.
Suarez et al., An. Asoc. Quim. Argent., **70**, 647 (1982); Chem. Abstr., **97** 107039z

A. auriculiformis.
FLAVANONES: 3,4′,7,8-tetrahydroxyflavanone, 4′,7,8-trihydroxyflavanone.
Drewes et al., Biochem. J., **98**, 493 (1966)

A. caven.
FLAVONOIDS: isorhamnetin, isovitexin, luteolin, quercetin, quercetin 3-O-glucoside.
Suarez et al., An. Asoc. Quim. Argent., **70**, 647 (1982); Chem. Abstr., **97** 107039z

A. concinna DC.
GLYCOSIDE: sonuside.
SAPONINS: sonunin A, sonunin B.
(Gly) Sharma et al., Curr. Sci., **46**, 382 (1977)
(Sap) Sharma et al., Pharmazie, **38**, 632 (1983)

A. concinna L.
TRITERPENOIDS: acacic acid, acacidiol, acacigenin A, acacigenin B, acaninin, machaerinic acid lactone.
Varshney et al., Ind. J. Chem., **14B**, 228 (1976)
Anjaneyulu et al., Phytochemistry **18**, 463 (1979)

A. constricta.
AMINE ALKALOID: N-methyl-β-phenethylamine.
NITRILE: acacipetalin.
(A) Camp, Norvell, Econ. Bot., **20**, 274 (1966)
(N) Seigler et al., Phytochemistry **15**, 219 (1976)

A. crombei.
BENZOPYRAN: crombeone.
Brandt et al., J. Chem. Soc., Chem. Commun., 116 (1971)

A. cultriformis Cunn. ex G.Don.
FLAVAN: (2R,3S,4S)-3′,4,7-trimethoxy-3,4-flavandiol.
Fourie et al., Phytochemistry **13**, 2573 (1974)

A. cyanophylla.
FLAVONOIDS: isorhamnetin-3-rutinoside, myricetin 4′-methylether-3-rhamnoside, quercetin-3-rutinoside.
PHENOLICS: (+)-catechol, gallic acid, (+)-gallocatechol.
Thieme et al., Pharmazie, **30**, 736 (1975)

A. dealbata.
GLUCOSIDE: 2′,3-dihydroxy-4,4′,6-trimethoxychalcone-6′-glucoside.
Imperato, Experientia, **38**, 67 (1982)

A. farnesiana.
FLAVONOIDS: isorhamnetin-3-rutinoside, myricetin 4′-methyl ether-3-rhamnoside, quercetin-3-rutinoside.
PHENOLICS: (+)-catechol, gallic acid, (+)-gallocatechol.
Thieme et al., Pharmazie, **30**, 736 (1975)

A. floribunda.
ISOCOUMARIN: floribundoside.
Paris et al., C. R. Acad. Sci., **231**, 72 (1950)

A. furcatispina.
FLAVONOIDS: isorhamnetin 3-O-glucoside, isovitexin, isovitexin rhamnoside, luteolin, quercetin 3-O-glucoside, vitexin.
Suarez et al., An. Asoc. Quim. Argent., **70**, 647 (1982); Chem. Abstr., **97** 107039z

A. greggei.
AMINE ALKALOID: N-methyl-β-phenethylamine.
Camp, Norvell, Econ. Bot., **20**, 274 (1966)

A. hebeclada.
NITRILE: acacipetalin.
Seigler et al., Phytochemistry **15**, 219 (1976)

A. horrida.
FLAVONOIDS: isorhamnetin-3-rutinoside, myricetin 4′-methyl ether-3-rhamnoside, quercetin-3-rutinoside.
PHENOLICS: (+)-catechol, gallic acid, (+)-gallocatechol.
Thieme et al., Pharmazie, **30**, 736 (1975)

A. jacquemonti.
DITERPENOIDS: acacia hemiacetals A and B.
Joshi et al., Tetrahedron, **35**, 1449 (1979)

A. kempearia.
FLAVANONE: 4′,7,8-trihydroxyflavanone.
Drewes et al., Biochem. J., **98**, 493 (1966)

A. leucophloea Willd.
SAPONIN: betulinic acid 3-O-maltoside.
DITERPENOIDS: leucophleol, leucophleoxol, leucoxol.
(Sap) Mishra et al., Indian J. Pharm. Sci., **47**, 154 (1985)
(T) Bansal et al., Phytochemistry, **19**, 1979 (1980)
(T) Perales et al., Tetrahedron Lett., 2843 (1980)

A. longifolia.
FLAVONOIDS: isorhamnetin-3-rutinoside, myricetin 4′-methyl ether-3-rhamnoside, quercetin-3-rutinoside.
PHENOLICS: (+)-catechol, gallic acid, (+)-gallocatechol.
Thieme et al., Pharmazie, **30**, 736 (1975)

A. mearnsii.
BENZOPYRONE: 6,13-bis(3,4-dihydroxyphenyl)-6a,7a,13a,14a-tetrahydro-6H,13H-p-dioxino-[2,3c 5,6-c′]-bis-[1]-benzopyran-3,11-diol.
FLAVANS: bileucofisetinidin, 3-O-methylfisetin.
(B) Drewes et al., J. Chem. Soc., C, 897 (1969)
(F) Drewes et al., J. Chem. Soc., Chem. Commun., 1246 (1968)

A. melanoxylon R.Br.
QUINONES: acamelin, 2,6-dimethoxy-1,4-benzoquinone.
Schmalle et al., Tetrahedron Lett., 149 (1980)

A. mellifera.
FLAVONOIDS: isorhamnetin-3-rutinoside, myricetin 4′-methyl ether-3-rhamnoside, quercetin-3-rutinoside.
PHENOLICS: (+)-catechol, gallic acid, (+)-gallocatechol.
Thieme *et al., Pharmazie,* **30,** 736 (1975)

A. nilotica.
FLAVONOIDS: isorhamnetin-3-rutinoside, myricetin 4′-methyl ether-3-rhamnoside, quercetin-3-rutinoside.
PHENOLICS: (+)-catechol, gallic acid, (+)-gallocatechol.
Thieme *et al., Pharmazie,* **30,** 736 (1975)

A. peuce.
FLAVANONES: fustin, fustin 3,3′,4′,7-tetramethyl ether, fustin 3,4′,7-trimethyl ether.
Weinger *et al., Justus Liebigs Ann. Chem.,* **627,** 229 (1959)

A. polyacantha subsp. camphylacantha.
FLAVONOIDS: isorhamnetin-3-rutinoside, myricetin 4′-methyl ether-3-rhamnoside, quercetin-3-rutinoside.
PHENOLICS: (+)-catechol, gallic acid, (+)-gallocatechol.
Thieme *et al., Pharmazie,* **30,** 736 (1975)

A. polystachya.
HISTAMINE ALKALOID: cinnamoylhistamine.
Johns, Lamberton, *Aust. J. Chem.,* **20,** 555 (1967)

A. praecox.
FLAVONOIDS: isorhamnetin, luteolin, luteolin 7-O-glucoside, quercetin, quercetin 3-O-galactoside, vitexin.
Suarez *et al., An. Asoc. Quim. Argent.,* **70,** 647 (1982); *Chem. Abstr.,* 97 107039z

A. retinoide.
FLAVONOID: 6-C-glucosylnaringenin.
Lorente *et al., Phytochemistry,* **21,** 1461 (1982)

A. rhodoxylon.
FLAVANONE: 4′,7,8-trihydroxyflavanone.
Clark-Lewis *et al., Aust. J. Chem.,* **25,** 1943 (1972)

A. rigida.
AMINE ALKALOID: N-methyl-β-phenethylamine.
Camp, Norvell, *Econ. Bot.,* **20,** 274 (1966)

A. roemeriana.
AMINE ALKALOID: N-methyl-β-phenethylamine.
Camp, Norvell, *Econ. Bot.,* **20,** 274 (1966)

A. saligna.
FLAVONOIDS: isorhamnetin-3-rutinoside, myricetin 4′-methyl ether-3-rhamnoside, quercetin-3-rutinoside.
PHENOLICS: (+)-catechol, gallic acid, (+)-gallocatechol.
Thieme *et al., Pharmazie,* **30,** 736 (1975)

A. saxatilis.
FLAVANS: (2R,3S,4R)-3′,4,7-trimethoxy-3,4-flavandiol, (2S,3S,4S)-3′,4,7-trimethoxy-3,4-flavandiol.
DuPreeze *et al., J. Chem. Soc., C,* 1800 (1970)

A. schottii.
AMINE ALKALOID: N-methyl-β-phenethylamine.
Camp, Norvell, *Econ. Bot.,* **20,** 274 (1966)

A. seyal.
FLAVONOIDS: isorhamnetin-3-rutinoside, myricetin 4′-methyl ether-3-rhamnoside, quercetin-3-rutinoside.
PHENOLICS: (+)-catechol, gallic acid, (+)-gallocatechol.
Thieme *et al., Pharmazie,* **30,** 736 (1975)

A. sieyberana
FLAVONOIDS: isorhamnetin-3-rutinoside, myricetin 4′-methyl ether-3-rhamnoside, quercetin-3-rutinoside.
NITRILE: acacipetalin.
PHENOLICS: (+)-catechol, gallic acid, (+)-gallocatechol.
(F, P) Thieme *et al., Pharmazie,* **30,** 736 (1975)
(N) Seigler *et al., Phytochemistry,* **15,** 219 (1976)

A. simplicifolia.
AMINE ALKALOIDS: N(b),N(b)-dimethyltryptamine, N-methyltryptamine.
CARBOLINE ALKALOID: 2-methyl-1,2,3,4-tetrahydro-β-carboline.
Poupat *et al., Phytochemistry,* **15,** 2019 (1976)

A. spirorbis.
AMINE ALKALOIDS: N(a)-trans-cinnamloylhistidine, hordenine.
Poupat *et al., Phytochemistry,* **14,** 1881 (1975)

A. taxensis.
AMINE ALKALOID: N-methyl-β-phenethylamine.
Camp, Norvell, *Econ. Bot.,* **20,** 274 (1966)

A. tortilis.
FLAVONOIDS: isorhamnetin-3-rutinoside, myricetin 4′-methyl ether-3-rhamnoside, quercetin-3-rutinoside.
PHENOLICS: (+)-catechol, gallic acid, (+)-gallocatechol.
Thieme *et al., Pharmazie,* **30,** 736 (1975)

A. willardiana.
AMINO ACID: willardiine.
Gmelin, *Acta Chem. Scand.,* **15,** 1188 (1961)

ACALYPHA

A. indica.
AMIDE: acalyphamide.
Talapatra *et al., Indian J. Chem.,* **20B,** 974 (1981)

ACANTHOPANAX (Araliaceae)

A. chiisanensis.
SAPONIN: chiisanoside.
Hahn *et al., Chem. Pharm. Bull.,* **32,** 1244 (1984)

A. trifoliatus.
NORTRITERPENOIDS:
24-nor-3α,11α-dihydroxylup-20(29)-en-25-oic acid,
24-nor-11α-hydroxy-3-oxolup-20(29)-en-28-oic acid.
Lischewski *et al., Phytochemistry,* **24,** 2355 (1985)

ACANTHOPHYLLUM

A. adenophorum.
DITERPENOID GLUCOSIDES: paniculides A, B, C, I, II, III, IV and V.
SAPONINS: acanthophyllum saponin, paniculatoside A.
Amanmuradov, Tanyurcheva, *Khim. Prir. Soedin.,* 326 (1969)

A. gypsophiloides.
SAPONINS: acanthophyllosides A, B, C and D.
Putieva *et al., Khim. Prir. Soedin.,* 806 (1977)

ACANTHOSPERMUM (Compositae)

A. australe.
SESQUITERPENOIDS: acanthamolide, acanthaustralide, acanthoglabrolide, acantholide, acanthospermal A, acanthospermal B, acanthospermolide.
Saleh *et al., J. Chem. Soc., Perkin I,* 1090 (1980)
Bohlmann *et al., Phytochemistry,* **20,** 1081 (1981)

A. glabratum (DC.) Willd.
SESQUITERPENE ALKALOID: acanthamolide.
SESQUITERPENOIDS: acanthoglabrolide, acantholide, dihydroacanthospermal A, glabratolide.
Saleh *et al., J. Chem. Soc., Perkin I,* 1090 (1980)

A. hispidum.
SAPONIN: acanthospermol galactosidopyranoside.
SESQUITERPENOIDS:
9-acetoxy-8-(2-methylbutanoyl)-14,15-dihydroxyacanthospermolide,
14-oxo-9-linolenyl-8-(2-methylbutanoyl)-15-hydroxyacanthospermolide,
14-oxo-9-linoleyl-8-(2-methylbutanoyl)-15-

hydroxyacanthospermolide,
14-oxo-8-(2-methylbutanoyl)-9,15-
dihydroxyacanthospermolide,
14-oxo-9-palmityl-8-(2-methylbutanoyl)-15-
hydroxyacanthospermolide,
14-oxo-9-stearyl-8-(2-methylbutanoyl)-15-
hydroxyacanthospermolide.
(Sap) Nair *et al.*, *Phytochemistry*, **15**, 1776 (1976)
(T) Bohlmann *et al.*, *ibid.*, **18**, 625 (1979))

ACANTHOSPORA (Lichenes)

A. chlorophana.
LACTONE ACIDS: acaranoic acid, (-)-acarenoic acid.
Santesson, *Acta Chem. Scand.*, **21**, 1993 (1980)

ACANTHOSYRIS

A. spinescens.
ACETYLENIC: 10,16-heptadecadien-8-ynoic acid.
Powell *et al.*, *Biochemistry*, **5**, 625 (1966)

ACER (Aceraceae))

A. caesium.
TRITERPENOID: acergenin.
Narayanan *et al.*, *Curr. Sci.*, **42**, 642 (1973)

A. negundo L.
SAPONIN: saponin P.
TRITERPENOIDS: acerocin, acerotin.
Kupchan *et al.*, *J. Chem. Soc., Chem. Commun.*, 969 (1970)

A. nikoense Maxim.
FLAVONOIDS: quercetin, quercitrin.
GLYCOSIDES: aceroside I, epi-rhododendrin.
PHENOLIC: ellagic acid.
STEROIDS: campesterol, β-sitosterol, β-sitosterol glucoside, stigmasterol.
TRITERPENOIDS: β-amyrin, β-amyrin acetate.
Inoue *et al.*, *J. Pharm. Soc., Japan*, **98**, 41 (1978)

A. saccharum Marsh.
FLAVONOID: quercetin.
PHENOLICS: caffeic acid, epi-coumaric acid, ferulic acid, vanillic acid.
Thakura, *J. Exptl. Bot.*, **28**, 795 (1977); *Chem. Abstr.*, **87** 197304c

ACHILLEA (Compositae)

A. ageratum L.
SESQUITERPENOIDS: ageratriol, agerol, β-elemen-9β-ol.
Garanti *et al.*, *Tetrahedron Lett.*, 1397 (1972)

A. biebersteinii.
COUMARINS: isoscopoletin, scopoletin, umbelliferone.
SESQUITERPENOIDS: bibsanthin, isororidentin.
(C) Shmatova *et al.*, *Khim. Prir. Soedin.*, 561 (1985)
(T) Yusupov *et al.*, *ibid.*, 580 (1979)

A. cartilaginea.
FLAVONOID: apigenin 7-O-β-D-ethylglucuronidopyranoside.
Zapesochnaya *et al.*, *Khim. Prir. Soedin.*, 814 (1976)

A. collina Becker.
MONOTERPENOIDS: caryophyllene, chamazulene, limonene, α-pinene, β-pinene, sabinene.
Cernaj *et al.*, *Biologia (Bratislava)*, **38**, 865 (1983); *Chem. Abstr.*, **100** 3548d

A. depressa.
SESQUITERPENOID: apressin.
Tsankova *et al.*, *Phytochemistry*, **20**, 1436 (1981)

A. filipendulina.
MONOTERPENOIDS: achillene, 2-bornanol, filipendulol.
Dembitskii, Yurina, *Khim. Prir. Soedin.*, 251 (1969)
Dembitskii *et al.*, *Izv. Akad. Nauk Kaz. SSR, Ser. Khim.*, 55 (1980)

A. lanulosa.
SESQUITERPENOIDS: achillin, hydroxyachillin, hydroxyachillin acetate.
White, Winter, *Tetrahedron Lett.*, 137 (1963)

A. millefolium L.
UNCHARACTERIZED ALKALOIDS: achilleine, achilletine.
FLAVONES: artemetin, casticin, 5-hydroxy-3,6,7,4'-tetramethoxyflavone.
SESQUITERPENOIDS: acetylbalchanolide, achillicin, achillin, millefin, millefolide.
(A) von Planta, *Justus Liebigs Ann. Chem.*, **58**, 21 (1846)
(F) Falk *et al.*, *J. Pharm. Sci.*, **64**, 1838 (1975)
(T) White, Winter, *Tetrahedron Lett.*, 137 (1963)
(T) Cuong *et al.*, *Phytochemistry*, **18**, 331 (1979)

A. moschata Wulf.
UNCHARACTERIZED ALKALOIDS: achilleine, achilletine, moschatine.
von Planta, *Justus Liebigs Ann. Chem.*, **155**, 153 (1870)

A. nobilis.
SESQUITERPENOID: anobin.
Adekenov *et al.*, *Khim. Prir. Soedin.*, 603 (1984)

A. santolina.
HALOGENATED SESQUITERPENOID: bibsanthin.
SESQUITERPENOID: sanchillin.
Mallabaev *et al.*, *Khim. Prir. Soedin.*, 793 (1984)

A. santolinoides.
MONOTERPENOIDS: camphene, camphor, 1,8-cineole, α-pinene.
Sanz *et al.*, *J. Nat. Prod.*, **48**, 993 (1985)

A. sibirica.
SESQUITERPENOID:
4β-hydroxy-3α,6,9-trimethyl-2,3,3a,4,5,7,9a,9b-
octahydroazuleno-[4,5b]-furan-2,7-dione.
Kaneko *et al.*, *Phytochemistry*, **10**, 3305 (1971)

ACHRAS

A. sapota.
TRITERPENOIDS: bassic acid, protobassic acid.
Heywood *et al.*, *J. Chem. Soc.*, 1124 (1939)
King *et al.*, *ibid.*, 4469 (1956)

ACHYRANTHES (Amaranthaceae)

A. fauriei Lev. et Van.
STEROID: inokosterone.
Takemoto *et al.*, *J. Pharm. Soc., Japan*, **87**, **305**, 1474 (1967)

A. rubrofusca.
STEROID: rubrosterone.
Hikino *et al.*, *Tetrahedron*, **32**, 3015 (1976)

ACHYROCLINE (Compositae))

A. satureoides var. albicans.
FLAVONES: alnustin, 7-hydroxy-3,5,8-trimethoxyflavone, 3,5,7,8-tetramethoxyflavone, 5,7,8-trimethoxyflavone.
Mesquita *et al.*, *Phytochemistry*, **25**, 1255 (1986)

A. tomentosa.
FLAVONOIDS: 3,5-dihydroxyy-7,8-dimethoxyflavone, 5,7,4'-trihydroxy-8,3'-dimethoxyflavone.
STEROID: β-sitosterol.
Ferraro *et al.*, *J. Nat. Prod.*, **48**, 817 (1985)

ACNISTUS

A. arborescens L.
STEROIDS: 3-ethoxy-2,3-dihydrowithacnistin, withacnistin, withaferin A.
Lavie, Yarden, *J. Chem. Soc.*, 2925 (1962)

A. breviflorus.
STEROIDS: acnistoferin, 4-hydroxy-5,6-epoxy-1-oxowithanolide.
Nittala *et al.*, *Phytochemistry*, **20**, 2735 (1981))

A. ramiflorum.
STEROIDS: acnistins A, B, C, D, E and F.
Usubillaga *et al.*, *J. Chem. Soc., Chem. Commun.*, 884 (1984)

ACOKANTHERA (Apocynaceae)

A. friesiorum.
STEROIDS: acoflorogenin, acofriosides F, L and M, acolongiflorosides E, G, H, J and K, acovenosides A and B.
Bally *et al.*, *Helv. Chim. Acta*, **34**, 1740 (1951)
Mohr *et al.*, *ibid.*, **37**, 403 (1954)

A. longiflora.
STEROIDS: acolongifloroside J, acovenoside A, acovenoside B.
Mohr *et al.*, *Helv. Chim. Acta*, **37**, 403 (1954)

A. lyciflora.
STEROID: acoflorogenin.
Hauschild-Rogat *et al.*, *Helv. Chim. Acta*, **50**, 2299 (1967)

A. ouabaio.
STEROID: ouabain.
Fieser *et al.*, *J. Biol. Chem.*, **114**, 705 (1936)

A. schimperi (A.DC.) Benth et Hook.
STEROIDS: acoschimperosides N, P, Q, S, T and U.
Bally *et al.*, *Helv. Chim. Acta*, **41**, 446 (1958)

A. venenata G.Don.
STEROIDS: acovenosides A, B and C.
von Euw, Reichstein, *Helv. Chim. Acta*, **33**, 485 (1950)
Mohr, Reichstein, *ibid.*, **34**, 1239 (1951)

ACONITUM (Ranunculaceae)

A. altaicum.
DITERPENE ALKALOIDS: aconitine, mesaconitine.
Monakhova *et al.*, *Khim. Prir. Soedin.*, 113 (1965)

A. anthora.
DITERPENE ALKALOID: atisine.
Edwards *et al.*, *Can. J. Chem.*, **32**, 465 (1954)

A. anthoroideum L.
DITERPENE ALKALOIDS: anthoroidine, condelfine.
Tel'nov *et al.*, *Khim. Prir. Soedin.*, 383 (1971)

A. balfourii Stapf.
DITERPENE ALKALOIDS: aconitine, pseudaconitine.
Dunstan, Carr, *J. Chem. Soc.*, **71**, 350 (1897)

A. barbatum var. puberulum.
DITERPENE ALKALOIDS: four aconitine alkaloids.
Yu, *Yaoxue Tongbao*, **17**, 301 (1981)

A. bullatifolium var. homostrictum.
DITERPENE ALKALOIDS: bullatines A, B, C, D, E, F and G.
Chu, Fang, *Hua Hsueh Hsueh Pao*, **31**, 222 (1965)

A. calliantum Koidz.
DITERPENE ALKALOIDS: hypaconitine, ignavine, mesaconitine.
Majima, Morio, *Justus Liebigs Ann. Chem.*, **476**, 171, 210 (1929)
Ochiai *et al.*, *J. Pharm. Soc., Japan*, **72**, 816 (1952)

A. carmichaeli.
DITERPENE ALKALOIDS: carmichaeline, Chuan-Wu base A, Chuan-Wu base B, fuziline, hokbusine, hokbusine B, senbusines A, B and C.

Konno *et al.*, *J. Nat. Prod.*, **45**, 128 (1982)
Hikino *et al.*, *ibid.*, **46**, 178 (1983)

A. chasmanthum Stapf.
DITERPENE ALKALOIDS: chasmaconitine, chasmanthine, chasmanthinine, homochasmanine, indaconitine.
Dunstan, Andrews, *J. Chem. Soc.*, **87**, 1620 (1905)
Achmatowicz, Marion, *Can. J. Chem.*, **43**, 1093 (1965)

A. columbianum.
DITERPENE ALKALOIDS: columbamine, columbianine, columbidine, 8-O-methyltalatisamine.
Pelletier *et al.*, *Heterocycles*, **23**, 331 (1984)

A. coreanum L. One aconitine alkaloid.
Monakhova *et al.*, *Khim. Prir. Soedin.*, 113 (1965)

A. crassicaule.
DITERPENE ALKALOIDS: crassicaulidine, crassicaulisine.
Wang, Fang, *Yaoxue Tongbao*, **17**, 300 (1982); *Chem. Abstr.*, **97** 212632k

A. czekanovskyi.
DITERPENE ALKALOIDS: hypaconitine, mesaconitine.
Zhapova *et al.*, *Khim. Prir. Soedin.*, 717 (1985)

A. deinorrhizum Stapf.
DITERPENE ALKALOIDS: aconitine, pseudaconitine.
Dunstan, Carr, *J. Chem. Soc.*, **71**, 250 (1897)

A. delphinifolium
DITERPENE ALKALOIDS: 14-O-acetylbrowniine, 14-O-acetylsachaconitine, 14-O-acetyltalatisamine, browniine, condelphine, delcosine, delphinifoline, isotalatizidine.
Aiyar *et al.*, *Phytochemistry*, **25.**, 973 (1986)

A. episcopale.
DITERPENE ALKALOIDS: episcopalidine, episcopalisine, episcopalisinine, episcopalitine.
Wang, Fang, *Yaoxue Tongbao*, **17**, 300 (1982)

A. excelsum Reichb.
DITERPENE ALKALOIDS: excelsine, excelsinine.
Tel'nov *et al.*, *Khim. Prir. Soedin.*, 129 (1973)

A. falconeri Stapf.
DITERPENE ALKALOIDS: falaconitine, falconerine, falconerine 8-acetate, mithaconitine.
Pelletier *et al.*, *J. Chem. Soc., Chem. Commun.*, 12 (1977)
Desai *et al.*, *Heterocycles*, **24** 1061 (1986)

A. fauriei L.
DITERPENE ALKALOID: mesaconitine.
Majima, Tamura, *Justus Liebigs Ann. Chem.*, **545**, 1 (1940)

A. ferox.
DITERPENE ALKALOIDS: diacetylpseudaconitine, pseudaconitine.
Purushothaman, Chandrasekharan, *Phytochemistry*, **13**, 1975 (1974)

A. finetianum Hand-Mazz.
DITERPENE ALKALOIDS: avadharidine, N-deacetylfinaconitine, N-deacetyllappaconitine, N-deacetylranaconitine, finaconitine, lycoctonine.
Jiang *et al.*, *Zhongcaoyao*, **13**, 5 (1982); *Chem. Abstr.*, **97** 107027u

A. fischeri Reichb.
DITERPENE ALKALOID: jesaconitine.
Makoshi, *Arch. Pharm.*, **247**, 243 (1909)

A. flavum.
DITERPENE ALKALOID: flavaconitine.
Lui *et al.*, *Yaoxue Tongbao*, **17**, 243 (1982)

A. foresti Stapf.
DITERPENE ALKALOIDS: 8-deacetylyunaconitine, foresaconitine, foresticine, forestine, liwaconitine, ludaconitine.
Chen *et al.*, *Yunnan Zhiwu Yanjiu*, **6**, 338 (1984)

A. franchetii Fin. et Gagn.
DITERPENE ALKALOIDS: franchetine, ludaconitine.
Chen, Song, *Huaxue Xuebao*, **41**, 843 (1983)

A. geniculatum Fletcher.
DITERPENE ALKALOIDS: chasmanine, crassicauline I,
8-deacetylyunaconitine, geniconitine, talatisamine,
vilmorrianine C.
Hao *et al., Zhiwu Xuebao*, **27**, 504 (1985); *Chem. Abstr.*, 104 31721f

A. gigas Lev. et Van.
DITERPENE ALKALOID: gigactonine.
Sakai *et al., J. Pharm. Soc., Japan*, **97**, 1376 (1978)

A. grossedentatum Nakai.
DITERPENE ALKALOIDS: hypaconitine, ignavine, mesaconitine.
Majima, Tomura, *Justus Liebigs Ann. Chem.*, **545**, 1 (1940)
Ochiai *et al., Chem. Pharm. Bull.*, **8**, 976 (1960)

A. hakusanense Nakai.
DITERPENE ALKALOIDS: hypaconitine, ignavine, mesaconitine.
Majima, Tomura, *Justus Liebigs Ann. Chem.*, **526**, 116 (1936)
Ochiai *et al., Chem. Pharm. Bull.*, **8**, 976 (1960)

A. hemsleyanum Pritz.
DITERPENE ALKALOIDS: guayewanines A, B and C,
yunaconitine.
Zhang *et al., Zhiwu Xuebao*, **24**, 259 (1982)

A. heterophylloides Stapf.
DITERPENE ALKALOID: heterophylloidine.
Pelletier *et al., Tetrahedron Lett.*, 313 (1981)

A. heterophyllum Wall.
DITERPENE ALKALOIDS: atidine, atisine, heteratisine,
heterophyllidine, heterophyllisine, hetidine, hetisine.
DITERPENOID: atisenol.
(A) Pelletier *et al., Tetrahedron Lett.*, 557 (1967)
(A) Jacobs, Craig, *J. Biol. Chem.*, **143**, 605 (1942)
(T) Pelletier *et al., J. Nat. Prod.*, **45**, 779 (1982)

A. ibukiense Nakai.
DITERPENE ALKALOIDS: ibukinamine, mesaconitine,
ryosenamine, ryosenaminol.
Sakai *et al., Chem. Pharm. Bull.*, **31**, 3338 (1983)

A. iliense.
DITERPENE ALKALOIDS: browniine, dehydrodelcorine,
delcoridine, delcorine, dictyocarpine, eldeline, ilidine,
lycoctonine.
Zhamerashvili *et al., Chem. Abstr.*, 97 107014n

A. japonicum Thunb.
DITERPENE ALKALOIDS: 11-acetylisohypognavine, aljesaconitine
A, aljesaconitine B, diacetylisohypognavine, higenamine,
15-hydroxyneoline, isohypognavine, takaosamine, Yakawo
Base I.
UNCHARACTERIZED ALKALOIDS: japaconitine A, japaconitine B.
Okomoto *et al., J. Pharm. Soc., Japan*, **76**, 550 (1956)
Sakai *et al., ibid.*, **99**, 647 (1979)
Bando *et al., Chem. Pharm. Bull.*, **33**, 4717 (1985)

A. japonicum var. ochiai No.21.
DITERPENE ALKALOID: Takawo Base I.
Okomoto *et al., J. Pharm. Soc., Japan*, **76**, 550 (1956)

A. jinyangense W.T.Wang.
DITERPENE ALKALOID: jynosine.
Chen, Song, *Yaoxue Xuebao*, **16**, 748 (1981)

A. kamtschaticum Nakai.
DITERPENE ALKALOIDS: hypaconitine, ignavine, mesaconitine.
Ochiai *et al., Chem. Pharm. Bull.*, **8**, 976 (1960)

A. karacolicum Rapaics.
DITERPENE ALKALOIDS: acetylnapelline, acetylnapelline
N-oxide, aconitine, 1-benzoylkarasamine, isoboldine,
karacolidine, karacoline, karasamine, napelline N-oxide,
songorine.
NAPHYLAMINE: phenyl-β-naphthylamine.
(A) Wiesner *et al., Chem. Ind.*, 173 (1957)
(A) Yunusov *et al., Khim. Prir. Soedin.*, 101 (1970)
(A) Sultankhodzhaev *et al., ibid.*, 207 (1986)
(N) Sultankhodzhaev *et al., ibid.*, 406 (1976)

A. kirinense Nakai.
DITERPENE ALKALOID: one aconitine base.
Monakhova *et al., Khim. Prir. Soedin.*, 113 (1965)

A. kongboense.
DITERPENOID ALKALOID: vilmorrianine A.
Wang *et al., Yaoxue Tongbao*, **17**, 395 (1982); *Chem. Abstr.*, 97
195756j

A. leucostomum.
DITERPENE ALKALOIDS: aksinatine, aksine, excelsine,
lappaconidine, lappaconitine, mesaconitine.
Furner *et al., J. Am. Chem. Soc.*, **82**, 5182 (1960)
Tel'nov *et al., Khim. Prir. Soedin.*, 129 (1973)

A. lucidusculum Nakai.
DITERPENE ALKALOIDS: ignavine, lucidusculine,
pseudokobusine.
Ochiai *et al., J. Pharm. Soc., Japan*, **75**, 976 (1955)
Natsume, *Chem. Pharm. Bull.*, **10**, 879 (1962)

A. lycoctonum L.
DITERPENE ALKALOIDS: lycaconitine, myoctonine.
Marion, Manske, *Can. J. Res.*, **24B**, 1 (1946)

A. majimae Nakai.
DITERPENE ALKALOIDS: isohypognavine, mesaconitine.
Ochiai *et al., J. Pharm. Soc., Japan*, **75** 638 (1955)

A. makchangense.
DITERPENE ALKALOID: mesaconitine.
Tamura, *J. Am. Chem. Soc.*, **58**, 1059 (1936)

A. mandschuricum.
DITERPENE ALKALOID: mesaconitine.
Morio, *Justus Liebigs Ann. Chem.*, **476**, 187 (1929)

A. masutum.
DITERPENE ALKALOID: one uncharacterized base.
Mura'eva, *Aktualnye Voprosy Farmatsii.*, **2**, 49 (1974)

A. miyabei Nakai.
DITERPENE ALKALOIDS: isodelphine, sachaconitine.
Pelletier *et al., Tetrahedron Lett.*, 4027 (1977)

A. monticola Steinb.
DITERPENE ALKALOIDS: acomonine, aconosine, norsongorine,
songoramine, songorine, songorine N-oxide.
Mura'eva *et al., Khim. Prir. Soedin.*, 128 (1972)

A. nagarum var. heterotrichum f. delsianum.
DITERPENE ALKALOID: nagarine.
Mody *et al., Heterocycles*, 91 (1982)

A. napellus L.
DITERPENE ALKALOIDS: isonapelline, luciculine, napelline,
neoline, neopelline.
Wiesner *et al., Chem. Ind.*, 173 (1957)
Wiesner *et al., Tetrahedron Lett.*, 17 (1960)

A. nasutum Fisch. et Rehd.
DITERPENE ALKALOID: aconosine.
AMINO PHENOL: N-methyladrenaline.
(A) Mura'eva *et al., Khim. Prir. Soedin.*, 128 (1972)
(Am) Samokish *et al., Chem. Abstr.*, 84 102283w

A. nemorosum Bieb.
DITERPENE ALKALOIDS: two uncharacterized bases.
Monakhova *et al., Khim. Prir. Soedin.,* 113 (1965)

A. nevadense Vechtr.
DITERPENE ALKALOIDS: nevadenine, nevadensine.
Gonzalez *et al., Heterocycles,* **23**, 2979 (1985)

A. orientale Mill.
DITERPENE ALKALOIDS: avadcharidine, avadcharine,
lappaconitine.
Kuzuvkov, Platonova, *J. Gen. Chem. USSR,* **29**, 2782 (1959)

A. palmatum.
DITERPENE ALKALOIDS: vakatidine, vakatisine, vakatisinine,
vakognavine.
Singh, Jaswal, *Tetrahedron Lett.,* 2219 (1968)

A. paniculatum Lam.
DITERPENE ALKALOIDS: paniculatine, panicutine.
Katz *et al., Helv. Chim. Acta,* **65**, 286 (1982)

A. pendulum.
DITERPENE ALKALOID: penduline.
Kiu *et al., Yaoxue Xuebao,* **18**, 39 (1983)

A. polyschistum Hand-Mazz.
DITERPENE ALKALOIDS: polyschistines A, B and C.
Wang *et al., Heterocycles,* **23**, 803 (1985).

A. ponticum.
DITERPENE ALKALOIDS: ponaconitines A, B and C,
pontaconitine.
Baytop, Tanker, *Istanbul Univ. Tip. Fak. Mecm.,* **25**, 157 (1962)
Rosenthaler, *Pharm. Acta Helv.,* **17**, 194 (1942)

A. puberulum.
DITERPENE ALKALOIDS: puberaconitidine, puberaconitine,
puberanidine, puberanine.
Yu, Das, *Planta Med.,* **49**, 85 (1983)

A. ranunculaefolium.
DITERPENE ALKALOID: ranaconitine.
Moller *et al., Tetrahedron Lett.,* 2189 (1969)

A. rotundifolium Kar. et Kir.
DITERPENE ALKALOIDS: two uncharacterized bases.
Platonova *et al., J. Gen. Chem. USSR,* **28**, 3126 (1958)

A. sachalinense Fr. Schmidt (Saglalien).
DITERPENE ALKALOIDS: jesaconitine, kobusine.
Suginome, Shimanouti, *Justus Liebigs Ann. Chem.,* **545**, 220 (1940)

A. sanyoense Nakai.
DITERPENE ALKALOIDS: hypognavine, hypognavinol.
Pelletier *et al., Tetrahedron Lett.,* 795 (1971)

A. saposhnikovii.
DITERPENE ALKALOIDS: 14-acetyldehydrotalatisamine,
14-acetyltalatisamine, isoboldine, talatisamine.
Sultankhodzhaev *et al., Khim. Prir. Soedin.,* 265 (1982)

A. senanense Nakai.
DITERPENE ALKALOIDS: hypaconitine, ignavine.
Ochiai *et al., Chem. Pharm. Bull.,* **8**, 976 (1960)

A. septentrionale Koelle.
DITERPENE ALKALOIDS: lappaconitine, septemtrionaline,
septentriodine.
Pelletier *et al., Heterocycles,* **12**, 377 (1979)

A. soongoricum Stapf.
DITERPENE ALKALOIDS: acetylsongoramin, acetylsongorine,
acetyltalitisamine, aconine, aconitine, luciculine, nemorine,
neoline, norsongorine, songoramine, songorine, talatisamine.
Yunusov *et al., Khim. Prir. Soedin.,* 101 (1970)
Wiesner *et al., Tetrahedron Lett.,* 867 (1971)

A. spicatum Stapf.
DITERPENE ALKALOID: bikhaconitine.
Tsuda, Marion, *Can. J. Chem.,* **41**, 3055 (1963)

A. stapfianum Hand-Mazz. var. pubipea Weng.
DITERPENE ALKALOID: dolaconine.
Lui, Chen, *Huaxue Xuebao,* **39**, 808 (1981)

A. stoerckianum Reichb.
DITERPENE ALKALOID: napelline.
Schultz, Berger, *Arch. Pharm.,* **262**, 553 (1924)

A. subcuneatum Nakai.
DITERPENE ALKALOIDS: 14-acetyldelcosine, 14-benzoylneoline,
deoxyjesaconitine, gomando bases A, B and C, ignavine,
jesaconitine, karakoline, neoline, penduline, tokoro bases A,
B, C and D.
Bando *et al., Heterocycles,* **16**, 1723 (1981)
Wada *et al., Chem. Pharm. Bull.,* **33**, 3658 (1985)

A. talassicum M. Pop.
DITERPENE ALKALOIDS: acetyltalatisamine, condelfine,
isotalatisidine, talatisamine, talatisidine, talatisine.
Khaimova *et al., Bulgarskoi Akademie Nauk,* **20**, 3 (1967)
Khaimova *et al., ibid.,* **20**, 193 (1967)
Pelletier *et al., J. Am. Chem. Soc.,* **89**, 4146 (1967)
Khaimova *et al., Tetrahedron,* **27**, 819 (1971)

A. tortuosum Willd.
DITERPENE ALKALOIDS: hypaconitine, ignavine, mesaconitine.
Ochiai *et al., Chem. Pharm. Bull.,* **8**, 976 (1971)

A. tranzschelii Steinb.
DITERPENE ALKALOIDS: isotalatisamine, talatisamine.
Tel'nov *et al., Khim. Prir. Soedin.,* 383 (1971)

A. umbrosum.
DITERPENE ALKALOID: umbrosine.
Tel'nov *et al., Khim. Prir. Soedin.,* 675 (1976)

A. varigatum.
DITERPENE ALKALOID: cammaconine.
Khaimova *et al., Tetrahedron,* **27**, 819 (1971)

A. vilmorrianum.
DITERPENE ALKALOIDS: vilmorrianine A, vilmorrianine B.
Chu, Chu, *Yao Hsueh Hsueh Pao,* **12**, 167 (1965)

A. yesoense Nakai.
DITERPENE ALKALOIDS: anisochasmaconitine,
ezochasmaconitine, ezochasmanine, lucidusculine,
pseudokobusine.
Takayama *et al., Heterocycles,* **15**, 403 (1981)

A. zeravschanicum Steinb.
DITERPENE ALKALOIDS: heteratisine, isotalatisidine,
talatisamine, zaravchanidine.
Przybylska, *Can. J. Chem.,* **41**, 291 (1963)
Aneja, Pelletier, *Tetrahedron Lett.,* 669 (1964)

A. zuccarini Nakai.
DITERPENE ALKALOIDS: hypaconitine, ignavine, mesaconitine.
Ochiai *et al., Chem. Pharm. Bull.,* **8**, 976 (1960)

ACORUS (Araceae)
A. calamus.
SESQUITERPENOIDS: acolamone, acoragermacrone, acorenol,
acorenone, acoric acid, acorone, aristolene,
(-)-cadala-1,4,9-triene, τ-cadinene, calacone, calacorene,
calameone, calamone, (+)-calamusenone, calarene,
(-)-β-curcumene, β-elemene, guaiene, isoacolamone,
isocaespitol, isocalamenediol, isocalamusenone,
isoshyobunone, preisocalamenediol, shyobunone,
epi-shyobunone.
Yamamura *et al., Tetrahedron,* **27**, 5419 (1971)
Iguchi *et al., Tetrahedron Lett.,* 2759 (1982)

ACOSMIUM
A. panamense.
SPARTEINE ALKALOID: (-)-4-hydroxysparteine.
Balondrin, Kinghorn, *Heterocycles*, **19**, 1931 (1982)

ACREMONIUM (Polyporaceae)
A. lazulae (Fuckel) Gams.
GLYCOSIDES: virescenosides D and H.
HALOGENATED TERPENOID: ascochlorin.
NORDITERPENOID: 19-norisopimara-7,15-dien-3-one.
(G) Cagnoli-Bellavita *et al.*, *J. Chem. Soc., Perkin* 1, 351 (1977)
(T) Tamura *et al.*, *J. Antibiotics*, **21**, 539 (1968)

ACRITOPAPPUS
A. confertus.
DITERPENOIDS: acritoconfertic acid, 8,16
16,17-bisepoxyacritoconfertic acid,
16,17-dihydroxy-7-acritoconfertenoic acid,
ent-16,17-dihydroxy-7,13-labdadien-15-oic acid,
ent-16,17-dihydroxy-7,13-labdadien-15-oic acid 16-acetate,
ent-16,17-dihydroxy-7,13-labdadien-15-oic acid methyl ester.
SESQUITERPENOIDS: 2,7(14)-bisaboladiene-10,11-diol,
2,7(14)-bisaboladien-10-one,
10,11-epoxy-2,7(14)-bisaboladiene.
Bohlmann *et al.*, *Phytochemistry*, **19**, 2695 (1980)
Bohlmann *et al.*, *ibid.*, **22**, 2243 (1983)
A. hagei.
DITERPENOIDS: acritopappus acid, acritopappusol,
16,18-dihydroxykolavenic acid, 16,18-dihydroxykolavenic
acid 16-acetate, 2,15-dihydroxykolavenic acid lactone,
16,18-dihydroxykolavenic acid methyl ester.
Bohlmann *et al.*, *Phytochemistry*, **19**, 2695 (1980)
A. prunifolius.
SESQUITERPENOIDS: 1-acripruninone, 7(11)-eudesmen-4-ol,
humulen-14-ol acetate, humulen-14-ol angelate.
Bohlmann *et al.*, *Phytochemistry*, **21**, 147 (1982)

ACRONYCHIA
A. baueri Schott.
ACRIDONE ALKALOIDS: acronidine, acronycine,
1,3-dimethoxy-10-methyl-9-acridone, melicopidine,
melicopine.
TERPENE ESTERS: p-geranylcinnamic acid methyl ester,
p-geranylferulic acid methyl ester.
TRITERPENOID: baurenol.
(A) Brown *et al.*, *Aust. J. Chem.*, A 2, 622 (1949)
(A) Lamberton, Price, *ibid.*, **6**, 66 (1953)
(T) Crow, Michael, *ibid.*, **8**, 129 (1955)
(T) Prager, Thredgold, *Aust. J. Chem.*, **19**, 451 (1966)
A. haplophylla.
FURANOQUINOLINE ALKALOIDS: acrophyllidine, acrophylline.
Lahey, McCamish, *Tetrahedron Lett.*, 1525 (1968)
A. muelleri.
TRITERPENOID: asarinin.
Beroza *et al.*, *J. Am. Chem. Soc.*, **78**, 1242 (1956)
A. vestita.
TRITERPENOID: acrovestone.
Banerji *et al.*, *Indian J. Chem.*, **11**, 693 (1973)

ACROPTILON
A. repens.
SESQUITERPENOIDS: acroptilin, acroptin, acrorepiolide.
Bohlmann *et al.*, *Phytochemistry*, **20**, 1152 (1981)
Serkerov *et al.*, *Khim. Prir. Soedin.*, 712 (1982)

ACTEA (Ranunculaceae)
A. racemosa Thunb.
GLYCOSIDE: actein.
TRITERPENOID: glycone A.
(Gly) Linde, *Arch. Pharm.*, **300**, 982 (1967)
(T) Piancatelli, *Gazz. Chim. Ital.*, **101**, 139 (1971)
A. rubra.
GLYCOSIDES: protoanemonin, ranunculin.
Linde. *Arch. Pharm.*, **300**, 982 (1967)

ACTIAS
A. silene.
CAROTENOID: 3-hydroxy-β,ε-caroten-3'-one.
Kayser *et al.*, *J. Comp. Physiol.*, **104** 27 (1975)

ACTINIDIA (Dilleniaceae/Actinidiaceae)
A. chinensis.
BENZOPYRAN: afzelechin.
King *et al.*, *J. Chem. Soc.*, 2948 (1955)
A. polygama (Sieb. et Zucc.) Maxim.
MONOTERPENOIDS: actinidiolide, actinidol, dehydroiridodial,
dehydroiridodial-β-D-gentiobioside, dihydroactinidiolide,
iridialo-β-D-gentiobioside, isodihydronepetalactone,
metabiether, neometatabiol, neonepetalactone.
Yoshihara *et al.*, *Chem. Lett.*, 433 (1978)
Murai, Tagawa, *Planta Med.*, **37**, 234 (1979)

ACTINODAPHNE
A. hookeri.
APORPHINE ALKALOID: actinodaphnine.
Krishna, Ghose, *J. Indian Chem. Soc.*, **9**, 429 (1932)
A. longifolia.
SESQUITERPENOIDS: longifolin, sesquirosefuran.
Hayashi *et al.*, *Chem. Ind. (London)*, 572 (1972)

ADANSONIA
A. digitata.
COUMARIN: scopoletin.
STEROID: β-sitosterol.
TRITERPENOIDS: bauerenol, bauerenol acetate, betulinic acid,
friedelin, lupeol, lupeol acetate, taraxerone.
Dan *et al.*, *Fitoterapia*, **57**, 445 (1986)

ADENIUM (Apocynaceae)
A. bohemianum.
STEROIDS: abobioside, abomonoside, echubioside, echujin,
somalin.
Schindler, Reichstein, *Helv. Chim. Acta*, **34**, 1732 (1951)
Striebel *et al.*, *ibid.*, **38**, 1001 (1955)
A. honghel.
SAPONINS: honghelosides A, B and C.
Hunger, Reichstein, *Helv. Chim. Acta*, **33**, 76, 1993 (1950)
A. obesum (Forsk.) Roem. et Schult.
STEROID: 16-anhydro-3-acetylgitoxigenin.
Vethaviyasar, John, *Planta Med.*, **44**, 123 (1982)
A. somalense Balf.
STEROID: somalin.
Hess, Hunger, *Helv. Chim. Acta*, **36**, 85 (1953)

ADENOCARPUS (Leguminosae)
A. commutatus.
DIPIPERIDYL ALKALOIDS: (-)-adenocarpine, (±)-adenocarpine, teidine.

Schopf, Kreibich, *Naturwissenschaften*, **41**, 335 (1954)
Ribas *et al.*, *Ann. Pharm. Fr.*, **51B**, 55 (1955)

A. complicatus Gay.
DIPIPERIDYL ALKALOID: teidine.
PYRIDINE ALKALOID: santiaguine.

Ribas *et al.*, *Ann. Pharm. Fr.*, **51B**, 55 (1955)

A. decorticans Boiss.
DIPIPERIDYL ALKALOIDS: decorticasine, N-depropionyldecorticasine, N-depropionyldecorticasine butyramide, N-depropionyldecorticasine isobutyramide, N-depropionyldecorticasine isovaleramide.

Clark, *J. Am. Chem. Soc.*, **54**, 3000 (1932)
Landa Velon, *Acta Cient. Compostelana*, **8**, 171 (1971)

A. foliosus.
DIPIPERIDYL ALKALOID: teidine.

Ribas *et al.*, *Ann. Pharm. Fr.*, **51B**, 55 (1955)

A. grandiflorus.
DIPIPERIDYL ALKALOIDS: adenocarpine, (-)-adenocarpine.

Ribas *et al.*, *Ann. Pharm. Fr.*, **51B**, 55 (1955)

A. hispanicus (Lam.) DC.
DIPIPERIDYL ALKALOID: (+)-adenocarpine.

Santos Ruiz, Llorente, *An. R. Soc. Esp. Fiz. Qhim.*, **37**, 624 (1941)

A. intermedius.
DIPIPERIDYL ALKALOIDS: (+)-adenocarpine, santiaguine.

Dominguez *et al.*, *An. R. Soc. Esp. Fiz. Qhim.*, **52B**, 133 (1956)

A. viscosus (Willd.) Webb et Benth.
DIPIPERIDYL ALKALOIDS: (+)-adenocarpine, teidine.

Ribas *et al.*, *Ann. Pharm. Fr.*, **51B**, 55 (1955)

ADENOCAULON
A. adhaerescens.
ACIDS: caffeic acid, caffeic acid methyl ester.
GLUCOSIDE: caffeoyl-1-β-D-glucopyranoside.

Kulesh *et al.*, *Khim. Prir. Soedin.*, 506 (1986)

ADENOCHLAENA (Euphorbiaceae)
A. siamensis Ridl.
DITERPENOIDS: chettaphanins I and II.

Sato *et al.*, *Tetrahedron Lett.*, 839 (1971)

ADENOSTEMMA
A. lavenia.
DITERPENOIDS: ent-11α,15α-dihydroxykaur-16-en-19-oic acid, ent-11α-hydroxy-15α-acetoxykaur-16-en-19-oic acid.

Hufford *et al.*, *Lloydia*, **42**, 183 (1979)

ADENOSTYLES
A. alliariae.
PYRROLIZIDINE ALKALOIDS: platyphylline, seneciphylline.
SESQUITERPENOIDS: adenostylone, 6-hydroxy-9-oxo-10(11)-furanoeremophilan, isoadenostylone, neoadenostylone.

(A) Yakhontova *et al.*, *Khim. Prir. Soedin.*, 122 (1976)
(T) Harmatha *et al.*, *Tetrahedron Lett.*, 1409 (1969)
(T) Harmatha *et al.*, *Collect. Czech. Chem. Commun.*, **34**, 1779 (1969)

ADERANTHERA
A. parvonica.
ALKALOID: one uncharacterized base.

Patel *et al.*, *Curr. Sci.*, **16**, 346 (1946)

ADHATODA
A. vasica Nees.
QUINAZOLINE ALKALOIDS: vasicine, vasicinone, vasicoline, vasicolinone.
SESQUITERPENOID: vasicol.

(A) Johne *et al.*, *Helv. Chim. Acta*, **54**, 826 (1971)
(T) Dhar *et al.*, *Phytochemistry*, **20**, 319 (1981)

ADIANTUM (Polypodiaceae)
A. capillus-veneris L.
TRITERPENOIDS: adiantone, adiantoxide, epoxyfilicane, 21-hydroxyadiantone.

Baddeley *et al.*, *J. Chem. Soc.*, 3891 (1961)
Berti *et al.*, *Tetrahedron*, **25**, 2939 (1969)

A. caudatum.
ALIPHATICS: 16-hentriacontanone, hentricontane.
STEROID: β-sitosterol.
TRITERPENOIDS: adiantone, isoadiantone.

Singh *et al.*, *Indian J. Pharm.*, **37**, 64 (1975)

A. monochlamys.
TRITERPENOIDS: adianene, adien-5-ene ozonide, ferna-7,9(11)-diene, fern-7-ene, fern-8-ene, 3-filicene, 21-hydroxyadiantone, ketohakonanol, neohopa-11,13(18)-diene, neohopane, neohopene, oxohakonanol.

Ageta *et al.*, *J. Chem. Soc., Chem. Commun.*, 1105 (1968)
Ageta *et al.*, *Tetrahedron Lett.*, 899 (1978)

A. pedatum.
TRITERPENOIDS: adipedatol, ferna-7,9(11)-diene, filicenal, isofernene, neohopa-11,13(18)-diene, neohopene.

Ageta, Iwata, *Tetrahedron Lett.*, 6069 (1966)
Ageta *et al.*, *J. Chem. Soc., Chem. Commun.*, 1105 (1968)

A. sulphureum.
FLAVONOIDS: galangen, isalpinin.

Wollenweber, *Phytochemistry*, **15**, 2013 (1976)

A. venestum Don.
TRITERPENOID: 21-hydroxyadiantone.

Zaman *et al.*, *Tetrahedron Lett.*, 3943 (1966)

ADINA (Rubiaceae)
A. cordifolia Roxb.
CARBOLINE ALKALOIDS: adifoline, cordifoline, 10-deoxyadifoline, deoxycordifoline.
CHROMONES: 5,7-dihydroxy-2-methylchromone, 5,7-dihydroxy-2-methylchromone 7-O-β-D-glucoside, 5,7-dihydroxy-2-methylchromone 7-O-pentosyl-β-D-glucoside.
FLAVONOIDS: chrysin, 7,4'-di-O-methyl-6-methylflavonone, gossypetin, kaempferol, noringenin 5-O-neohesperidoside.

(A) Cross *et al.*, *J. Chem. Soc.*, 2714 (1961)
(A) Brown, Row, *J. Chem. Soc., Chem. Commun.*, 453 (1967)
(C) Brown *et al.*, *J. Chem. Soc., Perkin I*, 1776 (1975)
(F) Srivastava *et al.*, *Indian J. Chem.*, **22B**, 1064 (1983)

A. rubescens Hemsl.
CARBOLINE ALKALOIDS: adirubine, anhydroadirubine, deoxicordifoline lactam, deoxycordifoline lactam methyl ester, macrolidine, 5-oxostrictosidine, rubescine.
GLUCOALKALOID: rubenine.
CHROMONES: noreugenin, noreugenin 7-β-D-glucoside.

(A) Brown, Fraser, *Tetrahedron Lett.*, 841 (1973)
(C) Brown *et al.*, *J. Chem. Soc., Perkin I*, 1776 (1975)

A. rubrostipulata K. Schumann.
SPIROINDOLONE ALKALOID: rubranidine.

Denis, Bull. Acad. R. Belg., Liege, (v), **23**, 174 (1937)

ADLUMIA

A. cirrhosa Rap.
OXOPROTOBERBERINE ALKALOID: α-allocryptopine.
Manske, Miller, *Can. J. Res.*, **18B**, 288 (1940)

A. fungosa Green.
ISOQUINOLINE ALKALOIDS: adlumidine, adlumine.
OXOQUINAZOLIDINE ALKALOID: protopine.
Schlotterbeck, *Am. Chem. J.*, **24**, 249 (1900)

ADONIS (Ranunculaceae)

A. aleppica Boiss.
STEROIDS: convallatoxol, convallotoxin, 3-epi-periplogenin, periplorhamnosin, strophanthidin, strophanthidin diginoside.
Junior *et al.*, *Dtsch. Apoth. Ztg.*, **125**, 1945 (1985); *Chem. Abstr.*, **104** 17632w

A. amurensis (L.) Regel et Redde.
STEROIDS: adonilide, adonitoxin, fukujusone, fukujusonorone, linoleon.
Shimizu *et al.*, *Lloydia*, **41**, 1 (1978)

A. annua.
CAROTENOID: 3-hydroxyechitenone.
Egger, *Phytochemistry*, **4**, 609 (1965)

A. chrysocyathus Hook f. ex Thoms. ex Hook.
STEROID: adonilidic acid.
Aitova *et al.*, *Khim. Prir. Soedin.*, 847 (1971)

A. mongolica Sim.
STEROIDS: adonitoxin, corchoside A, erysimoside, glucoolitoriside, K-strophanthoside, K-strophanthoside-β.
Thieme *et al.*, *Pharmazie*, **31**, 565 (1976)

A. tienschanicus.
COUMARINS: cymarin, scololetin, umbelliferone.
STEROIDS: strophanin-β, strophanthidin.
Komissarenko *et al.*, *Khim. Prir. Soedin.*, 287 (1977)

A. vernalis L.
AMINES: choline, cimarine.
SAPONINS: adonitoxigenin 3-[O-α-L-(2'-O-acetyl)-rhamnosido]-β-D-glucoside, adonitoxigenin 3-[O-α-L-(3'-O-acetyl)-rhamnosido]-β-D-glucoside, adonitoxigenin 3-O-α-L-rhamnosido-β-D-glucoside, strophanthidin digitaloside.
STEROIDS: acetyladonitoxin, 3-acetylstrophadogenin, adonitoxin, adonitoxol, linoleon, 3-epi-periplogenin, strophadogenin, k-strophanthidin, vernadigin.
(Am) Polkova *et al.*, *Chem. Ind.*, 1766 (1963)
(Sap) Winkler *et al.*, *Planta Med.*, 68 (1986)
(S) Wichtl, Junior, *Phytochemistry*, **19**, 2193 (1980)

A. wolgensis.
SAPONINS: adonitoxin, k-strophanthoside.
Yatsyuk, Komissarenko, *Khim. Prir. Soedin.*, 672 (1976)

ADOXA

A. moschatellina.
IRIDOID: adoxide.
Jensen *et al.*, *Biochem. Syst. Ecol.*, **7**, 103 (1979)

AECHMEA

A. glomerata.
ANTHOCYANINS: cyanidin, delphinidin, poenidin.
FLAVONOIDS: apigenin, luteolin, quercetin, tricin.
Asktakala *et al.*, *J. Am. Soc. Hortic. Sci.*, **100**, 546 (1975); *Chem. Abstr.*, 84 56511u

AEGICERAS

A. corniculatum.
TRITERPENOID: primulagenin A.
Margot, Reichstein, *Pharm. Acta Helv.*, **17**, 113 (1942)

A. major.
TRITERPENOIDS: aegecerin, aegiceradienol, aegiceradiol, aegigerin.
Barton, Brooks, *J. Chem. Soc.*, 257 (1951)
Rao *et al.*, *Tetrahedron*, **20**, 973 (1964)

AEGILOPS

A. kotschyi.
DITERPENOID: abscisic acid.
Wurzburger *et al.*, *Phytochemistry*, **15**, 225 (1976)

A. ovata.
BENZOPYRONE: aegicin.
Cooper *et al.*, *Isr. J. Chem.*, **16**, 12 (1977)

AEGINETIA

A. indica L.
POLYENES: polyene D, polyene E, polyene F.
SESQUITERPENE BENZOFURANONES: aeginetic acid, aeginetin, aeginetolide.
(P) Dighe *et al.*, *Indian J. Chem.*, **15B**, 550 (1977)
(T) Dighe *et al.*, *ibid.*, **12**, 413 (1974)
(T) Eschenmoser *et al.*, *Helv. Chim. Acta*, **65**, 353 (1982)

AEGLE (Rutaceae)

A. marmelos Correa.
AMIDE ALKALOIDS: aegeline, marmeline.
FURANOQUINOLINE ALKALOID: α-fagarine.
MONOTERPENE COUMARIN: (+)-marmin.
(A) Chatterjee, Bose, *J. Indian Chem. Soc.*, **29**, 425 (1952)
(A) Sharma *et al.*, *Phytochemistry*, **20**, 2606 (1981)
(T) Chatterjee, Bhattacharyya, *J. Chem. Soc.*, 1922 (1959)

AEGLOPSIS (Rutaceae)

A. chevalieri Swing.
PYRIDINE ALKALOID: halfordinol-3,3-dimethylallyl ether.
Dreyer, *J. Org. Chem.*, **33**, 3658 (1968)

AEGOPORDON

A. berarioides.
STEROIDS: sitosterol 3-O-glucoside, taraxasterol.
SESQUITERPENOID: 15-chloro-14-hydroxy-4,15-dihydroxycynaropicrin.
TRITERPENOID: lupeol.
Izaddoost *et al.*, *Fitoterapia*, **56**, 275 (1985)

AESCHRION

A. crenata Vell.
CARBOLINE ALKALOIDS: crenatidine, crenatine, isoparaine, paraine.
(A) Sanchez, Comin, *Phytochemistry*, **10**, 2155 (1971)
(T) Vitagliano, Comin, *ibid.*, **11**, 807 (1972)

AESCHYNOMENE (Leguminosae)

A. indica.
FLAVONOID: reynoutrin.
Morita *et al.*, *Chem. Abstr.*, 88 148947b

AESCULUS (Hippocastanaceae)

A. hippocastanum L.
ACIDS: linoleic acid, palmitic acid, stearic acid.
FLAVONES: 3,5-dihydroxy-3',4',7-trimethoxyflavone, myricetin 3',4',7-trimethyl ether.
QUINONE: plastoquinone 8.
SAPONINS: aescin, aesculin.
STEROID: phytosterol.
TRITERPENOIDS: barringtogenol, escigenin.
   (Ac, Sap, S) Wulff, Tschesche, *Tetrahedron*, **25**, 415 (1969)
   (F) Wollenweber *et al.*, *Tetrahedron Lett.*, 1601 (1970)
   (Q) Whistance *et al.*, *Phytochemistry*, **9**, 737 (1970)
   (T) Tschesche *et al.*, *Justus Liebigs Ann. Chem.*, **669**, 171 (1963)

A. indica.
SAPONIN: aesculuside A.
   Singh *et al.*, *Planta Med.*, 409 (1986)

A. parviflora.
AMINO ACID: cis-α-amino-2-carboxycyclopropane acetic acid.
   Fowden *et al.*, *Phytochemistry*, **8**, 437 (1969)

A. punduana.
ALIPHATIC: hentriacontanol.
TRITERPENOIDS: β-amyrin, barringtogenol C, barringtogenol C 21-tiglate, protoescigenin.
   Prakash, Nigam, *J. Indian Chem. Soc.*, **60**, 806 (1983)

A. turbinata Blume.
SAPONIN: japoaescinin.
TRITERPENOID: 16-deoxybarringtogenol C.
   (Sap) Matsukawa, *Chem. Zentralbl.*, *II*, 1719 (1935)
   (T) Yosioka *et al.*, *Tetrahedron Lett.*, 2577 (1967)

AFRAMOMUM (Zingiberaceae)

A. danielii.
DITERPENOID: (E)-8α,17-epoxylab-12-ene-15,16-dial.
   Kimbu *et al.*, *J. Chem. Soc., Perkin I*, 1303 (1979)

A. giganteum.
PHENOLICS: dehydrozingiberone, emodin, syringaldehyde, syringic acid.
   de Bernardi *et al.*, *Phytochemistry*, **15**, 1785 (1976)

AFZELIA

A. bella.
AMINO ACIDS: trans-4-carboxy-L-proline, trans-4-hydroxy-L-proline.
   Welter *et al.*, *Phytochemistry*, **17**, 131 (1978)

AGANOSMA

A. caryophyllata.
FLAVONOIDS: hyperin, isoquercetin, quercetin, quercetin 3-O-arabinoside, rutin.
PHENOLICS: ferulic acid, vanillic acid.
STEROID: β-sitosterol.
TRITERPENOID: ursolic acid.
   Sekhar *et al.*, *Fitoterapia*, **56**, 174 (1985)

AGAPANTHUS (Amaryllidaceae)

A. africanus L.
STEROIDS: agapanthagenin, 7-dehydroagapanthagenin, 8(14)-dehydroagapanthagenin, 9(11)-dehydroagapanthagenin, 27-hydroxyruscogenin, 7-spirostene-2,3,5-triol.
   Gonzalez *et al.*, *Phytochemistry*, **14**, 2259 (1975)

AGARICUS (Basidiomycetae)

A. amethystina.
STEROID: 22-dihydroergosterol.
   Endo *et al.*, *Chem. Abstr.*, *71499q*

A. bisporus.
HYDRAZIDE: agaritine.
   Daniels *et al.*, *J. Org. Chem.*, **23**, 3229 (1962)

A. campestris.
AMIDE: (S)-agaridoxin.
   Szent-Gyorgyi, *J. Org. Chem.*, **41**, 1603 (1976)

A. nebularis Batsch.
PURINE ALKALOID: nebularine.
   Szent-Gyorgyi, *J. Org. Chem.*, **41**, 1603 (1976)

A. placomyces.
STEROID: ergosterol.
   Endo *et al.*, *Tokyo Gakugei Daigaku Kiyo*, **27**, 124 (1975); *Chem. Abstr.*, *71499q*

A. silvaticus.
NITRO COMPOUNDS: β-nitraminoalanine, N-nitroethylenediamine.
   Chilton *et al.*, *Phytochemistry*, **14**, 2291 (1975)

A. xanthoderma.
PIGMENTS: agaricon, xanthodermin.
   Hilbig *et al.*, *Angew. Chem.*, **97**, 1063 (1985)

AGASTACHYS (Proteaceae)

A. odorata R.Br.
ALKALOID: one uncharacterized base.
   Bick *et al.*, *Aust. J. Chem.*, **32**, 2071 (1979)

AGATHIS (Araucariaceae)

A. australis.
DITERPENOIDS: agathalic acid, agatharesinol, agathic acid, agathic acid 19-methyl ester, araucarenolone, araucarol, araucarolone, araucarone, cis-communic acid, (-)-kaurene, isokaurene, isopimara-7,15-dien-3β-ol, sclarene.
   Enzell, Thomas, *Tetrahedron Lett.*, 2795 (1966)
   Carman, Marty, *Aust. J. Chem.*, **21**, 1923 (1968)

A. lanceolata.
NORDITERPENOID: noranticopalic acid.
   Dokhac *et al.*, *Phytochemistry*, **18**, 1839 (1979)

A. microstachya.
DITERPENOIDS: agathic acid, agathic acid methyl ester.
   Ruzicka, Hosking, *Helv. Chim. Acta.*, **13**, 1402 (1930)

A. robusta.
DITERPENOIDS: agathalic acid, trans-communic acid, 13β-hydroxylabda-7,14-dien-19-oic acid, 13β-hydroxylabda-8,14-dien-19-oic acid, 13β-hydroxylabda-8(17),14-dien-19-oic acid.
FLAVONE: robustaflavone.
   (F) Lin *et al.*, *Phytochemistry*, **13**, 1617 (1974)
   (T) Carman *et al.*, *Aust. J. Chem.*, **26**, 209 (1973)

AGATHOSMA

A. puberula.
COUMARIN: puberulin.
   Bohlmann *et al.*, *Chem. Ber.*, **108**, 2955 (1975)

AGAURIA

A. salicifolia.
TRITERPENOID: agaurilol.
   Dussy, Sosa, *Bull. Soc. Chim. Biol.*, **33**, 1672 (1951)

AGAVE (Agavaceae)

A. americana L.
SAPONINS: agavesaponins C, C', D, E and H, agavosides A, B and C.
STEROID: hecogenin.
DITERPENOID: piscidic acid.
(S) Marker et al., J. Am. Chem. Soc., **65**, 1199 (1943)
(Sap) Kintya et al., Khim. Prir. Soedin., 751 (1975)
(T) Nordal et al., Acta Chem. Scand., **20**, 1431 (1966)

A. asperriima Jacobi.
STEROID: hecogenin.
Marker et al., J. Am. Chem. Soc., **65**, 1199 (1943)

A. atrovirens Otto.
STEROID: manogenin.
Marker et al., J. Am. Chem. Soc., **65**, 1199 (1943)

A. attenuata Baker.
STEROID: sarsasapogenin.
Marker et al., J. Am. Chem. Soc., **65**, 1199 (1943)

A. bracteosa Wats.
STEROID: manogenin.
Marker et al., J. Am. Chem. Soc., **65**, 1199 (1943)

A. cantala Roxb.
SAPONINS: cantalanins A and B, cantalasaponin-2, cantalasaponin-4.
Pant et al., Phytochemistry, **25**, 1491 (1986)

A. chisoensis.
STEROID: manogenin.
Marker et al., J. Am. Chem. Soc., **65**, 1199 (1943)

A. crassipina Trel.
STEROID: manogenin.
Marker et al., J. Am. Chem. Soc., **65**, 1199 (1943)

A. crysantha Peebles.
STEROID: hecogenin.
Marker et al., J. Am. Chem. Soc., **65**, 1199 (1943)

A. deserti Engelm.
STEROID: hecogenin.
Marker et al., J. Am. Chem. Soc., **65**, 1199 (1943)

A. endlichiana Trel.
STEROID: hecogenin.
Marker et al., J. Am. Chem. Soc., **65**, 1199 (1943)

A. expansa.
STEROID: hecogenin.
Marker et al., J. Am. Chem. Soc., **65**, 1199 (1943)

A. ferox Wats.
STEROID: manogenin.
Marker et al., J. Am. Chem. Soc., **65**, 1199 (1943)

A. fourcroydes Baker.
STEROID: hecogenin.
Marker et al., J. Am. Chem. Soc., **65**, 1199 (1943)

A. funkiana Koch-Bouche.
STEROID: smilagenin.
Marker et al., J. Am. Chem. Soc., **65**, 1199 (1943)

A. gracilipes Trel.
STEROIDS: gitogenin, hecogenin, rockogenin.
Marker et al., J. Am. Chem. Soc., **69**, 2167 (1947)

A. havardiana Trel.
STEROID: manogenin.
Marker et al., J. Am. Chem. Soc., **65**, 1199 (1943)

A. heterocantha Zucc.
STEROID: smilagenin.
Marker et al., J. Am. Chem. Soc., **65**, 1199 (1943)

A. huachucensis Baker.
SAPONIN: agavosaponin.
STEROIDS: agavogenin, gitogenin, manogenin.
Marker et al., J. Am. Chem. Soc., **69**, 2167 (1947)

A. lechequilla Torr.
STEROIDS: gitogenin, ruizgenin, smilagenin.
Marker et al., J. Am. Chem. Soc., **69**, 2167 (1947)

A. lehmannii.
STEROID: manogenin.
Marker et al., J. Am. Chem. Soc., **65**, 1199 (1943)

A. lophantha.
STEROIDS: manogenin, smilagenin, tigogenin.
Marker et al., J. Am. Chem. Soc., **65**, 1199 (1943)

A. melliflua Trel.
STEROID: β-sitosterol.
Marker et al., J. Am. Chem. Soc., **65**, 1199 (1943)

A. mescal Koch.
STEROID: gitogenin.
Marker et al., J. Am. Chem. Soc., **65**, 1199 (1943)

A. mirabilis Trel.
STEROID: manogenin.
Marker et al., J. Am. Chem. Soc., **65**, 1199 (1943)

A. mitraeformis Trel.
STEROID: manogenin.
Marker et al., J. Am. Chem. Soc., **65**, 1199 (1943)

A. murpheyi Gibson.
STEROID: hecogenin.
Marker et al., J. Am. Chem. Soc., **65**, 1199 (1943)

A. palmaris Trel.
STEROID: β-sitosterol.
Marker et al., J. Am. Chem. Soc., **65**, 1199 (1943)

A. palmeri Engelm.
STEROID: hecogenin.
Marker et al., J. Am. Chem. Soc., **65**, 1199 (1943)

A. parassanas Trel.
STEROID: manogenin.
Marker et al., J. Am. Chem. Soc., **65**, 1199 (1943)

A. parviflora Torr.
STEROID: hecogenin.
Marker et al., J. Am. Chem. Soc., **65**, 1199 (1943)

A. quoitefera Trel.
STEROID: manogenin.
Marker et al., J. Am. Chem. Soc., **65**, 1199 (1943)

A. roezliana Baker.
STEROIDS: mexogenin, neomexogenin, neosamogenin, sarsasapogenin.
Marker, Lopez, J. Am. Chem. Soc., **69**, **2373**, 2383 (1947)

A. salmiana Otto.
STEROID: manogenin.
Marker et al., J. Am. Chem. Soc., **65**, 1199 (1943)

A. scabra.
STEROID: manogenin.
Marker et al., J. Am. Chem. Soc., **65**, 1199 (1943)

A. shawii Engelm.
STEROID: hecogenin.
Marker et al., J. Am. Chem. Soc., **65**, 1199 (1943)

A. sisalana Perrine.
STEROIDS: barbourgenin, hainangenin, hecogenin, tigogenin.
Blunden et al., J. Nat. Prod., **49**, 687 (1986)

A. striata Zucc.
STEROIDS: manogenin, neomanogenin.
Marker, Lopez, J. Am. Chem. Soc., **69**, 2375 (1947)

A. stricta subsp. glauca.
STEROID: gitogenin.
Marker et al., J. Am. Chem. Soc., **65**, 1199 (1943)

A. stricta subsp. nana.
STEROID: gitogenin.
Marker et al., J. Am. Chem. Soc., **65**, 1199 (1943)

A. stricta subsp. purpurea.
STEROID: tigogenin.
Marker et al., J. Am. Chem. Soc., **65**, 1199 (1943)

A. stricta subsp.
ROSEA, STEROID: gitogenin.
Marker et al., J. Am. Chem. Soc., **65**, 1199 (1943)

A. toumeyana Trel.
STEROID: hecogenin.
Marker et al., J. Am. Chem. Soc., **69**, 2167, 2375 et seq. (1947)

A. utahensis Trel.
STEROID: manogenin.
Marker et al., J. Am. Chem. Soc., **65**, 1199 (1943)

AGENATUM

A. conyzoides.
BENZOPYRAN: conyzorigin.
Adesogan et al., J. Chem. Soc., Chem. Commun., 152 (1978)

AGERATINA (Compositae/Eupatorieae)

A. aschenbornia.
CHROMENE: cepatoriochromene.
SESQUITERPENOID: ageroborniol.
Bohlmann, Fiedler, Phytochemistry, **17**, 566 (1978)

A. adenophora.
SESQUITERPENOID: 7,11-dehydro-12-hydroxy-8-oxoageraphone.
Bohlmann et al., Phytochemistry, **20**, 1432 (1981)

A. dendroides.
CHROMENES:
6-[1-(isobutyryloxy)ethyl]-7-methoxy-2,2-dimethylchromene,
7-methoxy-2,2,-dimethyl-6-[1-(2-methacrylacyloxy)ethyl]-chromene.
DITERPENOIDS: epi-9-angeloyloxy-10,11-dehydrobrickelliol,
dendrodinic acid, hebeclinolide,
7-hydroxy-7-hydro-6-dehydro-3,4(Z)-α-farnesene.
Bohlmann, Grenz, Chem. Ber., **110**, 1321 (1977)

A. espinosera (Gray) King et Rob.
MONOTERPENOID: 10-acetoxy-8,9-epoxy-6-methoxythymol
isobutyrate, 7-acetoxythymohydroquinone dimethyl ether,
6-hydroxythymohydroquinone dimethyl ether,
7-hydroxythymohydroquinone dimethyl ether.
Bohlmann et al., Chem. Ber., **110**, 301 (1977)

A. exertovenosa (Klatt) King et Rob.
CHROMENES:
6-[1-(isobutyryloxy)ethyl]-7-methoxy-2,2-dimethylchromene,
7-methoxy-2,2-dimethyl-6-[1-(2-methylacrylacyloxy)ethyl]-chromene.
DITERPENOIDS: epi-9-angeloyloxy-10,11-dehydrobrickelliol,
dendroidinic acid, hebeclinolide,
7-hydroxy-7-hydro-6-dehydro-3,4(Z)-α-farnesene.
Bohlmann, Grenz, Chem. Ber., **110**, 1321 (1977)

A. glabrata.
MONOTERPENOIDS: 10-acetoxy-8,9-epoxy-6-methoxythymol
isobutyrate, 10-benzoyloxy-8,9-epoxy-6-hydroxythymol
isobutyrate, 10-benzoyloxy-8,9-epoxy-6-methoxythymol
isobutyrate, 10-benzoyloxy-8,9-epoxythymol isobutyrate,
10-cinnamoyloxy-8,9-epoxy-6β-hydroxythymolnisobutyrate,
10-isobutyryloxy-8,9-dehydrothymol isobutyrate, isocostic
acid, 10-(2-methylbutyryloxy)-8,9-dehydrothymol

isobutyrate, 3-oxoisocostic acid.
Bohlmann et al., Chem. Ber., **110**, 301 (1977)

A. petiolare.
MONOTERPENOID: 10-acetoxy-8,9-epoxy-6-methoxythymol
isobutyrate.
SESQUITERPENOIDS: 3-acetoxy-4,7(11)-muuroladien-8-one,
2-acetoxy-3,4,6,11-tetrahydrocadinan-7-one, petiolaride.
Bohlmann et al., Chem. Ber., **110**, 301 (1977)

A. pichinchensis.
CHROMENES:
6-[1-(isobutyryloxy)ethyl]-7-methoxy-2,2-dimethylchromene,
7-methoxy-2,2,-dimethyl-6-[1-(2-methacrylacyloxy)ethyl]-chromene.
DITERPENOIDS: epi-9-angeloyloxy-10,11-dehydrobrickelliol,
dendroidinic acid, hebeclinolide,
7-hydroxy-7-hydro-6-dehydro-3,4(Z)-α-farnesene.
Bohlmann, Grenz., Chem. Ber., **110**, 1321 (1977)

A. riparia (Regel) King et Rob.
CHROMENES: ageratoriparin, 13-feruloyloxyeupatoriochromene,
13-feruloyloxyvanillochromene,
13-hydroxyacetovanillochromene,
6-(1-hydroxyethyl)-eupatoriochromene B,
6-hydroxyeupatoriochromene B,
13-hydroxymethylripariochromene,
13-hydroxyripariochromene.
Banerjee et al., Phytochemistry, **24**, 2688 (1985)

A. scorodinioides.
CHROMENE: acetovanillochromene.
Bohlmann et al., Chem. Ber., **110**, 301 (1977)

A. tomentella.
FLAVONOIDS: 6-methoxyapigenin, 6-methoxylutein,
6-methoxylutein 7,3'-dimethyl ether, 6-methoxylutein
7,4'-dimethyl ether, 6-methoxylutein 3'-methyl ether,
6-methoxylutein 7,3',4'-trimethyl ether.
Fang et al., J. Nat. Prod., **49**, 737 (1986)

AGERATUM (Compositae)

A. conyzoides.
CHROMENES: ageratochromene, demethoxyageratochromene.
Pham Truong Thi Tho et al., Tap Chi Hoat Dong Khoa Hoc, **14**, 29 (1976); Chem. Abstr., **86** 2364w

A. fastigiatum.
SESQUITERPENOID: fastigiolide.
Bohlmann et al., Phytochemistry, **22**, 983 (1983)

A. houstonianum.
BENZOFURAN: ageratone.
Anthonsen et al., Acta Chem. Scand., **24**, 721 (1970)

A. mexicanum Sims.
FLAVONOIDS: kaempferol, kaempferol 3,7-diglucoside,
kaempferol 3-(O-rhamnosylglucoside), quercetin, quercetin
3,7-diglucoside, quercetin 3-(O-rhamnosylglucoside).
Mionskowski et al., Acta Pol. Pharm., **32**, 633 (1975); Chem. Abstr., **85** 189221s

AGLAIA (Meliaceae)

A. odorata Lour.
AMIDE ALKALOIDS: odorine, odorinol.
TRITERPENOIDS: aglaiol, aglaiondiol, (24R)-aglaitriol,
(24Z)-aglaitriol.
(A) Shiengthong et al., Tetrahedron Lett., 2249 (1979)
(T) Shiengthong et al., Tetrahedron., **30**, 2211 (1974)

A. roxburghiana.

TRITERPENOIDS: roxburghiadiols A and B.

Purushothaman *et al.*, *Indian Drugs*, **23**, 260 (1986); *Chem. Abstr.*, 105 75902r

AGONIS

A. abnormis.

SESQUITERPENOID: α-humulene.

Chapman, *J. Chem. Soc.*, 785 (1928)

AGRIANTHUS

A. pungens.

SESQUITERPENOID: agriantholide.

Bohlmann *et al.*, *Phytochemistry*, **19**, 1873 (1980)

AGRIMONIA (Rosaceae)

A. pilosa.

ISOCOUMARIN: agrimonolide.

Yamato *et al.*, *J. Pharm. Soc., Japan*, **76**, 1086 (1958)

AGROSTEMMA (Caryophyllaceae)

A. githago L.

SAPONINS: githagenin, githagin, githagoside.

Wedekind, Schicke, *Z. Physiol. Chem.*, **190**, 1 (1930)

AILANTHUS (Simaroubaceae)

A. altissima (Mill.) Swingle.

CARBOLINE ALKALOIDS: canthin-6-one, 1-hydroxycanthin-6-one, 1-(2'-hydroxyethyl)-4-methoxy-β-carboline, methoxycanthin-6-one, methoxycanthin-6-one-3-N-oxide.

COUMARIN: scopoletin.

STEROID: β-sitosterol.

DITERPENOIDS: shinjulactones A, B, C, D, E, F, G, H, I, J and K.

(A) Varga *et al.*, *Fitoterapia*, **52**, 183 (1981)

(C, S) Szendrei *et al.*, *Herba Hung.*, **16**, 15 (1977); *Chem. Abstr.*, 88 166719h

(T) Naora *et al.*, *Bull. Chem. Soc., Japan*, **56**, 3694 (1983)

(T) Furuno *et al.*, *ibid.*, **57**, 2484 (1984)

A. excelsa.

CARBOLINE ALKALOID: 8-hydroxycanthin-6-one.

UNCHARACTERIZED ALKALOID: malanthine.

DITERPENOIDS: 13,18-dehydroexcelsin, 13,18-dehydroglaucarubol 15-isovalerate, excelsin.

TRITERPENOID: (3S,24S,25-trihydroxytirucall-7-ene.

(A) Cordell *et al.*, *Lloydia*, **41**, 166 (1978)

(T) Khan *et al.*, *Phytochemistry*, **19**, 2484 (1980)

A. glandulosa.

DITERPENOID: amaralide.

Stocklin *et al.*, *Tetrahedron Lett.*, 2399 (1970)

A. malabarica DC.

TRITERPENOIDS: ailanthol, epoxymalabaricol, malabaricanediol, malabaricol.

Joshi *et al.*, *Tetrahedron Lett.*, 1273 (1985)

AINSLEAEA

A. aciflora.

SESQUITERPENOID: taraxinic acid.

Jiu, *J. Pharm. Soc., Japan*, **102**, 911 (1982)

AJANIA

A. fastigiata.

FLAVONOID: apigenin 6,8-diglucopyranoside.

SESQUITERPENOIDS: ajadin, ajafin, ajanin, ajaphyllin, chlorfastin, isochrysartemin B, rupicolin B, rupicolin B oxide.

(F) Chumbalov *et al.*, *Khim. Prir. Soedin.*, 661 (1976)

(T) Yusupov *et al.*, *ibid.*, 579 (1979)

(T) Yusupov *et al.*, *ibid.*, 390 (1983)

A. fruticulosa.

FLAVONOIDS: artemiselin, axillarin, 5,3',4'-trihydroxy-3,6,7-trimethoxyflavone.

Belenovskaya *et al.*, *Khim. Prir. Soedin.*, 575 (1977)

AJUGA (Labiatae)

A. chamaepitys.

DITERPENOID: ajugapitin.

Hernandez *et al.*, *Phytochemistry*, **21**, 2909 (1982)

A. decumbens Thunb.

STEROID: ajugalactone.

Imai *et al.*, *Japanese Patent*, **08**, 640 (1973)

A. incisa.

STEROIDS: ajugasterones A, B and C.

Imai *et al.*, *J. Chem. Soc., Chem. Commun.*, **82**, 546 (1969)

A. iva.

STEROIDS: ivains I, II, III and IV.

Campsel *et al.*, *Chem. Lett.*, 1053 (1982)

A. nipponensis.

POLYSACCHARIDE: ajugose.

DITERPENOIDS: ajugamarin, dihydroajugamarin.

(Sacc) Herissey *et al.*, *C. R. Acad. Sci.*, **259**, 824 (1954)

(T) Shimomura *et al.*, *Chem. Pharm. Bull.*, **31**, 2192 (1983)

A. remota.

DITERPENOIDS: 6-acetoxyajugarin II, ajugarins I, II, III, IV and V.

Kubo *et al.*, *J. Chem. Soc., Chem. Commun.*, 618 (1982)

Kubo *et al.*, *Chem. Lett.*, 223 (1983)

A. reptans.

IRIDOIDS: ajugol, ajugoside, harpagide, harpagide acetate.

STEROIDS: ajugalactone, ajugasterone, β-ecdysone, 29-norcyasterone, 29-norsengosterone, polypodine B.

DITERPENOIDS: ajugareptansin, ajugareptansone A, ajugareptansone B, reptioside.

(I, T) Camps *et al.*, *Chem. Lett.*, 1093 (1981)

(S) Camps *et al.*, *ibid.*, 1313 (1982)

A. spectabilis Nakai.

IRIDOID: jaranidoside.

Chung *et al.*, *Saenyak Hakhoe Chi*, **11**, 15 (1980)

A. turkestanica.

STEROIDS: ajugalactone, cyasterone, ecdysterone, turkesterone.

Usmanov *et al.*, *Khim. Prir. Soedin.*, 710 (1977)

AKANIA

A. lucens.

TRITERPENOID: betulonic acid.

Robinson *et al.*, *Phytochemistry*, **9**, 907 (1970)

AKEBIA (Lardizabalaceae)

A. lagerocemora.

TRITERPENOID: 3-O-acetyloleanolic acid.

Braga *et al.*, *An. Acad. Bras. Cienc.*, **39**, 249 (1967)

A. quinata (Houtt.) Decne.

SAPONINS: akeboasides St-a, St-b, St-c, St-d, St-e, St-f, St-g1, St-g2, St-h, St-j, St-k and St-r.

NORTRITERPENOID: norarjunolic acid.

TRITERPENOID: 3-0-acetyloleanolic acid.

(Sap) Fujita *et al.*, *J. Pharm. Soc., Japan*, **94**, 194 (1974)

(T) Higuchi *et al.*, *Chem. Pharm. Bull.*, **24**, 1314 (1978)

A. trifoliata.

ANTHOCYANINS: chrysanthemin, cyanidin 3-p-coumarylxylosylglucoside, cyanidin 3-xylosylglucoside.

TRITERPENOID: 3-O-acetyloleanolic acid.

(An) Ishikura *et al.*, *Phytochemistry*, **15**, 442 (1976)

(T) Braga *et al.*, *An. Acad. Bras. Cienc.*, **39**, 249 (1967)

ALAFIA

A. multiflora.

PYRROLIZIDINE ALKALOID: alafine.

Paris *et al.*, *Ann. Pharm. Fr.*, **29**, 57 (1971)

ALANGIUM (Alangiaceae)

A. chinense.

PYRIDINE ALKALOID: anabasine.

Guo *et al.*, *Yaoxue Tongbao*, **17**, 390 (1982); *Chem. Abstr.*, 97 195755h

A. chinense var. pauciflorum.

PYRIDINE ALKALOID: anabasine.

Guo *et al.*, *Yaoxue Tongbao*, **17**, 390 (1982); *Chem. Abstr.*, 97 195755h

A. lamarckii Thw.

BENZOQUINOLIZIDINE ALKALOIDS: alancine, alangicine, alangide, alangimarckine, demethylcephaeline, demethylpsychotrine, demethyltubulosine, isotuberosine, marckiine.

PROTOBERBERINE ALKALOID: bharatamine.

QUINOLINE ALKALOID: ankorine.

UNCHARACTERIZED ALKALOIDS: alangine, alangine A, alangine B.

UNCHARACTERIZED ALKALOIDS: lamarchinine, marckidine.

STEROID: 24R-stigmasta-5,22,25-trien-3β-ol.

TRITERPENOIDS: alangidiol, isoalangidiol.

(A) Pakrashi *et al.*, *Experientia*, 1081 (1970)

(A) Chattopadhyay *et al.*, *Heterocycles*, **22**, 1965 (1984)

(S) Pakrashi, Achari, *Tetrahedron Lett.*, 365 (1971)

(T) Pakrashi, Ali, *ibid.*, 2143 (1967)

(T) Kapil *et al.*, *J. Chem. Soc., Chem. Commun.*, 904 (1971)

A. salviifolium.

PYRIDINE ALKALOID: anabasine.

Guo *et al.*, *Yaoxue Tongbao*, **17**, 390 (1982); *Chem. Abstr.*, 97 195755h

A. villosum.

NORTRITERPENOIDS: deoxyemmolactone, isodeoxyemmolactone.

Burbage, Jewers, *J. Chem. Soc., D*, 814 (1970)

ALATUS

A. striatus.

SESQUITERPENOID: alatol

Sigiura *et al.*, *Tetrahedron*, **35**, 3465 (1982)

ALBATRELLUS (Basidiomycetae)

A. ovinus.

POLYPRENOIDS: grifolin, neogrifolin, scutigeral, E,E-5-methyl-2-(3,7,11-trimethyl-2,6,10-dodecatrienyl)-1,3-dimethoxybenzene, E,E-5-methyl-4-(3,7,11-trimethyl-2,6,10-dodecatrienyl)-1,3-dimethoxybenzene, E,E-5-methyl-2-(3,7,11-trimethyl-2,6,10-dodecatrienyl)-1-hydroxy-3-methoxybenzene.

Besl *et al.*, *Chem. Ber.*, **110**, 3770 (1977)

Vrkoc *et al.*, *Phytochemistry*, **16**, 1409 (1977)

A. subrubrescens.

POLYPRENOIDS: neogrifolin, scutigeral.

Besl *et al.*, *Chem. Ber.*, **110**, 3770 (1977)

ALBIZIA (ALBIZZIA) (Leguminosae)

A. amara.

PHENOLICS: melacacidin, (-)-2,3-cis-3,4-cis-3-O-methylmelacacidin, 3′-O-methylmelanoxetin.

Deshpande *et al.*, *Indian J. Chem.*, **15B**, 201 (1977)

A. anthelmintica.

SAPONINS: deglucomusennin, musennin.

TRITERPENOID: acacic acid.

(Sap) Tschesche, Kammerer, *Justus Liebigs Ann. Chem.*, **724**, 183 (1969)

(T) Varshney, Shamsuddin, *Tetrahedron Lett.*, 2055 (1964)

A. julibrissin (Willd.) Durazz.

STEROID: albiside.

TRITERPENOID: acacic acid.

(S) Sergienko *et al.*, *Khim. Prir. Soedin.*, 708 (1977)

(T) Varshney, Shamsuddin, *Tetrahedron Lett.*, 2055 (1964)

A. lanata.

Four flavonoid glycosides.

Zadorozhnii *et al.*, *Khim. Farm. Zh.*, **20**, 855 (1986); *Chem. Abstr.*, 107 151179u

A. lebbek.

FLAVONOID: vicenin II.

PHENOLICS: melacacidin, (-)-2,3-cis-3,4-cis-3-O-methylmelacacidin, melanoxetin, 3′-O-methylmelanoxetin.

SAPONIN: lebbekanin A.

TRITERPENOIDS: acacic acid, albigenic acid, albigenin.

(F) Morita *et al.*, *Chem. Abstr.*, 88 148947b

(P) Deshpande *et al.*, *Indian J. Chem.*, **15B**, 201 (1977)

(Sap) Varshney *et al.*, *ibid.*, **11**, 1094 (1973)

(T) Barua, Raman, *Tetrahedron*, **18**, 155 (1962)

(T) Varshney, Shamsuddin, *Tetrahedron Lett.*, 2055 (1964)

A. procera.

ISOFLAVONES: biochanin A, daidzein, formononetin, genistein.

SAPONIN: proceranin A.

TRITERPENOID: proceragenin A.

(I) Deshpande *et al.*, *Indian J. Chem.*, **15B**, 201 (1977)

(S, T) Varshney *et al.*, *Arch. Pharm.*, **305**, 280 (1972)

ALCHORNEA (Euphorbiaceae)

A. cordifolia.

ACID: (+)-cis-14,15-epoxy-cis-11-eicosenoic acid.

Kleinan *et al.*, *Lipids*, **12**, 610 (1977)

A. floribunda.

IMIDAZOLE ALKALOID: alchorneinone.

IMIDAZOPYRIMIDINE ALKALOIDS: alchorneine, isoalchorneine.

Khuong-Huu-Laine *et al.*, *Tetrahedron*, **28**, 5207 (1972)

A. hirtella Benth.

IMIDAZOPYRIMIDINE ALKALOID: isoalchorneine.

Khuong-Huu-Laine *et al.*, *Tetrahedron*, **28**, 5207 (1972)

A. javanensis (Bl.) Muell.-Arg.

IMIDAZOPYRIMIDINE ALKALOIDS: alchornidine, alchornine.

Hart *et al.*, *J. Chem. Soc., Chem. Commun.*, 1484 (1969)

ALEANTARA
A. ekmaniana.
SESQUITERPENOID: epi-ilicic acid.
Bohlmann et al., *Phytochemistry*, **21**, 456 (1982)

ALECTORIA (Usneaceae)
A. fremontii.
STEROID: fremontol.
Solberg, *Acta Chem. Scand.*, **29B** 145 (1975)
A. nigricans.
ACID: alectorialic acid.
Solberg, *Z. Naturforsch.*, **22B**, 777 (1967)
A. ochroleuca (Hoffm.) Massal.
ACID: barbatic acid.
Robertson et al., *J. Chem. Soc.*, 1675 (1932)

ALEURIA (Ascomycetae)
A. aurantica.
CAROTENOID: aleuriaxanthin.
Kjosen, Liaaen-Jensen, *Acta Chem. Scand.*, **27**, 2495 (1973)

ALEURITES
A. fordii Hemsl.
DITERPENOIDS: 13-acetyl-16-hydroxyphorbol,
16-hydroxyphorbol,
12-O-palmitoyl-4-deoxy-16-hydroxylumiphorbol-13-acetate,
12-O-palmitoyl-4-deoxy-16-hydroxyphorbol-13-acetate,
12-O-palmitoyl-16-hydroxyphorbol-13-acetate,
1-O-palmitoyl-4,16-dihydroxyphorbol-13-acetate.
STEROIDS: campesterol, β-sitosterol, stigmasterol.
(T) Okuda et al., *Chem. Pharm. Bull.*, **22**, 971 (1974)
(T) Hiroto, Kushimizu, *Agr. Biol. Chem.*, **43**, 2523 (1979)
(S) Okuda et al., *Phytochemistry*, **14**, 2513 (1975)
A. montana.
DITERPENOIDS: 13-acetyl-16-hydroxyphorbol, aleuritolic acid.
Misra, Khastgir, *J. Indian Chem. Soc.*, **46**, 1063 (1969)

ALHAGI
A. kirghisorum.
FLAVONOIDS: isorhamnetin 3-β-D-galactopyranoside,
isorhamnetin 3-O-galactopyranosyl-6-rhamnopyranoside,
isorhamnetin 3-β-D-glucopyranoside, isorhamnetin
3-β-D-glucopyranoside-2-β-D-glucofuranoside, isorhamnetin
3-α-L-rhamnofuranoside-6-β-D-glucopyranoside.
Burasheva et al., *Khim. Prir. Soedin.*, 280 (1977)

ALISMA (Alismataceae)
A. plantago-aquatica L.
SESQUITERPENOIDS: alismol, alismoxide.
Oshima et al., *Phytochemistry*, **22**, 183 (1983)
A. plantago-aquatica var. orientalis.
TRITERPENOIDS: acetylalisol A, acetylalisol B, alisol A,
epi-alisol A, alisol B, alisol C, alisol C 23-acetate.
Myrata et al., *Tetrahedron Lett.*, 103 (1968)
A. rhizoma.
TRITERPENOIDS: alisol C, alisol C 23-acetate.
Murata et al., *Chem. Pharm. Bull.*, *18*, **1347**, 1369 (1970)

ALKANNA
A. hirsutissima.
PIGMENT: arnebin VII.
Dhar et al., *Indian J. Chem.*, **11**, 528 (1973)
A. orientalis.
FLAVONES: kaempferol, kaempferol-3,6-dimethylether,
kaempferol 3-glucoside, kaempferol 7-glucoside,
kaempferol 3-rutinoside, quercetin, quercetin 3-glucoside,
quercetin 3-rutinoside.
Mansour et al., *J. Nat. Prod.*, **49**, 355 (1981)

ALLAMANDA (Apocynaceae)
A. cathartica L.
SESQUITERPENOIDS: allamandicin, allamandin, allamdin.
Kupchan et al., *J. Org. Chem.*, **39**, 2477 (1974)
A. neriifolia.
IRIDOID: protoplumericin.
Yamauchi et al., *Chem. Pharm. Bull.*, **29**, 3051 (1981)

ALLANBLACKIA (Guttiferae/Clusiodaea)
A. floribunda Oliv.
XANTHONES:
4',5'-dihydro-1,6,7-trihydroxy-4',4',5'-
trimethylfuranoxanthone, 1,5-dihydroxyxanthone,
1,7-dihydroxyxanthone,
1,3,7-trihydroxy-2-(1,1,-dimethylprop-2-enyl)-xanthone,
1,3,7-trihydroxy-2-(3-methylbut-2-enyl)-xanthone.
Bandaranayake et al., *Phytochemistry*, **14**, 265, 1878 (1975)

ALLAUDIOPSIS
A. marnieriana.
FLAVONOID:
5,7,3'-trihydroxy-6,8-di-C-methyl-4',5'-dimethoxyflavanone.
Rabesa et al., *Phytochemistry*, **22**, 2092 (1983)

ALLIUM (Liliaceae)
A. albanum.
STEROID: tigogenin.
Ismailov et al., *Khim. Prir. Soedin.*, 550 (1976)
A. cepa.
AMINO ACIDS: τ-L-glutamy-S-(2-carboxypropyl)-cysteinglycine,
τ-glutamylisoleucine, τ-glutamyl-S-methylcysteine,
τ-glutamylvaline, τ-leucine, τ-methionine, τ-phenylalanine.
SAPONINS: alliofuroside A, alliospiroside A.
(Am) Virtanen et al., *Z. Physiol. Chem.*, **322**, 8 (1960)
(Sap) Kracets et al., *Khim. Prir. Soedin.*, 188 (1986)
A. cepa var. agrogatum.
SULPHIDES: dimethyltrisulphide, d-(1-propyl)-disulphide,
methylpropenyldisulphide, methylpropenyltrisulphide,
methylpropyltrisulphide.
Jiang et al., *Chem. Abstr.*, 105 206275p
A. erubescens.
STEROIDS: eruboside A, eruboside B.
Chincharadze et al., *Khim. Prir. Soedin.*, 509 (1979)
A. giganteum.
STEROIDS: aginoside, gantogenin, neoagigenin, neoascigenin.
Kel'ginbaev et al., *Khim. Prir. Soedin.*, 480 (1976)
A. karavatiense.
STEROIDS: karavatigenin A, karavatigenin B, karavatigenin B
3-O-β-D-glucopyranoside, karavatosides A, B, C, D and E.
Kristulas et al., *Khim. Prir. Soedin.*, 530 (1974)
Vollerner et al., *ibid.*, 69 (1984)

A. narcissiflorum.
STEROIDS: alliumosides A, B, C, D and E.
Krokhmalyuk, Kintya, *Khim. Prir. Soedin.*, 184 (1976)

A. nerinifolium Bak.
SULPHIDES: dimethyltrisulphide, di-(1-propyl)-disulphide,
methylpropenyldisulphide, methylpropenyltrisulphide,
methylpropyltrisulphide.
Jiang *et al.*, *Chem. Abstr.*, 105 206275p

A. rubellum.
STEROID: tigogenin.
Ismailov *et al.*, *Khim. Prir. Soedin.*, 550 (1976)

A. sativum.
AMINO ACIDS: τ-glutamyl-S-allylcysteine,
τ-glutamyl-S-(2-carboxy-1-propyl)-cysteinglycine.
STEROIDS: alliin, allicin.
(Am) Matikkala *et al.*, *Acta Chem. Scand.*, 16, 261 (1962)
(S) Cavallito *et al.*, *J. Am. Chem. Soc.*, 66, 1950 (1944)

A. schoenoprasum.
ALIPHATICS: dotriacontanol, octacosanol, triacontanol.
AMINO ACID: τ-glutamyl-S-allylcysteine.
(Al) Isono *et al.*, *J. Pharm. Soc., Japan*, 96, 86 (1976)
(Am) Matikkala *et al.*, *Acta Chem. Scand.*, 16, 2461 (1962)

A. turcomanicum.
SAPONINS: turosides A, B and C, turoside C 6-O-benzoate.
STEROIDS: alliogenin, neoagigenin, neoagigenin 6-O-benzoate,
neoagigenone, neoalliogenin, neoascigenin 6-O-benzoate,
yuccagenin.
(Sap) Pirtskhaleva *et al.*, *Khim. Prir. Soedin.*, 514 (1979)
(S) Pirtskhaleva *et al.*, *ibid.*, 534 (1977)

ALMEIDEA
A. guyanensis.
FURANOQUINOLINE ALKALOID: almeine.
Moulis *et al.*, *Phytochemistry*, 22, 2095 (1983)

ALNUS (Betulaceae)
A. glutinosa (L.) Gaertn.
STEROIDS: brassinolide, castasterone.
TRITERPENOIDS: alnusenone, taraxerol, taraxerone.
(S) Plattner *et al.*, *J. Nat. Prod.*, 49, 540 (1986)
(T) Chapon, David, *Bull. Soc. Chim.*, 333 (1953)

A. hirsuta.
PHENOLICS: hannokinin, hannokinol.
Nomura *et al.*, *Phytochemistry*, 20, 1097 (1981)

A. incana (L.) Moench.
TRITERPENOIDS: alnincanone, taraxerol, taraxerone.
Ryabinin *et al.*, *J. Gen. Chem. USSR*, 32, 2056 (1962)

A. pendula.
FLAVONE: alnustinol.
TRITERPENOID: 12-deoxyalnustic acid.
(F) Asakawa *et al.*, *Bull. Chem. Soc., Japan*, 44, 2761 (1971)
(T) Suga *et al.*, *Phytochemistry*, 25, 1243 (1986)

A. serrulatoides (Ait.) Willd.
TRITERPENOIDS: alnuselide, alnuseric acid, alnuserol,
alnuserrudiolone, alnuserrutriol,
(20S,24R)-20,24-epoxy-11α-hydroxydammaran-3-one
acetate,
(20S,24R0-20,24-epoxy-11α-hydroxy-24-methyldammaran-
3-one.
Hirata *et al.*, *Chem. Lett.*, 711 (1977)
Hirata *et al.*, *Bull. Chem. Soc., Japan*, 55, 369 (1982)

A. sieboldiana.
FLAVONE: strobopinin.
HEPTANOIDS: yashabushidiol A, yashabushidiol B,
yashabushiketodiol A, yashibushiketodiol B,
yashabushitriol.
TRITERPENOID: alnustic acid.
(F) Asakawa *et al.*, *Bull. Chem. Soc., Japan*, 44, 2761 (1971)
(H) Hashimoto *et al.*, *Chem. Pharm. Bull.*, 34, 1846 (1986)
(T) Suga *et al.*, *Bull. Chem. Soc., Japan*, 52, 1698 (1979)

ALOE (Liliaceae)
A. arborescens.
CAROTENOIDS: β-carotene, cryptoxanthin. lutein violaxanthin,
β-zeaxanthin.
Kudritskaya *et al.*, *Khim. Prir. Soedin.*, 573 (1985)

A. barbadensis.
ACID: aloetinic acid.
Hoffenberg *et al.*, *Seifen Oele Fette Wachse*, 105, 499 (1979)

A. saponaria.
ANTHRACENES: aloesaponols I, II, III and IV, aloesaponol IV
8-O-β-D-glucoside.
ANTHRAQUINONES: asphodalin, deoxyerytholaccin,
3,6,8-trihydroxy-1-methylanthraquinone 2-carboxylic acid
methyl ester.
PHENOLIC GLYCOSIDES: aloenin, aloesaponol I
6-O-β-glucopyranoside, aloesaponol II
6-O-β-glucopyranoside, aloesaponol III
8-O-β-glucopyranoside, aloesin, isoeleutherol
5-O-β-D-glucopyranoside.
SAPONINS: aloesaponarin I, aloesaponarin II.
(Anthracenes) Yagi *et al.*, *Chem. Pharm. Bull.*, 25, 1771 (1977)
(An) Makino *et al.*, *ibid.*, 26, 1111 (1978)
(P) Yagi *et al.*, *Chem. Pharm. Bull.*, 25, 1771 (1977)
(Sap) Cameron *et al.*, *J. Chem. Soc., Chem. Commun.*, 688 (1978)

ALPHITONIA
A. excelsa Boiss.
TRITERPENOIDS: alphitexolide, alphitolic acid, betulinic acid,
ceanothic acid.
Branch *et al.*, *Aust. J. Chem.*, 25, 2209 (1972)

A. petriei Braid et White.
TRITERPENOIDS: alphitolic acid, alphitonin, betulinic acid,
ceanothic acid, lupeol.
Branch *et al.*, *Aust. J. Chem.*, 25, 2209 (1972)

A. whitei Braid.
TRITERPENOIDS: alphitolic acid, betulinic acid, ceanothic acid.
Branch *et al.*, *Aust. J. Chem.*, 25, 2209 (1972)

ALPINIA (Zingiberaceae)
A. flabellata.
TESPENOID: aflabene.
Mori *et al.*, *Tetrahedron Lett.*, 2297 (1978)

A. japonica (Thunb.) Miq.
SESQUITERPENOIDS: α-agarofuran, alpinenone, alpiniol,
alpinolide, alpinolide peroxide, dihydroagarofuran,
9,10-eremophilene-11-ol, β-eudesmol, 10-epi-τ-eudesmol,
furopelargone A, furopelargone B, hanalpinol, hanalpinone,
hanamyol, 9-hydroxyalpinolide,
4α-hydroxydihydroagarofuran, 3α,4α-oxidoagarofuran,
3β,4β-oxidoagarofuran, pogostol.
Itokawa *et al.*, *Chem. Abstr.*, 105 75863d

A. officinarum L.
FLAVONE: 3,5,7-trihydroxyflavone.
Kimura *et al.*, *J. Pharm. Soc., Japan*, 55, 229 (1935)

A. speciosa.
BISNORDITERPENOID: 15,16-bisnor-8(17),11-labdadien-13-one.
Itokawa et al., Chem. Pharm. Bull., **28**, 3452 (1980)

ALSEODAPHNE (Lauraceae)
A. semicarpifolia Nees.
APORPHINE ALKALOID: srilankine.
Smolnycki et al., Tetrahedron Lett., 4617 (1978)

ALSOPHILA
A. spinulosa.
FLAVONOIDS: hegaflavone A, hegaflavone B.
Wada et al., Chem. Pharm. Bull., **33**, 4182 (1985)

ALSTONIA (Apocynaceae)
A. boonei de Wild.
MONOTERPENOID: boonein.
Marini-Bettolo et al., Tetrahedron, **39**, 323 (1983)

A. boulinaensis.
ESTER ALKALOIDS: 10-hydroxylanciferine, 10-methoxylanciferine.
Lewin et al., J. Indian Chem. Soc., **55**, 1096 (1978)

A. congensis.
CARBAZOLE ALKALOID: echitamidine.
CARBOLINE ALKALOIDS:
3,4,5-O-trimethoxycinnamoylvincamajine,
3,4,5-O-trimethoxyquebrachidine.
PYRROLINOINDOLE ALKALOID: echitamine.
UNCHARACTERIZED ALKALOIDS: porphyrine, porphyrosine.
Goodson et al., J. Chem. Soc., 2628 (1932)
Birch et al., Proc. Chem. Soc. (London), 62 (1961)

A. constricta F.Muell.
CARBOLINE ALKALOIDS: alstonidine, alstonilidine, alstoniline, alstonine, 3,4,5-O-trimethoxybenzoylquebrachidine, 3,4,5-trimethoxycinnamoylvincamajine.
Hawkins, Elderfield, J. Org. Chem., **7**, 573 (1942)
Crow et al., Aust. J. Chem., **23**, 2489 (1970)

A. deplanchei.
DIMERIC CARBOLINE ALKALOIDS: pleiocorine, pleiocraline.
Das et al., J. Org. Chem., **42**, 2785 (1977)

A. lanceolifera.
CARBOLINE ALKALOID:
10,11-dimethoxy-1-methyldeacetylpicraline-3',4',4'-trimethoxybenzoate.
Peitfrere-Auvray et al., Phytochemistry, **20**, 1987 (1981)

A. legouxiae.
CARBOLINE ALKALOID: N(a)-demethylquaternine.
Lewin et al., Ann. Pharm. Fr., **39**, 273 (1981)

A. lenormandii var. lenormandii.
CARBAZOLE ALKALOIDS: akuammiline,
10,11-dimethoxy-1-methyldeacetylpicraline benzoate,
10,11-dimethoxy-1-methyldeacetylpicraline 3',4',5'-trimethoxybenzoate,
10,11-dimethoxy-1-methylpicraline, lochnericine,
11-methoxyakuammicine, 11-methoxycompactinervine,
12-methoxycompactinervine, picraline.
Legseir et al., Phytochemistry, **25**, 1735 (1986)

A. lenormandii var. minutifolia.
CARBAZOLE ALKALOIDS: akuammiline,
10,11-dimethoxy-1-methyldeacetylpicraline benzoate,
10,11-dimethoxy-1-methyldeacetylpicraline 3',4',5'-trimethoxybenzoate,
10,11-dimethoxy-1-methylpicraline, lochnericine,
11-methoxyakuammicine, 11-methoxycompactinervine,
12-methoxycompactinervine, picraline.
Legseir et al., Phytochemistry, **25**, 1735 (1986)

A. macrophylla Wall.
CARBOLINE ALKALOIDS: alstophylline, O-benzoylvincamajine, macroalhine, picralstonine.
DIMERIC CARBOLINE ALKALOIDS: macralstonidine, macralstonine, macrocarpamine, villalstonine.
Kishi et al., Helv. Chim. Acta, **48**, 1349 (1965)
Mayerl, Hesse, ibid., **61**, 337 (1978)

A. muelleriana.
CARBOLINE ALKALOIDS: alstonerine, alstonisine.
DIMERIC CARBOLINE ALKALOID: alstonisidine.
Elderfield, Am. Scientist, **48**, 193 (1960)
Cook et al., J. Chem. Soc., D, 1306 (1969)

A. odontophora.
DIMERIC CARBOLINE ALKALOID: N(1')-demethylplieocorine.
Vercauteren et al., Phytochemistry, **18**, 1729 (1979)

A. plumosa var. communis Boiteau f. glabra Boiteau.
DIMERIC CARBOLINE ALKALOID: plumocraline.
Massiot et al., C. R. Acad. Sci., **292**, 191 (1981)

A. quaternata.
CARBOLINE ALKALOIDS: quaternatine, quaternidine, quaternine, quaternoxine.
PROTOBERBERINE ALKALOID: 11-methoxy-epi-3α-yohimbine.
Mamatos-Kalamaras et al., Phytochemistry, **14**, 1489 (1975)

A. scholaris R.Br.
CARBAZOLE ALKALOID: scholarine.
CARBOLINE ALKALOIDS: echitamidine, picralinal, picrinine.
INDOLE ALKALOID: nareline.
GLUCOALKALOID: venoterpine glucoside.
UNCHARACTERIZED ALKALOID: ditamine.
Hesse, Justus Liebigs Ann. Chem., **176**, 326 (1875)
Morita et al., Helv. Chim. Acta, **60**, 1419 (1977)

A. somersetensis F.M.Bailey.
DIMERIC CARBOLINE ALKALOIDS: macralstonidine, villalstonine.
Sharp, J. Chem. Soc., 1227 (1934)

A. spectabilis R.Br.
UNCHARACTERIZED ALKALOIDS: alstonamine, ditamine.
Hesse, Justus Liebigs Ann. Chem., **176**, 326 (1875)
Hesse, Ber. Dtsch. Chem. Ges., **11**, 1546 (1878)

A. venenata R.Br.
CARBOLINE ALKALOIDS: venenatine, 16-epi-venenatine, veneserpine, venoxidine.
IMIDAZOLIDINOCARBAZOLE ALKALOIDS: 5,20-dioxokopsane, echitoserpine, echitovenaldine, echitovenidine, echitoveniline, echitovenine, (-)-11-methoxyechitovenicine, (-)-11-methoxyechitovenidine, minovincinine.
PROTOBERBERINE ALKALOIDS: alstovenine, isovenatine, 16-epi-venenatine.
Chatterjee, Ray, J. Indian Chem. Soc., **41**, 638 (1964)
Chatterjee et al., Phytochemistry, **20**, 1981 (1981)

A. villosa.
DIMERIC CARBOLINE ALKALOID: villalstonine.
Sharp, J. Chem. Soc., 1227 (1934)

A. yunnanensis.
CARBOLINE ALKALOIDS: 17-acetylsarpagine, corynanthine, deacetylpicraline, deacetylpicraline 3,4,5-trimethoxybenzoate, normacusine B, perakine, picrinine, pseudoakuammigine, sarpagine, 16-epi-sarpagine, tetrahydroalstonine, vallosimine, vinorine.

IMIDAZOLIDINOCARBAZOLE ALKALOIDS: lochnerinine,
  11-methoxy-19-hydroxytabersonine,
  3-0x0-11-methoxytabersonine.
  ESTER: methyl-3,4,5-trimethoxycinnamate.
  Chen et al., Yaoxue Xuebao, **20**, 906 (1985); Chem. Abstr., 105 3520s

ALTERNARIA (Polyporaceae)
A. kikuchiana.
STEROIDS: ergosterol, lanosterol,
  24-methylene-24,25-dihydrolanosterol.
  Starratt et al., Phytochemistry, **15**, 2002 (1976)
A. porri.
PIGMENT: 1,2,8-trihydroxy-6-methoxy-3-methylanthraquinone.
  Stoessl et al., Can. J. Chem., **47**, 767 (1969)
A. solani.
PIGMENT: 1,2,8-trihydroxy-6-methoxy-3-methylanthraquinone.
  Stoessl et al., Can. J. Chem., **47**, 767 (1969)
A. tenuis.
TOXINS: altenariol, altenuic acids I, II and III, altenusin,
  altenusol, altertenuol.
  Piero et al., Biochem. Biophys. Acta, **230**, 170 (1971)
  Roselt et al., Biochem. J., **67**, 390 (1957)
A. zinnae.
SESQUITERPENOID: zinniol.
  White, Starratt, Can. J. Bot., **45**, 2087 (1967)

ALTHAEA (Malvaceae)
A. nudiflora.
PHENOLICS: p-coumaric acid, guaiacolic acid, syringic acid.
  Geronikaki et al., Khim. Prir. Soedin., 685 (1976)
A. officinalis var. 'Rusalka'.
ACIDS: chlorogenic acid, p-coumaric acid, ferulic acid,
  p-hydroxybenzoic acid, vanillic acid.
TROPANE ALKALOID: scopoline.
  Koleva et al., Farmatsiya (Sofia), **36**, 15 (1986); Chem. Abstr., 105 94531k
A. rosea.
PHENOLICS: p-coumaric acid, guaiacolic acid, syringic acid.
  Geronikaki et al., Khim. Prir. Soedin., 685 (1976)

ALUIS
A. balsamea.
SESQUITERPENOID: canadene.
  Shaw, J. Am. Chem. Soc., **73**, 2859 (1951)

ALYSSUM (Cruciferae)
A. minimum.
FLAVONES: kaempferol 4'-methyl
  ether-5-glucoside-7-glucuronide, quercetin 3',4'-dimethyl
  ether-7-glucuronide, quercetin 4'-methyl
  ether-7-glucuronide.
  Afsharypuor, Lockwood, J. Nat. Prod., **49**, 844 (1986)

ALYXIA
A. lucida.
COUMARINS: coumarin, 3-hydroxycoumarin,
  5-hydroxycoumarin, 8-hydroxycoumarin, scopoletin.
  Sadavongvivad et al., Phytochemistry, **16**, 1451 (1977)

AMANITA
A. bisporigera.
AMOTOXIN: amanin. Uncharacterized amotoxins and
  phallotoxins.
  Yocum et al., Lloydia, **40**, 178 (1977)
A. citrina.
INDOLE ALKALOID: bufotenine.
  Fiussello, Romano, Allionia, **25**, 77 (1982); Chem. Abstr., 100 48598m
A. mappa.
INDOLE ALKALOID: bufotenine.
  Wieland et al., Justus Liebigs Ann. Chem., **513**, 10 (1934)
A. muscaria (L. ex Fr.) Hooker.
CARBOLINE ALKALOID:
  1,2,3,4-tetrahydro-1-methyl-β-carboline-3-carboxylic acid.
OXOVANADIUM COMPLEX: amavadine.
  (A) Matsumoto et al., Helv. Chim. Acta, **52**, 716 (1969)
  (V) Bemsky et al., Chem. Abstr., 105 222722n
A. pantherina (DC.) Fr.
ISOXAZOLE: pantherine.
  Konda et al., Chem. Pharm. Bull., **33**, 1083 (1985)
A. phalloides (Bull.) Quel.
TOXINS: amanin, α-amanitin, β-amanitin, amanullinic acid,
  antamanide, phallacidin, phallacin, phallicin, phallisacin,
  phalloidin, phalloin.
  Faulstich et al., Justus Liebigs Ann. Chem., 2324 (1976)
  Yocum et al., Biochemistry, **17**, 3786, 3790 (1978)
A. verna.
AMOTOXIN: amanin. Uncharacterized amotoxins and
  phallotoxins.
  Yocum et al., Lloydia, **40**, 178 (1977)
A. virgineoides. Bas.
AMINO ACID: cyclopropylalanine.
  Ohto et al., Chem. Lett., 511 (1986)
A. virosa.
TOXINS: amanin, phallacidin, phalloidin.
  Malak, Planta Med., **29**, 80 (1976)
  Yocum et al., Lloydia, **40**, 178 (1977)

AMARANTHUS (Amarantaceae)
A. albus.
FLAVONOID: rutin.
  Bech et al., Farm. Zh. (Kiev), 90 (1977); Chem. Abstr., 87 197309h
A. flavus.
FLAVONOID: rutin.
  Bech et al., Farm. Zh. (Kiev), 90 (1977); Chem. Abstr., 87 197309h
A. retroflexus L.
ALCOHOLS: cis-3-hexen-1-ol, trans-3-hexen-1-ol.
  Flath et al., J. Agric. Food Chem., **32**, 92 (1984)
A. spinosus.
FLAVONOID: rutin.
SAPONINS:
  β-D-glucopyranosyl-β-D-glucopyranosyl-β-D-
  glucopyranosyloleanolic acid.
  (F) Bech et al., Farm. Zh. (Kiev), 90 (1077); Chem. Abstr., 87 197309h
  (Sap) Banerji, Indian J. Chem., **17B**, 180 (1978)
A. viridis.
STEROID: amasterol.
  Roy et al., Phytochemistry, **21**, 2417 (1982)

AMARORIA (Simaroubaceae)
A. soulameoides A.Gray.
CARBOLINE ALKALOIDS: amaroridine, amarorine.
  Clark et al., J. Chem. Soc., Perkin I, 1614 (1980)

AMARYLLIS (Amaryllidaceae)

A. bella-donna L.
AZAPHENANTHRENE ALKALOIDS: anhydrolycorine, anhydrolycorinone, bellamarine, lycorine.
Haensel, Thober, *Arch. Pharm.*, **315**, 767 (1982)

A. parkeri Worsley.
AMINE ALKALOID: parkamine.
AZAPHENANTHRENE ALKALOID: parkacine.
Boit, Doepke, *Naturwissenschaften*, **46**, 228 (1959)
Doepke, *ibid.*, **50**, 645 (1963)

A. vittata.
NORBELLADINE ALKALOID: ryllistine.
Ghosal, Razdan, *J. Chem. Res. Synop.*, 412 (1984)

AMBERBOA

A. lippi.
SESQUITERPENOID: amberboin.
Gonzalez *et al.*, *An. Real. Soc. Espan. Fiz. Chim.*, **63B**, 956 (1967)

AMBROSIA (Compositae)

A. ambrosioides.
SESQUITERPENOID: damsinic acid.
Doskotch, Hufford, *J. Pharm. Sci.*, **58**, 186 (1969)

A. artemisiifolia.
SESQUITERPENOIDS: 8-acetoxy-3-oxopseudoguaian-6,12-olide, ambrosic acid, artemidifolin, 4-hydroxy-3-oxopseudoguaian-6,12-olide.
Inayama *et al.*, *Chem. Pharm. Bull.*, **22**, 1435 (1974)

A. canescens.
SESQUITERPENOID: canambrin.
Romo, Rodriguez-Hahn, *Phytochemistry*, **9**, 1611 (1970)

A. chamissonis.
SESQUITERPENOID: chamissonin.
Geissman *et al.*, *J. Org. Chem.*, **31**, 2269 (1966)

A. confertiflora.
SESQUITERPENOIDS: chihuahuin, confertiflorin, confertin, dihydrotamaulipin A, α-epoxysantamarin, tamaulipins A and B.
Renold *et al.*, *J. Org. Chem.*, **35**, 4264 (1970)
Yoshioka *et al.*, *Phytochemistry*, **9**, 823 (1970)

A. cordifolia.
SESQUITERPENOID: cordilin.
Herz *et al.*, *Phytochemistry*, **12**, 1415 (1973)

A. cumanensis.
SESQUITERPENOIDS: altamisin, cumambrins A, B and C, 11-hydroxyguaien, isoguaiene.
Bohlmann *et al.*, *Phytochemistry*, **16**, 575 (1977)
Borges *et al.*, *Tetrahedron Lett.*, 1513 (1978)

A. deltoides.
SESQUITERPENOIDS: damsin, psilostachyin C.
Abu-Shady *et al.*, *J. Am. Pharm. Assoc.*, **42**, 387 (1953)
Miller *et al.*, *Tetrahedron Lett.*, 3397 (1965)

A. dumosa.
NORSESQUITERPENOID: dumosin.
SESQUITERPENOID: ambrosiol.
Mabry *et al.*, *J. Org. Chem.*, **31**, 681 (1966)
Seaman, Mabry, *Rev. Latinoam. Quim.*, **10**, 85 (1979)

A. elatior L.
SESQUITERPENOIDS: 6α-hydroxy-4(5)-ene-9β-O-anisate, 1β-hydroxyeudesa-4,11(13)-dien-12-oic acid.
Ahmoto *et al.*, *Chem. Pharm. Bull.*, **35**, 2272 (1987)

A. grayi.
FLAVONE: 4',5,7-trihydroxy-3,6,8-trimethoxyflavone.
Herz *et al.*, *Tetrahedron*, **31**, 1577 (1975)

A. hispida.
FLAVONE: hispidulin.
SESQUITERPENOID: damsin.
(F) Doherty *et al.*, *J. Chem. Soc.*, 5577 (1963)
(T) Abu-Shady, Soine, *J. Am. Pharm. Assoc.*, **42**, 387 (1953)

A. ilicifolia.
SESQUITERPENOID: costic acid, ilicic acid.
Herz *et al.*, *J. Org. Chem.*, **31**, 1632 (1966)

A. maritima.
SESQUITERPENOIDS: ambrosin, damsin.
Abu-Shady, Soine, *J. Am. Pharm. Assoc.*, **42**, 387 (1953)
Suchy *et al.*, *Collect. Czech. Chem. Commun*, **28**, 2257 (1963)

A. muricata.
SESQUITERPENOID: muricatin.
Gonzalez *et al.*, *An. Quim.*, **69**, 1333 (1973)

A. peruviana.
SESQUITERPENOIDS: peruvin, peruvinin.
Joseph-Nathan, Romo, *Tetrahedron*, **22**, 1723 (1966)
Romo *et al.*, *ibid.*, **23**, 529 (1967)

A. polystachya.
SESQUITERPENOIDS: granilin, ivaspensin.
Vichnewski *et al.*, *Phytochemistry*, **15**, 1531 (1976)

A. psilostachya.
SESQUITERPENOIDS: ambrosin, ambrosiol, 3-hydroxydamsin, isabelin, psilostachyins A, B and C.
Abu-Shady, Soine, *J. Am. Pharm. Assoc.*, **42**, 387 (1953)
Miller, Mabry, *J. Org. Chem.*, **32**, 2929 (1967)

A. tenuifolia.
SESQUITERPENOID: confertin.
Yoshioka, Mabry, *Tetrahedron*, **25**, 4767 (1969)

AMELLUS

A. strigosus.
MONOTERPENOIDS: (2E,8Z)-2,8-decadien-4,6-diynoic acid, (2E,8Z)-2,8-decadien-4,6-diynoic acid methyl ester.
Bohlmann *et al.*, *Chem. Ber.*, **100**, 611 (1967)

AMIANTHUM (Liliaceae)

A. muscaetoxicum Gray.
UNCHARACTERIZED ALKALOID: amianthine.
Meuss, *J. Am. Chem. Soc.*, **75**, 2772 (1933)

AMMI (Umbelliferae)

A. majus L.
COUMARINS: angenomalin, majurin, marmesin.
Abu-Mustafa *et al.*, *Tetrahedron Lett.*, 1657 (1971)

A. visnaga.
NORSESQUITERPENOID: kellactone.
Spath *et al.*, *Can. J. Chem.*, **31**, 715 (1953)

AMMOCHARIS

A. coranica (Ker.Gawl.) Herb.
BENZOISOQUINOLINE ALKALOID: (+)-epi-buphanisine.
UNCHARACTERIZED ALKALOID: coranicine.
Hauth, Stauffacher, *Helv. Chim. Acta*, **45**, 1305 (1962)

AMMODENDRON (Leguminosae)

A. argenteum O.Kutz.
BISPIPERIDINE ALKALOID: ammodendrine.
BIS-SPARTEINE ALKALOIDS: argentine.
LUPANINE ALKALOID: (-)-lupanine.

SPARTEINE ALKALOIDS: argentamine, N-methylcytisine, pachycaprine.
UNCHARACTERIZED ALKALOID: argentinamine.
Ngok *et al.*, *Khim. Prir. Soedin.*, 111 (1970)

A. conollyi Bge.
BISPIPERIDINE ALKALOIDS: ammodendrine, isoammodendrine.
SPARTEINE ALKALOIDS: anagyrine, pachycarpine, (+)-sparteine.
UNCHARACTERIZED ALKALOID: connolline.
Proskurnina, Merlis, *J. Gen. Chem. USSR*, **19**, 1396 (1949)
Ngok *et al.*, *Khim. Prir. Soedin.*, 111 (1970)

AMMOTHAMNUS (Leguminosae)
A. lehmanni Bge.
MATRINE ALKALOIDS: matrine N-oxide, sophocarpine.
SPARTEINE ALKALOIDS: ammothamnine, pachycarpine.
CHALCONE: ammothamnidin.
FLAVONOIDS: luteolin, quercetin.
(C,F) Sattikulov *et al.*, *Uzb. Khim. Zh.*, 11 (1983); *Chem. Abstr.*, 100 82724j
(T) Sadykov, Proskurnina, *J. Gen. Chem. USSR*, **13**, 314 (1943)
(T) Kushmuradov *et al.*, *Khim. Prir. Soedin.*, 377 (1975)

A. songoricus (Schrenk) Lipsky.
MATRINE ALKALOIDS: matrine N-oxide, sophocarpine.
Orekhov, Proskurnina, *Ber. Dtsch. Chem. Ges.*, **67**, 77 (1934)

AMOMUM (Zingiberaceae)
A. medium.
INDENES: 1H-indene-2,3-dihydro-4-carboxaldehyde, 1H-indene-2,3-dihydro-5-carboxaldehyde.
MONOTERPENOIDS: 1,8-cineole, p-cymene, limonene, myrcene, α-phellandrene, α-pinene, β-pinene.
Takido *et al.*, *Phytochemistry*, **17**, 327 (1978)

A. melegueta.
PHENOLIC: shogoal.
Connelly *et al.*, *Aust. J. Chem.*, **23**, 369 (1970)

AMOORA
A. rohituka (Aphanamixis polystachys).
PIPERIDINE ALKALOID: rohikutine.
LIMONOID: amoorinine.
(A) Harmon *et al.*, *Tetrahedron Lett.*, 721 (1979)
(T) Angihotri *et al.*, *Planta Med.*, **53**, 298 (1987)

AMORPHA (Leguminosae)
A. canescens.
BENZOPYRAN: amorpholone.
Piatak *et al.*, *Phytochemistry*, **14**, 1391 (1975)

A. fruticosa L.
BENZOPYRAN: amorphigenin.
FLAVONOIDS: amorilin, amorisin, amoritin.
GLYCOSIDES: amorphin, amorphol.
ISOCOUMARINS: amoradicin, amoradin, amoradinin.
SESTERTERPENOIDS: 5,7-dihydroxy-6-geranylflavan-4-one, muurolene.
ROTENOID: 6,12-dehydro-α-toxicarol.
(B, Gly) Kasymov *et al.*, *Khim. Prir. Soedin.*, 464 (1974)
(F) Rozsa *et al.*, *Heterocycles*, **19**, 1793 (1982)
(I) Rozsa *et al.*, *Phytochemistry*, **23**, 1818 (1984)
(R) Reisch *et al.*, *Arch. Pharm.*, **509**, 152 (1976)
(R) Reisch *et al.*, *Phytochemistry*, **15**, 234 (1976)

AMPELOCERA
A. ruizii.
PYRIDINE ALKALOID: 1-methyl-5-carbomethoxy-α-pyridone.
Burnell *et al.*, *Lloydia*, **38**, 444 (1975)

AMPHICOME (Bignoniaceae)
A. emodi.
IRIDOID: amphicoside.
Kapoor *et al.*, *Tetrahedron Lett.*, 2839 (1971)

AMPHIDINIUM (Gymnodiniaceae)
A. carteri.
STEROID: amphisterol.
Withers *et al.*, *Phytochemistry*, **18**, 899 (1979)

AMSINKIA (Boraginaceae)
A. hispida (Ruiz et Pav.) Johnst.
PYRROLIZIDINE ALKALOIDS: intermedine, lycopsamine.
Culvenor, Smith, *Aust. J. Chem.*, **19**, 1955 (1966)

A. intermedia Fisch et Mey.
PYRROLIZIDINE ALKALOIDS: echiumine, intermedine, lycopsamine, sincamidine.
Culvenor, Smith, *Aust. J. Chem.*, **19**, 1955 (1966)

A. lycopsoides Lehm.
PYRROLIZIDINE ALKALOIDS: intermedine, lycopsamine.
Culvenor, Smith, *Aust. J. Chem.*, **19**, 1955 (1966)

AMSONIA (Apocynaceae)
A. angustifolia.
CARBOLINE ALKALOIDS: antirhine methochloride, (+)-eburamanine, (+)-eburnamonine, (-)-eburnamonine, isoeburnamine.
IMIDAZOLIDINOCARBAZOLE ALKALOID: tabersonine.
STEROID: β-sitosterol, β-sitosterol β-D-glucoside.
TRITERPENOID: β-amyrin.
(A) Bartlett *et al.*, *C. R. Acad. Sci.*, **249**, 1259 (1959)
(A) Sakai *et al.*, *J. Pharm. Soc., Japan*, **93**, 483 (1973)S
(S, T) Tomczyk *et al.*, *Herba Pol.*, **23**, 25 (1977); *Chem. Abstr.*, 88 71497v

A. elliptica.
IMIDAZOLIDINOCARBAZOLE ALKALOID: 3-oxotabersonine, tabersonine N(b)-oxide.
UNCHARACTERIZED ALKALOID: amsonine.
Aimi *et al.*, *Chem. Pharm. Bull.*, **26**, 1182 (1978)

A. illustris.
CARBOLINE ALKALOIDS: (-)-eburmonine, (+)-eburnamenine, (+)-eburnamonine, isoeburnamine.
IMIDAZOLIDINOCARBAZOLE ALKALOID: tabersonine.
Bartlett *et al.*, *C. R. Acad. Sci.*, **249**, 1259 (1959)
Janot *et al.*, *Bull. Soc. Chim. Fr.*, 707 (1954)

A. tabernaemontana Walt. (A. salicifolia Pichon).
CARBOLINE ALKALOIDS: (+)-eburnamine, (-)-eburnamonine, isoeburnamine.
IMIDAZOLIDINOCARBAZOLE ALKALOIDS: (+)-1,2-dihydroaspidospermidine, pseudotabersonine, tabersonine, (+)-vincadine.
INDOLE ALKALOIDS: (+)-6,7-dehydrovincadine, 6,7-dehydro-epi-vincadine, (±)-vincadine, (-)-vincadine, epi-vincadine.
Bycroft *et al.*, *Helv. Chim. Acta*, **47**, 1147 (1964)
Zsadon *et al.*, *Chem. Ind.*, **5**, 229 (1973)
Zsadon *et al.*, *Acta Chem. Acad. Sci. Hung.*, **96**, 167 (1978)

AMYRIS
A. balsamifera.
COUMARIN: balsamiferone.
STEROID: balsoxin.
Burke *et al.*, *Phytochemistry*, **18**, 1073 (1979)

A. elemifera DC.
COUMARINS: bergapten,
5-hydroxy-7-(3,3-dimethylallyl)-coumarin, marmesin,
psoralen, scopoletin, subarenol, ulopterol.
Laguna, *Rev. Cubana Farm.*, **19**, 5-8 (1985); *Chem. Abstr.*, 104 48745q

A. madrensis.
COUMARIN: dehydrogeyerin.
Dominguez *et al.*, *Phytochemistry*, **5**, 1096 (1977)

A. simplicifolia.
COUMARIN: 3-(3,3-dimethylallyl)-xanthyletin.
Cordova *et al.*, *Phytochemistry*, **13**, 758 (1974)

ANABASIS (Chenopodiaceae)
A. aphylla L.
PYRIDINE ALKALOIDS: anabasamine, anabasine.
SPARTEINE ALKALOIDS: (+)-aphyllidine, aphylline, aphylline
methyl ether, oxoaphyllidine, oxoaphylline.
Orekhov, Men'shikov, *Ber. Dtsch. Chem. Ges.*, **65**, 234 (1932)
Tsuda, Marion, *Can. J. Chem.*, **42**, 764 (1964)

A. jaxartica (Bge.) Benth.
AMINE ALKALOID: N-methyl-4-oxo-α-phenethylamine.
UNCHARACTERIZED ALKALOIDS: jaksartine, jaksartinine.
Platonova *et al.*, *J. Gen. Chem. USSR*, **28**, 3128 (1958)

A. salsa (SAM) Benth.
PIPERIDINE ALKALOID: 2,6-dimethylpiperidine.
Kuzuvkov, Men'shikov, *J. Gen. Chem. USSR*, **20**, 15, 24 (1950)

ANACYCLUS (Compositae)
A. pyrethrum DC.
AMIDE: 2,4-decadienoic acid N-isobutylamide.
Burden *et al.*, *J. Chem. Soc., C*, 2477 (1969)

A. radiatus.
SESQUITERPENOID: mariolin.
Collado *et al.*, *Phytochemistry*, **24**, 2447 (1985)

ANACYSTIS (Chroccoccaceae)
A. nidulans.
CAROTENOIDS: caloxanthin, β,β-carotene-2,3,3'-triol,
nostoxanthin.
Stransky, Hager, *Arch. Mikrobiol.*, **71**, 132-164 (1970)
Smallidge, Quackenbush, *Phytochemistry*, **12**, 2481 (1973)

ANADENANTHERA
A. peregrina (L.) Speg.
CARBOLINE ALKALOIDS:
1,2,3,4-tetrahydro-6-methoxy-2,9-dimethyl-β-carboline,
1,2,3,4-tetrahydro-6-methoxy-2-methyl-β-carboline.
Filho *et al.*, *Phytochemistry*, **14**, 821 (1975)

ANAGALLIS (Primulaceae)
A. arvensis L.
TRITERPENE GLUCOSIDES: arvenins I and II.
Yamada *et al.*, *Tetrahedron Lett.*, 2099 (1977)

ANAGRAPHIS
A. echioides.
FLAVONE: echiodinin.
Govindachari *et al.*, *Tetrahedron*, **21**, 2633 (1965)

ANAGYRIS (Leguminosae/Papilionaceae)
A. foetida L.
SPARTEINE ALKALOIDS: anagyrine, cytisine, (+)-sparteine.
Gerrard, *Pharm. J.*, *17*, **109**, 227 (1886)
Ing, *J. Chem. Soc.*, 2195 (1931)

ANAMIRTA
A. cocculus.
PROTOBERBERINE ALKALOID: 8-oxotetrahydropalmatine.
BITTER PRINCIPLE: picrotoxin.
  (A) Verpoorte *et al.*, *Lloydia*, **44**, 221 (1981)
  (B) Clark, *J. Am. Chem. Soc.*, **57**, 1111 (1935)

A. paniculata Colebr.
UNCHARACTERIZED ALKALOIDS: menispermin,
paramenispermine.
Pelletier, Couerbe, *Justus Liebigs Ann. Chem.*, **10**, 198 (1834)

ANANAS (Bromeliaceae)
A. comosus (Stickm.) Merrill.
AMINE ALKALOID: 5-hydroxytryptamine.
STEROIDS: ergosterol peroxide, stigmast-5-ene-3β,7α-diol.
  (A) Regula, *Acta Bot. Croat.*, **36**, 83 (1977); *Chem. Abstr.*, 88 186093c
  (S) Pakrashi *et al.*, *Indian J. Chem.*, **13**, 755 (1975)

A. comosus var. cayense.
TRITERPENOID: ananasic acid.
Takata, Scheuer, *Tetrahedron*, **32**, 1077 (1976)

ANAPHALIS
A. adnata DC.
PYRONE: 4'-hydroxydehydrokawain.
Talapatra *et al.*, *Indian J. Chem.*, **14B**, 300 (1976)

A. contorta Hooker.
FLAVONOIDS: tiliroside, tiliroside acetate.
Lin, *J. Chin. Chem. Soc.*, **22**, 383 (1975); *Chem. Abstr.*, 84 147653h

A. subumbellata. Clarke.
AMIDE: (-)-anabellamide.
STEROID: β-sitosterol β-D-glucoside.
Talapatra *et al.*, *Chem. Abstr.*, 97 159472e

ANAPTYCHIA (Physciaceae)
A. heterochroa.
TRITERPENOID: zeorin.
Paterno, *Gazzetta*, **6**, 113, 129, 130 (1876)

A. hypoglauca.
TRITERPENOID: zeorin.
Huneck, *Chem. Ber.*, **94**, 614 (1961)

A. obscurata.
PIGMENTS: 2-chloro-1,3,8-trihydroxy-6-methylanthraquinone,
flavoobscurin A, flavoobscurin B.
Fox *et al.*, *Tetrahedron Lett.*, 919 (1969)

A. speciosa.
TRITERPENOID: zeorin.
Huneck, *Chem. Ber.*, **94**, 614 (1961)

ANASTREPTA
A. orcadensis.
SESQUITERPENOID: anastreptene.
DITERPENOID: anadensin.
Andersen *et al.*, *Tetrahedron*, **33**, 617 (1977)
Huneck *et al.*, *Tetrahedron Lett.*, 3787 (1983)

ANCHUSA (Boraginaceae)
A. officinalis L.
UNCHARACTERIZED ALKALOIDS: consolicine, consolidine,
cynoglossine,
SAPONINS: anchusosides 1, 1a, 2, 3, 4, 5, 7 and 7a,
quercilioside A.
  (A) Greiner, *Arch. Pharm.*, **238**, 505 (1900)
  (S) Romussi *et al.*, *ibid.*, **319**, 549 (1986)

ANCISTROCLADUS

A. congolensis.
ISOQUINOLINE ALKALOIDS: ancistrocongine, ancistrocongolensine, ancistroealaelsine.
Foucher *et al.*, *Plant Med. Phytother.*, **9**, 87 (1975)

A. ealaensis.
ISOQUINOLINE ALKALOIDS: ancistrocladonine, ancistroealaelsine.
Foucher *et al.*, *Phytochemistry*, **13**, 1253 (1974)

A. hamatus.
ISOQUINOLINE ALKALOID: hamatine.
Govindachari *et al.*, *Indian J. Chem.*, **13**, 641 (1975)

A. heyneanus Wall.
ISOQUINOLINE ALKALOIDS: ancistrocladine, ancistrocladinine, ancistrocladisine.
Govindachari *et al.*, *Indian J. Chem.*, **10**, 1117 (1972)

ANDIRA

A. inermis.
AMINE ALKALOID: geoffroyine.
SESQUITERPENOID: demethyopterocarpin.
(A) Goldschmeidt, *Monatsh. Chem.*, **33**, 1379 (1912)
(T) Harper *et al.*, *J. Chem. Soc., Chem. Commun.*, 310 (1965)

A. retusa.
AMINE ALKALOID: geoffroyine.
Goldschmeidt, *Monatsh. Chem.*, **33**, 1379 (1912)

ANDRACHNE (Euphorbiaceae)

A. rotundifolia SAM.
UNCHARACTERIZED ALKALOID: andrachnine.
Vil'yams *et al.*, *Khim. Prir. Soedin.*, 257 (1966)

ANDROCYMBIUM (Wurmbaeoideae)

A. burkei Baker.
COLCHICINE ALKALOID: colchicine.
Pijewska *et al.*, *Collect. Czech. Chem. Commun.*, **32**, 158 (1967)

A. capense (L.) Krause.
COLCHICINE ALKALOIDS: colchicine, β-lumicolchicine, τ-lumicolchicine.
Pijewska *et al.*, *Collect. Czech. Chem. Commun.*, **32**, 158 (1967)

A. eucomoides (Jacq.) Willd.
COLCHICINE ALKALOIDS: colchicine, 2-demethylcolchicine, N-formyl-N-deacetylcolchicine.
Pijewska *et al.*, *Collect. Czech. Chem. Commun.*, **32**, 158 (1967)

A. longipes Bak.
COLCHICINE ALKALOIDS: colchicine, 2-demethylcolchicine.
Pijewska *et al.*, *Collect. Czech. Chem. Commun.*, **32**, 158 (1967)

A. melanthioides.
ISOQUINOLINE ALKALOID: melanthioidine.
Santavy, Reichstein, *Helv. Chim. Acta.*, **33**, 1606 (1050)

A. melanthioides var. stricta Baker.
BISBENZYLISOQUINOLINE ALKALOID: melanthoidine.
COLCHICINE ALKALOIDS: androcymbine, colchicine, cornigerine, demecolcine, 2-demethylcolchicine, 3-demethylcolchicine, 2-demethyldelecolcine, 3-demethyldemecolcine, N-formyl-N-deacetylcolchicine, β-lumicolchicine, τ-lumicolchicine, β-lumicornigerine, O-methylandrocymbine.
Potesilova *et al.*, *Planta Med.*, 344 (1985)

ANDROGRAPHIS (Acanthaceae)

A. echioides Nees.
FLAVONOIDS: echinioidin, echinioidinin.
Balmain, Connolly, *J. Chem. Soc., Perkin I*, 1247 (1973)

A. paniculata Nees.
DITERPENOIDS: andrographiside, andrographolide, 14-deoxyandrographolide, 14-deoxy-11,12-didehydroandrographolide, 14-deoxy-11-oxoandrographolide, neoandrographolide.
Gorter, *Recl. Trav. Chim. Pays-Bas*, **30**, 151 (1911)
Balmain, Connolly, *J. Chem. Soc., Perkin I*, 1247 (1973)
Hu *et al.*, *Zhongcaoyao*, **12**, 531 (1981)

A. wightiana Arn. ex Nees.
FLAVONES: echioidinin, wightin, wightionolide.
Dutta *et al.*, *Indian J. Chem.*, **7**, 110 (1969)

ANDROMEDA (Ericaceae)

A. polifolia L.
IRIDOID: andromedoside.
Chung *et al.*, *Arch. Pharm.*, **313**, 702 (1980)

ANDROPOGON (Gramineae)

A. jwarancusa.
MONOTERPENE: (-)-4-carene.
Simonsen, *J. Chem. Soc.*, **121**, 2792 (1922)

A. schoenanthus.
MONOTERPENOID: geraniol.
Forster, Cardwell, *J. Chem. Soc.*, **103**, 1342 (1913)

ANDROSACE (Primulaceae)

A. saxifragifolia.
SAPONINS: saxifragifolin A, saxifragifolin B.
TRITERPENOID: androsacenol.
(Sap) Waltho *et al.*, *J. Chem. Soc., Perkin I*, 1527 (1986)
(T) Pal *et al.*, *Phytochemistry*, **23**, 1475 (1984)

A. septemtrionalis.
SAPONINS: androseptosides A, B, C, C1, D, D1 and F.
Pirozhkova, Kintya, *Khim. Prir. Soedin.*, 658 (1982)

ANDROSTACHYS

A. johnsonii Prain.
DITERPENOID: ent-3α-hydroxybeyer-15(16)-ene-2,12-dione.
Pegel *et al.*, *J. Chem. Soc., Chem. Commun.*, 1346 (1971)

ANDRYALA (Compositae)

A. canariensis.
STEROID: taraxasterol.
Dietrich, Jeger, *Helv. Chim. Acta*, **33**, 715 (1950)

A. pinnatifida.
SESQUITERPENOID: 8α-hydroxydehydrozaluzanin C.
Bohlmann *et al.*, *Phytochemistry*, **21**, 2029 (1982)

ANEMIA (Schizaeaceae)

A. hirsuta.
One norditerpenoid.
Zanno *et al.*, *Naturwissenschaften*, **59**, 512 (1972)

A. phyllitidis L. Sw.
NORDITERPENOIDS: antheridiogenin-An.
Nakanishi *et al.*, *J. Am. Chem. Soc.*, **93**, 5579 (1971)

ANEMONE (Ranunculaceae)

A. chinensis.
SAPONIN: anemosapogenin.
Chou, Chu, *Hua Hsueh Hsueh Pao*, 126 (1962)

A. narcissiflora.
SAPONINS: narcissifloridine, narcissiflorine, narcissiflorinine.
Masood et al., *Phytochemistry,* **20** 1675 (1981)

A. nemorosa L.
SAPONINS: obtusilobin, obtusilobinin.
Masood et al., *Phytochemistry,* **18,** 1539 (1979)

A. obtusiloba.
SAPONINS: obtusilobin, obtusilobinin.
Masood et al., *Phytochemistry,* **18,** 1539 (1979)

A. pratensis.
ANTIBIOTIC: anemonin.
Bekurts, *Arch. Pharm.,* **230,** 182 (1892)

A. pulsatilla.
ANTIBIOTIC: anemonin.
Bekurts, *Arch. Pharm.,* **230,** 182 (1892)

A. raddeana Regel.
SAPONINS: raddeanins A, B, C and D.
Wu, Chu, *Huaxue Xuebao,* **43,** 82 (1985)

ANETHUM (Apiaceae)

A. graveolens L.
COUMARIN:
6,7-dihydro-8,8-dimethyl-2H,8H-benzo-[1,2-b:5,4-']-dipyran-2,6-dione.
XANTHONE: dillanoside.
(C) Aplin et al., *J. Chem. Soc.,C,* 2593 (1967)
(X) Kozawu et al., *Chem. Pharm. Bull.,* **24,** 220 (1976)

A. sowa Roxb.
ALDEHYDE: dillapional.
PHTHALIDES: butylphthalide, Z-ligustilide, neocnidilide, senkyunolide.
(Al) Tomar et al., *Indian J. Chem.,* **20B,** 723 (1981)
(Ph) Gijbels et al., *Sci. Pharm.,* **50,** 158 (1982)

ANEURA (Hepaticae)

A. pinguis (L.) Dum.
SESQUITERPENOID: pinguisone.
Benesova et al., *Collect. Czech. Chem. Commun.,* **34,** 1810 (1969)

ANGELICA (Umbelliferae)

A. apaensis.
FURANOCOUMARIN: apaensin.
Snu et al., *Yunnan Zhiwu Yanjiu,* **3,** 279 (1981)

A. archangelica.
BENZOPYRONE: angelicain.
COUMARINS: bergapten, 8-hydroxy-5-methoxypsoralen, imperatorin, isoimperatorin, isopimpinellin, umbelliprenin.
MONOTERPENOIDS: trans-1(7),5-p-menthadien-2-yl acetate, 2-nitro-1,5-p-menthadiene, cis-6-nitro-1(7),2-p-menthadiene, trans-6-nitro-1(7),2-p-menthadiene.
SESQUITERPENOID: cis-α-copaen-8-ol.
TRITERPENOIDS: α-amyrin, lupeol acetate.
(B) Kozawa et al., *Chem. Pharm. Bull.,* **31,** 64 (1983)
(C) Patra et al., *Indian J. Chem.,* **14B,** 816 (1976)
(T) Escher et al., *Helv. Chim. Acta.,* **62,** 2061 (1979)

A. archangelica subsp. littoralis.
COUMARIN: archangelicin.
Nielsen et al., *Acta Chem. Scand.,* **18,** 932 (1964)

A. dahurica.
COUMARINS: biacangelicin, imperatorin, isoimperatorin, oxypeucedanin, oxypeucedanin hydrate.
Shlyin'ko et al., *Khim. Prir. Soedin.,* 280 (1977)

A. decursiva.
COUMARINS: andelin, decursidin, decursin.
SESQUITERPENOID: (S)-angelinol.
(C) Sano et al., *Chem. Pharm. Bull.,* **23,** 20 (1975)
(T) Hata et al., *J. Pharm. Soc., Japan,* **89,** 549 (1962)

A. glabra.
BENZOPYRAN: phellopterin.
COUMARINS: byakangelicin, byakangelicol.
(B) Chatterjee et al., *Arch. Pharm.,* **295,** 248 (1962)
(C) Kawanami et al., *Chem. Pharm. Bull.,* **22,** 1227 (1974)

A. glauca.
SESQUITERPENOIDS: isoimperiatorin, prangolarin.
Kapoor et al., *Phytochemistry,* **11,** 475 (1972)

A. keiskei Koidz.
CHALCONES: xanthoangelol, xanthoangenol.
Kozawa et al., *J. Pharm. Soc., Japan,* **98,** 210 (1978)

A. komarovii.
COUMARINS: biacangelicin, imperatorin, phellopterin, umbelliprenin.
Zorin et al., *Khim. Prir. Soedin.,* 644 (1983)

A. longeradiata.
COUMARIN: angeladin.
Hata et al., *Chem. Pharm. Bull.,* **21,** 518 (1977)

A. mori.
SESQUITERPENOID: isosamidin.
Hata et al., *Chem. Pharm. Bull.,* **22,** 957 (1978)

A. pubescens Maxim.
COUMARINS: angelols A, B, C, D, E, F, G and H.
Baba et al., *Chem. Pharm. Bull.,* **30,** 2025 (1986)

A. putrescens.
COUMARIN: angelol.
Hata et al., *Tetrahedron Lett.,* 4557 (1965)

A. sachalinensis.
COUMARIN: sachalinin.
Nikonov et al., *Khim. Prir. Soedin.,* 623 (1970)

A. silvestris L.
SESQUITERPENOIDS: angelic acid, bisabolangelone, phellandrene.
Novotny et al., *Tetrahedron Lett.,* 3541 (1966)

A. totaomae.
COUMARINS: anomalin, biakyangelicin.
Zorin et al., *Khim. Prir. Soedin.,* 644 (1983)

A. ursina Maxim.
BENZOPYRAN: pteryxin.
COUMARINS: bergapten, coumurrayin, 8-geranoxypsoralen, glabralactone, glabralactone hydrate, heraclenin, heraclenol, imperatorin, tert-O-methylheraclenol, osthole, xanthotoxin.
TRITERPENOIDS: ursinin, ursonic acid.
(B) Bohlmann et al., *Tetrahedron Lett.,* 3947 (1968)
(C) Baba et al., *J. Pharm. Soc., Japan,* **103,** 1091 (1983)
(T) Nikonov et al., *Khim. Prir. Soedin.,* 421 (1970)

ANGELONIA

A. grandiflora.
FLAVONE: 3',5,7-trihydroxy-3,4'-dimethoxyflavone.
Kupchan et al., *Phytochemistry,* **10,** 664 (1971)

ANGOPHORA
A. subvalentina.
TRITERPENOID: 24-methylcycloartenol.
Ohta, Shimizu, *Chem. Pharm. Bull.*, **6**, 325 (1958)

ANHALONIUM
A. fissuratum.
AMINE ALKALOID: hordenine.
Spath, *Ber. Dtsch. Chem. Ges.*, **75**, 1558 (1942)
A. lewinii Hennings.
AMINE ALKALOIDS: N-acetylmescaline, mescaline, N-methylmescaline.
ISOQUINOLINE ALKALOIDS: anhalamine, anhalidine, anhalinine, anhalonidine, anhalonine, mescalotam, O-methyl-(+)-anhalonidine, peyoglutam, 1,2,3,4-tetrahydro-6,7,8-trimethoxy-1-methylisoquinoline.
Kapadia, Fales, *J. Pharm. Sci.*, **57**, 2017 (1968)
Kapadia *et al.*, *J. Chem. Soc., Chem. Commun.*, 1688 (1968)
A. williamsii.
AMINE ALKALOID: mescaline.
ISOQUINOLINE ALKALOID: pellotine.
Heffter, *Ber. Dtsch. Chem. Ges.*, **27**, 2975 (1894)

ANIBA (Lauraceae)
A. burchellii.
NEOLIGNAN: burchellin.
Lima *et al.*, *Phytochemistry*, **11**, 2031 (1972)
A. coco.
PYRIDINE ALKALOID: anibine.
Mors *et al.*, *J. Am. Chem. Soc.*, **79**, 4507 (1957)
A. duckei.
PYRIDINE ALKALOID: duckein.
Correa, Gottlieb, *Phytochemistry*, **14**, 271 (1975)
A. gardneri.
DIMERIC PYRONE:
ret-(1R,6S,7S,8S)-5-methoxy-7-phenyl-8-[6-(4-methoxy-2-pyronyl]-1-(E)-styryl-2-oxabicyclo[4,2,0]-octa-4-en-3-one.
Mascarenhas *et al.*, *Phytochemistry*, **16**, 801 (1977)
A. gigantifolia.
PYRAN: 5,6,7,8-tetrahydroyangonin.
Achenbach *et al.*, *Chem. Ber.*, **104**, 2688 (1971)
A. guianensis.
SESQUITERPENOID: guianin.
Buchi *et al.*, *J. Am. Chem. Soc.*, **99**, 8073 (1977)
A. lancifolia.
PHENOLIC: 4-hydroxy-5-allylveratrole.
Gottlieb *et al.*, *Phytochemistry*, **11**, 1861 (1972)
A. mas.
PYRONES: 6-(3',4'-methylenedioxyphenyl)-4-methoxy-2-pyrone, 6-(3',4'-methylenedioxystyryl)-4-methoxy-2-pyrone.
De Diaz *et al.*, *Acta Amazonica*, **7**, 41 (1971); *Chem. Abstr.*, 88 3067e
A. roseodora.
SESTERTERPENOID: (-)-rubranin.
De Allelenia *et al.*, *Phytochemistry*, **17**, 517 (1978)
A. santalodora.
CARBOLINE ALKALOID: cecelin.
Aguiar *et al.*, *Phytochemistry*, **19**, 1859 (1980)

ANIGOZANTHOS (Haemodoraceae)
A. rufus. Labill.
PIGMENTS: anigorubone, anigozanthin, dihydroxyanigorubone, hydroxyanigorubone.
Cooke, Thomas, *Aust. J. Chem.*, **28**, 1053 (1975)

ANISOCYCLEA (Menispermaceae)
A. grandidieri H.Bn.
BISBENZYLISOQUINOLINE ALKALOIDS: 12-demethyltrilobine, 12-O'-demethyltrilobine, (-)-epi-stephanine.
Schittler, Weber, *Helv. Chim. Acta*, **55**, 2061 (1972)

ANISODORIS
A. nobilis.
PURINE ALKALOID: doridosine.
Kim *et al.*, *Lloydia*, **44**, 206 (1981)

ANISODUS (Solanaceae)
A. luridus Link et Otto.
BISPYRROLE ALKALOID: cuskhygrine.
TROPANE ALKALOID: hyoscyamine.
Tuppy, Faltaous, *Monatsh. Chem.*, **91**, 167 (1960)
Bremner, Cannon, *Aust. J. Chem.*, **21**, 1369 (1968)

ANISOMELES (Labiatae)
A. evindica.
DITERPENOID: ovatodiolide.
Ho Dac An *et al.*, *Bull. Soc. Chim. Fr.*, 1192 (1963)
A. indica O.Kuntze.
ACIDS: arachic acid, behenic acid, cerotic acid, heneicosanoic acid, lignoceric acid, palmitic acid, pentacosanoic acid, stearic acid, tricosanoic acid.
DITERPENOID: isoovatodiolide.
STEROIDS: β-sitosterol, stigmasterol.
(Ac) Chen *et al.*, *Chem. Abstr.*, 86 2376b
(S) Chen *et al.*, *Chem. Abstr.*, 84 102343r
(T) Manchand, Blount, *J. Org. Chem.*, **42**, 3824 (1977)
A. malabarica.
DITERPENOIDS: 2-acetoxymalabaric acid, anisomelic acid, anisomelol, anisomemyl acetate, malabaric acid, ovatodiolide.
Devi *et al.*, *Indian J. Chem.*, **16B**, 441 (1978)
Purushothaman *et al.*, *ibid.*, **13**, 1357 (1975)
A. ovata R.Br.
FLAVONOIDS: anisofolin A, anisofolin B, 5,6-dimethoxy-7,3',4'-trihydroxyflavone.
DITERPENOID: ovatodiolide.
(F) Rao *et al.*, *Phytochemistry*, **22**, 1522 (1983)
(T) Purushothaman *et al.*, *Indian J. Chem.*, **13**, 1357 (1975)

ANISOTES
A. sessiliflorus C.B.Cl.
QUINAZOLINE ALKALOIDS: aniflorine, anisessine, anisotine, deoxyaniflorine.
Arndt *et al.*, *Tetrahedron*, **23**, 3521 (1967)

ANISUM (Umbelliferae)
A. vulgare Gaertn.
SESQUITERPENOID: τ-himalchalene.
Tabacci *et al.*, *Helv. Chim. Acta*, **57**, 849 (1974)

ANNONA (Annonaceae)
A. bullata.
SESQUITERPENOID: bullatantriol.
Kutschabsky *et al.*, *Phytochemistry*, **24**, 2724 (1985)
A. cherimolia.
APORPHINE ALKALOIDS: anonaine, liriodenine, michelalbine, (+)-reticuline.
Cassels, *Rev. Latinoam. Quim.*, **8**, 133 (1977)

A. coriacea.

DITERPENOIDS: annonalide,
3,20-epoxy-3α-hydroxy-13-hydroxyacetyl-13-methyl-7-
podocarpen-19,6-olide.

Ferrari *et al.*, *Phytochemistry*, **10**, 3267 (1972)

Mussini *et al.*, *J. Chem. Soc., Perkin I*, 2551 (1973)

A. crassiflora Mart.

APORPHINE ALKALOIDS: anonaine, asimilobine.
BENZYLISOQUINOLINE ALKALOID: (+)-retuculine.
OXOAPORPHINE ALKALOID: liriodenine.

Hocquemiller *et al.*, *Plant Med. Phytother.*, **16**, 4 (1982)

A. montana.

CARBOLINE ALKALOIDS: annomontine, methoxyannomontine.

Lebouef *et al.*, *Plant Med. Phytother.*, **16**, 169 (1982)

A. muricata L.

ISOQUINOLINE ALKALOIDS: anomuricine, anomurine, muricine,
muricinine.

Lebouef *et al.*, *Planta Med.*, **42**, 37 (1981)

A. reticulata.

BENZYLISOQUINOLINE ALKALOID: (+)-reticuline.
DIFURAN: 14-hydroxy-25-deoxyrollinicin.

(A) Stermitz *et al.*, *Tetrahedron Lett.*, 1601 (1967)

(F) Etse, Waterman, *J. Nat. Prod.*, **49**, 684 (1986)

A. squamosa.

SESQUITERPENOIDS: α-caryophyllene, β-caryophyllene.
DITERPENOIDS: 17-acetoxy-19-kauranal,
17-hydroxykauran-19-al, kaur-16-en-19-al, kaur-16-en-19-ol
acetate.

Ramage, Simonsen, *J. Chem. Soc.*, 1208 (1938)

Bohlmann, Rao, *Chem. Ber.*, **106**, 841 (1973)

ANODENDRON (Apocynaceae)

A. affine Druce.

PYRROLIZIDINE ALKALOIDS: alloanodendrine, anodendrine.
STEROIDS: affinogenins D-I, D-II, D-III, D-IV and D-V,
affinosides A, B, C, D, E, F, G, La, Lb, Lc, Ld and Le.

(A) Sasaki, Hirata, *Tetrahedron Lett.*, 4065 (1969)

(S) Abe *et al.*, *Chem. Pharm. Bull.*, **30**, 1183 (1982)

(S) Abe *et al.*, *ibid.*, **33**, 3662 (1985)

A. paniculatum (Roxb.) A.DC.

STEROIDS: anodendrosides A, B, C, D, E-1, E-2, F and G.

Polonia *et al.*, *Helv. Chim. Acta.*, **53**, 1253 (1970)

ANOGEISSUS

A. latifolia.

BENZOPYRAN: 3,3′,4-tri-O-methylflavellagic acid.

Deshpande *et al.*, *Indian J. Chem.*, **14B**, 641 (1976)

ANOPTERUS

A. glandulosus Labill.

ESTER ALKALOID: anopterine.

Denne *et al.*, *Tetrahedron Lett.*, 2727 (1972)

A. macleayanus F.Muell.

ESTER ALKALOIDS: anopterine, hydroxyanopterine.

Denne *et al.*, *Tetrahedron Lett.*, 2727 (1972)

Hart *et al.*, *Aust. J. Chem.*, **29**, 1319 (1976)

ANTHELIA (Musci)

A. julacea (L.) Dum.

DITERPENOID: (-)-16-hydroxykaurane, 16β-kauran-16α-ol.

Huneck, Vevle, *Z. Naturforsch.*, **25B**, 277 (1970)

A. juratzkana (Limpr.) Trev.

DITERPENOID: (-)-16-hydroxykaurane.

Huneck, Vevle, *Z. Naturforsch.*, **25B**, 277 (1970)

ANTHEMIS (Compositae/Anthemideae)

A. austriaca.

SESQUITERPENOID: 8-acetoxy-4-oxonerolidol.

Bohlmann *et al.*, *Chem. Ber.*, **107**, 1074 (1974)

A. cotula.

One sesquiterpenoid.

Bohlmann *et al.*, *Tetrahedron Lett.*, 2417 (1966)

A. fuscata Brot.

ISOCOUMARIN: artemidin.

Bohlmann *et al.*, *Chem. Ber.*, **103**, 2856 (1970)

A. montana.

MONOTERPENOID: dehydronerol isovalerate.

Bohlmann, Kapteyn, *Tetrahedron Lett.*, 2065 (1973)

A. nobilis.

BENZOPYRAN: apiin.
MONOTERPENOID: 4-isopropenylbenzaldehyde.
SESQUITERPENOIDS: 3-dehydronobilin, 1,10-epoxynobilin,
hydroxynobilin, nobilin, epi-nobilin.
TRITERPENOID: taraxasterol.

(B) Holyalkar *et al.*, *Can. J. Chem.*, **43**, 2085 (1965)

(T) Holub, Samek, *Collect. Czech. Chem. Commun.*, **42**, 1053 (1977)

(T) Thomas *et al.*, *Helv. Chim. Acta*, **64**, 2393 (1981)

ANTHOCEPHALUS

A. cadamba.

CARBOLINE ALKALOIDS: cadambine, cadamine,
3α-dihydrocadambine, 3β-dihydrocadambine,
isodihydrocadambine.
SAPONINS: anthocephalus saponins A, B, C and D.
TRITERPENOID: cadambagenic acid.

(A) Brown *et al.*, *Tetrahedron Lett.*, 1629 (1976)

(Sap) Banerji *et al.*, *J. Indian Chem. Soc.*, **55**, 275 (1978)

(T) Sahu *et al.*, *Indian J. Chem.*, **12**, 284 (1974)

ANTHOCERIS (Solanaceae)

A. albicans.

TROPANE ALKALOID: 3α-n-butyryloxytropane.

Evans, Ramsey, *Phytochemistry*, **20**, 497 (1981)

A. genistoides.

TROPANE ALKALOID: aponorhyoscine.

Evans, Ramsey, *Phytochem*, **20**, 497 (1981)

A. littorea.

TROPANE ALKALOID: littorine.

Cannon *et al.*, *Aust. J. Chem.*, **22**, 221 (1969)

ANTHOCLEISTA

A. ambesiaca.

MONOTERPENOIDS: erythrocentaurine, sweroside.

Capelle, *Phytochemistry*, **12**, 1191 (1973)

A. djalonensis.

PHTHALIDE: djalonensin.
XANTHONE: lichexanthone.

(Ph) Okorie, *Phytochemistry*, **15**, 1799 (1976)

(X) Rubei *et al.*, *Phytochemistry*, **15**, 1093 (1976)

A. grandiflora.

IRIDOID: grandifloroside.

Chapelle, *Phytochemistry*, **15**, 1305 (1976)

A. procera.

PYRIDINE ALKALOID: gentianine.

Govindachari *et al.*, *J. Chem. Soc.*, 551, 2725 (1957)

A. vogelii.

AMIDE: fagaramide Iridoids secologanic acid, sweroside, vogeloside.

XANTHONES: decussatin, fagaramide.

(Am) Goodson, *Biochem. J.*, **15**, 123 (1921)

(I) Chapelle, *Planta Med.*, **29**, 268 (1976)

(X) Okorie, *Phytochemistry*, **15**, 1799 (1976)

ANTHOSPERMA

A. moschatum Labill.

OXONORAPORPHINE ALKALOID: antherospermidine.

Bick *et al.*, *Aust. J. Chem.*, **9**, 111 (1956)

ANTHOSTOMA

A. avocetta.

SESQUITERPENOID: avocettin.

Arigoni, *Pure Appl. Chem.*, **41**, 219 (1975)

ANTHOTROCHE

A. myoporoides.

TROPANE ALKALOID: aponoratropine.

Evans, Ramsey, *Phytochemistry*, **20**, 497 (1981)

A. pannosa.

TROPANE ALKALOID: (-)-hyoscyamine.

Ladenberg, *Justus Liebigs Ann. Chem.*, **206**, 282 (1880)

A. walcottii.

TROPANE ALKALOID: aponoratropine.

Evans, Ramsey, *Phytochemistry*, **20** 497 (1981)

ANTHRISCUS

A. sylvestris Hoffm.

ACID: (Z)-2-angeloyloxymethyl-2-butenoic acid.

DITERPENOIDS: anthricin, anthricinol methyl ether.

Kozawa *et al.*, *Chem. Pharm. Bull.*, **30**, 2885 (1982)

ANTHYLLIS (Papilionaceae/Fabaceae)

A. montana.

ANTHOCYANINS: cyanadin 3-galactoside, delphinidin 3-galactoside, poenidin 3-galactoside.

Sterk *et al.*, *Acta Bol. Neerl.*, **26**, 349 (1977); *Chem. Abstr.*, 87 197258r

A. onobrychioides.

FLAVONES: isorhamnetin 3-O-β-D-galactopyranoside, kaempferol 3-O-β-D-galactopyranoside, quetcetin 3-O-β-D-galactopyranoside, rhamnazin 3-O-galactoside, rhamnazin 3-O-galactoside-4'-O-glucoside, rhamnetin 3-O-β-D-galactopyranoside-3',4'-di-O-β-D-glucopyranoside, rhamnocitrin 3-O-β-D-galactopyranoside.

Barbera *et al.*, *Phytochemistry*, **25**, 2361 (1986)

A. vulneraria.

FLAVONE: fisetin.

PHYTOALEXINS: demethylvestitol, isovestitol.

(F) Gornet *et al.*, *Phytochemistry*, **11**, 2313 (1972)

(P) Ingham, *Phytochemistry*, **16** 1279 (1977)

A. vulneria L.

ANTHOCYANINS: cyanidin 3-galactoside, delphinidin 3-galactoside, paeonidin 3-galactoside.

Sterk *et al.*, *Acta Bologica Neerlandica*, **26**, 349 (1977); *Chem. Abstr.*, 87 197258r

ANTIARIS (Moraceae)

A. toxicaria (Pers.) Lesch.

STEROIDS: antiarigenin, α-antiarin, β-antiarin, antiogenin, strophalloside.

Tschesche, Haupt, *Ber. Dtsch. Chem. Ges.*, **69**, 1377 (1936)

Juslen *et al.*, *Helv. Chim. Acta*, **45**, 2285 (1962)

ANTIDESMA

A. bunias.

TRITERPENOID: dammar-20,24-dien-3β-ol.

Hui, Sung, *Aust. J. Chem.*, **21**, 2137 (1968)

A. menasu.

TRITERPENOID: antidesmanol.

Rizvi *et al.*, *Experientia*, **36**, 146 (1980)

A. pentandrum.

TRITERPENOID: lupeolactone.

Kikuchi *et al.*, *Chem. Lett.*, 603 (1983)

ANTIRHEA (Rubiaceae)

A. putaminosa (F.Muell.) Bail.

CARBOLINE ALKALOID: antirhine.

Johns *et al.*, *Aust. J. Chem.*, **20**, 1463 (1967)

ANTIRRHINUM (Scrophulariaceae)

A. hispanicum.

NAPHTHYRIDINE ALKALOID: 4-methyl-2,6-naphthyridine.

Brooker, Harkness, *Planta Med.*, **26**, 305 (1974)

A. majus.

NAPHYRIDINE ALKALOID: 4-methyl-2,6-naphthyridine.

AMINO ACIDS: alanine, aspartic acid, cystine, glutamic acid, glutamine, glycine, hydroxyproline, isoleucine, lysine, proline, serine, threonine, tryptophan, tyramine, valine.

BENZOFURAN: aureusidin.

GLUCOSIDE: aureusin.

IRIDOID: 5-o-β-glucosylantirrhinoside.

(A) Harkness, *Phytochemistry*, **10**, 2849 (1971)

(Am) Pande *et al.*, *Planta Med.*, **30**, 317 (1976)

(B, G) Geissman *et al.*, *J. Am. Chem. Soc.*, **72**, 5725 (1950)

(I) Guiso, Scarpati, *Gazz. Chim. Ital.*, **99**, 800 (1969)

A. molles.

NAPHTHYRIDINE ALKALOID: 4-methyl-2,6-naphthyridine.

Brooker, Harkness, *Planta Med.*, **26**, 305 (1974)

A. mollisum.

NAPHYRIDINE ALKALOID: 4-methyl-2,6-naphthyridine.

Brooker, Harkness, *Planta Med.*, **26**, 305 (1974)

A. orontium.

NAPHTHYRIDINE ALKALOID: 4-methyl-2,6-naphthyridine.

Harkiss, *Planta Med.*, **20**, 108 (1971)

A. orontium L. var. calycinum (Lam.).

IRIDOID: calycinoside.

Bianco *et al.*, *Gazz. Chim. Ital.*, **109**, 561 (1979)

A. tortuosum.

IRIDOIDS: antirrhide 5-O-β-glucosylantirrhinoside.

Scarpati, Guiso, *Gazz. Chim. Ital.*, **99**, 807 (1969)

ANTITOXICUM

A. funebre (Boiss Kotschy) Pobed.

PHENANTHROINDOLIZIDINE ALKALOID: antofine.

Platonova *et al.*, *J. Gen. Chem. USSR*, **28**, 3131 (1958)

A. scandens.
FLAVONOIDS: kaempferol, quercetin, quercetin
3-β-D-glucosido-7-α-L-rhamnoside.
PHENOLICS: caffeic acid, chlorgenic acid, ferulic acid, sinapic
acid.
Fursa *et al.*, *Khim. Prir. Soedin.*, 416 (1977)

ANVILLEA
A. garcini.
SESQUITERPENOIDS: 9β-hydroxy-1β,10α-epoxyparthenolide,
9α-hydroxyparthenolide, 9β-hydroxyparthenolide.
Rustaiyan *et al.*, *Phytochemistry*, **25**, 1229 (1986)

AOTUS (Leguminosae/Podalyriae)
A. subglauca Blackley et McKee.
INDOLE ALKALOID: (S)(4)-N(b)-dimethyltryptophan methyl
ester.
Johns *et al.*, *Aust. J. Chem.*, **24**, 439 (1971)

APHANAMIXIS
A. grandiflora.
LIMONOIDS: amoorastatin, amoorastatone, hydroxyamoorastatin.
Polonsky *et al.*, *Experientia*, **35**, 987 (1979)

A. polystachya.
LIMONOIDS: aphanamixin, aphanamixinin, aphanamixol,
polystachin, rohikutin.
Chandrasekharan *et al.*, *Sci. Cult.*, 363 (1968)
Mulholland *et al.*, *J. Chem. Res. Synop.*, 294 (1979)

APHANIZOMENON (Cyanophyceae)
A. flos-aquae.
CAROTENOIDS: aphanicin, aphanin, aphanizophyll.
UNCHARACTERIZED CAROTENOID: flavazin.
Tischer *et al.*, *Z. Physiol. Chem.*, **251**, 109 (1938)

APHELANDRA (Acanthaceae)
A. squarrosa Nees.
MACROCYCLIC ALKALOID: aphelandrine.
Daetwyler *et al.*, *Helv. Chim. Acta.*, **61**, 2646 (1978)

APHLOIA
A. madagascariensis.
XANTHONE: aphloiol.
Adjangba *et al.*, *Bull. Soc. Chim. Fr.*, 376 (1964)

APIOS (Leguminosae)
A. tuberosa Moench. (A. americana Medic).
BENZOPYRAN: apiocarpin.
Ingham *et al.*, *Phytochemistry*, **21**, 1409 (1982)

APIUM (Umbelliferae)
A. graveolens.
BENZOPYRAN: apiumetin.
COUMARIN: rutaretin.
Garg *et al.*, *Phytochemistry*, **17**, 2135 (1978)

APLOPAPPUS
A. heterophyllus.
BENZOFURAN: toxol.
STEROIDS: androstane-3β,16α,17α-triol,
stigmasta-8(14),22-3β-ol.
(B) Zalkow *et al.*, *Tetrahedron Lett.*, 2873 (1972)
(S) Zalkow *et al.*, *ibid.*, 217 (1964)

APOCYNUM
A. lancifolium.
FLAVONOIDS: neoisorutin, quercetin.
STEROIDS: cimarin, k-strophanthidin-β.
Murzagaliev *et al.*, *Chem. Abstr.*, *86 185911q*

APTOSIMUM (Scrophulariaceae)
A. spinescens Thunb.
FURANS: aptosimol, aptosimone (+)-piperitol, (+)-sesamin,
(+)-spinescin.
Brieskorn *et al.*, *Tetrahedron Lett.*, 2221 (1976)

APULEIA
A. leiocarpa.
FLAVONES: apuleidin, apuleisin, apuleitrin,
5-O-demethylapulein, oxyanin A, oxyanin B.
Braz, Filho, *Phytochemistry*, **10**, 2433 (1971)

AQUILLARIA
A. agallocha.
BENZOPYRAN: agarotetrol.
SESQUITERPENOIDS: α-agarofuran, β-agarofuran, agarol,
agarospirol, β-dihydroagarofuran,
3,4-dihydroxydihydroagarofuran,
4-hydroxydihydroagarofuran, norketoagarofuran.
(B) Yoshii *et al.*, *Tetrahedron Lett.*, 3921 (1978)
(T) Jain, Bhattacharyya, *ibid.*, 13 (1959)
(T) Varma *et al.*, *Tetrahedron*, **21**, 115 (1965)

A. malacacensis Wrod.
SESQUITERPENOIDS: jinkoh-eremol, jinkohol.
Nakanishi *et al.*, *J. Chem. Soc., Perkin I*, 601 (1983)

A. sinensis (Lour.) Gilg.
SESQUITERPENOIDS: baimuxinol, dehydrobaimuxinol.
Yang *et al.*, *Yaoxue Xuebao*, **21**, 576 (1986); *Chem. Abstr.*, 105
187604b

A. suensis.
SESQUITERPENOIDS: baimuxinal, baimuxinic acid.
Yang, Chen, *Yaoxue Xuebao*, **2**, 329 (1954)

AQUILEGIA (Ranunculaceae)
A. atrata.
BENZOFURANS: aquilegiolide, epi-aquilegiolide.
Guerriero *et al.*, *Phytochemistry*, **23**, 2394 (1984)

A. karelini (Baker) O.et B. Fedtsch.
QUATERNARY ALKALOID: magnoflorine iodide.
Nakano, *Chem. Pharm. Bull.*, **2**, 329 (1954)

A. olympica.
PROTOBERBERINE ALKALOID: berberine.
QUATERNARY ALKALOID: magnoflorine.
Efimova *et al.*, *Aktual. Vopr. Farm.*, **2**, 34 (1974); *Chem. Abstr.*, 84
147617z

ARACHIOIDES
A. standsii.
DITERPENOIDS: dryocrassol methyl ether, ryomenin.
Tanaka *et al.*, *Chem. Pharm. Bull.*, 27, 2874) (1979)

ARACHIS (Leguminosae)
A. hypogaea L.
ACIDS: behenic acid, lignoceric acid, myristic acid, palmitic
acid, stearic acid.
UNCHARACTERIZED ALKALOID: arachine.
CHROMENE: 5,7-dihydroxychromone Flavonoids chrysoeriol
daucosterin, esculetin, isoquercitrin, luteolin, pratensein,
quercetin, rutin.

PHENOLICS: caffeic acid, p-coumaric acid, ferulic acid, p-hydroxybenzoic acid, phloretic acid, protocatechuic acid, α-resorcylic acid, salixylic acid, vanillic acid.

STEROID: β-sitosterol.

(A) Mooser, *Landwirtsch. Vers. Stn.*, **60**, 321 (1904)

(F, P, S) Matsuura *et al.*, *J. Pharm. Soc., Japan*, **103**, 997 (1983)

ARACHNIODES

A. standishii.

PYRROLIDINE ALKALOID: 3,4-dihydroxy-2-hydroxymethylpyrrolidine.

Furukawa *et al.*, *Phytochemistry*, **24**, 593 (1985)

ARALIA (Araliaceae)

A. cordata.

DITERPENOIDS: grandifloric acid, 7α-hydroxy-(-)-pimara-8(14),15-dien-19-oic acid, 7β-hydroxy-(-)-pimara-8(14),15-dien-19-oic acid 7-oxo-(-)-pimara-8(14),15-dien-19-oic acid, (-)-pimara-8(14),15-dien-19-al, (-)-pimara-8(14),15-dien-19-oic acid.

Yahara *et al.*, *Chem. Pharm. Bull.*, **22**, 1639 (1974)

A. japonica.

TRITERPENOID: oleanolic acid.

Van der Haar, *Recl. Trav. Chim. Pays-Bas*, **46**, 775, 793 (1927)

A. mandschurica Rupr. et Maxim.

SAPONINS: aralosides A, B and C, oleanosides A, B, C, D, E, F, G, H and I.

Elyakov *et al.*, *Izv. Akad. Nauk SSSR, Otd. Khim. Nauk*, 1605 (1962)

Lutomski *et al.*, *Herba Pol.*, **23**, 183 (1977); *Chem. Abstr.*, 88 148979p

A. schmidtii.

SAPONIN: araloside A.

Kochetkov *et al.*, *J. Gen. Chem. USSR*, **31**, 658 (1961)

ARALIDIUM

A. pinnatifidum.

IRIDOID: aralidioside.

Jensen *et al.*, *Phytochemistry*, **19**, 2685 (1980)

ARALIOPSIS

A. soyauxii.

FURANOQUINOLINE ALKALOIDS: flindersamine, maculine, (+)-platydesmine, (-)-ribalinine, skimmianine.

QUINOLONE ALKALOID: araliopsine.

Vaquette *et al.*, *Phytochemistry*, **15**, 743 (1976)

A. tabouensis.

FUROQUINOLINE ALKALOIDS: halfordinine, isoplatydesmine, skimmianine.

INDOLOQUINAZOLINE ALKALOIDS: evodiamine, rhetsinine.

QUATERNARY ALKALOID: N-methylplatydesminium.

Fish *et al.*, *Planta Med.*, **29**, 318 (1976)

ARALIORHAMNUS

A. vaginatus Perrier.

PEPTIDE ALKALOIDS: aralionine, aralionine B, hysodricanine A, mauritine H, scutianine F.

Tschesche *et al.*, *Phytochemistry*, **16**, 1025 (1977)

ARARIBA

A. rubra.

CARBOLINE ALKALOID: harman.

UNCHARACTERIZED ALKALOID: aribine.

Rieth, Wohler, *Justus Liebigs Ann. Chem.*, **120**, 247 (1861)

Spath *et al.*, *Monatsh. Chem.*, **41**, 401 (1920)

ARAUCARIA (Araucariaceae)

A. angustifolia.

DITERPENOIDS: 7,13-abietadien-18-al, 7,13-abietadien-18-oic acid, 13α-acetoxy-8(17),13-labdadien-19-al, 15-acetoxy-8E,13-labdadien-19-al, 15-acetoxy-8(17),13-labdadien-19-oic acid, 19-acetoxy-8(E),13-labdadien-15-oic acid methyl ester, 13β-hydroxylabda-8,14-dien-19-oic acid methyl ester, 13β-hydroxylabda-8(17),14-dien-19-oic acid methyl ester, 15-hydroxylabda-8(17),13-dien-19-oic acid methyl ester, labda-8(17)E,13-dien-15,19-dioic acid dimethyl ester, labda-8,13-dien-15,19-diol, labda-8,14-dien-13α,19-diol, labda-8(17),14-dien-13α,19-diol, 8(17),12,14-labdatrien-19-oic acid methyl ester.

Ruzicka *et al.*, *Helv. Chim. Acta*, **24**, 504 (1941)

Caputo *et al.*, *Gazz. Chim. Ital.*, **105**, 639 (1975)

A. araucaria (Mol.) K.Koch.

DITERPENOIDS: (-)-atisirene, (-)-trachylobane.

Kapadi *et al.*, *Tetrahedron Lett.*, 2729 (1965)

A. bidwillii.

DITERPENOIDS: ent-4(18),13-clerodien-15-oic acid, ent-labda-8,13-dien-15-ol, ent-labda-8,13-dien-15-ol-15-acetate.

Caputo *et al.*, *Phytochemistry*, **15**, 1401 (1976)

A. cooki.

DITERPENOIDS: abietinal, 13β-hydroxylabda-8,14-dien-19-al, 13β-hydroxylabda-8(20),14-dien-19-al, 15-hydroxylabda-8,13-dien-19-al, 19-hydroxylabda-8,13-dien-15-oic acid, labda-8,14-diene-13α,19-diol.

Caputo *et al.*, *Phytochemistry*, **13**, 471 (1974)

A. cunninghamii.

DITERPENOIDS: 15-hydroxylabda-8(20),13-dien-19-oic acid, labda-8,13-diene-15,19 diacetate, labda-8,13-diene-15,19-diol 15-acetate.

FLAVONES: amentoflavone, amentoflavone 4″,7-dimethyl ether.

(F) Beckmann *et al.*, *Phytochemistry*, **10**, 2465 (1971)

(T) Caputo *et al.*, *Gazz. Chim. Ital.*, **13**, 475 (1974)

A. excelsa.

DITERPENOIDS: 7,13-abietadien-18-ol, 19-acetoxy-7,13-abietadiene, 13-acetoxy-8(17),14-labdadien-19-oic acid methyl ester, 19-norlabda-8(17),14-diene-4,13-diol.

Chirkova *et al.*, *Izv. Sib. Otd. Akad. Nauk SSSR*, **2**, 99 (1966)

Caputo *et al.*, *Gazz. Chim. Ital.*, **104**, 491 (1974)

A. hunsteinii.

DITERPENOIDS: ent-4(18),13-clerodien-15-ol, 8,15-labd-13-enediol.

Caputo, Mangoni, *Phytochemistry*, **13**, 467 (1974)

A. imbricata.

DITERPENOIDS: acetoxyimbricatolic acid, methyl 13-acetoxylabda-8(17),14-dien-19-oate, 15-acetoxylabd-8(17)-en-19-al, 15-acetoxylabd-8(20)-en-19-al, acetoxyimbricatolic acid, 16-acetoxylabd-8(17)-ene, methyl 15-acetoxylabd-8(17)-en-19-oate, 3β-hydroxylabd-8(20)-en-15-al, 15-hydroxylabd-8(17)-en-19-al, 15-hydroxylabd-8(20)-en-19-al, 15-hydroxylabd-8(17)-ene, methyl 15-hydroxylabd-8(17)-en-19-oate, methyllabd-8(17)-en-15-oate, 3β-hydroxylabd-8(20)-en-15-oic acid, imbricatadiol, imbricatolal, imbricatolic acid, labd-8(17)-ene-15,19-diol, labd-8(20)-ene-3β,15-diol.

Bruns, Weissman, *Tetrahedron Lett.*, 1901 (1966)

Caputo *et al.*, *Gazz. Chim. Ital.*, **106**, 1119 (1976)

ARBUTUS (Ericaceae)

A. andrachne L.

TRITERPENOID: ursolic acid.

Hassan et al., Khim. Prir. Soedin., 119 (1975)

A. unedo L.

ALCOHOL: dotriacontanol.

IRIDOIDS: unedide, unedoside.

(Al) Gascard, C. R. Acad. Sci., 159, 258 (1914)

(I) Davini et al, Phytochemistry, 20, 1583 (1981)

ARCHANGELICIA (Menispermaceae)

A. flava L. Merr.

BISBENZYLISOQUINOLINE ALKALOIDS: homoaromoline, limacine.

PROTOBERBERINE ALKALOIDS: berberine, columbamine, dehydrocorydaline, hydroxyberberine, jatrorrhizine, palmatine, pycnarrhine, thalifendine.

UNCHARACTERIZED ALKALOID,: shobakunine.

Verpoorte et al., J. Nat. Prod., 45, 582 (1982)

A. tschimganica.

COUMARINS: biacangelicin, biacangelicol, osthole, phellapterin, umbelliprenin, xanthotoxol.

Saidkhodzhaev et al., Khim. Prir. Soedin., 96 (1976)

ARCTEGUIETIA

A. pseudoarborea.

DITERPENOID: 13-labd-7-en-2α,15-diol.

Bohlmann et al., Justus Liebigs Ann. Chem., 2127 (1983)

ARCTIUM (Compositae)

A. lappa L.

SESQUILIGNANS: arctigenin, arctigenin 4-β-D-glucopyranoside, arctiin, lappaol C, lappaol D, lappaol E, matairesinol, sesquilignan AL-D, sesquilignan AL-F.

SESQUITERPENOID: dehydrofukinone.

(Sl) Takaoka et al., Bull. Chem. Soc., Japan, 50, 2821 (1977)

(Sl) Ichihara et al., Agr. Biol. Chem., 41, 1813 (1977)

(T) Naya et al., Chem. Lett., 235 (1972)

A. minus.

SESQUITERPENOID: arctiopicrin.

Suchy et al., Croat. Chem. Acta, 29, 247 (1953)

A. tomentosum.

SESQUITERPENOID: arctiopicrin.

Suchy et al., Collect. Czech. Chem. Commun., 24, 1542 (1959)

ARCTODIAPTOMUS

A. salinus.

CAROTENOID: crustaxanthin.

Bodea et al., Omajiu Taluca Ripan., 153 (1966)

ARCTOSTAPHLOS (Ericaceae)

A. uva-ursi (L.) Spreng.

STEROID: β-sitosterol.

TRITERPENOIDS: α-amyrin, betulinic acid, lupeol, oleanolic acid, taraxerol, ursolic acid, uvaol.

Huzii, Osumi, J. Pharm. Soc., Japan, 59, 176 (1939)

Kawai et al., Chem. Abstr., 88 34499b

ARCTOTIS (Ericaceae/Compositae))

A. grandis.

SESQUITERPENOIDS: 12-acetoxysesquisabinene, acetyltabarin, 8β-angeloyloxy-senoxyris-4-en-3-one, tabarin.

Bohlmann et al., Phytochemistry, 17, 1666, 1669 (1978)

Gonzalez et al., ibid., 22, 1509 (1983)

A. revoluta.

SESQUITERPENOIDS: 4β,14-dihydro-3-dehydrozaluzanin C, 11-hydroxyguaiene, isoguaiene, 4β,14,11α,13-tetrahydro-3-dehydrozaluzanin C, 4β,14,11β,13-tetrahydro-3-dehydrozaluzanin C.

NORDITERPENOID: gynurone.

DITERPENOIDS: ent-15-hydroxykaurene, ent-15α-hydroxykaurene, ent-kauran-15-one.

Bohlmann et al., Phytochemistry, 16, 575 (1977)

ARDISIA (Myrsinaceae)

A. cornudentata.

QUINONE: cornudentanone.

Tian et al., Phytochemistry, 26, 2361 (1987)

A. quinquegona.

PHENOLIC: 4-acetoxy-2-(2-acetoxypentadecyl)-methoxyphenol.

QUINOL: ardisianol.

QUINONES: 2-(2-acetoxypentadecyl)-6-methoxy-1,4-benzoquinone, ardisianone.

Kusami et al., Bull. Chem. Soc., Japan, 51, 943 (1978)

A. sieboldii.

QUINONES: ardisiaquinones A, B and C, (Z)-ardisiaquinone A.

Natori et al., Tetrahedron Lett., 1387, 3107 (1968)

A. solanacea.

TRITERPENOIDS: α-amyrin, β-amyrin, bauerenol.

Ahmad et al., Planta Med., 32, 162 (1977)

ARECA (Arecaceae)

A. catechu.

PIPERIDINE ALKALOIDS: guvacine, guvacoline, isoguvacine.

PYRIDINE ALKALOIDS: arecaidine, arecolidine, arecoline.

Winterstein, Weinhagen, Z. Physiol. Chem., 85, 372 (1913)

von Euler et al., Helv. Chim. Acta, 27, 382 (1944)

ARENICOLA

A. marina.

PIGMENT: arenichromene.

Duijn et al., Recl. Trav. Chim. Pays-Bas, 71, 585, 595 (1952)

ARGEMONE (Papaveraceae)

A. albiflora Hornem.

QUATERNARY ALKALOID: (-)-β-scoulerine methohydroxide.

Haisova et al., Collect. Czech. Chem. Commun., 38, 3312 (1973)

A. gracilenta Greene.

PAVINANE ALKALOIDS: (-)-argemonine methohydroxide, (-)-argemonine N-oxide, (-)-isonorargemonine, munitagine, platycerin.

Stermitz, McMurtry, J. Org. Chem., 34, 555 (1969)

A. hispida.

PAVINANE ALKALOID: norargemonine.

Soine, Gisvold, J. Am. Pharm. Assoc., 33, 185 (1944)

A. mexicana L.

ACID: argemonic acid.

PAVINANE ALKALOID: (-)-argemonine.

PROTOBERBERINE ALKALOIDS: α-allocryptopine. berberine protopine Flavone isorhamnetin 3-β-D-glucoside.

(A) Manske, Miller, Can. J. Res., 18B 288 (1940)

(A) Martell et al., J. Am. Chem. Soc., 85, 1022 (1963)

(Acid) Gunstone et al., Chem. Phys. Lipids, 20, 331 (1977)

(F) Rahman, J. Org. Chem., 27, 153 (1962)

A. munita Dur. et Hilg.

PAVINANE ALKALOIDS: 2,9-dimethoxypavinane, munitagine.

Coomes et al., J. Org. Chem., 38, 2701 (1973)

A. munita Dur. et Hilg. subsp. rotunda (Rydb.)
G.W.Ownb.
OXOPROTOBERBERINE ALKALOID: muramine.
PAVINANE ALKALOID: munitagine.
TROPANE ALKALOID: 2,9-dimethoxy-3-hydroxytropane.
　Coomes *et al., J. Org. Chem.,* **38**, 3701 (1973)

A. platyceras Link et Otto.
PAVINANE ALKALOID: platycerine.
QUATERNARY ALKALOIDS: argemonine methohydroxide,
　(-)-α-canadine methohydroxide, cyclanoline
　methohydroxide, magnoflorine, platycerine methohydroxide,
　(-)-α-stylopine methohydroxide.
　Slavik, Slavikova, *Collect. Czech. Chem. Commun.,* **41**, 285 (1976)

A. polyanthemos.
QUATERNARY ALKALOIDS: argemonine methohydroxide,
　(-)-α-canadine methohydroxide, cyclanoline
　methohydroxide, magnoflorine, platycerine methohydroxide,
　(-)-α-stylopine methohydroxide.
　Slavik, Slavikova, *Collect. Czech. Chem. Commun.,* **41**, 285 (1976)

A. subfusiformia var. subfusiformis.
ALKALOIDS: berberine, chelerylline, protopine, sanguinarine.
　Bandoni *et al., Phytochemistry,* **14**, 1785 (1975)

A. subfusiformis var. subinermis.
ALKALOIDS: berberine, chelerylline, protopine, sanguinarine.
　Bandoni *et al., Phytochemistry,* **14**, 1785 (1975)

ARGYLIA (Bignoniaceae)
A. radiata.
IRIDOIDS: argylioside, catalpol, 7-deoxy-8-epi-loganic acid.
　Bianco *et al., Phytochemistry,* **25**, 946 (1986)
　Bianco *et al., Planta Med.,* 55 (1986)

ARGYREIA
A. speciosa.
TRITERPENOIDS: epi-friedelinol, epi-friedelinol acetate.
　Sahu, Chakravarti, *Phytochemistry,* **10**, 1949 (1971)

ARIOCARPUS (Cactaceae)
A. fissuratus K.Schum. var. fissuratus.
AMINE ALKALOID: N-methyl-3,4-dimethoxy-β-phenethylamine.
　Norquist, McLaughlin, *J. Pharm. Sci.,* **59**, 1840 (1970)

A. kotschoubeyanus.
AMINE ALKALOID: N-methyltyramine.
　Keller *et al., J. Pharm. Sci.,* **62**, 408 (1973)

A. retusa.
AMINE ALKALOIDS: N-methyl-3,4-dimethoxy-β-phenethylamine,
　N-methyl-4-methoxy-β-phenethylamine.
　Norquist, McLaughlin, *J. Pharm. Sci.,* **59**, 1840 (1970)
　Horemann *et al., ibid.,* **61**, 41 (1972)

A. scapharostrus.
AMINE ALKALOIDS:
　N,N-dimethyl-3,4-dimethoxy-β-phenethylamine, hordenine,
　N-methyl-3,4-dimethoxyphenethylamine, N-methyltyramine.
　Bruhn, *Phytochemistry,* **14**, 2509 (1975)

ARISAEMA
A. tortuosum.
STEROIDS: campesterol, cholesterol, β-sitosterol, stigmasterol.
　Miglani *et al., Ind. J. Pharm.,* **40**, 24 (1978)

ARISTEGUIETIA
A. pseudoarborea.
DITERPENOID: 13-labd-7-en-2α,15-diol.
　Bohlmann *et al., Justus Liebigs Ann. Chem.,* 2126 (1983)

ARISTIDA (Graminaeae)
A. contorta.
AMINO ACIDS: alanine, glutamic acid, leucine, proline.
　Ghiglione *et al., C. R. Acad. Sci.,* **281D**, 451 (1975)

A. ramosa.
AMINO ACIDS: alanine, glutamic acid, leucine, proline.
　Ghiglione *et al., C. R. Acad. Sci.,* **281D**, 451 (1975)

A. rhiniochloa.
AMINO ACIDS: alanine, glutamic acid, leucine, proline.
　Ghiglione *et al., C. R. Acad. Sci.,* **281D**, 451 (1975)

ARISTOLOCHIA (Aristolochiaceae)
A. argentina Gris.
ALKALOIDS: aristolochine
　1-dimethylamino-3-hydroxy-4-methoxyphenanthrene.
LACTONE: argentilactone.
　(A) Castille *et al., J. Pharm. Belg.,* **4**, 569 (1922)
　(A) Crohare *et al., Phytochemistry,* **13**, 1957 (1974)
　(L) Priestap *et al., ibid.,* **16**, 1579 (1977)

A. clematis L.
ALKALOID: aristolochine.
　Castille *et al., J. Pharm. Belg.,* **4**, 569 (1922)

A. debilis.
ALKALOID: aristolochine.
SESQUITERPENOIDS: 1(10)-aristolen-15-al, aristolone,
　1(10)-aristolone, debilone, 3-oxoishwarane.
　(A) Rosenmund, Reichstein, *Pharm. Acta Helv.,* **18**, 243 (1943)
　(T) Ruecker *et al., Phytochemistry,* **23**, 1647 (1984)

A. indica L.
ALKALOID: aristolochine.
SESQUITERPENOIDS: ishwarane, ishwarol, ishwarone,
　ristolochene, (12S)-7,12-secoishwaran-12-ol,
　5βH,7β,10α-selina-4(14),11-diene.
　(A) Castille *et al., J. Pharm. Belg.,* **4**, 569 (1922)
　(T) Pakrashi *et al., J. Org. Chem.,* **45**, 4765 (1980)

A. maurorum.
PHENANTHRENE: aristolochic acid.
　Kery *et al., Fitoterapia,* **52**, 210 (1981)

A. mollissima Hance.
PHENANTHRENE: aristolochic acid.
STEROID: β-sitosterol.
SESQUITERPENOIDS: aristolactone, mollislactone.
　Liu *et al., Yaoxue Xuebao,* **18**, 684 (1983)

A. moupinensis Franch.
AMIDE ALKALOID: moupinamide.
　Xu, Sun, *Yaoxue Xuebao,* **19**, 48 (1984)

A. peduncularis.
SPIROINDOLONE ALKALOID: aristolarine.
　Kyburz *et al., Helv. Chim. Acta,* **67**, 804 (1984)

A. reticulata.
SESQUITERPENOID: aristolactone.
　Stenlake, Williams, *J. Chem. Soc.,* 2114 (1955)

A. rotunda L.
ALKALOID: aristolochine.
　Rosenmund, Reichstein, *Pharm. Acta Helv.,* **18**, 243 (1943)

A. silpho.
ALKALOID: aristolochine.
　Rosenmund, Reichstein, *Pharm. Acta Helv.,* **18**, 243 (1943)

A. taliscana.
PHENANTHRENE ALKALOID: taliscanine.
LIGNAN: dehydrodiisoeugenol.
　(A) Maldonado *et al., Ciencia (Mexico),* **24**, 237 (1966)
　(L) Ionescu *et al., Pharm. Sci.,* **66**, 1489 (1977)

A. tuberosa.
APORPHINE ALKALOIDS: tuberosinone, tuberosinone
N-β-D-glucoside.
Zhu *et al.*, *Heterocycles.*, 345 (1982)

ARISTOTELIA (Elaeocarpaceae)
A. chilensis (Mol.) Stuntz.
CARBAZOLE ALKALOID: aristotelinine.
INDOLE ALKALOID: aristonine.
SPIROINDOLE ALKALOID: aristone.
Bittner *et al.*, *J. Chem. Soc., Chem. Commun.*, 79 (1978)

A. fruticosa.
INDOLE ALKALOID: isopeduncularine.
Bick *et al.*, *Tetrahedron*, **41**, 3127 (1985)

A. peduncularis (Labill.) Hook f.
CARBAZOLE ALKALOIDS: aristoserratine, aristoserratone.
INDOLE ALKALOIDS: hobartine, isopeduncularine, peduncularine.
Hai *et al.*, *Helv. Chim. Acta*, **63**, 2130 (1980)
Kyburz *et al.*, *ibid.*, **67**, 804 (1984)

A. serrata (J.R.et G.Forst) W.R.B.Oliver.
CARBAZOLE ALKALOIDS: aristomakine, aristoserratine,
aristoserratone, aristoteline, aristotelinone, serratenone,
serratoline.
INDOLE ALKALOIDS: isopeduncularine, makomakine, makonine.
Bick *et al.*, *Tetrahedron Lett.*, 545 (1980)
Hai *et al.*, *Helv. Chim. Acta*, **63**, 2130 (1980)
Bick *et al.*, *Heterocycles*, **20**, 667 (1983)

ARMILLARIA (Tricholometaceae)
A. mellea.
SESQUITERPENOIDS: armillaridin, armillarin, melledonal,
melledonol, melleolide.
Donnelly *et al.*, *Tetrahedron Lett.*, 5343 (1985)

ARNEBIA (Boraginaceae)
A. decumbens.
BENZOQUINONES: shikonin, shikonin acetate, shikonin
isovalerate.
NAPHTHOQUINONE:
5,8-dihydroxy-2-(4-methylpent-3-enyl)-1,4-naphthoquinone.
STEROID: stigmasterol.
Afzal *et al.*, *Agr. Biol. Chem.*, **50**, 759, 1651 (1986)

A. euchroma.
SESQUITERPENOID: arnebinone.
Yao *et al.*, *Tetrahedron Lett.*, 3247 (1983)

A. nobilis.
NAPHTHOQUINONES:
2-(acetoxy-4-hydroxy-4-methylpent)-5,8-dihydroxy-1,4-naphthoquinone,
5,8-dihydroxy-2-(4-hydroxy-4-methylpent)-1,4-naphthalenedione.
PIGMENTS: arnebins I, II, III, IV, V, VI and VII.
(N) Bhakuni *et al.*, *Phytochemistry*, **10**, 1909 (1971)
(P) Dhar *et al.*, *Indian J. Chem.*, **11**, 528 (1973)

ARNICA (Compositae)
A. amplexicaulis.
Three terpenoids.
Bohlmann, Zdero, *Tetrahedron Lett.*, 2827 (1972)

A. chamissonis.
MONOTERPENOID: thymol isobutyrate.
Rinn, *Planta Med.*, **18**, 147 (1970)

A. chamissonis subsp. genuina Maguire.
SESQUITERPENOID: chamissonolide.
Willuhn *et al.*, *Tetrahedron*, **37**, 773 (1981)

A. foliosa.
SESQUITERPENOIDS: arnifolin, carabrone.
Evstratova *et al.*, *Khim. Prir. Soedin.*, 270 (1971)

A. longifolia.
SESQUITERPENOIDS: carabrone, helenalin, xantholongin.
Guenter, Hermann, *Planta Med.*, **37**, 325 (1979)

A. montana L.
SESQUITERPENOIDS: arnicolides A, B, C, D and E, arnifolin,
helenalin methacrylate.
TRITERPENOID: faradiol.
Zimmermann, *Helv. Chim. Acta.*, **26**, 642 (1943)
Poplawski *et al.*, *Collect. Czech. Chem. Commun.*, **36** 2189 (1971)
Hermann *et al.*, *Planta Med.*, **34**, 299 (1978)

ARTABOTRYS (Annonaceae)
A. suaveolens Blume.
APORPHINE ALKALOIDS: artabotrine, artabotrinine, isocorydine,
suaveoline.
Maranon, *Philipp. J. Sci.*, **38**, 259 (1928)

A. uncinatus.
SESQUITERPENOID: yinghaosu A.
Liang *et al.*, *Hua Hsueh Hsueh Pao*, **37**, 215 (1979)

A. venustus.
CATECHOL ALKALOID: artavenustine.
Cave *et al.*, *J. Nat. Prod.*, **49**, 602 (1986)

ARTEMISIA (Compositae/Anthemideae)
A. abrotanum L.
UNCHARACTERIZED ALKALOID: abrotine.
Giacosa, *Merck's Jahresblatt*, 1356 (1883)

A. absinthium L.
LIGNAN: lirioresinol A.
NORSESQUITERPENOIDS: 3,6-dihydrochamazulene,
5,6-dihydrochamazulene.
DITERPENOID: absinthin.
SESQUITERPENOIDS: anabsin, anabsinin, artabsin, artabsinolides
A, B, C and D, artemolin-a, artemolin-b,
2,3-diepi-artabsinolide C, hydroxypelenolide,
ketopelenolide, ketopelenolide a, ketopelenolide b.
(L) Briggs *et al.*, *J. Chem. Soc., C*, 3042 (1968)
(T) Beauhaire *et al.*, *J. Chem. Soc., Perkin I*, 861 (1982)

A. annua L.
ALIPHATIC: octacosanol.
FLAVONOID: 3,5-dihydroxy-6,7,3',4'-tetramethoxyflavone.
STEROIDS: β-sitosterol, stigmasterol.
SESQUITERPENOIDS: arteannuin A, arteannuin B, artemisic acid,
artemisilactone, artemisinic acid, quinghao acid, quinghaosu
I, II, III and IV.
(Al, F, S) Tu *et al.*, *Zhongyao Tongbao*, **10**, 419 (1985); *Chem. Abstr.*,
104 31868b
(T) Tu *et al.*, *Planta Med.*, **44**, 143 (1982)
(T) Zhu *et al.*, *Huaxue Xuebao*, **42**, 937 (1984)

A. arborescens.
SESQUITERPENOID: arborescin.
Meisels, Weizmann, *J. Am. Chem. Soc.*, **75**, 3865 (1953)

A. arbuscula.
SESQUITERPENOIDS: arbusculins A, B, C, D and E, badgerin,
tatridin A, tatridin B.
Irwin, Geissman, *Phytochemistry*, **8**, 2411 (1969)
Shafizadeh, Bhadane, *ibid.*, **12**, 857 (1973)

A. argentea.
SESQUITERPENOIDS: argentiolide A, argentiolide B.
El-Emary, Bohlmann, *Phytochemistry,* **19**, 845 (1980)

A. ashurbajevii.
SESQUITERPENOIDS: canin, chrysartemin B, granilin, hanphyllin.
Sakirov *et al., Khim. Prir. Soedin.,* 573 (1985)

A. austriaca.
SESQUITERPENOIDS: austricin, dehydroaustricin.
Rybalko, *J. Gen. Chem. USSR,* **33**, 2734 (1963)

A. balchanorum.
SESQUITERPENOIDS: balchanin, balchanolide, costunolide, hydroxybalchanolide, isobalchanolide.
Suchy *et al., Collect. Czech. Chem. Commun.,* **27**, 2925 (1962)

A. barrelieri.
SESQUITERPENOID: barrelin.
Villar *et al., Phytochemistry,* **22**, 777 (1983)

A. biglovii.
SESQUITERPENOID: arbiglovin.
Herz, Santhaman, *J. Org. Chem.,* **30**, 4340 (1965)

A. campestris.
CHROMENE: artemisenol.
De Pascual Teresa *et al., Phytochemistry,* **22**, 2587 (1983)

A. cana Pursh.
SESQUITERPENOIDS: artecanin, canin, viscidulin A.
Shafizadeh, Bhadane, *J. Org. Chem.,* **37**, 3168 (1972)

A. cana Pursh. subsp. cana.
SESQUITERPENOID: deacetylmatricarin.
Lee *et al., Phytochemistry,* **8**, 1515 (1969)

A. canariensis.
SESQUITERPENOID: tabarin.
Gonzalez *et al., An. Quim.,* **69**, 667 (1973)

A. capillaris.
ACETYLENICS: capillanol, norcapillene.
ISOCOUMARIN: capillarin.
PHENOLICS: capillartemisin A, capillartemisin B.
SESQUITERPENOID: capillarisin.
(Ac) Miyazawa *et al., Phytochemistry,* **14**, 1874 (1975)
(I) Bohlmann *et al., Chem. Ber.,* **95**, 39, 602 (1962)
(P) Kitagawa *et al., Chem. Pharm. Bull.,* **31**, 352 (1983)
(T) Komiya *et al., ibid.,* **23**, 1387 (1975)

A. carruthii.
SESQUITERPENOIDS: 11,13-dihydroludartin, ludartin.
Bohlmann *et al., Phytochemistry,* **20**, 2375 (1981)

A. caucasica.
SESQUITERPENOID: grossmizin.
Konovalova *et al., Khim. Prir. Soedin.,* 741 (1971)

A. chasarica.
SESQUITERPENOIDS: α-santonin, tauremisin.
Serkerov *et al., Khim. Prir. Soedin.,* 657 (1975)

A. cina.
SESQUITERPENOID: artemisin.
Cory, *J. Am. Chem. Soc.,* **77**, 1044 (1955)

A. compacta.
COUMARINS: dihydrocoumarin, β-santonin, scopoletin, umbelliferone.
Usynina *et al., Khim. Prir. Soedin.,* 809 (1976)

A. confertiflora.
SESQUITERPENOID: deacetylconfertiflorin.
Romo *et al., Can. J. Chem.,* **46**, 1535 (1968)

A. cretacea Fiori.
SESQUITERPENOID: taurin.
Callieri *et al., Acta Cryst., C* **39**, 1115 (1983)

A. cylindrica var. koenigii.
TRITERPENOID: fernenol.
Kundu *et al., Aust. J. Chem.,* **21**, 1931 (1968)

A. cylindrica var. media.
TRITERPENOID: fernenol.
Kundu *et al., Aust. J. Chem.,* **21**, 1931 (1968)

A. douglasiana.
SESQUITERPENOIDS: arglanin, arteglasins A and B.
Lee *et al., Phytochemistry,* **10**, 405 (1971)

A. dracunculus.
ISOCOUMARINS: (E)-artemidin, (Z)-artemidin, artemidinol, capillarin.
SESQUITERPENOID: spathulenol.
(I) Bohlmann *et al., Chem. Ber.,* **103**, 2856 (1970)
(I) Mallabaev *et al., Khim. Prir. Soedin.,* 811 (1976)
(T) Bowyer, Jefferies, *Chem. Ind.,* 1245 (1963)

A. fasciculata.
PHENOLIC: α-santonin.
Aleskerova *et al., Chem. Abstr.,* 107 4301z

A. feddei Lev. et Van.
MONOTERPENOIDS: yomogi alcohol, yomogi alcohol A, yomogiartenin.
SESQUITERPENOID: meridianone.
Matsueda *et al., J. Pharm. Soc., Japan,* **100**, 615 (1980)

A. filifolia.
MONOTERPENOIDS: filifolide A, filifolide B, (+)-filifolone, 4,5-seco-3,5-longibornanedione, 3,4-seco-4-longibornen-3-al.
SESQUITERPENOID: 3-longipinanone.
Torrance *et al., J. Org. Chem.,* **39**, 1068 (1973)
Bohlmann *et al., Phytochemistry,* **22**, 503 (1983)

A. finita.
SESQUITERPENOID: finitin.
Kitagawa *et al., Rep. Inst. Sci. Res. Manchukuo,* **6**, 124 (1942)

A. fragrans.
SESQUITERPENOIDS: alhanol, artapshin, fragranol, fragranol acetate, fragranol isobutyrate, fragranol isopropionate, fragranol methyl ester, fragranol sec-butyrate, jeiranbatanolide, shonachalin.
Serkerov *et al., Khim. Prir. Soedin.,* 787 (1985)

A. fragrans var. erivanica.
SESQUITERPENOID: erivanin.
Evstratova *et al., Khim. Prir. Soedin.,* 239 (1969)

A. franserioides.
SESQUITERPENOID: artefransin.
Lee, Geissman, *Phytochemistry,* **10**, 205 (1971)

A. frigide.
SESQUITERPENOID: hydroxyachillin.
White, Winter, *Tetrahedron Lett.,* 137 (1963)

A. glabella.
SESQUITERPENOID: arglabin.
Adekenov *et al., Khim. Prir. Soedin.,* 655 (1982)

A. glutinosa.
ACETOPHENONES: dehydroespeletone, glutinosol.
FLAVONOIDS: dihydroquercetin 7,3'-dimethyl ether, naringenin, palmatin, rhamnetin, 5,3',4'-trihydroxy-7-methylflavanone.
Gonzalez *et al., Phytochemistry,* **22**, 1515 (1983)

A. granatensis.
SESQUITERPENOIDS:
1-hydroxy-6β,7α,11β-H-eudesm-4-en-6,12-olide,
1-keto-6α,7β,11β-eudesmen-6,12-olide, tauramisin.
Gonzalez *et al., An. Quim.,* **70**, 231 (1974)

A. hanseniana var. phylostachus.
SESQUITERPENOIDS: arsubin, artemin, isoerivanin, tauremisin, taurin.
Serkerov et al., Khim. Prir. Soedin., 665 (1976)

A. herba.
SESQUITERPENOIDS:
3β,8α-dihydroxy-4(15),9-germacradien-12,6-olide,
3β,8α-dihydroxy-6αH,7H,11βH-germacran-5(14),9(11)-dien-12-olide,
3β,8α-dihydroxy-6αH,7H,11βH-germacran-5(14),9(11)-dien-3-oxolide.
Gordon et al., J. Nat. Prod., 44, 432 (1981)

A. herba alba.
SESQUITERPENOIDS: herbolides A, B, C, D, E and F.
Segal et al., Phytochemistry, 23, 2954 (1984)

A. herba alba subsp. valentina.
SESQUITERPENOID: torrentin.
Gomis et al., Phytochemistry, 18, 1523 (1979)

A. inculta.
SESQUITERPENOID: arteincultone.
Khafagy et al., Phytochemistry, 22, 1821 (1983)

A. ireyniana.
COUMARIN: scopoletin.
Konovalova et al., Khim. Prir. Soedin., 97 (1976)

A. japonica.
COUMARIN: scopoletin.
Konovalova et al., Khim. Prir. Soedin., 97 (1976)

A. klotzchiana.
FLAVONE: 3,5-dihydroxy-6,7,8-trimethoxyflavone.
SESQUITERPENOIDS: chrysartemin A, deacetylmatricarin.
(F) Dominguez et al., Phytochemistry, 14, 2511 (1975)
(T) Romo et al., Phytochemistry, 9, 1615 (1970)

A. koidzumii.
SESQUITERPENOID: koidzumiol.
Bohlmann, Zdero, Phytochemistry, 19, 149 (1980)

A. lactiflora.
SESQUITERPENOID: lactiflorenol.
Zeng et al., Zhongcaoyao, 13, 259 (1982)

A. leucodes.
LACTONE: artein.
SESQUITERPENOIDS: anhydroaustricin, artelin, austricin.
(L) Mallabaev et al., Khim. Prir. Soedin., 46 (1986)
(T) Saitbaeva et al., ibid., 115 (1986)

A. longiloba.
SESQUITERPENOID: longilobol.
Shafizadeh, Bhadane, Tetrahedron Lett., 2171 (1973)

A. ludoviciana.
SESQUITERPENOIDS: (1R,3R)-chrysanthemol, ludalbin, ludovicins A, B and C.
Lee, Geissman, Phytochemistry, 9 403 (1970)
Alexander, Epstein, J. Org. Chem., 40, 2576 (1975)

A. maritima L.
SESQUITERPENOIDS: artemin, artemisin, davanone, 1β-hydroxy-6β,7α,11βH-selen-4-en-6,12-olide.
Gonzalez et al., Phytochemistry, 16, 1836 (1977)
Gonzalez et al., J. Chem. Soc., Perkin I, 1243 (1978)G

A. maritima subsp. gallica.
SESQUITERPENOID: maritimin.
Gonzalez et al., Phytochemistry, 20, 2367 (1981)

A. maritima subsp. maritima.
SESQUITERPENOID: hydroxydavanone.
Jork, Nachtrab, Arch. Pharm., 312, 923 (1979)

A. mesatlantica.
FLAVONOID: 5,4'-dihydroxy-6,7,3',5'-tetramethoxyflavone.
Bouzid et al., Phytochemistry, 21, 803 (1982)

A. messerschmidtiana Bess.
ACIDS: caprylic acid, isovaleric acid.
Stefanov, Popov, Izv. Khim., 8, 518 (1975); Chem. Abstr., 84 176718p

A. mexicana.
SESQUITERPENOIDS: 8α-acetoxyarmexifolin, armefolin, armexifolin, artemorin, chrysartemin A, estafietin.
Mata et al., Phytochemistry, 23, 1665 (1984)

A. mexicana subsp. angustifolia.
SESQUITERPENOIDS: 8α-acetoxyarmexifolin, armefolin, armexifolin, armexin, artimexifolin, α-epoxyludalbin.
Romo de Vivar et al., Rev. Latinoam. Quim., 8, 127 (1977)
Mata et al., Phytochemistry, 23, 1665 (1984)

A. minus.
SESQUITERPENOID: arctiopicrin.
Suchy et al., Collect. Czech. Chem. Commun., 24, 1542 (1959)

A. molinieu.
SESQUITERPENOID: 7,8-dehydrobisabolol oxide A acetate.
Bohlmann, Zdero, Chem. Ber., 108, 2153 (1975)

A. monogyna.
SESQUITERPENOID: mibulactone.
Kariyone et al., J. Pharm. Soc., Japan, 69, 310 (1949)

A. monosperma.
ACETOPHENONE: artemispermol.
Bohlmann, Ehlers, Phytochemistry, 16, 1450 (1977)

A. nachitschevanica.
PHENOLIC: α-santonin.
Aleskerova et al., Chem. Abstr., 107 4301z

A. nilegarica.
AMINO ACIDS: arginine, glutamic acid, glycine, isoleucine, leucine, threonine.
Saxena et al., Acta Cientifica Indica Chemica, 11, 179 (1985); Chem. Abstr., 107 151193u

A. ochroleuca.
COUMARIN SESQUITERPENOIDS: drimanchone, secodrial, secodriol.
Gregor et al., Phytochemistry, 22, 1997 (1983)

A. pallens.
PYRONE:
2-(2-methyl-2-vinyltetrahydrofuran-5-yl)-2,6,6-trimethyldihydropyr-3-one.
SESQUITERPENOIDS: artemone, davana ether, davanafuran, davanone, isodavanone.
(P) Thomas, Helv. Chim. Acta, 57, 2081 (1964)
(T) Thomas, Pilton, ibid., 54, 1890 (1971)
(T) Thomas, Dubini, ibid., 57, 2066 (1974)

A. pauciflora.
SESQUITERPENOID: artepaulin.
Adekenov et al., Khim. Prir. Soedin., 238 (1983)

A. pontica.
COUMARIN SESQUITERPENOID:
7-o-(3-hydroxydrimen-11-yl)-isofraxidin.
Bohlmann et al., Chem. Ber., 107, 644 (1974)

A. princeps Pamp.
SESQUITERPENOIDS: τ-humulene, yomogin.
Geissman, J. Org. Chem., 31, 2523 (1966)
Yano, Nishijima, Phytochemistry, 13, 1207 (1974)

A. pygmaea.
SESQUITERPENOID: pygmol.
Irwin, Geissman, Phytochemistry, 12, 849 (1973)

A. rothrockii.

FLAVONE: chrysoplenol D.

SESQUITERPENOIDS: rothin A, rothin B.

(F) Ghisalberti et al., Aust. J. Chem., **20**, 1049 (1967)

(T) Irwin et al., Phytochemistry, **10**, 637 (1971)

A. santolina.

SESQUITERPENOIDS: arsanin, arsantin, artesin.

Moiseeva et al., Khim. Prir. Soedin., 167 (1973)

A. santolinifolia.

FLAVONOIDS: kaempferol, kaempferol 3-β-D-glucopyranosyl-6-β-L-rhamnopyranoside, luteolin, quercetin, rutin, scopolin.

Chumbalev, Fadeeva, Prikl. Teor. Khim., **5**, 71 (1974); Chem. Abstr., **85** 119568m

A. schmidtiana.

THIOPHENE: schmidtiol.

Jakupovic et al., Phytochemistry, **25**, 1663 (1986)

A. sieversiana.

SESQUITERPENOIDS: sieversin, sieversinin.

Nazarenko, Leont'eva, Khim. Prir. Soedin, 399 (1966)

A. silvatica.

COUMARIN: herniarin.

Konovalova et al., Khim. Prir. Soedin., 97 (1976)

A. tolonifera.

COUMARIN: herniarin.

Konovalova et al., Khim. Prir. Soedin., 97 (1976)

A. spicagera.

PHENOLICS: α-santonin, β-santonin.

Aleskerova et al., Chem. Abstr., **107** 4301z

A. sublessingiana.

SESQUITERPENOID: arsubin.

Tarasov et al., Khim. Prir. Soedin., 745 (1971)

A. taurica.

FLAVONOID: axillaroside.

SESQUITERPENOID: taurin.

(F) Oganesyan et al., Khim. Prir. Soedin., 599 (1976)

(T) Kochatova et al., Khim. Prir. Soedin., 205 (1968)

A. tilesii.

SESQUITERPENOID: matricarin.

Cekan et al., Collect. Czech. Chem. Commun., **24**, 1554 (1959)

A. transiliensis.

FLAVONES: 3',4',5,7-tetrahydroxy-3-methoxyflavone, transilin, transilitin.

Chumbalov et al., Khim. Prir. Soedin., **236**, 439 (1969)

A. tridentata.

MONOTERPENOIDS: artemeseole, 1,2-epoxy-2,5-dimethyl-3-vinyl-4-hexene, santolinic acid, santolinic acid methyl ester, santolinosides A, B, B' and C, 1,6,6-trimethyl-4-ethenyl-exo-2-oxabicyclo-[3,1,0]-hexane.

COUMARIN SESQUITERPENOID: tripartol.

SESQUITERPENOID: artevasin.

Gregor et al., Phytochemistry, **22**, 1997 (1983)

A. tridentata rothrockii.

MONOTERPENOID: rothrockene.

Epstein et al., J. Org. Chem., **47**, 175 (1982)

A. tridentata var. tridentata.

MONOTERPENOID: methyl santalinate.

Shaw et al., J. Chem. Soc., Chem. Commun., 590 (1975)

A. tripartita.

SESQUITERPENOID: colartin.

Irwin, Geissman, Phytochemistry, **8**, 2411 (1969)

A. tripartita subsp. rupicola.

SESQUITERPENOID: novanin.

Irwin, Geissman, Phytochemistry, **12**, 875 (1973)

A. umbelliformis Lam.

SESQUITERPENOIDS: 5-desoxy-5-hydroperoxytelekin, 5-desoxy-5-hydroperoxy-5-epi-telekin, umbellifolide.

Appendino et al., J. Chem. Soc., Perkin I, 2705 (1983)

A. vachanica.

UNCHARACTERIZED TERPENOID: vachanic acid.

Ketchakova et al., Khim. Prir. Soedin., 306 (1965)

A. verlatorum.

SESQUITERPENOIDS: anhydroverlatorin, artemorin, verlatorin.

Geissman, Phytochemistry, **9**, 2377 (1970)

A. vestita.

SESQUITERPENOID: aromadendrine.

Dolejs et al., Collect. Czech. Chem. Commun., **25**, 1483 (1960)

A. viscidula.

SESQUITERPENOIDS: viscidulins A, B and C.

Shafizadeh, Bhadane, J. Org. Chem., **37**, 3168 (1972)

A. vulgaris.

MONOTERPENOID: vulgarole.

SESQUITERPENOIDS: spathulenol, vulgarin.

TRITERPENOIDS: α-amyrin, α-amryin acetate, fernenol.

Nishimoto et al., Tetrahedron, **24**, 735 (1968)

Nano et al., Planta Med., **30**, 211 (1976)

ARTHONIA

A. impolita.

ACID: arthonioic acid.

Fehlhaber et al., Z. Naturforsch., **25**, 49 (1970)

ARTHRAXON

A. hispidus.

BENZOPYRAN: (-)-arthraxin.

Kaneta et al., Bull. Chem. Soc., Japan, **42**, 2084 (1969)

ARTHROPHYTUM

A. leptocladum M.Pop. (Hammada leptoclada).

AMINE ALKALOID: N-methyl-β-phenethylamine.

CARBOLINE ALKALOIDS: leptocladine, N-methyltetrahydro-β-carboline, 2-methyltetrahydro-β-carboline.

INDOLE ALKALOID: dipterine.

Yurushevskii, J. Gen. Chem. USSR, **9** 545 (1939)

ARTOCARPUS

A. chpalasha.

BENZOPYRAN: chaplashin.

TRITERPENOID: isocycloartenol.

(B) Rao et al., Indian J. Chem., **10**, 905, 989 (1972)

(T) Mahato et al., Phytochemistry, **10**, 1351 (1971)

A. heterophyllus.

FLAVONES: artoflavanone 2',4',5,7-tetrahydroxyflavone.

Dayal et al., Indian J. Chem., **12**, 895 (1974)

A. integrifolia.

DIOXOCIN: (+)-cyanomaclurin.

PIGMENT.: cycloartocarpin.

STEROID: artostenone.

(D) Appel et al., J. Chem. Soc., 752 (1935)

(P) Dave et al., Tetrahedron Lett., 9 (1962)

(S) Nath, Z. Physiol. Chem., **247**, 9 (1937)

A. pithecogallus C.Y.Wu.

FLAVONOIDS: morin, morin calcium chelate.

Mu et al., Dhiwu Xuebao, **24**, 147 (1982); Chem. Abstr., **97** 107043w

ARUM (Araceae)

A. maculatum L.

CYANO COMPOUNDS: isotriglochinin, triglochinin.
PRENOL: betulaprenol.

(Cy) Nahrstadt et al., Phytochemistry, **14**, 1870 (1975)
(P) Wellburn et al., Nature, **212**, 1364 (1966)

ARUNCUS (Rosaceae)

A. dioicus.

ACIDS: caffeic acid, ferulic acid.
GLUCOSIDE: caffeoyl-1-β-D-glucoside.

Kulesh et al., Khim. Prir. Soedin., 506 (1986)

ARUNDO (Gramineae)

A. caspicus.

TRITERPENOID: arundoin.

Eglinton et al., Tetrahedron Lett., 2323 (1964)

A. conspicua.

TRITERPENOID: arundoin.

Ohmoto et al., J. Chem. Soc., Chem. Commun., 601 (1969)

A. donax L.

BISINDOLE ALKALOID:
3,3′-bis(indolylmethyl)-dimethylammonium hydroxide.
INDOLE ALKALOIDS: donaxarine, donaxiridine, gramine.

Ubaidullaev et al., Khim. Prir. Soedin., 553 (1976)

ASA

A. foetida.

COUMARIN SESQUITERPENOIDS: farnisiferols A, B and C.

Caglioti et al., Helv. Chim. Acta, **41**, 2278 (1958)

ASAEMIA

A. axillaris.

Two terpenoids.

Bohlmann, Zdero, Chem. Ber., **107**, 1071 (1974)

ASANTHUS

A. solidaginifolia.

11 flavonoids.

Yu et al., J. Nat. Prod., **49**, 553 (1986)

A. thyrsiflora.

FLAVONES: 6-methoxyluteolin, quercetin, quercetin
3-O-β-D-galactoside, quercetin 3-O-β-D-rhamnogalactoside.

Yu et al., J. Nat. Prod., **49**, 553 (1986)

ASARUM (Aristolochiaceae)

A. caulescens.

FURANONE: (+)-aiofuranone.
SESQUITERPENOIDS: cauleslactone, caulesyl acetate,
caulolactone A, caulolactone B, dihydrofuranocaulesone,
furanocaulescones A, B and C, germacrone-4,5-epoxide,
4-hydroxy-β-bulnesene, selina-3,7(11)-dien-8-one,
selina-4(14),7(11)-dien-8-one.

Endo et al., Chem. Pharm. Bull., **27**, 275 (1979)

A. europeum L.

ACID: aconitic acid.
UNCHARACTERIZED ALKALOID: asarine.
MONOTERPENOIDS: bornyl acetate, methyleugenol.
SESQUITERPENOID: furopelargone A.

(Acid) Krogh, Acta Chem. Scand., **23**, 2923 (1969)
(A) Abdul'menev, Farmatsiya, **8**, 39 (1954)
(T) Jork, Biering, Arch. Pharm., **312**, 681 (1979)

A. taitonense.

KETONE: asatone.

Chen et al., Tetrahedron Lett., 1607 (1972)

ASCARINA

A. lucida.

FLAVONOIDS: isoorientin, isovitexin, kaempferol, kaempferol
3-O-diglucoside, kaempferol 3-O-glucoside, orientin,
quercetin, quercetin 3-O-diglucoside, quercetin
3-O-glucoside, vitexin.

Soltis et al., J. Nat. Prod., **45**, 415 (1982)

ASCHINEA

A. chrysantha.

ANTHRAQUINONE: funiculosin.

Howard et al., Biochem. J., **44**, 227 (1949)

ASCLEPIAS (Asclepiadaceae)

A. albicans.

SAPONIN: 3-O-glucopyranosyl-β-D-glucopyranoside.

Koike et al., Chem. Pharm. Bull., **28**, 401 (1980)

A. cornuti.

TRITERPENOID: β-amyrin.

Hormann, Firzlaff, Arch. Pharm., **268**, 64 (1930)

A. curassavica.

SAPONIN: asclepin.
STEROID: calotropin.

(Sap) Singh et al., Phytochemistry, **11**, 757 (1972)
(S) Kupchan et al., Science, **146**, 1685 (1964)

A. eriocarpa.

STEROIDS: eriocarpin, labriformidin, labriformin.

Sieber et al., Phytochemistry, **17**, 967 (1978)

A. fruticosa R.Br.

STEROIDS: afroside, 3′-epi-afroside, 3′-epi-afroside 3′-acetate,
asclepin, calactin, 19-deoxyuscharin, 3′-didehydroafroside,
gomphoside, 3-epi-gomphoside, 3-epi-gomphoside
3′-acetate, 4′β-hydroxygomphoside, uscharidin, uscharin.

Cheung et al., J. Chem. Soc., Perkin I, 2827 (1983)

A. humistrata.

TERPENOID: humistratin.

Nishio et al., J. Org. Chem., **47**, 2154 (1982)

A. labriformis.

STEROIDS: eriocarpin, labriformidin, labriformin.

Sieber et al., Phytochemistry, **17**, 967 (1978)

A. syriaca L.

PYRIDINE ALKALOID: nicotine.
STEROIDS: syriobioside, syriogenin.
TRITERPENOID: β-amyrin.

(A) Marion, Can. J. Res., **23B**, 165 (1945)
(S) Rashkas, Abubakirov, Khim. Prir. Soedin., 615 (1974)
(T) Mendive, J. Org. Chem., **5**, 235 (1940)

A. tuberosa.

COUMARINS: isorhamnetin, kaempferol, quercetin, rutin.
PHENOLICS: caffeic acid, chlorogenic acid.
STEROID: β-sitosterol.
TRITERPENOIDS: α-amyrin, β-amyrin, friedelin, lupeol.

Pagani, Boll. Chim. Farm., **114**, 450 (1975); Chem. Abstr., 84 56478p

ASCOCHYTA

A. viciae.

ANTIBIOTICS: ascochlorin, ascofuranone.

Tamura et al., J. Antibiotics, **21**, 539 (1968)

ASER

A. umbellatus.

SESQUITERPENOID: 4(15),7,(11)8-selinatriene.

Bohlmann et al., Phytochemistry, **19**, 435 (1980)

ASIMINA (Rosaceae/Annonaceae)

A. triloba Dun.

APORPHINE ALKALOIDS: anolobine, asimilobine.

GENIN: asimicin.

(A) Tomita, Kozuka, J. Pharm. Soc., Japan, **85**, 77 (1965)

(G) Rupprecht et al., Heterocycles, **24**, 1197 (1986)

ASPALATHUS

A. linearis.

GLUCOSIDE: aspalathin.

Koeppen et al., Biochem. J., **99**, 604 (1966)

ASPARAGOPSIS
(Bonnemaisoniaceae/Rhodophyceae)

A. armata.

STEROID: cholesta-5,25-diene-3β,24-diol.

Morisaki et al., Chem. Pharm. Bull., **24**, 3214 (1976)

ASPARAGUS (Liliaceae)

A. broussonetii Facq.

STEROIDS: penogenin, 25S-ruscogenin, sarsasapogenin.

Panova et al., Probl. Farm., **11**, 54 (1983); Chem. Abstr., 100 20435g

A. cochin-chinensis.

AMINO ACIDS: alanine, arginine, asparagine, aspartic acid, citrulline, glutamic acid, glycine, histidine, isoleucine, leucine, lycine, methionine, phenylalanine, proline, serine, threonine, tyrosine, valine.

Tomoda et al., Chem. Abstr., 85 30677y

A. officinalis L.

CAROTENOIDS: α-carotene, β-carotene, τ-carotene, lutein, zeaxanthin.

SAPONINS: asparagosides A, B, C, D, E, F, G, H and I., asparasaponins I and II.

(C) Meliksetyan et al., Izv. S.-KH. Nauk, **19**, 68 (1976); Chem. Abstr., 86 117589r

(Sap) Kintya et al., Khim. Prir. Soedin., **400**, 762 (1976)

(Sap) Goryanu, Kintya, ibid., 810 (1977)

A. sprengeri.

STEROIDAL GLYCOSIDES: sprengerinin A, B, C and D.

D. Sharma et al., Phytochemistry, **22**, 2259 (1983)

A. tenuifolium Lam.

STEROIDS: sarsasaponin, β-sitosterol, 25β-spirostan-3,5-diene, yamogenin.

Panova et al., Farmatsiya (Sofia), **25**, 39 (1975)

A. umbellatus Link.

STEROID: hispidogenin.

Maiti, Mukherjee, Chem. Ind., 1653 (1965)

ASPHODELINE

A. globifera.

FLAVONOIDS: acacetin, apigenin, apigenin 7-O-glucoside, asphodalin, chrysoeriol, chrysophanol, esculetin, luteolin, 6-methoxyluteolin, microcarpin, scutellarein.

STEROID: β-sitosterol.

SESQUITERPENOIDS:

1β-acetoxyeudesman-4(15),7(11)-dien-2α,12-olide, 1β-acetoxy-8β-hydroxyeudesman-4(15),7(11)-dien-2α,12-olide.

Ulubelen et al., Phytochemistry, **24**, 2923 (1985)

ASPHODELUS (Liliaceae)

A. albus.

STEROIDS: campesterol, fucosterol, β-sitosterol, stigmasterol.

TRITERPENOID: β-amyrin.

Abdel-Gawad et al., Fitoterapia, **47**, 111 (1976)

A. microcarpus.

ANTHRAQUINONE: asphodelin.

Makino et al., Chem. Pharm. Bull., **36**, 1111 (1978)

A. tenuifolius.

AMINO ACIDS: alanine, glycine, serine, valine.

Saxema et al., J. Inst. Chem. Calcutta, **48**, 18 (1976); Chem. Abstr., 85 14773r

ASPICILIA (Lecanoraceae)

A. calcarea (L.) Korb.(Lecanora calcarea (L.) Sommerf.).

MACROCYCLIC ALKALOID: aspicilin.

Huneck et al., Tetrahedron, **29**, 3687 (1973)

A. gibbosa (Ach.) Korb.(Lecanora gibbosa (Ach.) Nyl.).

MACROCYCLIC ALKALOID: aspicilin.

Huneck et al., Tetrahedron, **29**, 3687 (1973)

ASPIDOSPERMA (Apocynaceae)

A. album (Vahl.) R.Bent.

IMIDAZOLIDINOCARBAZOLE ALKALOIDS: N-acetyl-N-depropionylaspidoalbine, aspidoalbine, N-propionyl-N-deacetylaspidolimidine.

Ferrari et al., Can. J. Chem., **42**, 2705 (1964)

A. auriculatum Mfg.

CARBOLINE ALKALOIDS: dihydrocorynantheol, reserpinine.

Gilbert et al., Tetrahedron, **21**, 1141 (1965)

A. australe Mull Agrov.

CARBAZOLE ALKALOID: (±)-gutambuine.

Schmitz et al., Helv. Chim. Acta, **40**, 1189 (1957)

A. campus-belus.

CARBAZOLE ALKALOID: olivacine.

Garcia et al., Phytochemistry, **15**, 1093 (1976)

A. carapanauba Pichon.

SPIROINDOLONE ALKALOID: carapanaubine.

Gilbert et al., J. Am. Chem. Soc., **85**, 1523 (1963)

A. chakense var. spegazzini.

IMIDAZOLIDINOCARBAZOLE ALKALOIDS: spegazzine, spegazzinidine.

Djerassi et al., Experientia, **18**, 113 (1962)

A. compactinervium Kuhl.

CARBAZOLE ALKALOIDS: N-acetyl-11-hydroxy-aspidosparmatine, 1-acetyl-16-methylaspidospermidine, compactinervine.

Gilbert et al., Tetrahedron, **21**, 1141 (1965)

A. cuspa.

CARBAZOLE ALKALOID: burnamine.

SPIROINDOLE ALKALOID: aspidodasycarpine.

Burnell, Medina, Phytochemistry, **7**, 2045 (1968)

A. cylindrocarpon.

IMIDAZOLIDINOCARBAZOLE ALKALOIDS: N-acetylcylindrocarpinol, N-acetyl-17-demethoxycylindrocarine, N-benzoyl-17-demethyl-20-oxocylindrocarine, cylindrocarine, cylindrocarpidine, cylindrocarpine, demethoxycylindrocarpidine, N-formylcylindrocarpinol, 20-hydroxycylindrocarine.

Djerassi et al., Tetrahedron, **16**, 212 (1961)

Milborrow, Djerassi, J. Chem. Soc., C, 417 (1969)

A. dasycarpon A.DC.

CARBAZOLE ALKALOIDS: dehydro-de-N-methyluleine, des-N-methyluleine.

CARBOLINE ALKALOIDS: akuammidine, polyneuridine, polyneuridine aldehyde.

INDOLE ALKALOIDS: (+)-apparicine, (-)-apparicine, aspidodasycarpine, dasycarpidol, dasycarpidone, de-N-methyldasycarpidone.

Gilbert et al., Tetrahedron, 21, 1141 (1965)
Joule et al., ibid., 21, 1717 (1965)

A. desmanthum.

IMIDAZOLIDINOCARBAZOLE ALKALOID: aspidoalbine.

Garcia et al., Phytochemistry, 15, 1093 (1976)

A. discolor A.DC.

CARBOLINE ALKALOIDS:
19,20-dehydro-10-methoxy-dihydrocorynantheol, dihydrocorynantheol, isoreserpiline, isoreserpiline pseudoindoxyl, 10-methoxydihydrocorynantheol, 11-methoxydihydrocorynantheol.

IMIDAZOLIDINOCARBAZOLE ALKALOIDS:
demethoxyaspidospermine, demethylaspidospermine.

Gilbert et al., Tetrahedron, 21, 1141 (1965)

A. dispermum Fr. Allen.

IMIDAZOLIDINOCARBAZOLE ALKALOIDS:
N-acetyl-11-hydroxyaspidospermidine, aspidodispermine, compactinervine, deoxyaspidodispermine.

Ikeda, Djerassi, Tetrahedron Lett., 5837 (1968)

A. duckei Hub.

IMIDAZOLIDINOCARBAZOLE ALKALOIDS: kopsanol, epi-kopsanol, epi-kopsanol-10-lactam, kopsanone.

Filho et al., J. Chem. Soc., C, 1260 (1966)

A. eburneum Fr. All.

IMIDAZOLIDINOCARBAZOLE ALKALOIDS:
demethoxyaspidospermine, demethylaspidospermine.

INDOLE ALKALOID: (-)-apparicine.

Witkop, Patrick, J. Am. Chem. Soc., 76, 5603 (1954)
Ferreira et al., Experientia, 19, 585 (1963)

A. exaltum Monachino.

CARBOLINE ALKALOID: harman-3-carboxylic acid.

IMIDAZOLIDINOCARBAZOLE ALKALOIDS:
17-methyl-21-oxoaspidoalbine, 21-oxoaspidoalbine.

Sanchez et al., An. Acad. Bras. Cienc., 43, 603 (1971)

A. excelsum.

PROTOBERBERINE ALKALOIDS: O-acetylyohimbine, excelsine, excelsinine.

Burnell, Medina, Can. J. Chem., 45, 725 (1967)

A. fendleri Woodson.

IMIDAZOLIDINOCARBAZOLE ALKALOIDS: fendleridine, fendlerine, fendlispermine.

Medina et al., Rev. Latinoam. Quim., 4, 73 (1973)

A. formosanum.

IMIDAZOLIDINOCARBAZOLE ALKALOID: (-)-aspidocarpine.

CARBAZOLE ALKALOIDS: 1,13-dihydro-13-hydroxyuleine, uleine, epi-uleine.

CAROTENOID: lichexanthone.

Garcia et al., Phytochemistry, 15, 1093 (1976)

A. gilberti.

One uncharacterized alkaloid.

Miranda et al., Chem. Ber., 113, 3245 (1980)

A. gomezianum A.DC.

INDOLE ALKALOID: (-)-apparicine.

Gilbert et al., Tetrahedron, 21, 1141 (1965)

A. limae Woodson.

CARBOLINE ALKALOIDS: N-acetyl-11-hydroxyaspidospermidine, compactinervine.

IMIDAZOLIDINOCARBAZOLE ALKALOIDS: aspidolimidine, aspidolimine, limaspermine, limatine, limatinine, limopodine, 3'-methoxylimapodine, 3'-methoxylimaspermine, 11-methoxylimatine, 11-methoxylimatinine.

Limar, Schmid, Helv. Chim. Acta, 50, 89 (1967)

A. macrocarpon Mart.

IMIDAZOLIDINOCARBAZOLE ALKALOIDS: kopsanol, epi-kopsanol, kopsanone.

Filho et al., J. Chem. Soc., C, 1260 (1966)

A. marcgravianum Woodson.

CARBOLINE ALKALOID: dihydrocorynantheol.

IMIDAZOLIDINOCARBAZOLE ALKALOID:
demethoxyaspidospermine. 46 alkaloids.

Gilbert et al., Tetrahedron, 21, 1141 (1965)
Robert et al., J. Nat. Prod., 46, 694 (1983)

A. megalocarpon Muell. Arg.

IMIDAZOLIDINOCARBAZOLE ALKALOID: aspidocarpine.

McLean et al., Can. J. Chem., 38, 1547 (1960)

A. melanocalyx Muell. Arg.

IMIDAZOLIDINOCARBAZOLE ALKALOID:
18,18'-bis(O-demethylaspidocarpine).

Miranda, Gilbert, Experientia, 25, 575 (1969)

A. multiflorum A.DC.

CARBOLINE ALKALOIDS: N-acetyl-11-hydroxyaspidospermidine, compactinervine.

INDOLE ALKALOID: (-)-apparicine.

Gilbert et al., Tetrahedron, 21, 1141 (1965)

A. neblinine Monachino.

IMIDAZOLIDINOCARBAZOLE ALKALOID: neblinine.

Brown, Djerassi, J. Am. Chem. Soc., 86 2451 (1964)

A. nigricans Handro.

CARBAZOLE ALKALOIDS: N-acetyl-11-hydroxyaspidospermidine, compactinervine, olivacine N-oxide.

Gilbert et al., Tetrahedron, 21, 1141 (1965)

A. oblongum A.DC.

CARBOLINE ALKALOID:
19,20-dehydro-10-methoxydihydrocorynantheol. 35 alkaloids.

Spiteller, Spiteller-Friedmann, Monatsh. Chem., 93, 795 (1962)
Robert et al., J. Nat. Prod., 46, 708 (1983)

A. obscurinervum Azembuja.

IMIDAZOLIDINOCARBAZOLE ALKALOIDS:
dihydroobscurinervidine, dihydroobscurinervine, obscurinervidine, obscurinervine.

Brown, Djerassi, J. Am. Chem. Soc., 86, 2451 (1964)

A. olivaceum Muell. Arg.

CARBOLINE ALKALOIDS: N-acetyl-11-hydroxyaspidospermidine, compactinervine.

INDOLE ALKALOID: (-)-apparicine.

Gilbert et al., Tetrahedron, 21, 1141 (1965)

A. peroba.

CARBOLINE ALKALOID: macusine B.

INDOLINE ALKALOID: alkaloid Q.

Quaisuddin, Bangladesh J. Sci. Ind. Res., 15, 35 (1980)

A. polyneuron Muell. Arg.

CARBOLINE ALKALOIDS: polyneuridine, tombozine.

IMIDAZOLIDINOCARBAZOLE ALKALOIDS:
deacetylaspidospermine, palosine.

Paladini et al., An. Assoc. Quim. Argent., 50, 352 (1962)

A. populifolium A.DC.

CARBAZOLE ALKALOID:
14,19-dihydro-11-methoxycondylocarpine.
CARBOLINE ALKALOIDS: N-acetyl-11-hydroxyaspidospermidine,
compactinervine.
IMIDAZOLIDINOCARBAZOLE ALKALOIDS:
N-formyl-16,17-dimethoxyaspidofractinine,
N-formyl-17-methoxyaspidofractine,
17-methoxyaspidofractine, 17-methoxyaspidofractinine.
Gilbert et al., Tetrahedron, 21, 1141 (1965)

A. pyrifolium.

IMIDAZOLIDINOCARBAZOLE ALKALOIDS: aspidofiline,
(+)-aspidospermine, 6-demethoxypyrifoline, pyrifoline.
Craviero et al., Phytochemistry, 22, 1526 (1983)

A. quebracho Schlecht.

IMIDAZOLIDINOCARBAZOLE ALKALOIDS:
deacetylaspidospermine, (-)-quebrachamine.
UNCHARACTERIZED ALKALOID: hypoquebrachine.
Ewins, J. Chem. Soc., 105, 2738 (1914)

A. quebracho blanco Schlecht.

IMIDAZOLIDINOCARBAZOLE ALKALOIDS: aspidospermine,
(+)-12-dehydroaspidospermidine.
Mill, Nyburg, J. Chem. Soc., 1458 (1960)

A. quebracho blanco Schlecht f.pendulae Speg.

CARBAZOLE ALKALOID: aspidospermatine.
IMIDAZOLIDINOCARBAZOLE ALKALOIDS: aspidospermidine,
aspidospermine.
UNCHARACTERIZED ALKALOIDS: aspidosamine,
aspidospermicine.
Floriani, Rev. Cent. Estud. Farm. Bioquim., 25, 373, 423 (1935)

A. quirandy Hassler.

IMIDAZOLIDINOCARBAZOLE ALKALOID: aspidospermine.
UNCHARACTERIZED ALKALOIDS: aspidosamine, haslerine.
Floriani, Rev. Cent. Estud. Farm. Bioquim., 25, 373, 423 (1935)

A. refractum Mart.

IMIDAZOLIDINOCARBAZOLE ALKALOIDS: aspidofractine,
aspidofractinine, refractidine, refractine.
Djerassi et al., Helv. Chim. Acta, 46, 742 (1963)

A. rhombeinsignatum.

IMIDAZOLIDINOCARBAZOLE ALKALOIDS: limaspermidine,
15,16,17-trimethoxy-21-oxoaspidoalbine.
Medina, Di Genova, Planta Med., 37, 165 (1967)
Medina, Genova, ibid., 37, 165 (1979)

A. rigidum.

CARBOLINE ALKALOID: deacetylpicraline.
Arndt et al., Phytochemistry, 6, 1653 (1967)

A. spegazzini.

CARBOLINE ALKALOID: spegatrine.
Orazi et al., Can. J. Chem., 44, 1523 (1966)

A. spruceanum Benth.

CARBOLINE ALKALOIDS: N-acetyl-11-hydroxyspermidine,
compactinervine.
IMIDAZOLIDINOCARBAZOLE ALKALOIDS:
N-acetyl-N-depropionylaspidoalbine, aspidoalbine.
Ferrari et al., Can. J. Chem., 41, 1531 (1963)
Gilbert et al., Tetrahedron, 21, 1141 (1965)

A. subincanum Mart.

CARBOLINE ALKALOIDS: N-acetyl-11-hydroxyaspidospermidine,
compactinervine.
CARBAZOLE ALKALOIDS: 2-epi-dasycarpone, ellipticine
methonitrate, N-methyltetrahydro- ellipticine, subincanine.
Gaskell, Joule, Chem. Ind., 90, 2699 (1968)

A. ulei Mfg.

CARBAZOLE ALKALOIDS: N-methyltetrahydroellipticine, uleine.
Burnell, Casa, Can. J. Chem., 45, 89 (1967)

A. vargasii.

CARBOLINE ALKALOIDS: 9-methoxyolivacine,
N-methyltetrahydroellipticine.
Burnell, Casa, Can. J. Chem., 45, 89 (1967)

A. verbascifolium.

IMIDAZOLIDINOCARBAZOLE ALKALOID: N-formylkopsanol.
Braekman et al., Bull. Soc. Chim. Belg., 78, 63 (1969)

A. No. RDJ 119070.

CARBOLINE ALKALOID: 10-methoxydihydrocorynantheol.
Dastour et al., Helv. Chim. Acta, 50, 213 (1967)

ASPILIA (Compositae)

A. kotschyi (Sch. Bip.).

TRITERPENOIDS: echinocytic acid, oleanolic acid.
Kapundu, Bull. Soc. R. Sci. Liege, 54, 129 (1985)

A. pluriseta.

DITERPENOIDS: 15β-angeloyloxy-ent-kaurenic acid,
15β-angeloyloxy-16,17-epoxy-ent-kaurenic acid,
(-)-laur-16-en-19-oic acid.
Lwande et al., Fitoterapia, 56, 126 (1985)

ASPLENIUM (Polypodiaceae/Aspleniaceae)

A. bulbiferum.

FLAVONOID: kaempferol 3,7-dirhamnoside.
Imperato, Chem. Ind. (London), 566 (1987)

A. laciniatum.

ACID: stearic acid.
ALIPHATIC: octatriacontane.
BENZOQUINONE: phthiocol.
NAPHTHOQUINONES: 3,3'-bis(2-methyl-1,4-naphthoquinone),
2-methyl-1,4-naphthoquinone.
STEROIDS: β-sitosterol, β-sitosterol D-O-glucoside.
Gupta et al., Curr. Sci., 45, 44 (1976)
Gupta et al., Indian J. Chem., 15B, 394 (1977)

A. montanum Willd.

XANTHONES: isomangiferin, mangiferin.
Smith, Harborne, Phytochemistry, 10, 2117 (1971)

A. viride.

FLAVONE: 3',4',5,7-tetrahydroxy-3-methoxyflavone.
Voirin et al., Phytochemistry, 13, 275 (1974)

ASTASIA (Euglenaceae/Euglenophyceae)

A. longa.

STEROID: 24-methylpollinasterol.
Rohmer, Brandt, Eur. J. Biochem., 36, 446 (1973)

ASTER (Compositae)

A. bakeranus.

One monoterpenoid.
Tsankova et al., Phytochemistry, 22, 1285 (1983)

A. salignus.

PHENOLICS: caffeic acid, chlorogenic acid, hydroxycinnamic
acid, isochlorogenic acid a, isochlorogenic acid b,
isochlorogenic acid c, neochlorogenic acid.
Sergeeva et al., Khim. Prir. Soedin., 655 (1975)

A. schreberi.

ACETYLENICS: 1,7,9,13-heptadecatetraen-11-yne,
1,7,9,13-heptadecatetraen-11-yn-3-ol,
2,7,9,13-heptadecatetraen-11-yn-1-ol,
4,8,10,16-heptadecatetraen-6-yn-3-ol,
8,10,14,16-heptadecatetraen-6-yn-3-one.

MONOTERPENOIDS: deca-2,6-dien-4-ynoic acid,
deca-2,6-dien-4-ynoic acid methyl ester.
Bohlmann, Zdero, *Chem. Ber.*, **102**, 1037 (1969)

A. tataricus.

TRITERPENOID: shionone.
Nakoaki, *J. Pharm. Soc., Japan*, **49**, 1169 (1928)

ASTERACANTHA (Asteracanthaceae)

A. longifolia (Hygrophila spinosa). One uncharacterized alkaloid.

STEROID: stigmasterol.
TRITERPENOIDS: betulin, lupeol.
(A) Basu, Lal, *Q. J. Pharmacol.*, **20**, 38 (1947)
(S, T) Gupta *et al.*, *J. Nat. Prod.*, **46**, 938 (1983)

ASTERIAS

A. amurensis.

STEROIDS: amuresterol, cholestane-3β,6α,15α,24-tetraol.
Kobayashi, Mitsuhashi, *Tetrahedron*, **30**, 2147 (1974)

A. rubens.

CAROTENOID: 7′,8′-tetrahydroastaxanthin.
Cheesman *et al.*, *Biol. Rev.*, **42**, 131 (1967)

ASTERICUS (Compositae)

A. aquaticus L.

SESQUITERPENOID: asteriscanolide.
San Feliciano *et al.*, *Tetrahedron Lett.*, 2369 (1985)

ASTILBE

A. rivularis.

COUMARIN: bergenin.
TRITERPENOIDS: acetyl-β-peltoboykinolic acid, astilbic acid, peltoboykinolic acid.
Sastry *et al.*, *Indian J. Chem.*, **15B**, 494 (1977)

A. thunbergii.

TRITERPENOID: astilbic acid.
Takahashi *et al.*, *Chem. Pharm. Bull.*, **20**, 2106 (1972)

ASTRAGALUS (Papilionaceae)

A. alexandrinus Boiss.

SAPONIN: furostene glycoside.
Khafagy *et al.*, *J. Pharm. Sci.*, **20**, 7 (1979)

A. angustifolius Lam.

FLAVONOIDS: isorhamnetin 3-α-L-arabinoside, isorhamnetin
3-α-L-arabinoside-α-L-rhamnoside, isorhamnetin
3-β-D-glucoside, kaempferol 3-α-L-arabinoside, nicotiflorin.
TRITERPENOID: 20S,24R-silversigenin
epoxylanost-9(11)-ene-3β,6α,16α,25-tetraol.
Panova *et al.*, *Farmatsiya*, **33**, 23 (1983)

A. brachycarpus.

COUMARINS: scopoletin, umbelliferone.
FLAVONOIDS: hyperin, quercetin.
Alaniya *et al.*, *Khim. Prir. Soedin.*, 813 (1976)

A. caucasicus.

FLAVONOL: ascaside.
Alaniya *et al.*, *Khim. Prir. Soedin.*, 351 (1975)

A. centralpinus.

FLAVONOIDS: 3,5-dihydroxyflavone,
5,4′-dimethoxy-7-hydroxyflavan-7-ol, isorhamnetin
3-O-glucobioside, isorhamnetin 3-O-glucorhamnoside,
isorhamnetin 3-O-glucoside, isorhamnetin 7-O-rhamnoside,
3,5,7,3′-tetramethoxy-4′-hydroxyflavone,
5,7,4′-trihydroxy-3,3′-dimethoxyflavone.

Pashkov, Marichkova, *Probl. Farm.*, **11**, 36 (1983); *Chem. Abstr.*, **100** 3517t

A. cicer.

FLAVONOIDS: isorhamnetin glycoside, kaempferol glycoside, quercetin glycoside.
Alaniya *et al.*, *Khim. Prir. Soedin.*, 528 (1983)

A. dasyanthus.

SAPOGENIN: dasyanthogenin.
Evstratova *et al.*, *Khim. Prir. Soedin.*, 102 (1981)

A. dipelta.

FLAVONOIDS: astragalin, populnin, rodinin, trifolin.
Luk'yanichkov *et al.*, *Khim. Prir. Soedin.*, 453 (1987)

A. galegiformis.

SAPONINS: cyclogaleginosides A and B,
TRITERPENOID: cyclogalegenin.
(Sap) Aleniya *et al.*, *Khim. Prir. Soedin.*, 477 (1984)
(T) Aleniya *et al.*, *ibid.*, 332 (1983)

A. glycyphyllos L.

STEROIDS: campesterol, sitosterol, soyasapogenol B, stigmasterol.
Elenga *et al.*, *Pharmazie*, **41**, 300 (1986)

A. membranaceus.

SAPONINS: astragalosides I, II, III, IV, V, VI and VII.
TRITERPENOID: astramembrangenin.
(Sap) Cao *et al.*, *Huaxue Xuebao*, **41**, 1137 (1983)
(T) Kitagawa *et al.*, *Chem. Pharm. Bull.*, **31**, 716 (1983)

A. orbiculatus.

STEROID: cycloorbigenin.
Agzamova *et al.*, *Khim. Prir. Soedin.*, 455 (1986)

A. pamirensis.

TRITERPENOIDS: cyclosiversigenin, cyclosiversigenin,
3-O-β-D-xylopyranoside.
Agzamova *et al.*, *Khim. Prir. Soedin.*, 117 (1986)

A. pterocephalus.

SAPONINS: cyclosiversiosides E and F.
STEROIDS: β-sitosterol, β-sitosterol-β-D-glucopyranoside.
TRITERPENOID: siversigenin.
Agzamova *et al.*, *Khim. Prir. Soedin.*, 117 (1986)

A. sieversianus.

SAPONINS: cyclosiverosides A, B, C, D, E, F, G and H.
Svechnikova *et al.*, *Khim. Prir. Soedin.*, **312**, 460 (1983)

A. sinicus.

FLAVONE: astragalin.
Takagi *et al.*, *J. Pharm. Soc., Japan*, **97**, 1369 (1977)

A. taschkendicus.

SAPONINS: askendosides A, B, C and D.
TRITERPENOID: cycloasgenins A and B.
(Sap) Israev *et al.*, *Khim. Prir. Soedin.*, **173**, 587 (1983)
(T) Israev *et al.*, *ibid.*, 572 (1981)

A. tibetanus Benth ex Bge. (A. laxmanii Bge.).

UNCHARACTERIZED ALKALOID: smirnovine.
Ryabinin *et al.*, *Dokl. Akad. Nauk SSSR*, **61**, 317 (1948)

A. trigonus.

TRITERPENOID: 6-oxocycloartan-3β,16β-di-O-glucopyranoside.
El-Sabakhy, Waterman, *Planta Med.*, 350 (1985)

ASTRANTIA (Umbelliferae/Saniculoidaea)

A. major.

LIMONOIDS: isoorientin, kaempferol, quercetin.
TRITERPENOID: astrantiagenin.
(L) Hiller *et al.*, *Pharmazie*, **30**, 809 (1975)
(T) Weitke *et al.*, *ibid.*, **33**, 541 (1978)

ASTROCASIA (Euphorbiaceae)

A. phyllantoides.
BISPIPERIDINE ALKALOIDS: astrocasine, astrophylline.
Lloyd, Tetrahedron Lett., 1761, 4537 (1965)

ATALANTIA (Rutaceae)

A. ceylanica.
BISCARBAZOLE ALKALOIDS: atalantine, ataline.
Fraser, Lewis, J. Chem. Soc., Chem. Commun., 615 (1973)

A. macrophylla.
CARBAZOLE ALKALOID: atalaphyllidine.
Busa, Experientia, 31, 1387 (1975)

A. monophylla Correa.
ACRIDONE ALKALOIDS: N-methylbicycloatalaphylline,
N-methyl-1,5-dihydroxy-2,3-dimethoxyacridin-9-one.
CARBAZOLE ALKALOIDS: atalaphylline, atalaphylline
3,5-dimethyl ether, atalaphyllinine, N-methylatalaphylline.
LIMONOIDS: atalantin, atalantolide, cyclo-epi-atalantin,
dehydroatalantin.
(A) Govindachari et al., Tetrahedron, 26, 2906 (1970)
(A) Kulkarni, Sabata, Phytochemistry, 20, 867 (1981)
(T) Dreyer et al., Tetrahedron, 32, 2367 (1976)

A. racemosa W. et A.
COUMARIN: racemol.
Purushothaman et al., Indian Drugs, 23, 487 (1986); Chem. Abstr., 105
168851p

ATHAMANTHA (Umbelliferae)

A. oreoselinum.
COUMARIN: athamantin.
Halpern et al., Helv. Chim. Acta, 40, 758 (1957)

ATHANASIA (Compositae)

A. dentata L.
Four furane sesquiterpenoids.
Bohlmann, Grenz, Chem. Ber., 108, 357 (1975)

A. dimorpha DC.
Three furane sesquiterpenoids.
Bohlmann, Grenz, Chem. Ber., 108, 357 (1975)

A. dregeana.
SESQUITERPENOIDS: athanadregolide, athanagrandione.
Bohlmann et al., Phytochemistry, 18, 995 (1979)

A. linifolia Harv.
Four furane sesquiterpenoids.
Bohlmann, Grenz, Chem. Ber., 108, 357 (1975)

A. punctata Berg.
Three furane sesquiterpenoids.
Bohlmann, Grenz, Chem. Ber., 108, 357 (1975)

ATHEROSPERMA
(Monimiaceae/Atherospermataceae)

A. moschatum Labill.
BISBENZYLISOQUINOLINE ALKALOIDS: atherospermoline,
berbamine, isotetrandrine.
OXONORAPORPHINE ALKALOIDS: atheroline, atherospermidine,
moschatoline.
UNCHARACTERIZED ALKALOID: atherospermine.
Zeyer, Vsjschr. Prakt. Pharm., 10, 504 (1861)
Bick, Douglas, Aust. J. Chem., 20, 1403 (1967)

ATHRIXIA

A. angustissima.
SESQUITERPENOIDS: 10-isobutyryloxyathrixol epoxide,
7,10-diisobutyryloxyathrixol epoxide.
Bohlmann, Zdero, Phytochemistry, 16, 1773 (1977)

A. elata.
DITERPENOID: secoathrixic acid.
Bohlmann et al., Phytochemistry, 21, 1806 (1982)

A. fontana.
DITERPENOIDS: 10-acetoxyathrixol epoxide,
4-carbomethoxyathrixanone.
Bohlmann, Zdero, Phytochemistry, 16, 1773 (1977)

A. phylicoides.
DITERPENOIDS: 4-carbomethoxyathrixanone,
16,17-dihydro-16,17-epoxyathrixianone,
4-formylathrixanone, 4-hydroxymethylathrixianone,
4-methoxyathrixanone.
Bohlmann, Zdero, Phytochemistry, 16, 1773 (1977)

ATHROTAXIS (Taxodiaceae)

A. selaginioides Don.
TRITERPENOIDS: athrotaxin, cubenol.
Talvitre et al., Finn. Chem. Lett., 93 (1979)

ATHYRIUM

A. mesosorum.
XANTHONES: 3-methoxy-1,6,7-trihydroxyxanthone,
6-methoxy-1,3,7-trihydroxyxanthone.
Owen et al., J. Chem. Soc., Perkin I, 1018 (1974)

ATLAS

A. cedar.
SESQUITERPENOIDS: (E)-10,11-dihydroatlantone,
(Z)-dihydroatlantone.
Irie et al., Chem. Pharm. Bull., 23, 1892 (1975)

ATRACTYLIS (ATRACTYLODES) (Compositae)

A. gummifera.
NORDITERPENOIDS: atractyligenin, 2-O-methylatractyligenin.
Piozzi et al., Gazz. Chim. Ital., 99 373 (1969)

A. japonica.
SESQUITERPENOIDS: 3β-acetoxyatractylon, (+)-atractylon,
3β-hydroxyactractylon, selina-4(14),7(11)-dien-8-one.
Nishikawa et al., J. Pharm. Soc., Japan, 96, 1089 (1976)

A. lancea.
SESQUITERPENOIDS: 3β-acetoxyatractylon, hinesol,
3β-hydroxyactrylon.
Nishikawa et al., J. Pharm. Soc., Japan, 96, 1089 (1976)

A. lancea De Candolle var. chinensis Kitamura.
ACETYLENICS: acetylatractylodinol, atractylodinol.
Nishikawa et al., J. Pharm. Soc., Japan, 96, 1322 (1976)

A. lancea var. simplicifolia.
SESQUITERPENOIDS: acetylatractylodinol, atractylinodinol,
atractylodin.
Nishikawa et al., J. Pharm. Soc., Japan, 96, 1089 (1976)

A. ovata.
SESQUITERPENOIDS: 3β-acetylatractylon, atractylon,
3β-hydroxyatractylon.
Nishikawa et al., J. Pharm. Soc., Japan, 96, 1089 (1976)

A. simplicifolia.
SESQUITERPENOID: 3β-acetoxyatractylon.
Nishikawa et al., J. Pharm. Soc., Japan, 96 1089 (1976)

ATRAPHAXIS

A. frutescens.
FLAVONOIDS: kaempferol 3-β-D-glucopyranoside, myricetin
3-β-D-glucopyranoside, quercetin 3-β-D-glucopyranoside.
Chumbalev et al., Khim. Prir. Soedin., 424 (1975)

A. purifolia.
FLAVONOIDS: 7-O-methylgossypetin 8-β-O-glucopyranoside,
7-O-methylgossypetin 3-O-α-L-rhamnopyranoside,
7-O-methylgossypetin
3-O-α-L-rhamnopyranoside-8-β-D-glucopyranoside.
Chumbalev et al., Khim. Prir. Soedin., 660 (1976)

ATROPA (Solanaceae)

A. bella-donna L.
TROPANE ALKALOIDS: apoatropine, atropamine, atropine,
belladonnine, bellaridine, (-)-hyoscyamine, scopolamine.
Hesse, Justus Liebigs Ann. Chem., 361, 87 (1891)
Gadamer, Arch. Pharm., 239, 294 (1901)

A. boetica.
TROPANE ALKALOID: (-)-hyoscyamine.
Marion, Thomas, Can. J. Chem., 32, 1116 (1954)

ATYLOSIA

A. trinerva.
ALCOHOL: atylosol.
Tripathi et al., Phytochemistry, 17, 2001 (1978)

AUBRIETIA

A. deltoidea L.
AMINO ACID: N-τ-glutamyltyramine.
Morris et al., Biochemistry, 1, 706 (1962)

AUCUBA (Cornaceae)

A. japonica Thunb.
IRIDOID: aucubin.
MONOTERPENOIDS: eucommioside
1-O-β-D-glucopyranosyl-eucommiol.
Bernini et al., Phytochemistry, 23, 1431 (1984)

AUSTROBAILEYA (Austrobaileyaceae)

A. maculata C.T.White.
LIGNANS: austrobailignans 1, 2, 3, 4, 5, 6 and 7.
Murphy et al., Aust. J. Chem., 28, 81 (1975)

A. scandens C.T.White.
LIGNANS: austrobailignans 1, 2, 3, 4, 5, 6 and 7.
Murphy et al., Aust. J. Chem., 28, 81 (1975)

AUSTROCEDRUS

A. chilensis.
DITERPENOID: 8,20-dihydroxy-9(11),13-abietadien-12-one.
Cairness et al., J. Nat. Prod., 46, 135 (1983)

AUSTROEUPATORIUM

A. chaparense.
DITERPENOIDS: 18-acetoxyaustrochaparol 7-acetate,
austrochaparol, austrochaparol acetate,
12-desoxyaustrofolin-12-ol, 15,16-dihydroaustrofolin-15α-ol,
15,16-dihydroaustrofolin-15β-ol, 18-hydroxyaustrochaparol
7-acetate, 18-oxoaustrochaparol 7-acetate.
Bolhmann et al., Phytochemistry, 19, 111 (1980)

A. inulaefolium.
DITERPENOIDS: austrofolin, austroinulin,
ent-15,16-epoxy-2β,3β,19-trihydroxy-18-nor-13(16),14-
labdadiene-7,12-dione.
Bohlmann et al., Chem. Ber., 110, 1034 (1977)
Oberti et al., Phytochemistry, 23, 2003 (1984)

AUSTROPLENEKIA

A. populnea.
Two terpenoids.
Prasad et al., Phytochemistry, 23, 1655 (1984)

AVENA (Gramineae)

A. sativa L.
FLAVONOIDS: isoorientin 2″-arabinoside, isoswertisin
2″-rhamnoside, isovitexin 2″-arabinoside, vitexin
2″-rhamnoside.
PIGMENT: O″-rhamnosylisoswertisin.
SAPONINS: avenacoside A, avenacoside B,
26-desglucoavenacoside A, 26-desglucoavenacoside B.
Chopin et al., Phytochemistry, 16, 2041 (1977)
(P) Nabeta, Kadota, Phytochemistry, 16, 1112 (1977)
(Sap) Tschesche, Schmidt, Z. Naturforsch., 21B, 896 (1966)

AVICENNIA

A. officinalis L.
IRIDOIDS: avicennioside, 7-cinnamoyl-8-epi-loganic acid,
2′-cinnamoylmussaenosidic acid, geniposidic acid.
Koenig et al., Phytochemistry, 26, 423 (1987)

AYAPANA (Compositae)

A. amygdalina.
DITERPENOID: ent-3β,7α-dihydroxy-8(17),13-labdadien-15-oic
acid.
SESQUITERPENOID: 2,6,10-bisabolatriene.
Bohlmann et al., Phytochemistry, 18, 1997 (1979)

A. ecuadorensis King et Rob.
NORSESQUITERPENOIDS:
cis-3-hydroxy-2,3-dihydroeuparin-acetoxymethyl-trans-
crotonate,
trans-3-hydroxy-2,3-dihydroeuparin-β-acetoxymethyl-trans-
crotonate, trans-3-hydroxy-2,3-dihydroeuparin angelate.
Bohlmann et al., Chem. Ber., 110, 1034 (1977)

AZADIRACHTA (Meliaceae).

A. indica A. Juss.
STEROID: vilasanin.
SESTERTERPENOIDS: 6-O-acetylnimbandiol, 17-epi-azadiradione,
nimbandiol, nimbinene, nimbolicinol, nimbidin, nimbolin A,
nimbolin B.
ESTER TERPENOID: salannolide.
TETRANORTRITERPENOIDS: isoazadirolide, isomargosinolide,
isonimbocinolide, margodunolide, margosinolide,
nimbocinolide.
TRITERPENOIDS: isonimbocinone, nimbocinome.
(S) Kraus et al., Justus Liebigs Ann. Chem., 181 (1981)
(T) Siddiqui et al., Tetrahedron, 42, 4849 (1986)
(T) Siddiqui et al., Phytochemistry, 25, 2183 (1986)

AZIMA (Salvadoraceae)

A. tetracantha Lam.
MACROCYCLIC ALKALOIDS: azcarpine, azimine.
Smalberger et al., Tetrahedron, 24, 6417 (1968)

AZORELLA (Mulineae)

A. trifurcata (Gaertn.) Pers.

SESQUITERPENOID:
2-cis-9-cis-pentadeca-2,9-diene-4,6-diyne-1,8-diol acetate.
Bohlmann *et al.*, *Chem. Ber.*, **104**, 1322 (1971)

AZUREOCEREUS (Cactaceae)

A. ayacuchensis.

AMINO ACID: tyramine hydrochloride.
Lee *et al.*, *Lloydia*, **38**, 366 (1975)

B

BACCHARIS (Compositae/Asteraceae)

B. articulata.
DITERPENOIDS: articulin, articulinol.
TRITERPENOID: barticulidiol.
Gianello, Giordano, *Rev. Latinoam. Quim.*, **13**, 76 (1982)

B. bigelovii.
FLAVONES: alnusin, alnusinol, 7-benzoylchrysin, 2α,5,7-trihydroxyflavone.
Arriaga-Giner *et al.*, *Z. Naturforsch.*, **41C**, 946 (1986)

B. concava Pers.
FLAVONOIDS: cirsimaritin, pectolinarigenin, salvigenin.
Zamorano *et al.*, *Bol. Soc. Chil. Quim.*, **32**, 101 (1987); *Chem. Abstr.*, 107 151212z

B. conferta.
COUMARINS: 2-[1-(acetoxymethyl)vinyl]-5-hydroxycoumaran, 6-acetyl-2-(1-formylvinyl)-5-hydroxycoumaran.
DITERPENOID: bacchofertin.
(C) Bohlmann, Zdero, *Chem. Ber.*, **109**, 1450 (1976)
(T) Guerro, Romo de Vivar, *Rev. Latinoam. Quim.*, **4**, 178 (1973)

B. cordifolia.
TRICHOTHECENES: isomiotoxin D, miotoxin D.
Habermehl *et al.*, *Ann. Chem.*, 633 (1985)

B. crispa. Spreng.
DITERPENOIDS: bacrispin, 1-deoxybacrispin.
Tonn *et al.*, *An. Assoc. Quim. Argentina*, **68**, 237 (1980)

B. eleagnoides.
FLAVAN: cirsimaritin.
Mesquita *et al.*, *Phytochemistry*, **25**, 1255 (1986)

B. genistelloides.
SESQUITERPENOID: palustrol.
Buchi *et al.*, *Tetrahedron Lett.*, 14 (1959)

B. halimifolia. L.
TRITERPENOID: baccharis oxide.
Anthonsen *et al.*, *Acta Chem. Scand.*, **24**, 2479 (1970)

B. heterophylla.
TRITERPENOIDS: maniladiol, oleanolic acid.
Arriaga-Giner *et al.*, *Phytochemistry*, **25**, 719 (1986)

B. incanum.
FLAVONOIDS: 5,4'-dihydroxy-3,6,7,8,3'-pentamethoxyflavone, 5,4'-dihydroxy-6,7,8,3'-tetramethoxyflavone.
DITERPENOID: bincantriol.
(F) Faini *et al.*, *J. Nat. Prod.*, **45**, 501 (1982)
(T) San Martin *et al.*, *Phytochemistry*, **25**, 264 (1985)

B. kingii.
DITERPENOID: kingidiol.
Bohlmann *et al.*, *Phytochemistry*, **23**, 1511 (1984)

B. linearis.
ALKALOIDS: three uncharacterized bases.
Montes *et al.*, *Rev. R. Acad. Cienc. Exactes, Fis. Nat. Madrid*, **65**, 499 (1971)

B. macraei.
DITERPENOIDS: bacchasmacranone, 2α-hydroxybacchasmacranone, 2β-hydroxybacchasmacranone.
Gambaro *et al.*, *Phytochemistry*, **25**, 2175 (1986)

B. magellanica.
COUMARIN: scopoletin.
FLAVONOID: hesperidin.
Cordano *et al.*, *J. Nat. Prod.*, **45**, 6531 (1982)

B. megapotamica Spreng.
NORTRITERPENOIDS: baccharin, baccharinol. 13 macrocyclic trichothecenes.
Kupchan *et al.*, *J. Org. Chem.*, **42**, 4221 (1977)
Jarvis *et al.*, *ibid*, **52**, 45 (1987)

B. minutiflora.
DITERPENOIDS: ent-16β-acetyl-17-nor-19-kauranol, ent-16β,17-epoxykaurane.
Bohlmann *et al.*, *Phytochemistry*, **21**, 399 (1982)

B. patens.
FLAVONOIDS: gardenin D, 5,3',4'-trihydroxy-6,7,8-trimethoxyflavone, xanthomicrol.
de Silva *et al.*, *J. Nat. Prod.*, **48**, 861 (1985)

B. rhomboidulus.
TRITERPENOID: brein.
Laird *et al.*, *J. Am. Chem. Soc.*, **82**, 4198 (1960)

B. sarothroides.
FLAVONES: 5,4'-dihydroxy-3,6,7,8-tetramethoxyflavone, 3',5,7-trihydroxy-3,4'-dimethoxyflavone, 5,7,4'-trihydroxy-3,6,8-trimethoxyflavone, 5,7,4'-trihydroxy-3,6,8-trimethoxyflavone 7-methylether.
DITERPENOIDS: hautriwaic acid, 2β-hydroxyhautriwaic acid.
TRITERPENOID: oleanolic acid.
(F) Wollenweber *et al.*, *Z. Naturforsch.*, **41C**, 87 (1986)
(T) Arriaga-Giner *et al.*, *Phytochemistry*, **25**, 719 (1986)

B. scandens.
NORSESQUITERPENOID: bacchascandone.
DITERPENOID: bacchalineol malonate.
Bohlmann *et al.*, *Phytochemistry*, **18**, 1993 (1979)

B. sternbergiana.
DITERPENOIDS: two ent-labdanes.
Bohlmann *et al.*, *Phytochemistry*, **23**, 1109 (1984)

B. tola.
DITERPENOIDS: ent-beyerene-18,19-diol, ent-beyeren-18-ol, 8,13-epoxy-14-labden-19-ol, neohibaenol.
San Martin *et al.*, *Phytochemistry*, **22**, 1461 (1983)

B. tricuneata.
DITERPENOIDS: bacchalineol, bacchatricuneatins A, B, C and D.
Wagner *et al.*, *J. Org. Chem.*, **43**, 3339 (1978)
Bohlmann *et al.*, *Phytochemistry*, **18**, 1993 (1979)

B. tricuneata var. tricuneata (L.f.) Pers.
DITERPENOIDS: bacchatricuneatins A, B, C and D.
Wagner *et al.*, *Tetrahedron Lett.*, 3039 (1977)

B. trimera.
FLAVONE: eupatorin.
Kupchan *et al.*, *J. Pharm. Sci.*, **54**, 929 (1965)

B. trinervis.
ACETYLENIC: 7,9,15-heptadecatriene-11,13-diynoic acid methyl ester.
DITERPENOIDS: 15-hydroxy-ent-labd-8(17)-en-19-oic acid, pinobanksin 3-acetate
(Acet) Bolhmann *et al.*, *Chem. Ber.*, **103**, 2327 (1970)
(T) Jakupovic *et al.*, *Pharmazie*, **41**, 157 (1986)

B. tucumaensis.
DITERPENOID: 7,13-labdadiene-15,18-dioic acid.
Tonn *et al., Phytochemistry,* **21,** 2599 (1982)

B. vaccinoides.
DITERPENOID: hautriwaic acid.
TRITERPENOIDS: maniladiol, oleanolic acid.
Arriaga-Giner *et al., Phytochemistry,* **25,** 719 (1986)

BACKHOUSIA
B. angustifolia.
MONOTERPENOID: angustione.
Gibson *et al., J. Chem. Soc.,* 1184 (1930)

BACOPA (Scrophulariaceae)
B. monniera.
TRITERPENOIDS: bacogenins A-1, A-2 and A-3.
Chandel, Kulshreshta, *Phytochemistry,* **16,** 141 (1977)

BAECKEA
B. crenulata.
ALCOHOL: baeckeol.
Hems *et al., J. Chem. Soc.,* 1208 (1940)

BAEOMETRA (Liliaceae/Wurmbaeoideae)
B. columellaris Salisb.
UNCHARACTERIZED ALKALOID: columellarine.
Pijewska *et al., Collect. Czech. Chem. Commun.,* **32,** 158 (1967)

BAERIA
B. chrysostoma.
SESQUITERPENOIDS: chrysostomalide, chrysostomalide acetate,
chrysostomalide isobutyrate.
Bohlmann, Zdero, *Phytochemistry,* **17,** 3022 (1978)

B. coronaria.
SESQUITERPENOID: chrysostomalide isobutyrate.
Bohlmann, Zdero, *Phytochemistry,* **17,** 3022 (1978)

BAHIA
B. oppositifolia.
SESQUITERPENOIDS: bahia I, bahia II, bahifolin.
Herz *et al., Phytochemistry,* **11,** 371 (1972)
Nelson, Aplund, *ibid,* **22,** 2755 (1983)

B. pringlei.
SESQUITERPENOIDS: bahia I, bahia II.
Romo de Vivar *et al., Can. J. Chem.,* **48,** 2849 (1969)

B. woodhousei (Gray) Gray. (Picradeniopsis woodhousei (Gray) Rydb.
SESQUITERPENOID: woodhousin.
Herz, Bhat, *J. Org. Chem.,* **37,** 306 (1972)

BAIKIAEA (Leguminosae)
B. plurijuga.
AMINOACID: 1,2,3,6-tetrahydro-2-pyridine carboxylic acid.
King *et al., J. Chem. Soc.,* 3590 (1950)

BAILEYA (Compositae)
B. multiradiata Harv. et Gray.
SULPHOXIDE:
(5E)-6-methylsulphinyl-1-phenyl-5-heptene-1,3-diyne.
SESQUITERPENOIDS: baileyin, baileyolin, multigilin,
multiradiatin, multistatin, pleniradin.
(Sul) Bohlmann, Zdero, *Chem. Ber.,* **109,** 1964 (1976)
(T) Dominguez *et al., Planta Med.,* **30,** 356 (1977)
(T) Pettit *et al., Lloydia,* **41,** 29 (1978)

B. pauciradiata Harv. et Gray.
SESQUITERPENOID: paucin.
Waddell, Geissman, *Tetrahedron Lett.,* 515 (1969)

B. pleniradiata Cov.
SESQUITERPENOIDS: baileyin, baileyolin, paucin, plenolin.
Waddell, Geissman, *Tetrahedron Lett.,* 515 (1969)
Herz *et al., J. Org. Chem.,* **44,** 1873 (1979)

BALAMEANDA
B. chinensis.
ISOFLAVONE GLUCOSIDE: tectoridin.
Khara *et al., Indian J. Chem.,* **16B,** 78 (1978)

BALANOCARPUS (Dipterocarpaceae)
B. zeylanicus (Trimen).
POLYPHENOL: balanocarpol.
Diyasena *et al., J. Chem. Soc., Perkin I,* 1807 (1985)

BALANOPS
B. australiana.
TRITERPENOID: friedelinol.
Drake *et al., J. Am. Chem. Soc.,* **58,** 1681 (1936)

BALDUINA (Compositae)
B. angustifolia (Pursh.) Robins.
SESQUITERPENOID: balduilin.
Herz *et al., J. Am. Chem. Soc.,* **81,** 6061 (1959)

B. uniflora.
SESQUITERPENOID: balduilin.
Herz *et al., J. Am. Chem. Soc.,* **81,** 6061 (1959)

BALFOURODENDRON (Rutaceae)
B. redelsamia.
FURANOQUINOLINE ALKALOID: flindersiamine.
Brown *et al., Aust. J. Chem.,* **7,** 181 (1954)

B. riedelianum (Engl.) Engl.
ACRIDONE ALKALOID: evoxanthine.
FURANOQUINOLINE ALKALOIDS: balfourodine,
(+)-isobalfourodine, ribalinium.
QUATERNARY ALKALOID: O(4)-methylbalfourodinium.
QUINOLONE ALKALOIDS: (+)-ribalidine, ribalinine.
Rapoport, Holden, *J. Am. Chem. Soc.,* **81,** 3738 (1959)
Jurd, Wong, *Aust. J. Chem.,* **36,** 1615 (1983)

BALIOSPERMUM
B. montanum.
PYRROLIZIDINE ALKALOID: heliospermine.
DITERPENOID: montanin.
TERPENE ESTER: baliospermin.
Ogura *et al., Planta Med.,* **33,** 128 (1978)

BALLONITES
B. roxburghii Planch.
STEROIDS: balanitisins A, B, C, D, E and F.
Varshney *et al., Indian J. Pharm.,* **39,** 125 (1977)

BALLOTA (Labiatae)
B. acetobulosa.
DITERPENOID: 18-hydroxyballonigrin.
Savona *et al., J. Chem. Soc., Perkin I,* 1271 (1978)

B. foetida Lamk.
FLAVONOIDS: apigenin 6,8-di-C-glucoside, apigenin
7-O-glucoside.
Darbour *et al., Pharmazie,* **41,** 605 (1986)

B. hispanica.
DITERPENOIDS: hispaninic acid, hispanolone, hispanonic acid
   methyl ester.
Lopez de Lerma *et al.*, *Tetrahedron Lett.*, 1273 (1980)

B. lanata.
DITERPENOID: 13-hydroxyballogrinolide.
Savona *et al.*, *Phytochemistry*, **17**, 2132 (1978)

B. nigra L.
DITERPENOIDS: ballonigrin, ballotenol, ballotinone.
Savona *et al.*, *J. Chem. Soc., Perkin I*, 1607 (1976)

B. rupestris.
DITERPENOIDS: ballonigrin, ballonigrinone, rupestralic acid.
Savona *et al.*, *J. Chem. Soc., Perkin I*, 322 (1977)

BALSAMITA
B. major.
SESQUITERPENOIDS: dehydroisoerivanin, isoerivanin.
Samek *et al.*, *Collect. Czech. Chem. Commun.*, **44**, 1468 (1979)

BALTIMORA
B. recta.
SESQUITERPENOIDS: encelin, haagenolide.
Herz, Kumar, *Phytochemistry*, **18**, 1743 (1979)
Ciccio *et al.*, *ibid*, **20**, 517 (1981)

BANGIA (Rhodophyceae)
B. fuscopurpurea.
CAROTENOIDS: α-carotene, β-carotene, α-cryptoxanthin,
   β-cryptoxanthin, fucoxanthin, lutein, zeaxanthin.
Bjoernland *et al.*, *Phytochemistry*, **15**, 291 (1976)

BANISTERIOPSIS
B. argentea.
CARBOLINE ALKALOIDS: 5-methoxytetrahydroharman,
   (-)-N(6)-methyltetrahydroharman.
INDOLE ALKALOID: dimethyltryptamine-N(6)-oxide.
Ghosal, Majumber, *Phytochemistry*, **10**, 2840 (1971)

B. caapi Spruce.
CARBOLINE ALKALOIDS: harmine, tetrahydroharmine.
PHTHALIDE-PYRROLIDINE ALKALOID: dihydroshihunine.
UNCHARACTERIZED ALKALOID: telepathine.
Spath, Lederer, *Ber.*, **63**, 120 (1930)
Kawanishi *et al.*, *J. Nat. Prod.*, **45**, 637 (1982)

B. lutea.
CARBAZOLE ALKALOID: harmine.
UNCHARACTERIZED ALKALOID: telepathine.
Bayon, *Quoted by Villalba, Bolletin Laboratory Samper-Martinez,
   Bogata* (1927)

B. metallicolor.
CARBAZOLE ALKALOID: harmine.
UNCHARACTERIZED ALKALOID: telepathine.
Spath, Lederer, *Ber.*, **63**, 120 (1930)

BAPTISIA (Papilionaceae)
B. australis.
SPARTEINE ALKALOIDS: cytisine, (+)-sparteine.
Govindachari *et al.*, *J. Chem. Soc.*, 3839 (1957)

B. lecontei.
FLAVONOID: fustin 3-O-β-glucoside.
ISOFLAVONOID: sphaerobioside.
(F) Markham *et al.*, *Tetrahedron*, **24**, 823 (1968)
(I) Rosler *et al.*, *Chem. Ber.*, **98**, 2193 (1965)

B. minor Lehm.
PAVINANE ALKALOID: baptifoline.
Marion, Turcotte, *J. Am. Chem. Soc.*, **70**, 3253 (1948)

B. perfoliata (L.) R.Br.
PAVINANE ALKALOID: baptifoline.
Bohlmann *et al.*, *Chem. Ber.*, **95**, 944 (1962)

B. sphaerocarpa.
ISOFLAVONOID: sphaerobioside.
Rosler *et al.*, *Chem. Ber.*, **98**, 2193 (1965)

B. tinctoria R. Br.
SPARTEINE ALKALOIDS: cytisine, (+)-sparteine.
Bohlmann *et al.*, *Chem. Ber.*, **95**, 944 (1962)

BARBACENIA
B. coniostigma.
Three unnamed triterpenoids.
Pinto *et al.*, *Anais. Acad. Bras. Cienc.*, **53**, 69 (1981)

BARBILOPHAZIA (Hepaticae)
B. attenuata (Mart.) Loeske.
SESQUITERPENOIDS: α-barbatene, β-barbatene.
DITERPENOIDS: barbilycopodin,
   18-acetoxy-3S,4S:7S,8S-diepoxydolabellan,
   10R,18-diacetoxy-3S,4S-epoxydolabell-7E-ene,
   18-hydroxydolabell-7E-en-3-β-ol.
Huneck *et al.*, *J. Chem. Soc., Perkin I*, 809 (1986)

B. barbata.
SESQUITERPENOIDS: α-barbatene, β-barbatene.
Andersen *et al.*, *Phytochemistry*, **12**, 2709 (1973)

B. floerkei (Web. et Mohr) Loeske.
SESQUITERPENOIDS: α-barbatene, β-barbatene.
DITERPENOID: barbilycopodin.
Andersen *et al.*, *Phytochemistry*, **12**, 2709 (1973)

B. hatcheri (Evans) Loeske.
SESQUITERPENOID: hercynolactone.
Huneck *et al.*, *Tetrahedron Lett.*, 3959 (1982)

B. lycopodioides (Wallr.) Loeske.
SESQUITERPENOIDS: α-barbatene, β-barbatene, hercynolactone.
DITERPENOIDS: 18-acetoxy-3S,4S:7S,8S-diepoxydolabellane,
   barbilycopodin,
   10R,18-diacetoxy-3S,4S-epoxydolabell-7E-ene,
   18-hydroxydolabell-7-en-3-one.
Huneck *et al.*, *J. Chem. Soc., Perkin I*, 809 (1986)

BARLERIA (Acanthaceae)
B. prionitis L.
IRIDOIDS: acetylbarlerin, barlerin.
Taneja, Tiwari, *Tetrahedron Lett.*, 1995 (1975)

BAROSMA (Rutaceae)
B. betulina Bartl.
MONOTERPENOIDS: p-menthane-8-thiol-1-one,
   2-epi-p-menthane-thiol-1-one.
Lamparsky, Schudel, *Tetrahedron Lett.*, 3323 (1971)

B. crenulata (L.) Hook.
MONOTERPENOIDS: p-menthane-8-thiol-1-one,
   2-epi-p-menthane-thiol-1-one.
Lamparsky, Schudel, *Tetrahedron Lett.*, 3323 (1971)

B. serratifolia Willd.
MONOTERPENOIDS: p-menthane-8-thiol-1-one,
   2-epi-p-menthane-8-thiol-1-one.
Lamparsky, Schudel, *Tetrahedron Lett.*, 3323 (1971)

BARRACENIA

B. bicolor.
TRITERPENOIDS: 3β,20(R)-dihydroxydammar-24-ene,
20(R)-hydroxydammar-24-en-3-one,
3β,20(R)-dihydroxydammar-24-ene.
Baker et al., Phytochemistry, 15, 795 (1976)

BARRINGTONIA (Lecythidaceae)

B. acutangula.
TRITERPENOIDS: acutangulic acid, barrigenic acid, barrigenols
A, B, C, D and E, barrinic acid, tanginol, tangulic acid.
Barua et al., J. Am. Pharm. Ass., 40, 937 (1961)
Narayan et al., Curr. Sci., 45, 518 (1976)

B. asiatica.
TRITERPENOID: A1-barrigenol.
Nozoe, J. Chem. Soc., Japan, 56, 689 (1935)

B. racemosa.
TRITERPENOID: R1-barrigenol.
Lin et al., J. Chin. Chem. Soc., 4, 77 (1957)

B. speciosa Forst.
TRITERPENOIDS: anhydrobartogenic acid, bartogenic acid,
19-epi-bartogenic acid.
Rao et al., Indian J. Chem., 25B, 113 (1986)

BARTLETTINA

B. karwinskiana.
SESQUITERPENOIDS: 3β-acetoxy-11β,13-epoxy-epi-tulupinolide,
11β,13-dihydroxy-epi-tulipinolide,
11β,13-epoxy-8β-acetoxy-α-cyclocostunolide,
11β,13-epoxy-epi-tulipinoide.
Gao et al., Phytochemistry, 25, 1231 (1986)

BARTSIA (Scrophulariaceae)

B. trixago.
IRIDOID: bartsioside.
Bianco et al., Gazz. Chim. Ital., 106, 725 (1976)

BASSIA (Chenopodiaceae)

B. butyracea.
TRITERPENOID: bassic acid.
Van der Hass, Recl. Trav. Chim. Pays-Bas, 48, 1155 et seq. (1929)

B. latifolia.
TRITERPENOID: bassic acid.
Van der Hass, Recl. Trav. Chim. Pays-Bas, 48, 1155 et seq. (1929)

B. longifolia.
TRITERPENOID: bassic acid.
Van der Hass, Recl. Trav. Chim. Pays-Bas, 48, 1155 et seq. (1929)

B. parkeri.
TRITERPENOID: bassic acid.
Van der Hass, Recl. Trav. Chim. Pays-Bas, 48, 1155 et seq. (1929)

BATHIORHAMNUS

B. cryptophorus.
PIPERIDINE ALKALOIDS: O-acetylcryptophorinine,
cryptophorinine.
Bruneton et al., Plant Med. Phytother., 9, 21 (1975)

BAUHINIA (Leguminosae)

B. candicans.
AMINE ALKALOIDS: choline, trigonelline.
FLAVONOIDS: kaempferol 3-O-β-rutinoside, kaempferol
3-O-β-rutinoside-7-O-α-rhamnopyranoside.

STEROIDS: campesterol, cholesterol, β-sitosterol, β-sitosterol
3-glucoside, stigmasta-3,5-dien-7-one, stigmasterol.
Iribarren et al., J. Nat. Prod., 46, 750 (1983)

B. purpurea.
FLAVONOIDS: apigenin, apigenin 7-O-glucoside, quercetin,
quercitrin, rutin.
Abd-el-Wahab et al., Herba Hung., 26, 27 (1987); Chem. Abstr., 107
151201v

B. racemosa Lamk.
PHENOLIC: racemosol.
Anjaneyulu et al., Tetrahedron, 42, 2417 (1986)

B. vahlii L.
FLAVONOIDS: agathoflavone, agathaflavone dimethyl ether,
agathoflavone methyl ether, quercetin, quercetin
3-O-β-glucoside.
STEROIDS: campesterol, β-sitosterol, stigmasterol.
TRITERPENOID: betulinic acid.
Sultana et al., J. Indian Chem. Soc., 62, 337 (1985)

B. variegata var. candida.
FLAVONOIDS: apigenin, apigenin 7-O-glucoside, quercetin,
quercitrin, rutin.
Abd-el-Wahab et al., Herba Hung., 26, 27 (1987); Chem. Abstr., 107
151201v

BAURELLA

B. simplicifolia.
Two uncharacterized alkaloids.
Tillequin et al., Lloydia, 43, 498 (1980)

B. simplicifolia subsp. neo-scotia.
ACRIDONE ALKALOIDS: 1,3-dimethoxyacridan-9-one,
1,2,3,4-tetramethoxyacridan-9-one.
Tillequin et al., Lloydia, 43, 498 (1980)

BAZZANIA (Hepaticae)

B. angustifolia.
SESQUITERPENOIDS: (+)-bazzanene, isobazzanene.
Wu, Liu, Tetrahedron, 39, 2657 (1983)

B. pompeana.
SESQUITERPENOIDS: (+)-bazzanene, bazzanenol,
(R)-(-)-cuparane, δ-cuparenol.
Matsui et al., Phytochemistry, 14, 1037 (1975)

B. tridens.
SESQUITERPENOID: tridensenone.
Toyota et al., Phytochemistry, 20, 2359 (1981)

B. trilobata (L.) Gray.
SESQUITERPENOIDS: α-bazzanene, β-bazzanene, 11-drimanol.
Andersen et al., Phytochemistry, 12, 1818 (1973)

BEAUMONTIA

B. grandiflora Wallich.
STEROIDS: beaumontoside, beauwalloside, wallichoside.
Krasso et al., Helv. Chim. Acta, 46, 1691 (1963)

BEDFORDIA

B. salicina.
DIMERIC SESQUITERPENOIDS: 9β,9'α-bis-dihydroligularenolide,
9β,9'β-bis-dihydroligularenolide.
SESQUITERPENOIDS: 1,9-dihydro-8-methoxyligularenone,
3β,4β-dimethyl-7β-(1-carbomethoxyethyl)-9,10-
dehydrodecal-8-one, 4-hydroxybedfordic acid,
isobedfordiaterpenalcohol, 4-oxobedfordic acid methyl ester.
DITERPENOIDS: bedfordiaterpenalcohol,
3α-hydroxy-ent-kaurenic acid methyl ester.
Bohlmann, Ngo Le Van, Phytochemistry, 17, 1173 (1978)

BEESIA (Ranunculaceae)
B. calthaefolia.
TERPENOIDS: beesioside I, beesioside II.
Sakurai *et al.*, *Chem. Pharm. Bull.*, **34**, 582 (1986)

BEFARIA
B. glutinosa.
FLAVONOIDS: kaempferol, quercetin, quercetin
3-O-α-L-rhamnoside.
TRITERPENOIDS: betulin, lupeol.
Torrenagra *et al.*, *Rev. Latinoam. Quim.*, **16**, 102 (1985)

BEGONIA (Begoniaceae)
B. tuberhybrida.
DITERPENOID: hexanocucurbitacin D.
Doskotch *et al.*, *Lloydia*, **32**, 115 (1969)

BEILSCHMEIDIA
B. elliptica White et Francis.
APORPHINE ALKALOID: laurelliptine.
Clezy *et al.*, *Experientia*, 19, (1963)
B. podogarica Kostermans.
APORPHINE ALKALOIDS: 2-hydroxy-1,9,10-trimethoxyaporphine,
10-hydroxy-1,2,9-trimethoxyaporphine,
2-hydroxy-1,9,10-trimethoxynoraporphine.
Johns *et al.*, *Aust. J. Chem.*, **23**, 363 (1970)

BEJARANOA
B. semistriata.
SESQUITERPENOIDS: α-hydroxybejaranolide,
β-hydroxybejaranolide.
DITERPENOIDS: 17,20-dihydroxygeranylnerol, semistriatin
methyl ether.
Bohlmann *et al.*, *Phytochemistry*, **20**, 1639 (1981)

BELAMEANDA
B. chinensis.
FLAVONOID: irigenin.
Hu *et al.*, *Zhongyao Tongbao*, **7**, 29 (1982); *Chem. Abstr.*, 97 107032s

BELLARDIA (Scrophulariaceae)
B. montana.
AMINE ALKALOIDS: bellendine,
tri-(α-methylene-τ-butyrolactonyl)amine.
Ralph *et al.*, *Phytochemistry*, **25**, 972 (1986)
B. trixago (L. All.) var flaviflora (Boiss.) Maire.
DITERPENOIDS: isotrixagol, isotrixagoyl oxide, trixagodiols A,
B and C, trixagoene, trixagotriol.
de Pascual Teresa *et al.*, *Tetrahedron*, **38**, 1837 (1982)
B. trixago (L. All.) var. versicolor (Willd.) P.Coutinho.
DITERPENOIDS: isotrixagol, isotrixagoyl oxide, trixagodiols A,
B and C, trixagoen trixagotriol
de Pascual Teresa *et al.*, *Tetrahedron*, **38**, 1837 (1982)

BELLIUM
B. bellidioides.
SESQUITERPENOID: matricariol 3-methylcrotonyl ester.
Bohlmann *et al.*, *Chem. Ber.*, **105**, 1919 (1972)

BERBERIS (Berberidaceae)
B. actinacantha.
AMINE ALKALOIDS: chiloenamines A and B, chiloenine.
BISBENZYLISOQUINOLINE ALKALOIDS: berbitine,
dihydroxytaxilamine.
Shamma *et al.*, *J. Chem. Soc., Chem. Commun.*, 799 (1983)
Rahimazadeh, *Planta Med.*, 339 (1986)
B. amurensis Rupr. var. japonica (Regel) Rehd.
BISBENZYLISOQUINOLINE ALKALOID: berbamunine.
Kametani *et al.*, *J. Pharm. Soc., Japan*, **88**, 1163 (1968)
B. aristata DC.
BISBENZYLISOQUINOLINE ALKALOID: taxilamine.
PROTOBERBERINE ALKALOID: karachine.
Blasko *et al.*, *J. Am. Chem. Soc.*, **104**, 2039 (1982)
B. baluchistanica Ahrendt.
BISBENZYLISOQUINOLINE ALKALOIDS: pakistanamine,
pakistanine.
BENZYLISOQUINOLINE ALKALOIDS: dihydrosecoquettamine,
quettamine, secoquettamine, gandharamine, obaberine,
oxyacanthine.
ISOQUINOLINE-BENZYLISOQUINOLINE ALKALOIDS:
baluchistanamine, baluchistine.
Miana *et al.*, *Experientia*, **35**, 1137 (1979)
Zarga *et al.*, *Heterocycles*, **18**, 63 (1982)
B. brandisiana.
APORPHINE ALKALOID: isoboldine.
BENZYLISOQUINOLINE ALKALOID: (+)-reticuline.
BISBENZYLISOQUINOLINE ALKALOIDS: (+)-berbamine,
(+)-berbamine 2′-β-N-oxide, (+)-isotetrandrine.
ISOQUINOLINE ALKALOID: thalifoline.
PROTOBERBERINE ALKALOIDS: berberine, palmatine.
Hussain *et al.*, *J. Nat. Prod.*, **49**, 538 (1986)
B. buxifolia Lam.
AMINE ALKALOIDS: chiloenamine, chiloenamine A,
chiloenamine B, chiloenine.
BISBENZYLISOQUINOLINE ALKALOIDS: calafatamine, calafatine,
calafatine 2-α-N-oxide, calafatine 2-β-N-oxide, curacutine,
(-)-osornine, (-)-talcamine.
Fejardo *et al.*, *Heterocycles*, **15**, 1137 (1981)
Leet *et al.*, *J. Chem. Soc., Perkin I*, 65 (1984)
B. calliobotrys Aitch. ex Bienert.
BISBENZYLISOQUINOLINE ALKALOID: khyberine.
Hussain *et al.*, *Tetrahedron Lett.*, 4573 (1980)
B. chilensis.
BISBENZYLISOQUINOLINE ALKALOIDS:
7-O-demethylisothalicberine, isothalicberine.
Torres *et al.*, *Gazz. Chim. Ital.*, **109**, 567 (1979)
B. cristata DC.
PROTOBERBERINE ALKALOID: karachine.
Blasko *et al.*, *J. Am. Chem. Soc.*, **104**, 2039 (1982)
B. Darwinii Hook.
BENZYLISOQUINOLINE ALKALOID: santiagonamine.
ISOQUINOLINE ALKALOID: (±)-nuevamine.
PROTOBERBERINE ALKALOID: berberine.
Awe, *Arch. Pharm.*, **284**, 352 (1951)
Valencia *et al.*, *Tetrahedron Lett.*, 599 (1984)
B. empetrifolia Lam.
ISOQUINOLINE ALKALOIDS: (±)-chilenine, coyhaiquinine,
natalimine, puntarenine.
PROTOBERBERINE ALKALOID: berberine.
Fajardo *et al.*, *Bol. Soc. Chil. Quim.*, **30**, 51 (1985)
B. floribunda G.Don.
PROTOBERBERINE ALKALOID: (-)-corypalmine.
Manske, *Can. J. Res.*, **20B**, 57 (1942)

B. gagnepainii.
FLAVONOIDS: hyperoside, isorhamnetin galactoside, luteolin, quercetroside, rutoside.
Dauguet et al., Plant Med. Phytother., **16**, 16 (1982)

B. glauca DC.
PROTOBERBERINE ALKALOID: berberine.
Awe, Arch. Pharm., **284**, 352 (1951)

B. hakeoides.
BISBENZYLISOQUINOLINE ALKALOIDS: (+)- pakastonamine, (+)-patagonine, valdiberine, (+)-valdivianine.
Urzua et al., Rev. Latinoam. Quim., **16**, 66 (1985)

B. heterobotris.
PROTOBERBERINE ALKALOID: berberine.
Awe, Arch. Pharm., **284**, 352 (1951)

B. heteropoda Schrenk.
BISBENZYLISOQUINOLINE ALKALOIDS: berbamine, oxyacanthine.
PROTOBERBERINE ALKALOIDS: berberine, columbianine, jatrorrhizine, palmatine.
Awe, Arch. Pharm., **284**, 352 (1951)
Kondo, Tomita, ibid., **268**, 549 (1930)

B. himaloica.
PROTOBERBERINE ALKALOID: corypalmine.
Manske, Can. J. Res., **20B**, 57 (1942)

B. iliensis M.Pop.
BISBENZYLISOQUINOLINE ALKALOIDS: berbamine, oxyacanthine.
PROTOBERBERINE ALKALOIDS: berberine, jatrorrhizine.
Tomita, Pharm. Bull., Japan, **1**, 101 (1953)

B. insignis Hook.
PROTOBERBERINE ALKALOID: umbellatine.
Orekhov, J. Am. Pharm. Ass., **30**, 247 (1941)

B. japonica.
BISBENZYLISOQUINOLINE ALKALOID: isotetrandrine.
PROTOBERBERINE ALKALOID: columbianine.
Chatterjee, Banerji, J. Indian Chem. Soc., **30**, 705 (1953)
Bick et al., Aust. J. Chem., **9**, 111 (1956)

B. lambertii.
PROTOBERBERINE ALKALOID: columbianine.
UNCHARACTERIZED ALKALOIDS: berlambine, lambertine.
Chatterjee, Banerji, J. Indian Chem. Soc., **30**, 705 (1953)

B. laurina Billb.
BISBENZYLISOQUINOLINE ALKALOIDS: belarine, espinidine, espinine.
UNCHARACTERIZED ALKALOID: lambertine.
Falco et al., J. Chem. Soc., Chem. Commun., 1056 (1971)

B. lyceum Royle.
UNCHARACTERIZED ALKALOIDS: berbenine, berbericine, berbericinine.
Ikram et al., Pakistan J. Sci. Ind. Res., **9**, 343 (1966)

B. maximowiczii.
PROTOBERBERINE ALKALOID: jatrorrhizine.
Tomita, Tani, J. Pharm. Soc., Japan, **61**, 83 (1941)

B. nervosa.
PROTOBERBERINE ALKALOID: berberine.
Monkovic, Spenser, Can. J. Chem., **43**, 2017 (1965)

B. obenga.
BISBENZYLISOQUINOLINE ALKALOID: 2'-N-methylberbamine.
Karimova et al., Khim. Prir. Soedin., 80 (1977)

B. orthobotrys (Bienert ex Aitch.) Schneid.
BISBENZYLISOQUINOLINE ALKALOIDS: chitraline, 1-O-methylpakistanine.
Hussain et al., J. Nat. Prod., **44**, 274 (1981)

B. ottowensis L.
PROTOBERBERINE ALKALOIDS: berberine, hydroxyberberine, palmatine.
Yusufbekov et al., Dokl. Akad. Nauk Tadzh. SSR, **28**, 712 (1985);
   Chem. Abstr., 105 3564j

B. pseudanubalata.
BISBENZYLISOQUINOLINE ALKALOIDS: O-methyloxyacanthine, oxyacanthine.
PROTOBERBERINE ALKALOIDS: berberine, palmatine.
STEROIDS: β-sitosterol, β-sitosterol β-D-glucoside.
Pant et al., Fitoterapia, **57**, 427 (1986)

B. tachamoskyana.
BISBENZYLISOQUINOLINE ALKALOID: obamegine.
Tomita, Kubo, J. Pharm. Soc., Japan, **79**, 317 (1959)

B. thunbergii DC.
BISBENZYLISOQUINOLINE ALKALOID: berbamine.
Bick et al., Aust. J. Chem., **9**, 111 (1956)

B. thunbergii DC. var. Maximowiczii.
PROTOBERBERINE ALKALOIDS: columbianine, jatrorrhizine.
UNCHARACTERIZED ALKALOID: shobakunine.
Orekhov et al., Arch. Pharm., **271**, 323 (1933)
Tomita, Tani, J. Pharm. Soc., Japan, **61**, 83 (1941)

B. umbellata Wall.
PROTOBERBERINE ALKALOID: umbellatine.
Orekhov, J. Am. Pharm. Ass., **20**, 247 (1941)

B. valdiviana Phil.
BISBENZYLISOQUINOLINE ALKALOID: (+)-temuconine.
Guinaudeau et al., Heterocycles, **19**, 1009 (1982)

B. vulgaris L.
BISBENZYLISOQUINOLINE ALKALOIDS: berbamine, oxyacanthine.
PROTOBERBERINE ALKALOIDS: berberine, columbianine, jatrorrhizine, palmatine.
UNCHARACTERIZED ALKALOID: bervulsine.
Cava et al., Lloydia, **28**, 73 (1965)

B. zabeliana Schneid.
DIMERIC ALKALOID: chitraline.
Hussain et al., Tetrahedron Lett., 4573 (1980)

BERKHEYA (Compositae/Arctotideae)
B. angustifolia (Houtt.) Merrill.
ACETYLENICS: cis-1-methylthio-1,4-di-(2-thienyl)-1-buten-3-yn, trans-1-methylthio-1,4-di-(2-thienyl)-1-buten-3-yn, 2-(3-[5-oxo-1-propenyl)-2-thienyl]-2-propynylidene)-3-thietanol, 2-(3-[5-oxo-1-propenyl)-2-thienyl]-2-propynylidene)-3-thietanone.
Bohlmann, Suwita, Chem. Ber., **108**, 515 (1975)

B. carduoides (Less.) Hutch ex Fourc.
ACETYLENICS: cis-1-methylthio-1,4-di(2-thienyl-1-buten-3-yn, trans-1-methylthio-1,4-di(2-thienyl)-1-buten-3-yn, 2-(3-[5-oxo-1-propenyl)-2-thienyl]-2-propynylidene)-3-thietanol, 2-(3-[5-oxo-1-propenyl)-2-thienyl]-2-propynylidene)-3-thietanone.
Bohlmann, Suwita, Chem. Ber., **108**, 515 (1975)

B. radula.
SESQUITERPENOID: berkheyaradulene
Bohlmann et al., Chem. Ber., **110**, 3777 (1977)

BERLANDIERA (Compositae)

B. pumila.
SESQUITERPENOIDS: 3,4-epoxypumilin, pumilin.
Fischer *et al.*, *J. Heterocycl. Chem.*, **19**, 181 (1982)

B. subacaulis (Nutt.) Nutt.
SESQUITERPENOIDS: berlandin, subacaulin.
Herz *et al.*, *J. Org. Chem.*, **37**, 2532 (1973)

B. tescana.
SESQUITERPENOIDS: 3,4-epoxypumilin, pumilin.
Fischer *et al.*, *J. Heterocycl. Chem.*, **19**, 181 (1982)

BERSAMA (Melianthaceae)

B. abyssinica Fresen.
STEROIDS: abyssinin, bersaldegenin 3-acetate, bersaldegenin 1,3,5-orthoacetate, bersamagenin, berschillogenin, 3-epi-berschillogenin, hellibrigenin 3-acetate, hellibrigenin 3,5-diacetate, 16α-hydroxybersaldegenin 3-acetate, 16α-hydroxybersaldegenin 1,3,5-orthoacetate, 1-hydroxybersamagenin 1,3,5-orthoacetate, scilliglaucosidin.
Kupchan *et al.*, *J. Org. Chem.*, **36**, 2611 (1971)
Kubo, Matsumoto, *Tetrahedron Lett.*, 4601 (1984)

BERTYA

B. cupressoidea.
DITERPENOID: bertyadionol.
Ghisalberti *et al.*, *Tetrahedron Lett.*, 4599 (1970)

B. sp. nova.
DITERPENOID: bertydionol.
Ghisalberti *et al.*, *Tetrahedron Lett.*, 4599 (1970)

BESHORNERIA

B. yuccoides.
SAPONINS: beschornin, beshornoside.
Kintya *et al.*, *Phytochemistry*, **21**, 1447 (1982)

BETA

B. vulgaris.
PIGMENTS: betanin, neobetanin.
Lard *et al.*, *Phytochemistry*, **24** 2383 (1985)

BETONICA (Labiatae)

B. macrantha.
IRIDOIDS: harpagide, harpagide acetate.
Komissarenko *et al.*, *Khim. Prir. Soedin.*, 109 (1976)

B. nivea.
IRIDOIDS: harpagide, harpagide acetate.
Komissarenko *et al.*, *Khim. Prir. Soedin.*, 109 (1976)

B. occidentalis.
IRIDOIDS: harpagide, harpagide acetate.
Komissarenko *et al.*, *Khim. Prir. Soedin.*, 109 (1976)

B. officinalis L. (Stachys betonica Benth.).
QUATERNARY ALKALOID: stachydrine.
IRIDOIDS: harpagide, harpagide acetate.
(A) Kung, *Z. Physiol. Chem.*, **85**, 217 (1913)
(I) Komissarenko *et al.*, *Khim. Prir. Soedin.*, 109 (1976)

B. wiesen.
QUATERNARY ALKALOID: stachydrine.
Kung, Trier, *Z. Physiol. Chem.*, **85**, 209 (1913)

BETULA (Betulaceae)

B. alba.
GLYCOSIDE: roseoside.
MONOTERPENOIDS: betalbusides A and B.
TRITERPENOID: betulin.
(Gl) Ciper *et al.*, *Phytochemistry*, **15**, 1990 (1976)
(T) Tschesche *et al.*, *Chem. Ber.*, **110**, 3111 (1977)

B. costata Trautv.
TRITERPENOIDS: betulafolientetraol oxide, betulafolientriol I, dammar-24-ene-3α,17α,20-triol.
Reichardt *et al.*, *J. Org. Chem.*, **46**, 4576 (1981)

B. ermanni Cham.
TRITERPENOIDS: 20,24-epoxydammarane-3β,6α,25-triol, 20,24-epoxydammarane-3β,6α,25-triol 6-acetate.
Malinovskaya *et al.*, *Khim. Prir. Soedin.*, 587 (1978)

B. lenta L.
SESQUITERPENOIDS: α-betulenol, β-betulenol.
Triebs, *Ber.*, **69**, 41 (1936)

B. mandschurica.
TRITERPENOID: 20,24-dihydroxydammar-25-en-3-one.
Malinovskaya *et al.*, *Khim. Prir. Soedin.*, 346 (1980)

B. nigra.
FLAVONOIDS: apigenin, apigenin 7,4'-dimethyl ether, cirsimaretin, combretol, exmanin, 5-hydroxy-3,3',4',5',7-pentamethoxyflavone, isokaempferide, isorhamnetin, kaempferol 3,7,4'-trimethyl ether, luteolin, pachypodol, quercetin 3,7,3',4'-tetramethyl ether, rhamnocitrin, rutin, salvigenin, 3',4',5,7-tetrahydroxy-3-methoxyflavone.
Chumbalov *et al.*, *Khim. Prir. Soedin.*, **236**, 439 (1969)
Wollenweber, *Phytochemistry*, **16**, 295 (1977)

B. platyphylla Sukatchev.
GLUCOSIDE: hannokinin glucoside.
TRITERPENOIDS: betulafolienpentaol, betulafolientetraols A and B, betulafolientriol, betulafolientriol oxide, betulin, 3-epi-ocotillol II.
(G) Nomura *et al.*, *Phytochemistry*, **20**, 1097 (1981)
(T) Han, Song, *ibid*, **16**, 1075 (1977)

B. platyphylla Sukatchev. var. japonica.
TRITERPENOIDS: betulafolientriol, dammar-24-ene-12β,20-(S)-diol-3-one, hydroxyhopanone.
Nagai *et al.*, *Tetrahedron Lett.*, 4239 (1968)

B. pubescens Ehrh.
SESQUITERPENOID: 14-hydroxycaryophyllen-4,5-oxide.
Pokhilo *et al.*, *Khim. Prir. Soedin.*, 598 (1984)

B. utilis.
TRITERPENOIDS: 3-0-acetyloleanolic acid, karachic acid.
Khan, Atta-ur-Rahman, *Phytochemistry*, **14**, 789 (1975)

B. verrucosa Ehrh. (B. pendula Roth.).
STEROIDS: citrostadienol, (Z)-sitosterol, α-sitosterol.
PRENOLS: betulaprenols 6, 7, 8, 9, 10, 11, 12 and 13.
TRITERPENOIDS: lupane-3β,20-diol, lupane-3β,20,28-triol.
(P) Wellburn *et al.*, *Nature*, **212**, 1364 (1966)
(S) Bates *et al.*, *Tetrahedron Lett.*, 6163 (1968)
(T) Lindgren, Svahn, *Acta Chem. Scand.*, **20**, 1720 (1966)

BEYERIA (Euphorbiaceae)

B. brevifolia Muell.-Arg.
DITERPENOID:
(15S)-15,16-dihydroxy-3,4-seco-enantio-pimara-4(18),7-dien-3-oic acid.
Chow, Jefferies, *Aust. J. Chem.*, **21**, 2529 (1968)

B. brevifolia Muell.-Arg. var. *brevifolia* Airy Shaw.
TRITERPENOID: lup-20(29)-ene-3β,16α,18-triol.
Errington *et al.*, *Aust. J. Chem.*, **29**, 1809 (1976)

B. calycina.
DITERPENOID: ent-18-hydroxy-3,4-secobeyer-15-ene-3,17-dioic acid.
Ghisalberti *et al.*, *Phytochemistry*, **17**, 1961 (1978)

B. leschenaultii (DC) Bail.
DITERPENOIDS: 6β-acetoxy-17-hydroxybeyer-15-en-3-one,
19-acetyl-3-oxo-16-kauran-17-oic acid, beyerol,
(-)-12β,17-dihydroxy-16α-kauran-18-oic acid,
17-hydroxy-3,4-secobeyera-4(18),15-dien-3-oic acid,
16α-kaurane-17,19-diol, 16α-kaurane-3α,17,19-triol,
(-)-kaur-16-ene-3β,18-diol.
TRITERPENOID: lup-2-ene-3β,16α-diol.
Baddeley *et al.*, *Tetrahedron*, **20**, 1983 (1964)
Baddeley *et al.*, *Aust. J. Chem.*, **17**, 578 (1974)

B. leschenaultii var. *drummondii.*
DITERPENOID: beyerol 17-cinnamate.
Jefferies *et al.*, *Aust. J. Chem.*, **15**, 521 (1962)

B. sp. nova.
DITERPENOIDS: (-)-16α,17-dihydroxy-16α-kauran-19-oic acid,
8α,13-oxidoperu-14-en-18-oic acid,
8β,13-oxidoperu-14-en-18-oic acid.
Jefferies *et al.*, *Aust. J. Chem.*, **18**, 1441 (1965)

BIDENS (Compositae)
B. aurea (Ait.) Sherff.
ESTER: 1'-(2-acetoxy-2-methylpropionoyloxy)-eugenol isobutyrate. Four unnamed polyenes.
Bohlmann, Zdero, *Chem. Ber.*, **108**, 440 (1975)

BIFURCARIA (Phaeophyceae/Cystosieraceae)
B. bifurcata Ross.
PHENOLIC ANTIBIOTICS: difucol hexaacetate, diphlorethol pentaacetate, heptafuhalol octadecaacetate, pentafuhalol tridecaacetate, tetrafuhalol undecaacetate, trifucol nonaacetate, trifuhalol octaacetate.
DITERPENOIDS: eleganediol, epoxyeleganolone,
12-(S)-hydroxygeranylgeraniol.
(P) Glombitza *et al.*, *Phytochemistry*, **15**, 1279 (1976)
(T) Baird *et al.*, *Tetrahedron Lett.*, 1849 (1980)

B. galapagensis (Piccone et Gruiow.) Womersley.
DITERPENOID: bifurcarenone.
Sun *et al.*, *Tetrahedron Lett.*, 3123 (1980)

BILLBERGIA (Bromeliaceae)
B. vittata.
ANTHOCYANINS: cyanidin, delphinidin, poenidin.
FLAVONOIDS: apigenin, luteolin, scutellarein,
scutellarein-β-D-glucoside, quercetin, tricin.
Ashtakala *et al.*, *J. Proc. Am. Hort. Soc.*, **100**, 546 (1975); *Chem. Abstr.*, 84 56511u

BINAPSIS
B. arvense.
FLAVONE: isorhamnerin 3-β-D-glucoside.
Rahman *et al.*, *J. Org. Chem.*, **27**, 153 (1962)

BIOTA
B. orientalis.
SESQUITERPENOIDS: α-biotol, cuparene, α-cuparenol,
β-cuparenol, curcumenether, α-isobiotol, isocuparenol.
Tomita *et al.*, *Tetrahedron Lett.*, 843 (1968)

BISHOFIA
B. javanica.
STEROID: β-sitosterol.
TRITERPENOIDS: friedelin, friedelinol, epi-friedelinol.
Chen *et al.*, *Zhongcaoyao*, **28**, 250 (1987); *Chem. Abstr.*, 107 172507q

BIXA (Bixaceae)
B. orellana L.
NORSESTERTERPENOID: bixin.
Karrer, Jucker, *Carotenoids, Basle*, (1976)

BLAINVILLEA
B. latifolia.
One germacranolide.
Rojarkar *et al.*, *J. Chem. Res. Synop.*, 272 (1986)

BLECHNUM (Blechnaceae)
B. japonicum.
STEROID: shidasterone.
Hikino *et al.*, *Chem. Pharm. Bull.*, **23**, 1458 (1975)

B. minus Ettingh.
STEROID: 2-deoxyecdysone.
Chong *et al.*, *J. Chem. Soc., Chem. Commun.*, 1217 (1970)

B. nipponicum.
STEROID: shidasterone.
Imai *et al.*, *J. Chem. Soc., Chem. Commun.*, 352 (1970)

B. procerum.
ANTHOCYANIN: 4',5,7-trihydroxyflavilium.
Crowden *et al.*, *Phytochemistry*, **13**, 1947 (1974)

BLEEKERIA
B. vitiensis.
CARBOLINE ALKALOID: bleekerine.
Sainsbury, Webb, *Phytochemistry*, **11**, 2337 (1972)

BLEPHARIS
B. sindica.
FLAVONOIDS: apigenin, blepharin, prunine-6''-O-coumarate,
terniflorin.
STEROID: β-sitosterol.
TRITERPENOID: oleanolic acid.
Ahmad *et al.*, *J. Chem. Soc., Pak.*, **6**, 217 (1984); *Chem. Abstr.*, 103 3663z

BLIGHIA (Sapindaceae)
B. sapida Koenig.
AMINIOACIDS: trans-α-amino-2-carboxycyclopropane acetic acid, glycylglycylglycine.
TRITERPENOID: hederagenin.
(Am) Fowden *et al.*, *Phytochemistry*, **8**, 437, (1969)
(Am) Fowden *et al.*, *ibid.*, **8**, 1043 (1969)
(T) Row *et al.*, *Indian J. Chem.*, **9**, 385 (1971)

BLUMEA
B. amplectens var. *arenaria.*
ACID: dihydrosterculic acid.
ACETYLENIC THIOPHENE: amplectol.
SESQUITERPENOID: fewrutinin.
TRITERPENOID: taraxeryl acetate.
Pathak *et al.*, *Planta Med.*, **53**, 103 (1987)

B. membranacea.
DITERPENOID: 5-hydroxy-p-cymene-2-oxybornylene.
Bokadia *et al.*, *Acta Cient. Indica*, **2**, 239 (1976)

B. mollis.
MONOTERPENOID: chrysanthenone.
Geda *et al.*, *Perfume Flavor*, **7**, 27 (1982)

BOCCONIA (Papaveraceae)
B. arborea.
OXOPROTOBERBERINE ALKALOIDS: α-allocryptopine, protopine N-oxide.
PHENANTHROISOQUINOLINE ALKALOIDS: chelerythrine, dihydrosanguinarine, 11-O-methyldihydrochelerythrine.
Manske, *Can. J. Res.*, **116**, 153 (1938)
MacLean *et al.*, *Can. J. Chem.*, **47**, 1951 (1969)

B. cordata Willd.
OXOPROTOBERBERINE ALKALOIDS: α-allocryptopine, protopine N-oxide.
PHENANTHROISOQUINOLINE ALKALOIDS: bocconine, bocconoline.
PROTOBERBERINE ALKALOID: dehydrocheilanthifoline.
UNCHARACTERIZED ALKALOID: heleritine.
Tani, Takao. J. Pharm. Soc., *Japan*, **82**, 755 (1962)
Iwasu *et al.*, *Phytochemistry*, **22**, 627 (1983)

B. frutescens L.
OXOPROTOBERBERINE ALKALOIDS: α-allocryptopine, β-allocryptopine.
PHENANTHROISOQUINOLINE ALKALOID: chelerythrine.
Jowett, Pyman, *J. Chem. Soc.*, **103**, 291 (1913)

B. macrocarpa Maxim.
OXOPROTOBERBERINE ALKALOIDS: α-allocryptopine, cryptopine, protopine.
PHENANTHROISOQUINOLINE ALKALOID: sanguinarine.
Spath, Kuffner, *Ber.*, **64**, 1123 (1931)
Spath, Kuffner, *ibid.*, **64**, 2034, (1931)

BOEHMERIA (Urticaceae)
B. cylindrica.
PHENANTHROINDOLIZIDINE ALKALOID: cryptopleurine. Three uncharacterized bases.
Hart *et al.*, *Aust. J. Chem.*, **22**, 1805 (1969)

B. holosericea.
ACIDS: caffeic acid, linoleic acid, palmitic acid, stearic acid.
FLAVONOIDS: astragalin, hyperin, kaempferol 3-rutinoside, rutin.
STEROIDS: camperesterol, β-sitosterol, stigmasterol.
Takemoto *et al.*, *Phytochemistry*, **14**, 2534 (1975)

B. platyphylla.
ACETOPHENONE ALKALOID:
3,4-dimethoxy-2-(2-piperidyl)-acetophenone.
Hart *et al.*, *Aust. J. Chem.*, **21**, 1397 (1968)

B. tricuspis.
ACIDS: caffeic acid, linoleic acid, palmitic acid, stearic acid.
FLAVONOIDS: astragalin, hyperin, kaempferol 3-rutinoside, rutin.
STEROIDS: campesterol, β-sitosterol, stigmasterol.
Takemoto *et al.*, *Phytochemistry*, **14**, 2534 (1975)

BOENNINGHAUSENA
B. albiflora.
COUMARINS: bhubaneswin, 3-(1,1-dimethylallyl)-xanthyletin, (E)-7-hydroxy-6-(3-hydroxy-3-methyl-1-butenyl)-2H-1-benzopyran-2-one, (Z)-7-hydroxy-6-(3-hydroxy-3-methyl-1-butenyl)-2H-1-benzopyran-2-one, micropubscin.
Basa *et al.*, *Heterocycles*, **22**, 333 (1984)

BOERHAAVIA (Nyctaginae)
B. diffusa L.
ALKALOID: punarnavine.
Basu, Sharma, *Q. J. Pharmacol.*, **20**, 41 (1947)

BOESENBERGIA
B. pandurata.
BENZOPYRANS: boesenbergin A, boesenbergin B.
FLAVONE: 5,7-dimethoxyflavone.
(B) Mahidol *et al.*, *Aust. J. Chem.*, **37**, 1739 (1984)
(C) Jaipetch *et al.*, *Phytochemistry*, **22**, 625 (1983)

BOHADSCHIA.
B. argus.
TRITERPENOIDS: holost-9(11)-ene-3β,17α-diol, holost-9(11)-en-3β-ol.
Stonik *et al.*, *Khim. Prir. Soedin.*, 790 (1982)

B. graeffei.
SAPONIN: echinoside A.
Takeda Chem. Ind. Ltd, *Japanese Patent, 212200* 1980

BOLBITIS
B. subcordata.
STEROIDS: ecdysterone, ponasterone A.
Hikino *et al.*, *Planta Med.*, **31**, 71 (1977)

BOLBOSTEMMA (Cucurbitaceae)
B. paniculatum.
SAPONIN: tubeimoside.
Kasai *et al.*, *Chem. Pharm. Bull.*, **34**, 3974 (1986)

BOLDEA
B. fragrans. Gray. (B. chilensis Juss.).
APORPHINE ALKALOID: boldine.
Merck, *Jahresb.*, **36**, 110 (1922)

BOLETUS
B. bovinus L. ex Fr.
QUINONE: bovinone.
Beaumont *et al.*, *J. Chem. Soc.*, C, 2393 (1969)

B. calopus Fr.
PHENOLIC: xercomic acid.
Edwards *et al.*, *J. Chem. Soc.*, 2582 (1971)

B. erythropus (Fr. ex Fr.) Secr.
PHENOLIC: xercomic acid.
Edwards *et al.*, *J. Chem. Soc.*, 2582 (1971)

BOLUSANTHUS
B. speciosus.
QUINOLIZINE ALKALOIDS: 5,6-dehydrolupanine, 6β-hydroxylupanine.
SPARTEINE ALKALOIDS: 11α-allylcytisine, cytisine, 13-hydroxyanagyrine, N-methylcytisine.
ISOFLAVONOIDS: biochanin A, bolusanthin, genistein, 3'-O-methylpratensein, orobol.
(A) Asres *et al.*, *Phytochemistry*, **25**, 1449 (1986)
(I) Asres *et al.*, *Z. Naturforsch*, **40C**, 617 (1985)

BOMBAX (Bombaceae)
B. malabaricum DC.
QUINONE: bombaxquinone.
　Seshadri et al., Curr. Sci., **40**, 630 (1971)
B. munguba Mart.
ESTERS: 1,3-dipalmitoyl-2-linoleoylglycerol,
　1,3-dipalmitoyl-2-oleoylglycerol,
　1,3-dipalmitoyl-2-sterculoylglycerol.
　Schuch et al., J. Am. Oil Chem. Soc., **63**, 778 (1986); Chem. Abstr., **105**
　39417c

BONAFOUSIA
B. tetrastachya (Humbolt Bompland et Kunth.) Markgraf.
BIS-CARBOLINE ALKALOIDS: bonafousine, isobonafousine.
BIS-HOMOCARBOLINE ALKALOID:
　bis-(hydroxy-11-coronaridinyl)-12.
　Danak et al., Bull. Soc. Chim. Fr., 490 (1980)

BONANNIA (Umbelliferae)
B. graeca (L.) Halacsy.
DITERPENOID: bonandiol.
　Bruno et al., Tetrahedron Lett., 4287 (1984)

BONNEMAISONIA (Rhodophyceae)
B. hamifera.
CAROTENOIDS: α-carotene, β-carotene, α-cryptoxanthin,
　β-cryptoxanthin, lutein, zeaxanthin.
　Bjoernland et al., Phytochemistry, **15**, 291 (1976)

BOOPHONE (Amaryllidaceae)
B. disticha (L.F.) Herb. (Haemanthus toxicaria Hab.).
BENZOISOQUINOLINE ALKALOIDS: 3-acetylnerbowdine,
　buphanamine, buphanisine, buphanitine.
UNCHARACTERIZED ALKALOIDS: buphanine, nerbowdine.
　Tutin, J. Chem. Soc., **99**, 1240 (1911)
　Renz et al., Helv. Chim. Acta, **38**, 1209 (1955)
B. fischerii Baker.
BENZOISOQUINOLINE ALKALOIDS: buphanamine, buphanidrine,
　buphanisine.
UNCHARACTERIZED ALKALOID: buphanine.
　Renz et al., Helv. Chim. Acta, **38**, 1209 (1955)

BORONIA (Rutaceae)
B. ternata Endl.
QUINOLONE ALKALOID: 1-acetoxymethyl-2-propyl-4-quinolone.
　Duffield, Jefferies, Aust. J. Chem., **16**, 292 (1963)

BORREIRA (Rubiaceae)
B. capitata.
INDOLE ALKALOIDS: borrecapine, borrecozine, borreline,
　dehydroborrecapine.
　Jossang et al., Planta Med., **43**, 301 (1981)

BOSCHNAKIA
B. rossica Hult.
PYRIDINE ALKALOIDS: boschniakine, boschnilactone,
　boschniaside.
　Sakan et al., Tetrahedron, **23**, 4635 (1967)
　Nurai et al., Planta Med., **46**, 45 (1982)

BOSISTOA (Rutaceae)
B. euodiformis F.Muell.
BENZOPYRANS: franklinene, franklinol.
TRITERPENOID: bosistoin.
　Croft et al., Aust. J. Chem., **28**, 2019 (1975)

BOSWELLIA
B. carteri.
DITERPENOIDS: incensole, incensole oxide,
　(S)-1-isopropyl-4,8,12-trimethylcyclotetradeca-3E,7E,11E-
　trien-1-ol.
TRITERPENOIDS: acetyl-β-boswellic acid, 4,23-dihydroroburic
　acid, 11-keto-α-boswellic acid.
　Savoir et al., Bull. Soc. Chim. Belg., **76**, 368 (1967)
　Pardhy, Bhattacharyya, Indian J. Chem., **16A**, 176 (1978)
B. serrata.
TRITERPENOIDS: 3β-acetoxytirucalla-8,24-dien-21-oic acid,
　3-ketotirucalla-8,24-dien-21-oic acid.
　Pardhey, Bhattacharyya, Indian J. Chem., **16B**, 174 (1978)
　Klein, Obermann, Tetrahedron Lett., 349 (1978)
B. serrate.
DITERPENOID: serratol.
　Pardhy, Bhattacharyya, Indian J. Chem., **16B**, 171 (1978)

BOTHRIOCHLOA
B. intermedia.
SESQUITERPENOIDS: acorenone B, intermedeol, neointermedeol.
　McClure et al., J. Chem. Soc., Chem. Commun., 1135 (1968)

BOTHRIOCLINE
B. laxa.
SESQUITERPENOIDS: bothrioclinin, 7-hydroxybothrioclinin.
　Bohlmann, Zdero, Phytochemistry, **16**, 1261 (1977)

BOTRYOCOCCUS (Chlorophyceae)
B. braunii.
CAROTENOID: botryococcene.
　Maxwell et al., Phytochemistry, **7**, 2157 (1968).

BOUCERIA
B. aucheriana.
STEROID: boucerin.
　Nikaido et al., Chem. Pharm. Bull., **15**, 725 (1967)

BOURRERIA
B. verticillata.
ISOQUINOLINE ALKALOID: emetine.
　Orazi, Rev. Fac. Cienc. Quim., **19**, 17 (1946)

BOUVARDIA (Rubiaceae)
B. ternifolia. Schlecht.
PEPTIDES: bouvardin, bouvardin 6-methyl ether,
　deoxybouvardin.
　Jolad et al., J. Am. Chem. Soc., **99**, 8040 (1977)

BOWDICHIA
B. nitida.
ISOFLAVONE: calycosin.
QUINONE: bowdichione.
　Brown et al., Justus Liebigs Ann. Chem., 1295 (1974)
B. virgilioides.
DITERPENOID: alcornol.
　Harwich, Dunnenberger, Arch. Pharm., **278**, 341 (1900)

BOWEIA
B. volubilis Harvey.
STEROIDS: bovogenins A, B, C, D and E, bovosides A, B, C and D.

Katz, *Helv. Chim. Acta.*, **33**, 1420 (1950)

BRACHIARIA
B. ruziziensis Germain et Evrard.
ACIDS: p-coumaric acid, ferulic acid, vanillic acid.

Renard *et al.*, *Chem. Abstr.*, 86 103117y

BRACHYLAENA (Compositae)
B. hutchinsii Hutch.
SESQUITERPENOIDS: brachylaenalones A and B.

Brooks, Campbell, *J. Chem. Soc., Chem. Commun.*, 630 (1969)

B. transvaalensis.
SESQUITERPENOIDS: brachylaenolide, dehydrobrachylaenolide.

Bohlmann *et al.*, *Phytochemistry*, **21**, 647 (1982)

BRACHYMERIS
B. montana.
SESQUITERPENOID: brachymerolide.

Bohlmann *et al.*, *Phytochemistry*, **21**, 1989 (1982)

BRACKENRIDGEA.
B. zanguebarica.
PHENOLIC: brackenin

Drewes *et al.*, *Phytochemistry*, **22**, 2823 (1983)

BRANDEGEA (Cucurbitaceae)
B. bigelovii Cogn.
TRITERPENOIDS: cucurbitacin O, cucurbitacin P, cucurbitacin Q.

Kupchan *et al.*, *J. Org. Chem.*, **35**, 2891 (1970)

BRASILIA
B. serjana.
ACID: serjanic acid.

Savoir *et al.*, *Tetrahedron Lett.*, 2129 (1969)

B. sickii.
SESQUITERPENOIDS: 15-acetylbrasilic acid methyl ester, eight unnamed sesquiterpenoids.

Bohlmann *et al.*, *Phytochemistry*, **22**, 1213 (1983)

BRASSICA (Brassicaceae)
B. campestris L.
STEROIDS: brassinolide, campesterol, castasterone.

Fernholz, MacPhillemy, *J. Chem. Soc.*, **63**, 1155 (1941)

Fernholz, MacPhillemy, *ibid.*, **63**, 1157 (1941)

Abe *et al.*, *Agric. Biol. Chem.*, **46**, 2609 (1982)

B. campestris L. subsp. pekinensis.
PHYTOALEXINS: brassinin, methoxybrassinin.

Takasugi *et al.*, *J. Chem. Soc., Chem. Commun.*, 1077 (1986)

B. juncea.
STEROID: 24-methylene-25-methylcholesterol.

Matsumoto *et al.*, *Phytochemistry*, **22**, 2619 (1983)

B. napus L.
FLAVONE: brassicoside.

ISOTHIOCYANATES: allylisothiocyanate, butenylisothiocyanate, pentenylisothiocyanate.

STEROIDS: $4\alpha,14\alpha,24$-trimethyl-5α-cholesta-8,25-dien-3β-ol, $4\alpha,1,24$-trimethyl-$9\alpha,19$-cyclocholest-24-en-3β-ol.

(F) Kuhn, *Chem. Ber.*, **81**, 363 (1948)

(I) Georgieva *et al.*, *Dokl. Bulg. Akad. Nauk*, **39**, 127 (1986)

(S) Itoh *et al.*, *Phytochemistry*, **16**, 1448 (1977)

B. napus L. var. oleifera.
FLAVONE: isorhamnetin 3β-D-glucoside.

Rahman *et al.*, *J. Org. Chem.*, **27**, 153 (1962)

B. nigra (L.) Koch.
ISOTHIOCYANATES: allylisothiocyanate, sinigrin.

Georgieva *et al.*, *Dokl. Bulg. Akad. Nauk*, **39**, 127 (1986)

B. oleracea L.
ACID: ascorbalamic acid.

ANTHOCYANINS: cyanidin 3-p-coumaryl-5-glucoside, cyanidin 3-(di-p-coumaryl)-5-glucoside, cyanidin 3-(disinapyl)-5-glucoside, cyanidin 3-(diferulyl)-5-glucoside, cyanidin 3-malonyl-5-glucoside, cyanidin 3-sianapyl-5-glucoside, cyanidin 3-sophoroside-5-glucoside.

(Ac) Couchman *et al.*, *Phytochemistry*, **12**, 707 (1973)

(An) Hrazdina *et al.*, *ibid.*, **16**, 297 (1977)

B. rapa L.
STEROIDS: brassicasterol, campesterol, $4\alpha,7\alpha,24$-trimethyl-5α-cholesta-8,24-dien-3β-ol.

Itoh *et al.*, *Phytochemistry*, **16**, 1448 (1977)

BREDEMEYERA
B. floribunda.
TRITERPENOID: bredemolic acid.

UNCHARACTERIZED TERPENOID: tenuifolic acid.

Tschesche, Gupta, *Chem. Ber.*, **93**, 1903 (1960)

BRENANIA
B. brieyi.
SAPONINS: oleanolic acid 3-O-β-D-galactoside, oleanolic acid 3-O-β-D-glucoside, oleanolic acid 3-O-β9-glycuronoside.

Menu *et al.*, *Ann. Pharm. Fr.*, **34**, 427 (1976)

BREYNIA
B. officinalis (L.) Hemsl.
SAPONIN: breynin A.

Sesaki *et al.*, *Tetrahedron Lett.*, 2439 (1973)

BRICKELLIA (Eupatoriaceae)
B. diffusa.
SESQUITERPENOID: 2-angeloylbrickellidiffusic acid.

Bohlmann *et al.*, *Phytochemistry*, **21**, 691 (1982)

B. eupatoreides.
DITERPENOIDS:
17-acetoxy-3α-angeloyloxy-12-cleistanthen-11-one, 17-acetoxy-15,16-epoxycleistanth-12-en-11-one, 17-acetoxy-14β-hydroxy-12-cleistanthen-11-one, 3α-angeloyloxy-12-cleistanthen-11-one, 8,9:15,16-diepoxy-11-oxo-12-cleistanthen-17-al, 15,16-epoxy-12-cleistanthen-11-one, 15,16-epoxy-12-cleistanthen-11-one 17-acetate, 15,16-epoxy-12-cleistanthen-11-one 17-aldehyde.

Bohlmann *et al.*, *Phytochemistry*, **21**, 181 (1982)

B. guatamaliensis.
SESQUITERPENOID:
3,7,11-trimethyl-1,4,6,10-dodecatetraen-3,8-diol.

Bohlmann *et al.*, *Tetrahedron Lett.*, 5109 (1969)

B. laciniata.
SESQUITERPENOID: 4,5-dehydronerolidol.

Bohlmann *et al.*, *Phytochemistry*, **17**, 763 (1978)

B. paniculata.
DITERPENOID: angeloyloxy-2α-hydroxycativic acid.

Gomez *et al.*, *Phytochemistry*, **22**, 1292 (1983)

B. scoparia.
FLAVONOIDS: 5,7-dihydroxy-3',4',6-trimethoxyflavone, 5-hydroxy-6,7,3',4'-tetramethoxyflavone, kaempferol 6,7-dimethyl ether, 6-methoxykaempferol 3-O-β-D-glucoside, quercetin 6-methyl ether, 5,7,3',4'-tetrahydroxy-3,6-dimethoxyflavone, 5,7,4'-trihydroxy-3,6-dimethoxyflavone, 5,7,3'-trihydroxy-3,6,4'-trimethoxyflavone.
Li *et al., J. Nat. Prod.,* **49**, 732 (1986)

B. veronicaefolia.
BENZOFURAN: 7-acetyl-2,3-dihydro-2-isopropenyl-3,6-benzofurandiol.
SESQUITERPENOID: 9-angeloyloxy-10,11-epoxy-3,7,11-trimethyl-1,4,6-dodecatrien-3-ol. Diterpenoid 3-oxocativic acid.
Bohlmann *et al., Chem. Ber.,* **109**, 1436 (1976)

BROCCHIA
B. cinerea.
COUMARIN TERPENOIDS: acetylpetachol B, petachol B.
Greyer *et al., Phytochemistry,* **24**, 85 (1985)

BROMELIA (Bromeliaceae)
B. pinguin.
DITERPENOID: 7,8-dihydroxy-15-pimaren-3-one.
Raffauf *et al., J. Org. Chem.,* **46**, 1094 (1981)

BROSIMUM
B. rubescens.
COUMARIN: xanthyletin.
Dasgupta *et al., J. Chem. Soc., C,* **33** (1969)

BROUSSONETIA (Moraceae)
B. kazinoki Sieb.
ISOPRENOIDS: kazinol J, kazinol L, kazinol M, kazinol N, kazinol P.
Kato *et al., Heterocycles,* **24**, 2141 (1986)
Kato *et al., Chem. Pharm. Bull.,* **34**, 2448 (1986)

B. papyrifera (L.) Vent.
FLAVONOIDS: 4',7-dihydroxyflavan, 7-hydroxy-4'-methoxyflavan, kazinol A, kazinol B.
PHENOLICS: broussonins A, B, C, D, E and F, 4'-O-methyldavidioside.
(F) Coxon *et al., Phytochemistry,* **19**, 889 (1980)
(F) Ikuta *et al., Heterocycles,* **23**, 2835 (1985)
(P) Alves de Lima *et al., ibid.,* **14**, 1831 (1975)

B. zeylandica.
BISQUINOLINE ALKALOID: broussonetine.
Gunatilaka *et al., Phytochemistry,* **23**, 929 (1984)

BRUCEA (Simaroubaceae)
B. amarissima.
QUASSINOIDS: bruceins D, E and F.
Polonsky *et al., C. R. Acad. Sci.,* **267C**, 1346 (1968)

B. antidysenterica.
CARBOLINE ALKALOIDS: canthin-6-one, 1,11-dimethoxycanthin-6-one, 11-hydroxycanthin-6-one.
QUASSINOIDS: bruceanols A and B, bruceantarin, bruceantin, bruceantinol, dehydrobruceantarin, dehydrobruceantin, dehydrobruceantol, dehydrobrucein B, isobrucein B.
(A) Fukamiya *et al., J. Nat. Prod.,* **49**, 428 (1986)
(T) Lin *et al., Shenyang Yaoxueyuan Xuebao,* **3**, 192 (1986); *Chem. Abstr.,* 105 222753y

B. sumatrana Roxb.
QUASSINOID: brusatol.
Sim *et al., J. Org. Chem.,* **33**, 429 (1968)

BRUGUIERA (Rhizophoraceae)
B. cylindrica.
DITHIOLANE: 4β-hydroxy-1,2-dithiolane.
Kato *et al., Phytochemistry,* **15**, 220 (1976)

B. exaristata Ding Hou.
TROPANE ALKALOIDS: tropine benzoate, tropine butyrate, tropine isobutyrate, tropine-1,2-dithiolane-3-carboxylic acid, tropine isovalerate, tropine propionate.
Loder, Russell, *Aust. J. Chem.,* **22**, 1271 (1969)

B. gymnorrhiza (L.) Lam.
No alkaloids present.
Loder, Russell, *Aust. J. Chem.,* **22**, 1271 (1969)

B. parviflora (Roxb.) W.et A. ex Graff.
No alkaloids present.
Loder, Russell, *Aust. J. Chem.,* **22**, 1271 (1969)

B. sexangula (Lour.) Poir.
TROPANE ALKALOIDS: tropine benzoate, tropine butyrate, tropine isobutyrate.
Loder, Russell, *Aust. J. Chem.,* **22**, 1271 (1969)

BRUNSDONNA
B. tubergenii.
UNCHARACTERIZED ALKALOID: coruscine.
Boit, Ehmke, *Chem. Ber.,* **90**, 369 (1957)

BRUNSVIGIA (Amaryllidaceae)
B. cooperi.
ISOQUINOLINE ALKALOIDS: brunsvigine, nerbowdine.
Inubushi *et al., J. Org. Chem.,* **25**, 2157 (1960)
Hauth, Stauffacher, *Helv. Chim. Acta,* **46**, 810 (1963)

B. rosea (Amaryllis belladonna).
BENZOISOQUINOLINE ALKALOIDS: amaryllisine, ambelline.
Burlinghame *et al., J. Am. Chem. Soc.,* **96**, 4976 (1964)

BRYONIA (Cucurbitaceae)
B. alba.
TRITERPENOIDS: cucurbitacin B, cucurbitacin D, cucurbitacin E, cucurbitacin I, cucurbitacin J, cucurbitacin K, cucurbitacin L, dihydrocucurbitacin B, dihydrocucurbitacin E, iso-23,24-dihydrocucurbitacin D, tetrahydrocucurbitacin I.
Pohlmann, *Phytochemistry,* **14**, 1587 (1975)
Panosyen *et al., Bioorg. Khim.,* **5**, 721 (1979)

B. dioica (Jacq.) Tutin.
TRITERPENOIDS: 25-o-acetylbryomaride, bryodulcosigenin, bryogenin, bryomaride, bryonolic acid, bryononic acid, bryosigenin, brusulcoside, cucurbitacin B, cucurbitacin D, cucurbitacin E, cucurbitacin I, cucurbitacin J, cucurbitacin K, cucurbitacin L, cucurbitacin S, dihydrocucurbitacin B, dihydrocucurbitacin E, elaterinide, 3-hydroxymultiflora-7,9(11)-dien-29-oic acid, elaterinide, tetrahydrocucurbitacin I.
Hylands, Salama, *Phytochemistry,* **15**, 559 (1976)
Ripperger *et al., Tetrahedron,* **32**, 1567 (1976)

BRYOPHYLLUM (Crassulaceae)
B. daigremontana.
GIBBERELLIN: gibberellin A-20,
Zeevaert, *Planta Med.,* **114**, 285 (1973)

BRYOPHYLLUM (Crassulaceae)

B. tubiflorum.

SAPONIN: bryotoxin,

Capon *et al.*, *J. Chem. Res. Synop.*, 333 (1985)

BRYOPSIS (Bryopsidaceae/Chlorophyceae)

B. corticulans.

CAROTENOID: ε-carotene,

Manchand *et al.*, *J. Chem. Soc.*, 2019 (1965)

BUCHENOVIA

B. capitata.

FLAVONE ALKALOIDS: capitavine,
2,3-dihydro-4'-hydroxycapitavine.

Ahond *et al.*, *Bull. Soc. Chim. Fr.*, 41 (1984)

B. macrophylla Eichl.

FLAVONE ALKALOIDS: N-bisdemethylbuchenavianine,
buchenavianine, O-demethylbuchenavianine,
N-demethylbuchenavianine, N-demethylcapitavine,
capitavine, 2,3-dihydrocapitacine.

Ahond *et al.*, *Bull. Soc. Chim. Fr.*, 41 (1984)

BUCKLEYA (Santalaceae)

B. lanceolata Miq.

ACIDS: cis-ferulic acid, trans-ferulic acid.
ESTER: stearyl-O-acetyl-trans-ferulate.
STEROID: β-sitosterol 3-β-D-glucoside.

Sashida *et al.*, *J. Pharm. Soc.*, Japan, **97**, 695 (1977)

BUCSERA

B. graveolens var. villosula.

TRITERPENOID: 2,3-olean-12-ene-2,3,28-trioic acid.

Ruzicka *et al.*, *Helv. Chim. Acta*, **21**, 1371 (1928)

BUDDLEJA (Buddlejaceae/Loganiaceae)

B. davidii Franch.

SESQUITERPENOIDS: buddledins A, B, C, D and E.

Yoshida *et al.*, *Chem. Pharm. Bull.*, **26**, 2543 (1978)

B. gloriosa.

IRIDOID: catalpol.

Inouye *et al.*, *Chem. Pharm. Bull.*, **19**, 1438 (1971)

B. madagascariensis.

STEROIDS: β-sitosterol, stigmasterol.
TRITERPENOIDS: α-amyrin, β-amyrin, 11-keto-β-amyrin.

Kapoor *et al.*, *Fitoterapia*, **32**, 235 (1982)

B. variabilis.

IRIDOID: catalpol.

Inouye *et al.*, *Chem. Pharm. Bull.*, **19**, 1435 (1971)

BUELLIA (Lichenes)

B. canescens (Dicks.) De Not. Phenolic

canescolide

Sala *et al.*, *J. Chem. Soc., Chem. Commun.*, 1041 (1978)

BULBOCAPNUS

B. cavus Bernh.

APORPHINE ALKALOID: bulbocapnine.

Kikkawa *et al.*, *J. Pharm. Soc., Japan*, **79**, 1245 (1959)

BULBOCODIUM

B. vernum.

SPIROAPORPHINE ALKALOID: bulbocodine.

Reichstein *et al.*, *Planta Med.*, **16**, 357 (1968)

BULBOPHYLLUM (Orchidaceae)

B. leopardium. Phenolic

bulbophyllanthrin

Majumder *et al.*, *Phytochemistry*, **24**, 2083 (1985)

BULNESIA

B. sarmienti.

SESQUITERPENOIDS: α-bulnesene, β-bulnesene, bulnesol,
α-guaiene, τ-guaiene, guaioside, guaioxiole, guanoxide,
α-patchoulene.

Takeda *et al.*, *Chem. Ind., (London)*, 1715 (1962)

BUPLEURUM (Umbelliferae)

B. chinensis DC.

SAPONIN:
3-O-[α-L-arabinopyranosyl-β-D-glucopyranosyl]-oleanolic
acid β-D-glucopyranosyl ester.

Seto *et al.*, *Agric. Biol. Chem.*, **50**, 939 (1986)

B. falcatum L.

COUMARIN: (-)-anomalin.
TRITERPENOIDS: saikogenins A, B, C, D, E, F and G.
SAPONINS: longispinogenin, saikosaponins a, b-1, b-2, b-3, b-4,
c, d, e, A, B, C and D, saikoside Ia, saikoside Ib.

Banerji *et al.*, *Indian J. Chem.*, **15B**, 293 (1977)

B. fruticosum L.

MONOTERPENOID: (+)-phellandrene.

Wallach *et al.*, *Justus Liebigs Ann. Chem.*, **240**, 1 (1905)

B. kummingense.

SAPONINS: 2''-O-acetylsaikosaponin A, 3''-O-acetylsaikosaponin
A, 4''-O-acetylsaikosaponin A, 2''-acetylsaikosaponin D,
16-epi-chikusaikoside I.

Seto *et al.*, *Agric. Biol. Chem.*, **50**, 943 (1986)

B. longeradiatum Turez.

SAPONINS: chikuakoside A, chikuakoside B.

Kimata *et al.*, *Chem. Pharm. Bull.*, **30**, 4373 (1982)

B. ranunculoides.

SESQUITERPENOIDS:
2-cis-9-cis-pentadec-2,9-diene-4,6-diyn-1-al,
cis-pentadeca-2,9-diene-4,6-diyn-1-ol,
trans-pentadeca-1,8,10-triene-4,6-diyn-1-al.

Bohlmann *et al.*, *Chem. Ber.*, **104**, **1322**, 2030 (1971)

B. rotundifolium L.

FLAVONOIDS: isorhamnetin, isorhamnetin 3-glucoside,
isorhamnetin 3-rutinoside, quercetin, quercetin 3-glucoside,
rutin.
SAPONINS: rotundoside E, rotundoside F, Triterpenoids
rotundiogenins A, B, C, D, E and F.

(F) Baeva *et al.*, *Khim. Prir. Soedin.*, 648 (1983)
(Sap, T) Kobayashi et al., Chem. Pharm. Bull., 29, 2222 (1981) (Sap, T)
Kobayashi *et al.*, *ibid.*, 2230 (1981)

BUPHTHALMUM (Compositae/Inuleae)

B. salicifolium L.

THIOPHEN: 5-methyl-5'-(but-3-en-1-yl)-bithenyl,
5-hydroxymethyl-5'-(but-3-en-1-yl)-2,2'-bithienyl.

Bohlmann, Berger, *Chem. Ber.*, **98**, 883 (1965)

BURASIA

B. madagascariensis.

PROTOBERBERINE ALKALOID: burasaine.

Resplandy, *Mem. Inst. Scient. Madagascar*, **10D**, 37 (1961)

BURCHARDIA

B. multiflora.
FLAVONOID: luteol
PHENOLICS: benzoic acid, salicylic acid.
Potesilova et al., Phytochemistry, **26**, 1031 (1987)

BURSERA (Burseraceae)

B. arida (Rose) Standl.
TRITERPENOID: benulin.
Ionescu et al., J. Org. Chem., **42**, 1627 (1977)

B. fagaroides.
BENZOPYRAN: 5'-demethoxy-α-peltatin A methyl ether.
Bianchi et al., Tetrahedron Lett., 2759 (1969)

B. graveolens.
NORTRITERPENOID: 2,3-seco-12-oleanene-2,3,28-trioic acid.
Crowley et al., J. Chem. Soc., 4254 (1964)

B. graveolens var. villulosa.
BENZOFURAN:
(+)-2,4,5,6,7,7-hexahydro-7-hydroxy-3,6-
dimethylbenzofuran.
Crowley et al., J. Chem. Soc., 4254 (1964)

B. morelensis.
NEOLIGNANS: deoxypodophyllotoxin, morelensin.
Jolad et al., J. Pharm. Sci., **66**, 892 (1977)

BUTEA

B. monosperma.
ALKALOID: monospermine.
ANTHRAQUINONE: 3,6,8-trihydroxy-1-methylanthraquinone
2-carboxylic acid.
(A) Mehta, Bokadia, Chem. Ind., (London), 98 (1981)
(An) Gadgil et al., Tetrahedron Lett., 2223 (1968)

BUXUS (Buxaceae)

B. arborescens Mill.
STEROIDAL ALKALOIDS: buxbarbarine,
O-tigloylcyclovirobuxeine A.
Mokry, Voticky, Chem. Zvesti, **38**, 101 (1984)

B. balearica (L.) Willd.
STEROIDAL ALKALOIDS: baleabuxine, baleabuxoxazine,
N-benzoylbaleabuxadiene F, N-benzoylbaleabuxidine,
N-3-benzoylcyclobuxidine F, buxamine E, buxaminol E,
buxamideine K, buxidine F, cyclobuxidine F,
cycloprotobuxine A, cycloprotobuxine D,
cyclomicrophylline B, N-isobutyrylbaleabuxadienine F,
N-isobutyrylbaleabuxaline F, N-isobutyrylcyclobuxidine F,
N-isobutyrylcyclobuxidine H, N-isobutyrylcyclobuxine F.
Khuong-Huu-Laine et al., Tetrahedron, **22**, 3321 (1966)
Kurakina et al., Khim. Prir. Soedin., 231 (1970)

B. hyrcana.
STEROIDAL ALKALOID: buxtauine.
Aliev et al., Khim. Prir. Soedin., 409 (1974)

B. koreana Nakai.
STEROIDAL ALKALOIDS: cyclokoreanines A and B.
Nakano et al., J. Chem. Soc., C, 1805 (1968)

B. madagascarica.
STEROIDAL ALKALOIDS: buxamine A, buxitrienine C,
cycloprotobuxine F, 16-deoxybuxidienine C.
Khuong-Huu-Laine et al., C. R. Acad. Sci., **273C**, 558 (1971)

B. malayana Ridl.
STEROIDAL ALKALOIDS: cyclovirobuxeine A, cyclovirobuxeine
B, cyclovirobuxeine D, O-vanilloylcyclovirobuxeine B.
Nakano et al., J. Chem. Soc., C, 1412 (1966)
Khuong-Huu-Laine et al., Ann. Pharm. Fr., **28**, 211 (1970)

B. microphylla Sieb. et Zucc.
STEROIDAL ALKALOIDS: cyclobuxomicreine, cyclobuxophylline,
cyclobuxophyllinine, cyclobuxosuffrine, cyclobuxoviridine,
cyclobuxoxazine, cyclomicrobuxeine, cyclomicrobuxine,
cyclomicrobuxinine, cyclomicrophyllidine A,
cyclomicrophyllines A, B and C, cyclomicrosine,
cycloprotobuxine C, cyclosuffrobuxine, cyclosuffrobuxinine,
dihydrocyclomicrophyllidine A, dihydrocyclomicrophylline
A, dihydrocyclomicrophylline F.
Nakano, Terao, J. Chem. Soc., 4512 (1965)

B. microphylla Sieb. et Zucc. var. suffruticosa Mak.
STEROIDAL ALKALOIDS: cyclobuxine D, cyclobuxomicreine,
cyclobuxophylline, cyclobuxosuffrine, cyclobuxoviridine,
cyclobuxoxazine, cyclomicrobuxeine, cyclomicrobuxine,
cyclomicrobuxinine, cyclomicrophyllidine A,
cyclomicrophyllines A, B and C, cyclomicrosine,
cyclomikuranine, cycloprotobuxine C, cyclosuffrobuxine,
cyclosuffrobuxinine, dihydrocyclomicrophyllidine A,
dihydrocyclomicrophylline A, dihydrocyclomicrophylline F.
Nakano, Terao, Tetrahedron Lett., 1034 (1964)
Nakano, Terao, ibid., 1045 (1964)
Nakano et al., J. Chem. Soc., C, 1412 (1966)

B. microphylla Sieb. et Zucc. var. suffruticosa Mak. f. major.
STEROIDAL ALKALOID: cyclomikuramine.
Nakano, Terao, J. Chem. Soc., C, 1412 (1966)
Nakano, Terao, ibid., 1805 (1966)

B. papilosa. C.K. Schm. Linn.
STEROIDAL ALKALOIDS: 30-acetoxy-N-benzoylbuxidienine,
N-benzoyl-16-acetylcyclobuxidine, buxpapamine,
buxpapine, harappamine, karachicine, moenjodaramine,
papilamine, papilicine, papilinine.
Atta-ur-Rahman et al., J. Chem. Soc., Perkin I, 919 (1986)
Atta-ur-Rahmen et al., Planta Med., **53**, 75 (1987)

B. rolfei.
STEROIDAL ALKALOIDS: cyclobuxoxazine A, cyclomethoxazine
B.
Khuong-Huu-Laine et al., Bull. Soc. Chim. Fr., 1216 (1966)

B. sempervirens L.
STEROIDAL ALKALOIDS: N-benzoyl-O-acetylbuxodienine E,
N-benzoylbuxodienine E, buxaline C, buxaltine, buxamine
E, buxaminol E, buxandrine, buxanine, buxarine, buxatine,
buxazidine B, buxazine, buxene, buxenine G, buxenone,
buxeridine, buxidine, buxiramine, buxocyclamine A,
buxpsiine, cyclobuxamine, cyclobuxine, cyclobuxine B,
cyclobuxine D, cyclomicrobuxinine, cyclomicrosine,
cycloprotobuxine C, cycloprotobuxine D, cyclovirobuxeine
A, cyclovirobuxeine B, cyclovirobuxines A, B, C and F,
deoxycyclobuxoxazine A, N-methylbuxene, norbuxamine,
pseudocyclobuxine D.
Kupchan, Ohta, J. Org. Chem., **31**, 608 (1966)
Khodzhaeva et al., Khim. Prir. Soedin., 718 (985) (985)

B. sempervirens L. var. argentea Hort ex Staud.
STEROIDAL ALKALOID: cyclobuxargentine C.
Kuchkova et al., Chem. Zvesti., **30**, 174 (1976)

B. sempervirens L. var. bullatus.
STEROIDAL ALKALOIDS: buxaminol B, cyclobullatine A.
Bauerova, Paulik, Collect. Czech. Chem. Commun., **40**, 3055 (1975)

B. wallichiana.
ALKALOIDS: two uncharacterized bases.
Vassova et al., Pharmazie, **25**, 363 (1970)

BYRSONIMA

B. crassifolia.
UNCHARACTERIZED TERPENOID: byrsonimol.
Heyl, Heil, Festschrift. Alax. Tschirch., 62 (1926)

C

CABRALEA

C. eichleriana.
TRINORTRITERPENOIDS: cabralea hydroxylactone, cabralealactone, cabraleone, eichlearialactone.
NORTRITERPENOIDS: cabralin, isocabralin.
TRITERPENOIDS: cabraleadiol, eichlerianic acid, ocotillone.
Rao et al., Phytochemistry, 14, 1071 (1975)

C. polytricha.
TRINORTRITERPENOIDS: cabralea hydroxlactone, cabralealactone, cabraleone.
TRITERPENOIDS: cabraleadiol, 3-epi-ocotillol, 3-epi-ocotillone.
Cason, Brown, Tetrahedron, 28, 315 (1974)
Rao et al., Phytochemistry, 14, 1071 (1975)

C. toona.
SESQUITERPENOIDS: cubenol, epi-cubenol.
Ohta, Hirose, Tetrahedron Lett., 2073 (1967)

CABUCALA

C. caudata Miq.
BISCARBOLINE ALKALOID: cabifiline.
Massiot et al., C. R. Acad. Sci., 294, 579 (1982)

C. fasciculata.
SPIROINDOLONE ALKALOIDS: caboxine A, caboxine B, isocaboxine A, isocaboxine B.
UNCHARACTERIZED ALKALOIDS: (-)-caberine, (-)-cabucraline, cabulatine.
Titeaux et al., Phytochemistry, 14, 564 (1975)

C. madagascariensis Pichon.
CARBOLINE ALKALOIDS: cabucine, 10-methoxyajmalicine.
SPIROINDOLONE ALKALOID: 10,11-dimethoxyisomitraphylline.
Douzoua et al., Ann. Pharm., 30, 199 (1972)

C. striolata.
CARBOLINE ALKALOIDS: ajmalicinine, 10,11-dimethoxyajmalicine.
Bombardelli et al., Fitoterapia, 45, 183 (1974)

C. torulosa.
CARBOLINE ALKALOIDS: (-)-aricine, (-)-cabucine, cabucraline, (+)-quebrachidine, (-)-tetrahydroalstonine, (-)-vincamagine.
Titeux et al., Phytochemistry, 14, 1648 (1975)

CACALIA (Compositae)

C. auriculata.
SESQUITERPENOID: cacalonol.
Takemoto et al., J. Pharm. Soc., Japan, 94, 1593 (1974)

C. bulbifera.
DITERPENOIDS: (-)-kauran-16β-ol, kaur-16-en-19.ol.
El-Emary et al., Phytochemistry, 14, 1660 (1975)

C. decomposita.
SESQUITERPENOIDS: cacalol, cacalone.
Romo, Joseph-Nathan, Tetrahedron, 20, 2331 (1975)

C. delfinifolia.
SESQUITERPENOIDS: cacalal, cacalolide, tetrahydromaturinone.
Naya et al., Chem. Lett., 1179 (1977)

C. hastata L.
PYRROLIZIDINE ALKALOID: hastacine.
SESQUITERPENOIDS: cacalolide, dehydrocacalohastin.
(A) Konovalova, Men'shikov, J. Gen. Chem. USSR, 15, 328 (1945)
(T) Naya et al., Chem. Lett., 73 (1976)

C. hastata var. tamakae L.
SESQUITERPENOID: tetrahydromaturinone.
Naya et al., Chem. Lett., 73 (1976)

C. robusta Tolm.
PYRROLIZIDINE ALKALOID: hastacine.
Konovalova, Men'shikov, J. Gen. Chem. USSR, 15, 328 (1945)

C. yatabei.
PYRROLIZIDINE ALKALOID: yamataimine.
Kikichi et al., Tetrahedron Lett., 767 (1978)

CACCINIA

C. glauca.
TRITERPENOIDS: caccigenin, caccigenin lactone, 23-deoxycaccigenin. One uncharacterized alkaloid.
(A) Siddiqui et al., Phytochemistry, 17, 2049 (1978)
(T) Tiwari et al., Indian J. Chem., 8, 593 (1970)

CADABA

C. farinosa.
MACROCYCLIC ALKALOID: cadabicine.
Ahmad et al., Phytochemistry, 24, 2709 (1985)

C. fruticosa.
QUATERNARY ALKALOID: cadabine.
Ahmad et al., Pakistan J. Sci. Ind. Res., 17, 109 (1975)

CADIA

C. ellisiana.
LUPANINE ALKALOID: 13-hydroxylupanine-2-pyrrolcarbonate.
Lindner et al., IRCS Libr. Compend., 2, 1550 (1974)

C. purpurea.
SPARTEINE ALKALOID: cadaine.
TWO UNCHARACTERIZED BASES. FLAVONOIDS: apigenin, apigenin 7-O-glucoside, chrysoeriol, 7,3'-dihydroxy-4'-methoxyflavone.
(A) van Eijk et al., Pharm. Weekbl., 33, 1865 (1982)
(F) Zweekhorst-Van Laer et al., Pharm. Weekbl., 111, 1289 (1976)

CAESALPINIA (Leguminosae/Caesalpiniaceae)

C. bonducella.
DITERPENOIDS: α-caesalpin, β-caesalpin, τ-caesalpin, ε-caesalpin, #87-ceasalpin.
Balmain et al., Tetrahedron Lett., 5027 (1967)

C. coriara.
TANNINS: chebulagic acid, chebulinic acid.
Haslam et al., J. Chem. Soc., C, 2381 (1967)

C. digyna Rotte.
MACROCYCLIC ALKALOID: caesalpinine A.
Mahato et al., J. Am. Chem. Soc., 105, 4441 (1983)

C. pulcherrima.
DITERPENOID: X-caesalpin.
Sengupta et al., Chem. Ind. (London), 534 (1970)

C. sappan L.
AMINO ACIDS: alanine, aspartic acid, glycine, leucine, norvaline, proline, threonine, valine.
Nigam et al., Indian J. Pharm., 39, 85 (1977)

C. spinosa.
GLUCOSIDE: tannin.
Armitage et al., J. Chem. Soc., 1842 (1961)

CAJANUS
C. cajan.
ISOFLAVONES: cajanin, ferreirin.
BENZOPYRAN: cajanone.
(B) Preston et al., Phytochemistry, 16, 143 (1977)
(I) Ingham, Z. Naturforsch., 31, 504 (1976)

CALAMINTHA
C. nepeta subsp. glandulosa.
MONOTERPENOIDS: limonene, piperitenone, piperitone oxide.
De Pooter et al., Phytochemistry, 25, 691 (1986)

CALCEOLARIA (Scrophulariaceae)
C. hypericina.
PHENYLPROPANOIDS: calceolarioside A, calceolarioside B.
Nicoletti et al., Gazz. Chim. Ital., 116, 431 (1986)
C. integrifolia.
PIGMENT: (-)-dunnione.
Price et al., J. Chem. Soc., 1552, 1940 (1939)

CALDARIELLA
C. acidophila.
QUINONE: caldariellaquinone.
De Rosa et al., J. Chem. Soc., Perkin I, 653 (1977)

CALEA
C. axillaris.
SESQUITERPENOID: calaxin.
Nabin et al., J. Org. Chem., 44, 1881 (1979)
C. jamaicensis.
SESQUITERPENOIDS: jamaicolides A, B, C and D.
Ober et al., Phytochemistry, 25, 877 (1986)
C. oxylepis.
SESQUITERPENOIDS:
1-acetoxy-8-O-(2-methylbutanoyl)-desacetylzacatechinolide,
1-hydroxy-8-O-(2-methylbutanoyl)-deacetylzacatechinolide,
1-oxo-8-O-(2-methylbutanoyl)-desacetylzacatechinolide.
Bohlmann et al., Phytochemistry, 21, 1164 (1982)
C. pinnatifolia.
SESQUITERPENOID: arucanolide.
Ferreira et al., Phytochemistry, 19, 1481 (1980)
C. reticulata.
SESQUITERPENOID: 4(15),5E,10(14)-germacratrien-1-one.
Fattorusso et al., Tetrahedron Lett., 4149 (1978)
C. solidaginea.
SESQUITERPENOIDS: solidaginolides A and B.
Ober et al., Phytochemistry, 24, 2728 (1985)
C. subcordata.
SESQUITERPENOIDS: subcordatolides A, B and C.
Ober et al., Phytochemistry, 23, 1289 (1984)
C. ternifolia.
SESQUITERPENOIDS: 8β-angeloyloxy-9α-acetoxyternifolin,
8α-angeloyloxy-9α-(2-methylbutanoyloxy)-ternifolin.
Lee et al., Phytochemistry, 21, 2313 (1982)
C. teucrifolia.
SESQUITERPENOIDS: acetoxycaleteucrin, acetoxynerolidol,
calefolione angelate, calefolione methylbutyrate,
caleteucrifolin, caleteucrin.
Bohlmann et al., Phytochemistry, 20, 1643 (1981)

C. trichomata.
SESQUITERPENOIDS: trichomatolides A, B, C, D and E.
Ober et al., Phytochemistry, 23, 1439 (1984)
C. urticifolia.
SESQUITERPENOIDS: calein D, caleurticolide acetate,
juanislamin, juanislamin-2,3-epoxide.
Borges-del-Castillo et al., J. Nat. Prod., 44, 348 (1981)
C. zactechichi.
CHROMENES: caleochromene A, caleochromene B,
Sesquiterpenoids 1α-acetoxyzacatechinolide, caleins A, B, C
and D, 1α-hydroxyzacatechinolide acetate,
1α-hydroxyzacatechinolide methyl ether,
1-oxozacatechinolide.
(C) Quijano et al., Rev. Latinoam. Quim., 8, 90 (1977)
(T) Bohlmann, Zdero, Phytochemistry, 16, 1065 (1977)
(T) Quijano et al., ibid., 18, 1745 (1979)

CALENDULA (Compositae)
C. arvensis.
FLAVONOIDS: isoquercitroside, narcissoside, rutoside.
Pinkas et al., C. R. Acad. Sci., 281D, 447 (1975)
C. officinalis L.
CAROTENOIDS: flavochrome, mutatochrome.
SAPONINS: calendulosides A, B, C, D, G and H.
STEROIDS: fucostanol,
4α-methyl-24-methylenecholest-7-en-3β-ol,
4α-methylstigmasta-7,24(28)-dien-3β-ol,
4β-methylstigmasta-7,24(28)-dien-3β-ol.
SESQUITERPENOID: calendrin.
TRITERPENOIDS: heliantriol C, ursadiol, 12-ursene-3,16,21-triol.
(C) Karrer et al., Helv. Chim. Acta, 28, 471 (1945)
(Sap) Verchenko et al., Khim. Prir. Soedin., 532 (1974)
(S) Sucrow et al., Pol. J. Chem., 53, 1071 (1979)
(T) St. Pyrek et al., ibid., 53, 2465 (1979)

CALIBANUS
C. hookeri.
STEROIDS: calibagenin, 5-cholestene-3,14,16,22-tetraol.
Giral, Rivera, Phytochemistry, 23, 2089 (1984)

CALLIANDRA (Leguminosae)
C. haematocephala Hassk.
ACIDS: (2S,4R,5S)-dihydroxypipecolic acid,
(2S,4S,5S)-dihydroxypipecolic acid. One uncharacterized
alkaloid.
(Acid) Marlier et al., Phytochemistry, 11, 2597 (1972)
(A) Marlier et al., ibid., 18, 479 (1979)
C. pittieri.
PHENOLICS: 2,4-cis-4,5-trans-4,5-dihydroxypipecolic acid,
cis-4-hydroxypipecolic acid.
Romeo et al., Phytochemistry, 22, 1615 (1983)

CALLICARPA (Verbenaceae)
C. arborea Roxb.
AMINO ACID: L(+)-α-amino-β-(p-methoxyphenyl)-propionic
acid.
STEROID: β-sitosterol.
TRITERPENOIDS: β-amyrin, lupeol acetate, maslinic acid,
oleanolic acid, oleanolic acid methyl ester acetate, ursolic
acid.
(Am) Zhang et al., Zhongcaoyao, 16, 556 (1985)
(S, T) Anjaneyulu et al., Curr. Sci., 46, 667 (1977)
C. candicans.
DITERPENOID: callicarpone.
Kawazu, Mitsui, Tetrahedron Lett., 3519 (1966)

C. formosana.
FLAVONOIDS: 5-hydroxy-3,4,7,4'-tetraethoxyflavone, 3,5,7,3',4'-pentamethoxyflavone, 3,5,7,4'-tetramethoxyflavone.
TRITERPENOIDS: 2α,-dihydroxyurs-12-en-28-oic acid, ursolic acid.
　　Chen et al., J. Chin. Chem. Soc., **33**, 329 (1987); Chem. Abstr., 107 194917a

C. integerrima.
ACIDS: salicylic acid, syringic acid, vanillic acid.
ALIPHATIC: triacontane.
FLAVONES: 5-hydroxy-3',4',6,7-tetramethoxyflavone, 3',4',5,7-tetramethoxyflavone.
STEROID: β-sitosterol.
　　Wang et al., Zhongcaoyao, **17**, 108 (1986); Chem. Abstr., 105 39359k

C. longifolia.
DITERPENOID: callicarpone acetate.
　　Chatterjee et al., Tetrahedron, **28**, 4319 (1972)

C. macrophylla Vahl.
AMINO ACID: L(+)-α-amino-β-(p-methoxyphenyl)-propionic acid.
STEROIDS: β-sitosterol, β-sitosterol β-D-glucoside.
DITERPENOIDS: acetylcallicarpenone, acetylcalliterpenone, callicarpenone, calliterpenone.
TRITERPENOIDS: crategolic acid, 2α-hydroxyursolic acid, ursolic acid.
　　(Am) Zhang et al., Zhongcaoyao, **16**, 556 (1985)
　　(S, T) Sengupta et al., J. Indian Chem. Soc., **53**, 218 (1976)
　　(T) Ahmad et al., Tetrahedron Lett., 2179 (1973)

C. maingayi.
DITERPENOID: maingaiyic acid.
　　Nishimo et al., Tetrahedron Lett., 1541 (1971)

CALLICHILIA
C. barteri Hook.
IMIDAZOLIDINOCARBAZOLE ALKALOIDS: beninine, callichiline.
INDOLE ALKALOID: 16-epi-aspidodasycarpine.
　　Naranjo et al., Helv. Chim. Acta, **53**, 749 (1970)

CALLIGONUM (Polygonaceae)
C. leucocladum (Schrenk.) Bunge.
ACIDS: caffeic acid, chlorogenic acid, p-coumaric acid, 5-p-cumaroquinic acid, 3,4-dimethoxycinnamic acid, ferulic acid, 5-feruloquinic acid, 2-hydroxy-4-methoxybenzoic acid, isoferulic acid, neochlorogenic acid, sinapic acid.
FLAVONOIDS: aromadendrin, hyperoside, kaempferol, quercetin.
　　(A) Vorovskii et al., Rastit. Resur., **12**, 76 (1976)
　　(F) Dubinin et al., Khim. Prir. Soedin., 427 (1975)

C. minimum Lipsky.
CARBOLINE ALKALOIDS: calligonidine, calligonine, eleagnine, harman, harman N-oxide, harmanine, oxyeleagnine.
　　Abdusalanov, Sadykov, Nauchn. Tr. Tashk. Gos. Uni., **286**, 76 (1966)

CALLILEPIS
C. laureola.
MONOTERPENOID: 7-oxo-10-isovaleryloxy-8,9-epoxythymol isovalerate.
　　Bohlmann, Zdero, Phytochemistry, **16**, 1854 (1977)

C. salicifolia.
SESQUITERPENOIDS: 12-isocomenal, 12-isocomenol.
　　Bohlmann et al., Phytochemistry, **21**, 139 (1982)

CALLISTEMON (Myrtaceae)
C. lanceolatus L.
PHLOROGLUCINOL: myrtucommulone.
TRITERPENOID: 2α-hydroxyuvaol.
　　(Ph) Lounasmaa et al., Phytochemistry, **16**, 1851 (1977)
　　(T) Younes, Aust. J. Chem., **28**, 221 (1975)

CALLITRIS (Cupressaceae)
C. columellaris.
DITERPENOIDS: 8,11,13-abietatrien-19-oic acid, 4-epi-dehydroabetic acid, 9(20),13(16),14-labdatrien-18-oic acid.
SESQUITERPENOIDS: callitrin, callitrisin, columellarin, dihydrocallitrisin, dihydrocolumellarin.
　　Bracknell, Carman, Tetrahedron Lett., 73 (1978)

C. glauca.
MONOTERPENOID: (-)-citronellic acid.
　　Clark, Chem. Ind., 450 (1952)

C. quadrivalis.
MONOTERPENOID: thymoquinone.
　　Liebermann, Iljinsky, Ber. Dtsch. Chem. Ges., **18**, 3194 (1885)
　　Liebermann, Iljinsky, ibid., **18**, 3220 (1885)

C. rhomboidea.
DITERPENOID: 4-epi-isocommunic acid.
　　Prasad, Krishnamurty, Phytochemistry, **16**, 801 (1977)

CALLITROPIS
C. araucarioides.
SESQUITERPENOID: τ-eudesmol.
　　McQuillan, Parrack, J. Chem. Soc., 2973 (1956)

CALLODONIA
C. triquetra.
COUMARIN SESQUITERPENOID: callodocin.
　　Borisov et al., Khim. Prir. Soedin., 247 (1975)

CALLUNA
C. vulgaris (L.) Hull.
ACIDS: caffeic acid, chlorogenic acid, p-coumaric acid, ferulic acid, gentisic acid, p-hydroxybenzoic acid, protocatechuic acid, sinapic acid, syringic acid, vanillic acid.
FLAVONOIDS: kaempferol, orcinol, orcinol β-D-glucoside, quercetin, scopoletin.
　　Mantilla et al., An. Edafol. Agrobiol., **34**, 765 (1975)

CALOCEPHALUS
C. brownii.
SESQUITERPENOIDS: calocephalin, pseudoivalin.
　　Batterham et al., Aust. J. Chem., **19**, 143 (1966)

CALOCYBE
C. gambosa.
PHENOXAZINES: 2-aminophenoxazone, phenoxazone.
　　Schlunegger et al., Helv. Chim. Acta, **59**, 1383 (1976)

CALODENDRUM (Rutaceae)
C. capense (L.f.) Thunb.
LIMONOID: rutaevin.
SESQUITERPENOID: calodendrolide.
　　Cassady et al., J. Chem. Soc., Chem. Commun., 86 (1977)

CALONECTRIA

C. nivalis.
SESQUITERPENOIDS: calonectrin, 15-deacetylcalonectrin.
Gardner et al., J. Chem. Soc., Perkin I, 2576 (1972)

CALONYCTION

C. aculeatum.
DITERPENOIDS: ent-6α,7α,1,17-tetrahydroxykauran-19-oic acid,
ent-7α,16α,17-trihydroxykauran-19-oic acid.
GIBBERELLINS: gibberellin A-30, gibberellin A-31, gibberellin
A-33, gibberellin A-34.
Murofushi et al., Tetrahedron Lett., 789 (1973)

CALOPHYLLUM (Guttiferae/Clusiaceae)

C. amoenum.
ALIPHATIC: hentriacontane.
STEROID: β-sitosterol.
TRITERPENOIDS: friedelan-3β-ol, friedelin.
Banerji et al., Q. J. Crude Drugs Res., 15, 133 (1977); Chem. Abstr., 88
133274f

C. apetalum Willd.
BENZOPYRAN: apetalic acid.
COUMARIN: apetalolide.
STEROID: apetalactone.
Govindachari et al., J. Chem. Soc., C, 1323 (1968)

C. australianum.
COUMARIN: calaustralin.
Bhushan et al., Indian J. Chem., 13, 746 (1975)

C. blancoi.
BENZOPYRAN: blancoic acid.
Sout, Sears, J. Org. Chem., 33, 4185 (1968)

C. bracteatum Thw.
BENZOPYRAN: calophyllolide.
STEROID: β-sitosterol.
TRITERPENOIDS: taraxerol, taraxerone.
XANTHONES: calabaxanthone, 6-deoxyjacareubin,
1,8-dihydroxy-2,3,7-trimethoxyxanthone,
1,5-dihydroxyxanthone, 1,7-dihydroxyxanthone, guanandin,
jacareubin, tovopyrifolin C,
1,3,5-trihydroxy-2-methoxyxanthone.
Somanathan et al., J. Chem. Soc., Perkin I, 1935 (1972)

C. brasiliense Camb.
ACIDS: (2R,3R)-brasilensic acid, (2R,3S)-brasilensic acid.
XANTHONES: dehydrocycloguanandin, 6-deoxyjacareubin,
3,7-dimethoxy-1-hydroxyxanthone, gentisin, guanandin,
4-hydroxyxanthone, isoguanandin, jacareubin,
osajaxanthone.
(Acid) Stout et al., Tetrahedron Lett., 3285 (1962)
(X) King et al., J. Chem. Soc., 3932 (1953)
(X) da Silva Pereira et al., An. Acad. Bras. Cienc., 38, 426 (1966)

C. calaba L.
DIPYRAN: isochapeleiric acid.
XANTHONES: buchanoxanthone, calabaxanthone,
calocalabaxanthone, 6-deoxyjacareubin,
1,6-dihydroxy-5-methoxyxanthone,
2,8-dihydroxy-1-methoxyxanthone, 1,7-dihydroxyxanthone,
1,8-dimethoxy-2-hydroxyxanthone, guanandin,
2-hydroxy-1,8-dimethoxyxanthone,
8-hydroxy-1,2-dimethoxyxanthone, jacareubin, scriblitifolic
acid, 1,2,8-trihydroxyxanthone, 1,5,6-trihydroxyxanthone.
(D) Guerriero et al., Phytochemistry, 10, 2139 (1971)
(X) Somanathan et al., J. Chem. Soc., Perkin I, 1935 (1972)
(X) Kumar et al., Phytochemistry, 21, 807 (1982)

C. canum Hook.
XANTHONES: jacareubin, osajaxanthone,
1,3,5,6-tetrahydroxy-2-(3-methylbut-2-enyl)-xanthone,
1,3,7-trihydroxy-2-(3-methylbut-2-enyl)-xanthone.
Carpenter et al., J. Chem. Soc., C, 486 (1969)

C. chapelleri.
BENZOPYRAN: chapelieric acid.
Guerriero et al., Phytochemistry, 10, 2139 (1971)

C. cordata-oblongum Thw.
DITERPENOID: cordatooblongic acid.
XANTHONES: buchanoxanthone, cordatooblonguxanthone,
3-hydroxy-4-methoxyxanthone, 2-hydroxyxanthone,
4-hydroxyxanthone, scriblitifolic acid,
1,5,6-trihydroxyxanthone.
(T) Gunasekera, Sultanbawa, J. Chem. Soc., Perkin I, 2215 (1975)
(X) Somanathan et al., ibid., 1935 (1972)

C. cuneifolium Thw.
XANTHONES: calabaxanthone, 6-deoxyjacareubin,
1,7-dihydroxyxanthone, gunandin, scriblitifolic acid,
thwaitesixanthone, trapezifolixanthin,
1,3,5-trihydroxy-2-(3-methyl-2-butenyl)-xanthone.
Gunasekera et al., J. Chem. Soc., Perkin I, 1505 (1977)

C. fragrans Ridley.
XANTHONES: buchanoxanthone, 6-deoxyjacareubin,
1,7-dihydroxyxanthone, 1,2-dimethoxy-8-hydroxyxanthone,
8-hydroxy-1,2-dimethoxyxanthone, jacareubin,
1,3,5,6-tetrahydroxy-2-(3-methylbut-2-enyl)-xanthone,
1,5,6-trihydroxyxanthone.
Locksley, Murray, J. Chem. Soc., C, 1567 (1969)

C. inophyllum L.
BENZOPYRAN: calophynic acid.
COUMARINS: calaustralin, calophylloside, ponnalide.
TRITERPENOIDS: calophyllal, calophyllic acid.
XANTHONES: buchanoxanthone, 6-deoxyjacareubin,
1,7-dihydroxy-3,6-dimethoxyxanthone,
1,7-dihydroxyxanthone, guanandin, jacareubin,
1,3,5,6-tetrahydroxy-2-(3-methylbut-2-enyl)-xanthone,
1,5,6-trihydroxyxanthone.
(B, C) Gautier et al., Tetrahedron Lett., 2715 (1972)
(T) Govindachari et al., Tetrahedron, 23, 1901 (1967)
(X) Jackson et al., Phytochemistry, 8, 927 (1979)

C. lankaensis.
ISOCOUMARIN: calozeylanic acid.
Samaraweera et al., Tetrahedron Lett., 5083 (1981)

C. neo-ebudicum Guill.
XANTHONES: 6-deoxyjacareubin, jacareubin,
1,3,5,6-tetrahydroxy-2-(3-methylbut-2-enyl)-xanthone,
1,3,7-trihydroxy-2-(3-methylbut-2-enyl)-xanthone.
Scheinmann et al., Phytochemistry, 10, 1331 (1971)

C. papuana.
BENZOPYRANS: isopapuanic acid, papuanic acid.
Stout et al., J. Org. Chem., 33, 4191 (1968)

C. pulcherrimum Wall ex Choisy.
XANTHONES: 1,5-dihydroxyxanthone, 1,7-dihydroxyxanthone.
Jackson et al., J. Chem. Soc., C, 178 (1966)

C. ramiflorum Schwarz.
XANTHONES: 1,7-dihydroxyxanthone,
1-hydroxy-6,7-dimethoxyxanthone,
1-hydroxy-3,5,6-trimethoxy-2-(3-methylbut-2-enyl)-
xanthone, jacareubin.
Carpenter et al., J. Chem. Soc., C, 2421 (1969)
Subramanian et al., Phytochemistry, 14 298 (1975)

C. recedens.
BENZOPYRANS: isocalolongic acid, recedensic acid, recedensolide.
Guerriero et al., Phytochemistry, **12**, 185 (1973)

C. sclerophyllum Vesq.
XANTHONES: 1,7-dihydroxyxanthone, jacareubin, 1-methoxy-3,5,6-trihydroxyxanthone, 1,3,5,6-tetrahydroxy-2-(3-methylbut-2-enyl)-xanthone, 1,3,5,6-tetrahydroxyxanthone.
Jackson et al., J. Chem. Soc., C, 178 (1966)

C. scriblitifolium Hems. et Wyatt Smith.
XANTHONES: 6-deoxyjacareubin, guanandin, jacareubin, scriblitifolic acid, 1,3,5-trihydroxy-2-(3-methylbut-2-enyl)-xanthone, 1,3,7-trihydroxy-2-(3-methylbut-2-enyl)-xanthone.
Samanathan et al., J. Chem. Soc., Perkin I, 1935 (1972)

C. soulattri Burm. f.
BENZOPYRAN: soulattrolide.
XANTHONES: buchanoxanthone, 6-deoxyjacareubin, 1,7-dihydroxyxanthone, scriblitifolic acid, 1,3,5-trihydroxy-2-(3-methylbut-2-enyl)-xanthone.
(B) Gunasekera et al., J. Chem. Soc., Perkin I, 1505 (1977)
(X) Samanathan et al., ibid., 1935 (1972)

C. thwaitesii Planch et Triana.
ISOCOUMARIN: calozeylanic acid.
TRITERPENOIDS: friedelinol, taraxerol.
XANTHONES: 1,5-dihydroxyxanthone, 1,7-dihydroxyxanthone, jacareubin, thwaitesixanthone.
(I) Samanathan et al., Tetrahedron Lett., 5083 (1981)
(T, X) Dahanayake et al., J. Chem. Soc., Perkin I, 2510 (1974)

C. tomentosum Wight.
COUMARINS: tomentolide A, tomentolide B.
STEROID: apetalactone.
XANTHONES: buchanoxanthone, 6-deoxyjacareubin, 1,5-dihydroxyxanthone, 1,7-dihydroxyxanthone, jacareubin, 1,3,5-trihydroxy-2-(3-methylbut-2-enyl)-xanthone.
(C) Nigam et al., Tetrahedron Lett., 2633 (1967)
(S) Govindachari et al., J. Chem. Soc., C, 1323 (1968)
(X) Karunanayake et al., Unpublished work cited in J. Chem. Soc., Perkin I, 2515 (1974)

C. trapezifolium Thw.
STEROID: β-sitosterol.
TRITERPENOIDS: simiarenol, taraxerol.
XANTHONES: buchanoxanthone, calabaxanthone, 6-deoxyjacareubin, 3,8-dihydroxy-1,2-dimethoxyxanthone, 1,5-dihydroxyxanthone, 1,7-dihydroxyxanthone, 2-hydroxyxanthone, toxyloxanthone, trapezifolixanthone.
Samanathan et al., J. Chem. Soc., Perkin I, 2515 (1974)

C. verticillatum L.
NEOFLAVONOID: calofloride.
Ramiandrasoa et al., Tetrahedron, **39**, 3923 (1983)

C. walkeri Wight.
ISOCOUMARIN: calozeylanic acid.
TRITERPENOIDS: simiarenol, taraxerol.
XANTHONES: buchanoxanthone, calabaxanthone, 1,5-dihydroxy-2,3-dimethoxyxanthone, 1,5-dihydroxyxanthone, 1,7-dihydroxyxanthone, guanandin, 1,3,5-trihydroxy-2-(3-methyl-2-butenyl)-xanthone.
(I) Samaraweera et al., Tetrahedron Lett., 5083 (1981)
(T, X) Dahanayake et al., J. Chem. Soc., Perkin I, 2510 (1974)

C. zeylanicum Kosterm.
XANTHONE: zeyloxanthone.
Karunanayake et al., Tetrahedron Lett., 4977 (1979)

CALOPLACA
C. ferruginea.
ANTHRAQUINONE: fallacinal.
Nakano et al., Phytochemistry, **11**, 3505 (1972)

CALOPOGONIUM
C. mucumoides.
FLAVONOIDS: 5,4'-dihydroxy-6'',6''-dimethylpyrano(3'',2'' 6,7)-isoflavone, 7-methoxy-3',4'-methylenedioxyisoflavone, 3',4'-methylenedioxy-6'',6''-dimethylpyrano(2'',3'' 7,8)-isoflavone, 7,2',4',5'-tetramethoxyisoflavone, 7,3',4'-trimethoxyisoflavone.
Vilain et al., Bull. Soc. R. Sci. Liege, **45**, 468 (1976); Chem. Abstr., 86 152677q

CALOSCYPHA
C. fulgens.
CAROTENOID: β,τ-carotene.
Weisgraber et al., Experientia, **27**, 1017 (1971)

CALOSTEMMA
C. purpurea.
BENZYISOQUINOLINE ALKALOID: crinine.
Boit, Ehmke, Chem. Ber., **90**, 369, 2203 (1957)

CALOTROPIS (Asclepiadaceae)
C. gigantea.
TRITERPENOIDS: α-amyrin, β-amyrin, α-amyrin benzoate, β-amyrin benzoate, giganteol, isogiganteol, taraasterol benzoate.
Row, Curr. Sci., **37**, 156 (1968)

C. procera Dryand. R.Br.
STEROIDS: calactin, calotoxin, calotropin, calotropogenin, voruscharin.
Kupchan et al., Science, **146**, 1685 (1964)

CALPURNIA (Leguminosae)
C. aurea (Ait.) Benth.
SPARTEINE ALKALOID: calpurmenine.
Radema et al., Phytochemistry, **18**, 2063 (1979)

C. aurea subsp. aurea.
QUINOLIZIDINE ALKALOIDS: calpurmenine, calpurmenine pyrrolecarboxylic acid, calpurnine, digittine, digittine aminoalcohol, 3β,4α-dihydroxy-13α-O-(2α-pyrrolylcarbonyl)-lupanine, 13-hydroxylupanine, 13-hydroxylupanine tiglate, 3β,4α,13α-trihydroxylupanine, virgiline, virgiline pyrrolecarboxylic acid.
Asres et al., Phytochemistry, **25**, 1443 (1986)

C. aurea subsp. sylvatica.
QUINLOIZIDINE ALKALOIDS: calpurmenine, calpurmenine pyrrolecarboxylic acid.
Radema et al., Phytochemistry, **18**, 2063 (1979)

C. subdecandra (L'Herit.) Schweickerdt.
QUINOLIZIDINE ALKALOID: calpurnine.
Goosen, J. Chem. Soc., 3067 (1963)

CALTHA (Ranunculaceae)
C. palustris L.
GLYCOSIDE: ranunculin.
NORTRITERPENOIDS: caltholide, epi-caltholide.
Rastogi et al., Phytochemistry, **23**, 1699 (1984)

CALVATIA (Basidiomycetae)
C. lilacina.
ACID: calvatic acid.
Gasco et al., Tetrahedron Lett., 3431 (1974)

CALYCANTHUS (Calycanthaceae)
C. floridus L.
DIMERIC PYRROLOINDOLE ALKALOIDS: calycanthidine, calycanthine, meso-calycanthine.
O'Donovan et al., J. Chem. Soc., C, 1570 (1966)
C. glaucus Willd.
DIMERIC PYRROLOINDOLE ALKALOID: calycanthine.
Eccles, Proc. Am. Pharm. Assoc., 84, 382 (1888)
C. occidentalis Hook et Arn.
DIMERIC PYRROLOINDOLE ALKALOIDS: calycanthine, folicanthine.
O'Donovan et al., J. Chem. Soc., C, 1570 (1966)

CALYCOPTERIS
C. floribunda.
FLAVONOIDS: calycopterin, gossypol, quercetin.
Rao et al., Indian J. Chem., 11, 403 (1973)
Kasim et al., Curr. Sci., 44, 888 (1975)

CALYCOTOME (Leguminosae)
C. spinosa Link.
UNCHARACTERIZED ALKALOIDS: calycotamine, calycotomine.
White, N. Z. J. Sci. Tech., 25, 93 (1943)

CALYPOGEIA (Hepaticae)
C. granulata Inouye.
SESQUITERPENOIDS: 3-acetoxy-2,3-bicyclogermacrenediol, 3α-acetoxy-2,3-bicyclogermacrenediol, 2-acetoxy-3-hydroxybicyclogermacrene, 3-acetoxy-2-hydroxybicyclogermacrene.
TRINORSESQUITERPENOID: trinoranastreptene.
Takeda et al., Bull. Chem. Soc., Japan, 56, 1265 (1983)

CALYTHRIX
C. tetragona.
KETONE: calythrone.
Penfold et al., J. Chem. Soc., 412 (1940)

CALYX
C. nicaaensis.
STEROID: calysterol.
Fattorusso et al., Tetrahedron, 31, 1715 (1975)

CAMELLIA (Theaceae/Camelliaceae)
C. irawadiensis.
PURINE ALKALOIDS: caffeine, theobromine.
Nagata et al., Phytochemistry, 24, 2271 (1985)
C. japonica L.
TRITERPENOIDS: camelledionol, camellediononol, camelliagenins A, B and C, 3β,18β-dihydroxy-28-nor-12-oleanen-16-one, 18-hydroxy-28-nor-3,16-oleanenedione.
Itokawa et al., Phytochemistry, 20, 2539 (1981)
C. oleifera.
SAPONIN: oleiferone.
Sokol'skii et al., Khim. Prir. Soedin., 102 (1975)

C. sasanqua Thunb.
MONOTERPENOID GLYCOSIDE: sasanquin.
TRITERPENOIDS: camelliagenin D, 22α-hydroxyerythrodiol, olean-12-ene-3β,22α,28-triol.
Ito, Ogina, Tetrahedron Lett., 1127 (1967)
Yosioka et al., Chem. Ind. (London), 745 (1968)
C. sinensis (L.) O.Kuntze.
PURINE ALKALOIDS: caffeine, theobromine.
TRITERPENOID: camelliagenin D.
Ito, Ogina, Tetrahedron Lett., 1127 (1967)
C. sinensis (L.) var. kolkhida.
FLAVONOIDS: apigenin 8-C-glucoside, apigenin 6,8-di-C-glucoside, dihydrokaempferol, kaempferol, kaempferol 3-O-glucoside, kaempferol 3-O-rhamnodiglucoside, kaempferol 3-O-rutinoside, myricetin 3-O-glucoside, naringenin, quercetin, quercetin 3-O-glucoside, quercetin 3-O-rhamnodiglucoside, quercetin 3-O-rutinoside, tricetin.
Chkhikvishvili, Prikl. Biokhim. Mikrobiol., 22, 410 (1986); Chem. Abstr., 105 57942z
C. taliensis.
PURINE ALKALOIDS: caffeine, theobromine.
Nagata et al., Phytochemistry, 24, 2271 (1985)

CAMOENSIS
C. brevicalyx.
ALKALOIDS: two lupinane bases.
Waterman, Faulkner, Phytochemistry, 21, 2401 (1975)
C. maxima.
SPARTEINE ALKALOIDS: camoensidine, camoensine, leontidine.
Santamaria, Khuong-Huu-Laine, Phytochemistry, 14, 2401 (1975)

CAMPANULA (Campanulaceae)
C. cephalates.
FLAVONOIDS: isoquercitrin, rhamnetin, rhamnetin 3-galactoside, rhamnetin 3-glucoside.
Teslov et al., Chem. Abstr., 86 185873d
C. glomerata L.
FLAVONOIDS: hyperoside, isoquercitrin, isorhamnetin, isorhamnetin 3-galactoside, isorhamnetin 3-glucoside, quercetin 3-glucuronide, rutin, trifolin.
PYRAN: 9-(tetrahydro-2-pyranyl)-2,8-nonadiene-4,6-diyn-1-ol.
Bentley et al., J. Chem. Soc., C, 830 (1969)
C. medium L.
PIPERIDONE ALKALOID: camperine.
PYRAN: 9-(tetrahydro-2-pyranyl)-1,8-nonadiene-4,6-diyn-3-ol.
(A) Dopke, Werner, Pharmazie, 25, 128 (1970)
(P) Teslov et al., Chem. Abstr., 86 185873d
(P) Badanyan et al., J. Chem. Soc., Perkin I, 145 (1973)
C. patula.
FLAVONOIDS: campanoside, graveobioside A.
Teslov et al., Khim. Prir. Soedin., 117 (1977)
C. pyramidalis.
PYRAN: 9-(tetrahydro-2-pyranyl)-1,8-nonadiene-4,6-diyn-3-ol.
Badanyan et al., J. Chem. Soc., Perkin I, 145 (1973)

CAMPHYLANDRA
C. aurantiaca.
STEROID: 25(27)-spirostene-1,2,3,4,5,6,7-heptaol.
Miyahara et al., Tetrahedron Lett., 83 (1980)

CAMPOVASSOURIA

C. bupleurifolia.
SESQUITERPENOID: 8β-tigloylguaigrazielolide.
Bohlmann *et al.*, *Phytochemistry*, **22**, 2860 (1983)

CAMPSIDIUM

C. valdivianum.
IRIDOIDS: plantarenaloside, stansioside.
Gambarino *et al.*, *J. Nat. Prod.*, **48**, 992 (1985)

CAMPSIS (Bignoniaceae)

C. chinensis (Lam.) Voss.
IRIDOIDS: campenoside, 5-hydroxycampenoside.
Kobayashi *et al.*, *Heterocycles*, **16**, 1475 (1981)

CAMPTOTHECA

C. acuminata Decne.
INDOLOQUINOLINE ALKALOIDS: camptothecin,
11-hydroxycamptothecin, 10-methoxycamptothecin.
Lin *et al.*, *Huaxue Xuebao*, **40**, 85 (1982)

CANANGA (Annonaceae)

C. odorata (Lam.) Hook. f. et Thoms.
ALKALOIDS: canangine, sampangine.
SESQUITERPENOID: ylangene.
(A) Rao *et al.*, *J. Nat. Prod.*, **49**, 346 (1986)
(T) Herout, Dimitrov, *Chem. Listy*, **46**, 432 (1952)

CANARIUM

C. muelleri.
TRITERPENOID: canaric acid.
Carman, Connelly, *Tetrahedron Lett.*, 627 (1964)

C. samoense.
SESQUITERPENOID: maaliol.
Stoll *et al.*, *Helv. Chim. Acta*, **40**, 1205 (1957)

C. strictum.
SESQUITERPENOID: canarone.
TRITERPENOIDS: canaric acid,
20-epi-pseudotaraxastane-3β,20-diol,
epi-pseudotaraxastanonol.
Wagh *et al.*, *J. Org. Chem.*, **29**, 2479 (1964)

CANAVALIA (Leguminosae)

C. gladiata.
GIBBERELLINS: canavalia gibberellin I, canavalia gibberellin II,
gibberellin A-21, gibberellin A-22, gibberellin A-59.
Takahashi *et al.*, *Tetrahedron Lett.*, 4861 (1967)

C. lineata.
FLAVONOID: rutin.
Morita *et al.*, *Chem. Abstr.*, **88** 148947b

CANELLA

C. winteriana.
SESQUITERPENOID: canellal.
El-Feraly *et al.*, *J. Chem. Soc., Chem. Commun.*, 75 (1978)

CANNABIS (Cannabaceae)

C. sativa L.
MACROCYCLIC ALKALOID: cannabisativine.
FLAVONOIDS: apigenol O-glucoside, cannflavins A and B,
isovitexin O-glucoside, orienti
(A) Loffer *et al.*, *Tetrahedron Lett.*, 2815 (1975)
(F) Barrett *et al.*, *Experientia*, **42**, 452 (1986)
(P) Boeren *et al.*, *Experientia*, **33**, 848 (1977)

(T) Lousberg *et al.*, *Phytochemistry*, **16**, 595 (1976)
(T) Crombie *et al.*, *Tetrahedron Lett.*, 661 (1979)

C. sativa (Czechoslavakian).
DITERPENOIDS: cannabioxic acid, tetrahydrocannabinolic acid.
Hanus *et al.*, *Acta Univ. Palacki. Olomuc. Fac. Med.*, **74**, 161 (1975)

C. sativa (Egyptian).
DITERPENOIDS: cannabidiolic acid, cannabinol, cannabinolic
acid.
Schultz, Huffner, *Z. Naturforsch.*, **14B**, 98 (1959)

C. sativa (Indian).
DITERPENOIDS: cannabinol, cannabinolic acid,
(±)-8,9-dihydroxy-6a(10a)-tetrahydrocannabinol,
(±)-9,10-6a(10a)-tetrahydrocannabinol.
Elsohly *et al.*, *Experientia*, **34**, 1127 (1978)

C. sativa (Lebanese).
DITERPENOIDS: cannabicitran, cannabicoumarone, cannabinol.
Grote, Spiteller, *Tetrahedron*, **34**, 3207 (1978)

C. sativa 'Meao' variant.
DITERPENOIDS: cannabinol, tetrahydrocannabivarinic acid.
Shoyama *et al.*, *Chem. Pharm. Bull.*, **25**, 2306 (1977)

C. sativa (South African).
DITERPENOIDS: cannabinol, cannabispirol.
Boeren *et al.*, *Experientia*, **35**, 1278 (1979)

C. sativa (Thai).
DITERPENOIDS: cannabichromene, cannabichromenic acid,
cannabichromevarin, cannabichromevarinic acid,
cannabidihydrophenanthene, cannabigerovarin,
cannabigerovarinic acid, cannabispiradienone.
Shoyama *et al.*, *Chem. Pharm. Bull.*, **23**, 1984 (1975)
Crombie *et al.*, *Tetrahedron*, 661 (1979)

C. sativa (Turkish).
DITERPENOIDS: cannabidiolic acid, cannabinol, cannabinolic
acid.
Schultz, Huffner, *Z. Naturforsch.*, **14B**, 98 (1959)

CANSCORA (Gentianaceae)

C. decussata (Schult.).
MONOTERPENOID: (-)-loliolide.
TRITERPENOID: canscoradione.
XANTHONES: 1,5-dihydroxy-3,7-dimethoxyxanthone,
3,5-dihydroxy-1-methoxyxanthone,
1,7-dihydroxy-3,5,6-trimethoxyxanthone,
2-C-β-D-glucopyranosyl-1,3,5,6-tetrahydroxyxanthone,
1-hydroxy-3,5-dimethoxyxanthone,
1-hydroxy-3,5,6,7-tetramethoxyxanthone,
1-methoxy-3,4-dihydroxyxanthone,
1-methoxy-3,5-dihydroxyxanthone 3-rutinoside,
1,3,5,6,7-pentamethoxyxanthone,
1,3,6,7,8-pentamethoxyxanthone,
3-rutinosyl-5-hydroxy-1-methoxyxanthone,
1,3,5-trihydroxy-5,6-dimethoxyxanthone,
1,3,7-trihydroxy-5,6-dimethoxyxanthone,
1,3,8-trihydroxy-3-methoxyxanthone,
1,5,6-trihydroxy-3-methoxyxanthone.
(T) Ghosal *et al.*, *J. Pharm. Sci.*, **65**, 1549 (1976)
(X) Locksley *et al.*, *Phytochemistry*, **10**, 3179 (1971)
(X) Ghosal *et al.*, *Phytochemistry*, **15**, 1041 (1976)

CANTHARELLUS (Basidiomycetae)

C. cinnabarinus.
CAROTENOID: canthaxanthin.
Haxo, *Bot. Gaz.*, **112**, 228 (1950)

CANTHIUM

C. dicoccum.
TRITERPENOID: 3-epi-betulin.
Das, *Chem. Ind.*, 1331 (1971)

C. euryoides.
MACROCYCLIC ALKALOID: canthiumine.
Boulvin *et al.*, *Bull. Soc. Chim. Belg.*, **78**, 583 (1969)

CANTLEYA

C. corniculata.
PYRIDINE ALKALOID: cantleyine.
Sevenet *et al.*, *Bull. Soc. Chim. Fr.*, 3120 (1970)

CAPPARIS (Capparadiaceae)

C. aegyptica.
ACIDS: citric acid, oxalic acid, tartaric acid.
QUATERNARY ALKALOID: stachydrine.
STEROIDS: β-sitosterol, β-sitosterol β-D-glucoside.
Hammouda *et al.*, *Pharmazie*, **30**, 747 (1975)

C. aphylla.
UNCHARACTERIZED ALKALOIDS: cappariline, capparine, capparinine.
Manzoor-i-Khuda, Jeelani, *Pakistan J. Sci. Ind. Res.*, **11**, 250 (1968)

C. decidua.
SPERMIDINE ALKALOIDS: capparidisine, capparisine.
Ahmad *et al.*, *Z. Naturforsch.*, **41B**, 1033 (1986)

C. deserti.
ACIDS: citric acid, oxalic acid, tartaric acid.
QUATERNARY ALKALOID: stachydrine.
STEROIDS: β-sitosterol, β-sitosterol β-D-glucoside.
Hammouda *et al.*, *Pharmazie*, **30**, 747 (1975)

C. formosana Hemsl.
ACID: caffeic acid.
COUMARIN: scopoletin.
STEROIDS: campesterol, β-sitosterol, stigmasterol.
Liu *et al.*, *Chem. Abstr.*, 88 60087k

C. leucophylla.
ACIDS: citric acid, oxalic acid, tartaric acid.
QUATERNARY ALKALOID: stachydrine.
STEROIDS: β-sitosterol, β-sitosterol β-D-glucoside.
Hammouda *et al.*, *Pharmazie*, **30**, 747 (1975)

C. spinosa L.
QUATERNARY ALKALOID: stachydrine.
FLAVONOIDS: kaempferol 3-rutinoside, quercetin 3-rutinoside.
(A) Cornforth, Henry, *J. Chem. Soc.*, 601 (1952)
(F) Thomas *et al.*, *Chem. Abstr.*, 86 185906s

C. tormentosa L.
QUATERNARY ALKALOID: stachydrine.
Schultz, Trier, *Z. Physiol. Chem.*, **67**, 59 (1910)

CAPRARIA

C. biflora.
ANTIBIOTIC: biflorine.
de Lima *et al.*, *Rev. Quim. Ind. (Rio de Janeiro)*, **22**, 14 (1953)

CAPSICODENDRON

C. dinisii.
SESQUITERPENOIDS: 6β-acetoxydrimenin, 6-acetoxyisodrimenin.
Mahmoud *et al.*, *J. Nat. Prod.*, **43**, 365 (1980)

CAPSICUM (Solanaceae)

C. annuum L.
CAROTENOIDS: capsanthin, capsanthin-5,6-epoxide, capsorubin, β-carotene, cryptoxanthin, α-cryptoxanthin, lutein, phytofluene, xanthophyll.
STEROIDS: capsicoside, lanostenol, lanosterol, 4α-methyl-5α-cholest-8(14)-en-3β-ol.
(C) Barber *et al.*, *J. Chem. Soc.*, 4019 (1967)
(C) Bovonat *et al.*, *Chem. Abstr.*, 84 132595s
(S) Matsumoto *et al.*, *Phytochemistry*, **22**, 2621 (1983)

C. frutescens.
SESQUITERPENOIDS: capsaicin, capsidiol, capsidol.
Gordon *et al.*, *Can. J. Chem.*, **51**, 749 (1973)

C. pendulum.
SESQUITERPENOID: capsaicin.
Gutsu *et al.*, *Chem. Abstr.*, 100 48599n

CAPURONETTA

C. elegans.
IMIDAZOLIDINOCARBAZOLE ALKALOIDS: 14,15-anhydrocapuronidine, 14,15-anhydrodihydrocapuronidine, capuronidine, capuronine, capuvosine.
Chardon-Loriaux, Husson, *Tetrahedron Lett.*, 1845 (1975)

CARAPA

C. guaianensis.
LIMONOIDS: 6α-acetoxyepoxyazadiradione, 6α-acetoxygedunin, 11β-acetoxygedunin, 7-deacetoxy-7-oxogedunin, 6α,11β-diacetoxygedunin, 6β,11β-diacetoxygedunin, epoxyazadiradione, 6α-hydroxygedunin.
TRINORTRITERPENOID: andirobin.
Lavie *et al.*, *Bioorg. Chem.*, **2**, 59 (1972)
Marcelle *et al.*, *Phytochemistry*, **14**, 2717 (1975)

C. procera.
BENZOPYRANS: carapolides A, B, C, D, E and F.
LIMONOIDS: 7-deacetoxy-7-oxogedunin, proceranolide, proceranone, procerin.
(B) Kimba *et al.*, *Tetrahedron Lett.*, 1613 (1984)
(L) Sondergam *et al.*, *Phytochemistry*, **20**, 173 (1981)

CARAUPA (Guttiferae/Clusiaceae)

C. densiflora.
XANTHOLIGNOIDS: cadensin A, cadensin B, kielcorin.
XANTHONE: 2,6-dihydroxy-7,8-dimethoxyxanthone.
(XL) Castelao *et al.*, *Phytochemistry*, **16**, 735 (1977)
(X) de Lima *et al.*, *An. Acad. Bras. Cienc.*, **42**, 133 (1970)

CARDARIA (Cruciferae)

C. draba (L.) Desv. (Lepidium draba L.).
One uncharacterized alkaloid.
Potapov, *Med. Zh. Uzb.*, 61 (1976)

CARDUNCELLUS

C. monspeliensium All.
FLAVONOIDS: luteolin, luteolin 7-glucoside, quercetin, quercetin 7-glucoside.
Proliac, *Fitoterapia*, **47**, 113 (1976)

CARDUUS (Compositae)

C. acanthoides L.
IMIDE ALKALOIDS: (+)-acanthoidine, (+)-acanthoine.
Frydman, Delofeu, *Tetrahedron*, **18**, 1063 (1962)

C. getulus Pomel.
TRITERPENOIDS: erythrodiol 3-acetate, taraxasterol, taraxasterol acetate.
Abdel-Salam *et al.*, *Inst. J. Crude Drug Res.*, **21**, 79 (1983); *Chem. Abstr.*, 100 3478f

CAREX (Cyperaceae)

C. brevicollis DC.
CARBOLINE ALKALOIDS: brevicarine, brevicolline, dihydrobrevicolline, harman, harmine, harmol.
Terent'eva *et al.*, *Khim. Prir. Soedin.*, 397 (1969)

CAREYA

C. arborea.
TRITERPENOIDS: carearborin, careyagenols D and E.
Tang, *Phytochemistry*, **18**, 651 (1979)

CARICA

C. papaya L.
MACROCYCLIC ALKALOIDS: carpaine, dehydrocarpamine I, dehydrocarpamine II, pseudocarpaine.
Tang, *Phytochemistry*, **18**, 651 (1979)

CARISSA

C. carandas.
LIGNAN: carinol.
TRITERPENOID: carindone.
(L) Pal *et al.*, *Phytochemistry*, **14**, 2302 (1975)
(T) Singh, Rastogi, *Phytochemistry*, **11**, 1707 (1972)

C. edulis.
FURAN: carissanol.
Achenbach *et al.*, *Phytochemistry*, **22**, 749 (1983)

C. lanceolata.
SESQUITERPENOID: carissone.
Mohr *et al.*, *Helv. Chim. Acta*, **37**, 462 (1954)

CARLINA (Compositae)

C. corymbosa L. var. globosa.
TRITERPENOIDS: 11α-hydroxy-β-amyrin, 30-norlupan-3β-ol-20-one.
Piozzi *et al.*, *Phytochemistry*, **14**, 1662 (1975)

CARNEGIA (Cactaceae)

C. gigantea (Engelm.) Britton et Rose.
ISOQUINOLINE ALKALOIDS: carnegine, dehydroheliamine, gigantine.
Ordez *et al.*, *Phytochemistry*, **22**, 2101 (1983)

CARPESIUM

C. abrotanoides.
SESQUITERPENOIDS: carabrol, carabrone, carpesiolin, granilin.
Maruyama *et al.*, *Phytochemistry*, **22**, 2773 (1983)

C. eximum.
SESQUITERPENOID: carabrone.
Konovalova *et al.*, *Khim. Prir. Soedin.*, 721 (1972)

CARPOLOBIA (Polygalaceae)

C. lutea.
One uncharacterized saponin.
Delaude, *Bull. Soc. R. Sci. Liege*, **44**, 495 (1975)

CARTEROTHAMNUS

C. anomalochaeta.
DITERPENOIDS: carterochaetic acid, carterochaetol, carterothamnotriol.
Bohlmann *et al.*, *Phytochemistry*, **18**, 621 (1979)

CARTHAMNUS (Compositae)

C. lanatus.
SESQUITERPENOID: α-bisabolol-β-D-fucopyranoside.
San Feliciano *et al.*, *Phytochemistry*, **21**, 2115 (1982)

C. oxyacantha.
SESQUITERPENOID: hinesol-β-D-fucopyranoside.
Rustaiyan *et al.*, *Chem. Ber.*, **109**, 3953 (1976)

C. tinctorius L.
ACETYLENICS: 3-cis-11-trans-trideca-1,3,11-triene-5,7,9-triyne, 3-trans-11-trans-trideca-1,3,11-triene-5,7,9-triyne.
AMIDE: serotobenine.
MONOTERPENOID: deca-4,6-diyn-1-ol.
PHYTOALEXIN: safynol.
QUINONE: ubiquinone-9.
(Acet) Kogiso *et al.*, *Agr. Biol. Chem.*, **40**, 2085 (1976)
(Am) Sato *et al.*, *ibid.*, **49**, 2969 (1985)
(P) Nakada *et al.*, *ibid.*, **41** 1761 (1977)
(Q) Hagimori *et al.*, *ibid.*, **42**, 499 (1978)
(T) Bohlmann, Zdero, *Chem. Ber.*, **103**, 2853 (1970)

CARUM (Umbelliferae)

C. copticum.
GLYCOSIDES: 3-galactosyloxy-5-hydroxytoluene, α-galactoside, 2-methyl-3-glucosyloxy-5-isopropylphenol.
Garg *et al.*, *Proc. Nat. Acad. Sci. India*, **55A**, 95 (1985)

C. roxburghianum.
BENZOPYRAN: seselin.
Gupta *et al.*, *J. Indian Chem. Soc.*, **51**, 904 (1974)

CARYA

C. pecan.
FLAVONOIDS: azaleatin 3-arabinoside, azaleatin 3-rutinoside, caryatin 3'-glucoside, quercetin 3-glucoside.
El-Ansari *et al.*, *Z. Naturforsch.*, **32C**, 444 (1977)

CARYOCAR

C. glabrum.
SESQUITERPENOID: caryocarsapogenin.
Ruzicka, van Veen, *Z. Physiol. Chem.*, **184**, 69 (1929)

CARYOPTERIS (Verbenaceae)

C. x cladonensis.
DITERPENOID: caryopterone.
Giles *et al.*, *J. Chem. Soc.*, *Perkin I*, 1632 (1969)

C. divaricata.
DITERPENOIDS: caryoptin, caryoptin hemiacetal, caryoptinol, dihydrocaryoptinol.
Harada *et al.*, *J. Am. Chem. Soc.*, **100**, 8022 (1978)

CASCARA

C. segrada.
ANTHRAQUINONE: frandulaemodin.
Sargent *et al.*, *J. Chem. Soc.*, C, 307 (1970)

CASEARIA (Flacourtiaceae)
C. ilicifolia Vent.
FLAVONOIDS: apigenin, kaempferol, quercetin.
STEROID: β-sitosterol.
TRITERPENOIDS: cycloartenol, 24-methylenecycloartenol.
　　Weniger et al., Planta Med., 32, 170 (1978)

CASELLA
C. atromarginata.
DITERPENOIDS:
　　2-hydroxy-17,19-diacetoxy-1,13(16),14-spongiatrien-3-one,
　　2,17,19-trihydroxy-1,13(16),14-spongiatrien-3-one.
　　de Silva et al., Heterocycles, 17, 167 (1982)

CASIMIROA
C. edulis Llave et Lejarza.
GLUCOSIDIC ALKALOID: casimiroedine.
PYRIMIDINE ALKALOID: zapotidine.
QUINOLINE ALKALOIDS: casimiroine, edulinine, edulitine.
LIMONOID: zapoterin.
　　(A) King et al., J. Chem. Soc., 4613 (1056) (1056)
　　(T) Dreyer et al., J. Org. Chem., 33, 3577 (1968)

CASSIA (Leguminosae)
C. absus.
QUATERNARY ALKALOID: chaksine, isochaksine.
　　Siddiqui, Ahmed, Proc. Indian Acad. Sci., 2, 421 (1935)
C. alata L.
ANTHROQUINONES: aloe-emodin, chrysophanol, emodin, rhein.
　　Villaroya et al., Asian J. Pharm., 3, 10, 17 (1976); Chem. Abstr., 86
　　　　40173r
C. alexandria.
ACIDS: linoleic acid, oleic acid, palmitic acid, stearic acid.
　　Osman et al., Nahrung, 20, 483 (1976); Chem. Abstr., 85 119625c
C. angustifolia.
ANTHRAQUINONES: meso-aloe-emodin, trans-aloe-emodin
　　dianthrone diglucoside.
SAPONINS: sennosides A, A', B, C, D, E and F.
　　(An) Nakajima et al., J. Pharm. Pharmacol., 37, 703 (1985)
　　(Sap) Stoll et al., Helv. Chim. Acta, 33, 313 (1950)
C. carnaval Speg.
PIPERIDINE ALKALOIDS: carnavaline, cassine, prosopinone.
　　Lythgoe, Vernengo, Tetrahedron Lett., 1133 (1967)
C. dentata.
FLAVONOIDS: piceatanol, resveratrol.
STEROID: β-sitosterol.
　　De Oliviera et al., Rev. Latinoam. Quim., 8, 82 (1977)
C. excelsa Schrad.
PIPERIDINE ALKALOID: cassine.
　　Highet, J. Org. Chem., 29, 471 (1966)
C. fistula.
FLAVONOIDS: (-)-epi-afzelechin, (+)-catechin, clitorin,
　　dihydrokaempferol, kaempferol, kaempferol β-D-glucoside,
　　kaempferol 3-neohesperidoside.
　　Patil et al., Indian J. Chem., 21B, 626 (1982)
C. garrettiana.
ANTHRONE: cassioloin.
　　Hata et al., Chem. Pharm. Bull., 24, 1688 (1976)
C. grandis.
ANTHROQUINONE: aloe-emodin.
　　Gritranapan et al., Chem. Abstr., 100 65011g
C. italica.
ACIDS: linoleic acid, oleic acid, palmitic acid, stearic acid.
　　Osman et al., Nahrung, 20, 483 (1976); Chem. Abstr., 85 119625c

C. javanica.
ACIDS: cerotic acid, isostearic acid, pentacosanoic acid, stearic
　　acid.
ALCOHOL: myricyl alcohol.
ALIPHATICS: ceryl alcohol, hentriacontane, hentriacontanol,
　　heptacosane, octacosane, octacosanol.
DITERPENOIDS: afzelechin, rimuene.
ESTER: ethylarachidate.
STEROIDS: β-sitosterol, β-sitosterol β-D-glucoside.
　　(Ac, E, T) Chaudhuri et al., Indian J. Pharm. Sci., 47, 172 (1985)
　　(Al, S) Joshi et al., Planta Med., 28, 190 (1975)
C. macrantha.
CHROMENES:
　　3,5-dihydroxy-8-isobutyryl-2-methyl-7-methoxychromone,
　　6-O-galactosylrubrofusarin, rubrafusarin.
STEROID: β-sitosterol.
　　De Oliviera et al., Rev. Latinoam. Quim., 8, 82 (1977)
C. marylandica L.
AMINE ALKALOID: N-methyl-β-phenethylamine.
　　Camp, Norvell, Econ. Bot., 20, 274 (1966)
C. multijuga.
ACID: stearic acid.
STEROIDS: β-sitosterol, β-sitosterone.
　　De Oliviera et al., Rev. Latinoam. Quim., 8, 82 (1977)
C. nodosa.
FLAVONOID: 5-O-α-L-rhamnopyranosylvelutin.
　　Tiwari et al., J. Indian Chem. Soc., 59, 526 (1982)
C. nomame.
ANTHRAQUINONES: emodin-9-anthrone, physcion, physcion
　　9-anthrone, physcion 10,10'-bianthrone.
　　Kitanaka et al., J. Nat. Prod., 48, 849 (1985)
C. obtusifolia L.
ANTHRAQUINONES: alaternin 1-O-β-D-glucopyranoside,
　　chrysoobtusin 8-O-β-D-glucopyranoside.
　　Kitanaka et al., Chem. Pharm. Bull., 33, 1274 (1985)
C. occidentalis.
ANTHRAQUINONES: bianthaquinone, chrysophanol, funiculosin.
FLAVONOIDS: jaceidin 7-rhamnoside, matteucinol
　　7-rhamnoside.
XANTHONE: casiollin.
　　(An, X) Howard et al., Biochem. J., 44, 227 (1949)
　　(F) Tiwari, Singh, Phytochemistry, 16, 1107 (1977)
　　(X) Tiwari et al., Planta Med., 32, 375 (1977)
C. petosiana.
NORDITERPENOIDS: colensanone, colensenone.
　　Mason, Luso, 3, 119 (1982)
C. renigera.
ANTHRAQUINONE:
　　1-hydroxy-3,8-dimethoxy-2-methylanthraquinone.
FLAVONOIDS: 5-hydroxy-4'-methoxyflavanone
　　7-α-L-rhamnopyranoside,
　　5-hydroxy-3',4',5,6,7-pentamethoxyflavanone
　　5-O-α-L-rhamnopyranoside, quercetin 3,6-dimethyl ether.
　　Tiwari et al., Phytochemistry, 16, 798 (1977)
C. roxburghii.
ANTHRAQUINONE: roxburghinol.
　　Ashok et al., Phytochemistry, 24, 2673 (1985)
C. senna L.
BIANTHRAQUINONE: sennoside A.
　　Christ et al., ArzneimForsch., 28, 225 (1978)

C. siamea Lamk.
CHROMONE ALKALOID: cassiadinine.
ISOQUINOLINE ALKALOIDS: siamine, siaminines A, B and C.
BENZOPYRAN· barakol.
CHROMONE: 5-acetonyl-7-hydroxy-2-methylchromone.
ISOCOUMARIN: (+)-6-hydroxymellein.
QUINONES: cassiamins A, B and C.
STEROID: β-sitosterol.
TRITERPENOIDS: betulin, cycloart-23-en-3β,25-diol, friedelin, lupeol, oleanolic acid.
 (A) El-Sayyed et al., J. Nat. Prod., 47, 708 (1984)
 (B) Bycroft et al., J. Chem. Soc., C, 1686 (1970)
 (Ch, I) Mallick, Phytochemistry, 25, 1727 (1986)
 (Q) Dutta et al., Tetrahedron Lett., 3023 (1964)
 (S, T) Varshney, Pal, J. Indian Chem. Soc., 54, 548 (1977)

C. sophora.
ANTHRAQUINONE: chrysophanol.
FLAVONOID: rhamnetin 3-O-β-D-glucoside.
 Grove et al., J. Chem. Soc., C, 230 (1966)
 Tiwarts et al., Planta Med., 28, 182 (1975)

C. speciosa.
ALIPHATIC: octacosanol.
ANTHRAQUINONES: chrysophanol, emodin, 8-O-methylchrysophanol.
STEROID: β-sitosterol.
TRITERPENOIDS: 3-O-acetylbetulinic acid, lupeol.
 de Oliviera et al., Rev. Latinoam. Quim., 8, 82 (1977)

C. tora L.
ANTHRAQUINONE: emodin.
 Pal et al., Indian J. Pharm., 39, 116 (1977)

C. torosa L.
ANTHRACENES: phlegmacin A1, phlegmacin A2, phlegmacin B1, phlegmacin B2.
ANTHRAQUINONES: chrysophanol, emodin, germichrysone, physcion, torosachrysone, xanthorin.
 (An) Takahashi et al., Phytochemistry, 16, 999 (1977)
 (Aq) Takido et al., Lloydia, 40, 191 (1977)

CASSINE
C. metabelica.
PIPERIDINE ALKALOIDS: cassine, cassinine.
 Wagner et al., Tetrahedron Lett., 125 (1977)

CASSINOPSIS
C. ilicifolia Kuntz.
BENZOISOQUINOLINE-CARBOLINE ALKALOID: (-)-deoxytubulosine.
 Montiero et al., J. Chem. Soc., Chem. Commun., 317 (1965)

CASSIPOUREA
C. gerrardii Alston.
THIOALKALOIDS: gerrardine, gerrardoline.
 Wright, Warren, J. Chem. Soc., C, 283 (1967)

C. gummiflua Tul. var. verticillata Lewis.
THIOALKALOID: cassipourine.
 Wright, Warren, J. Chem. Soc., C, 283 (1967)

CASSYTHA (Lauraceae)
C. americana. (C. filiformis L.)
APORPHINE ALKALOID: launobine.
OXOAPORPHINE ALKALOIDS: cassamedine, cassemedinine.
 Cava et al., J. Org. Chem., 33, 2243, 2443 (1968)

C. glabella R.Br.
APORPHINE ALKALOID: cassythicine.
 Tomita, Kikkawa, J. Pharm. Soc., Japan, 52, 3833 (1958)

C. melantha R.Br.
APORPHINE ALKALOID: cassythicine.
 Tomita, Kikkawa, J. Pharm. Soc., Japan, 52, 3833 (1958)

C. pubescens R.Br.
APORPHINE ALKALOID: laurelliptine.
 Johns et al., Aust. J. Chem., 19, 2331 (1966)

C. racemosa Nees.
APORPHINE ALKALOIDS:
 1,2-dimethoxy-9,10-methylenedioxyaporphine,
 1,2-dimethoxy-9,10-methylenedioxyoxoaporphine.
 Johns et al., Aust. J. Chem., 20, 1457 (1967)

CASTANEA
C. caudata.
ALIPHATICS: ditriacontane, heptacosane, hexacosane, monotriacontane, nonacosane, octacosane, pentacosane, triacontane, tritriacontane.
STEROID: β-sitosterol.
TRITERPENOIDS: friedelanol, epi-friedelanol, friedelin, glutinol, lupeol.
 Tachi et al., J. Pharm. Soc., Japan, 98, 349 (1978)

CASTANOPSIS (Fagaceae)
C. cuspidata.
ALIPHATICS: ditriacontane, heptacosane, hexacosane, monotriacontane, nonacosane, octacosane, pentacosane, triacontane, tritriacontane.
STEROID: β-sitosterol.
TRITERPENOIDS: friedelanol, epi-friedelanol, friedelin, glutinol, lupeol.
 Tachi et al., J. Pharm. Soc., Japan, 98, 349 (1978)

C. indica.
TRITERPENOIDS: castanopsin, castanopsol, castanopsone.
 Pant, Rastogi, Phytochemistry, 17, 575 (1978)

C. lamontii.
TRITERPENOID: 3α,4α-epoxyfriedelane.
 Hui, Li, Phytochemistry, 15, 1313 (1976)

CASTANOSPERMUM
C. australe.
ISOFLAVONOID: koparin.
TRITERPENOID: castanogenol.
 (I) Berry et al., Aust. J. Chem., 30, 1827 (1977)
 (T) Rao et al., Indian J. Chem., 7, 1203 (1969)

CASTELA
C. nicholsonii Hook.
BITTER PRINSIPLES: amarolide, castelanolide.
 Mitchell et al., Phytochemistry, 10, 411 (1971)

C. tweedii.
QUASSINOID: castelanone.
 Polonsky et al., C. R. Acad. Sci., 288C, 267 (1979)

CASUARINA (Casuarinaceae)
C. cunninghamiana Miq.
FLAVONOIDS: kaempferol, kaempferol glucoside, quercetin, quercetin glucoside.
PHENOLIC: ellagic acid.
 Natarajan et al., Phytochemistry, 10, 1283 (1971)

C. equisetifolia L.
FLAVONOIDS: kaempferol, kaempferol glycoside, quercetin, quercetin glycoside, trifolin.
PHENOLICS: (+)-catechin, ellagic acid, (-)-gallocatechin.
STEROID: β-sitosterol.
Neelankantan et al., Fitoterapia, 57, 120 (1986)

C. junghuniana Miq.
FLAVONOIDS: kaempferol, kaempferol glycoside, quercetin, quercetin glycoside, trifolin.
STEROID: β-sitosterol.
Neelankantan et al., Fitoterapia, 57, 120 (1986)

C. rigida Miq.
FLAVONOIDS: kaempferol, kaempferol glycoside, quercetin, quercetin glycoside, trifolin.
STEROID: β-sitosterol.
Neelankantan et al., Fitoterapia, 57, 120 (1986)

C. sticta Ait.
FLAVONOID: hinokiflavone.
ELLAGITANNINS: casuarictin, casuarinin, casuriin, pedunculagin, stachyurin, stictinin, tellimagrandin I.
(F) Sawada et al., J. Pharm. Soc., Japan, 78, 1023 (1958)
(T) Okuda et al., J. Chem. Soc., Perkin I, 1765 (1983)

CATALPA (Bignoniaceae)
C. ovata G.Don.
NAPHTHOPYRAN: 4,9-dihydroxy-α-lapachone.
SESQUITERPENOIDS: catalpalactone, catalponol.
(N) Hayashi et al., Chem. Pharm. Bull., 23, 384 (1975)
(T) Inoue et al., ibid, 28, 1224 (1980)

C. speciosa.
IRIDOID: specioside.
DITERPENOID: specionin.
Chang, Nakanishi, J. Chem. Soc., Chem. Commun., 605 (1983)

CATANANCHE
C. caerulea.
BENZOPYRAN: carlinoside.
PROLIAC, REYNAUD, PLANTA MED., 32, 68 (1977):

CATHA
C. cassinoides.
STEROID: iguesterin.
Gonzalez et al., An. Quim., 70, 376 (1974)

C. edulis.
SESQUITERPENE ALKALOID: cathidin D.
ALKAMINE: merucanthine.
(A) Cais et al., Tetrahedron, 31, 2727 (1975)
(Al) Brenneisen et al., Planta Med., 50, 531 (1984)

CATHARANTHUS (Apocynaceae)
C. lanceus Boj.
CARBOLINE ALKALOID: pericycline.
IMIDAZOLIDINOCARBAZOLE ALKALOIDS: cathanneine, hoerhammericine, hoerhammerinine.
UNCHARACTERIZED ALKALOID: cathalanceine.
Aynilian et al., Tetrahedron Lett., 89 (1972)

C. longifolius.
BIS-IMIDAZOLIDINOCARBAZOLE ALKALOID: vindiolicine.
Rabaron et al., Plant Med. Phytother., 7, 53 (1973)

C. ovalis Markgraf.
IMIDAZOLIDINOCARBAZOLE ALKALOIDS: cathovaline, kitraline, kitramine, vincolicine, vincovaline, vincovalinine.
Langlois et al., Helv. Chim. Acta, 63, 773 (1980)

C. roseus G.Don.
CARBAZOLE ALKALOID: gomaline.
IMIDAZOLIDINOCARBAZOLE ALKALOIDS: catharanthamine, deacetylvindoline, 17-desacetoxyleurosine, desacetylvinblastine, 21'-oxoleurosine, 16-epi-19-vindolenine, 16-epi-19-vindolenine N-oxide.
ANTHOCYANINS: hirsutidin, malvidin, petunidin.
(A) Bruyn et al., Bull. Soc. Chim. Belg., 91, 75 (1982)
(A) Atta-ur-Rahman et al., Phytochemistry, 22, 1021 (1983)
(An) Carew, Kruegar, ibid., 15, 442 (1976)

C. trichophyllus.
IMIDAZOLIDINOCARBAZOLE ALKALOIDS: cathaphylline, trichophylline.
Mukhapadhyay et al., Tetrahedron, 39, 3639 (1983)

CAUCALIS (Umbelliferae)
C. scabra.
SESQUITERPENOIDS: caucalol acetate, caucalol diacetate.
Sasaki et al., Tetrahedron Lett., 623 (1966)

CAULERPA
C. brownii.
DITERPENOIDS: caulerpol, caulerpol diacetate.
Blackman, Wells, Tetrahedron Lett., 2729 (1976)

C. flexilis.
SESQUITERPENOID: flexilin.
Blackman, Wells, Tetrahedron Lett., 3063 (1978)

C. prolifera Lamaroux.
SESQUITERPENOIDS: caulerpenyne, furocaulerpin.
de Napoli et al., Experientia, 37, 1132 (1981)

C. trifaria.
DITERPENOID: ent-15,16-diacetoxy-7,13(16),14-labdatriene.
Blackman, Wells, Tetrahedron Lett., 3063 (1978)

CAULOPHYLLUM
C. robustum.
SAPONINS: cauloside A, cauloside B, cauloside C, cauloside D, cauloside G.
TRITERPENOID: caulophyllogenin.
Strigina et al., Phytochemistry, 14, 1583 (1975)

C. thalictroides (L.) Michx.
QUINOLIZIDINE ALKALOIDS: caulophylline, N-methylcytisine.
Power, Salway, J. Chem. Soc., 104, 194 (1913)

CEANOTHUS (Rhamnaceae)
C. americanus L.
PEPTIDE ALKALOIDS: americine, ceanothines A, B, C, D, E and F.
TRITERPENOIDS: ceanothenic acid, ceanothic acid.
(A) Servis et al., J. Am. Chem. Soc., 91, 5619 (1969)
(T) Eade et al., Aust. J. Chem., 24, 621 (1971)

C. integerrimus Hook et Arn.
PEPTIDE ALKALOIDS: integerrenine, integerressine, integerrine.
Tschesche et al., Tetrahedron Lett., 1311 (1968)

C. velutinus (Hook) Dougl.
One uncharacterized alkaloid.
Richards, Lynn, J. Am. Pharm. Assoc., 28, 332 (1934)

CEDRELA (Meliaceae)

C. angustifolia.
LIMONOIDS: angustidienolide, angustinolide.
Lavie et al., Tetrahedron, **26**, 219 (1970)

C. fissilis.
LIMONOID: fissinolide.
Zelnik, Rosita, Tetrahedron Lett., 6441 (1966)

C. glaziovii.
TRITERPENOIDS: mexicanolide, odoratol, 24-epi-odoratol, odoratone.
Connolly et al., Tetrahedron Lett., 3449 (1967)

C. mexicana.
LIMONOID: 7,11-diacetoxydihydronomilin.
TRITERPENOIDS: mexicanol, mexicanolide, odoratol.
Marcelle et al., Tetrahedron Lett., 505 (1981)

C. odorata.
TRITERPENOIDS: angolensic acid methyl ester, mexicanolide, odoratol, odoratone, photogedunin.
Akisanya et al., J. Chem. Soc., 3827 (1960)
Bevan et al., ibid., 890 (1963)

C. odorata var. brasiliensis.
SESQUITERPENOID: cedrelanol.
Smalders, Ingegheria Chimica, **49**, 12 (1967)

C. toona.
PHENOL: siderin.
SESQUITERPENOIDS: cedrelone, T-muurolol.
(P) Nagasampagi et al., Phytochemistry, **14**, 1673 (1975)
(T) Parihar, Dutt, J. Indian Chem. Soc., **27**, 77 (1950)

CEDRELOPSIS

C. grevei.
ALDEHYDE: 3-acetoxy-2,6-dihydroxy-4-methoxybenzaldehyde.
CHROMENES: alloevodionol, greveichromenol, greveiglycol, ptaeroxylin.
(A) Schulte et al., Arch. Pharm., **306**, 857 (1973)
(C) Dean, Robinson, Phytochemistry, **10**, 3221 (1971)

CEDRUS (Pinaceae)

C. atlantica Manetti.
SESQUITERPENOIDS: cis-atlantone, trans-atlantone, epi-cubenol.
Adams et al., J. Chem. Soc., Perkin I, 1502 (1975)

C. atlantica Manetti cv. glauca.
FLAVONOIDS: isorhamnetin, kaempferol, laricitrin, myricetin, quercetin, syringetin.
Niemann, Z. Naturforsch., **32C**, 1015 (1977)

C. deodara (Roxb.) G.Don.
MONOTERPENOID: limonene carboxylic acid.
SESQUITERPENOIDS: allohimachalol, atlantolone, atlantone, isocentdarol, isohimachalone, longiborneal.
Krisnappa et al., Phytochemistry, **17**, 599 (1978)

C. lebanotica.
SESQUITERPENOID: cis-atlantone.
Pfau, Helv. Chim. Acta, **15**, 1481 (1932)

CELASTRUS (Celastraceae)

C. orbiculatus Thunb.
SESQUITERPENOIDS: isocelorbicol, isocelorbicol 1-cinnamate-2,9-diacetate.
Smith et al., J. Org. Chem., **41**, 3249 (1976)

C. paniculatus.
SESQUITERPENE ALKALOIDS: celapagine, celapaginine, celapanigine, celapanine.
SESQUITERPENOIDS: malkangunin, malkanguniol, paniculatadiol.
(A) Wagner et al., Tetrahedron, **31**, 1949 (1975)
(T) den Hertog et al., Tetrahedron Lett., 845 (1973)

C. scandens L.
CAROTENOID: celaxanthin.
TRITERPENOID: celastrol.
LeRosen, Zechmeister, Arch. Biochem., **1**, 17 (1942)

CELSIA

C. coromandeliana Vahl.
STEROIDS: celsianol, 5,9(11)-stigmastadien-3β-ol.
Sen,Chaudhury, J. Indian Chem. Soc., **47**, 1063 (1970)

CELTIS

C. laevigata.
STEROIDS: β-sitosterol, stigmasterol.
TRITERPENOID: moretenol.
Santa-Cruz et al., Phytochemistry, **14**, 2532 (1975)

CENTAUREA (Compositae)

C. aggregata.
FLAVONOID: euparotin.
Kupchan et al., J. Pharm. Sci., **54**, 929 (1965)

C. arbutifolia.
SESQUITERPENOID: arbutifolin.
Gonzalez et al., Phytochemistry, **20**, 1895 (1981)

C. breviceps.
UNCHARACTERIZED ALKALOID: brevicepsine.
Kurmaz et al., Farm. Zh. (Leningrad), **17**, No. **2**, 40 (1962)

C. cachanahuen.
FLAVONOIDS: decussatin, swerchirin, swertiaperenin.
TRITERPENOID: oleanolic acid.
Versluys et al., Experientia, **38**, 771 (1982)

C. canariensis.
SESQUITERPENOIDS: aguerin A, aguerin B, subexpinnatin.
Gonzalez et al., Tetrahedron Lett., 895 (1982)

C. canariensis var. subpinnata.
SESQUITERPENOID: aguerin B.
Gonzalez et al., Phytochemistry, **17**, 995 (1978)

C. cineraria L.
FLAVONOID: 5,7,4'-trihydroxy-8-methoxyflavone.
STEROID: β-sitosterol.
TRITERPENOIDS: α-amyrin, β-amyrin.
El-Emary et al., Fitoterapia, **54**, 133 (1983)

C. clementei.
FLAVONOIDS: 7,4'-di-O-methylscutellarein, hispidulin, isokaempferide, neglectin.
SESQUITERPENOID: clementein.
(F) Gonzalez et al., J. Nat. Prod., **48**, 819 (1985)
(T) Massonet et al., Tetrahedron Lett., 1641 (1983)

C. collina L.
FLAVONOIDS: 6-methoxyquercetin, quercetin, rhamnetin.
Kananzi et al., Pharmazie, **37**, 454 (1982)

C. erythraea.
XANTHONE: 1,6,8-trihydroxy-3,5,7-trimethoxyxanthone.
Veshta et al., Khim. Prir. Soedin., 258 (1982)

C. hyrcanica.
SESQUITERPENOID: repin.
Evstratova et al., Khim. Prir. Soedin., 451 (1972)

C. hyssopifolia.
SESQUITERPENOIDS: hyssopifolins A, B, C, D and E, vahlenin.
Gonzalez et al., Phytochemistry, 13, 1193 (1974)

C. janeri.
SESQUITERPENOIDS: chlorojanerin, janerin.
Gonzalez et al., An. Quim., 73, 86 (1977)

C. kandavanensis.
TERPENOIDS: 9β-hydroxykandavanolide, kandavanolide.
Rustaiyan, Ardebili, Planta Med., 50, 364 (1984)

C. linifolia.
SESQUITERPENOIDS: aguerin, linichlorins A, B and C, vahlenin.
Gonzalez et al., Phytochemistry, 17, 997 (1978)

C. littorale.
XANTHONES: 1,8-dihydroxy-3,5-dimethoxyxanthone,
1,8-dihydroxy-3,5,6,7-tetramethoxyxanthone,
1-hydroxy-3,5,6,7,8-pentamethoxyxanthone,
1-hydroxy-3,5,8-trimethoxyxanthone.
Van der Sluis et al., Phytochemistry, 24, 2601 (1985)

C. macrocephala Muss.-Puschk.
PYRAN:
2-acetoxy-6-(1,3,9-undecatriene-5,7-diynyl)-tetrahydropyran.
Bohlmann et al., Chem. Ber., 103, 2100 (1970)

C. melitensis L.
FLAVONOIDS: homoorientin, isovitexin, luteolin 7-O-glucoside,
6-methoxyluteolin 7-O-glucoside, 6-methoxyquercetin
7-O-glucoside, orientin, schaftoside, vicenin 2.
SESQUITERPENOIDS: melitensin, melitensin
8-hydroxyisobutyrate.
(F) Kamanzi et al., Plant Med. Phytother., 17, 45 (1983)
(T) Gonzalez et al., Phytochemistry, 14, 2039 (1975)

C. nigrescens.
FLAVONOID: demethoxycentaureidin.
Bohlmann et al., Tetrahedron Lett., 3239 (1967)

C. polypodiiflora.
ACID: 1-hydroxybenzoic acid.
Rasulov et al., Khim. Prir. Soedin., 524 (1983)

C. repens.
SESQUITERPENOID: repdiolide.
Stevens et al., Phytochemistry, 21, 1093 (1982)

C. salonitana.
AMINE: N-phenyl-β-naphthylamine.
SESQUITERPENOID: salonitenolide.
(Am) Evstratova et al., Khim. Prir. Soedin., 582 (1977)
(T) Suchy et al., Collect. Czech. Chem. Commun., 32, 2016 (1967)

C. scabiosa L.
SESQUITERPENOID: scabiolide.
Suchy et al., Collect. Czech. Chem. Commun., 27, 1905 (1962)

C. seridis.
SESQUITERPENOID: 15-acetylartemisitolin.
Gonzalez et al., Phytochemistry, 12, 2997 (1973)

C. solstitalis L.
FLAVONOIDS: homoorientin, orientin, quercetin, quercetin
3-O-glucoside, quercetin 7-O-glucoside, schaftoside.
SESQUITERPENOIDS: acetylsolstitalin, solstitalin.
(F) Kamanzi et al., Plant Med. Phytother., 10, 78 (1976)
(T) Zarghami, Heina, Chem. Ind. (London), 1556 (1969)

C. sventenii.
SESQUITERPENOID: aguerin B.
Gonzalez et al., Phytochemistry, 17, 955 (1978)

CENTELLA
C. asiatica.
SAPONINS: isothankuniside, thankuniside.
TRITERPENOIDS: brahmic acid, isobrahmic acid, isothankunic
acid, madasiatic acid, medicassic acid, thankunic acid.
(Sap) Dutta, Basu, Bull. Nat. Inst. Sci. India, 37, 178 (1968)
(T) Rastogi et al., J. Sci. Ind. Res., 19B, 252 (1960)

CENTRANTHUS (Valerianaceae)
C. ruber (L.) DC.
VALPOTRIATE: AHD-valtrate.
Handjieva et al., Phytochemistry, 17, 561 (1978)

CENTRARIA
C. nivalis.
TRITERPENOID: epi-friedelinol.
Dominguez et al., Phytochemistry, 12, 224 (1973)

C. sanguinea.
ACID: anziaic acid.
Asahina et al., Ber. Dtsch. Chem. Ges., 70, 1826 (1937)

CENTRATHERUM
C. punctatum.
SESQUITERPENOID: centratherin.
Ohno et al., Phytochemistry, 18, 681 (1979)

CEPHALIS
C. acuminata Karsten.
ISOQUINOLINE ALKALOID: emetine.
Battersby, Garratt, J. Chem. Soc., 3512 (1959)

C. ipecacuanha. (Brot.) Rich.
ISOQUINOLINE ALKALOIDS: cephaeline, emetine, ipecamine,
isoemetine.
Battersby, Garratt, J. Chem. Soc., 3512 (1959)

CEPHALARIA (Dipsacaceae)
C. coriaceae.
PHENOLIC: swertidin.
Zemtsova et al., Khim. Prir. Soedin., 705 (1977)

C. gigantea.
PHENOLIC: swertidin.
Zemtsova et al., Khim. Prir. Soedin., 705 (1977)

C. kotschyi.
FLAVONOIDS: cynaroside, hyperoside, quercimeritrin.
Movsumov, Aliev, Khim. Prir. Soedin., 804 (1975)

C. nachiczevanica.
FLAVONOIDS: cynaroside, hyperoside, quercimeritrin.
Movsumov, Aliev, Khim. Prir. Soedin., 804 (1975)

C. transylvanica.
TRITERPENOIDS: hederagenin, oleanic acid.
Tagiev et al., Khim. Prir. Soedin., 822 (1976)

C. uralensis.
GLYCOSIDE: swertiajaponin.
Zemtsova et al., Khim. Prir. Soedin., 705 (1977)

CEPHALOSPHAERA
C. usambarensis.
TRITERPENOID: 24-methylenecycloartenol.
Taylor, Ritchie, Aust. J. Chem., 14, 473 (1961)

CEPHALOSPORIUM

C. aphidicola.
DITERPENOID: aphidicolin.
Dalziel et al., J. Chem. Soc., Perkin I, 2841 (1973)

C. caerulens.
SESTERTERPENOIDS: cephalonic acid, ophiobolin D.
Irai et al., Tetrahedron Lett., 4111 (1967)
Nozoe et al., ibid., 4113 (1967)

CEPHALOTAXUS (Cephalotaxaceae)

C. drupacea.
HOMOISOQUINOLINE ALKALOID: cephalotaxine.
Paundler et al., J. Org. Chem., 28, 2194 (1963)

C. fortuni Hook.
HOMOISOQUIBOLINE ALKALOIDS: acetylcephalotaxine,
cephalotaxine.
Paundler et al., J. Org. Chem., 38, 2110 (1973)

C. hainanensis.
HOMOISOQUINOLINE ALKALOID: Homoisoquinoline alkaloid,
hainanensine.
Sun, Liang, Yao Hsueh Hsueh Pao, 15, 39 (1980)

C. harringtonia. (Forbes) K.Koch.
HOMOISOQUINOLINE ALKALOIDS: deoxyharringtonine, drupacine,
harringtonine.
Powell et al., J. Org. Chem., 39, 676 (1974)

C. harringtonia (Forbes) K.Koch var. drupacea.
HOMOISOQUINOLINE ALKALOID: harringtonine.
Powell et al., Tetrahedron Lett., 4081 (1969)

CERAMIUM (Rhodophyceae)

C. rubrum.
CAROTENOIDS: α-carotene, β-carotene, α-cryptoxanthin,
β-cryptoxanthin, fucoxanthin, lutein, zeaxanthin.
Bjoernland et al., Phytochemistry, 15, 291 (1976)

CERASTIUM

C. arvense.
FLAVONOIDS: ceravensin, ceravensin 7-O-glucoside, isovitexin
7,2″-di-O-glucoside.
Dubois et al., Phytochemistry, 21, 1141 (1982)

CERATOCYSTIS

C. minor.
ISOCOUMARINS: 6,8-dihydroxy-3-hydroxymethylisocoumarin,
6,8-dihydroxy-3-methylisocoumarin, scytalone.
McGraw et al., Phytochemistry, 16, 1315 (1977)

CERATOPETALUM

C. apetalum.
TRITERPENOID: friedelanone.
Drake, Jacobsen, J. Am. Chem. Soc., 57, 1570, 1854 (1935)

CERBERA

C. manghes.
ACIDS: cerberic acid, cerberinic acid.
MONOTERPENOID: cerbinal.
Abe et al., Chem. Pharm. Bull., 25 3422 (1977)

CERBERTA

C. dilatata.
SAPONIN: cerbertin.
Cable et al., Aust. J. Chem., 18, 1079 (1965)

C. floribunda.
SAPONIN: cerbertin.
Cable et al., Aust. J. Chem., 18, 1079 (1965)

CERCIDIPHYLLUM

C. japonica.
FLAVONOID: aromadendrol.
Gripenberg, Acta Chem. Scand., 6, 1152 (1952)

CERCOSPORA

C. taiwanensis.
PYRONES: cis-3S,4S-4-hydroxymellein, mellein, taiwapyrone.
Camarda et al., Phytochemistry, 15, 537 (1976)

CEREUS (Cactaceae)

C. grandiflorus.
FLAVONOIDS: cacticin, narcissin.
Rahman et al., J. Org. Chem., 27, 153 (1962)

C. pecten aboriginum.
ISOQUINOLINE ALKALOID: carnegine.
Heyl, Arch. Pharm., 239, 451 (1901)

C. validus.
AMINE ALKALOID: 3-nitro-4-hydroxyphenethylamine.
Neme et al., Phytochemistry, 16, 277 (1977)

CEROPEGIA

C. dichotoma.
TRITERPENOIDS: guimarenol, lup-18-en-3β-ol.
Gonzalez et al., An. Quim., 69, 921 (1973)

CERURA

C. vinula.
CAROTENOID: 3-hydroxy-β,ε-caroten-3′-one.
Kayser, J. Compar. Physiol., 104, 27 (1975)

CETRARIA

C. nivalis.
TRITERPENOID: 22α-hydroxystictan-3-one.
Wilkins, Phytochemistry, 16, 608 (1977)

C. richardsonii.
STEROID: ergosterol peroxide.
Sviridonov, Strigina, Khim. Prir. Soedin., 669 (1976)

CHAENOMELES (Rosaceae)

C. japonica.
MONOTERPENOIDS: betalbuside A, betalbuside B.
Tschesche et al., Chem. Ber., 110, 3111 (1977)

CHAENORHINUM (Scrophulariaceae)

C. minus (L.) Lange.
MACROCYCLIC ALKALOID: O-methylorantine.
Datwyler et al., Helv. Chim. Acta, 62, 2712 (1979)

C. organifolium.
MACROCYCLIC ALKALOID: chaenorhine.
Hesse, Schmid, Izv. Otd. Khim. Nauk, Bulg. Akad. Nauk, 5, 279 (1972)

C. rubrifolium.
MACROCYCLIC ALKALOID: (+)-chaenorhine.
Bosshardt et al., Pharm. Acta Helv., 51, 371 (1976)

C. villosum.
MACROCYCLIC ALKALOID: O-methylorantine.
Bosshardt *et al., Pharm. Acta Helv.,* **51,** 371 (1976)

CHAITURUS
C. marrubiastrum.
DITERPENOID: deoxymarrubialactone.
Popa, Salei, *Khim. Prir. Soedin.,* 399 (1976)

CHAMAECYPARIS (Cupressaceae)
C. formosensis Maxium.
41 norsesquiterpenoids and sesquiterpenoids.
Feng, *J. Chin. Chem. Soc.,* **33,** 265 (1986); *Chem. Abstr.,* 107 4282a

C. lawsoniana (A.Murr.) Parl.
SESQUITERPENOIDS: canadene, α-cadinol.
Shaw, *Can. J. Chem.,* **31,** 193 (1953)

C. nootkatensis (D.Don.) Spach.
DITERPENOID: 6-epi-manool oxide.
MONOTERPENOID: chaminic acid.
SESQUITERPENOIDS: α-alaskene, β-alaskene, cis-calamenene, trans-calamenene, calameone, calamone, (+)-calamusenone, chanootin, nootkatene, nootkatin, (+)-nootkatone.
Andersen *et al., Tetrahedron Lett.,* 905 (1972)

C. obtusa (Sieb. et Zucc.) Endl.
TRITERPENOIDS: chamaecydin, 6-hydroxychamaecydin.
Hirose *et al., J. Chem. Soc., Chem. Commun.,* 1535 (1983)

C. pisifera.
DITERPENOIDS: pisiferal, pisiferdiol, pisiferic acid, pisiferin, pisiferol.
Yatagi *et al., Phytochemistry,* **19,** 1149 (1980)

C. pisifera var. plumosa.
DITERPENOIDS: O-methylpisiferic acid, pisiferic acid.
Kobayashi *et al., Agr. Biol. Chem.* **50,** 240 (1986)

C. thyoides (L.) Britt.
SESQUITERPENOIDS: cuparene, cuparenic acid, hinokiic acid.
Matsui *et al., Phytochemistry,* **14,** 1037 (1975)

CHAMAEDAPHNE
C. calyculata.
FLAVONOID: guaijaverin.
Glyzin *et al., Khim. Prir. Soedin.,* 515 (1975)

CHAMAENERION (Onagraceae)
C. angustifolium (L.) Scop. (Epilobium angustifolium L.).
PHENOLICS: chamaeneric acid I, chamaeneric acid II.
TRITERPENOID: 2-hydroxyursolic acid.
(P) Sasov *et al., Khim. Prir. Soedin.,* 106 (1986)
(T) Glen *et al., J. Chem. Soc., C,* 510 (1967)

CHAMONIXIA (Basidiomycetae)
C. caespitosa.
PIGMENTS: chamonixin, gyrocyanin, gyrosporin.
Steglich *et al., Z. Naturforsch.,* **32C,** 46 (1977)

CHARTOLEPIS
C. intermedia.
SESQUITERPENOID: grossheimin.
Samek *et al., Tetrahedron Lett.,* 4775 (1971)

CHASMANTHERA (Menispermaceae)
C. dependens.
MORPHINANDIENONE ALKALOID: pallidine.
PAVINE ALKALOID: bisnorargeminine.
PROTOBERBERINE ALKALOIDS: coreximine, govanine.
DITERPENOID: 8-hydroxycolumbin.
(A) Ohiri *et al., Planta Med.,* **49,** 17 (1983)
(T) Oguakwa *et al., ibid.,* 198 (1986)

CHEILANTHES (Cruciferae)
C. albomarginata.
FLAVONOIDS: genkwanin, kumatukenin, rhamnocitrin.
Wollenweber, *Phytochemistry,* **15,** 2013 (1976)

C. allionii.
SAPONINS: digitoxigenin glucodigulomethyloside, digitoxigenin gulomethyloside.
STEROID: alliotoxigenin.
Makarevich *et al., Khim. Prir. Soedin.,* 754 (1975)

C. bullosa.
FLAVONOIDS: acacigenin, apigenin, apigenin dimethyl ether.
Wollenweber. *Phytochemistry,* **15,** 2013 (1976)

C. farinosa.
STEROIDS: cheilanthone A, cheilanthone B.
SESTERTERPENOIDS: cheilanthatriol, cheilarinosin.
Iyer *et al., Indian J. Chem.,* **10,** 482 (1972)

C. fragrans.
FLAVONOID: quercetin 3-O-glucosylgalactosylrhamnoside.
Imperato *et al., Chem. Ind.,* 709 (1985)

C. grisea.
FLAVONOIDS: apigenin 7,4'-dimethyl ether, apigenin 7-methyl ether, kaempferol, kaempferol 7,4'-dimethyl ether, kaempferol 3,7,4'-trimethyl ether.
Wollenweber, *Phytochemistry,* **15,** 2013 (1976)

C. marantae.
TRITERPENOID: 6β,22-dihydroxyhopane.
Gonzalez *et al., Phytochemistry,* **15,** 1996 (1976)

C. rufa.
FLAVONOIDS: genkwanin, kumatukenin, rhamnocitrin.
Wollenweber, *Phytochemistry,* **15,** 2013 (1976)

C. scoparius Brouss.
STEROIDS: 16β-acetoxystrophanthidin, arguayoside, corchoroside A, 16-dehydrostrophanthidin, helveticoside, strophanthidin, taucidoside.
Gonzalez *et al., Farmaco Nueva, 41, 264, 267,* 273 (1976)

C. tenuifolia.
STEROIDS: cheilanthone A, cheilanthone B.
Faux *et al., J. Chem. Soc., Chem. Commun.,* 243 (1970)

CHELIDONIUM (Papaveraceae)
C. majus L.
DIMERIC ALKALOID: chelidimerine.
OXOPROTOBERBERINE ALKALOIDS: α-allocryptopine, β-allocryptopine.
OXOQUINOLIZIDINE ALKALOID: protopine.
PHENANTHROISOQUINOLINE ALKALOIDS: chelerythrine, chelidonine, oxychelidonine, sanguinarine.
PROTOBERBERINE ALKALOID: berberine.
SPARTEINE ALKALOID: (-)-sparteine.
UNCHARACTERIZED ALKALOIDS: chelidamine, chelirubine, corysamine.
Platonova *et al., J. Gen. Chem. USSR,* **26,** 188 (1965)
Kun *et al., J. Pharm. Sci.,* **56,** 372 (1969)

CHENOPODIUM (Chenopodiaceae)

C. ambrosioides L.
DITERPENOID: aritasone.

Nakajima *et al.*, *J. Pharm. Soc., Japan*, **82**, 1278 (1962)

C. atrovirens.
FLAVONOIDS: isorhamnetin 3-O-glycoside, quercetin 3-O-glycoside.

Crawford, *Brittonia*, **27**, 279 (1975); *Chem. Abstr.*, 84 56525b

C. botrys L.
SESQUITERPENOIDS: 11-acetoxy-cis-guai-10(14)-en-4α-ol, 4α-acetoxyselinane-6α,11-diol, 6α-acetoxyselin-4(15)-en-11-ol, 6α-acetoxyselin-3-en-11-ol-2-one, chenopodic acid, α-chenopodiol, α-chenopodiol 6-acetate, β-chenopodiol, cis-guai-10-en-4α,11-diol, selinane-3α,4α,6α,11-tetraol, selinane-3β,,6α,11-tetraol, selin-3-en-6α,11β-dihydroxy-2-one, selin-4(15)-en-3β,11-diol, selin-4-en-3α,6α,11-triol, selin-4(11)-en-3α,6α,11-triol, selin-4(15)-en-3β,6α,11-triol.

de Pascual Teresa *et al.*, *Tetrahedron*, **36**, 371 (1980)

C. dessicatum var.
DESSICATUM, FLAVONOIDS: isorhamnetin 3-O-glycoside, quercetin 3-O-glycoside.

Crawford, *Brittonia*, **27**, 279 (1975); *Chem. Abstr.*, 84 56525b

C. dessicatum var. leptophylloides.
FLAVONOIDS: isorhamnetin 3-O-glycoside, quercetin 3-O-glycoside.

Crawford, *Brittonia*, **27**, 279 (1975); *Chem. Abstr.*, 84 56525b

C. hians.
FLAVONOIDS: isorhamnetin 3-O-glycoside, quercetin 3-O-glycoside.

Crawford, *Brittonia*, **27**, 279 (1975); *Chem. Abstr.*, 84 56525b

C. leptophyllum.
FLAVONOIDS: isorhamnetin 3-O-glycoside, quercetin 3-O-glycoside.

Crawford, *Brittonia*, **27**, 279 (1975); *Chem. Abstr.*, 84 56525b

CHICORUM

C. endiva.
ACID: chicoric acid.

Scarpati *et al.*, *Tetrahedron*, **4**, 43 (1958)

C. intybus.
ACID: chicoric acid.

Scarpati *et al.*, *Tetrahedron*, **4**, 43 (1958)

CHIDLOWIA

C. sanguinea.
PURINE ALKALOID: triacanthine.

Monseur, Adriaens, *J. Am. Chem. Soc.*, **84**, 2148 (1962)

CHILOSCYPHUS (Hepaticae)

C. polyanthus.
SESQUITERPENOIDS: chiloscyphone, ent-5β-hydroxydiplophyllin, ent-3-oxodiplophyllin, α-selinene.

Asakawa *et al.*, *Phytochemistry*, **18**, 1007 (1979)

CHIMONANTHUS (Calycanthaceae)

C. fragrans Lindl. (Meratia praecox Rehd. et Wils.).
DIMERIC ALKALOID: (-)-chimonanthine.
STEROID: isoallochionographalone.

(A) Hodgson, Robinson, *Proc. Chem. Soc.*, 465 (1961)
(S) Sauer *et al.*, *Justus Liebigs Ann. Chem.*, **742**, 152 (1970)

CHIONOGRAPHIS

C. japonica Maxim.
STEROIDS: allochionographolone, chionogralactone, chionographolone, chionograsterol A, chionograsterol B, chionograsterone, kryptogenin.

Takeda *et al.*, *J. Pharm. Soc., Japan*, **88**, 104 (1968)
Sauer *et al.*, *Justus Liebigs Ann. Chem.*, **742**, 152 (1970)

CHISOCHETON

C. paniculatus.
TETRANORTRITERPENOIDS: 6α-acetoxyazadirone, 6α-acetoxy-16-oxoazadirone.
TRITERPENOIDS: chisochetons A, B and C.

Saika *et al.*, *Indian J. Chem.*, **16B**, 1042 (1978)
Connolly *et al.*, *J. Chem. Soc., Perkin I*, 2959 (1979)

CHLORANTHUS

C. glaber.
SESQUITERPENOIDS: chlorantholactone A, chlorantholactone B.

Uchida *et al.*, *Heterocycles*, **9**, 139 (1978)

C. japonicus Sieb.
SESQUITERPENOIDS: chlorantholactone, shizukacoradienol, shizukanolide.

Kawabata *et al.*, *Agr. Biol. Chem.*, **48**, 713 (1984)

CHLORODESMIS (Chlorophyceae)

C. fastigiata.
TERPENOIDS: chlorodesmin, dihydrochlorodesmin.

Wells, Barrow, *Experientia*, **35**, 1544 (1979)

CHLOROGALUM

C. pomeridianum.
SAPONIN: amolonin.
STEROIDS: chlorogenin, tigogenin.

Tschesche *et al.*, *Justus Liebigs Ann. Chem.*, **699**, 212 (1966)

CHLOROPHORA (Moraceae)

C. tinctoria Gand.
TERPENOID: 6-prenylpinocembrin.
XANTHONE: 1,3,6,7-tetrahydroxyxanthone.

Gottlieb *et al.*, *Phytochemistry*, **14**, 1674 (1975)

CHLOROXYLON

C. swietenia (L.) DC.
FURANOQUINOLINE ALKALOIDS: chloroxylonine, skimmianine.
QUINOLINE ALKALOIDS: swietenidine A, swietenidine B.
COUMARINS: alloxanthoxyletin, demethylluvangetin, 8-prenylnodokenetin, swietenocoumarins A, B, C, D, E and F, swietenol, swietenone.
PHENOLIC: 2,4-dihydroxy-5-prenylcinnamic acid.

(A) Auld, *J. Chem. Soc.*, **95**, 964 (1909)
(C) Bhide *et al.*, *Indian J. Chem.*, **15B**, 440 (1977)
(P) Majumder *et al.*, *ibid.*, **15B**, 200 (1977)

CHOANEPHRA

C. trispora.
SESQUITERPENOIDS: anhydrotrisporone, trisporone.

Cainelli *et al.*, *Chem. Ind. (London)*, **49**, 628, 740 (1967)

CHOISYA (Rutaceae)

C. arizonica.
COUMARIN: ramosin.
Dreyer et al., Phytochemistry, 11, 705 (1972)

C. mollis.
COUMARIN: ramosin.
Dreyer et al., Phytochemistry, 11, 705 (1972)

C. ternata H.B.K.
FURANOQUINOLINE ALKALOIDS: choisyine, evoxine, skimmianine.
FLAVONOIDS: astragalin, kaempferol 3-O-rhamnoside, quercetol 3-O-rutinoside.
(A) Frolova et al., Med. Prom., 35 (1958)
(F) Tea et al., Plant Med. Phytother., 9, 187 (1975)

CHONDRIA (Rhodophyceae)

C. californica.
CYCLIC POLYSULPHIDES: 4-dioxo-1,2,4,6-tetrathiepane, 1-oxo-1,2,4-trithiolane, 4-oxo-1,2,4-trithiolane, 1,2,3,5,6-pentathiepane, 1,2,4,6-tetrathiepane, 1,2,4-trithiolane.
Wratten, Faulkner, J. Org. Chem., 41, 2465 (1976)

C. oppositiclada.
SESQUITERPENOIDS: cycloeudesmol, isocycloeudesmol.
Suzuki et al., Chem. Lett., 1267 (1980)

CHONDROCOCCUS (Rhodophyceae)

C. hornemannii.
HALOGENATED MONOTERPENOIDS: chondrocolactone, chondrocole A, chondrocole B.
Wollard et al., Tetrahedron Lett., 2367 (1978)

CHONDRODENDRON

C. limacifolium.
ALKALOIDS: two uncharacterized bases.
Barltrop, Jeffreys, J. Chem. Soc., 159 (1954)

C. platyphyllum (St.Hill.) Miers.
BISBENZYLISOQUINOLINE ALKALOIDS: chondrofoline, isochondrodendrine.
UNCHARACTERIZED ALKALOIDS: bebeerine B, α-bebeerine.
King, J. Chem. Soc., 737 (1940)

C. tomentosum.
BISBENZYLISOQUINOLINE ALKALOIDS: chondrocurarine chloride, (+)-chondrocurine, (-)-curine, (+)-tomentocurine.
King, J. Chem. Soc., 1472 (1937)
Dutcher, J. Am. Chem. Soc., 74, 2221 (1952)

CHONEMORPHA

C. macrophylla.
STEROIDAL ALKALOIDS: chonemorphine, funtumufrine C, japindine.
Banerji et al., Indian J. Chem., 11, 1056 (1973)

C. penangensis.
STEROIDAL ALKALOID: chonemorphine.
Chatterjee, Das, Chem. Ind. (London), 1445 (1959)

CHORISIA

C. insignis.
STEROIDS: brassicasterol, campesterol, cholesterol, 24-ethylcholesta-1,3,6-triene, β-sitosterol, stigmasterol, stigmasta-3,5-dien-7-one, stigmasta-4,6-dien-3-one, stigmast-4-en-3,6-dione, stigmast-4-en-3-one.
Becerra et al., Rev. Latinoam. Quim., 14, 92 (1983)

C. speciosa.
STEROIDS: brassicasterol, campesterol, cholesterol, 24-ethylcholesta-1,3,6-triene, stigmasta-3,5-dien-7-one, stigmasta-4,6-dien-3-one, stigmast-4-en-3,6-dione, stigmast-4-en-3-one, stigmasterol.
Becerra et al., Rev. Latinoam. Quim., 14, 92 (1983)

CHRESTA

C. sphaerocephala.
SESQUITERPENOID: chrestanolide.
Bohlmann et al., Phytochemistry, 20, 518 (1981)

CHROMOLAENA

C. collina.
DITERPENOID: 7β-acetoxy-trans-communic acid.
Bohlmann, Fiedler, J. Indian Chem. Soc., 55, 1161 (1978)

C. glaberrina.
SESQUITERPENOID: chromolaenin.
Bohlmann, Lothar, Chem. Ber., 111, 408 (1978)

C. laevigata.
SESQUITERPENOID: chromolaenin.
Bohlmann, Lothar, Chem. Ber., 111, 408 (1978)

C. meridensis.
FLAVONOIDS: eriodactyol 7-methyl ether, eupatolitin, rhamnetin, velutin.
Amaro et al., Rev. Latinoam. Quim., 14, 86 (1983)

C. pseudoinsignis.
SESQUITERPENOID: 7,12-dihydroxy-4-amorphen-3-one.
Bohlmann et al., Phytochemistry, 21, 371 (1982)

CHRYSANTELLUM

C. procumbens Rich.
SAPONINS: chrysantellin A, chrysantellin B.
Becchi et al., Eur. J. Biochem., 108, 27 (1980)

CHRYSANTHEMUM (Compositae)

C. anethifolium Brous.
STEROID: β-sitosterol.
TRITERPENOID: α-amyrin.
Sayed et al., Bull. Fac. Pharm. Cairo Univ., 12, 101 (1976); Chem. Abstr., 85 43709k

C. boreale.
ESTER: buranoic acid 3-methyl-2-(2,4-dihexadiynylidene)-4-hydroxy-1,6-dioxaspiro-[4,5]-dec-3-yl ester.
Matsuo et al., Tetrahedron Lett., 1885 (1974)

C. cineriaefolium Bocc.
UNCHARACTERIZED ALKALOID: chrysanthemine.
NORTRITERPENOID: pyrethrol.
SESQUITERPENOID: pyrethrosin.
UNCHARACTERIZED TERPENOIDS: jasmolin I, jasmolin II.
(A) Yoshimura et al., Z. Physiol. Chem., 77, 290 (1912)
(T) Castille, Louvain, An. R. Acad. Farm., 32, 721 (1966)
(T) Godin et al., J. Chem. Soc., C, 332 (1966)

C. coronarium.
CAROTENOID: mutatochrome.
SESQUITERPENOID: dihydrocumambrin A.
El-Masry et al., Phytochemistry, 23, 2983 (1984)

C. floculosum.
SESQUITERPENOID: α-bisabolone.
Bohlmann, Rao, Tetrahedron Lett., 1295 (1972)

C. indicum.
SESQUITERPENOID: yehujalactone.
Chien, Chen, Yao Hsueh Hsueh Pao, 10, 129 (1963)

C. japonense.
MONOTERPENOIDS: nojigiku acetate, nojigiku alcohol.
Matsuo et al., Tetrahedron Lett., 4219 (1974)

C. leucanthemum L.
ACETYLENIC:
4-acetoxy-8,9-epoxy-7-(2-hexadiynylidene)-1,6-dioxaspiro-[4,4]-non-2-ene.
FLAVONOIDS: acacetin, scopoletin.
PHENOLIC: 2,6-dimethoxy-p-benzoquinone.
(Ac) Bohlmann et al., Justus Liebigs Ann. Chem., 668, 51 (1963)
(F, P) Waddell et al., J. Tenn. Acad. Sci., 56, 53 (1983); Chem. Abstr., 100 48584d

C. morifolium Ramat.
HALOGENATED SESQUITERPENOID: chlorochrymorin.
Osawa et al., Tetrahedron Lett., 5135 (1973)

C. morifolium Ramat. var. sinense Makino f. esculentum Makino.
ALIPHATICS: heptacosane, pentocosane.
FLAVONOID: apigenin.
Takahashi et al., Chem. Abstr., 97 159570k

C. nauseatum.
DITERPENOID: 16-succinyloxygrindelic acid.
Rose, Phytochemistry, 19, 2689 (1980)

C. ornatum var. spontaneum.
SESQUITERPENOID: angeloylcumambrin B.
Haruno et al., Phytochemistry, 20, 2583 (1981)

C. parthenium (L.) Bernh.
SESQUITERPENOIDS: chrysartenin A, chrysartenin B.
Romo et al., Phytochemistry, 9, 1615 (1970)

C. poteriifolium.
SESQUITERPENOID: cis, cis-2α-hydroxycostunolide.
Bohlmann, Ehlers, Phytochemistry, 16, 137 (1977)

C. sinense.
MONOTERPENOID: chrysanthenone.
Penfold et al., J. Chem. Soc., 1496 (1939)

C. vulgare (L.) Bernh. (Tanacetum vulgare L.).
MONOTERPENOID: (+)-filifolide.
SESQUITERPENOIDS: (+)-vulgarone A, (+)-vulgarone B.
Uchio et al., Int. Congr. Ess. Oils, 20, 1049 (1967)

CHRYSOSPLENIUM (Saxafragaceae)
C. alternifolium.
FLAVONOIDS: 3,7-di-O-methylquercetagenin, penduletin, 3,6,7,3'-O-tetramethylquercetagenin, 3,6,7-O-trimethylquercetagenin.
Jay et al., Phytochemistry, 15, 517 (1976)

C. oppositifolium.
FLAVONOIDS: 3,6,7,3'-O-tetramethylquercetagenin, 3,7,3'-O-trimethylquercetagenin, 3,6,7-O-trimethylquercetagenin.
Jay et al., Phytochemistry, 15, 517 (1976)

C. tetrandum.
FLAVONOID: penduletin.
Jefferies et al., Aust. J. Chem., 20, 1049 (1967)

CHRYSOTHAMNUS (Compositae)
C. paniculatus.
DITERPENOIDS: chrysolic acid, chrysothane, 12 norsesquiterpenoids and diterpenoids.
Timmermann et al., Phytochemistry, 22, 523 (1983)

C. nauseosus.
DITERPENOID: 1-hydroxygrindelic acid.
Rose et al., Phytochemistry, 19, 2689 (1980)

C. viscidiflorus.
FLAVONOIDS: quercetagenin 3,5,6,3'-tetramethyl ether, 5,7,4'-trihydroxy-3,6,8,3'-tetramethoxyflavone.
Urbatsch et al., Phytochemistry, 14, 2279 (1975)

CHUKRASIA
C. tabularis A Juss.
ACID: tannic acid.
FLAVONOIDS: quercetin, quercetin 3-galactoside.
LIMONOIDS: 6-acetoxychukrasin, 12α-acetoxyphragmalin 3,30-diisobutyrate, 12α-acetoxyphragmalin 30-propionate, chukrasins A, B, C, D and E, phragmalin 3,30-diisobutyrate, phragmalin 3-isobutyrate-30-propionate, tabularin.
(Ac, F) Purushothaman et al., J. Res. Indian Med., 10, 25 (1975); Chem. Abstr., 85 2561n
(T) Ragettli, Tam, Helv. Chim. Acta, 61, 1814 (1978)

CHYSIS
C. bractescens.
PYRROLIZIDINE ALKALOIDS: chysin A, chysin B, (+)-1-(methoxycarbonyl)-pyrrolizidine.
Luning, Trankner, Acta Chem. Scand., 22, 2324 (1968)

CICAS (CYCAS)
C. circinalis.
ALKALOID: cycasin.
Riggs et al., Chem. Ind., 926 (1956)

C. revoluta.
ALKALOID: cycasin.
Riggs et al., Chem. Ind. (London), 926 (1956)

CIMICIFUGA (Ranunculaceae)
C. acerina.
SAPONIN: 25-O-methylcimigenoside.
TRITERPENOIDS: acerinol, acerionol, isodahurinol.
Kusano et al., J. Pharm. Soc., Japan, 96, 321 (1976)

C. acerina var. peltata.
TRITERPENOIDS: 24-O-acetylhydroshengmanol xyloside, acetylshengmanol, acetylshengmanol xyloside, shengmanol xyloside.
Kimura et al., J. Pharm. Soc., Japan, 103, 293 (1983)

C. dahurica. Two uncharacterized alkaloids.
TRITERPENOID: dahurinol.
(A) Hata et al., Chem. Pharm. Bull., 26 2279 (1978)
(T) Sakurai et al., J. Pharm. Soc., Japan, 92, 724 (1972)

C. japonica.
SAPONIN: 25-O-methylcimigenoside.
TRITERPENOIDS: acerinol, acerionol, 24-O-acetyldihydroshangmanol xyloside.
Sakurai et al., Chem. Pharm. Bull., 29, 955 (1981)

C. simplex.
TRITERPENOIDS: acerinol, acerionol, cimicifugenol, cimicifugin, cimicifugoside, cimigol, isodahurinol.
Takemoto et al., J. Pharm. Soc., Japan, 90, 68 (1970)
Kusano et al., Chem. Pharm. Bull., 25, 3182 (1977)

CINCHONA
C. amygdalifolia.
QUINOLINE ALKALOIDS: dihydroquinidine, quinidine, epi-quinidine, quinine, epi-quinine.
Forst, Bohringer, Ber. Dtsch. Chem. Ges., 15, 520, 1656 (1882)

C. calisaya Ledger.
QUINOLINE ALKALOIDS: dihydroquinidine, homocinchonidine, quinidine, epi-quinidine, quinine, epi-quinine.
Forst, Bohringer, Ber. Dtsch. Chem. Ges., 15, 520, 1656 (1882)

C. calisaya Ledger var. javanica.
UNCHARACTERIZED ALKALOID: javanine.
QUINOLINE ALKALOIDS: dihydroquinidine, quinidine,
epi-quinidine, quinine, epi-quinine.
Hesse, *Ber. Dtsch. Chem. Ges.,* **10**, 2162 (1877)

C. ledgeriana.
QUINOLINE ALKALOIDS: alkaloid CL-I, alkaloid CL-II, alkaloid
Cl-III, alkaloid Cl-IV, alkaloid CL-V, alkaloid CL-VI,
cinchophyllamine, conquinamine, isocinchophyllamine,
quinamine, quinine.
Potier *et al., Bull. Soc. Chim. Fr.,* **7**, 2309 (1966)
Zeches *et al., Phytochemistry,* **19**, 2451 (1980)

C. micrantha.
QUINOLINE ALKALOIDS: cinchonine, quinidine, epi-quinidine,
quinine, epi-quinine.
Martinotti, *Ind. Chim. Rome,* **6**, 395 (1931)

C. pelletierana Wedd.
QUINOLINE ALKALOID: quinine.
UNCHARACTERIZED ALKALOIDS: conchairamidine,
conchairamine, cuscamidine, cuscamine.
Hesse, *Justus Liebigs Ann. Chem.,* **200**, 304 (1880)

C. pitayensis.
QUINOLINE ALKALOIDS: dihydroquinidine, quinidine,
epi-quinidine, quinine, epi-quinine.
Forst, Bohringer, *Ber. Dtsch. Chem. Ges.,* **15**, 520, 1656 (1882)

C. rosulenta.
QUINOLINE ALKALOID: dicinchonine.
Hesse, *Justus Liebigs Ann. Chem.,* **227**, 154 (1885)

C. succirubra Pav.
QUINOLINE ALKALOIDS: cinchamidine, conquinamine,
dicinchonine, diconquinine, homocinchonidine, quinamine,
epi-quinidine, quinine.
UNCHARACTERIZED ALKALOID: succirubine.
Khuda *et al., Sci. Res., (Dacca),* **2**, 1 (1965)

C. tucujensis.
QUINOLINE ALKALOID: cinchonidine.
Goutarel *et al., Helv. Chim. Acta,* **33**, 150 (1950)

CINERARIA

C. albicans.
SESQUITERPENOID: 1-oxocinalbicol.
Abraham, *Phytochemistry,* **17**, 1629 (1978)

C. decipiens.
SESQUITERPENOID: 1-oxocinalbicol.
Abraham, *Phytochemistry,* **17**, 1629 (1978)

C. fruticulorum.
MONOTERPENOID: 2-pinene-4,10-diol.
SESQUITERPENOID: 5,7,11-cineratrien-9-one.
TRITERPENOID: biscineradienone.
Bohlmann *et al., Phytochemistry,* **21**, 2531 (1982)

C. lydrata.
SESQUITERPENOIDS: cinalyatol acetate, 1,10-dihydrocinalyatol
acetate.
Abraham, *Phytochemistry,* **17**, 1629 (1978)

CINNAMOMUM (Lauraceae)

C. camphora Sieb.
LIGNANS: cinnamonol, dimethylmatairesinol,
dimethylsecoisolariciresinol, hinokinin, kusunokinin.
MONOTERPENOIDS: (+)-camphene, (-)-camphene, (+)-camphor,
3,7-dimethylocta-1,5-dien-3,7-diol,
3,7-dimethylocta-1,7-dien-3,6-diol.

SESQUITERPENOIDS: camphorenol, epi-camphorenol,
camphorenone.
(L) Takaoka *et al., Bull. Chem. Soc., Japan,* **50**, 2821 (1977)
(T) Hikino *et al., Tetrahedron Lett.,* 5069 (1967)

C. cassia Blume.
PHENOLICS: (-)-epi-catechin, procyanidins B1, B2, B5, B7 and
C1.
TANNINS: cinnamtannins A1, A2, A3 and A4.
DITERPENOIDS: cinncassiols A, B, C, C1, C3, D1, D2, D3 and
D4, cinncassiol D1 glucoside, cinncassiol D2 glucoside.
(P, Tan) Morimoto *et al., Chem. Pharm. Bull.,* **34**, 633 (1986)
(T) Nohara *et al., ibid.,* **29**, 2451 (1981)

C. japonicum.
MONOTERPENOIDS: cis-yabunikkeol, trans-yabunikkeol.
Fujita *et al., Bull. Chem. Soc., Japan,* **43**, 1599 (1970)

C. kanahirai.
CYCLOPENTANE: 1,2,3,3-tetramethylcyclopentane.
Saito *et al., J. Chem. Soc., Japan,* **64**, 1024 (1943)

C. micranthum.
ALIPHATICS: octatriacontane, tetracosanol, triacontanoic acid.
STEROIDS: cholesterol, β-sitosterol.
Lin *et al., Chem. Abstr.,* 85 189222t

C. obtusifolium Nees.
PHENOLICS: (-)-epi-catechin, procyanidins B2, B5, B7 and C1.
TANNINS: cinnamtannins A2, A3 and A4.
Morimoto *et al., Chem. Pharm. Bull.,* **34**, 633 (1986)

C. sieboldii Meisner.
PHENOLICS: (+)-catechin, (-)-epi-catechin, proanthocyanidins
B1, B2 and B5.
Morimoto *et al., Chem. Pharm. Bull.,* **33**, 4338 (1985)

C. zeylanicum.
DITERPENOIDS: cinnzeylanol, cinnzeylanol 1-acetate.
Isogai *et al., Agr. Biol. Chem.,* **40**, 2305 (1976)

CINNAMOSMA

C. fragrans.
SESQUITERPENOIDS: bemadienolide, bemarivolide.
Canonica *et al., Tetrahedron,* **25**, 3903 (1969)

CIRSIUM

C. arisanense Kitamura.
FLAVONOID: linarin.
Morita *et al., Chem. Abstr.,* 88 34562s

C. arvense.
FLAVONOID: linarin.
Rendyuk *et al., Khim. Prir. Soedin.,* 282 (1977)

C. ferum Kitamura.
ACID: fumaric acid.
FLAVONOID: pectolinarin.
Morita *et al., Chem. Abstr.,* 88 34562s

C. kawakamii.
FLAVONOID: luteolin 7-glucoside.
Lin, *J. Chin. Chem. Soc.,* **22**, 275 (1975); *Chem. Abstr.,* 84 28055e

C. setosum Willd.
FLAVONOID: linarin.
Rendyuk *et al., Acta Pharm. Jugosl.,* **27**, 135 (1977); *Chem. Abstr.,* 88
117796x

C. wallichii.
ACID: fumaric acid.
Lin, *J. Chin. Chem. Soc.,* **22**, 275 (1975); *Chem. Abstr.,* 84 28055e

CISSAMPELOS

C. insularis Makino.
BISBENZYLISOQUINOLINE ALKALOID: insularine.
Kondo, Yano, *J. Pharm. Soc., Japan, No.* **548,** 107 (1927)

C. pareira L.
BISBENZYLISOQUINOLINE ALKALOIDS: cissampareine, (++)-curine-4″-methyl ether, (±)-hayatidine, hayatine, tetrandrine, tetrandrine N-2′-oxide.
PROTOBERBERINE ALKALOID: cyclanoline.
Kupchan *et al., J. Pharm. Sci.,* **54,** 580 (1965)
Dahmen *et al., Arch. Pharm.,* **310,** 95 (1977)

CISSUS

C. pallida.
DIMER: pallidol.
Khan *et al., Phytochemistry,* **25,** 1945 (1986)

CISTANCHE (Orobanchaceae)

C. salsa.
IRIDOIDS: 6-deoxycatalpol, mussaenosidic acid.
Kobayashi *et al., Chem. Pharm. Bull.,* **33,** 3645 (1985)

CISTUS (Cistaceae)

C. bourgeanus.
TRITERPENOID: casasequic acid.
de Pascual Teresa *et al., An. Quim.,* **75,** 131 (1979)

C. clusii.
FLAVONOIDS: apigenin 7,4′-dimethyl ether, luteolin 7,3′-dimethyl ether, luteolin 7,3′,4′-trimethyl ether, quercetin 3,7,3′,4′-tetramethyl ether, quercetin 7,3′,4′-trimethyl ether.
MONOTERPENOIDS: borneol, bornyl acetate, camphene, limonene, α-pinene, β-pinene.
SESQUITERPENOIDS: caryophyllene oxide, ent-13-epi-manoyl oxide.
TRITERPENOID: (-)-betuligenol.
de Pascual Teresa *et al., An. Quim.,* **73,** 40 (1983)

C. hirsutus.
ALIPHATICS: hentriacontane, nonacosane.
Kaehlen *et al., Z. Pflanzenphysiol.,* **77,** 99 (1975); *Chem. Abstr.,* **84** 71470y

C. ladaniferus.
ALIPHATICS: hentriacontane, nonacosane.
DITERPENOIDS: 8-abieten-18-oic acid, 6β-acetoxy-7-oxo-8-labden-15-oic acid, ambreinolide, dihydropalustric acid, labdane-8α,15-diol, labdanolic acid, labdan-8α,15,19α-triol, 12-norambreinolide, 8-epi-norambreinolide.
(Al) Kaehlen *et al., Z. Pflanzenphysiol.,* **77,** 99 (1975); *Chem. Abstr.,* **84** 71470y
(T) de Pascual Teresa *et al., An. Quim.,* **75,** 335 (1979)

C. laurifoluus L.
ALIPHATICS: hentriacontane, nonacosane.
DINORDITERPENOIDS: salmantic acid, salmantic acid methyl ester, salmantidiol.
DITERPENOID: 6β-acetoxy-8,13S-epoxy-15-methoxy-ent-labdane.
(Al) Kaehlen *et al., Z. Pflanzenphysiol.,* **77,** 99 (1975); *Chem. Abstr.,* **84** 71470y
(T) de Pascual Teresa *et al., Phytochemistry,* **22,** 2783 (1983)

C. monspeliensis L.
ALIPHATICS: hentriacontane, nonacosane.
DITERPENOIDS: cistadoic acid, cistodiol.
(Al) Kaehlen *et al., Z. Pflanzenphysiol.,* **77,** 99 (1975); *Chem. Abstr.,* **84** 71470y
(T) Berti *et al., Tetrahedron Lett.,* 1401 (1970)

C. palhinhae.
DITERPENOIDS: three labdanes and one clerodane.
de Pascual Teresa *et al., Phytochemistry,* **22,** 2805 (1983)

C. populifolius L.
DITERPENOIDS: ent-cleroden-2-acetoxy-15-oic acid, isodehydropopulifolic acid, oxopopulifolic acid, populifolic acid.
de Pascual Teresa *et al., An. Quim.,* **74,** 476 (1978)

C. rheifolia.
QUINOLIZIDINE ALKALOID: kayawongine.
Kelly *et al., J. Nat. Prod.,* **46,** 353 (1983)

C. salviifolius.
ALIPHATICS: hentriacontane, nonacosane.
Kaehlen *et al., Z. Pflanzenphysiol.,* **77,** 99 (1975); *Chem. Abstr.,* **84** 71470y

C. symphytifolius Lam.
DITERPENOIDS: cistadienic acid, cistenolic acid.
Calabuig *et al., Phytochemistry,* **20,** 2255 (1981)

CITHAREXYLUM

C. subserratum.
FLAVONOID: 5,4′-dihydroxy-7,3′-dimethoxyflavone 6-glucoside.
SAPONINS: durantoside I, durantoside II.
(F) Mathuram *et al., Phytochemistry,* **15,** 838 (1976)
(Sap) Ganapathy, Rao, *Fitoterapia,* **54,** 13 (1983)

CITRULLUS (Cucurbitaceae)

C. battich.
AMINO ACIDS: citrulline, α,β-diaminopropionic acid.
INDOLE: indolacetic acid.
TERPENOID: p-hydroxymethylanisole.
Inatomi, *Meiji Daigaku Nogakubu Kenkyu Hokoku,* **34,** 7 (1975); *Chem. Abstr.,* **84** 71479h

C. colocynthis (L.) Schrad.
TRITERPENOIDS: citrullone, colocynthin.
Yankov, Khusain, *Dokl. Bulg. Akad. Nauk,* **28,** 1641 (1975)

C. ecirrhosus.
TRITERPENOIDS: cucurbitacin J, cucurbitacin K, cucurbitacin L.
Enslin *et al., J. Sci. Food Agr.,* **8,** 673 (1957)

C. lanatus var. sitroides.
STEROID: stigmast-7-en-3β-ol β-D-glucopyranoside.
Seifert *et al., Pharmazie,* **32,** 125 (1977)

CITRUS (Rutaceae/Zygophyllaceae)

C. aurantium.
COUMARINS: aurapten, (+)-auraptenal.
GLYCOSIDE: rhoifolin.
NORTRITERPENOID: isolimonic acid.
(C) Lundin *et al., Tetrahedron,* **21,** 89 (1965)
(Gl) Coussio *et al., Experientia,* **20,** 562 (1964)
(T) Bennett *et al., Phytochemistry,* **19,** 2417 (1980)

C. chinensis.
SESQUITERPENOIDS: α-sinensal, β-sinensal.
Stevens *et al., J. Org. Chem.,* **30,** 1690 (1965)

C. grandis.
ACRIDONE ALKALOIDS: grandisimine, grandisine.
QUATERNARY ALKALOID: stachydrine.
Wu *et al., Phytochemistry,* **22,** 1493 (1983)

C. hassaku Hort.
PHENYLPROPANOIDS: citrusins A, B and C, coniferin, syringin.
Sawabe *et al., Nippon Nogei Kagaku Kaishi,* **60,** 593 (1986); *Chem. Abstr.,* 105 222728u

C. ichangensis.
BITTER PRINCIPLE: ichangin.
Dreyer, *J. Org. Chem.*, **31**, 2279 (1966)

C. jambhiri Lush.
COUMARINS: hesperidin, neohesperidin, tangeretin.
Chaliha et al., *Tetrahedron*, **21**, 1441 (1965)

C. junos.
MONOTERPENOIDS: camphene, citral, citronellol, limonene, linalool, nerol, pinene.
SESQUITERPENOID: bicyclogermacrene.
Nishimura et al., *Tetrahedron Lett.*, 3097 (1969)

C. laurifolius.
DITERPENOID: acetyllaurifolic acid. Monoterpenoids limonene, linalool, nerol.
de Pascual Teresa et al., *An. Quim.*, **74**, 1540 (1978)

C. limon.
FLAVONOIDS: citronetin, isolimocitrol, isolimocitrol 3-β-D-glucoside, isolimocitrol 3,7,4′-trimethyl ether, 2″-O-xylosylvitexin.
Bentili et al., *Tetrahedron*, **20**, 2313 (1964)
Tomas et al., *Chem. Abstr.*, **84** 102365z

C. medica L. var. sarcodactylis (Noot.) Swingle.
FLAVONOIDS: 3,5,6-trihydroxy-7,4′-dimethoxyflavone, 3,5,6-trihydroxy-7,3′,4′-trimethoxyflavone.
He et al., *Yaoxue Xuebao*, **20**, 433 (1985); *Chem. Abstr.*, **104** 17641y

C. orlando x tangelo x clementine.
CAROTENOID: β-citraurinene.
Leuenberger, Stewart, *Phytochemistry*, **15**, 227 (1976)

C. paradisi L.
STEROID: 5,6-dihydroergosterol.
LIMONOIDS: nomilin, nomilinic acid.
SESQUITERPENOIDS: 3,7(11)-eudesmadien-2-one, (+)-nootkatone.
(S) Barton, Cox, *J. Chem. Soc.*, 1357 (1948)
(T) Demole et al., *Helv. Chim. Acta*, **66**, 1381 (1983)

C. reticulata.
CAROTENOID: reticulataxanthin.
STEROIDS: calamin, cyclocalamin, retrocalamin.
(C) Yokoyama et al., *J. Org. Chem.*, **30**, 2412 (1965)
(S) Bennett, Hasegawa, *Tetrahedron*, **37**, 17 (1981)

C. reticulata cv Austera x Fortunella sp.
LIMONOID: 6-keto-7-deacetylnominol.
Hasegawa et al., *Phytochemistry*, **25**, 1984 (1986)

C. sinensis Osbeck.
PHENYLPROPANOIDS: citrusins A, B and C.
Sawabe et al., *Nippon Nogei Kagaku Kaishi*, **60**, 539 (1986); *Chem. Abstr.*, 105 222728u

C.
SUDACHI FLAVONOIDS: demethoxysudachitin, 4′-β-D-glucosylsudachitin 7-O-(3-hydroxy-3-methylglutalate), sudachitin.
Kumamoto et al., *Agr. Biol. Chem.*, **49**, 2797 (1985)

C. trifoliata.
FLAVONOID: citrifoliol.
Asahina et al., *Chem. Abstr.*, 23 3475

C. unshiu Marc.
FLAVONOID: hesperidin.
MONOTERPENOIDS: β-elemene, limonene, linalool, myrcene, τ-terpinene.
(F) Okuyama et al., *Chem. Abstr.*, 87 81233z
(T) Kekelidze et al., *Khim. Prir. Soedin.*, 572 (1985)

CITRUS-PONCIRUS hybrid.
DINORTRITERPENOIDS;: 1-abeo-7-acetoxyisoobacun-3,10-olide;, 1-abeo-7-acetoxy-9(11)-obacunene;. (11)-obacunene;.
Bennett et al., *Phytochemistry*, **21**, 2349 (1982)

CLADONIA (Cladoniaceae)
C. bellidiflora (Ach.) Schaerer.
ACID: squamatic acid.
Asahina et al., *Ber. Dtsch. Chem. Ges.*, 70, **62**, 64 (1937)

C. crispata (Ach.) Flotow.
ACID: squamatic acid.
Asahina et al., *Ber. Dtsch. Chem. Ges.*, 70, **62**, 64 (1937)

C. cryptochlorophaeca.
ACID: cryptochlorophaeic acid.
Shibata, *Phytochemistry*, **4**, 133 (1965)

C. deformis (L.) Hoffm.
TRITERPENOIDS: 14-taraxerene, zeorin.
Huneck, *Chem. Ber.*, **94**, 614 (1961)

C. incrassata Floerke.
BENZOPYRAN: rhodocladonic acid.
Baker et al., *Can. J. Chem.*, **47**, 2733 (1969)

C. pleurota (Floerke) Schaerer.
CAROTENOIDS: β-carotene, lutein, neozeaxanthin, violaxanthin.
PIGMENTS: rhodocladonic acid, usnic acid.
Asperges et al., *Bull. Soc. R. Bot. Belg.*, **110**, 260 (1977); *Chem. Abstr.*, 88 101632e

C. polycarpoides.
BISANHYDRIDE: homohevesdride.
Archer et al., *Phytochemistry*, **26**, 2117 (1987)

C. polydactyla (Floerke) Spreng.
PIGMENT: rhodocladonic acid.
Baker et al., *Can. J. Chem.*, **47**, 2733 (1969)

CLADOPHORA (Chlorophyceae)
C. glomerata Kutz.
CAROTENOIDS: β-carotene, τ-carotene, α-cryptoxanthin, flavoxanthin, isocryptoxanthin, torularhodin.
Mantiasevic, Horgas, *Chem. Abstr.*, 88 85996u

CLADRASTIS (Leguminosae)
C. amurensis Benth.
QUINOLIZIDINE ALKALOID: cytisine.
Govindachari et al., *J. Chem. Soc.*, 3839 (195)

C. platycarpa.
FLAVONOIDS: bayin, cladrastin 7-O-β-D-glucoside, fujikinin, platycarpanetin 7-O-β-D-glucoside, pseudobaptigenin.
Ohashi et al., *Phytochemistry*, **15**, 354 (1976)

CLARISIA (Moraceae)
C. racemosa R. et P.
STEROID: β-sitosterol.
STILBENE: 3,5-dihydroxy-4′-methoxystilbene.
XANTHONE: 1,3,6,7-tetrahydroxyxanthone.
(S, St) Gottlieb et al., *Phytochemistry*, **14**, 1674 (1975)
(X) Cotterill, Scheinmann, *J. Chem. Soc., Chem. Commun.*, 664 (1975)

CLAUSENA
C. anisata (Willd.) Oliv.
CARBAZOLE ALKALOID: clausanitine.
COUMARINS: chalepin, coumurrayin, imperatorin, osthole, pimpinellin, scopoletin, umbelliferone, xanthotoxin, xanthotoxol.

MONOTERPENOIDS: limonene, methylchavicol, myrcene.
(A) Okorie et al., *Phytochemistry*, **14**, 2720 (1975)
(C, T) Reisch et al., *Sci. Pharm.*, **53**, 153 (1985)

C. dentata.
COUMARIN: nordentatin.
Govindachari et al., *Tetrahedron*, **24**, 753 (1968)

C. heptaphylla Wt. et Arn.
CARBAZOLE ALKALOIDS: clausenapine, heptaphylline, heptazolidine, heptazoline.
Chakraborty et al., *Chem. Ind. (London)*, 303 (1974)

C. indica Oliv.
CARBAZOLE ALKALOID: 6-methoxyheptaphylline.
PYRIDINE ALKALOID: indicaine.
BENZOPYRAN: clausanidin.
SESQUITERPENOID: clausantalene.
(A) Joshi et al., *Indian J. Chem.*, **10**, 1123 (1972)
(B) Kuffner et al., *Monatsh. Chem.*, **104**, 911 (1973)
(T) Joshi et al., *Experientia*, **31**, 138 (1975)

C. pentaphylla (Roxb.) DC.
COUMARINS: clausarin, clausenidin, clausenin, clausmarin A, clausmarin B, dentatin.
SESTERTERPENOID: clausenolide.
STEROID: β-sitosterol.
TRITERPENOID: O-methylclausenol.
(C) Anwer et al., *Experientia*, **33**, 412 (1977)
(T) Chakraborty et al., *J. Chem. Soc., Chem. Commun.*, 246 (1979)

C. wildenovii.
DITERPENOIDS: diclausenan A, diclausenan B.
Rao et al., *Tetrahedron Lett.*, 1019 (1976)

CLAVICEPS (Clavicepidaceae/Hypocraceae)

C. paspali.
ERGOT ALKALOIDS: paspalicine, paspaline, paspalinine.
Springer, Clardy, *Tetrahedron Lett.*, 231 (1981)

C. purpurea Tul.
ERGOT ALKALOIDS: chanoclavine, ergoclavine, ergocornine, ergocorninine, ergocristine, ergocristinine, α-ergocryptine, β-ergocryptine, ergocryptinine, ergometrine, ergometrinine, ergomollinine, ergomonamine, ergosecalinine, ergosine, ergosinine, ergostine, ergotamine, ergotaminine, ergotoxine, ergovalide.
Stoll et al., *Helv. Chim. Acta*, **28**, 1283 (1945)
Ban'kovskaya et al., *Khim. Prir. Soedin.*, 134 (1973)

CLAVULARIA

C. inflata.
SESQUITERPENOIDS: 12-acetoxysinularene, cyclosinularene.
Braekman et al., *Tetrahedron*, **37**, 179 (1981)

C. koellikeri.
MONOTERPENOIDS: clavularin A, clavularin B.
Endo et al., *J. Chem. Soc., Chem. Commun.*, 322 (1983)

CLEISTANTHUS (Euphorbiaceae)

C. collinus (Roxb.) Benth. et Hook f.
GLYCOSIDES: cleistanthins A, B and C, diphyllin-4-O-[β-2,3-di-O-methylxylopyranosyl]-β-D-glucopyranoside.
NAPHTHOFURAN: collinusin.
Anjaneyulu et al., *Indian J. Chem.*, **15B**, 10 (1977)

C. schlechteri.
DITERPENOID: cleistanthol.
McGarry et al., *J. Chem. Soc., Chem. Commun.*, 1074 (1969)

CLEISTOPHOLIS (Annonaceae)

C. patens.
OXOAPORPHINE ALKALOID: isomoschatoline.
SESQUITERPENOIDS: methyl-(-)-trans, trans-10,11-dihydroxyfarnesoate, methyl-(+)-10-hydroxy-6,11-cyclofarnes-7(14)-enoate.
(A) Abd-El Atti et al., *J. Nat. Prod.*, **45**, 476 (1982)
(T) Waterman et al., *Phytochemistry*, **24**, 523 (1985)

CLEMATIS (Ranunculaceae)

C. apiifolia.
TRITERPENOID: 3-O-acetyloleanolic acid.
Fujita et al., *J. Pharm. Soc., Japan*, **94**, 189 (1974)

C. biternata.
TRITERPENOID: 3-O-acetyloleanolic acid.
Fujita et al., *J. Pharm. Soc., Japan*, **94**, 189 (1974)

C. chinensis Osbeck.
SAPOGENINS: prosapogenin CP-0, prosapogenin CP-2a, prosapogenin CP-3a.
SAPONINS: prosaponins A, B, C, CP-2, CP-6, CP-7 and CP-8.
Kizum Tomimori, *Chem. Pharm. Bull.*, **27**, 2388 (1979)

C. japonica.
TRITERPENOID: 3-O-acetyloleanolic acid.
Fujita et al., *J. Pharm. Soc., Japan*, **94**, 189 (1974)

C. patens.
TRITERPENOID: 3-O-acetyloleanolic acid.
Fujita et al., *J. Pharm. Soc., Japan*, **94**, 189 (1974)

C. recta var. mandschurica.
SAPONINS: clematosides A, A', B and C.
Chirva, Konyhukov, *Khim. Prir. Soedin.*, 60 (1969)

C. songarica.
SAPONINS: saponins A, B and B'.
Krokhmalyuk et al., *Khim. Prir. Soedin.*, 600 (1975)

C. stans.
TRITERPENOID: 3-O-acetyloleanolic acid.
Fujita et al., *J. Pharm. Soc., Japan*, **94**, 189 (1974)

C. stans var. austrojaponensis.
TRITERPENOID: 3-O-a.cetyloleanolic acid.
Fujita et al., *J. Pharm. Soc., Japan*, **94**, 189 (1974)

C. uncinata var. ovatifolia.
TRITERPENOID: 3-O-acetyloleanolic acid.
Fujita et al., *J. Pharm. Soc., Japan*, **94**, 189 (1974)

C. vitalba.
SAPONINS: vitalbosides A, B, C, D, E, F, G and H.
Chirva et al., *Dokl. Akad. Nauk SSSR*, **217**, 969 (1974)

CLEOME (Capparidaceae)

C. icosandra.
DITERPENOID: cleomeolide.
Mahato et al., *J. Am. Chem. Soc.*, **101**, 4720 (1979)

C. viscosa.
DITERPENOIDS:
(3E,7Z,11Z)-17,20-dihydroxycembra-3,7,11,15-tetraen-19-oic acid,
(1R,3E,7Z,12R)-20-hydroxycembra-3,7,15-trien-19-oic acid.
Koselan et al., *Aust. J. Chem.*, **38**, 1365 (1985)

CLEONIA

C. lusitanica.
DITERPENOID: cleonioic acid.
Garcia-Alvarez et al., *Phytochemistry*, **18**, 1835 (1979)

CLERODENDRON (Verbenaceae)

C. calamitosum.
DITERPENOIDS: caryoptenol, caryoptin, 3-epi-caryoptin.
　　Hosozawa et al., Phytochemistry, 13, 308, 1019 (1974)

C. campbellii.
STEROIDS: 4α-methylstigmasta-7,25-dien-3β-ol,
　　stigmasta-5,22E,24S-trien-3β0-ol,
　　24S-stigmasta-5,22,25-trien-3β-ol.
　　Bolger et al., Tetrahedron Lett., 3043 (1970)
　　Pakrashi et al., ibid., 365 (1971)

C. inerme.
FLAVONOIDS: 4'-methylscutellarein, pectolinarigenin.
　　Vandantham et al., Phytochemistry, 16, 294 (1977)

C. infortumatum.
STEROID: 5,25-stigmastadien-3β-ol.
DITERPENOID: clerodin.
TRITERPENOIDS: clerodolone, clerodone. (S) Manzoor-i-Khuda
　　et al., Tetrahedron, 22, 2377 (1966)
　　(T) Manzoor-i-Khuda, Sarela, ibid., 21, 979 (1965)

C. neriifolium.
DITERPENOID: nerifolinol.
　　Purushothaman et al., Indian Drugs, 29, 640 (1980); Chem. Abstr., 106
　　　30005e

C. neutens.
STEROIDS: cholesterol, clerodolone,
　　(24S)-ethylcholesta-5,22,25-trien-3β-ol, β-sitosterol,
　　stigmasterol.
　　Nair et al., Curr. Sci., 45, 391 (1976)

C. phlomoides.
FLAVONOID: scutellarein.
　　Arisawa et al., Chem. Pharm. Bull., 18, 916 (1970)

C. serratum.
FLAVONOIDS: luteolin, luteolin 7-O-β-D-glucuronide,
　　scutellarein.
PHENOLIC: (±)-catechin.
STEROIDS: α-spinasterol, stigmasterol.
TRITERPENOIDS: queretaroic acid, serratagenic acid.
　　(F, P, S) Nair et al., Curr. Sci., 45, 391 (1976)
　　(T) Rangaswami, Sarayan, Tetrahedron, 25, 3701 (1969)

C. splendens.
STEROIDS: clerodolone, (24S)-ethylcholesta-5,22,25-trien-3β-ol.
TRITERPENOIDS: β-amyrin, friedelan-3β-ol, friedelanone.
　　Joshi et al., J. Indian Chem. Soc., 62, 409 (1985)

C. thomsonae Balf.
IRIDOID: 8-O-acetylmioporoside.
　　Lemmel, Rimpler, Z. Naturforsch., 36C, 708 (1981)

C. trichotomum Thunb.
DIMERIC CARBOLINE ALKALOIDS:
　　N:N'-diglucopyranosyltrichotomine, trichotomine,
　　trichotomine G1.
TERPENOID: clerodendrin A.
　　(A) Iwadare et al., Tetrahedron, 30, 4105 (1974)
　　(T) Kato et al., J. Chem. Soc., Chem. Commun., 1632 (1971)

C. trichotomum Thunb. var. fargesii Rehd.
GLYCOSIDE: clerodendroside.
　　Morita et al., J. Pharm. Soc., Japan, 97, 976 (1977)

C. ugandense.
IRIDOID: ugandoside.
　　Jacke, Rimpler, Phytochemistry, 22, 1729 (1983)

C. uncinatum Schinz.
HYDROQUINONE: uncinatone.
　　Dorsaz et al., Helv. Chim. Acta, 68, 1605 (1985)

CLETHRA (Clethraceae)

C. barbinervis Sieb. et Zucc.
TRITERPENOIDS: barbinervic acid, clethric acid.
　　Takahashi, Takani, Chem. Pharm. Bull., 26, 2689 (1978)

CLIBADIUM

C. glomeratum.
One diterpenoid.
　　Czerson et al., Phytochemistry, 18, 257 (1979)

CLIMACOPTERA

C. transoxana.
SAPONINS: copterosides A, B, C, D, E, F, G and H.
　　Annaev et al., Khim. Prir. Soedin., 60 (1984)

CLITOCYBE (Agaricaceae)

C. acromelalga.
PYRIDINE ALKALOIDS: acromelic acid A, acromelic acid B.
QUATERNARY ALKALOID: clithioneine.
　　Konno et al., Phytochemistry, 23, 1003 (1984)

C. fasciculata.
PYRROLIZIDINE ALKALOID: lepistine.
　　Laing et al., Tetrahedron Lett., 269 (1975)

C. illudens (Schw.) Sacc.
ACID: atromentic acid.
SESQUITERPENOIDS: illudin M, illudin S, illudoic acid, illudol.
　　(A) Anchel et al., Phytochemistry, 10, 3259 (1971)
　　(T) Anchel et al., Proc. Nat. Acad. Sci., USA, 36, 300 (1950)

C. nebularis (Batsch ex Fr.) Kummer.
STEROID: portensterol.
　　Regerat et al., Ann. Pharm. Fr., 34, 323 (1976)

C. subilludens.
BENZOFURAN: thelephoric acid.
　　Gripenberg, Tetrahedron, 10, 135 (1960)

CLITORIA

C. ternatea L.
ACIDS: arachidic acid, behenic acid, lignoceric acid, linoleic
　　acid, linolenic acid, oleic acid, palmitic acid, stearic acid.
FLAVONOIDS: clitorin, kaempferol 3-glucoside, kaempferol
　　3-neohesperidoside, kaempferol 3-rutinoside.
STEROID: stigmast-4-ene-3,6-dione.
　　(Ac) Debnath et al., J. Inst. Chem. Calcutta, 47, 253 (1975); Chem.
　　　Abstr., 84 147724g
　　(F) Morita et al., J. Pharm. Soc., Japan, 97, 649 (1977)
　　(S) Ripperger, Pharmazie, 33, 82 (1978)

CLIVIA (Amaryllidaceae)

C. miniata (Regel.) Benth.
PERHYDROINDOLE ALKALOIDS: clivacetine, clivatine, clivealine,
　　cliviahaksine, cliviamartine, cliviasine, clivicosine,
　　clividine, clivimine, clivionine, clivojuline, clivonidine,
　　lycorine.
　　Dopke, Rosham, Z. Chem., 20, 374 (1980)
　　Dopke et al., ibid., 23, 102 (1983)

CLUSIA

C. congestiflora.
BENZOPHENONE: clusianone.
　　McCandlesh et al., Acta Crystallogr., 32B, 1793 (1976)

C. ellipticifolia Cuatr.
TRITERPENOIDS: friedelin, epi-friedelinol.
　　Salama, Chem. Abstr., 105 206298y

CLYTIA

C. richartiana.
TERPENOID: saudin.

Mossa et al., J. Org. Chem., 50, 916 (1985)

CNEORUM (Cneoraceae)

C. pulverulentum Vent.
CNEORINS: cneorins A, B, C, D, E, F, H, K and L, cneorum-chromones C, E and G, cneorum-coumarin B.
STEROID: 5-stigmastene-3,7,20-triol.

(C) Mondon, Epe, Tetrahedron Lett., 1273 (1976)

(S) Mondon et al., Chem. Ber., 108, 1989 (1975)

C. tricoccum.
BENZODIOXIN: (E)-bethancorol.
DITERPENOIDS: cneorubines U, W, X and Y.
SESTERTERPENOIDS: tricoccines R0, R1, R2, R3, R4, R5, R6, S1, S2, S3, S4, S5, S6, S7, S8, S10, S13, S16, S22, S27, S32.

Trautmann et al., Chem. Ber., 113, 3848 (1981)

CNESTUS (Connaraceae)

C. glabra.
AMINO ACID: methionine sulfoximine.

Jeannoda et al., Phytochemistry, 24, 854 (1985)

C. polyphylla.
AMINO ACID: methionine sulfoximine.

Jeannoda et al., Phytochemistry, 24, 854 (1985)

CNICOTHAMNUS

C. lorentzii.
SESQUITERPENOID: cnicothamnol.

Bohlmann et al., Phytochemistry, 18, 95 (1979)

CNICUS (Compositae)

C. benedictus L.
LIGNANS: 2-acetylnortracheloside, arctigenin, nortracheloside, salonitenolide, trachelogenin.
SESQUITERPENOID: cnicin.

(L) Vanhaelen et al., Phytochemistry, 14, 2709 (1975)

(T) Suchy et al., Chem. Ber., 93, 2449 (1960)

CNIDIUM (Umbelliferae)

C. officinale L.
BENZOFURANS: cnidilide, ligustilide.

Mitsuhashi et al., Tetrahedron, 20, 1971 (1964)

COCCIFERA

C. plurota.
TRITERPENOID: zeorin.

Barton, Bruun, J. Chem. Soc., 1683 (1952)

COCCOLOBA

C. excoriata.
STEROID: β-sitosterol. Triterpenoids betulinic acid, friedelin, lupeol, lupeol acetate, taraxerone, ursolic acid.

Dan et al., Fitoterapia, 57, 445 (1986)

COCCULUS (Menispermaceae)

C. carolinus.
PHENANTHRENE ALKALOID: carococculine.
QUATERNARY ALKALOIDS: magnoflorine, palmatine.
FLAVONOIDS: (-)-bivurnitol, (+)-quercitrol.

Elsohly et al., J. Pharm. Sci., 65, 132 (1976)

C. diversifolius DC.
BISBENZYLISOQUINOLINE ALKALOID: tetrandrine.
UNCHARACTERIZED ALKALOID: kukoline.

Ohta, Ber. Ges. Physiol., 33, 352 (1925)

C. japonicus.
BISBENZYLISOQUINOLINE ALKALOID: tetrandrine.

Chen, Chen, J. Biol. Chem., 109, 681 (1935)

C. laurifolius DC.
APORPHINE ALKALOID: laurifoline.
BENZYLISOQUINOLINE ALKALOIDS: (±)-coclaurine, erythroculine, erythroculinol.
BIPHENYL ALKALOIDS: laurifine, laurifinine, laurifonine.
ERYTHRINA ALKALOIDS: coccoline, coccolinine, coccudinone, cocculimine, cocculitine, cocculitinine, cocculitrine, coccuvine, coccuvinine.

Inubushi et al., Tetrahedron Lett., 153 (1969)

Bhakuni, Jain, Tetrahedron, 36, 3107 (1980)

C. leaebe.
BISBENZYLISOQUINOLINE ALKALOIDS: cocsoline, penduline.

Ikram et al., Planta Med., 45, 253 (1982)

C. pendulus Diels.
BISBENZYLISOQUINOLINE ALKALOIDS: cocsoline, cocsolinine, penduline.

Jishi et al., Indian J. Chem., 12, 517 (1974)

C. sarmentosus Diels.
BRIDGED BISBENZYLISOQUINOLINE ALKALOIDS: isotrilobine, menisarine, trilobine.

Tomita, Furukawa, J. Pharm. Soc., Japan, 83, 190 (1963)

C. trilobus (Thunb.) DC.
BISBENZYLISOQUINOLINE ALKALOIDS: coclobine, daphnoline, isotrilobine, trilobine.

Wada et al., Tetrahedron Lett., 5179 (1966)

Ito et al., J. Pharm. Soc., Japan, 89, 1163 (1969)

COCHLEARIA (Cruciferae)

C. arctica Schlecht.
PYRROLE ALKALOIDS: hygrine, hygroline.
UNCHARACTERIZED ALKALOID: cochlearine.

Platonova, Kuzovkov, Med. Prom., 17, 19 (1963)

COCHLIOBOLUS

C. heterostrophus.
TETRANORTRITERPENOIDS: geranylnerilidol, ophiobola-7,18-dien-20-ol, ophiobolins A, B and C.

Nozoe et al., Tetrahedron Lett., 4457 (1968)

C. lunata.
ANTIBIOTICS: lunatoic acid A, lunatoic acid B.

Nukina, Marumo, Tetrahedron Lett., 2603 (1977)

C. miyabeanus.
NORSESTERTERPENOIDS: cochliobolins A, B and C.
TRITERPENOIDS: cochlioquinone A, cochlioquinone B, ophiobolin A.

Orsenigo et al., Phytopathology, 2, 29, 189 (1957)

Carruthers et al., J. Chem. Soc., Chem. Commun., 164 (1971)

C. setariae.
SESQUITERPENOIDS: cis-sativenediol, trans-sativenediol.

Nukina et al., J. Am. Chem. Soc., 97, 2542 (1975)

COCHLOSPERMUM

C. gillivraei.
FLAVONOIDS: (+)-afzelechin, apigenin, naringenin

Cook et al., Phytochemistry, 14, 2510 (1975)

COCSINIUM (Menispermaceae)

C. blumeanum Miers.

PROTOBERBERINE ALKALOIDS: berberine, jatrorrhizine, palmatine.

Awe, *Arch. Pharm.*, **284**, 352 (1951)

C. fenestratum Colebr.

PROTOBERBERINE ALKALOID: berberine.

Awe, *Arch. Pharm.*, **284**, 352 (1951)

CODIUM (Chlorophyceae)

C. bursa.

STEROID: (24S)-24-methylcholesta-5,25-dien-3β-ol.

Romeo *et al.*, *Chem. Abstr.*, 97 141766u

C. fragile.

STEROID: codisterol.

Rubinstein, Goad, *Phytochemistry*, **13**, 481 (1974)

CODONOCARPUS (Lobeliaceae)

C. australis A.Cunn.

MACROCYCLIC ALKALOIDS: codonocarpine, N-methylcodonocarpine.

Doskotch *et al.*, *Tetrahedron*, **30**, 3229 (1974)

CODONOCEPHALUM (Compositae)

C. grane Schrenk. (Inula grandis Schrenk.).

One uncharacterized alkaloid.

Zabolotnaya, Safronic, Vilp., XI, 152 (1959), *given in S. Yu Yunusov (ed.)* (1974) The Alkaloids (University of Tashkent)

CODONOPSIS (Lobeliaceae)

C. clematidea (Schrenk) Clarke.

PYRROLE ALKALOIDS: codonopsine, codonopsinine.

Matkhalikova *et al.*, *Khim. Prir. Soedin.*, 607 (1969)

C. lanceolata.

STEROIDS: α-spinasterol, stigmasterol.

TRITERPENOIDS: albigenic acid, echinocystic acid, oleanolic acid.

(S, T) Yans *et al.*, *Chem. Abstr.*, 84 102371y

(T) Han *et al.*, *Chem. Abstr.*, 86 40163n

C. pilosula.

STEROIDS: δ-spinasterol, δ-spinasterol glucoside, stigmast-7-en-3β-ol, stigmast-7-en-3β-ol glucoside.

TRITERPENOIDS: friedelin, taraxerol, taraxeryl acetate.

Wong *et al.*, *Planta Med.*, **49**, 60 (1983)

COELIDIUM

C. fourcadei.

PIPERIDINE ALKALOIDS: α-aldotripiperideine, isotripiperideine.

Arndt, De Plessis, *J. S. Afr. Chem. Inst.*, **21**, 54 (1968)

COELOCLINE (Annonaceae)

C. polycarpa DC.

PROTOBERBERINE ALKALOID: berberine.

Awe, *Arch. Pharm.*, **284**, 352 (1951)

COFFEA (Rubiaceae)

C. arabica.

PURINE ALKALOID: caffeine.

DITERPENOID: ent-16-kauren-19-ol.

(A) Charrier, Berthand, Chem. Abstr., *84 118445m* 118445m

(T) Wohlberg *et al.*, *Phytochemistry*, **14**, 1677 (1975)

C. canephora.

PURINE ALKALOID: caffeine.

Charrier, Berthand, *Chem. Abstr.*, 84 118445m

C. dewevrei.

ACIDS: methoxytetramethyluric acid, 1,3,7,9-tetramethyluric acid, O(2),1,9-trimethyluric acid.

Petermann *et al.*, *Phytochemistry*, **16**, 620 (1977)

C. dewevrei var. aruwimiensis.

ACIDS: methoxytetramethyluric acid, 1,3,7,9-tetramethyluric acid, O(2),1,9-trimethyluric acid.

Petermann *et al.*, *Phytochemistry*, **16**, 620 (1977)

C. dewevrei var. excelsa.

ACIDS: methoxytetramethyluric acid, 1,3,7,9-tetramethyluric acid, O(2),1,9-trimethyluric acid.

Petermann *et al.*, *Phytochemistry*, **16**, 620 (1977)

C. eugenioides.

PURINE ALKALOID: caffeine.

Charrier, Berthand, *Chem. Abstr.*, 84 118445m

C. liberica.

ACIDS: methoxytetramethyluric acid, 1,3,7,9-tetramethyluric acid, O(2),1,9-trimethyluric acid.

Petermann *et al.*, *Phytochemistry*, **16**, 620 (1977)

C. stenophylla.

PURINE ALKALOID: caffeine.

Charrier, Berthand, *Chem. Abstr.*, 84 118445m

C. vanneyi.

TRITERPENOID: mascaroside.

Ducroiux *et al.*, *J. Chem. Soc., Chem. Commun.*, 396 (1975)

COIX

C. lacryma-jobi.

ANTINEOPLASTIC AGENT: coixenolide.

Lee, *Chem. Abstr.*, 87 130483s

COLA

C. acuminata.

PURINE ALKALOID: theobromine.

Blout *et al.*, *J. Am. Chem. Soc.*, **72**, 479 (1950)

COLCHICUM (Liliaceae)

C. autumnale L.

COLCHICINE ALKALOIDS: O-acetylcolchiceine, N-acetyldemecolcine, 2-acetyl-2-demethylcolchicine, 3-acetyl-3-demethylcolchicine, autumnaline, colchicine, colchifoline, demecolcine, 2-demethylcolchicine, 3-demethylcolchicine.

Santavy, Reichstein, *Helv. Chim. Acta*, **33**, 1606 (1950)

Potesilova *et al.*, *Collect. Czech. Chem. Commun.*, **32**, 141 (1967)

C. cornigerum Tachk. et Drar.

COLCHICINE ALKALOIDS: autumnaline, colchicine, cornigerine, 3-demethylcolchicine, α-lumicornigerine, β-lumicornigerine, α-lumi-3-demethylcolchicine, 2-methylcolchicine, 3-methylcolchicine.

El-Hamidi, Santavy, *Collect. Czech. Chem. Commun.*, **27**, 2111 (1962)

Saleh *et al.*, *ibid.*, **28**, 3413 (1963)

Battersby *et al.*, *J. Chem. Soc., C*, 3514 (1971)

C. doerfleri.

COLCHICINE ALKALOIDS: autumnaline, colchicine, colchiculine, 2-demethylcolchicine, N-formyl-N-deacetylcolchicine.

Gasic *et al.*, *Planta Med.*, **32**, 368 (1977)

C. kesselringii Rgl.

COLCHICINE ALKALOIDS: allocolchicine, colchiceine, colchicine, 2-demethylcolchicine, 3-demethylcolchicine, 2-demethyllumicolchicine, 3-demethyl-α-lumicolchicine, N-formyldesacetylcolchicine, α-lumicolchicine, regelamine, regelidine, regelinine.

ISOQUINOLINE ALKALOIDS: crociflorinone, kesselridine, kesselringine.

UNCHARACTERIZED ALKALOIDS: crocifloridine, crociflorine, crociflorinine.
Turdikulov *et al.*, *Khim. Prir. Soedin.*, 541 (1971)
Yusupov *et al.*, *Izd-vo Farmatsericheskii Uzbekistanskii SSR,* 181 (1973)

C. latifolium S. et S.
COLCHICINE ALKALOIDS: autumnaline, colchiciline, 3-demethyl-N-formyl-N-deacetylcolchicine, 2,3-didemethylcolchicine, 10,11-epoxycolchicine.
Potesilova *et al.*, *Collect. Czech. Chem. Commun.*, 41, 3146 (1976)

C. luteum Baker.
COLCHICINE ALKALOIDS: colchiceine, colchicine, 3-demethylcolchamine, 2-demethylcolchicine, 3-demethylcolchicine, 3-demethyl-α-lumicolchicine, N-formyldesacetylcolchicine, lumicolchicine.
ISOQUINOLINE ALKALOID: kesselringine.
SPIROISOQUINOLINE ALKALOIDS: luteidine, luteine.
Mukhamed'yarova *et al.*, *Khim. Prir. Soedin.*, 354 (1976)
Yusupov *et al.*, ibid., 359 (1976)

C. speciosum Stev.
COLCHICINE ALKALOIDS: colchamine, colchiceine, colchicine, specionine, speciosine.
Masinova, Santavy, *Collect. Czech. Chem. Commun.*, 19, 1283 (1954)
Kiselev, *J. Gen. Chem. USSR*, 26, 3218 (1956)

C. szovitsii.
APORPHINE ALKALOID: szovitsinine. Dibenzocycloheptanone alkaloids szovitsamine, szovitsidine.
Yusupov *et al.*, *Khim. Prir. Soedin.*, 271 (1975)

C. vasiani.
ISOQUINOLINE ALKALOID: autumnaline.
Potesilova *et al.*, *Collect. Czech. Chem. Commun.*, 34, 3540 (1969)

COLEUS (Labiatae/Lamiaceae)
C. amboinicus Louriero.
TRITERPENOIDS: crategolic acid, 2α,3α-dihydroxyoleanolic acid, euscaphic acid, oleanolic acid, pomolic acid, 2α,3α,19α,23-tetrahydroxyursolic acid, tomentic acid, ursolic acid.
Brieskorn, Riedel, *Arch. Pharm.*, 310, 910 (1977)

C. aquaticus.
DITERPENOIDS: coleon C, coleon D.
Ruedi, Eugster, *Helv. Chim. Acta*, 55, 1736 (1972)

C. barbatus.
DITERPENOIDS: barbatusin, barbatusol, coleon E, coleon F.
Kelecom *et al.*, *Phytochemistry*, 23, 1677 (1984)

C. blumei.
ANTHOCYANIN: cyanidin 3,5-diglucoside.
FLAVONOID: dihydrokaempferol.
Lamprecht *et al.*, *Phyton (Buenos Aires)*, 33, 157 (1975); *Chem. Abstr.*, 84 147738q

C. caerulescens.
DITERPENOID: coleon Y.
Grob *et al.*, *Helv. Chim. Acta*, 61, 871 (1978)

C. carnosus.
DITERPENOIDS: 7α-acetoxy-6β,20-dihydroroyleanone, 7α-acetoxyroyleanone.
Yoshizaka *et al.*, *Helv. Chim. Acta*, 62, 2754 (1979)

C. forskohlii.
DITERPENOIDS:
7β-acetoxy-8,13-epoxy-6β-hydroxylabd-14-en-11-one, 6β-acetoxy-8,13-epoxy-1α,7β,9-trihydroxylabd-14-en-11-one, coleonols A, B, C, D, E and F, coleosol.
Bhat *et al.*, *Tetrahedron Lett.*, 1669 (1977)
Tandon *et al.*, *Indian J. Chem.*, 16B, 341 (1978)

C. garckeanus.
DITERPENOIDS: 16-O-acetylcoleon W, 16-O-acetylcoleon X, coleon X, coleon Z.
Miyase *et al.*, *Helv. Chim. Acta*, 62, 2374 (1979)

C. ignarius.
DITERPENOIDS: coleon A, coleon B.
Eugster *et al.*, *Helv. Chim. Acta*, 46, 530 (1963)

C. kilimandschari.
DITERPENOIDS: coleon E, coleon F.
Ruedi, Eugster, *Helv. Chim. Acta*, 56, 1129 (1973)

C. somaliensis.
DITERPENOIDS: coleon G, coleon H, coleon I, coleon J, coleon K, coleon O.
Moir *et al.*, *Helv. Chim. Acta*, 56, 2534, 2539 (1973)

COLLADONIUM
C. triquestra.
COUMARIN: badrakemin.
Saidkhodzhaev, Nikonov, *Khim. Prir. Soedin.*, 490 (1973)

COLLETIA (Rhamnaceae)
C. spinosissima Gmel.
BENZYLISOQUINOLINE ALKALOID: colletine.
Sanchez, Comin, *Tetrahedron*, 23, 1139 (1963)

COLLETOTRICHUM
C. capsici.
NORSESQUITERPENOID: colletodiol.
MacMillan *et al.*, *J. Chem. Soc., Perkin I*, 1487 (1983)

C. nicotianae.
NORTRITERPENOIDS: colletotrichins A, B and C.
Grove *et al.*, *J. Chem. Soc., C*, 230 (1966)

COLLYBIA
C. maculata.
SESQUITERPENOIDS: collybolide, isocollybolide.
Bui *et al.*, *Tetrahedron*, 30, 1327 (1974)

COLONA
C. auriculata.
BENZOPYRAN: tiliroside.
Lin, *J. Chin. Chem. Soc.*, 22, 383 (1975); *Chem. Abstr.*, 84 147653

COLUBRINA
C. faralaotra.
UNCHARACTERIZED ALKALOID: faralaotrine.
Guinaudeau *et al.*, *Planta Med.*, 37, 304 (1975)

C. faralaotra subsp. trichocarpa.
ACIDS: p-coumaric acid, ferulic acid, p-hydroxybenzoic acid, syringic acid, vanillic acid.
FLAVONOIDS: kaempferol, myricetol, quercetol.
Guinaudeau *et al.*, *Planta Med.*, 39, 54 (1976)

C. granulosa.
TRITERPENOIDS: colubrinic acid, granulosic acid.
Rotman, Jurd, *Phytochemistry*, 17, 491 (1978)

C. texensis.
PEPTIDE ALKALOID: texensine.
Wani *et al.*, *Tetrahedron Lett.*, 4675 (1973)

COLYTHIS
C. elliptica.
TRITERPENOID: colysanoxide.
Ageta et al., Tetrahedron Lett., 4349 (1982)
C. pothifolia.
TRITERPENOID: colysanoxide.
Ageta et al., Tetrahedron Lett., 4349 (1982)

COMBRETIUM
C. apiculatum.
PHENANTHROID: 4,6,7-trihydroxy-2,3-dimethoxyphenanthrene.
Letcher et al., J. Chem. Soc., C, 3070 (1971)
C. hereroense.
PHENANTHROID: 2,3,5,7-tetrahydroxyphenanthrene-5,7-dimethyl ether.
Letcher et al., J. Chem. Soc., Perkin I, 2941 (1972)
C. micranthum G.Don.
QUATERNARY ALKALOIDS: choline, 4-hydroxystachydrine, stachydrine.
UNCHARACTERIZED ALKALOID: combretine.
Lehmann, Heil. U. Gewuerz-Pflan., 22, 1 (1943)
Bassene et al., Ann. Pharm. Fr., 44, 191 (1986)
C. molle.
TRITERPENOIDS: mollic acid, mollic acid 3-O-glucoside, mollic acid 3-O-xyloside.
Pegel et al., J. Chem. Soc., Perkin I, 1711 (1985)
C. psidioides S.
PHENANTHROID: 2,3,5,7-tetrahydroxyphenanthrene-3,5,7-trimethyl ether.
Letcher et al., J. Chem. Soc., Perkin I, 1179 (1973)
C. zeyheri.
ALKALOID: N-methyl-L-tyrosine.
Mwauluka et al., Phytochemistry, 14, 1657 (1975)

COMMELINA
C. benghalensis.
ALIPHATICS: dotriacontanol, octacosanol, triacontanol.
STEROIDS: campesterol, β-sitosterol, stigmasterol.
Pandey, Gupta, J. Res. Indian Med., 10, 79 (1975); Chem. Abstr., 85 2563q
C. communis.
ANTHOCYANIN: malonylawobanin.
Goto et al., Tetrahedron Lett., 4863 (1983)
C. undulata.
TRITERPENOIDS: dammar-12,25-dien-3β-acetate, dammar-12,25-dien-3β-ol.
Sharma et al., Phytochemistry, 21, 2420 (1982)

COMMIPHORA(Burseraceae)
C. dalzielii.
TRITERPENOIDS: β-amyrin, cabraleadiol, cabraleadiol 3-acetate, cabraleone, isofourquierone, lupeol, epi-lupeol.
Waterman, Ampofo, Phytochemistry, 24, 2925 (1985)
C. molmol.
SESQUITERPENOIDS: 2-acetoxyfuranodiene, 4,5-dihydrofuranodien-6-one, 1(10)Z,4Z-furanodien-6-one, furanoeudesma-1,3-diene, furanoeudesma-1,4-diene-3,5-dione, furanoeudesma-1,4-dien-6-one, 2-methoxyfuranodiene, 2-methoxyfuranoguaia-9-en-8-one.
Brieskorn et al., Phytochemistry, 22, 187 (1983)
Brieskorn et al., ibid., 22, 1207 (1983)

C. mukul.
STEROIDS: 6-acetoxy-3,5-cholestanediol, guggulsterols I, II, III, IV and V, E-guggulsterone, Z-guggulsterone.
DITERPENOIDS: α-camphorene, mukulol.
(S) Purushothaman, Chandrasekharan, Indian J. Chem., 14B, 802 (1976)
(T) Patil et al., Tetrahedron, 28, 2341 (1972)
C. myrrha.
SESQUITERPENOID: comiferin.
Mincione, Iaverone, Chim. Ind. (Milan). 54, 525 (1972)
C. pyracanthoides.
TRITERPENOIDS: commic acid C, commic acid D, commic acid E.
UNCHARACTERIZED TERPENOIDS: commic acid A, commic acid B.
Thomas, Miller, Experientia, 76, 62 (1960)

COMMITHECA (Rubiaceae)
C. liebrechtsiana.
ANTHRAQUINONES: lucidin, rubiadin.
Hocquemiller et al., Plant Med. Phytother., 10, 248 (1976)

COMULARIA
C. camerunensis.
INDOLE ALKALOIDS: eburnamenine, eburnamine, isoeburnamine, pleiocarpine, pleiocarpinine.
Bruneton, Planta Med., 46, 58 (1982)

CONANDRON
C. ramoidioides.
GLYCOSIDES: acteoside, conandroside.
Nonaka et al., Phytochemistry, 16, 1265 (1977)

CONDURANGO
C. cortex.
STEROIDS: condurangoglycoside A0, condurangoglycoside C0, 2,22-dideoxy-20-hydroxyecdysone.
Ikekawa et al., J. Chem. Soc., Chem. Commun., 448 (1980)
Hayashi et al., Chem. Pharm. Bull., 38, 1954 (1980)

CONIUM (Umbelliferae)
C. maculatum L.
PIPERIDINE ALKALOIDS: N-acetyl-(+)-coniine, N-acetyl-(-)-coniine, conhydrine, α-conhydrine, τ-coniceine, (+)-coniine, N-methyl-(+)-coniine, N-methyl-(-)-coniine, pseudoconhydrine.
Giesecke, Arch. Pharm., 20, 97 (1827)
Gabriel, Ber. Dtsch. Chem. Ges., 42, 4059 (1900)

CONOCARPUS
C. erectus.
BENZOFURAN: conocarpin.
Hayashi et al., Phytochemistry, 14, 1085 (1975)

CONOCEPHALUM
C. conicum.
FLAVONOIDS: apigenin 7-O-glucuronide, apigenin 7-O-glucuronide-4-O-glucuronide, chrysoeriol, chrysoeriol 7-O-glucuronide-4-O-glucuronide, lucenin-2, luteolin, luteolin 7,3'-di-O-glucuronide, luteolin 7-O-glucuronide-4-O-glucuronide, luteolin 7-O-glucuronide-3',4'-di-O-rhamnoside, vicenin-2.
SESQUITERPENOIDS: 8-acetoxyzaluzanin C, 8-acetoxyzaluzanin D.
(F) Markham et al., Phytochemistry, 15, 147 (1976)
(T) Asakawa et al., ibid., 18, 285 (1979)

CONOCLINIOPSIS
C. prasiifolia.
SESQUITERPENOIDS: conoprasiolide, conoprasiolide 5'-O-acetate, two bejaranolides and two furoheliangolides.
Bohlmann *et al., Phytochemistry*, **23**, 1509 (1984)

CONOPHARYNGIA
C. durissima Stapf.
CARBOLINE ALKALOIDS: akuammiline, anhydrovobasinediol.
DIMERIC ALKALOIDS: conoduramine, conodurine.
HOMOCARBOLINE ALKALOIDS: conopharyngine, coronaridine hydroxyindolenine, isovoacangine.
Gorman *et al., J. Am. Chem. Soc.*, **82**, 1142 (1960)
Das *et al., C. R. Acad. Sci.*, **264C**, 1765 (1967)

C. johnstonii Stapf.
CARBAZOLE ALKALOID: tubotaiwine N-oxide.
Pinar *et al., Helv. Chim. Acta*, **55**, 2972 (1972)

C. jollyana Stapf.
HOMOCARBOLINE ALKALOIDS: 20-hydroxy-19-oxocoronaridine, jollyanine, 19-oxocionopharyngine.
Hootele, Petcher, *Chimia*, **22**, 245 (1968)

C. longiflora Stapf.
IMIDAZOLIDINOCARBAZOLE ALKALOID: conoflorine.
Dugan *et al., Helv. Chim. Acta*, **50**, 60 (1967)

CONSOLIDA
C. orientalis Schroed.
ANTHOCYANIN: delphinidin 3-rutinoside-5-glucoside.
Sulyok *et al., Chem. Abstr.*, **105** 168914m

CONVALLARIA (Liliaceae)
C. keisukei.
SAPONINS: convallosaponins A, B, C and D, glucoconvallosaponin A, glucoconvallosaponin B.
STEROIDS: convallogenin A, convallogenin B.
Kimura *et al., Chem. Pharm. Bull.*, **15**, 1204 (1967)

C. majalis L.
CARDENOLIDES: peripalloside, periplogenin, strophalloside, strophanolloside, strophanthidin, strophanthidol.
FLAVONOIDS: isorhamnetin 3-galactoside, isorhamnetin 3-galactodirhamnoside, isorhamnetin 3-galactorhamnoside, kaempferol 3-galactodirhamnoside, kaempferol 3-galactorhamnoside, kaempferol 3-galactoside, quercetin 3-galactodirhamnoside, quercetin 3-galactorhamnoside, quercetin 3-galactoside.
SAPONINS: convalloside, neoconvallotoxide, toloside.
STEROIDS: canescin, convallamarin, convallamarogenin, convallarin, convallatoxin, myalin.
(C) Kubelka *et al., Pharm. Acta Helv.*, **50**, 353 (1975)
(F) Malinowski *et al., Acta Pol. Pharm.*, **33**, 767 (1976); *Chem. Abstr.*, **87** 180762c
(Sap) Buchvarov *et al., Farmatsiya*, **29**, 36 (1979)
(S) Tschesche *et al., Chem. Ber.*, **106**, 3010 (1973)

CONVOLVULUS (Convolvulaceae)
C. erinaceus Ldt.
HYGRINE ALKALOID: cuskhygrine.
Spath, Tuppy, *Monatsh. Chem.*, **79**, 119 (1948)

C. glomeratus.
GLUCOSIDE: 2-C-methylerythritol.
Anthonsen *et al., Acta Chem. Scand.*, B **30**, 91 (1976)

C. hamadae V.Petrov.
HYGRINE ALKALOIDS: cuskhygrine, hygrine.
Spath, Tuppy, *Monatsh. Chem.*, **79**, 119 (1948)

C. krauseanus.
TROPANE ALKALOID: convolisine.
Atripova, Yunusov, *Khim. Prir. Soedin.*, 527 (1979)

C. lineatus L.
TROPANE ALKALOIDS: convolvamine, convolvine.
Yunusov *et al., Dokl. Akad. Nauk Uzb. SSR*, 17 (1958)

C. microphyllus Sieb.
ALIPHATICS: dotriacontanol, hexacosanol, octacosanol, triacontanol.
STEROIDS: β-sitosterol, ε-sitosterol.
Bisht *et al., Chem. Abstr.*, **88** 133284j

C. pluricaulis Chois.
UNCHARACTERIZED ALKALOID: sankhpuspine.
Basu, Dandiya, *J. Amer. Pharm. Assoc.*, **37**, 27 (1948)

C. pseudocanthabrica Schrenk.
TROPANE ALKALOIDS: convolvamine, convolvicine, convolvidine, convolvine.
Orekhov, Konovalova, *Arch. Pharm.*, **271**, 145 (1933)

C. subhirsutus Rgl. et Schmalh.
TROPANE ALKALOIDS: confoline, convolvamine, convolvine.
Orekhov, Konovalova, *Arch. Pharm.*, **27**, 145 (1933)
Sharova *et al., Khim. Prir. Soedin.*, 672 (1980)

CONYZA (Compositae)
C. podocephala.
DITERPENOIDS: conycephaloside, conypododiol.
Bohlmann *et al., Phytochemistry*, **21**, 1693 (1982)

C. sticta.
FLAVONOIDS: conyzatin, 5,7-dihydroxy-3,8,4'-trimethoxyflavone.
DITERPENOIDS: conyzic acid, stictic acid.
(F) Tandon, Rastogi, *Phytochemistry*, **16**, 1455 (1977)
(T) Tandon, Rastogi, *Indian J. Chem.*, **18B**, 529 (1979)

C. ulnifolia.
SESQUITERPENOID: oxoselinene.
Bohlmann, Jakupovic, *Phytochemistry*, **18**, 1367 (1979)

COOPERIA (Amaryllidaceae)
C. drummondii Herb.
BENZOISOQUINOLINE ALKALOID: lycorine.
Greathouse, Rigler, *Am. J. Bot.*, **28**, 702 (1941)

C. pedunculata.
BENZOISOQUINOLINE ALKALOIDS: lycorine, pseudolycorine.
Greathouse, Rigler, *Am. J. Bot.*, **28**, 702 (1941)

COPAIFERA
C. multijuga.
DITERPENOID: copaiferolic acid.
SESQUITERPENOID: α-multijugenol.
Delle Monache *et al., Ann. Chim., (Rome)*, **60**, 233 (1970)

C. officinalis L.
DITERPENOID: (-)-hardwickiic acid.
Cocker *et al., Tetrahedron Lett.*, 1983 (1965)

COPRIMUS
C. rotundus.
NORSESQUITERPENOID: isokobusone.
Hikino *et al., Chem. Pharm. Bull.*, **17**, 1390 (1969)

COPRINARIUS (Agaricales)
C. cotoneus.
PIGMENT: leprocybin.
Kopanski *et al., Ann. Chem.*, 1280 (1982)

COPROSMA (Rubiaceae)

C. acerosa.
PIGMENT: soranjidiol.
Bowie et al., Aust. J. Chem., 15, 332 (1962)

C. areolata.
ANTHRAQUINONE: copreolatin.
Briggs et al., J. Chem. Soc., 1718 (1952)

C. australis.
ANTHRAQUINONE: morindone.
Roberts et al., Aust. J. Chem., 30, 1553 (1977)

C. linariifolia Hook. f.
PIGMENTS: anthragallol 2-ethyl ether,
3-hydroxy-2-methylanthraquinone, nordamnacanthal,
rubiadin, rubiadin 1-methyl ether.
Gambarino et al., J. Nat. Prod., 48, 992 (1985)

C. lucida.
PIGMENT: 1,3-dihydroxy-2-hydroxymethylathraquinone.
Briggs et al., J. Chem. Soc., Perkin I, 1789 (1976)

C. propinqua A.Cunn.
COUMARIN: scopoletin.
IRIDOID: asperuloside.
Gambarino et al., J. Nat. Prod., 48, 992 (1985)

C. pyrifolium.
IRIDOIDS: asperuloside, deacetylasperuloside.
Gambarino et al., J. Nat. Prod., 48, 992 (1985)

C. robusta Raoul.
ANTHRAQUINONES: rubiadin, rubiadin 1-methyl ether.
COUMARIN: scopoletin.
IRIDOID: asperuloside.
Gambarino et al., J. Nat. Prod., 48, 992 (1985)

C. rotundifolia A.Cunn.
ANTHRAQUINONES: lucidin, rubiadin, rubiadin 1-methyl ether.
COUMARIN: scopoletin.
Gambarino et al., J. Nat. Prod., 48, 992 (1985)

C. tenuicaulis Hook. f.
ANTHRAQUINONES: 1,2-dihydroxy-6-methylanthroquinone,
3-hydroxy-2-methylanthraquinone, rubiadin, rubiadin
1-methyl ether.
COUMARIN: scopoletin.
PIGMENT: 6-methylalizarin.
(A, C) Briggs et al., J. Chem. Soc., Perkin I, 1789 (1976)
(P) Brew et al., J. Chem. Soc., C, 2001 (1971)

COPTIS (Ranunculaceae)

C. groenlandica.
PROTOBERBERINE ALKALOID: isocoptisine.
UNCHARACTERIZED ALKALOID: groenlandicine.
Cooper et al., Planta Med., 19, 23 (1971)

C. japonica Makino.
PROTOBERBERINE ALKALOIDS: berberine, columbamine,
coptisine, worenine.
FLAVONOIDS: coptiside I, coptiside II.
(A) Kitasato, J. Pharm. Soc., Japan, 30, 705 (1953)
(F) Fujiwara et al., Chem. Pharm. Bull., 24, 407 (1976)

C. occidentalis.
PROTOBERBERINE ALKALOID: berberine.
Awe, Arch. Pharm., 284, 352 (1951)

C. teeta Wall.
UNCHARACTERIZED ALKALOID: coptine.
Gross, Am. J. Pharm., 14, 193 (1873)

C. trifolia.
PROTOBERBERINE ALKALOID: berberine.
Awe, Arch. Pharm., 284, 352 (1951)

CORALLIA (Rhizophoraceae)

C. brachiata (Lour.) Merr.
INDOLE ALKALOID: (+)-hygrine.
Fitzgerald, Aust. J. Chem., 18, 589 (1965)

CORALLINA

C. buchordii.
TRITERPENOID: 3,4-seco-20(29)-lupen-3-oic acid.
Corbett et al., J. Am. Chem. Soc., 102, 1171 (1980)

CORBICULA

C. leana.
STEROID: corbisterol.
Toyama et al., Bull. Chem. Soc., Japan, 25, 355 (1952)

CORCHORUS

C. capsularis.
TRITERPENOIDS: capsugenin, corosin.
Manzoor-i-Khuda et al., Pakistan J. Sci. Ind. Res., 14, 49 (1971)

C. olitorius.
STEROIDS: coroloside, deglucocoroloside.
TRITERPENOID: corosin.
(S) Masslenikova et al., Khim. Prir. Soedin., 535 (1975)
(T) Manzoor-i-Khuda et al., Pakistan J. Sci. Ind. Res., 14, 49 (1971)

C. trilocularis.
STEROID: trilocularin.
Rao, Rao, Phytochemistry, 14, 533 (1975)

CORDIA

C. alliodora.
BENZOPYRAN: cordiaquinol C.
HOMOSESQUITERPENOIDS: alliodorin, alliodorol, allioquinol C,
cordiachromen A, cordillinol.
TRITERPENOIDS: 3,29-dihydroxyolean-12-en-27-oic acid,
3,29-dioxoolean-12-en-27-oic acid,
3-hydroxyolean-12-ene-27,28-dioic acid,
3-hydroxyolean-12-en-27-oic acid,
3-hydroxy-29-oxoolean-12-en-27-oic acid,
3-oxoolean-12-en-27-oic acid.
Manners, Jurd, J. Chem. Soc., Perkin I, 405 (1977)
Chen et al., J. Org. Chem., 48, 3525 (1983)

C. millenii.
QUINONES: cordiachromes A, B, C, D, E and F.
Moir, Thomson, J. Chem. Soc., Perkin I, 1352 (1973)

C. obliqua L.
TRITERPENOIDS:
5,7-dimethoxytaxifolin-3-)-α-L-rhamnopyranoside,
lup-20(29)-ene-3-O-β-maltoside.
Chuahen, Srivastava, Phytochemistry, 17, 1005 (1978)

C. verbenacea.
TRITERPENOIDS: cordialin A, cordialin B.
Velde et al., J. Chem. Soc., Perkin I, 2697 (1982)

CORDYCEPS (Ascomycetae)

C. ophioglossipides.
ESTER ALKALOID: ophiocardin.
Koenig et al., Chem. Ber., 113, 2221 (1980)

CORDYLA

C. africana.
ISOFLAVONOIDS: 6,7,3',4'-tetramethoxyisoflavone,
5,6,7,8-tetramethoxy-3',4'-methylenedioxyisoflavone.
Braz Filho et al., Phytochemistry, 10, 2834 (1971)

CORDYLINE (Agavaceae)

C. australis Hook. f.
STEROID: australigenin.
> Blundon *et al.*, *J. Nat. Prod.*, **47**, 246 (1984)

C. cannabina.
STEROIDS: brisbagenin, brisbagenone.
> Jewers *et al.*, *Steroids*, **24**, 203 (1974)

C. cannifolia.
STEROID: cordylagenin.
> Jewers *et al.*, *Tetrahedron Lett.*, 1475 (1974)

C. indivisa.
STEROID: pollinasterol.
> Itoh *et al.*, *Phytochemistry*, **16**, 1081 (1977)

C. stricta.
STEROIDS: cordylagenin, 18-hydroxycrabbogenin, pompeygenin, strictagenin.
> Jewers *et al.*, *Tetrahedron Lett.*, 1475 (1974)
> Blunden *et al.*, *Tetrahedron*, **37**, 2911 (1981)

COREOPSIS

C. grandiflora.
FLAVONOID: 7,8,3',4'-tetrahydroxyflavanone.
> King *et al.*, *J. Chem. Soc.*, 569 (1951)

C. martema.
CHALCONE: marein.
> Harborne *et al.*, *J. Am. Chem. Soc.*, **78**, 829 (1956)

C. neucensis.
CHALCONES: lanceolin, marein.
> Harborne, *Phytochemistry*, **16**, 927 (1977)

C. tinctoria.
CHALCONE: marein.
> Harborne *et al.*, *J. Am. Chem. Soc.*, **78**, 829 (1956)

CORIARIA (Coriariaceae)

C. japonica A.Gray.
SESQUITERPENOIDS: coryamyrtin, corianin.
> Kariyone *et al.*, *J. Pharm. soc., Japan*, **50**, 106 (1930)

CORIOLUS (Basidiomycetae)

C. consors.
NORSESQUITERPENOIDS: coriolins A, B and C.
SESQUITERPENOID: hirsutene.
> Takahashi *et al.*, *Tetrahedron Lett.*, 1637 (1970)

CORNULACA

C. monacantha Del.
TRITERPENOIDS: azizic acid, manevalic acid.
> Dawidar *et al.*, *Chem. Pharm. Bull.*, **27**, 2938 (1979)

CORNUS (Cornaceae)

C. controversa Hemsl.
SESQUITERPENOID: cornusol.
> Kurihara, Kikuchi, *J. Pharm. Soc., Japan*, **98**, 969 (1978)

C. femina.
GLUCOSIDE: cornoside.
> Hensen *et al.*, *Acta Chem. Scand.*, **27**, 367 (1973)

C. florida L.
TRITERPENOID: betulinic acid.
> Ruzicka *et al.*, *Helv. Chim. Acta*, **21**, 1706 (1938)

CORONILLA (Leguminosae)

C. glauca.
STEROID: alloglaucotoxigenin.
> Stoll *et al.*, *Helv. Chim. Acta*, **32**, 293 (1949)

C. hyrcana.
STEROIDS: deglucohyrcanoside, hyrcanoside.
> Bagirov, Komissarenko, *Khim. Prir. Soedin.*, 251 (1967)

C. varia L.
LACTOSE: 6-deoxy-D-mannono-1,4-lactone.
STEROIDS: coronillin, deglucohyrcanoside.
> (L) Opletal *et al.*, *Pharmazie*, **41**, 605 (1986)
> (S) Bagirov, Komissarenko, *Khim. Prir. Soedin.*, 251 (1967)

CORTEX

C. chinae.
SAPONINS: quinovinglycosides A, B and C.
> Hlasiwatz, *Justus Liebigs Ann. Chem.*, **111**, 182 (1859)

C. condurango.
ALKALOIDS: condurangamine A, condurangamine B.
> Pailer, Ganzinger, *Monatsh. Chem.*, **106**, 37 (1975)

CORTICUM (Hymenomycetae)

C. caeruleum.
PYRAN: cortalcerone.
> Baute *et al.*, *Phytochemistry*, **15**, 1753 (1976)

CORTINARIUS (Agaricaceae)

C. brunneus.
AMINO ACID: L-4-hydroxy-3-methoxyphenylalanine.
> Dardenne *et al.*, *Phytochemistry*, **16**, 1822 (1977)

C. infractus.
CARBOLINE ALKALOID: infractin.
> Steglich *et al.*, *Tetrahedron Lett.*, 2341 (1984)

C. odorifer.
PIGMENT: anhydrophlegmacin.
> Steglich *et al.*, *Z. Naturforsch.*, **27B**, 1286 (1972)

C. orellanus.
TOXIN: orellanine.
> Antkowiak *et al.*, *Bull. Acad. Pol. Sci.*, **23** 729 (1975); *Chem. Abstr.*, **84** 71474c

CORYDALIS (Papaveraceae/Fumariaceae)

C. ambigua Cham. et Schltd.
BENZOPHENANTHRIDINE ALKALOID: ambinine.
PROTOBERBERINE ALKALOIDS: corybulbine, corydaline, corydaline I, dehydrocorydaline, (±)-tetrahydrocoptisine.
> Freund, Josehpi, *Ber. Dtsch. Chem. Ges.*, **25**, 2411 (1892)
> Cui *et al.*, *Yaoxue Xuebao*, **19**, 904 (1985); *Chem. Abstr.*, 103 19853x

C. aurea Willd.
BENZYLISOQUINOLINE ALKALOID: bicucine.
ISOQUINOLINE ALKALOID: corypalline.
ISOQUINOLINE LACTONE ALKALOIDS: bicuculline, cordrastine.
OXOPROTOBERBERINE ALKALOID: α-allocryptopine.
PROTOBERBERINE ALKALOIDS: capauridine, capaurine, corydaline, (-)-tetrahydropalmatine.
UNCHARACTERIZED ALKALOID: aurotensine.
> Manske, *Can. J. Res.*, **9**, 436 (1933)

C. bulleyana.
ISOQUINOLINE ALKALOIDS: (+)-6-acetonylcorynoline, acetylcorydamine, (+)-acetylcorynoline, (+)-acetylisocorynoline, allocryptopine, bulleyanine, (+)-bulleyanoline, (-)-cheilanthifoline, (+)-conspermine, corycavamine, corydamine, (+)-corynoline, (+)-corynoloxine, dihydrosanguinarine, (+)-12-formyloxycorynoline, (+)-12-hydroxycorynoline,

(+)-isoboldine, (+)-isocorynoline, (+)-norjuziphine, (+)-6-oxoacetylcorynoline, (+)-scoulerine, (±)-stylopine.
Hong et al., Planta Med., 193 (1986)

C. caseana A. Gray.
ISOQUINOLINE LACTONE ALKALOID: bicuculline.
OXOPROTOBERBERINE ALKALOID: α-allocryptopine.
PROTOBERBERINE ALKALOIDS: caseamidine, caseamine, (±)-corypalmine, (-)-scoulerine, (-)-tetrahydropalmatine, (±)-tetrahydropalmatine.
Manske, Miller, Can. J. Res., 16B, 153 (1938)
Yu et al., Can. J. Chem., 49, 124 (1971)

C. cava.
APORPHINE ALKALOID: predicentrine.
Johns et al., Aust. J. Chem., 22, 1277 (1969)

C. cheilantheifolia Hemsl.
OXOPROTOBERBERINE ALKALOID: α-allocryptopine.
PROTOBERBERINE ALKALOIDS: berberine, (-)-canadine, cheilanthifoline, (-)-corypalmine.
Manske, Can. J. Res., 16B, 438 (1938)

C. clarkei.
PROTOBERBERINE ALKALOIDS: caseanadine, clarkeanidine.
Rothera et al., J. Nat. Prod., 48, 802 (1985)

C. claviculata (L.) DC.
DIBENZAZONINE ALKALOID: crassifolazonine.
ISOQUINOLINE ALKALOIDS: culacorine, cularine, limousamine, oxocularine, ribasidine.
Boente et al., Tetrahedron Lett., 2295 (1983)
Boente et al., Heterocycles, 23, 1069 (1985)

C. consperma.
NAPHTHOISOQUINOLINE ALKALOID: conspermine.
Fang et al., Planta Med., 50, 25 (1984)

C. crystallina Engelm.
ISOQUINOLINE LACTONE ALKALOID: bicuculline.
Manske, Can. J. Res., 9, 436 (1933)

C. decumbens Pers.
APORPHINE ALKALOID: bulbocapnine.
PROTOBERBERINE ALKALOID: dehydrocorydaline.
Kikkawa et al., J. Pharm. Soc., Japan, 79, 1244 (1959)

C. delavayi Franch.
NAPHTHOISOQUINOLINE ALKALOID: 6-acetylcorynoline.
Luo, Yunnan Zhiwu Yanjiu, 4, 69 (1982)

C. fedtshencoana Rgl.
APORPHINE ALKALOID: bulbocapnine.
NAPHTHOISOQUINOLINE ALKALOID: sanguinarine.
OXOQUINOLIZIDINE ALKALOID: protopine.
Manske, Can. J. Res., 21B, 117, 240 (1943)

C. fimrillifera Korsh.
OXOQUINOLIZIDINE ALKALOID: protopine.
Haworth, Perkin, J. Chem. Soc., 1769 (1926)

C. gigantea Trautv. et Mey.
ISOQUINOLINE ALKALOIDS: (-)-adlumidine, (-)-adlumine, (-)-cheilanthifoline, dihydroxysanguinarine, (-)-scoulerine.
ISOQUINOLINE LACTONE ALKALOID: bicuculline.
NAPHTHOISOQUINOLINE ALKALOID: sanguinarine.
OXOQUINOLIZIDINE ALKALOID: protopine.
Manske, Can. J. Res., 15B, 274 (1937)
Margvelashvili et al., Khim. Prir. Soedin., 832 (1976)

C. glaucescens Rgl.
APORPHINE ALKALOID: bulbocapnine.
OXOQUINOLIZIDINE ALKALOID: protopine.
PROTOBERBERINE ALKALOID: corydaline.
Mottus et al., Can. J. Chem., 31, 1144 (1953)

C. gortschakovii.
APORPHINE ALKALOIDS: bracteoline, corytuberine, domesticine, isoboldine.
BENZYLISOQUINOLINE ALKALOIDS: corgoine, gorchacoine, sendaverine.
ISOQUINOLINE LACTONE ALKALOIDS: (+)-adlumine, (+)-bicuculline.
OXOPROTOBERBERINE ALKALOID: cryptopine.
OXOQUINOLIZIDINE ALKALOID: protopine.
PROTOBERBERINE ALKALOID: isocorydaline.
Tomita, Kitamura, J. Pharm. Soc., Japan, 79, 1092 (1959)
Irgashev et al., Khim. Prir. Soedin., 127 (1977)

C. govaniana.
PROTOBERBERINE ALKALOID: govanine.
Mehra et al., J. Chem. Soc., B, 14, 58 (1976)

C. incisa Pers.
ISOQUINOLINE ALKALOIDS: corydamine, N-formylcorydamine.
NAPHTHOISOQUINOLINE ALKALOIDS: acetylcorynoline, acetylisocorynoline, corynoline, (+)-14-epi-corynoline, corynoloxine.
PROTOBERBERINE ALKALOID: (-)-corypalmine.
Chatterjee et al., J. Indian Chem. Soc., 29, 921 (1952)
Yakao, Chem. Pharm. Bull., 19, 247 (1971)

C. koidzumiana.
PROTOBERBERINE ALKALOIDS: corydalidzine, corynoxidine, epi-corynoxidine.
Tani et al., Tetrahedron Lett., 803 (1973)

C. ledebouriana K. et K.
NAPHTHOISOQUINOLINE ALKALOID: sanguinarine.
OXOPROTOBERBERINE ALKALOID: α-allocryptopine.
OXOQUINOLIZIDINE ALKALOID: protopine.
SPIROISOQUINOLINE ALKALOID: ledeborine.
Israilov et al., Khim. Prir. Soedin., 268 (1975)

C. lutea (L.) DC.
OXOPROTOBERBERINE ALKALOID: ochrobirine.
PROTOBERBERINE ALKALOIDS: (-)-tetrahydrocolumbamine, (-)-,tetrahydropalmatine.
Manske, Can. J. Res., 18B, 75 (1940)

C. marshalliana.
APORPHINE ALKALOIDS: bulbocapnine, corydine, domesticine, isobolbine.
ISOQUINOLINE LACTONE ALKALOIDS: adlumidine, bicuculline.
OXOQUINOLIZIDINE ALKALOID: protopine.
Manske, Can. J. Res., 8, 210, 404 (1933)

C. micrantha (Engelm) Gray.
PROTOBERBERINE ALKALOIDS: capauridine, capaurine, (-)-scoulerine, (-)-tetrahydropalmatine.
Manske, Can. J. Res., 16, 81 (1938)

C. montana. (Engelm) Britton.
PROTOBERBERINE ALKALOIDS: capauridine, capaurimine, capaurine, (-)-scoulerine, (±)-tetrahydropalmatine.
Manske, Can. J. Res., 16, 81 (1938)

C. nobilis Pers.
APORPHINE ALKALOID: corytuberine.
ISOQUINOLINE LACTONE ALKALOIDS: bicuculline, corlumine.
OXOPROTOBERBERINE ALKALOID: cryptopine.
PROTOBERBERINE ALKALOID: isocorypalmine.
Manske, Can. J. Res., 14B, 325, 347 (1936)

C. ochotensis Turez.
OXOPROTOBERBERINE ALKALOIDS: cryptocavine, ochrobirine.
PROTOBERBERINE ALKALOIDS: corytenchine, corytenchirine, lienkonine.
UNCHARACTERIZED ALKALOID: aurotensine.
Lu, Wang, Phytochemistry, 21, 809 (1982)

C. ochotensis Turez. var. raddeana.
AMINE ALKALOID: aobamidine.
BENZYLISOQUINOLINE ALKALOID: aobamine.
SPIROISOQUINOLINE ALKALOIDS: raddeanamine, raddeanidine,
   raddeanine, raddeanone.
   Kametani et al., Heterocycles, **4**, 723 (1976)

C. ochroleuca Koch.
ISOQUINOLINE LACTONE ALKALOID: bicuculline.
OXOPROTOBERBERINE ALKALOID: ochrobirine.
PROTOBERBERINE ALKALOIDS: (-)-corypalmine,
   (-)-tetrahydrocolumbamine, (-)-tetrahydropalmatine.
   Manske, Can. J. Res., **18B**, 75 (1940)

C. ophiocarpa (Hook.) Makino.
OXOPROTOBERBERINE ALKALOIDS: α-allocryptopine,
   cryptocavine.
PROTOBERBERINE ALKALOIDS: carpoxidine,
   13-hydroxystylopine, ophiocarpine.
UNCHARACTERIZED ALKALOIDS: bebeerine, (-)-cryptopalmine.
   Manske, Marion, J. Am. Chem. Soc., **62**, 2042 (1940)
   Tani et al., J. Pharm. Soc., Japan, **98**, 1243 (1978)

C. paczoskii N. Busch.
APORPHINE ALKALOID: bulbocapnine.
PROTOBERBERINE ALKALOID: corydaline.
SPIROISOQUINOLINE ALKALOID: corpaine.
   Kikkawa et al., J. Pharm. Soc., Japan, **79**, 1244 (1959)

C. pallida Pers.
ISOQUINOLINE ALKALOID: corypalline.
PROTOBERBERINE ALKALOIDS: capauridine, capaurimine,
   capaurine, palmatine chloride, (±)-tetrahydropalmatine.
   Kametani et al., J. Pharm. Soc., Japan, **95**, 1103 (1975)

C. pallida Pers. var. tenuis Yatabe.
PROTOBERBERINE ALKALOID: kikemanine.
   Kametani, J. Chem. Soc., C, 1060 (1970)

C. persica Cham. et Schlecht.
NAPHTHOISOQUINOLINE ALKALOIDS: chelerythrine, sanguinarine.
OXOQUINOLIZIDINE ALKALOID: protopine.
   Manske, Can. J. Res., **21B**, 140 (1943)

C. platycarpa Makino.
ISOQUINOLINE LACTONE ALKALOID: bicuculline.
PROTOBERBERINE ALKALOIDS: corybulbine,
   dehydrocapaurimine, (-)-tetrahydrocolumbamine,
   (-)-tetrahydropalmatine.
   Manske, Can. J. Res., **16**, 81 (1938)
   Tani et al., J. Pharm. Soc., Japan, **90**, 903 (1970)

C. popovii Nevski.
APORPHINE ALKALOID: bulbocapnine.
PROTOBERBERINE ALKALOID: corydaline.
   Kikkawa, J. Pharm. Soc., Japan, **79**, 1244 (1959)

C. pseudoadunca M.Pop.
APORPHINE ALKALOID: adukaine.
ISOQUINOLINE ALKALOID: (-)-adlumidine.
ISOQUINOLINE LACTONE ALKALOIDS: (+)-bicuculline,
   (±)-bicuculline, 2-hydrastine.
NAPHTHOISOQUINOLINE ALKALOID: sanguinarine.
OXOQUINOLIZIDINE ALKALOID: protopine.
PROTOBERBERINE ALKALOIDS: coramine, (-)-scoulerine.
   Yunusov et al., Khim. Prir. Soedin., 340 (1966)

C. rosea Leych.
ISOQUINOLINE ALKALOIDS: (-)-adlumidine, (±)-adlumidine,
   (-)-adlumine.
OXOQUINOLIZIDINE ALKALOID: protopine.
   Manske, Can. J. Res., **16B**, 89 (1938)

C. scouleri H.K.
OXOPROTOBERBERINE ALKALOID: α-allocryptopine.
PROTOBERBERINE ALKALOID: (-)-scoulerine.
   Manske, Can. J. Res., **20B**, 49 (1942)

C. sempervirens (L.) Pers.
AMINE ALKALOID: adlumiceine.
ISOQUINOLINE LACTONE ALKALOIDS: bicuculline, capnoidine.
LACTONE ALKALOID: adlumiceine enol-lactone.
OXOPROTOBERBERINE ALKALOID: cryptopine.
   Preininger et al., Phytochemistry, **12**, 2513 (1973)

C. sewerzovii Rgl.
ISOQUINOLINE LACTONE ALKALOIDS: (-)-bicuculline, corlumine.
NAPHTHOISOQUINOLINE ALKALOID: sanguinarine.
OXOPROTOBERBERINE ALKALOIDS: α-allocryptopine, cryptopine.
OXOQUINOLIZIDINE ALKALOID: protopine.
PROTOBERBERINE ALKALOID: corydaline.
   Yunusov et al., Khim. Prir. Soedin., 61 (1968)

C. sibirica Pers.
ISOQUINOLINE LACTONE ALKALOIDS: bicuculline, corlumine.
OXOPROTOBERBERINE ALKALOID: cryptopine.
PROTOBERBERINE ALKALOID: (-)-scoulerine.
SPIROISOQUINOLINE ALKALOID: sibiricine.
   Manske, Can. J. Chem., **47**, 3585 (1969)

C. solida Sm.
APORPHINE ALKALOID: bulbocapnine.
PROTOBERBERINE ALKALOID: solidaline.
SPIROBENZYLISOQUINOLINE ALKALOID: (±)-corysolidine.
   Manske et al., Can. J. Chem., **56**, 383 (1978)
   Rohimizadeh et al., Phytochemistry, **25**, 2245 (1986)

C. stewartii.
UNCHARACTERIZED ALKALOIDS: corycidine, corydicine,
   corydinine.
   Ikram et al., Pakistan J. Sci. Ind. Res., **9**, 34 (1966)

C. stricta Steph.
ISOQUINOLINE ALKALOIDS: (+)-α-hydrastine,
   N-methylcorypalline.
NAPHTHOISOQUINOLINE ALKALOID,: sanguinarine.
OXOQUINOLIZIDINE ALKALOID: protopine.
   Irgashev et al., Khim. Prir. Soedin., 490 (1983)

C. tashiroi.
PROTOBERBERINE ALKALOID: dehydrodiscratamine chloride.
   Tani et al., Planta Med., **41**, 403 (1981)

C. ternata Makino.
APORPHINE ALKALOID: (-)-corydine.
OXOPROTOBERBERINE ALKALOID: α-allocryptopine.
PROTOBERBERINE ALKALOIDS: (+)-tetrahydrocoptisine,
   (±)-tetrahydrocoptisine, (-)-tetrahydrocoptisine.
   Arumugam et al., Chem. Ber., **91**, 40 (1958)

C. thalictrifolia Franch.
ISOQUINOLINE ALKALOID: adlumidine.
PROTOBERBERINE ALKALOIDS: (-)-corypalmine, thalictrifoline.
   Chatterjee et al., J. Indian Chem. Soc., **29**, 921 (1952)

C. tuberosa DC. (C. cava Schwg.).
APORPHINE ALKALOIDS: bulbocapnine, (-)-corydine,
   corytuberine.
OXOPROTOBERBERINE ALKALOIDS: corycavamine, corycavidine,
   corycavine.
PROTOBERBERINE ALKALOIDS: (+)-canadine, corybulbine,
   corydaline, (+)-corypalmine, dehydrocorydaline,
   isocorybulbine, isocorypalmine, (+)-thalictricavine.
   Spath et al., Ber. Dtsch. Chem. Ges., **56**, 877 (1923)
   Gadamer, Knorck, Apoth. Ztg., **41**, 928 (1926)

C. vaginans.
　SPIROISOQUINOLINE ALKALOID: 1-O-methylcorpaine.
　　Margvelashvili *et al.*, *Khim. Prir. Soedin.*, 123 (1976)

CORYLUS
C. avellana.
　STEROIDS: avenasterol, campesterol, citrostadienol, gramisterol,
　　obtusifoliol, β-sitosterol, stigmasterol.
　　Van Dijck *et al.*, *Rev. Fr. Corps Gras*, **22**, 619 (1975); *Chem. Abstr.*, **84**
　　161775f

C. maxima.
　STEROIDS: avenasterol, campesterol, citrostadienol, gramisterol,
　　obtusifoliol, β-sitosterol, stigmasterol.
　　Van Dijck *et al.*, *Rev. Fr. Corps Gras*, **22**, 619 (1975); *Chem. Abstr.*, **84**
　　161775f

CORYNANTHE
C. mayumbensis.
　CARBOLINE ALKALOID: 2-isorauniticine.
　　Strombon, Bruhn, *Acta Pharm. Suec.*, **15**, 127 (1978)

C. yohimbe.
　CARBOLINE ALKALOIDS: alloyohimbine, 19-dehydroyohimbine,
　　yohimbine.
　　Arndt, Jerassi, *Experientia*, **21**, 566 (1965)

CORYPHANTHA
C. bumamma.
　AMINE ALKALOID: N-methyl-4-methoxyphenethylamine.
　　Bruhn *et al.*, *Acta Pharm. Suec.*, **12**, 199 (1975)

C. calipensis.
　AMINE ALKALOIDS: (-)-calipamine, macromerine methyl ether.
　　Bruhn, Agurell, *J. Pharm. Sci.*, **63**, 574 (1974)

C. cornifera var. echinus.
　AMINE ALKALOID: 4-methoxy-α-hydroxy-β-phenethylamine.
　　Hornemann *et al.*, *J. Pharm. Sci.*, **61**, 41 (1972)

C. greenwoodii.
　AMINE ALKALOIDS: corypanthine,
　　N,N-dimethyl-3,4-dimethoxy-α-methoxyphenethylamine.
　　Mayer *et al.*, *J. Nat. Prod.*, **46**, 688 (1983)

C. macromeria (Engelm.) Lem. var. runyonii B. et R.
　AMINE ALKALOID: N-formylnormacromerine.
　　Keller *et al.*, *J. Pharm. Sci.*, **62**, 408 (1973)

C. runyonii B. et R.
　AMINE ALKALOID: macromerine.
　　Hodgkins *et al.*, *Tetrahedron Lett.*, 1321 (1967)

CORYSPERMUM
C. leptopyrum.
　ISOQUINOLINE ALKALOID: salsoline.
　　Spath *et al.*, *Ber. Dtsch. Chem. Ges.*, **67**, 1214 (1934)

COSMOS
C. bipinnatus.
　BENZOPYRAN: cosmosiin.
　　Nakaoki *et al.*, *J. Pharm. Soc., Japan*, **55**, 967 (1935)

COSTUS
C. speciosus.
　ALIPHATICS: 14-oxoheptacosanoic acid, 15-oxooctacosanoic
　　acid, 14-oxotricosanoic acid,
　　tetradecyl-13-methylpentadecanoate,
　　tetradecyl-11-methyltridecanoate, triacontanoic acid,
　　triacontanol.
　FLAVONOID: diosgenin.

SAPONINS: costicoside I, costicoside J.
STEROIDS: β-sitosterol, 5α-stigmast-9(11)-en-3β-ol.
　　(Al, F, S) Gupta *et al.*, *Phytochemistry*, **25**, 1899 (1986)
　　(Sap) Singh *et al.*, *ibid.*, **25**, 911 (1986)

COTULA (Compositae/Asteraceae)
C. hispida.
　SESQUITERPENOID: isovaleryloxycostunolide.
　　Bohlmann, Zdero, *Phytochemistry*, **18**, 336 (1979)

COTYLEDON (Crassulaceae)
C. orbiculata.
　BUFADIENOLIDES: orbicusides A, B and C.
　　Steyn *et al.*, *J. Chem. Soc., Perkin I*, 1633 (1986)

C. wallichii Harv.
　SAPONIN: cotyledoside.
　　Van Wyk *et al.*, *J. S. Afr. Chem. Inst.*, **28**, 281 (1975)

COURBONIA
C. virgata.
　QUATERNARY ALKALOID: stachydrine ethyl ester.
　　Henry, King, *J. Chem. Soc.*, 2866 (1950)

COUROUPITA
C. guianensis.
　ALKALOIDS: couroupitine A, couroupitine B.
　　Sen *et al.*, *Tetrahedron Lett.*, 609 (1974)

CRASPIDOSPERMUM
C. verticillatum Boj.
　CARBOLINE ALKALOIDS: 14-vincine, 16-epi-14-vincine.
　　Kan-Fan *et al.*, *C. R. Acad. Sci.*, **272C**, 1431 (1971)

C. verticillatum Boj. var. petiolare.
　CARBOLINE ALKALOID: andrangine.
　HOMOCARBOLINE ALKALOIDS: andranginine,
　　deformylstemmadine.
　　Kan-Fan *et al.*, *Bull. Soc. Chim. Fr.*, 2839 (1974)

CRATAEGUS (Rosaceae)
C. monogyna L.
　FLAVONOIDS: 4‴-O-acetylvitexin 2″-O-rhamnoside,
　　homoorientin rhamnoside, isoorientin 2″-O-rhamnoside,
　　isoschaftoside, 8-methoxykaempferol
　　3-O-β-D-glucopyranoside, orientin rhamnoside, rutin,
　　saponaretin rhamnoside, schaftoside, spiraeoside, vicenin-1,
　　vicenin-2, vicenin-3, vitexin 2″-O-rhamnoside.
　　Nikolov, *Khim. Prir. Soedin.*, 422 (1975)
　　Nikolov *et al.*, *Chem. Abstr.*, **97** 212665y

C. oxyacantha L.
　TRITERPENOIDS: acantholic acid, trans-crategolic acid.
　UNCHARACTERIZED TERPENOID: neotaegolic acid.
　　Tschesche, Popper, *Chem. Ber.*, **92**, 320 (1959)

C. phaenopyrum.
　FLAVONOID: 5,7,3′,4′-tetrahydroxyflavanone 5.3′-diglucoside.
　　Kowalewski, *Planta Med.*, **19**, 311 (1971)

CRATOXYLON (Guttiferae/Clusiaceae)
C. celibicum Blume.
　XANTHONE: celibaxanthone.
　　Stout *et al.*, *Tetrahedron*, **19**, 667 (1963)

CREMASTOSPERMA (Annonaceae)

C. polyphlebum.

BISBENZYLISOQUINOLINE ALKALOID: phlebicine.

Cava et al., J. Org. Chem., 39, 3588 (1974)

CREPIS (Compositae)

C. conyzaefolia.

ACIDS: (6Z,9Z,12R,13S)-12,13-epoxy-6,9-octadecadienoic acid, (6E,9Z,12R,13S)-12,13-epoxy-6,9-octadecadienoic acid.

LACTONE: (-)-(S,S)-12-hydroxy-13-octadec-cis-9-enolide.

Spencer et al., Phytochemistry, 16, 282, 764 (1977)

C. japonica.

SESQUITERPENOIDS: crepisides A, B, C, D, E, F, G, H and I, glucozaluzanin C.

Miyase et al., Chem. Pharm. Bull., 33, 4451 (1985)

C. napifera (Franch) Babc.

TRITERPENOID: taraxeryl acetate.

Liang, Chen, Yaoxue Tongbao, 17, 242 (1982); Chem. Abstr., 97 159491k

C. pygmaea.

SESQUITERPENOIDS: 4,5-dihydro-11-nor-hydroxy-7,11-santonin, 8-epi-grossheimin, 11-hydroxy-13-nor-3-oxo-1,7(11)-eudesmadien-12,6-olide.

Casinovi et al., Planta Med., 44, 186 (1982)

C. virens.

SESQUITERPENOID: 8-epi-grossheimin.

Barbetti et al., Collect. Czech. Chem. Commun., 44, 3123 (1979)

CRESSA

C. cretica L.

FLAVONOID: quercetin glycoside.

Purushothaman, J. Res. Indian Med., 9, 109 (1974); Chem. Abstr., 85 2560m

CRINODENDRON (Elaeocarpaceae)

C. hookerianum.

TRITERPENOID: cucurbitacin H.

Holzepfel, Enslin, J. S. Afr. Chem. Inst., 17, 142 (1964)

CRINUM (Amaryllidaceae)

C. angustum.

CARBOLINE ALKALOID: angustine.

ISOQUINOLINE ALKALOIDS: 6α-hydroxybuphanisine, 6β-hydroxybuphanisine.

Ali et al., Phytochemistry, 20, 1121, 1731 (1981)

C. asiaticum L.

ISOQUINOLINE ALKALOIDS: criasbetaine, crinamine, hamayne, lycorine, lycoriside, palmilycorine, ungeremine.

Ghosal et al., Phytochemistry, 24, 2703 (1985)

Ghosal et al., J. Chem. Res. Synop., (3), 112 (1986)

C. asiaticum L. var. japonicum Bak.

PHENANTHRIDINE ALKALOID: 4,5-etheno-8,9-methylenedioxyphenanthridine.

ESTER: methyl linoleate.

STEROID: stigmasterol.

TRITERPENOIDS: cycloartenol, cyclolaudenol, 31-norcyclolaudenol.

Takagi et al., J. Pharm. Soc., Japan, 97, 1155 (1977)

C. augustum.

AMINE ALKALOID: N-(3-hydroxy-4-methoxybenzylidene)-4'-hydroxyphenethylamine.

Ghosal et al., J. Chem. Res. Synop., (1), 28 (1986)

C. erubescens.

ISOQUINOLINE ALKALOIDS: flexinine, 6-hydroxycrinamine.

Boit, Ehmke, Ber. Dtsch. Chem. Ges., 90, 369 (1957)

C. fimbriatum.

ISOQUINOLINE ALKALOID: 6-hydroxycrinamine.

Fales et al., Chem. Ind., 1415 (1959)

C. gigantum Andr.

ISOQUINOLINE ALKALOID: lycorine.

Greathouse, Rigler, Am. J. Bot., 28, 702 (1941)

C. latifolium.

AMINE ALKALOID: latisoline.

ISOQUINOLINE ALKALOIDS: pratorimine, pratosine.

C. macrantherum Engl.

AMINE ALKALOIDS: O-acetylmacranthine, diacetylmacranthine, macranthine.

ISOQUINOLINE ALKALOID: acetylcaranine.

LACTONE ALKALOID: macronine.

Hauth, Stauffacher, Helv. Chim. Acta, 47, 185 (1964)

C. moorei Hook. f.

ISOQUINOLINE ALKALOIDS: crinamidine, crinine.

Boit, Ehmke, Chem. Ber., 87, 1704 (1954)

C. oligantum.

ISOQUINOLINE ALKALOID: oliganine.

Dopke et al., Z. Chem., 23, 101 (1983)

C. ornatum.

ALKALOIDS: ornazamine, ornazidine.

Onyriuka, Jackson, Isr. J. Chem., 17, 185 (1978)

C. powellii Hort.

ISOQUINOLINE ALKALOIDS: cherylline, precriwelline.

UNCHARACTERIZED ALKALOIDS: cinalbine, crinosine.

Wildman, Bailey, J. Org. Chem., 83, 3749 (1968)

C. powellii Hort. var. harlamense.

ISOQUINOLINE ALKALOIDS: crispaline, powellamine.

Dopke et al., Arch. Pharm., 295, 868 (1962)

C. pratense Herb.

ISOQUINOLINE ALKALOID: lycorine.

Greathouse, Rigler, Am. J. Bot., 28, 702 (1941)

C. scabrum.

ISOQUINOLINE ALKALOID: lycorine.

Greathouse, Rigler, Am. J. Bot., 28, 702 (1941)

C. zeanicum.

ISOQUINOLINE ALKALOID: 6-hydroxycrinamine.

Goosen et al., J. Chem. Soc., 1088 (1960)

C. zeylanicum.

ISOQUINOLINE ALKALOID: zeylamine.

Dopke et al., Z. Chem., 26, 438 (1986)

CRIOCERAS

C. dipladeniiflorus.

CARBOLINE ALKALOID: 12-methoxy-14-vincamine.

DIMERIC IMIDAZOLIDINOCARBAZOLE ALKALOID: criophylline.

INDOLE ALKALOID: 12-methoxyvoaphylline.

Bruneton et al., Phytochemistry, 13, 1963 (1974)

C. longiflorus Pierre.

CARBOLINE ALKALOIDS: 14,15-dehydrovincamine, 16epi-14,15-dehydrovincamine.

Cave et al., C. R. Acad. Sci., 272C, 1387 (1971)

CRITONIA (Compositae/Eupatorieae)

C. daleoides.
BENZOFURAN: tremetone.
: Bohlmann *et al., Phytochemistry,* **16**, 1973 (1977)

C. morifolia (Mill.) King et Rob.
SESQUITERPENOIDS: critonilide, isocritonilide.
: Bohlmann *et al., Chem. Ber.,* **110**, 301 (1977)

CROCUS (Iridaceae)

C. albiflorus.
CAROTENOID: crocetin.
: Jucker, Karrer, *Carotenoids,* 242 (1968)

C. luteus.
CAROTENOID: crocetin.
: Jucker, Karrer, *Carotenoids,* 242 (1968)

C. sativa L.
CAROTENOIDS: crocetin, crocetin di-β-glucoside, crocetin
β-D-gentiobioside, crocetin-β-D-glucoside,
crocetin-β-D-glucoside-D-gentiobioside, crocin, crocin-2,
crocin-3, crocin-4.
: Dhingra *et al., Indian J. Chem.,* **13**, 339 (1975)
: Pfander *et al., Pure Appl. Chem.,* **47**, 121 (1976)

CRONQUISTIANTHUS (Compositae)

C. chachapoyensis.
SESQUITERPENOIDS: 8β-angeloyloxycronquistianthusic acid,
cronquistianthusic acid, elemacronquistianthusic acid,
8β-tigloyloxyconquistianthusic acid.
: Bohlmann *et al., Justus Liebigs Ann. Chem.,* 240 (1984)

CROOMIA

C. heterosepala Okuyama.
ALKALOID: croomine.
: Noro *et al., Chem. Pharm. Bull.,* **27**, 1495 (1979)

C. nana Burm.
PYRROLIZIDINE ALKALOIDS: cronaburmine, crotananine.
: Siddiqui *et al., Phytochemistry,* **17**, 2143 (1978)

CROPTILON

C. divaricatum.
DITERPENOID: norjulslimdioline.
: Dominguez *et al., Rev. Latinoam. Quim.,* **3**, 177 (1973)

CROSSOPTERYX

C. kotschyana Fenzl.
UNCHARACTERIZED ALKALOID: crossoptine.
: Blaise, *Les Crossopteryx Africains Etude Botanique, Thesis, Paris*
(1932)

CROSSOSTYLIS

C. multiflora.
TROPANE ALKALOID: tropan-3α-yl ferulate.
: Gneccio *et al., J. Nat. Prod.,* **46**, 398 (1983)

C. sebertii.
TROPANE ALKALOIDS: tropan-3α-yl cinnamate, tropan-3α-yl
ferulate.
: Gnaccio *et al., J. Nat. Prod.,* **46**, 398 (1983)

CROTALARIA

C. albida.
PYRROLIZIDINE ALKALOID: croalbidine.
: Sawhney *et al., Indian J. Chem.,* **11**, 88 (1973)

C. anagyroides.
PYRROLIZIDINE ALKALOIDS: anacrotine 1, methylpyrrolizidine.
: Atal *et al., Tetrahedron Lett.,* 537 (1966)

C. aridicola Domin.
PYRROLIZIDINE ALKALOIDS:
7β-acetoxy-1-methoxymethyl-1,2-dehydro-8-pyrrolizidine,
1β,2β-epoxy-1α-methoxymethyl-8-pyrrolizidine,
7α-hydroxy-1-methoxymethyl-1,2-dehydro-8-pyrrolizidine,
7β-hydroxy-1-methoxymethyl-1,2-dehydro-8-pyrrolizidine.
: Culvenor *et al., Aust. J. Chem.,* **20**, 757 (1967)

C. axillaris Ait.
PYRROLIZIDINE ALKALOIDS: axillaridine, axillarine.
: Crout, *J. Chem. Soc., C,* 1379 (1969)

C. barbata.
PYRROLIZIDINE ALKALOID: crobarbatine.
: Puri *et al., Experientia,* **29**, 390 (1973)

C. burha.
PYRROLIZIDINE ALKALOIDS: croburhine, crotalarine.
BENZOPYRAN: sumatrol.
: (A) Rao *et al., Indian J. Chem.,* **13**, 835 (1975)
: (B) Kerry *et al., J. Chem. Soc.,* 1601 (1939)

C. californicus.
DITERPENOID: barbasoic acid methyl ester.
: Wilson *et al., J. Am. Chem. Soc.,* **98**, 3669 (1976)

C. candicans.
PYRROLIZIDINE ALKALOIDS: crocandine, cropodine.
: Haksar *et al., Indian J. Chem.,* **21B**, 492 (1982)

C. crassipes Hook.
PYRROLIZIDINE ALKALOID: retusamine.
: Culvenor *et al., Aust. J. Chem.,* **20**, 801 (1967)

C. crispata F.Muell.
PYRROLIZIDINE ALKALOIDS: crispatine, fulvine.
: Culvenor, Smith, *Aust. J. Chem.,* **16**, 239 (1963)

C. dura Wood et Evans.
PYRROLIZIDINE ALKALOID: crotaline.
: Marais, *Onderstepoort Journal,* **20**, 61 (1944)

C. fulva Roxb.
PYRROLIZIDINE ALKALOID: fulvine.
: Schoental, *Aust. J. Chem.,* **16**, 233 (1963)

C. globifer Mey.
PYRROLIZIDINE ALKALOIDS: crotaline, globiferine.
: Marais, *Onderstepoort Journal,* **20**, 61 (1944)
: Brown *et al., Phytochemistry,* **23**, 457 (1984)

C. goreensis Guill. et Perr.
PYRROLIZIDINE ALKALOID:
7β-hydroxy-1-methylene-8-pyrrolizidine.
: Culvenor, Smith, *Aust. J. Chem.,* **14**, 284 (1961)

C. grahamiana R.Wight et Walk. Arm.
PYRROLIZIDINE ALKALOID: grahamine.
: Atal *et al., Aust. J. Chem.,* **22**, 1773 (1969)

C. grantiana.
PYRROLIZIDINE ALKALOID: grantianine.
: Adams *et al., J. Am. Chem. Soc.,* **84**, 571 (1942)

C. incana.
PYRROLIZIDINE ALKALOID: anacrotine.
: Mattocks, *J. Chem. Soc., C,* 235 (1968)

C. intermedia.
PYRROLIZIDINE ALKALOIDS: integerrimine, usaramine.
: Suri *et al., Indian J. Pharm.,* **37**, 96 (1975)

C. laburnifolia L.
PYRROLIZIDINE ALKALOIDS: crotalaburnine, hydroxysenkirkine.
Crout, *J. Chem. Soc., Perkin I*, 1602 (1972)

C. laburnifolia L. subsp. eldomae.
PYRROLIZIDINE ALKALOIDS: crotafoline, madurensine.
Crout, *J. Chem. Soc., Perkin I*, 1602 (1972)

C. madurensis.
PYRROLIZIDINE ALKALOIDS: cromadurine, madurensine.
BENZOPYRANS: crotmadine, crotmarine.
(A) Atal *et al., Tetrahedron Lett.*, 537 (1966)
(B) Bhakuni *et al., J. Nat. Prod.*, 47, 585 (1984)

C. mitchelii Benth.
PYRROLIZIDINE ALKALOID: retusamine.
Culvenor *et al., Aust. J. Chem.*, 20, 801 (1967)

C. mucronata Desv.
PYRROLIZIDINE ALKALOIDS: mucronitine, mucronitinine.
Bhacca, Sharma, *Tetrahedron*, 24, 6319 (1968)

C. nana Burns.
PYRROLIZIDINE ALKALOIDS: crotaburnine, crotananine.
Siddiqui *et al., Phytochemistry*, 17, 2143 (1978)

C. novae-hollandiae DC.
PYRROLIZIDINE ALKALOID: retusamine.
Culvenor *et al., Aust. J. Chem.*, 20, 801 (1967)

C. retusa L.
PYRROLIZIDINE ALKALOID: retusamine.
Culvenor *et al., Aust. J. Chem.*, 10, 464 (1957)

C. semperflorus Vent.
PYRROLIZIDINE ALKALOID: crosemperine.
Atal *et al., Aust. J. Chem.*, 20, 205 (1967)

C. spectabilis.
PYRROLIZIDINE ALKALOID: retronecanol.
Manske, *Can. J. Res.*, 5, 651 (1931)

C. stricta.
PYRROLIZIDINE ALKALOID: crotastrictine.
Ghandhi *et al., Curr. Sci.*, 37, 285 (1968)

C. tetragona.
AMINO ACID: 2-amino-5-hydroxyhexanoic acid.
Suri *et al., Indian J. Pharm.*, 37, 96 (1975)

C. trifoliastrum Willd.
PYRROLIZIDINE ALKALOIDS:
1β,2β-epoxy-1α-hydroxymethyl-8-purrolizidine,
1β,2β-epoxy-1α-methoxymethyl-8-pyrrolizidine,
7α-hydroxy-1-methoxymethyl-1,2-dehydro-8-pyrrolizidine,
7β-hydroxy-1-methoxymethyl-1,2-dehydro-8-pyrrolizidine.
Culvenor *et al., Aust. J. Chem.*, 20, 757 (1967)

C. usaramoensis E.G.Baker.
PYRROLIZIDINE ALKALOIDS: usaramine, usaramoensine.
Culvenor, Smith, *Aust. J. Chem.*, 19, 2127 (1966)

C. verrucosa.
FLAVONOID: kaempferol 7,4'-di-O-L-rhamnopyranoside.
Rao *et al., Fitoterapia*, 56, 175 (1985)

C. virgulata subsp. grantiana.
PYRROLIZIDINE ALKALOIDS: grantaline, grantianine.
Smith, Culvenor, *Phytochemistry*, 23, 473 (1984)

C. walkeri Arnott.
PYRROLIZIDINE ALKALOIDS: O-acetylcrotaverrine, crotaverrine.
STEROID: β-sitosterol.
Suri *et al., Indian J. Chem.*, 14B, 471 (1976)

CROTON (Euphorbiaceae)

C. argyrophylloides.
DITERPENOID: terpenoid Aa-ICM.
Monte *et al., J. Nat. Prod.*, 47, 55 (1984)

C. balsamifera Jacq.
PHENANTHRENE ALKALOID: norsinoacutine.
Chambers *et al., J. Chem. Soc., Chem. Commun.*, 449 (1966)

C. bonplandianum.
PHENANTHRENE ALKALOID:
3-methoxy-4,6-dihydroxymorphanandien-7-one.
Tiwari *et al., Phytochemistry*, 20, 863 (1981)

C. californicus.
DITERPENOID: barbascic acid methyl ester.
Wilson *et al., J. Am. Chem. Soc.*, 98, 3669 (1976)

C. californicus var. tenuis.
XANTHONE: 1,2,3,4,6,7-hexamethoxyxanthone.
Tammami *et al., Phytochemistry*, 16, 2040 (1977)

C. caudatus.
NORDITERPENOIDS: crotocaudin, isocrotocaudin.
Chatterjee *et al., Tetrahedron*, 33, 2407 (1977)

C. columnaris. (C. joufra).
DITERPENOID: plaunol, swassin.
Rosengsumran *et al., J. Nat. Prod.*, 45, 772 (1982)

C. corylifolius.
DITERPENOIDS: crotofolin A, crotofolin E.
Chan *et al., J. Am. Chem. Soc.*, 97, 4037 (1975)

C. diasii.
DITERPENOID: diasin.
De Alvarenga *et al., Phytochemistry*, 17, 1773 (1978)

C. discolor Willd.
PHENANTHRENE ALKALOID: 8,14-dihydrosalutaridine.
SPIRONORAPORPHINE ALKALOID: discolorine.
Stuart *et al., J. Chem. Soc., C*, 1228 (1970)

C. draco.
DITERPENOID: draconin.
Rodriguez-Hahn *et al., Rev. Latinoam. Quim.*, 6, 123 (1975)

C. eleuteria.
DITERPENOIDS: cascarilladiene, cascarillin, cascarillin A, cascarillone.
Claude-Lefontaine *et al., Bull. Soc. Chim. Fr.*, 88 (1976)

C. flavans L.
PHENANTHRENE ALKALOIDS: flavinantine, flavinine, norsinoacutine.
Stuart, Chambers, *Tetrahedron Lett.*, 2879 (1967)

C. glabellus.
FLAVONOID: ayanin.
De Garcia *et al., Chem. Abstr.*, 105 206297x

C. humilis.
AMINE ALKALOID:
N-N(2)-methylbutyryl-α-glutamyl)-2-phenethylamine.
Kutney *et al., Tetrahedron Lett.*, 3263 (1971)

C. labdaniferus.
DITERPENOID: 6β-acetoxy-7-oxo-8-labdan-15-oic acid.
de Pascual Teresa *et al., Phytochemistry*, 21, 899 (1982)

C. linearis Jacq.
PHENANTHRENE ALKALOID: dihydrosalutaridine.
SPIROISOQUINOLINE ALKALOIDS: crotonosine, jaculadine, jacularine.
Stuart *et al., J. Chem. Soc., C*, 1228 (1970)

C. lucidus.
NORDITERPENOID: crotonin.
Chan *et al., J. Chem. Soc., Chem. Commun.*, 191 (1967)

C. macrostachys.
DITERPENOID: crotepoxide.
Kupchan et al., J. Am. Chem. Soc., 90, 2982 (1968)

C. nitens.
DITERPENOID: crotonitenone.
Burke et al., J. Chem. Soc., Perkin I, 2666 (1981)

C. niveus.
DITERPENOID: nivenolide.
Rojas, Rodriguez-Hahn, Phytochemistry, 17, 574 (1978)

C. oblongifolius.
DITERPENOIDS: 11-dehydro-(-)-hardwickiic acid, oblongifoliol, oblongifolic acid.
Aiyar, Seshadri, Phytochemistry, 11, 1473 (1972)

C. penduliflorus.
DITERPENOID: penduliflaworosin.
Adesogan, J. Chem. Soc., Perkin I, 1151 (1981)

C. poilanei.
DITERPENOID: poilaneic acid.
Sato et al., Phytochemistry, 20, 1715 (1981)

C. rhamnofolius.
DITERPENOID:
20-acetoxy-9-hydroxy-13,15-seco-4-tiglia-1,6,10-triene-3,13-dione.
Stuart, Barratt, Tetrahedron Lett., 2399 (1969)

C. salutaris.
PHENANTHRENE ALKALOID: salutaridine.
Barton et al., J. Chem. Soc., 2423 (1965)

C. sellowii.
PHENOLIC: sellovicine B.
Bohlmann et al., Chem. Ber., 102, 1691 (1969)

C. sonderianus.
DITERPENOIDS: 3,4-seco-sonderianol, sonderianin, sonderianol.
Craviero et al., Phytochemistry, 21, 2571 (1982)

C. sparsiflorus Morong.
APORPHINE ALKALOID: sparsiflorine.
SPIROISOQUINOLINE ALKALOIDS: crotsparinine, isocrotsparinine.
DITERPENE ESTER: croton factor S.
(A) Chatterjee et al., Tetrahedron Lett., 1539 (1965)
(T) Upadhyay, Hecker, Phytochemistry, 15, 1070 (1976)

C. sublyratus.
DITERPENOIDS: plaunol, plaunol A, plaunol B, plaunolide.
Takahashi et al., Phytochemistry, 22, 302 (1983)

C. tiglium L.
DITERPENOIDS: crotofolin A, ingenol 3,20-dibenzoate, phorbol, phorbol 12-tiglate-13-decanoate.
TOXIN: crotin.
Chan et al., J. Am. Chem. Soc., 97, 4437 (1975)

C. turumquirensis Stayerm.
QUATERNARY ALKALOID: turumquirensine.
Burnell, Casa, Nature, 203, 297 (1964)

C. verreauxii Baill.
DITERPENOIDS: croverin, dihydrocroverin.
Fujita et al., J. Chem. Soc., Chem. Commun., 920 (1980)

C. wilsonii.
ALKALOIDS: five aporphine bases.
Stuart, Chambers, Tetrahedron Lett., 4135 (1967)

CRYPTOCARYA

C. angulata C.T.White.
PHENANTHRENE ALKALOID:
N,N-dimethylaminoethyl-3,4-dimethoxyphenanthrene.
Yunusov et al., Bull. Soc. Chim., 7, 70 (1946)

C. bourdilloni.
PYRANONE: cryptocaryalactone.
Govindachari et al., Indian J. Chem., 10, 149 (1972)

C. bowiei.
ISOQUINOLINE ALKALOIDS: cryptaustaline, cryptowoline.
Ewing et al., Aust. J. Chem., 6, 78 (1953)

C. chinensis.
ISOQUINOLINE ALKALOIDS: (-)-caryachine, (+)-caryachine, crychine.
Lu, Lan, J. Pharm. Soc., Japan, 83, 177 (1966)

C. konishii.
BENZYLISOQUINOLINE ALKALOID: crykonisine.
Lu, J. Pharm. Soc., Japan, 87, 1278 (1967)

C. longifolia Kosterm.
BENZYLISOQUINOLINE ALKALOIDS: longifolidine, longifolonine.
Bick et al., Aust. J. Chem., 34, 195 (1981)

C. odorata.
APORPHINE ALKALOID: cryptodorine.
Bick et al., Bull. Soc. Chim. Fr., 12, 4596 (1972)

C. pleurosperma White et Francis.
PHENANTHROINDOLIZIDINE ALKALOIDS: cryptopleuridine, cryptopleurine, cryptopleurospermine.
PIPERIDINE ALKALOID: pleurospermine.
Johns et al., Aust. J. Chem., 23, 353 (1970)

CRYPTOLEPIS

C. buchanani.
PYRIDINE ALKALOID: buchananine.
Dutta et al., Phytochemistry, 17, 2047 (1978)

C. sanguinolenta Sch.
UNCHARACTERIZED ALKALOID: cryptolepine.
Clinquart, Bull. Acad. Med. Belg., 9, 627 (1929)

CRYPTOMERIA (Taxodiaceae)

C. japonica D.Don.
DITERPENOIDS: cryptojaponol, cryptomeriol, cryptomerione, cryptometone, isokaurene, phyllocladan-16α-ol (+)-sugiol.
Nagahama et al., Bull. Chem. Soc., Japan, 37, 1029 (1964)

CRYPTOMONAS

C. ovata.
CAROTENOID: alloxanthin.
Haxo, Fork, Nature, 184, 1051 (1952)

C. ovata var. palustris.
CAROTENOID: ε-carotene.
Chapman, Haxo, Plant Cell Physiol., 4, 57 (1963)

CRYPTOSTEGIA

C. grandiflora.
TERP ENOID: cryptograndoside.
Krasso et al., Helv. Chim. Acta, 55, 1352 (1972)

CRYPTOSTYLIS (Orchidaceae)

C. erythroglossa Hayata.
ISOQUINOLINE ALKALOIDS:
(-)-1,2,3,4-tetrahydro-6,7-dimethoxy-2-methyl-1-(3,4-methylenedioxyphenyl)-isoquinoline,
(-)-1,2,3,4-tetrahydro-6,7-dimethoxy-2-methyl-(3,4,5-trimethoxyphenyl)-isoquinoline.
Agurell et al., Acta Chem. Scand., 28B, 239 (1974)

C. fulva Schltr.
ISOQUINOLINE ALKALOIDS: cryptostylines I, II and III.
Leander et al., Acta Chem. Scand., 27, 710 (1973)

CRYSTOPHORA

C. moniliformis.
NORDITERPENOID: 8,13-epoxy-14,15-dinor-12-labden-3β-ol.
Ravi *et al.*, *Aust. J. Chem.*, **35**, 171 (1982)

CUCUBALUS

C. baccifer.
FLAVONOIDS: homoorientin, isosaponarin, orientin, saponaretin, saponaretin 6″-O-galactoside, vitexin.
Darmograi *et al.*, *Sb. Nauchn. Tr. Ryazan. Med. Inst.*, **50**, 33 (1975); *Chem. Abstr.*, 84 56522y

CUCUMARIA

C. frondosa.
TRITERPENOID: frondogenin.
Findlay *et al.*, *J. Nat. Prod.*, **47**, 320 (1984)

CUCUMIS (Cucurbitaceae)

C. angolensis.
TRITERPENOID: cucurbitacin F.
Enslin *et al.*, *J. Sci. Food Agr.*, **8**, 673 (1957)

C. hirsutus.
TRITERPENOID: cucurbitacin G.
Enslin *et al.*, *J. Sci. Food Agr.*, **8**, 673 (1957)

C. prophetarum.
TRITERPENOID: cucurbitacin Q-1.
Atta-ur-Rahman *et al.*, *Phytochemistry*, **12**, 2741 (1973)

C. sativus.
STEROIDS: 24-ethyl-5α-cholesta-7,22-dien-3β-ol, 24-ethylcholest-5-en-7β-ol, 24-methylenepollinasterol, 24(R)-14α-methyl-24-ethyl-5α-cholest-9(11)-en-3β-ol, 7,22-stigmastadien-3β-ol, 7-stigmasten-3β-ol.
Matsumoto *et al.*, *Phytochemistry*, **22**, 2622 (1983)
Akahisa *et al.*, *Lipids*, **21**, 491 (1986)

CUCURBITA (Cucurbitaceae)

C. lundelliana.
TRITERPENOID: isomultiflorenone.
Sengupta, Khastgir, *Tetrahedron*, **19**, 123 (1963)

C. pepo L.
GIBBERELLINS: gibberellin A-39, gibberellin A-48, gibberellin A-49.
STEROIDS: 24α-ethyllathosterol, spinasterol, 5α-stigmasta-7,25-dien-3β-ol.
DINORSESQUITERPENOIDS: cucurbic acid, cucurbic acid glucoside, cucurbic acid methyl ester.
DITERPENOIDS: (+)-cis-abscisic acid, (+)-trans-abscisic acid, (+)-dehydrovomifoliol, (+)-vomifoliol.
(S) Nes *et al.*, *Lipids*, **12**, 511 (1977)
(T) Fukui *et al.*, *Agr. Biol. Chem.*, **41**, 175, 181 (1977)

CUDRANIA (Moraceae)

C. cochinchinensis.
BENZOPHENONE: cudranone.
Ottersen *et al.*, *Acta Chem. Scand.*, **31B**, 434 (1977)

C. cochinchinensis var. gerontogea.
BENZOPHENONE: cudranone.
Chang *et al.*, *J. Pharm. Sci.*, **66**, 908 (1977)

C. javanensis.
XANTHONE: cudranixanthone.
Murty *et al.*, *Phytochemistry*, **11**, 2089 (1972)

C. tricuspidata.
FLAVONOIDS: cudraflavone A, cudraflavone B.
Fujimoto *et al.*, *Planta Med.*, **50**, 161 (1984)

CULLUMIA (Compositae/Arctotideae)

C. squarrosa (L.) R.Br.
Thienyl and thietane polyacetylenes.
Bohlmann, Suwita, *Chem. Ber.*, **108**, 515 (1975)

C. sulcata Less.
Four thienyl polyacetylenes.
Bohlmann, Suwita, *Chem. Ber.*, **108**, 515 (1975)

CUNNINGHAMIA

C. lanceolata Hook. f.
FLAVONOID: robustaflavone.
Ansari *et al.*, *J. Indian Chem. Soc.*, **62**, 406 (1985)

CUNURIA

C. balsamona Cham. et Schl.
STEROIDS: β-sitosterol, stigmasterol.
TRITERPENOID: friedelin.
Lin *et al.*, *Chem. Abstr.*, 87 81231x

CUPRESSUS (Cupressaceae)

C. cashmeriana.
FLAVONOIDS: amentoflavone, cupressuflavone, hinokiflavone, isocryptomerin, quercetin 3-O-(6″-O-α-rhamnopyranosyl)-β-D-glucopyranoside, sequoiaflavone.
Khabir *et al.*, *J. Nat. Prod.*, **50**, 511 (1987)

C. dupreziana.
SESQUITERPENOID: α-duprezianene.
Piovetti *et al.*, *Phytochemistry*, **16**, 103 (1977)

C. funebris Endl.
SESQUITERPENOID: α-funebrene.
Motl, Paknikar, *Collect. Czech. Chem. Commun.*, **33**, 1939 (1968)

C. goveniana var. abramasiana.
NORDITERPENOID: cupressol.
Jolad *et al.*, *J. Nat. Prod.*, **47**, 983 (1984)

C. lusitanica.
FLAVONOID: podocarpusflavone A.
Rizvi *et al.*, *Phytochemistry*, **13**, 1990 (1974)

C. macrocarpus.
DITERPENOID: (+)-isophyllocladene.
SESQUITERPENOID: juniperol.
Naffa, Ourisson, *Bull. Soc. Chim. Fr.*, 382 (1954)

C. pygmaea (Lemm.) Sarg.
MONOTERPENOID: α-thujaplicinol.
TROPOLONES: 3-isopropenyl-7-hydroxytropolone, pygmaein.
Zavarin, *J. Org. Chem.*, **27**, 3368 (1962)

C. semopervirens L.
FLAVONOIDS: cupressuflavone, hinokiflavone, isocryptomerin, quercetin, quercetin 3-O-α-L-rhamnopyranoside.
DITERPENOIDS: trans-communic acid, cupressic acid, neocupressic acid, (+)-sempervirol.
(F) Khabir *et al.*, *J. Nat. Prod.*, **50**, 511 (1987)
(T) Balansard *et al.*, *Trans. Soc. Pharm. Montpellier*, **33**, 367 (1973)

C. torulosa.
FLAVONOID: cupressuflavone.
DITERPENOID: toruloasal.
(F) Nakazawa *et al.*, *Chem. Pharm. Bull.*, **10**, 1036 (1962)
(T) Enzell, *Acta Chem. Scand.*, **15**, 1303 (1961)

C. tricuspidata (Carr) Lav.
MONOTERPENOIDS: methyl-4-trans-dehydrogeranate, α-pinene, β-thujaplicin, τ-thujaplicin.
SESQUITERPENOID: nootkatin.
Senter *et al.*, *Phytochemistry*, **14**, 2233 (1975)

CURARIA

C. candicans.

BISBENZYLISOQUINOLINE ALKALOIDS: (+)-candicusine,
(-)-limacine N-oxide.
Lavault et al., J. Chem. Res. Synop., 248 (1985)

CURCULIGO (Amaryllidaceae)

C. orchioides.

ALIPHATIC: 3-methoxy-5-acetyl-31-tritriacontene.
GLYCOSIDES: curculigine A,
5,7-dimethoxy-dihydromyricetin-3-O-α-L-xylopyranosyl-4-
O-β-D-glucopyranoside.
(Al) Mehta et al., Acta Cienc. Indica, 7, 174 (1982); Chem. Abstr., 97
141652d
(Gl) Tiwari et al., Planta Med., 29, 291 (1976)

CURCUMA

C. aromatica.

MONOTERPENOID: 2-bornanol.
SESQUITERPENOID: curcumene.
Stothers et al., Can. J. Chem., 54, 1211 (1976)

C. longa.

SESQUITERPENOIDS: curlone, α-turmerone, β-turmerone.
Kuso et al., Phytochemistry, 22, 596 (1983)

C. xanthorrhiza.

SESQUITERPENOID: xanthorrhizol.
Rimpler et al., Z. Naturforsch., 256, 995 (1970)

C. zedoaria.

SESQUITERPENOIDS: curculone, curcumadiol curcumenol,
curcumol, curdione, curzerenone, ent-curzerenone,
isocurcumenol, isofuranodienone, procurcumenol,
procurzerenone, zederone, zedoarone.
Hikino et al., Chem. Pharm. Bull., 19, 93 (1971)

CUSPIDUM

C. filix.

BISQUINONE: filixic acid.
Roehm, Justus Liebigs Ann. Chem., 318, 253 (1901)

CUTLERIA

C. multifida.

HOMOMONOTERPENOID: multifidene.
Jaenicke et al., J. Am. Chem. Soc., 96, 3324 (1974)

CYANOPHYTA (Cyanophyceae)

C. sp.

CAROTENOIDS: isocryptoxanthin, mutatochrome.
Francis, Halfen, Phytochemistry, 11, 2347 (1972)

CYANOSTEGIA

C. angustifolia Turez.

FLAVONOIDS: 5,7,4'-trihydroxy-3,8-dimethoxyflavone,
5,7,4'-trihydroxy-3,8,3'-trimethoxyflavone.
Ghisalberti et al., Aust. J. Chem., 20, 1049 (1967)

C. microphylla.

FLAVONOIDS: chrysosplenol,
5,7,4'-trihydroxy-3,8-dimethoxyflavone,
5,7,4'-trihydroxy-3,8,3'-trimethoxyflavone.
Ghisalberti et al., Aust. J. Chem., 20, 1049 (1967)

CYATHOCLINE

C. lyrata.

MONOTERPENOID: lyratol.
Dergan et al., Tetrahedron Lett., 5337 (1967)

C. purpurea.

SESQUITERPENOID: isoivangustin.
Nayasampagi et al., Phytochemistry, 20, 2034 (1981)

CYATHULA

C. bulleri Brodie strain 6680a (ATCC 38351).

SESQUITERPENOIDS: cybrodal, cybrodic acid, cybrodol.
Ayer, McGaskill, Tetrahedron Lett., 1917 (1980)

C. capitata Monquim-Tandon.

STEROIDS: amarasterone A, amarasterone B, cyasterone,
poststerone, precyasterone, sengosterone.
Takemoto et al., Tetrahedron Lett., 3191, 5136 (1967)
Hikino et al., Tetrahedron, 26, 887 (1970)

CYATHUS (Basidiomycetae)

C. africanus.

DITERPENOIDS: cyafrin A-4, cyafrin A-5, cyafrin B-4.
Ayer et al., Can. J. Chem., 56, 2113 (1978)

C. helenae.

DITERPENOIDS: cyathin A-3, cyathin A-4, cyathin A5, cyathin
B-3, cyathin C-3, cyathin C-5.
Ayer et al., Can. J. Chem., 56, 717 (1978)

C. striatus.

ANTIBIOTICS: striatins A, B and C.
Hecht et al., J. Chem. Soc., Chem. Commun., 665 (1978)

CYCLACHAENA

C. xanthifolia.

SESQUITERPENOIDS: dehydrocoronopilin, ivoxanthin.
Nomikov et al., Khim. Prir. Soedin., 375 (1969)

CYCLAMEN (Primulaceae)

C. europeum.

SAPOGENINS: cyclamigenins A-1, A-2, B, C and D,
cyclamiretin, cyclamiretin A.
SAPONIN: cyclamin.
Dorchai et al., Tetrahedron, 24, 1377, 5649 (1968)

C. graecum Lam.

SAPOGENIN: cyclamiretin C.
Harvala, Hylands, Planta Med., 33, 180 (1978)

C. hederifolium Aiton.

SAPOGENIN: cyclamiretin C.
Harvala, Hylands, Planta Med., 33, 180 (1978)

CYCLEA

C. barbata.

BISBENZYLISOQUINOLINE ALKALOID: homoaromoline.
Tomita et al., J. Pharm. Soc., Japan, 87, 1012 (1967)

C. burmanni.

UNCHARACTERIZED ALKALOIDS: burmannaline, burmannine.
Saradamma, Bull. Cent. Res. Inst. Univ. Travancore, 3, 55 (1954)

C. hainanensis.

PROTOBERBERINE ALKALOID: α-hainanine.
Zhang et al., Chih Wu Hsueh Pao, 23, 216 (1981)

C. insuklaris (Mak.) Diels.

ALKALOIDS: cycleanine, insulanoline.
Fujita, Murai, J. Pharm. Soc., Japan, 71, 1043 (1951)

98

C. peltata.
BISBENZYLISOQUINOLINE ALKALOIDS: cycleacurine, cycleadrine, cycleahomine, cycleanine, cycleanorine, cycleapeltine, fangchinoline.
Kupchan *et al.*, *J. Org. Chem.*, **38**, 1846 (1973)

CYCLOBIUM
C. clausseni.
QUINONE: claussequinone.
Gottlieb *et al.*, *Phytochemistry*, **12**, 2495 (1973)
C. vecchi.
QUINONE: claussequinone.
Gottlieb *et al.*, *Phytochemistry*, **12**, 2495 (1973)

CYDONIA
C. oblongata.
GLYCOSIDES: marmelolactine A, marmelolactine B, roseoside.
Ishihara *et al.*, *Agr. Biol. Chem.*, **47**, 2121 (1983)

CYLICODISCUS
C. gabunensis.
CHALCONE: okanin.
Russell *et al.*, *J. Chem. Soc.*, 1506, 1940 (1934)

CYMARA
C. sibthorpiana.
SESQUITERPENOID: sibthorpine.
Omar *et al.*, *Phytochemistry*, **23**, 2381 (1984)

CYMBALARIA (Scrophulariaceae)
C. muralis Gaertn. Meyer et Scherb.
IRIDOID: antirrhinoside.
Kapoor *et al.*, *Phytochemistry*, **13**, 1018 (1974)

CYMBELLA
C. cymbiformis Kutz.
CAROTENOIDS: β-carotene, fucoxanthin, lutein.
Illyes *et al.*, *Stud. Cercet. Biochim.*, **18**, 109 (1975); *Chem. Abstr.*, **85** 17083g

CYMBIDIUM
C. giganteum.
SAPONIN: cymbidioside.
Dalmen, Leander, *Phytochemistry*, **17**, 1975 (1978)

CYMBOPOGON
C. citratus.
TRITERPENOIDS: cymbopogone, cymbopogonol.
Crawford *et al.*, *Tetrahedron Lett.*, 3099 (1975)
Hanson *et al.*, *Phytochemistry*, **15**, 1074 (1976)
C. flexuosus.
ALIPHATICS: triacontane, triacontanol.
STEROID: β-sitosterol.
TRITERPENOID: arundoin.
Thappa *et al.*, *Indian J. Chem.*, **13**, 1108 (1975)
C. senarensis.
MONOTERPENOID: (+)-piperitone.
Read, Walker, *J. Chem. Soc.*, 308 (1934)
C. winterianus.
MONOTERPENOID: hydroxycitronellic acid.
Sethi *et al.*, *Indian Perfum.*, **23**, 167 (1979)

CYMOPHOLIA (Chlorophyceae)
C. barbata.
SESQUITERPENOIDS: cyclocymopol, cyclocymopol methyl ether, isocymopolone.
McConnell *et al.*, *Phytochemistry*, **21**, 2139 (1982)

CYMOPTERUS
C. watsonii.
FUROCOUMARINS: byakangelicin, heraclenol.
Frank *et al.*, *Phytochemistry*, **14**, 1681 (1975)

CYNANCHUM
C. africanum.
STEROIDS: cynafoside A, cynafoside B.
Tsukamoto *et al.*, *Tennen Yuki Kagobutsu Toronkai Kown Yoshishu*, 27th 529 (1985)
C. caudatum Max.
STEROIDAL ALKALOIDS: gagamine, gagaminine.
STEROIDS: caudatin, 12-O-cinnamoyl-20-O-acetylglycosarcostin, 20-O-cinnamoylikemagenin, 12-O-cinnamoylikemagenol, 20-O-cinnamoylsarcostin, cynanchogenin, 5,6-epoxycaudatin, glycocynanchogenin, glycopenupogenin, ikemagenin, isoikemagenin, penupogenin.
(A) Yamagishi *et al.*, *Chem. Pharm. Bull.*, **20**, 2289 (1972)
(S) Bando *et al.*, *ibid.*, **28**, 2258 (1980)
C. glaucescens Hand.-Mazz.
SAPOGENINS: glaucogenins A, B and C.
SAPONINS: glaucogenin C 3-O-β-D-thevetoside, glaucoside F, glaucoside G.
Nakagawa *et al.*, *Chem. Pharm. Bull.*, **31**, 879 (1983)
C. otophyllum.
STEROID: quinyangshengenin.
Mu *et al.*, *Yaoxue Xuebao*, **18**, 356 (1983)
C. sibiricum.
STEROIDS: sibiricoside D, sibiricoside E.
Tursunova *et al.*, *Khim. Prir. Soedin.*, 171 (1975)
C. vincetoxicum.
AZAPHENANTHRENE ALKALOID: vincetene.
INDOLIZIDINE ALKALOID: 6-hydroxy-2,3-dimethoxyphenanthro-[9,10-f]-indolizidine.
ISOQUINOLINE ALKALOID: 6-hydroxy-2,3-dimethoxy-9,11,12,13,13a,14-hexahydrodibenzo-[fmh-pyrrolo-1,2-b]-isoquinoline.
Weigrabe *et al.*, *Justus Liebigs Ann. Chem.*, **721**, 154 (1969)
Budzikiewicz *et al.*, *Justus Liebigs Ann. Chem.*, 1212 (1980)
C. wilfordi Hemsl.
STEROIDS: 12-O-cinnamoyl-20-O-ikemanoylsarcostin, 12-O-cinnamoyl-20-O-tigloylsarcostin, 12-O-cumanoyl-20-O-tigloylsarcostin, wilforine.
Hayashi, Mitsuhashi, *Chem. Pharm. Bull.*, **23**, 139 (1975)

CYNARA (Compositae)
C. cardunculus.
SESQUITERPENOID: cynaratriol.
Bernhard *et al.*, *Helv. Chim. Acta*, **62**, 1288 (1979)
C. scolymus L.
PIGMENTS: cyanidol 3-caffeylglucoside, cyanidol caffeylsophoroside, cyanidol dicaffeylsophoroside, cyanidol diglucoside, cyanidol glucoside, cyanidol 5-glucoside-3-caffeylsophoroside, cyanidol sophoroside.
SESQUITERPENOIDS.: cynarapicrin, cynaratriol, cynarolide, isoamberboin.
(P) Aubert *et al.*, *Chem. Abstr.*, 97 195777s
(T) Bernhard *et al.*, *Helv. Chim. Acta*, **62**, 1288 (1979)

CYNESPORA

C. melonis.
SESQUITERPENOIDS: 2-hydroxyguaioxide, 3-hydroxyguaioxide.
Takeda, *Pure Appl. Chem.*, **21**, 181 (1970)

CYNOGLOSSUM (Boraginaceae)

C. amabile Stapf. et Drummond.
PYRROLIZIDINE ALKALOIDS: amabiline, echinatine.
Culvenor, Smith, *Aust. J. Chem.*, **20**, 2499 (1967)

C. glochidiatum.
PYRROLIZIDINE ALKALOID: amabiline.
Suri *et al.*, *Indian J. Pharm.*, **37**, 69 (1975)

C. lanceolatum.
PYRROLIZIDINE ALKALOIDS: cynaustine, cynaustraline.
Suri *et al.*, *Indian J. Pharm.*, **37**, 69 (1975)

C. latifolium R.Br.
PYRROLIZIDINE ALKALOID: latifoline.
Crowley, Culvenor, *Aust. J. Chem.*, **15**, 139 (1962)

C. officinale L.
PYRROLIZIDINE ALKALOID: heliosupine.
UNCHARACTERIZED ALKALOIDS: consolicine, consolidine,
cynoglossine.
Culvenor *et al.*, *Aust. J. Chem.*, **9**, 512 (1956)

C. pictum Ait.
PYRROLIZIDINE ALKALOIDS: echinatine, echinatine N-oxide,
heliosupine, heliosupine N-oxide.
UNCHARACTERIZED ALKALOID: pictumine.
Culvenor, *Aust. J. Chem.*, **9**, 512 (1956)
Man'ko, Marchenko, *Khim. Prir. Soedin.*, 655 (1972)

C. viridiflorum Pall.
PYRROLIZIDINE ALKALOIDS: heliosupine, heliosupine N-oxide,
viridiflorine.
Men'shikov *et al.*, *J. Gen. Chem. USSR*, **18**, 1736 (1948)

CYNOMETRA

C. ananta.
IMIDAZOLE ALKALOIDS: anantine, cynodine, cynometrine.
Khuong-Huu-Laine *et al.*, *Tetrahedron Lett.*, 1757 (1973)

C. hankei.
IMIDAZOLE ALKALOIDS: cynodine, cynometrine,
N(1)-demethylcynodine, N(1)-demethylcynometrine.
Waterman, Faulkner, *Phytochemistry*, **20**, 2765 (1981)

CYNOSEIRA

C. stricta.
TRINORTRITERPENOIDS:
2-(10,11-dihydroxygeranylgeranyl)-6-methyl-1,4-
benzenediol,
2-(10,11-dihydroxygeranylgeranyl)-6-methyl-1,4-
benzoquinone.
Amico *et al.*, *Phytochemistry*, **21**, 421 (1982)

CYNURA

C. japonica Makino.
STEROIDS: 3-epi-neoruscogenin,
spirosta-5,25(27)-diene-1,3-diol.
Takehira *et al.*, *Tetrahedron Lett.*, 3647 (1977)

CYPERUS (Cyperaceae)

C. articulus.
SESQUITERPENOID: cyperotundone.
Naves, Ardizio, *Bull. Soc. Chim. Fr.*, 372 (1954)

C. brevibracteatus.
BENZOPYRAN: (-)-breverin.
QUINONE: breviquinone.
Allan *et al.*, *Tetrahedron Lett.*, 7 (1973)

C. conicus.
QUINONE: conicaquinone.
MacLeod *et al.*, *Tetrahedron Lett.*, 241 (1972)

C. corymbosus.
SESQUITERPENOID: isocorybolene.
Gambarino *et al.*, *Phytochemistry*, **24**, 2726 (1985)

C. rotundus L.
ACIDS: p-coumaric acid, ferulic acid, p-hydroxybenzoic acid,
protocatechuic acid, vanillic acid.
SESQUITERPENOIDS: caryophyllene, cyperene, cyperenone,
cyperol, cyperolone, cyperotundone, β-elemene,
α-humulene, isocyperol, kobusone, mustakone, rotundene,
rotundenol, α-rotundol, β-rotundol, rotundone, β-selinene.
Hikino *et al.*, *Tetrahedron*, **27**, 4831 (1971)
Komai *et al.*, *Chem. Abstr.*, **97** 123942u

C. rotundus L. (Indian).
SESQUITERPENOID: (+)-copadiene.
Kapadia *et al.*, *Tetrahedron Lett.*, 4661 (1967)

C. scaber.
BENZOPYRAN: scaberin.
QUINONE: scabequinone.
SESQUITERPENOID: scabediol.
Allan *et al.*, *Tetrahedron Lett.*, 3 (1973)

C. scariosus.
SESQUITERPENOIDS: cyperenol, isopatchoula-3,5-diene,
patchoulenol, rotundene.
Nerali, Chakravarti, *Tetrahedron Lett.*, 2447 (1967)

CYPHOLOPHUS (Urticaceae)

C. friesianus.
ALKALOIDS: acetylcypholophine, cypholophine.
Hart *et al.*, *J. Chem. Soc., D*, 441 (1976)

CYPHOMANDRA

C. betaceae Sendt.
AMIDE ALKALOID: solacaproine.
ANTHOCYANINS: malvinidin 3-diglucoside, paeonidin
3-diglucoside, pelargonidin 3-diglucoside.
(A) Evans *et al.*, *J. Chem. Soc., Perkin I*, 2017 (1972)
(An) Bobbio *et al.*, *Food Chem.*, **12**, 189 (1983); *Chem. Abstr.*, **100**
63936a

CYRTANTHUS (Amaryllidaceae)

C. pallidus Sims.
ISOQUINOLINE ALKALOID: lycorine.
Greathouse, Rigler, *Am. J. Bot.*, **28**, 702 (1941)

CYSTOSIERA (Phaeophyceae)

C. balearica.
SPIROPYRAN: cystoketal.
Amico *et al.*, *J. Nat. Prod.*, **47**, 947 (1984)

C. caespitosa.
Two polyenes.
Amico *et al.*, *J. Chem. Res. Synop.*, 262 (1962)

C. crinita.

DITERPENOIDS: crinitol,
1,7-dihydroxy-3,7,11,15-tetramethylhexadecatetraen-13-one,
oxocrinitol acetate.

Fattorusso et al., Tetrahedron Lett., 937 (1976)
Amico et al., Phytochemistry, **20**, 1085 (1981)

C. elegans.

DINORTRITERPENOIDS:
2,12-dihydroxy-16-(2-hydroxy-5-methoxy-3-methylphenyl)-
2,6,10,14-tetramethyl-2,10,14-hexadecatrien-4-one,
2,12-dihydroxy-16-(2-hydroxy-5-methoxy-3-methylphenyl)-
2,6,10,14-tetramethyl-3,10,14-hexadecatrien-5-one.

DITERPENOID: eleganolone.

Francisco et al., Phytochemistry, **17**, 1003 (1978)
Banaigs et al., Tetrahedron Lett., 3271 (1982)

C. fimbriata (Desf.) Bory.

ACIDS: arachidic acid, lauric acid, myristic acid, oleic acid,
palmitic acid.

Sayed et al., J. Pharm. Sci., **20**, 199 (1979)

C. mediterranea.

DITERPENOIDS: cystoseirols A, B, C, D and E, mediterraneol.

Francisco et al., J. Org. Chem., **51**, 2707 (1986)

C. stricta.

DITERPENOIDS: cystoseirols A, B, C, D and E.

Francisco et al., J. Org. Chem., **51**, 2707 (1986)

C.

TAMARISCIFOLIA, DITERPENOIDS: cystoseirols A, B, C, D and E.

Frabcisco et al., J. Org. Chem., **51**, 2707 (1986)

CYTINUS (Rafflesiaceae)

C. hypocistis.

PIGMENT: isoterchebin.

Furstenwerth et al., Ann. Chem., 112 (1976)

CYTISUS (Leguminosae)

C. caucasicus Grossh.

QUINOLIZIDINE ALKALOIDS: (+)-lupanine, pachycarpine,
(+)-sparteine.

Clemo et al., J. Chem. Soc., 429 (1931)

C. hirsutus L.

QUINOLIZIDINE ALKALOID: 7-hydroxysparteine.

Mollov et al., Dokl. Bolg. Akad. Nauk, SSR **24**, 1657 (1971)

C. laburnum L.

IMIDAZOLIDINE ALKALOID: cytisine.
PYRROLIZIDINE ALKALOIDS:
1-hydroxymethyl-8Hβ-pyrrolizidine, laburnine.

Galinovsky et al., Monatsh. Chem., **80**, 550 (1949)

C. monspessulanus L.

LUPINANE ALKALOID: monspessulanine.

White, N. Z. J. Sci. Tech., **27B**, 335, 339, 474, 478 (1946)

C. nigricans L. var. elongatus Willd.

IMIDAZOLIDINE ALKALOID: cytisine.
UNCHARACTERIZED ALKALOID: calycotomine.

White, N. Z. J. Sci. Tech., **25B**, 152 (1944)

C. proliferus L.f.

IMIDAZOLIDINE ALKALOID: cytisine.
UNCHARACTERIZED ALKALOID: calycotomine.

White, N. Z. J. Sci. Tech., **25B**, 152 (1944)

C. ratisbonensis Schaeff.

QUINOLIZIDINE ALKALOIDS: (-)-lupanine, (-)-sparteine.

Bellavita, Farm. Sci., **3**, 424 (1948)

C. ruthenicus Fisch.

QUINOLIZIDINE ALKALOID: sparteine.

Bellavita, Farm. Sci., **3**, 424 (1948)

C. scoparius (L.) Link.

QUINOLIZIDINE ALKALOIDS: genisteine, hydroxylupanine,
isosparteine, lupanine, (-)-sparteine.
UNCHARACTERIZED ALKALOID: sarothamnine.
FLAVONOID: scoparin.
GIBBERELLIN: gibberellin A-35.

(A) Valeur, J. Pharm. Sci., **8**, 573 (1913)
(F) Beyersdorff et al., Naturwissenschaften, **17**, 392 (1962)
(G) Yamane et al., Agr. Biol. Chem., **38**, 649 (1974)

D

DACRYDIUM (Podocarpaceae)

D. bidwillii.
DITERPENOIDS: isopimara-7,15-dien-18-al, isopimara-7,15-dien-18-ol, isopimara-7,15-diene, 18-norisopimaradien-4α-ol.
Church, Ireland, *J. Org. Chem.*, **28**, 17 (1963)
Grant *et al.*, *Aust. J. Chem.*, **20**, 969 (1967)

D. biforme.
NORDITERPENOID: 18-norisopimara-4(19),7,15-triene. Diterpenoids biformene, manool, sclarene.
Carman, Dennis, *Aust. J. Chem.*, **20**, 157 (1967)

D. colensoi.
DITERPENOIDS: 2α-carboxycolensen-2β-ol, colensenone, 2,3-dicarboxy-2,3-secomanoyl oxide, 2β-hydroxycolensen-2α-oic acid, 2β-hydroxy-1α-hydroxymethylcolensene-2α-carboxylic acid lactone, 2α-hydroxymanoyl oxide, 18-hydroxy-2-oxomanoyl oxide, 8β-hydroxysandaracopimar-15-ene, manoyl oxide, 2-oxomanoyl oxide, sandaracopimaradiene-3β,19-diol, sandaracopimara-8(14),16-diene-2α,3β,18,19-tetraol, sandaracopimara-8(14),16-diene-2α,18,19-triol, sandaracopimara-8(14),16-diene-3β,18,19-triol, sandaracopimaradien-19-ol.
Grant *et al.*, *Aust. J. Chem.*, **22**, 1265 (1969)
Grant, McGrath, *Tetrahedron*, **26**, 1619 (1970)

D. cupressinum.
SESQUITERPENOID: juniperol.
DITERPENOIDS: laurenene, rimuene, (+)-sugiol.
Corbett *et al.*, *J. Chem. Soc., Perkin I*, 1774 (1979)

D. elatum.
DITERPENOID: cedrol.
Stork, Breslow, *J. Am. Chem. Soc.*, **75**, 3291 (1953)

D. franklini.
SESQUITERPENOID: dacrinol.
Baggaley *et al.*, *Acta Chem. Scand.*, **21**, 2247 (1967)

D. intermedium.
STEROIDS: dacrysterone, β-ecdysone 2-cinnamate, β-ecdysone 3-cinnamate, β-ecdysone 3-p-coumarate, makisterone, polypodine D 2-cinnamate, ponasterone 2-cinnamate, makisterone A.
Russell *et al.*, *Aust. J. Chem.*, **26**, 1805 (1973)

D. kirkii.
DITERPENOIDS: 2β,3β-dihydroxymanoyl oxide, 7α-hydroxymanool.
Cambie *et al.*, *Aust. J. Chem.*, **22**, 1691 (1969)

DACRYMYCES (Basidiomycetae)

D. stillatus.
CAROTENOID: phytofluene.
Goodwin, *Biochem. J.*, **53**, 538 (1953)

DACTYLARIA

D. lutea.
PIGMENTS: altersolanols A, B and C.
Stoessl *et al.*, *Can. J. Chem.*, **47**, 767 (1969)
Stoessl *et al.*, *ibid.*, 777 (1969)

DACTYLINA

D. arctica.
STEROIDS: 5α,8α-epidioxy-5α-ergosta-6,22-dien-3β-ol, usnic acid.
Strigina, Sviridov, *Phytochemistry*, **17**, 327 (1978)

DACTYLOCAPNOS

D. macrocarpos Hutch.
OXOPROTOBERBERINE ALKALOID: α-allocryptopine.
OXOQUINOLIZIDINE ALKALOID: protopine.
Manske, Miller, *Can. J. Res.*, **18B**, 288 (1940)

DAEDAELIA (Basidiomycetae)

D. confragosa.
ACIDS: dotriacontanoic acid, palmitic acid.
STEROID: ergosterol.
TRITERPENOID: epi-lupeol.
Merijanian *et al.*, *Rev. Latinoam. Quim.*, **14**, 73 (1983)

D. quercina.
TRITERPENOIDS: carboxyacetylquercinic acid, quercinic acid.
Adam *et al.*, *Tetrahedron Lett.*, 1461 (1967)

DAEMONOROPS

D. draco.
SESQUITERPENOID: pterocarpal.
Nasini *et al.*, *Phytochemistry*, **20**, 514 (1981)

DAHLIA (Compositae)

D. scapigera.
POLYENE: 1-acetoxy-4,5-epoxy-tetradecadiene-8,10-diyne.
Lam *et al.*, *Phytochemistry*, **10**, 1877 (1971)

D. tenuicaulis.
ACETYLENICS: 1,3-diacetoxytetradeca-4,6-diene-8,10,12-triyne, 1-acetoxytetradeca-4,6-diene-8,10,12-trinye.
CHALCONES: 4,2',4'-trihydroxychalcone, 3,2',4'-trihydroxy-4-methoxychalcone, 2'-hydroxy-4,4',6-trimethoxychalcone.
FLAVANS: 5-hydroxy-7,4'-dimethoxyflavanone, 5,7,4'-trimthoxyflavan-4-ol, 5,7,4'-trimethoxyflavanone.
Lam *et al.*, *Phytochemistry*, **14**, 1621 (1975)

DALBERGIA (Leguminosae)

D. barretoana.
SESQUITERPENOID: (S)-methoxydalbergione.
TRITERPENOID: 3-o-acetyloleanolic acid.
Braga *et al.*, *An. Acad. Bras. Cienc.*, **39**, 249 (1967)
Fujita *et al.*, *J. Pharm. Soc., Japan.*, **94**, 189 (1974)

D. inundata.
ISOFLAVONE: orobol 3'-methyl ether.
Markham *et al.*, *Phytochemistry*, **7**, 791 (1968)

D. latifolia Roxb.
BENZOPYRAN: dalbinol.
FLAVONOIDS: daliriodain, 7,4'-dihydroxyflavan.
XANTHONE: 2,7-dihydroxy-3-methoxyxanthone.
(B) Chibber *et al.*, *Phytochemistry*, **17**, 1442 (1978)
(F, X) Donnelly *et al.*, *Proc. R. Ir. Acad.*, **83B**, 39 (1983);
Chem. Abstr., *100 20402u*

D. nitidula Welw. ex Bak.
BENZOPYRANS: hemileiocarpin, heminitidulan.
Brandt et al., J. Chem. Soc., Perkin I, 137 (1978)

D. paniculata.
BENZOPYRANONES: dalpanin, dalpanol.
Adinarayana et al., Phytochemistry, 12, 2543 (1973)

D. retusa.
ISOFLAVONE: retusin.
Jurd et al., Tetrahedron Lett., 2149 (1972)

D. sericea.
ISOFLAVONE:
7-hydroxy-6-methoxy-3',4'-methylenedioxyisoflavone.
TRITERPENOIDS: pseudobaptigenin, betulin, erythrodiol, glutinol, taraxerol, tirucallol acetate.
Parthasarathy et al., Phytochemistry, 15, 226 (1976)

D. sissoo.
COUMARIN: dalbergin.
ISOFLAVONE: psitectorigenin.
(C) Mukherjee et al., Tetrahedron, 27, 799 (1971).
(I) Dhingra et al., Indian J. Chem., 12, 1118 (1974)

D. spinosa.
ISOFLAVONOIDS: dalspinin, dalspinosin.
Dosen et al., Indian J. Chem., 21B, 385 (1982)

D. stevensonii Standl.
COUMARIN: stevenin.
ISOFLAVONE: tectorigenin. tcw(C) Ahmad et al., J. Chem. Soc., Perkin 1, 1737 (1973)
(I) Khara et al., Indian J. Chem., 16B, 78 (1978)

D. volubilis.
ALIPHATIC: triacontane.
COUMARIN: 7-hydroxy-4-methoxycoumarin.
FLAVONOIDS: biochanin A, claroin, dalbergin, formononetin, (+)-medicarpin, 7-O-methyldalbergin, tectoridin, tectorigenin, volubilin.. Steroid β-sitosterol.
Chawla et al., Phytochemistry, 15, 235 (1976)
Khara et al., Indian J. Chem., 16B, 78 (1978)

DALEA (Fabaceae)
D. frutescens.
AMINE ALKALOID: N-methyl-β-phenethylamine.
Camp, Norvell, Econ. Bot., 20, 274 (1966)

D. tuberculata.
COUMARIN: coumarin.
FLAVONOIDS: saltilin, torreonin.
STEROID: β-sitosterol.
TRITERPENOIDS: β-amyrin, friedelin, friedelanol, epi-friedelanol.
Dominguez et al., Rev. Latinoam. Quim., 13, 39 (1982)

DAMNACANTHUS (Rubiaceae)
D. major.
PIGMENTS: damnacanthal, 1,3,5-trihydroxyanthraquinone-2-carboxaldehyde.
Thomson et al., J. Chem. Soc., C, 2001 (1971)

D. subspinosus Hand-Mazz.
PIGMENT: 5-hydroxydamnacanthol ethyl ether.
Li et al., Yaoxue Xuebao, 21, 303 (1986); Chem. Abstr., 105 94503c

DANAIS (Rubiaceae)
D. fragrans Gaertn.
ANTHROQUINONE: 1-hydroxydimethylanthraquinone.
FLAVONOIDS: kaempferol 3-O-rhamnodiglucoside, kaempferol 3-O-rhamnoglucoside, quercetin 3-O-rhamnoglucoside, rubiadin, rubiadin xyloglucoside.
Andre et al., Plant Med. Phytother., B 10, 110 (1976)

DANIELLA (Leguminosae)
D. ogea.
DITERPENOIDS: daniellic acid, ozic acid, ozol.
Bevan et al., J. Chem. Soc., C, 1063 (1968)

D. oliveri.
DITERPENOIDS: daniellic acid, oliveric acid.
Hauser et al., Tetrahedron, 26, 3461 (1970)

DAPHNANDRA (Monimiaceae)
D. apatela.
BISBENZYLISOQUINOLINE ALKALOID: apateline.
Bick, Sotheeswaran, Aust, J. Chem., 31, 2077 (1978)

D. aromatica.
BISBENZYLISOQUINOLINE ALKALOID: aromoline.
Bick, Whalley, Univ. Queensland Papers, Dept. Chem., 1, No. 33 (1948)

D. dielsii Perkins.
BISBENZYLISOQUINOLINE ALKALOID: repanduline.
Bick, Whalley, Univ. Queensland Papers, Dept. Chem., No. 1, 28 (1946)

D. johnsonii.
BISBENZYLISOQUINOLINE ALKALOIDS: johnsonine, N-methylapateline, N-methylnorapateline.
Bick, Loew, Aust. J. Chem, 31, 2539 (1978)

D. micrantha Benth.
BISBENZYLISOQUINOLINE ALKALOIDS: daphnandrine daphnoline.
Pyman, J. Chem. Soc., 105, 1679 (1914)

D. repandula.
BISBENZYLISOQUINOLINE ALKALOIDS: O-methylrepandine, repandine, repandinine, repanduline.
Bick et al., J. Chem. Soc., 692 (1953)

DAPHNE (Thymelaeaceae)
D. genkwa Sieb. et Zucc.
FLAVONOIDS: apigenin, genkwanin, 3'-hydroxygenkwanin, yuenkanin.
DITERPENOIDS: yuanhaufin, yuanhuacium ester, yuanhaudin, yuanhuapine.
(F) Li et al., Zhongcaoyao, 14, 392 (1983); Chem. Abstr., 100 64929a
(T) Wang et al., Huaxue Xuebao, 40, 835 (1982)

D. odora Thunb.
COUMARINS: daphneticin, daphnetin, daphnoretin, odoracin, umbelliferone.
FLAVONOIDS: daphnodorin A, daphnodorin B, daphnodorin C.
PHENOLIC: daphneolone.
(C) Kosigo et al., Agric. Biol. Chem., 40, 2119 (1976)
(F) Baba et al., Chem. Pharm. Bull., 34, 595 (1986)
(P) Kosigo et al., Phytochemistry, 13, 2332 (1974)

D. sericea.
FLAVONOIDS: apigenin, apigenin 7-β-D-glucoside, isovitexin, luteolin 7-methyl ether, luteolin 7-methyl ether 5-β-D-glucoside.
Ulubelen et al., Phytochemistry, 21, 801 (1982)

D. tangutica.
COUMARINS: daphneticin, daphnetin, daphnoretin, 7-hydroxy-8-methoxycoumarin.
LIGNANS: (-)-dihydrosesamin, (-)-lariciresinol, (-)-pinoresinol, (+)-syringaresinol.
DITERPENOID: gniditrin.
Zhuang et al., Planta Med., 45, 172 (1982)

DAPHNIPHYLLUM (Daphniphyllaceae)

D. bancanum Kurz.
UNCHARACTERIZED ALKALOID: daphniphylline.
Irikawa *et al.*, *Tetrahedron*, 24, 5691 (1968)

D. gracile.
ALKALOIDS: daphnicine, daphnigraciline, daphnigracine, hydroxydaphigraciline, oxodaphnigraciline, epi-oxodaphnigraciline, oxodaphnigracine.
Yamamura *et al.*, *Chem. Lett.*, 923 (1978)
Yamamura *et al.*, *ibid.*, 393 (1980)

D. humile.
ALKALOIDS: deoxyyuzurianine, methyl homodaphniphyllate.
Toda *et al.*, *Tetrahedron*, 30, 2683 (1979)

D. macropodum Miq.
ALKALOIDS: codaphniphyllidine, daphmarine, daphmacropodine, daphnilactone A, daphnilactone B, macrodaphnine, macrodaphniphyllamine, macrodaphniphyllidine, methyl homodaphniphyllate, neodaphniphylline, neoyuzurimine, secodaphniphylline, yuzurimine A, yuzurimine B.
IRIDOIDS: asperuloside, daphylloside.
 (A) Irikawa *et al.*, *Tetrahedron Lett.*, 5363 (1966)
 (A) Toda *et al.*, *Tetrahedron*, 30, 2683 (1974)
 (I) Inouye *et al.*, *J. Pharm. Soc., Japan*, 86, 943 (1966)

D. teijsmanni.
ALKALOIDS: daphnijasmine, daphniteijsmannine, daphniteijsmine, deacetyldaphnijsmine, dihydrodaphnilactone, methyl homodaphniphyllate.
Yamamura, Hirata, *Tetrahedron Lett.*, 2849 (1974)
Yanamura, Hirata, *ibid.*, 3673 (1974)
Toda *et al.*, *Tetrahedron*, 30, 2683 (1974)

DARLINGIA (Proteaceae)

D. darlingiana (F.Muell.) L.A.S Johnson.
PIPERIDINE ALKALOIDS: bellendine, darlingine.
PYRROLE ALKALOIDS: darlingianine, tetrahydrodarlingianine.
Bick *et al.*, *Aust. J. Chem.*, 32, 2523 (1979)

D. ferrugina.
PIPERIDINE ALKALOID: (+)-ferruginine.
Bick *et al.*, *Aust. J. Chem.*, 32, 2537 (1979)

DARWINIA (Myrtaceae)

D. grandiflora.
TERPENOID: baeckeol.
Hems *et al.*, *J. Chem. Soc.*, 1208 (1940)

DATISCA (Datiscaceae)

D. cannabina.
FLAVONOIDS: cannabin, datinate-CM, datin, datinoside, datiscanin, datiscetin, datiscin, galangin, galanginoside, isalpinin.
Zapesochnaya *et al.*, *Rastit. Resur.*, 12, 237 (1976); *Chem. Abstr.*, 85 59588f
Zapesochnayi *et al.*, *Khim. Prir. Soedin.*, 176 (1982)

D. glomerata.
TRITERPENOIDS: datiscacin, datiscoside, datiscoside C.
Eddy *et al.*, *J. Chem. Soc., Perkin I*, 1333 (1983)

DATURA (Solanaceae)

D. alba.
TROPANE ALKALOID: (+)-hyoscyamine.
Bremner, Canno, *Aust. J. Chem.*, 21, 1369 (1968)

D. arborea L.
TROPANE ALKALOIDS: daturine, (-)-hyoscyamine, scopolamine.
Marion, Thomas, *Can. J. Chem.*, 33, 1853 (1955)

D. candida.
TROPANE ALKALOIDS: 6β,7β-dihydroxylittorine, hyoscine, hyoscyamine, meteloidine, norhyoscine.
Griffen *et al.*, *Aust. J. Chem.*, 29, 2329 (1976)

D. ceratocaula.
TROPANE ALKALOID: 6α-(2-methylbutanoyloxy)tropan-3β-ol.
Beresford, Woolley, *Phytochemistry*, 13, 2511 (1974)

D. cornigera Hook.
TROPANE ALKALOID: 6α-tigloyltropane.
Eade, Schartner, *Helv. Chim. Acta*, 18, 344 (1935)

D. fastuosa. (D. metel).
TROPANE ALKALOIDS: fastudine, fastudinine, fastunine, fastusine, fastusinine, (-)-hyoscyamine.
TRITERPENOID: daturadiol.
Khaleque *et al.*, *Sci. Res. (Dacca)*, 3, 212 (1966)

D. ferox.
TROPANE ALKALOID: meteloidine.
Sheehan, Bissel, *J. Org. Chem.*, 19, 270 (1954)

D. innoxia Mill.
TRITERPENOIDS: daturadiol, daturanolone.
Kocor *et al.*, *J. Org. Chem.*, 38, 3685 (1973)

D. metel L.
TROPANE ALKALOID: norhyoscyamine.
STEROID: withametelin.
 (A) Carr, Reynolds, *J. Chem. Soc.*, 101, 946 (1912)
 (S) Oshima *et al.*, *Tetrahedron Lett.*, 2025 (1987)

D. metel L. var. fastuga.
TROPANE ALKALOID: 3β-acetoxytropane.
Evans, Major, *J. Chem. Soc., C*, 1621 (1966)

D. meteloides DC.
TROPANE ALKALOIDS: apohyoscine, (-)-3β,6α-ditigloyloxytropane, 3β,6α-ditigloyloxytropane-7β-ol, (-)-hyoscyamine, meteloidine, norhyoscyamine.
Evans, Woolley, *J. Chem. Soc.*, 4936 (1965)

D. quercifolia.
STEROIDS: daturalactone-3, daturalactone-4, 12-oxowithanolide.
TROPANE ALKALOID: (-)-hyoscyamine.
 (A) Marion, Spenser, *Can. J. Chem.*, 32, 1116 (1954)
 (S) Qurishi *et al.*, *Phytochemistry*, 18, 283, 1756 (1979)

D. sanguinea Ruiz. et Pav.
TROPANE ALKALOIDS: 3β-acetoxytropane, 6α-isovaleryloxy-3-tigloyloxytropane-7β-ol, (-)-3α-(2,3-dihydroxy-2-phenylpropionoyloxy)-6,7-epoxytropane.
Evans, Major, *J. Chem. Soc., C*, 2775 (1968)
Van Eijk *et al.*, *Planta Med.*, 28, 139 (1975)

D. stramonium L.
TROPANE ALKALOIDS: atropine, (-)-hyoscyamine, scopolamine.
STEROID: vitastramonolide.
SESQUITERPENOIDS: capsidol, 4-hydroxylubimin.
 (A) Gadamer, *Arch. Pharm.*, 239, 294 (1901)
 (S) Masslenikova *et al.*, *Tezisy Dokl. - Sov. Indiiskii Simp. Khim. Prir. Soedin.*, 53 (1978)
 (T) Stoessl *et al.*, *J. Chem. Soc., Chem. Commun.*, 431 (1975)

D. stramonium x D. discolor.
TROPANE ALKALOIDS: apoatropine, apohyoscine, hyoscine, hyoscyamine, meteloidine, noratropine, pseudotropine, tropine.
Al-Yahya *et al.*, *J. Pharm. Pharmacol.*, 27, 87P (1975)

D. stramonium var. stramonium.
TROPANE ALKALOIDS: flurodaturatine, homoflurodaturatine.
Maier *et al.*, *Monatsh. Chem.*, 112, 1425 (1981)

DATURICARPA

D. elliptica Stapf.

INDOLE ALKALOIDS: ibogaine, ibogaline, ibophyllidine,
iboxygaine, voacangine.

Bruneton et al., Plant Med. Phytother., 10, 20 (1976)

DAUCUS (Umbelliferae)

D. carota L.

UNCHARACTERIZED ALKALOID: daucine.

AMINOACIDS: alanine, threonine, tyrosine, valine.

CAROTENOIDS: α-carotene, β-carotene, #87-carotene.

PHYTOALEXIN: 6-methoxymellein.

STEROID: diosgenin.

SESQUITERPENOIDS: α-bergamotene, carotol, (-)-β-bisabolene,
(-)-daucene, daucol.

(A) Pictet, Court, Bull. Soc. Chim., 1, 1001 (1907)

(Am) Ali et al., J. Indian Chem. Soc., 64, 230 (1987)

(C) Karrer et al., Helv. Chim. Acta, 14, 1083 (1931)

(Ph) Korusaki, Nishi, Phytochemistry, 22 669 (1983)

(T) Larsen et al., J. Am. Chem. Soc., 99, 8015 (1977)

DAVALLIA

D. divericata var. chinense.

TRITERPENOID: davallic acid.

Nakanishi et al., Tetrahedron Lett., 1451 (1963)

DAVIDSONIA

D. pruriens.

FLAVONOIDS: kaempferol, kaempferol 3-O-rhamnoside,
kaempferol 3-O-rhamnoside sulphate, luteolin 7-O-xyloside,
myricetin, myricetin 3-O-rhamnoside, myricetin
3-O-rhamnoside sulphate, quercetin, quercetin
7-O-rhamnoside, quercetin 7-O-rhamnoside sulphate.

PHENOLICS: epi-catechin gallate, (-)-epi-gallocatechin gallate,
gallylproanthocyanidin.

TRITERPENOIDS: 3β-hydroxybauer-7-en-28-oic acid, methyl
3-oxobauer-7-en-28-oate.

(F, P) Wilkins et al., Phytochemistry, 16, 144 (1977)

(T) Meksuriyen et al., ibid., 25, 1685 (1986)

DECACHAETA

D. haenkeana.

FLAVONOIDS: quercetagenin 6,7,4'-trimethyl ether,
quercetagenin 6,7,4'-trimethyl ether 6-sulphate.

Miski et al., Phytochemistry, 24, 3078 (1985)

D. ovalifolia.

FLAVONOIDS: 6-methoxyacacetin, 6-methoxyacigenin.

SESQUITERPENOIDS:
4α,15-dihydro-7α-hydroxy-3-deoxyzaluzanin C,
7α-hydroxycostunolide, 7α-hydroxyreynosin,
7α-hydroxysantamarin pinnatifidin.

De Luengo et al., Phytochemistry, 25, 1917 (1986)

D. thieleana.

TRITERPENOIDS: decathieleanolide, thieleanine.

Castro et al., Ann. Chem., 974 (1983)

DECALEPSIS

D. hamiltonii.

TRITERPENOIDS: α-amyrin, β-amyrin.

Simpson, J. Chem. Soc., 283, 286 (1944)

DECODON

D. verticillata L.

MACROCYCLIC ALKALOIDS: decaline, decamine, decinine,
decodine, 12-O-desmethyldecaline, desmethylvertaline,
dihydrolythrine, dihydrodecamine, vertaline.

UNCHARACTERIZED ALKALOID: verticillatine.

Ferris, J. Org. Chem., 27, 2985 (1962)

DELESSERIA

D. sanguinea.

HYDROQUINONE: jacaranone.

Yvin et al., J. Nat. Prod., 48, 814 (1985)

DELIA

D. emoryi.

PIGMENT: dalrubone.

Dreyer et al., Tetrahedron, 31, 287 (1975)

D. tinctoria.

PIGMENT: dalrubone.

Dreyer et al., Tetrahedron, 31, 287 (1975)

DELISEA (Rhodophyceae/Bonnemaisoniaceae)

D. fimbriata. (D. pulchra).

HALOGENATED MONOTERPENOIDS: acetoxyfimbrolides a, b, c, d,
e, f and g.

Kaslauskas et al., Tetrahedron Lett., 37, 41 (1977)

DELONIX

D. regia.

ACIDS: oxaloacetic acid, pyruvic acid.

AMINOACIDS: alanine, threonine, tyrosine, valine.

(Ac) Mukherjee et al., Phytochemistry, 14, 1915 (1975)

(Am) Ali et al., J. Indian Chem. Soc., 64, 230 (1987)

DELPHINIUM (Ranunculaceae)

D. ajacis L. (Consolida ambigua).

DITERPENE ALKALOIDS: ajacine, ajacinine, ajacinidine,
ajacinoidine, ajaconine, ajacusine, ajadine, delcosine,
trimethylacetyldelcosine.

Goodson, J. Chem. Soc., 245 (1947)

Pelletier et al., Heterocycles, 9, 463 (1978)

D. araraticum.

One uncharacterized diterpene alkaloid.

Samatov, Aptechn. Delo, 14, 26 (1965)

D. barkeyi.

DITERPENE ALKALOIDS: anthranoyllycoctonine, deltaline.

Couch, J. Am. Chem. Soc., 58, 684 (1936)

D. bicolor Nutt.

DITERPENE ALKALOIDS: delphocurarine. Two uncharacterized
alkaloids.

Jones, Benn, Tetrahedron Lett., 4351 (1972)

D. biternatum Huth.

DITERPENE ALKALOIDS: 10-benzoylbrowniine,
10-benzoyliliensine, delphatine, delbine, delbiterine,
iliensine.

Salimov et al., Khim. Prir. Soedin., 106 (1978)

D. bonvalotii.

DITERPENE ALKALOIDS: bonvalol, bonvalone, bonvalotine.

Jiang et al., Heterocycles, 22, 2429 (1984)

D. brownii Rybd.

DITERPENE ALKALOID: browniine.

Benn et al., Can. J. Chem., 41, 477 (1963)

D. cardinale.
DITERPENE ALKALOIDS: dehydrobrowniine, hetisinone.
> Benn, Can. J. Chem., 44, 1 (1966)
> Aplin et al., ibid, 46, 2635 (1968)

D. cardiopetalum DC.
DITERPENE ALKALOIDS: 15-acetylcardiopetaline, 14-acetyldihydrogadesine, acetylhetisinone, 14-benzoyldihydrogadesine, 14-benzoylgadesine, cardiopetamine, cardiopetalidine, cardiopetaline, dihydrogadesine, karacoline.
> Gonzalez et al., Heterocycles, 24, 1513 (1986)

D. carolinianum Walt.
DITERPENE ALKALOID: delcaroline.
FLAVONOIDS:
quercetin-(3''-benzoyl-2''-glucosyl)-3-glucoside-7-rhamnoside, quercetin-(glucosylbenzoyl)-glucoside, quercetin-(zylosylbenzoyl)-glucoside.
> (A) Pelletier et al., Heterocycles, 16, 747 (1981)
> (F) Warnock et al., Phytochemistry, 22, 1834 (1983)

D. cashmirianum.
DITERPENE ALKALOID: (+)-cashmiradelphine.
> Shamma et al., J. Nat. Prod., 42, 615 (1979)

D. confusa M.Pop.
DITERPENE ALKALOID: condelfine.
> Rabinovitch, Konovalova, J. Gen. Chem. USSR, 12, 329 (1942)

D. consolida L.
DITERPENE ALKALOIDS: consolidine, delcosine, delsoline, delsonine, lycoctonine.
> Marion, Edwards, J. Am. Chem. Soc., 69, 2010 (1947)
> Goodson, J. Chem. Soc., 245 (1947)

D. corymbosum Rgl.
DITERPENE ALKALOIDS: anthranoyllycoctonine, delcorine, delpyrine, deoxydelcorine, deoxylycorine, dictyocarpine, eldelidine, lycoctonine, methyllycaconitine, methyllycoctonine.
> Narzullaev et al., Khim. Prir. Soedin., 412 (1974)

D. denudatum.
DITERPENE ALKALOIDS: condelfine, delnudine, hetisinone, isotalatizidine.
> Goetz, Wiesner, Tetrahedron Lett., 5335 (1969)

D. dictyocarpum DC.
DITERPENE ALKALOIDS: dictyocarpine, delectine, delphirine, eldeline, eldelidine, lycoctonine, methyllycaconitine.
> Narzullaev et al., Khim. Prir. Soedin., 498 (1972)
> Samilov et al., ibid, 665 (1975)

D. elatum L.
DITERPENE ALKALOIDS: eldeline, delpheline, delatine, delphelatine, eldelidine, methyllycaconitine.
> Feofilaktov, Alekseeva, J. Gen. Chem. USSR, 24, 739 (1954)

D. elisabethae.
DITERPENE ALKALOIDS: anthranoyllococtonine, lycoctonine, methyllycaconitine.
> Beshitaishvili et al., Soobshch. Akad. Nauk Gruz. SSR, 79, 617 (1975); Chem. Abstr., 84 86723g

D. flexuosum Bieb. (MB).
DITERPENE ALKALOIDS: anthranoyllycoctonine, delflexine, delpyrine, methyllycoctonine.
> Brutko, Massagetov, Khim. Prir. Soedin., 21 (1967)

D. foetidum.
Three uncharacterized alkaloids.
> Zolonitskaya, Dokl. Akad. Nauk Arm. SSR, 37, 95 (1963)

D. freynii Conrath.
DITERPENE ALKALOIDS: anthranoyllycoctonine, delfrenine, delflexine, delpyrine, methyllycoctonine.
> Brutko, Massagetov, Khim. Prir. Soedin., 21 (1967)

D. glaucescens Rydb.
DITERPENE ALKALOIDS: glaucedine, glaucenine, glaucephine, glaucerine.
> Pelletier et al., J. Org. Chem., 46, 3284 (1981)

D. gracile.
DITERPENE ALKALOID: graciline.
> Gonzalez et al., Heterocycles, 22, 667 (1984)

D. grandiflorum L.
DITERPENE ALKALOIDS: anthranoyllycaconitine, delpyrine, methyllycaconitine.
> Strzelecka, Diss. Pharm. Pharmacol., 20, 325 (1968)
> Savchenko, Aktual. Vopr. Farm., 2, 22 (1974); Chem. Abstr., 84 147613v

D. iliense Huth.
DITERPENE ALKALOID: ilidine.
> Zhamietashvili et al., Khim. Prir. Soedin., 876 (1977)

D. menziesii DC.
UNCHARACTERIZED DITERPENE ALKALOID: delphocurarine.
> Hely, Sudd, Apoth. Ztg., 43, 28 (1903)

D. nelsonii Gr.
UNCHARACTERIZED DITERPENE ALKALOID: delphocurarine.
> Hely, Sudd, Apoth. Ztg., 43, 28 (1903)

D. nudicaule Torr. et Gray.
DITERPENE ALKALOIDS: 2-dehydrohetisine, 6-deoxydelcorine, dictyocarpine, dihydrogadesine, hetisine, lycoctonine, methyllycaconitine, nudicaulamine, nudicaulidine, nudicauline.
> Kilanthaivel et al., Heterocycles, 23, 2515 (1985)

D. nuttallianum Pritz.
DITERPENE ALKALOID: hetisine 13-O-acetate.
> Benn et al., Heterocycles, 24, 1605 (1986)

D. occidentale.
DITERPENE ALKALOID: deltaline.
> Couch, J. Am. Chem. Soc., 58, 684 (1936)

D. oreophilum Huth.
DITERPENE ALKALOIDS: anthranoyllycoctonine, delpyrine, delsine, demissine, methyllycoctonine, oreodine, oreoline.
> Khazlikhim et al., Khim. Prir. Soedin., 869 (1977)

D. orientale J.Gay.
DITERPENE ALKALOIDS: anthranoyllycoctonine, delcoline, delorine, delsoline, methyllycoctonine.
> Platonova, Kuzuvkov, Med. Prom. SSSR, 4, 17 (1963)
> Platonova, Kuzuvkov, ibid., 4, 19 (1963)

D. pentagynum Lam.
DITERPENE ALKALOIDS: 14-acetylgadesine, gadeline, gadesine, gadenine, pentagydine.
> Gonzalez et al., Tetrahedron Lett., 959 (1983)

D. puniceum Pall.
DITERPENE ALKALOIDS: anthranoyllycoctonine, delpyrine, methyllycoctonine.
> Brutko, Massagetov, Rastit. Resur., 6, 243 (1970)

D. pyramidatum Albov.
DITERPENE ALKALOIDS: delpyrine, methyllycoctonine.
> Brutko, Massagetov, Khim. Prir. Soedin., 21 (1967)

D. rotundifolium Afan.
DITERPENE ALKALOIDS: browniine, delsemine, methyllycaconitine.
> Yunusov, Abubakirov, Dokl. Akad. Nauk Uzb. SSR, 8, 21 (1949)

D. rugulosum.
One uncharacterized diterpene alkaloid.
> Namedov, Aptech. Delo., 14, 26 (1903)

D. scopulorum Gr.
UNCHARACTERIZED DITERPENE ALKALOID: delphocurarine.
> Heyl, Sudd, Apoth. Ztg., 43, 28 (1903)

D. semibarbatum Bienert.
DITERPENE ALKALOIDS: anthranoyllycoctonine, delsemidine, delsemine, delpyrine, lycoctonine, methyllycaconitine.
Yunusov, Abubakirov, *J. Gen. Chem. USSR.*, **22**, 1461 (1952)

D. staphisagria L.
DITERPENE ALKALOIDS: delphidine, delphinine, delphinoidine, delphirine, delphisine, staphidine, staphimine, staphinine, staphirine, staphisagnine, staphisagrine, staphisagroine.
Marquis, *Ann. Chim. Phys.*, **52**, 35 (1953)
Pelletier *et al.*, *Phytochemistry*, **16**, 404 (1977)

D. tatsiense Franch.
DITERPENE ALKALOIDS: deacetylambiguine, tatsiensine, tatsinine.
Pelletier *et al.*, *Heterocycles*, **20**, 1347 (1983)

D. ternatum.
DITERPENE ALKALOID: dehydroelidine.
Matveev *et al.*, *Khim. Prir. Soedin.*, 131 (1985)

D. tricorne.
DITERPENE ALKALOID: tricornine.
Pelletier, Bhattacharyya, *Phytochemistry*, **16**, 1464 (1977)

D. vestitum Wall.
DITERPENE ALKALOIDS: delvestridine, delvestrine.
Desai *et al.*, *Heterocycles*, **23**, 2483 (1985)

DENDRANTHEMA (Compositae/Anthemideae)
D. maximowiczii (Kort.) Tzvel.
SESQUITERPENOID: dendranthemol acetate.
Bohlmann *et al.*, *Chem. Ber.*, **108**, 735 (1975)

D. nactongense.
SESQUITERPENOID: dendranthemol acetate.
Bohlmann *et al.*, *Chem. Ber.*, **108**, 735 (1975)

DENDROBIUM (Orchidaceae)
D. aduncum.
SESQUITERPENOID: aduncin.
Gawell, Leander, *Phytochemistry*, **15**, 1991 (1976)

D. amoenum.
SESQUITERPENOIDS: amoenin, amotin.
Dahnen, Leander, *Phytochemistry*, **17**, 1949 (1978)

D. anosmum Lindl.
PYRIDOAMIDAZOLE ALKALOID: di(tetrahydropyridoimidazolium).
Leander, Luning, *Tetrahedron Lett.*, 905 (1968)

D. chrysanthemum Wall.
PYRROLIDINE ALKALOID: (-)-hygrine.
Liebermann, Khuling, *Ber.*, **24**, 407 (1891)

D. crassinoides B. et Rf.
GLUCOSIDES: cis-crassinodine, trans-crassinodine.
Dahmen *et al.*, *Acta Chem. Scand.*, **30B**, 297 (1976)

D. crepidatum.
DIMERIC ALKALOID: dendrocrepine.
INDOLIZIDINE ALKALOIDS: crepidamine, crepidine.
Eleander *et al.*, *Acta Chem. Scand.*, **27**, 1907 (1973)

D. findleyanum.
LACTONE ALKALOID: 2-hydroxydendrobine.
Granelli *et al.*, *Acta Chem. Scand.*, **24**, 1209 (1970)

D. hildebrandii Rolfe.
TERTIARY AMINE ALKALOIDS: 6-hydroxynobiline, N-isopentyl-6-hydroxydendroxinium chloride.
Eleander, Leander, *Acta Chem. Scand.*, **25**, 717 (1971)
Hedman, Leander, *ibid*, **26**, 3177 (1972)

D. nobile Lindl.
LACTONE ALKALOIDS: dendramine, dendrine, dendrobine, dendrobine N-oxide, dendroxine, N-isopentenyldendrobium bromide, N-isopentenyldendroxine, N-isopentenyldendroxinium chloride, N-isopentenyl-6-hydroxydendroxinium chloride nobiline, nobilonine.
PHENANTHROQUINONE: denbinobin.
STEROIDS: β-sitosterol, β-sitosterol β-D-glucoside.
SESQUITERPENOID: nobilomethylene.
(A) Inubishi, Nakano, *Tetrahedron Lett.*, 2723 (1965)
(A) Hedman, Leander, *Acta Chem. Scand.*, **26**, 3177 (1972)
(Ph) Talapatra *et al.*, *Indian J. Chem.*, **21B**, 386 (1982)
(T) Okamoto *et al.*, *Chem. Pharm. Bull.*, **20**, 418 (1972)

D. ochreatum Lindl.
SAPONINS: dendrosteroside, ochreasteroside, 16-epi-ochreasteroside.
Behr, Leander, *Phytochemistry*, **15**, 1403 (1976)

D. parishii.
PYRIDOIMIDAZOLE ALKALOID: di(tetrahydropyridoimidazolium).
Leander, Luning, *Tetrahedron Lett.*, 905 (1968)

D. pierardii Roxb.
LACTONE ALKALOID: pierardine.
Eleander *et al.*, *Acta Chem. Scand.*, **23**, 2177 (1969)

D. primulinum Lindl.
IMIDAZOLE ALKALOID: dendroprimine.
PYRROLIDINE ALKALOID: (-)-hygrine.
Blomqvist *et al.*, *Acta Chem. Scand.*, **27**, 1439 (1973)

D. superbum Rchb. f.
25 ketones and acetates.
Flath *et al.*, *J. Agric. Food Chem.*, **30**, 841 (1982); *Chem. Abstr.*, **97** 106997y

DENNSTAEDTIA
D. scabra.
SESQUITERPENOID: 4-hydroxypterosin A, pterosin V.
Murakami *et al.*, *Chem. Pharm. Bull.*, **23**, 1630 (1975)

D. wilfordii.
SESQUITERPENOIDS: pterolactones A and B.
Murakami *et al.*, *Chem. Pharm. Bull.*, **28**, 1869 (1980)

DENTARIA (Cruciferae)
D. enneaphylos.
LACTAM ALKALOID: sisymbrin.
Turemitch *et al.*, *Sci. Pharm.*, **53**, 163 (1985)

DERMOCYBE
D. sanguinea (Wulf. ex Fr.) Wunche.
ANTHRAQUINONES: dermoglaucin, dermocybin.
Steglich *et al.*, *Chem. Ber.*, **105**, 2928 (1972)

DERRIS
D. eclipta.
PISCICIDE: rotenone.
Carlson *et al.*, *Tetrahedron*, **29**, 2731 (1973)

D. elliptica.
BENZOFURAN: tubaic acid.
BENZOPYRAN: β-tubaic acid.
Obara *et al.*, *Agric. Biol. Chem.*, **40**, 1245 (1976)

D. glabrescens.
COUMARIN: derrusnin.
East *et al.*, *J. Chem. Soc.*, C, 365 (1969)

D. malaccensis.
BENZOPYRAN: α-toxicarol.
ISOFLAVONE: toxicarolisoflavone.
Cahn *et al., J. Chem. Soc.,* **513,** 734 (1938)

D. obtusa.
AURONATE: derriobtusone.
ALIPHATIC: heptacosanol.
STEROID: β-sitosterol.
Do Nascimento *et al., Phytochemistry,* **15,** 1553 (1976)

D. robusta.
COUMARINS: derrusnin, robustic acid, robustin, robustone.
Jain *et al., Tetrahedron,* **30,** 2485 (1974)

D. scandens.
BENZOPYRANONE: chandalone.
COUMARINS: scandenin, scandenone, scandinone.
(B) Falshaw *et al., J. Chem. Soc., C,* 374 (1969)
(C) Pelter *et al., Tetrahedron Lett.,* 1209 (1964)
(c) Pelter *et al., ibid.,* 2817 (1967)

D. trifoliata.
FLAVONOID: rhamnetin 3-O-neohesperidoside.
Nar *et al., J. Nat. Prod.,* **49,** 710 (1986)

D. uruau.
SAPONIN: derrisaponin.
Parente, Mors, *Anais Acad. Bras. Cienc.,* **52,** 503 (1980)

DESMARESTIA (Phaeophyceae)
D. ligulata.
ACIDS: citric acid, lactic acid, oxalic acid, succinic acid.
Sato *et al., Chem. Abstr.,* 84 102368c

DESMIA (Rhodophyceae)
D. hornemanni.
HALOGENATED MONOTERPENOIDS: 3-bromo-7-chloromyrcene,
7-bromo-10E-chloromyrcene, 7-bromo-10Z-chloromyrcene,
7-bromomyrcene, 10E-bromomyrcene, 10Z-bromomyrcene,
9-chloromyrcene, 3-chloro-7,10(Z)-dibromomyrcene,
7-chloro-10(E)-bromomyrcene,
7-chloro-10(Z)-bromomyrcene,
3,7-dibromo-10(Z)-chloromyrcene.
Naya *et al., Chem. Lett.,* 839 (1976)

D. japonicus. (Chondrococcus japonicus).
HALOGENATED MONOTERPENOIDS:
(e)-10-bromo-3-chloromyrcene,
(Z)-10-bromo-3-chloromyrcene.
Naya *et al., Chem. Lett.,* 839 (1976)

DESMODIUM (Leguminosae)
D. pulchellum Bent. ex Baker.
INDOLE ALKALOIDS: 5-methoxy-N,N-dimethyltryptamine,
5-methoxy-N,N-dimethyltryptamine N-oxide,
5-methoxy-N-methyltryptamine.
Culvenor *et al., Aust. J. Chem.,* **17,** 1301 (1964)
Ghosal, Mukherjee, *Chem. Ind. (London),* 793 (1965)

D. styracifolium Merr.
FLAVONOIDS: schaftoside, vicenins 1, 2 and 3.
Yasukawa *et al., J. Pharm. Soc., Japan,* **106,** 517 (1986)

D. tiliifolium.
ISOQUINOLINE ALKALOID: salsoline.
Spath *et al., Ber.,* **67,** 1214 (1934)

DESMONCUS (Palmae)
D. polyacanthos.
CATECHINS: (+)-afzelechin, (+)-catechin, (+)-epi-catechin.
Marletti *et al., Phytochemistry,* **15,** 443 (1976)

DESMOS
D. dasymachalus.
APORPHINE ALKALOID: dasymachaline.
Chan, Toh, *Phytochemistry,* **25,** 1999 (1986)

DEUTZIA (Philadelphaceae)
D. scabra Thunb.
IRIDOIDS: deutzioside, scabroside, scasbrosidol.
Plouvier, *C. R. Acad. Sci.,* **261,** 4248 (1965)
Esposito *et al., J. Nat. Prod.,* **46,** 614 (1983)

DIALIUM (Caesalpinaceae)
D. ovoideum.
ALKALOID: albizziine.
Peiris *et al., Phytochemistry,* **16,** 1821 (1977)

DIANELLA (Liliaceae)
D. ensifolia Redoute.
CHROMONES: 5,7-dihydroxy-2,8-dimethyl chromone,
5,7-dihydroxy-2,6,8-trimethylchromone.
PHENOLICS: dianellidin, methyl
2,4-dihydroxy-3,6-dimethylbenzoate, methyl
2,4-dihydroxy-6-methylbenzoate, methyl
2,4-dihydroxy-3,5,6-trimethylbenzoate.
Lojanapiwatna *et al., J. Sci. Soc., Thailand,* **8,** 95 (1982); *Chem. Abstr.,* 97 178728j

D. nigra.
PHENOLICS: dianellidin, dianellin, dianellinone.
Cooke *et al., Aust. J. Chem.,* **18,** 218 (1965)
Briggs *et al., New Zealand J. Sci.,* **18,** 559 (1975); *Chem. Abstr.,* 84 132647k

D. revoluta R.Br.
CHROMONE: 5,7-dihydroxy-6-methyl-2-nonacosylchromone.
Cooke *et al., Tetrahedron Lett.,* 1039 (1970)

DIANTHUS (Caryophyllaceae)
D. barbatus L.
PYRAN: barbapyroside.
SAPONINS: barbatoside A, barbatoside B.
(P) Plouvier *et al., Phytochemistry,* **25,** 546 (1986)
(Sap) Cordell *et al., Lloydia,* **40,** 361 (1977)

D. caryophyllus.
ANTHOCYANIN: pelargonidin 3-malylglucoside.
Terahara *et al., Phytochemistry,* **25,** 1715 (1986)

D. deltoidea.
ANTHOCYANIN: cyanidin 3-malylglucoside.
PYRAN: barbapyroside.
(An) Terahara *et al., Phytochemistry,* **25,** 1715 (1986)
(P) Plouvier *et al., Phytochemistry,* **25,** 546 (1986)

D. superbus var.longicalycinus.
SAPONINS: dianoside A, dianoside B.
Oshima *et al., Planta Med.,* **50,** 40 (1984)

DICENTRA (Fumariaceae/Papaveraceae)
D. canadensis (Goldie) Walp.
APORPHINE ALKALOIDS: (-)-corydine, isocorydine.
ISOQUINOLINE ALKALOID: cancentrine.
Clark *et al., J. Am. Chem. Soc.,* **92,** 4998 (1973)

D. chrysantha Walp.
ISOQUINOLINE LACTONE ALKALOID: bicuculline.
OXOPROTOBERBERINE ALKALOIDS: cryptocavine, cryptopine.
UNCHARACTERIZED ALKALOID: chrycentrine.
Manske, Marion, *J. Am. Chem. Soc.,* **62,** 2042 (1940)

D. cucullaria (L.) Bernh.
ISOQUINOLINE LACTONE ALKALOIDS: bicuculline, corlumine.
ISOQUINOLINE ALKALOID: cularine.
OXOPROTOBERBERINE ALKALOID: cryptopine.
> Manske, *Can. J. Res.*, **18**, 288 (1940)

D. eximia (Ker.) Torr.
APORPHINE ALKALOIDS: (+)-corydine, (-)-corydine, dicentrine, eximidine.
OXOPROTOBERBERINE ALKALOID: coreximine.
> Tomita *et al.*, *J. Pharm. Soc., Japan*, **87**, 880 (1967)

D. formosa (Andr.) Walp.
APORPHINE ALKALOIDS: (-)-corydine, corytuberine, dicentrine.
ISOQUINOLINE ALKALOID: cularine.
> Manske, *Can. J. Res.*, **10**, 521 (1934)

D. ochroleuca Engelm.
ISOQUINOLINE ALKALOID: bicuculline.
OXOPROTOBERBERINE ALKALOID: cryptopine.
> Manske, *Can. J. Res.*, **9**, 436 (1933)

D. oregana Eastwood.
APORPHINE ALKALOIDS: (-)-corydine, dicentrine.
ISOQUINOLINE ALKALOID: cularine.
OXOPROTOBERBERINE ALKALOID: α-allocryptopine.
PROTOBERBERINE ALKALOID: (-)-corypalmine.
> Kametani, Fukumoto, *Chem. Ind. (London)*, 291 (1963)

D. pusilla Sieb. et Zucc.
APORPHINE ALKALOID: dicentrine.
> Manske, *Can. J. Res.*, **14B**, 348 (1936)

D. spectabilis.
OXOQUINOLIZIDINE ALKALOID: protopine.
> Manske, *Can. J. Res.*, **15B**, 274 (1937)

DICHAPETCHUM
D. cymosum.
PYRIDINE ALKALOID: trigonelline.
> Holtz *et al.*, *Z. Biol.*, **8**, 57 (1924)

DICHROA (Saxifragaceae)
D. febrifuga Lour.
QUINAZOLINE ALKALOIDS: febrifugine, isofebrifugine.
UNCHARACTERIZED ALKALOIDS: dichroidine, dichroine A, dichroine B, dichroine-α, dichroine-β, dichroine-τ,
> Koepfli *et al.*, *J. Am. Chem. Soc.*, **69**, 1837 (1947)

DICHROSTACHYS
D. cinerea Maab.
STEROID: β-sitosterol.
TRITERPENOIDS: β-amyrin, friedelen-3β-ol, friedelan-3-one.
> Joshi *et al.*, *J. Indian Chem. Soc.*, **54**, 649 (1977)

DICOMA
D. anomala.
SESQUITERPENOIDS: 14-acetoxydicomanolide, dicomanolide, 9-hydroxydehydrozaluzanin C, 24-oxodicomanolide.
> Bohlmann *et al.*, *Phytochemistry*, **17**, 570 (1978)

DICORIA
D. canescens.
MONOTERPENOID: (-)-cis-chrysanthemol-β-D-glucopyranoside.
> Miyakado *et al.*, *Phytochemistry*, **13**, 2881 (1974)

DICRANOSTIGMA (Fumariaceae)
D. franchetianum (Prain.) Fedde.
NAPHTHOISOQUINOLINE ALKALOID: chelidonine.
OXOQUINOLIZIDINE ALKALOID: protopinc.
> Manske, *Can. J. Res.*, **20B**, 53 (1942)

DICTAMNUS (Rutaceae)
D. albus L.
FURANOISOQUINOLINE ALKALOIDS: dictamnine, isomaculosidine.
QUINOLONE ALKALOID: preskimmianine.
COUMARIN: auraptene.
NORSESQUITERPENOID: fraxinellone.
> (A) Storer, Young, *Tetrahedron*, **29**, 1217 (1973)
> (C) Reisch *et al.*, *Planta Med.*, **15**, 320 (1967)
> (T) Coggon *et al.*, *J. Chem. Soc., B*, 1521 (1970)

D. angustifolia G.Don.
FURANOISOQUINOLINE ALKALOID: dictamine.
QUINOLINE ALKALOIDS: dubamine, dubamidine, evoxine, skimmianine.
> Sidyakin *et al.*, *Uzb. Khim. Zh.*, **6**, 56 (1962)

D. caucasicus Fisch.
FURANOISOQUINOLINE ALKALOIDS: dictamine, dimethoxyisodictamine, α-fagarine, isodictamine, 6α-methoxyisodictamine, robustine, skimmianine.
> Asahina *et al.*, *Ber.*, **63**, 2045 (1930)
> Asatiani *et al.*, *Soobshch. Akad. Nauk Gruz SSR*, **64**, 1 (1971)

D. hispanicus Webb.
COUMARINS: bergapten, psoralen, xanthotoxin.
STEROID: β-sitosterol.
TRITERPENOIDS: α-amyrin, β-amyrin.
> Gonzalez *et al.*, *An. Quim.*, **73**, 430 (1977)

DICTYOPHLEBA
D. lucida.
STEROIDAL ALKALOIDS: distyolucidamine, dictyolutidine.
> Janot *et al.*, *Bull. Soc. Chim. Fr.*, **11**, 3472 (1966)

DICTYOPTERIS (Phaeophyceae)
D. australis.
MONOTERPENOIDS: (+)-butylcyclohepta-2,6-dione, (+)-6-butylcyclohepta-2,4-dione.
> Moore, Yost, *J. Chem. Soc., Chem. Commun.*, 937 (1973)

D. divaricata.
SESQUITERPENOID: epi-cubenol.
> Suzuki *et al.*, *Bull. Chem. Soc.,Japan*, **54**, 2366 (1981)

D. plagiogramma.
MONOTERPENOIDS: (+)-(R)-butylcyclohepta-2,5-dione, (+)-6-butylcyclohepta-2,4-dione.
> Moore, Yost, *J. Chem. Soc., Chem. Commun.*, 937 (1973)

D. undulata.
DITERPENOIDS: chromazonarol, isochromazonarol, yakuzunol, zonaroic acid.
> Ravi *et al.*, *Pure Appl. Chem.*, **51**, 1893 (1978)

D. zonarioides.
QUINONE: cyclospongiaquinone.
> Ravi *et al.*, *Pure Appl. Chem.*, **51**, 1893 (1978)

DICTYOSTELUM
D. discoideum.
PURINE ALKALOID: discadenine
> Obata *et al.*, *Agric. Biol. Chem.*, **37**, 1989 (1973)

DICTYOTA (Phaeophyceae/Dictyotaceae)

D. acutiloba.
DITERPENOIDS: dictyolene, dictyoxepin.
> Sun et al., J. Am. Chem. Soc., 99, 3516 (1977)

D. binghamiae.
DITERPENOIDS: dictyol G acetate, dictyoxide A, dictyotriol A diacetate.
> Pathirana et al., Can. J. Chem., 62, 1666 (1984)

D. cervicornis Kuetzing.
DITERPENOIDS: cervicol, dolostane, 3-secodolastene.
> Texiera et al., Bull. Soc. Chim. Belg., 95, 263 (1986)
> Texiera et al., J. Nat. Prod., 49, 570 (1986)

D. crenulata.
NORDITERPENOIDS: acetoxycrenulatin, crenulatin.
DITERPENOIDS: 4-acetoxycrenulide, crenulide.
> Sun et al., J. Org. Chem., 48, 1903 (1983)
> Sun et al., ibid., 1907 (1983)

D. dentata.
DITERPENOID: dictyol H.
> Alvarado et al., J. Nat. Prod., 48, 132 (1985)

D. dichotoma (Hudson) Lemaroux.
SESQUITERPENOIDS: tricyclodictyofuran A, tricyclodictyofuran B, tricyclodictyofuran C.
DITERPENOIDS: dictyoacetal, dictyofuran C, dictyofuran T, dictyol B acetate, dictyol C, dictyol F, epi-dictyol F, dictyone, dictydiol, dictytriene A, dictytriene B, dictyodiol, dictyotadiol, 3,4-epoxy-7,18-dollabelladiene, 3,4-epoxy-14-hydroxy-7,18-dollabelladiene, 3,4-epoxy-14-oxo-7,18-dollabelladiene, 14-oxo-7,18-dollabelladiene, 14-oxo-3,7,18-dollabellatriene.
> Enoki et al., Chem. Lett., 1399 (1983)
> Enoki et al., ibid., 1627 (1983)
> Enoki et al., Tetrahedron Lett., 1731 (1985)

D. dichotoma var. implexa.
DITERPENOIDS: dictyol B, dictyol I acetate, pachydictyol A.
> De Rosa et al., Phytochemistry, 25, 2179 (1986)

D. divaracata.
DITERPENOIDS:
4(S)-acetoxy-14(S)-hydroxydolast-1(15),7,9-triene, 7(S)-acetoxy-14(S)-dihydroxydolast-1(15),8-diene, 7(S)-acetoxy-4(S),14(S)-dihydroxydolast-1(15),8-diene, 4(S),7S)-diacetoxy-14(S)-hydroxydolast-1(15),8-diene, 7(S)-acetoxy-14(S)-hydroxydolast-1(15),8-diene.
> Sun et al., Tetrahedron, 37, 1237 (1981)

D. flabellata.
DITERPENOID: pachydictyol A epoxide.
> Robertson, Fenical, Phytochemistry, 16, 1071 (1977)

D. indica.
DITERPENOIDS: dictyotriol A, dictyotriol B.
> Li, Xue, Hydrobiologia, 116, 168 (1984)

D. linearis.
DITERPENOIDS: amijidictyol, 14-deoxyamijiol, isoamijiol.
> Ochi et al., Chem. Lett., 1253 (1980)

D. masonii.
DITERPENOID: hydroxydilophol.
> Sun, Fenical, J. Org. Chem., 44, 1354 (1979)

D. prolificans.
DITERPENOIDS:
18-acetoxy-19-oxo-1(9),6,13-xenicatrien-17,18-olide, 1(9)E,6E,13-xenicatriene-17,18:19,18-diolide.
> Ravi et al., Aust. J. Chem., 35, 121 (1982)

DIDYMOCARPUS

D. aurantiaca.
CHALCONES: aurentiacin, aurentiacin A, 5,6-dehydrokawain, 7,8-epi-5,6-dehydrokawain.
> Rezende et al., Phytochemistry, 10, 3167 (1971)
> Narayan et al., ibid., 15, 229 (1976)

D. oblonga.
DITERPENOIDS: didymooblongin, (-)-ent-16α-kauranol, (-)-ent-16α-kauran-19-oic acid.
> Mitra et al., Indian J. Chem., 19B, 79 (1980)

D. pedicellata.
UNCHARACTERIZED ALKALOIDS: pedicelline, pedicine, pedicinine.
PYRANONE: 5,6-dehydrokawain.
SESQUITERPENOID: α-humulene.
> (A) Grippa, Rend. Accad. Naz., 4, 18, (1968)
> (A) Grippa, ibid., 317 (1968)
> (P) Rezende et al., Phytochemistry, 10, 3167 (1971)
> (T) Clemo, Harris, J. Chem. Soc., 665 (1952)

DIDYMOSALPINX

D. abbeokutae.
DITERPENOID: abbeokutone.
> Taylor, J. Chem. Soc., 1360 (1967)

DIGITALIS (Scrophulariaceae)

D. canariensis. L.
STEROID: canarigenin.
> Studer et al., Helv. Chim. Acta, 116, 23 (1963)

D. canariensis L. var. isabelliana (Webb) Lindinger.
SAPONIN: digitoxigenin allomethyloside.
> Kaiser et al., Experientia, 21, 575 (1965)

D. ciliata.
COUMARIN: cynaroside.
SAPONIN: gitaloxin.
TRITERPENOID: deacetyllanostroside.
> (C) Gvazova et al., Khim. Prir. Soedin., 818 (1976)
> (S, T) Kamertelizde et al., ibid., 824 (1976)

D. grandiflora Mill. (D. ambigua Murr.)
ANTHRAQUINONE: digitolutein.
FLAVONOIDS: apigenol glucuronide, chrysoeriol glucuronide, kaempferol heterobioside, luteolol glucuronide, luteolol 7-glucoside.
> Boudouin et al., C. R. Acad. Sci., 283D, 1177 (1976)

D. lanata Ehrh.
ANTHRAQUINONE: 3-methylpurpurin.
SAPONINS: digalonin, digifolein, diginatin, digitalonin, digitonin, digitoxigenin allomethyloside, digitoxigenin digilanidobioside, gitorin, neoglucodigifucoside, tigonin.
STEROIDS: campesterol, cholesterol, digitogenin, 28-isofucosterol, lanatigonin, lanatosides A, B, C and D, neodigalogenin, neodigitalogenin, neodigitogenin.
TRITERPENOIDS: cycloartenol, 24-methylenecycloartanol.
> (An) Furuya et al., Phytochemistry, 11, 1073 (1972)
> (Sap) Kaiser, Experientia, 21, 575 (1965)
> (S) Tschesche et al., Tetrahedron, 18, 959 (1962)
> (S, T) Helmbold et al., Planta Med., 33, 185 (1978)

D. orientalis.
PIGMENTS: digitoemodin, 1-hydroxyanthraquinone-3-carboxylic acid, α-hydroxy-β-hydroxymethylanthraquinone, isochrysophenol, 3-methylantharufin.
SAPONIN: origidin.
> (P) Imre et al., Phytochemistry, 13, 681 (1974)
> (Sap) Mannich, Schneider, Arch. Pharm., 279, 223 (1941)

D. purpurea L.

SAPONINS: digacetinin, digifolein, digitonin, F-gitonin, gitoroside, gitoxoside, purpronin, purpureagitoside, strospeside, tigonin.

STEROIDS: desglucodigitonin, digalonin, diginin, 14α-digipronin, digipurpurogenin I, digipurpurogenin II, digitalin, digitalonin, digitalogenin, digitogenin, digitalin, gitaloxigenin, gitaloxin, gitorin, gitostin, neodigitogenin, neogitostin, glucoverodoxin, neodigitalogenin, purpnin, tigogenin.

BENZOPYRAN: digicitrine.

(B) Meier et al., Helv. Chim. Acta, **45**, 232 (1965)
(Sap) Furuya et al., Chem. Pharm. Bull., **18**, 1080 (1970)
(S) Tschesche et al., Justus Liebigs Ann. Chem., **606**, 160 (1957)
(S) Wada, Satoh, Chem. Pharm. Bull., **12**, 752 (1964)

D. schischkinii.

ANTHRAQUINONE: 3-methylpurpurin.

PIGMENTS: digitoemodin, isochrysophenol, 3-methylantharufin, methylquinizarin.

STEROID: glucoacetyldigoxoside.

(A) Furuya et al., Phytochemistry, **11**, 1073 (1972)
(P) Sanderman et al., Naturwissenschaften, **52**, 262 (1965)
(S) Imre, Ersoy, Experientia, **34**, 1254 (1978)

D. thapsi.

FLAVONE: thapsin.

Rao et al., Indian J. Chem., **11**, 403 (1973)

D. trojana.

ANTHRAQUINONES: digitopurpone-1-methyl ether, hydroxypachybasin, hydroxyziganein.

Imre et al., Phytochemistry, **15**, 317 (1976)

D. viridiflora.

PIGMENTS: digitoemodin, isochrysophenol.

Bick et al., Biochem. J., **98**, 112 (1966)

DILLENIA (Dilleniaceae)

D. indica.

TRITERPENOIDS: 3-β-hydroxylupan-13,28-lactone.
FLAVANONE: 3,5,7,4'-tetrahydroxy-3'-methoxyflavone.

(F) Paranasasivam et al., J. Chem. Soc., Perkin I, 612 (1975)
(T) Banerji et al., Phytochemistry, **14**, 1447 (1975)

D. pentagyna.

DITERPENOID: dipoloic acid.

Srivastava et al., Curr. Sci., **53**, 646 (1984)

DILOPHUS (Phaeophyceae/Dictyotaceae)

D. dichotoma.

DITERPENOID: dictyol D.

Denise et al., Experientia, **33**, 413 (1977)

D. fasciola.

SESQUITERPENOIDS: (1R,4R,10R)-1,4-epoxycadinane, 4(15),5E,10(14)-germacratrien-1α-acetate, (2S,8R)-germacra-1(11),E6-trien-2-ol acetate.
LIPID: (-)-(R)-1-O-geranylgeranylglycerol.

Fattorusso et al., Gazz. Chim. Ital., **109**, 589 (1979)

D. ligulatus.

DITERPENOID: dictyoxide.

Amica et al., Phytochemistry, **18**, 1895 (1979)

D. okamuri.

DITERPENOIDS: fukurinal, fukurinolol.

Ochi et al., Chem. Lett., 1927 (1982)

D. prolificans.

DITERPENOIDS: 3α-acetoxydilopholone, 3β-acetoxydilopholone, dilopholone, epoxydilophone.

Kazlauskas et al., Tetrahedron Lett., 4155 (1978)

DIMELAENA

D. oriena.

TRITERPENOID: zeorin.

Huneck, Lehn, Bull. Soc. Chim., 1702 (1963)

DIMEROSTEMMA

D. asperatum.

SESQUITERPENOID: dimerostemmolide.

Bohlmann et al., Phytochemistry, **23**, 1802 (1984)

D. bishopii.

SESQUITERPENOID: dimerostemmolide.

Bohlmann et al., Phytochemistry, **23**, 1802 (1984)

D. brasilianum.

SESQUITERPENOIDS:
1-O-(4-acetoxyangeloyloxy)-dimerostemmolide, 1-O-(4-acetoxyangeloyloxy)-isodimerostemmolide, 1-O-(5-acetoxyangeloyloxy)-isodimerostemmolide, 1-O-(4-acetoxytiglyl)-dimerostemmolide, 1-O-(2,3-epoxy-2-methylpronaloyl)-dimerostemmolide, dimerostemmobrasiolide, dimerostemmolide, 1-O-(2-hydroxymethylpropanoyl)-dimerostemmolide, 1-O-(2-hydroxymethylpropanoyl)-isodimerostemmolide, 1-O-(2-hydroxymethylpropanoyl)-8-(2-methylpropanoyl)-dimerostemmolide, 1-O-tiglyldimerostemmolide.

DITERPENOID: dimerobrasiolide.

Bohlmann et al., Phytochemistry, **21**, 1343 (1982)

D. lippoides.

Three unnamed sesquiterpenoids.

Bohlmann et al., Phytochemistry, **20**, 838 (1981)

DIMORPHOTHECA (Calendulaceae)

D. aurantiaca.

DITERPENOID: 3α-tigloyloxy-9,11-dehydrostachenoic acid.

Bohlmann et al., Phytochemistry, **16**, 1073 (1977)

D. pluvalis Moench.

DITERPENOIDS: 3β-acetoxysandaracopimar-7,15-dien-20-oic acid, 3β-acetoxysandaracopimar-8(14),15-dien-20-oic acid, 20-cupressenoic acid, sandaracopimar-7,15-dien-20-oic acid, sandaracopimar-8(14),15-dien-20-oic acid.

Bohlmann, Ngo Le Van, Chem. Ber., **109**, 1446 (1976)

DIOSCOREA (Dioscoreaceae)

D. batatus.

PHENANTHRENES: batatasin I, batatasin II, batatasin III, batatasin IV.

Hashimoto et al., Phytochemistry, **17**, 1179 (1978)

D. bulbifera.

STEROIDS: diosbulbins A, B and C diosgenin.

Komori et al., Chem. Ber., **101**, 3096 (1968)

D. bulbifera forma spontanea.

STEROIDS: diosbulbins D, E, F and G.
SAPONINS: diosbulbinoside D, diosbulbinoside F.

Komori et al., Chem. Ber., **101**, 3096 (1968)

D. capillaris.

STEROID: diosgenin.

Marker et al., J. Am. Chem. Soc., **65**, 1199 (1943)

D. chiapasensis Matuda.

STEROIDS: chiapagenin, isochiapagenin.

Harrison et al., J. Org. Chem., **26**, 155 (1961)

D. colletii Hook.

STEROIDS: 3,5-deoxyneotigogenin, 3,5-deoxytigogenin, diosgenin, diosgenin palmitate, isoparthogenin, sarsasapogenin, epi-sarsasapogenin, β-sitosterol, smilagenin, epi-smilagenin, smilagenone, yamogenin, yamogenin β-D-glucoside, yamogenin palmitate.

Minghe *et al., Planta Med.,* **49,** 38 (1983)

Liu *et al., Yaoxue Xuebao,* **20,** 143 (1985); *Chem. Abstr.,* 103 3692h

D. colletii Hook var. hypoglauca (Palibin) Pei et Ting.

SAPONINS: gracillin, progracillin.

Tang *et al., Zhiwu Xuebao,* **28,** 453 (1986); *Chem. Abstr.,* 105 222756b

D. composita.

STEROID: diosgenin.

Marker *et al., J. Am. Chem. Soc.,* **65,** 1199 (1943)

D. cyphocarpa.

STEROID: diosgenin.

Marker *et al., J. Am. Chem. Soc.,* **65,** 1199 (1943)

D. deltoidea Wall.

SAPONIN:

diosgenin-3-O-glucopyranosly-O-β-D-glucopyranosyl-β-D-glucopyranoside.

STEROIDS: campesterol, deltofolin, deltonin, diosgenin, β-sitosterol, spirostan-3,5-dione, stigmasterol.

Stohs *et al., Planta Med.,* **28,** 101 (1975)

Passeshnichenko *et al., Dokl. Akad. Nauk SSSR,* **246,** 742 (1981)

D. dugessi.

STEROID: diosgenin.

Marker *et al., J. Am. Chem. Soc.,* **65,** 1199 (1943)

D. dumetorum Pax.

TROPANE ALKALOID: dioscine.

Pinter, *J. Chem. Soc.,* 2236 (1952)

D. galeottiana.

STEROID: diosgenin.

Marker *et al., J. Am. Chem. Soc.,* **65,** 1199 (1943)

D. gracillum.

STEROIDS: gracilin, protodioscin.

Tsukumoto, Kawasaki, *J. Pharm. Soc., Japan,* **74,** 1127 (1954)

D. grandiflora.

STEROID: diosgenin.

Marker *et al., J. Am. Chem. Soc.,* **65,** 119 (1943)

D. hirsuta Blume.

TROPANE ALKALOID: dioscorine.

STEROID: diosgenin.

(A) Boorsma, *Meded. Lds Pttuin,* 13 (1984)

(S) Marker *et al., J. Am. Chem. Soc.,* **65,** 1199 (1943)

D. hirsuticaulis.

STEROID: diosgenin.

Marker *et al., J. Am. Chem. Soc.,* **65,** 1199 (1943)

D. jaliscana.

STEROID: diosgenin.

Marker *et al., J. Am. Chem. Soc.,* **65,** 1199 (1943)

D. lobata.

STEROID: diosgenin.

Marker *et al., J. Am. Chem. Soc.,* **65,** 1199 (1943)

D. macrostachya.

STEROIDS: diosgenin, ricogenin.

Marker *et al., J. Am. Chem. Soc.,* **71,** 3856 (1949)

D. mexicana.

STEROIDS: diosgenin, neokammagenin.

Marker *et al., J. Am. Chem. Soc.,* **69,** 2375 (1947)

D. militaris.

STEROID: diosgenin.

Marker *et al., J. Am. Chem. Soc.,* **65,** 1199 (1943)

D. minima.

STEROID: diosgenin.

Marker *et al., J. Am. Chem. Soc.,* **65,** 1199 (1943)

D. multinervis.

STEROID: diosgenin.

Marker *et al., J. Am. Chem. Soc.,* **65,** 1199 (1943)

D. paranensis Kunth.

STEROID: yamogenin.

Marker *et al., J. Am. Chem. Soc.,* **69,** 2184 (1947)

D. platycalpata.

STEROID: diosgenin.

Marker *et al., J. Am. Chem. Soc.,* **65,** 1199 (1943)

D. plumifera.

STEROID: diosgenin.

Marker *et al., J. Am. Chem. Soc.,* **65,** 1199 (1943)

D. prazeri.

SAPONINS: diosgenin
3-O-α-L-rhamnopyranosyl-O-di-β-D-glucopyranoside,
diosgenin 3-O-α-L-rhamnopyranosyl-β-D-glucopyranoside,
prazerigenin A
3-O-α-L-rhamnopyranosyl-O-di-β-D-glucopyranoside,
prazerigenin A
3-O-L-α-rhamnopyranosyl-β-D-glucopyranoside.

STEROIDS: prazerigenin A, prazerigenin A glycopyranoside, prazerigenin B, prazerigenin C.

PHENANTHRENES:
5,6-dihydroxy-2,4-dimethoxy-9,10-dihydrophenanthrene,
5,6-dihydroxy-1,3,4-trimethoxy-9,10-dihydrophenanthrene.

(P) Rajamaran *et al., Indian J. Chem.,* **13,** 1137 (1975)

(S, Sap) Rajamaran *et al., Indian J. Chem.,* **14B,** 735 (1976)

(Sap) Wij *et al., ibid.,* **15B,** 451 (1977)

D. pringlei.

STEROID: diosgenin.

Marker *et al., J. Am. Chem. Soc.,* **65,** 1199 (1943)

D. remotiflora.

STEROID: diosgenin.

Marker *et al., J. Am. Chem. Soc.,* **65,** 1199 (1943)

D. sanzibarensis.

TROPANE ALKALOID: dioscine.

Pinder *et al., J. Chem. Soc.,* 2236 (1952)

D. sativa.

SAPONIN: trillin.

Marker *et al., J. Am. Chem. Soc.,* **62,** 1548 (1940)

Marker *et al., ibid.,* **62,** 3349 (1940)

D. septemloba.

SAPONINS: kikuba saponin, methylprotogracilin, protogracilin.

Kawasaki *et al., Chem. Pharm. Bull.,* **22,** 2164 (1974)

D. spiculiflora.

STEROIDS: botogenin, neobotogenin.

Wahlens *et al., J. Org. Chem.,* **22,** 182 (1957)

D. subtomentosa.

STEROID: diosgenin.

Marker *et al., J. Am. Chem. Soc.,* **65,** 1199 (1943)

D. tenuipes Franch et Sarat.

STEROIDS: 2-acetoxy-3,4-dihydroxy--pregn-16-en-20-one,
4-acetoxy-2,3-dihydroxy-5α-pregn-16-en-20-one,
2-acetoxy-2,3,4-trihydroxy--pregn-16-en-20-one,
4-acetoxy-2,3,4-trihydroxy-5α-pregn-16-en-20-one,
diotigenin, glycosides S-3a, S-4, U-3, 3-methoxaldiotigenin
4-acetate, 2,3,4-trihydroxy-5β-pregn-16-en-20-one.

Kiyosawa, Kawasaki, *Chem. Pharm. Bull.,* **25,** 163 (1977)

Kiyosawa, Kawasaki, *Tetrahedron Lett.,* 4599 (1977)

D. testudinaria.

STEROID: yamogenin.

Marker *et al., J. Am. Chem. Soc.,* **65,** 1199 (1943)

D. tokoro Makino.

SAPONINS: prototokorin, tokoronin, yononin.

STEROIDS: dioscin, diosgenin, 3-epi-diosgenin, tokorogenin, yonogenin.

(Sap) Kawasaki, Yamauchi, *J. Pharm. Soc., Japan,* **83,** 757 (1963)

(Sap) Tomita, Uomori, *Phytochemistry*, **13**, 729 (1974)

(S) Uomori *et al.*, *Phytochemistry*, **22**, 203 (1983)

D. tryphida.

ANTHOCYANINS: malvidin 3,5-diferulate, malvidin 3,5-diglucoside, peonidin 3,5-diglucoside.

Carreno-Diaz *et al.*, *J. Food Sci.*, **42**, 615 (1977); *Chem. Abstr.*, 86 185971j

D. ulinei.

STEROID: diosgenin.

Marker *et al.*, *J. Am. Chem. Soc.*, **65**, 1199 (1943)

D. urceolata.

STEROID: diosgenin.

Marker *et al.*, *J. Am. Chem. Soc.*, **65**, 1199 (1943)

D. zingiberensis Wright.

SAPONIN: protozingiberensis saponin.

STEROIDS: diosgenin palmitate, gracillin, protogracillin, β-sitosterol.

Liu *et al.*, *Zhiwa Xuebao*, **27**, 68 (1985); *Chem. Abstr.*, 103 3668e

DIOSCOREOPHYLLUM

D. cummingsii.

TRITERPENOIDS: columbin, isocolumbin.

Cava *et al.*, *J. Am. Chem. Soc.*, **78**, 5317 (1950)

DIOSPYROS (Ebenaceae)

D. canaliculata.

CHROMENE: canaliculatin.

Waterman *et al.*, *J. Chem. Res. Synop.*, 2 (1985)

D. celebica.

FLAVONOIDS: dihydrodiosindigo B, diosindigo B.

QUINONES: celebaquinone, diomeloquinone A, isocelebaquinone.

Maiti *et al.*, *J. Chem. Soc., Perkin I*, 675 (1986)

D. discolor.

ANTHRAQUINONE: 1,3,5,6-tetrahydroxy-2-methylanthraquinone-8-O-β-D-glucopyranoside.

Srivastava *et al.*, *Curr. Sci.*, **54**, 998 (1985)

D. galpini.

NAPHTHOQUINONES: isoxylospyrin, whyteone.

Van der Vijver *et al.*, *Pharm. Weekbl.*, **111**, 1273 (1976)

D. heterotricha.

NAPHTHOQUINONES: 8'-hydroxyisodiospyrin, mamegakinone.

Alves *et al.*, *An. Fac. Farm. Porto*, **33**, 5 (1973); *Chem. Abstr.*, 17097q

D. ismailii.

CHROMENE: ismailin.

Waterman *et al.*, *J. Chem. Res. Synop.*, 2 (1985)

D. japonica.

NAPHTHALENONE: (R)-shinanolone.

Kuroyanagi *et al.*, *Chem. Pharm. Bull.*, **19**, 2314 (1971)

D. kaki.

CAROTENOIDS: β-carotene, cryptoxanthin, lycopene.

LIGNAN: (-)-divanillyltetrahydrofuran ferulate.

NAPHTHALENONE: (R)-shinanolone.

(C) Kolesnik *et al.*, *Khim. Prir. Soedin.*, 456 (1977)

(L) Matsuura *et al.*, *Phytochemistry*, **24**, 626 (1985)

(N) Kuroyanagi *et al.*, *Chem. Pharm. Bull.*, **19**, 2314 (1971)

D. lotus.

PIGMENT: bisisodiospyrin.

Natori *et al.*, *Chem. Pharm. Bull.*, **19**, 2308 (1971)

D. lycioides.

NAPHTHOQUINONES: 8'-hydroxyisodiospyrin, mamegakinon.

Alves *et al.*, *An. Fac. Farm. Porto*, **33**, 5 (1973); *Chem. Abstr.*, 17097q

D. mannii.

QUINONES: diospyrin, 3'-methoxydiospyrin, 7-methyljuglone.

TRITERPENOIDS: betulin, betulinic acid, lupeol.

Jeffrey *et al.*, *Phytochemistry*, **22**, 1832 (1983)

D. melanoxylon.

PIGMENTS: 3,3'-bi[6-hydroxy-5-methoxy-2-methylnaphthoquinone], 4,11-dihydroxy-5-methoxy-2,9-dimethyldinaphtho-[1,2-b 2',3'-d]-furan-7,12-quinone.

Sankaram *et al.*, *Phytochemistry*, **20**, 1093 (1981)

D. mollis.

NAPHTHALENES: 1,8-dihydroxy-3-methylnaphthalene, 4,5,8-trimethoxy-2-naphthaldehyde.

Borsub *et al.*, *Tetrahedron Lett.*, 105 (1976)

D. montana.

TRITERPENOID: allobetulin.

Lillie *et al.*, *J. Chem. Soc., Perkin I*, 2155 (1976)

D. natalensis.

NAPHTHOQUINONES: isoxylospyrin, rotundiquinone, whyteone.

Van der Vijver *et al.*, *Pharm. Weekbl.*, **111**, 1273 (1976)

D. quiloensis.

NAPHTHALENES: 4,5,6,8-tetrahydroxy-2-naphthaldehyde, 4,5,8-trihydroxy-2-naphthaldehyde.

Harper *et al.*, *J. Chem. Soc., C*, 626 (1970)

D. whyteana.

NAPHTHOQUINONES: isoxylospyrin, whyteone.

Van der Vijver *et al.*, *Pharm. Weekbl.*, **111**, 1273 (1976)

DIOTIS

D. maritima.

ACETYLENIC: 9,16-heptadecadiene-4,6-diyn-3-ol.

de Pascual Teresa *et al.*, *An. Quim.*, **73**, 1525 (1977)

DIPHASIA

D. klaineana.

PHENOLIC: tecleanone.

Casey *et al.*, *Tetrahedron Lett.*, 401 (1975)

DIPHYSA

D. robinioides.

BENZOPYRONES: (-)-4'-O-methylglabridin, (-)-4'-O-methylpreglabridin.

FLAVONOIDS: diphysolone, diphysolone 4'-methyl ether, ferreirin, kievitone.

STILBENE: ent-3,5,3',4',5'-pentahydroxystilbene.

(B, S) Castro *et al.*, *J. Nat. Prod.*, **49**, 1080 (1986)

(F) Ingham *et al.*, *Z. Naturforsch.*, **38C**, 899 (1983)

DIPIDAX (Liliaceae/Wurmbaeoideae)

D. triqueta (Jacq.) Baker.

COLCHICINE ALKALOID: cornigerine.

Pihewska *et al.*, *Collect. Czech. Chem. Commun.*, **32**, 158 (1967)

DIPLAZIUM (Athyriaceae)

D. subsinuatum.

TRITERPENOIDS: 17,24-dihydroxyhopan-28,22-olide, 17-hydroxyhopan-28,22-olide-24-glucopyranoside.

Tanaka *et al.*, *Chem. Pharm. Bull.*, **30**, 3632 (1982)

DIPLOCLISIA

D. glaucescens.

STEROIDS: 24(28)-dehydromakisterone A, 20-hydroxyecdysone, makisterone, 24-epi-makisterone, pterosterone.

Miller *et al.*, *Planta Med.*, 40 (1985)

DIPLODIA
D. macrospora.
INDOLE ALKALOID: chaetoglobosine A.
Springer et al., Tetrahedron Lett., 1905 (1980)
D. pinea.
MONOTERPENOIDS: diplodialides A, B and C.
DITERPENOID: diplodiatoxin.
Wada, Ishida, J. Chem. Soc., Perkin I, 1154 (1979)

DIPLOKNEMA
D. butyracea.
FLAVONOIDS: myricetin, myricetin 3-O-rhamnoside, quercetin, quercetin 3-O-rhamnoside.
Khetwal et al., Fitoterapia, 57, 128 (1986)

DIPLOPAPPUS
D. fruticosus.
SESQUITERPENOIDS: 2,8-decadiene-4,6-diyn-1,10-dial, cis,cis-matricarianol, trans,trans-matricarianol, cis,cis-matricarianol acetate.
Bohlmann et al., Chem. Ber., 105, 1919 (1972)

DIPLOPHYLLUM (Hepaticae/Scapaniaceae)
D. albicans (L.) Dum.
SESQUITERPENOIDS: acetoxydiplophyllin, albicanol, anastreptene, 11-dihydrodiplophyllin, diploalbicanol, diplophyllin, diplophyllolide, diplophyllolide B, ent-selina-4,11-diene.
Ohta et al., Tetrahedron, 33, 617 (1977)
Asakawa et al., Phytochemistry, 18, 1007 (1979)
D. taxifolium (Wahl.) Dum.
SESQUITERPENOIDS: acetoxydiplophyllin, albicanol, diploalbicanol, diplophyllin, diplophyllolide, ent-selina-4,11-diene.
Ohta et al., Tetrahedron, 33, 617 (1977)
Asakawa et al., Phytochemistry, 18, 1007 (1979)

DIPLOPTERYGIUM
D. glaucum.
TRITERPENOID: hopan-22-ol.
Baddeley et al., J. Chem. Soc., 3891 (1961)

DIPLORRHYNCHUS
D. condylocarpon (Muell) Arg.
CARBAZOLE ALKALOIDS: mossambine, norfluorocurarine.
IMIDAZOLIDINOCARBAZOLE ALKALOIDS: condyfoline, condylocarpine.
INDOLE ALKALOID: stemmadenine.
Yuldashev, Yunusov, Khim. Zh. Uzbeksk., 7, 44 (1963)
Schumann, Schmid, Helv. Chim. Acta, 46, 1996 (1963)
D. condylocarpon (Muell) Arg. subsp. mossambicensis (Benth.) Duvign.
CARBAZOLE ALKALOIDS: mossambine, norfluorocurarine, tombozine.. Indole alkaloid stemmadenine.
Schumann, Schmid, Helv. Chim. Acta, 46, 1996 (1963)

DIPSACUS (Dipsacaceae)
D. azureus Schrenk.
PYRIDINE ALKALOIDS: cantleyine, gentianine.
Sevenet et al., Bull. Soc. Chim. Fr., 8, 3120 (1970)
D. sylvestris Huds. (D. fullonum L.).
IRIDOIDS: sylvestrosides I, II, III and IV.
Jensen et al., Phytochemistry, 18, 273 (1979)

DIPTEROCARPUS (Dipterocarpaceae)
D. dyeri.
SESQUITERPENOIDS: α-gurjunene, β-gurjunene, ent-β-gurjunene, τ-gurjunene, gurjuresene.
Asakawa et al., Phytochemistry, 17, 457 (1978)A
D. elatus.
SESQUITERPENOIDS: 5β,10α-selina-3,11-diene, 10α-selina-4,11-diene.
Lund et al., Phytochemistry, 9, 2419 (1970)
D. gracilis.
SESQUITERPENOID: apitonene-1.
TRITERPENOIDS: gracilols A, B and C.
Kitao, Ikeda, Zairyo, 16, 848 (1967)
Ikeda, Kitao, Mokuzai Gakkaishi, 20, 460 (1974)
D. hispidus.
TRITERPENOIDS: ocotillone, ocotillone I, ocotillone II.
Bisset et al., Phytochemistry, 5, 865 (1966)
D. macrocarpus.
SESQUITERPENOIDS: dipterol, dipterone.
Krishnamurty et al., Indian J. Chem., 12, 520 (1974)
D. pilosus.
SESQUITERPENOID: humulene epoxide III.
DITERPENOID: hollongdione.
TRITERPENOIDS: 2-acetylasiatic acid methyl ester, caryophyllenol I caryophyllenol II.
Gupta, Dev, Tetrahedron, 27, 635 (1971)
Gupta, Dev, ibid., 27, 823 (1971)

DIRCA (Thymelaeaceae)
D. occidentalis.
ESTERS: dircin, pimelea factor P2.
LIGNANS: (-)-lariciresinol, (-)-medioresinol, (+)-syringaresinol.
Badawi et al., J. Pharm. Sci., 72, 1285 (1983)

DISCARIA
D. crenata.
CARBOLINE ALKALOID: crenatine A.
Pucheco et al., Phytochemistry, 12, 954 (1973)
D. febrifuga Mart.
PEPTIDE ALKALOIDS: discarines A, B, C and D.
Digel et al., Z. Physiol. Chem., 364, 1641 (1983)

DISTEMONANTHUS
D. benthamianus.
FLAVONES: ayanin, oxyayanin A, oxyayanin B.
King et al., J. Chem. Soc., 4587 (1954)
Kupchan et al., Phytochemistry, 10, 664 (1971)

DISYNAPHIA
D. halimifolia.
SESQUITERPENOIDS: disyfolide, disyhamifolide.
Bohlmann et al., Phytochemistry, 20, 1077 (1981)

DITTRICHIA
D. viscosa.
SESQUITERPENOIDS: 6α-hydroxy-9-desacetylineupatorolide, 2-methylpropanoyldittricholide.
Bohlmann, Gupta, Phytochemistry, 21, 1443 (1982)

DOCYNIOPSIS
D. tschonoski.
FLAVONE: toringin.
Marsh, Biochem. J., 59, 58 (1955)

DODONAEA (Sapindaceae)

D. alternata.
TRITERPENOID: lup-20(29)-ene-3β,11β-diol.
Ghisalberti et al., Phytochemistry, 12, 1125 (1973)

D. attenuata.
DITERPENOID:
ent-17-acetoxy-15,16-epoxy-18-hydroxy-3,13(16),14-clerodatrien-18-oic acid.
Jefferies, Payne, Tetrahedron Lett., 4777 (1967)

D. attenuata var. linearis.
DITERPENOID: huatriwaic acid.
Jefferies, Payne, Tetrahedron Lett., 4777 (1967)

D. boroniaefolia.
Three diterpenoids.
Jefferies et al., Aust. J. Chem., 26, 2199 (1973)

D. lobulata.
DITERPENOIDS: 6β,8β-dihydroxy-enantio-labdan-15-oic acid,
7α,8β-dihydroxy-enantio-labdan-15-oic acid,
enantio-labdanoic acid, enantio-labdanolic acid.
Dawson et al., Aust. J. Chem., 19, 2113 (1966)

D. microzyga.
DITERPENOID: ent-2α,17-dihydroxylabdan-17,13-dien-15-oic acid.
Jefferies et al., Aust. J. Chem., 27, 1097 (1974)

D. petiolaris.
DITERPENOIDS:
ent-3β-acetoxy-8(17),13(16),14-labdatrien-15-oic acid,
ent-3β-hydroxy-15,16-epoxy-8(17),13(16),14-labdatrien-18-oic acid.
Jefferies et al., Aust. J. Chem., 34, 1001 (1981)

D. viscosa.
DITERPENOIDS: dodonic acid, hautriwaic acid. Eight flavonoids.
(F) Sachdev et al., Phytochemistry, 22, 1253 (1983)
(T) Jefferies, Payne, Tetrahedron Lett., 4777 (1967)

DOLICHOS (Papilionaceae)

D. falcata.
TRITERPENOIDS: medicagenic acid, medicagenic acid 3-O-β-D-glucoside.
Feng et al., Zhongcaoyao, 16, 47 (1985)

D. lablab.
GIBBERELLIN: 3-O-glucopyranosylgibberellin A-4.
Yokoya et al., Agric. Biol. Chem., 42, 1811 (1978)

DOLICHOTHELE (Cactaceae)

D. longimamma.
AMINE ALKALOID: ubine.
ISOQUINOLINE ALKALOIDS: longimammadine, longimammanine, longimammidine, longimammine, longimammosine.
Tomina et al., Izv. Akad. Nauk SSSR, 2181 (1971)
Ranieri, McLaughlin, J. Org. Chem., 41, 319 (1976)

D. sphaerica R. et R.
ALKALOID: dolichotheline.
Rosenberg, Paul, Tetrahedron Lett., 1039 (1969)

DOREMA

D. ammonaicum.
BENZOPYRAN: ammoresinol.
SESQUITERPENOID: ferulene.
(B) Kunz et al., J. Prakt. Chem., 141, 350 (1934)
(T) Semmler et al., Ber., 50, 1826 (1917)

D. hyrcanum.
GLYCOSIDES: hyrcanoside, pleoside.
Nurmukhamedova et al., Khim. Prir. Soedin., 101 (1976)

DORONICUM (Compositae)

D. grandiflorum Lam.
FLAVONOIDS: 5,7,4'-trihydroxy-8,3'-dimethoxyflavone, 5,7,4'-trihydroxy-8-methoxyflavone.
Reynaud et al., Pharmazie, 38, 628 (1983)

D. microphyllum.
SAPONINS: doronicosides A, B, C and D
Alieva et al., Khim. Prir. Soedin., 658 (1977)

D. pardalianches L.
PYRROLIZIDINE ALKALOID: otosenine.
Rajagopalan et al., Indian J. Chem., 24B, 882 (1985)

DORSTENIA (Moraceae)

D. contraferva.
STEROID: syriogenin.
Okada, Anjyo, Chem. Pharm. Bull., 23, 2039 (1975)

DORYPHORA (Monimiaceae)

D. sassafras.
ISOQUINOLINE ALKALOIDS: corypalline, doryphorinine.
UNCHARACTERIZED ALKALOIDS: doryafrine, doryanine, doryflavine, doryphorine.
Chen et al., Lloydia, 37, 493 (1974)
Francisca et al., Indian J. Chem., 15B, 182 (1977)

DRACAENA (Liliaceae)

D. australis.
STEROID: smilagenin.
Marker et al., J. Am. Chem. Soc., 65, 1199 (1943)

D. draco.
STEROID: dracogenin.
Gonzalez et al., Rev. Latinoam. Quim., 3, 8 (1972)

DRACOCEPHALUM

D. nutans.
FLAVONOID: leuteolin 5-O-β-L-galactoside.
Shamyrina et al., Khim. Prir. Soedin., 577 (1977)

DRACONTOMELUM (Anacardiaceae)

D. mangiferum Bl.
QUINOLIZIDINE ALKALOID:
(-)-1,2,3,4,6,7-hexahydro-12H-indolo-[2,3-a]-quinolizidine.
Johns et al., Aust. J. Chem., 19, 1951 (1966)

DREGEA (Asclepiadaceae)

D. sinensis var. corrugata.
STEROID: drevogenin.
Jin et al., Yunnan Zhiwu Yanjiu, 9, 227 (1987); Chem. Abstr., 107 194915w

D. volubilis Benth.
STEROID: drevogenin P.
Sauer et al., Helv. Chim. Acta, 48, 857 (1965)

DRIMYS (Winteraceae)

D. leucolata.
SESQUITERPENOID: polygodial.
Barnes, Loder, Aust. J. Chem., 15, 322 (1962)
Barnes, Loder, ibid., 15, 389 (1962)

D. winteri L.
SESQUITERPENOIDS:
11,12-bis(7-drimen-11-oxy)-11,12-epoxy-7-drimene, confertifolin, drimenin, drimenol, fuegin, futronolide, isodrimenin, valdiviolide.

Kato et al., Tetrahedron Lett., 4257 (1972)
Aasen et al., Acta Chem. Scand., 31, 51 (1977)

DROSERA (Droseraceae)
D. ramentacea.
DIQUINONE: biramentaceone.
Krishnamoorthy et al., Phytochemistry, 8, 1591 (1969)

D. whittakeri.
NAPHTHOQUINONES: hydroxydroserone, droserone.
Lugg et al., J. Chem. Soc., 1457 (1936)

DROSOPHILA
D. subatrata.
ACETYLENIC: drosophilin C.
Jones et al., J. Chem. Soc., 2257 (1960)

D. vestita.
ACETYLENIC: drosophilin C.
Jones et al., J. Chem. Soc., 2257 (1960)

DRYADODAPHNE
D. novo-guineensis.
BISBENZYLISOQUINOLINE ALKALOIDS: dryadine, dryadadaphnine.
Bick et al., J. Chem. Soc., C, 1627 (1967)

DRYMARIA (Caryophyllaceae)
D. cordata (L.) Willd.
ACIDS: capric acid, caproic acid, caprylic acid, lauric acid, linoleic acid, linolenic acid, myristic acid, oleic acid, palmitic acid, stearic acid.
ALKALOID: cordatanine.
STEROID: α-spinasterol.
(Ac, S) Hu et al., Zhongcaoyao, 13, 343 (1982)
(A) Chen et al., Zhiwu Xuebao, 28, 450 (1986); Chem. Abstr., 106 15732w

D. drummondii.
TRITERPENOID: isoursenol.
Dominguez et al., Phytochemistry, 14, 815 (1975)

DRYOBALANOPS (Dipterocarpaceae)
D. aromatica Gaertn.
TRITERPENOIDS: dryobalanolide, dryobalanone, alphitolic acid, 11-oxoasiatic acid.
Cheung et al., J. Chem. Soc., C, 1047 (1968)
Cheung et al., ibid., 2686 (1968)

DRYOPHANTA
D. divisa.
CYCLOHEPTANE: purpurogallin.
Barltrop et al., J. Chem. Soc., 116 (1948)

DRYOPTERIS (Aspidiaceae/Dryopteridaceae)
D. austriaca.
PHENOLIC: desaspidin.
Aebo et al., Helv. Chim. Acta, 40, 572 (1957)

D. crassirhizoma.
TRITERPENOIDS: diploptene, dryocrassol, dryocrassyl acetate, 9(11)-fernene.
Ghisalberti et al., Tetrahedron Lett., 3297 (1975)

D. erythrosora.
PHLOROGLUCINOLS: aspidin AB, methylene-bis-aspidinol, trisdesaspidin BBB, trisflavaspidic acid.
Widen et al., Planta Med., 28, 144 (1975)

D. filix-mas (L.) Schott.
DIMER: desaspidin.
Aeib et al., Helv. Chim. Acta, 40, 266 (1957)

D. fragrans.
KETONE: fragrinol.
Molodozhnikova et al., Khim. Farm. Zh., 5, 32 (1971); Chem. Abstr., 75 11585x

D. noveboracensis.
STEROIDS: 24-ethyllathosterol, β-sitosterol, spinasterol.
Nes et al., Lipids, 12, 511 (1977)

DUBOISIA (Solanaceae)
D. hopwoodii F. Muell.
PYRIDINE ALKALOIDS: nicotine, (+)-nornicotine.
Spath, Biniecki, Ber., 72, 1809 (1939)

D. leuchhardtii F. Muell.
TROPANE ALKALOIDS: butropine, (±)-hyoscine, (-)-hyoscyamine, 1-phenyl-1,2,3,4-tetralin-1,4-dicarboxylic acid discopine ester, 1-epi-1,2,3,4-tetralin-1,4-dicarboxylic acid discopine ester, α-scopadonnine, β-scopadonnine, valtropine.
Rosenblum, Taylor, J. Pharm. Pharmacol., 6, 410 (1954)
Kagai et al., J. Pharm. Soc., Japan, 100, 216 (1980)

D. myoporoides R. Br.
TROPANE ALKALOIDS: (-)-hyoscyamine, isoporoidine, norhyoscyamine, poroidine, scopolamine, tigloidine, valeroidine.
Berger et al., J. Chem. Soc., 1685 (1938)

DUGALDIA
D. hoopesii.
Two sesquiterpenoids.
Bohlmann et al., J. Nat. Prod., 47, 658 (1984)

DUGESIA (Compositae)
D. mexicana Gray.
SESQUITERPENOID: dugesualactone.
Bohlmann, Zdero, Chem. Ber., 109, 2651 (1976)

DUGUETIA
D. obovata.
APORPHINE ALKALOIDS: (-)-duguevanine, (-)-N-formylduguevanine, (-)-N-formylxylopine.
ISOQUINOLINE ALKALOID: probovatine.
Roblot et al., J. Nat. Prod., 46, 862 (1983)

D. spixiana.
APORPHINE ALKALOID: duguexpixin.
Desbourger et al., J. Nat. Prod., 48, 310 (1983)

DULACEA (Olacaceae)
D. guainensis (Engl.) O. Kuntze.
SESQUITERPENE ALKALOIDS: manicolines A and B.
SESQUITERPENOID: manicol.
(A) Polonsky et al., J. Chem. Soc., Chem. Commun., 731 (1981)
(T) Polonsky et al., J. Chem. Soc., Perkin I, 2065 (1980)

DUMORIA
D. heckelii.
TRITERPENOID: bassic acid.
King et al., J. Chem. Soc., 1338 (1955)

DUNALIA (Solanaceae)

D. australis (Griseb.) Sleum.

SAPONINS: dunawithanins A and B.

STEROIDS: withanolide,
4,7,20-trihydroxy-1-oxowitha-2,5,24-trienolide.

(Sap) Adam et al., Int. Conf. Chem. Biotechnol. Biol. Act. Nat. Prod., **3**, 1919 (1981)

(T) Adam, Hesse, Tetrahedron Lett., 1199 (1971)

DURANTA (Verbenaceae)

D. plumieri.

FLAVONOIDS: 4'-methoxyscutellarein, pectolinarigenin, pectolinarigenin 7-O-rutinoside, scutellarein, scutellarein 7-O-rhamnoside.

STEROID: β-sitosterol.

TRITERPENOID: α-amyrin.

Makboul et al., Fitoterapia, **52**, 219 (1981)

D. repens.

IRIDOIDS: durantosides I, II, III and IV.

Rimpler, Timm, Z. Naturforsch., **29C**, 11 (1974)

DYERA

D. lowii.

SACCHARIDE: dambonitol.

Comollo et al., J. Chem. Soc., 3319 (1953)

DYNAMENA

D. pumila.

STEROIDS: cholesterol, fucostanol.

Ando et al., Biochem. Syst. Ecol., **3**, 247 (1975); Chem. Abstr., **84** 118460n

DYSOXYLON (Meliaceae)

D. acutangulum.

SESQUITERPENOID: (+)-8-hydroxycalamenene.

Nishizawa et al., Phytochemistry, **22**, 2083 (1983)

D. alliceum.

SESQUITERPENOIDS: bicalamenene, (+)-8-hydroxycalamenene.

Nishizawa et al., Tetrahedron Lett., 1535 (1985)

D. binectariferum.

TRITERPENOID: dysobinin.

Singh et al., Phytochemistry, **15**, 2001 (1976)

D. frazerianum.

SESQUITERPENOIDS: (±)-cadinene, dysoxylonene, δ-elemene.

Gough et al., Tetrahedron Lett., 763 (1961)

Herout et al., Collect. Czech. Chem. Commun., **19**, 118 (1954)

D. lenticullure.

HOMOERYTHRINA ALKALOID: dyshomerythrine.

Aladesenmi et al., J. Chem. Res. Synop., 108 (1984)

DYSSODIA

D. anthemidifolia.

ACETYLENICS:
5-(4-acetoxy-1-butynyl)-5'-methyl-2,2'-bithiophene,
5-(3,4-diacetoxy-1-butynyl)-5'-methyl-2,2'-bithiophene,
5-(4-hydroxy-1-butynyl)-5'-methyl-2,2'-bithiophene,
5-(3-hydroxy-4-isovaleryloxy-1-butynyl)-2,2'-bithiophene
3'-methoxy-2,2' 5',2''-terthiophene, 5-methyl-2,2'
5',2''-terthiophene.

Bohlmann, Zdero, Chem. Ber., **109**, 901 (1976)

D. setifolia.

ACETYLENICS:
5-(4-acetoxy-1-butynyl)-5'-methyl-2,2'-bithiophene,
5-(3,4-diacetoxy-1-butynyl)-5'-methyl-2,2'-bithiophene,
5-(4-hydroxy-1-butynyl)-5'-methyl-2,2'-bithiophene,
5-(3-hydroxy-4-isovaleryloxy-1-butynyl)-2,2'-bithiophene,
3'-methoxy-2,2' 5',2''-terthiophene 5-methyl-2,2'
5',2''-terthiophene.

Bohlmann, Zdero, Chem. Ber., **109**, 901 (1976)

E

ECBALLIUM (Cucurbitaceae)
E. elaterium (L.) A.Rich.
NAPHTHOPENTANOFURANONES: elateric acid, isoelateric acid.
HEXANORTRITERPENOIDS: 16-deoxy-16-hexanorcucurbitacin O, hexanorcucurbitacin I.
TRITERPENOIDS: anhydro-22-oxo-epi-isocucurbitacin D, cucurbitacin E, cucurbitacin R.
Rao *et al., J. Chem. Soc., Perkin I*, 2552 (1974)

ECHEVERIA
E. elegans.
ACID: phorbic acid.
Kringstad, *Phytochemistry*, **14**, 2710 (1975)

ECHINACEA (Asteraceae)
E. angustifolia.
GLYCOSIDE: echinacoside.
Becker *et al., Z. Naturforsch.*, **37C**, 351 (1986)

ECHINOCEREUS (Cactaceae)
E. blanckii.
AMINES: N',N'-dimethylhistamine, 3,4-dimethoxyphenethylamine.
Wagner *et al., Planta Med.*, **45**, 95 (1982)
E. cinerascens.
AMINE ALKALOIDS:
N,N-dimethyl-3,4-dimethoxyphenethylamine, N-methyl-3,4-dimethoxyphenylamine.
Bruhn *et al., Phytochemistry*, **16**, 622 (1977)
E. merkeri.
AMINE ALKALOID: N,N-dimethyl-3,4-dimethoxyphenethylamine.
Agurell *et al., J. Pharm. Sci.*, **58**, 1413 (1969)

ECHINOCHLOA
E. crusgalli. (L.) Beauv.
TRITERPENOID: sawamilletin.
Ito, *J. Chem. Soc., Japan*, **59**, 274 (1938)

ECHINOCYSTIS (Cactaceae/Cistaceae)
E. macrocarpa.
GIBBERELLIN: gibberellin A-43.
Beeley *et al., Phytochemistry*, **14**, 779 (1975)

ECHINODONTIUM (Polyporaceae)
E. tinctorium Ellis et Ev.
TRITERPENOID: echinodol.
Bond *et al., J. Am. Chem. Soc.*, **88**, 3882 (1966)
E. tsugicola.
TRITERPENOIDS: deacetoxyechinodol, deacetoxy-3-epi-echinodol, deacetoxyechinodone, echinodol, 3-epi-echinodol, echinodone.
Kanamatsu *et al., Chem. Pharm. Bull.*, **20**, 1993 (1972)

ECHINOPS (Compositae)
E. echinatus.
FLAVANONES: echinacin, echinaticin.
Singh *et al., Chem. Ind. (London)*, 713 (1986)
E. ritro L.
QUINOLINE ALKALOIDS: 4-aminoquinoline oxymethylate, echinopsine, echinorine.
UNCHARACTERIZED ALKALOIDS: echinopseine, echinopsidine, pseudoechinopsine.
FLAVONOIDS: apigenin, apigenin 7-glucuronide, apigenin 7-rhamnoglucoside, chrysoeriol, chrysoeriol 7-glucoside, luteolin, luteolin 7-glucoside, quercetin, rutin.
(A) Greshoff, *Recl. Trav. Chim. Pays-Bas*, **10**, 360 (1900)
(F) Chevrier *et al., Fitoterapia*, **47**, 115 (1976)
E. sphaerocephalus L.
QUINOLINE ALKALOIDS: 4-aminoquinoline oxymethylate, echinopsidine, echinopsine, echinorine.
Greshoff, *Recl. Trav. Chim. Pays-Bas*, **19**, 360 (1900)

ECHITES
E. hirsuta.
COUMARIN: fraxetin.
Wesseley *et al., Ber.*, **62**, 120 (1929)

ECHIUM (Boraginaceae)
E. plantagineum L.
PYRROLIZIDINE ALKALOID: echiumine.
Culvenor, Smith, *Aust. J. Chem.*, **19**, 1955 (1966)
E. vulgare L.
UNCHARACTERIZED ALKALOIDS: consolicine, consolitine.
Greiner, Arch. Pharm. (Weinheim, Ger.), **238**, 505 (1900)

ECHUDOSOPHORA
E. koreensis.
SPARTEINE ALKALOID: N-(3-oxobutyl)-cytisine.
Murakoshi *et al., Phytochemistry*, **16**, 1460 (1977)

ECKLONIA (Phaeophyceae)
E. maxima.
PHLOROTANNINS: 7,9'-bieckol, 7,7'-bieckol-9,9'-bieckol, furodehydroeckol, phloroeckol A, phloroeckol B, tetraphloroethol C.
Glombitza *et al., Planta Med.*, 308 (1985)

ECLIPTA (Compositae)
E. alba.
PYRIDINE ALKALOID: nicotine.
Marion, *Can. J. Res.*, **23B**, 165 (1945)
E. erecta.
THIOPHENE:
2-(4-acetoxy-3-hydroxybutenyl)-5-(1,3-pentadiynyl)-thiophene.
Bohlmann *et al., Chem. Ber.*, **103**, 834 (1970)

EDGEWORTHIA (Thymelacaceae)
E. papyrifera.
STEROID: 24-methylcholesta-5,8,17(20)-trien-3β-ol.
Sugiyama *et al., Chem. Abstr.*, 70 862

EGERIA

E. densa.

ANTHOCYANIN: 5-O-methylcyanidin-3-glucoside.

Mamose *et al., Phytochemistry,* **16**, 1321 (1977)

EHRETIA (Boraginaceae/Ehretiaceae)

E. aspera.

PYRROLIZIDINE ALKALOID: ehretinine.

Suri *et al., Phytochemistry,* **19**, 1273 (1980)

E. microphylla Lamk. (Carmosa retusa (Vahl.) Masam.).

NAPHTHOQUINONE: microphyllone.

TRITERPENOID: bauerenol.

Agarwal *et al., Tetrahedron,* **36**, 1435 (1980)

EKEBERGIA

E. capensis.

COUMARIN: ekebergin.

Taylor, *Phytochemistry,* **20**, 2263 (1981)

ELAEAGNUS (Elaeagnaceae)

E. angustifolia L.

CARBOLINE ALKALOIDS: dihydroharman, elaegnine, harman, 2-methyl-1,2,3,4-tetrahydro-β-carboline, N-methyltetrahydroharmol, tetrahydroharmol.

Massagetov, *J. Gen. Chem. USSR.,* **16**, 129 (1946)

Nikolaeva, *Khim. Prir. Soedin.,* 638 (1970)

ELAEIS

E. guineensis.

ACIDS: linoleic acid, myristic acid, oleic acid, palmitic acid, stearic acid.

PHENOLICS: o-cresol, 2,6-dimethoxyphenol, guaicol, 1-ethyl-3,5-dimethoxy-4-hydroxybenzene, 1-methyl-3,5-dimethoxy-4-hydroxybenzene.

(Ac) Wuidart *et al., Oleagineux,* **30**, 401 (1975); *Chem. Abstr.,* **84** 147627c

(P) Chan *et al., Malays. J. Sci.,* **6**, 139 (1980); *Chem. Abstr.,* **97** 123920k

ELAEOCARPUS (Elaeocarpaceae)

E. archboldianus.

CARBOLINE ALKALOID: elaeocarpidine.

Johns *et al., J. Chem. Soc., Chem. Commun.,* 410 (1968)

E. densiflorus Kunth.

CARBOLINE ALKALOID: elaeocarpidine.

Johns *et al., Aust. J. Chem.,* **22**, 793 (1969)

Johns *et al., ibid.,* 801 (1969)

E. dolichostylus Schl.

ALKALOIDS: elaeocarpiline, isoelaeocarpiline.

HEXANORTRITERPENOID: hexanorcucurbitacin F.

(A) Johns *et al., J. Chem. Soc., Chem. Commun.,* 1324 (1968)

(T) Fang *et al., J. Nat. Prod.,* **47**, 988 (1984)

E. ganitrus.

ALKALOID: rudrakine.

Ray *et al., Phytochemistry,* **18**, 700 (1979)

E. kaniensis Schltr.

ALKALOIDS: elaeokanidines A, B and C, elaeokanines A, B, C, D and E.

Hart *et al., J. Chem. Soc., Chem. Commun.,* 460 (1971)

E. lanceofolius.

FLAVONOID: 4'-methylmyricetin.

Ray *et al., Phytochemistry,* **15**, 1797 (1976)

E. polydactylus Schl.

ALKALOIDS: elaeocarpidine, elaeocarpine, isoelaeocarpicine, isoelaeocarpiline, isoelaeocarpine.

Johns *et al., Aust. J. Chem.,* **22**, 793 (1969)

Johns *et al., ibid.,* 801 (1969)

E. sphaericus Schl.

ALKALOIDS: epiallo-elaeocarpiline, epi-elaeocarpiline, isoelaeocarpiline.

Johns *et al., J. Chem. Soc., Chem. Commun.,* 804 (1970)

ELAEODENDRON (Celastraceae)

E. balae.

TRITERPENOIDS: 8,29-dihydroxyfriedelan-3-one, D:B-friedoolean-5-ene-3α,29-diol.

Weeratunga *et al., Phytochemistry,* **24**, 2369 (1985)

E. glauca Pers.

SAPONINS: elaeodendrosides A, B, C, D, E, H, I and J.

TRITERPENOIDS: 25,28-dihydroxy-3-friedelanone, 25-hydroxy-3-friedelanone-28-aldehyde.

(Sap) Shimada *et al., Heterocycles,* **15**, 335 (1981)

(T) Weeratunga *et al., Aust. J. Chem.,* **36**, 1067 (1983)

ELAEOSELENUM

E. foetidum.

DITERPENOID: foetidin.

Pinar *et al., Phytochemistry,* **22**, 2775 (1983)

ELEAGIA (Restionaceae)

E. utilis.

TRITERPENOIDS: (20S)-dammar-24-ene-3β,20,26-triol, isofourquierol.

Biftu, Stevenson, *J. Chem. Soc., Perkin I,* 360 (1978)

ELEOCHARIS (Cyperaceae)

E. tuberosa.

ANTIBIOTIC: puchiin.

Chen *et al., Nature,* **156**, 234 (1945)

ELEPHANTOPUS (Compositae)

E. carolianus.

SESQUITERPENOID: deoxyelephantopin.

Kurokawa *et al., Tetrahedron Lett.,* 2863 (1970)

E. elatus.

SESQUITERPENOIDS: elephantin, elephantopin.

Kupchan *et al., J. Am. Chem. Soc.,* **88**, 3674 (1966)

E. mollis.

SESQUITERPENOIDS: 1-hydroxy-15-senecioyloxy-α-curcumene, molephantin, molelephantinin, phantomolin, phantomolin epoxide.

Lee *et al., J. Pharm. Sci.,* **64**, 1077 (1975)

Bohlmann, Zdero, *Chem. Ber.,* **109**, 3956 (1976)

E. nudatus.

SESQUITERPENOID: nudaphantin.

Haruna *et al., J. Nat. Prod.,* **48**, 93 (1985)

E. scaber.

SESQUITERPENOIDS: deoxyelephantopin, isodeoxyelephantopin.

Govindachari *et al., Isr. J. Chem.,* **8**, 762 (1970)

E. tomentosus.

SESQUITERPENOID: dihydroelephantopin.

Rustaiyan *et al., Rev. Latinoam. Quim.,* **9**, 200 (1978)

ELEUTHERINE (Iridaceae)

E. americana Merr. et Heyne.
NAPHTHALENE: hongconin.
> Chen et al., Chem. Pharm. Bull., **34**, 2743 (1986)

E. bulbosa.
NAPHTHOPYRAN: eleutherin.
SESQUITERPENOID: eleutherinol.
> (N) Schmid et al., Helv. Chim. Acta, **33**, 1751 (1950)
> (T) Ebnother et al., ibid., **35**, 910 (1952)

E. subaphylla Gagnep.
BENZOQUINONES: eleutherin, isoeleutherin.
PHENOLIC: eleutherol.
> Institut de Matiere Medicale, Rev. Med. (Hanoi), 157 (1975); Chem. Abstr., 85 106637h

ELMERILLIA (Magnoliaceae)

E. papuana (Scultr.) Dandy.
APORPHINE ALKALOID: elmerrillicine.
> Cleaver et al., Aust. J. Chem., **29**, 2003 (1976)

ELODEA (Hydrochariaceae)

E. canadensis (Michx.) L.C.H.Richard. Carotenoid.
eloxanthin Diterpenoids abieta-8(14)-en-18-oic acid 9α,13α-endoperoxide, abieta-8(14)-en-18-oic acid 9β,13β-endoperoxide.
> (Car) Hay, Biochem. J., **31**, 532 (1937)
> (T) Monaco et al., Tetrahedron Lett., 4609 (1987)

ELSHOLTZIA (Labiatae)

E. densa.
MONOTERPENOID: elsholtzidol.
> Vashist, Atal, Experientia, **26**, 817 (1970)

ELVIRA

E. biflora.
PHENOLIC: 2-(1,5-dimethylhex-4-enyl)-4-methylphenol.
SESQUITERPENOID: elvirol.
> Bohlmann, Grenz, Tetrahedron Lett., 1005 (1969)

EMBELIA

E. ribes.
ACID: embelic acid.
BENZOQUINONE: vlangin.
> (A) Joshi et al., J. Chem. Soc., Perkin I, 327 (1975)
> (B) Rao et al., J. Org. Chem., **24**, 4529 (1961)

E. tsjeriam.
ACID: embelic acid.
> Joshi et al., J. Chem. Soc., Perkin I, 327 (1975)

EMILIA (Compositae)

E. flammea.
PYRROLIZIDINE ALKALOID: emiline.
> Tomczyk, Kohlmuenzer, Herba Pol., **17**, 226 (1971)

EMMENOSPERMUM (Rhamnaceae)

E. alphitonioides.
DINORTRITERPENOIDS: deoxyemmolactone, emmolactone, isodeoxyemmolactone.
NORTRITERPENOIDS: ceanothic acid, jungullic acid.
TRITERPENOIDS: ebelin lactone, emmolic acid.
> Eade et al., J. Chem. Soc., Chem. Commun., 579 (1969)
> Burbage, Jewers, J. Chem. Soc., D, 814 (1970)

E. pancherianum Baill.
TRITERPENOIDS: 24,25-dihydroebelin lactone, 24,25-dehydroebelin lactone.
> Baddeley et al., Aust. J. Chem., **33**, 2071 (1980)

EMMOTUM

E. nitens.
SESQUITERPENOIDS: emmotins A, B, C, D, E, F, G and H.
> de Oliviera et al., Phytochemistry, **15**, 1267 (1976)

ENANTIA

E. pilosa.
APORPHINE ALKALOIDS: oliveridine, oliverine.
OXOAPORPHINE ALKALOIDS: lanugosine, liriodenine, N-oxyoliveridine, N-oxyoliverine.
> Nieto et al., Lloydia, **39**, 350 (1976)

E. polycarpa.
BENZYLISOQUINOLINE ALKALOID: polycarpine.
> Jossang et al., C. R. Acad. Sci., **284C**, 467 (1977)

ENCELIA (Compositae/Heliantheae)

E. farinosa Gray.
CHROMENE: eupatoriochromene.
SESQUITERPENOIDS: encelin, farinosin.
> (C) Anthonsen, Acta Chem. Scand., **23**, 3605 (1969)
> (T) Geissman, Mukherjee, J. Org. Chem., **33**, 656 (1968)

E. virginensis.
SESQUITERPENOID: virginin.
> Sims, Berryman, Phytochemistry, **11**, 44 (1972)

ENCELIOPSIS

E. covillei.
BENZOFURANS:
2-acetyl-5-(1-hydroxymethyl)-6-methoxybenzofuran, 2-isopropylene-5-(1-methoxyethyl)-6-methoxybenzofuran.
> Mitsakos et al., Phytochemistry, **15**, 2243 (1976)

ENDIANDRA (Lauraceae)

E. introrsa.
DITERPENOIDS: endiandric acid, endiandric acid lactone.
> Bandaranayake et al., Aust. J. Chem., **34**, 1655 (1981)

ENGLERASTRUM (Labiatae)

E. scandens Alston.
DITERPENOIDS: rastronols A, B, C, D, E, F, G and H.
> Nomoto et al., Helv. Chim. Acta, **59**, 772 (1976)

ENHYDRA (Compositae/Heliantheae)

E. fluctuans Lour.
DITERPENOIDS: 15α-angeloyloxykaur-16-en-19-oic acid, enhydrin, 17-hydroxy-(-)-kaur-15-en-19-oic acid.
SESQUITERPENOIDS: fluctuadin, fluctuanin.
> Ali et al., Indian J. Chem., **9**, 84 (1971)
> Ali et al., Tetrahedron, **28**, 2285 (1972)

ENICOSTEMMA (Gentianaceae)

E. hyssopifolium.
AMINE ALKALOID: enicoflavine.
> Chaudhuri et al., Chem. Ind. (London), 127 (1975)

E. littorale Bl.
PYRIDINE ALKALOID: gentianine.
> Plat et al., Bull. Soc. Chim., 1302 (1963)

ENKIANTHUS (Ericaceae)

E. cernuus (Sieb. et Zucc.) Makino.
STEROID: β-sitosterol.
TRITERPENOIDS: α-amyrin, betulinic acid, 6-hydroxyursolic acid, lupeol, oleanolic acid, taraxerol, ursolic acid.
Sakakibara et al., Phytochemistry, 22, 2553 (1983)

E. perulatus.
FLAVONOIDS: koaburanin, lyoniside.
Ogawa et al., Shoyakugaku Zasshi, 30, 24 (1976); Chem. Abstr., 87 2389d

E. subsessilis (Miq.) Makino subsp. nudipes (Honda) Kitamura.
FLAVONOIDS: astragalin, chrysin 7-glucoside, hyperoside, maltol, quercetin.
Agawa et al., Shoyakugaku Zasshi, 30, 24 (1976); Chem. Abstr., 87 2389d

E. subsessilis (Miq.) Makino subsp. subsessilis.
FLAVONOIDS: chrysin 7-glucoside, hyperoside, nudiposide.
Ogawa et al., Shoyakugaku Zasshi, 30, 24 (1976); Chem. Abstr., 87 2389d

ENTADA

E. phaseoloides.
AMIDE: entadamide A.
TRITERPENOID: entagenic acid.
(Am) Ikegami et al., Chem. Pharm. Bull., 33, 5153 (1985)
(T) Chakravarti et al., Curr. Sci., 20, 199 (1954)

E. pusalthia.
TRITERPENOID: entagenic acid.
Chakravarti et al., Curr. Sci., 20, 199 (1954)

E. scandens.
STEROIDS: prosapogenin, β-sitosterol.
TRITERPENOID: lupeol.
Hariharan, Indian J. Pharm., 37, 67 (1975)

ENTANDROPHRAGMA (Meliaceae)

E. angolense.
LIMONOID: gedunin.
NORTRITERPENOID: angolensic acid methyl ester.
TRITERPENOID: entandrolide.
Okorie et al., Phytochemistry, 16, 2029 (1977)

E. bussei Harms.
LIMONOIDS: busseins A, B, C, D, E, F, G, H, I, J, K, L and M, entandrophragmin.
Guex, Tamm, Helv. Chim. Acta, 67, 885 (1984)

E. caudatum.
LIMONOIDS: bussein A, bussein B, entandrophragmin, phragmalin.
Taylor, J. Chem. Soc., C, 3495 (1965)

E. cylindricum Sprague.
LIMONOIDS: entandrophragmin, sapelins A, B, C, D, E and F.
Chan et al., J. Chem. Soc., C, 311 (1970)

E. spicata.
TRITERPENOID: spicatin.
Connolly et al., Phytochemistry, 20, 2596 (1981)

E. utile.
TRITERPENOID: utilin.
Akisanya et al., J. Chem. Soc., 3827 (1960)

EPANORA

E. lecanora.
LEUCINE ESTER: (S)-epanorin.
Zopf et al., Justus Liebigs Ann. Chem., 313, 330 (1900)

EPERUA

E. falcata.
DITERPENOID: eperuic acid.
King, Jones, J. Chem. Soc., 658 (1955)

E. purpurea.
DITERPENOIDS: eperuic acid, eperuol.
Medina et al., J. Nat. Prod., 46, 462 (1983)

EPHEDRA (Ephedraceae)

E. ciliata SAM.
AMINE ALKALOIDS: (+)-ephedrine, (-)-ephedrine, (+)-pseudoephedrine.
Mitchell, J. Chem. Soc., 1153 (1940)

E. distachya L.
AMINE ALKALOIDS: (-)-ephedrine, N-methylephedrine, (-)-norephedrine, (+)-pseudoephedrine.
Mitchell, J. Chem. Soc., 1153 (1940)

E. equisetina Bge.
AMINE ALKALOIDS: (-)-ephedrine, (+)-pseudoephedrine.
Mitchell, J. Chem. Soc., 1153 (1940)
Hoover, Hass, J. Org. Chem., 12, 506 (1947)

E. foliata Boiss.
AMINE ALKALOID: ephedrine.
Schmidt, Bumming, Arch. Pharm. (Weinheim, Ger.), 247, 141 (1909)

E. gerardiana (Stapf.) Wall.
AMINE ALKALOID: ephedrine.
Mitchell, J. Chem. Soc., 1153 (1940)

E. helvetica C.A.Meyer.
AMINE ALKALOIDS: ephedrine, N-methylephedrine, (-)-norephedrine.
Mitchell, J. Chem. Soc., 1153 (1940)
Hoover, Hass, J. Org. Chem., 12, 506 (1947)

E. intermedia Schrenk.
AMINE ALKALOIDS: (-)-ephedrine, (+)-pseudoephedrine.
MACROCYCLIC ALKALOIDS: ephedradines A, B and C.
Konno et al., Phytochemistry, 18, 697 (1979)

E. lomatolepis Schrenk.
AMINE ALKALOIDS: (-)-ephedrine, (+)-pseudoephedrine.
Chen, Schmidt, Proc. Soc. Exp. Biol. Med., 21, 351 (1924)

E. monosperma SAM.
AMINE ALKALOIDS: (-)-ephedrine, (+)-pseudoephedrine.
Schmidt, Arch. Pharm. (Weinheim, Ger.), 244, 239 (1906)
Schmidt, ibid., 244, 251 (1906)

E. nebrodensis Tineo.
AMINE ALKALOID: ephedrine.
Emde, Arch. Pharm. (Weinheim, Ger.), 245, 662 (1907)

E. pachyclada Boiss.
AMINE ALKALOID: ephedrine.
Mitchell, J. Chem. Soc., 1153 (1940)

E. procera var. chrysocarpa.
AMINE ALKALOIDS: (-)-ephedrine, pseudoephedrine.
Mitchell, J. Chem. Soc., 1153 (1940)

E. sinica Stapf. (E. flava Porter Smith).
AMINE ALKALOIDS: ephedrine, N-methylephedrine, (-)-norephedrine. Mercks Uber. Neuer. Geb. Pharmakother., 1 (1888)
Hoover, Hass, J. Org. Chem., 12, 506 (1947)

EPIGAEA (Ericaceae)
E. asiatica Maxim.
STEROID: β-sitosterol.

TRITERPENOIDS: α-amyrin, betulinic acid, lupeol, lupeol taraxerol ursolic acid

Sando, *J. Biol. Chem.*, **56**, 457 (1973)

Kawai *et al.*, *Chem. Abstr.*, 88 34499b

EPILOBIUM (Onagraceae)
E. hirsutum L.
TRITERPENOID: 23-hydroxytormentic acid.

de Pascual Teresa *et al.*, *An. Quim.*, **75**, 135 (1979)

EPIMEDIUM
E. koreanum.
FLAVONOIDS: epimedin A, epimedin B, epimedin C.

Oshima *et al.*, *Heterocycles*, **26**, 935 (1987)

EPITHELANTHA
E. micromeris B. et R.
TRITERPENOIDS: epithelanthic acid, epithelanthic acid methyl ester.

West, McLaughlin, *Lloydia*, **40**, 499 (1977)

EQUISETUM (Equisetaceae)
E. arvense L.
ACIDS: caffeic acid, p-coumaric acid, ferulic acid, gallic acid, p-hydroxybenzoic acid, vanillic acid.

FLAVONOIDS: dihydrokaempferol, dihydroquercetin, equisetin, naringenin.

(A) Syrchina *et al.*, *Khim. Prir. Soedin.*, 416 (1975)

(F) Syrchina *et al.*, *ibid.*, 424 (1975)

E. palustre L.
MACROCYCLIC ALKALOID: palustridine.

UNCHARACTERIZED ALKALOID: palustrine.

Eugster *et al.*, *Helv. Chim. Acta*, **36**, 1387 (1953)

ERANTHIS (Ranunculaceae)
E. pinnatifida.
SESQUITERPENOID: norcimifugin.

Wada *et al.*, *Phytochemistry*, **13**, 297 (1974)

ERECHTITES
E. hieracifolia (L.) Raf.
PYRROLIZIDINE ALKALOID: hieracifoline.

Manske, *Can. J. Res.*, **17B**, 8 (1939)

E. quadridentata DC.
PYRROLIZIDINE ALKALOID: retrorsine.

Warren *et al.*, *J. Am. Chem. Soc.*, **72**, 1421 (1950)

EREMANTHUS (Compositae)
E. elaeagnus Schultz.-Bip.
ALKALOID: eremanthine.

SESQUITERPENOID: eremanthin.

NORDITERPENOIDS: eremantholides A and C.

DITERPENOID: eremantholide B.

(A) Raffauf *et al.*, *J. Am. Chem. Soc.*, **97**, 6884 (1975)

(T) LeQuesne *et al.*, *J. Chem. Soc., Perkin I*, 1572 (1978)

E. glomeratus.
SESQUITERPENOID: ereglomerulide.

Bohlmann *et al.*, *Phytochemistry*, **20**, 1609 (1981)

E. goyazensis.
NORDITERPENOID: eremantholide C.

SESQUITERPENOIDS: eregoyazidin, eregoyazin, goyazensolide.

Vichnewski *et al.*, *J. Org. Chem.*, **42**, 3910 (1977)

EREMOCARPUS (Euphorbiaceae)
E. setigerus (Hook.) Benth.
DITERPENOID: eremone.

Jolad *et al.*, *J. Org. Chem.*, **47**, 1356 (1982)

EREMOPHILA (Myoporaceae)
E. abietina.
DITERPENOIDS: cembr-4-ene-15,19,20-triol, (1R,3S,4S,8R,11Z)-3,15-epoxycembr-11-ene-18,19-dioic acid.

Ghisalberti *et al.*, *Aust. J. Chem.*, **36**, 1187 (1983)

E. cuneifolia.
DITERPENOIDS: 5,14-dihydroxy-3-visciden-20-oic acid, 5-hydroxy-14-oxo-3-visciden-20-oic acid.

Ghisalberti *et al.*, *Aust. J. Chem.*, **37**, 635 (1984)

E. decipiens.
DITERPENOID: 18-hydroxydecipia-2(4),14-dien-1-oic acid.

Ghisalberti *et al.*, *Tetrahedron Lett.*, 1775 (1975)

E. dempsteri.
DITERPENOIDS: (1R,3Z,7E,11Z)-15-hydroxycembra-3,7,11-trien-19-oic acid, (1R,3Z,8,11Z)-15-hydroxycembra-3,11-dien-19-oic acid.

Ghisalberti *et al.*, *Aust. J. Chem.*, **39**, 1703 (1986)

E. drummondii.
DITERPENOID: 7,8,16-trihydroxy-19-serrulatonoic acid.

Croft *et al.*, *Aust. J. Chem.*, **34**, 1951 (1981)

E. exilifolia.
DITERPENOIDS: 2-(9-aldehyde-4,8-dimethyl-3,7-nonadienyl)-6-methyl-2,6-octadiendioic acid, 2-(9-hydroxy-4,8-dimethyl-3,7-nonadienyl)-6-methyl-2,6-octadienedioic acid.

Ghisalberti *et al.*, *Aust. J. Chem.*, **34**, 1491 (1981)

E. fraseri.
FLAVONES: 3',5,5'-trihydroxy-2 3,4',6,7-tetramethoxyflavone.

DITERPENOIDS: cembr-4-ene-15,19,20-triol, (1R,3S,4R,7E,11Z)-3,15-epoxy-19-oxocembra-7,11-dien-18-ol.

(F) Jefferies *et al.*, *Aust. J. Chem.*, **15**, 532 (1962)

(T) Ghisalberti *et al.*, *ibid.*, **39**, 1703 (1986)

E. freelingi.
DITERPENOID: eremolactone.

SESQUITERPENOIDS: freelingnite, freelingyne.

Knight, Pattenden, *Tetrahedron Lett.*, 1115 (1975)

E. georgei.
DITERPENOID: 3,15-epoxy-4-hydroxycembra-7(Z),11(Z)-diene.

SESQUITERPENOIDS: 7β-hydroxyprezizaene, 2,6,6,8-tetramethyltricyclo-[6,2,1,0]-undecan-7β-ol.

Carrol *et al.*, *Phytochemistry*, **15**, 777 (1976)

Ghisalberti *et al.*, *Tetrahedron*, **23**, 3301 (1978)

E. glutinosa.
DITERPENOIDS: 14S,15-dihydro-6E-formyl-2E,10Z,14Z-trimethyl-2,6,10,14-hexadecatetraenedioic acid, 6E-formyl-2E,10Z, 14Z-trimethyl-2,6,10,14-hexadecatetraenedioic acid.

Ghisalberti *et al.*, *Aust. J. Chem.*, **34**, 1491 (1981)

E. granatica.
DITERPENOIDS:
(1R,3S,4S,7E,11Z)-3,15-epoxy-19-oxocembra-7,11-dien-18-
oic acid,
(1R,3S,4S,7Z,11Z)-3,15-epoxy-19-oxocembra-7,11-dien-19-
oic acid, 14-serrulatene-2α,7,8,20-tetraol,
14-serrulatene-7,8,20-triol.
Ghisalberti *et al.*, *Aust. J. Chem.*, **36**, 1187 (1983)

E. latrobei.
SESQUITERPENOIDS: (-)-10,11-dehydromyoporone, (-)-ngaione.
Blackburne, Sutherland, *Aust. J. Chem.*, **25**, 1787 (1972)

E. maculata (Ker) F.Muell.
SESQUITERPENOID: (-)-10,11-dehydromyoporone.
Blackburne, Sutherland, *Aust. J. Chem.*, **25**, 1787 (1972)

E. mitchelli.
SESQUITERPENOIDS: 7αH-eremophila-1,11-dien-9-one,
eremophilone, allo-eremophilone,
8α-hydroxy-7αH-eremophila-1,11-dien-9-one,
8α-hydroxy-7αH-eremophila-1(10),11-dien-9-one,
2-hydroxyeremophilone.
Chetty *et al.*, *Tetrahedron Lett.*, 307 (1969)

E. platycalyx.
DITERPENOID:
(1R,3S,4R,7E,11Z)-3,15-epoxy-19-oxocembra-7,11-dien-18-
ol.
Ghisalberti *et al.*, *Aust. J. Chem.*, **39**, 1703 (1986)

E. rotundifolia.
DINORSESQUITERPENOID: eremoacetal.
Dimitriadis, Massy-Westropp, *Aust. J. Chem.*, **32**, 2003 (1979)

E. scoparia.
SESQUITERPENOIDS: 8,11-dihydroxy-3-eudesman-2-one,
8,11-dihydroxy-3-eudesman-2-one 8-acetate,
(4aR,6S,7S,8aS)-7-hydroxy-6-(1-hydroxy-1-methylethyl)-
4,8a-dimethyl-4a,5,6,7,8,8a-hexahydronaphthalene-2-(1H)-
one,
(4aS,6S,7S,8aS)-7-hydroxy-6-(1-hydroxy-1-methylethyl)-
4,8a-dimethyl-4a,5,6,7,8,8a-hexahydronaphthalene-2-(1H)-
one,
(2S,3S,4aR,8aS)-3-(1-hydroxy-1-methylethyl)-5,8a-
dimethyl-7-oxo-1,2,3,4,4a,7,8,8a-octahydronaphthalene-2-yl-
acetate.
Babidge *et al.*, *Aust. J. Chem.*, **37**, 629 (1984)

E. serrulata.
DITERPENOID: dihydroserrulatic acid.
Babidge *et al.*, *Aust. J. Chem.*, **37**, 629 (1984)

EREMOSPARTON (Leguminosae)
E. flaccidum Litv.
AMINE ALKALOID: sphaerophysine.
Merlis, *J. Gen. Chem. USSR*, **22**, 347 (1952)

EREMOSTACHYS (Labiatae)
E. molucelloides.
HALOGENATED ACIDS: 9,10-tetrabromostearic acid,
9,10,12,13-tetrabromostearic acid.
Gusakova, Umarov, *Khim. Prir. Soedin.*, 717 (1976)

E. speciosa Rupr.
QUATERNARY ALKALOID: stachydrine.
Pulatova, *Khim. Prir. Soedin.*, 62 (1969)

EREMURUS (Liliaceae)
E. fuscus.
AMINE ALKALOIDS: hordenine, hordenine methyl ether.
Akramov, *Dokl. Akad. Nauk SSSR*, **2**, 34 (1961)

E. luteus Baker.
AMINE ALKALOIDS: hordenine, hordenine methyl ether.
Sheveleva *et al.*, *Mat. Nauk Conf.*, 107 (1970)

E. regelii Vved.
AMINE ALKALOID: hordenine.
Wilkinson, *J. Chem. Soc.*, 2079 (1958)

E. transchanica Vved.
AMINE ALKALOIDS: hordenine, hordenine methyl ether.
Akramov, *Dokl. Akad. Nauk SSSR*, **2**, 34 (1961)

ERICA (Ericaceae)
E. arborea L.
TRITERPENOIDS: betulin, friedelin, lupeol, ursolic acid.
Hassan *et al.*, *Khim. Prir. Soedin.*, 119 (1977)

E. cinerea L.
TRITERPENOID: ursolic acid.
Fujii, Shimada, *J. Pharm. Soc., Japan*, **58**, 1687 (1938)

E. scoparia.
ACID: 2-hydroxyphenylacetic acid.
NITRILE: 2,4-dihydroxyphenylacetonitrile.
Ballester *et al.*, *Phytochemistry*, **14**, 1667 (1975)

ERICAMERIA
E. diffusa.
FLAVONE: 3',5,7-trihydroxy-3,4'-dimethoxyflavone.
FLAVONOIDS: kaempferol 3,5,7-trimethyl ether, quercetin
3,3',4'-trimethyl ether.
Kupchan *et al.*, *Phytochemistry*, **10**, 664 (1971)
Urbatsch *et al.*, *ibid.*, **15**, 440 (1976)

ERIGERON (Compositae)
E. khorossanicus.
DITERPENOID: erigerolide.
Orezkurdyev *et al.*, *Khim. Prir. Soedin.*, 396 (1981)

E. philadelphicus.
DITERPENOID: erigerol.
Waddell *et al.*, *J. Org. Chem.*, **48**, 4450 (1983)

ERIODERMA
E. chilense.
DEPSIDONES: norargopsin, norpannarin.
Elix *et al.*, *Aust. J. Chem.*, **39**, 719 (1986)

E. physcioides.
ALDEHYDE: eriodermin.
Connolly *et al.*, *Phytochemistry*, **23**, 857 (1984)

ERIODICTYON
E. californicum.
FLAVONE: 3',4',5,7-tetrahydroxyflavone.
Horhammer *et al.*, *Tetrahedron Lett.*, 5133 (1966)

E. glutinosum.
FLAVONE: 3',4',5,7-tetrahydroxyflavone.
Horhammer *et al.*, *Tetrahedron Lett.*, 5133 (1966)

ERIOLEANA
E. hookeriana.
STEROID: β-sitosterol.
TRITERPENOIDS: α-amyrin, β-amyrin acetate, lupeol.
Samiuddin *et al.*, *Indian J. Pharm.*, **39**, 17 (1977)

ERIOPHYLLUM (Compositae)

E. confertiflorum.
SESQUITERPENOIDS: eriofertopin, eriofertopin 2-acetate.
Kupchan et al., Phytochemistry, 16, 1834 (1977)

E. lanatum (Pursh.) Forbes.
SESQUITERPENOIDS: 2α-acetoxy-8-epi-ivangustin,
1-acetyl-14-(methylpropanoyl)-6-hydroxyeriolanolide,
6-acetyl-14-(methylpropanoyl)-1-hydroxyeriolanolide,
1-acetoxy-14-(methylpropenoyloxy)-6-hydroxyeriolanolide,
6-acetoxy-14-(methylpropenoyloxy)-1-hydroxyeriolanolide.
Kupchan et al., J. Chem. Soc., Chem. Commun., 842 (1973)
Bohlmann et al., Phytochemistry, 20, 2239 (1981)

E. steachadifolium.
One sesquiterpenoid.
Bohlmann et al., Phytochemistry, 20, 2239 (1981)

ERIOSTEMON (Rutaceae)

E. brucei.
COUMARINS: 1-deoxybruceol, eriobrucinol, hydroxybruceol.
Jefferies, Worth, Tetrahedron, 29, 903 (1973)

E. buxifolia.
FLAVONE: eriostemin.
Lassak, Aust. J. Chem., 25, 2517 (1972)

E. crowei.
MONOTERPENOID: croweacin.
Penfold et al., J. Chem. Soc., 756 (1938)

E. hispidulus.
FLAVONE: eriostemin.
Lassak, Aust. J. Chem., 25, 2517 (1972)

E. myoporoides.
SESQUITERPENOID: maaliol.
Hellyer, Aust. J. Chem., 15, 157 (1962)

E. trachyphyllus.
COUMARIN: trachyphyllin.
Lassak et al., Aust. J. Chem., 22, 2175 (1969)

ERLANGEA (Compositae)

E. cordifolia S. Moore.
SESQUITERPENOIDS: cordifene, cordifene oxide, cordifolia 31,
cordifolia P2, 4,15-epoxy-4,15-dihydrocordifene.
Begley et al., J. Chem. Soc., Perkin I, 819 (1984)

E. inyangama.
SESQUITERPENOIDS: EI-I and EI-II.
Bohlmann, Czerson, Phytochemistry, 17, 568 (1978)

E. remifolia.
SESQUITERPENOIDS: glaucolide I, 19-hydroxyglaucolide.
Bohlmann, Czerson, Phytochemistry, 17, 1190 (1978)

ERVATAMIA

E. coronaria.
INDOLE ALKALOIDS: dregamine, ervaticine, hecubine,
hyderabadine, 19-(2-ketopropyl)-coronaridine, mehranine.
IMIDAZOLIDINOCARBAZOLE ALKALOID: hecubine.
Atta-ur-Rahman et al., Z. Naturforsch., 38B, 1310 (1983)
Atta-ur-Rahman et.al., ibid., 38B, 1700 (1983)
Atta-ur-Rahman et al., Heterocycles, 23, 2975 (1985)

E. hainanensis.
BISINDOLE ALKALOIDS: ervahanines A, B and C.
Feng et al., J. Nat. Prod., 44, 670 (1981)

E. heyneana.
ISOQUINOLINE ALKALOID: (-)-heyneatine.
Gunasekera et al., Phytochemistry, 19, 1212 (1980)

E. orientalis.
INDOLE ALKALOIDS: 19,20-dehydroervatamine, ervatamine,
20-epi-ervatamine.
Knox, Slobbe, Tetrahedron Lett., 2149 (1971)

E. pandacaqui.
INDOLE ALKALOID: ervafoline.
Henriques et al., Tetrahedron Lett., 3707 (1978)

ERYMANTHUS

E. goyazensis.
SESQUITERPENOIDS: eregoyazidin, eregoyazin.
Vichnewski et al., J. Org. Chem., 42, 3910 (1977)

ERYNGIUM (Umbelliferae)

E. amethystinum.
MONOTERPENOIDS: O-(2-angeloylmethyl)-cis-crotonyl-ferulol,
O-(2-angeloylmethyl)-cis-crotonyl-isoferulol,
O-(2-isovaleryloxy-cis-crotonyl)-ferulol,
O-2-(2-methylbutyryloxymethyl)-cis-crotonyl-ferulol,
O-2-(2-methylbutyryloxymethyl)-cis-crotonyl-isoferulol.
Bohlmann, Zdero, Chem. Ber., 104, 1957 (1971)

E. bromeliifolium.
SAPONINS: saponins A, B, C, D, E, F, G and H.
TRITERPENOIDS: betulic acid-3-O-D-glucoside, betulic
acid-3-O-D-glucoside-β-D-glucopyranoside.
Hiller et al., Pharmazie, 31, 891 (1976)

E. campestre.
COUMARINS: aegelinol, aegelinol benzoate, agasyllin,
grandivittin.
Erdelmeier, Sticher, Planta Med., 407 (1985)

E. giganteum.
TRITERPENOIDS: giganterumgenins A, B, C, D, E, F, G, H and
K.
Hiller et al., Pharmazie, 30, 105 (1975)

E. ilicifolium.
COUMARINS: deltoin, (-)-marmesin, prantschimgin.
Pinar, Galam, J. Nat. Prod., 48, 853 (1985)

E. planum.
FLAVONOID: kaempferol 3,7-di-O-rhamnoside.
TRITERPENOIDS: A-1-barrigenol, erynginol A.
(S) Zarnack et al., Z. Chem., 17, 445 (1977)
(T) Hiller et al., Pharmazie, 28, 409 (1973)

E. serbicum.
MONOTERPENOIDS: 9-(2-angeloylmethyl-cis-crotonyl)-ferulol,
9-O-(2-angeloylmethyl-cis-crotonyl)-isoferulol,
9-O-(2-isovaleryloxymethyl-cis-crotonyl)-isoferulol.
Bohlmann, Zdero, Chem. Ber., 104, 1957 (1971)

ERYSIMUM (Cruciferae)

E. arkansanum.
SULPHONE: cheiroline.
Schlagdenhauffer, Reeb, C. R. Acad. Sci., 131, 953 (1900)

E. aureum.
SULPHONE: cheiroline.
Schlagdenhauffer, Reeb, C. R. Acad. Sci., 131, 953 (1900)

E. canescens.
STEROIDS: eryscenoside, erysimine.
Sauer et al., Tetrahedron Lett., 1703 (1966)

E. carniolicum.
FLAVONOIDS: isorhamnetin kaempferol, quercetin, rhamnetin,
rhamnocitrin.
STEROIDS: campesterol, cholesterol, β-sitosterol.
(F) Kozjek et al., Circ. Farm., 35, 8 (1977); Chem. Abstr., 87 51057y
(S) Umek, Korbar-Smid, Fitoterapia, 52, 277 (1982)

E. cuspidatum.
CARDENOLIDE: cheirotoxin.
STEROID: glucolocundijoside.
- (C) Nagata et al., Helv. Chim. Acta, 40, 41 (1957)
- (S) Masslenikova et al., Khim. Prir. Soedin., 166 (1975)

E. diffusum.
STEROID: sinapoylerysimoside.
- Navruzoca et al., Khim. Prir. Soedin., 750 (1973)

E. marschallianum.
STEROID: sinapoylglucoerysimoside.
- Matsuyina, Khim. Prir. Soedin., 603 (1975)

E. perofskianum.
STEROIDS: erycorchoside, eryperoside.
- Kwalewski et al., Helv. Chim. Acta, 43, 957 (1960)

E. repandum.
CARDENOLIDES: glucoperiplorhamnoside, periplorhamnoside.
- Bochvarov et al., Khim. Prir. Soedin., 581 (1977)

E. suffruticosa.
STEROID: bipindogulomethyloside.
- Makarevitch, Khim. Prir. Soedin., 178 (1974)

ERYTHRAEA (Gentianaceae)
E. centaurium (L.) Pers.
PYRIDINE ALKALOIDS: gentianidine, gentianine.
BITTER PRINCIPLE: centapicrin.
XANTHONES: 1,8-dihydroxy-2,3,4,6-tetramethoxyxanthone, 1,8-dihydroxy-3,4,6-trimethoxyxanthone, methylbellidifolin, methylswetianin.
- (A) Xioa-tian et al., Sci. Sinica, (Peking), 14 (1965)
- (B) Sakina, Aota, J. Pharm. Soc., Japan, 96, 683 (1976)
- (X) Takagi et al., J. Pharm. Soc., Japan, 102, 546 (1982)

ERYTHRINA
E. americana Mill.
ERYTHRINA ALKALOIDS: erysodine, erythroidine.
- Folkers et al., J. Am. Chem. Soc., 62, 1677 (1940)

E. arborescens Roxb.
ERYTHRINA ALKALOIDS: erysotramidine, erytharbine, erythrascine, erysopinophorine.
STEROIDS: campesterol, β-sitosterol, stigmasterol.
- (S) Singh, Chawla, Indian J. Chem., 14B, 388 (1976)
- (A) Ali et al., J. Pharm. Soc., Japan, 93, 1617 (1973)
- Tiwari, Masood, Phytochemistry, 18, 2069 (1979)

E. bidwillii.
ALKALOID: erybidine.
- Ito et al., Chem. Pharm. Bull., 19, 1509 (1971)

E. brucei.
ISOQUINOLINE ALKALOID: 8-oxoerythrinine.
- Dagna, Steglich, Phytochemistry, 23, 449 (1984)

E. costaricensis Michel.
ERYTHRINE ALKALOID: erysonine.
- Folkers et al., J. Am. Chem. Soc, 63, 1544 (1941)

E. crista-galli L.
BENZOPYRAN: erycristagallin.
BENZYLISOQUINOLINE ALKALOID: cristadine.
ERYTHRINA ALKALOIDS: crystamidine, erysodine.
- (B) Mitscher et al., Heterocycles, 22, 1673 (1984)
- (A) Juichi et al., ibid., 19, 849 (1982)

E. excelsa.
FLAVONE: 4',5,7-trihydroxy-6,8-diprenylisoflavone.
- Fomum et al., Planta Med., 341 (1986)

E. falcata.
ERYTHRINA ALKALOID: erythratidine.
- Deylofeu, Ber, 85, 620 (1952)

E. folkersii.
ERYTHRINA ALKALOIDS: erysosaline, erysotinone, erythravine.
- Millington et al., J. Am. Chem. Soc., 96, 1909 (1972)

E. glauca Willd.
ERYTHRINA ALKALOIDS: erysothiovine, erysothropine, erythratine, erythratine.
- Folkers et al., J. Am. Chem. Soc., 66, 1083 (1944)

E. indica Lam.
ERYTHRINA ALKALOID: erythrinine.
- Ito et al., J. Chem. Soc., D, 1076 (1970)

E. lithosperma Blume.
ERYTHRINA ALKALOIDS: erythratinidone, erythrinine, 3-demethoxyerythratidinone.
BENZYLISOQUINOLINE ALKALOIDS: N-norprotosinomenine, protosinomenine.
- Barton et al., J. Chem. Soc., Perkin I, 874 (1973)

E. lysistemon Hutchins.
ERYTHRINA ALKALOIDS: erysotrine, erystemine, erythristemine, 11-hydroxyerysovine, 11-methoxyerythraline.
- Ito et al., J. Pharm. Soc., Japan, 93, 1617 (1973)

E. melanacantha.
ERYTHRINA ALKALOID: erymelanthine.
- Dagne, Steglich, Tetrahedron Lett., 5067 (1983)

E. poeppigiana (Walp.) Skeels.
BENZYLISOQUINOLINE ALKALOID: (+)-nororientaline.
- Barton et al., J. Chem. Soc., Perkin I, 874 (1973)

E. salviiflora.
ERYTHRINA ALKALOIDS: erysosaline, erysotinone, erythravine.
- Millington et al., J. Am. Chem. Soc., 96, 1909 (1972)

E. sandwicensis Deg.
ERYTHRINA ALKALOID: erythramine.
- Folker et al., J. Am. Chem. Soc., 61, 1232 (1939)

E. senegalensis.
FLAVONES: erythrisenegalone, 4',5,7-trihydroxy-6,8-diprenylisoflavone.
- Fomum et al., Planta Med., 341 (1986)

E. steyermarkii.
ERYTHRINA ALKALOID: erythravine.
- Hargreaves et al., Lloydia, 37, 569 (1974)

E. sudumbrans (Hassk.) Merrill. (E. hypaphorus).
ERYTHRINA ALKALOID: erythramine.
INDOLE ALKALOID: hypaphorine.
- Folkers et al., J. Am. Chem. Soc., 61, 1232 (1939)

E. thollonia.
One uncharacterized glycoalkaloid.
- Lapiere, Bull. Soc. Chim. Biol., 31, 862 (1949)

E. variegata.
ERYTHRINA ALKALOID: erythrartine.
ISOFLAVONES: alpinumisoflavone, dihydrooxyresveratrol, erythrinin A, erythrinin B, erythrinin C, osajin, oxyresveratrol.
- (A) El-Olemy et al., Lloydia, 41, 343 (1978)
- (I) Deshpande et al., Indian J. Chem., 15B, 205 (1977)

ERYTHROCHITON (Rutaceae)
E. brasilensis Nees et Mart.
FURANOQUINOLINE ALKALOID: τ-fagarine.
- Johne et al., Pharmazie, 32, 415 (1977)

ERYTHRONIUM

E. sibiricum.
FLAVONOIDS: isorhamnetin, kaempferol, myricetin, quercetin.
Astankovich et al., Aktual. Vopr. Bot. Resur. Sib., 168 (1976); Chem. Abstr., 85 189200j

ERYTHROPHLEUM

E. chlorostachys.
DITERPENE ALKALOIDS: 3β-acetoxyerythrosuamine, 3β-acetylnorerythrostachaldine, N-2-hydroxyethyl-N-methylcinnamide, norcassamidine, norerythrostachaldine.
Griffen et al., Phytochemistry, 10, 2793 (1971)
Loder, Nearn, Tetrahedron Lett., 2497 (1975)

E. couminga.
NORDITERPENE ALKALOID: norerythrosuamide.
DITERPENE ALKALOIDS: coumingidine, coumingine, 19-nor-4-dehydrocassaidine.
Cronlund, Okagawa, Acta Pharm. Suec., 12, 467 (1975)

E. fordii.
DITERPENOIDS: erythrolic acid, erythrophleadienolic acid.
Arya, J. Sci. Ind. Res. (India), 21B, 381 (1962)

E. guineense E.Don.
DITERPENE ALKALOIDS: cassaidine, cassaine, dehydronorerythrosuamine, erythrophlamine, erythrophleguine, erythrophleine, homophleine, norcassamidide, norcassamine, norerythrosuamine.
Dalma, Ann. Chim. Appl., 25, 569 (1935)
Friedrich et al., Chem. Ber., 104, 3535 (1971)

E. ivorense.
DITERPENE ALKALOIDS: cassaide, cassamide, erythrophlamide, 19-hydroxycassaine, ivorine.
Cronlund, Sandberg, Acta Pharm. Suec., 8, 351 (1971)

E. lasianthum.
DITERPENE ALKALOID: erythrophleine.
Laborde, Ann. Mus. Col. Marseilles, 5, 305 (1907)

E. quincense G.Don.
DITERPENE ALKALOIDS: cassaidine, cassaine, cassamine.
DITERPENOIDS: cassamic acid, cassaminic acid.
(A) Dalma, Helv. Chim. Acta, 22, 1497 (1939)
(T) Chapman et al., J. Chem. Soc., 4010 (1963)

E. suaveolens.
DITERPENE ALKALOID: 3β-acetoxy-7α-hydroxy-N-(2-hydroxyphenyl)-N-methyl-(E)-cass-13(15)-en-16-amide.
Cronlund, Acta Pharm. Suec., 10, 373 (1973)

E. worense.
DITERPENE ALKALOID: 3-(3-methylcrotonyl)-cassamine.
Cronlund, Acta Pharm. Suec., 10, 407 (1973)

ERYTHROXYLON (Erythroxylaceae)

E. areolatum.
TROPANE ALKALOIDS: cinnamoylcocaine, β-cocaine.
Perrot, Bull. Sci. Pharmacol., 42, 266 (1935)

E. australe.
TROPANE ALKALOID: meteloidine.
DITERPENOIDS: 15,16-epoxystachan-1-one, 2-hydroxystacha-2,15-dien-1-one, 2β-hydroxystach-15-en-1-one, 19-hydroxystach-15-en-1-one, stach-15-en-1-one.
(A) Johns, Lamberton, Aust. J. Chem., 20, 1301 (1967)
(T) Connolly, Harding, J. Chem. Soc., Perkin I, 1996 (1972)

E. coco Lam.
PYRROLE ALKALOIDS: (-)-dihydrocuscohygrine, hygroline.
TROPANE ALKALOIDS: cinnamoylcocaine, β-cocoaine.
Turner et al., Phytochemistry, 20, 1403 (1981)

E. dekindtii.
Two tropane esters.
Al-Yahya et al., J. Chem. Soc., Perkin I, 2130 (1979)

E. laurifolium.
TROPANE ALKALOIDS: cinnamoylcocaine, β-cocaine.
Chemnitius, J. Prakt. Chem., 116, 285 (1927)

E. mamacoca.
TROPANE ALKALOID: nortropacocaine.
El-Iman et al., Phytochemistry, 24, 2285 (1985)

E. monogynum Roxb.
TROPANE ALKALOIDS: cinnamoylcocaine, β-cocaine.
DITERPENOIDS: (+)-devadarene, (+)-devadarool, erythroxydiol A, erythroxydiol, erythroxydiol Y, erythroxydiol Z, erythroxylol A, erythroxylol A acetate epoxide, erythroxylol A epoxide, erythroxylol B, erythroxytriol P, erythroxytriol Q, (+)-hibaene epoxide, hydroxydevadarool, 4α-hydroxy-18-norhibaene, 4β-hydroxy-18-norhibaene, isoatisirene, (-)-pimara-8(14),15-diene, (+)-stach-15-ene-17,19-diol, (+)-stach-15-ene-17,19-diol epoxide, (+)-stach-15-en-17-ol, (+)-stach-15-en-19-ol, (+)-stach-15-en-19-ol epoxide, (+)-stach-15-en-19-ol epoxyacetate.
Martin, Murray, J. Chem. Soc., 2529 (1968)

E. montanum.
TROPANE ALKALOIDS: cinnamoylcocaine, β-cocaine.
Perrot, Bull. Sci. Pharmacol., 42, 266 (1935)

E. novogranatense.
FLAVONOIDS: (+)-catechin-3-rhamnoside, ombuin-3-O-rutoside.
DITERPENOID: erythrodiol.
(F) Bonefeld et al., Phytochemistry, 25, 1205 (1986)
(T) Power, Tutin, J. Chem. Soc., 93, 891 (1908)

E. ovatum.
TROPANE ALKALOIDS: cinnamoylcocaine, β-cocaine.
Kovacs et al., Helv. Chim. Acta, 37, 892 (1954)

E. pulchrum.
TROPANE ALKALOIDS: cinnamoylcocaine, β-cocaine.
Kovacs et al., Helv. Chim. Acta, 37, 892 (1954)

E. retusum.
TROPANE ALKALOIDS: cinnamoylcocaine, β-cocaine.
Kovacs et al., Helv. Chim. Acta, 37, 892 (1954)

E. truxillense Rusby.
PYRROLE ALKALOID: cuskhygrine.
TROPANE ALKALOIDS: cinnamoylcocaine, β-cocaine, (+)-pseudococaine, α-truxilline, β-truxilline.
Willstatter et al., Annalen, 434, 266 (1923)

E. vacciniifolium.
TROPANE ALKALOIDS: catuabines A, B and C.
Graf, Lude, Arch. Pharm. (Weinheim, Ger.), 311, 139 (1978)

ESCHSCHOLTZIA

E. californica.
CAROTENOID: eschscholtziaxanthin.
Strain, J. Biol. Chem., 123, 425 (1938)

ESCOBEDIA

E. laevis.
CAROTENOID: azafrin.
Kuhn *et al., Ber.,* **64**, 333 (1931)

E. scabrifolia.
CAROTENOID: azafrin.
Kuhn *et al., Ber.,* **64**, 333 (1931)

ESPELETIA (Compositae)

E. grandiflora.
DITERPENOID: grandiflorolic acid.
Piozzi *et al., Gazz. Chim. Acta,* **98**, 907 (1968)

E. schultzii.
DITERPENOIDS: 15-acetoxykaur-16-en-19-oic acid,
15α-hydroxykaur-16-en-19-oic acid,
kaura-9(11),16-dien-19-oic acid.
Brieskorn, Pohlmann, *Tetrahedron Lett.,* 5661 (1968)

E. timotensis.
DITERPENOID: 15α-hydroxykaur-16-en-19-oic acid.
Brieskorn, Pohlmann, *Tetrahedron Lett.,* 5661 (1968)

E. weddellii.
DITERPENOIDS: 15α-hydroxykaur-16-en-19-oic acid,
(-)-kaur-15-en-19-al.
Karle, *Acta Crystallogr.,* **28B**, 588 (1972)

ESPELETIOPSIS

E. guacharaca.
SESQUITERPENOIDS: copacamphor, 6-siliphilperfolen-5-one.
Bohlmann *et al., Phytochemistry,* **19**, 2399 (1980)

ESPOSTOA

E. huanucensis.
AMINE ALKALOIDS: hordenine, N-methyltyramine, tyramine.
Mata *et al., Lloydia,* **39**, 461 (1976)

ETHULIA

E. conyzoides.
COUMARINS: cycloethuliacoumarin, ethuliacoumarin.
Bohlmann, Zdero, *Phytochemistry,* **16**, 1092 (1977)

E. conyzoides var. gracilis Asch. et Schweinf.
COUMARINS: ethuliacoumarin A, ethuliacoumarin B,
isoethuliacoumarin A, isoethuliacoumarin B,
5-methyl-4-hydroxycoumarin.
Mahmoud *et al., Pharmazie,* **38**, 486 (1983)

EUCALYPTUS (Myrtaceae)

E. agglomerata.
KETONE: agglomeratone.
Hellyer, *Aust. J. Chem.,* **17**, 1419 (1967)

E. caesia.
PHENOLIC: torquatone.
Bowyer *et al., Aust. J. Chem.,* **12**, 442 (1959)

E. calophylla.
BENZOPYRAN: (+)-afzelechin.
delle Monache *et al., Phytochemistry,* **11**, 2333 (1972)

E. citriodora.
STEROID: β-sitosterol.
SESQUITERPENOIDS: aromadendrin, 7-O-methylaromadendrin.
TRITERPENOIDS: betulinic acid, ursolic acid.
Dayal, *J. Indian Chem. Soc.,* **59**, 1008 (1982)

E. decorticans.
NORSESQUITERPENOID: flavescone.
Bick *et al., J. Chem. Soc.,* 3690 (1965)

E. dives.
MONOTERPENOIDS: (+)-piperitone, (-)-piperitone.
Simonsen, *J. Chem. Soc.,* **119**, 1646 (1921)

E. globulus Labill.
BENZOPYRANS: euglobals Ia, Ib, Ic, IIa, IIb, IIc, III, IVa, IVb,
V, VI and VII.
FLAVONOIDS: quercetol, quercetol 3-glucoside, quercitrin,
hyperoside, rutin.
DITERPENOIDS: cineole, pinocarveol, (-)-trans-pinocarveol,
(-)-pinocarvone.
SESQUITERPENOIDS: alloaromadendrine, α-eudesmol,
β-eudesmol, globulol.
(B) Kozuka *et al., Chem. Pharm. Bull.,* **30**, 1952 (1982)
(F) Boukef *et al., Plant Med. Phytother.,* **10**, 30 (1976)
(T) Sawada *et al., ibid.,* **28**, 2546 (1980)

E. grandis.
PHENOLIC: grandinol.
Crow *et al., Tetrahedron Lett.,* 1073 (1977)

E. macathuri.
MONOTERPENOID: eucalyptol.
SESQUITERPENOIDS: aromadendrine, carissone, α-eudesmol,
β-eudesmol.
McQuillan *et al., J. Chem. Soc.,* 2973 (1956)

E. mackliana.
KETONE: agglomerone.
MONOTERPENOID: eucalyptol.
SESQUITERPENOID: aromadendrine.
Hellyer *et al., Aust. J. Chem.,* **17**, 1419 (1967)

E. microcorys.
TRITERPENOID: cycloeucalenol
Itoh *et al., Phytochemistry,* **16**, 1448 (1977)

E. robusta.
BENZOPYRANS: robustadial A, robustadial B.
Xu *et al., J. Am. Chem. Soc.,* **106**, 734 (1984)

E. spathulata var. grandiflora.
SESQUITERPENOID: spathulenol.
Bowyer, Jefferies, *Chem. Ind. (London),* 1254 (1963)

E. tereticornis.
TRITERPENOIDS: 2α-hydroxyursolic acid, ursolic acid, ursolic
acid lactone.
Dayal *et al., Curr. Sci.,* **56**, 670 (1987)

E. torquata.
PHENOLIC: torquatone.
Bowyer *et al., Aust. J. Chem.,* **15**, 1448 (1962)

EUCHARIS (Amaryllidaceae)

E. grandiflora Blanck.
ISOQUINOLINE ALKALOID: lycorine.
Reichert, *Arch. Pharm. (Weinheim, Ger.),* **226**, 328 (1938)

EUCHRESTA

E. horsfeldtii.
MATRINE ALKALOID: (+)-5,9-dihydroxymatrine.
QUINOLIZIDINE ALKALOID: cytisine.
Ohmiya *et al., Phytochemistry,* **18**, 645 (1979)

E. japonica.
FLAVANONE: euchrestaflavanone.
Shirataki *et al., Phytochemistry,* **21**, 2959 (1982)

EUCLEA

E. natalensis.

NAPHTHAQUINONES: birameutaceone, diospyrin, isodiospyrin, mamegakinone, 7-methyljugone, 2-methylnaphthazarin, xylospyrin.

Ferreira et al., An. Fac. Farm. Porto, 33, 61 (1973); Chem. Abstr., 85 2553m

EUCOMIS (Liliaceae)

E. autumnalis (Mill.) Chittenden.

BENZOPYRANS: autumnariniol, autumnariol.
STEROID: eucosterol.

Fritz et al., Helv. Chim. Acta, 54, 207 (1971)

E. bicolor.

TRITERPENOID:
(23S)-17,23-epoxy-3β,16β,31-trihydroxy-27-nor-5α-lanost-8-ene-15,24-dione.

Ziegler, Tamm, Helv. Chim. Acta, 59, 1997 (1976)

E. punctata.

STEROID: eucosterol.

Sidwell et al., J. Am. Chem. Soc., 97, 3518 (1975)

EUCOMMIA (Eucommiaceae)

E. ulmoides Oliv.

IRIDOID: ulmoside.
LIGNANS: eucommin A,
(+)-1-hydroxypinoresinol-4,4'-glucopyranoside,
(+)-medioresinol di-O-β-D-glucopyranoside, (-)-olivil
4',4''-di-O-β-D-glucopyranoside.
MONOTERPENOID: eucommiol.

(I, T) Bianco et al., Gazz. Chim. Ital., 108, 17 (1978)
(L) Deyama, Chem. Pharm. Bull., 33, 3651 (1985)

EUGENIA (Myrtaceae)

E. caryophyllata Thunb.

SESQUITERPENOIDS: caryophylla-3(12),6-dien-4-ol, caryophylla-3(12),7(13)-dien-6α-ol, caryophyllene oxide.

Iwamuro et al., Agric. Biol. Chem., 47, 2099 (1983)

E. crebrinervis (Syzygium crebrinervis).

TRITERPENOIDS: arjunolic acid, asiatic acid, ellagic acid.

Bannon et al., Aust. J. Chem., 29, 1135 (1976)

E. gustavioides (Cleistocalyx gustavioides).

TRITERPENOIDS: arjunolic acid, asiatic acid crategolic acid.

Bannon et al., Aust. J. Chem., 29, 1135 (1976)

E. jambolana.

PHENOLICS: 2,6-dihydroxy-4-methoxyacetophenone, methylxanthoxyletin.
SAPONIN: eugenin.

(P) Linde, Arch. Pharm. (Weinheim, Ger.), 316, 971 (1983)
(Sap) Sengupta, Bikram, J. Indian Chem. Soc., 42, 266 (1965)

E. jambos L.

UNCHARACTERIZED ALKALOID: jambosine.

Gerrard, Chem. Abstr., 48 396 (1885)

E. pseudomata.

MONOTERPENOID: verbenone.

Banthorpe et al., Chem. Ber., 66, 643 (1966)

EUGLENA (Euglenophyceae/Euglenaceae)

E. helicorubescens.

CAROTENOID: astaxanthin.

Tischer, Z. Physiol. Chem., 240, 191 (1941)

EUMORPHA

E. prostrata.

SESQUITERPENOIDS: eumorphistonol, 10-oxolasiosperman.

Bohlmann, Zdero, Phytochemistry, 17, 1155 (1978)

E. sericea.

SESQUITERPENOIDS: cycloeumorphinone, eumorphinone, seven furan sesquiterpenoids.

Bohlmann, Zdero, Phytochemistry, 17, 1155 (1978)

EUONYMUS (Celastraceae)

E. alatus.

STEROIDS: 6α-hydroxystigmast-4-en-3-one, β-sitosterol, stigmast-4-ene-3,6-dione, stigmast-4-en-3-one.
SESQUITERPENOIDS: alatol, euolalin.

(S) Chen et al., Zhongcaoyao, 14, 385 (1983); Chem. Abstr., 100 3500g
(T) Sigiura et al., Tetrahedron Lett., 2307 (1975)

E. alatus f. alatus.

FLAVONOIDS: euonymin, kaempferol 3-rhamnosylxyloside-7-glucoside, kaempferol 3-xyloside.

Ishikura et al., Chem. Abstr., 87 114584v

E. alatus f. ciliato-dentatus.

FLAVONOID: kaempferol 3-rhamnosylxyloside.

Ishikura et al., Chem. Abstr., 87 114584v

E. alatus f. striatus.

MACROCYCLIC SESQUITERPENE ALKALOIDS: alatusamine, alatusinine, neolatamine.
SESQUITERPENOID: euolalin.
UNCHARACTERIZED TERPENOID: alatolin.

(A) Ishiwata et al., Phytochemistry, 22, 2839 (1983)
(T) Sigura et al., Tetrahedron Lett., 2307 (1975)

E. bungeanus.

TRITERPENOIDS: benulin, bungeanic acid, 3β,22α-dihydroxyolean-12-en-29-oic acid, moronic acid, oleanolic acid, wilforlide A, wilforlide B.

Zhou et al., Chem. Abstr., 107 151211y, Chem. Abstr., 107 194892p

E. europaeus L.

MACROCYCLIC SESQUITERPENE ALKALOIDS: 2,5-bisdiacetylevonine, 2-deacetylevonine, 4-deoxyevonine, evoeuropine, evoevoline, evomine, evonine, evonoline, evozine, isoevonine.
STEROIDS: euomonoside, euonoside.
SESQUITERPENOIDS: five uncharacterized sesquiterpene esters.

(A) Budziekiewicz et al., Z. Naturforsch., 27B, 800 (1972)
(A) Crombie, Whiting, Phytochemistry, 12, 703 (1973)
(S) Reichstein et al., Helv. Chim. Acta, 35, 87 (1952)
(T) Budziekiewicz, Romer, Tetrahedron, 31, 1761 (1975)

E. medirossica.

STEROID: euonoside.

Kislichenko, Makarevich, Khim. Prir. Soedin., 804 (1973)

E. sieboldiana Blume.

MACROCYCLIC SESQUITERPENE ALKALOIDS: euonymine, evonine, neoeuonymine, neoevonine.

Sigiura et al., Tetrahedron Lett., 2659, 2733 (1971)

E. tingens.

TRITERPENOIDS: 20-hydroxytingenone, tingenone.

Brown et al., J. Chem. Soc., Perkin I, 2721 (1973)

E. verrucosus.

SESQUITERPENE ESTERS: ever-1, ever-2, ever-4, ever-6, ever-7, ever-8, ever-10.

Begley et al., J. Chem. Soc., Perkin I, 535 (1986)

EUPATORIADELPHUS

E. fistulosus.
TRITERPENOID: 3α-tigloyloxy-6β-hydroxytrametone.
Bohlmann et al., Phytochemistry, 17, 2101 (1978)

EUPATORIUM (Compositae)

E. adenophorum.
SESQUITERPENOID: 5,8-epoxy-4-hydroxy-3-amorphanone.
Shukla et al., Chem. Ind. (London), 863 (1983)

E. album.
SESQUITERPENOID: eupatoralbin.
DITERPENOIDS: ent-11α,15α-dihydroxykaur-16-en-19-oic acid,
eupatalbin, ent-11α-hydroxy-15-oxokaur-16-en-19-oic acid,
ent-11,12,15-trihydroxykaur-16-en-19-oic acid.
Herz et al., J. Org. Chem., 44, 2999 (1979)

E. altissimum.
One sesquiterpenoid.
Herz et al., Phytochemistry, 17, 1953 (1978)

E. azureum.
TRITERPENOIDS: epi-friedelinol, taraxasterol acetate.
Nomamura, J. Pharm. Soc., Japan, 75, 80 (1955)

E. bupleurifolium.
MONOTERPENOIDS: limonene, α-pinene, β-pinene.
Bauer et al., Trib. Farm., 42, 37 (1974); Chem. Abstr., 85 119567k

E. cannabinum L.
PYRROLIZIDINE ALKALOIDS: echinatine, supinine.
TRITERPENOIDS: dammaradienyl acetate, eucannabinolide,
eupatoropicrin, taraxasterol.
(A) Pedersen, Phytochemistry, 14, 2086 (1975)
(T) Drozdz et al., Collect. Czech. Chem. Commun., 37, 1546 (1972)
Talapatra et al., Aust. J. Chem., 27, 1137 (1974)

E. chinense var. hakonense.
SESQUITERPENOIDS: eupachifolins A, B, C, D and E,
eupahakonensin, eupahakonin A, eupahakonin B,
eupasimplicin A, eupasimplicin B, peroxyeupahakonin A.
Ito et al., Phytochemistry, 21, 715 (1982)

E. chinense var. simplicifolium.
SESQUITERPENOIDS: eupachifolins A, B, C, D and E.
Ito et al., Chem. Lett., 1473 (1979)

E. cuneifolium.
SESQUITERPENOIDS: eupacunolin, eupatocunin, eupatocunoxin.
Kupchan et al., J. Org. Chem., 38, 2189 (1973)

E. formosanum.
SESQUITERPENOIDS: eupaformonin, eupaformosanin, eupatolide.
Lee et al., Phytochemistry, 16, 1068 (1977)

E. glechnophyllum Less.
CHROMENES: ethylencecalol, methylencecalol.
Becerra et al., Rev. Latinoam. Quim., 14 92 (1983)

E. hookerianum.
FLAVONOIDS: acacetin, apigenin 7-rhamnoside, kaempferitrin,
kaempferol 3-glucoside-7-rhamnoside, kaempferol
7-rhamnoside.
Ferraro et al., An. Asoc. Quim. Argent., 71, 327 (1983) Chem. Abstr.,
100 20437j

E. hyssopifolium.
SESQUITERPENOIDS: eupahyssopin, eupassofilin, eupassopilin,
eupassopin.
Lee et al., Tetrahedron Lett., 1051 (1976)

E. japonicum.
SESQUITERPENOID: euponin.
Nakajima et al., Agric. Biol. Chem., 44, 2893 (1980)

E. jhanii.
DITERPENOIDS: jhanic acid, jhanidiol, jhanidiol acetate,
jhanilactone, jhanol, jhanol acetate.
Gonzalez et al., An. Quim., 75, 128 (1978)

E. inulaefolium.
FLAVONOIDS: pedalitin,
5,6,3'-trihydroxy-7,4'-dimethyloxyflavone.
Ferraro et al., Phytochemistry, 16, 1618 (1977)

E. laevigatum.
NORSESQUITERPENOID: laevigatin.
de Oliviera et al., Tetrahedron Lett., 2653 (1978)

E. ligustrinum DC.
SESQUITERPENOID: ligustrin.
Romo et al., Tetrahedron, 24, 5087 (1968)

E. lindleyanum.
SESQUITERPENOIDS: eupalinins A, B, C and D.
Ito et al., Chem. Lett., 1469 (1979)

E. macrocephalum.
MONOTERPENOID: p-menth-1-ene-3,6-diol.
Gonzalez et al., An. Quim., 68, 319 (1972)

E. maculatum L.
PYRROLIZIDINE ALKALOIDS: echinatine, trachelantimidine.
Tsuda, Marion, Can. J. Chem., 41, 1919 (1963)

E. micranthum.
FLAVONOIDS: 7-methylgaramodendrin, rhamnocitrin.
Sagareishvili et al., Khim. Prir. Soedin., 710 (1985)

E. microphyllum.
FLAVONOIDS: 5-hydroxy-6,7,3',4'-tetramethoxyflavone,
quercetin, rutin.
Torrenegra et al., Rev. Latinoam. Quim., 16, 64 (1985)

E. mikanioides.
SESQUITERPENOIDS: deacetyleupaserrin,
2,8-dihydroxy-1(10),4,11(13)-germacratrien-12,6-olide,
mollisorin A.
Herz et al., J. Org. Chem., 45, 489 (1980)

E. mohrii.
SESQUITERPENOID: anomalide.
Cox et al., Tetrahedron Lett., 3569 (1979)

E. odoratum.
CHALCONE: 2',3',4,4',6'-pentahydroxychalcone.
TRITERPENOID: epoxylupeol.
Bose et al., Phytochemistry, 12, 667 (1973)
Talapatra et al., Indian J. Chem., 15B, 806 (1977)

E. perfoliatum.
SESQUITERPENOIDS: 8β-angeloyloxyeupatundin,
11,13α-dihydroeuperfolide, eufoliatin, eufoliatorin,
euperfolide, euperfolin, euperfolitin,
7-hydroxy-7-hydro-6-dehydro-3,4(Z)-α-farnesene.
TRITERPENOID: urs-20-en-3β-ol.
Bohlmann et al., Phytochemistry, 16, 1973 (1977)
Herz et al., J. Org. Chem., 42, 2264 (1977)

E. recurvens.
SESQUITERPENOID: eupecurvin.
Herz et al., J. Org. Chem., 43, 3559 (1978)

E. riparium.
CHROMENES: acetovanillochromene, ripariochromene A.
TRITERPENOID: taraxasterol palmitate.
(C) Tsukayama et al., Bull. Chem. Soc., Japan, 48, 80 (1975)
(C) Bohlmann et al., Chem. Ber., 110, 301 (1977)
(T) Talapatra et al., J. Indian Chem. Soc., 55, 296 (1978)

E. rotundifolium.
SESQUITERPENOIDS: eupachlorin, eupachloroxin, euparotin, euparotin acetate, eupatoroxin, 10-epi-eupatoroxin, eupatundin.
Kupchan et al., Tetrahedron Lett., 3517 (1968)

E. sachalinense.
SESQUITERPENOIDS: hiyodorilactones A, B, C, D, E and F, peroxysachalinin, sachalin, sachalinin.
Ito et al., Chem. Lett., 1503 (1979)
Takahashi et al., Chem. Pharm. Bull., 27, 2539 (1979)

E. semiserratum DC.
FLAVONES: eupatilin, eupatorin.
SESQUITERPENOIDS: deacetyleupaserrin, eupaserrin.
(F) Kupchan et al., Tetrahedron, 25, 1603 (1969)
(T) Kupchan et al., J. Org. Chem., 38, 1260 (1973)

E. serotinum.
SESQUITERPENOIDS: 8β-angeloyloxy-15-hydroxyparthenolide, 7β-hydroxy-1-oxolongipinene, 8β-(3-methylpent-2E-enoyloxy)-elemacroquistanthus acid, 8β-(3-methylpent-2E-enoyloxy)-15-hydroxyparthenolide, 8β-tigloyloxy-15-hydroxyparthenolide, euserotin.
Herz et al., J. Org. Chem., 44, 2784 (1979)
Bohlmann et al., Planta Med., 76 (1985)

E. sessilifolium.
SESQUITERPENOIDS: eupasessifolides A and B.
Bohlmann et al., Phytochemistry, 18, 1401 (1979)

E. tashiroi Hayata.
MONOTERPENOID: eupatriol.
Wu et al., Chem. Pharm. Bull., 33, 4025 (1985)

E. tinifolium H.B.K.
MONOTERPENOIDS: tinifoline, tinifolinediol.
delle Monache et al., Farmaco Ed. Sci., 36, 960 (1981)

E. villosum.
DITERPENOID: evillosin.
Manchand et al., J. Org. Chem., 44, 1322 (1979)

EUPHORBIA (Euphorbiaceae)

E. acaulis.
DITERPENOID: caudicifolin.
Satti et al., Phytochemistry, 25, 1441 (1986)

E. antiquorum.
DITERPENOID: 3-O-angeloylingenol.
Adolf et al., J. Sci. Soc. Thailand, 9, 81 (1983)

E. atoto.
PIPERIDINE ALKALOID:
(+)-aza-1-methylbicyclo-[3,3,1]-nonan-3-one.
Beecham et al., Aust. J. Chem., 20, 2291 (1967)

E. balsamifera.
TRITERPENOID: germanicol.
Russel et al., Phytochemistry, 15, 1933 (1976)

E. biglandulosa Desf.
DITERPENOIDS:
20-O-acetyl-13-O-isobutyryl-12-O-(2E)-2,4-octadienyl-4-deoxyphorbol,
20-O-acetyl-12-(Z,E)-2,4-octadienyl-13-O-propionyl-4-deoxyphorbol,
1,20-O-diacetyl-12-O-(Z,E)-2,4-octadienyl-4-deoxyphorbol.
Falcone, Crea, Justus Liebigs Ann. Chem., 1116 (1979)

E. caudicifolia.
DITERPENOID: candicifolin.
Ahmad et al., Phytochemistry, 16, 1844 (1977)

E. characais L.
DITERPENOIDS: characiol, characiol-5,6-epoxide.
Seip et al., Phytochemistry, 23, 1689 (1984)

E. coerulescens.
DITERPENOIDS: 12-deoxy-4β-hydroxyphorbol 13-dodecanoate-20-acetate, 12-deoxy-4-hydroxyphorbol 13-undecanoate-20-acetate.
Evans, Kinghorn, Phytochemistry, 14, 1669 (1975)

E. condylocarpon.
FLAVONOID: trifolin.
Roshchin, Khim. Prir. Soedin., 576 (1977)

E. cooperi.
SESQUITERPENOID:
4β,9α,13α,16,20-pentahydroxytiglia-1,6-dien-3-one 16-isobutyrate-13-(2-methylbut-2-enoate).
Gschwendt, Hecker, Tetrahedron Lett., 567 (1970)

E. corollata.
TRITERPENOID: corollatadiol.
Piatak, Riemann, Tetrahedron Lett., 4525 (1972)

E. cyparissias L.
DITERPENOIDS: 12-deoxy-16,19-dihydroxyingenol, 13-hydroxyingenol.
TRITERPENOID: 5-gluten-3α-ol.
Hecker, Pure Appl. Chem., 49, 1423 (1977)

E. echinus.
TRITERPENOID: obtusifoldienol.
Gonzalez, Breton, An. Quim., 55B, 93 (1959)

E. esula L.
DITERPENOID: ingenol-3,20-dibenzoate.
Kupchan et al., Science, 191, 571 (1976)

E. fischeriana.
AMIDE ALKALOID: aurantiamide O-acetate.
Maiti et al., Experientia, 32, 1166 (1976)

E. fortissima.
DITERPENOIDS: 12-deoxy-4-hydroxyphorbol 13-dodecanoate-20-acetate, 12-deoxy-4β-hydroxyphorbol 13-dodecenoate 20-acetate, 12-deoxy-4β-hydroxyphorbol 13-octenoate 20-acetate.
Evans, Kinghorn, Phytochemistry, 14, 1669 (1975)

E. geniculata.
FLAVONOIDS: kaempferol, kaempferol 3-rutinoside, quercetin, quercetin 3-rhamnoside.
Ismail et al., Pharmazie, 32, 538 (1977)

E. helioscopa L.
DITERPENOIDS: euphohelins A, B, C and D, euphohelioscopin A, euphohelioscopin B, euphoscopins A, B and C, helioscopinolides A, B and C.
Shizuri et al., Chem. Lett., 65 (1983)
Kosemura et al., Bull. Chem. Soc., Japan, 58, 3112 (1985)

E. hermentiana.
DITERPENOIDS: four methoxyingenol esters.
Lin et al., Phytochemistry, 22, 2795 (1983)

E. heterophylla.
AMINOACIDS: alanine, aspartic acid, cysteine, glutamic acid, methionine, proline, serine.
Kumar et al., J. Inst. Chem. (India), 48, 192 (1976); Chem. Abstr., 86 68376h

E. indica.
STEROIDS: campesterol, β-sitosterol, stigmasterol.
TRITERPENOID: taraxerol.
Rizk et al., Planta Med., 32, 177 (1977)

E. ingens E. Mey ex Boiss.
DITERPENOIDS: 16-hydroxyingenol, ingol 3,7,12-triacetate-8-nicotinate.
Opferkuch, Hecker, Tetrahedron Lett., 3611 (1973)

E. jacquionontii.
AMINOACIDS: alanine, aspartic acid, glutamic acid, leucine, norvaline.
Kumar et al., J. Inst. Chem. (India), **48,** 192 (1976); Chem. Abstr., 86 68376h

E. jaxartica.
DITERPENOID: euphorbol hexacozonate.
Azimov et al., Khim. Prir. Soedin., 136 (1970)

E. jolkini.
DITERPENOIDS: ingenol-2,4,6,8,10-tetradecapentaenoate, jolkinols A, B, C and D, jolkinolides A, B, C, D and E.
Uemura et al., Tetrahedron Lett., 4593 (1976)

E. kamerunica.
Six diterpenoids.
Abo et al., Planta Med., **43,** 392 (1981)

E. kansui.
DITERPENOIDS: 20-deoxyingenol, 20-deoxyingenol 3-benzoate, 20-deoxyingenol 5-benzoate, 13-hydroxyingenol, 13-oxyingenol 13-dodecanoate-20-pentanoate.
Uemura et al., Tetrahedron Lett., 2527 (1974)

E. kansui lioii.
DITERPENOIDS: kansuine A, kansuine B.
Adolf, Hecker, Experientia, **27,** 1393 (1971)

E. lactea Ham.
DITERPENOIDS: 12-deoxyphorbol 13,20-diacetate, euphorbol hexacosonate, tinyatoxin.
TRITERPENOID: 24-methylenecycloartenol.
Baslas et al., Herba Hung., **75,** 49 (1986); Chem. Abstr., 107 151204y

E. lathyris L.
DITERPENOIDS:
6,20-epoxylathyrol-5,10-diacetate-3-phenylacetate, euphorbiasteroid, 16-hydroxyingenol, 7-hydroxylathyrol, lathyrol.
Sahai et al., Phytochemistry, **20,** 1660 (1981)

E. maddeni.
DITERPENOID: euphorinin.
Sahai et al., Phytochemistry, **20,** 1660 (1981)

E. marginata.
DITERPENOID: 16-hydroxyingenol.
Hecker, in: Handbook of General Pathology, New York: Springer Verlag, 1975, vol 6, p 651.

E. millii Ch. des Moulins.
DITERPENOID ALKALOIDS: milliamines A, B, C, D, E, F, G, H and I.
Marston et al., Planta Med., **50,** 319 (1984)

E. nerifolia.
TRITERPENOID: glut-6(10)-en-1-one.
Anjeneyulu et al., Tetrahedron, **29,** 3909 (1973)

E. obtusifolia.
TRITERPENOID: obtusifoldienol.
Gonzalez, Breton, An. Quim., **55B,** 93 (1959)

E. paralias.
STEROID: β-sitosterol.
TRITERPENOIDS: β-amyrin, betulin, oleanolic acid, ursolic acid, uvaol.
Khafagy et al., Planta Med., **29,** 301 (1976)

E. poisonii.
SESQUITERPENOID: euphorbia compound D.
DITERPENOIDS: candletoxin A, candletoxin B, resiniferotoxin, tingatoxin.
Schmidt, Evans, Experientia, **33,** 1197 (1977)
Fakunle et al., Tetrahedron Lett., 2119 (1978)

E. polyacantha.
DITERPENOIDS: 12-deoxy-4β-hydroxyphorbol 13-dodecanoate-20-acetate, 12-deoxy-4β-hydroxyphorbol 13-dodecenoate-20-acetate, 12-deoxy-4β-hydroxyphorbol 13-octenoate-20-acetate.
Evans, Kinghorn., Phytochemistry, **14,** 1669 (1975)

E. prostrata.
FLAVONOIDS: cosmosiin, kaempferol, quercetin, quercetin 3-rhamnoside, rhamnetin 3-galactoside.
Ismail et al., Pharmazie, **32,** 538 (1977)

E. pulcherrima.
ACID: deca-2,4,6-trienoic acid.
ANTHRAQUINONE: 2-methylanthraquinone.
ESTERS: germanicyl acetate, epi-germanicyl acetate, germanicyl behenate, germanicyl tetracosanoate.
STEROID: β-sitosterol.
TRITERPENOID: pulcherrol.
(Ac) Warnaar, Lipids, **12,** 707 (1977)
(An, E) Mahato et al., J. Indian Chem. Soc., **54,** 388 (1977)
(T) Dominguez, Bruer, Rev. Latinoam. Quim., **1,** 68 (1970)

E. regis-tubae.
TRITERPENOID: obtusifoldienol.
Gonzalez, Breton, An. Quim., **55B,** 93 (1959)

E. resinifera.
DITERPENOIDS: 12,20-dideoxyphorbol-13-angelate, 12,20-dideoxyphorbol 13-isobutyrate, α-euphol, preresiniferotoxin, resiniferol, resiniferotoxin.
Hargenhahn et al., Tetrahedron Lett., 1595 (1975)

E. royleana.
TRITERPENOIDS: cycloroylenol, cycloroylenol acetate, glut-5-en-3β-ol.
Bhat et al., Tetrahedron Lett., 5207 (1982)

E. segieriana.
DITERPENOID: ingenol.
Upadhyay et al., Planta Med., **30,** 32 (1976)

E. splendens.
MACROLIDE: lasiodiplodin.
Lee et al., Phytochemistry, **21,** 1119 (1982)
Matsunaga et al., J. Chem. Soc., Chem. Commun., 1128 (1984)

E. supina.
TRITERPENOID: spirosupinanonediol.
Matsunaga et al., J. Chem. Soc., Chem. Commun., 1128 (1984)

E. thymifolia.
Three uncharacterized alkaloids.
Jabber Khan, Pakistan J. Sci. Ind. Res., **8,** 293 (1965)

E. tirucalli L.
ANTHOCYANINS: cyanidin 3,5-diglucoside, tulipandin 3-rhamnoglucoside.
PHENOLIC: 3,4,3′-tri-O-methylellagic acid.
STEROID: β-sitosterol.
DITERPENOIDS:
20-O-acetylresiniferonol-9,13,14-orthophenylacetate, cycloeuphorbol, 4-deoxyphorbol, 12-deoxyphorbol 13-isobutyrate-20-acetate, euphorbinol, euphorbia factors Ti-1, Ti-2, Ti-3, Ti-4, αTi-1 and αTi-2.
TRITERPENOIDS: euphorbol, euphorbol hexacosonoate, taraxasterol, taraxerone, tinyatoxin, tirucallol.
(An, T) Baslas et al., Herba Hung., **22,** 35 (1983); Chem. Abstr., 100 64928z
(T) Furstenberger, Hecker, Planta Med., **22,** 241 (1972)
(T) Afza et al., Pakistan J. Sci. Ind. Res., **22,** 173 (1979)

E. triaculeata.
DITERPENOID: triaculetin.
Ahmad et al., Arab Gulf J. Sci. Res., **1,** 99 (1983)

E. unispina.

DITERPENOID: resiniferotoxin.

Hergenhahn *et al.*, *Tetrahedron Lett.*, 1595 (1975)

EUPHRASIA (Scruphulariaceae)

E. rostkoviana Hayne.

GLYCOSIDE: eukovoside.

IRIDOID: erostoside.

(Gl) Sticher *et al.*, *Helv. Chim. Acta*, **65**, 1538 (1982)

(I) Salama *et al.*, *Planta Med.*, **47**, 90 (1983)

E. salisbergensis Funck.

IRIDOID: euphroside.

Sticher *et al.*, *Helv. Chim. Acta*, **64**, 78 (1981)

EUPOMATIA (Eupomatiaceae)

E. laurina R.Br.

AZAPHENANTHRENE ALKALOID: eupolauramine.

Bowden *et al.*, *J. Chem. Soc., Perkin 2*, 658 (1976)

EUPTELEA (Eupteleaceae)

E. polyandra (Sieb. et Zucc.)

SAPONINS: eupteleosides A and B.

Goto *et al.*, *Chem. Pharm. Bull.*, **12**, 516 (1964)

EURYCLES (Amaryllidaceae)

E. sylvestris Salisb.

ISOQUINOLINE ALKALOID: lycorine.

Greathouse, Rigler, *Amer. J. Bot.*, **28**, 702 (1941)

EURYCOMA

E. longifolia.

DITERPENOIDS: 3,4-dihydroeurycomalactone, eurycomalactone, eurycomanol, eurycomanone, laurycolactones A and B.

Darise *et al.*, *Phytochemistry*, **21**, 2091 (1981)

EURYOPS (Compositae)

E. abrotanifolius.

SESQUITERPENOIDS:

6α-angeloyloxy-1,10-epoxy-4-hydroxyeremophilane, 17,18-epoxy-17,18-dihydroneoadenostylone, 4α-hydroxy-6β-isovaleryloxy-9-oxofuranoeremophilane, 10β-hydroxy-2β-(2-methylacryloyloxy)-furanoeremophilane, 4α-hydroxy-6β-senecioyloxy-9-oxofuranoeremophilane.

Bohlmann *et al.*, *Chem. Ber.*, **107**, 2730 (1974)

E. acraeus.

SESQUITERPENOIDS:

3β-angeloyloxy-1,10-epoxufuranoeremophilane, 10β-hydroxy-1α-(3-methylpent-2c-enoyloxy)-furanoeremophilane, 10β-hydroxy-1α-(3-methylpent-2t-enoyloxy)-furanoeremophilane, 3α-tigloyloxy-1β,10β-epoxyfuranoeremophilane.

Bohlmann, Zdero, *Phytochemistry*, **17**, 1135 (1978)

E. annae.

SESQUITERPENOIDS:

9α-acetoxy-8α-angeloyloxy-2-oxo-10βH-eremophil-11(12)ene, 8α-angeloyloxy-9α-hydroxy-2-oxo-10βH-eremophil-11(12)-ene, 9α-angeloyloxy-8α-hydroxy-2-oxo-10βH-eremophil-11(12)-ene, 8α-hydroxy-2-oxo-10βH-eremophil-11(12)-ene, 6β-(2-methyl-2,3-epoxybutyryloxy)-10αH-furanoeremophil-9-one, 9α-(3-methylpent-3-enoyloxy)-10αH-furanoeremophilone.

Bohlmann, Zdero, *Phytochemistry*, **17**, 1135 (1978)

E. brevilobus.

SESQUITERPENOIDS:

1α-angeloyloxy-6β-acetoxy-10β-hydroxyfuranoeremophilane, 3α-3-methylpent-2-cis-enoyloxy)-9α-hydroxyeremophil-8-on-11(12)-ene.

Bohlmann, Zdero, *Phytochemistry*, **17**, 1135 (1978)

E. brevipappus.

SESQUITERPENOIDS: 6β-acetoxyligularenolide, 1β,10β-epoxy-6β-tigloyloxyfuranoeremophilane, 13-hydroxycacalol methyl ether, 9α-hydroxy-6β-(2-methylacryloyloxy)-euryopsin, 9α-hydroxy-6β-senecioyloxyeuryopsin 6β-(2-methylacrylyl)-9-euryopsin, 6β-senecioyloxyeuryopsin, 6β-tigloyloxyeuryopsin.

Bohlmann, Zdero, *Phytochemistry*, **17**, 1135 (1978)

E. chrysanthemoides.

SESQUITERPENOIDS:

6β-angeloyloxy-4,5-didehydro-5,6-secoeuryopsin, 10β-angeloyloxy-2β-hydroxyfuranoeremophilane, 2,9-dioxoeuryopsin.

Bohlmann *et al.*, *Chem. Ber.*, **107**, 2730 (1974)

E. empetrifolius.

SESQUITERPENOIDS: 3β-acetoxy-6β-angeloyloxyeuryopsin, 3β-acetoxy-6β-senecioyloxyeuryopsin, 3β-acetoxy-6β-tigloyloxyeuryopsin, 3β-hydroxy-6β-angeloyloxyeuryopsin, 3β-hydroxy-6β-(3-methylpent-2-enoyloxy)-euryopsin, 3β-hydroxy-6β-senecioyloxyeuryopsin, 3β-hydroxy-6β-tigloyloxyeuryopsin, 3β-(2-methylacryloxy)-bicyclogermacrene, 6β-tigloyloxyeuryopsin.

Bohlmann, Zdero, *Phytochemistry*, **17**, 1135 (1978)

E. evansi.

SESQUITERPENOID:

1β,10β-epoxy-6β-tigloyloxufuranoeremophilane.

Bohlmann, Zdero, *Phytochemistry*, **17**, 1135 (1978)

E. floribundus.

SESQUITERPENOIDS: euryopsol, euryopsonol, 4α-hydroxy-6β-(2-methylacryloyloxy)-1β,10β-epoxyfuranoeremophilane,

Rivett *et al.*, *Tetrahedron*, **23**, 2431 (1967)

Bohlmann, Zdero, *Phytochemistry*, **17**, 1135 (1978)

E. galpinii.

SESQUITERPENOID: 6β-isobutyryloxy-10-furanoeremophil-9-one.

Bohlmann, Zdero, *Phytochemistry*, **17**, 1135 (1978)

E. hebecarpus.

SESQUITERPENOIDS:

6-angeloyloxy-4,5-didehydro-5,6-secoeuryopsin, angeloyloxyeuryopsonol 6β-angeloyloxy-4β-hydroxyeuryopsin, 6β-angeloyloxy-4α-hydroxy-9-oxofuranoeremophilane, 3α-angeloyloxy-9-oxofuranoeremophilane, 1,4:3,14-diepoxy-4-hydroxy-5,6,9,10-tetrahydro-4,5-secofuranoeremophilane, 1,10-epoxy-4α-hydroxy-6β-isobutyryloxyfuranoeremophilane, 6-hydroxy-4,5-didehydro-5,6-secoeuryopsin-2-methacrylate, 4α-hydroxy-6β-isobutyryloxyeuryopsin, 4α-hydroxy-6β-isobutyryloxy-9-oxofuranoeremophilane, 4α-hydroxy-6β-isovaleryloxyeuryopsin, 9-oxoeuryopsin, 4-oxo-5,6,9,10-tetrahydro-4,5-secofuranoeremophilane-5,1-carbolactone, 5,6,9,10-tetrahydro-1,4 3,14-diepoxy-4-hydroxy-4,5-secofuranoeremophilane.

Bohlmann *et al.*, *Chem. Ber.*, **107**, 2730 (1974)

E. imbricatus.

SESQUITERPENOID: lateriflorol-9-(2-methylpent-2-cis-enoate), lateriflorol-9-(2-methylpent-2-trans-enoate).
Bohlmann, Zdero, *Phytochemistry*, **17**, 1135 (1978)

E. lateriflorus.

SESQUITERPENOIDS:
3β-angeloyloxy-6-isobutyryloxyeuryopsin-9-one,
3β-angeloyloxy-6β-propionyloxyeuryopsin-9-one,
3β,6β-dipropionyloxyeuryopsin-9-one,
3β-hydroxy-6β-angeloyloxyeuryopsin,
lateriflorol-9-angelate-2-acetate, lateriflorol-9-senecionate,
lateriflorol-2-one-8,9-ditiglate.
Bohlmann, Zdero, *Phytochemistry*, **17**, 1135 (1978)

E. linifolius.

BENZOFURAN: 1,2-dihydrocacalohastin.
SESQUITERPENOIDS:
1,10-epoxy-6β-(2-methylacryloyloxy)-9-oxofuranoeremophilane, 2-methylacrylyloxyeuryopsinol.
Bohlmann *et al.*, *Chem. Ber.*, **107**, 2730 (1974)

E. multifidus.

SESQUITERPENOIDS:
9α-hydroxy-6β-angeloyloxy-1β,11β-epoxyfuranoeremophilane,
9α-hydroxy-6β-2-methacryloyloxy)-1β,10β-epoxyfuranoeremophilane,
9α-hydroxy-6β-senecioyloxy-1β,10β-epoxyfuranoeremophilane,
6β-(2-methylacrylyloxy)-10βH-furanoeremophil-9-one,
6β-senecioyloxy-10βH-furanoeremophilan-9-one,
6β-tigloyloxy-10βH-furanoeremophilan-9-one.
Bohlmann, Zdero, *Phytochemistry*, **17**, 1135 (1978)

E. oligoglossus.

SESQUITERPENOID: 1α,10α-epoxy-tigloyloxyfuranoeremophilane,
Bohlmann, Zdero, *Phytochemistry*, **17**, 1135 (1978)

E. oligoglossus subsp. oligoglossus.

SESQUITERPENOIDS:
3β-acetoxy-6-(2,3-epoxy-2-methylbutyryloxy)-10αH-furanoeremophil-9-one,
9α-hydroxy-6β-angeloyloxy-1β,11β-epoxyfuranoeremophilane,
9α-hydroxy-6β-isobutyryloxy-1β,10β-epoxyfuranoeremophilane,
9α-hydroxy-6β-isovaleryloxy-1β,10β-epoxyfuranoeremophilane,
9α-hydroxy-6β-(3-methylpent-3t-enoyloxy)-1β,10β-epoxyfuranoeremophilane,
9α-hydroxy-6β-senecioyloxy-1β,10β-epoxyfuranoeremophilane,
9α-hydroxy-6β-tigloyloxy-1β,10β-epoxyfuranoeremophilane.
Bohlmann, Zdero, *Phytochemistry*, **17**, 1135 (1978)

E. othonnoides.

SESQUITERPENOIDS: 6β-angeloyloxy-1α,10α-epoxyeuryopsin,
6-angeloyloxyeuryopsin,
6β-angeloyloxy-9-oxo-1,10-epoxyeuryopsin,
10-hydroxy-1-angeloyloxyfuranoeremophilane,
3-hydroxy-4-isobutyryloxy-9-oxoeuryopsin,
3-hydroxy-6-isovaleryloxy-9-oxoeuryopsin,
10-hydroxy-1-senecioyloxyfuranoeremophilane,
6β-isobutyryloxy-9-oxo-1,10-epoxyeuryopsin,
6β-isovaleryloxy-9-oxo-1,10-epoxyeuryopsin.
Bohlmann *et al.*, *Chem. Ber.*, **105**, 3523 (1972)

E. pectinatus.

SESQUITERPENOIDS:
6β-(2,3-epoxy-2-methylbutyryloxy)-1β,10β-furanoeremophil-9-one,
4α-hydroxy-6β-isovaleryloxy-10αH-furanoeremophilan-9-

one.
Bohlmann, Zdero, *Phytochemistry*, **17**, 1135 (1978)

E. pectinatus x chrysanthemoides.

SESQUITERPENOID: 3β-angeloyloxyeuryopsin.
Bohlmann, Zdero, *Phytochemistry*, **17**, 1135 (1978)

E. pedunculatus.

SESQUITERPENOIDS: 3β-acetoxy-6β-angeloyloxyeuryopsin,
3α-angeloyloxy-1β,10β-epoxyfuranoeremophilane,
anhydrooplopanone,
3β-hydroxy-6β-(3-methylpentanoyloxy)-1β,10β-epoxyfuranoeremophilane,
3β-(3-methylpentanoyloxy)-euryopsin,
3α-(3-methylpent-2-cis-enoyloxy)-1β,10β-epoxyfuranoeremophilane,
3α-(3-methylpent-2-trans-enoyloxy)-1β,10β-epoxyfuranoeremophilane,
3α-3-methylpent-2-cis-enoyloxy)-euryopsin.
Bohlmann, Zdero, *Phytochemistry*, **17**, 1135 (1978)

E. rehmanni.

SESQUITERPENOIDS: 9α-hydroxy-6β-angeloyloxyeuryopsin,
3β-(2-methylacrylyloxy)-6β-angeloyloxy-10αH-furanoeremophil-9-one,
3β-(2-methylacrylyloxy)-6β-(2-methylacrylyloxy)-10αH-furanoeremophilan-9-one,
3β-(2-methylacrylyloxy)-6β-tigloyloxy-10αH-furanoeremophilan-9-one.
Bohlmann, Zdero, *Phytochemistry*, **17**, 1135 (1978)

E. rupestris var. rupestris.

SESQUITERPENOIDS: 2-dehydrorupestrol-9-angelate, eurupestrol 9-angelate,
Bohlmann, Zdero, *Phytochemistry*, **17**, 1135 (1978)

E. spathaceus.

SESQUITERPENOIDS:
6β-angeloyloxy-3β-hydroxy-9-oxoeuryopsin,
6β-angeloyloxy-3β-(2-methacryloxy)-9-oxoeuryopsin,
6β-angeloyloxy-9-oxo-2,3-didehydroeuryopsin,
3β,6β-bis(2-methylacryloxy)-9-oxoeuryopsin,
3β-hydroxy-6β-(2-methylacryloxy)-9-oxoeuryopsin,
3β-hydroxy-6β-(2-methyl-2,3-epoxy-butyryloxy)-euryopsin-9-one,
6β-(2-methylacrylyloxy)-9-oxo-2,3-didehydroeuryopsin,
6β-(2-methylacrylyloxy)-9-oxoeuryopsin,
6β-(2-methylacrylyloxy)-9-oxofuranoeremophilane.
Bohlmann *et al.*, *Chem. Ber.*, **107**, 2730 (1974)

E. speciosissimum.

SESQUITERPENOIDS:
6β-angeloyloxy-9-oxo-1α,10α-epoxyeuryopsin,
2-hydroxymethylcrotonyleuryopsonol, euryopsonol isobutyrate, euryopsonol isovalerate,
6β-isobutyryloxy-9-oxo-1,10-epoxyeuryopsin,
6β-isovaleryloxy-9-oxo-1,10-epoxyeuryopsin.
Bohlmann *et al.*, *Chem. Ber.*, **105**, 3523 (1972)

E. sulcatus.

SESQUITERPENOIDS: 3β-acetoxy-6β-angeloyloxyeuryopsin,
4α-hydroxy-6β-tigloyloxyeuryopsin.
Bohlmann, Zdero, *Phytochemistry*, **17**, 1135 (1978)

E. tenuisissimus.

SESQUITERPENOIDS:
6β-angeloyloxy-4,5-didehydro-5,6-secoeuryopsin,
6β-angeloyloxy-4β-hydroxyeuryopsin,
1,10-epoxy-4α-hydroxy-6β-isobutyryloxyfuranoeremophilane.
Bohlmann *et al.*, *Chem. Ber.*, **107**, 2730 (1974)

E. tysonii.
 SESQUITERPENOIDS: 6β-acetoxy-12-hydroxy-8,12H
  1(10),9-dehydrofuranoeremophilane,
  10β-hydroxy-1α-tigloyloxyfuranoeremophilane,
  6β-(2-methylacryloyloxy)-12-hydroxy-8,12
  1(10),9-furanoeremophilane,
  6β-tigloyloxy-10αH-furanoeremophilan-9-one.
  Bohlmann, Zdero, *Phytochemistry*, **17**, 1135 (1978)

E. virgineus.
 SESQUITERPENOIDS:
  4α-hydroxy-9-oxo-6β-tigloyloxyfuranoeremophilane,
  4α-hydroxy-6β-tigloyloxyeuryopsin.
  Bohlmann *et al.*, *Chem. Ber.*, **107**, 2730 (1974)
  Bohlmann *et al.*, *ibid.*, **107**, 2912 (1974)

E. wagneri.
 SESQUITERPENOID:
  4α-hydroxy-6β-(2-methyl-2,3-epoxybutyryloxy)-10αH-
  furanoeremophil-9-one.
  Bohlmann, Zdero, *Phytochemistry*, **17**, 1135 (1978)

EUSCAPHIS (Staphyleaceae)

E. japonica.
 TRITERPENOID: euscaphic acid.
  Takahashi *et al.*, *Chem. Pharm. Bull.*, **22**, 650 (1974)

EUSTOMA

E. grandiflorum (Raf.) Shinners.
 XANTHONES: eustomin, 1-hydroxy-3,7-dimethoxyxanthone,
  1-hydroxy-3,5,6,7,8-pentamethoxyxanthone,
  1-hydroxy-3,7,8-trimethoxyxanthone.
  Sullivan *et al.*, *J. Pharm. Sci.*, **66**, 828 (1977)

EUTREPTIELLA

E. gymnastica.
 CAROTENOID: eutreptiellanone.
  Fiksdahl *et al.*, *Phytochemistry*, **23**, 649 (1984)

EUXYLOPHORA (Rutaceae)

E. paraensis Hub.
 CARBOLINE ALKALOIDS: euxylophoricines A, B and C,
  euxylophorine A, euxylophorine B, paraensine.
 QUINOLONE ALKALOIDS: paraensidimerines A, B, C, D, E, F
  and G.
  Danieli *et al.*, *Phytochemistry*, **11**, 1833 (1972)
  Jurd, Wong, *Aust. J. Chem.*, **34**, 1635 (1981)

EVERNIA (Lichenes)

E. prunastri (L.) Ach.
 DEPSIDE: evernin.
  Nicollier *et al.*, *Helv. Chim. Acta*, **59**, 2979 (1976)

EVODIA (Rutaceae)

E. alata.
 ACRIDONE ALKALOIDS: arborinine, evoprenine, normelicopidine.
 FURANOQUINOLINE ALKALOID: evolatine.
  Gell *et al.*, *Aust. J. Chem.*, **8**, 114 (1955)

E. celata F.Muell.
 ACRIDONE ALKALOID: melicopine.
  Gell *et al.*, *Aust. J. Chem.*, **8**, 114 (1955)

E. evelleryana F.Muell.
 FURANOQUINOLINE ALKALOID: evellerine.
  Johns *et al.*, *Aust. J. Chem.*, **24**, 1897 (1968)

E. floribunda.
 DITERPENOIDS: floribundic acid, floridiolic acid, floridiolide A,
  floridiolide B.
  Billet *et al.*, *Tetrahedron Lett.*, 3825 (1975)
  Billet *et al.*, *J. Chem. Res.*, 1475 (1978)

E. fraxinifolia.
 TRITERPENOID: isobauerenol.
  Lahey, Leeding, *Proc. Chem. Soc. (London)*, 342 (1958)

E. grauca.
 NORTRITERPENOIDS: graucins A, B and C.
  Nakatani *et al.*, *Tennen Yuki Kagobutsu Toronkai Koen Yoshishu*, **26**, 9
  (1983)

E. hortensis.
 MONOTERPENOID: evodone.
  Stetter *et al.*, *Chem. Ber.*, **93**, 603 (1960)

E. hupehensis.
 COUMARIN: xanthotoxol.
  Spath *et al.*, *Ber.*, **69**, 767 (1936)

E. littoralis.
 FURANOQUINOLINE ALKALOID: evolitrine.
  Cooke, Haynes, *Aust. J. Chem.*, **7**, 273 (1954)

E. meliifolia Benth.
 PROTOBERBERINE ALKALOID: berberine.
  Gupta, Spenser, *Can. J. Chem.*, **43**, 133 (1965)

E. rutaecarpa Benth. et Hook.
 CARBOLINE ALKALOIDS: dehydroevodiamine, evodiamine,
  rhetsinine, rutaecarpine.
 QUINOLINE ALKALOIDS: evocarpine, wuchuyine.
 LIMONOID: rutaevin.
  (A) Nakasato *et al.*, *J. Pharm. Soc., Japan*, **82**, 619 (1962)
  (T) Dreyer, *J. Org. Chem.*, **32**, 3442 (1967)

E. xanthoxyloides F.Muell.
 ACRIDONE ALKALOIDS: evoxanthidine, evoxanthine, melicopine,
  xanthoxoline.
 FUROQUINOLINE ALKALOIDS: anhydroevoxine, evoxine,
  evoxoidine.
 UNCHARACTERIZED ALKALOID: xanthevolide.
  Hughes *et al.*, *Austral. J. Sci. Res.*, **5A**, 401 (1952)

EXCOECARIA (Euphorbiaceae)

E. agallocha L.
 One nortriterpenoid.
  Ohigashi *et al.*, *Agric. Biol. Chem.*, **38**, 1093 (1974)

F

FABIANA (Solanaceae)
F. imbricata Hort.
QUINOLINE ALKALOID: fabianine.
Edwards, Elmore, *Can. J. Chem.*, **40**, 256 (1962)

FAGARA (Fagaraceae)
F. angolensis.
NAPHTHOISOQUINOLINE ALKALOID: angoline.
Palmer, Paris, *Ann. Pharm. Fr.*, **13**, 657 (1955)

F. coca (Gill) England.
FURANOQUINOLINE ALKALOID: τ-fagarine.
OXOPROTOBERBERINE ALKALOIDS: α-fagarine, fagarine II, fagarine III.
QUATERNARY ALKALOID: N-methylisocorydine.
Giacopello *et al.*, *Tetrahedron*, **20**, 2971 (1964)

F. hymelis (St.Hill) England.
AMIDE ALKALOIDS: (±)-tembamide, (+)-tembamide.
Albonico *et al.*, *J. Chem. Soc., C*, 1327 (1967)

F. lepreurii.
ACRIDONE ALKALOID: 1-hydroxy-2,3-dimethoxy-N-acridone.
NAPHTHOISOQUINOLINE ALKALOID: angoline.
Calderwood, Fish, *J. Pharm. Pharmacol.*, **18**, 1195 (1966)

F. macrophylla.
BENZOPHENANTHRIDINE ALKALOID: xanthofagarine.
Torto *et al.*, *Ghana J. Sci.*, **8**, 3 (1969)

F. mayu. One naphthoisoquinoline alkaloid.
COUMARINS: bergapten, isopimpinellin, psoralen, xanthotoxin.
(A) Assem *et al.*, *Phytochemistry*, **18**, 511 (1979)
(C) Torres *et al.*, *An. Asoc. Quim. Argent.*, **63**, 187 (1975); *Chem. Abstr.*, 85 139724n

F. naranjillo (Griseb.) Engl.
APORPHINE ALKALOID: tembetarine.
Albonico *et al.*, *Chem. Ind. (London)*, 1580 (1964)

F. nigrescens Fries.
APORPHINE ALKALOID: (+)-N-methylcorydine.
Kuck, *Chem. Ind. (London)*, 118 (1966)

F. officinalis L.
PROTIBERBERINE ALKALOIDS: (±)-sinactine, (-)-sinactine.
Goto, Kitasato, *J. Chem. Soc.*, 1234 (1930)

F. parvifolia A.Chev.
UNCHARACTERIZED ALKALOID: parvifagarine.
Paris *et al.*, *Ann. Pharm. Fr.*, **6**, 409 (1948)

F. rhoifolia Lam.
OXOPROTOBERBERINE ALKALOID: α-cryptopine.
PROTOBERBERINE ALKALOIDS: (-)-canadine, N-methylcanadine.
Calderwood, Fish, *Chem. Ind. (London)*, **6**, 237 (1966)

F. rubescens.
ACRIDONE ALKALOIDS:
1-hydroxy-2,3-dimethoxy-N-methylacridone,
1-hydroxy-3-methoxy-10-methylacridone.
AMIDE ALKALOID: rubesamide.
Dadson *et al.*, *J. Chem. Soc., Perkin I*, 146 (1976)

F. semiarticulata.
NAPHTHOISOQUINOLINE ALKALOID: chelerythrine.
Manske, *Can. J. Res.*, **21B**, 140 (1943)

F. tessmannii.
AMIDE ALKALOID: fagaramide.
Okorie, *Phytochemistry*, **15**, 1799 (1976)

F. tinguassoiba Hoehn.
APORPHINE ALKALOID:
1-hydroxy-2,9,10-trimethoxy-N-methylaporphine.
Riggs *et al.*, *Can. J. Chem.*, **39**, 1330 (1961)

F. xanthoxyloides.
ISOQUINOLINE ALKALOID: fagaronine.
LIGNAN: fagarol.
(A) Messmer *et al.*, *J. Pharm. Sci.*, **61**, 1858 (1972)
(L) Abe *et al.*, *Chem. Pharm. Bull.*, **22**, 2650 (1974)

FAGAROPSIS
F. glabra.
LIMONOID: isofraxinellone.
Blaise *et al.*, *Phytochemistry*, **24**, 2379 (1985)

FAGONIA (Zygophyllaceae)
F. alba.
FLAVONOIDS: herbacetin, herbacetin 8-methylether
3,7-diglucoside, herbacetin 8-methyl ether 3-rutinoside,
herbacetin 8-methyl ether 3-rutinoside-7-glucoside,
isorhamnetin 3-glucoside, isorhamnetin 3-rutinoside.
El-Negoumy *et al.*, *Phytochemistry*, **25**, 2423 (1986)

F. cretica.
TRITERPENOIDS: chinovic acid, fagogenin, fagogenin A,
fagogenin B.
Hussein *et al.*, *Pakistan J. Sci. Ind. Res.*, **9**, 269 (1966)
Ali, Hameed, *ibid.*, **10**, 140 (1967)

F. indica.
SAPOGENIN: nahogenin.
Atta-ur-Rahman *et al.*, *Heterocycles*, **19**, 217 (1982)

F. taeckholmiana.
FLAVONOIDS: herbacetin, herbacetin 8-methyl ether
3,7-diglucoside, herbacetin 8-methyl ether 3-rutinoside,
herbacetin 8-methyl ether 3-rutinoside-7-glucoside,
isorhamnetin 3-glucoside, isorhamnetin 3-rutinoside.
El-Negoumy *et al.*, *Phytochemistry*, **25**, 2423 (1986)

FAGOPYRUM (Polygonaceae)
F. esculentum Moench.
PIPERIDINE ALKALOID: 3,4-dihydroxy-2-piperidinomethanol.
PIGMENT: fagopyrin.
(A) Koyama, Sakamura, *Agric. Biol. Chem.*, **38**, 1111 (1974)
(P) Brockmann *et al.*, *Tetrahedron Lett.*, 1575 (1979)

FAGUS (Fagaceae)
F. crenata.
HYDROCARBONS: ditriacontane, heptacosane, hexacosane,
monotriacontane, nonacosane, octacosane, pentacosane,
triacontane, tritriacontane.
STEROID: β-sitosterol.
TRITERPENOIDS: friedelanol, epi-friedelanol, friedelin, glutinol.
Tachi *et al.*, *J. Pharm. Soc., Japan*, **98**, 349 (1978)

F. japonica.
HYDROCARBONS: ditriacontane, heptacosane, hexacosane, monotriacontane, nonacosane, octacosane, pentacosane, triacontane, tritriacontane.

STEROID: β-sitosterol.

TRITERPENOIDS: friedelanol, epi-friedelanol, friedelin, glutinol.

Tachi et al., J. Pharm. Soc., Japan, **98**, 349 (1978)

FALCARIA
F. vulgaris.
ACETYLENIC: 1,9-heptadecadiene-4,6-diyn-3-one.

COUMARINS: scopoletin, umbelliferone.

PHENOLICS: caffeic acid, chlorogenic acid, p-coumaric acid, ferulic acid, p-hydroxybenzoic acid, p-hydroxyphenylacetic acid, protocatechuic acid, salicylic acid, syringic acid, vanillic acid.

MONOTERPENOIDS: anethole, bisabolene, borneol, camphene, carvene, 1,8-cineole, p-cymene, fenchol, fenchone, isoborneol, isobornyl acetate, limonene, α-phellandrene, α-pinene, β-pinene.

(Acet) Bohlmann et al., Chem. Ber., **94**, 958 (1961)

(A, C, T) Gudej et al., Acta Pol. Pharm., 34, **299**, 305 (1977); Chem. Abstr., 88 60130u

FALKENBERGIA
(Rhodophyceae/Bonnemaisoniaceae)
F. rufolansa (Harv.) Smith.
STEROID: cholesta-5,25-diene-3β,24-diol.

Francisco et al., Steroids, **34**, 163 (1979)

FARADAYA (Verbenaceae)
F. splendida F.Muell.
TRITERPENOID: faradoic acid.

Eade et al., Aust. J. Chem., **27**, 2289 (1974)

FARAMEA
F. cyanea.
NAPHTHOPYRANS: faramol, 7-methoxyfaramol.

Ferrari et al., Phytochemistry, **24**, 2753 (1985)

FARFUGIUM (Compositae)
F. hiberniflorum (Makino) Kitam.
ACIDS: linoleic acid, linolenic acid, palmitic acid.

STEROID: β-sitosterol.

SESQUITERPENOIDS: 6β-acetoxyfuranoeremophilan-3α-yl angelate, 3-acetyl-6-angeloyloxyfuranofukinol, 6β-hydroxyfuranoeremophilan-3α-ol.

(Ac, S) Ito et al., J. Pharm. Soc., Japan, **97**, 1374 (1977)

(T) Nagano, Takahashi, Bull. Chem. Soc., Japan, **45**, 1935 (1972)

F. japonicum Kitamura.
DIMERIC SESQUITERPENOID: 12,12′-bi(3-angeloyloxyfuranoeremophila-7,11-diene).

Sesquiterpenoids 3β-angeloyloxy-8β,10β-dihydroxyeremophilenolide, 3β-angeloyloxyfuranoeremophilane, 3β-angeloyloxy-9β-hydroxyfuranoeremophilane, 3β-angeloyloxy-10β-hydroxyfuranoeremophilane, 3β-angeloyloxy-10β-hydroxy-9β-senecioyloxyfuranoeremophilane, 3β-angeloyloxy-9β-senecioloyloxyfuranoeremophilane, farfugin A, farfugin B, 8β-hydroxyeremophilenolide. 6β-senecioyloxyfuranoeremophilane

Kurihara et al., J. Pharm. Soc., Japan, **101**, 35 (1981)

F. japonicum Kitamura var. formosanum.
SESQUITERPENOIDS: farformolide A, farformolide B.

Ito et al., J. Pharm. Soc., Japan, **98**, 1592 (1978)

F. japonicum Kitamura var. luchnense.
SESQUITERPENOID: 8β-hydroxyeremophilenolide.

Ito et al., J. Pharm. Soc., Japan, **100**, 69 (1980)

FATSIA (Araliaceae)
F. japonica.
SAPONINS: fatsiaside A, fatsiaside B, fatsiaside C, fatsiaside D, fatsiaside E, fatsiaside F, fatsiaside G, α-fatsin, β-fatsin, β-2-fatsin, β-3-fatsin, β-4-fatsin, 3-O-[β-glucopyranosyl-β-L-arabinopyranosyl]-oleanic acid, 3-O-[β-glucopyranosyl-β-D-glucopyranosyl]-hederagenin.

Tanamura, Takamura, J. Pharm. Soc., Japan, **95**, 1 (1975)

Aoki et al., Phytochemistry, **15**, 781 (1976)

Gabadadze et al., Izv. Akad. Nauk Gruz SSR, **3**, 243 (1977); Chem. Abstr., 88 148969k

FELICIA
F. uliginosa.
ESTER: tetradeca-5,7,9,11-tetraenoic acid methyl ester.

Bohlmann, Zdero, Phytochemistry, **15**, 1318 (1976)

F. wrightii.
ISOCOUMARINS: eugenol isovalerate, 3-propylisocoumarin.

Bohlmann, Zdero, Phytochemistry, **15**, 1318 (1976)

FERETIA
F. apodanthera.
IRIDOIDS: 10-ethylapodanthoside, apodanthoside, 11-methylixoside, ixoside.

Bailleul et al., Planta Med., **37**, 316 (1979)

FERREIRA
F. spectabilis.
PIGMENT: chrysarobin.

Grundon et al., J. Chem. Soc., 4580 (1952)

FERREYANTHUS
F. verbascifolium.
SESQUITERPENOID: ferreyanthuslactone.

Bohlmann et al., Phytochemistry, **16**, 285 (1977)

FERULA (Umbelliferae)
F. aitchinsonii.
COUMARINS: karatavic acid, karatavikinol, tavicone, umbelliprenin.

Veselovskaya et al., Khim. Prir. Soedin., 397 (1982)

F. akitschkensis.
SESQUITERPENOIDS: akiferidin, akiferidinin, akiferinin, akitschenin, akitschenol.

Kushmuradov et al., Khim. Prir. Soedin., **137**, 725 (1978)

F. alliacea.
BENZOPYRAN: phellopterin.

Chatterjee et al., Arch. Pharm. (Weinheim, Ger.), **295**, 248 (1962)

F. asafoetida.
COUMARIN SESQUITERPENOID: kamolonol.

Hafer et al., Monatsh. Chem., **115**, 1207 (1984)

F. badkhysi.
SESQUITERPENOIDS: badkhysidin, badkhysinin.

Serkerov et al., Khim. Prir. Soedin., 176 (1972)

F. ceratophylla.
SESQUITERPENOIDS: fecerol, fecerin, ferocin, ferolinin.

Golivina, Saidkhodzhaev, Khim. Prir. Soedin., 796 (1977)

F. communis.
COUMARIN: ferulenol.
SESQUITERPENOID: α-ferulene.
Carboni *et al., Tetrahedron Lett.,* 3017 (1965)

F. communis subsp. communis.
ESTER: fercomin.
LACTONE: fercolide.
Miski, Mabry, *Phytochemistry,* **25,** 1673 (1986)

F. communis var. genuina.
COUMARINS: ferulenol, hydroxyferulenol.
Lamnaouer *et al., Phytochemistry,* **26,** 1613 (1987)

F. conocaula.
COUMARIN SESQUITERPENOIDS: cauferinin, conferdione,
   conferin, conferol, conferone, dihydroconferin, ferocolidin,
   ferocolin, ferocolinin, kauferidin, kauferin.
Vandychev *et al., Khim. Prir. Soedin.,* 658 (1974)
Kuliev *et al., ibid.,* 151 (1979)

F. diversivittata.
COUMARINS: diversine, diversinine, diversoside, umbelliferone.
COUMARIN SESQUITERPENOIDS: pheselol, pheselol acetate.
SESQUITERPENOID: diversolide.
STEROID: β-sitosterol.
Nabiev *et al., Khim. Prir. Soedin.,* 517 (1978)

F. dshaudshamyr.
KETONE: dshamirone.
Kamilov, Nikonov, *Khim. Prir. Soedin.,* 817 (1976)

F. fedtschenkoana.
SESQUITERPENOID: federin.
Kamilov *et al., Khim. Prir. Soedin.,* 527 (1974)

F. feganensis.
COUMARINS: chimganidin, ferolin.
Kadyrov *et al., Khim. Prir. Soedin.,* 704 (1977)

F. foetidissima.
COUMARINS: conferol, conferone.
COUMARIN SESQUITERPENOID: ferukrinone.
Kir'yanova, Skylar, *Khim. Prir. Soedin.,* 652 (1984)

F. foliosa.
COUMARINS: kamolol, kamolone.
COUMARIN SESQUITERPENOID: foliferin.
(C) Kadyrov *et al., Khim. Prir. Soedin.,* 704 (1977)
(T) Kadyrov *et al., ibid.,* **178,** 518 (1978)

F. galbaniflora.
SESQUITERPENOID: 10-epi-junenol.
Thomas *et al., Tetrahedron,* **32,** 2261 (1976)

F. gigantea.
SESQUITERPENOIDS: feguolide, giferolide.
Savina *et al., Khim. Prir. Soedin.,* 490 (1979)

F. grigorjevii.
DITERPENOID: grilactone.
Kir'yalova *et al., Khim. Prir. Soedin.,* 446 (1972)

F. iliensis.
MONOTERPENOIDS: bisabolene, β-calocorene, camphene,
   copaene, ar-curcumene, β-myrcene, cis-ocimene,
   trans-β-ocimene, α-pinene, β-pinene.
COUMARIN SESQUITERPENOID: ferilin.
SESQUITERPENOIDS: α-bisabolol, α-cadinol.
Veselovskya, Skylar, *Khim. Prir. Soedin.,* 387 (1984)
Dembitskii *et al., Izv. Akad. Nauk Kaz. SSR,* 63 (1985); *Chem. Abstr.,*
   103 19866d

F. involucrata.
ESTERS: involucrin, involucrinin, ugaferin, ugaferinin.
Potapov *et al., Chem. Abstr.,* 87 2357s

F. jaeschkeana.
SESQUITERPENOIDS: jaeschferin, jaeschkeanadiol.
Srivaman *et al., Tetrahedron,* **29,** 985 (1973)

F. juniperina.
SESQUITERPENOIDS: juniferdin, juniferin, juniferidin, juniferinin.
PHENOLICS: juniperin, juniperinin, juperin.
(P) Sagitdinova *et al., Khim. Prir. Soedin.,* 547 (1976)
(T) Sagitdinova, Saidkhodzhaev, *ibid.,* 790 (1977)

F. karatavica.
COUMARIN SESQUITERPENOID: karatavikinol.
SESQUITERPENOID: karatavin.
Bagirov, *Khim. Prir. Soedin.,* 655 (1978)

F. kelleri.
COUMARIN SESQUITERPENOIDS: ferukrin, kellerin.
Adrianova *et al., Khim. Prir. Soedin.,* 795 (1978)

F. kokanica.
COUMARIN SESQUITERPENOID: kokanidin.
Nabiev *et al., Khim. Prir. Soedin.,* 578 (1982)

F. kopetdagensis.
COUMARIN SESQUITERPENOIDS: acetylferukrin, fekolin,
   fekolone, kopedin, kopeoside, kopetdaghin, kopeolone.
Nabiev *et al., Khim. Prir. Soedin.,* 48 (1982)

F. korshinskyi.
COUMARIN SESQUITERPENOID: feroside.
SESQUITERPENOIDS: fekorin, fekorin A.
Kadyrov *et al., Khim. Prir. Soedin.,* 152 (1975)

F. krylovii.
COUMARIN SESQUITERPENOIDS: fekrol, ferukrin.
Veselovskaya *et al., Khim. Prir. Soedin.,* 851 (1979)

F. kumistanica.
SESQUITERPENOID: ferutidin.
Saidkhodzdaev, Nikonov, *Khim. Prir. Soedin.,* 525 (1974)

F. lancerottensis.
SESQUITERPENOIDS: epoxyjaeschkeanadiol p-hydroxybenzoate,
   jaeschkeanadiol p-hydroxybenzoate, lancerodiol, lancerodiol
   p-hydroxybenzoate, lancerodiol p-methoxybenzoate,
   lancerotriol, lancerotriol p-hydroxybenzoate.
Fraga *et al., Phytochemistry,* **24,** 501 (1985)

F. lapidosa.
SESQUITERPENOIDS: lapidolin, lapiferinin, lapidin.
Golovina *et al., Khim. Prir. Soedin.,* 318 (1981)

F. lehmanni.
UNCHARACTERIZED TERPENOIDS: lehmferidin, lehmferin.
Sagitdinova *et al., Khim. Prir. Soedin.,* 709 (1983)

F. linkiis.
SESQUITERPENOIDS: felikiol 3-angelate, ferutriol 5-isovalerate,
   lancerotol p-hydroxybenzoate, lancerotol
   p-methoxybenzoate, lancerotol veratrate, linkiol, webiol
   angelate, webiol epoxyangelate.
Diaz *et al., Phytochemistry,* **25,** 1161 (1986)

F. loscosii.
COUMARINS: coladin, coladonin, isovalerylcoladonin,
   umbelliprenin.
Pinar *et al., Phytochemistry,* **16,** 1987 (1977)

F. marmarica L.
COUMARIN SESQUITERPENOID: marmaricin.
Dawidar *et al., Chem. Pharm. Bull.,* **27,** 3153 (1979)

F. microcarpa.
COUMARIN SESQUITERPENOIDS: fecarpin, microferin,
   microferinin.
Golovina *et al., Khim. Prir. Soedin.,* 566 (1978)

F. microloba.
COUMARIN SESQUITERPENOIDS: galbanic acid methyl ester, microlobin.
Nabiev et al., Khim. Prir. Soedin., 781 (1983)

F. mogoltarica.
COUMARIN SESQUITERPENOIDS: mogoltadin, mogoltadone, mogoltavin, mogoltavinin, mogoltavitsin, mogoltin, mogolton.
Khasanov et al., Khim. Prir. Soedin., 10, 25 (1974)

F. moschata.
COUMARIN SESQUITERPENOID: conferol.
Vandyshev et al., Khim. Prir. Soedin., 670 (1972)

F. nevskii.
COUMARIN SESQUITERPENOID: nevskone.
Bagirov et al., Khim. Prir. Soedin., 652 (1978)

F. olgae.
IRIDOID: lamioside.
SESQUITERPENOIDS: laferin, laferinin, olgin, olgoferin, talasin A, talasin B.
Konovalova et al., Khim. Prir. Soedin., 590 (1975)

F. oopoda.
COUMARIN SESQUITERPENOIDS: feropolol, feropolidin, feropolin.
SESQUITERPENOIDS: badkhysidin, badkhysinin, 11,13-dihydroopodin, grillactone, feropodin, oopodin, semopodin.
Serkerov et al., Khim. Prir. Soedin., 392 (1976)
Malone et al., J. Chem. Soc., Perkin 2, 1683 (1980)

F. pallida.
SESQUITERPENOIDS: angrendiol, ferolin, palliferidin, palliferin, palliferinin, pallinin.
UNCHARACTERIZED TERPENOIDS: taufedin, taufedinin, tauferin, taulin.
Kushmuradov et al., Khim. Prir. Soedin., 53 (1986)

F. penninerva.
COUMARIN SESQUITERPENOIDS: kamolol, kamolone.
SESQUITERPENOID: ferolide.
Nurukhamedova et al., Khim. Prir. Soedin., 533 (1980)

F. persica.
COUMARIN: farnesiferol C.
Caglioti et al., Helv. Chim. Acta, 42, 2557 (1959)

F. polyantha.
COUMARIN SESQUITERPENOIDS: polyanthin, polyanthinin.
Khasanov et al., Khim. Prir. Soedin., 517 (1974)

F. pseudooreoselinum.
COUMARIN SESQUITERPENOIDS: fezelol, epi-fezelol, fezelone.
LACTONES: α-reolone, β-reolone.
SESQUITERPENOID: roseolin.
(C, T) Kir'yamlov, Buk'reeva, Khim. Prir. Soedin., 425 (1973)
(L) Bukreeva et al., ibid., 30 (1977)

F. rigidula.
KETONE: crocatone.
Sadykov et al., Khim. Prir. Soedin., 107 (1973)

F. rubroarenosa.
MONOTERPENOID: rubaferin.
SESQUITERPENOIDS: rubaferidin, rubaferinin.
Golovina et al., Khim. Prir. Soedin., 712 (1978)

F. samarcandia.
SESQUITERPENOID: mogolton.
Khasanov et al., Khim. Prir. Soedin., 617 (1973)

F. schschurowskiana.
COUMARIN SESQUITERPENOID: feshurin.
Kadyrov et al., Khim. Prir. Soedin., 228 (1979)

F. stylosa.
MONOTERPENOID: stylosidin.
Bagirov, Aliev, Khim. Prir. Soedin., 99 (1981)

F. syreitschikowii.
COUMARINS: imperatorin, oxypeucedanin, oxypeucedanin hydrate, pranchimgin, saxalin.
Bizhanova, Nikonov, Khim. Prir. Soedin., 278 (1977)

F. szovitsiana DC.
COUMARINS: aurapten, farnesiferol C, umbelliferone.
Caglioti et al., Helv. Chim. Acta, 42, 2557 (1959)
Gashimov et al., Chem. Abstr., 85 189214s

F. tadshikorum.
COUMARIN SESQUITERPENOIDS: tadshiferin, tadshikorin.
Perel'son et al., Khim. Prir. Soedin., 593 (1976)

F. tenuisecta.
MONOTERPENOID: feroferin.
SESQUITERPENOIDS: ferolin, ferolinicin, fertenidin, ferutin, ferutinin, fetinin, teferidin, teferin, tenuferidin, tenuferin, tenuferinin.
TERPENE ESTER: ferutinine.
Potapov et al., Izv. Akad. Nauk Kaz. SSR., 63 (1980)

F. tetterina.
COUMARIN SESQUITERPENOID: feterin.
Perel'son et al., Khim. Prir. Soedin., 318 (1978)

F. tingitana.
SESQUITERPENOIDS: acetyldesoxodihydrolaserpilin, acetyltingitanol, 14-p-anisoyloxydauca-4,8-diene, 4β-hydroxy-6α-p-hydroxybenzoyloxy-10α-angeloyloxydauc-8-ene, tingitanol.
Ulubelen et al., Tetrahedron, 40, 5197 (1984)
Miski, Mabry, J. Nat. Prod., 49, 657 (1986)

F. tschatcalensis.
SESQUITERPENOID: chatferin.
Sagitdinova et al., Khim. Prir. Soedin., 721 (1983)

F. tschimganica.
SESQUITERPENOIDS: angrendiol, ferolin.
Saidkhodzhaev et al., Khim. Prir. Soedin., 519 (1977)

F. ungamica.
KETONE: crocatone.
Serkerov et al., Khim. Prir. Soedin., 539 (1975)

F. xeromorpha.
COUMARINS: 3'-(α-methylbutyryloxy)-4'-acetoxydihydroseselin, xanthogallol.
SESQUITERPENOIDS: fecseridin, fecserinin, fecserin, xeroferin.
(C) Aminov et al., Khim. Prir. Soedin., 283 (1977)
(T) Bizhanova et al., ibid., 576 (1978)

FERULAGO
F. turcomanica.
COUMARINS: furocoumarin, isoimperatorin, isooxypeucedanin, osthole, oxypeucedanin.
Adrianova et al., Khim. Prir. Soedin., 514 (1975)
Serkerov et al., ibid., 94 (1976)

FESTUCA (Gramineae)
F. arundinacea Schreb.
PYRROLIZIDINE ALKALOID: festucine.
Yates, Tookey, Aust. J. Chem. (Weinheim, Ger.), 18, 53 (1965)

FIBRAURIA (Menispermaceae)

F. chloroleuca Miers.
PROTOBERBERINE ALKALOIDS: jatrorrhizine, palmatine.
DITERPENOIDS: fibleucin, fibraurin, 6-hydroxyfibraurin.
(A) Orekhov, *Arch. Pharm. (Weinheim, Ger.)*, **271**, 323 (1933)
(T) Ito, Furukawa, *J. Chem. Soc., Chem.Commun.*, 653 (1969)

F. tinctoria.
UNCHARACTERIZED ALKALOIDS: fibraminine, fibranine.
Chu *et al.*, *Hua Hsueh Hsueh Pao.*, **28**, 59 (1962)

FICARIA

F. verna Hudson.
TRITERPENOIDS: hederagenin, oleanolic acid.
Figurkin, Figurkina, *Rastit. Resur.*, **12**, 557 (1976); *Chem. Abstr.*, **86** 27654d

FICUS (Moraceae)

F. bengalensis.
AMINE ALKALOID: (R)-synephrine.
COUMARIN: bergapten.
LEUCOANTHOCYANINS: delphinidin 3-O-α-L-rhamnoside, leucocyanidin 3-O-β-D-galactosylcellobiose, pelargonidin 3-O-α-L-rhamnoside.
STEROID: β-sitosterol.
(A) Stewart *et al.*, *J. Biol. Chem.*, **239**, 930 (1964)
(C, S) Ahmad, *J. Indian Chem. Soc.*, **53**, 1165 (1976)
(L) Subramanian *et al.*, *Indian J. Chem.*, **15B**, 762 (1977)

F. cunninghamii.
COUMARINS: herniarin, isopimpinellin, marmesin, umbelliferone.
El-Khrisy *et al.*, *Fitoterapia*, **56**, 184 (1985)

F. eriobotryoides.
COUMARINS: bergapten, marmesin.
STEROID: β-sitosterol.
TRITERPENOID: β-amyrin.
El-Khrisy *et al.*, *Fitoterapia*, **56**, 184 (1985)

F. lacor.
AMINOACIDS: alanine, asparagine, threonine, tyrosine, valine.
Ali *et al.*, *J. Indian Chem. Soc.*, **64**, 230 (1987)

F. macrophyllus.
TRITERPENOID: moretenol
Galbraith *et al.*, *Aust. J. Chem.*, **18**, 226 (1965)

F. nitida.
TRITERPENOIDS: acetoxynitidiol, nitidiol.
Elgamel *et al.*, *Naturwissenschaften*, **62**, 486 (1975)

F. palmata.
COUMARIN: bergapten.
STEROID: β-sitosterol.
Ahmad *et al.*, *J. Indian Chem. Soc.*, **53**, 1165 (1976)

F. pantoniana King.
COUMARIN ALKALOIDS: ficine, isoficine.
Johns *et al.*, *Tetrahedron Lett.*, 1987 (1965)

F. racemosa.
TRITERPENOID: lupeol.
Agrawal, *Rocz. Chem.*, **51**, 1265 (1977)

F. religiosa.
AMINOACIDS: alanine, asparagine, threonine, tyrosine, valine.
Ali *et al.*, *J. Indian Chem. Soc.*, **64**, 230 (1987)

F. septica.
PHENANTHROINDOLIZIDINE ALKALOIDS: septicine, tylocrebine, tylophorine.
Russell, *Naturwissenschaften*, **50**, 443 (1963)

F. sycomorus.
COUMARINS: bergapten, marmesin, psoralen.
El-Khrisy *et al.*, *Fitoterapia*, **56**, 184 (1985)

FILIFOLIUM

F. sibiricum (L.) Kitam.
FLAVONOIDS: 3,6-dimethoxyquercetagenin, eriodictyol, filifolin.
Wang, *Yaoxue Xuebao*, **19**, 441 (1984); *Chem. Abstr.*, 103 3679j

FILIREA

F. media.
IRIDOIDS: deoxysyringoxide, hydroxynuezhenide, igustroside, nuezhenide, oleuropein, syringoxide.
Marekov *et al.*, *Khim. Ind., (Sofia)*, **58**, 132 (1986); *Chem. Abstr.*, 105 160854s

FIMBRISTYLIS (Cyperaceae)

F. dichotoma.
FURANOQUINONE: dihydrocyperoquinone.
Allen *et al.*, *Tetrahedron Lett.*, 4669 (1969)

FISSISTIGMA

F. oldhami.
APORPHINE ALKALOID: fisostigmine A.
Zu *et al.*, *Zhongcaoyao*, **14**, 148 (1983)

FITZROYA (Cupressaceae)

F. patagonica Hook.F.
FLAVONOIDS: isocryptomerin, podocarpusflavone.
Naqvi *et al.*, *Curr. Sci.*, **56**, 480 (1987)

FLACOURTIA (Flacourtiaceae)

F. jangomas.
SESQUITERPENOID: jangomolide.
Ahmad *et al.*, *Phytochemistry*, **23**, 1269 (1984)

FLAVERIA

F. bidentis.
FLAVONOID: isorhamnetin 3,7-disulphate.
Cabrera *et al.*, *Phytochemistry*, **16**, 400 (1977)

F. brownii.
FLAVONOID: 6-ethoxykaempferol 3-O-glucoside.
Al-Khubaizi *et al.*, *Phytochemistry*, **17**, 163 (1978)

F. chloraefolia.
FLAVONOIDS: patuletin 3-sulphate, quercetin 3-sulphate.
Barron *et al.*, *Phytochemistry*, **25**, 1719 (1986)

FLEISCHMANNIA (Compositae/Eupatoreae)

F. pycnocephala.
LIGNAN: 2-epi-fargesin.
Kakisawa *et al.*, *Phytochemistry*, **11**, 2289 (1972)

F. sinclairia.
DITERPENOIDS: 13,14H-kolaveninic acid, 13,14Z-kolaveninic acid.
Bohlmann *et al.*, *Phytochemistry*, **17**, 2101 (1978)

F. viscidipes.
Two caryophyllane terpenoids.
Bohlmann *et al.*, *Rev. Latinoam. Quim.*, **15**, 1 (1984)

FLEMINGIA (MOGHANIA) (Leguminosae)

F. chappar.
PTEROCARPINS: flemichapparin A, flemichapparin B.
Adityachaudhury, *J. Indian Chem. Soc.,* **47,** 1023 (1970)

F. congesta.
BENZOPYRANS: flemingins A, B, C, D, E and F.
FLAVONOID: myricitrin.
(B) Cardillo *et al., Phytochemistry,* **12,** 2027 (1973)
(F) Morita *et al., Chem. Abstr.,* **88** 148947b

F. macrophylla.
FLAVONOID: fleminone.
Rao *et al., Phytochemistry,* **22,** 2287 (1983)

F. rhodocarpa.
BENZOPYRANS: flemingins A, B, C, D, E and F.
Cardillo *et al., Phytochemistry,* **12,** 2027 (1973)

F. stricta.
BENZOPYRANS: flemistrictins A, B, C and D.
Rao, *Indian J. Chem.,* **14B,** 339 (1976)

F. wallichi.
PIGMENTS: flemichins A, B and C, flemiwallichin A.
Rao *et al., Indian J. Chem.,* **13,** 1000 (1975)

FLINDERSIA (Rutaceae)

F. australis.
QUINOLINE ALKALOID: flindersine.
Matthews, Schreiber, *Ber. Dtsch. Pharm. Ges.,* **24,** 385 (1914)

F. bourjotiana.
TRITERPENOIDS: bourjotinolones A, B and C, bourjotone.
Breen *et al., Aust. J. Chem.,* **19,** 455 (1966)

F. brassii.
LIGNANS: brassilignan, flinderbrassin.
Nimgirawath *et al., Aust. J. Chem.,* **30,** 451 (1977)

F. brayleana.
COUMARINS: brayleyanin, braylin.
Anet *et al., Aust. J. Sci. Res.,* **2A,** 608 (1949)

F. dissosperma.
FURANOQUINOLINE ALKALOID: maculine.
Brown *et al., Aust. J. Chem.,* **7,** 180 (1954)

F. fournieri Panch. et Seb.
DIMERIC INDOLE ALKALOIDS: borreverine,
15'-hydroxy-14',15'-dihydroborreverine, isoborreverine.
Tilliquin, Koch, *Phytochemistry,* **18,** 1559 (1979)
Tilliquin, Koch, ibid., 2066 (1979)

F. ifflaiana.
FURANOQUINOLINE ALKALOID: ifflaiamine.
TRITERPENOID: ifflaionic acid.
(A) Brown *et al., Aust. J. Chem.,* **16,** 480 (1963)
(T) Brown *et al., ibid.,* **16,** 491 (1963)

F. laevicarpa C.T.White et Francis.
ACID: salicylic acid.
CHROMENES: carpachromene, flindersiachromanone.
FURANOQUINOLINE ALKALOID: skimmianine.
BENZOPYRANS: flindercarpins 1, 2 and 3.
PENTANES: (-)-erythro-1,5-diphenylpentane-1,3-diol,
1,5-diphenylpentan-3-ol.
STEROIDS: hesperidin, β-sitosterol.
TRITERPENOID: lupeol.
Roy, *Indian J. Chem.,* **16B,** 463 (1970)
Picher *et al., Aust. J. Chem.,* **29,** 2023 (1976)

F. maculosa.
FURANOQUINOLINE ALKALOIDS: flindersiamine, maculine, maculosidine.
Anet *et al., Aust. J. Sci. Res.,* **5A,** 412 (1952)
Brown *et al., Aust. J. Chem.,* **7,** 180 (1954)

F. pimenteliana.
BENZOPYRAN: seselin.
Gupta *et al., J. Indian Chem. Soc.,* **51,** 904 (1974)

FLOURENSIA (Compositae)

F. campestris.
FLAVONOID: galetin 3,6-dimethyl ether.
Dillon *et al., Phytochemistry,* **16,** 1318 (1977)

F. cernua DC.
ACETOPHENONE: 4'-hydroxy-3'-isovalerylacetophenone.
CHROMENE:
6-acetyl-2,2-dimethyl-8-(3-methyl-2-butenyl)-2H-1-benzopyran.
SESQUITERPENOIDS: flourensadiol, flourensic acid.
(Acet, C) Bohlmann *et al., Chem. Ber.,* **108,** 26 (1975)
(T) Kingston *et al., Phytochemistry,* **14,** 2033 (1975)

F. hirsuta.
FLAVONOIDS: axillarin, jaceidin, kaempferol 3-methyl ether, pinocembrin.
Dillon *et al., Phytochemistry,* **16,** 1318 (1977)

F. ilicifolia.
FLAVONES: galetin 3,6-dimethyl ether, galetin 6-methyl ether, jaceidin, kaempferol 3-methyl ether, pinocembrin.
Dillon *et al., Phytochemistry,* **16,** 1318 (1977)

F. macrophylla.
ACIDS: 8β,9β-dihydroxycostic acid, 1β,9β-dihydroxycostic acid, 9α-hydroxy-4(15),11(13)-eudesmadien-13-oic acid, 9β-hydroxy-4(15),11(13)-eudesmadien-13-oic acid, 9-oxocostic acid.
Bowekar *et al., Tetrahedron,* **21,** 1521 (1965)

F. oolepis.
SESQUITERPENOID: ilicol.
Guerreiro *et al., Phytochemistry,* **18,** 1235 (1979)

F. resinosa.
FLAVONES: chrysin, 5,7-dihydroxyflavone.
TRITERPENOID: resinone.
(F) Asakawa *et al., Bull. Chem. Soc., Japan,* **44,** 2761 (1971)
(T) Rodriguez-Hahn, Rodriguez, *Rev. Latinoam. Quim.,* **3,** 148 (1973)

F. retinophylla.
FLAVONOIDS: kumatakenin, pinocembrin.
Dillon *et al., Phytochemistry,* **16,** 1318 (1977)

FLUEGGEA

F. microcarpa.
ALIPHATIC: hexacosane.
CARBOLINE ALKALOID: N(b)-methyltetrahydro-β-carboline.
FLAVONOIDS: bergenin, quercetin.
PHENOLIC: ellagic acid.
STEROIDS: β-sitosterol, β-sitosterol β-D-glucoside.
TRITERPENOIDS: friedelanol, friedelin.
Kumar *et al., Planta Med.,* 466 (1985)

FOENICULUM (Umbelliferae)

F. vulgare L.
COUMARINS: bergapten, umbelliferone.
Abyshev *et al., Azerb. Med. Zh.,* **53,** 34 (1976); *Chem. Abstr.,* **86** 2357w

FOKIENIA (Cupressaceae)
F. hodginsii.
SESQUITERPENOIDS: cryptomeriol, fokienol.

Vig et al., Indian J. Chem., 14B, 926 (1976)

FOMES
F. fomentarius.
BENZOTROPOLONES: anhydrodehydrofomentariol, anhydrofomentariol.

Favre-Bonvin et al., Phytochemistry, 16, 495, 1852 (1977)

FOMITOPSIS (Basidiomycetae)
F. insularis.
Two sesquiterpenoids.

Nozoe et al., Tetrahedron Lett., 3125 (1971)

FONTANESIA (Oleaceae)
F. phillyreoidea Labill.
PYRIDINE ALKALOID: fontaphilline.

Budziekiewicz et al., Chem. Ber., 100, 2798 (1967)

FORSYTHIA (Oleaceae)
F. japonica.
PHTHALIDE: 3-ethyl-7-hydroxyphthalide.

Kameoka et al., Phytochemistry, 14, 1676 (1975)

F. koreana.
LIGNANS: arctiin, phyllygenin, phyllyrin, (+)-pioresinol.
COUMARIN: rutin. Triterpenoids betulinic acid, oleanolic acid.

Nishibe et al., J. Pharm. Soc., Japan, 97, 1134 (1977)

F. suspensa.
COUMARIN: rutin.
LIGNANS: phyllygenin, phyllyrin, (+)-pioresinol.
TRITERPENOIDS: betulinic acid, oleanolic acid.

Nishibe et al., J. Pharm. Soc., Japan, 97, 1134 (1977)

F. viridissima.
IRIDOIDS: forsythide, forsythide methyl ester.

Inouye, Nishioka, Chem. Pharm. Bull., 21, 497 (1973)

FOUQUIERA (Fouquieriaceae)
F. splendens.
TRITERPENOIDS: fourquierol, isofourquierol, ocotillol II.

Butruille, Dominguez, Tetrahedron Lett., 639 (1974)

FRANSERA (Compositae)
F. ambrosioides Cav.
SESQUITERPENOID: franserin.

Romo et al., Can. J. Chem., 46, 1535 (1968)

FRASERA (Gentianaceae)
F. albicaulis Dougl. et Griseb.
XANTHONES: 1-hydroxy-2,3,5-trimethoxyxanthone, 2-hydroxy-1,3,4,7-tetramethoxyxanthone, 2-hydroxy-1,3,7-trimethoxyxanthone, 1,2,3,4,5-pentamethoxyxanthone, swerchirin, 1,2,3,5-tetrahydroxyxanthone, 1,2,3,7-tetramethoxyxanthone, 1,3,4,5-tetramethoxyxanthone, 1,3,4,7-tetramethoxyxanthone, 1,3,5-trimethoxyxanthone, 1,3,7-trimethoxyxanthone.

Stout et al., Tetrahedron, 25, 1947 (1969)
Stout et al., ibid., 25, 1961 (1969)
Ghosal et al., Phytochemistry, 15, 1041 (1976)

F. caroliniensis Walt.
XANTHONES: 1,3-dihydroxy-4,5-dimethoxyxanthone, 1-hydroxy-2,3,4,5-tetramethoxyxanthone, 1-hydroxy-2,3,5-trimethoxyxanthone, 1-hydroxy-2,3,7-trimethoxyxanthone, 1,2,3,5-tetrahydroxyxanthone, 1,2,3,7-tetrahydroxyxanthone.

Stout et al., Tetrahedron, 25, 1947 (1969)
Stout et al., ibid., 25, 1961 (1969)

FRAXINUS (Oleaceae)
F. excelsior L.
COUMARINS: fraxidin 8-O-β-D-glucoside, fraxin 7-methyl ether.
IRIDOIDS: deoxysyringoxide, 10-hydroxyligustroside, hydroxynuezhenide, ligustroside, nuezhenide, oleuropein, syringoxide.

(C) de Almeida et al., Phytochemistry, 13, 1225 (1974)
(C) Jensen et al., Phytochemistry, 15 221 (1976)
(I) Marekov et al., Khim. Ind. (Sofia), 58, 132 (1986); Chem. Abstr., 105 168854s

F. griffithii.
GLUCOSIDE: syringin.

Rao et al., Lloydia, 41, 56 (1978)

F. malacophylla Hems.
QUINOLOZIDINE ALKALOID: sinine.

Tonkin, Work, Nature, 156, 630 (1945)

F. ornus.
COUMARIN: fraxetin.

Wessely et al., Ber, 62, 120 (1929)

FRITILLARIA (Liliaceae)
F. camtschatcensis (L.) Ker.
STEROIDAL ALKALOIDS: camtschatcanidine, hapepunine.

Kaneko et al., Phytochemistry, 20, 327 (1981)

F. collicola.
UNCHARACTERIZED ALKALOID: peimunine.

Li, J. Chin. Pharm. Assoc., 2, 235 (1940)

F. delavayi.
STEROIDAL ALKALOID: chuanbeinone.

Kaneko et al., Tetrahedron Lett., 2387 (1986)

F. harenlinii.
STEROIDAL ALKALOIDS: harepermine, hareperminiside.

Min et al., Phytochemistry, 25, 2008 (1986)

F. hupehensis (Hsaio et K.C. Hsia).
STEROIDAL ALKALOIDS: hepehenine, hupehenisine.

Wu et al., Yaoxue Xuebao, 21, 546 (1986); Chem. Abstr., 106 15699r

F. imperialis L.
STEROIDAL ALKALOIDS: cevacine, cevanine, imperialine, isobaimonidine.

Suri, Ram, Indian J. Chem., 7, 1057 (1969)
Masterova et al., Arch. Pharm. (Weinheim, Ger.), 315, 157 (1982)

F. raddeana.
UNCHARACTERIZED ALKALOIDS: alvanidine, alvanine, raddeanine.

Aslanov, Sadykov, J. Gen. Chem. USSR, 26, 579 (1956)

F. roylei Hook.
STEROIDAL ALKALOIDS: fritimine, peimidine, peimine, peiminine, peimiphine, peimisine, peimitidine, peimitine.

Chou, Chen, Chin. J. Physiol., 7, 41 (1933)
Chou, J. Am. Pharm. Assoc., 36, 215 (1947)

F. sewerzowii (Korolkowia sewerzowii).
UNCHARACTERIZED ALKALOID: alginine.

Yunusov et al., J. Gen. Chem. USSR, 9, 1911 (1939)

F. thunbergii Miq.

UNCHARACTERIZED ALKALOID: peimunine.

DITERPENOIDS: ent-(16)-atisan-13,17-oxide,
ent-15,16-epoxykauran-17-ol,
ent-16-hydroxykaura-17-yl-ent-15,16-epoxykauran-17-ol,
ent-16-hydroxykaura-17-7l-ent-laur-15-en-17-oate,
isopimaracran-19-ol.
    (A) Li, *J. Chin. Pharm. Assoc.,* **2,** 235 (1940)
    (T) Kitajima *et al., Chem. Pharm. Bull.,* **30,** 3922 (1982)

F. ussuriensis Maxim.

STEROIDAL ALKALOIDS: pingpeimines A and B,
sipeimine-3-β-D-glucoside.
    Xu, *Zhongcaoyao,* **14,** 55 (1983)

F. verticillata. Willd.

STEROIDAL ALKALOID: fritillarizine.
    Kaneko *et al., Chem. Pharm. Bull.,* **28,** 3711 (1980)

F. verticillata Willd. var. thunbergii Bak.

STEROIDAL ALKALOIDS: fritillarine, peimunine, verticine,
verticinone, vertine.
UNCHARACTERIZED ALKALOID: verticilline.
    Chou, Chen, *Chin. J. Physiol.,* **6,** 265 (1932)
    Chi *et al., J. Am. Chem. Soc.,* **62,** 1778 (1944)

F. walujewii.

STEROIDAL ALKALOIDS: imperialine, sinpeinine A.
    Liu *et al., Yaoxuye Xuebao,* **19,** 894 (1985); *Chem. Abstr.,* 103 19851v

FRITSCHIELLA
(Chlorophyceae/Chaetophoraceae)

F. tuberosa.

CAROTENOID: fritschiellaxanthin.
    Buchecker *et al., Helv. Chim. Acta,* **61,** 1962 (1978)

FRUCTUS

F. momordicae.

SAPONINS: mogrosides I, II, III and IV.
TERPENOID: mogrol.
    Takemoto *et al., J. Pharm. Soc., Japan,* **103,** 1155, (1983)
    Takemoto *et al., ibid.,* **103,** 1167 (1983)

FRULLANIA (Hepaticae)

F. brittoniae subsp. truncetifolia.

STILBENES: brittonin A, brittonin B.
    Asakawa *et al., Phytochemistry,* **15,** 1057 (1976)

F. brotheri Stapf.

NORSESQUITERPENOID: (+)-brothenolide.
SESQUITERPENOID: (+)-α-frullanolide.
    Takeda *et al., Bull. Chem. Soc., Japan,* **56,** 1120 (1983)

F. dilatata (L.) Dum.

FLAVONOIDS: 6-hydroxyluteolin 7,4'-diglucoside,
6-hydroxyluteolin 7-glucoside, luteolin, luteolin
7-diglucoside, luteolin 7,4'-diglucoside, luteolin 7-glucoside,
pedalitin.
SESQUITERPENOIDS: cis-α-cyclocostumolide,
dihydroeremofrullanolide, eremofrullanolide,
(+)-frullanolide, (-)-frullanolide, oxofrullanolide.
    (F) Mues *et al., Cryptogram.: Bryol. Lichenol.,* **4,** 111 (1983); *Chem. Abstr.,* 100 48524j
    (T) Asakawa *et al., Bull. Soc. Chim. Fr.,* 1465 (1977)

F. tamarisci (L.) Dum.

FLAVONOIDS: acylatedluteolin 7-diglucoside, acylatedluteolin
7-glucoside, luteolin.
SESQUITERPENOIDS: (+)-frullanolide, tamariscol.
    (F) Mues *et al., Cryptogram.: Bryol. Lichenol.,* **4,** 111 (1983); *Chem. Abstr.,* 100 48524j
    (T) Connolly *et al., Tetrahedron Lett.,* 1401 (1984)

F. tamarisci (L.) Dum. var. obscura.

SESQUITERPENOID: cis-α-cyclocostunolide.
    Asakawa *et al., Tetrahedron Lett.,* 3957 (1975)

FUCHSIA (Onagraceae)

F. boliviana var. luxurians.

FLAVONOIDS: kaempferol 3-galactoside, quercetin
3-galactoside.
    Williams *et al., Phytochemistry,* **22,** 1953 (1983)

F. excorticata.

FLAVONOIDS: apigenin 7-glucuronide, apigenin
7-glucuronidesulphate, diosmetin 7-glucuronide, eriodictyol
7-glucoside, luteolin 7-glucuronide, luteolin
7-glucuronidesulphate.
    Williams *et al., Phytochemistry,* **22,** 1953 (1983)

F. fulgens.

FLAVONOIDS: quercetin 3-glucuronide, quercetin 3-rhamnoside,
quercetin 3-rutinoside.
    Williams *et al., Phytochemistry,* **22,** 1953 (1983)

F. procumbens.

FLAVONOIDS: apigenin 7-glucuronide, apigenin
7-glucuronidesulphate, luteolin 7-glucuronide, luteolin
7-glucuronidesulphate.
    Williams *et al., Phytochemistry,* **22,** 1953 (1983)

F. splendens.

FLAVONOID: luteolin 7-glucoside.
    Williams *et al., Phytochemistry,* **22,** 1953 (1983)

F. triphylla.

PHENOLIC: galloylglucoside.
    Williams *et al., Phytochemistry,* **22,** 1953 (1983)

FUCUS (Phaeophyceae/Fucaceae)

F. vesiculosus (L.).

BIPHENYL:
4-(2'',4'',6''-trihydroxyphenyl)-4'',6''-dihydroxyphenoxy-
2,2',4',6,6'-pentahydroxybiphenyl,
4-(2'',4'',6''-trihydroxyphenoxy)-2,2',4',6,6'-
pentahydroxybiphenyl.
    Craigie *et al., Can. J. Chem.,* **55,** 1575 (1977)

FULGENSIA

F. fulgida (Nyl.) Szat.

ANTHRAQUINONE: fragilin.
DEPSIDONES: fulgidin, fulgoicin.
    Mahandru, Gilbert, *J. Chem. Soc., Perkin I,* 2249 (1983)

FUMARIA (Fumariaceae/Papaveraceae)

F. indica.

AMINE ALKALOIDS: narceimine, narlumidine.
NAPHTHOISOQUINOLINE ALKALOID:
(-)-8-methoxydihydrosanguinarine.
    Seth *et al., Chem. Ind. (London),* 744 (1979)
    Pandey *et al., Phytochemistry,* **18,** 695 (1979)

F. judaica.

UNCHARACTERIZED ALKALOID: fumjudaine.
    Saleh, Gabr, *J. Pharm. Sci. Arab Repub.,* **6,** 61 (1965)

F. kralikii.

SPIROISOQUINOLINE ALKALOIDS: fumaritine N-oxide,
norfumaritine.
    Colton *et al., J. Nat. Prod.,* **48,** 846 (1985)

F. micrantha Lag.

OXOQUINOLIZIDINE ALKALOIDS: fumaramine, protopine.
    Platonova *et al., J. Gen. Chem. USSR,* **26,** 181 (1956)

F. microcarpa.
HOMOISOQUINOLINE ALKALOID: fumarofine.
> Blasko *et al., Tetrahedron Lett.,* 3175 (1981)

F. officinalis L.
OXOPROTOBERBERINE ALKALOIDS: cryptocavine, cryptopine.
OXOQUINOLIZIDINE ALKALOID: protopine.
PROTOBERBERINE ALKALOIDS: N-methylsinactine, (-)-sinactine.
SPIROISOQUINOLINE ALKALOIDS: dihydrofumariline, fumaricine, fumariline, fumarophycine.
UNCHARACTERIZED ALKALOIDS: aurotensine, fumaritine.
> Mollov *et al., C. R. Acad. Bulg. Sci.,* **20,** 557 (1967)
> Madirossica *et al., Phytochemistry,* **22,** 759 (1983)

F. parviflora Lam.
AMINE ALKALOIDS: fumariflorine, fumariflorine ethyl ester.
HOMOISOQUINOLINE ALKALOIDS: lahoramine, lahorine.
ISOQUINOLINE LACTONE ALKALOIDS: (-)-adlumine, (+)-bicuculline, hydrastine.
OXOPROTOBERBERINE ALKALOIDS: cryptopine, izmirine.
OXOQUINOLIZIDINE ALKALOIDS: fumaramine, fumaridine, protopine, Spiroisoquinoline alkaloids parfumidine, parfumine, parviflorine.
> Israilov *et al., Khim. Prir. Soedin.,* 588 (1970)
> Guinaudeau *et al., J. Nat. Prod.,* **46,** 934 (1983)

F. schleicheri Soy-Will.
OXOQUINOLIZIDINE ALKALOIDS: fumaramine, fumaridine, protopine.
SPIROISOQUINOLINE ALKALOID: fumaritrine.
TERTIARY AMINE: (+)-fumschleicherine.
UNCHARACTERIZED ALKALOIDS: fumarinine, fumaritine.
> Israilov *et al., Khim. Prir. Soedin.,* 588 (1970)
> Kir'yakov *et al., Phytochemistry,* **19,** 2507 (1980)

F. schrammii.
AMINE ALKALOIDS: bucucullinidine, (+)-fumschleicherine.
> Kir'yakov *et al., Phytochemistry,* **20,** 1721 (1981)

F. vaillantii Loisl.
ISOQUINOLINE ALKALOIDS: egerine, norjuziphine.
ISOQUINOLINE LACTONE ALKALOIDS: adlumine, bicuculline, hydrastine.
OXOQUINOLIZIDINE ALKALOIDS: fumaramine, fumaridine, fumvailine, protopine, vaillantine.
> Alimova *et al., Khim. Prir. Soedin.,* 874 (1979)
> Gozler *et al., Tetrahedron,* **39,** 577 (1983)

FUNARIA
F. hygrometrica.
PIGMENT: bracteatin.
> Wetz *et al., Phytochemistry,* **16,** 1108 (1977)

FUNKIA (Liliaceae)
F. ovata.
SAPONINS: funkiosides A, B, C, D, E, F, G and L.
> Kintya *et al., Khim. Prir. Soedin.,* 69 (1977)

FUNTUMIA (Apocynaceae)
F. africana.
STEROIDAL ALKALOIDS: funtumufrine B, funtuphyllamines A, B and C.
> Janot *et al., C. R. Acad. Sci.,* **250,** 2445 (1960)

F. elastica (Preuss) Stapf.
STEROIDAL ALKALOIDS: irehamine, irehidiamine A, irehidiamine B, irehine.
STEROID: cyclofuntumenol.
> (A) Truong-Ho *et al., Bull. Soc. Chim. Fr.,* **594,** 2332 (1963)
> (S) Mukam *et al., Tetrahedron Lett.,* 2779 (1973)

F. latifolia Stapf.
STEROIDAL ALKALOIDS: funtadienine, funtessine, funtuline, funtumine, funtumufrine C, latifoline, latifolinine.
STEROIDS: 5,24-cholestadien-3-ol, 4α-methyl-5α-cholesta-7,24-dien-3β-ol, 29-norlanosterol.
> (A) Janot *et al., Bull. Soc. Chim. Fr.,* 2169 (1964)
> (S) Charles *et al., C. R. Acad. Sci.,* **268,** 2105 (1968)

FURCELLARIA (Rhodophyceae)
F. fastigiata.
STEROIDS: cholesterol, 22-trans-dehydrocholesterol, desmosterol, fucosterol, isofucosterol, 24-methylenecholesterol β-sitosterol.
> Chandler *et al., Can. J. Pharm. Sci.,* **12,** 92 (1977); *Chem. Abstr.,* **88** 133242u

FURCRAEA
F. elegans Tod.
STEROIDS: hecogenin, tigogenin.
> Dixit *et al., Indian J. Chem.,* **15B,** 582 (1977)

F. selloa.
STEROID: furcogenin.
> Marker *et al., J. Am. Chem. Soc.,* **65,** 1199 (1943)

FUSARIUM
F. culmorium.
SESQUITERPENOID: cyclonerotriol.
> Hanson *et al., J. Chem. Soc., Perkin I,* 1586 (1975)

F. diversispermum.
SESQUITERPENOIDS: diacetoxyscirpenol, diacetoxyscirpenone.
> Flury *et al., J. Chem. Soc., Chem. Commun.,* **26,** 27 (1965)

F. moniliforme.
AZA-ANTHRAQUINONE ALKALOID: 8-O-methylbostrycoidine.
NORDITERPENOID: gibberellenic acid.
> (A) Steyn *et al., Tetrahedron,* **35,** 1551 (1979)
> (T) Gerzon *et al., Experientia,* **13,** 487 (1957)

F. nivale.
SESQUITERPENOIDS: nivalenol, novalenol 4,15-diacetate.
> Tatsumo *et al., Chem. Pharm. Bull.,* **16,** 2519 (1968)

F. roseum.
SESQUITERPENOID: 4-acetoxyscirpendiol.
> Ishii *et al., J. Agric. Food Chem.,* **26,** 649 (1978)

F. sambusinum.
SESQUITERPENOIDS: diacetoxyscirpenol, diacetoxyscirpenone.
> Flury *et al., J. Chem. Soc., Chem. Commun.,* **26,** 27 (1965)

F. tricinctum.
SESQUITERPENOID: 4,8-diacetoxy-12,13-epoxytrichothec-9-ene-2,15-diol.
> Ilus *et al., Phytochemistry,* **16,** 1839 (1977)

FUSCUS
F. evanescens.
STEROIDS: stigmasta-5,24(25)-diene-3β,7α-diol, stigmasta-5,24(25)-diene-3β,7β-diol, stigmasta-3,5,24(28)-trien-7-one.
> Itakawa *et al., Phytochemistry,* **11,** 2317 (1972)

F. vesiculosus.
ETHYLENIC: 1,6,9,12,15,18-henicosahexaene.
CAROTENOIDS: fucoxanthin, neoxanthin, zeaxanthin.
> (C) Galasko *et al., J. Chem. Soc., C,* 1264 (1969)
> (E) Halsall *et al., J. Chem. Soc., Chem. Commun.,* 448 (1971)

FUSIDIUM (Polyporaceae)

F. coccineum K. Tubaki.

NORTRITERPENOID: 9,11-dehydro-12-hydroxyfusidic acid.

Godtfredson *et al.*, *Tetrahedron,* **35,** 2419 (1979)

G

GABNERIA

G. fulgens.
PIGMENT: gesnerin.

Crowden et al., *Phytochemistry*, **13**, 1947 (1974)

GABUNIA (Apocynaceae)

G. eglandulosa Stapf.
HOMOCARBOLINE ALKALOID: 19-hydroxycoronaridine.
INDOLE ALKALOIDS: eglandine, eglandulosine,
(-)-19-hydroxyisovoacangine.

Agwada et al., *Helv. Chim. Acta*, **58**, 1001 (1978)

G. odoratissima Stapf.
DIMERIC INDOLE ALKALOID: gabunine.
CARBOLINE ALKALOID: pericyclivine,

Cava et al., *Tetrahedron Lett.*, 931 (1965)

GAILLARDIA (Compositae)

G. amblyodon.
SESQUITERPENOIDS: amblyodin, amblyodiol,
deacetylgailladipinnatin.

Herz et al., *Phytochemistry*, **13**, 1187 (1974)
Herz et al., *ibid*, 1189 (1974)

G. aristata Pursh.
FLAVONOIDS: apigenin, apigenin
6-hexoside-7-glucoarabinoside-8-rhamnoside,
6-C-glucoarabinoglucosyl-5,3'4'-trihydroxy-7-
methoxyflavone, isoquercitrin, swertisin.
MONOTERPENOID: 5-hydroxymethyl-2-isopropyltenol
diisobutyrate.
SESQUITERPENOID: 3-epi-isotelekin.

(C) Zielinska-Stasiak, Gill, *Rocz. Chem.*, **51**, 921 (1977)
(T) Bohlmann et al., *Chem. Ber.*, **102**, 864 (1969)

G. arizonica.
SESQUITERPENOIDS: fastigilins A, B and C, gaillardilin.

Herz et al., *Tetrahedron*, **22**, 693 (1966)

G. fastigiata.
SESQUITERPENOIDS: bigelovin, fastigilins A, B and C.

Parker et al., *J. Org. Chem.*, **27**, 4127 (1962)
Herz et al., *Tetrahedron*, **22**, 1907 (1966)

G. jasminoides forma grandiflora.
IRIDOIDS: gardenoside, gardoside.

Inouye et al., *Phytochemistry*, **13**, 2219 (1974)

G. mexicana.
SESQUITERPENOID: neoleonine.

Dominguez et al., *Rev. Latinoam. Quim.*, **1**, 136 (1970)

G. pinnatifida.
SESQUITERPENOIDS: gaillardilin, gaillardpinnatin, mexicanin I.

Herz et al., *Tetrahedron*, **22**, 693 (1966)

G. pulchella Foug.
BENZOPYRAN: swertisin.
SESQUITERPENE ALKALOID: neopulchellidine, pulchellidine.
SESQUITERPENOIDS: gaillardin, isogaillardin, neogaillardin,
pulchellins A, B, C, D, E and F, pulchelloids A, B and C.

(A) Yamagita et al., *Tetrahedron Lett.*, 3007 (1970)
(B) Wagner et al., *Phytochemistry*, **11**, 851 (1972)
(B) Wagner et al., *ibid.*, **11**, 1857 (1972)
(T) Inayama et al., *Heterocycles*, **17**, 219 (1982)

G. pulchella Foug. var. japonica.
SESQUITERPENOID: neogaillardin.

Inayama et al., *Phytochem.*, **12**, 1741 (1973)

G. spathulata.
SESQUITERPENOID: spathulin.

Herz et al., *J. Org. Chem.*, **32**, 1042 (1967)

GALANTHUS (Amaryllidaceae)

G. caucasicus (Bak.) Grossh.
BENZINDOLE ALKALOIDS: demethylhomolycorine, galanthamine,
galanthine, galanthinine, galanthusine, lycorine, tazettine.
FLAVONOIDS: hyperoside, rutin.

(A) Taskadze et al., *Khim. Prir. Soedin.*, 773 (1970)
(F) Taskadze et al., *ibid*, 116 (1977)

G. elwesii Hook.
BENZINDOLE ALKALOID: galanthamine.
BENZISOQUINOLINE ALKALOIDS: elwesine, epi-elwesine.

Boit, Dopke, *Naturwissenshaften*, **48**, 406 (1961)

G. nivalis.
BENZIOSQUINOLINE ALKALOID: lycorine.
DIPHENYL ALKALOIDS: galanthamine, narvedine, nivalidine,
tazettine.
METHYLINDOLE ALKALOID: hippeastrine.

Kalashnikov et al., *Khim. Prir. Soedin.*, 380 (1970)

G. nivalis var. gracilis.
DIBENZOAZAOCTANE ALKALOID: nivalidine.

Kalashnikov et al., *Khim. Prir. Soedin.*, 380 (1970)

G. woronowii Losinsk.
DIPHENYL ALKALOIDS: galanthamidine, galanthamine,
galanthidine, galanthine.
BENZISOQUINOLINE ALKALOID: lycorine.

Barton, Kirby, *J. Chem. Soc.*, 806 (1962)
Yakovleva et al., *J. Gen. Chem.*, USSR, **33**, 1691 (1963)

GALEGA (Leguminosae)

G. officinalis L.
GUANIDINE ALKALOID: galegine.
QUINAZOLINE ALKALOID: (±)-peganine.

Tanret, *Acad.Sci.*, **158**, 1182 (1914)
Tanret, *Acad.Sci.*, **158**, 1182 (1914)

GALEOPSIS (Labiatae)

G. agallocha.
SESQUITERPENOID: pregaleopsin.

Rodriguez, Savona, *Phytochemistry*, **19**, 1805 (1980)

G. angustifolia Hoffm.
SESQUITERPENOIDS: galeolone, galeopsin, galeopsinolone,
galeopsitirione, pregaleopsin.

G. pubescens.
IRIDOIDS: glucoside VII, gluroside.

Rodriguez, Savona, *Phytochemistry*, **19**, 1805 (1980)
Perez-Sirvent et al., *ibid*, **22**, 527 (1983)

G. reuteri.
SESQUITERPENOIDS: galeuterone, pregaleuterone.

Savona et al., *Phytochemistry*, **23**, 2958 (1984)

G. tetrahit L.
IRIDOIDS: galiridoside, gluroside.
Sticher, *Tetrahedron Lett.,* 3197 (1970)
Sticher, Weisflog, *Pharm. Acta Helv.,* **50,** 394 (1975)

GALIPEA
G. officinalis Hancock (Cusparia trifoliata Engl.).
QUINOLINE ALKALOIDS: cuspareine, cusparine, galipidine, galipoidine, galipoline.
Schlager, Leeb, *Monatsh. Chem.,* **81,** 714 (1950)

GALIUM (Rubiaceae)
G. album.
IRIDOID: secogalioside.
Bock *et al., Acta Chem. Scand.,* **30B,** 743 (1976)

G. asparine L.
IRIDOID: asperuloside.
Briggs *et al., J. Chem. Soc.,* 3940 (1954)
Briggs *et al., ibid.,* 4182 (1954)

G. glaucum.
IRIDOID: monotropein.
Bridel, *Acad. Sci.,* **176,** 1742 (1923)

G. mollugo L.
FURAN: furomollugin.
IRIDOIDS: galioside, gardenosidic acid, mollugoside.
(F) Schildknecht *et al., Justus Liebigs Ann. Chem.,* 1772 (1976)
(I) Iavarone *et al., Phytochemistry,* **22,** 175 (1983)

G. verum L.
IRIDOIDS: iridoids V-1, V-2 and V-3.
Boejthe-Horvath *et al., Phytochemistry,* **21,** 2917 (1982)

GANARIUM
G. schweinfurthii.
TRITERPENOIDS: elemadienolic acid, elemadienonic acid.
Ruzicka *et al., Helv. Chim. Acta,* **25,** 439 (1942)

GANODERMA (Polyporaceae)
G. applanatum.
STEROIDS: ergosta-7,22-dien-3β-ol, ergosta-7,22-dien-3-one, ergosterol peroxide.
Ripperger *et al., Phytochemistry,* **14,** 2297 (1975)
Sviridonov, Strigina, *Khim. Prir. Soedin.,* 669 (1976)

G. australe. (G. adspermum).
STEROIDS: 5,8-epidioxyergosta-6,22-dien-3β-ol,
5,8-epidioxyergosta-6,9(11),22-trien-3β0-ol,
ergosta-7,22-dien-3β-ol, ergosta-7,22-dien-3-one,
ergosta-4,6,8(14),22-tetraen-3-one, ergost-7-en-3β-ol.
Gonzalez *et al., An. Quim.,* **82C,** 149 (1986)

G. lucidum (Fr.) Karst.
TRITERPENOIDS:
3,7-dihydroxy-11,15,23-trioxo-8-lanosten-26-oic acid,
7,15-dihydroxy-3,11,23-trioxo-8-lanosten-26-oic acid,
ganoderic acids, A, B, C, D, E, F, T, V, W, X, Y and Z,
ganoderic acid C2, ganoderic acid E, ganoderic acid I,
ganoderic acid K, lucidenic acid F.
Hirotani, Furuya, *Phytochemistry,* **25,** 1189 (1986)
Kikuchi *et al., Chem. Pharm. Bull.,* **34,** 3695 (1986)

GARCINIA (Guttiferae)
G. atroviridis.
ACID: garcinic acid.
Lewis *et al., Phytochemistry,* **4,** 619 (1965)

G. conrauana.
Lactones and biflavonoids.
Hussain, Waterman, *Phytochemistry,* **21,** 1393 (1982)

G. cowa.
XANTHONE:
1,3,6-trihydroxy-7-methoxy-8-(3,7-dimethyl-2,6-octadienyl)-xanthone.
Lee *et al., Phytochemistry,* **16,** 2038 (1977)

G. echinocarpa.
FLAVONES: norelloflavone, volkensiflavone.
STEROID: β-sitosterol.
Bandaranayake *et al., Phytochemistry,* **14,** 1878 (1975)

G. eugeniifolia.
XANTHONES: 1,4,7-trihydroxy-3-methoxyxanthone,
1,6,7-trihydroxyxanthone.
Jackson *et al., J. Chem. Soc., C,* 2201 (1969)

G. homobroniana.
TERPENOID: bronianone.
Ollis *et al., J. Chem. Soc., Chem. Commun.,* 879 (1969)

G. indica.
ACID: garcinic acid.
FLAVONES: morelloflavone, volkensiflavone.
XANTHONE: 1,7-dihydroxyxanthone.
(Ac) Lewis *et al., Phytochemistry,* **4,** 619 (1965)
(F, X) Cotterill *et al., ibid.,* **16,** 148 (1977)

G. livingstonii.
FLAVONE: podocarpusflavone A.
Rizvi *et al., Phytochemistry,* **13,** 1990 (1974)

G. mangostana L.
ANTHOCYANINS: cyanidin 3-glucoside, cyanidin sophoroside.
FLAVONES: norelloflavone, volkensiflavone.
STEROID: β-sitosterol.
XANTHONE:
1,5-dihydroxy-3-(3-methyl-2-butenyl)-3-methoxyxanthone.
(An) Du *et al., J. Food Sci.,* **42,** 1667 (1977); *Chem. Abstr.* 87 180740u
(F,S) Bandaranayake *et al., Phytochemistry,* **14,** 1878 (1975)
(X) Sen *et al., Phytochemistry,* **20,** 183 (1981)

G. morella.
TERPENOIDS: α-gambogic acid, β-gambogic acid, isomorellin, morreollin, morellin.
Rao *et al., Proc. Indian Acad. Sci.,* **87A,** 75 (1978)

G. multiflora.
BENZOPYRAN: xanthochymusside.
Konoshima *et al., Tetrahedron Lett.,* 4203 (1970)

G. nervosa.
XANTHONES: isocowanin, isocowanol, nervosaxanthone.
Ampofo *et al., Phytochemistry,* **25,** 2351 (1986)

G. pedunculata.
XANTHONE: 1,3,5,7-tetrahydroxyxanthone.
Rao *et al., Phytochem,* **13,** 1241 (1974)

G. polyantha.
XANTHONES: isocowamin, isocowanol, nervosaxanthone.
Ampofo *et al., Phytochemistry,* **25,** 2351 (1986)

G. pyrifera.
XANTHONES: isocowanin, isocowanol, nervosaxanthone.
Ampofo *et al., Phytochemistry,* **25,** 2351 (1986)

G. spicata.
DIPYRAN: isochapelieric acid.
Gunatilaka *et al., Phytochemistry,* **23,** 323 (1984)

G. terpnophylla.
FLAVONES: norelloflavone, volkensiflavone.
STEROID: β-sitosterol.
Bandaranayake *et al., Phytochemistry,* **14,** 1878 (1975)

G. xanthochymus.

BENZOPYRAN: xanthochymusside.

SESQUITERPENOIDS: isoxanthochymol, xanthochymol.

    (B) Konoshima et al., Tetrahedron Lett., 4203 (1970)

    (T) Karanjgoaker et al., ibid., 4977 (1973)

GARDENIA (Rubiaceae)

G. gummifera.

TRITERPENOID: 19α-hydroxyerythrodiol.

    Reddy et al., Planta Med., 32, 206 (1977)

G. jasminoides.

GLUCOSIDE: crocin, picrocrocinic acid.

IRIDOIDS: genipin 1-β-D-glucoside, geniposide, scandoside methyl ether.

    (G) Takeda et al., Chem. Pharm. Bull., 24, 2644 (1976)

    (I) Inouye et al., Phytochemistry, 13, 2219 (1974)

G. jasminoides forma grandiflora.

IRIDOID: methyl deacetylasperulosate.

    Inouye et al., J. Pharm. Soc., Japan, 94, 577 (1974)

G. latifolia.

TRITERPENOIDS: 3-O-acetylspinosic acid methyl ester, 3-epi-siaresinolic acid,

    Reddy et al., Phytochemistry, 14, 307 (1975)

G. shimada.

Three uncharacterized alkaloids.

    Haginawa et al., J. Pharm. Soc., Japan, 90, 219 (1970)

GARDNERIA (Loganiaceae)

G. multiflora Mak.

HYDROINDOLE ALKALOIDS: demethylgarneramine, gardfloramine, gardmultine.

SPIROINDOLONE ALKALOID: chitosenine.

    Haginawa et al., J. Pharm. Soc., Japan, 90, 219 (1970)

    Sakai et al., Tetrahedron Lett., 715 (1975)

G. nutans Sieb. et Zucc.

HYDROINDOLE ALKALOIDS: gardmultine, gardneramine, gardnerine, gardnutine, hydroxygardnutine.

    Sakai et al., Tetrahedron Lett., 1485 (1969)

GARRYA (Garryaceae)

G. laurifolia Hartw.

DITERPENOID ALKALOIDS: cuachicicine, garryfoline.

    Djerassi et al., J. Am. Chem. Soc., 77, 4801 (1955)

    Djerassi et al., ibid., 77, 6633 (1955)

G. longifolia.

DITERPENE ALKALOID: garryfoline.

    Djerassi et al., J. Am. Chem. Soc., 77, 4801 (1955)

G. ovata.

DITERPENE ALKALOID: lingheimerine.

    Pelletier et al., Heterocycles, 9, 1409 (1978)

G. ovata var. lingheimeri.

DITERPENE ALKALOIDS: lingheimerine, ovatine.

    Pelletier et al., Heterocycles, 9, 1409 (1978)

G. veatchii.

DITERPENE ALKALOIDS: garryine, veatchine.

    Oneto, J. Am. Pharm. Assoc., 35, 204 (1946)

GARULEUM (Calendulaceae)

G. bipinnatifidum DC.

DITERPENOIDS: 14β-acetoxy-9,11-didehydroatisiren-20-oic acid, 6β-acetoxysandaracopimar-15-ene-8β,11α-diol, 11α-acetoxysandaracopimar-15-ene-8β,12β-diol, 6β-acetoxysandaracopimar-15-en-8β-ol, 12β-acetoxysandaracopimar-15-en-8β-ol, 12β-(p-hydroxycinnamoyloxy)-sandaracopimar-15-ene-8β,11α-diol, sandaracopimar-15-en-8β,11α-diol, sandaracopimar-15-en-6β,8β,11α-triol, sandaracopimar-15-en-8β,11α,12β-triol 12-p-hydroxycinnamate.

    Bohlmann, Grenz, Ber., 111, 1509 (1978)

G. fruticosum.

DITERPENOIDS: sandaracopimar-15-en-8β,12α-diol, sandaracopimar-15-en-8β,12β-diol, sandaracopimar-15-en-8β,18-diol, sandaracopimar-15-en-8β,11α,12β-triol, sandaracopimar-15-en-6β,8β,11α-triol-6-acetate, sandaracopimar-15-en-8β,11α,12β-triol-12-acetate.

    Bohlmann et al., Chem. Ber., 106, 826 (1973)

G. sonchifolium.

DITERPENOIDS: 11α-acetoxyatisiren-20-oic acid, 7α-acetoxy-9,11-didehydroatisiren-20-oic acid, 14β-acetoxy-9,11-didehydroatisiren-20-oic acid.

    Bohlmann, Grenz, Ber., 111, 1509 (1978)

GASCARDIA

G. madagascariensis.

SESTERTERPENOID: gascardic acid.

    Boekman et al., J. Am. Chem. Soc., 101, 5060 (1979)

GASTROLOBIUM

G. callistachys.

INDOLE ALKALOID: S(+)N(b)-methyltryptophan methyl ester.

    Johns et al., Aust. J. Chem., 24, 439 (1971)

G. calycinum Benth.

ALKALOID: cygnine.

    Mann, Ince, Proc. R. Soc. London, 79B, 488 (1907)

GAULTHERIA (Ericaceae)

G. adenothrix.

STEROID: β-sitosterol.

TRITERPENOIDS: α-amyrin, betulinic acid, lupeol, oleanolic acid, taraxerol, ursolic acid.

    Kawai et al., Chem. Abstr. 88 34499b

G. miqueliana.

STEROID: β-sitosterol.

TRITERPENOIDS: α-amyrin, betulinic acid, lupeol, oleanolic acid, taraxerol, ursolic acid.

    Kawai et al., Chem. Abstr. 88 34499b

GAZANIA (Compositae)

G. krebsiana.

SESQUITERPENOIDS: gazaniolide, isovaleryloxygazaniolide.

    Bohlmann, Zdero, Phytochemistry, 18, 322 (1979)

G. rigens (L.) Gaertn.

CAROTENOID: gazaniaxanthin.

    Schon, Biochem. J., 32, 1566 (1938)

GEIJERA

G. africana.

SESQUITERPENOIDS: dihydrogrisein, gafrinin, griesenin, vermeerin.

    de Villiers et al., J. Chem. Soc., 2049 (1961)

    De Kock et al., Tetrahedron, 24, 6037 (1968)

    De Kock et al., ibid., 24, 6045 (1968)

G. aspera.

SESQUITERPENOIDS: aspergeijeric acid methyl ester, geijerin, geijerinin, vermeerin.

    Anderson et al., Tetrahedron, 23, 4153 (1967)

G. balansae.

ALKALOID: O-acetylbalansine.

Ahond *et al., Phytochemistry,* **18,** 1415 (1979)

G. parviflora.

COUMARINS: dehydrogeijerin, dihydrogeiparvarin, geiparvarin.

MONOTERPENOIDS: cogeigerene, pregeigerene.

(C) Dreyer *et al., Phytochemistry,* **11,** 763 (1972)

(T) Gough *et al., Tetrahedron Lett.,* 763 (1961)

G. salicifolia Schott.

FURANOISOQUINOLINE ALKALOIDS: acetylplatydesmine, platydesmine.

Johns, Lamberton, *Aust. J. Chem.,* **19,** 1991 (1966)

GEISSOSPERMUM

G. laeve (Vellezo) Baill.

CARBOLINE ALKALOID: flavopereirine.

Bejar *et al., Acad. Sci.,* **244,** 2066 (1957)

G. vellosii.

CARBOLINE ALKALOIDS: flavopereirine, geissolosimine, geissoschizine, geissospermine, geissovelline, pereirine, tombozine, velbsimine, vellosine.

CARBAZOLE ALKALOID: geissoschizoline.

Moore, Rapoport, *J. Org. Chem.,* **38,** 215 (1973)

GELONIUM

G. multiflorum.

TRITERPENOIDS: 7-bauren-3-ol, multiflorenol, epi-multiflorenol.

Fukuoka *et al., Chem. Pharm. Bull.,* **20,** 974 (1971)

GELSEMIUM (Loganiaceae)

G. elegans Benth.

INDOLONE ALKALOIDS: humantendine, humantenidine, humantenine, humantenirine, humantenmine, koumine, kouminicine, kouminidine, kouminine.

Yang, Chen, *Yaoxue Xuebao,* **17,** 633 (1982)

Yang, Chen, *ibid.,* **18,** 104 (1983)

G. sempervirens (L.) Ait.

SPIROINDOLONE ALKALOIDS: gelsedine, gelsemicine, gelsemidine, gelsemine, gelseminine, gelseverine, 1-methoxygelsemine.

PYRIDOINDOLE ALKALOID: sempervirine.

COUMARIN: fabiatin.

IRIDOIDS: gelsemide, gelsemide 7-glucoside, gelsemiol, gelsemiol 1-glucoside, gelsemiol 3-glucoside, 9-hydroxysemperoside, semperoside.

(A) Schwartz, Marion, *Can. J. Chem.,* **31,** 958 (1953)

(A) Wichtl *et al., Monatsh. Chem.,* **104,** 87 (1973)

(C,I) Jensen *et al., Phytochemistry,* **26,** 1725 (1987)

GENIPA

G. americana.

IRIDOID: geniposidic acid.

Guarneccia *et al., Tetrahedron Lett.,* 5125 (1972)

GENISTA (Leguminosae)

G. abchasica DC.

QUINOLINE ALKALOID: cytisine.

QUINOLIZIDINE ALKALOID: pachycarpine.

Bostogonashvili, *Farm. An. Gruz SSR,* **10,** 196 (1967)

G. aetnensis DC.

QUINOLINE ALKALOID: cytisine.

QUINOLIZIDINE ALKALOID: retamine.

Bohlmann *et al., Ber.,* **98,** 659 (1965)

G. cinerea (Vill.) DC.

QUINOLIZIDINE ALKALOIDS: cinegalleine, cinegalline, cinevanine, cineverine, formylcinegalline.

Faugeras, Paris, *Acad. Sci.,* **270D,** 203 (1970)

G. hystrix.

DIPYRIDYL ALKALOID: hystrine.

Steinegger *et al., Phytochemistry,* **7,** 849 (1968)

G. ispanica L.

QUINOLINE ALKALOID: cytisine.

Sckiladze, Tbilsk. Med., 22, *400 (1065)*

G. lucida.

QUINOLIZIDINE ALKALOIDS: anagyrine, cineverine, cytisine, 13-hydroxylupanine, lupanine, N-methylcytisine, retamine, sparteine.

ACIDS: caffeic acid, p-coumaric acid, gallic acid, p-hydroxybenzoic acid, protocatechuic acid, salicylic acid, vanillic acid.

Adzet *et al., Trans. Soc. Pharm. Montpellier,* **35,** 237 (1975); *Chem. Abstr.* 84 102282v

G. lusitanica.

UNCHARACTERIZED ALKALOID: lusitanine.

Steinegger, Vicky, *Pharm. Acta Helv.,* **40,** 610 (1965)

G. lydia.

One uncharacterized alkaloid.

Ulubelen, Dogue, *Planta Med.,* **26,** 338 (1974)

G. ramossissima.

QUINOLIZIDINE ALKALOIDS: anagyrine, cineverine, cytisine, 13-hydroxylupanine, lupanine, N-methylcytisine, retamine, sparteine.

ACIDS: caffeic acid, p-coumaric acid, gallic acid, p-hydroxybenzoic acid, protocatechuic acid, salicylic acid, vanillic acid.

Adzet *et al., Trans. Soc. Pharm. Montpellier,* **35,** 237 (1975); *Chem. Abstr.* 84 101181v

G. sparioides.

QUINOLIZIDINE ALKALOIDS: anagyrine, cineverine, cytisine, 13-hydroxylupanine, lupanine, retamine, sparteine.

ACIDS: caffeic acid, p-coumaric acid, gallic acid, p-hydroxybenzoic acid, protocatechuic acid, salicylic acid, vanillic acid.

Adzet *et al., Trans. Soc. Pharm. Montpellier,* **35,** 237 (1975); *Chem. Abstr.* 84 102282v

G. tinctoria L.

QUINOLINE ALKALOIDS: cytisine, methylcytisine, tinctorine.

LUPINE ALKALOID: anagyrine.

Knoefel, Schuette, *J. Prakt. Chem.,* **312,** 887 (1971)

G. transcaucasica Schischk.

QUINOLINE ALKALOIDS: cytisine, methylcytisine.

LUPINE ALKALOID: anagyrine.

Bostogonashvili, *Farm. Akad. Nauk Gruz. SSR,* **10,** 196 (1967)

G. valentina.

QUINOLIZIDINE ALKALOIDS: anagyrine, cineverine, cytisine, 13-hydroxylupanine, lupanine, N-methylcytisine, retamine, sparteine, Acids caffeic acid, p-coumaric acid, gallic acid, p-hydroxybenzoic acid, protocatechuic acid, salicylic acid, vanillic acid.

Adzet *et al., Trans. Soc. Pharm. Montpellier,* **35,** 237 (1975); *Chem. Abstr.* 84 102282v

GENTIANA (Gentianaceae)

G. argentea.

FLAVONOIDS: apigenin O-glucoside, isoorientin, isovitexin, luteolin, luteolin O-glucoside.

Chulia *et al., Plant Med. Phytother.,* **11,** 112 (1977)

G. asclepiadea.
PYRIDINE ALKALOIDS: gentianidine, gentiolutine.

FLAVONOIDS: isoorientin, isoorientin 4'-O-glucoside, isoorientin
2''-O-glucoside, isovitexin, isovitexin 4'-O-glucoside,
isovitexin 2''-O-glucoside, mangiferin, mangiferin
6-O-β-D-glucoside, mangiferin 7-O-β-D-glucoside,
saponarin.

(A) Marekov, *Acad. Bulg. Sci.*, **23**, 803 (1970)

(F) Goetz et al., *Helv. Chim. Acta*, **60**, 2104 (1977)

G. bavarica L.
FLAVONOID: gentiabavarurinoside.

Hostettmann et al., *Helv. Chim. Acta*, **59**, 2592 (1976)

G. bellidifolia.
XANTHONES: bellidifolin, swerchirin,
1,3,5,8-tetrahydroxyxanthone.

Markham et al., *Tetrahedron*, **21**, 1449 (1965)

G. bulgarica L.
One uncharacterized alkaloid.

Marekov et al., *Acad. Bulg. Sci.*, **18**, 999 (1965)

G. burseri Lapeyr.
FLAVONOIDS: trans-feruloyl-2''-isoorientin,
trans-feruloyl-2''-isoorientin 4'-O-β-D-glucoside.

Luong Minh Duc et al., *Helv. Chim. Acta*, **59**, 1294 (1976)

G. campestris (L.) Borner.
FLAVONES: corymbiferin, mangiferin, swertisin.

GLUCOSIDE: campestroside.

XANTHONES: 4,5-dimethoxy-3,8-dihydroxyxanthone
1-O-β-D-glucopyranoside,
1,3,5,8-tetrahydroxyxanthone-9-O-β-D-glucoside.

(F) Kaldas et al., *Helv. Chim. Acta*, **58**, 2188 (1975)

(G) Kaldas et al., *Phytochemistry*, **17**, 295 (1978)

(X) Kaldas et al., *Helv. Chim. Acta*, **57**, 2557 (1974)

G. caucasica MB.
PYRIDINE ALKALOIDS: gentianine, gentinamine.

Rakhmatullaev et al., *Khim. Prir. Soedin.*, 32 (1969)

G. cruciata.
PYRIDINE ALKALOID: gentianine.

Kurinnaya, *Farmatsiya (Kiev)*, 197 (1956)

G. depressa.
FLAVONOIDS: apigenin O-glucoside, isoorientin, isovitexin,
luteolin, luteolin O-glucoside.

IRIDOID: depressoside.

(F) Chulia et al., *Plant Med. Phytother.*, **11**, 112 (1977)

(I) Chulia et al., *J. Nat. Prod.*, **48**, 54 (1985)

G. elwesii.
FLAVONOIDS: apigenin O-glucoside, isoorientin, isovitexin,
luteolin, luteolin O-glucoside.

Chulia et al., *Plant Med. Phytother.*, **11**, 112 (1977)

G. kaufmanniana Rgl. et Schamlh.
PYRIDINE ALKALOIDS: gentianamine, gentianine, gentianinine.

Rakhmatullaev et al., *Khim. Prir. Soedin.*, 32 (1969)

G. germanica (Willd.) Borner.
BENZOPYRAN: swertiajaponin.

Bouillant et al., *Phytochemistry*, **11**, 1858 (1972)

G. kirilowii Tur.
PYRIDINE ALKALOID: gentianine.

Proskernina, *J. Gen. Chem. USSR*, **14**, 1148 (1944)

G. lutea.
XANTHONES: gentisein, gentisin,
1-hydroxy-3,7-dimethoxyxanthone.

Atkinson et al., *Tetrahedron*, **25**, 1509 (1969)

G. olgae Rgl. et Schmalh.
PYRIDINE ALKALOIDS: gentianadine, gentiananine.

Rakhmatullaev et al., *Khim. Prir. Soedin.*, 32 (1969)

G. olivieri Griseb.
PYRIDINE ALKALOIDS: gentianaine, gentiananine, gentianidine,
gentianine, gentioflavine, gentiotibetine, oliveramine,
oliveridine, oliverine.

Rakhmatullaev et al., *Khim. Prir. Soedin.*, 608 (1969)

G. panonica Scop.
GLUCOSIDES: amaropanin, amaroswerin, desoxymarogentin.

Wagner et al., *Phytochemistry*, **13**, 615 (1974)

G. pedicellata.
FLAVONOID: apigenin O-glucoside, isoorientin, isovitexin,
luteolin, luteolin O-glucoside.

IRIDOIDS: 2'-caffeoylloganin, 2'-p-coumaroylloganin,
2'-feruloylloganin, loganin.

(F) Chulia et al., *Plant Med. Phytother.*, **11**, 112 (1977)

(I) Garcia et al., *Planta Med.*, 327 (1986)

G. pneumonanthe L.
No alkaloids present.

Kurinnaya, *Aptechn. Delo*, 4 (1954)

G. prolata.
FLAVONOIDS: apigenin O-glucoside, isoorientin, isovitexin,
luteolin, luteolin O-glucoside.

Chulia et al., *Plant Med. Phytother.*, **11**, 112 (1977)

G. punctata L. Four uncharacterized alkaloids.
FLAVONOIDS: 2''-trans-caffeoylisoorientin 4'-O-β-D-glucoside,
2''-trans-feruloylisoorientin 4'-O-β-D-glucoside,
2:-trans-feruloylisovitexin, 2:-trans-isoorientin
4'-O-β-D-glucoside, isoscoparin.

IRIDOID: gentioflavoside.

(A) Marekov et al., *Akad. Bulg. Sci.*, **18**, 999 (1965)

(F) Luong et al., *Helv. Chim. Acta*, **60**, 2099 (1977)

(I) Popov, Marekov, *Chem. Ind. (London)*, 655 (1971)

G. purpurea L.
MONOTERPENOID: gentiolactone.

Suhr et al., *Phytochemistry*, **17**, 135 (1978)

G. pyrenaica L.
FLAVONOIDS: isocytisoside 7-O-glucoside, isopyrenin,
isopyrenin 7-O-glucoside, isoquercetin, isoscoparin
7-O-glucoside, isovitexin 7-O-glucoside.

Marston et al., *Helv. Chim. Acta*, **59**, 2596 (1976)

G. ramosa.
BENZOPYRAN: swertiajaponin.

Bouillant et al., *Phytochemistry*, **11**, 1858 (1972)

G. scabra var. buergeri.
DIPYRAN: gentiopicriside.

Hayashi et al., *J. Pharm. Soc., Japan*, **96**, 679 (1976)

G. sikkimensis.
FLAVONOIDS: apigenin O-glucoside, isoorientin, isovitexin,
luteolin, luteolin O-glucoside.

Chulia et al., *Plant Med. Phytother.*, **11**, 112 (1977)

G. tibetica King.
PYRIDINE ALKALOID: gentialutine.

Rulko et al., *Pol. J. Pharmacol. Pharm.*, **26**, 561 (1974)

G. triflora var. japonica.
DIPYRAN: gentiopicrin.

MONOTERPENE GLYCOSIDE: trifloroside.

(D) Hayashi et al., *J. Pharm. Soc., Japan*, **96**, 679 (1976)

(T) Inouye et al., *Tetrahedron*, **30**, 571 (1974)

G. turkestanorum Gand.
PYRIDINE ALKALOIDS: gentianaine, gentianamine, gentiananine,
gentianidine, gentianine. Samatov et al., *Khim. Prir.
Soedin.*, 182 (1967)

G. verna L.
XANTHONE: 1-hydroxy-3-methoxy-7-glucosylxanthone.

Hostettmann et al., *Helv. Chim. Acta*, **57**, 1155 (1974)

GERAEA
G. viscida.
SESQUITERPENOID: gerin.
> Rodriguez et al., Phytochemistry, **18**, 1741 (1979)

GERANIUM (Geraniaceae)
G. bourbon.
MONOTERPENOIDS: (+)-β-citronellol, (-)β-citronellol.
SESQUITERPENOIDS: α-bourbonene, β-bourbonene, furopelargones A, B, C and D, guaia-6,9-dione, 11-norbourbonan-1-one, norbourbonone.
> Krepinsky et al., Tetrahedron Lett., 359 (1966)
> Gianotti, Schwang, Bull. Soc. Chim. Fr., 2452 (1968)

G. macrorrhizum L.
FLAVONOIDS: acetovanillon, ermanin, isokaempferide, kumatakenin, quercetin, retusin, rhamnosin.
ACID: gallic acid, Phenolics glucogallin.
STEROIDS: β-sitosterol, β-sitosterol-β-D-glucoside.
SESQUITERPENOIDS: germacrone, germazone.
> (F, P, S) Ivancheva et al., Dokl. Bolg. Akad. Nauk., **29**, 205 (1976); Chem. Abstr. 85 30659u
> (T) Ognayanov et al., Collect. Czech. Chem. Commun., **23**, 2033 (1958)
> (T) Tsankova, Ognyanov, Tetrahedron Lett., 3833 (1976)

G. phaeum.
FLAVONOID: quercetin.
PHENOLICS: caffeic acid, chebulagic acid, chlorogenic acid, ellagic acid, gallic acid.
> Lufertova et al., Chem. Abstr. 86 152629a

G. thunbergii.
TANNINS: elaeocarpusin, geraniin.
PHENOLS: brevifolin, corilagin, ellagic acid, ellagitannin, gallic acid, pyrogallol.
> (P) Okuda et al., Phytochemistry, **14**, 1877 (1975)
> (T) Nonaka et al., Chem. Pharm. Bull., **34**, 941 (1986)

GEOFFROEA
G. surinamensis.
TYROSINE ALKALOID: geoffroyine.
> Goldschmeidt, Monatsh. Chem., **34**, 659 (1913)

GERBERA (Compositae/Arctotideae)
G. crocea.
SESQUITERPENOID: isogerberacumarin.
> Boglmann et al., Chem. Ber., **106**, 382 (1973)

G. jamesonii var. hybrida.
COUMARINS: 4β-D-glucopyranosyloxy-5-methylcoumarin, 5-methyl-4-rutinosylcoumarin, prunasin.
GLYCOSIDES: amygdalin, vicianin.
> Nagumo et al., Chem. Pharm. Bull., **33**, 4803 (1985)

G. piloselloides Cass.
CHROMENE: 6-acetyl-2,2-dimethyl-8-(3-methyl-2-butenyl)-2H-1-benzopyran.
> Bohlmann et al., Chem. Ber., **108**, 26 (1975)

GEUM (Rosaceae)
G. japonicum Thunb.
ELLAGITANNINS: gemin A, gemin B, gemin C.
> Yoshida et al., J. Chem. Soc., Perkin I, 315 (1985)

GIGARTINA (Rhodophyceae)
G. stellata.
CAROTENOIDS: α-carotene, β-carotene, α-cryptoxanthin, β-cryptoxanthin, lutein, zeaxanthin.
> Bjoernland et al., Phytochemistry, **15**, 291 (1976)

GILMANIELLA
G. humicola Barron.
NORSESQUITERPENOID: gilmacolin.
> Chexal et al., Helv. Chim. Acta, **62**, 1129 (1979)

GINKGO (Ginkgoaceae)
G. biloba L.
ESTER: 6-(pentadec-8-enyl)-2,4-dihydroxybenzoic acid.
FLAVONES: bilobetin, ginkgetin.
SESQUITERPENOIDS: bilobalide A, bilobanone, 10,11-(E)-dihydroatlantone, 10,11-(Z)-dihydroatlantone.
DITERPENOIDS: ginkgolides A, B, C and M.
STEROIDS: 24α-ethyllathosterol, β-sitosterol, spinasterol.
> (E) Gellerman et al., Phytochemistry, **15**, 1959 (1976)
> (F) Beckmann et al., Phytochemistry, **10**, 2465 (1971)
> (S) Nes et al., Lipids, **12**, 511 (1977)
> (T) Maruyama et al., Tetrahedron Lett., 299 et seq. (1967)
> (T) Plattier et al., Recherche, **19**, 131 (1974)

GINSENG
G. radix.
GLYCOSIDES: malonylginsenoside Rb-1, malonylginsenoside Rb-2, malonylginsenoside Rc, malonylginsenoside Rd.
> Kitagawa et al., Chem. Pharm. Bull., **31**, 3353 (1983)

GIRGENSOHNIA (Chenopodiaceae)
G. diptera Bge.
INDOLE ALKALOID: dipterine.
PIPERIDINE ALKALOID: N-methylpiperidine.
> Juraschevski, Stepanova, J. Gen. Chem., USSR, **9**, 203 (1939)

G. oppositifolia (Pall.) Fenzl.
PIPERIDINE ALKALOIDS: girgensohnin, N-methylpiperine.
> Juraschevski, Stepanova, J. Gen. Chem., USSR, **16**, 141 (1946)

GLAUCIDIUM (Podophyllaceae)
G. palmatum Sieb. et Zucc.
SESQUITERPENOID: glaupalol.
> Itie et al., J. Chem. Soc., Chem. Commun., 547 (1967)

GLAUCIUM (Papaveraceae)
G. controruplicatum Boiss.
APORPHINE ALKALOID: sinoacutine.
> Tin-Wa et al., J. Pharm. Sci., **65**, 755 (1976)

G. corniculatum (L.) Curt.
APORPHINE ALKALOID: norbracteoline.
BENZOQUINOLIZIDINE ALKALOID: protopine.
NAPHTHOISOQUINOLINE ALKALOID: (-)-chelidonine.
PROTOBERBERINE ALKALOID: α-allocryptopine.
> Israilov et al., Khim. Prir. Soedin., 751 (1983)

G. corniculatum (L.) Curt. subsp. refractum.
APORPHINE ALKALOIDS: bulbocapnine, corydine, dicentrine, glaucine, isocorydine.
OXOPROTOBERBERINE ALKALOIDS: α-allocryptopine, protopine.
PHENANTHRENE ALKALOID: (-)-methylflavinantine.
> Shafiee et al., J. Nat. Prod., **48** 855 (1985)

G. elegans Fisch et Mey.
APORPHINE ALKALOID: glaucine.
NAPHTHOISOQUINOLINE ALKALOID: dihydrochelerythrine.
NORAPORPHINE ALKALOID: O-methylatheroline.
Sonnet, Jacobson, *J. Pharm. Sci.*, **60**, 1254 (1971)

G. fimbrilligerum Boiss.
APORPHINE ALKALOIDS: (-)-corydine, corytuberine, epi-glaufidine, glaufine, glaufinine, glaunidine, glaunine, isocorydine.
BENZOQUINOLIZIDINE ALKALOID: protopine.
NAPHTHOISOQUINOLINE ALKALOIDS: α-allocryptopine, chelerythrine, sanguinarine.
Konovalova *et al.*, *J. Gen. Chem., USSR*, **9**, 1939 (1939)
Karimova, Yunusov, *Khim. Prir. Soedin.*, 250 (1986)

G. flavum Crantz.
APORPHINE ALKALOIDS: aurotensine, cataline, dehydroglaucine, glaucine, isoboldine, isocorydine.
BENZOQUINOLIZIDINE ALKALOID: protopine.
NORAPORPHINE ALKALOID: O-methylatheroline.
PROTOBERBERINE ALKALOID: sinactine.
Yakhontova *et al.*, *Khim. Prir. Soedin.*, 214 (1972)

G. flavum Crantz var. leucarpum.
One aporphine alkaloid.
Kiryakov, Panov, *Dokl. Akad. Nauk Bolg.*, **22**, 1019 (1969)

G. flavum Crantz var. vestitum.
NAPHTHOISOQUINOLINE ALKALOID: 6-iminosanguinarine.
OXOAPORPHINE ALKALOIDS: arosine, arosinine, corunnine.
Castedo *et al.*, *Heterocycles*, **16**, 533 (1981)

G. grandiflorum Boiss. et Huet.
APORPHINE ALKALOIDS: glaucine, glauvine, isoboldine, thalicmidine.
BENZOQUINOLIZIDINE ALKALOID: protopine.
NAPHTHOISOQUINOLINE ALKALOID: sanguinarine.
OXOAPORPHINE ALKALOID: pontevedrine.
NORAPORPHINE ALKALOID: O-methylatheroline.
Ribas *et al.*, *Tetrahedron Lett.*, 3093 (1971)

G. nana.
IRIDOID: lytanthosalin.
Chaudhuri *et al.*, *Helv. Chim. Acta*, **64**, 2401 (1981)

G. serpieri Heldr.
APORPHINE ALKALOIDS: aurotensine, glaucine.
Manske, *Can. J. Res.*, **16**, 81 (1938)

G. squamigerum K. et K.
APORPHINE ALKALOID: corydine.
BENZOQUINOLIZIDINE ALKALOID: protopine.
NAPHTHOISOQUINOLINE ALKALOIDS: α-allocryptopine, chelerythrine, sanguinarine.
Platonova *et al.*, *J. Gen. Chem., USSR*, **26**, 173 (1956)

G. vitellinum.
APORPHINE ALKALOID: 4α-hydroxybulbocapnine.
Shafiel *et al.*, *Lloydia*, **42**, 174 (1979)

GLEDITSIA (Leguminosae)
G. japonica Miquel.
FLAVONOIDS: homoorientin, isoorientin, isoquercitrin, isovitexin, luteolin 7-glucoside, vitexin.
Yoshizaki *et al.*, *Chem. Pharm. Bull.*, **25**, 3408 (1977)

G. sinensis Lam.
FLAVONOIDS: homoorientin, isoorientin, isoquercitrin, isovitexin, luteolin 7-glucoside, orientin, vitexin.
Yoshizaki *et al.*, *Chem. Pharm. Bull.*, **25**, 3408 (1977)

G. triacanthos L.
PURINE ALKALOID: triacanthine.
TRITERPENE GLYCOSIDES: gleditschiosides A, B, C, D and E, triacanthoside A, triacanthoside C, triacanthoside G.
(A) Belikov *et al.*, *J. Gen. Chem., USSR*, **24**, 919 (1954)
(G) Badelbaeva *et al.*, *Khim. Prir. Soedin.*, 635 (1973)

GLIRICIDIA
G. sepium.
FLAVONOIDS: 2'-O-methylsepiol, robinetin, sepiol, 7,3',4'-trihydroxyflavanone.
Jurd, Manners, *J. Agric. Food Chem.*, **25**, 723 (1977); *Chem. Abstr.* 87 35858j

GLOBULARIA (Globulariaceae)
G. alypum.
IRIDOIDS: globularicisin, globularidin, globularimin, globularin, globularinin.
Chaudhuri *et al.*, *Helv. Chim. Acta*, **64**, 3 (1981)

G. cordifolia.
IRIDOID: globularifolin.
Chaudhuri, Sticher, *Helv. Chim. Acta*, **63**, 117 (1980)

G. virescens.
COLCHICINE ALKALOIDS: O-acetylcolchicine, N-acetyldemecolcine, 2-acetyl-2-demethylcolchicine, 3-acetyl-3-demethylcolchicine, N-acetylisodemecolcine.
Potesilova *et al.*, *Collect. Czech. Chem. Commun.*, **32**, 141 (1967)

GLOCHIDON (Euphorbiaceae)
G. eriocarpum.
TRITERPENOID: glochidol.
Hui *et al.*, *Phytochemistry*, **15**, 561 (1976)

G. ferdinandii Muell.-Arg.
No alkaloids present.
Johns, Lamberton, *Aust. J. Chem.*, **20**, 555 (1967)

G. hohenhackeri.
TRITERPENOIDS: glochidiol, glochidone, epi-lupeol, lup-20(29)-ene-1β,3β-diol, lup-20(29)-ene-1,3-dione.
Ganguly *et al.*, *Tetrahedron*, **22**, 1513 (1966)
Hui, Li, *Phytochemistry*, **17**, 156 (1978)

G. macrophylla.
TRITERPENOIDS: glochilocudiol, lup-20(29)-ene-1β,3β-diol, lup-20(29)-ene-3β,23-diol, lup-20(29)-ene-1,3-dione, methylbetulinate.
Hui, Lee, *Phytochemistry*, **17**, 156 (1978)

G. multiloculare.
TRITERPENOIDS: glochidiol, glochilocudiol.
Ganguly *et al.*, *Tetrahedron*, **22**, 1513 (1966)

G. philippicum (Cav.) C.B.Rob.
ALKALOIDS: cinnamoylhistidine, glochidicine, glochidine, N(a)-(4'-oxodecanoyl)-histamine.
Johns, Lamberton, *Aust. J. Chem.*, **20**, 555 (1967)

G. puberum.
STEROID: β-sitosterol.
TRITERPENOIDS: friedelan-3β-ol, friedelin, glochidiol, glochidone, glochidonol, lup-20(29)-ene-1β,3β-diol, lup-20(29)-ene-1,3-dione, lupenone, lupeol.
Hui *et al.*, *Phytochemistry*, **17**, 156 (1978)

G. wrightii.
TRITERPENOID: glochidonol.
Hui, Lee, *J. Chem. Soc., C*, 1710 (1969)

GLOMERIS
G. marginata.
QUINAZOLONE ALKALOIDS: glomerine, homoglomerine.
Schildknecht et al., Z. Naturforsch., 21B, 121 (1966)

GLORIOSA (Liliaceae)
G. superba L.
COLCHICINE ALKALOIDS: 2-demethyl-β-lumicolchicine,
N-formyl-deacetyl-β-lumicolchicine,
N-formyldeacetyl-β-lumicolchinine,
N-formyldeacetyl-O(2)-demethyl-β-lumicolchinine.
Canonica et al., Chem. Ind. (Milan), 49, 1304 (1967)
Dvorackova et al., Collect. Czech. Chem. Commun., 49, 1536 (1984)

G. virescens.
COLCHICINE ALKALOIDS: O-acetylcolchicine,
N-acetyldemecolcine, 2-acetyl-2-demethylcolchicine,
3-acetyl-3-demethylcolchicine, N-acetylisodemecolcine.
Potesilova et al., Collect. Czech. Chem. Commun., 32, 141 (1967)

GLOSSOCALYX
G. brevipes.
APORPHINE ALKALOID: (+)-N-methyllaurotetanine β-N-oxide.
Montgomery et al., J. Nat. Prod., 48, 833 (1985)

GLOSSOGYNE
G. ternuifolia (Labill.) Cass.
FLAVONOIDS: luteolin, luteolin 7-O-β-D-glucopyranoside.
Lin et al., Chem. Abstr. 86 13812a

GLYCINE (Leguminosae)
G. max Merrill.
SAPONINS: glyceollins I, II, III and IV, soyasaponin I, II, III
and A1, A2.
Lyne et al., Tetrahedron Lett., 3127 (1978)
Kitagawa et al., Chem. Pharm. Bull., 33, 1069 (1985)

GLYCOSMIS (Rutaceae)
G. arborea (Roxb.) DC.
QUINAZOLONE ALKALOIDS: arborine, glycorine, glycosmicine,
glycosminine.
ACRIDONE ALKALOID: arborinine.
TRITERPENOIDS: arborinol, isoarborinol.
(A) Chakravarti et al., Tetrahedron, 16, 224 (1961)
(T) Vorbruggen et al., Justus Liebigs Ann. Chem., 668, 57 (1963)

G. bilocularis.
ACRIDONE ALKALOID: hydroxyarborinine.
Bowen et al., Phytochemistry, 19, 1566 (1980)

G. citrifolia (Willd.) Lindl. (G. cochinchinensis Pierre).
ACRIDONE ALKALOIDS: furofolines I and II, glycobismine A,
glycocitrine I, glycocitrine II, glycofoline, glyfoline,
3-O-methylglycocitrine I, 3-O-methylglycocitrine II,
pyranofoline.
Wu et al., J. Chem. Soc., Perkin I, 1681 (1983)

G. cyanocarpa.
FURANOACRIDONE ALKALOID: glycarpine.
Sarkar et al., Phytochemistry, 17, 2145 (1978)

G. mauritiana.
One acridone and one quinolone alkaloid.
Rastogi et al., Phytochemistry, 9, 945 (1980)

G. pentaphylla (Retz.) Correa.
ACRIDONE ALKALOIDS: N-desmethylnoracronycine,
des-N-methylacronycine, noracronycine.
AMIDE ALKALOID: glycomide.
CARBAZOLE ALKALOIDS: glycozolidine, glycozoline.

QUINAZOLINE ALKALOID: glycophymoline.
QUINAZOLONE ALKALOID: glycophymine.
QUINOLONE ALKALOID: glycosolone.
Sarkar, Chakraborty, Phytochemistry, 18, 694 (1979)

GLYCYRRHIZA (Papilionaceae)
G. glabra L.
CHALCONE: licochalcone.
COUMARIN: liqcoumarin.
FLAVONOIDS: 7-acetoxy-2-methylisoflavone,
2,7-dihydroxy-3',4'-dimethoxyisoflavone,
7,4'-dihydroxyflavone, echinatin, formononetin, glabridin,
glabrol, glycyrol, glyzarin, 7-hydroxy-2-methylisoflavone,
isoliquiritigenin, kumatakenin, licoflavonol,
licoisoflavanone, licoisoflavone A, licoisoflavone B,
licoricone, 7-methoxy-2-methylisoflavone.
TRITERPENOIDS: deoxoglabrolide, glabralonic acid, glabrolide,
glycyrrhetic acid, glycyrrhetol,
24-hydroxy-11-deoxyglycyrrhetic acid,
18α-hydroxyglycyrrhetic acid, 23-hydroxyglycyrrhetic acid,
28-hydroxyglycyrrhetic acid, 21α-hydroxyisoglabrolide,
24-hydroxyliquiretic acid, isoglabrolide, liquiridiolic acid,
liquoric acid.
(C) Saito, Tetrahedron Lett., 4661 (1975)
(Cou) Bhardwaj et al., Phytochemistry, 15, 1182 (1976)
(F) Bhardwaj et al., Phytochemistry, 15, 352 (1976)
(F) Saito et al., Chem. Pharm. Bull., 24, 752 (1976)
(F) Saito et al., ibid., 24 1242 (1976)
Ingham, Phytochemistry, 16, 1457 (1977)
(I) Bhardwaj et al., ibid., 16, 402 (1977)
(T) Canonica et al., Gazz. Chim. Ital., 98, 712 (1968)
(T) Elgamel, El-Tawil, Planta Med., 27, 159 (1975)

G. macedonica.
SAPONINS: saponins A and B.
TRITERPENOIDS: isomacedonic acid, macedonic acid.
(S) Senenchenko et al., Aktual. Vopr. Farm., 2, 65 (1974); Chem. Abstr.
84 132598v
(T) Kir'yalov, Naugol'naya, J. Gen. Chem., USSR, 33, 697 (1963)

G. pallidiflora.
TRITERPENOID: meriestrotrophic acid.
Kir'yalov, Naugol'naya, J. Gen. Chem., USSR, 33, 694 (1963)

G. uralensis.
TRITERPENOID: uralenic acid.
Kir'yalov, Naugol'naya, J. Gen. Chem., USSR, 34, 2814 (1964)

GMELINA
G. arborea.
LIGNAN: 4,8-dihydroxysesamin.. Sesquiterpenoids arboreol,
arboreol 1-ethyl ether, arboreol 1-methyl ether, gmelofuran.
(L) Anjaneyulu et al., Tetrahedron, 33, 133 (1977)
(T) Joshi et al., Tetrahedron Lett., 4719 (1978)

G. leuchhardtii.
LIGNAN: gmelinol.
Birch et al., Aust. J. Chem., 7, 83 (1954)

GNAPHALIUM (Compositae)
G. affine.
FLAVONOIDS: apigenin, apigenin 4'-β-D-glucoside,
dehydro-p-asebolin, luteolin, luteolin 4'-β-D-glucoside,
quercetin, quercetin 4'-β-D-glucoside.
Itakura et al., Agric. Biol. Chem., 39, 2237 (1975)

G. pellitum.
FLAVONOID: 5-hydroxy-7,8-dimethoxyflavone.
STEROID: β-sitosterol.
Escarria et al., Phytochemistry, 16, 1618 (1977)

G. undulatum.
DITERPENOID: 8,13-gnaphalenediol.
Bohlmann *et al., Phytochemistry,* **19,** 71 (1980)

GNETUM (Gnetaceae)

G. gnemon.
BENZOPYRAN: swertijaponin.
Vallant *et al., Phytochemistry,* **17,** 2136 (1978)

G. scandens.
ACIDS: malvalic acid, sterculic acid.
Mustafa *et al., J. Am. Oil Chem. Soc.,* **63,** 1191 (1986); *Chem. Abstr.*
105 206282p

G. ula.
STILBENES: gnetin, 2,3',5',6-stilbenetetraol.
Zyman *et al., Indian. J. Chem.,* **22B,** 101 (1983)
Prakash *et al., Phytochemistry,* **24,** 622 (1985)

GNIDIA (Thymelaeaceae)

G. lamprantha.
DITERPENOIDS: gnidicin, gnididin, gniditrin.
Kupchan *et al., J. Am. Chem. Soc,* **97,** 672 (1975)

G. latifolia Gilg.
DITERPENOIDS: glndiglaucin, gnidilatin.
SESQUITERPENOID: gnididione.
Kupchan *et al., J. Org. Chem.,* **42,** 348 (1977)

GNOXYS

G. psilophylla.
One sesquiterpenoid.
Bohlmann, Zdero, *Phytochemistry,* **18,** 339 (1979)

G. sancto-antonii.
SESQUITERPENOIDS: 6β-acetoxy-3-angeloyloxyeremophilane,
    6β-acetoxy-3β-senecioyloxyfuranoeremophilane,
    6β-acetoxy-3β-tigloyloxyfuranoeremophilane,
    6β-angeloyloxy-3β-senecioyloxyfuranoeremophilane,
    3β,6β-diangeloyloxyfuranoeremophilane.
Bohlmann *et al., Phytochemistry,* **16,** 774 (1977)

GOCHNATIA

G. paniculata.
NORDITERPENOIDS:
    17α-acetoxy-6α,15-dihydroxy-3-kolavene-18-ol,
    17α-acetoxy-3-kolavene-6α,15,18-triol,
    17-phenylacetoxy-6α,15-dihydroxy-3-kolaven-18-ol,
    17-phenylacetoxy-3-kolavene-6α,15,18-triol.
TRITERPENOIDS: gochnatiolides A and B, gochnatol 17-acetate.
Bohlmann *et al., Phytochemistry,* **22,** 191 (1983)

G. polymorpha.
Two bisabolenes, four guaianolides and two dimeric guaianolides.
Bohlmann *et al., Phytochemistry,* **25,** 1175 (1986)

G. smithii.
Two guaianolides.
Ortega *et al., Phytochemistry,* **23,** 1507 (1984)

GOBBELIA

G. pachycarpa.
QUINOLIZIDINE ALKALOID: sophorbenzamine.
Abdusalamov *et al., Khim. Prir. Soedin.,* 71 (1976)

GOMPHIDIUS (Basidiomycetae)

G. glutinosus.
PHENOLIC: xercomic acid.
Edwards *et al., J. Chem. Soc.,* 2582 (1971)

G. rutilus.
PHENOLIC: xercomic acid.
Edwards *et al., J. Chem. Soc.,* 2582 (1971)

GOMPHOCARPUS (Asclepiadaceae)

G. fruticosus R.Br.
STEROIDS: frugoside, gofruside, gomphogenin.
Lardon *et al., Helv. Chim. Acta,* **52,** 1940 (1969)

GONIOMA

G. kamasii.
UNCHARACTERIZED ALKALOID: rhazidine.
Markey *et al., Tetrahedron Lett.,* 157 (1967)

G. malagasy.
ALKALOID: goniomine.
Chaironi *et al., J. Am. Chem. Soc.,* **102,** 6920 (1980)

GONIOTHALAMUS (Annoniaceae)

G. macrophyllus (Bl.) Hook. f.et Thomas.
PHENOLICS: goniothalamin, goniothalamin oxide.
Sam *et al., Tetrahedron Lett.,* 2541 (1987)

GONOCARYUM

G. pyriforme.
CAROTENOIDS: δ-carotene, τ-carotene.
Winterstein, *Z. Physiol. Chem.,* **219,** 249 (1933)

GONOCLINIOPSIS

G. prasiifolia.
Four sesquiterpenoids.
Bohlmann *et al., Phytochemistry,* **23,** 1509 (1984)

GONYAULAX (Pyrrhophyceae/Gonyaulacaceae)

G. foliaceum.
STEROID: 4-methylgorgostanol.
Alam *et al., J. Org. Chem.,* **44,** 4466 (1979)

G. tamarensis.
STEROID: dinosterol.
Shimizu *et al., J. Am. Chem. Soc.,* **98,** 1059 (1976)

GOODENIA (Goodeniaceae)

G. ramelii.
DITERPENOIDS: 8β,13-epoxyperu-14,15,18-triol,
    19-(β-carboxypropionyloxy)-(-)-kaur-16-en-3β-ol.
Coates *et al., Tetrahedron,* **24,** 795 (1968)

G. strophiolata.
DITERPENOID:
    19-(β-carboxypropionyloxy)-(-)-kaur-16-en-3β-ol.
Coates *et al., Tetrahedron,* **24,** 795 (1968)

GOSSYPIUM (Malvaceae)

G. barbadense.
FLAVONOIDS: astragalin, isoquercitrin.
PHENOLICS: caffeic acid, chlorogenic acid.
SESQUITERPENOIDS: helioside A, helioside B,
    6-deoxyhemigossypol, hemigossypol, para-hemigossypol.

TRITERPENOIDS: 6,6'-dimethoxygossypol, 6-methoxygossypol.
  (F, P) Ishak *et al., Egypt. J. Chem.,* **17,** 135 (1974); *Chem. Abstr.* 86 117611s
  (T) Stipanovic *et al., Phytochemistry,* **14,** 1077 (1975)
  (T) Bell *et al., ibid.,* **17,** 1297 (1978)

G. hirsutum.
SESQUITERPENOIDS: (+)(S)-abscicic acid, bisabolene oxide, lacinilenes A, B and C.
NORDITERPENOID: strigol.
SESTERTERPENOIDS: heliocides H-1, H-2, H-3 and H-4.
TRITERPENOIDS: 6,6'-dimethoxygossypol, 6-methoxygossypol.
  Stipanovic *et al., J. Agric. Food Chem.,* **28,** 115 (1978)
  Stipanovic *et al., Phytochemistry,* **17,** 151 (1978)

G. raimondii.
SESQUITERPENOID: raimondal.
  Stipanovic *et al., Phytochemistry,* **19,** 1735 (1980)

GOYAZIANTHUS
G. tetrastichus.
DITERPENOID: 3,4-epoxy-13-kolavene-2,15-diol.
  Bohlmann *et al., Phytochemistry,* **21,** 939 (1982)

GRANGEA
G. maderaspatana.
DITERPENOID: strictic acid.
  Iyer *et al., Indian J. Chem.,* **18B,** 529 (1979)

GRATIOLA (Scrophulariaceae)
G. officinalis L.
SAPONIN: gratioside.
TRITERPENOIDS: gratiogenin, gratiogenin glucoside, 16-hydroxygratiogenin.
  Tschesche *et al., Justus Liebigs Ann. Chem.,* **674,** 196 (1964)

GRAZELIS
G. intermedia.
DITERPENOID: grazielic acid.
  Bohlmann *et al., Phytochemistry,* **20,** 1069 (1981)

GREVILLEA (Proteaceae)
G. robusta.
MACROCYCLIC: robustol.
PHENOLICS: bis-norstriatol, norstrictol.
  (M) Cannon *et al., Aust. J. Chem.,* **26,** 2257 (1973)
  (M) Cannon *et al., ibid.,* **26,** 2277 (1973)
  (P) Varma *et al., Phytochemistry,* **15,** 1418 (1976)

GREWIA (Tiliaceae)
G. asiatica.
BENZOPYRAN: tetrahydro-6-(2-hydroxy-16,19-dimethylhexacosyl)-4-methyl-2H-pyran-2-one.
KETOALCOHOL: grewinol.
  (B) Lakshmi *et al., Phytochemistry,* **15,** 1397 (1976)
  (K) Lakshni *et al., Lloydia,* **39,** 372 (1976)

G. tomentosa.
BENZOPYRAN: tilirioside.
  Lin, *J. Chin. Chem. Soc.,* **22,** 383 (1975); *Chem. Abstr.* 84 147653

GRIFFONIA
G. simplicifolia.
GLYCOSIDE: griffonin.
LACTONE: griffonilide.
  Dwuma-Badu *et al., Lloydia,* **39,** 385 (1976)

GRIFOLIA (Basidiomycetae)
G. confluens.
PHENOLIC: grifolin.
  Goto *et al., Tetrahedron,* **19,** 2079 (1963)

GRINDELIA (Compositae/Astereae)
G. acutifolia.
DITERPENOIDS: 6α-hydroxy-17-acetoxygrindelic acid, 6α-hydroxy-17-isovaleroxygrindelic acid, 6α-hydroxy-17-(α-methylbutyroxy)-grindelic acid, 6-oxo-17-acetoxygrindelic acid, 6-oxo-17-isobutyroxygrindelic acid.
  Timmermann *et al., Phytochemistry,* **26,** 467 (1987)

G. aphanactis Rydb.
FLAVONOIDS: luteolol, 3-methylquercetol, quercetol.
PHENOLIC: vanillic acid.
  Torck *et al., Plant Med. Phytother.,* **10,** 188 (1976)

G. camporum.
12 diterpenoids.
  Timmermann *et al., Phytochemistry,* **22,** 523 (1983)

G. havardii.
ACIDS: havardic acids A, B, C, D, E and F.
DIOL: havardiol.
  Jolad *et al., Phytochemistry,* **26,** 483 (1987)

G. humilis.
DITERPENOID: 6-hydroxygrindelic acid.
  Rose *et al., Phytochemistry,* **20,** 2249 (1981)

G. paludosa.
TERPENOIDS: 1,2-dihydropalutropone, palutropone.
  Bohlmann *et al., Phytochemistry,* **21,** 167 (1982)

G. robusta.
DITERPENOIDS: 7,8-epoxy-7,8-dihydrogrindelic acid, grindelic acid, 6-oxogrindelic acid.
  Panizzi *et al., Tetrahedron Lett.,* 376 (1961)

G. squarrosa.
FLAVONOIDS: luteolol, 3-methylkaempferol, 3-methylquercetol, quercetol.
DITERPENOIDS: grindelic acid, oxygrindelic acid.
  (F) Torck *et al., Acad. Sci.,* **282,** 1453 (1976)
  (T) Bruun *et al., Acta Chem. Scand.,* **76,** 1675 (1962)

G. stricta.
DINORDITERPENOID: grindelistrictoic acid.
DITERPENOID: strictanonoic acid.
  Bohlmann *et al., Phytochemistry,* **21,** 167 (1982)

GRISELINIA (Cornaceae)
G. littoralis.
IRIDOID: griselinoside.
  Jensen *et al., Phytochemistry,* **19,** 2685 (1980)

GROSSHEIMIA
G. macrocephala.
SESQUITERPENOIDS: cynaropicrin, 4'-deoxycynaropicrin, grossheimin, grossheimol, 4'-nor-2'-methoxycynaropicrin.
  Daniewski *et al., Collect. Czech. Chem. Commun.,* **47,** 3160 (1982)
  Barbetti *et al., Farmaco Ed. Sci.,* **40,** 755 (1985)

GROSSWEILERODENDRON
G. balsamiferum.
DITERPENOIDS: agbanindiol A, agbanindiol B, agbaninol, kolavic acid methyl ester.
  Ekong, Okogun, *J. Chem. Soc., C,* 2153 (1969)

GUAIACUM (Zygophyllaceae)
G. officinale L.
FURAN: furoguaiacidin.
PHENOLIC: furoguaiaoxidin.
SAPOGENIN: 24-O-β-D-glucopyranosylofficigenin.
NORTRITERPENOID: 3β,20-dihydroxy-30-norolean-12-en-28-oic acid.
TRITERPENOID: officigenin.
  (F, P) Majumder et al., Chem. Ind. (London), 77 (1974)
  (Sap) Ahmad et al., J. Nat. Prod., **48**, 826 (1985)
  (T) Ahmad et al., Phytochemistry, **23**, 2613 (1984)

GUAREA (Meliaceae)
G. cedrata.
STEROID: 3,4-seco-4(28),7,24-tirucallatriene-3,21-dioic acid methyl ester.
  Connolly et al., J. Chem. Res. Synop., 256 (1978)
G. glabra.
LIMONOIDS: glabretal, glabretal acrylate, glabretal angelate, glabretal α-hydroxyisovalerate, glabretalone, glabretal tiglate.
  Ferguson et al., J. Chem. Soc., Chem. Commun., 159 (1973)
G. thompsonii.
LIMONOIDS: 6,12β-diacetoxyangolensic acid methyl ester, dihydrogedunin.
SESQUITERPENOID: angolensic acid methyl ester.
SESTERTERPENOIDS: fuarealactones B and C.
  Connolly et al., J. Chem. Res. Synop., 256 (1978)
G. trichilioides.
TERPENOID: fissinolide.
  Zelnik, Rosito, Tetrahedron Lett., 6441 (1966)

GUATTERIA (Annonaceae)
G. dielsiana.
AZAFLUORENONE ALKALOIDS: dielsine, dielsinol, 6-methoxyonychine, onychine.
OXOAPORPHINE ALKALOIDS: isomoschatoline, liriodenine, O-methylmoschatoline.
QUINONE: dielsiquinone.
  Goulart et al., Phytochemistry, **25**, 1691 (1986)
G. discolor.
NORAPORPHINE ALKALOIDS: guadiscine, guadiscoline.
  Hocquemiller et al., Tetrahedron Lett., 4247 (1982)
G. elata.
APORPHINE ALKALOID: puterine.
  Hsu et al., Lloydia, **40**, 505 (1977)
G. gaumeri.
BIZBENZYLISOQUINOLINE ALKALOID: guattagaumerine.
  Dehaussy et al., Planta Med., **49**, 25 (1983)
G. megalophylla Diels.
BISBENZYLISOQUINOLINE ALKALOID: (R,R)-(-,-)-O,O-dimethylcurine.
  Galeffi et al., Gazz. Chim. Ital., **105**, 1207 (1975)
G. melosma.
APORPHINE ALKALOIDS: 3-hydroxynornuciferine, melosmidine, melosmine.
OXOAPORPHINE ALKALOIDS: isomoschatoline, oxoanolobine.
  Abd-el-Atti et al., J. Nat. Prod., **45**, 476 (1982)
G. ouregou.
APORPHINE ALKALOID: pentouregine.
  Cortes et al., Phytochemistry, **24**, 2776 (1985)
G. psilopus.
OXOAPORPHINE ALKALOID: atherospermidine.
  Bick, Douglas, Aust. J. Chem., **18**, 1997 (1965)

G. scandens.
APORPHINE ALKALOIDS: guattescidine, guattescine.
  Hocquemiller et al., J. Nat. Prod., **46**, 335 (1983)
G. schomburgkiana.
APORPHINE ALKALOIDS: belemine, dehydroguattescine.
  Cortes et al., J. Nat. Prod., **48**, 254 (1985)
G. subsessilis.
OXOAPORPHINE ALKALOID: subsessiline.
  Hasegawa et al., Acta Cient. Venez., **23**, 165 (1972)

GUAZUMA
G. tomentosa Kunth.
STEROID: β-sitosterol.
TRITERPENOIDS: friedelin 3α-acetate, friedelin-3β-ol.
  Anjaneyulu et al., Curr. Sci., **46**, 776 (1977)

GUETTARDA (Rubiaceae)
G. angelica.
TRITERPENOID: 3,23-dihydroxy-12-ursen-28-oic acid.
  Sousa et al., Phytochemistry, **23**, 2589 (1984)
G. eximia.
One carboline alkaloid.
  Kan-Fan, Husson, J. Chem. Soc., Chem. Commun., 618 (1978)
G. heterosepala.
INDOLE ALKALOID: guettardine.
  Brillanceau et al., Tetrahedron Lett., 2767 (1984)

GUEVARIA
G. sodiroi.
SESQUITERPENOID: guevariolide.
  Bohlmann et al., Phytochemistry, **20**, 1144 (1981)

GUIBORTIA
G. coleosperma.
FLAVAN: guibourtacacidin.
  Roux et al., Biochem. J., **87**, 439 (1963)

GUILLONEA
G. scabra.
SESQUITERPENOIDS: guillonein, shairidin.
  Pinar et al., Phytochemistry, **22**, 987 (1983)

GUMMIPHORA
G. mukul.
DITERPENOID: allylcembrol.
  Ruecker et al., Arch. Pharm. (Weinberg, Ger.), **305**, 486 (1972)

GUTIERREZIA
G. dracunuloides.
DITERPENOID: gutierolide.
  Cruse et al., J. Chem. Soc., Chem. Commun., 1278 (1971)
G. gymnospermoides.
MONOTERPENOID: 2-trans-6-trans-8-trans-2,6,8-decatriene-4-yn-1-ol.
  Bohlmann et al., Chem. Ber., **98**, 369 (1965)
G. microcephala.
FLAVONOIDS: 5,2'-dihydroxy-3,6,7,8,4',5'-hexamethoxyflavone, 5,7,2',4'-tetrahydroxy-3,6,8,5'-tetramethoxyflavone, 5,7,2',5'-tetrahydroxy-3,6,8,4'-tetramethoxyflavone, 5,7,2',4'-tetrahydroxy-3,8,5'-trimethoxyflavone, 5,7,2'-trihydroxy-3,6,8,4',5'-pentamethoxyflavone, 5,7,2'-trihydroxy-3,6,4',5'-tetramethoxyflavone.
  Fang et al., Phytochemistry, **24**, 3029 (1985)

G. sarotirae.
DITERPENOIDS: articulin, articulin acetate, gutierrezial.
Bohlmann et al., Phytochemistry, 23, 2007 (1984)

G. wrightii.
FLAVONOIDS: chrysoeriol,
5,7-dihydroxy-3,4,6,8-tetramethoxyflavone, isorhamnetin,
isorhamnetin 3-O-β-D-glucoside, kaempferol, kaempferol
3-O--D-glucoside, luteolin, quercetin, quercetin
3-O-α-D-glucuronide, quercetin 3-O-β-D-glucoside,
3,5,7,3',4'-pentahydroxy-6-methoxyflavone,
3,5,7,4'-tetrahydroxy-6,3'-dimethoxyflavone,
5,7,4'-trihydroxy-3,6-dimethoxyflavone,
5,7,4'-trihydroxy-3,6,8,3'-tetramethoxyflavone,
5,7,4'-trihydroxy-3,6,8-trimethoxyflavone.
Fang et al., J. Nat. Prod., 49, 738 (1986)

GYMINDA
G. costarricensis Standl.
DITERPENOIDS: gymindone,
14-hydroxyabieta-8,11,13-trien-19-oic acid,
14-hydroxyabieta-8,11,13-trien-7-one,
hydroxymethoxyabieta-8,11,13-trien-19-oic acid.
Castro et al., Ing. Cienc. Quim., 10, 5 (1986); Chem. Abstr. 106 30032m

GYMNACRANTHERA (Myristacaceae)
G. paniculata (A.DC.) var. zippeliana (Miq.) Sinclair.
INDOLE ALKALOID: 3-demethylaminoethyl-1,5-dimethoxyindole.
Johns et al., Aust. J. Chem., 30, 1737 (1967)

GYMNEMA (Asclepiadaceae)
G. sylvestre R.Br.
QUINOLINE ALKALOID: gymnamine.
TRITERPENOIDS: gymnemagenin, gymnemic acid, gymnemic
acid A-1, gymnestrogenin.
(A) Rao et al., Chem. Ind. (London), 537 (1972)
(T) Stocklin et al., Helv. Chim. Acta, 51, 1235 (1968)

GYMNOCARPIUM
G. robertianum Newn. (G. jessoense Koidz.).
XANTHONES: gymnocarposide, mangiferin,
1,3,6,7-tetrahydroxyxanthone.
Murakami et al., J. Pharm. Soc., Japan, 106, 378 (1986)

GYMNOCLADA (Leguminosae)
G. dioica.
STEROID: 8(14),22-stigmastadien-3-ol.
Abramson, Su Kim, Phytochemistry, 12, 951 (1973)

GYMNOCOLEA (Musci)
G. inflata (Hudson) Dum.
DITERPENOID: gymnocolin.
Huneck et al., Tetrahedron Lett., 115 (1983)

GYMNOGONGRUS (Rhodophyceae)
G. flabelliformis.
COUMARIN: gongrine.
Ito et al., Agric. Biol. Chem., 29, 832 (1965)

GYMNOMITRION
G. obtusa.
SESQUITERPENOIDS: gynomitrene, gynomitrol, gynomitrol
acetate,
8-hydroxymethyl-1,2,6-trimethyltricyclo-[5,3,1,0]-undec-8-
en-11-yl acetate,
1,2,6-trimethyl-8-methylenetricyclo-[5,3,1,0]-undecane-5,11-
diyl diacetate,
1,2,6-trimethyl-8,8-oxymethylenetricyclo-[5,3,1,0]-
undecane-11-yl acetate.
Connolly et al., J. Chem. Soc., Perkin I, 2847 (1974)

GYMNOSPERMA
G. glutinosa.
DITERPENOID: gymnospermin.
TRITERPENOID: olean-12-ene-3,11-dione.
Miyakado et al., Phytochemistry, 13, 189 (1974)

GYMNOSPERMIUM
G. alberti (Regel) Takht.
QUINOLIZIDINE ALKALOIDS: darmasine, darvasamine,
N-methylcytisine.
Kurbanov et al., Rastit. Resur., 18, 372 (1982)

GYMNOSPORIA
S. rothiana.
TRITERPENOIDS;: gymnosporol olean-12-ene-3,11-dione.
Govindachari et al., Indian J. Chem., 8, 395 (1970)

G. wallichiana.
TRITERPENOIDS: gymnosporic acid, wallichianic acid,
wallichianol.
Kulshreshtha, Phytochemistry, 16, 1783 (1977)

GYMNOSTEMMA (Cucurbitaceae)
G. pentaphyllum Makino.
SAPONINS: gymnosaponins E, G, I and K, gypenosides XV,
XVI, XVII, XVIII, XIX, XX, XXI, XXXVI, XXXVIII,
XLII, XLIII, XLIV, XLV, XLVI, LIII and LIV.
STEROIDS: (24E)-dimethyl-5α-cholesta-7,22-dien-3β-ol,
24,24-dimethyl-5α-cholesta-7,25-dien-3β-ol,
24,24-dimethyl-5α-cholest-7-en-3β-ol,
14α-methyl-5α-ergosta-9(11),24(28)-dien-3β-ol.
(Sap) Takemoto et al., J. Pharm. Soc., Japan, 104, 939, 1043 (1984)
(S) Akihara et al., Phytochemistry, 26, 2412 (1987)

GYMNOSTROMA
G. missouriensis.
MONOTERPENOID: nectriapyrone.
Nair, Carey, Tetrahedron Lett., 1655 (1975)

GYNANDROPSIS
G. adenocarpa.
GLUCOSINOLATES: glucobrassicin, glucocapparine, glucoiberine,
neoglucobrassicin.
Saleh et al., Pharmazie, 31, 818 (1976)

G. grandiflora.
GLUCOSINOLATES: glucobrassicin, glucocapparine, glucoiberine,
neoglucobrassicin.
Saleh et al., Pharmazie, 31, 818 (1976)

G. gynandra.
GLUCOSINOLATES: glucobrassicin, glucocapparine, glucoiberine,
neoglucobrassicin.
Saleh et al., Pharmazie, 31, 818 (1976)

G. macrophylla.
GLUCOSINOLATES: glucobrassicin, glucocapparine, glucoiberine,
neoglucobrassicin.
Saleh et al., Pharmazie, 31, 818 (1976)

G. speciosa.
GLUCOSINOLATES: glucobrassicin, glucocapparine, glucoiberine
neoglucobrassicin.
Saleh et al., Pharmazie, 31, 818 (1976)

G. tricopus.
> GLUCOSINOLATES: glucobrassicin, glucocapparine, glucoiberine, neoglucobrassicin.
>> Saleh *et al., Pharmazie,* **31,** 818 (1976)

GYNOCARDIA (Flacourtiaceae)
G. odorata.
> CARBONITRILE: gynocardin.
> TRITERPENOIDS: O-acetylodollactone, odolactone, odollactone.
>> (C) Coburn *et al., J. Org. Chem.,* **31,** 4312 (1966)
>> (T) Pradhan *et al., Tetrahedron Lett.,* 865 (1984)

GYNURA (Compositae)
G. crepioides.
> COUMARIN TERPENOID: gynurone.
>> Bohlmann *et al., Phytochemistry,* **16,** 494 (1977)

G. japonica Makino.
> SAPONIN: ophiopogonin.
> STEROIDS: 3-epi-diosgenin-3-O-β-D-glucopyranoside, 3-epi-ruscogenin, 3-epi-sceptrumgenin-3-O-β-glucopyranoside.
>> Takahara *et al., Tetrahedron Lett.,* 3647 (1977)

GYROMITRA (Ascomycetae)
G. esculenta (Pers.) Fr.
> TOXINS: gynomitrin, hexylidene, 3-methylbutylidene, pentylidene.
>> List *et al., Arch. Pharm. (Weinreim, Ger),* **38,** 294 (1968)
>> Pyysalo, *Naturwissenshaften,* **62,** 395 (1975)

GYPSOPHILA (Caryophyllaceae)
G. acutifolia.
> SAPONIN: acutifolioside.
>> Chirva *et al., Khim. Prir. Soedin.,* 330 (1974)

G. pacifica.
> SAPONIN: gypsoside.
> TRITERPENOIDS: 12,13-dihydro-13-hydroxygypsogenin, gypsogenin.
>> (Sap) Kochetkov *et al., Tetrahedron Lett.,* 477 (1963)
>> (T) Khorlin *et al., J. Gen. Chem., USSR,* **32,** 782 (1962)

G. paniculata.
> SAPONIN: gypsoside.
>> Kochetkov *et al., Tetrahedron Lett.,* 477 (1963)

G. patrinii.
> SAPONIN: philoside B.
>> Bikharov *et al., Khim. Prir. Soedin.,* 603 (1974)

G. struthium.
> STEROIDS: 24R-ergosta-6,22-dien-5α-acetoxy-3-β-D-glucoside, 24R-ergosta-5,22-dien-7α-ol 3-β-D-O-glucoside, spinasterol, spinasteryl-3-β-O-D-glucoside, 24-stigmast-22-en-3β-ol, 24R-stigmast-22-en-3β-β-D-O-glucoside.
> TRITERPENOIDS: betulinic acid, oleanolic acid, 3β-hydroxylup-20(29)-en-27-oic acid.
>> Del Castillo *et al., Fitoterapia,* **57,** 61 (1986)

GYROCARPUS (Hernandiaceae)
G. americanus Jacq.
> QUATERNARY ALKALOID: (+)(S)-magnocurarine.
>> McKenzie, Price, *Aust. J. Chem.,* **6,** 180 (1953)

H

HABRANTHUS (Amaryllidaceae)
H. brachyandrus.
BENZHOMOISOQUINOLINE ALKALOID: habranthine.
Wildman, Brown, *Tetrahedron Lett.*, 4573 (1968)

HABROPETALUM
H. dawei.
TRINORSESQUITERPENOID: isoshinanolone.
Hanson *et al.*, *Phytochemistry*, **20**, 1161 (1981)

HAEMANTHUS (Amaryllidaceae)
H. albomaculatus.
LACTONE ALKALOID: albomaculine.
Briggs *et al.*, *J. Am. Chem. Soc.*, **78**, 2899 (1956)

H. amarylloides Jacq.
HOMOACRIDINE ALKALOID: manthine.
Wildman, Kaufman, *J. Am. Chem. Soc.*, **77**, 1248 (1955)

H. katherinae.
BENZOISOQUINOLINE ALKALOID: epi-haemanthamine.
Nogueiras *et al.*, *Tetrahedron Lett.*, 2743 (1971)

H. montanus Baker.
HOMOACRIDINE ALKALOID: montanine.
Wildman, Kaufman, *J. Am. Chem. Soc*, **77**, 1248 (1955)

H. natalensis.
BENZOISOQUINOLINE ALKALOID: epi-haemanthidine.
Goosen *et al.*, *J Chem. Soc.*, 1088 (1960)

H. puniceus x katherinae.
ISOQUINOLINE ALKALOIDS: haemanthamine, haemanthidine.
Fales *et al.*, *J. Am. Chem. Soc.*, **82**, 197 (1960)
Fales *et al.*, *ibid.*, **82**, 3368 (1960)

H. tigrinus.
ANTHRACENE ALKALOID: coccinine.
Wildman, Kaufman, *J. Am. Chem. Soc.*, **77**, 1248 (1955)

HAEMODORUM (Haemodoraceae)
H. distichophyllum (Hook).
PIGMENT: haemodorin.
Bick *et al.*, *Aust. J. Chem.*, **26**, 1377 (1973)

HAEMATOMMA (Lecanoraceae)
H. coccineum (Dicks.) Korb.
TRITERPENOID: zeorin.
Huneck, Lehn, *Bull. Soc. Chim.*, 1702 (1963)

H. erythromma (L.) Massal.
PHENOLIC: thamnolic acid.
Wachtmeister *et al.*, *Acta Chem. Scand.*, **9**, 1395 (1955)

H. ventosum (L.) Massal.
STEROID: 5,8-peroxyergosterol divaricatinate.
Bruun, Motzfeldt, *Acta Chem. Scand.*, **29**, 274 (1975)

HAEMATOXYLON
H. compechianum.
INDENOPYRAN: haematoxylin.
Parkin *et al.*, *J. Chem. Soc.*, **93**, 489 (1908)

HALFORDIA (Rutaceae)
H. kendack.
ISOQUINOLINE ALKALOID: halfordamine.
COUMARIN: halkendin.
(A) Crow, Hodgkin, *Aust. J. Chem.*, **21**, 2075 (1968)
(C) Nakayama *et al.*, *ibid.*, **24**, 209 (1971)

H. scleroxyla.
PYRIDINE ALKALOIDS: halfordanine, halfordine, halfordinine, halfordinol, halfordisone, N-methylhalfordine.
Crow, Hodgkin, *Aust. J. Chem.*, **21**, 3075 (1971)

HALIDONA (Rhodophyceae)
H. longley.
STEROID: haliclonasterol.
Bergman *et al.*, *J. Org. Chem.*, **14**, 1078 (1949)

HALIDRYS (Phaeophyceae)
H. siliquosa.
DINORTRITERPENOIDS:
(6'E,11'R,12'R)-12-hydroxy-5,13-dioxohalidrol,
(6'Z,11'R,12'R)012-hydroxy-5,13-dioxohalidrol,
(6'E,11'R,12'S)-12-hydroxy-5,13-dioxohalidrol.
Higgs *et al.*, *Tetrahedron*, **37**, 3209 (1981)

HALIMEDA (Chlorophyceae/Codiaceae) Halimeda sp. (unclassified).
DITERPENOIDS: halimedalactone halimestrial
Paul *et al.*, *Science*, **221**, 747 (1983)

HALIMIUM (Cistaceae)
H. viscosum.
DITERPENOIDS: O-acetylhydrohaliminic acid,
15-acetoxy-7,13E-labdadien-17-oic acid,
14,15-dinor-13-oxo-7-labden-17-oic acid, hydrohalimic acid,
2β-hydroxyhydrohalimic acid,
15-hydroxy-7,13E-labdadien-17-oic acid.
de Pascual Teresa *et al.*, *Phytochemistry*, **25**, 711 (1986)

HALLERIA
H. lucida.
QUINONES: halleridone, hallerone.
Messana *et al.*, *Phytochemistry*, **23**, 2617 (1984)

HALMIUM
H. umbellatum.
Two diterpenoids.
de Pascual Teresa *et al.*, *An. Quim.*, **71**, 112 (1975)

HALOSTACHYS (Chenopodiaceae)
H. caspica (Pall.) SAM.
AMINE ALKALOID: halostachine.
Man'shikov, Rubinshtein, *J. Gen. Chem.*, *USSR*, **13**, 801 (1943)

HALOZYLON (Chenopodiaceae)

H. articulatum.
ISOQUINOLINE ALKALOID: N-methylisosalsoline.
Carling, Sandberg, *Acta Pharm. Suec.*, **7**, 285 (1970)

H. salicornicum.
PIPERIDINE ALKALOID: halosaline.
ISOQUINOLINE ALKALOID: carnegine. Two uncharacterized alkaloids.
Michel *et al.*, *Acta Pharm. Suec.*, **4**, 97 (1967)
Brown *et al.*, *J. Org. Chem.*, **37**, 1825 (1972)

H. wakhanicum Eug. Kor. (Arthrophytum wakhanicum Pauls. Eug. Kor).
AMINE ALKALOID: N-methyl-β-phenethylamine.
Camp, Norvell, *Econ. Bot.*, **20**, 274 (1966)

HAMAMELIS (Hamamelidaceae)

H. virginiana L.
TANNIN: hamamelitannin.
Mayer *et al.*, *Justus Liebigs Ann. Chem.*, **688**, 232 (1965)

HAMELIA

H. patens Jacq.
OXINDOLE ALKALOIDS: isomaruquine, isopteropodine, maruquine, palmirine, pteropodine, rumberine, seneciophylline.
STEROID: stigmas-4-ene-3,6-dione.
(A) Nroges *et al.*, *Tetrahedron Lett.*, 3197 (1979)
(A) Borges del Castillo *et al.*, *Chem. Abstr.*, 97 123849e
(S) Ripperger, *Pharmazie*, **33**, 82 (1978)

HAMMARBYA (Orchidaceae)

H. paludosa (L.) Kuntze. (Malaxis paludosa L.).
PYRROLIZIDINE ALKALOIDS: hammarbine, paludosine.
Lindstrom, Luning, *Acta Chem. Scand.*, **26**, 2963 (1972)

HANDELIA

H. trichophylla.
SESQUITERPENOIDS: chrysartenin B, handelin, hanphyllin.
Tarasov *et al.*, *Khim. Prir. Soedin.*, 667 (1976)

HANNOA

H. klaineana.
DITERPENOIDS: chaparrinone, glaucarubolone, kleineanone.
Polansky, Bourguignon-Zylber, *Bull. Soc. Chim. Fr.*, 2793 (1965)
Moron, Polansky, *Tetrahedron Lett.*, 385 (1968)

H. undulata (Guillen et Perr.) Planch.
QUASSINOIDS: ailanthinone, undulatone.
Wani *et al.*, *Tetrahedron*, **35**, 17 (1979)

HAPLOCARPHA

H. lyrata.
SESQUITERPENOIDS: 13-acetoxysesquisabinene, 12-sesquisabienal, 12-sesquisabienol, 12-sesquisabienol acetate.
Bohlmann *et al.*, *Phytochemistry*, **21**, 1157 (1982)

H. scaposa.
SESQUITERPENOIDS: 13-acetoxysesquisabinene, 12-sesquisabienal, sesquisabienol, sesquisabienol acetate.
Bohlmann *et al.*, *Phytochemistry*, **21**, 1157 (1982)

HAPLOPAPPUS (Compositae)

H. angustifolius.
DITERPENOID: haplopappic acid.
Silva, Sammes, *Phytochemistry*, **12**, 1755 (1973)

H. ciliatus.
DITERPENOID: haplociliatic acid.
Bittner *et al.*, *Phytochemistry*, **17**, 1797 (1978)

H. foliosus.
FLAVONOIDS: esculetin, isorhamnetin, isorhamnetin 3-O-β-D-glucoside, kaempferol, kaempferol 3-O-β-D-glucoside, kaempferol methyl ether, quercetin, quercetin 3-O-β-D-galactoside, quercetin 3-O-β-D-glucoside, quercetin methyl ether.
DITERPENOID: haplopappic acid.
(F) Ulubelen *et al.*, *J. Nat. Prod.*, **45**, 363 (1982)
(T) Silva, Sammes, *Phytochemistry*, **12**, 1755 (1973)

H. freemontii.
SESQUITERPENOIDS: 1β-hydroxy-β-cyperone, 12(13)-hydroxy-β-cyperone, 8-oxo-β-cyperone.
Jakupovic *et al.*, *Planta Med.*, 411 (1986)

H. myrtifolium.
LIGNANS: (-)-haplomyrfolin, haplopyrtin.
Evcum *et al.*, *Phytochemistry*, **25**, 1949 (1986)

H. obtusifolium.
COUMARINS: obtusidin, obtusiprenin, scopoletin.
Matkarimov *et al.*, *Khim. Prir. Soedin.*, 173 (1982)

H. tenuisectus.
SESQUITERPENOID: isosesquicarene.
Bohlmann *et al.*, *Phytochemistry*, **18**, 1749 (1979)

HAPLOPHYLLUM (Rutaceae)

H. acutifolium.
ISOQUINOLINE ALKALOID: acutine.
FURANOQUINOLINE ALKALOID: skimmianine.
Razzakova *et al.*, *Khim. Prir. Soedin.*, 206 (1973)

H. alberti-regelii.
LIGNAN: diphyllin.
Nesmelova *et al.*, *Khim. Prir. Soedin.*, 643 (1983)

H. bucharicum Litv.
FURANOQUINOLINE ALKALOIDS: bucharamine, dictamnine, haplopine, robustine, skimmianine.
PROTOBERBERINE ALKALOID: α-fagarine.
QUINOLONE ALKALOIDS: bucharaine, bucharidine, folifidine.
PYRANOQUINOLINE ALKALOID: haplobucharine.
LIGNAN: diphyllin.
(A) Sharafutdinova, Yunusov, *Khim. Prir. Soedin.*, 394 (1969)
(A) Nasmalova *et al.*, *ibid.*, 815 (1975)
(L) Nesmelova *et al.*, *Khim. Prir. Soedin.*, 643 (1983)

H. bungei Trautv.
FURANOQUINOLINE ALKALOIDS: dictamnine, skimmianine.
QUINOLONE ALKALOID: robustinine.
Kurbanov, Yunusov, *Khim. Prir. Soedin.*, 289 (1967)

H. cimicidum.
DIMERIC ALKALOID: haplophytine.
IMIDAZOLIDINOCARBAZOLE ALKALOIDS: haplocidine, haplocine.
Yates *et al.*, *J. Am. Chem. Soc.*, **95**, 7842 (1973)

H. dauricum.
COUMARIN: dauroside D.
Vdovin *et al.*, *Khim. Prir. Soedin.*, 441 (1983)

H. dubium Eug. Kor.

DIMERIC ALKALOID: haplophytine.
FURANOQUINOLINE ALKALOIDS: dubinidine, dubinine, evoxine, skimmianine.
QUINOLINE ALKALOIDS: dubamine, foliosidine, foliosine.
QUINOLONE ALKALOIDS: graveoline, norgraveoline.
　　Yunusov, Sidyakov, *Dokl. Akad. Nauk SSSR*, **12**, 15 (1950)
　　Razzakova *et al.*, *Khim. Prir. Soedin.*, 810 (1979)

H. foliosum Vved.

FURANOQUINOLINE ALKALOIDS: dubinidine, folifinine, foliminine, folizine, haplofoline.
PYRANOQUINOLINE ALKALOIDS: folifine, haplopholine.
QUINOLINE ALKALOIDS: folifosidine, folimidine, folimine, foliosidine, foliosidine acetonide, foliozidine, graveoline, robustinine.
　　Razzakova *et al.*, *Khim. Prir. Soedin.*, **133**, 755 (1972)
　　Bessonova *et al.*, *ibid.*, 52 (1974)

H. hispanicum.

FURANOQUINOLINE ALKALOID: haplatine.
　　Gonzalez *et al.*, *An. Quim.*, **68**, 1133 (1972)

H. latifolium.

AMIDE ALKALOID: haplamidine.
FURANOQUINOLINE ALKALOID: haplatine.
　　Masmalova *et al.*, *Khim. Prir. Soedin.*, 666 (1975)
　　Nesmalova *et al.*, *ibid.*, 427 (1977)

H. obtusifolium Ldb.

FURANOQUINOLINE ALKALOIDS: evoxine, skimmianine.
　　Ubaidullaev *et al.*, *Khim. Prir. Soedin.*, 343 (1972)

H. pedicellatum Bge.

FURANOQUINOLINE ALKALOIDS: α-fagarine, robustine, skimmianine.
　　Ubaidullaev *et al.*, *Khim. Prir. Soedin.*, 343 (1972)

H. perforatum (MB.) Kar. et Kir.

FURANOQUINOLINE ALKALOIDS: evoxine, haplophylline, haplophytine, haplopine, perforine, skimmianine, triacetylglycoperine.
QUINOLONE ALKALOID: haplamine.
LIGNAN: diphyllin.
　　(A) Bessonova *et al.*, *Khim. Prir. Soedin.*, 360 (1968)
　　(A) Abdullaev *et al.*, *ibid.*, 425 (1977)
　　(L) Nesmelova *et al.*, *ibid.*, 643 (1983)

H. popovii Eug. Kor.

FURANOQUINOLINE ALKALOIDS: evoxine, skimmianine.
QUINOLONE ALKALOID: hapovine.
　　Razzakova *et al.*, *Khim. Prir. Soedin.*, 528 (1981)

H. ramosissimum Vved.

FURANOQUINOLINE ALKALOIDS: dictamnine, evoxine, skimmianine.
　　Kurbanov *et al.*, *Khim. Prir. Soedin.*, **67**, 289 (1967)

H. robustum Bge.

FURANOQUINOLINE ALKALOIDS: dictamnine, α-fagarine, haplopine, robustine.
QUINOLONE ALKALOID: robustinine.
　　Narasimhan *et al.*, *Tetrahedron*, **27**, 1351 (1971)

H. tuberculatum.

QUINOLONE ALKALOID:
3-dimethylallyl-4-dimethylallyloxy-2-quinolone.
　　Lavie *et al.*, *Tetrahedron*, **23**, 3011 (1968)

HAPLOPHYTON (Apocynaceae)

H. cimicidum A.DC.

CARBOLINE ALKALOIDS: eburnamine, methyleburnamine.
DIMERIC ALKALOIDS: cimiciphytine, cimilophytine, haplophytine, methyleburnamine, norcimiciphytine.
IMIDAZOLIDINOCARBAZOLE ALKALOIDS: cimicidine, cimicine, haplocidine, haplocine.
　　Bartlett *et al.*, *C. R. Acad. Sci.*, **249**, 1259 (1959)
　　Adesomoju *et al.*, *J. Org. Chem.*, **48**, 3015 (1983)

HAPLORMOSIA

H. monophylla.

ISOFLAVONES: dehydroferreirin, ferreirin.
　　King *et al.*, *J. Chem. Soc.*, 4580 (1952)
　　King et al. ibid., *ibid.*, 4752 (1952)

HARDWICKIA

H. pinnata Roxb.

SESQUITERPENOID: τ-cadinene.
DITERPENOIDS: (-)-hardwickiic acid, kolavelool, kolavenic acid, kolavenol, kolavenolic acid, kolavonic acid, kolovic acid.
　　Misra *et al.*, *Tetrahedron*, **35**, **979**, 985 (1979)

HARPAGOPHYTUM (Pedaliaceae)

H. procumbens (Burchell) DC.

IRIDOIDS: harpagoside, procumbide.
　　Lichti, Wartburg, *Tetrahedron Lett.*, 835 (1964)

HARPEPHYLLUM

H. caffrum.

FLAVONOIDS: apigenin 7-glucoside, kaempferol 3-galactoside, kaempferol 3-rhamnoside, quercetin 3-arabinoside, quercetin 3-glucoside, quercetin 3-rhamnoside.
PHENOLICS: gallic acid, protocatechuic acid.
　　Sherbeiny *et al.*, *Planta Med.*, **29**, 129 (1976)

HARPULLIA (Sapindaceae)

H. pendula Planch.

TRITERPENOID: A-1-barrigenol 22-angelate.
　　Lewis, Khong, *Aust. J. Chem.*, **29**, 1351 (1976)

HARRINGTONIA

H. abyssinica.

CHROMONE: 2-hydroxymethylalloptaeroxylin.
　　Balde *et al.*, *Phytochemistry*, **26**, 2415 (1987)

HARRISONIA

H. abyssinica.

TETRANORTRITERPENOIDS: harrisonin, obacunone, pedonin.
　　Kubo *et al.*, *Heterocycles*, **5**, 485 (1976)
　　Hassanali *et al.*, *Phytochemistry*, **26**, 573 (1987)

HARTWRIGHTIA

H. floridana.

DITERPENOID: hartwrightic acid.
　　Bohlmann *et al.*, *Phytochemistry*, **20**, 843 (1981)

HARUNGANA (Guttiferae/Clusiaceae)

H. madagascariensis.

ANTHRONE: haronginanthrone.
TRITERPENOID: harunganin.
XANTHONE: 1,7-dihydroxyxanthone.
　　Ritchie *et al.*, *Tetrahedron Lett.*, 1431 (1964)

HAZUNTA

H. costata.
DIMERIC CARBOLINE ALKALOID:
(19R)-19-hydroxy-tabernaelegantine A.
Urrea et al., Bull. Soc. Chim. Fr., 147 (1981)

H. modesta var. brevituba.
INDOLE ALKALOIDS: 16-epi-methuenine, methuenine.
Bui et al., Phytochemistry, 16, 703 (1977)

H. modesta var. divaricata.
INDOLE ALKALOIDS: 16epi-methuenine, methuenine.
Bui et al., Phytochemistry, 18, 1329 (1979)

H. modesta var. modesta subvar. divaricata.
DIMERIC CARBAZOLE ALKALOID: hazuntiphylline.
Bui et al., J. Nat. Prod., 49, 321 (1986)

H. silicicola Pichon.
INDOLE ALKALOID: (±)-oxo-6-silicine.
Reis et al., Tetrahedron Lett., 1085 (1976)

HEBECLINIUM

H. macrophyllum.
DITERPENOID: hebeclinolide.
Bohlmann, Grenz, Chem. Ber., 110, 1321 (1977)

HEBELOMA

H. crustuleniforme.
TRITERPENOIDS:
3β-acetyl-2α-(3'-hydroxy-3'-methylglutamyl)-crustulinol,
crustulinol.
de Bernardi et al., Tetrahedron Lett., 1630 (1983)

H. sinapizens.
TRITERPENOID: crustulinol.
de Bernardi et al., Tetrahedron Lett., 1630 (1983)

H. vinosophyllum.
SAPONINS: hebevinosides I, II and III.
Fujimoto et al., Tennen Yuki Kagobutsu Toronkai Koen Yoshishu, 26, 62 (1983)

HECHTIA

H. texensis.
STEROID: bacogenin.
Callow et al., Chem. Ind. (London), 699 (1951)

HEDERA (Araliaceae)

H. helix L.
SAPONIN: hederin. Van der Haar, Arch. Pharm. (Weinheim, Ger.) 250, 424 (1912)

H. nepalensis K.Koch.
GLYCOSIDES: campesterol β-D-glucoptranoside, hederagenin
β-D-glucoside, hederagenin 3-O-β-D-glucuronopyranoside,
oleanolic acid
3-O-α-L-rhamnopyranosyl-α-L-arabinopyranoside,
β-sitosterol β-D-glucoside, stigmasterol β-D-glucoside.
SAPONINS: HN-saponins A, D, D-1, D-2, E, F, H, I, K, M, N
and P.
Kizu et al., Chem. Pharm. Bull., 33, 3324 (1985)

H. rhombea.
SAPONINS: kizuta saponin K-8, kizuta saponin K-11.
Kizu et al., Chem. Pharm. Bull., 33, 3473 (1985)

HEDERANTHERA (Apocynaceae)

H. barteri (Hook.f.) Pichon.
DIIMIDAZOLIDINOCARBAZOLE ALKALOIDS: amataine,
demethylvobtusine, goziline, owerreine.
IMIDAZOLIDINOCARBAZOLE ALKALOIDS: hederantherine,
17-hydroxyhederantherine.
Agwada et al., Helv. Chim. Acta, 53, 1567 (1970)
Narango et al., ibid., 55, 1849 (1972)

HEDLOMA

H. pulegoides.
MONOTERPENOID: (+)-pulegone.
Baeyer, Henrich, Ber., 28, 652 (1895)

HEDWIGIA

H. ciliata.
FLAVONOIDS: lucenin-2, luteolin
7-O-neohesperidoside-4'-sophoroside, vicenin-2.
Osterdahl et al., Acta Chem. Scand., B 32, 93 (1978)

HEDYCARYA

H. angustifolia.
SESQUITERPENOID: hedycaryol.
Jones, Sutherland, J. Chem. Soc., Chem. Commun., 1229 (1968)

HEDYCHIUM (Zingeriberaceae)

H. spicatum Buch.-Ham. ex Sm.
DITERPENOIDS: 7-hydroxyhedychenone, hedychenone.
Sharma et al., Phytochemistry, 15, 827 (1976)

HEDYOTIS

H. acutangula.
TRITERPENOIDS: olean-12-ene-3β,28,29-triol,
olean-12-ene-3β,28,29-triacetate.
Hui, Li, J. Chem. Soc., Perkin I, 23 (1976)

H. auriculata L.
UNCHARACTERIZED ALKALOIDS: auricularine, hedyotine.
Ratnagiriswaran, Venkatachalam, J. Indian Chem. Soc., 19, 389 (1942)

H. diffusa.
IRIDOIDS: 6-O-p-courmaroylscandoside methyl ester,
6-O-feruloylscandoside methyl ester,
6-O-p-methoxycinnamoylscandoside methyl ester.
Nishihama et al., Planta Med., 43, 28 (1981)

HEDYSARUM (Leguminosae)

H. caucasicum.
FLAVONOIDS: antioside, kaempferol, quercetin.
Alaniya, Khim. Prir. Soedin., 646 (1983)

H. coronarium.
ANTHOCYANINS: malvidin 3,5-diglucoside, paeonidin
3,5-diglucoside, paeonidin 3-glucoside.
Chriki, Harborne, Phytochemistry, 22, 2322 (1983)

H. denticulatum.
XANTHONES: isomangiferin, mangiferin.
Glyzin et al., Khim. Prir. Soedin., 283 (1977)

H. polybotrys Hand-Mazz.
ACIDS: lignoceric acid, stearic acid.
ESTER: lignoceryl ferulate.
PTEROCARPAN: 3-hydroxy-9-methoxypterocarpan.
STEROID: β-sitosterol.
TRITERPENOID: ursolic acid.
Pan et al., Chem. Abstr., 105 187579x

H. sericeum.
FLAVONOIDS: antioside, hyperoside, isoquercitrin, kaempferol, mangiferin, quercetin.
Alaniya, *Khim. Prir. Soedin.*, 646 (1983)

HEIMIA (Lythraceae)
H. myrtifolia Cham. et Schlecht.
QUINOLIZIDINE ALKALOID: lythridine.
Douglas *et al.*, *Lloydia*, **27**, 25 (1964)

H. salicifolia Link et Otto.
QUINOLIZIDINE ALKALOIDS: dehydrodecodine, demethyllasubin I, demethyllasubine II, heimine, lyfoline, lythridine, lythrine, nesodine, sinicuichine, vertine.
Horhammer *et al.*, *Z. Naturforsch.*, **26B**, 970 (1971)
Rother, *J. Nat. Prod.*, **48**, 33 (1985)

HELENIUM (Compositae)
H. alternifolium.
SESQUITERPENOIDS: alternilin, brevilin A, linifolin A, linifolin B.
Herz *et al.*, *J. Org. Chem.*, **33**, 2780 (1969)
Herz *et al.*, *J. Am. Chem. Soc.*, **81**, 1481 (1969)

H. amarum.
SESQUITERPENOIDS: amaralin, aromaticin, heleniamarin.
Romo *et al.*, *Tetrahedron*, **20**, 79 (1964)
Ottersen *et al.*, *Acta Chem. Scand.*, **32B**, 79 (1978)

H. arenarium (L.) Moench.
α-PYRONES: arenol, homoarenol.
Vrkoc *et al.*, *Tetrahedron Lett.*, 247 (1971)

H. arizonicum.
SESQUITERPENOID: isotenulin.
Herz *et al.*, *J. Org. Chem.*, **27**, 4043 (1962)

H. aromaticum.
SESQUITERPENOIDS: aromaticin, aromatin, mexicanin I.
Romo *et al.*, *Chem. Ind. (London)*, 1839 (1963)

H. autumnale L.
CAROTENOID: helenien.
SESQUITERPENOIDS: akihalin, autumnolide, carolenalin, carolenin, carolenalone, dihydroflorilenalin, florilenalin, halshalin, helenalin, helenalin lactone, 2-methoxydihydrohelenin, picrohelenin, plenolin, sulferalin, 4-O-tigloyl-11,13-dihydroautumnolide.
Kondo *et al.*, *Tetrahedron Lett.*, 2155 (1977)
Furukawa *et al.*, *Chem. Pharm. Bull.*, **26**, 1335 (1978)

H. brevifolium.
SESQUITERPENOID: brevilin A.
Herz *et al.*, *J. Am. Chem. Soc.*, **81**, 1481 (1969)

H. coclinium.
SESQUITERPENOID: neohelenin.
Herz *et al.*, *J. Org. Chem.*, **27**, 4043 (1962)

H. elegans.
SESQUITERPENOIDS: helenalin, tenulin.
Herz *et al.*, *J. Org. Chem.*, **27**, 4043 (1962)

H. flexuosum.
SESQUITERPENOIDS: flexuosin A, flexuosin B.
Herz *et al.*, *J. Am. Chem. Soc.*, **82**, 2276 (1960)

H. laciniatum.
SESQUITERPENOID: helenalin.
Herz *et al.*, *J. Org. Chem.*, **27**, 4043 (1962)

H. linifolium.
SESQUITERPENOIDS: linifolin A, linifolin B.
Herz *et al.*, *J. Org. Chem.*, **27**, 4043 (1962)

H. mexicanum.
SESQUITERPENOIDS: helenalin, mexicanins A, B, C, D, E, F, G, H and I.
Romo *et al.*, *Tetrahedron Lett.*, 579 (1985)

H. microcephalum.
SESQUITERPENOIDS: helenalin, isohelenol, mexicanin E, microhelenins A, B and C, microlenin.
Lee *et al.*, *Phytochemistry*, **16**, 393 (1977)
Imakura *et al.*, *J. Pharm. Sci.*, **67**, 1228 (1978)

H. pinnatifidum.
SESQUITERPENOID: pinnatifidin.
Herz *et al.*, *J. Am. Chem. Soc.*, **81**, 1481 (1958)

H. puberulum.
SESQUITERPENOIDS: desacetylisobigelovin, helepuberinic acid, puberolide.
Bohlmann, Jakupovic, *Phytochemistry*, **18**, 131 (1979)

H. quadridentatum.
SESQUITERPENOIDS: quadridenin, quadridentin.
Watanabe *et al.*, *J. Agric. Food Chem.*, **33**, 83 (1985).

H. tenuifolium.
SESQUITERPENOID: tenulin,
Clark, *J. Am. Chem Soc.*, **61**, 1836 (1939)

H. thurberi.
SESQUITERPENOID: thurberilin.
Herz *et al.*, *Tetrahedron*, **21**, 1711 (1965)

H. virginicum.
SESQUITERPENOID: virginolide,
Herz, Santhaman, *J. Org. Chem.*, **32**, 507 (1967)

HELIABRAVOA (Cactaceae)
H. chende.
TRITERPENOID: oleanic aldehyde.
Shamma, Rosenstock, *J. Org. Chem.*, **24**, 726 (1959)

HELIANTHELLA (Compositae)
H. uniflora.
CHROMENES: 6-acetyl-2,2-dimethyl-2H-1-benzopyran, eupatoriochromene.
Bohlmann *et al.*, *Phytochemistry*, **17**, **566**, 1677 (1978)

HELIANTHUS (Compositae)
H. annuus L.
STEROID: 7,24-stigmastadien-3β-ol.
CAROTENOID: lutein.
SESQUITERPENOID: annuithrin.
DITERPENOIDS: (-)-kaur-16-en-19-oic acid, trachyloban-19-oic acid.
TRITERPENOIDS: grandiflorolic acid, heliantriols B-0, B-1 and B-2, heliantriols C and F.
(C) Karrer, Eugster, *Helv. Chim. Acta*, **33**, 300 (1950)
(S) Hamberg *et al.*, *Phytochemistry*, **12**, 1767 (1973)
(T) Spring *et al.*, *ibid.*, **20**, 1883 (1981)
(T) Mitscher *et al.*, *J. Nat. Prod.*, **46**, 745 (1983)

H. argophyllus.
SESQUITERPENOIDS: argophyllins A and B, argophyllone B.
Stipanovic *et al.*, *Phytochemistry*, **24**, 358 (1985)

H. ciliaris.
DITERPENOID: ciliaric acid.
SESQUITERPENOIDS: calaxin, ciliarin.
Nabin *et al.*, *J. Org. Chem.*, **44**, 1831 (1979)

H. debilis.
DITERPENOID: tetrachrysin.
Ohno *et al.*, *Phytochemistry*, **18**, 1687 (1979)

H. decapetalus.
DITERPENOID: ent-11β-acetoxy-16-atisen-19-oic acid.
Bohlmann *et al.*, *Phytochemistry*, **19**, 863 (1980)

H. heterophyllus.
BENZOQUINONE: 2,6-dimethoxybenzoquinone.
CAROTENOID: lutein.
COUMARIN: (-)-8-methoxyobliquin.
DINORSESQUITERPENOIDS: dehydrovomifoliol,
9-hydroxy-4-megastigmen-3-one, (±)-vomifoliol.
SESQUITERPENOIDS: 2′,3′-dihydroleptocarpin,
2′,3′-dihydronivensin.
(B, C, T) Herz, Bruno, *Phytochemistry*, **25**, 1913 (1986)
(Car.) Karrer, Eugtser, *Helv. Chim. Acta*, **33**, 300 (1950)

H. hirsutus.
Two pimaranes.
Herz *et al.*, *Phytochemistry*, **23**, 1453 (1984)

H. mollis.
SESQUITERPENOIDS: mollisorin A, mollisorin B.
Ohno, Mabry, *Phytochemistry*, **18**, 1003 (1979)

H. niveus subsp. canescens.
SESQUITERPENOIDS: niveusins A, B and C.
Ohno, Mabry, *Phytochemistry*, **19**, 609 (1980)

H. occidentalis.
DITERPENOID: occidentalic acid.
Herz *et al.*, *Phytochemistry*, **22**, 2021 (1983)

H. occidentalis var. dowellianus.
MONOTERPENOID: 10-hydroxyverbenone.
Bohlmann *et al.*, *Planta Med.*, **50**, 202 (1984)

H. schweinitzii.
SESQUITERPENOID: 4,15-isoatripliciolide tiglate.
Gershenzon, Mabry, *Phytochemistry*, **23**, 2557 (1984)

H. strumosus.
Two ent-pimaranes and one heliangolide.
Herz *et al.*, *Phytochemistry*, **23**, 1453 (1984)

H. tuberosus L.
DITERPENOID: ent-12-acetoxy-16-kauren-19-oic acid.
SESQUITERPENOIDS: 2,7(14),9-bisabolatrien-11-ol, heliangine.
Miyazawa *et al.*, *Phytochemistry*, **22**, 1040 (1983)

HELICHRYSUM (Compositae)

H. arenarium.
BENZOFURAN: arenopitialide.
ISOCOUMARIN: floribundaside.
PIGMENT: arenol.
PYRAN: helipyrone diacetate.
(B, P) Budesinsky *et al.*, *Phytochemistry*, **14**, 1383 (1985)
(B, P) Budesinsky *et al.*, *ibid.*, **14**, 1845 (1985)
(I) Hansel, *Arch. Pharm.* (Weinheim, *Ger.*) **292**, 398 (1959)
(Pyran) Hansel *et al.*, *Tetrahedron Lett.*, 3369 (1970)

H. aureo-nitens.
TERPENOID: aureonitol.
Bohlmann *et al.*, *Phytochemistry*, **18**, 664 (1979)

H. bracteatum (Vent.) Andr.
FLAVONOIDS: eriodictyol, naringenin,
5,7,3′,4′,5′-pentahydroxyflavanone.
DINORSESQUITERPENOID: 4-formyl-1-methoxycarboxyazuleine.
(F) Forkmann, *Z. Naturforsch.*, **38C**, 891 (1983)
(T) Bohlmann *et al.*, *Chem. Ber.*, **106**, 1337 (1973)

H. caespititium.
PHLOROGLUCINOL: caespitin.
Dekker *et al.*, *S. Afr. J. Chem.*, **36**, 114 (1983)

B. chionosphaerum.
DITERPENOIDS;: 19-helifulvanoic acid;, 19-helifulvanol;.
SESQUITERPENOIDS;: helifulvanic acid;, humulen-14-ol;.
Bohlmann *et al.*, *Phytochemistry*, **19**, 869 (1980)

H. dendroioideum.
DITERPENOIDS: ent-kaur-15-ene-17,19-diol,
(-)-kaur-16-ene-3,18-diol, erythroxydiol A,
stach-15-ene-3α,17-diol, stach-15-ene-3β,17-diol,
stach-15-ene-3α,19-diol, stach-15-ene-17,19-diol.
Lloyd, Fales, *Tetrahedron Lett.*, 4891 (1967)

H. fulvum.
DITERPENOID: helifulvanolic acid.
Bohlmann *et al.*, *Phytochemistry*, **18**, 1359 (1979)

H. graveolens.
FLAVONES: apigenin, apigenin-4′-glucoside,
apigenin-7-glucoside, astragalin,
3,5-dihydroxy-6,7,8-trimethoxyflavone, galangin 3-methyl
ether, helichrysin B, luteolin, luteolin-4′-glucoside,
luteolin-7-glucoside, kaempferol, isosalipurposide,
naringenin, naringenin-4′-glucoside.
Cubukcu, Damadyan, *Fitoterapia*, **57**, 124 (1986)

H. heterolasium.
DITERPENOIDS: 14,15-dihydro-16-helicallenoic acid,
16-helicallenal.
Bohlmann *et al.*, *Phytochemistry*, **18**, 889 (1979)

H. italicum.
PYRAN: helipyrone.
Hansel *et al.*, *Tetrahedron Lett.*, 3369 (1970)

H. kraussii.
BENZOPYRAN: helichrysoside.
Candy *et al.*, *Tetrahedron Lett.*, 1211 (1975)

H. orientale (L.) Gaertn.
FLAVONOIDS: astrogalin,
3,5-dihydroxy-6,7,8-trimethoxyflavone, kaempferol,
quercetin, scopaletin, tiliroside.
Cubukcu, *Plant Med. Phytother.*, **10**, 44 (1976)

H. splendidum.
SESQUITERPENOID: helisplendiolide
Bohlmann *et al.*, *Phytochemistry*, **18**, 885 (1979)

HELICIA

H. erratica.
GLYCOSIDE: helicide.
Wei-Shin *et al.*, *Justus Liebigs Ann. Chem.*, 1893 (1981)

HELICOBASIDIUM

H. mompa.
PIGMENTS: helicobasidin, helicobasin.
Nishikawa *et al.*, *Agric. Biol. Chem.*, **26**, 696 (1962)

HELIETTA

H. longifoliata.
QUINOLONE ALKALOID: isodictamnine.
NORDITERPENOID: heliettin.
(A) Mammarella, Comin, *An. Asoc. Quim. Argentina*, **59**, 239 (1971)
(T) Piozzi *et al.*, *Tetrahedron*, **23**, 1129 (1967)

H. parviflora.
FURANOQUINOLINE ALKALOID: heliparvifoline.
Chang *et al.*, *J. Pharm. Sci.*, **65**, 561 (1975)

HELIOPSIS (Compositae)
H. buphthalmoides.
SESQUITERPENOIDS: dehydroheliobuphthalmin, heliobuphthalmin.
Bohlmann *et al.*, *Phytochemistry*, **17**, 330 (1978)
H. scabra Dunal.
LIGNAN: helianthoidin.
NAPHTHODIOXOLONE: helioxanthin.
(L) Burden *et al.*, *Tetrahedron Lett.*, 1035 (1968)
(N) Burden *et al.*, *J. Chem. Soc., C*, 693 (1969)

HELIOTROPIUM (Boraginaceae)
H. acutiflorum Kit. et Kar.
PYRROLIZIDINE ALKALOIDS: heliotrine, heliotrine N-oxide.
Culvenor *et al.*, *Aust. J. Chem.*, **7**, 277 (1954)
H. arguzoides Kit. et Kar.
PYRROLIZIDINE ALKALOIDS: trichodesmine, trichodesmine N-oxide.
Akramov *et al.*, *Dokl. Akad. Nauk SSSR*, **4**, 30 (1961)
H. curassavicum.
PYRROLIZIDINE ALKALOIDS: 7-angelylheliotridine, 7-angelylheliotrine N-oxide, heliotrine, heliotrine N-oxide, lasiocarpine.
Rajagapalan *et al.*, *Indian J. Chem.*, **15B**, 494 (1977)
H. dasycarpum Ldb.
PYRROLIZIDINE ALKALOIDS: heliotrine, heliotrine N-oxide.
Rajagapalan *et al.*, *Indian J. Chem.*, **15B**, 494 (1977)
H. eichwaldii.
PYRROLIZIDINE ALKALOID: 7-angeloylheliotrine.
Suri *et al.*, *Indian J. Chem.*, **13**, 505 (1975)
H. europeum.
PYRROLIZIDINE ALKALOIDS: acetyllasiocarpine, heliotrine N-oxide, lasiocarpine N-oxide.
UNCHARACTERIZED ALKALOID: isoheliotrine.
Culvenor *et al.*, *Aust. J. Chem.*, **28**, 2319 (1975)
H. indicum.
PYRROLIZIDINE ALKALOIDS: acetylindicine, indicine, indicine N-oxide, indicinine.
STEROIDS: campesterol, chalinasterol, β-sitosterol, stigmasterol.
(A) Mattocks, *J. Chem. Soc., C*, 329 (1967)
(S) Andhiwal *et al.*, *Indian Drugs*, **22**, 567 (1985); *Chem. Abstr.*, **104** 3416s
H. lasiocarpum F. et M.
PYRROLIZIDINE ALKALOIDS: heliotrine, heliotrine N-oxide, lasiocarpine, lasiocarpine N-oxide.
Yunusov, Sidyakin, *Dokl. Akad. Nauk SSSR*, **1**, 3 (1950)
H. olgae Bge.
PYRROLIZIDINE ALKALOIDS: heliotrine, heliotrine N-oxide, incanine, incanine N-oxide, lasiocarpine, lasiocarpine N-oxide.
Kiyamitdinova *et al.*, *Khim. Prir. Soedin.*, 411 (1967)
H. ovalifolium.
PYRROLIZIDINE ALKALOID: heliofoline.
Subramanian *et al.*, *Phytochemistry*, **20**, 1991 (1981)
H. ramosissimum.
PYRROLIZIDINE ALKALOID: heliotrine.
Habib, *Bull. Fac. Sci. Riyadh Univ.*, **7**, 67 (1975); *Chem. Abstr.*, **85** 119597v
H. strigosum.
PYRROLIZIDINE ALKALOID: strigosine.
Mattocks, *J. Chem. Soc.*, 1974 (1964)
H. supinum L.
PYRROLIZIDINE ALKALOIDS: heliosupine, supinine.
Men'shikov *et al.*, *J. Gen. Chem., USSR*, **19**, 1382 (1949)

H. transoxanum Bge.
PYRROLIZIDINE ALKALOIDS: heliotrine, heliotrine N-oxide.
Akramov *et al.*, *Khim. Prir. Soedin.*, 258 (1968)

HELLEBORUS (Ranunculaceae)
H. dumetorum Waldst. et Kit.
STEROID: 14-hydroxy-3-oxo-1,4,20,22-bufatetraenolide.
Wissner, *Planta Med.*, **24**, 201 (1973)
H. dumetorum var. atrorubens (Waldst. et Kit.) Merxm. et Podlech.
STEROID: helleborogenone.
Petricis *et al.*, *Acta Pharm. Jugosl.*, **220**, 158 (1972)
H. niger L.
STEROID: spirosta-5,25(27)-diene-1,3,11-triol.
Linde *et al.*, *Helv. Chim. Acta*, **54**, 1703 (1971)
H. occidentalis.
STEROIDS: 14-hydroxy-3-oxo-1,4,20,22-bufatetraenolide, helleborogenone.
Wissner, *Planta Med.*, **24**, 201 (1973)
H. odorus (L.) Waldst. et Kit.
STEROIDS: bufalone, 11α-hydroxydesglucohellebrin, 14-hydroxy-3-oxo-5,14-bufo-20,22-dienolide, spirosta-5,25(27)-diene-1,3,11-triol.
Hauser *et al.*, *Helv. Chim. Acta*, **55**, 2625 (1972)
Kissmer *et al.*, *Planta Med.*, 152 (1986)
H. viridis L.
UNCHARACTERIZED ALKALOIDS: sprintillamine, sprintilline.
STEROIDS: helleborogenone, 14-hydroxy-3-oxo-1,4,20,22-bufotetraenolide.
(A) Keller, Schobel, Arch. Pharm. (Weinheim, *Ger.*) **265**, 238 (1927)
(S) Wissner, *Planta Med.*, **24**, 201 (1973)
H. viridis L. var. occidentalis (Reuter) Schiffner.
STEROID: 14-hydroxy-3-oxo-1,4,20,22-bufotetraenolide.
Wissner, *Planta Med.*, **24**, 201 (1973)

HELMINTHOSPORIUM
H. laersii.
TRITERPENOID: alboleersin.
Ashley, Raistrick, *Biochem. J.*, **32**, 449 (1938)
H. oryzae.
SESTERTERPENOIDS: ophiobolins A and B.
Canonica *et al.*, *Tetrahedron Lett.*, 2211 (1966)
H. sativum.
SESQUITERPENOIDS: helminthogermacrene, helminthosporal, helminthosporol.
Winter *et al.*, *J. Org. Chem.*, **45**, 4786 (1980)
H. siccans.
ACID: presiccanochromenic acid.
Nozoe *et al.*, *Tetrahedron Lett.*, 3643 (1968)
H. zizaniae.
SESTERTERPENOIDS: ophiobolins A and B.
Nozoe *et al.*, *J. Am. Chem. Soc.*, **87**, 4968 (1967)

HELMINTHOSTACHYS
H. zeylanica.
BENZOPYRANS: ugonins A, B and C.
Murakami *et al.*, *Chem. Pharm. Bull.*, **21**, 1849 (1973)
Murakami *et al.*, *ibid.*, **21**, 1851 (1973)

HELVELLA
H. esculenta.
One uncharacterized alkaloid.
Awe, Arch. Pharm. (Weinheim, *Ger.*) **271**, 537 (1933)

HEMANDIA

H. ovigera.

BENZODIOXOLE: epi-aschantin.

Nishino *et al., Tetrahedron Lett.,* 335 (1973)

HEMEROCALLUS (Liliaceae)

H. fulva.

ALDEHYDES: dotriacontanal, octacosanal, triacontanal.
GLUTAMINE: pinnatanine.

(Al) Isono *et al., J. Pharm. Soc., Japan,* **96,** 86 (1976)

(G) Grove *et al., Tetrahedron,* **29,** 2715 (1973)

H. minor Mill.

FLAVONOIDS: chrysophanol, hemerocallin, hemerocallone, rhein.
STEROIDS: β-sitosterol, τ-sitosterol.

Xiu *et al., Zhongcaoyao,* **13,** 1 (1982); *Chem. Abstr.,* **97** 107026t

HEMIARIA

H. glabra.

FLAVONE: narcissin.

Rahman *et al., J. Org. Chem.,* **27,** 153 (1962)

HEMIDESMUS

H. indicus.

SAPONIN: desinine.
TRITERPENOIDS: α-amyrin, β-amyrin.

(Sap) Oberai *et al., Phytochemistry,* **24,** 2395 (1985)

(T) Ruzicka, Wurz, *Helv. Chim. Acta,* **22,** 948 (1939)

HEMISELMIS (Cryptophyceae/Hemiselmidaceae)

H. virescens.

CAROTENOIDS: crocoxanthin, alloxanthin.

Chapman, *Phytochemistry,* **5,** 1331 (1966)

HEMIZONIA

H. congesta.

SESQUITERPENOID: 11-hydroxycubebol.

Bohlmann *et al., Phytochemistry,* **20,** 2383 (1981)

H. lutescens.

DITERPENOID: 14,15-epoxy-8(17)-labden-15-ol.

Bohlmann *et al., Phytochemistry,* **20,** 2383 (1981)

HEMSLEYA

H. graciliflora.

TRITERPENOIDS: cucurbitacin Ia, 23,24-dihydrocucurbitacin F, 23,24-dihydrocucurbitacin F 25-acetate.

Meng *et al., Yaoxue Xuebao,* **20,** 446 (1985); *Chem. Abstr.,* **104** 17642z

HERACLEUM (Umbelliferae/Pastinaceae)

H. asperum.

COUMARINS: bergapten, byakangelicin, heraclesol, imperatorin, isobergapten, isopimpinellin, phellopterin, pimpinellin, xanthotoxin.

Komissarenko *et al., Khim. Prir. Soedin.,* 446 (1987)

H. candicans.

BENZOPYRONE: 8-geranylpsoralen.
COUMARIN: prangenin.
DITERPENOID: candicopimaric acid.

(B) Stanley, Vannier, *J. Am. Chem. Soc.,* **79,** 3488 (1957)

(C) Kidwai *et al., Tetrahedron,* **20,** 87 (1964)

(T) Bandopadhyay *et al., Indian J. Chem.,* **11,** 1097 (1973)

H. canescens.

COUMARINS: 8-geranoxypsoralin, heraclemin, imperatorin, isobergapten, osthole, xanthotoxin.

Kumar *et al., Planta Med.,* **30,** 291 (1976)

H. dissectum.

COUMARINS: bergapten, imperatorin, isobergapten, 7-isopentenyloxycoumarin, isopimpinellin, pimpinellin, sphondin, umbelliferone.

Belenocskaya *et al., Khim. Prir. Soedin.,* 574 (1977)

Gusak *et al., Chem. Abstr.,* **86** 68378k

H. granatense Boiss. et Elench.

COUMARINS: bergapten, byakangelicin, byakangelicin acetonide, 8-(3-chloro-2-hydroxy-3-methylbutyryloxy)psoralen, heraclenol, heraclenol acetonide, 8-(2-hydroxy-3-butenyloxy)psoralen, imperatorin, isobergapten, isoheraclenin, isopimpinellin, isorhamnetin 7-rhamnoglucoside, phellopterin, scopoletin, sphondin, umbelliferone, umbelliprenin, xanthotoxin.
STEROIDS: β-sitosterol, β-sitosterol β-D-glucoside.
TRITERPENOID: ursolic acid.

Kutney *et al., Tetrahedron,* **28,** 5091 (1972)

Gonzalez *et al., An. Quim.,* **73,** 858 (1977)

H. grandiflorum.

COUMARINS: bergapten, isobergapten, isopimpinellin, pimpinellin, xanthotoxin.

Kerimov *et al., Khim. Prir. Soedin.,* 253 (1976)

H. lehmannianum Bunge.

COUMARINS: angelicin, antoside, bergapten, flavantaside, isobergapten, isopimpinellin, pimpinellin, epi-rutin, sphondin.

Temirbekov *et al., Rastit. Resur.,* **13,** 342 (1977); *Chem. Abstr.,* **87** 2382w

H. levskovii.

CHROMENE: psoralen.

Austin *et al., Phytochemistry,* **12,** 1657 (1973)

H. mantegazzanium Sommier et Levier.

BENZOPYRANS: apterin, phellopterin.

Fischer *et al., Phytochemistry,* **15,** 1079 (1976)

H. moellendorffii Hance.

COUMARINS: bergapten, byakangelicin, heraclesol, imperatorin, isobergapten, isopimpinellin, phellopterin, pimpinellin, xanthotoxin.

Komissarenko *et al., Khim. Prir. Soedin.,* 446 (1987)

H. moellendorffii Hance var. paucivitatum.

COUMARINS: bergapten, byakangelicin, heraclesol, imperatorin, isobergapten, isopimpinellin, pimpinellin, xanthotoxin.

Wu *et al., Yaoxue Xuebao,* **21,** 599 (1986); *Chem. Abstr.,* **105** 18724h

H. nepalense.

COUMARINS: bergapten, isobergapten, isopimpinellin, sphondin.

Gupta *et al., Phytochemistry,* **14,** 2533 (1975)

H. pinnatum.

COUMARINS: 8-geranoxypsoralin, heraclenin, imperatorin, isobergapten, osthole, xanthotoxin.

Jumar *et al., Planta Med., B* **30,** 291 (1976)

H. ponticum.

COUMARINS: bergapten, byakangelicin, heraclesol, imperatorin, isobergapten, isopimpinellin, phellopterin, pimpinellin, xanthotoxin.

Komissarenko *et al., Khim. Prir. Soedin.,* 446 (1987)

H. pyrenaicum Lam.

COUMARINS: apiol
8-(3-chloro-2-hydroxy-3-methylbutyloxy)-5-
metoxypsoralen,
8-(3-chloro-2-hydroxy-3-methylbutyryloxy)-psoralen,
isoheraclenin, isopimpinellin, scopoletin,

Gonzalez *et al., An. Quim.,* **72,** 584 (1976)

Sethi *et al., Phytochemistry,* **15,** 1773 (1976)

H. sibiricum.

COUMARINS: bergapten, imperatorin, isobergapten,
isopimpinellin, pimpinellin, sphondin, umbelliferone.

Gusak *et al., Chem. Abstr.,* 86 68378k

H. sosnowskyi.

COUMARINS: angelicin, bergapten, isobergapten, pimpinellin,
sphondin, umbelliferone.

Muradyan *et al., Biol. Zh. Arm.,* **29,** 97 (1976); *Chem. Abstr.,* 86 68363b

H. sphondylum L.

COUMARINS: bergapten, byakangelicin, heraclesol, imperatorin,
isobergapten, isopimpinellin, phellopterin, pimpinellin,
xanthotoxin.

Komissarenko *et al., Khim. Prir. Soedin.,* 446 (1987)

H. thomsoni.

COUMARINS: apterin, 5-hydroxyangelicin,
5-hydroxy-8-(1',1'-dimethylallyl)psoralen, vaginidiol.

Fischer *et al., Phytochemistry,* **15,** 1079 (1976)

Gupta *et al., Indian J. Chem.,* **16B,** 38 (1978)

H. trachyloma.

COUMARINS: angelicin, bergapten, isobergapten, pimpinellin,
sphondin, umbelliferone.

Muradyan *et al., Biol. Zh. Arm.,* **29,** 97 (1976); *Chem. Abstr.,* 86 68363b

H. wallichii.

BISBENZYLISOQUINOLINE ALKALOID: cycleanine.

COUMARINS: bergapten, columbianetin, isobergapten,
isopimpinellin, marmesin, sphondin, vaginidiol.

STEROID: stigmasterol.

Gupta *et al., Phytochemistry,* **15,** 576 (1976)

H. woroschiklowii.

COUMARINS: bergapten, byakangelicin, heraclesol, imperatorin,
isobergapten, isopimpinellin, phellopterin, pimpinellin,
xanthotoxin.

Komissarenko *et al., Khim. Prir Soedin.,* 446 (1987)

HERBA

H. rutae.

COUMARIN: rutaretin.

Garg *et al., Phytochemistry,* **17,** 2135 (1978)

HERNANDIA (Lauraceae/Hernandiaceae)

H. bivalvis Benth.

APORPHINE ALKALOID: hernandine.

Greenhalgh, Lahey, *in: Current Trends in Heterocyclic Chemistry*
London, Butterworth, 100 (1966)

H. nymphaefolia.

APORPHINE ALKALOID: hernagine.

PYRIDINE ALKALOID: 3-cyano-4-methoxypyridine.

Yakushijn *et al., Phytochemistry,* **19,** 161 (1980)

H. ovigera L.

APORPHINE ALKALOIDS: hernandeline, hernovine,
N-methylhernangerine, N-methylhernovine,
N-methylnandigerine, N-methylovigerine, nandigerine,
ovigerine.

OXOAPORPHINE ALKALOID: hernandonine.

Ito, Furukawa, *Tetrahedron Lett.,* 3023 (1970)

H. peltata.

BISBENZYLISOQUINOLINE ALKALOIDS: (+)-malekulatine,
(+)-vanuatine, (+)-vateamine.

Bruneton *et al., J. Org. Chem.,* **48,** 3957 (1983)

HERNIARIA (Caryophyllaceae/Illecebraceae)

H. glabra L.

SAPONINS: hernaria saponins A and B.

Bucharov, Scherbak, *Khim. Prir. Soedin.,* 307 (1970)

HESPERATHUSA

H. crenulata (Roxb.) M.Roem.

QUINOLONE ALKALOID: 4-methoxy-1-methyl-2-quinolone.

COUMARINS: 6-formyl-7-methoxycoumarin,
7-methoxycoumarin, suberosin.

(A) Nayar *et al., Phytochemistry,* **10,** 2843 (1971)

(C) Ubaev *et al., Khim. Prir. Soedin.,* 248 (1974)

(C) Basa *et al., J. Nat. Prod.,* **45,** 503 (1982)

HESPERIS (Cruciferae)

H. matronalis L.

QUATERNARY ALKALOID: hesperaline.

Gmelin, *Mohrrie, Arch. Pharm. (Weinheim, Ger.)* **300,** 176 (1967)

HETEROCHORDARIA

H. abietina.

IMINOACID: L-1,4-thiazane-3-carboxylic acid.

Kawauchi *et al., Chem. Abstr.,* 88 71433w

HETEROCONDYLUS

H. vitalbe.

SESQUITERPENOID:
epi-9-angeloyloxy-10,11-epoxydihydrobrickelliol.

Bohlmann, Grenz, *Chem. Ber.,* **110,** 1321 (1977)

HETERODEA (Lichenes)

H. beaugleholei.

DEPSIDE: nordivaricatic acid.

Elix *et al., Aust. J. Chem.,* **30,** 2333 (1977)

HETERODENDRON (Sapindaceae)

H. oleaefolium.

CYANOGLYCOSIDES: cardiospermin, dihydroacacipetalin.

Huebel *et al., Phytochemistry,* **14,** 2723 (1975)

HETEROSMILAX

H. japonica Kunth.

ACIDS: palmitic acid, stearic acid.

STEROID: β-sitosterol.

Chen *et al., Chem. Abstr.,* 86 27676n

HETEROTHALAMUS

H. spartioides.

STEROIDS: sitosterol, stigmasterol.

UMBELLIFERONES: aurapten, aurapten-6',7'-epoxide,
5'-ketoaurapten
5'-oxo-2',3',6',7'-tetrahydroaurapten-6',7'-epoxide.

Boeker *et al., Rev. Latinoam. Quim.,* **17,** 47 (1986)

HETEROTHECA (Compositae/Asteraceae)

H. grandiflora.
SESQUITERPENOIDS: 1(10),4-cadinadien-9-one, α-calacoren-3-ol, iso-epi-cubenol.
Bohlmann et al., *Phytochemistry*, **18**, 1675 (1979)

H. inuloides Cass.
SESQUITERPENOID: dihydro-7-hydroxycadalene.
Bohlmann, Zdero, *Chem. Ber.*, **109**, 2021 (1976)

H. latifolia.
SESQUITERPENOIDS: 3-acetoxy-1(10),4-cadinadien-15-oic acid, 10-hydroxycalamenen-15-oic acid.
Bohlmann et al., *Phytochemistry*, **21**, 2982 (1982)

H. subaxillaris.
SESQUITERPENOIDS: 1(10),4-cadinadien-9ol 9-angelate, 3,9-calamenediol, 3,14-calamenediol.
Bohlmann et al., *Phytochemistry*, **18**, 1185 (1979)

HETEROTROPHA (Aristolochiaceae)

H. curvistigma.
SESQUITERPENOIDS: asatone, elemicin, heterocurvistone, safrol.
Niwa et al., *Phytochemistry*, **20**, 1137 (1981)

H. hexaloba.
SESQUITERPENOIDS: asatone, elemicin, safrol.
Yamamura et al., *Phytochemistry*, **15**, 426 (1976)

H. nipponica.
SESQUITERPENOIDS: asatone, elemicin, safrol.
Yamamura et al., *Phytochemistry*, **15**, 426 (1976)

H. takaoi.
NEOLIGNAN: heterotropanone.
SESQUITERPENOIDS: asatone, elemicin, safrol.
Niwa et al., *Tetrahedron Lett.*, 4891 (1978)

HEVEA (Euphorbiaceae)

H. brasiliensis.
CHROMAN: plastochromanol 8.
Whittle et al., *Biochem. J.*, **96**, 17 (1965)

HEXALOBUS

H. crispiflorus.
APORPHINE ALKALOIDS: N-carbamoylanonaine, N-carbamoylasimilobine, N-formylanonaine, 3-hydroxy-6α,7-dehydronuciferine, 3-hydroxynornuciferine.
Achenbach et al., *Justus Liebigs Ann. Chem.*, 1623 (1982)

HEYNEA

H. trijuga Roxb.
HOMOTRITERPENOID: heynic acid.
Purushothaman et al., *Indian J. Chem.*, **22B**, 820 (1983)

HIBISCUS (Malvaceae)

H. abelomoschus.
SESQUITERPENOID: trans-trans-farnesol.
Bates et al., *J. Org. Chem.*, **28**, 1086 (1963)

H. cannabinus.
ACID: hibiscus acid.
FLAVONOIDS: kaempferol 3-β-C-rhamnopyranoside-7-α-L-rhamnofuranoside, myricetin 3'-α-D-glucoside.
PHENOLICS: proanthocyanidin B-5, proanthocyanidin B-6.
(Ac) Lewis et al., *Phytochemistry*, **4**, 619 (1975)
(F) Padukina et al., *Khim. Prir. Soedin.*, **257**, 388 (1976)
(P) Pham Van Thinh et al., *ibid*, 336 (1982)

H. elatus.
SESQUITERPENOIDS: hibiscones A, B, C and D, hibiscoquinones A, B, C and D.
Ferreira et al., *J. Chem. Soc., Perkin I*, 249 (1980)
Ferreira et al., *ibid.*, 257, (1980)

H. furcatus.
ACID: hibiscus acid.
Lewis et al., *Phytochemistry*, **4**, 619 (1975)

H. mutabilis f. versicolor.
FLAVONOIDS: guaijaverin, hyperin, isoquercitrin, quercetin 3-sambubioside.
Ishikura et al., *Agric. Biol. Chem.*, **46**, 1705 (1982)

H. rosa-sinensis L.
ACETYLENIC ESTERS: methyl-8-oxo-9-octadecynoate, methyl-10-oxo-11-octadecynoate.
ALIPHATIC: hentriacontane.
ANTHOCYANIN: cyanin chloride.
FLAVONOID: quercetin.
(Ac) Nakatani et al., *Phytochemistry*, **25**, 449 (1986)
(An, Al, F) Srivastava et al., *J. Res. Indian Med.*, **9**, 103 (1974); *Chem. Abstr.*, 85 2557r

H. subdariffa.
SAPONIN: β-sitosterol-β-D-galactopyranoside.
Osman et al., *Aust. J. Chem.*, **28**, 217 (1975)

H. tiliaceus.
SESQUITERPENOIDS: hibiscones A, B, C and D, hibiscoquinones A, B, C and D.
Ferreira et al., *J. Chem. Soc., Perkin* 1, 249 (1980)
Ferreira et al., *ibid.*, 257, (1980)

HIERACIUM (Compositae)

H. irazuensis.
SESQUITERPENOID: irazunolide.
Hasbun et al., *J. Nat. Prod.*, **45**, 749 (1982)

HIMANTANDRA (Magnoliaceae)

H. baccata (Galbulimina baccata).
ALKALOIDS: himandreline, himandridine, himandrine, himbacine, himbadine, himbosine, himgaline, himgravine.
FURAN: galbacin.
NAPHTHALENES: (-)-galbulin, galcalin.
(A) Brown et al., *Aust. J. Chem.*, **9**, 283 (1956)
(F, N) Tanaoka et al., *Bull. Chem. Soc., Japan*, **49**, 3564 (1976)

H. belgraveana F. Muell. (Galbulimina belgraveana).
ALKALOIDS: himandravine, himandrine, himbacine, himbeline, himdreline, himgaline, himgrine.
Brown et al., *Aust. J. Chem.*, **9**, 283 (1956)

HIMANTHALIA (Phaeophyceae/Himanthaliaceae)

H. elongata.
PHLOROTANNIN: bisfucopentaphlorethol A.
Grosse-Damiues et al., *Phytochemistry*, **22**, 2043 (1983)

HINTERHUBERA (Compositae/Asteraceae)

H. imbricata Cuatr.
DITERPENOIDS: 18-angeloyloxy-3-hydroxyimbricatol, 18-isovaleryloxy-3-hydroxyimbricatol, 18-methylbutyryloxy-3-hydroxyimbricatol.
Bohlmann et al., *Chem. Ber.*, **106**, 2479 (1975)

HIPPEASTRUM (Amaryllidaceae)

H. ananua Phil.
BENZOISOQUINOLINE ALKALOID: hippeastidine.
LACTONE ALKALOID: epi-homolycorine.
  Pacheco et al., Rev. Latinoam. Quim., **9**, 28 (1978)

H. aulicum.
BENZOISOQUINOLINE ALKALOID: galanthine.
  Proskurnina et al., J. Gen. Chem., USSR, **26**, 172 (1956)

H. candicum.
ALKALOID: candimine.
  Dopke, Arch. Pharm. (Weinheim, Ger.) **295**, 920 (1962)

H. caubium var. robustum.
ALKALOID: chlidanthine.
  Boit, Chem. Ber., **89**, 1129 (1956)

H. punecium.
BENZOISOQUINOLINE ALKALOID: cavinine.
  Samuel, Org. Mass Spectra, **10**, 427 (1975)

H. vittatum.
INDANONE ALKALOIDS: hippacine, hippadine, hippafine, hippagine, hippeastrine, lycorine, tazettine, vittatine.
  El Mohgazi et al., Planta Med., **28**, 336 (1975)

HIPPOCRATEA

H. mucronata.
FLAVONOIDS: 5-hydroxy-7,3',4'-trimethoxyflavone, isorhamnetin 3-O-α-D-glucoside, kaempferol, kaempferol 7-O-α-D-glucoside.
  Alyas et al., Niger. J. Pharm. Sci., **2**, 99 (1986); Chem. Abstr., 105 222749b

HIPPOMANE (Euphorbiaceae)

H. mancinella.
DITERPENOID: mancinellin.
  Adolf, Hecker, Tetrahedron Lett., 1587 (1975)

HIPPOPHAE (Elaeagnaceae)

H. rhamnoides.
ACIDS: gallic acid, linoleic acid, linolenic acid, palmitic acid, palmitoleic acid.
INDOLE ALKALOID: serotonin.
FLAVONOIDS: isorhamnetin, kaempferol, myricetin, quercetin.
  (Ac) Zham'yanson, Khim. Prir. Soedin., 133 (1978)
  (A) Petrova, Men'shikov, J. Gen. Chem., USSR, **31**, 2413 (1961)
  (A, F) Chumbalov et al., Khim. Prir. Soedin., 663 (1976)

HIRSCHFELDIA

H. incana.
FLAVONOIDS: kaempferol 3-O-glucoside, quercetin 3,7-di-O-glucoside.
PHENOLIC: caffeic acid rhamnoglucoside.
  Souleles, Philianos, J. Nat. Prod., **46**, 751 (1983)

HISTIOPTERIS

H. incisa.
NORSESQUITERPENOIDS: pterosin Q, pterosin Q 3β-L-arabinopyranoside.
  Murakami et al., Chem. Pharm. Bull., **23**, 936 (1975)

HODGKINSONIA (Rubiaceae)

H. frutescens F.Muell.
TETRAMERIC ALKALOIDS: quadrigemine A, quadrigemine B.
TRIMERIC ALKALOID: hodgkinsine.
  Perry, Smith, J. Chem. Soc., Perkin I, 1671 (1978)

HOLARRHENA

H. africana A.DC.
STEROIDAL ALKALOIDS: conessine, holafrine, holarrhetine.
  Rostock, Seebeck, Helv. Chim. Acta, **41**, 11 (1958)

H. antidysenterica Wall.
STEROIDAL ALKALOIDS: conamine, conarrhimine, conessidine, conessimine, conessine, holadysamine, holadysine, holantosines A, B, C, D, E and F, holarosine A, holarrhessimine, holarrhidine, holarrhimine, holarrhine, holonamine, 12-hydroxyconessine, hydroxyconessine, 7α-hydroxyconessine, isoconessimine, kurchiline, kurchine, kurchiphyllamine, kurchiphylline, norconessine, N,N,N',N'-tetramethylholarrhenine.
UNCHARACTERIZED ALKALOID: lettocine.
STEROID: 5,23-stigmastadien-3-ol.
  (A) Janot et al., Tetrahedron, **26**, 1695 (1970)
  (A) Khuong-Huu et al., Bull. Soc. Chim. Fr., 804 (1971)
  (S) Steiner et al., Helv. Chim. Acta, **60**, 475 (1977)

H. congolensis Stapf.
STEROIDAL ALKALOIDS: conessine, holarrhenine.
  Stork et al., J. Am. Chem. Soc., **84**, 2018 (1962)

H. crassifolia.
STEROIDAL ALKALOID: N-formylconkurchine.
  Einhorn et al., Phytochemistry, **11**, 769 (1972)

H. curtisii King et Gamble.
STEROIDAL ALKALOIDS: N-demethylholacurtine, holacurtine.
  Cannon et al., J. Chem. Soc., Thailand, **6**, 81 (1980)

H. febrifuga Klotsch.
STEROIDAL ALKALOIDS: conessine, dimethylfuntumine, holafebrine, N-methylfuntumine, methylholamine.
  Cave et al., Phytochemistry, **12**, 923 (1973)

H. floribunda (G.Don.) Dur. et Schinz.
STEROIDAL ALKALOIDS: dihydroholaphyllamine, holaline, holamine, holaphyllamine, holaphylline, holarrhesine.
PURINE ALKALOID: triacanthine.
STEROID: holathurin B-1.
  (A) Hoyer et al., Planta Med., **34**, 47 (1978)
  (S) Kuznetsova et al., Khim. Prir. Soedin., 482 (1982)

H. mitis.
STEROIDAL ALKALOID: N(3)-methylholarrhimine.
  Wannigama et al., Ann. Pharm. Fr., **30**, 535 (1972)

H. wulfsbergii.
STEROIDAL ALKALOID: conessine.
  Marshall, Johnson, J. Am. Chem. Soc., **84**, 1485 (1962)

HOLODEIMA

H. grisea.
TRITERPENOID: griseogenin.
  Tursch et al., Tetrahedron, **23**, 761 (1967).

HOLOPTELEA

H. integrifolia.
TRITERPENOID: 2α,3α-dihydroxyolean-12-en-28-oic acid.
  Cheung, Yan, J. Chem. Soc., Chem. Commun., 369 (1970)

HOLOSTACHYS (Chenopodiaceae)

H. caspica (Pall.) SAM.
AMINE ALKALOID: holostachine.
  Men'shikov, Rubinshtein, J. Gen. Chem., USSR, **13**, 801 (1943)

HOLOTHURIA

H. leucospilota.
SAPONINS: holothurin A holothurin B.
Matsuma, Yamanouchi, *Nature*, **191**, 75 (1961)

H. lubrica.
SAPONINS: holothurins A and B.
Matsuma, Yamanouchi, *Nature*, **191**, 75 (1961)

H. polii.
TRITERPENOID: 17-deoxyholothurinogenin.
Hebermehl, Volkwein, *Justus Liebigs Ann. Chem.*, **731**, 53 (1970)

H. vagabunda.
TRITERPENOID: holothurinogenin.
Yasumoto *et al.*, *Agric. Biol. Chem.*, **31**, 7 (1967)

HOMALIUM (Flacourtiaceae)

H. pornyense Guill.
SPERMINE ALKALOIDS: homaline, hopromalinol, hopromine, hoprominol.
Pais *et al.*, *Tetrahedron*, **29**, 1001 (1973)

HOMERIA

H. glauca (Wood et Evans).
STEROID: $1\alpha,2\alpha$-epoxyscillaroside.
Maude, Potgeister, *J. S. Afr. Vet. Med. Assoc.*, **37**, 73 (1966)

HOMOGYNE (Compositae)

H. alpina (L.) Cass.
SESQUITERPENOID: 3β-tigloyloxybakkenolide A.
Harmatha *et al.*, *Collect. Czech. Chem. Commun.*, **41**, 2047 (1976)

HOPEA (Dipterocarpaceae)

H. jucunda Thw.
POLYPHENOL: balanocarpol.
Diyasena *et al.*, *J. Chem. Soc.*, *Perkin I*, 1807 (1985)

H. parviflora.
One sesquiterpenoid.
Purushothaman *et al.*, *Indian Drugs*, **21**, 516 (1984)

HORDEUM (Gramineae)

H. vulgare.
QUATERNARY AMINE: candicine.
Rabtzsch, *Planta Med.*, **6**, 103 (1958)

HORTIA

H. arborea.
BENZOPYRAN: clausinidin.
Joshi *et al.*, *J. Chem. Soc.*, *Perkin I*, 1561 (1974)

H. badinii.
CARBOLINE ALKALOIDS: hortiacine, rutaecarpine.
COUMARIN: coumurrayin.
PHENOLICS:
methyl-3-[2,4-dimethoxy-3-prenylphenyl]-propionate,
methyl-3-[2-methoxy-6',6'-dimethylpyrano-(2',3'
3,4-phenyl)]-propionate.
(A, P) Correa *et al.*, *Phytochemistry*, **14**, 2059 (1975)
(C) Ramstad *et al.*, *Tetrahedron Lett.*, 811 (1968)

H. longifolia.
COUMARIN: coumurrayin.
Ramstad *et al.*, *Tetrahedron Lett.*, 811 (1968)

H. regia.
TETRANORTRITERPENOID: guyanin.
Jacobs *et al.*, *Tetrahedron Lett.*, 1453 (1986)

HOSTA (Liliaceae)

H. kiyosumiensis F.Maek.
STEROIDS: 25,27-gitogenin, 5,22-spirost-25(27)-ene-$2\alpha,3\beta$-diol.
Takeda *et al.*, *Tetrahedron*, **21**, 2089, 2742 (1965)

H. undulata.
ALDEHYDES: dotriacontanal, octacosanal, triacontanal.
Isono *et al.*, *J. Pharm. Soc.*, *Japan*, **96**, 86 (1975)

HOVEA (Leguminosae)

H. linearis.
ALKALOIDS: (+)-16-epi-ormosanine, (±)-piptanthine.
Lamberton *et al.*, *Tetrahedron Lett.*, 3875 (1975)

HOVENIA (Rhamnaceae)

H. dulcis Thunb.
PEPTIDE ALKALOID: hovenine A.
SAPONINS: hovenosides A, B, C, D, G and I.
TRITERPENOIDS: hovenalactone, hovenic acid.
(A) Yakai *et al.*, *Phytochemistry*, **12**, 2985 (1973)
(Sap) Kawai *et al.*, *ibid.*, **13**, 2829 (1974)
(T) Kimura *et al.*, *J. Chem. Soc.*, *Perkin I*, 1923 (1981)

H. tomentilla.
PEPTIDE ALKALOID: hovenine A.
Yakai *et al.*, *Phytochemistry*, **12**, 2985 (1973)

HOYA (Asclepiadaceae)

H. australis.
NORTRITERPENOID: 3,4-seco-3-norolean-12-en-1-ol.
Baas, *Z. Naturforsch.*, **38**, 487 (1983)

HUMBERTIA

H. madagascariensis.
SESQUITERPENOID: himbertiol.
Raulais, *C. R. Acad. Sci.*, **256**, 3369 (1963)

HUMIRIANTHERA

H. rupestris.
TERPENOIDS: humirianthenolides A, B, C, D, E and F.
Zoghbi *et al.*, *Phytochemistry*, **20**, 1669 (1981)

HUMULUS (Cannabinaceae)

H. lupulus L.
PHENOLIC: xanthohumol.
SESQUITERPENOIDS: α-coracalene, τ-calocorene, humulol.
DITERPENOIDS: adhulupone, adhumulone, adlupulone,
cohumulone, colupdox a, colupdox b, colupdox c, colupox
b, colupulone, 4-deoxyhumulone, helupone, hulupinic acid,
humuladienone, humulene epoxide I, humulene epoxide II,
humulinone, humulone, metacamphorene,
selina-4(14),7(11)-diene, selina-3,7(11)-diene,
4,5-epi-thiocaryophyllene.
TRITERPENOIDS: luparenol, luparol, luparone.
(P) Verzele *et al.*, *Bull. Soc. Chim. Belg.*, **66**, 452 (1957)
(T) Spetsig *et al.*, *J. Inst. Brewing*, **65**, 413 (1960)
(T) Naya *et al.*, *Bull. Chem. Soc.*, *Japan*, **42**, 2083 (1969)
(T) Sharpe, Peppard, *Chem. Ind. (London)*, 664 (1977)

H. lupulus var. Hersbrucken.
SESQUITERPENOID: aromadendrene epoxide.
Tressl *et al.*, *J. Agric. Food Chem.*, **31**, 892 (1983)

HUNNEMANNIA

H. fumariaefolia Sweet.
ISOQUINOLIME ALKALOID: oxyhydrastinine.
OXOPROTOBERBERINE ALKALOID: hunnemannine.
OXOQUINOLIZIDINE ALKALOID: protopine.
El-Shanawany *et al., J. Nat. Prod.,* **46,** 753 (1983)

HUNTERIA (Apocynaceae)

H. attenuata.
CARBOLINE ALKALOID: eppiallo-corynantheine.
Phillipson, Hemingway, *Phytochemistry,* **14,** 1855 (1975)

H. congolense.
ALKALOIDS: 17-hydroxypseudoakammicine,
17-hydroxypseudoakuammigine, lanceomigine.
LeMen *et al., Tetrahedron Lett.,* 2871 (1981)

H. corymbosa.
INDOLE ALKALOID: corymine.
Kiang, Smith, *Proc. Chem. Soc. London,* 298 (1962)

H. eburnea Pichon.
CARBOLINE ALKALOIDS: antirhine methochloride,
eburnamenine, eburnamine, eburnamonine,
(±)-eburnamonine, eburnaphylline, hunterburnine,
hunterburnine α-methochloride, hunterburnine
β-methochloride, hunteriamine, hunterine, isoeburnamine,
pleiocarpamine.
IMIDAZOLIDINOCARBAZOLE ALKALOIDS: eburcine, eburine.
INDOLE ALKALOID: ericinine.
OXOCARBOLINE ALKALOID: burnamicine.
QUATERNARY ALKALOID: hunteracine.
Bartlett *et al., J. Org. Chem.,* **28,** 2197 (1963)
Bartlett, Taylor, *J. Am. Chem. Soc.,* **85,** 1203 (1963)

H. elliotii.
CARBOLINE ALKALOID: hunterburnine α-methochloride.
Soendergaard, Nartney, *Phytochemistry,* **15,** 1322 (1976)

H. umbellata (K.Schum.) Stapf.
INDOLE ALKALOIDS: erinicine, erinine, eripine, isocorymine.
DIMERIC INDOLE ALKALOID: umbellamine.
Morita *et al., Helv. Chim. Acta,* **51,** 138 (1968)

H. zeylanica.
INDOLE ALKALOID: 3-epi-dihydrocorymine 17-acetate.
Levaud *et al., Phytochemistry,* **21,** 445 (1982)

HUPERZIA (Lycopodiaceae)

H. selago.
FLAVONE: selagin.
Voirin *et al., Phytochemistry,* **15,** 840 (1976)

H. serrata (Thunb.) Trev.
LYCOPODIUM ALKALOIDS: huperzine A, huperzine B.
Liu *et al., Can. J. Chem.,* **64,** 837 (1986)

HURA

H. crepitans.
TOXIN: huratoxin.
Sabata *et al., Agric. Biol. Chem.,* **35,** 2113 (1971)

HYBANTHUS (Violaceae)

H. enneasperma.
AMIDE ALKALOID: aurantiamide.
Banerji *et al., Indian J. Chem.,* **13,** 1234 (1975)

HYDNOCARPA

H. octandra.
TRITERPENOIDS: octandrolal, octandrolic acid, octandrolol,
octandronal, octandronic acid, octandronol, polpunonic acid.
Gunasekera, Sultanbawa, *Chem. Ind. (London),* 790 (1973)

HYDRANGEA (Hydrangeaceae/Saxifragaceae)

H. heteromalla.
STEROID: β-sitosterol.
TRITERPENOID: ursolic acid.
Talapatra *et al., J. Indian Chem. Soc.,* **52,** 1222 (1975)

H. macrophylla Ser.
IRIDOIDS: hydrangenosides A, B, C, D, E, F, G and H.
Uesato *et al., Chem. Pharm. Bull.,* **29,** 3421 (1981)

H. macrophylla Ser. var. thunbergii Makino.
COUMARINS: hydrangeol, umbelliferone.
GLYCOSIDE: phyllodulcin 8-O-β-D-glucoside.
(C) Suzuki *et al., Agric. Biol. Chem.,* **41,** 205 (1977)
(Gly) Suzuki *et al., ibid.,* **41,** 1815 (1977)

H. vestita.
FLAVONOIDS: daphnetin 8-methyl ether, fraxetin, umbelliferone.
Talapatra *et al., J. Indian Chem. Soc.,* **52,** 1222 (1975)

HYDRASTIS

H. canadensis L.
ISOQUINOLINE ALKALOIDS: canadaline, hydrastidine,
(-)-hydrastine.
PROTOBERBERINE ALKALOIDS: berberastine, (-)-canadine.
Gleye *et al., Phytochemistry,* **13,** 675 (1974)
Messana *et al., Gazz. Chim. Ital.,* **110,** 539 (1980)

HYDROCOTYLE (Umbelliferae)

H. vulgaris L.
TRITERPENOIDS: hydrocotylegenins A, B, C, D, E and F.
Kieper *et al., Pharmazie,* **30,** 619 (1975)

HYELLA (Cyanophyceae)

H. caespitosa.
CARBAZOLE ALKALOIDS: chlorohyellazole, hyellazole.
Cordellina II *et al., Tetrahedron Lett.,* 4915 (1979)

HYAENANCHE

H. globosa.
SESQUITERPENOIDS: dihydroisohyenanchin, hyenanchin,
isodihydrohyenanchin.
DINORDITERPENOID: capensin.
SPIROTERPENOID: lambicin.
Corbella *et al., Tetrahedron,* **25,** 4835 (1969)

HYGROPHOROPSIS

H. aurantiaca (Wulfen et Fr.).
ESTERS: methylvariegatate, methylvariegatate pentamethyl
ether.
Edwards, *J. Chem. Res. Synop.,* 276 (1977)

HYLOCOMIUM

H. splendens (Hedw.) Br.
FLAVONOID: apigenin 7-rhamnoglucoside.
Vandekerkhove, *Z. Pflanzenphysiol.,* **85,** 135 (1977); *Chem. Abstr.,* **87**
197317j

HYMENAEA

H. coubarril.
DITERPENOIDS: copalic acid, eperua-7,13-dien-15-oic acid,
labdan-8β-ol-15-oic acid, labd-13-en-8β-ol-15-oic acid.
Cunningham *et al.*, *Phytochemistry*, **13**, 294 (1974)

H.oblongifolia.
DITERPENOIDS: guamaic acid, enantio-pinifolic acid,
(-)-pinifolic acid.
Cunningham *et al.*, *Phytochemistry*, **12**, 633 (1973)

H. verrucosa.
DITERPENOID: 8(17),13(16),14-labdatrien-18-oic acid.
Martin, Langenheim, *Phytochemistry*, **13**, 523 (1974)

HYMENOCALLIS (Amaryllidaceae)

H. amanacea.
BENZOISOQUINOLINE ALKALOID: narcissidine.
Boit, Stender, *Chem. Ber.*, **87**, 624 (1954)

H. arenicola.
BENZOISOQUINOLINE ALKALOIDS: caribine, havanine, zaidine.
Dopke *et al.*, *Z. Chem.*, **20**, 26 (1980)

HYMENOCARDIA

H. acida Tul.
PEPTIDE ALKALOID: hymenocardine.
Pais *et al.*, *Bull. Soc. Chim. Fr.*, **29**, 79 (1968)

HYMENOCLEA

H. salsola.
SESQUITERPENOIDS: neoambrosin, salsolin.
Geissman, Toribio, *Phytochemistry*, **6**, 1563 (1967)

H. subola.
SESQUITERPENOID: ambrosin.
Torrence *et al.*, *J. Pharm. Sci.*, **64**, 887 (1975)

HYMENODICTYON (Rubiaceae)

H. excelsum.
PIGMENTS: 6-methylalizarin, morindone, nordamnacanthal.
Roberts *et al.*, *Aust. J. Chem.*, **30**, 1553 (1977)

HYMENOXYS

H. anthemoides.
SESQUITERPENOIDS: anthemoidin, themoidin.
Herz *et al.*, *J. Org. Chem.*, **35**, 2611 (1970)

H. grandiflora (T. et G.) Parker.
SESQUITERPENOIDS: hymenoflorin, hymenograndin.
DITERPENOID: perigrandin.
Herz *et al.*, *J. Org. Chem.*, **39**, 2013 (1974)

H. greenei.
SESQUITERPENOIDS: greenein, psilotropin.
de Silva, Geissman, *Phytochemistry*, **9**, 59 (1970)

H. insignis.
SESQUITERPENOID: acetylhymenograndin, hymenograndin.
Herz *et al.*, *J. Org. Chem.*, **45**, 493 (1980)

H. odorata.
SESQUITERPENOIDS: hymenolide, hymenolone, hymenovin,
hymenoxon, hymenoxynin, odoratin.
Petterson, Kim, *J. Chem. Soc., Perkin 2*, 1399 (1976)

H. richardsonii.
SESQUITERPENOID: psilotropin.
de Silva, Geissman, *Phytochemistry*, **9**, 59 (1970)

H. scaposa.
FLAVONE: scaposin.
Mabry *et al.*, *Tetrahedron*, **24**, 3675 (1968)

HYOSCYAMUS (Solanaceae)

H. albus L.
TROPANE ALKALOID: (-)-hyoscyamine.
Bremner, Cannon, *Aust. J. Chem.*, **21**, 1369 (1968)

H. muticus.
TROPANE ALKALOID: (-)-hyoscyamine.
Marion, Thomas, *Can. J. Chem.*, **33**, 1853 (1955)

H. niger L.
TROPANE ALKALOIDS: (-)-hyoscyamine, scopolamine.
Marion, Thomas, *Can. J. Chem.*, **33**, 1853 (1955)

H. orientalis.
TROPANE ALKALOIDS: atropamine, belladonnine, cuskhygrine,
hyoscyamine, scopolamine.
AMINES: betaine, choline.
Telezhko, *Aktual. Vopr. Farm.*, **2**, 45 (1974); *Chem. Abstr.*, **84** 102349x

H. reticulatus.
TROPANE ALKALOID: (-)-hyoscyamine.
Marion, Thomas, *Can. J. Chem.*, **33**, 1853 (1955)

HYPECOUM (Papaveraceae)

H. erectum L.
SPIROISOQUINOLINE ALKALOIDS: hypecoridine, hypecorine,
hypecorinine, protopine.
Yakhontova *et al.*, *Khim. Prir. Soedin.*, 624 (1972)

H. parviflorum L.
AMINE ALKALOID: (-)-peshwarine.
Shamma *et al.*, *Tetrahedron*, **34**, 635 (1978)

H. pendulum L.
OXOQUINOLIZIDINE ALKALOID: protopine.
Mottus *et al.*, *Can. J. Chem.*, **31**, 1144 (1953)

H. procumbens L.
OXOQUINOLIZIDINE ALKALOID: protopine.
ISOQUINOLINE ALKALOID: (-)-corydalisol.
SPIROALKALOID: turkiyenine.
Gozler *et al.*, *J. Am. Chem. Soc.*, **106**, 6101 (1984)

H. trilobum Trautv.
NAPHTHOISOQUINOLINE ALKALOIDS: chelerythrine, sanguinarine.
OXOQUINOLIZIDINE ALKALOID: protopine.
Mottus *et al.*, *Can. J. Chem.*, **31**, 1144 (1953)

HYPERICUM (Guttiferae/Hypericaceae)

H. androsaemum L.
XANTHONE: toxyloxanthone B.
Deshpande *et al.*, *Indian J. Chem.*, **71**, 518 (1973)

H. chinense L.
PHENOLICS: chinesin I, chinesin II.
Nagai *et al.*, *Chem. Lett.*, 1337 (1987)

H. japonicum Thunb.
FLAVONOIDS: 3,5,7,3',4'-pentahydroxyflavone-7-O-rhamnoside,
quercetin.
STEROID: sitosterol.
Chen *et al.*, *Chem. Abstr.*, 86 27678q
Gu *et al.*, *Zhongcaoyao*, **14**, 347 (1983); *Chem. Abstr.*, 100 3473a

H. mysorense Heyne.
PYRANS:
3-(1,1-dimethyl-2-propenyl)-6-phenyl-4H-pyran-4-one,
3-(1,1,-dimethyl-2-propenyl)-6-phenyl-2,4(3H)-dioxopyran,
hyperenone, mysorenone.
Vishwakarma *et al.*, *Indian J. Chem.*, **25B**, 466 (1986)

H. perforatum.
FLAVONOIDS: hyperoside, isoquercitrin, rutin.
Dorosiev, *Pharmazie*, **40**, 585 (1985)

H. sampsonii.
XANTHONES: 2-hydroxy-3,4-dimethoxyxanthone, hyperxanthone, isomangiferin, mangiferin, toxyloxanthone B.
Chen et al., Heterocycles, 23, 2543 (1985)

H. uliginosum.
PHENOLICS: uliginosin A, uliginosin B.
Parker et al., J. Am. Chem. Soc., 90, 4716 (1968)
Parker et al., ibid., 90, 4723 (1968)

HYPHOLOMA (Agaricales)
H. fascicuklare.
PIGMENTS: fasciculin A, fasciculin B, hypholomin A, hypholomin B.
Fiasson et al., Chem. Ber., 110, 1047 (1977)

HYPNEA (Rhodophyceae)
H. japonica.
STEROIDS: cholesta-5,22-dien-3β-ol, cholesta-5,25-dien-3β-ol, 22-dehydrocholesterol.
Tsuda et al., Chem. Pharm. Bull., 7, 747 (1959)

HYPOCHOERIS (Compositae)
H. cretensis.
SESQUITERPENOIDS: hypocretenoic acid methyl ester, hypocretenolide.
Bohlmann et al., Phytochemistry, 21, 2119 (1982)

H. retosus.
SESQUITERPENOID: hypochaerin.
Gonzalez et al., Phytochemistry, 15, 991 (1976)

HYPODEMATIUM
H. crenata Kuhn.
FLAVONOIDS: iriflophenone 3-C-β-D-glucoside, kaempferol 3-C-β-D-glucoside.
XANTHONE: mangiferin.
Murakami et al., J. Pharm. Soc., Japan, 106, 378 (1986)

H. fauriei Tagawa.
FLAVONOIDS: iriflophenone 3-C-β-D-glucoside, kaempferol 3-C-β-D-glucoside.
XANTHONE: mangiferin.
Murakami et al., J. Pharm. Soc., Japan, 106, 378 (1986)

HYPOESTES (Acanthaceae)
H. rosea.
DITERPENOIDS: hypoestoxide, roseadione, roseanolone.
TETRATERPENOID: dihypoestoxide.
Adesomoju et al., J. Nat. Prod., 47, 308 (1984)

HYPOGYMNIA (Hepaticae)
H. vittata.
ACID: vittatolic acid.
Hirayama et al., Chem. Pharm. Bull., 23, 693 (1975)

HYPOLEPIS
H. punctata (Thunb.) Mett.
INDANONES: 3S-pteroside D, 3R-pterosin D, 3S-pterosin D, 2R,3R-pterosin L 2'-O-β-D-glucoside, 2S,3R-pterosin L 2'-O-β-D-glucoside.
SESQUITERPENOIDS: hypolepins A, B and C.
(I) Murakami et al., Chem. Pharm. Bull., 24, 2241 (1976)
(T) Hayashi et al., Chem. Lett., 63 (1973)

HYPOXIS (Amaryllidaceae/Hypoxidaceae)
H. nyasica Bak.
GLUCOSIDES: hypoxoside, nyasoside.
Marini-Bettolo et al., Tetrahedron, 41, 665 (1985)

H. obtusa Busch.
GLYCOSIDE: hypoxoside.
Marini-Bettolo et al., Tetrahedron, 38, 1683 (1982)

HYPOXYLON (Xylariaceae)
H. fragiforme.
COUMARIN: isoochracein.
Anderson et al., J. Chem. Soc., Perkin I, 2185 (1983)

H. haematostroma.
COUMARINS: isoochracein, mellein, ramulosin.
Anderson et al., J. Chem. Soc., Perkin I, 2185 (1983)

H. howeianum.
COUMARIN: 4-hydroxyisoochracein.
Anderson et al., J. Chem. Soc., Perkin I, 2185 (1983)

H. mammatum (Vahl.) Miller.
DITERPENOID: hymatoxin A.
Bodo et al., Tetrahedron Lett., 2355 (1987)

H. venustuissimum.
COUMARINS: isoochracein, mellein, ramulosin.
Anderson et al., J. Chem. Soc., Perkin I, 2185 (1983)

HYSSOPUS (Labiatae)
H. ferganensis.
ACIDS: caffeic acid, p-coumaric acid, ferulic acid, gallic acid, sinapic acid, vanillic acid.
Zotov et al., Chem. Abstr., 88 71448e

H. officinalis L.
MONOTERPENOID: (1S,2R,5R)-3-pinanone.
Banthorpe et al., Chem. Rev., 66, 643 (1966)

H. seravshamicus.
ACIDS: caffeic acid, p-coumaric acid, ferulic acid, gallic acid, sinapic acid, vanillic acid.
Zotov et al., Chem. Abstr., 88 71448e

HYPTIS (Labiatae)
H. fruticosa.
DITERPENOIDS: hyptol, 14-methoxytaxadione.
Marlette et al., Gazz. Chim. Ital., 106, 119 (1976)

H. suaveolens.
DITERPENOIDS: suaveolic acid, suaveolol.
Manchand et al., J. Org. Chem., 39, 2706 (1974)

HYSTERIOPSIS
H. incisa.
GLUCOSIDE: pteroside Q.
Kurakami et al., Chem. Pharm. Bull., 22, 2758 (1974)

I

IANTHELLA
I. ardis.
HALOGENATED TERPENOIDS: (-)-aeroplysinin-1,
  (±)-aeroplysinin-2.
  Fattorusso *et al.*, *J. Chem. Soc., Chem. Commun.*, 751 (1970)
I. basta.
CAROTENOID: bastaxanthin.
  Herzberg *et al.*, *Acta Chem. Scand.*, B 27, 267 (1983)

IBERIS (Cruciferae)
I. amara L.
THIO COMPOUNDS: (R)-3-methylsulfinylpropylamine,
  3-methylthiopropylamine.
TRITERPENOIDS: cucurbitacin E, cucurbitacin I.
  (T) Nielsen *et al.*, *Phytochemistry*, 16, 1519 (1977)
  (Th) Dalgaard *et al.*, *ibid.*, 16, 931 (1977)

IBOZA (Labiatae)
I. riparia (Hochst.) N.E.Br.
DITERPENOIDS: ibozol, 8(14),15-isopimaradiene-7,18-diol,
  8(14),15-sandaracopimaradiene-7α,18-diol.
  Zelnik *et al.*, *Phytochemistry*, 17, 1795 (1978)
  Van Puyvelde *et al.*, *J. Org. Chem.*, 47, 3628 (1982)

ICACINA (Icacinaceae)
I. chaessensis.
DITERPENOID: icacinol.
  Onokoko *et al.*, *Tetrahedron*, 41, 745 (1985)
I. guesfeldtii.
STEROIDAL ALKALOIDS: icaceine, icacine.
  Onokoko, Vanhaelen, *Phytochemistry*, 19, 303 (1980)
I. mannii.
DITERPENOID: icacenone.
  On'okoko *et al.*, *Phytochemistry*, 24, 2452 (1985)

ICHTHYOTHERE
I. terminalis.
DITERPENOIDS: ent-3α-acetoxy-9(11),16-kauradien-19-oic acid,
  9,15-dihydroxy-16-kauren-19-oic acid,
  ent-2-hydroxy-9(11),16-kauradien-19-oic acid,
  ent-3-hydroxy-9(11),16-kauradien-19-oic acid,
  7-hydroxy-9(11),16-kauradien-19-oic acid,
  ichthyotherminolide.
TRITERPENOID: terminolic acid.
  Bohlmann *et al.*, *Phytochemistry*, 21, 2317 (1982)
I. ulei.
DITERPENOIDS: 19-acetoxyichthyoluteolide, ichthyoluteolide.
  Bohlmann *et al.*, *Phytochemistry*, 21, 2317 (1982)

IGNATIA
I. amara L.
STRYCHNINE ALKALOID: brucine.
  Robinson, *Prog. Org. Chem.*, 1, 1 (1952)

ILEX (Aquifoliaceae)
I. asprella Champ. Numerous alkanes.
  Chen *et al.*, *Chem. Abstr.*, 86 27677p
I. cornuta.
SAPONIN:
  3-O-β-D-glucopyranosyl-α-L-arabinopyranosylpomolic acid.
  Otsuka Pharmaceuticals Ltd, Japanese Patent, 83 146 600
I. goshiensis.
TRITERPENOID: neoilexonol.
  Yagashita, Nishimura, *Agric. Biol. Chem.*, 25, 517 (1961)
I. hainanensis.
TRITERPENE GLUCOSIDE: hainanenside.
  Min *et al.*, *Yaoxue Xuebao*, 19, 691 (1984)
I. oldhami.
TRITERPENE GLUCOSIDE: pedunculoside.
  Hase *et al.*, *Nippon Kogaku Kaishi*, 778 (1973)
I. paraguairensis.
PURINE ALKALOID: theophylline.
  Blout *et al.*, *J. Am. Chem. Soc.*, 72, 479 (1950)
I. pedunculosa.
TRITERPENE GLUCOSIDE: pedunculoside.
  Hase *et al.*, *Nippon Kogaku Kaishi*, 778 (1973)
I. perado.
PURINE ALKALOID: theobromine.
  Bohinc *et al.*, *Planta Med.*, 28, 374 (1975)
I. pubescens Hook et Arn.
TRITERPENE GLYCOSIDES: ilexasaponin, ilexoside A.
  Qin *et al.*, *Huaxue Xuebao*, 45, 249 (1987); *Chem. Abstr.*, 107 4283v
I. rotunda Thunb.
TRITERPENOID: rotundic acid.
TRITERPENE GLUCOSIDE: pedunculoside.
  Hase *et al.*, *Nippon Kogaku Kaishi*, 778 (1973)

ILLICIUM (Illiciaceae)
I. anisatum L.
SESQUITERPENOIDS: anisatin, anisotin, neoanisatin,
  pseudoanisitin.
  Yamada *et al.*, *Tetrahedron Lett.*, 4797 (1965)
I. religiosum.
ACID: shikimic acid.
SESQUITERPENOIDS: anisatin, neoanisatin.
  Hamada *et al.*, *Tetrahedron Lett.*, 4797 (1965)
I. verum Hook.
MONOTERPENOID: trans-anethole.
  Li *et al.*, *Chem. Abstr.*, 104 3421q

ILLIGERA
I. pentaphylla.
PHENANTHRENE ALKALOID: thaliporphinemethine.
  Ross *et al.*, *J. Nat. Prod.*, 48, 835 (1985)

IMPATIENS (Balsaminaceae)
I. balsamina.
TRITERPENOID: hosenkol A.
  Shoji *et al.*, *J. Chem. Soc., Chem. Commun.*, 871 (1983)

IMPERATA (Gramineae)

I. cylindrica Beauv.
TRITERPENOIDS: arborinol methyl ether, arundoin, 9(11)-fernen-3-ol.
    Ogunkoya *et al., Phytochemistry*, **16**, 1606 (1977)

I. cylindrica Beauv. var. koenigii Durand et Schinz.
TRITERPENOIDS: arborinione, arundoin, cylindrin.
    Hui, Lam, *Phytochemistry*, **4**, 333 (1965)
    Nishimoto *et al., Tetrahedron*, **24**, 735 (1968)

I. cylindrica var. media.
TRITERPENOID: arundoin.
    Hui, Lam, *Phytochemistry*, **4**, 333 (1965)

INCARVILLEA (Bignoniaceae)

I. olgae Rgl.
PYRIDINE ALKALOIDS: indicaine, (-)-plantagonine.
    Danilova, Konovalova, *J. Gen. Chem., USSR*, **22**, 2237 (1952)

INDIGOFERA (Leguminosae)

I. suffruticosa.
TERPENOID: louisfieserone.
    Dominguez *et al., Tetrahedron Lett.*, 429 (1978)

I. tetrantha.
FLAVONOIDS: acacetin, apigenin, isoquercitrin.
STEROID: β-sitosterol.
TRITERPENOID: ursolic acid.
    Thusoo *et al., J. Indian Chem. Soc.*, **59**, 1007 (1982)

INERTIA

I. hypoleuca.
TRITERPENOID: α-amyrenone.
    Lou-Cam *et al., Phytochemistry*, **12**, 475 (1973)

INEZIA

I. integrifolia.
SESQUITERPENOID: 1α,2α-diacetoxyalantalactone.
    Bohlmann *et al., Phytochemistry*, **21**, 2743 (1982)

INOLOMA (Basidiomycetae)

I. traganum.
ANTIBIOTIC: inolomin.
    Fragner, *Experientia*, **5**, 167 (1949)

INONOTUS (Basidiomycetae)

I. obliquus.
TRITERPENOIDS: 3β-hydroxylanosta-8,24-dien-21-al, inotodiol, trametenolic acid.
    Kahlos *et al., Planta Med.*, **50**, 197 (1984)

INULA (Compositae)

I. aschersoniana.
SESQUITERPENOID: 8β-hydroxydihydroeremanthin.
    Papano *et al., Phytochemistry*, **19**, 152 (1980)

I. bifrons.
MONOTERPENOID: 2-methoxythymol isobutyrate.
    Bohlmann *et al., Phytochemistry*, **17**, 1163 (1978)

I. britannica.
MONOTERPENOIDS: 7,10-diisobutyryloxythymol isobutyrate, 10-hydroxy-8,9-epoxythymol isobutyrate.

SESQUITERPENOIDS: 12-acetoxyeremophilene britannin, 15-desoxy-cis, cis-artemisiifolin, 3-hydroxy-2-senecioyloxy-isoalantolactone, methyleremophilen-12-oic acid.
    Bohlmann *et al., Phytochemistry*, **16**, 1243 (1977)
    Bohlmann *et al., ibid.*, **16**, 1302 (1977)

I. caspica.
SESQUITERPENOID: incaspin.
    Adekenov *et al., Khim. Prir. Soedin.*, 797 (1984)

I. crithmoides L.
SESQUITERPENOID: inucrithniolide.
    Mahmoud *et al., Phytochemistry*, **20**, 735 (1981)

I. cuspidata.
SESQUITERPENOIDS: 4(15),5E,10(14)-germacratrien-1-ol, incapitolide A.
    Bohlmann *et al., Phytochemistry*, **21**, 157 (1981)

I. eupatorioides.
SESQUITERPENOIDS: ineupatolde, ineupatorolide A, ineupatorolide B.
THIOPHENE: ineupatoriol.
    Baruch *et al., Phytochemistry*, **21**, 665 (1982)

I. germanica.
SESQUITERPENOIDS: germanin A, germanin B.
    Konovalova *et al., Khim. Prir. Soedin.*, 578 (1974)
    Baruch *et al., J. Org. Chem.*, **45**, 4838 (1980)

I. grandis Schrenk.
ALKALOID: one uncharacterized base. Flavone 4',5,6-trihydroxy-3,7-dimethoxyflavone.
SESQUITERPENOIDS: alantolactone, granilin, grandicin, grandulin, igalan.
    (A) Zabolotnaya *et al., Tr. Vilr. XI*, 152 (1959)
    (F) Nikonova *et al., Khim. Prir. Soedin.*, 96 (1975)
    (T) Maruyama *et al., Phytochemistry*, **14**, 2247 (1975)

I. graveolens.
SESQUITERPENOID: graveolide.
    d'Alcantres *et al., Gazz. Chim. Ital.*, **103**, 239 (1973)

I. helenium L.
COUMARINS: scopoletin, umbelliferone.
SESQUITERPENOIDS: 4α-H-confertin, 1-deoxy-8-epi-ivangustin, dihydroalantolactone, dihydroisoalantolactone, 1β,10α-epoxy-1,10-H-cis-inunolide, 4β,5α-epoxy-4,5-cis-inunolide, germacrene D-lactose, 1β-hydroxyalantolactone, 9β-hydroxycostunolide, isoalantolactone, 9-isobutyryloxy-costunolide, 4-epi-isoinuviscolide, 8-epi-isoivangustin, 9-isovaleryloxycostunolide, 9β-(2-methylbutyryloxy)-costunolide, 2-oxoalantolactone, 4H-tomentosin, 8-epi-tomentosin.
TRITERPENOID: dammaradienol.
    (C) Khvorost, Komissarenko, *Khim. Prir. Soedin.*, 820 (1976)
    (T) Bohlmann *et al., Phytochemistry*, **17**, 1163 (1978)

I. heterolepis.
SESQUITERPENOIDS: 1,10-epoxyhaageanolide, 9-hydroxyartemorin.
    Bohlmann *et al., Phytochemistry*, **21**, 1166, 1357 (982) **21**, 116, 1357 (1982)

I. indica.
SESQUITERPENOID: inuolide.
    Dhaneshwar *et al., Acta Crystallogr., C* **39**, 462 (1983)

I. japonica.
SESQUITERPENOID: deacetylinulicin.
    Evstratova *et al., Khim. Prir. Soedin.*, 730 (1974)

I. racemosa.

SESQUITERPENOIDS: alloalantolactone, dihydroinunolide, inunolide, neotsantolactone.

Bhandar, Rastogi, *Indian J. Chem.*, **22B**, 286 (1983)

I. royleana.

UNCHARACTERIZED ALKALOID: royline.

SESQUITERPENOIDS: dehydroroyleanone, 1-deoxy-8-epi-ivangustin, 4β,5α-epoxy-4,5-cis-inunolide, 2α-hydroxyalantolactone, 9β-hydroxycostunolide, inuroyleanone, 4-epi-isoinuvicsolide, 8-epi-isoivangustin, ivalin acetate, 8-epi-ivangustin, 7-ketoroyleanone, 9β-(2-methylbutyryloxy)-costunolide, 2-oxoalantolactone, 3β-propionoyloxycostunolide, royleanone, 4H-tomentosin.

(A) Chopra *et al.*, *Indian J. Med. Res.*, **33**, 139 (1945)
(T) Bhat *et al.*, *Tetrahedron*, **31**, 1001 (1975)
(T) Bohlmann *et al.*, *Phytochemistry*, **17**, 1163 (1978)

I. spiraefolia.

MONOTERPENOID: 7,10-diisobutyryloxythymol isobutyrate.

Bohlmann, Zdero, *Phytochemistry*, **16**, 1243 (1977)

I. viscosa.

SESQUITERPENOID: inuviscolide.

Bohlmann *et al.*, *Chem. Ber.*, **110**, 1330 (1977)

IPHIGENIA (Wurmbaeoideae)

I. stellata.

COLCHICINE ALKALOIDS: colchicine, colchicoside, cornigerine, 3-demethylcolchicine, 3-demethyl-N-deacetylcolchicine, N-formyl-N-deacetylcolchicine, β-lumicolchicine.

HOMOAPORPHINE ALKALOIDS: S-(+)-bechuanine, kreysiginine, multifloramine.

Potesilova *et al.*, *Planta Med.*, 72 (1985)

IPHIONA

I. mucronata.

Thymol derivatives.

Metwally *et al.*, *Z. Naturforsch.*, **40B**, 1597 (1985)

I. scabra.

SESQUITERPENOIDS: iphionane, isoiphionane.

El-Ghazouly *et al.*, *Phytochemistry*, **26**, 439 (1987)

IPOMOEA (Convolvulaceae)

I. alba L.

INDOLIZIDINE ALKALOIDS: dihydroipalbidine, ipalbidine.

Gourley *et al.*, *J. Chem. Soc., Chem. Commun.*, 709 (1969)

I. calonyction.

STEROIDS: calonysterone, muristerone, muristerone A.

Canonica *et al.*, *J. Chem. Soc., Chem. Commun.*, 1060 (1972)

I. dichroa.

GLYCOSIDES: dichrosides A, B, C and D

Harrison *et al.*, *Carbohydr. Res.*, **143**, 207 (1985); *Chem. Abstr.*, 104 65842r

I. hildebrandii Vatke.

ERGOT ALKALOID: cycloclavine.

Stauffacher *et al.*, *Tetrahedron*, **25**, 5879 (1969)

I. muricata.

GLYCOALKALOID: ipomine.

GLYCOSIDES: glycosides Mb-1, Mb-2, Mb-3, Mb-4 and Mb-5

(A) Dawidar *et al.*, *Tetrahedron*, **33**, 1733 (1977)
(Gly) Noda *et al.*, *Tennen Yuki Kagobutsu Toronkai Koen Yoshishu*, **27**, 427 (1985)

I. orizabensis.

GLYCOSIDES: glycosides Oc-1, Oc-2, Oc-3 and Oc-4

Noda *et al.*, *Tennen Yuki Kagobutsu Toronkai Koen Yoshishu*, 27th, 427 (1985)

I. tricolor.

ANTHOCYANIN: heavenly blue anthocyanin.

Kondo *et al.*, *Tetrahedron Lett.*, 2273 (1987)

I. violacea.

ERGOT ALKALOID: chanoclavine 1-carboxylic acid.

Choong, Shough, *Tetrahedron Lett.*, 3137 (1977)

IRESINE (Amarantaceae)

I. celosioides.

SESQUITERPENOIDS: dihydroiresin, dihydroiresone, iresin, iresone, isoiresin.

Djerassi *et al.*, *J. Am. Chem. Soc.*, **76**, 2966 (1954)
Rabbe *et al.*, *Bull. Soc. Chim. Belg.*, **67**, 632 (1958)

IRIS (Iridaceae)

I. florentina.

ISOFLAVONES: iridin, irifloroside, irigenin.

NORSESQUITERPENOIDS: irone, β-irone, τ-irone.

XANTHONES: irisxanthone, isomangiferin, isoswertiajaponin, isoswertisin, mangiferin, swertisin.

(I) Arisawa *et al.*, *Chem. Pharm. Bull.*, **21**, 2323 (1973)
(T) Ruzicka *et al.*, *Helv. Chim. Acta*, **30**, 1807 (1947)
(T) Ruzicka *et al.*, *ibid.*, **30**, 1810 (1947)
(X) Fujita *et al.*, *Chem. Pharm. Bull.*, **30**, 2342 (1982)

I. germanica L.

ISOFLAVONES: irilone, 4′,5,7-trihydroxy-3′,6-dimethoxyisoflavone, 3′,4′,5′-trihydroxy-6,7-dimethoxyisoflavone, 3′,4′,5′-trihydroxy-6,7-methylenedioxyisoflavone.

PHENOLICS: acetovanillone, irigenin, irisolidone.

STEROID: β-sitosterol.

TRITERPENOIDS: α-amyrin, β-amyrin, iridogermanal, α-iridogermanal, τ-iridogermanal, δ-iridogermanal.

(I) Pailer *et al.*, *Monatsh. Chem.*, **104**, 1394 (1973)
(I) Dhar *et al.*, *Phytochemistry*, **12**, 734 (1973)
(P, S) Ali *et al.*, *ibid.*, **22**, 2061 (1983)
(T) Marner *et al.*, *J. Org. Chem.*, **47**, 2531 (1982)

I. hollandica.

ISOPRENOID QUINONES: 3-demethylplastoquinone 8, 3-demethylplastoquinone 9, phylloquinone, plastochromanol 8, plastoquinone 9.

STEROID: tocopherol.

Etman-Gervais, *C. R. Acad. Sci.*, **282**, 1171 (1976)
Etman-Gervais *et al.*, *Nouv. J. Chim.*, **1**, 323 (1977); *Chem. Abstr.*, 87 180695h

I. hookeriana.

ISOFLAVONOIDS: iridin, irigenin, irisflorentin, junipigenin A.

Shawl *et al.*, *J. Nat. Prod.*, **48**, 849 (1985)

I. kashmiriana.

FLAVONE: irisolidone.

Dhar *et al.*, *J. Indian Chem. Soc.*, **52**, 784 (1975)

I. kumanonensis.

ISOFLAVONE: irigenin.

Dhar *et al.*, *Phytochemistry*, **11**, 3097 (1972)

I. pallida.

TERPENOIDS: iriflorental, iripallidal.

Krick *et al.*, *Z. Naturforsch.*, **38C**, 179 (1983)

I. tectorum.

ISOFLAVONES: iristectorigenin, iristectorin A, tectoridin.

Khara *et al.*, *Indian J. Chem.*, **16B**, 78 (1978)

I. tingitana.

AMINOACIDS: τ-glutamyl-β-alanine, τ-glutamyl-β-aminoisobutyric acid.

Larsen, *Acta Chem. Scand.*, **16**, 1511 (1962)

I. unguiclaris Poir.

FLAVONES: irigenin, iristectorigenin A, kanzakiflavone 1, kanzakiflavone 2.

Arisawa *et al.*, *Chem. Pharm. Bull.*, **24**, 815 (1976)

Arisawa *et al.*, *ibid.*, **24**, 1609 (1976)

IRYANTHERA

I. grandis.

CHALCONE: 2',4'-dihydroxy-4,6'-dimethoxydihydrochalcone.

CHROMAN: 2,8-dimethyl-2-(4,12-dimethyl-8-carboxy-3,7,11-tridecatrienyl)-6-chromanol.

FLAVANS: (±)-7,4'-dihydroxy-3'-methoxyflavan, (+)-5,7-dimethoxy-4'-hydroxyflavan.

PHENOLICS: 1-(2'-hydroxy-4',6'-dimethoxyphenyl)-3-(3'-methoxy-4"-methylenedioxyphenyl)-propane, 1-(2'-hydroxy-4',6'-dimethoxyphenyl)-3-(3",4"-methylenedioxyphenyl)-propane.

(Chal, Chr) Vieira *et al.*, *Phytochemistry*, **22**, 2281 (1983)

(F, P) Diaz *et al.*, *ibid.*, **25**, 1395 (1986)

I. juruensis.

LACTONE: juruenolide.

Vieira *et al.*, *Phytochemistry*, **22**, 711 (1983)

I. ulei.

LACTONE: juruenolide.

Vieira *et al.*, *Phytochemistry*, **22**, 711 (1983)

ISMENE (Amaryllidaceae)

I. calithina.

ERYTHRINA ALKALOID: pretazettine.

Wildman *et al.*, *J. Am. Chem. Soc.*, **89**, 5514 (1967)

ISOCARPHA

I. atriplicifolia.

SESQUITERPENOIDS: atripliciolide isobutyrate, atripliciolide isovalerate, atripliciolide methacrylate, atripliciolide tiglate.

Bohlmann *et al.*, *Phytochemistry*, **17**, 471 (1978)

I. oppositifolia.

AGERATONE: dihydroageratone-O-methyl ether.

Bohlmann *et al.*, *Phytochemistry*, **16**, 768 (1977)

ISOCOMA

I. wrightii. (Haplopappus heterophyllus).

SESQUITERPENOIDS: comene, isocomene, 1,2,3,4-tetrahydro-1,1,5,6-tetramethylnaphthalene.

TRITERPENOIDS: friedelan-3α-ol, friedelin.

Zalkow *et al.*, *Lloydia*, **41**, 96 (1979)

ISODON (Labiatae)

I. amethystoides.

DITERPENOID: wangzoazine A.

Wang *et al.*, *Zhongcaoyao*, **13**, 11 (1982)

I. emarginatus.

TERPENOID: emarginatoside.

Row, Rukamini, *Indian J. Chem.*, **4**, 149 (1966)

I. inflexus.

DITERPENOID: inflexin.

Kubo *et al.*, *Chem. Lett.*, 99 (1977)

I. japonicus Hara.

DITERPENOIDS: enmein, enmein 3-acetate, isodoacetal, isodocarpin, isodonal, isodopodin, nodosin, epi-nodosin, nodosinin, nodosinol, epi-nodosinol, odonicin, oridonin,

ponicidin.

Kubo *et al.*, *Tetrahedron*, **30**, 615 (1974)

I. kameba.

DITERPENOIDS: kamebamacetals A and B, kamebanin.

Kubo *et al.*, *Chem. Lett.*, 1289 (1977)

I. lasiocarpus.

DITERPENOIDS: isolasiokaurin, lasiodonin, lasiokaurin, lasiokaurinol.

Fujita *et al.*, *Chem. Pharm. Bull.*, **22**, 280 (1974)

I. okuyama.

DITERPENOID: mebadonin.

Isobe *et al.*, *Nippon Kagaku Kaishi*, 2143 (1972)

I. shikokianus var. intermedius.

DITERPENOIDS: isodomedin, shikokianin.

Kubota, Kubo, Bull. Chem. Soc., *Japan*, **42**, 1778 (1969)

I. trichocarpus.

DITERPENOIDS: ememodin, ememofin, ememogin, ememonin, ememosin, enmedol, enmenol, isodocarpin, isodonol, isodotricin, nodosin, oridonin.

Kubo *et al.*, *Tetrahedron*, **30**, 615 (1974)

ISOLONA

I. pilosa.

APORPHINE ALKALOID: isopiline.

Hocquemiller *et al.*, *J. Chem. Soc., Chem. Commun.*, 447 (1977)

I. zenkeri.

APORPHINE ALKALOID: zenkerine.

Hocquemiller *et al.*, *Planta Med.*, **50**, 23 (1984)

ISOMOPSIS

I. aggregata.

TRITERPENOID: 3-epi-isocucurbitacin B.

Arisawa *et al.*, *J. Pharm. Sci.*, **73**, 411 (1984)

ISOPLEXIS (Scrophulariaceae)

I. isabelliana.

STEROID: gitoroside.

Freitag *et al.*, *Helv. Chim. Acta.*, **50**, 1335 (1967)

I. sceptrum (L.) Steud.

STEROIDS: (24R)-funchaligenin, isoplexigenins A, B, C and D, sceptrumgenin.

Gonzalez *et al.*, *An. Quim.*, **69**, 1031 (1973)

ISOPYRUM

I. biternatum L.

UNCHARACTERIZED ALKALOID: isopyroine.

Frankforter, *J. Am. Chem. Soc.*, **25**, 99 (1903)

I. thalictroides L.

QUATERNARY PROTOBERBERINE ALKALOIDS: dehydropseudocheilanthifoline, pseudoberberine, pseudocolumbamine, pseudocoptisine.

UNCHARACTERIZED ALKALOIDS: isopyrine, pseudoisopyrine.

Moulis *et al.*, *Phytochemistry*, **16**, 1283 (1977)

ISOTOMA (Campanulaceae)

I. longifera.

PIPERIDINE ALKALOID: (-)-cis-8,10-diphenyllobelidiol.

Arthur, Chen, *J. Chem. Soc.*, 750 (1963)

ITEA (Escalloniaceae)
I. ilicifolia.
CARBOHYDRATE: allitol.

Wolfrom *et al.*, *J. Am. Chem. Soc.*, **68**, 1443 (1946)

I. virginica.
CARBOHYDRATE: allitol.

Steiger *et al.*, *Helv. Chim. Acta*, **19**, 184 (1936)

IVA (Compositae)
I. acerosa.
FLAVONES: acerosin.

Farkas *et al.*, *Tetrahedron*, **23**, 3557 (1967)

I. ambrosiaefolia.
SESQUITERPENOID: apachin.

Yoshioka *et al.*, *Phytochemistry*, **10**, 401 (1971)

I. angustifolia.
SESQUITERPENOIDS: ivangulin, ivangustin.

Herz *et al.*, *J. Org. Chem.*, **32**, 3658 (1967)

I. annua.
SESQUITERPENOIDS: ivanuol, ivanuol acetate.

Bohlmann, Zdero, *Phytochemistry*, **18**, 2034 (1979)

I. asperifolia.
SESQUITERPENOID: asperilin.

Herz, Viswanathan, *J. Org. Chem.*, **29**, 1022 (1964)

I. axillaris.
SESQUITERPENOIDS: anhydroivaxillarin, axivalin, ivaxillarin, ivaxillin.

Herz *et al.*, *J. Org. Chem.*, **31**, 3232 (1966)

I. dealbata.
SESQUITERPENOIDS: ivalbatin, ivalbin.

Herz *et al.*, *J. Org. Chem.*, **32**, 682 (1967)
Chikamatsu, Herz, *ibid.*, **38**, 585 (1973)

I. frutescens.
SESQUITERPENOID: ifrutescin.

Herz *et al.*, *Phytochemistry*, **11**, 1829 (1972)

I. imbricata.
SESQUITERPENOID: ivalin.

Herz, Hogenaufer, *J. Org. Chem.*, **27**, 905 (1962)

I. macrocephala.
SESQUITERPENOID: dihydropseudoivalin.

Herz *et al.*, *J. Org. Chem.*, **30**, 118 (1965)

I. microcephala.
SESQUITERPENOIDS: ivalin, microcephalin.

Herz *et al.*, *J. Org. Chem.*, **29**, 1700 (1964)

I. texensis.
SESQUITERPENOID: asperilin.

Herz, Viswanathan, *J. Org. Chem.*, **29**, 1022 (1964)

I. xanthofolia.
SESQUITERPENOID: cyclachaenin.

Bohlmann *et al.*, *Phytochemistry*, **18**, 1892 (1979)

IXORA
I. chinensis.
IRIDOIDS: ixoroside, ixoside.

Takeda *et al.*, *Phytochemistry*, **14**, 2647 (1975)

I. parviflora.
TERPENOID: ixorol.

Ali, Kapadia, *Pakistan J. Sci. Ind. Res.*, **11**, 12 (1968)

J

JABOROSA (Solanaceae)

J. integrifolia Lam.
LACTONES: jaborosalactones A, B, C, D, E and F.
Tschesche *et al., Tetrahedron,* **29,** 1909 (1972)

J. leucotricha.
WITHANOLIDE: jaborosalactone L.
Lavie *et al., Phytochemistry,* **25,** 1765 (1986)

JACARANDA (Bignoniaceae)

J. caucana.
TRITERPENOIDS: jacarandic acid, jacoumaric acid.
NORMONOTERPENOID: jacaranone.
Ogura *et al., Phytochemistry,* **16,** 286 (1977)
Ogura *et al., Lloydia,* **40,** 157 (1977)

JACQUINIA (Theophrastaceae)

J. angustifolia.
DITERPENOIDS: crocetin aldehyde, crocetin half-aldehyde.
Eugster *et al., Helv. Chim. Acta,* **52,** 806 (1969)

J. armillaris.
TRITERPENOID: armillarigenin.
De Mahaas *et al., Bull. Soc. Chim. Fr.,* 226 (1966)

J. pungens.
TRITERPENOIDS: jacquinic acid, primulagenin A.
Khan *et al., Tetrahedron,* **21,** 1735 (1965)

JAMESONIA

J. scammanae.
SESQUITERPENOID: jamesonin.
Satake *et al., Chem. Pharm. Bull.,* **32,** 4620 (1984)

JASMINUM (Oleaceae)

J. auriculatum.
ACIDS: linoleic acid, linolenic acid, malvalic acid, palmitic acid, stearic acid.
TRITERPENOID: jasminol.
(Ac) Srivastava *et al., J. Appl. Chem. Biotechnol.,* **27,** 55 (1977); *Chem. Abstr.,* **87** 18958z
(T) Deshpande, Upadhyay, *Experientia,* **26,** 10 (1970)

J. grandiflorum.
MACROCYCLIC TERPENOID: 5'-hydroxyjasmonic acid lactone.
MONOTERPENOID: (-)-jasmine lactone.
Winter *et al., Helv. Chim. Acta,* **45,** 1250 (1962)

J. primulinum.
DITERPENE GLUCOSIDE: jasminin.
Kubota *et al., J. Chem. Soc., Japan,* **89,** 62 (1968)

JASONIA (Compositae/Inulae)

J. glutinosa DC.
SESQUITERPENOID: kudtdiol.
de Pascual Teresa *et al., Tetrahedron Lett.,* 4141 (1978)

J. tuberosa.
FLAVONES: dihydroquercetin 3',7-dimethyl ether, hispidulin, penduletin.
SESQUITERPENOID: 2β-hydroxy-T-cadinol.
(F) Bohm *et al., Phytochemistry,* **16,** 1205 (1977)
(F) Gonzalez *et al., An. Quim.,* **73,** 460 (1977)
(T) de Pascual Teresa *et al., An. Quim.,* **74,** 1536 (1978)

JATEORHIZA (Menispermaceae)

J. palmata (Lam.) Miers. (J. columba).
PROTOBERBERINE ALKALOIDS: columbamine, jateorrhizine, palmatine.
DIMERIC PROTOBERBERINE ALKALOID: bisjatrorrhizine.
DITERPENOIDS: chasmanthin, isojateorinyl glucoside, jateorin, jaterorinyl glucoside, palmarin.
TRITERPENOIDS: columbin, columbinyl glucoside.
(A) Carvalhas, *J. Chem. Soc., Perkin I,* 327 (1972)
(T) Barton *et al., J. Chem. Soc.,* 4809 (1962)
(T) Itokawa *et al., Planta Med.,* **53,** 271 (1987)

JATROPHA (Euphorbiaceae)

J. curcus.
ACIDS: linoleic acid, oleic acid, palmitic acid, stearic acid.
DITERPENOIDS: curcusones A, B, C and D.
(Ac) Ferrao *et al., Rev. Port. Bioquim. Appl.,* **4,** 17 (1982); *Chem. Abstr.,* 97 195753f
(T) Naengchomnong *et al., Tetrahedron Lett.,* 2439 (1986)

J. dioica.
DITERPENOID: riolozatrione.
Dominguez *et al., Phytochemistry,* **19,** 2478 (1980)

J. gossypifolia.
FURANONE: gadain.
DITERPENOIDS: jatropholone A, jatropholone B, jatrophone.
(F) Banerji *et al., Phytochemistry,* **23,** 2323 (1984)
(T) Purushothaman *et al., Tetrahedron Lett.,* 972 (1979)

J. macrorhiza.
DITERPENOID: jatrophatrione.
TRITERPENOID: aleuritolic acid.
Torrance *et al., J. Pharm. Sci.,* **66,** 1348 (1977)

JAVA

J. citronella.
SESQUITERPENOID: sesquicitronellene.
Naves *et al., Helv. Chim. Acta,* **49,** 1029 (1966)

JUGLANS (Juglandaceae)

J. mandshurica.
NAPHTHOQUINONES: 3,3'-bijuglone, cyclotrijuglone, juglone.
Kirakawa *et al., Phytochemistry,* **25,** 1494 (1986)

J. regia.
NAPHTHOQUINONES: 3,3'-bisjuglone, juglone, trisjuglone.
BENZOPYRAN: avicularin.
STEROID: β-sitosterol.
(N, S) Sidhu *et al., Indian J. Chem.,* **13,** 749 (1975)
(B) Babu *et al., Phytochemistry,* **17,** 2042 (1978)

JULBERNARDIA

J. paniculata.
PIPERIDINE ALKALOID: carboxy-4(R),5(R)-dihydroxypiperidine.
Shewry *et al., Phytochemistry,* **15,** 1981 (1976)

JULOCROTON (Euphorbiaceae)

J. montevidensis (L.) Klotzsch.
PIPERIDINEDIONE ALKALOID: julocrotine.
Nakano *et al., J. Org. Chem.,* **26,** 1184 (1961)

JUNCUS

J. effusus.
COUMARIN: 7,8-dihydroxycoumarin.

Williams *et al.*, *Biochemical Systematics and Ecology*, **3**, 181 (1975)

JUNGERMANNIA (Hepaticae)

J. infusca.
DITERPENOIDS: ent-15β-hydroxykaurene, ent-kaur-16-en-15-ol, ent-kaur-16-en-15-one.

Matsuo *et al.*, *Phytochemistry*, **16**, 489 (1977)

J. rosulans (Stapf.) Stapf.
SESQUITERPENOIDS: cuprenenol, rosulantol.

Matsuo *et al.*, *Tetrahedron Lett.*, 3821 (1977)

J. sphaerocarpa.
DITERPENOIDS: ent-11α-hydroxykaur-16-en-15-one, ent-11α-hydroxykauren-15α-yl acetate, ent-kaur-16-ene-11,15-diol.

Connolly, Thornton, *J. Chem. Soc., Perkin I*, 736 (1973)

J. thermarum Stapf.
DITERPENOIDS: ent-8(14),15-pimaradien-19-oic acid, ent-pimarene-8,19-diol, (-)-thermarol.

Matsuo *et al.*, *Tetrahedron Lett.*, 2451 (1976)

J. torticalyx.
DITERPENOID: jungermanool.

Matsuo *et al.*, *Experientia*, **32**, 966 (1976)

JUNGIA (Compositae/Mutisieae)

J. malvaefolia Muschler.
SESQUITERPENOIDS: iso-α-cedren-15-al, iso-α-cedrene-14,15-dial, jungianol.

Bohlmann, Zdero, *Chem. Ber.*, **112**, 427 (1979)

J. spectabilis.
ACETYLENICS: pentadeca-8,10-diene-2,4,6-triyne, pentadeca-2,8,10-triene-4,6-diyne.

Bohlmann, Zdero, *Phytochemistry*, **16**, 239 (1977)

J. stuebelli.
SESQUITERPENOIDS: 9-acetoxy-iso-α-cedrene-15,14-olide, 7-oxo-14-norisoα-cedren-15-al.

Bohlmann *et al.*, *Phytochemistry*, **22**, 1201 (1983)

JUNIPERUS (Cupressaceae)

J. arizonica.
DITERPENOID: trans-communic acid.

Ahoud *et al.*, *Bull. Soc. Chim. Fr.*, 347 (1964)

J. californica.
SESQUITERPENOID: widdringtonic acid.

Arya, *J. Sci. Ind. Res. (India)*, **21B**, 504 (1962)

J. communis L.
BENZOPYRAN: afzelechin.
PIGMENT: xanthoperol.
DITERPENOID: communic acid.
MONOTERPENOIDS: 4-acetyl-1,4-dimethyl-1-cyclohexene, campholic aldehyde.
SESQUITERPENOIDS: α-cadinol, (-)-α-cadinol, β-elemen-7α-ol, (+)-junenol, junicedrol, juniperol, longifolene.

(B) Hillis *et al.*, *Aust. J. Chem.*, **13**, 390 (1960)

(P) Brendenberg *et al.*, *Acta Chem. Scand.*, **10**, 1511 (1956)

(T) de Pascual Teresa *et al.*, *An. Quim.*, **73**, 463 (1977)

(T) de Pascual Teresa *et al.*, *ibid.*, **73**, 568 (1977)

J. conferta.
FURANONE: savinin.
DITERPENOID: neoisodextropimaric acid.

J. contorta.
SESQUITERPENOID: longifolan-3α,7α-oxide.

Doi *et al.*, *Tetrahedron Lett.*, 4003 (1977)

SESQUITERPENOIDS: 3α,7α-epoxylongifolane, longifolol-7(15)-en-5-ol.

Doi *et al.*, *Tetrahedron Lett.*, 4003 (1977)

J. foetidissima Willd.
SESQUITERPENOIDS: 8,14-cedranediol, 8,14-cedranolide, 8,14-cedranoxide, 8-cedren-13-ol.

Baggaley *et al.*, *Tetrahedron*, **24**, 3399 (1968)

J. formosana Hayata.
LIGNAN: detetrahydroconidendrin.
DITERPENOIDS: 7α-methoxydeoxycryptojaponol, 7β-methoxydeoxycryptojaponol, sugiol.
SESQUITERPENOID: isocedrolic acid.

(L, T) Kim *et al.*, *J. Chin. Chem. Soc.*, **29**, 213 (1982)

(T) Kuo *et al.*, *J. Chin. Chem. Soc.*, **31**, 417 (1984)

J. horizontalis.
FLAVONES: cupressusflavone, sciadopitysin.

Beckmann *et al.*, *Phytochemistry*, **10**, 2465 (1971)

J. indica (J. pseudosabina).
FLAVONOIDS: amentoflavone, cupressuflavone, hinokiflavone, isocryptogenin.

Khatoon *et al.*, *J. Indian Chem. Soc.*, **62**, 410 (1985)

J. macropoda.
FLAVONOIDS: armentoflavone, hinokiflavone, isocryptomarin, junipegenin C, kaempferol 3-O-β-D-glucoside, quercetagenin 3-O-α-L-rhamnoside, quercetin 3-O-α-L-rhamnoside.

Sethi *et al.*, *Phytochemistry*, **20**, 341 (1981)

J. oxycedrus L.
DITERPENOIDS: 11β-hydroxymanoyl oxide, methyl cis-communate, methyl-trans-communate, methylimbricatalate, methylimbricatolate, myrecommunic acid, myrecommunic acid dimer, myrecommunic acid methyl ester, ylangene.
SESQUITERPENOIDS: 1-bulgarene, α-bulgarene, τ-cadinene.

de Pascual Teresa *et al.*, *An. Quim.*, **70**, 1015 (1974)

de Pascual Teresa *et al.*, *ibid.*, **73**, 1527 (1978)

J. phoenicea L.
DITERPENOIDS: 7,13-abietadien-19-oic acid, 8,13-abietadien-19-oic acid, 7,13-abietadien-3-one, 4-epi-abietic acid, (12R)-12-acetoxymyreocommunic acid, 7α-hydroxy-4-epi-dehydroabietal, 12-hydroxy-8(20)-labden-19-oic acid, 6α-hydroxysandaracopimaric acid, 4-epi-palustric acid.
SESQUITERPENOIDS: α-acoradiene, β-acoradiene, α-acorenol, β-acorenol, α-cadinol methyl ether.

Tabacek, Poisson, *Bull. Soc. Chim. Fr.*, 3264 (1969)

de Pascual Teresa *et al.*, *An. Quim.*, **74**, 459 (1978)

J. procera.
CYCLOHEPTANE: procerin.

Runeberg, *Acta Chem. Scand.*, **15**, 645 (1961)

J. recurva Buch.
SESQUITERPENOIDS: 8-cedren-13-ol, thujopsene.

Oda *et al.*, *Agric. Biol. Chem.*, **41**, 201 (1977)

J. rigida Sieb. et Zucc.
DITERPENOIDS: 3β-hydroxysandaracopimaric acid, nootkatinol, norisodextropimaric acid.
SESQUITERPENOIDS: α-acoradiene, β-acoradiene, δ-acoradiene, α-acorene, β-acorene, cadinenol, calacorene.

Naya, Kotake, Bull. Chem. Soc., *Japan*, **42**, 2088 (1969)

Tomita *et al.*, *Tetrahedron Lett.*, 1371 (1970)

Kuzo Doi, Kawamura, *Phytochemistry*, **11**, 841 (1972)

J. sabina L.
FURANONE: savinin. Diterpenoid palustradiene.
MONOTERPENE: (+)-sabinene, sabinol.
SESQUITERPENOID: oplopenone.
  (F) Doi, *Phytochemistry,* **11,** 1175 (1972)
  (T) de Pascual Teresa *et al., An. Quim.,* **74,** 680 (1978)

J. silicicola.
DITERPENOID: trans-communic acid.
  Ahond *et al., Bull. Soc. Chim. Fr.,* 348 (1964)

J. squamata (Lamb) D.Don.
SESQUITERPENOID: isocedrolic acid, 4-ketocedrol.
  Kuo *et al., Experientia,* **32,** 827 (1976)

J. thurifera.
SESQUITERPENOID: 8-acetoxyelemol.
  de Pascual Teresa *et al., An. Quim.,* **73,** 151 (1977)

J. utahensis.
CYCLOHEPTANE: utahin.
SESQUITERPENOID: cedrolic acid.
  Runeberg, *Acta Chem. Scand.,* **14,** 797 (1960)

J. virginiana L.
SESQUITERPENOID: cedrol.
TERPENOID GLUCOSIDE: podophyllotoxin-β-D-glucoside.
  Kupchan *et al., J. Pharm. Sci.,* **54,** 459 (1965)

JURINEA (Compositae)

J. alata.
SESQUITERPENOID: alatolide.
  Drozdz *et al., Collect. Czech. Chem. Commun.,* **38,** 727 (1973)

J. albicaulis.
SESQUITERPENOIDS: albicolide, juricanolide.
  Todorova, Ognyanov, *Planta Med.,* **50,** 452 (1984)

J. carduiformis.
SESQUITERPENOID: desacetylrepin.
  Rustaiyan *et al., Phytochemistry,* **20,** 1154 (1981)

J. maxima.
SESQUITERPENOIDS: jurmolide, maximolide, salonitenolide, salonitolide.
  Zakirov *et al., Khim. Prir. Soedin.,* 656 (1975)

J. suffruticosa Rgl.
ALKALOIDS: jurinine. Two uncharacterized bases.
  Yunusov *et al., Dokl. Akad. Nauk SSSR,* **11,** 29 (1959)
  Toda *et al., Tetrahedron,* **28,** 1477 (1972)

JUSTICIA (Acanthaceae)

J. gendarussa Burm.
STEROID: β-sitosterol.
  Wahi *et al., J. Res. Indian Med.,* **9,** 65 (1974); *Chem. Abstr.,* 85 17102n

J. hayati.
NAPHTHOFURANS: justicidins A and B.
  Okigawa *et al., Tetrahedron,* **26,** 4301 (1970)

J. procumbens L. var. decumbens Honda.
LACTONE: justicidin F.
NAPHTHOFURANS: diphyllin, justicidin A, justicidin B.
  (L) Horii *et al., Chem. Pharm. Bull.,* **25,** 1803 (1977)
  (N) Govindachari *et al., Tetrahedron,* **25,** 2815 (1969)

K

KADSURA (Schisandraceae)

K. coccinea.
LIGNANS: isokadsuranin, kadsuranin, kadsutherin.
Li et al., Planta Med., 297 (1985)

K. japonica.
CYCLIC DIBENZOOCTANES: angeloylbinankadsurin A, capreoylbinankadsurin A, norkadsurin A.
SESQUITERPENOIDS: α-elemol, epi-α-elemol, germacrene C.
TRITERPENOID: kadsuric acid.
(C) Ookawa et al., Chem. Pharm. Bull., 29, 123 (1981)
(T) Morikawa, Hirose, Tetrahedron Lett., 1299 (1969)

KALANCHOE (Crassulaceae)

K. daigremontiana.
STEROIDS: clerosterol, 25(27)-dehydroporiferasterol.
Nes et al., Lipids, 12, 511 (1977)

K. spathulata.
STEROIDS: campesterol, stigmasterol.
TRITERPENOIDS: friedelin, glutinol, taraxerol.
Gaind et al., Phytochemistry, 15, 1999 (1976)

KALLSTROEMIA

K. maxima.
STEROID: 25S-spirost-4-ene-3,12-dione.
Dominguez et al., Planta Med., 534 (1985)

K. pubescens.
SAPONINS: kallstroemins A, B, C, D and E.
Mahato et al., Indian J. Chem., 15B, 445 (1977)

KALMIA (Ericaceae)

K. latifolia L.
STEROIDS: 24α-ethyllathosterol, β-sitosterol, spinasterol.
TOXINS: kalmitoxins I, II, III, IV, V and VI.
(S) Nes et al., Lipids, 12, 511 (1977)
(To) El-Naggar et al., J. Nat. Prod., 43, 617 (1980)

K. leucanthoe.
TOXIN: grayanitoxin I.
Wood et al., J. Am. Chem. Soc., 76, 5689 (1954)

KALOPANAX (Araliaceae)

K. septemlobus.
SAPONINS: kalopanax septemlobus saponin A, kalopanax septemlobus saponin B.
Khorlin et al., Izv. Akad. Nauk SSSR, 1588 (1966)

KAUNEA

K. arbuscularis.
MONOTERPENOID: 10-acetoxy-8,9-dihydroxythymol.
SESQUITERPENOIDS:
2β-hydroxy-3α,4α-epoxy-3,4-dihydrokauniolide, kauniolide, novanin.
Bohlmann et al., Planta Med., 50, 284 (1984)

K. ignorata.
SESQUITERPENOIDS: kauniolide 2-oxoludartin.
Bohlmann et al., Phytochemistry, 20, 2375 (1981)

K. saltensis.
MONOTERPENOID: 3β-hydroxycarvotanactone.
Bohlmann et al., Phytochemistry, 20, 2375 (1981)

KAYEA (Guttiferae/Clusiaceae)

K. stylosa Thw.
XANTHONES: kayeaxanthone, 10-O-methylmacluraxanthone.
Gunasekera, Sultanbawa, Unpublished work.

KENTROPHYLLUM

K. lanatum (DC.).
FLAVONOIDS: eriodictyol, luteolin, quercetin.
Reynaud et al., Plant Med. Phytother., 10, 170 (1976)

KHAYA (Meliaceae)

K. anthotheca (Welw.) C.DC.
LIMONOIDS: 11α-acetoxyazadirone, 11β-acetoxyazadirone, anthothecol, deacetylanthothecol, 3-deacetylkhivorin, deoxyhavanensin, deoxyhavanensin acetate, 1α,11:14β,15-diepoxy-6-hydroxymeliaca-5,9,20,22-tetraene-3,7-dione, 14α,15α-epoxy-16-oxo-1,3,7-triacetoxymeliacin, havanensin, havanensin acetate, khayanthone, khayasin, khayasin B, khayasin T, khivorin, 3-isobutyryloxy-1-oxo-meliacate.
Adesogan et al., J. Chem. Soc., C, 205 (1970)
Halsall, Troke, J. Chem. Soc., Perkin I, 1758 (1975)

K. grandifoliola.
STEROID: 20-hydroxypregn-4-en-3-one acetate.
TERPENOIDS: 6-acetoxyangolensic acid methyl ester, fissinolide, grandifolienone, 6-hydroxyangolensic acid methyl ester.
(S) Adesogan, Taylor, Chem. Ind., (London), 1365 (1967)
(T) Connolly et al., Tetrahedron, 23, 4035 (1967)

K. ivorensis.
COUMARINS: esculetin, scoparone, scopoletin, umbelliferol.
STEROIDS: campesterol, β-sitosterol, stigmasterol.
TERPENOIDS: 6-deoxyswietenolide, ivorensic acid.
(C, S) Adesina, Fitoterapia, 54, 141 (1983)
(T) Connolly et al., J. Chem. Soc., Chem. Commun., 162 (1965)

K. madagascariensis Jumelle et Perrier.
LIMONOIDS: 11α-acetoxykhivorin, 3-deacetylkhivorin, fissinolide, methyl 3β-acetoxy-2-hydroxy-1-oxomeliacate.
Taylor, J. Chem. Soc., Chem. Commun., 1172 (1968)

K. nyasica.
TERPENOIDS: 11-acetoxykhivorin, khayasin, nyasin.
Taylor, J. Chem. Soc., Chem. Commun., 1172 (1968)

K. senegalensis L.
COUMARINS: esculetin, scoparone, scopoletin, umbelliferol.
LIMONOIDS: 7-deacetoxy-7-oxokhivorin, 7-deoxy-7-oxokhivorin, 6-epi-detigloyloxyswietenin acetate, 6-hydroxyangolensic acid methyl ester, 12β-hydroxy-6-deoxydestigloylswietenin acetate, khayanthone, khayasin, khayasin B, khayasin T, methyl-6-acetylangolensate, methyl 6-hydroxyangolensate, methylsenegalensate.
POLYSACCHARIDE: polysaccharide A.
STEROIDS: campesterol, β-sitosterol, stigmasterol.
(C, S) Adesina, Fitoterapia, 54, 141 (1983)
(P) Aspinall, Bhattacharyya, J. Chem. Soc., C, 365 (1970)
(T) Adesogan, Taylor, ibid, 1974 (1968)

KIBATALIA

K. arborea.
STEROIDAL ALKALOID: holafebrine.
Janot et al., Bull. Soc. Chim. Fr., 285 (1962)

K. gitingensis Woods.
STEROIDAL ALKALOIDS: gitingensine, kibataline.
Cave et al., Bull. Soc. Chim., 2415 (1964)
Bernal-Santos, Phillipp. J. Sci., 96, 411 (1967)

KICKXIA

K. spuria.
IRIDOID: kickxioside.
Nicoletti et al., Planta Med., 53, 295 (1987)

KIELMEYERA (Guttiferae/Clusiaceae)

K. candidissima.
XANTHONES: 1,7-dihydroxyxanthone,
1,3-dimethoxy-5-hydroxyxanthone,
2,3-dimethoxy-4-hydroxyxanthone.
Ferreira et al., Phytochemistry, 11, 1512 (1972)

K. coriacea Mart.
LIGNOIDS: cadensin A, cadensin B, kielcorin, kielcorin B.
XANTHONES: 6-deoxyjacareubin,
1,5-dihydroxy-3-methoxyxanthone, 3
4-dihydroxy-2-methoxyxanthone,
2,3-dimethoxy-4-hydroxyxanthone,
2,3-dimethoxy-5-hydroxyxanthone,
2,3-methylenedioxy-4-methoxyxanthone,
2-methoxyxanthone, osajaxanthone.
(L) Pinto et al., Phytochemistry, 26, 2045 (1987)
(X) Lopes et al., Phytochemistry, 16, 1101 (1977)

K. corymbosa (Spr.) Mart.
XANTHONES: 1,3-dimethoxy-5-hydroxyxanthone,
1-hydroxy-7-methoxyxanthone, 2-methoxyxanthone,
osajaxanthone.
de Barros Correa et al., An. Acad. Bras. Cienc., 36, 239 (1964)

K. excelsa Camb.
XANTHONES: 2,8-dihydroxy-1-methoxyxanthone,
1,7-dihydroxyxanthone, gentisin,
1-hydroxy-7-methoxyxanthone.
Antonaccio et al., An. Acad. Bras. Cienc., 37, 229 (1965)

K. ferruginea A.P.Duarte.
XANTHONES: 6-deoxyjacareubin,
1,3-dimethoxy-5-hydroxyxanthone,
2,3-dimethoxy-4-hydroxyxanthone,
2,4-dimethoxy-3-hydroxyxanthone, jacareubin.
Gottlieb et al., Phytochemistry, 8, 665 (1969)

K. petiolaris (Spr.) Mart.
XANTHONE: 2,8-dihydroxy-1-methoxyxanthone.
Gottlieb et al., Tetrahedron, 24, 1601 (1968)

K. rubriflora Camb.
XANTHONES: 1,7-dihydroxy-2,3,8-trimethoxyxanthone,
2,3-dimethoxy-4-hydroxyxanthone,
2,4-dimethoxy-3-hydroxyxanthone,
2,3-methylenedioxy-4-hydroxyxanthone,
2,3-methylenedioxy-4-methoxyxanthone.
Gottlieb et al., Phytochemistry, 10, 2253 (1971)

K. rupestris A.P.Duarte.
XANTHONES: 1,3-dimethoxy-5-hydroxyxanthone,
3-hydroxy-1,5,6-trimethoxyxanthone.
de Barros Correa et al., Phytochemistry, 9, 2537 (1970)

K. speciosa St. Hill.
XANTHONES: 6-deoxyjacareubin,
1,3-dimethoxy-5-hydroxyxanthone,
2,4-dimethoxy-3-hydroxyxanthone,
2,3-methylenedioxy-4-hydroxyxanthone,
2,3-methylenedioxy-4-methoxyxanthone.
De Oliviera et al., An. Acad. Bras. Cienc., 38, 421 (1971)

KIGELIA (Bignoniaceae)

K. pinnata.
DITERPENOID: pinnatal.
NORMONOTERPENOID: norviburtinal.
Joshi et al., Tetrahedron, 38, 2703 (1982)

KINGIELLA

K. taenialis (Lindl.) Rolfe.
PYRROLIZIDINE ALKALOID: phalaenopsine.
Brandage, Granelli, Acta Chem. Scand., 24, 354 (1970)

KLEINIA (Compositae)

K. fulgens Hook f. (Senecio fulgens).
SESQUITERPENOID: 4,5-dihydroxykleinifulgin.
Bohlmann et al., Phytochemistry, 20, 251 (1981)

K. kleinioides.
PYRROLIZIDINE ALKALOID: dehydroisosenaetnine.
Bohlmann, Knoll, Phytochemistry, 17, 599 (1978)

KNAUTIA (Dipsacaceae)

K. arvensis.
SAPONINS: knautioside A, knautioside B.
Surkova et al., Khim. Prir. Soedin, 661 (1975)

K. montana.
GLYCOSIDES: swertiajaponin, swertisin.
Zamtsova et al., Khim. Prir. Soedin., 705 (1977)

KNEMA (Myristicaceae)

K. attenuata (Wall.) Warb.
TETRALIN: (-)-attenuol.
Joshi et al., Tetrahedron, 35, 1665 (1979)

KNIGHTIA

K. deplanchei.
Four tropane alkaloids.
Kan-Fan, Lounasmaa, Acta Chem. Scand., 27, 1039 (1973)

K. strobilina.
TROPANE ALKALOIDS: knightalbinol, knightolamine, strobamine,
strobolamine.
Lounasmaa et al., Phytochemistry, 19, 949 (1980)

KOANOPHYLLON

K. admantium.
DITERPENOIDS: koanoadmantic acid methyl ester, koanophyllic
acids A, B, C and D.
Bohlmann, Phytochemistry, 20, 1903 (1981)

K. conglobatum.
DITERPENOIDS: koanophyllic acids A, B and C.
Bohlmann, Phytochemistry, 20, 1903 (1981)

KOELREUTERIA (Sapindaceae)

K. paniculata.
SAPONINS: koelreuteriasaponin A, koelreuteriasaponin B.
Chirva et al., Khim. Prir. Soedin., 328 (1970)

KOKOONA (Celastraceae)

K. zeylanica Thwaites.

TRITERPENOIDS: friedelin, kokoondiol, kokoonol, kokoononol, kokzeylanol, kokzeylanonol, zeylandiol, zeylanol, zeylononol.

STEROID: zeylasterone.

(S) Kamal *et al.*, *Tetrahedron Lett.*, 4749 (1980)

(T) Gunatilaka *et al.*, *J. Chem. Soc., Perkin I*, 2459 (1983)

KOMAROVIA

K. anisospermum.

COUMARINS: isoimperatorin, 5-methoxy-8-geranylhydroxypsoralen, phellopterin.

Sokolova *et al.*, *Khim. Prir. Soedin.*, 166 (1976)

KOPSIA (Apocynaceae/Plumerioideae)

K. fruticosa.

IMIDAZOLIDINOCARBAZOLE ALKALOIDS: fruticosine, kopsamine, kopsidine, kopsine.

Battersby *et al.*, *J. Chem. Soc., Chem. Commun.*, 786 (1966)

K. jasminiflora Pitard.

IMIDAZOLIDINOCARBAZOLE ALKALOIDS: jasminiflorine, kopsijasmine.

Ruangrungsi *et al.*, *Tetrahedron Lett.*, 3679 (1987)

K. longiflora.

IMIDAZOLIDINOCARBAZOLE ALKALOIDS: kopsiflorine, kopsilongine, kopsinine.

Crow, Michael, *Aust. J. Chem.*, **8**, 129 (1955)

K. officinalis L.

IMIDAZOLIDINOCARBAZOLE ALKALOIDS: chanofruticosine methyl ester, (+)-5,22-dioxokopsane, (-)-isoburnamine, (-)-kopsinine, kopsoffine, de-N-methoxycarbonylchanofruticone methyl ester, 6,7-methylenedioxychanofruticosine methyl ester, (-)-N-methoxycarbonyl-11,12-dimethoxykopsinaline, (-)-N-methoxycarbonyl-12-methoxykopsinaline, (-)-N-methoxycarbonyl-11,12-methylenedioxykopsinaline, (-)-12-methoxykopsinaline, (-)-11,12-methylenedioxykopsinaline, (-)-quebrachamine, (-)-tetrahydroalstonine.

Feng *et al.*, *J. Nat. Prod.*, **47**, 117 (1984)

K. singapurensis.

IMIDAZOLIDINOCARBAZOLE ALKALOIDS: kopsapurine, kopsingine.

Thomas *et al.*, *J. Am. Chem. Soc.*, **89**, 3325 (1967)

KOROLKOWIA (Liliaceae)

K. sewerzowii (Rgl.) Rgl. (Fritillaria sewerzowii Rgl.).

STEROIDAL ALKALOIDS: algamine, alginine, korseveramine, korseveridine, korseveriline, korseverine, korseverinine, korseversinine, korsevinine, korsidine, korsiline, korsinamine, korsine, korsine N-oxide, korsinine, sevcorine, seveline, seveline N-oxide, severine, severine N-oxide.

Samikov *et al.*, *Khim. Prir. Soedin.*, 233 (1978)

Samikov *et al.*, *ibid*, 251 (1986)

KRAMERIA (Krameriaceae)

K. ramosissima.

NORLIGNANS: ramosissin, 2-(2,3,4,6-tetramethoxyphenyl)-5-(E)-propenylbenzofuran.

Achenbach *et al.*, *Phytochemistry*, **26**, 2041 (1987)

K. triandra.

TYROSINE ALKALOID: geoffroyine.

Goldschmeidt, *Monatsh. Chem.*, **34**, 659 (1913)

KREYSIA

K. multiflora Reichb.

HOMOAPORPHINE ALKALOIDS: kreysigine, kreysiginine, multifloramine.

SPIROAPORPHINE ALKALOIDS: dihydrokreysiginone, kreysiginone.

Battersby *et al.*, *J. Chem. Soc., Chem. Commun.*, 934 (1967)

KYDIA

K. calycina.

PHENOLICS: hibiscone C, hibisoquinone B, isohemigossypol 1-methyl ether.

Joshi *et al.*, *Planta Med.*, **49**, 127 (1983)

L

LABURNUM (Leguminosae)

L. alpinum.
ISOFLAVONE: alpinusisoflavone.
Jackson *et al., J. Chem. Soc., C,* 3389 (1971)

L. anagyroides.
QUINOLIZIDINE ALKALOIDS: cytisine, methylcytisine,
(-)-N-(3-oxobutyryl)-cytisine.
Gray *et al., J. Pharm. Pharmacol.,* 95P (1981)

LACHENALIA

L. unifolia.
FLAVONOIDS: diosmetin 7,3′-disulphate, diosmetin 3′-sulphate,
luteolin 7,3′-disulphate, luteolin 3′-sulphate, tricetin
7,3′-disulphate, tricetin 3′-sulphate.
Williams *et al., Phytochemistry,* **15,** 349 (1976)

LACHNANTHES (Hamamodoraceae)

L. tinctoria Ell.
PIGMENTS: N-(1-carboxy-2-methylbutyryl)-lachnanthopyridone,
lachnanthocarpone, lachnanthoside, lanthnanthoflurone.
PYRAN: 3,4-dihydroxy-5-phenylnaphthalic aldehyde.
(P) Weiss, Edwards, *Tetrahedron Lett.,* 4325 (1969)
(Py) Bazan *et al., Phytochemistry,* **15,** 1413 (1976)

LACTARIUS (Russulaceae)

L. blennius.
SESQUITERPENOIDS: blennins A, B, C and D.
de Bernardi *et al., Phytochemistry,* **19,** 99 (1980)

L. camphoratus.
SESQUITERPENOIDS: 1,10-epoxyxaryophyllen-12-ol,
12-hydroxycaryophyllene-4,5-epoxide.
Daniewski *et al., Phytochemistry,* **20,** 2733 (1981)

L. deliciosus.
SESQUITERPENOIDS: 14-hydroxyguai-1,3,5,9,11-pentaene,
lactarazulene, lactarofulvene, lactaroviolin,
14-stearylguai-1,3,5,9,11-pentaene.
Vokac *et al., Collect. Czech. Chem. Commun.,* **35,** 1296 (1970)
Daniewski *et al., Rocz. Chem.,* **50,** 2095 (1976)

L. germanicum.
TRITERPENOID: germanicol.
Russel *et al., Phytochemistry,* **15,** 1933 (1976)

L. necator.
SESQUITERPENOIDS: anhydrolactarorufin A, 3-deoxylactarorufin
A, lactaronecatorin.
ALKALOID: necatorone.
(A) Fugmann *et al., Tetrahedron Lett.,* 3575 (1984)
(T) Daniewski *et al., Rocz. Chem.,* **51,** 1395 (1977)

L. pergameus.
SESQUITERPENOIDS: lactaral, velleral,
Magnusson *et al., Tetrahedron,* **29,** 1621 (1973)

L. rufus.
SESQUITERPENOIDS: isolactarorufin, lactarorufins A, B, C, D
and N.
Daniewski, Kocor, *Bull. Acad. Pol. Sci., Ser. Sci. Chim.,* **18,** 585 (1970)

L. scrobiculatus.
Four sesquiterpenoids.
de Bernardi *et al., Chem. Ind. (London),* **58,** 177 (1976)

L. theiogalus.
ALIPHATIC: lactarinic acid.
Belova *et al., Khim. Prir. Soedin.,* 507 (1986)

L. torminosus (Schaeff. ex Fries) S.F.Gray.
SESQUITERPENOID: hydroxyblennin A.
Seppa, *Phytochemistry,* **18,** 1226 (1979)

L. uvidus Fries.
SESQUITERPENOIDS: uvidins A, B, C, D and E.
de Bernardi *et al., J. Chem. Soc., Perkin I,* 2739 (1983)

L. vellereus.
SESQUITERPENOIDS: lactaral, velleral.
Magnusson *et al., Tetrahedron,* **29,** 1621 (1973)

L. velutinus.
SESQUITERPENOID: stearylvelutinal.
Favre-Bonvin *et al., Tetrahedron Lett.,* 1907 (1982)

LACTUCA (Compositae/Cichioideae)

L. chinensis Makino.
ESTERS: stearylarachidate, stearyl palmitate, stearyl stearate.
TRITERPENOID: lupenyl acetate.
Chen *et al., Chem. Abstr.,* 86 27679r

L. lacianata.
SESQUITERPENOIDS: 11,13-dihydroglucozaluzanin C,
dihydrosantamarin, glucozaluzanin C,
9-hydroxy-11,13-dihydrozaluzanin C, 9-hydroxyzaluzanin
C, lactucopicriside, lactulide A, lactuside A, lactuside B.
Nishimura *et al., Phytochemistry,* **25,** 2375 (1986)

L. plumieri.
MONOTERPENOIDS: (R)-deca-4,6,8-triyn-1,3-diol,
dec-1-ene-4,6,8-triyn-3-one.
Bentley *et al., J. Chem. Soc.,* 1096 (1969)

L. sativa L.
CAROTENOID: lactucaxanthin.
Siefermanntharmis *et al., Phytochemistry,* **20,** 85 (1981)

L. serriola.
SESQUITERPENOID: 8-deoxylactucin.
St. Pyrek, *Rocz. Chem.,* **51,** 2165 (1977)

L. virosa L.
SESQUITERPENOID: lactucin.
Barton, Narayan, *J. Chem. Soc.,* 963 (1958)

LACTUCARIUM

L. germanicum.
TRITERPENOID: germanicol.
Barton, Brooks, *J. Chem. Soc.,* 257 (1951)

LADYGINIA

L. bucharica.
SAPONINS: ladyginosides A, B, C, D, E and F.
Patkhullaeva *et al., Khim. Prir. Soedin.,* 733 (1973)

LAETREA

L. thelypteris Bosy.
STEROID: pterosterone.
Takemoto *et al., Chem. Pharm. Bull.,* **15,** 1816 (1967)

LAGASCEA

L. rigida.
CHROMENE: eupatoriochromene.

Bohlmann et al., Phytochemistry, 17, 566 (1978)

Bohlmann et al., ibid., 17, 1677 (1978)

LAGENARIA (Cucurbitaceae)

L. leucantha.
GIBBERELLINS: gibberellin A-50, gibberellin A-52.

Fukui et al., Agric. Biol. Chem., 42, 1571 (1978)

L. siceraria.
TRITERPENOID: 22-deoxocucurbitacin D.

Enslin et al., J. Chem. Soc., C, 964 (1967)

LAGERSTROEMIA (Lythraceae)

L. indica L.
QUINAZOLINE ALKALOIDS: dihydroverticillatine, lagerine.

Ferris et al., J. Am. Chem. Soc., 93, 2958 (1971)

L. lancasteri.
TRITERPENOIDS: lagerenol, lagerenoyl acetate.

Talapatra et al., Phytochemistry, 22, 2559 (1983)

L. subcostata.
QUINOLIZIDINE ALKALOIDS: lasubine I, lasubine II, subcosine I, subcosine II.

Fuji et al., Chem. Pharm. Bull., 26, 2515 (1978)

LAGGERA

L. aurita (Schultz).
SESQUITERPENOID: laggerol.

Zutshi, Bakadia, Indian J. Chem., 14B, 64 (1976)

LAGOCHILUS (Labiatae)

L. hirsutissimum.
TERPENOID: lagochirzidin.

Nurmatova et al., Khim. Prir. Soedin., 788 (1979)

L. hirtus F. et M.
QUATERNARY ALKALOID: stachydrine.

Proskurnina, Utkin, Med. Prom., 9, 30 (1960)

L. inebrians Bge.
QUATERNARY ALKALOID: stachydrine.

DITERPENOIDS: lagochilin, lagochilin 3-acetate.

(A) Pulatova et al., Khim. Prir. Soedin., 62 (1969)

(T) Islamov et al., ibid., 404 (1978)

L. platycalyx Schrenk.
QUATERNARY ALKALOID: stachydrine.

Proskurnina, Utkin, Meditsinskaya Promyshlennost SSSR 9, 30 (1960)

L. pubescens Vved.
QUATERNARY ALKALOID: stachydrine.

DITERPENOIDS: 16-O-acetyl-3,18-isopropylidenelagochilin, 15-acetyllagochilin, 16-acetyllagochilin.

(A) Pulatova et al., Khim. Prir. Soedin., 62 (1969)

(T) Mavlankulova et al., ibid., 82 (1978)

L. setulosus Vved.
ALKALOIDS: lagochilline, stachydrine.

DITERPENOID: lagochilin.

(A) Pulatova et al., Khim. Prir. Soedin., 62 (1969)

(T) Mavlankulova et al., ibid., 82 (1978)

LALLEMANTIA (Labiatae)

L. peltata (L.) F. et M.
UNCHARACTERIZED ALKALOID: lallemancine

Platonova et al., Meditsinskaya Promyshlennost SSSR 2, 14 (1962)

LAMIASTRUM (Labiatae)

L. galeobdolon flavidum (L.) Ehrend et Polatschek.
IRIDOIDS: 8-O-acetylharpagide, 10-deoxymelittoside, harpagide.

Bianco et al., Phytochemistry, 25, 1981 (1986)

LAMINARIA

L. ochroleuca.
PHENOLICS: diphlorethol pentaacetate, phloroglucinol triacetate, triphlorethol heptaacetate.

Glombitza et al., Phytochemistry, 15, 1082 (1976)

LAMIUM (Labiatae)

L. album L.
QUATERNARY ALKALOID: stachydrine.

TERPENOID: lamalbid.

(A) Pulatova et al., Khim. Prir. Soedin., 62 (1969)

(T) Brieskorn, Ahlborn, Tetrahedron Lett., 4037 (1973)

L. amplexicaule L.
IRIDOIDS: 5-deoxylamioside, 6-deoxylamioside, ipolamiide, ipolamiidioside, lamiol, lamioside.

Agostini et al., Gazz. Chim. Ital., 112, 9 (1982)

LAMPROLOBIUM

L. fruticosum.
QUINOLIZIDINE ALKALOID: lamprolobine.

Hart et al., J. Chem. Soc., Chem. Commun., 302 (1968)

LANSIUM (Meliaceae)

L. annamalayanum.
SESQUITERPENOIDS: α-chigadmarene, lansene, lansol.

Somasekar et al., J. Indian Chem. Soc., 29, 604 (1952)

Somasekar et al., ibid., 29, 620 (1952)

L. domesticum Jack. var. duku.
DINORTRITERPENOID: lansic acid.

TETRANORTRITERPENOIDS: dukunolides A, B and C.

Nishizawa et al., J. Org. Chem., 50, 5487 (1985)

LANTANA (Verbenaceae)

L. achyranthifolia.
FLAVONOIDS: chrysoplenetin, penduletin.

NAPHTHOQUINONE: diodantunezone.

Dominguez et al., Planta Med., 49, 63 (1983)

L. camara L.
MONOTERPENOID: geraniol.

SESQUITERPENOIDS: caryophyllene, cedrene, farnesol, linalool, cis-nerolidol, trans-nerolidol.

TRITERPENOIDS: lantadene A, lantadene B, lantanilic acid, lantanolic acid, lantic acid, 22β-[(S)-2-methylbutanoyloxy]-3-oxoolean-12-en-28-oic acid.

Barua et al., Phytochemistry, 15, 987 (1976)

Johns et al., Aust. J. Chem., 36, 1895 (1983)

L. camara var. ('Common Pink').
TRITERPENOID: lantabetulic acid.

Hart et al., Aust. J. Chem., 29, 655 (1976)

L. camara var. ('Townsville Prickly Orange').
TRITERPENOID: icterogenin.

Barton, Mayo, J. Chem. Soc., 887 (1954)

L. hybrida.
GLYCOSIDES: 1-caffeylrhamnose, 1-(3-glucosyloxy-4-hydroxycinnamoyl)-glucoside.

Imperato et al., Phytochemistry, 15, 1786 (1976)

L. trilaefolia.
> TRITERPENOID: 24-hydroxy-3-oxours-12-en-28-oic acid.
>> Johns et al., Aust. J. Chem., **36**, 2537 (1983)

LANUROCERASUS
L. officinalis.
> STEROIDS: cholesterol, β-sitosterol, stigmasterol.
>> Tushishvili et al., Khim. Prir. Soedin., 709 (1977)

LAPPULA (Boraginaceae)
L. intermedia (Ldb.) M.Pop.
> PYRROLIZIDINE ALKALOID: lasiocarpine.
>> Man'ko et al., Khim. Farm., **26**, **5**, 166 (1968)

LARIX (Pinaceae)
L. darurild.
> CATECHINS: (-)-epi-afzelechin, (+)-catechin, (-)-epi-catechin.
>> Polyakova, Altual'nye Voprosy Botanicheskogo Resursovedeniya u Sibiri 190 (1976); Chem. Abstr., 85 189201k

L. decidua Miller.
> CATECHINS: (-)-epi-afzelechin, (+)-catechin, (-)-epi-catechin.
> DITERPENOIDS: isolacuresinol, (-)-13-epi-manool.
>> (C) Polyakova, Altual'nye Voprosy Botanicheskogo Resursovedeniya u Sibiri 190 (1976); Chem. Abstr., 85 189201k
>> (T) Briggs et al., Tetrahedron Lett., 14 (1959)

L. europaea.
> DITERPENOIDS: 6-acetyllarixol, larixyl acetate.
>> Weinhaus et al., Chem. Ber., **93**, 2625 (1960)

L. gmelini.
> DITERPENOIDS: larixol, larixyl acetate, epi-manool.
> TRITERPENOIDS: asqualene, cycloartenone.
>> D'yachenko et al., Khim. Prir. Soedin., 56 (1986)

L. leptolepis.
> FLAVONOIDS: dihydroquercetin, kaempferol 3-(p-coumaroylglucoside, naringenin 7-glucoside.
>> Niemann, Acta Botanica Neerlandica, **25**, 349 (1976); Chem. Abstr., 86 27673j

L. olgensis.
> CATECHINS: (-)-epi-afzelechin, (+)-catechin, (-)-epi-catechin.
>> Polyakova, Altual'nye Voprosy Botanicheskogo Resursovedeniya u Sibiri 190 (1976); Chem. Abstr., 85 189201k

L. sibirica.
> CATECHINS: afzelechin, (-)-epi-afzelechin, (+)-catechin, (-)-epi-catechin.
> DITERPENOID: larixol.
> MONOTERPENOIDS: 3-carene, β-phellandrene, α-pinene, β-pinene.
>> (C) King et al., J. Chem. Soc., 2948 (1955)
>> (C) Polyakova, Altual'nye Voprosy Botanicheskogo Resursovedeniya u Sibiri 190 (1976); Chem. Abstr., 85 189291k
>> (T) Weinhaus, Angew. Chem., **59**, 148 (1947)

LARREA (Zygophyllaceae)
L. divaricata.
> NORTRITERPENOID: larreagenin A.
>> Hebermehl, Moller, Justus Liebigs Ann. Chem., 169 (1974)

L. tridentata.
> FLAVONOIDS: 2,6-di-C-glucopyranosylapigenin, 6,8-di-C-glucopyranosylchrysoeriol, gossypetin 3,7-dimethyl ether, 5,8,4'-trihydroxy-3,7,3'-trimethoxyflavone.
> DITERPENOID: 3'-demethoxyisoguaiacin.
>> (F) Sakakibara et al., Phytochemistry, **15**, 727 (1976)
>> (T) Fronczek et al., J. Nat. Prod., **50**, 497 (1987)

LASALLIA
L. papulosa.
> ANTHRAQUINONE: 2-chloro-1,3,8-trihydroxy-6-methylanthraquinone.
>> Fox et al., Tetrahedron Lett., 919 (1969)

LASER (Umbelliferae)
L. trilobum.
> DITERPENOIDS: isolaserolide, laserolide.
>> Holub et al., Collect. Czech. Chem. Commun., **38**, 731 (1973)

LASERPITIUM (Umbelliferae)
L. archangelica.
> SESQUITERPENOID: archangelolide.
>> Holub, Samek, Collect. Czech. Chem. Commun., **43**, 2444 (1978)

L. halleri Crantz subsp. halleri.
> SESQUITERPENOIDS: hallerin, hallerol.
>> Appendino, Gariboldi, J. Chem. Soc., Perkin 1, 2017 (1983)

L. latifolium.
> SESQUITERPENOID: laserpitin.
>> Feldman, Justus Liebigs Ann. Chem., **135**, 276 (1965)

L. prutenicum.
> SESQUITERPENOIDS: 4-acetoxy-isoprunetin-10-one, prutenin.
>> Bohlmann, Zdero, Chem. Ber., **104**, 1611 (1971)

L. siler.
> SESQUITERPENOIDS: isomontanolide, isomontanolide acetate, polhovolide, tarolide, tribolide.
>> Holub et al., Collect. Czech. Chem. Commun., **43**, 2471 (1978)
>> Nawrot et al., Biochemical Systematics and Ecology, **11**, 243 (1983)

LASIANTHERA
L. austrocaledonica.
> ALKALOID: tetrahydrocantleyine.
>> Sevenet et al., Phytochemistry, **15**, 576 (1976)

L. fruticosa.
> SESQUITERPENOID: lasiodiol acetate.
>> Wiemar, Ales, J. Org. Chem., **46**, 5449 (1981)

L. podocephala.
> SESQUITERPENOID: podocephalol.
>> Bohlmann, Lonitz, Chem. Ber., **111**, 843 (1978)

LASIOCORYS
L. capensis.
> DITERPENOID: lasiocoryin.
>> Gafner, Kruger, J. Chem. Soc., Chem. Commun., 249 (1974)

LASIODISCUS (Rhamnaceae)
L. marmoratus.
> PEPTIDE ALKALOIDS: lasiodine A, lasiodine B.
>> Manchand et al., Tetrahedron, **25**, 937 (1969)

LASIOLAENA
L. morii.
> SESQUITERPENOIDS: 8,9-dihydroxy-3,4-epoxy-2-oxolasiolaenin, 9-hydroxy-8-(2-acetoxymethyl-2-butenoyl)-2-oxolasiolaenin, 9-hydroxy-8-tigloyloxy-2-oxolasiolaenin.
> DITERPENOIDS: 19-acetoxy-12,20-dihydroxy-14-methylenegeranylnerol, 12,19,20-trihydroxy-14-methylenegeranylnerol.
>> Bohlmann et al., Phytochemistry, **21**, 161 (1982)

L. santosii.
> SESQUITERPENOID: lasiolaenolide-9-al.
>> Bohlmann et al., Phytochemistry, **20**, 1613 (1981)

LASIOSPERMUM

L. radiatum.
SESQUITERPENOIDS: dehydrolasiosperman,
cis-dihydrophymaspermone, trans-dihydrophymaspermone,
lasiosperman.
Boronowski, *Tetrahedron*, **277**, 4101 (1971)
Rao *et al.*, *Tetrahedron Lett.*, 1039 (1972)

LASTREA

L. tenericaulis.
AMINOACIDS: arginine, aspartic acid, cysteine, glycine,
histidine, hydroxyproline, isoleucine, leucine, lycine,
methionine, proline, serine.
Patil *et al.*, *J. Polynol.*, **12**, 149 (1976); *Chem. Abstr.*, **88** 148993p

LATHAEA

L. thelypteris.
STEROID: pterosterone.
Takemoto *et al.*, *Chem. Pharm. Bull.*, **15**, 1816 (1971)

LATHYRUS (Leguminosae)

L. japonicus.
ACID: cis-5-hydroxy-L-pipecolic acid.
Hatanaka *et al.*, *Phytochemistry*, **16**, 1041 (1977)

L. odoratus L.
ISOXAZOLINE ALKALOID: 2-(2-cyanoethyl)-3-isoxazolin-5-one.
Van Rompuy *et al.*, *Biochem. Biophys. Res. Commun.*, **56**, 199 (1974)

L. tingitanus L.
ALKALOID: lathyrine.
Bell, *Biochem. Biophys. Acta*, **47**, 602 (1961)

LAUNAEA

L. nudicaulis.
STEROIDS: β-sitosterol, β-sitosterol acetate.
TRITERPENOID: taraxerol.
Majumder, Laha, *J. Indian Chem. Soc.*, **59**, 881 (1982)

LAUREALIA (Atherospermataceae)

L. novae-zelandae.
APORPHINE ALKALOIDS: laureline, laurepukine.
Aston, *J. Chem. Soc.*, **97**, 1381 (1910)
Aston, *ibid.*, **97**, 1386 (1910)

L. philippiana.
APORPHINE ALKALOIDS: 4-hydroxyanonaine,
4-hydroxynantenine.
Urzua, Cassels, *Phytochemistry*, **21**, 773 (1978)

L. sempervirens.
One aporphine alkaloid.
Bianchi *et al.*, *Gazz. Chim. Ital.*, **92**, 818 (1962)

LAURENCIA (Rhodophyceae)

L. alata.
ACETYLENE: alatenyne.
Hall *et al.*, *Aust. J. Chem.*, **39**, 1401 (1986)

L. brongniantii.
Four brominated indole alkaloids.
Carter *et al.*, *Tetrahedron Lett.*, 4479 (1978)

L. caespitosa.
HALOGENATED SESQUITERPENOIDS: caespitol, furocaespitone.
Gonzalez *et al.*, *Tetrahedron Lett.*, 2391 (1973)
Gonzalez *et al.*, *ibid.*, 3625 (1973)

L. caribaea.
HALOGENATED SESQUITERPENOID: carabaical.
Izac *et al.*, *Tetrahedron Lett.*, 1799 (1981)

L. chilensis.
KETAL: tetracyclinpolyketal.
MONOTERPENOID: chilenone A.
(K) Bittner *et al.*, *Tetrahedron Lett.*, 4031 (1987)
(T) San Martin *et al.*, *ibid.*, 4063 (1983)

L. concinna.
HALOGENATED DITERPENOID: concinndiol.
Sims *et al.*, *J. Chem. Soc., Chem. Commun.*, 470 (1973)

L. elata.
HALOGENATED SESQUITERPENOID: elatol.
Sims *et al.*, *Tetrahedron Lett.*, 3487 (1974)

L. filiformis.
SESQUITERPENOIDS: isolaurene, laurene.
HALOGENATED SESQUITERPENOIDS: allolaurinterol, austradiol
acetate, austradiol 4,6-diacetate,
4-bromo-α-chamigren-8-one, 4-bromo-β-chamigren-8-one,
dihydrolaurene, filiformin, filiforminol, heterocladol,
6-hydroxyaplysistatin, laurencin, spirolaurenone.
Polonsky *et al.*, *J. Am. Chem. Soc.*, **100**, 2575 (1978)
Brennan *et al.*, *J. Org. Chem.*, **47**, 3917 (1982)

L. glandulifera Keutzing.
SESQUITERPENOIDS: isolaurene, laurene.
HALOGENATED SESQUITERPENOIDS,:
4-bromo-α-chamigren-8-one, 4-bromo-β-chamigren-8-one,
4-bromo-8,9-epoxy-α-chamigrene, laurencin, spirolaurenone.
Suzuki *et al.*, *Tetrahedron Lett.*, 821 (1974)

L. hybrida.
DITERPENOID: hybridalactone.
Higgs *et al.*, *Tetrahedron*, **37**, 4259 (1981)

L. imbricata.
HALOGENATED SESQUITERPENOIDS:
5-acetoxy-2S,10R-dibromo-3-S-chloro-7S-chamigranol,
acetoxyintricatol, bermudenyol, bermudenyol acetate.
McMillan *et al.*, *Tetrahedron Lett.*, 2039 (1974)
Cordellina *et al.*, *Can. J. Chem.*, **60**, 2675 (1982)

L. intermedia.
SESQUITERPENOID: debromolaurinterol.
HALOGENATED SESQUITERPENOIDS: isolaurinterol, laurinterol.
Irie *et al.*, *Tetrahedron Lett.*, 1837 (1966)

L. irieii.
HALOGENATED DITERPENOIDS: irieol, irieols A, B, C, D, E, F
and G, noeirieone.
Howard *et al.*, *Tetrahedron Lett.*, 3847 (1982)

L. johnstonii.
HALOGENATED SESQUITERPENOIDS: johnstonol, prepacifenol
epoxide.
Sims *et al.*, *Tetrahedron Lett.*, 195 (1972)

L. nana.
HALOGENATED SESQUITERPENOIDS:
10-bromo-7,12-dihydroxy-2,3-laurene,
10-bromo-7-hydroxy-11-iodolaurene.
Izac, Sims, *J. Am. Chem. Soc.*, **101**, 6136 (1979)

L. nidifica.
SESQUITERPENOID: laurequinone.
HALOGENATED SESQUITERPENOIDS: isomaneonene A,
isomaneonene B, maneonene, menaonines A, B, C and D,
nidificene, nidifidiene, nidifidienol.
Shizuri *et al.*, *Phytochemistry*, **23**, 2672 (1984)

L. nipponica Yamada.
HALOGENATED SESQUITERPENOIDS: bromoacetal A, bromoacetal
B, 10-bromo-7,8-epoxy-2,9-chamigradiene-5,15-diolol,
laurallene, laureatin, isolaureatin, laurefucin, laurefucin
acetate, isodihydrolaurenol, isodihydrolaurenol acetate,
isoprelaurefucin, (-)-obtusone, spironippol.
Suzuki *et al.*, *Bull. Chem. Soc., Japan*, **55**, 1561 (1982)

Furasaki *et al., ibid.,* **56,** 2501 (1983)

Kurata *et al., Chem. Lett.,* 561 (1983)

L. obtusa (Hudson) Lemaroux.

KETAL: neoobtusin.

HALOGENATED DITERPENOIDS: laurencianol, obtusadiol.

SESQUITERPENOIDS: brasilenol, epi-brasilenol, brasilenyl acetate.

HALOGENATED SESQUITERPENOIDS:
2-bromo-4-(4-bromo-3,3-dimethylcyclohexyl)-1-methyl-7-oxobicyclo-[2,2,1]-heptane, bromo-8-epi-caparrapi oxide, isoobtusol, laurencianyne, obtusallene, obtusane, obtusenol, obtusenyne, obtusin, obtusol, α-snyderol, β-snyderol.

(K) Caccamese, Toscano, *Gazz. Chim. Ital.,* **116,** 177 (1986)

(T) Imre *et al., Phytochemistry,* **20,** 833 (1981)

(T) Gonzalez *et al., Tetrahedron Lett.,* 4143 (1983)

L. okamurii.

HALOGENATED SESQUITERPENOIDS: deoxyokamurallene, isoaplysin, johnstonol, okamurallene.

Suzuki *et al., Chem. Lett.,* 289 (1982)

L. oppositiclada.

HALOGENATED SESQUITERPENOID: rhodophytin.

Fenical *et al., J. Am. Chem. Soc.,* **96,** 5580 (1974)

L. pacifica.

HALOGENATED SESQUITERPENOIDS: kylinone, prepacifenol.

Selover, Crews, *J. Org. Chem.,* **45,** 69 (1980)

L. cf. palisada Yamada.

HALOGENATED SESQUITERPENOIDS: 5-acetoxypalisadin B, 12-hydroxypalisadin B, palisadin A, palisadin B, palisol.

Paul, Fenical, *Tetrahedron Lett.,* 2787 (1980)

L. perforata.

HALOGENATED SESQUITERPENOIDS: perforatone, perforenol, perforenone A, perforenone B.

Gonzalez *et al., Tetrahedron Lett.,* 2499 (1975)

L. pinnata Yamada.

STEROID: pinnasterol.

HALOGENATED DITERPENOID: pinnaterpene A.

HALOGENATED SESQUITERPENOIDS: isopaurepinnacin, laurepinnacin, (3E)-pinnatifidenyne, (3Z)-pinnatifidenyne.

(S) Fukuzawa *et al., Tetrahedron Lett.,* 4085 (1981)

(T) Gonzalez *et al., Tetrahedron,* **38,** 1009 (1982)

L. poetei.

SESQUITERPENOIDS: dactyol, poetediene, poeteol.

Fenical *et al., J. Org. Chem.,* **43,** 3629 (1978)

Wright *et al., Tetrahedron Lett.,* 4649 (1983)

L. snyderiae.

HALOGENATED DITERPENOID: neoconcinndiol hydroperoxide.

Howard *et al., J. Am. Chem. Soc.,* **99,** 6444 (1976)

L. snyderiae var. guadalupensis.

SESQUITERPENOIDS: guadalupol, epi-guadalupol, isoconcinndiol.

Howard, Fenical, *Phytochemistry,* **19,** 2774 (1980)

L. subopposita.

SESQUITERPENOIDS: 7-hydroxylaurene, laurene, oplopanone.

HALOGENATED SESQUITERPENOIDS: acetyllaurefucin, allolaurinterol, 10-bromo-7-laurene, isoprelaurefucin, laurefucin, oppositol. Kazlauskas et al., Aust. J. Chem., **29,** 2533 (1976)

Wratten, Faulkner, *J. Org. Chem.,* **42,** 3343 (1977)

L. venusta Yamada.

HALOGENATED SESQUITERPENOIDS: venustin A, venustin B.

TRITERPENOID: venustatriol.

Sakemi *et al., Tetrahedron Lett.,* 4287 (1986)

LAURUS (Lauraceae)

L. nobilis L.

MONOTERPENOID: dehydro-1,8-cineole.

SESQUITERPENOID: laurenobiolide.

Takeda *et al., Chem. Pharm. Bull.,* **24,** 667 (1976)

LAVANDULA (Labiatae)

L. canariensis.

STEROID: sitosterol-3-β-D-gluvcoside.

TRITERPENOIDS: α-amyrin, β-amyrin, 2,3-dihydroxyursolic acid, oleanolic acid, 2,3,19,23-tetrahydroursolic acid, ursolic acid.

Breton *et al., J. Nat. Prod.,* **49,** 941 (1986)

L. gibsonii.

MONOTERPENOIDS: 3-hydroxy-α,4-dimethylstyrene, 3-hydroxy-α,α,4-trimethylbenzyl alcohol, 3-hydroxy-α,α,4-trimethylbenzyl methyl ether.

TRITERPENOID: ursolic acid.

Patwardhan, Gupta, *Phytochemistry,* **22,** 2080 (1983)

L. stoechas L.

MONOTERPENOID: (+)-fenchone.

TRITERPENOID: lavanol.

Manzoor-i-Khuda *et al., Pakistan J. Sci. Ind. Res.,* **10,** 164 (1967)

LAWSONIA

L. alba, (L. inermis).

NORTRITERPENOID: 30-norlupan-3β-ol-20-one.

Thompson, Brown, *Phytochemistry,* **7,** 845 (1968)

L. inermis.

XANTHONES: laxanthone I, laxanthone II.

Bhardioaj *et al., Phytochemistry,* **16,** 1616 (1977)

LECANORA (Lecanoraceae)

L. gangleoidess Nyl.

BISANTHRAQUINONE: skyrin.

DEPSIDES: atramorin, chloroatramorin.

DEPSIDONES: gangleoidin, leonidin.

Devlin *et al., J. Chem. Soc., Perkin I,* 1491 (1986)

L. reuteri.

XANTHONE: norlichexanthone.

Broadbent *et al., Phytochemistry,* **14,** 2082 (1975)

L. rubina.

ACID: placcodiolic acid.

Huneck, *Tetrahedron,* **28,** 4011 (1972)

LECCENORA

L. sulphurella.

DEPSIDE: 3,5-dichloro-2'-O-methylanziaiac acid.

Huneck *et al., Phytochemistry,* **16,** 995 (1977)

LECCINUM

L. scabrum.

PIGMENTS: gyrosporin, xerocomorubin.

Besl *et al., Chem. Abstr.,* 88 117773n

LECIDEA (Lecideaceae)

L. confluens.

ACID: confluentic acid.

Huneck, *Chem. Ber.,* **95,** 328 (1962)

L. fuscoatra.

ESTER: confluentic acid methyl ester.

Huneck, *Phytochemistry,* **13,** 221 (1974)

LEDEBOURIELLA

L. seseloides.

BENZOPYRANS: 3'-angeloylhamaudol, 3'-angeloylhamaudol
3'-O-β-D-glucoside, hamaudol, hamaudol 3'-acetate,
ledebouriellol.
Sasaki *et al., Chem. Pharm. Bull.,* **30**, 3555 (1980)

LEDUM (Ericaceae)

L. columbianum.

SESQUITERPENOID: ledol.
Buchi *et al., Tetrahedron Lett.,* 14 (1959)

L. groenlandicum.

SESQUITERPENOID: ledol.
Graham *et al., Aust. J. Chem.,* **13**, 372 (1960)

L. palustre L.

FLAVONE: 5-hydroxy-7,4'-dimethoxy-6-methylglavone.
MONOTERPENOIDS: lepalene, myrcene.
SESQUITERPENOIDS: alloaromadendrene, ledol, palustrol.
Nikhaelova *et al., Khim. Prir. Soedin.,* 322 (1979)

LEGNEPHORA

L. moorea.

APORPHINE ALKALOID: isocorydine.
Kikkawa, *J. Pharm. Soc., Japan,* **78**, 1006 (1958)

LEMAIREOCEREUS (Cactaceae)

L. beneckei.

TRITERPENOID: queretaroic acid.
Djerassi *et al., J. Am. Chem. Soc.,* **78**, 3785 (1956)

L. chichips.

TRITERPENOIDS: chichipigenin, olean-12-ene-3β-16,22-triol.
Khong, Lewis, *Aust. J. Chem.,* **28**, 165 (1975)

L. dumorteri.

TRITERPENOID: dumorterigenin.
Djerassi *et al., J. Am. Chem. Soc.,* **76**, 2969 (1954)

L. longispinus.

TRITERPENOIDS: erythrodiol, longispinogenin.
Djerassi *et al., J. Am. Chem. Soc.,* **75**, 5940 (1953)

L. queretaroensis.

TRITERPENOID: queretaroic acid.
Djerassi *et al., J. Am. Chem. Soc.,* **78**, 3785 (1956)

L. stellatus.

TRITERPENOID: stellatogenin.
Djerassi *et al., J. Am. Chem. Soc.,* **75**, 2254 (1953)

L. thurberi.

STEROIDS: thurberin, thurberogenin.
Djerassi *et al., J. Am. Chem. Soc.,* **75**, 2254 (1953)

LEMMAPHYLLUM

L. microphyllum.

TRITERPENOIDS: 8(26),14(27)-onaceradienol, polypodatetraene,
α-polypodatetraene.
Shiojima *et al., Tetrahedron Lett.,* 3733 (1983)

L. microphyllum var. microphyllum.

TRITERPENOIDS: α-onaceradiene, serratene.
Ageta *et al., Chem. Pharm. Bull.,* **30**, 2272 (1982)

L. microphyllum var. obovatum.

TRITERPENOIDS: 12,21-baccadiene, 20,24-dammaradiene,
onoceranoxide.
Masuda *et al., Chem. Pharm. Bull.,* **31**, 2530 (1983)

LEMNA (Lemnaceae)

L. perpusilla.

GIBBERELLIN: allogibberic acid.
Pryce, *Phytochemistry,* **12**, 1745 (1973)

LENS

L. esculenta.

FLAVONOIDS: kaempferol
7-O-β-D-glucoside-O-α-L-rhamnosyl-3-O-α-L-rhamnoside,
kaempferol
3-O-α-L-rhamnoside-β-D-glucosyl-7-O-α-L-rhamnoside,
kaempferol 3-rutinosyl-7-O-β-D-glucoside, kaempferol
3-rutinosyl-7-O-α-L-rhamnoside.
PHENOLICS: p-coumaric acid, ferulic acid.
El-Negoumy *et al., Rev. Latinoam. Quim.,* **18**, 88 (1987)

LENTINUS

L. dactyloides.

HOMOTRITERPENOID: eburicoic acid.
Holker *et al., J. Chem. Soc.,* 2422 (1953)

L. edodes Sing. (Cortinellus shiitake P. Henn.).

ALIPHATICS: octanol, 1-octen-3-ol, cis-2-octen-1-ol.
PURINE ALKALOIDS: deoxyeritadenine, eritadenine.
CYCLIC THIO DERIVATIVE: 1,2,4,6-tetrathiepane,
1,2,4-trithiolane.
(A) Kamiya *et al., Tetrahedron,* **28**, 899 (1972)
(C) Morita *et al., Chem. Pharm. Bull,* **15**, 998 (1967)

LENZITES (Basidiomycetae)

L. trabea.

TRITERPENOID: 15-hydroxytrametenolic acid.
Lawrie *et al., J. Chem. Soc., C,* 1776 (1976)

LEONOTIS (Labiatae)

L. dysophylla.

DITERPENOIDS: 8β-hydroxymarrubin, leonotin.
Kaplan *et al., J. Chem. Soc., C,* 1656 (1970)

L. leonitis.

DITERPENOID: leonitin.
Eagle *et al., J. Chem. Soc., Perkin* 1, 994 (1978)

L. leonurus (L.) R.Br.

DITERPENOID: marubiin.
Castine *et al., Chem. Ind.,* 1832 (1961)

L. nepetaefolia (L.) Ait. f.

ALLENE: labellenic acid.
COUMARIN: 4,6,7-trimethoxy-5-methylcoumarin.
DITERPENOIDS: leonotin, leonotinin, methoxynepaetefolin,
nepetaefolin, nepetaefolinol, nepetaefuran, newpetaefurinol.
(Al) Bagbey *et al., Chem. Ind.,* 1861 (1964)
(C) Purushothaman *et al., J. Chem. Soc., Perkin I,* 2594 (1976)
(T) Kaplan *et al., J. Chem. Soc., C,* 1656 (1970)
(T) Manchand, *Tetrahedron Lett.,* 1907 (1973)

LEONTICE (Berberidaceae)

L. albertii Rgl.

ALKALOID: thaspine.
MATRINE ALKALOIDS: albertamine, albertidine, albertine,
leontalbamine, leontalbine, leontalbinine, leontidine,
leontine, matrine, (+)-sophoramine.
PIPERIDINE ALKALOID: anabasine.
Iskandarov *et al., Khim. Prir. Soedin.,* 628 (1972)
Sadykov *et al., ibid.,* 6 (1973)

L. darvasica Rgl.

ALKALOID: thaspine.
MATRINE ALKALOIDS: darvasine, darvasoline, leontine.
QUINOLIZIDINE ALKALOIDS: (-)-lupanine, N-methylcytisine.
Iskandarov et al., Khim. Prir. Soedin., 132 (1969)
Yunusov et al., ibid., 1 (1974)

L. ewersmannii Bge.

ALKALOID: thaspine.
BENZYLISOQUINOLINE ALKALOID: petaline.
MATRINE ALKALOID: leontine.
QUINOLIZIDINE ALKALOID: (+)-lupanine.
SPARTEINE ALKALOIDS: leontamine, leontidine, pachycarpine.
SAPONINS: leontosides A, B, C, D and E.
(A) Platonova et al., J. Gen. Chem., USSR, 23, 880 (1953)
(A) Ryblko, Proskurnina, ibid., 31, 308 (1961)
(Sap) Mzhel'skaya et al., Khim. Prir. Soedin., 421 (1966)

L. leontopetalum L.

BENZYLISOQUINOLINE ALKALOID: petaline.
QUINOLIZIDINE ALKALOIDS: leontoformidine, leontoformine.
Panov et al., C. R. Acad. Bulg. Sci., 25, 55 (1972)

L. smirnovii Trautv.

ALKALOID: thaspine.
MATRINE ALKALOIDS: isoleontalbine, leontisimine, sophocarpine.
QUINOLIZIDINE ALKALOIDS: 2-argemonine, leontamine, leontidine, (-)-lupanine, N-methylcytisine.
Tkashalashvili et al., Khim. Prir. Soedin., 807 (1975)

LEONURUS (Labiatae)

L. cardiaca L.

IRIDOIDS: ajugol, ajugoside, gluridoside, leonuride.
DITERPENOID: leocardin.
(I) Guiso et al., Gazz. Chim. Ital., 104, 25 (1974)
(I) Buzogamy et al., Clujul Medical, 56, 385 (1983); Chem. Abstr., 100 65051v
(T) Malakov et al., Phytochemistry, 24, 2341 (1985)

L. marrubiastrum.

DITERPENOIDS: aldehydomarrubialactone, demethylmarrubiaketone, marrubialactone, marrubiaside, marrubiastrol.
Tschesche et al., Chem. Ber., 111, 2130 (1978)

L. quinquelobatus Gilib.

QUATERNARY ALKALOIDS: choline, stachydrine.
IRIDOIDS: ajugol, ajugoside, gluridoside.
(A) Kazlova, Farmatsiya, 6, 23 (1967)
(I) Buzogamy et al., Clujul Medical, 56, 385 (1983); Chem. Abstr., 100 65051v

L. sibirica L.

GUANIDINE ALKALOID: leonurine.
DITERPENOIDS: isoleosibirin, leosibiricin, leosibirin.
(A) Goto et al., Tetrahedron Lett., 545 (1962)
(T) Savona et al., Phytochemistry, 21, 2699 (1982)

L. turkestanicus V.Kreez et Kupr.

QUATERNARY ALKALOID: stachydrine.
Pulatova, Khim. Prir. Soedin., 62 (1969)

LEPIDOCORYPHANTHA

L. runyonii.

AMINE ALKALOID: N-methyl-4-methoxyphenethylamine.
Agurell, Experientia, 25, 1132 (1969)

LEPIDOPHORUM

L. grisleyi.

SESQUITERPENOID: lepidophorone.
Bohlmann, Zdero, Chem. Ber., 106, 379 (1973)

LEPIDOSPARTIUM

L. squamatum.

SESQUITERPENOID: 1,10-epoxydecompositin.
Bohlmann et al., Chem. Ber., 105, 3523 (1972)

LEPIDOTRICHILIA

L. volensii.

SESQUITERPENOID: volenol.
Hoffman et al., J. Org. Chem., 43, 1254 (1978)

LEPIDOZIA (Hepaticae)

L. reptans.

SESQUITERPENOID: eudesm-3-ene-6β,7α-diol.
Connolly et al., Phytochemistry, 25, 1745 (1986)

L. vitrea.

SESQUITERPENOIDS: (-)-isobicyclogermacrenal, lepidozienal.
Nakayama et al., Chem. Lett., 1097 (1981)

LEPISTA

L. glaucocana.

ESTER: 2,8-decadiene-4,6-diynedioic acid dimethyl ester.
Hearn et al., J. Chem. Soc., Perkin I, 2335 (1974)

LEPRA (Lichenes)

L. candelairis.

PIGMENT: calycin.
Letcher et al., Tetrahedron Lett., 3541 (1967)

LEPRARIA (Fungi Imperfecti)

L. flava.

DITERPENOID: pinastric acid.
Asano et al., Ber., 68, 1565 (1935)

L. latebrarum.

TRITERPENOID: zeorin.
Huneck, Chem. Ber., 94, 714 (1961)

LEPTACTINIA

L. densiflora.

ISOQUINOLINE ALKALOID: tetrahydroharman.
UNCHARACTERIZED ALKALOID: leptaflorine.
Paris, Blond, C. R. Acad. Sci., 241, 241 (1955)

LEPTADENIA

L. pyrotechnica F.

AMINOACIDS: (-)-alanine, (-)-arginine, (-)-isoleucine, (-)-lysine, (-)-methionine, (-)-threonine.
DIPEPTIDES: (±)-alanyl-(-)-alanine, glycyl-(-)-alanine.
Dhawan et al., Curr. Sci., 45, 198 (1976)

LEPTINOTARSA

L. decenlineata Say.

STEROIDAL ALKALOID: leptinidine.
Kuhn, Low, Angew. Chem., 69, 236 (1957)

LEPTOLAENA (Chlaenaceae)

L. diospyroidea cavaco var. tampoketsensis.

ACIDS: caffeic acid, p-coumaric acid, gallic acid, gentisic acid, vanillic acid.
FLAVONOIDS: kaempferol-3-rhamnoside, myricetol-3-rhamnoside, myricitrin, quercetin-3-rhamnoside.
Paris et al., Plant Med. Phytother., 9, 230 (1975)

L. pauciflora.

ACIDS: caffeic acid, p-coumaric acid, gallic acid, gentisic acid, vanillic acid.

FLAVONOIDS: kaempferol-3-rhamnoside, myricetol-3-rhamnoside, myricitrin, quercetin-3-rhamnoside.

Paris *et al., Plant Med. Phytother.,* **9,** 230 (1975)

LEPTORHABDOS (Scrophulariaceae)

L. parviflora Benth.

MATRINE ALKALOIDS: isosophoramine, sophocarpine, (-)-sophoridine.

PYRIDINE ALKALOID: leptorhabine.

UNCHARACTERIZED ALKALOID: aloperine.

Bocharnikova, Massagetov, *J. Gen. Chem., USSR,* **34,** 1025 (1964)

Kadyrov *et al., Khim. Prir. Soedin.,* 683 (1974)

LEPTOSPERMUM (Myrtaceae)

L. flavescens Sm.

CYCLOHEXENE: grandiflorone.

NORSESQUITERPENOID: flavescone.

(C) Hellyer *et al., J. Chem. Soc., C,* 1496 (1966)

(T) Bick et al., *J. Chem. Soc.,* 3690 (1965)

L. flavescens var. grandiflora.

CYCLOHEXENE: grandiflorone.

Hellyer, Pinhey, *J. Chem. Soc., C,* 1496 (1966)

L. lanigerum.

CYCLOHEXENE: grandiflorone.

Hellyer, Pinhey, *J. Chem. Soc., C,* 1496 (1966)

L. lehmannii.

BENZOPHENONE: 2,4,6-trihydroxy-2-methylbenzophenone.

McGookin *et al., J. Chem. Soc.,* 2021 (1951)

L. liversidgei.

MONOTERPENOID: (+)-isopulegol.

Pickard *et al., J. Chem. Soc.,* **117,** 1252 (1920)

LEPTOSTIGMA (Rubiaceae)

L. arnothianum.

FLAVONOIDS: isorhamnetin 3-O-rutinoside, isorhamnetin 3-O-xylosylrutinoside, quercetin 3-O-rutinoside.

Nunes-Alarcon *et al., Rev. Latinoam. Quim.,* **18,** 9 (1987)

LESPEDEZA (Leguminosae)

L. bicolor var. japonica.

INDOLE ALKALOID: lespedamine.

Morimoto, Oshio, *Justus Liebigs Ann. Chem.,* **682,** 212 (1965)

L. capitata.

FLAVONOIDS: corlinoside, isocorlinoside, isoshaftoside, neocorlinoside, neoshaftoside, shaftoside.

Linard *et al., Phytochemistry,* **21,** 797 (1982)

L. hedysaroides.

FLAVONOIDS: bioquercetin, homoorientin, kaempferol 3-O-robinoside, orientin, saponaretin, vitexin.

Lachman *et al., Khim. Prir. Soedin.,* 136 (1978)

L. sericea.

TRITERPENOID: 3β,22β,24-trihydroxyolean-12-ene.

Steffens *et al., J. Am. Chem. Soc.,* **105,** 1669 (1983)

LESQUERELLA (Cruciferae)

L. densipila.

ACIDS: densipolic acid, ricinoleic acid.

Smith *et al., J. Org. Chem.,* **27,** 3112 (1962)

L. lasiocarpa.

ACID: lesquerolic acid.

Smith *et al., J. Org. Chem.,* **27,** 3112 (1962)

LETHARIA (Lichenes)

L. vulpina (L.) Hue.

ACID: usnic acid.

Dean *et al., J. Chem. Soc.,* 1250 (1953)

LEUCAENA (Leguminosae)

L. glauca (L.) Benth.

PYRIDONE ALKALOID: leucaenine.

FLAVONOIDS: kaempferol 3-arabinoside, kaempferol xyloside, quercetin 3-arabinoside, quercetin galactoside, quercetin rhamnoside.

Mascre, *C. R. Acad. Sci.,* **204,** 890 (1937)

Morita *et al., Chem. Abstr.,* 88 148947b

LEUCANTHEMELLA

L. serotina.

SESQUITERPENOID: 8-hydroxybalchanin.

Holub *et al., Collect. Czech. Chem. Commun.,* **47,** 2927 (1982)

LEUCANTHEMOPSIS

L. pallida subsp. flaveola.

PHENOLICS: 4-hydroxy-5-propionyl-1,3-di-O-methylpyrogallol, 4-hydroxy-4-propionyl-2-O-isopentenylpyrogallol, 5-(1'-isovalerianoyloxy)-ethyl-13-di-O-methyl-2-O-isopentenylpyrogallol.

Bellido *et al., Phytochemistry,* **24,** 2758 (1985)

L. pulverulenta.

SESQUITERPENOID: hispanolide.

de Pascual Teresa *et al., Phytochemistry,* **22,** 1985 (1983)

LEUCANTHEMUM

L. vulgare.

FLAVONOID: apigenin 7-O-β-D-glucuronide.

Sagareishvili *et al., Khim. Prir. Soedin.,* 647 (1983)

LEUCOCARPUS

L. perfoliatus.

IRIDOID: boschnaloside.

Ozaki *et al., Helv. Chim. Acta,* **62,** 1708 (1979)

LEUCOCENDRON

L. adscendens.

SESQUITERPENOID: leucodrin.

Rapson *et al., J. Chem. Soc.,* 282 (1938)

L. concinnum.

SESQUITERPENOID: leucodrin.

Rapson, *J. Chem. Soc.,* 1085 (1939)

L. stokei.

SESQUITERPENOID: leucodrin.

Rapson, *J. Chem. Soc.,* 1085 (1939)

LEUCOJUM (Amaryllidaceae)

L. aestivum L.

DIPHENYL ALKALOIDS: galanthamine, leucotamine, O-methylleucotamine.

ERYTHRINA ALKALOIDS: isotazettine, tazettine.

ISOQUINOLINE ALKALOID: lycorine.

UNCHARACTERIZED ALKALOID: aestivine.

Kobayashi *et al., Heterocycles,* **19,** 1219 (1982)

L. vernum L.

DIPHENYL ALKALOID: galanthamine.

INDOLE ALKALOID: narcipoetine.

Boit, Stender, *Ber.,* **87,** 681 (1954)

Proskurnina, Yakovleva, *J. Gen. Chem., USSR,* **25,** 1035 (1955)

LEUCOPAXILLUS

L. tricolor.
PIGMENTS: asperflavin, endocrocin, phlegmacin.
Besl *et al.*, *Chem. Abstr.*, 88 117773n

LEUCOSCEPTRUM

L. japonicum (Miq.) Kitamura et Murata.
GLYCOSIDES: acteoside, leucosceptoside A, leucoseptoside B, martynoside.
Miyase *et al.*, *Chem. Pharm. Bull.*, **30**, 2732 (1982)

LEUCOSPERMUM (Proteaceae)

L. conocarpodendron.
SESQUITERPENOIDS: conocarpic acid, conocarpin.
Kruger, Perold, *J. Chem. Soc., C*, 2127 (1970)

L. reflexum Buek. ex Meisner.
SESQUITERPENOIDS: conocarpic acid, conocarpin, reflexin.
Perold *et al.*, *J. Chem. Soc., Perkin 1*, 2450, 2457 (1972)

LEUCOTHOE (Ericaceae)

L. grayana Max.
STEROID: β-sitosterol.
DITERPENOIDS: leucothols A, B, C and D,
3S,6R,14,16R-tetrahydroxy-5-oxo-5,10-seco-ent-kaur-10-ene.
TRITERPENOIDS: α-amyrin, betulinic acid, lupeol, oleanolic acid, taraxerol, ursolic acid.
TOXINS: grayanotoxins I, II, III, IV, V, VI, VII, VIII, IX, X, XI, XII, XIII, XIV, XV, XVI, XVII, XVIII, XIX.
(S) Kawai *et al.*, *Chem. Abstr.*, 88 34499b
(T) Furusaki *et al.*, *Bull. Chem. Soc., Japan*, **54**, 657 (1981)
(T) Katai *et al.*, *Chem. Lett.*, 443 (1985)

L. keiskei.
FLAVONE: poriolide.
Ogiso *et al.*, *Tetrahedron Lett.*, 3071 (1972)

LEVISTICUM (Umbelliferae)

L. officinale.
COUMARINS: bergapten, coumarin, umbelliferone.
Albulescu *et al.*, *Farmacia*, **23**, 159 (1975); *Chem. Abstr.*, 84 147632a

LIABUM

L. eggersii.
SESQUITERPENOIDS: 19-acetoxymodheptene, liabinolide.
Bohlmann *et al.*, *Phytochemistry*, **23**, 1800 (1984)

L. floribundum.
SESQUITERPENOIDS: 5,6:7,8-bisepoxyeudesmene, liabinolide.
Bohlmann *et al.*, *Phytochemistry*, **23**, 1800 (1984)

LIATRIS (Compositae)

L. chapmanii.
DITERPENOIDS: chapliatrin, isochapliatrin, liatrin.
Herz *et al.*, *Phytochemistry*, **14**, 1803 (1975)

L. cylindracea.
SESQUITERPENOIDS: hydroxyliacylindrolide, liacylindrolide.
Bohlmann, Dutta, *Phytochemistry*, **18**, 847 (1979)

L. elegans.
DITERPENOIDS: ligantrol, ligantrol acetate.
Smirnova *et al.*, *Khim. Prir. Soedin.*, 63 (1969)

L. gracilis.
DITERPENOID: acetylchapliatrin.
Herz *et al.*, *Phytochemistry*, **14**, 1803 (1975)

L. graminifolia.
DITERPENOID: deoxygraminiliatrin.
Herz *et al.*, *J. Org. Chem.*, **40**, 199 (1975)

L. laevigata.
DITERPENOIDS: ent-8,15R-epoxy-3-oxopimara-12α,16-diol, ent-8,15R-epoxy-3β,12,16-pimaranetriol, ent-8,15S-epoxy-3α,12,16-pimaranetriol.
TRITERPENOIDS: lupenyl acetate, lupeol.
Herz *et al.*, *Phytochemistry*, **22**, 715 (1983)

L. microcephala.
NORTRITERPENOID: 30-nortaraxasterene-3β-acetate.
Herz *et al.*, *Phytochemistry*, **22**, 1457 (1983)

L. platylepis.
SESQUITERPENOID: 2,3-seco-1,3,11(13)-eudesmatrien-12,8-olide.
Bohlmann *et al.*, *Phytochemistry*, **18**, 1228 (1979)

L. provincialis.
SESQUITERPENOID: provincialin.
Herz, Wahlberg, *J. Org. Chem.*, **38**, 2485 (1973)

L. punctata.
SESQUITERPENOIDS: liatripunctin, punctatin.
Herz, Wahlberg, *Phytochemistry*, **12**, 1421 (1973)

L. pycnostachya.
SESQUITERPENOIDS: epoxyspicatin, pycnolide, spicatin.
Herz *et al.*, *J. Org. Chem.*, **40**, 199 (1975)

L. secunda.
SESQUITERPENOIDS: liatripunctin, liscundin, liscunditrin.
Herz, Sharma, *Phytochemistry*, **14**, 1561 (1975)

L. spicata (L.) Kuntze.
SESQUITERPENOID: spicatin.
Herz *et al.*, *J. Org. Chem.*, **40**, 199 (1975)

LIBANOTIS

L. buchtorimensis (Fisch.) DC.
SESQUITERPENOID: selinene.
Wang *et al.*, *Zhiwu Xuebao*, **28**, 192 (1986); *Chem. Abstr.*, 105 75889s

L. intermedia.
STEROIDS: campesterol, β-sitosterol, stigmasterol.
MONOTERPENOIDS: β-elemene, terpinen-4-ol.
Motl *et al.*, *Collect. Czech. Chem. Commun.*, **42**, 2815 (1977)

L. transcaucasica.
SESQUITERPENOID: (-)-selinene.
Pigulevskii, Borovkov, *J. Gen. Chem., USSR*, **32**, 3106 (1962)

LIBOCEDRUS (Cupressaceae)

L. bidwillii.
PHENOLIC: β-peltatin A methyl ether.
DITERPENOID: sugiol.
TOXIN: deoxypodophyllotoxin.
Russell, Phytochemistry, 14, *2708 (197)* (197)

L. formosana.
MONOTERPENOID: shonanic acid.
NORDITERPENOID: shonanol.
Lin, Lin, *J. Chin. Chem. Soc.*, **12**, 51 (1968)

LIBOTHAMNUS

L. granatesianum.
SESQUITERPENOID: 4,11-selinadiene-12,15-dial.
Bohlmann *et al.*, *Phytochemistry*, **19**, 1145 (1980)

LICARIA

L. aritu.
DITERPENOID: licarin A.
> Aiba *et al.*, *Phytochemistry*, **12**, 1163 (1973)

L. canella.
DITERPENOIDS: canellins A, B and C.
> Giesbrecht *et al.*, *Phytochemistry*, **13**, 2285 (1974)

L. macrophylla.
DITERPENOID: macrophyllin.
> Franca *et al.*, *Phytochemistry*, **13**, 2839 (1974)

LIDBECKIA (Compositae/Anthemideae)

L. pectinata.
SESQUITERPENOIDS: dehydroleucodin, epoxyisoestafiatin.
> Bohlmann, Zdero, *Tetrahedron*, **28**, 621 (1972)

LIGULARIA (Compositae)

L. brachyphylla Hand-Mazz.
PYRROLIZIDINE ALKALOIDS: angeloyloxyligularinine, ligularine.
SESQUITERPENOIDS: liguhodgsonal, ligularinone A.
> (A) Klasek *et al.*, *Collect. Czech. Chem. Commun.*, **36**, 2205 (1971)
> (T) Bohlmann *et al.*, *Chem. Ber.*, **110**, 2640 (1977)

L. calthaefolia.
SESQUITERPENOID: ligucalthaefolin.
> Bohlmann *et al.*, *Chem. Ber.*, **110**, 2640 (1977)

L. clivorum Maxim.
PYRROLIZIDINE ALKALOID: clivorine.
SESQUITERPENOIDS: 8-angeloyloxyligularinone, liguhodgsonal.
> (A) Klasek *et al.*, *Collect. Czech. Chem. Commun.*, **32**, 2512 (1967)
> (T) Bohlmann *et al.*, *Chem. Ber.*, **110**, 2640 (1977)

L. dentata (A.Gray) Hara.
PYRROLIZIDINE ALKALOIDS: ligudentine, ligularine, ligularinine, ligularizine.
SESQUITERPENOID: liguhodgsonal.
> (A) Asada, Furuya, *Chem. Pharm. Bull.*, **32**, 475 (1984)
> (T) Bohlmann *et al.*, *Chem. Ber.*, **110**, 2640 (1977)

L. dictyoneura.
BENZDIPYRONE: yentianmasu.
> Zhao *et al.*, *Zhongcaoyao*, **13**, 10 (1982); *Chem. Abstr.*, 97 107023q

L. elegans Cass.
PYROLLIZIDINE ALKALOID: ligularine.
> Klasek *et al.*, *Collect. Czech. Chem. Commun.*, **36**, 2205 (1971)

L. fischeri.
SESQUITERPENOIDS: 1,10-epoxyfuranoeremophilan-9-one, ligularone, liguloxide, liguloxidol, liguloxidol acetate, 6-oxofuranoeremophilane.
> Morigama *et al.*, *Chem. Lett.*, 637 (1972)
> Yamakawa *et al.*, *Chem. Pharm. Bull.*, **25**, 2535 (1977)

L. hodgsonii Hook f.
SESQUITERPENOIDS: 8α-angeloyloxyligularinone, 8β-angeloyloxyligularinone, liguhodgsonal.
> Bohlmann *et al.*, *Chem. Ber.*, **110**, 2640 (1977)

L. japonica.
SESQUITERPENOIDS: liguhodgsonal, ligujapone.
> Bohlmann *et al.*, *Chem. Ber.*, **110**, 2640 (1977)

L. macrophylla.
SESQUITERPENOID: ligolide.
> Nikonova, Nikonov, *Khim. Prir. Soedin.*, 742 (1976)

L. sibirica.
SESQUITERPENOIDS: ligularenolide, ligularone, 6-oxofuranoeremophilane, petasalbine, petasalbine methyl ether.
> Takahashi *et al.*, *Tetrahedron Lett.*, 3739 (1968)
> Yamakawa *et al.*, *Chem. Pharm. Bull.*, **25**, 2535 (1977)

L. speciosa.
SESQUITERPENOID: isofukinone.
> Bohlmann, Fritz, *Phytochemistry*, **19**, 2471 (1980)

L. tussilaginea Makino.
SESQUITERPENOIDS: 3β-angeloyloxyeremophilenolide, 3β-angeloyloxy-6β-hydroxyeremophilenolide, 3β-angeloyloxy-8β-hydroxyeremophilenolide, tsuwaburinonol.
> Kurihawa, Suzuki, *J. Pharm. Soc., Japan*, **98**, 1441 (1978)

L. vorobierii.
SESQUITERPENOID: 6-acetoxy-3-angeloyloxy-1,10-epoxyfuranoeremophilane.
> Bohlmann, Suwita, *Chem. Ber.*, **110**, 1759 (1977)

LIGUSTICUM (Umbelliferae)

L. acutilobum.
TERPENOIDS: ligustilide, riligustilide.
> Meng *et al.*, *Lanzhou Daoxue Xuebao, Ziran Kexueban*, 76 (1983)

L. chuanxing.
PHENOLICS: 8-oxabicyclo-[3,2,1]-oct-3-en-2-one-5-hydroxymethyl-6-endo-3'-methoxy-4'-hydroxyphenyl, 3-n-butyl-3-hydroxy-4,5,6,7-tetrahydro-6,7-dihydroxyphthalide.
> Wen *et al.*, *Zhongcaoyao*, **17**, 109 (1986); *Chem. Abstr.*, 105 175884m

L. elatum.
TERPENOID: (-)-anomalin.
> Kapoor *et al.*, *Phytochemistry*, **11**, 477 (1972)

L. lucidum.
DITERPENOID GLUCOSIDE: oleuropein.
> Panizzi *et al.*, *Gazz. Chim. Ital.*, **90**, 1449 (1960)

L. mucronatum.
MONOTERPENOID: O-angeloyloxyepoxyisoeugenol.
> Bohlmann, Zdero, *Chem. Ber.*, **104**, 2033 (1971)

L. pyrenaicum.
BENZOPYRAN: 9-acetoxy-O-isovaleryldihydrooreselol.
> Bohlmann *et al.*, *Chem. Ber.*, **102**, 1963 (1969)

L. wallichii.
BENZOFURAN: diligustilide.
PHTHALIDE: (Z)-ligustidiol.
> (B) Kaouadji *et al.*, *Tetrahedron Lett.*, 4677 (1983)
> (P) Kaouadji *et al.*, *ibid.*, 4675 (1983)

LIGUSTRUM (Oleaceae)

L. japonicum L.
SECOIRIDOIDS: ligustaloside A, ligustaloside B, oleuropein.
> Inouye *et al.*, *Phytochemistry*, **21**, 2305 (1982)

L. lucidum.
DITERPENOIDS: ligustrosidic acid.
IRIDOIDS: ligustaloside A, ligustaloside B.
SECOIRIDOID: oleuropein.
> Inouye *et al.*, *Phytochemistry*, **21**, 2305 (1982)
> Kikuchi, Yamauchi, *J. Pharm. Soc., Japan*, **1055**, 142 (1985)

L. nilghirense var. obovata C.B.Cl.
FLAVONOID: kaempferitrin.
> Joshi *et al.*, *Proc. Indian Acad. Sci.*, **86A**, 41 (1977)

LILIUM (Liliaceae)

L. humboltii.
GENIN: lilagenin.
> Knight *et al.*, *J. Chromatogr.*, **133**, 222 (1977)

L. rubrum magnificum.
GENIN: lilagenin.
> Knight *et al.*, *J. Chromatogr.*, **133**, 222 (1977)

L. tigrinum Ker.

CAROTENOIDS: cis-antheraxanthin, trans-antheraxanthin.

Karrer, Oswald, *Helv. Chim. Acta,* **18,** 1303 (1935)

L. tigrinum Ker. cv. ('Red Night').

CAROTENOIDS: 6-epi-kerpozanitin, lilizanthin.

Fischer, Eugster, *Helv. Chim. Acta,* **68,** 1708 (1985)

LIMACIA

L. cuspidata (Miers.) Hook f. Thom.

BISBENZYLISOQUINOLINE ALKALOIDS: cuspidapine, limacine, limacusine.

Tomita, Furukawa, *Tetrahedron Lett.,* 4293 (1966)

LINACIOPSIS

L. loangensis.

BISBENZYLISOQUINOLINE ALKALOID: 2'-norisotetrandrine.

Cava *et al., Planta Med.,* **35,** 31 (1979)

LINANTHEMUM

L. humboldtianum.

UNCHARACTERIZED ALKALOID: linantenine.

Ribieri, Machado, *Engenharia et Quimica,* **3,** 152 (1951)

LINARIA (Schophulariaceae)

L. cymbalarica (L.) Mill.

MONOTERPENOID: 7-hydroxy-8-epi-iridodial.

Bianco *et al., Planta Med.,* **46,** 38 (1982)

L. japonica.

DITERPENOIDS: linaridial, linarienone.

IRIDOID: linarioside.

Kitagawa *et al., Tetrahedron Lett.,* 23 (1975)

L. muratus.

IRIDOID: linaride.

Bianco *et al., Gazz. Chim. Ital.,* **107,** 83 (1977)

L. popovii Kupr.

QUINAZOLINE ALKALOID: peganine.

Yunusov *et al., Dokl. Akad. Nauk SSSR,* **11,** 25 (1956)

L. saxatilis.

DITERPENOIDS:
15-acetoxy-11,15-epoxy-ent-cleroda-4(18),12-dien-16-ol, isolinaridial, isolinaridiol, isolinaridiol diacetate, isolinaritriol, isolinaritriol 12,15-diacetate, isolinaritriol 12,16-diacetate, isolinaritriol 15,16-diacetate, isolinaritriol triacetate.

San Feliciano *et al., Tetrahedron,* **41,** 671 (1985)

L. transiliensis Kupr.

QUINAZOLINE ALKALOIDS: deoxyvasicine, peganine, vasicinone.

Plechanova *et al., Frunze Izdaniya-Vo Ilim* 54 (1965)

L. vulgariformis E.Nik.

QUINAZOLINE ALKALOID: peganine.

Plechanova *et al., Frunze Izdaniya-Vo Ilim* 54 (1965)

L. vulgaris Mill.

QUINAZOLINE ALKALOID: peganine.

Men'shikov *et al., J. Gen. Chem., USSR,* **29,** 3846 (1959)

LINDBERGIA

L. indica.

SAPONINS:
α-L-rhamnopyranosyl-α-L-arabinopyranosyl-sitosterol,
α-L-rhamnopyranosyl-α-L-rhamnopyranosylsitosterol,
α-L-rhamnopyranosyl-β-glucopyranosyl-sitosterol.

Tiwari, Chaudhury, *Phytochemistry,* **18,** 2044 (1979)

LINDELOFIA (Boraginaceae)

L. anchusoides Bunge. (L. macrostyla Bge. M.Pop.).

PYRROLIZIDINE ALKALOIDS: lindelofamine, lindelofine, lindelofine N-oxide.

Shirul'nikova *et al., J. Gen. Chem., USSR,* **32,** 2705 (1962)

L. angustifolia.

PYRROLIZIDINE ALKALOIDS: amabiline, echinatine.

Suri *et al., Indian J. Pharm.,* **37,** 69 (1975)

L. olgae (Rgl. et Schmalh.) Brand.

PYRROLIZIDINE ALKALOIDS: viridiflorine, viridiflorine N-oxide.

Akramov *et al., Dokl. Akad. Nauk SSSR,* **10,** 29 (1962)

L. pterocarpa M.Pop.

PYRROLIZIDINE ALKALOIDS: viridiflorine, viridiflorine N-oxide.

Akramov *et al., Dokl. Akad. Nauk SSSR,* **6,** 28 (1964)

L. spectabilis.

PYRROLIZIDINE ALKALOID: 7-acetylechinatine.

Suri *et al., Indian J. Chem.,* **13,** 503 (1975)

L. stylosaa (Kar. et Kir.) Brand.

PYRROLIZIDINE ALKALOIDS: echinatine, echinatine N-oxide, lindelofine, lindelofine N-oxide, viridiflorine, viridiflorine N-oxide.

Akramov *et al., Dokl. Akad. Nauk SSSR,* **6,** 35 (1961)
Kirmitdinova *et al., Khim. Prir. Soedin.,* 411 (1967)

L. tschimganica (Lipsky) M.Pop.

PYRROLIZIDINE ALKALOIDS: echinatine, viridiflorine.

Akramov *et al., Dokl. Akad. Nauk SSSR,* **4,** 35 (1965)

LINDERA (Lauraceae)

L. citriodora (Sieb. et Zucc.) Hemsl.

ACIDS: decanoic acid, cis-decenoic acid, cis-4-decenoic acid, dodecanoic acid, eicosenoic acid, linoleic acid, octadecanoic acid, octadecenoic acid, octanoic acid, oleic acid, undecanoic acid.

Furukawa *et al., Yakagaku Zasshi,* **25,** 296 (1976); *Chem. Abstr.,* **85** 59590a

L. erythrocarpa.

FLAVONOIDS: kanakugin, kanakugiol, linderone, lucidone, methyllinderone, methyllucidone.

Liu *et al., J. Pharm. Soc., Japan,* **95,** 1114 (1975)

L. glauca.

SESQUITERPENOIDS: aciphylline, aciphyllol, glaucic acid, glaucic acid methyl ester, glaucyl alcohol.

Nii *et al., Nippon Nogei Kagaku Kaishi,* **57,** 725 (1983)
Nii *et al., ibid.,* **57,** 733 (1983)

L. lucida.

FLAVONES: lucidin, lucidin dimethyl ether.

SESQUITERPENOID: lucidone.

(F) Lee *et al., J. Chem. Soc.,* 2743 (1965)
(T) Lee, *Tetrahedron Lett.,* 4243 (1968)

L. obtusiloba.

PYRONE: C-19-obtusilactone dimer.

Niwa *et al., Chem. Lett.,* 581 (1977)

L. oldhami.

BISBENZYLISOQUINOLINE ALKALOID: lindoldhamine.

Lu, Chen, *Heterocycles,* **4,** 1073 (1976)

L. strichynifolia.

SESQUITERPENOIDS: isogermafurene, isogermafurenolide, isolinderalactone, isolinderoxide, lindenene, lindenenol, lindenenone, linderalactone, linderane, linderene, linderene acetate, linderoxide, lindestrene, lindestrenolide, neosericenyl acetate.

Takeda *et al., J. Chem. Soc., C,* 2786 (1969)
Wada *et al., ibid.,* 1070 (1971)

L. triloba.

SESQUITERPENOID: 1-acetoxydelobanone.

Takeda *et al.*, *Tetrahedron*, **27**, 6049 (1971)

LINDHEIMERA

L. texana.

TRITERPENOIDS:
(16S,23,24)-cycloartan-3-one-16,23,24,25-tetraol,
(3S,16S,23R)-16,23-epoxycycloart-24-en-3-ol,
(3S,16S,23S)-16,23-epoxycycloart-24-en-3-ol,
(16S,23R)-epoxycycloart-24-en-3-one,
(16S,23S)-epoxycycloart-24-en-3-one,
(16S,23R)-16,23-epoxy-23,25-epidioxycycloartan-3-one,
(16S)-23,24,25,26,27-pentanorcycloartan-3-one-16,22-olide.

Herz *et al.*, *Phytochemistry*, **24**, 2645 (1985)

LINEARIS

L. vulgariformis.

QUINAZOLINE ALKALOID: peganine.

Plekhanova *et al.*, *Frunze Izdaniya-Vo Ilim*, 54 (1965)

L. vulgaris.

QUINAZOLINE ALKALOID: peganine.

Men'shikov *et al.*, *J. Gen. Chem., USSR*, **29**, 3846 (1969)

LIPARIS (Orchidaceae)

L. auriculata Blume.

UNCHARACTERIZED ALKALOID: auriculatine.

Nishikawa, Hirata, *Tetrahedron Lett.*, 6289 (1968)

L. bicallosa Schltr.

PYRROLIZIDINE ALKALOID: malaxine.

Leander, Luning, *Tetrahedron Lett.*, 3477 (1967)

L. hachrijoensis.

PYRROLIZIDINE ALKALOID: malaxine.

Leander, Luning, *Tetrahedron Lett.*, 3477 (1967)

L. keitaoensis Hay.

PYRROLIZIDINE ALKALOIDS: keitaoine, keitine.

Lindstrom, Luning, *Acta Chem. Scand.*, **26**, 2963 (1972)

L. kumokiri F.Maekawa.

GLUCOALKALOID: kuramerine.
PYRROLIZIDINE ALKALOID: kumokirine.

Nishikawa *et al.*, *Tetrahedron Lett.*, 2597 (1967)

L. kurameri French et Sav.

GLUCOALKALOID: kuramerine.

Nishihawa *et al.*, *Tetrahedron Lett.*, 2597 (1967)

L. nervosa.

PYRROLIZIDINE ALKALOID: nervosine.

Nishikawa, Hirata, *Tetrahedron Lett.*, 2591 (1967)

LIPPIA (Verbenaceae)

L. citriodora.

SESQUITERPENOID: caryophyllane-2,6β-oxide.

Kaiser *et al.*, *Helv. Chim. Acta*, **59**, 1803 (1976)

L. dulcis.

SESQUITERPENOID: hernandulin.

Compadre *et al.*, *Science*, 227(4685), 417 (1984)

L. integrifolia.

SESQUITERPENOIDS: africanone, 1,5-bicyclohummulenedione.

Catalan *et al.*, *Phytochemistry*, **22**, 1507 (1983)
Martayet *et al.*, *ibid.*, **23**, 688 (1984)

L. nodiflora.

FLAVONES: nepetin, nodifloretin.

Brieskorn *et al.*, *Tetrahedron Lett.*, 3447 (1968)

L. rehmannii.

TRITERPENOID: rehmannic acid.

Barton, de Mayo, *J. Chem. Soc.*, **887**, 900 (1954)

LIQUIDAMBER (Hamamelidaceae)

L. formosana.

ELLAGITANNIN: liquidambin.
MONOTERPENOID: (±)-terpineol.
TRITERPENOIDS: forucosolic acid, liquidambonic acid, liquidambrovonic acid.

(E) Okuda *et al.*, *Phytochemistry*, **26**, 2053 (1987)
(T) Phom Truong Thi Tho, *Tap. San. Hoa-oc.*, **133**, 34 (1975)
(T) Yankov *et al.*, *Dokl. Bolg. Akad. Nauk*, **33**, 75 (1980)

L. orientalis Mill.

TRITERPENOID: 3-epi-oleanic acid.

Huneck, *Tetrahedron*, **19**, 479 (1967)

LIRIODENDRON (Magnoliaceae)

L. tulipfera L.

APORPHINE ALKALOIDS: (+)-caaverine, lirinidine, lirinine, lirinine N-oxide, O-methyllirine.
LIGNANS: (+)-medioresinol, (+)-pinoresinol, (+)-synringaresinol.
SESQUITERPENOIDS: lipiferolide, liriodenolide, peroxyferolide, epi-tulipdienolide, tulipinolide, epi-tulipinolide, epi-tulipinolide diepoxide.
NORAPORPHINE ALKALOID: liriodenine.

(A) Zirev *et al.*, *Khim. Prir. Soedin.*, **67**, 505 (1973)
(A) Doskotch *et al.*, *J. Org. Chem.*, **45**, 1441 (1980)
(L) Fujimoto *et al.*, *Chem. Abstr.*, 87 197244h
(T) Doskotch *et al.*, *Phytochemistry*, **14**, 769 (1975)

LITCHIA

L. chinensis.

AMINOACIDS: alanine, asparagine, threonine, tyrosine, valine.

Ali *et al.*, *J. Indian Chem. Soc.*, **64**, 230 (1987)

LITHOCARPUS (Fagaceae)

L. attenuata.

TRITERPENOIDS: 3-methoxyoleana-2,12-dien-1-one, olean-12-ene-1,3-dione.

Hui, Li, *Phytochemistry*, **14**, 785 (1975)

L. correa.

TRITERPENOIDS: 29-acetoxytaraxer-14-en-3-one, 14-hydroxytaraxer-3-one, 29-hydroxytaraxer-14-en-3-one, lup-20(29)-ene-3β,27-diol, myricadiol, taraxerane-3β,13α-diol, taraxer-14-ene-3β,29-diacetate, taraxer-14-ene-3β,29-diol.

Hui *et al.*, *J. Chem. Soc., Perkin I*, 23 (1976)

L. irwinii.

TRITERPENOID: lithocarpic lactone.

Hui *et al.*, *J. Chem. Soc., Perkin I*, 617 (1975)

L. lithioides.

TRITERPENOID: lithocarpic lactone.

Hui *et al.*, *J. Chem. Soc., Perkin I*, 617 (1975)

L. litseifolius.

CHALCONES: phlorizin, trilobatin.

Rui-Lin *et al.*, *Agric. Biol. Chem.*, **46**, 1933 (1982)

L. polystachya.

TRITERPENOIDS: lithocarpdiol, lithocarpolone, 24-methylenecycloartan-3β,21-diol.

Arthur, Ko, *Phytochemistry*, **13**, 2551 (1974)

LITHOSPERMUM (Boraginaceae)

L. erythrorhizon Sieb. et Zucc.
PIGMENTS: acetylshikonin, deoxyshikonin,
β-hydroxyisovalerylshikonin, β-hydroxyshikonin,
isobutyrylshikonin, isovalerylshikonin, lithospermidin A,
lithospermidin B, (α-methylbutyryl)-shikonin, shikizarin,
(R)-shikonin.
Krivoshchekova *et al.*, *Khim. Prir. Soedin.*, 726 (1976)
Hisamichi *et al.*, *Chem. Abstr.*, 97 178739p

L. euchromum.
PIGMENT: (R)-shikonin.
Raudnitz *et al.*, *Justus Liebig's Ann. Chem.*, **68**, 1479 (1935)

L. officinale L.
PIGMENTS: acetylshikonin, deoxyshikonin, β-hydroxyshikonin,
isobutyrylshikonin, isovalerylshikonin, shikonin.
Krivoshchekova *et al.*, *Khim. Prir. Soedin.*, 726 (1976)

L. purpureo-caeluleum.
CYANOGLYCOSIDE: lithospermoside.
Sosa *et al.*, *Phytochemistry*, **16**, 707 (1977)

LITSEA (Lauraceae)

L. chrysocoma.
APORPHINE ALKALOID: laurotetanine.
Greshoff, *Ber.*, **23**, 3537 (1890)

L. citrata.
APORPHINE ALKALOIDS: laurotetanine, N-methyllaurotetanine.
Barger *et al.*, *Ber.*, **66**, 450 (1933)

L. cubeba (Lour.) Pers.
APORPHINE ALKALOID: laurotetanine.
MONOTERPENOIDS: camphene, citronellal, citronellol, linalool,
α-pinene, β-terpineol, δ-terpineolneol.
(A) Barger *et al.*, *Ber.*, **66**, 450 (1933)
(T) Jian *et al.*, *Chem. Abstr.*, 106 15731v

L. dealbata.
TRITERPENOID: taraxerol.
Barton *et al.*, *J. Chem. Soc.*, 2131 (1955)

L. japonica.
APORPHINE ALKALOID: laurolitsine.
Nakasato, Nomura, *Chem. Pharm. Bull.*, **7**, 780 (1959)

L. laeta.
APORPHINE ALKALOIDS: laetanine, laetine.
Rastogi, Borthakur, *Phytochemistry*, **19**, 998 (1980)

L. nitida.
APORPHINE ALKALOID: litsedine.
Gopinath *et al.*, *Indian J. Chem.*, **13**, 197 (1975)

L. salicifolia.
APORPHINE ALKALOID: laetine.
Rastogi, Bothakur, *Phytochemistry*, **19**, 998 (1980)

L. sebifera.
APORPHINE ALKALOID: litseferine, sebiferine.
Sivakumaran, Gopinath, *Indian J. Chem.*, **14B**, 151 (1976)

L. tomentosa.
TRITERPENOID: litsomentol.
Govindachari, *J. Chem. Soc., Chem. Commun.*, 665 (1971)

LITTORIA

L. modesta.
COLCHICINE ALKALOIDS: O-acetylcolchiceine,
N-acetyldemecolcine, 2-acetyl-2-demethylcolchicine,
3-acetyl-3-demethylcolchicine, N-acetylisodemecolcine.
Potesilova *et al.*, *Collect. Czech. Chem. Commun.*, **32**, 141 (1967)

LOBARIA (Stictaceae)

L. amplissima (Scop.) Forss.
FURAN: xantholamine.
Solberg, *Lichenologist*, **4**, 271 (1970); *Chem. Abstr.*, 75 126657m

L. isidiosa var. subisidiosa.
SESTERTERPENOID: retigeranic acid.
Rao *et al.*, *Curr. Sci.*, **34**, 9 (1965)

L. kazawoensis.
TRITERPENOID: retigeric acid B.
Takahashi *et al.*, *Phytochemistry*, **11**, 2039 (1972)

L. linita (Ach.) Rabenh.
ORSELLINATES: methylevernate, methylgyrophorate,
methyllecanorate, methylorsellinate, tenuiorin.
Maass, *Bryologist*, **78**, 178 (1975)

L. pulmonaria (L.) Hoffm. 16 fatty acids.
PHENOLIC: tenuiorin.
STEROIDS: ergosterol, ergosterol acetate, fecosterol, fecosterol
acetate.
(Ac, S) Catalano *et al.*, *Phytochemistry*, **15**, 221 (1976)
(P) Asahina *et al.*, *Ber.*, **66**, 1910 (1933)

L. retigera.
SESTERTERPENOID: retigeranic acid.
TRITERPENOIDS: retigeradiol, retigeric acid, retigeric acid A.
Takahashi *et al.*, *Phytochemistry*, **11**, 2039 (1972)

L. sachalinensis.
TRITERPENOIDS: retigeric acid A, retigeric acid B.
Takahashi *et al.*, *Phytochemistry*, **11**, 2039 (1972)

L. subretigera. (L. pseudopulmonaria).
TRITERPENOIDS: retigeradiol, retigeranic acid A, retigeranic
acid B.
SESTERTERPENOID: retigeranic acid.
Takahashi *et al.*, *Phytochemistry*, **11**, 2039 (1972)

LOBELIA (Lobeliaceae/Campanulaceae)

L. cardinalis L.
ISOQUINOLINE ALKALOID: lobinaline.
Manske, *Can. J. Res.*, **16B**, 445 (1938)

L. inflata L.
LOBELINE ALKALOIDS: isolobinine, (±)-lelobanidine,
(+)-lelobanidine, (-)-lelobanidine, (-)-lelobanidine II,
lobelanidine, (±)-lobeline, (-)-lobeline, lobinanidine,
lobinine, norlobelanidine, norlobelanine.
Thoma, *Justus Liebigs Ann. Chem.*, **540**, 99 (1939)
Wieland *et al.*, *ibid.*, **540**, 103 (1939)

L. portoricensis.
One lobeline alkaloid.
Melendez *et al.*, *J. Pharm. Sci.*, **56**, 1677 (1967)

L. pyramidalis.
One lobeline alkaloid.
Jain, Nigam, Madhye Brahati, *Part II*, **9A**, 29 (1960)

L. salicifolia.
UNCHARACTERIZED ALKALOID: salicilobine.
Steinegger, Ochsner, *Pharm. Acta Helv.*, **31**, 89 (1956)

L. sessilifolia Lamb.
LOBELINE ALKALOIDS: lobelanidine, lobelanine, (-)-lobeline.
Wieland, Dragendorff, *Justus Liebigs Ann. Chem.*, **473**, 53 (1929)

L. syphilitica L.
ISOQUINOLINE ALKALOIDS: syphilobines A, B and F.
LOBELINE ALKALOIDS: (-)-dehydro-trans-8,10-diethyllobelidiol,
(-)-cis-8,10-diethylnorlobelionol, 8,10-diethylnorlobelidone,
cis-8,10-diethylnorlobelionol, lophilacrine.
UNCHARACTERIZED ALKALOID: lophiline.
Tschesche *et al.*, *Chem. Ber.*, **74**, 3327 (1961)

LOBELIA (Lobeliaceae/Campanulaceae)

L. urens.

LOBELINE ALKALOIDS: lobelanidine, (-)-lobeline.
Steinegger, Grutter, *Pharm. Acta Helv.,* **46,** 71 (1973)

LOBESIA

L. botrana.

TERPENOID: 7(E0,9(Z)-dodecadienyl acetate.
Roelofs *et al., Mitt. Schweiz. Entomol. Ges.,* **46,** 71 (1973)

LOBIVIA (Cactaceae)

L. backenbergii.

AMINE ALKALOIDS: hordenine, N-methyltyramine, tyramine.
Follas *et al., Phytochemistry,* **16,** 1459 (1977)

L. binghamiana.

AMINE ALKALOIDS: hordenine, N-methyltyramine, tyramine.
Follas *et al., Phytochemistry,* **16,** 1459 (1977)

L. pentlandii.

AMINE ALKALOIDS: hordenine, N-methyltyramine, tyramine.
Follas *et al., Phytochemistry,* **16,** 1459 (1977)

LOCHNERA (Apocynaceae)

L. lancea Boj. (ex A.DC.) K.Schum.

UNCHARACTERIZED ALKALOID: lanceine.
Janot *et al., Ann. Pharm. Fr.,* **15,** 474 (1957)

L. rosea (L.) Reichb. f. (Catharanthus roseus).

CARBOLINE ALKALOID: lochnerine.
Janot, LeMen, *C. R. Acad. Sci.,* **243,** 1789 (1956)

LOLIUM (Gramineae)

L. cuneatum Nevski.

PYRROLIZIDINE ALKALOIDS: lolidine, loline, lolinine, lolinidine, norloline.
Yunusov, Akramov, *J. Gen. Chem. USSR,* **25,** 1765, 1813 (1955)
Yunusov, Akramov, *ibid.,* **25,** 1813 (1955)

L. multiflorum Lam.

ALKALOID: annuloline.
Axelrod, Belize, *J. Org. Chem.,* **23,** 919 (1958)

L. perenne L.

DIAZAPHENANTHRENE ALKALOIDS: perlolidine, perloline, perlolyrine.
MONOTERPENOID: loliolide.
(A) Jeffreys, *J. Chem. Soc., C,* 1091 (1970)
(T) Manske, *Can. J. Res., B* **16,** 438 (1938)

LOMARTIUM (Proteaceae/Umbelliferae)

L. columbianum.

COUMARINS: columbianadin, cis-p-coumaroyllomatin, trans-p-coumaroyllomatin, (+)-dihydrooroselol, lomatin, selinidin.
Nielsen *et al., Phytochemistry,* **15,** 1049 (1976)

L. ferrugineum.

QUINONE: valdivione.
Mechendale *et al., Phytochemistry,* **14,** 801 (1975)

L. macrocarpum.

SESQUITERPENOID: macrocarpin.
Steck, *Phytochemistry,* **12,** 2283 (1973)

L. nuttallii.

NORSESQUITERPENOID: lomatin.
Willette, Soine, *J. Pharm. Sci.,* **51,** 149 (1962)

LOMATHOGONIUM

L. carinthiacum (Wulfen) Rchb.

IRIDOID: erythrocentaurin.
XANTHONES: 1,8-dihydroxy-3,5-dimethoxyxanthone, 1,8-dihydroxy-3,7-dimethoxyxanthone, 1-hydroxy-3,7-dimethoxyxanthone, 1-hydroxy-3,7,8-trimethoxyxanthone, 1-hydroxy-4,6,8-trimethoxyxanthone.
Sorig *et al., Pharmazie,* **32,** 803 (1977)

LONCHOCARPUS

L. costaricensis.

FLAVONOIDS: 8-(3,3-dimethylallyloxy)-5,7-dimethoxyflavone, 7-(3,3-dimethylallyloxy)-8-(3,3-dimethylallyl)-5-methoxyflavone, 7-(3,3,-dimethylallyloxy)-8-(3-hydroxy-3-methyl-trans-but-1-enyl)-flavanone, dimethylprecansone B.
Waterman, *Phytochemistry,* **24,** 571 (1985)

L. laxiflorus.

ISOFLAVAN: lonchocarpin.
PTERAN: philenopteran 9-methyl ether.
Pelter, Amenechi, *J. Chem. Soc., C,* 887 (1969)

LONICERA (Caprifoliaceae)

L. alpigena.

IRIDOID: alpigenoside.
Bailleul *et al., J. Nat. Prod.,* **44,** 573 (1981)

L. caerulea.

IRIDOID: periclymenosidic acid.
Calis *et al., J. Nat. Prod.,* **48,** 108 (1985)

L. morrowii.

IRIDOIDS: kingiside, loniceroside, morrowonoside.
Souza, Mitsuhashi, *Tetrahedron Lett.,* 2725 (1969)

L. periclymenum L.

IRIDOID: periclymenoside.
Calis *et al., Helv. Chim. Acta,* **67,** 160 (1984)

L. ruprechtiana.

CAROTENOIDS: loniceraxanthin, rhodoxanthin.
Rahman, Egger, *Z. Naturforsch.,* **28C,** 434 (1973)

L. webbiana.

CAROTENOID: loniceraxanthin.
Rahman, Egger, *Z. Naturforsch.,* **28C,** 434 (1973)

L. xylosteum L.

THIOALKALOIDS: loxylostosidine A, loxylostosidine B, xylostosidine.
Chaudhuri *et al., Tetrahedron Lett.,* 559 (1981)

LOPHOCEREUS (Cactaceae)

L. schottii.

ALKALOIDS: lophocerine, lophocine, piloceredine, pilocerine.
STEROID: lophenol.
(A) Wani *et al., J. Chem. Res. Synop.,* 15 (1980)
(S) Djerassi *et al., J. Am. Chem. Soc.,* **80,** 1005 (1958)
(S) Djerassi *et al., ibid.,* **80,** 6084 (1958)

LOPHOLAENA

L. dregeana.

SESQUITERPENOID: 6β-acetoxy-10βH-furanoeremophilane.
Bohlmann *et al., Phytochemistry,* **16,** 1769 (1977)

LOPHOMYRTUS

L. bullata.

TERPENOID: bullatenone.
Brandt *et al., J. Chem. Soc.,* 3245 (1954)

LOPHOPHORA (Cactaceae)

L. diffusa.

ISOQUINOLINE ALKALOID: O-methylpellotine.

Bruhn, Agurell, *Phytochemistry,* **14,** 1442 (1975)

L. williamsii (Lem. et Salm. Dyek.) Coult.

AMINE ALKALOID: mescaline.

ISOQUINOLINE ALKALOIDS: anhalamine, anhalenine, lophophorine, O-methylpeyoruvic acid, O-methylpeyoxylic acid, pellotine, peyocactine.

PYRROLE ALKALOID: peyonine.

Kapadia *et al., J. Am. Chem. Soc.,* **92,** 6943 (1970)

Rao, *J. Pharm. Pharmacol.,* **22,** 544 (1970)

LORANTHUS (Loranthaceae)

L. falcatus L.f.

STEROIDS: β-sitosterol, stigmasterol.

TRITERPENOIDS: β-amyrin acetate, oleanolic acid, oleanolic acid methyl ester acetate.

Anjaneyulu *et al., Curr. Sci.,* **46,** 850 (1977)

L. grewinkii.

TRITERPENOID: loranthol.

Atta-ur-Rahman *et al., Phytochemistry,* **12,** 3004 (1973)

LOROGLOSSUM

L. hircinum.

PHENANTHRENOID: loroglossol.

Hardegger *et al., Helv. Chim. Acta,* **46,** 1171 (1963)

LOROSTEMON

L. coelhoi Paula.

XANTHONE: lorostemin.

Braz Filho *et al., Phytochemistry,* **12,** 947 (1973)

L. negrensis Fross.

XANTHONE: lorostemin.

Braz Filho *et al., Phytochemistry,* **12,** 947 (1973)

LOTUS (Leguminosae)

L. cornicuklatus.

PHYTOALEXINS: sativan, vestitol.

Ingham, *Phytochemistry,* **16,** 1279 (1977)

L. uliginosus.

PHYTOALEXINS: demethylvestitol, vestitol.

Ingham, *Phytochemistry,* **16,** 1279 (1977)

LUDWIGIA (Onagraceae)

L. adscendens.

FLAVONOIDS: kaempferol 3-O-glucoside, myricetin 3-O-galactoside, myricetin 3-O-rhamnoside, quercetin, quercetin 3-O-galactoside, quercetin 3-O-glucoside.

Huang, *Chem. Abstr.,* 104 31779f

L. epilobioides var. epilobioides.

FLAVONOIDS: orientin, isoorientin, isovitexin, vitexin.

Huang, *Chem. Abstr.,* 104 31779f

L. epilobioides var. greatrexii.

FLAVONOIDS: orientin, isoorientin, isovitexin, vitexin.

Huang, *Chem. Abstr.,* 104 31779f

L. hyssopifolia.

FLAVONOIDS: orientin, isoorientin, isovitexin, vitexin.

Huang, *Chem. Abstr.,* 104 31779f

L. octovalvis.

FLAVONOIDS: orientin, isoorientin, isovitexin, vitexin.

Huang, *Chem. Abstr.,* 104 31779f

L. ovalis.

FLAVONOIDS: orientin, isoorientin, isovitexin, vitexin.

Huang, *Chem. Abstr.,* 104 31779f

L. peploides var. stipulacea.

FLAVONOIDS: kaempferol 3-O-glucoside, quercetin, quercetin 3-O-galactoside, quercetin 3-O-glucoside.

Huang, *Chem. Abstr.,* 104 31779f

L. perennis.

FLAVONOIDS: kaempferol 3-O-glucoside, quercetin, quercetin 3-O-galactoside, quercetin 3-O-glucoside.

Huang, *Chem. Abstr.,* 104 31779f

LUFFA (Cucurbitoideae)

L. aegyptica Mill.

SAPONINS: aegyptinin A, aegyptinin B.

Varshney *et al., Natl. Acad. Sci. Lett. (India),* **5,** 403 (1982)

L. amara.

TRITERPENOID: amarinin.

Mikherjee *et al., Plant Cell Physiol.,* **27,** 935 (1986); *Chem. Abstr.,* 105 168857v

L. echinata.

TRITERPENOIDS: 2-epi-cucurbitacins A, B, C and D, isocucurbitacin B.

Lavie *et al., J. Chem. Soc.,* 3259 (1962)

L. operciliata.

TRITERPENOID: isocucurbitacin B.

Abrew Matos, Gottlieb, *An. Acad. Bras. Cienc.,* **39,** 245 (1967)

LUFFARIELLA

L. variabilis.

ANTIBIOTIC: manoalide.

de Silva *et al., Tetrahedron Lett.,* 1611 (1980)

LUNARIA (Cruciferae)

L. annua L.

AMINOACIDS: τ-glutamyl-β-alanine, τ-glutamyl-τ-aminobutyric acid.

Morris *et al., J. Biol. Chem.,* **237,** 2180 (1962)

L. biennis Moench.

MACROCYCLIC AKLALOIDS: lunariamine, lunaridine, lunarine, numismine.

Poupat *et al., C. R. Acad. Sci.,* **269,** 335 (1969)

LUNASIA

L. amara Blanco.

QUINOLONE ALKALOIDS: (-)-hydroxylunine, isobalfourodine, lunine.

Clarke, Grundon, *J. Chem. Soc.,* 4196 (1964)

L. amara var. repanda (Leuterb. et K.Schum.)

QUINOLONE ALKALOIDS: lunidine, lunidonine.

Ruegger, Stauffacher, *Helv. Chim. Acta,* **46,** 2329 (1963)

L. costata Miq.

QUINOLONE ALKALOIDS: lunacridine, lunacrine, lunamarine, lunaridine.

Boorsmaa, *Meded. Lds. Pttuin,* **13,** 126 (1899)

Boorsmaa, *ibid.,* **31,** 126 (1899)

L. quercifolia (Warb.) Leuterb. et K. Schum.

FURANOQUINOLONE ALKALOIDS: lunasine, lunine.

Amelink, *Pharmaceutisch Weekblad,* **69,** 1390 (1932)

L. temulentum.

UNCHARACTERIZED ALKALOID: temuline.

Hofmeister, *Archiv fuer Experimentelle Pathologie und Pharmacologie,* **30,** 203 (1892)

LUNULARIA (Hepaticae)

L. cruciata.
ACID: dihydrohydrangeaic acid.

Huneck et al., Z. Naturforsch., **26B**, 738 (1971)

LUPINUS (Leguminosae)

L. albus L.
LUPANINE ALKALOID: (+)-lupanine.

QUINAZOLINE ALKALOID: albine.

SPARTEINE ALKALOIDS: 5-dehydro-13-hydroxymultiflorone, 13-hydroxymultiflorone.

UNCHARACTERIZED ALKALOID: dehydroalbine.

Wiewiorowski et al., Bulleton de l'Academie Polonaise des Sciences, Serie des Sciences, **12**, 217 (1964)

L. angustifolius L.
LUPANE ALKALOIDS: angustifoline, (+)-13-trans-cinnamoylhydroxylupanine, (+)-lupanine.

Wiewiorowski et al., Monatsh. Chem., **88**, 663 (1957)

L. argenteus var. stenophyllus.
SPARTEINE ALKALOIDS: anagyrine, 5-dehydrolupanine, α-isolupanine, lupanine, sparteine, thermopsine.

Keller et al., J. Pharm. Sci., **67**, 430 (1978)

L. barbiger.
LUPINANE ALKALOID: trilupine.

QUINOLIZIDINE ALKALOID: sparteine.

Winterfield, Nitzsche, Arch. Pharm. (Weinheim, Ger.), **278**, 393 (1940)

L. caudatus.
SPARTEINE ALKALOID: thermopsine.

Orekhov et al., Ber., **66**, 625 (1933)

L. corymbosus Heller.
SPARTEINE ALKALOID: thermopsine.

Orekhov et al., Ber., **66**, 625 (1933)

L. cosentinii.
One uncharacterized alkaloid.

Beck et al., Lloydia, **42**, 385 (1979)

L. formosus.
DIPIPERIDYL ALKALOIDS: acetylhystrine, (+)-methylammodendrine.

Fitch et al., J. Org. Chem., **39**, 2974 (1974)

L. hartwegii.
QUINOLIZIDINE ALKALOIDS: aphylline, epi-aphylline, 10,17-dioxosparteine, gramine, α-isolupanine, nuttaline, virgiline.

Anderson et al., J. Org. Chem., **41**, 3441 (1976)

L. kingsii Wats.
LUPANE ALKALOID: (+)-lupanine.

Ranedo, Anales Fisica Quimica, **20**, 527 (1922)

L. laxiflorus var. silvicola.
SPARTEINE ALKALOID: anagyrine.

Ing, J. Chem. Soc., 504 (1933)

L. laxus Rydb.
LUPANE ALKALOIDS: (+)-lupanine, lupilaxine, trilupine.

SPARTEINE ALKALOID: (-)-sparteine.

Marion et al., Can. J. Chem., **31**, 181 (1953)

L. lindenianus.
SPARTEINE ALKALOID: lindenianine.

Nakano et al., J. Org. Chem., **39**, 3584 (1974)

L. luteus.
QUINOLIZIDINE ALKALOIDS: lupanine, (-)-sparteine, (-)-trans-4'-hydroxycinnamoyllupinine.

GIBBERELLINS: gibberellin A-18, gibberellin A-23, gibberellin A-28.

(A) Baumert et al., Ber., **15**, 1951 (1882)

(A) Baumert et al., ibid., **631**, 1951 (1882)

(A) Murakoshi et al., Phytochemistry, **14**, 2714 (1975)

(Gib) Yokota et al., Planta, **87**, 180 (1969)

L. luteus cv Barpine.
CHROMONE: lupilutin.

ISOFLAVONOIDS: barpisoflavone A, barpisoflavone B, barpisoflavone C, 2,3-dehydrokievitone hydrate, lupiwighteone, lupiwighteone hydrate, 5-O-methylderrone, 5-O-methyllupiwighteone.

Tahara et al., Agric. Biol. Chem., **50**, 1797 (1986)

Tahara et al., ibid., **50**, 1809 (1986)

L. macounii Rydb.
LUPINANE ALKALOID: rhombinine.

Manske, Marion, Can. J. Res., **21B**, 144 (1943)

L. niger.
QUINOLIZIDINE ALKALOID: lupinine.

SPARTEINE ALKALOID: (-)-sparteine.

Baumert et al., Ber., **15**, **631**, 1951 (1882)

L. nuttallii.
SPARTEINE ALKALOID: nuttaline.

Goldberg, Balthis, J. Chem. Soc., D, 660 (1969)

L. palmeri Wats.
QUINOLIZIDINE ALKALOID: lupinine.

UNCHARACTERIZED ALKALOIDS: pentalupine, tetralupine.

Sadykov, J. Gen. Chem., USSR, **19**, 143 (1948)

L. perennis.
LUPANE ALKALOIDS: 13-hydroxylupanine, (+)-lupanine.

SPARTEINE ALKALOID: angustifoline.

Ranedo, Anales Fisica Quimica, **20**, 527 (1922)

L. pilosus.
PYRROLIZIDINE ALKALOID: (+)-isolupinine.

FLAVONOIDS: apigenin, genistein, 3',4'-methylenedioxyorobol.

(A) Winterfield, Holschneider, Ber., **64**, 141 (1931)

(F) Lemeshev et al., Chem. Abstr., 86 68364c

L. polyphyllus Lindl.
LUPANE ALKALOID: (+)-lupanine.

QUINOLIZIDINE ALKALOID: lupinine.

SPARTEINE ALKALOID: sparteine.

Clemo et al., J. Chem. Soc., 432 (1931)

L. sericeus Pursh.
SPARTEINE ALKALOID: (-)-isosparteine.

UNCHARACTERIZED ALKALOIDS: (-)-isolupanine, lupilaxine, spathulatine.

Marion et al., Can. J. Chem., **31**, 181 (1951)

L. sericeus var. flexuosus C.P.Smith.
UNCHARACTERIZED ALKALOID: octalupine.

Couch et al., J. Am. Chem. Soc., **61**, 1523 (1939)

L. spathulatus.
UNCHARACTERIZED ALKALOID: spathulatine.

Couch et al., J. Am. Chem. Soc., **62**, 554 (1940)

L. termis.
LUPANE ALKALOID: (±)-lupanine.

Ranedo, Anales Fisica Quimica, **20**, 527 (1922)

L. wyethii.
KUPANE ALKALOID: 13-hydroxylupanine.

Deulofeu et al., J. Org. Chem., **10**, 179 (1945)

LYCEUM

L. euriopeum.
UNCHARACTERIZED ALKALOID: lyceamine.

Manzoor-i-Khuda, Sultana, *Pakistan J. Sci. Ind. Res.*, **11**, 247 (1968)

LYCHIUM

L. chinense.
PEPTIDE ALKALOID: kukoamine A.

STEROIDS: withanolide I, withanolide II.

(A) Furuyama *et al.*, *Tetrahedron Lett.*, 1355 (1980)

(S) Haansel *et al.*, *Arch. Pharm. (Weinheim, Ger.)* **308**, 653 (1975)

LYCHNOPHORA

L. affinis.
SESQUITERPENOIDS: lychnopholic acid, lychnophorolide A, lychnophorolide B.

LeQuesne *et al.*, *J. Org. Chem.*, **47**, 1519 (1981)

L. bahiensis.
SESQUITERPENOID:
15-hydroxy-16-[1(Z)-methyl-1-propenyl]-eremantholide.

Bohlmann *et al.*, *Phytochemistry*, **21**, 1087 (1982)

L. blanchei.
SESQUITERPENOIDS:
2,3-dihydro-3-hydroxy-8-(2-methylpropenoyl)-8-deacetyleremoglomerulide,
2,3-dihydro-3-methoxy-8-(2-methylpropenoyl)-8-deacetylereglomerulide,
8-(2-methylpropenoyl)-8-deacetylereglomerulide.

Bohlmann *et al.*, *Phytochemistry*, **21**, 1087 (1982)

L. columnaris.
SESQUITERPENOIDS: 11-acetoxy-14-lychnenoic acid,
1,10-epoxy-α-humulen-14-ol, α-humulen-14-aldehyde,
α-humulen-14-ol acetate, lychnocolumnic acid.

Bohlmann *et al.*, *Phytochemistry*, **21**, 1087 (1982)

L. martiana.
SESQUITERPENOIDS: acetyllychnofolic acid, lychnofolic acid.

Vichnewski *et al.*, *Phytochemistry*, **19**, 685 (1980)

L. passerina.
Two sesquiterpenoids.

Bohlmann *et al.*, *Angew. Chem.*, **93**, 280 (1981)

L. phylicaefolia.
SESQUITERPENOID: lychnofolide.

Bohlmann *et al.*, *Phytochemistry*, **19**, 2381 (1980)

L. salicifolia.
SESQUITERPENOIDS: 2-epi-lychnosalicifolide, lychnosalicifolide.

Bohlmann *et al.*, *Phytochemistry*, **21**, 1087 (1982)

L. sellowii.
SESQUITERPENOIDS:
5α-hydroxy-6α-angeloyloxy-4(15)-isogoyazensenolide,
5α,6α-diangeloyloxy-4(15)-isogoyazensenolide.

Bohlmann *et al.*, *Phytochemistry*, **21**, 1087 (1982)

LYCICHITUM

L. camtschatiense var. japonicum.
One uncharacterized alkaloid.

Katsui *et al.*, *Tetrahedron Lett.*, 6257 (1966)

LYCIUM (Solanaceae)

L. chinense Mill.
STEROIDS:
6,7-epoxy-5,20-dihydroxy-1-oxowitha-2,24-dienolide,
6,7-epoxy-5-hydroxy-1-oxowitha-2,24-dienolide.

Haansel *et al.*, *Arch. Pharm. (Weinheim, Ger.)* **308**, 563 (1975)

LYCOPERSICON (Solanaceae)

L. esculentum.
STEROIDAL GLYCOALKALOID: tomatine.

CAROTENOIDS: lycopene, lycophyll, lycoxanthin.

FLAVONOIDS: kaempferol, quercetin.

SAPONIN:
5α-furostane-3β,22,26-triol-3-O-β-D-glucopyranosyl-β-D-glucopyranosyl-α-galactopyranosyl-α-D-glucopyranosyl-26-O-β-D-glucopyranoside.

(A) Fontaine *et al.*, *Arch. Biochem.*, **27**, 461 (1950)

(F) Koizumi *et al.*, *Phytochemistry*, **15**, 342 (1976)

(Sap) Sato, Sakamura, *Agric. Biol. Chem.*, **37**, 225 (1973)

(T) Karrer *et al.*, *Helv. Chim. Acta*, **14**, 43 (1931)

L. esculentum var. (Tangella).
CAROTENOID: prolycopene.

Zechmeister *et al.*, *Proc. Natl. Acad. Sci. USA*, **27**, 468 (1941)

L. hirsutum.
GLYCOSIDIC STEROIDAL ALKALOID: tomatine.

Kuhn *et al.*, *Ber.*, **90**, 203 (1957)

L. peruvianum.
STEROIDAL GLYCOALKALOID: tomatine.

Kuhn *et al.*, *Angew. Chem.*, **68**, 212 (1956)

L. peruvianum var. chutatum.
STEROIDAL GLYCOALKALOID: tomatine.

Kuhn *et al.*, *Angew. Chem.*, **68**, 212 (1956)

L. pimpinellifolium. (Solanum racemigerum).
STEROIDAL GLYCOALKALOID: tomatine.

Fontaine *et al.*, *Arch. Biochem.*, **18**, 467 (1948)

L. putatum.
STEROIDAL GLYCOALKALOID: tomatine.

Fontaine *et al.*, *J. Am. Chem. Soc.*, **73**, 878 (1951)

LYCOPODIUM (Lycopodiaceae)

L. acrifolium.
LYCOPODIUM ALKALOID: acrifoline.

Manske, Marion, *J. Am. Chem. Soc.*, **69**, 2126 (1947)

L. alopecuroides.
LYCOPODIUM ALKALOIDS: acetyldibenzoylalopecurine,
alolycopine, alopecurine, lycopecunine, lycopodine.

Ayer *et al.*, *Can. J. Chem.*, **49**, 524 (1971)

L. annotinum L.
DIAZAPHENANTHRENE ALKALOIDS: α-obscurine, β-obscurine.

LYCOPODIUM ALKALOIDS: annofoline, annopodine, annotine,
annotinine, isolycopodine, α-lofoline, β-lofoline, lycodine,
lycodoline lycofawcine lycofoline, lyconnotine, lycopodine.

Anet, Khan, *Can. J. Chem.*, **37**, 1589 (1959)

Ayer *et al.*, *Tetrahedron Lett.*, 4587 (1968)

L. annotinum var. acrifolium Fern.
LYCOPODIUM ALKALOIDS: acrifoline, alkaloid L-28, alkaloid
L-30, alkaloid L-31.

Manske, Marion, *J. Am. Chem. Soc.*, **69**, 2126 (1947)

L. carolinianum L.
UNCHARACTERIZED ALKALOID: carolinianine.

Miller *et al.*, *Bull. Soc. Chim. Belg.*, **80**, 629 (1971)

L. carolinianum L. var. affine.
LYCOPODIUM ALKALOID: anhydrolycocernuine.

UNCHARACTERIZED ALKALOID: carolinianine.

Miller *et al.*, *Bull. Soc. Chim. Belg.*, **80**, 629 (1971)

L. cernuum.
DIAZAPHENANTHRENE ALKALOIDS: cernuine, lycocernuine.

TRITERPENOIDS: lycocernuic acid A, lycocernuic acid B.

(A) Marion, Manske, *Can. J. Res.*, **26B**, 1 (1948)

(T) Raffauf *et al.*, *J. Am. Chem. Soc.*, **100**, 7437 (1978)

L. clavatum L.

LYCOPODIUM ALKALOID: fawcettimine.

TRITERPENOIDS: lyclaninol, lyclanitin, lyclavanol, lyclavinin, lyclavatol, lycocryptol, 16-oxoserrat-14-ene-3α,21β-diol, 16-oxoserrat-14-ene-3β,21α-diol, 16-oxoserrat-14-ene-3β,21β-diol, serrat-14-ene-3α,20β,21,24-tetraol, serrat-14-ene-3β,21β,24-triol, serrat-14-ene-3α,21β,24-triol-16-one.

  (A) Burnell, Mootoo, *Can. J. Chem.*, **39**, 1090 (1961)
  (T) Tsuda *et al.*, *Chem. Pharm. Bull.*, **23**, 1336 (1975)

L. clavatum var. europeus.

LYCOPODIUM ALKALOIDS: clavanoline, clavatine, clavatoxine, de-N-methyl-α-obscurine.

  Achmatowicz, Uzieblo, *Rocz. Chem.*, **18**, 88 (1938)

L. clavatum var. megastachyon Fern. et Bissel.

LYCOPODIUM ALKALOIDS: acetyllycoclavine, lucidoline, lycoclavine, lycopodine, methyllycoclavine.

  Ayer, Law, *Can. J. Chem.*, **40**, 2088 (1962)

L. complanatum.

UNCHARACTERIZED ALKALOID: complanatine.
TRITERPENOID: tohogenol.

  (A) Manske, Marion, *Can. J. Res.*, **20B**, 87 (1942)
  (T) Inubushi *et al.*, *Chem. Pharm. Bull.*, **13**, 450 (1965)

L. densum Labill.

LYCOPODIUM ALKALOIDS: alkaloid L-34, alkaloid L-35.

  Manske, *Can. J. Res.*, **31**, 894 (1953)

L. fawcettii Lloyd.

LYCOPODIUM ALKALOIDS: acetylfawcettiine, diacetyllycofoline, fawcettiine, lycodine, lycofawcine, lycofoline.

  Anet, Rao, *Tetrahedron Lett.*, **20**, 9 (1960)

L. flabelliforme Fernald.

LYCOPODIUM ALKALOIDS: flabellidine, flabelliformine, flabelline, de-N-methyl-α-obscurine, α-obscurine, β-obscurine.

  Young, McLean, *Can. J. Chem.*, **41**, 2731 (1963)

L. gnidioides.

LYCOPODIUM ALKALOIDS: gnidioidine, lycognidine.

  Nyembo *et al.*, *Bull. Soc. Chim. Belg.*, **85**, 595 (1976)

L. inundatum L.

LYCOPODIUM ALKALOIDS: dehydrolycopecurine, (-)-3-dehydro-trans-9-methyl-10-ethyllobelidiol, inundatine, isoinundatine, lycopodine.

  Braekman *et al.*, *Bull. Soc. Chim. Belg.*, **80**, 83 (1971)

L. lucidulum.

ALKALOID: luciduline.

  Ayer *et al.*, *Can. J. Chem.*, **46**, 3631 (1968)

L. lucidum Michx.

LYCOPODIUM ALKALOIDS: dihydrolycolucine, 6-hydroxylycopodine, 12-hydroxylycopodine, lucidoline, lycolucine, lycopodine.
DITERPENOID: lycoxanthol.

  (A) Ayer *et al.*, *Can. J. Chem.*, **57**, 1105 (1979)
  (T) Burnell, Moinas, *J. Chem. Soc., Chem. Commun.*, 897 (1971)

L. magellanicum.

ALKALOIDS: 5-dehydromagellanine, magellanine.

  Loyola *et al.*, *Phytochemistry*, **18**, 1721 (1979)

L. megastacyum.

TRITERPENOID: serratenediol.

  Inubushi *et al.*, *J. Chem. Soc., C*, 3109 (1971)

L. obscurum var. dendroideum (Michx.) D.C.Eaton.

DIAZAPHENANTHRENE ALKALOIDS: α-obscurine, β-obscurine.

  Moore, Marion, *Can. J. Chem.*, **31**, 952 (1953)

L. paniculatum.

LYCOPODIUM ALKALOID: paniculine.

  Morales *et al.*, *Phytochemistry*, **18**, 1719 (1979)

L. phlegmara.

TRITERPENOIDS: phlegmanols A, B, C, D, E and F, phlegmaric acid, serratenediol.

  Inubushi *et al.*, *Chem. Pharm. Bull.*, **19**, 2640 (1972)

L. saururus Lam.

LYCOPODIUM ALKALOID: sauroxine.
UNCHARACTERIZED ALKALOIDS: pillijanine, sausurine.

  Deulofeu, DeLanghe, *J. Am. Chem. Soc.*, **64**, 968 (1942)

L. selago L.

ISOQUINOLINE ALKALOIDS: pseudoselagine, selagine.

  Achmatowicz, Rodewalk, *Rocz. Chem.*, **30**, 233 (1956)

L. serratum.

LYCOPODIUM ALKALOID: serratidine.
TRITERPENOIDS: 16-oxolycoclavinol, 16-oxoserratenetriol, tohogeninol, tohogenol.

  (A) Inubushi *et al.*, *J. Pharm. Soc., Japan*, **84**, 1108 (1964)
  (T) Tsuda *et al.*, *Chem. Pharm. Bull.*, **33**, 1290 (1975)

L. serratum var. serratum forma intermedium.

LYCOPODIUM ALKALOID: lycopodine. Uncharacterized alkaloids lycobergine, lycoserramine, lycoserrine, lycothuine.

  Inubushi *et al.*, *J. Pharm. Soc., Japan*, **87**, 1394 (1967)

L. serratum var. serratum forma serratum.

UNCHARACTERIZED ALKALOID: lycoserrine.

  Inubushi *et al.*, *J. Pharm. Soc., Japan*, **87**, 1394 (1967)

L. serratum Thunb. var. thunbergii Mak.

TRITERPENOIDS: serrat-14-ene-3β,21α-diol, serrat-14-ene-3β,21β-diol, serrat-14-ene-3β,21α,24-triol.
ALKALOIDS: serratidine, serratinidine, serratinine.

  (A) Yasui *et al.*, *Tetrahedron Lett.*, 3967 (1966)
  (T) Tsuda *et al.*, *ibid.*, 1279 (1964)
  (T) Inubushi *et al.*, *Chem. Pharm. Bull.*, **13**, 104 (1965)

L. verticillatum.

LYCOPODIUM ALKALOID: lycoverticine.

  Nyumbo *et al.*, *Bull. Soc. Chim. Belg.*, **85**, 595 (1976)

L. wrightianum.

TRITERPENOID: 3,14,20,21,24-serratenepentaol.

  Tsuda *et al.*, *Chem. Pharm. Bull.*, **28**, 3275 (1980)

LYCORIS

L. radiata Herb.

ACID: i2-(4-hydroxybenzoyl)-maleic acid. Benzindolizidine alkaloid lycorine.
INDOLE ALKALOIDS: 9-demethylhomolycorine, O-2-demethyllycoramine, galanthamine, lycorenine.
ISOQUINOLINE ALKALOIDS: norpluviine, pluviine.
FLAVAN: 7,3'-dihydroxy-4'-methoxy-8-methylflavan.

  (Ac) Koizumi *et al.*, *Phytochemistry*, **15**, 342 (1976)
  (A) Kobayashi *et al.*, *Chem. Pharm. Bull.*, **28**, 3433 (1980)
  (F) Numata *et al.*, *ibid.*, **31**, 2146 (1983)

LYGODIUM

L. flexuosum.

ANTHRAQUINONE: tectoquinone.
STEROIDS: α-sitosterol, stigmasterol.
TRITERPENOIDS: O-p-coumaryldryocrassol, dryocrassol.

  Achari *et al.*, *Planta Med.*, 329 (1986)

LYNGBYA
L. majuscula.
HALOGENATED PYRROLE ALKALOIDS: malyngamide A,
   malyngamide B.
PYRROLE ALKALOIDS: pukeleimides A, B, C, D, E, F and G.
DEPSIPEPTIDE: majusculamide C.
   (A) Cordellina II, Moore, *Tetrahedron Lett.,* 2007 (1979)
   (A) Cordellina II *et al., J. Am. Chem. Soc.,* **101,** 240 (1979)
   (D) Carter *et al., J. Org. Chem.,* **49,** 236 (1984)

LYONIA (Ericaceae)
L. ovalifolia var. elliptica.
DITERPENOIDS: lyoniols A, B, C and D.
TRITERPENOIDS: O-p-hydroxy-trans-cinnamoylmaslinic acid,
   lyofolic acid.
   Yasue *et al., J. Pharm. Soc., Japan,* **91,** 1200 (1971)

LYSICHITON (Araceae)
L. camtschatcense var. japonicum Makino.
OXOAPORPHINE ALKALOID: lysicamine.
   Katsui *et al., Tetrahedron Lett.,* 6257 (1966)

LYTHRUM (Lythraceae)
L. anceps Makino.
PIPERIDINE ALKALOID: lythramidine.
MACROCYCLIC ALKALOIDS: lythracines I, II, III, IV, V, VI and
   VII, lythramine, lythrancepines I, II and III, lythranidine,
   lythranine.
   Fujita, Saeki, *J. Chem. Soc., Perkin I,* 306 (1973)

LYTTA
L. vesicatoria.
MONOTERPENOID: catharidin.
   Coffey, *Recl. Trav. Chim. Pays-Bas,* **42,** 1026 (1923)

M

MAACKIA (Leguminosae)

M. amurensis Rupr. et Maxim.
LUPANE ALKALOID: (-)-lupanine.
QUINOLIZIDINE ALKALOID: cytisine.
Tsarev, C. R. Acad. Sci. USSR, 42, 122 (1944)

MACARANGA

M. denticulata.
TRITERPENOID: 3-epi-taraxerol.
Bose, Khastgir, Indian J. Chem., 12, 11 (1973)

M. peltata Muell.
TRITERPENOID: cyclopeltenyl acetate.
Anjaneyulu, Reddy, Indian J. Chem., 20B, 1033 (1981)

M. tanarius.
DITERPENOIDS:
6,20-epoxylathyrol-5,10-diacetate-3-phenylacetate, macarangonol.
Adolf et al., Tetrahedron Lett., 2241 (1970)

MACFADYENA

M. cynanchoides.
IRIDOIDS: 5,7-bisdeoxycynanchoside, cynanchoside.
Adrian et al., Phytochemistry, 21, 231 (1982)

MACHAERANTHERA (Compositae)

M. scabrella.
COUMARINS: 5'-oxo-3-aurapten, 5'-oxoaurapten.
Bohlmann, Grenz, Chem. Ber., 109, 1584 (1976)

M. tanacetifolia (H.B.K.) Nees.
SESQUITERPENOID: (+)-eudesmol-α-L-arabinopyranoside.
Yoshioka et al., J. Org. Chem., 34, 3697 (1969)

MACHAERIUM

M. acutifolium.
ISOFLAVONE: duartin.
Ollis et al., Phytochemistry, 17, 1401 (1978)

M. opacum.
ISOFLAVONE: duartin.
Ollis et al., Phytochemistry, 17, 1401 (1978)

M. scleroxylon.
PHENOLICS: dalbergin, 3,4-dimethoxydalbergione, O-methyldalbergin.
TRITERPENOID: O-acetyloleanolic aldehyde.
Mukherjee et al., Tetrahedron, 27, 799 (1971)
Yoshimoto et al., Mokkuzai Gakkaishi, 21, 686 (1975) Chem. Abstr., 84 147705b

M. villosum.
ISOFLAVONE: duartin.
Ollis et al., Phytochemistry, 17, 1401 (1978)

MACHAEROCEREUS (Cactaceae)

M. gummosus.
TRITERPENOIDS: gummosogenin, macheric acid, macherinic acid.
Djerassi et al., J. Am. Chem. Soc., 77, 1825 (1955)

MACHILUS (Lauraceae)

M. acuminatissimum.
One uncharacterized alkaloid.
Lu, J. Pharm. Soc., Japan, 87, 1278 (1967)

M. glaucescens Wight.
OXOAPORPHINE ALKALOID: machigline.
Talapatra et al., J. Indian Chem. Soc., 59, 1364 (1982)

M. japonica. Sieb. et Zucc.
LIGNANS: (+)-galbacin, (+)-galbelgin, (-)-licarin-B, machilusin.
Hughes et al., Aust. J. Chem., 7, 104 (1954)
Takeoka et al., Bull. Chem. Soc., Japan, 49, 3564 (1976)

M. kusanoi.
BENZOISOQUINOLINE ALKALOID: L-(-)-norarmepavine.
Tomita et al., J. Pharm. Soc., Japan, 83, 15 (1963)

MACKLINAYA

M. macrosciadia (F.Muell.) F.Muell.
QUINAZOLINE ALKALOIDS:
6,7,8,9-tetrahydro-11H-pyrido-[2,1-b]-quinazoline,
6,7,8,9-tetrahydro-11H-pyrido-[2,1-b]-quinazolin-11-one.
Spath, Platzer, Ber. Dtsch. Chem. Ges., 68, 2221 (1935)

M. subulata Philipson.
QUINAZOLINE ALKALOIDS:
6,7,8,9-tetrahydro-11H-pyrido-[2,1-b]-quinazoline,
6,7,8,9-tetrahydro-11H-pyrido-[2,1-b]-quinazolin-11-one.
Johns, Lamberton, J. Chem. Soc., Chem. Commun., 267 (1965)

MACLURA (Moraceae)

M. pomifera.
SESTERTERPENOID: pomiferin.
TRITERPENOID: lupane-3β,20-diol.
XANTHONE: toxyloxanthone.
(T) Gearien, Klein, J. Pharm. Sci., 64, 104 (1975)
(X) Deshpande et al., Indian J. Chem., 11, 518 (1973)

MACOWANIA

M. corymbosa.
DITERPENOID: 3,4-dihydro-cis-clerodan-15-oic acid.
Bohlmann, Zdero, Phytochemistry, 16, 1583 (1977)

MACROCYSTIS (Phaeophyceae)

M. pyrifer.
AMINO ACIDS: alanine, arginine, aspartic acid, cystine, glutamic acid, glycine, histidine, isoleucine, leucine, lycine, methionine, ornithine, phenylalanine, proline, serine, threonine, valine.
Mateus et al., Bot. Mar., 19, 155 (1976); Chem. Abstr., 85 119623a

MACROPIPER

M. excelsum.
LIGNANS: demethoxyexcelsin, syringiaresinol.
Russell et al., Phytochemistry, 12, 1799 (1973)

MACRORUNGIA (Acanthaceae)
M. longistrobus C.B.Cl.
QUINOLINE ALKALOIDS: dehydroisolongistrobine, isolongistrobine, isomacrorine, longistrobine, macrorine, macrorungine, normacrorine.

Arndt *et al.*, *Tetrahedron*, **25**, 2767 (1969)

Wuonolo, Woodward, *ibid.*, **32**, 1085 (1976)

MACROTOMIA (Boraginaceae)
M. echinoides (L.) Boiss. (Aipyanthus echioides).
PYRROLIZIDINE ALKALOID: macrotomine.

Men'shikov, Petrova, *J. Gen. Chem. USSR*, **22**, 1457 (1952)

MACROZAMIA
M. riedlei.
GLYCOALKALOID: macromazine.

Lythgoe, Riggs, *J. Chem. Soc.*, 2716 (1949)

MACUNA
M. deeringiana.
QUINOLINE ALKALOID:
1,2,3,4-tetrahydro-6,7-dihydroxy-1-methyl-3-isoquinoline carboxylic acid.

Daxenbichler *et al.*, *Tetrahedron Lett.*, 1801 (1972)

M. mutisiana.
ISOQUINOLINE ALKALOID:
1,2,3,4-tetrahydro-6,7-dihydroxyisoquinoline carboxylic acid.

Bell *et al.*, *Phytochemistry*, **10**, 2191 (1971)

M. pruriens.
UNCHARACTERIZED ALKALOIDS: prurienine, prurieninine.

Santra, Majumder, *Indian J. Pharm.*, **15**, 62 (1953)

MADHUCA
M. longifolia.
TRITERPENOID: protobassic acid.

SAPONINS: MI-saponins A, B and C.

Kitagawa *et al.*, *Chem. Pharm. Bull.*, **26**, 1100 (1978)

MAESA
M. chisia.
STEROID: 7,22-stigmastadien-3-ol-β-D-glucopyranoside.

Nigam *et al.*, *Indian J. Chem.*, **5**, 395 (1967)

MAGNOLIA (Magnoliaceae)
M. acuminata (L.) L.
DITERPENOID: acuminatin.

Aiba *et al.*, *Phytochemistry*, **12**, 1163 (1973)

M. coco.
BISBENZYLISOQUINOLINE ALKALOID: magnococline.

Yan, Lui, *J. Chin. Chem. Soc.*, **18**, 91 (1971)

M. fargesii.
LIGNAN: fargesin.

Kakisawa *et al.*, *Bull. Chem. Soc., Japan*, **43**, 3631 (1970)

M. fuscata Andr.
BISBENZYLISOQUINOLINE ALKALOIDS: magnolamine, magnoline.

Proskurnina, Orekhov, *Bull. Soc. Chim.*, **5**, 1357 (1938)

M. grandiflora L.
QUATERNARY AMINES: candicine, salicifoline.

GLUCOSIDES: lirioresinol A, syringin.

SESQUITERPENOIDS: magnolialide, peroxycostunolide, peroxyparthenolide.

(A) Erspamer, *Arch. Biochem. Biophys.*, **82**, 431 (1959)

(G) Briggs *et al.*, *J. Chem. Soc., C*, 3042 (1968)

(T) El-Feraly *et al.*, *Phytochemistry*, **18**, 881 (1979)

M. kachirachira.
BENZYLISOQUINOLINE ALKALOID: D-(+)-norarmepavine.

Yanu, Lu, *J. Pharm. Soc., Japan*, **83**, 22 (1963)

M. kobus.
SESQUITERPENOIDS: kobusunin A, kobusunin B.

Ida *et al.*, *Phytochemistry*, **21**, 673 (1982)

M. liliiflora Desr.
Two uncharacterized alkaloids.

Nakuno *et al.*, *Pharm. Bull.*, **1**, 29 (1953)

M. obovata.
ACID: palmitic acid.

APORPHINE ALKALOID: obovaine.

ESTERS: ethyl palmitate, ethyl stearate.

MONOTERPENOIDS: bornyl acetate, camphene, caryophyllene, caryophyllene epoxide, limonene, β-pinene.

Ito, Asai, *J. Pharm. Soc., Japan*, **94**, 729 (1974)

Sashida *et al.*, *ibid*, **96**, 218 (1976)

M. officinalis L.
TERPENOID: bornylmagnolol.

Tsumura, Juntendo Ltd, *Japanese Patent, 84 95, 229*

M. x soulangeana Soulange-Bodin.
ALKALOID: oxylauranine.

SESQUITERPENOIDS: soulangianolides A and B.

(A) Ziyaev *et al.*, *Khim. Prir. Soedin.*, 528 (1975)

(T) El-Feraly *et al.*, *Phytochemistry*, **22**, 2239 (1983)

M. sprengeri.
AMINE: magnosprengerine.

Cao *et al.*, *Zhongcaoyao*, **16**, 386 (1985); *Chem. Abstr.*, 104 17645c

M. wilsonii Rehd.
ISOQUINOLINE ALKALOID: magnocurarine.

Wang, *Yaoxue Fenxi Zazhi*, **2**, 95 (1982); *Chem. Abstr.*, 97 195751d

MAGUEY (Agavaceae)
M. cacaya.
STEROIDS: cacogenin, macogenin.

Marker, *J. Am. Chem. Soc.*, **69**, 2399 (1947)

M. canasto.
STEROID: manogenin.

Marker *et al.*, *J. Am. Chem. Soc.*, **65**, 1199 (1943)

M. canon de abra.
STEROID: tigogenin.

Marker *et al.*, *J. Am. Chem. Soc.*, **65**, 1199 (1943)

M. cenisto.
STEROID: manogenin.

Marker *et al.*, *J. Am. Chem. Soc.*, **65**, 1199 (1943)

M. cimmaron.
STEROID: manogenin.

Marker *et al.*, *J. Am. Chem. Soc.*, **65**, 1199 (1943)

M. cuchacamba.
STEROID: manogenin.

Marker *et al.*, *J. Am. Chem. Soc.*, **65**, 1199 (1943)

M. el ojital.
STEROID: hecogenin.

Marker *et al.*, *J. Am. Chem. Soc.*, **65**, 1199 (1943)

M. espadin.
STEROID: hecogenin.

Marker *et al.*, *J. Am. Chem. Soc.*, **65**, 1199 (1943)

M. estrella.
STEROID: sarsasapogenin.
Marker *et al.*, *J. Am. Chem. Soc.*, **65**, 1199 (1943)

M. tequilla manso.
STEROID: hecogenin.
Marker *et al.*, *J. Am. Chem. Soc.*, **65**, 1199 (1943)

MAGYDARIS (Umbelliferae)

M. panacifolia (Vahl.) Lange.
DITERPENOIDS: magydardiendiol,
magydar-2,10(20),13-trien-17-acetate,
magydar-2,10(20),13-trien-17-ol.
de Pascual Teresa *et al.*, *Tetrahedron Lett.*, 4563 (1978)

MAHONIA (Berberidaceae)

M. aquifolia (Pursh.) Nutt. (Berberis aquifolium Pursh.)
APORPHINE ALKALOIDS: corydine, corytuberine, isoboldine,
isocorydine, magnoflorine.
BISBENZYLISOQUINOLINE ALKALOID: berbamine.
QUATERNARY ALKALOIDS: columbamine, jatrorrhizine,
palmatine.
PROTOBERBERINE ALKALOIDS: berberine, tetrahydrojatrorrhizine,
tetrahydroberberine.
Slavik *et al.*, *Chem. Pap.*, **39**, 547 (1985); *Chem. Abstr.*, 104 31708g

M. nepalensis DC. (Berberis nepalensis Spreng.).
BENZYLISOQUINOLINE ALKALOIDS: neprotine, umbellatine.
Chatterjee, *Sci. Cult.*, **7**, 619 (1942)

M. philippinensis Nutt.
PROTOBERBERINE ALKALOIDS: jatrorrhizine, shobakunine.
Santos, *Univ. Philipp. Nat. Appl. Sci. Bull.*, **1**, 183 (1931)

M. trifoliolata Fedde.
PROTOBERBERINE ALKALOID: berberine.
Awe, *Arch. Pharm.*, **284**, 352 (1951)

MAJIDEA

M. fosteri Radlk.
TRITERPENOID GLYCOSIDE: majideagenin.
Kapundu *et al.*, *Bull. Soc. Chim. Belg.*, **93**, 497 (1984)

MAJORANA

M. hortensis.
FLAVANONE: majoranin.
Mabry *et al.*, *Curr. Sci.*, **41**, 202 (1972)

MAKINOA (Dilenaenaceae)

M. crispata.
SESQUITERPENOID: crispatanolide.
Asakawa *et al.*, *J. Chem. Soc., Chem. Commun.*, 1232 (1980)

MALABAILA (Pastinaceae)

M. dasycarpa.
COUMARINS: bergapten, imperatorin, isobergapten,
isopimpinellin, pimpinellin, sphondin, umbelliferone.
Gusak *et al.*, *Chem. Abstr.*, 86 68378k

M. graveolens.
COUMARINS: bergapten, imperatorin, isobergapten,
isopimpinellin, pimpinellin, sphondin, umbelliferone.
Gusak *et al.*, *Chem. Abstr.*, 86 68378k

MALACOCARPUS (Zygophyllaceae)

M. crithmofolius.
PIPERIDINE ALKALOID: anabasine.
Spath, Kesztler, *Ber. Dtsch. Chem. Ges.*, **70**, 72 (1937)

MALAPROUNEA

M. africana.
TRITERPENOID: malaprounic acid.
Wani *et al.*, *J. Nat. Prod.*, **46**, 537 (1983)

MALAXIS (Orchidaceae)

M. congesta.
PYRROLIZIDINE ALKALOID: malaxine.
Leander, Luning, *Tetrahedron Lett.*, 3477 (1967)

M. grandifolia Schltr.
PYRROLIZIDINE ALKALOID: grandiflorine.
Lindstrom *et al.*, *Acta Chem. Scand.*, **25**, 1900 (1971)

MALLOTUS (Euphorbiaceae)

M. hookerianus.
TRITERPENOIDS: 3β,28-dihydroxyurs-12-en-27-oic acid,
3-oxours-12-ebe-27,28-dioic acid.
Hui, Li, *Phytochemistry*, **15**, 985 (1976)

M. japonicus.
PHENOLICS: 4-O-galloylbergenin, 11-O-galloylbergenin,
11-O-galloyldemethylbergenin.
Yoshida *et al.*, *Phytochemistry*, **21**, 1180 (1982)

M. paniculatus.
TRITERPENOID: 29-nor-21H-hopane-3,22-dione.
Hui *et al.*, *Phytochemistry*, **8**, 819 (1969)

M. repandus.
ISOCOUMARIN: bergenin.
TERPENOIDS: mallotucins A and B.
TRITERPENOID: 3β-hydroxy-13-ursan-28,12-olide benzoate.
(I) Tomizawa *et al.*, *Phytochemistry*, **15**, 328 (1976)
(T) Kawashima *et al.*, *Heterocycles*, **5**, 277 (1976)

M. stenanthus.
TRITERPENOID: mallotin.
Pal *et al.*, *Phytochemistry*, **14**, 2263 (1975)

MALOUETIA (Apocynaceae)

M. arborea Miers.
STEROIDAL ALKALOIDS: malarboreine, malarborine.
Soti *et al.*, *Tetrahedron Lett.*, 1437 (1967)

M. bequaertiana.
STEROIDAL ALKALOIDS: funtumufrine C, funtuphyllamine B,
malouetine, malouphyllamine, malouphylline.
Janot *et al.*, *Bull. Soc. Chim. Fr.*, 648 (1962)

MALUS (Rosaceae)

M. communis.
FLAVANONE: 3',4',5,7-tetrahydroxyflavanone-7-glucoside.
Kowakewski, *Planta Med.*, **19**, 311 (1971)

M. silvestris Mill.
FLAVONOIDS: quercetin 3-O-α-L-arabinofuranoside, quercetin
3-O-α-D-galactoside, quercetin 3-O-β-D-glucoside,
quercetin 3-O-rhamnoside, quercetin 3-O-xyloside, rutin.
Teuber Wuenscher *et al.*, *Chem. Abstr.*, 88 148955c

MALVA (Malvaceae)

M. sylvestris.
FLAVONOIDS: gossypin 3-sulphate, hypoletin
8-O-β-D-glucoside 3'-sulphate.
Nawwar *et al.*, *Phytochemistry*, **16**, 145 (1977)

MAMMEA (Guttiferae/Clusiaceae)

M. acuminata.
XANTHONES: 2-hydroxyxanthone, 2-methoxyxanthone.
Karunanayake et al., Unpublished work. Given in J. Chem. Soc., Perkin 1, 2241, 2259 (1972)

M. africana G.Don.
XANTHONES: 1,5-dihydroxyxanthone, 1,5,6-trihydroxyxanthone, 1,6,7-trihydroxyxanthone.
Jackson et al., J. Chem. Soc., C, 2201, 2421 (1969)

M. americana L.
COUMARINS:
4-[1-(acetoxypropyl)]-5,7-dihydroxy-6-(3-ethyl-2-butenyl)-8-(2-methyl-1-oxobutyl)-2H-1-benzopyran-2-one, cyclomammein, cyclomammeisin, cycloneomammein.
XANTHONES: 1,5-dihydroxyxanthone, 2-hydroxyxanthone, 4-hydroxyxanthone, 2-methoxyxanthone.
(C) Crombie et al., J. Chem. Soc., Perkin I, 2241, (1972)
(C) Crombie et al., ibid., 1, 2259 (1972)
(X) Finnegan et al., Tetrahedron Lett., 6087 (1966)

M. indica.
STEROID: β-sitosterol.
TRITERPENOIDS: friedelin, lupeol, taraxerol, taraxerone.
Arjaneyulu et al., Indian J. Pharm. Sci., 44, 58 (1982)

M. longifolia.
COUMARIN: surangin A.
Buchiet et al., J. Chem. Soc., C, 542 (1966)

MANDRAGORA (Solanaceae)

M. scopoliae.
TROPANE ALKALOID: (-)-hyoscyamine.
Bremner, Cannon, Aust. J. Chem., 21, 1369 (1968)

M. vernalis.
TROPANE ALKALOIDS: hyoscine, norhyoscyamine.
King et al., J. Chem. Soc., 476 (1919)

MANFREDA (Agavaceae)

M. gigantea.
STEROID: gitogenin.
Marker et al., J. Am. Chem. Soc., 65, 1199 (1943)

M. maculosa Hook.
STEROIDS: hecogenin, manogenin.
Marker et al., J. Am. Chem. Soc., 85, 1199 (1943)

M. tigrina Engelm.
STEROID: manogenin.
Marker et al., J. Am. Chem. Soc., 65, 1199 (1943)

M. virginica.
STEROID: gitogenin.
Marker et al., J. Am. Chem. Soc., 65, 1199 (1943)

MANGIFERA (Anacardiaceae)

M. indica L.
TRITERPENOIDS: ambolic acid, ambonic acid, isomangiferolic acid, mangiferolic acid, mangiferonic acid.
Corsano, Mincione, Tetrahedron Lett., 2377 (1965)

MANILA

M. elemi.
TRITERPENOID: brein.
Laird et al., J. Am. Chem. Soc., 82, 4108 (1960)

MANILKARA

M. hexandra.
TRITERPENOID: triandrin.
Pant, Rastogi, Indian J. Chem., 15B, 911 (1977)

MANNIA

M. fragrans.
SESQUITERPENOID: (R)(-)-cuparenone.
Benesova, Collect. Czech. Chem. Commun., 41, 8812 (1976)

MANSONIA (Sterculiaceae)

M. altissima Chev.
SESQUITERPENOIDS: mansonones A, B, C, D, E, F, G and H.
Marini-Bettolo et al., Tetrahedron Lett., 4857 (1965)
Tanaka et al., ibid., 2767 (1966)

MAPPIA (Olacaceae)

M. foetida Miers.
ALKALOIDS: camptothecin, mappicine, 9-methoxycamptothecin.
Govindachari et al., J. Chem. Soc., Perkin I, 1215 (1974)

MARAH (Cucurbitaceae)

M. macrocarpus (Greene) Greene.
TRITERPENOIDS: echinocystic acid methyl ester, maragenins I, II and III.
Hylands, Salama, Tetrahedron, 35, 417 (1979)

M. oreganus.
FLAVONOIDS: kaempferol 3-O-diglucoside, kaempferol 3,7-di-O-glucoside, kaempferol 3-O-glucoside, kaempferol 3-O-glucoside-7-O-rhamnoside, quercetin 3-O-diglucoside, quercetin 3,7-di-O-glucoside, quercetin 3-O-glucoside, quercetin 3-O-glucoside-7-O-rhamnoside.
TRITERPENOID: dihydrocucurbitacin B.
(F) Nicholls et al., J. Nat. Prod., 45, 453 (1982)
(T) Kupchan et al., Phytochemistry, 17, 767 (1978)

MARASMIUS (Basidiomycetae)

M. alliaceus.
SESQUITERPENOIDS: alliacide, alliacols A and B, alliacolides I, II and III, 11-hydroxyalliacolide, 12-hydroxyalliacolide, 12-noralliacolide.
Anke et al., J. Antibiotics, 34, 1271 (1981)
Farrell et al., J. Chem. Soc., Perkin I, 1790 (1981)

M. conigenus.
SESQUITERPENOID: marasmic acid.
Kavanagh et al., Proc. Nat. Acad. Sci., 35, 843 (1949)

MARCHANTIA (Hepaticae)

M. polymorpha L.
SESQUITERPENOIDS: (S)-2-hydroxycuparene, prelunaric acid.
Ohta et al., Phytochemistry, 23, 1607 (1984)

MARGINOSPORUM (Rhodophyceae)

M. aberrans.
AMIDE: dichloroacetamide.
HALOGENATED DITERPENOID: aplysin 20.
SESQUITERPENOID: 3,5-dinitroguaiacol.
HALOGENATED SESQUITERPENOID: aplysinal.
(Am) Ohta et al., Phytochemistry, 16, 1085 (1977)
(T) Ohta, Takagi, ibid., 16, 1062 (1977)

MARGOTIA

M. gummifera.
DITERPENOIDS: gummiferolic acid, isomargotianoic acid, margotianin, margotianoic acid.
Rodriguez, Pinar, An. Quim., 75, 936 (1979)

MARKHAMIA

M. hildebrandtii (Baker) Sprague.
NAPHTHOQUINONES: dehydro-α-lapachone, 2-isopentylfurano-1,4-naphthoquinone, lapachol.
Chen et al., Hua Hsueh, **44**, 61 (1986); Chem. Abstr., 105 168912j

M. stipulata.
NAPHTHOPYRAN: tectol.
SESQUITERPENOID: dehydro-α-lapachone.
(N) Joshi et al., Planta Med., **34**, 219 (1978)
(T) Manners et al., Phytochemistry, **15**, 225 (1976)

MARRUBIUM (Labiatae)

M. allyson.
DITERPENOIDS: 6-acetylpremarrubenol, premarrubenol.
Savona et al., Phytochemistry, **18**, 859 (1979)

M. alternidens Rech.
QUATERNARY ALKALOID: stachydrine.
Paudler, Wagner, Chem. Ind., 1693 (1963)

M. catariifolium.
DITERPENOID: peregrinol.
Salei et al., Khim. Prir. Soedin., 301 (1967)

M. incanum.
DITERPENOID: marrubiin-3-one.
Salei et al., Khim. Prir. Soedin., 249 (1966)

M. leonurioides.
DITERPENOIDS: marrubiin-3-one, peregrinol.
Salei et al., Khim. Prir. Soedin., 301 (1967)

M. peregrinum.
DITERPENOIDS: marrubiin-3-one, peregrinol.
Salei et al., Khim. Prir. Soedin., 301 (1967)

M. praecox.
DITERPENOID: peregrinol.
Salei et al., Khim. Prir. Soedin., 301 (1967)

M. sericeum.
DITERPENOIDS: 6-acetylmarrubenol, 19-acetylmarrubenol, 6-acetylpremarrubenol, premarrubenol.
Savona et al., Phytochemistry, **18**, 859 (1979)

M. supinum.
DITERPENOIDS: 6-acetylmarrubenol, 19-acetylmarrubenol, 6-acetylpremarrubenol, premarrubenol.
Savona et al., Phytochemistry, **18**, 859 (1979)

M. vulgare L.
QUATERNARY ALKALOID: betonicine.
MONOTERPENOIDS: camphene, p-cymene, p-fenchene, limonene, α-pinene, sabinene, α-terpinolene.
DITERPENOIDS: marrubenol, peregrinol.
(A) Cocker, J. Chem. Soc., 2540 (1953)
(T) Salei et al., Khim. Prir. Soedin., 301 (1967)
(T) Karryev et al., Izv. Akad. Nauk Turkm. SSR, 86 (1976); Chem. Abstr., 86 2355u

MARSDENIA (Asclepiadaceae)

M. condurango Rchb.
STEROIDAL ALKALOIDS: condurangamines A and B.
STEROIDS: condurangoglycosides A, B and C.
(A) Pailer, Ganzinger, Monatsh. Chem., **106**, 37 (1975)
(S) Saner et al., Helv. Chim. Acta, **53**, 221 (1970)

M. erecta R.Br.
STEROIDS: marsdenin, marsectohexol, marsectohexol-11,12-diacetate.
Saner et al., Helv. Chim. Acta, **53**, 221 (1970)

M. flavescens.
STEROIDS:
12-O-acetyl-3β,8β,12β,14β,20-pentahydroxypregn-5-en-10-one, flavescin.
Duff et al., Phytochemistry, **12**, 2943 (1973)

M. formosana.
STEROIDS: glycosides MF-A, MF-B, MF-C and MF-D.
TRITERPENOIDS: marsformol, marsformosanone, marsformoxides A and B, 9(11),12-ursadien-3-one.
(S) Ito et al., Chem. Pharm. Bull., **26**, 3189 (1978)
(T) Lai, Ito, ibid., **27**, 2248 (1979)

M. rostrata.
STEROIDAL ALKALOIDS: dihydrorostratamine, rostratamine.
Gellert, Simmons, Aust. J. Chem., **27**, 919 (1974)

M. sinensis.
ALIPHATIC: hentriacontane.
TRITERPENOIDS: α-amyrin acetate, lupenyl acetate.
Du et al., Zhongcaoyao, **17**, 290 (1986); Chem. Abstr., 105 168930p

M. tenacissima (Roxb.) Wight et Arn.
STEROIDS: drevogenin Q, tenacigenin, tenacigenins A, B and C, tenascogenin, tenacissigenin.
Singhal et al., Indian J. Chem., **19B**, 178 (1980)
Luo et al., Huaxue Xuebao, **40**, 321 (1982); Chem. Abstr., 97 123892c

M. tomentosa Decne.
STEROIDAL ALKALOID: tomentomine.
STEROIDS: deacetyldehydrotomentodin, deacetylkidjoladinin, deacetyltomentosin, dehydrotomentosin, kidjoladinin, kidjolanin, penupogenin 20-O-acetate, tomentin, tomentogenin, tomentogenin 12-O-acetate, tomentogenin-12-(2-methyl-2-butenoate)-20-O-acetate, tomentogenin-12-(2-methyl-2-propenoate)-20-acetate, tomentonin, tomentosin.
(A) Sato et al., Chem. Pharm. Bull., **25**, 876 (1977)
(S) Mitsuhashi et al., Chem. Pharm. Bull., **13**, 267 (1965)
(S) Sato et al., ibid., **25**, 611 (1977)

MARSILIA (Marsiliaceae)

M. minuta L.
TRITERPENOID: marsilagenin A.
Chakravarti et al., Tetrahedron, **31**, 1781 (1975)

MARSUPELLA (Hepaticae)

M. emarginata (Ehrh.) Dum. subsp. tubulosa.
SESQUITERPENOIDS: (+)-marsupellol, (-)-marsupellol.
Matsuo et al., Chem. Lett., 73 (1979)

MARTENSIA (Rhodophyceae)

M. fragilis.
INDOLE ALKALOIDS: fragilamide, martensine A, 10-epi-martensine A, martensine B.
Kirkup, Moore, Tetrahedron Lett., 2087 (1983)

MATRICARIA (Compositae)

M. caucasica Poir.
Acetylenic compounds.
Bohlmann et al., Chem. Ber., **100**, 611 (1967)

M. chamomilla L.
SESQUITERPENOIDS: (-)-bisaboloxides A, B and C.
Sorm et al., Collect. Czech. Chem. Commun., **16**, 626 (1951)
Stilcher et al., Planta Med., **23**, 132 (1973)

M. discoidea DC.
COUMARINS: herniarin, umbelliferone.
FLAVONOIDS: cynaroside, luteolin.
Provoskii et al., Khim. Prir. Soedin, 712 (1985)

M. inodora L.
MONOTERPENOID: (2E,8Z)-2,8-decadiene-4,6-diynoic acid.
Bohlmann *et al.*, *Chem. Ber.*, **100**, 611 (1967)

M. oreades Boiss.
Acetylenic compounds.
Bohlmann *et al.*, *Chem. Ber.*, **100**, 611 (1967)

M. radix.
SESQUITERPENOID: chamomillol.
Reichling *et al.*, *Z. Naturforsch.*, **38C**, 159 (1983)

M. suffruticosa.
SESQUITERPENOID: 9-acetoxyparthenolide.
Bohlmann *et al.*, *Chem. Ber.*, **108**, 437 (1975)

M. zuurbergensis.
SESQUITERPENOID: zuurbergenin.
Bohlmann, Zdero, *Phytochemistry*, **16**, 136 (1977)

MATTEUCIA (Athyriaceae)

M. orientalis.
BENZOPYRAN: matteucinol.
Arthur *et al.*, *J. Chem. Soc.*, 3197 (1960)

MATTHIOLA (Cruciferae)

M. incana R.Br.
FLAVONOIDS: kaempferol 7-glucoside, kaempferol
3-rhamnoarabino-7-rhamnoside, kaempferol
rhamnogluco-7-rhamnoside, kaempferol 7-rhamnoside.
Matkawska *et al.*, *Herba Pol.*, **22**, 138 (1976); *Chem. Abstr.*, 86 68367f

MAYACA (Mayacaceae)

M. fluviatilis.
FLAVONOIDS: kaempferol 3-O-glucoside, luteolin
5-O-glucoside, quercetin 3-O-glucoside, quercetin
3-O-rutinoside.
Roberts, Haynes, *Phytochemistry*, **24**, 3077 (1985)

MAYTENUS (Celastraceae)

M. arbutifolia.
MACROCYCLIC ALKALOIDS: celacinnine, celallocinnine.
Kupchan *et al.*, *J. Chem. Soc., Chem. Commun.*, 329 (1974)

M. buchananii.
MACROCYCLIC ALKALOIDS: maysenine, maysine, maytanvaline,
normaysine.
Kupchan *et al.*, *J. Org. Chem.*, **42**, 2349 (1977)

M. buxifolia.
MACROCYCLIC ALKALOIDS: N(1)-acetyl N(1)-deoxymayfoline,
mayfoline.
Diaz, Ripperger, *Phytochemistry*, **21**, 255 (1982)

M. chuchuhuasca.
TRITERPENOIDS: pristimerin, tingenone.
Martinod *et al.*, *Phytochemistry*, **15**, 562 (1976)

M. dispermus.
DITERPENOIDS: dispermol, dispermone, maytenoquinone.
Martin, *Tetrahedron*, **29**, 2553 (1973)

M. diversifolia.
TRITERPENOIDS: 3,22-dihydroxy-12-oleanen-29-oic acid,
maytenfolic acid, maytenfolioln, maytensifolin A.
Lac *et al.*, *Tetrahedron Lett.*, 707 (1984)

M. guianensis.
PHENOLICS: 4'-O-methyl-(-)-epigallocatechin, proanthocyanidin
A.
STEROIDS: β-sitostenone, β-sitosterol.
TRITERPENOID: friedelan-3,7-dione.
De Sousa *et al.*, *Phytochemistry*, **25**, 1776 (1986)

M. hookeri Loes.
MACROCYCLIC ALKALOID: maytanbutine.
Zhou *et al.*, *Huaxue Xuebao*, **39**, 933 (1982)

M. ilicifolia.
TRITERPENOIDS: maitenin, maytensin.
Delle Monache *et al.*, *Gazz. Chim. Ital.*, **102**, 317 (1972)

M. mossambicensis.
MACROCYCLIC ALKALOIDS: hydroxyisocelabenzine,
isocyclocelabenzine.
Wagner, Burghes, *Helv. Chim. Acta*, **65**, 739 (1982)

M. myrsinoides.
MACROCYCLIC ALKALOIDS: acetylmaymyrsine, maymyrsine.
Baudouin *et al.*, *Heterocycles*, **22**, 2221 (1984)

M. ovatus Loes.
MACROCYCLIC ALKALOIDS: maytine, maytoline.
Kupchan *et al.*, *J. Am. Chem. Soc.*, **92**, 6667 (1970)

M. rigida.
TRITERPENOIDS: 11-hydroxy-20(29)-lupen-3-one, rigidenol.
Marta *et al.*, *Gazz. Chim. Ital.*, **106**, 61 (1979)

M. senegalensis.
TRITERPENOID: maytenonic acid.
Abraham *et al.*, *J. Pharm. Sci.*, **64**, 1085 (1971)

M. serrata.
PERHYDRONAPHTHALENE ALKALOID: maytolidine.
MACROCYCLIC ALKALOIDS: maytanbutacine, maytanbutine,
maytanprine, maytansine.
Kupchan, Smith, *J. Org. Chem.*, **42**, 115 (1977)
Kupchan, Smith, *ibid.*, **42**, 2349 (1977)

MECANOPSIS (Papaveraceae)

M. cambrica (L.) Vig.
APORPHINE ALKALOID: mecambroline.
PROTOBERBERINE ALKALOID: mecambridine.
SPIROISOQUINOLINE ALKALOIDS: fungipavine, mecambrine.
Slavik, Slavikova, *Collect. Czech. Chem. Commun.*, **28**, 1720 (1963)

MEDICAGO (Leguminosae)

M. rugosa L.
FLAVONOIDS: (-)-isosativan, (-)-medicarpin, (-)-vestitol,
(±)-vestitone.
PHYTOALEXIN: isosativanone.
Ingham, *Planta Med.*, **45**, 46 (1982)

M. sativa L.
COUMARIN: sativol.
TRITERPENOID: medicagenic acid.
(C) Kalra *et al.*, *Tetrahedron Lett.*, 2153 (1967)
(T) Walter *et al.*, *J. Am. Chem. Soc.*, **76**, 2271 (1954)

MEDICOSMA

M. cunninghamii.
FURANOQUINOLINE ALKALOID: medicosmine.
CHROMENE: alloevodionol.
(A) Lamberton, Price, *Aust. J. Chem.*, **6**, 173 (1953)
(C) Barnes *et al.*, *ibid.*, **17**, 975 (1964)

MEIOCARPUM

M. lepidatum.
PHENANTHRENE ALKALOIDS: methoxyatherospermine,
methoxyatherospermine N-oxide.
Leboeuf *et al.*, *Plant Med. Phytother.*, **11**, 284 (1973)

MELALEUCA

M. alternifolia.
SESQUITERPENE: viridiflorene.
Swords, Hunter, *J. Agric. Food Chem.*, **26**, 734 (1978)

M. cuticularis.
TRITERPENOID: malaleucic acid.
Arthur *et al.*, *Chem. Ind. (London)*, 926 (1956)

M. linearifolia.
SESQUITERPENOID: melalinol.
Davenport *et al.*, *Chem. Abstr.*, 44 3677

M. raphniphylla.
TRITERPENOID: malaleucic acid.
Arthur *et al.*, *Chem. Ind. (London)*, 926 (1956)

M. viminea.
TRITERPENOID: malaleucic acid.
Arthur *et al.*, *Chem. Ind. (London)*, 926 (1956)

M. viridiflora.
SESQUITERPENOID: viridiflorol.
Baker, Smith, *Proc. Roy. Soc., N.S.W.*, 205 (1913)

MELAMPODIUM (Compositae/Heliantheae)

M. americanum.
SESQUITERPENOIDS: 9-deacetylmelampodinin A, leucanthinin, melampodinin, melampodinins A, B and C.
Malcolm *et al.*, *Phytochemistry*, **21**, 151 (1982)

M. argophyllum.
SESQUITERPENOIDS: cinerenin, melampodin C.
Perry, Fischer, *J. Org. Chem.*, **40**, 3480 (1975)

M. cinereum.
SESQUITERPENOIDS: cinerenin, 4,5-dihydromelampodin B, melcanthins A, B and C, melnerin.
Klimash *et al.*, *Phytochemistry*, **20**, 840 (1981)

M. diffusum.
SESQUITERPENOID: melfusin.
Quijano *et al.*, *J. Nat. Prod.*, **44**, 266 (1985)

M. heterophyllum.
SESQUITERPENOID: 11,13-dihydromelampodin.
Dominguez *et al.*, *Phytochemistry*, **20**, 1431 (1981)

M. leucanthum.
SESQUITERPENOIDS: 9-acetoxymelnerin A, 9-acetoxymelnerin B, leucanthins A, B and C, leucanthinin, melampodin, melampodin B, melampodin C, melampodinin.
Perry, Fischer, *J. Org. Chem.*, **40**, 3480 (1975)
Oliver *et al.*, *ibid.*, **45**, 4028 (1980)

M. linearilobum.
SESQUITERPENOIDS: linearilobins A, B, C, D, E, F, G and H.
Seaman *et al.*, *Phytochemistry*, **19**, 849 (1980)

M. longicorne.
SESQUITERPENOIDS: longicornins A, B, C and D.
Malcolm *et al.*, *J. Chem. Soc., Chem. Commun.*, 2759 (1983)

M. longipilum.
SESQUITERPENOIDS: longipilin, longipin.
Seaman, Fischer, *Phytochemistry*, **17**, 2131 (1978)

M. perfoliatum.
DITERPENOID: 18-angeloyloxykaurenoic acid.
Bohlmann, Zdero, *Chem. Ber.*, **109**, 1670 (1976)

M. rosei.
SESQUITERPENOID: melrosin A.
Fischer *et al.*, *J. Chem. Soc., Chem. Commun.*, 1243 (1982)

MELAMPYRUM (Scrophulariaceae)

M. sylvaticum L.
TERPENOID: melampyroside.
Ahn, Pacholy, *Tetrahedron*, **30**, 4049 (1974)

MELANDRUM

M. album.
FLAVONOIDS: apigenin, homoorientin, isosaponarin, luteolin.
Zykova *et al.*, *Farm. Zh. (Kiev)*, 61 (1976); *Chem. Abstr.*, 85 156487e

M. firmum.
PHENOLICS: linarin, melandrin.
STEROID: α-spinasterylglucoside.
Woo *et al.*, *Phytochemistry*, **26**, 2099 (1987)

MELANOSELENIUM (Umbelliferae)

M. decipiens.
SESQUITERPENOIDS: decipienins A, B, C, D, E, F, G and H.
Gonzalez *et al.*, *Rev. Latinoam. Quim.*, **7**, 37 (1976)

MELANOXYLON

M. braunia.
ANTHRAQUINONE:
1,2,6,7-tetrahydroxy-8-methoxy-3-methylanthraquinone.
Gottlieb *et al.*, *Phytochemistry*, **10**, 1379 (1971)

MELASMA

M. arvense.
GLUCOSIDE: melasmoside.
Yang *et al.*, *Yunnan Zhiwu Yanjiu*, **5**, 213 (1983)

MELASTOMA (Melastomaceae)

M. bathricum.
TRITERPENOID: melastomic acid.
Manzoor-i-Khuda *et al.*, *J. Bangladesh Acad. Sci.*, **5**, 55 (1981)

M. intermedium Dunn.
PHENOLICS: gallic acid, pyromucic acid.
STEROIDS: β-sitosterol, sitosterol β-D-glucoside.
TRITERPENOID: oleanolic acid.
Yuan, *Zhongcaoyao*, **13**, 12 (1982); *Chem. Abstr.*, 97 212669c

MELIA (Meliaceae)

M. azadirach L.
UNCHARACTERIZED ALKALOIDS: azaridine, margosine, paraisine.
TERPENOIDS: 1-cinnamylmelianolone, kilactone, kulinone, kulolactone, meldenin, melianin A, melianin B, meliandiol, melianone, methylkulonate, minbolidin A, minbolidin B, minbolin A, minbolin B, ohchinin, ohchinin acetate, ohchinolal, ohchinolide A, ohchinolide B, solannin.
TOXINS: meliatoxins A-1, A-2, B-1 and B-2.
(A) Carratala, *Rev. Assoc. Med. Argent.*, **53**, 338 (1939)
(T) Chang *et al.*, *Tetrahedron*, **29**, 1911 (1973)
(T) Lee *et al.*, *Tetrahedron Lett.*, 3543 (1987)
(Tox) Kraus, Bokel, *Chem. Ber.*, **114**, 267 (1981)

M. azadirach L. var. japonica Makino.
LIMONOIDS: ohchinal, ohchinin acetate, ohchinolide A, ohchinolide B, sendalactone, sendanal, sendanin.
MONOTERPENOIDS: τ-cadinene, δ-cadinene, α-cadinol, δ-cadinol, camphene, citronellol, o-cresol, m-cresol, p-cresol, p-cymene, guaiacol, limonene, linalool, 1-p-menthene, α-pinene, β-pinene, terpinolene.
PHENOLIC: phenol.
Kameoka *et al.*, *Nippon Nogei Kagaku Kaishi*, **50**, 127 (1976); *Chem. Abstr.*, 85 2551j
Ochi *et al.*, *Chem. Lett.*, 1137 (1979)

MELIA (Meliaceae)

M. dubia.
 TETRANORTRITERPENOID: compositolide.
  Purushothaman et al., Phytochemistry, 23, 135 (1984)

M. indica.
 LIMONOIDS: nimbidinin, nimbin, nimbinin, nimbolide, salannic
  acid.
  Mitra et al., Phytochemistry, 10, 857 (1971)

M. toosenden.
 TERPENOIDS: 21-O-acetylisosendantriol, chuanliansu.
  Nakanishi et al., Chem. Lett., 69 (1986)

MELIANTHUS

M. comosus Vahl.
 STEROIDS: melianthugenin, melianthusigenin.
  Anderson, Koekenoer, J. S. Afr. Chem. Inst., 22, 5119 (1969)

M. major.
 ESTER: queretaroic caffeate.
 STEROIDS: β-sitosterol, β-sitosterol-β-D-glucoside.
  Agarwal, Rastogi, Phytochemistry, 15, 430 (1976)

MELICOPE (Rutaceae)

M. fareana F.Muell.
 FURANOQUINOLINE ALKALOID: acronycidine.
 QUINOLONE ALKALOIDS: melicopicine, melicopidine,
  melicopine.
  Price, Aust. J. Chem., A 2, 249 (1949)

M. leptococca.
 ACRIDONE ALKALOIDS: acronycidine, acronycine, acronydine,
  kokusaginine, melicopicine, melicopidine.
 CARBOLINE ALKALOIDS: 5-methoxy-N,N-dimethyltryptamine,
  6-methoxy-2-methyltetrahydro-β-carboline.
 INDOLE ALKALOID: 3-dimethylaminoacetyl-5-methoxyindole.
  Skaltsioumis et al., J. Nat. Prod., 46, 732 (1983)

M. octandra.
 PYRANS: melisimplexin, meliternatin, octandrenalone.
 STEROID: β-sitosterol.
 TERPENOID: multiflorenol.
  Free et al., Aust. J. Chem., 29, 695 (1976)

M. perspicuinervia.
 FURANOQUINOLINE ALKALOID: halfordinine.
  Murphy et al., Aust. J. Chem., 27, 187 (1974)

MELILOTUS

M. alba Colebr.
 PTEROCARPANS: melilotocarpans A, B, C, D and E.
  Miyase et al., Chem. Pharm. Bull., 30, 1986 (1982)

MELINIS (Gramineae)

M. multiflora.
 ACIDS: arachidic acid, behenic acid, lauric acid, linoleic acid,
  linolenic acid, myristic acid, oleic acid, palmitic acid,
  stearic acid.
 CAROTENOID: β-carotene.
 STEROIDS: campesterol, cholesterol, β-sitosterol, stigmasterol.
 TRITERPENOID: lupeol.
  Calle Alvarez et al., Chem. Abstr., 105 206296w

MELISSA

M. officinalis.
 MONOTERPENOIDS: caryophyllene, citronellal, geranyl acetate,
  linalool.
  Gasic et al., Arch. Farm., 35, 93 (1985); Chem. Abstr., 107 194945h

MELITTIS (Labiatae)

M. melissophyllum L.
 IRIDOIDS: 8-O-acetylharpagide, ajugol, ajugoside, melittoside.
  Guiso et al., Gazz. Chim. Ital., 104, 25 (1974)

MELOCHIA (Sterculiaceae)

M. corchorifolia.
 PEPTIDE ALKALOID: frangufoline.
 PYRIDINE ALKALOID: 6-methoxy-3-propenyl-2-pyridine
  carboxylic acid.
 FLAVONOIDS: hibifolin, trifolin.
 GLYCOSIDE: melocorin.
  (A) Tschesche, Reutel, Tetrahedron Lett., 3817 (1968)
  (A) Bhakuni et al., Chem. Ind. (London), 464 (1986)
  (F, G) Nair et al., Indian J. Chem., 15B, 1045 (1977)

M. pyramidata.
 PYRIDINE ALKALOID: melochinine.
 UNCHARACTERIZED ALKALOID: melochine.
  Medina, Spiteller, Chem. Ber., 112, 376 (1979)

M. tomentosa L.
 INDOLONE ALKALOIDS: melosatines A, B and C.
 PEPTIDE ALKALOIDS: melonovine A, melonovine B.
 QUINOLONE ALKALOIDS: melochinone, melovinone.
  Kapadia et al., Tetrahedron, 36, 2441 (1980)

MELODINUS (Apocynaceae)

M. aureus.
 INDOLE ALKALOID: (+)-20-epi-ibogamine.
  Naassou et al., Phytochemistry, 17, 1449 (1978)

M. australis (F.Muell.) Pierre.
 IMIDAZOLIDINOCARBAZOLE ALKALOIDS: 6β-hydroxykopsinine,
  14α-hydroxykopsinine, (+)-17-methoxyquebrachamine,
  venalstonidine.
 INDOLE ALKALOID: stemmadenine.
  Linde, Helv. Chim. Acta, 48, 1822 (1965)

M. balansae Baill.
 IMIDAZOLIDINOCARBAZOLE ALKALOIDS: baloxine,
  11-hydroxy-(-)-tabersonine, melobaline.
  Mehri et al., Bull. Soc. Chim. Fr., 3291 (1972)

M. balansae var. paucivenosus.
 BISIMIDAZOLIDINOCARBAZOLE ALKALOID: paucivenine.
  Mehri et al., Phytochemistry, 17, 1451 (1978)

M. celastroides.
 IMIDAZOLIDINOCARBAZOLE ALKALOID: (-)-buxomeline.
  Rabaron et al., Phytochemistry, 17, 1452 (1978)

M. insulae-pinorum Boiteau.
 INDOLE ALKALOIDS: 15α-hydroxykopsinine,
  19β-hydroxykopsinine, 19β-hydroxyvenalstonidine,
  insulopinine, isositsirikine, venalstonidine, (+)-venalstonine,
  14-vincanol, 16-epi-14-vincanol.
  Batchely et al., Ann. Pharm. Fr., 43, 359 (1985)

M. monogynus.
 TRITERPENOID: lupane-3β,20-diol.
  Chatterjee et al., J. Sci. Ind. Res. (India), 18B, 262 (1959)

M. panifera. Triterpenoid,
 lupane-3β,20-diol
  Chatterjee et al., J. Sci. Ind. Res. (India), 18B, 262 (1959)

M. scandens Forst.
 IMIDAZOLIDINOCARBAZOLE ALKALOIDS: meloscandonine,
  (+)-meloscine, (±)-epi-meloscine, epi-meloscine N-oxide,
  scandine.
 BISIMIDAZOLIDINOCARBAZOLE ALKALOIDS: scandomeline,
  epi-scandomeline, scandomelonine, epi-scandomelonine.
  Daudon et al., J. Org. Chem., 41, 3275 (1976)

MENEGAZZIA (Hepaticae)

M. asahinae (Yas. ex Zahlbr.) Sant.
DEPSIDONES: atranorin, constictic acid, menegassaic acid, stictic acid.
PHENOLIC: chloroatranorin.
Hirayama et al., Chem. Pharm. Bull., 24, 2340 (1976)

M. dispora.
DEPSIDONE: acetylhypoconsticic acid.
Elix et al., Aust. J. Chem., 38, 1735 (1985)

M. terebrata (Hoffm.) Massal.
DEPSIDONES: atranorin, constictic acid, menegazzaic acid, stictic acid.
PHENOLIC: chloroatronorin.
Hirayama et al., Chem. Pharm. Bull., 24, 2340 (1976)

MENISPERMUM (Menispermaceae)

M. canadense L.
BISBENZYLISOQUINOLINE ALKALOIDS: acutumine, dauricine.
Doskotch, Knapp, Lloydia, 34, 292 (1971)

M. dauricum DC.
BISBENZYLISOQUINOLINE ALKALOIDS: dauricine, dauricinoline, dauricoline, daurinoline, tetrandrine.
APORPHINE ALKALOID: dauriporphine.
HALOGENATED ALKALOIDS: acutumidine, acutuminine.
OXOISOAPORPHINE ALKALOIDS: bianfugecine, bianfugedine, bianfugenine.
PHENANTHRENE ALKALOID: sinomenine.
Tomita et al., J. Pharm. Soc., Japan, 90, 1178 (1970)
Takani et al., Chem. Pharm. Bull., 31, 3091 (1983)
Hou, Xue, Yaoxue Xuebao, 20, 112 (1985); Chem. Abstr., 103 3691g

MENTHA (Labiatae)

M. aquatica L.
FLAVONOIDS: acacetin, acacetin 7-rutinoside, apigenin, apigenin 7-rutinoside, apigenin 7-β-D-glucopyranoside, eriodictyol 7-β-D-glucopyranoside, eriodictyol 7-rutinoside, hesperetin 7-β-D-glucopyranoside, hesperitin, luteolin, luteolin 7-β-D-glucopyranoside, luteolin 7-rutinoside.
Burzanska-Hermann et al., Rocz. Chem., 51, 701 (1977)

M. arvensis L. var. piperascens.
MONOTERPENOID: (-)-2,5-trans-p-menthanediol.
Hashizue et al., Tetrahedron Lett., 3355 (1967)

M. citrata.
MONOTERPENOID: 1-vinylmenth-4(8)-ene.
Singh et al., Phytochemistry, 20, 19 (1981)

M. gattefosse.
MONOTERPENOIDS: (+)-menthone, (-)-menthone, (+)-pulegone.
Tuttle, Wallach, Justus Liebigs Ann. Chem., 277, 157 (1899)

M. gentilis.
MONOTERPENOID: (+)-1-acetoxy-p-menth-3-one.
Nagasawa et al., J. Agr. Chem. Soc., Japan, 51, 81 (1977)

M. piperiae.
ACIDS: caffeic acid, p-coumaric acid.
FLAVONOIDS: apigenin-7-glycoside, luteolin.
Kohlmunzer et al., Herba Pol., 21, 130 (1975); Chem. Abstr., 84 14683w

M. x piperita L.
MONOTERPENOIDS: (-)-menthol, (-)-menthone.
DITERPENOIDS: β-betulenol, (-)-bicycloelemene, mint sulphide.
Yoshida et al., J. Chem. Soc., Chem. Commun., 512 (1979)

M. piperita L. var. japonica.
MONOTERPENOID: 1-menthyl-β-D-glucoside.
Sakata, Mitsui, Agr. Biol. Chem., 39, 1329 (1975)

M. piperita L. var. kubamskaya.
FLAVONOIDS: dimethylsudachitin, 5-hydroxy-6,7,3',4'-tetramethoxyflavone.
Zakharova et al., Khim. Prir. Soedin., 645 (1983)

M. piperita L. var. prilwkskaya.
FLAVONOIDS: 5,7-dihydroxy-6,8,4'-trimethoxyflavone, dimethoxysudachitin, hymenotoxin, menthocubanone.
Zakharova et al., Khim. Prir. Soedin., 645 (1983)

M. piperita L. var. vulgaris.
MONOTERPENOID: menthofuran.
Weinhaus, Dequein, Angew. Chem., 47, 415 (1934)

M. pulegium L.
MONOTERPENOIDS: isopulegone, (-)-menthone, piperitenone.
Brunel, C. R. Acad. Sci., 140, 793 (1905)

M. rotundifolium (L.) Hudson.
MONOTERPENOID: rotundifolone.
Shimizu et al., Agr. Biol. Chem., 30, 89 (1966)

M. sylvestris.
MONOTERPENOID: (+)-pulegone.
Kon, J. Chem. Soc., 1616 (1930)

MENTZELIA

M. decapetala.
IRIDOIDS: decapetaloside, loasaside, mentzeloside, strictoside.
El-Naggar et al., J. Nat. Prod., 45, 539 (1982)

MENYANTHES (Menyanthaceae)

M. trifoliata L.
UNCHARACTERIZED ALKALOID: gentiatibetine.
FLAVONOIDS: hyperoside, rutin.
GLUCOSIDE: menthiafolin.
(A) Rulko et al., Rocz. Chem., 43, 1831 (1969)
(F) Mel'chakova et al., Khim. Prir. Soedin., 106 (1976)
(G) Battersby et al., J. Chem. Soc., Chem. Commun., 1277 (1968)

MERATIA

M. praecox Rehd. et Wils.
QUINOLINE ALKALOID: calycanthine.
Takamizawa et al., Tetrahedron, 23, 2959 (1967)

MERCURIALIS (Euphorbiaceae)

M. leiocarpa.
DIMERIC ALKALOID: 3,3'-bis(1,1'-dimethyl-2,2'-dioxo-4,4'-dimethoxy-5,5'-dihydroxy-5,5'-dimethoxycarbonyl-3-pyrroline).
Masui et al., Phytochemistry, 25, 1470 (1986)

M. perennis L.
CHROMOGEN: hermidin.
Swan, J. Chem. Soc., Perkin I, 1757 (1985)

MERENDERA (Liliaceae)

M. bulbocodium Ram.
COLCHICINE ALKALOID: colchicine.
FLAVONOID: multiflorin.
Pijewska et al., Collect. Czech. Chem. Commun., 32, 158 (1967)

M. jolanta E.Czerniak.
COLCHICINE ALKALOIDS: colchameine, colchamine, 2-demethylcolchicine, 3-demethylcolchicine, α-lumicolchicine, jolantamine, jolantidine.
Chommadov et al., Khim. Prir. Soedin., 790 (1983)

M. raddeana Rgl.

COLCHICINE ALKALOIDS: colchameine, colchamine, colchicine,
2-demethylcolchicine, 3-demethylcolchicine,
2-demethyl-α-lumicolchicine, 3-demethyl-α-colchicine,
N-formyldeacetylcolchicine, α-lumicolchicine, meradinine,
merenderine.
Canonica et al., Chem. Ind., (Milan), **49**, 1304 (1967)

M. robusta Bge.

COLCHICINE ALKALOIDS: colchameine, colchamine, colchiceine,
colchicine, deacetylcolchiceine, deacetylcolchicine,
3-demethylcolchamine, 2-demethylcolchicine,
3-demethylcolchicine, 3-demethyl-α-lumicolchicine,
N-formyldeacetylcolchicine, α-lumicolchicine,
β-lumicolchicine, N-methylcolchamine.
Salem et al., Collect. Czech. Chem. Commun., **28**, 3413 (1963)

M. trigina.

ISOQUINOLINE ALKALOID: merenderine.
Badger et al., J. Chem. Soc., 445 (1960)

MERIANDRA (Labiatae)

M. benghalensis (Roxb.) Benth.

FLAVONE: penduletin.
MONOTERPENOID: camphor.
SESQUITERPENOIDS: benghalensin A, benghalensin B,
benghalensitriol.
TRITERPENOIDS: 2α,3α-dihydroxyolean-12-en-28-oic acid,
2α,3β-dihydroxyolean-12-en-28-oic acid,
2α,3α-dihydroxyurs-12-en-28-oic acid,
2α,3β-dihydroxyurs-12-en-28-oic acid, oleanolic acid,
ursolic acid.
Perales et al., J. Org. Chem., **48**, 5318 (1983)

MERILLIA (Rutaceae)

M. caloxylon.

CHALCONE: 2′,3-dihydroxy-4,4′,6′-trimethoxychalcone.
Sesquiterpenoid eupatorin.
(C) Fraser et al., Phytochemistry, **13**, 1561 (1974)
(T) Adams, Lewis, Planta Med., **32**, 86 (1977)

MERISTROTROPIS

M. triphylla.

TRITERPENOID: meristrotropic acid.
Kir'yalev, Nangol'naya, J. Gen. Chem., USSR., **33**, 694 (1963)

MESEMBRYANTHEMUM (Aizoiaceae)

M. anatomicum Harv.

INDOLONE ALKALOID: mesembrine.
Popelak et al., Naturwissenschaften, **47**, 156 (1960)

M. expansum L.

INDOLONE ALKALOID: mesembrine.
Popelak et al., Naturwissenschaften, **47**, 156 (1960)

M. myrtifolia.

TRITERPENOID: myrtifolic acid.
Gunasekera, Sultanbawa, J. Chem. Soc., Perkin I, 6 (1977)

M. tortuosum L.

INDOLONE ALKALOID: mesembrine.
QUINOLONE ALKALOID: mesembrinine.
Bodendorf, Krieger, Arch. Pharm., **290**, 441 (1957)

MESPILUS (Rosaceae)

M. germanica L.

FLAVONOIDS: isoquercitrin, quercetin, quercetin
3-α-L-(4-O-acetyl)-rhamnopyranosido-7-α-L-
rhamnopyranoside, quercetin 3,7-di-α-L-rhamnopyranoside,
quercetin 3-β-D-glucopyranosido-7-α-L-rhamnopyranoside,
quercitrin.
Nikolev et al., Chem. Abstr., 105 168913k

MESUA (Guttiferae/Clusiaceae)

M. ferrea L.

COUMARIN: ferruol A.
XANTHONES: 1,5-dihydroxyxanthone,
1,3,6-trihydroxy-7,8-dimethoxyxanthone,
1,5,6-trihydroxyxanthone.
(C) Govindachari et al., Tetrahedron, **23**, 4161 (1967)
(X) Locksley et al., J. Chem. Soc., C, 1332 (1971)

M. salicina Planch et Triana.

XANTHONES: 1,5-dihydroxyxanthone,
3,7-dihydroxy-1,5,6-trimethoxyxanthone,
5,6-dihydroxy-1,3,7-trimethoxyxanthone,
1,5,6-trihydroxyxanthone.
Gunasekera et al., J. Chem. Soc., Perkin I, 2447 (1975)

M. thwaitesii Planch et Triana.

COUMARIN: cyclomammeisin.
XANTHONES: 1,5-dihydroxyxanthone, 1,5,6-trihydroxyxanthone.
(C) Crombie et al., J. Chem. Soc., Perkin I, 2248 (1972)
(X) Locksley et al., J. Chem. Soc., C, 1332 (1971)
(X) Bandaranake et al., Phytochemistry, **14**, 265 (1975)

METANARTHECIUM (Liliaceae)

M. luteo-viride Maxim.

STEROIDS:
2β-acetoxy-3α-hydroxy-11α-(tri-O-acetyl-α-L-
arabinopyranosyl)oxy-5β-25R-spirostane,
2β-acetoxy-3β-hydroxy-11α-(tri-O-acetyl-α-L-
arabinopyranosyl)oxy-5β-25R-spirostane,
11-O-α-L-arabinopyranosylprotometeogenin,
11--galactosylnogiragenin, luvigenin, metagenin,
3-epi-metagenin, meteogenin, neonogiragenin, nogiragenin.
Yosioka et al., Tetrahedron, **30**, 2283 (1974)
Kitagawa et al., Tetrahedron Lett., 1885 (1976)

METAPLEXIS (Asclepiadaceae)

M. japonica Makino.

STEROIDS: gagaimol dibenzoate, gagaimol 7-methyl ether,
7-hydroxy-12-O-benzoyldeacetylmetaplexigenin,
3β-methoxysarcostin, 7α-methylsarcostin.
Nomura et al., Planta Med., **41**, 206 (1981)

METASEQUOIA (Taxodiaceae)

M. glyptostroboides Hu et Cheng.

FLAVONES: amentoflavone, amentoflavone-4′,7-dimethyl ether,
amentoflavone-4″,7″-dimethyl ether,
amentoflavone-4′,4″,7″-trimethyl ether,
2,3-dihydrohinokiflavone, ginkgetin, sciadopitysin,
sotetsuflavone.
LIGNANS: agatharesinol, athrotaxin, hydroxyathrotaxin,
hydroxymetasequirin A, metasequirin A, metasequirin B.
DITERPENOID: 3-acetoxylabd-8(20),13-dien-15-oic acid.
(F) Beckmann et al., Phytochemistry, **10**, 1465 (1971)
(L) Enoki et al., Chem. Abstr., 88 71421r, 71422s
(T) Braun, Breiterbach, Tetrahedron, **33**, 145 (1977)

METROSIDEROS (Myrtaceae)

M. umbellata.
SESQUITERPENOIDS: α-metrosiderene, β-metrosiderene.
Corbett, Hanger, *J. Chem. Soc.,* 1179 (1954)

METZGERIA

M. furcata.
FLAVONOID: isofurcatain.
Markham *et al., Z. Naturforsch.,* **37C**, 562 (1982)

MICARIA (Lichenes)

M. prasina L.
DEPSIDE: prasinic acid.
PHENOLICS: methoxymicareic acid, micareic acid.
Elix *et al., Aust. J. Chem.,* **37**, 2346 (1984)

MICHELIA (Magnoliaceae)

M. alba.
APORPHINE ALKALOIDS: norushinsunine, ushinsunine.
Tomita, Furukawa, *J. Pharm. Soc., Japan,* **82**, 925 (1962)

M. champaca L.
APORPHINE ALKALOID: ushinsunine.
Tomita, Furukawa, *J. Pharm. Soc., Japan,* **82**, 925 (1962)

M. compressa.
APORPHINE ALKALOID: shinsunine.
QUATERNARY ALKALOID: michepressine.
SESQUITERPENOIDS: compressanolide, michelenolide, micheliolide.
(A) Tomita, Furukawa, *J. Pharm. Soc., Japan,* **82**, 925 (1962)
(T) Ogura *et al., Phytochemistry,* **17**, 957 (1978)

M. lanuginosa Wall.
APORPHINE ALKALOID: michelangine.
OXOAPORPHINE ALKALOID: lanuginosine.
SESQUITERPENOID: lanuginolide.
(A) Talapatra *et al., Chem. Ind.,* **31**, 1056 (1969)
(T) Talapatra *et al., J. Chem. Soc., Chem. Commun.,* 1534 (1970)

MICRANDROPSIS

M. scleroxylon.
PHENANTHROID: micrandrol C.
de Alvarenga *et al., Phytochemistry,* **13**, 1283 (1974)

MICROCIONA

M. toxistyla.
SESQUITERPENOIDS: microcionins 1, 2, 3 and 4, toxistylide A, toxistylide B.
Cimino *et al., Tetrahedron Lett.,* 3723 (1975)
Cimino *et al., ibid.,* 3619 (1979)

MICROGLOSSA (Compositae)

M. zeylandica.
FLAVONE: 5,4'-dihydroxy-6,7,8,3'-tetramethoxyflavone.
SESQUITERPENOIDS: caryophyllen-1,10-epoxide, farnesene.
DITERPENOIDS: dihydromicroglossic acid, microglossic acid.
TETRATERPENOID: squalene.
TRITERPENOID: dammadienyl acetate.
Gunatilaka *et al., Phytochemistry,* **26**, 2408 (1987)

MICROLEPIS

M. marginata.
DITERPENOID GLUCOSIDES: microlepin, 6-epi-microlepin.
Tanaka *et al., Chem. Pharm. Bull.,* **24**, 2891 (1976)

MICROMELUM (Rutaceae)

M. minutum (Forst.f.) Seem.
SESQUITERPENOID: micromelin.
Lamberton *et al., Aust. J. Chem.,* **20**, 973 (1967)

M. zeylanicum.
PYRIDINE ALKALOIDS: O-methylhalfordine, O-methylhalfordinol.
Bowen, Perera, *Phytochemistry,* **21**, 433 (1982)

MICROMERIS (Labiatae)

M. benthami.
TRITERPENOIDS: 3,19-dihydroxyurs-12-en-28-oic acid, micromeric acid.
Brieskorn, Wunderer, *An. Quim.,* **64**, 175 (1968)

M. chamaissonis.
FLAVONE: xanthomicrol.
Stout *et al., Tetrahedron.,* **14**, 296 (1961)

MICROPLUMERIA

M. anomala.
IMIDAZOLIDINOCARBAZOLE ALKALOID: anomaline.
Reis Luiz *et al., Phytochemistry,* **22**, 2301 (1983)

MIKANIA (Compositae)

M. alvimii.
DITERPENOID: ent-14-oxo-15-nor-8(17),12-labdadien-18-oic acid.
Bohlmann *et al., Phytochemistry,* **21**, 173 (1982)

M. banisteriae.
DITERPENOIDS: 18-acetoxy-ent-kaurene, 4β,19-epoxy-18-nor-ent-kaurene, 18-hydroxy-16α,17-epoxy-ent-kaurene, ent-kaur-16-en-18-al.
Castro, Jakupovic, *Phytochemistry,* **24**, 2450 (1985)

M. batatifolia.
FLAVONE: batatifoline.
Herz *et al., Chem. Ber.,* **103**, 1822 (1970)

M. cordata.
FLAVONE: mikanin.
Kiang *et al., Phytochemistry,* **7**, 1035 (1968)

M. cordifolia.
SESQUITERPENOID: micordilin.
Herz *et al., J. Org. Chem.,* **42**, 1720 (1977)

M. goyazensis.
SESQUITERPENOIDS: 4,5-epoxy-1-desoxodeacetylchrysanolide, 1,10-epoxy-6-hydroxyinunolide.
Bohlmann *et al., Phytochemistry,* **21**, 1349 (1982)

M. micrantha.
SESQUITERPENOID: mikanokryptin.
Herz *et al., Phytochemistry,* **14**, 233 (1975)

M. monogasensis.
SESQUITERPENOID: dihydromikanolide.
Herz *et al., Tetrahedron Lett.,* 3111 (1967)

M. oblongifolia.
DITERPENOID: cinnamoylgrandifloric acid.
Vichnewski *et al., Phytochemistry,* **16**, 2028 (1977)

M. pohlii.
SESQUITERPENOIDS: 1,10-epoxydesacetyllaurenobiolide, 1,10-epoxy-6-hydroxyinunolide.
Bohlmann *et al., Phytochemistry,* **21**, 1349 (1982)

M. purpurascens.
MONOTERPENOID: 9-acetoxy-8,10-epoxy-2-methoxythymol tiglate.
SESQUITERPENOID: purpurascenolide.
    Bohlmann et al., *Phytochemistry*, **21**, 705 (1982)

M. scandens.
SESQUITERPENOIDS: dihydromikanolide, mikanolide, miscandenin.
    Herz et al., *J. Org. Chem.*, **35**, 1453 (1970)

MILLETIA (Leguminosae)

M. auriculata.
FLAVONES: auriculasin, auriculatin, auriculin, aurmillone, isoauriculasin.
ISOFLAVONE: psitectorigenin.
ROTENOID: summatrol.
    (F, R) Kapoor et al., *Tetrahedron*, **32**, 749 (1976)
    (I) Dhingra et al., *Indian J. Chem.*, **12**, 1118 (1974)

M. dura.
BENZOPYRAN: tephrosin.
ISOFLAVONES: durlettone, durmillone, milldurone.
    (B) Olis et al., *Tetrahedron*, **23**, 4741 (1967)
    (I) Campbell et al., *J. Chem. Soc., C*, 1787 (1969)

M. ovalifolia.
PIPERIDINE ALKALOID: ovaline.
CHALCONES: ovalichalcone A, ovalitenin A, ovalitenin B, ovalitenone.
FLAVANONES: 7-hydroxy-6,8-di-C-prenylflavanone, 7-hydroxy-8-C-prenylflavanone.
    (A) Gupta, Krishnamurti, *Phytochemistry*, **18**, 2021 (1979)
    (C) Gupta, Krishnamurti, *Indian J. Chem.*, **17B**, 291 (1979)
    (F) Gupta et al., *Phytochemistry*, **15**, 832 (1976)

M. pachycarpa.
ROTENOIDS: cis-12α-hydroxyrotenone, cis-12α-hydroxyrot-2-enonic acid, rotenone, rot-2'-enonic acid.
    Singhal et al., *Phytochemistry*, **21**, 949 (1982)

M. pendula.
QUINONE: claussequinone.
    Gottlieb et al., *Phytochemistry*, **14**, 2495 (1975)

MIMOSA (Mimosaceae/Leguminosae)

M. caesalpiniaefolia.
GLYCOSIDE: 3-O-arabinosylmorolic acid.
    De Alencar et al., *Rev. Latinoam. Quim.*, **7**, 44 (1976)

M. pudica.
PYRIDINE ALKALOID: mimosine.
    Renz, *Zeit. Physiol. Chem.*, **244**, 153 (1936)

M. rubicaulis.
STEROID: β-sitosterol.
TRITERPENOIDS: β-amyrin, friedelin.
    Kumar et al., *Curr. Sci.*, **44**, 889 (1975)

MIMULUS (Scrophulariaceae)

M. guttatus DC.
CAROTENOIDS: deepoxyneoxanthin, mimulaxanthin.
    Nitsche, *Z. Naturforsch.*, **28C**, 481 (1973)

MIMUSOPS (Sapotaceae)

M. elengi.
TRITERPENOID: bassic acid.
    King et al., *J. Chem. Soc.*, 1338 (1958)

M. heckelii.
SAPONIN: makore saponin.
TRITERPENOID: bassic acid.
    King et al., *J. Chem. Soc.*, 1338 (1958)

M. littoralis Kurz.
FLAVONOIDS: quercetin, quercetol.
TRITERPENOIDS: β-amyrin, bassic acid.
    Dixit, Srivastava, *Indian J. Pharm.*, **39**, 85 (1977)

MINNEOLA

M. tangor. (Citrus reticulaya x C. sinensis).
CAROTENOID: reticulataxanthin.
    Yokoyama et al., *J. Org. Chem.*, **30**, 2482 (1965)

MIOPORUS

M. insulare.
IRIDOIDS: acetylharpagide, mioporoside.
    Bianco et al., *Gazz. Chim. Ital.*, **105**, 175 (1975)

MIRABILIS (Nyctaginaceae)

M. jalapa L.
ALIPHATICS: n-hexacosanol, 12-triacosanone.
AMINO ACIDS: alanine, leucine, glycine, tryptophan, valine.
STEROIDS: β-sitosterol, β-sitosterol β-D-glucoside.
TRITERPENOIDS: β-amyrin, β-amyrin-α-L-rhamnosyl-O-β-D-glucoside.
    Behari et al., *Collect. Czech. Chem. Commun.*, **41**, 295 (1975)

MITRAGYNA (Rubiaceae)

M. ciliata Aubr. et Pellegr.
CARBOLINE ALKALOIDS: ciliaphylline, mitraciliatine.
TRITERPENOID: quinovic acid.
    (A) Soliman, *J. Chem. Soc.*, 1760 (1939)
    (T) Barton, de Mayo, *ibid.*, 3111 (1953)

M. diversifolia Hook f.
INDOLONE ALKALOID: rhyncophylline.
UNCHARACTERIZED ALKALOID: mitraversine.
    Field, *J. Chem. Soc.*, **119**, 887 (1921)
    Kondo et al., *J. Pharm. Soc., Japan*, **57**, 237 (1937)

M. hirsuta Havil.
CARBOLINE ALKALOID: hirsutine.
    Shellard et al., *J. Pharm. Pharmacol.*, **18**, 553 (1966)

M. inermis Kuntze. (M. africana Korth.)
INDOLONE ALKALOIDS: isorhyncophylline N-oxide, rhyncophylline, rhyncophylline N-oxide.
TRITERPENOID: quinovic acid.
    (A) Shellard et al., *Phytochemistry*, **10**, 2505 (1971)
    (T) Soliman, *J. Chem. Soc.*, 1760 (1939)

M. japonica var. microphylla.
INDOLONE ALKALOID: javaphylline.
    Shellard, Alam, *J. Chromatogr.*, **32**, 472 (1968)

M. javanica Koord.
CARBOLINE ALKALOIDS: mitrajavanine, mitrajavine.
INDOLONE ALKALOID: javaphylline.
    Shellard et al., *J. Pharm. Pharmacol.*, **18**, 533 (1966)

M. rotundifolia (Roxb.) Kuntze.
INDOLONE ALKALOIDS: isorhyncophylline N-oxide, mitragynol, rhyncophylline, rhyncophylline N-oxide, rotundifoline.
    Barger et al., *J. Org. Chem.*, **4**, 418 (1939)
    Shellard et al., *Phytochemistry*, **10**, 2505 (1971)

M. rubrostipulata.
INDOLONE ALKALOIDS: (-)-mitraphylline, (+)-mitraphylline, rotundifoline, anti-rotundifoline N-oxide, vinylrotundifoline.
TRITERPENOID: quinovic acid.
(A) Beaton *et al.*, *Can. J. Chem.*, **36**, 1031 (1958)
(A) Lala, *J. Inst. Chem.*, *(India)*, **53**, 30 (1981)
(T) Barton, de Mayo, *J. Chem. Soc.*, 3111 (1953)

M. speciosa Korth.
CARBOLINE ALKALOIDS: mitragynine, paynantheine, speciociliatine, speciogynine.
INDOLONE ALKALOIDS: isomitrophylline, mitragynine, speciofoline, speciophylline, uncarine D.
UNCHARACTERIZED ALKALOID: mitraspecine.
Beckett *et al.*, *Planta Med.*, **14**, 277 (1966)
Hemingway, Phillipson, *J. Pharm. Pharmacol.*, **24**, 169P (1972)

M. stipulosa Kuntze. (M. macrophylla Hiern.).
INDOLONE ALKALOID: rhynvcophylline.
Raymond-Hamet *et al.*, *C. R. Acad. Sci.*, **199**, 587 (1934)

MNIOBRYUM

M. vahlenbergii var. glacialis.
STEROIDS: campesterol, cholesterol, β-sitosterol, stigmasterol.
Salberg, *Chem. Abstr.*, 100 20441f

MOCHIA

M. velutina.
SESQUITERPENOID: arbusculin A.
Irwin *et al.*, *Phytochemistry*, **8**, 2411 (1969)

MOGHANIA

M. macrophylla.
ANTHOCYANIN: procyanidin.
ISOFLAVONE: 5,7,2′,4′-tetrahydroxyisoflavone.
STEROIDS: β-sitosterol, β-sitosterol glucoside.
TRITERPENOIDS: α-amyrin, lupeol.
Prasad *et al.*, *Phytochemistry*, **16**, 1120 (1977)

M. strobilifera.
STEROIDS: β-sitosterol, stigmasterol.
Cheb *et al.*, *J. Chin. Chem. Soc.*, **23**, 111 (1976); *Chem. Abstr.*, **85** 74977a

MOLLUGO

M. hirta.
TRITERPENOIDS: mollugogenols A, B, C, D, E and F.
TRITERPENOID GLYCOSIDE: mollugocin A.
Chaudhury, Chakrabarti, *Indian J. Chem.*, **13**, 947 (1975)

M. spergula.
TRITERPENOIDS: spergulagenic acid, spergulagenin, spergulagenol, spergulatriol.
Kitagawa *et al.*, *Tetrahedron Lett.*, 2327 (1976)

MOMORDICA (Cucurbitaceae)

M. charantia L.
STEROIDS: elasterol, 5α-stigmasta-7,25-dien-3β-ol, 5α-stigmasta-7,22,25-trien-3β-ol.
TRITERPENOIDS: momordicosides A, B, C, D, E, F-1, F-2, G and I.
(S) Sucrow *et al.*, *Chem. Ber.*, **103**, **745**, 750 (1970)
(T) Okabe *et al.*, *Tetrahedron Lett.*, 77 (1982)

M. cochinochinensis Sprenger.
SAPONINS: momordins I, II and III.
Iwamoto *et al.*, *Chem. Pharm. Bull.*, **33**, 1 (1985)

MONACHOSORUM (Pteridaceae)

M. arakii Tagawa.
DINORSESQUITERPENOIDS: monachosorin A, monachosorin B, monachosorin C. mukagolactone.
Satake *et al.*, *Chem. Pharm. Bull.*, **33**, 4175 (1985)

MONARDA (Labiatae)

M. fistulosa L.
MONOTERPENOID: thymoquinone.
Liebermann, Iljinski, *Ber. Dtsch. Chem. Ges.*, **3194**, 3220 (1885)

MONASCUS

M. anka.
PIGMENT: ankaflavin.
Manchand *et al.*, *Phytochemistry*, **12**, 2531 (1973)

MONECHMA

M. debile.
TERPENOIDS: monechmol, monechmol 3-O-β-D-glucoside.
Ayoub *et al.*, *Planta Med.*, **50**, 520 (1984)

MONNIERA

M. cuneifolia Michx. (Herpestris monniera HB).
UNCHARACTERIZED ALKALOID: herpestrine.
Basu, Walia, *Indian J. Pharm.*, **6**, 85, (1944)
Basu, Walia, *ibid.*, **6**, 91 (1944)

M. trifolia.
FURANOQUINOLINE ALKALOIDS: delbine, montrifoline.
Bhattacharyya, Serur, *Heterocycles*, **16**, 371 (1981)

MONOCLEA

M. forsteri.
FLAVONE: 3′,4′-dihydroxywogonin.
Markham *et al.*, *Phytochemistry*, **11**, 2047 (1972)

MONODORA (Annonaceae)

M. angolensis.
APORPHINE ALKALOID: wilsonirine.
Johns *et al.*, *Aust. J. Chem.*, **23**, 363 (1970)

M. myristica.
INDOLE ALKALOID: 6-(3,3-dimethylallyl)-indole.
INDOLES: isomonodoroindole, monodoroindole.
(A) Waterman, Thomas, *Fitoterapia*, **57**, 58 (1986)
(I) Muhammad *et al.*, *J. Bangladesh Acad. Sci.*, **11**, 1 (1987); *Chem. Abstr.*, 107 172452t

M. tenuifolia.
INDOLE: 3-dimethylallylindole.
Adeoye *et al.*, *J. Nat. Prod.*, **49**, 534 (1986)

MONOGYNE

M. alpina.
SESQUITERPENOID: 2-(angeloyloxy)-bakkenolide A.
Harmatha *et al.*, *Collect. Czech. Chem. Commun.*, **41**, 2047 (1976)

MONOMORIUM

M. pharaonis.
TERPENOID: farnal.
Ritter *et al.*, *Tetrahedron Lett.*, 2617 (1977)

MONOTROPA (Pyrrolaceae)

M. hypopitys L.
IRIDOID: monotropein.
Bridel, *C. R. Acad. Sci.*, **176**, 1742 (1923)

MONTANOA

M. frutescens.
SESQUITERPENOID: montefrusin.

Quijano *et al.*, *Phytochemistry*, **18**, 843 (1979)

M. leucantha subsp. leucantha.
SESQUITERPENOID: leucanthanolide.

Oshima *et al.*, *J. Nat. Prod.*, **49**, 313 (1986)

M. speciosa.
SESQUITERPENOID: 1,2-dehydro-3-oxocostic acid

Seaman *et al.*, *Phytochemistry*, **24**, 607 (1985)

M. tomentosa.
DITERPENOIDS: (-)-kaur-9(11),16-dien-19-oic acid, kaurenoic acid, monogynoic acid, montanol, pretomentol, pretomexanthol, prezoapatanol, tomentol, tomexanthol, zoapatanol, zoapatanolide A, zoapatanolide C, zoapatanolide D.

Quijano *et al.*, *Phytochemistry*, **24**, 2337, (1985)
Quijano *et al.*, *ibid.*, **24**, 2741 (1985)

M. tomentosa subsp. xanthiifolia.
SESQUITERPENOIDS:
8α-hydroxy-9β-senecioyloxy-trans-germa-1(10),4-dien-cis-6,12-olide,
9β-hydroxy-8α-senecioyloxy-trans-germa-1(10),4-dien-cis-6,12-olide.

Castro *et al.*, *Phytochemistry*, **24**, 2449 (1985)

MOQUINEA

M. velutina.
SESQUITERPENOID: cyclocostunolide.

Tomassini, Gilbert, *Phytochemistry*, **11**, 1177 (1972)

MORA

M. excelsa Benth.
TRITERPENOID: morolic acid.

Barton, Brooks, *J. Chem. Soc.*, 257 (1951)

MORAEA (Iridaceae)

M. graminicola Oberm.
STEROID: 3-dehydro-16β-hydroxyobovogenin.

van Wyk, *J. S. Afr. Chem. Inst.*, **25**, 82 (1972)

M. polystachya Ker.
STEROID: 3-dehydro-16β-hydroxybovogenin.

Van Wyk, *J. S. Afr. Chem. Inst.*, **25**, 82 (1972)

MORARDA

M. didyma.
FLAVANONE: didymin.

Kariyone *et al.*, *J. Pharm. Soc., Japan*, **74**, 363 (1954)

MORCHELLA

M. esculenta.
AMINO ACID: cis-3-amino-L-proline.

Hatanaka *et al.*, *Mushroom Sci.*, **9**, 809 (1976); *Chem. Abstr.*, **86** 103099u

MORINDA (Rubiaceae)

M. angustifolia.
ANTHRAQUINONE: morindone.

Roberts *et al.*, *Aust. J. Chem.*, **30**, 1553 (1977)

M. citrifolia.
PIGMENTS: damnacanthol, nordamnacanthol, soranjidiol.

Bowie *et al.*, *Aust. J. Chem.*, **15**, 332 (1962)
Murti *et al.*, *Indian J. Chem.*, **10**, 246 (1972)

M. lucida.
PIGMENT: soranjidiol.
TERPENOID: oruwacin.

(P) Bowie *et al.*, *Aust. J. Chem.*, **15**, 332 (1962)
(T) Adesogan, *Phytochemistry*, **18**, 175 (1979)

M. tinctoria var. tomentosa.
PIGMENTS: alizarin-1-methyl ether, ibericin, morindone, morindone 6-primeveroside, rubiadin, soranjidiol.
TRITERPENOID: ursolic acid.

Rao *et al.*, *Indian J. Chem.*, **15B**, 497 (1977)
Rao *et al.*, *J. Indian Chem. Soc.*, **60**, 585 (1983)

M. umbellata.
PIGMENTS: 6-methylxanthopurpurin, soranjidiol.

Stoessl, *Can. J. Chem.*, **47**, 767 (1969)

MORITHAMNUS

M. crassus.
Three sesquiterpenoids.

Bohlmann *et al.*, *Phytochemistry*, **19**, 2769 (1980)

MORONOBEA (Guttiferae/Clusiaceae)

M. pulchra.
BENZOPHENONE: marupone.

Dias *et al.*, *Phytochemistry*, **13**, 1953 (1974)

MORTONIA

M. greggii.
SESQUITERPENOIDS: glucopyranosyldesacetylmortonol B, mortonin, mortonins A, B, C and D, mortonol A, mortonol B.

Martinez *et al.*, *Phytochemistry*, **23**, 1651 (1984)

MORUS (Moraceae)

M. alba L.
ACIDS: cis-5-hydroxy-L-pipecolic acid, trans-5-hydroxy-L-pipecolic acid.
BENZOPYRANS: albafuran C, albanin F, albanin G, albanol A, albanol B.
CHROMENES: mulberrin, mulberrochromin, mulberrofuran M, mulberrofuran Q.
FLAVONOIDS: kuwanons A, B, C, D, J, Y and Z.

(Ac) Hatanaka *et al.*, *Phytochemistry*, **16**, 1041 (1977)
(B) Takasugi *et al.*, *Chem. Lett.*, 1577 (1980)
(B) Rao *et al.*, *Tetrahedron Lett.*, 3013 (1983)
(C) Nomura *et al.*, *Heterocycles*, **9**, 1593 (1978)
(F) Uedaa *et al.*, *Chem. Pharm. Bull.*, **30**, 3042 (1982)

M. australis.
BENZOFURAN: mulberrofuran D.
FLAVONOIDS: cyclomorusin, kuwanon C, kuwanon G, kuwanon H, morusin, oxydihydromorusin.

Nomura *et al.*, *Planta Med.*, **49**, 90 (1983)

M. laevigata.
AMINO ACIDS: asparagine, tyrosine.

Ali *et al.*, *J. Indian Chem. Soc.*, **64**, 230 (1987)

M. lhou Koidz.
STILBENES: kuwanon P, kuwanon S, kuwanon T, kuwanon X, mulberrofuran H.

Kirakura *et al.*, *Chem. Pharm. Bull.*, **33**, 1088 (1985)
Fukai *et al.*, *ibid.*, **33**, 4288 (1985)

M. rubra.
FLAVONES: rubraflavone A, rubraflavone B.

Deshpande *et al.*, *Indian J. Chem.*, **12**, 431 (1974)

MOSTEA
M. brunonis Didr. var. brunonis.
CARBOLINE ALKALOID: mostueine.
INDOLONE ALKALOID: 20-(N-4)-dehydrogelsemicine.
Onanga et al., C. R. Acad. Sci., 291C, 191 (1980)

MUCUNA (Leguminosae)
M. pruriens.
UNCHARACTERIZED ALKALOIDS: mucunadine, mucunine.
Santra, Majumder, Indian J. Pharm., 15, 60 (1953)

MUEHLENBECKIA (Polygonaceae)
M. tamnifolia.
ANTHRAQUINONES: chrysophanic acid, emodin, rhein.
Martinod et al., Politecnica, 3, 111 (1973)

MUNCUNA
M. sloanei.
AMINE: L-dopa.
Rai et al., Curr. Sci., 46, 778 (1977)

MUNDULEA
M. sericea.
DIPYRAN: sericetin.
IMIDAZOLE: 2-aminoimidazole.
(D) Burrows et al., Proc. Chem. Soc. (London), 177 (1960)
(I) Fellows et al., Phytochemistry, 16, 1399 (1977)
M. suberosa.
DIPYRAN: sericetin.
Burrows et al., Proc. Chem. Soc. (London), 177 (1960)

MUNNOZIA
M. maronii.
SESQUITERPENOID: maroniolide.
Bohlmann, Grenz, Phytochemistry, 18, 334 (1979)

MUNTAFERA
M. sessilifolia.
HOMOCARBOLINE ALKALOID: (-)-3,6-oxidoisovoacangine.
INDOLE ALKALOID: (-)-6-hydroxy-3-oxoisovoacangine.
Panas et al., Phytochemistry, 14, 1120 (1975)

MURRAYA (Rutaceae)
M. elongata.
COUMARONE: murralongin.
Talapatra et al., Tetrahedron Lett., 5005 (1973)
M. euchrestifolia Hayata.
CARBAZOLE ALKALOIDS: bismurrayafoline A, bismurrayafoline B, (±)-murrafoline, (-)-murrafoline, murrayaline, murrayastine, pyrayafoline.
QUINONE ALKALOIDS: murrayaquinone A, murrayaquinone B, pyrayaquinone A, pyrayaquinone B.
Wu et al., Heterocycles, 20, 1267 (1983)
Furukawa et al., Chem. Pharm. Bull., 34, 2672 (1986)
M. exotica.
CARBAZOLE ALKALOID: exozoline.
CAROTENOIDS: semi-α-carotenone, semi-β-carotenone.
COUMARIN: murrayatin.
(A) Ganguly, Sarker, Phytochemistry, 17, 1816 (1978)
(C) Barik, Dey, ibid., 22, 2273 (1983)
(T) Buchecker et al., Helv. Chim. Acta, 53, 1210 (1970)

M. koenigii Spreng.
CARBAZOLE ALKALOIDS: bicyclomahanimbicine, bicyclomahanimbine, currayangine, cyclomahanimbine, girimbine, isomahanimbine, koenigine, koenimbidine, koenimbine, koenine, mahanimbicine, mahanimbidine, mahanimbine, mahanimbinine, mahanimbinol, mahanine, mukonidine, mukonine, mukonal, mukonicine, mukulodine, murrayacine, murrayanine, murrayazolidine, murrayazolinine.
Chakraborty et al., Phytochemistry, 17, 834 (1978)
Roy et al., J. Indian Chem. Soc., 59, 1369 (1982)
M. paniculata (L.) Jacq.
BISINDOLE ALKALOID: yuehchukine.
COUMARINS: coumurrayin, isomexoticin, meranzin hydrate, murpanicin, murpanidin, murralongin, murragatin, phebalosin.
(A) Kong et al., Jiegou Huaxue, 4, 30 (1985); Chem. Abstr., 104 31728p
(C) Raj et al., Phytochemistry, 15, 1787 (1976)
(C) Yang et al., Yaoxue Xuebao, 18, 760 (1983); Chem. Abstr., 100 64986s

MUSA (Musaceae)
M. x paradisiaca L.
STEROID:
(24R)-4,14,24-trimethyl-5α-cholesta-8,25(27)-dien-3β-ol.
Dutta et al., Phytochemistry, 22, 2563 (1983)
M. x sapientum L.
STEROIDS:
(24S)-14α,24-dimethyl-9β,19-cyclo-5α-cholest-25-en-3β-ol, methylenepollinasterol.
Knapp et al., Phytochemistry, 11, 3497 (1972)
Akihisa et al., Lipids, 21, 494 (1986)

MUSANGA
M. cecropioides.
TRITERPENOIDS: 2-acetyltormentic acid, 3-acetyltormentic acid, euscaphic acid, tormentic acid.
Ojinnaka et al., J. Nat. Prod., 48, 337 (1983)

MUSCARI (Liliaceae)
M. comosum (L.) Mill.
SAPONIN: muscaroside A.
TRITERPENOID:
(23S,24S)-17,23-epoxy-24,31-dihydroxy-27-nor-5α-lanost-8-ene-3,15-dione.
(Sap) Adinolfi et al., J. Nat. Prod., 46, 559 (1983)
(T) Adinolfi et al., Can. J. Chem., 61, 2633 (1983)

MUSSAENDA (Rubiaceae)
M. parviflora.
IRIDOIDS: mussaenoside, shanzhiside methyl ester.
Takeda et al., Phytochemistry, 16, 1401 (1977)
M. shikokiana.
IRIDOIDS: mussaenoside, shanzhiside methyl ester.
Takeda et al., Phytochemistry, 16, 1401 (1977)

MUTISIA (Compositae)
M. homoeantha.
SESQUITERPENOID: mitusianthol.
Bohlmann et al., Phytochemistry, 18, 99 (1979)

MYCENA

M. megaspora.
METHYL ETHER: drosophilin A methyl ether.
Van Eijk et al., *Phytochemistry,* **14**, 2506 (1975)

M. pura.
AMINO ACID: L-τ-propylideneglutamic acid.
Hatanaka et al., *Mushroom Sci.,* **9**, 809 (1976); *Chem. Abstr.,* **86** 103099m

MYCOCALIA (Basidiomycetae)

M. reticulata.
SESQUITERPENOIDS: 6α,7β-dihydroxydrimenin, 7-ketodihydrodrimenin, 7-oxodihydrodrimenin.
Ayer, Fung, *Tetrahedron,* **33**, 2771 (1977)

MYLIA (Hepaticae)

M. taylori (Hook.) Gray.
DITERPENOID: (15S,16S)-2β,16-epoxyverrucosan-16-ol.
SESQUITERPENOIDS: (-)-dihydromylione A, myliol, taylorione.
Matsuo et al., *Experientia,* **33**, 991 (1977)
Harrison et al., *J. Chem. Res. Synop.,* 212 (1986)

M. verrucosa.
DITERPENOIDS: (-)-9-acetoxy-2β,7α-dihydroxyverrucosane,
(-)-1-acetoxy-2β-hydroxyverrucosane,
(-)-2β-acetoxy-11-hydroxyverrucosane,
(-)-9-acetoxy-2β-hydroxyverrucosane,
(-)-11-acetoxy-2β-hydroxyverrucosane,
(-)-2β-9α-diacetoxyverrucosane,
(-)-2β-hydroxy-9-oxoverrucosane,
(-)-2β-hydroxyverrucosane, 2β,7α-verrucosanediol,
2,11-verrucosanediol, 2-verrucosanol.
Takaoka et al., *J. Chem. Soc., Perkin I,* 2711 (1979)

MYOPORUM (Myoporaceae)

M. acutinatum.
SESQUITERPENOID: (-)-10,11-dehydromyoporone.
Blackburne, Sutherland, *Aust. J. Chem.,* **25**, 1787 (1972)

M. betcheanum.
SESQUITERPENOID: (-)-10,11-dehydromyoporone.
Blackburne, Sutherland, *Aust. J. Chem.,* **25**, 1787 (1972)

M. crassifolium.
SESQUITERPENOID: anymol.
Engelmann, Bauer, *Naturwissenschaften,* **40**, 363 (1953)

M. deserti A. Cunn.
IRIDOIDS: (+)-(1R)-1-acetoxymyodesert-3-ene,
(-)-(1S)-1-acetoxymyodesert-3-ene,
(1R)-1-methoxymyodesert-3-ene.
SESQUITERPENOIDS: (-)-10,11-dehydromyoporone,
myodesmone, myoporone, (-)-ngaione, (+)-ngaione.
Grant et al., *Aust. J. Chem.,* **33**, 853 (1980)

M. lactum.
SESQUITERPENOIDS: myodesmone, myoporone, (-)-ngaione.
Birch et al., *Aust. J. Chem.,* **6**, 385 (1953)

MYRCIA

M. acris.
MONOTERPENOID: myrcene.
Power, Kleber, *Pharm. Rundschau,* **13**, 60 (1895)

MYRIANTHUS (Myrianthaceae)

M. arboreus.
PEPTIDE ALKALOIDS: myrianthines A, B and C.
STEROIDS: β-sitosterol, β-sitosterol 3-O-β-D-glucopyranoside, Triterpenoids euscaphic acid, myrianthic acid, tomentic acid, ursolic acid.
(A) Manchand et al., *Ann. Pharm. Fr.,* **26**, 771 (1968)
(S, T) Ojinnaka et al., *J. Nat. Prod.,* **48**, 1002 (1985)

MYRICA (Myricaceae)

M. gale L.
MACROCYCLICS: galeon, porson.
Malterud et al., *Tetrahedron Lett.,* 3069 (1976)

M. nagi.
DITERPENOIDS: myricanol, myricanone.
TRITERPENOIDS: myricardilol, myricolal.
Campbell et al., *J. Chem. Soc., Chem. Commun.,* 1206 (1970)

MYRICARIA (Tamaricaceae)

M. alopecurioides.
PHENOLICS: dehydrodigallic acid, dehydrotrigallic acid.
Chumbalov et al., *Khim. Prir. Soedin.,* 131 (1976)

MYRISTICA (Myristaceae)

M. fragrans.
PHENOLICS:
2-(4-allyl-2,6-dimethoxyphenoxy)-1-(4-hydroxy-3-methoxyphenyl)-1-propanol,
2-(4-allyl-2,6-dimethoxyphenoxy)-1-(3,4-methylendioxyphenyl)-1-propanol,
2-(4-allyl-2,6-dimethoxyphenoxy)-1-(3,4,5-trimethoxyphenyl)-propane.
Isogai et al., *Agr. Biol. Chem.,* **37**, 1479 (1973)

M. otoba.
DITERPENOID: otobain.
Stevenson, *Chem. Ind. (London),* 270 (1962)

M. simarum A. DC.
LIGNAN KETONE: 1-oxootobain.
Kuo et al., *Experientia,* **1**, 451 (1971)

MYROXYLON

M. balsamum.
TRITERPENOIDS: 6-hydroxy-3-oxoolean-12-en-28-oic acid,
(20R)-24-ocotillone.
Wahlberg, Enzell, *Acta Chem. Scand.,* **25**, 352 (1971)

MYRRHIS (Umbelliferae)

M. octodecimguttata.
QUINOLIZIDINE ALKALOID: myrrhine.
Tursch et al., *Tetrahedron,* **31**, 1541 (1975)

MYRTILLOCACTUS (Cactaceae)

M. cochal.
TRITERPENOIDS: chichipigenin, cochalic acid.
Sandoral et al., *J. Am. Chem. Soc.,* **79**, 4468 (1957)

M. eichlamii.
TRITERPENOID: chichipigenin.
Djerassi et al., *J. Am. Chem. Soc.,* **79**, 3525 (1957)

M. geometrizans.
TRITERPENOID: chichipigenin.
Djerassi et al., *J. Am. Chem. Soc.,* **79**, 3525 (1957)

M. nova.
TRITERPENOID: myrtillogenic acid.
Djerassi et al., *J. Am. Chem. Soc.,* **79**, 3525 (1957)

M. schenckii.
TRITERPENOID: chichipigenin.

Djerassi *et al., J. Am. Chem. Soc.,* **79**, 3525 (1957)

MYRTOPSIS

M. microcarpa.
FURANOQUINOLINE ALKALOID: 8-methoxyflindersine.

Hifnawy *et al., Phytochemistry,* **16**, 1035 (1977)

M. myrtoidea.
AMINE ALKALOID: N-benzyltryptamine.

Hifnawy *et al., Phytochemistry,* **16**, 1035 (1977)

M. sellingii.
FURANOQUINOLINE ALKALOID: myrtopsine.
COUMARINS: myrsellin, myrsellinol.

(A) Hifnawy *et al., Planta Med.,* **26**, 346 (1976)
(C) Hifnawy *et al., Phytochemistry,* **16**, 1035 (1977)

MYRTUS (Myrtaceae)

M. communis.
ANTIBIOTIC: myrtucummulone A.

Rotstein, Lifshitz, *Antimicrobial Agents & Chemotherapy,* **6**, 539 (1974)

MYXOPHYTA (Cyanophyceae/Myxophyceae)

Myxophyta spp.
CAROTENOIDS: myxoxanthin, myxoxanthophyll.

Heilbron, Lythgoe, *J. Chem. Soc.,* 1376 (1936)

MYZODENDRON (Myzodendraceae)

M. punctulatum.
PHENYLBUTANONE: myzodendrone.

Reyes *et al., J. Nat. Prod.,* **49**, 318 (1986)

N

NAEMATOLOMA (Scrophulariaceae)

N. fasciculare (Fr.) Karst.

TRITERPENOIDS: fasciculols A, B, C, D, E, F, G, fasciculol C
2-depsipeptide, fasciculol C 3-depsipeptide.

Kubo *et al.*, *Chem. Pharm. Bull.*, **33**, 3821 (1983)

NANDINA (Berberidaceae)

N. domestica Thunb.

APORPHINE ALKALOIDS: dehydroisoboldine, dehydronantenine,
4,5-dioxodehydronantenine, domesticine, isoboldine,
isodomesticine, nandinine, nantenine.

PROTOBERBERINE ALKALOID: berberine.

QUATERNARY ALKALOID: nandazurine.

Kunimoto, Murakami, *Shyakugaku Zasshi*, **33**, 84 (1979)

N. domestica Thunb. var. leucocarpa.

APORPHINE ALKALOIDS: dehydroisoboldine, hydroxynantenine.

Kunimoto *et al.*, *J. Pharm. Soc., Japan*, **94**, 1149 (1974)

NANOPHYTON (Chenopodiaceae)

N. erinaceum (Pall.) Bge.

PIPERIDINE ALKALOIDS: 2,6-dimethylpiperidine,
1,2,6-trimethylpiperidine.

Yunusov *et al.*, *Nauchn. Tr. Tashk. Gos. Univ.*, **262**, 16 (1964)

NAPOLEONAEA (Lecythidaceae/Napoleonaceae)

N. vogelii.

TRITERPENOIDS: napoleogenin A, napoleogenin B.

Kapundu *et al.*, *Bull. Soc. Chim. Belg.*, **89**, 1001 (1980)

NARCISSUS (Amaryllidaceae)

N. folli.

ISOQUINOLINE ALKALOIDS: lycorine, tazettine.

Abdu'azimov, Yunusov, *Khim. Prir. Soedin.*, 64 (1967)

N. incomparabilis.

ISOQUINOLINE ALKALOID: galanthine.

Fales *et al.*, *J. Am. Chem. Soc.*, **78**, 4145 (1956)

N. jonquilla.

ISOQUINOLINE ALKALOID: jonquilline.

Dopke, Dalmer, *Naturwissenshaften* (1965)

N. kristalli.

DIPHENYL ALKALOID: galanthamine.

ISOQUINOLINE ALKALOIDS: lycorine, tazettine.

UNCHARACTERIZED ALKALOIDS: (+)-narvedine, (±)-narvedine.

Smirnova *et al.*, *Dokl. Akad. Nauk SSSR*, **154**, 171 (1964)

N. papyraceus Ker.-Gawl.

ISOQUINOLINE ALKALOID: papyramine.

Hung *et al.*, *Huaxue Xeubao*, **39**, 529 (1981)

N. poeticus L.

INDOLE ALKALOID: narcipoetine.

ISOQUINOLINE ALKALOIDS: lycorine, tazettine.

LUPINANE ALKALOID: pancracine.

PHENOLIC ACID: piscidic acid.

(Acid) Nordal *et al.*, *Acta Chem. Scand.*, **18**, 1979 (1964)

(A) Bopke, *Z. Chem.*, **14**, 57 (1974)

N. poeticus L. var. ornatus.

UNCHARACTERIZED ALKALOID: poetaricine.

Dopke, *Naturwissenshaften*, **50**, 595 (1963)

N. pseudonarcissus L.

HOMOISOQUINOLINE ALKALOID: narcissamine.

ISOQUINOLINE ALKALOID: galanthine.

Fales *et al.*, *J. Am. Chem. Soc.*, **78**, 445 (1956)

N. tazetta L.

ISOQUINOLINE ALKALOIDS: lycorine, maritidine,
O-methylmaritidine, pancratine, suisenine, tazettine.

Tani *et al.*, *Chem. Pharm. Bull.*, **29**, 3381 (1981)

N. tazetta L. var. chinensis Roem.

ISOQUINOLINE ALKALOIDS: epi-galanthamine,
6α-hydroxy-3-O-methyl-epi-maritidine,
6β-hydroxy-3-O-methyl-epi-maritidine,
maritidine-O-lycoramine, 3-O-methylmaritidine.

Ma *et al.*, *Heterocycles*, **24**, 2089 (1986)

NARDIA (Hepaticae)

N. scalaris (Schrad.) Gray.

DITERPENOIDS: ent-hydroxy-9(11),16-kauradien-6-ol, nardiin.

TRITERPENOID: (+)-3-oxo-21-methoxyserrat-14-ene.

Benes *et al.*, *Collect. Czech. Chem. Commun.*, **47**, 1873 (1982)

NARDOSMIA (Compositae)

N. laevigata DC. (Petasites laevigatus Reichb.).

PYRROLIZIDINE ALKALOIDS: platyphylline, renardine,
senecionine, senkirkine.

Wunderlich, *Chem. Ind. (London)*, 2089 (1962)

Briggs *et al.*, *J. Chem. Soc.*, 2492 (1965)

NARDOSTACHYS (Valerianaceae)

N. chinensis.

SESQUITERPENOIDS: aristola-1,9-diene,
aristola-1(10),8-dien-2-one, aristol-9-en-1-ol, aristolene,
isonardosinone, narchinol A, nardosinone.

Rucker *et al.*, *Justus Liebigs Ann. Chem.*, **748**, 211, 214 (1971)

N. jatamsi.

SESQUITERPENOIDS: angelicin, jatamanshic acid, lomatin,
maaliol, nardol, nardosiacione, norseychellanone,
224-patchoulene, β-patchoulene, patchouli alcohol.

Stanbhag *et al.*, *Tetrahedron*, **21**, 3591 (1965)

Ruecker *et al.*, *Phytochemistry*, **15**, 224 (1976)

NARTHECIUM (Liliaceae)

N. ossifragum (L.)Hudson.

SAPONINS: narthecin, xylosin.

STEROIDS: cholest-5-ene-3β,22-diol, cholest-5-ene-5,22-diol,
cholest-5-ene-3β,24-diol.

(Sap) Ceh, Hauge, *Acta Vet. Scand.*, **22**, 391 (1982)

(S) Stabursvik, *Acta Chem. Scand.*, **7**, 1220 (1953)

NAUCLEA (Rubiaceae)

N. diderrichii.

CARBOLINE ALKALOIDS: alkaloid ND-305B, alkaloid ND-363C,
alkaloid ND-370, nauclechine, naucledine, naucleonidine,
naucleonine.

HOMOCARBOLINE ALKALOID: nauclederine.

INDOLE ALKALOID: nauclexine.

PYRIDINE ALKALOID: 3-methoxycarbonyl-5-vinylpyridine.

SECOIRIDOID: diderroside.

(A) Murray, McLean, *Can. J. Chem.,* **50,** 1478, (1972)

(A) Murray, McLean, *ibid.,* **50,** 1496 (1972)

(A) Dimitrienko *et al., Tetrahedron Lett.,* 2599 (1974)

(I) Adouye *et al., Phytochemistry,* **22,** 975 (1983)

N. latifolia Sm.

CARBOLINE ALKALOIDS: nauclefine, naucletine, naulafine, laufoline.

Hotellier *et al., Phytochemistry,* **14,** 1403 (1975)

N. parva.

CARBOLINE ALKALOID: parvine.

Sainsbury, Webb, *Phytochemistry,* **14,** 2691 (1975)

NAUMBERGIA (Primulaceae)

N. thyrsifolia (L.) Reichb. (Lysimachia thyrsiflora L.)

SAPONINS: saponins A and B.

Karpova *et al., Khim. Prir. Soedin.,* 364 (1975)

NAVANAX

N. inermis.

SESQUITERPENOIDS: narvenones A, B and C.

Sleeper, Fenical, *J. Am. Chem. Soc.,* **99,** 2367 (1977)

NAVICULA (Haptophyceae)

N. torquetum.

CAROTENOID: ε-carotene.

Chapman, Haxo, *Plant Cell Physiol.,* **4,** 57 (1963)

NECTANDRA (Lauraceae)

N. elaiophora.

SESQUITERPENOIDS: (-)-curcumene, (+)-curcumene, (-)-α-curcumene.

Naves, *Bull. Soc. Chim.,* 987 (1951)

N. rodioei Hook.

UNCHARACTERIZED ALKALOID: chondrodine.

Scholtz, *Ber. Dtsch. Chem. Ges.,* **29,** 2054 (1896)

N. rubra (Mez.) C.K.Allen.

FURANS: rubrenolide, rubrynolide.

Franca *et al., J. Chem. Soc., Chem. Commun.,* 514 (1972)

NEISOSPERMA

N. kilneri (F.V.Muell.) Fosberg et Sachet.

CARBOLINE ALKALOIDS: angustine, 1-carbamoyl-β-carboline, 1-carbamoyl-7-methoxy-β-carboline, cathafoline, descarbomethoxygambirtannine, isoreserpiline, 10-methoxygeissoschizol, ochrolifuanine A, reserpiline.

Batchily *et al., Ann. Pharm. Fr.,* **43,** 513 (1985)

NELUMBO (Nymphaceae)

N. lutea (Willd.) Pers.

APORPHINE ALKALOIDS: N-norarmepavine, (-)-N-nornuciferine.

Kupchan *et al., Tetrahedron,* **19,** 227 (1963)

N. nucifera Gaertn.

APORPHINE ALKALOIDS: (±)-armepavine dehydronuciferine, dehydroroermerine, 2-hydroxy-1-methoxyaporphine isoliensine, isoliensinine, 4'-methyl-N-methylcoclaurine, neferine, roermerine.

ISOQUINOLINE ALKALOIDS: isoliensinine, liensinine, lotusine, O-methylcorypalline.

BISBENZYLISOQUINOLINE ALKALOID: neferine.

Nishibe *et al., J. Nat. Prod.,* **49,** 548 (1986)

NEMALION (Rhodophyceae)

N. helminthoides.

CAROTENOIDS: α-carotene, β-carotene, α-cryptoxanthin, β-cryptoxanthin, fucoxanthin, lutein, zeaxanthin.

Bjoernland *et al., Phytochemistry,* **15,** 291 (1976)

NEMUARON (Monomiaceae)

N. vaillardii.

APORPHINE ALKALOIDS: laurotetanine, N-methyllaurotetanine, norisocorydine.

BISBENZYLISOQUINOLINE ALKALOID: nemuarine.

OXOAPORPHINE ALKALOID: atheroline.

Bick *et al., Aust. J. Chem.,* **20,** 1403 (1967)

NEOCALLITROPIS

N. araucarioides.

SESQUITERPENOID: araucariol.

Rudloff, *Chem. Ind. (London),* 2126 (1964)

NEOCHAMAELEA

N. pulverulenta.

CHROMENES: pulverin, spatheliabischromene.

STEROID: β-sitosterol.

DITERPENOID: pulverin.

SESTERTERPENOIDS: cneorin G, cneorin NP-32, cneorin NP-36, cneorin NP-38, cneorin R.

(C, S) Gonzalez *et al., Phytochemistry,* **14,** 1656 (1975)

(T) Mondon *et al., Tetrahedron Lett.,* 2015 (1981)

NEOLITSEA (Lauraceae)

N. aciculata (Blume) Koidz.

ACIDS: dodecanoic acid, cis-4-decenoic acid, cis-4-dodecenoic acid, cis-4-tetradecenoic acid.

SESQUITERPENOIDS: ,i.isolinderalactone, linderadine, linderalactone, linderane, linderine, litseaculane, litsealactone, neoliacine, neolinderane, pseudolinderane, zeylanane, zeylanine.

(Acid) Furukawa *et al., Yukagaku,* **25,** 496 (1976); *Chem. Abstr.,* **85** 119617b

(T) Takeda *et al., J. Chem. Soc., C,* 973 (1970)

(T) Nozaki *et al., J. Chem. Soc., Chem. Commun.,* 1107 (1983)

N. aurata (Hay.) Koidz.

SPIROISOQUINOLINE ALKALOIDS: anonaine, N-methyllitsericine, roemerine.

Lu *et al., J. Chin. Chem. Soc.,* **22,** 349 (1975); *Chem. Abstr.,* **84** 132659r

N. buisanensis Yamamoto et Kamikoti.

ISOQUINOLINE ALKALOIDS: laurolitsine, litsericine.

Lu *et al., J. Chin. Chem. Soc.,* **22,** 349 (1975); *Chem. Abstr.,* **84** 132659r

N. daibuensis Kamikoti.

BENZYLISOQUINOLINE ALKALOID: (S)-(+)-reticuline.

Lu *et al., Chem. Abstr.,* **88** 34561r

N. fischeri.

SESQUITERPENOIDS: fischeric acid, isofischeric acid.

Joshi, Kamal, *Indian. J. Chem.,* **9,** 80 (1971)

N. pulchella.

APORPHINE ALKALOIDS: neolitsine, neolitsinine.

TRITERPENOIDS:

3β-methoxy-24,25-dimethyllanosta-9(11)-en-24-ol, 3-methoxy-24,25-dimethyllanosta-9(11)-en-24-ol methyl ether, 24-methoxy-24,25-dimethyllanost-9(11)-en-3-one.

(A) Hui *et al., J. Chem. Soc.,* 2285 (1965)

(T) Chan, Hui, *J. Chem. Soc., Perkin I,* 490 (1973)

N. sericea (Blume) Koidz.
ACIDS: dodecanoic acid, cis-4-decenoic acid, cis-4-dodecenoic acid, cis-4-tetradecenoic acid.
APORPHINE ALKALOID: laurolitsine.
SESQUITERPENOIDS: isosericenin, neosericenin, sericealactone, sericenic acid, sericenin.
  (Acid) Furukawa *et al., Yukagaku,* **25,** 496 (1976); *Chem. Abstr.,* **85** 119617b
  (A) Kozuka *et al., J. Pharm. Soc., Japan,* **82,** 1567 (1962)
  (T) Takeda *et al., J. Chem. Soc., C,* 1547 (1970)

N. zeylanica.
SESQUITERPENOIDS: linderalactone, neolinderane, zeylanicin, zeylanidine, zeylanin.
  Joshi *et al., Tetrahedron,* **23,** 261 (1967)
  Joshi *et al., ibid.,* **23,** 273 (1967)

NEOLLOYDIA
N. texensis.
TRITERPENOID: ocotillol II.
  Warnhoff, Walls, *Can. J. Chem.,* **43,** 3311 (1965)

NEORAUTANENIA (Leguminosae/Papilionaceae)
N. amboensis.
PHYTOALEXINS: ambanol, edudiol, edudiol methyl ether, edulane, (6aR,11aR)-edunol, (6aS,11aS)-edunol.
ROTENOIDS: 12α-hydroxyisomillettone, neobanol, neobanone, rotenonone.
  (P) Brink *et al., Phytochemistry,* **16,** 273 (1977)
  (P) Oberholzer *et al., Tetrahedron Lett.,* 1165 (1977)
  (R) Oberholzer *et al., Phytochemistry,* **15,** 1283 (1976)

N. edulis.
FURANOFLAVONE: pachyrrhizin.
ISOFLAVONES: dehydroneotenone, neodulin, neotenone.
PHYTOALEXINS: edudiol, edudiol methyl ether, edulane.
  Brink *et al., Phytochemistry,* **16,** 273 (1977)

N. ficifolia.
BENZOPYRANS: folinin, folitenol.
PTEROCARPAN: ficifolinol.
  Brink *et al., J. S. Afr. Chem. Inst.,* **23,** 24 (1970)

N. pseudopachyrrhiza Harms.
FURANOFLAVONE: pachyrrhizin.
ISOFLAVONES: neotenone, nepseudin.
ROTENOID: dolineone.
  Crombie, Whiting, *J. Chem. Soc.,* 1569 (1963)

NEPETA (Labiatae)
N. cataria L.
MONOTERPENOIDS: 5,9-dehydronepetalactone, (1R)-nepetalic acid.
  Sastry *et al., Phytochemistry,* **11,** 453 (1972)

N. granatensis.
DITERPENOID: 7,18-dihydroxy-14-abietene.
  Gonzalez *et al., An. Quim.,* **72,** 65 (1976)

N. hindostana.
SESQUITERPENOIDS: nehepediol, nehepetol, nepeticin, nepetidin.
TRITERPENOID: 2α,3β,23-trihydroxyurs-12-en-28-oic acid.
  Ahmad *et al., Phytochemistry,* **25,** 1487 (1986)

N. mussini.
MONOTERPENOIDS: cis-nepetalactone, 4aα,7β,7aα-nepetalactone.
  Eisenbaum *et al., Int. Congr. Essent. Oils, 6th,* 149 (1974)

N. taydea.
TERPENOIDS: two uncharacterized terpenoids.
  Gonzalez *et al., An. Quim.,* **69,** 1059 (1973)

NERINE (Amaryllidaceae)
N. bowdenii W.C.R.Wats.
ISOQUINOLINE ALKALOIDS: crinine, 6-hydroxybuphanidrine, 6-hydroxypowelline.
  Slabaugh, Wildman, *J. Org. Chem.,* **36,** 3202 (1971)

N. corusca.
UNCHARACTERIZED ALKALOID: coruscine.
  Boit, Ehmke, *Chem. Ber. Dtsch. Chem. Ges.,* **90,** 369 (1957)

N. crispa Hort.
UNCHARACTERIZED ALKALOID: crispinine.
  Dopke, *Naturwissenshaften,* **49,** 469 (1962)

N. falcata.
ISOQUINOLINE ALKALOID: falcatine.
  Wildman, Kaufman, *J. Am. Chem. Soc.,* **77,** 4807 (1955)

N. flexuosa.
ISOQUINOLINE ALKALOID: flexinine.
  Fales, Wildman, *J. Org. Chem.,* **26,** 181 (1961)

N. flexuosa alba.
ISOQUINOLINE ALKALOID: undulatine.
  Wernhoff, Wildman, *J. Am. Chem. Soc.,* **82,** 1472 (1960)

N. krigeii.
INDOLE ALKALOIDS: krigenamine, neronine.
  Garbett *et al., Chem. Ind. (London),* 468 (1961)

N. laticonia.
ISOQUINOLINE ALKALOID: falcatine.
  Fales, Wildman, *J. Am. Chem. Soc.,* **80,** 4395 (1958)

N. powelii.
ALKALOID: criwelline.
  Boit, Ehmke, *Chem. Ber.,* **89,** 2093 (1956)

N. sarniensis (L.) Herbert.
INDOLE ALKALOID: nerinine.
  Boit, *Chem. Ber.,* **87,** 1704 (1954)

N. undulata.
UNCHARACTERIZED ALKALOID: crispine.
  Boit, *Chem. Ber.,* **89,** 1129 (1956)

NERIUM (Apocynaceae)
N. indicum Mill.
FLAVANOID: kaempferol-3-glycoside.
SAPONIN: oleandrin.
STEROIDS: β-sitosterol, β-sitosterol glucoside.
TRITERPENOIDS: oleanolic acid, ursolic acid.
  (F, S, T) Lin *et al., Chem. Abstr.,* 84 118469x
  (Sap) Neumann, *Ber. Dtsch. Chem. Ges.,* **70,** 1547 (1937)

N. odorum Sol.
PIGMENTS: neriumosides A-1. A-2, B-1, B-2 and C-1.
SAPONINS: 16-dehydroadynerigenin-α-diginoside, 16-dehydroadynerigenin-α-D-digitaloside, neriaside, neridienone A, neridienone B, odorosides A, B, C, D, E and F.
  (P) Yamauchi *et al., Tetrahedron Lett.,* 1115 (1976)
  (Sap) Yamauchi *et al., Chem. Pharm. Bull.,* **22,** 1680 (1974)

N. oleander L.
SAPONINS: adigoside, adynerin, neriantin, oleandrin.
TRITERPENOID: ursolic acid.
  (Sap) Jeger *et al., Helv. Chim. Acta,* **42,** 977 (1959)
  (T) Hassan *et al., Khim. Prir. Soedin.,* 119 (1977)

NERVILIA (Orchidaceae)
N. purpurea Schlechter.
STEROIDS: cyclohomonervilol, cyclonervilasterol, 24-epi-cyclonervilasterol, cyclonervilasterone, dehydronervilasterol, dihydrocyclonervilasterol, 24-epi-dihydrocyclonervilasterol, nervilasterol, nervilasterone, nervisterol.

Kikuchi et al., Chem. Pharm. Bull., 34, 3183 (1986)

NEUROLAENA
N. lobata.
SESQUITERPENOIDS: lobatin A, lobatin B, neurolenin A, neurolenin B.

Borges-del-Castillo et al., J. Nat. Prod., 45, 762 (1982)

NEUROSPORA
N. crassa.
STEROID: 4α-methyl-5α-ergosta-8,24(28)-dien-3β-ol.

SUBDEN, RENAUD, STEROIDS: 34, 643 (1979)

NEWCASTELIA (Verbenaceae)
N. viscida E. Pritzel.
DITERPENOID: isopimar-9(11),15-diene-3β,19-diol.

Jefferies, Ratajczak, Aust. J. Chem., 26, 173 (1973)

NICANDRA (Solanaceae)
N. physaloides.
STEROIDS: Nic-1, Nic-2, Nic-3, Nic-7 Nic-10, Nic-11, Nic-12, Nic-17, Nic-1 lactone, nicandrenone.

Begley, Morehead, J. Chem. Soc., Chem. Commun., 125 (1974)
Glotter et al., Phytochemistry, 15, 1317 (1976)

N. physaloides var. albiflora.
STEROIDS: nicalbin A, nicalbin B.

Begley, Morehead, J. Chem. Soc., Chem. Commun., 125 (1974)

NICOTIANA (Solanaceae)
N. affinis.
PYRIDINE ALKALOIDS: cis-nicotine-1'-N-oxide, trans-nicotine-1'-N-oxide.

Phillipson, Handa, Phytochemistry, 14, 2683 (1975)

N. debneyi.
SESQUITERPENOIDS: cyclodebneyol, debneyol.

Burden et al., Phytochemistry, 25, 1607 (1986)

N. glauca.
PYRIDINE ALKALOID: anabasine.

Orekhov, C. R. Acad. Sci., 189, 945 (1929)

N. glutinosa L.
DITERPENOIDS: (13Z)-labda-8,14-diene-2,13-diol, (13R)-labda-8,14-dien-13-ol, (12Z)-labda-12,14-dien-8α-ol, (13Z)-labda-8,14-dien-13-ol-2-one, (13R)-labd-14-ene-8α,13-diol, (13S)-labd-14-ene-8α,13-diol.

Colledge, Reid, Chem. Ind. (London), 570 (1975)

N. plumboginifolia.
STEROIDAL ALKALOID: solaplumbine.

Singh et al., Phytochemistry, 13, 2020 (1974)

N. setchellii.
DITERPENOID: 7,13-labdadien-15-ol.

Suzuki et al., Phytochemistry, 22, 1294 (1983)

N. sylvestris.
PYRIDINE ALKALOIDS: cis-nicotine-1'-N-oxide, trans-nicotine-1'-N-oxide.

Phillipson, Handa, Phytochemistry, 14, 2683 (1975)

N. tabacum L.
PYRIDINE ALKALOIDS: (-)-anatabine, (±)-anatabine, anatelline, lathreine, (-)-N-methylanabasine, (-)-N-methylanatabine, 5-methyl-2,3'-bipyridine, myosmine, nicotelline, nicotianamine, nicotimine, nicotine, nicotyrine, nornicotine.

UNCHARACTERIZED ALKALOIDS: α-socratine, β-socratine, τ-socratine.

DIKETONE: prenylnorsolanadione.

PHYTOALEXIN: glutinosone.

POLYPRENOID: norsolenesene.

PYRAN: spiroxabovalide.

DITERPENOIDS: 10-isopropyl-3,7,13-trimethyltetradeca-2,6,11,13-tetraen-1-al.

NORSESQUITERPENOIDS: isonordrimenone, 3-oxo-α-ionol, solanofuran, (+)-solanone.

SESQUITERPENOIDS: 1-oxo-α-cyperone, phytuberol, solanascone, 3,7,13-trimethyl-10-isopropyl-2,6,11,13-tetradecatetraen-1-al, 4-(2,2,6-trimethyl-6-vinylcyclohexyl)-2-butanone.

TOXINS: caffeoylputrescine, caffeoylspermidine, dicaffeoylspermidine.

(A) Spath et al., Ber. Dtsch. Chem. Ges., 69, 393 (1936)
(A) Kisaki et al., Phytochemistry, 7, 323 (1968)
(Ph) Murai et al., Bull. Chem. Soc., Japan, 53, 1045 (1980)
(Po) Enzell et al., Tetrahedron Lett., 1983 (1971)
(Py) Demole et al., Helv. Chim. Acta, 56, 265 (1973)
(T) Takagi et al., Agr. Biol. Chem., 43, 2395 (1979)
(Tox.) Deletang, Ann. Tab. Sect., 2, 123 (1974); Chem. Abstr., 84 147656m

N. tabacum Burley.
DINORSESQUITERPENOIDS: (E)-5-isopropyl-6,7-epoxy-8-hydroxy-8-methylnonan-2-one, (E)-5-isopropyl-8-hydroxy-8-methylnon-6-en-2-one, 5-isopropyl-8-hydroxynonan-2-one, (E)-5-isopropyl-8-hydroxynon-6-en-2-one, (E)-3-isopropyl-6-methylhepta-4,6-dien-1-ol, 5-isopropylnonane-2,8-diol, (E)-5-isopropylnon-3-ene-2,8-diol, 3,3,5-trimethyl-8-isopropyl-4,9-dioxabicyclo-[3,3,1]-nonan-2-ol.

MONOTERPENOIDS: exo-1-(1-methyl-4-isopropyl-7,8-dioxabicyclo-[3,2,1]-oct-6-yl)-ethanol, exo-1-(1-methyl-4-isopropyl-7,8-dioxabicyclo-[3,2,1]-ocy-6-yl)-methylketone, endo-2-(1-methyl-4-isopropyl-7,8-oxabicyclo-[3,2,1]-oct-6-yl)-propen-2-ol.

Demole, Demole, Helv. Chim. Acta, 58, 1867 (1975)

N. tabacum L. (Greek).
PYRIDINE ALKALOIDS: anabasine, nicotine, nicotyrine, nornicotine.

DINORSESQUITERPENOIDS: (3S,6R,7E,9R)-megastigmadiene-3,9-diol, (3S,6R,7E,9S)-megastigmadiene-3,9-diol, megastigma-5(13),7-diene-6,9-diol, megastigma-4,6,8-trien-3-one, 6-megastigma-4,7,9-trien-3-one.

NORDITERPENOID: 15-nor-8α-hydroxy-12E-labden-14-ol.

NORSESQUITERPENOID: 11-nor-8-hydroxy-9-drimenone.

(A) Spath et al., Ber. Dtsch. Chem. Ges., 69, 393 (1936)
(T) Aasen et al., Acta Chem. Scand., 26, 2573 (1972)
(T) Behr et al., ibid, 32B, 391 (1978)
(T) Wahlberg et al., ibid, 36B, 573 (1982)

N. tabacum PB. (Bergerac).
PYRIDINE ALKALOIDS: anabasine, nicotine, nicotyrine.
DITERPENOIDS: (13E)-15-acetoxy-13-labden-8-ol,
　　(11E,13R)-labda-11,14-diene-8,13-diol,
　　(11E,13S)-labda-11,14-diene-8,13-diol.
　　(A) Kasaki et al., Phytochemistry, 7, 323 (1968)
　　(T) Colledge, Reid, Chem. Ind. (London), 570 (1975)

N. tabacum L. (Turkish).
PYRIDINE ALKALOIDS: isonicoteine, nicotine, nicotoine.
MONOTERPENOIDS:
　　18-methyl-3-isopropylcyclopentane-1,2-dicarboxylic acid,
　　2-methyl-5-isopropyl-1-cyclopentene-1-carboxylic acid.
DITERPENOIDS: α-levantanolide, α-levantenolide, β-levantenolie.
　　(A) Noga, Tachl. Mitt. Oesterr. Takakregie, (1914)
　　(T) Giles et al., Tetrahedron, 9, 107 (1963)
　　(T) Chuman et al., Agr. Biol. Chem., 42, 203 (1978)

N. tabacum L. (Virginian).
NORSESQUITERPENOIDS:
　　3-isopropenyl-5-methyl-1,2-dihydronaphthalene,
　　cis-2-isopropenyl-8-methyl-1,2,3,4-tetrahydro-1-
　　naphthalenol, megastigma-5,8-dien-4-one.
　　Demole et al., Helv. Chim. Acta, 62, 67 (1979)

NIDORELLA
N. anomala.
DITERPENOIDS: ent-3β-hydroxy-15-beyeren-19-oic acid,
　　3α-hydroxy-19-stachenoic acid.
BISDITERPENOID: nidoanomalin.
　　Bohlmann et al., Phytochemistry, 21, 1175 (1982)

N. hottentotica.
BISSESQUITERPENOIDS: nidhottin, 6α-nidhottinol, 6β-nidhottinol,
　　6-nidhottinone.
DITERPENOIDS: 6-O-angeloyloxyseconidorellalactone,
　　seconidorellalactone.
　　Bohlmann et al., Phytochemistry, 21, 1109 (1982)

NIGELLA (Ranunculaceae)
N. damascena L.
AMINE ALKALOIDS: damascenine, damascinine.
SESQUITERPENOID: (-)-α-selinene.
　　(A) Schneider, Pharm. Zentralbl., 31, 173 (1890)
　　(A) Dopke, Fritsch, Pharmazie, 25, 69 (1970)
　　(T) Tillequin et al., Planta Med., 30, 59 (1976)

NITRARIA (Zygophyllaceae)
N. komakovii.
CARBOLINE ALKALOID: komarovidine.
　　Sulyaganov et al., Khim. Prir. Soedin., 732 (1980)

N. schoberi L.
QUINOLINE ALKALOID: nitramine.
CARBOLINE ALKALOIDS: nitrarine,
　　tetramethyotetrahydro-β-carboline.
　　Novgorodova et al., Khim. Prir. Soedin., 196 (1973)

N. sibirica.
INDAZOLE ALKALOID: nitrabirine.
　　Ibragimov et al., Khim. Prir. Soedin., 213 (1983)

NOLINA (Cactaceae)
N. erumpens.
STEROID: diosgenin.
　　Marker et al., J. Am. Chem. Soc., 65, 1199 (1943)

N. greeni.
STEROID: diosgenin.
　　Marker et al., J. Am. Chem. Soc., 65, 1199 (1943)

NOTHOFAGUS (Fagaceae)
N. dombeyi.
FLAVANONE: aromadendrin.
　　Gripenberg, Acta Chem. Scand., 6, 1152 (1952)

NOTHOLAENA
N. aschenborniana.
FLAVONOIDS:
　　5,4'-dihydroxy-3,7,3'-trimethoxy-8-acetoxyflavone,
　　5-hydroxy-3,7,2',3',4'-pentamethoxy-8-acetoxyflavone.
　　Jay et al., Z. Naturforsch., 37C, 721 (1982)

N. californica.
FLAVONOIDS: apigenin 7-methyl ether, kaempferol 3,7-dimethyl
　　ether, quercetin 3,7-dimethyl ether, quercetin
　　3,7,4'-trimethyl ether.
　　Wollenweber et al., Chem. Abstr., 97 212663w

N. candida.
FLAVONOIDS: apigenin, apigenin 7-dimethyl ether,
　　5-hydroxy-3,3',4',5',7-pentamethoxyflavone.
　　Farkas et al., Chem. Ber. Dtsch. Chem. Ges., 100, 2296 (1967)
　　Wollenweber, Phytochemistry, 15, 2013 (1976)

N. neglecta.
FLAVONOIDS: 5-hydroxy-7-methoxy-8-acetoxyflavanone,
　　7-O-methyl-8-acetoxygalangin.
　　Wollenweber et al., J. Nat. Prod., 45, 216 (1982)

N. schaffneri.
FLAVONOID: combretol.
　　Wollenweber, Phytochemistry, 15, 2013 (1976)

N. standleyi.
FLAVONOIDS: kaempferol, kaempferol 3,7-dimethyl ether,
　　kaempferol 7,4'-dimethyl ether, kaempferol 3-methyl ether,
　　kaempferol 4'-methyl ether.
　　Wollenweber, Phytochemistry, 15, 2013 (1976)

NOTONIA
N. petraea.
SESQUITERPENOID: notonipetrone.
　　Bohlmann, Zdero, Phytochemistry, 18, 1063 (1979)

NUMMULARIA (Xylariaceae)
N. broomeiana.
ANTIBIOTIC: pyrenophorin.
　　Anderson et al., J. Chem. Soc., Perkin 1, 2185 (1983)

NUPHAR (Nymphaeaceae)
N. japonicum.
PIPERIDINE ALKALOID: anhydronupharamine.
　　Arata et al., J. Pharm. Soc., Japan, 87, 1094 (1967)

N. luteum (L.) Sm.
BISQUINOLIZIDINE ALKALOIDS: allothiobonupharidine,
　　6,6'-dihydroxyneobinupharidine, neothiobinupharidine,
　　nupshleine, pseudothiobinupharidine, thiobinupharidine,
　　thiobidesoxynupharidine, thiobinupharidine sulphoxide,
　　thionupharoline, thionuphleine sulphoxide,
　　epi-thionuphleine sulphoxide.
QUINOLIZIDINE ALKALOIDS: 7-epi-deoxynupharidine,
　　nupharoline, nupharolutine.
　　Achmatowicz, Wrobel, Tetrahedron Lett., 927 (1964)
　　Wrobel et al., Bull. Acad. Pol. Sci. Ser. Sci. Chem., 24, 99 (1976)

N. luteum L. subsp. macrophyllum.
BISQUINOLIZIDINE ALKALOIDS: bis-6,7-oxidodeoxynupharidine,
　　6,6'-dihydroxythionuphlutines A and B.

QUINOLIZIDINE ALKALOIDS: deoxynupharidin-7α-ol,
7-epi-deoxynupharidin-7α-ol.
Lalonde *et al.*, *J. Am. Chem. Soc.*, **94**, 8522 (1972)

N. luteum L. subsp. variegatum.
PIPERIDINE ALKALOIDS: 3-epi-nuphamine, 3-epi-nupharamine.
Forrest, Ray, *Can. J. Chem.*, **49**, 1774 (1971)

N. pumila (Timm.) DC.
QUINOLIZIDINE ALKALOID: nupharopumiline.
Reura, Lounasmaa, *Phytochemistry*, **16**, 1122 (1977)

N. variegatum Engelm.
PIPERIDINE ALKALOIDS: 3-epi-nupharamine, nuphenine.
Forrest, Ray, *Can. J. Chem.*, **49**, 1774 (1971)

NYCTANTHES (Nyctanthaceae)
N. arbor-tristes.
FLAVONOIDS: astragalin, nicotiflorin.
IRIDOID: nyctanthoside.
STEROID: nycosterol.
TRITERPENOID: nyctanthic acid.
(F) Gupta, Bokadia, *Chem. Abstr.*, 87 18988j
(S) Vasista, *J. Benares Hindu Univ.*, **2**, 348 (1938)
(T) Rimpler, Junghanns, *Tetrahedron Lett.*, 2423 (1975)

NYMANIA (NYMPHAEA) (Nymphaeaceae)
N. alba L.
UNCHARACTERIZED ALKALOID: nympheine.
HOMOSESTERTERPENOIDS: nymania 1, nymania 2, nymania 3.
(A) Bures, Pizak, *Cas. Cesk. Lek.*, **15**, 223, 243 (1935)
(T) MacLachlin *et al.*, *Phytochemistry*, **21**, 1700 (1982)

O

OCHROCARPUS (Guttiferae/Clusiaceae)

O. odoratus.
XANTHONES: 1,5-dihydroxyxanthone, 1,3,5,6-tetrahydroxyxanthone, 1,3,6,7-tetrahydroxyxanthone, 1,5,6-trihydroxyxanthone, 2,3,4-trihydroxyxanthone.
Quillinan *et al., J. Chem. Soc., Perkin I,* 1382 (1972)

OCHROSIA (Apocynaceae)

O. balansae Guill.
CARBAZOLE ALKALOID: 1,2-dihydro-9-methoxyellipticine.
PROTOPINE ALKALOID: dimethoxypicraphylline.
Bruneton *et al., Ann. Pharm. Fr.,* **30,** 629 (1972)

O. elliptica Labill.
CARBAZOLE ALKALOIDS: ellipticine, elliptimine.
INDOLE ALKALOIDS: apparicine, cathenamine, nephrosine, norfluorocurarine, pleiocarpamine, tetrahydroalstonine.
Goodwin *et al., J. Am. Chem. Soc.,* **81,** 1903 (1959)
Pawekla *et al., Z. Naturforsch.,* **41C,** 381 (1986)

O. lifuana Guill.
BISCARBOLINE ALKALOIDS: dehydro-3-ochrolifuanine, ochrolifuanine, ochrolifuanine A.
CARBOLINE ALKALOID: decarboxydihydrogambertannine.
Peube-Locou *et al., Phytochemistry,* **12,** 199 (1973)
Preaux *et al., ibid.,* **13,** 2607 (1974)

O. moorei.
CARBOLINE ALKALOIDS: 10,11-dimethoxy-19,20-dihydro-16S,20R-sitsirikine, reserpiline N-oxide.
INDOLONE ALKALOID: 3S,7S-ochropposinine oxindole.
Ahond *et al., Lloydia,* **44,** 193 (1981)

O. nakaiana.
CARBOLINE ALKALOID: 10-methoxycorynantheol methoperchlorate.
Sakai *et al., J. Pharm. Soc., Japan,* **94,** 1274 (1974)

O. oppositifolia.
CARBAZOLE ALKALOID: 9-methoxyelliptisine.
CARBOLINE ALKALOIDS: ochropposine, ochropposinine.
Peube-Locou et al. *Phytochemistry,* **11,** 2109 (1972)

O. poweri Bailey.
CARBOLINE ALKALOID: poweridine.
INDOLE ALKALOIDS: ochropamine, ochropine.
UNCHARACTERIZED ALKALOIDS: elliptamine, poweramine, powerchrine, powerine.
Doy, Moore, *Aust. J. Chem.,* **15,** 548 (1962)
Douglas *et al., ibid.,* **17,** 246 (1964)

O. sandwicensis Gray.
CARBAZOLE ALKALOID: 9-methoxyelliptisine.
CARBOLINE ALKALOID: ochrosandwine.
Jordan, Scheuer, *Tetrahedron,* **21,** 3731 (1965)

OCHTODES (Rhodophyceae)

O. crockeri
MONOTERPENOIDS: (6S*,E)-2,4-ochtodiene-1,6-diol (6S*,Z)-2,4-ochtodiene-1,6-diol
Paul *et al., J. Org. Chem.,* **45,** 3401 (1980)

O. secundiramea.
HALOGENATED MONOTERPENOIDS: ochtodene, ochtodiol.
McConnell, Fenical, *J. Org. Chem.,* **43,** 4238 (1978)

OCIMUM (Labiatae)

O. basilicum.
MONOTERPENOID: ocimene.
SESQUITERPENE: 1-epi-bicyclosesquiphellandrene.
Terhune *et al., Phytochemistry,* **13,** 1183 (1974)

O. gratissimum.
SESQUITERPENOID: gratissimene.
Dembitskii *et al., Khim. Prir. Soedin.,* 398 (1982)

O. spicatum.
STEROIDS: campesterone, β-sitosterol, stigmasterol.
TRITERPENOIDS: maslinic acid, epi-maslinic acid, oleanolic acid.
Xaasen *et al., J. Nat. Prod.,* **46,** 936 (1983)

OCOTEA (Lauraceae)

O. aciphylla.
NEOLIGNANS: ferrearin, 3'-methoxyburchellin, oxaguianin.
Felicio *et al., Phytochemistry,* **25,** 1707 (1986)

O. acutangula (Mez.).
PHENANTHRENE ALKALOIDS: (S)(-)-O-methylpallidine, pallidine, pallidinine.
Vecchietti *et al., J. Chem. Soc., Perkin I,* 578 (1981)

O. barcellensis.
SESQUITERPENOIDS: (-)-curcumene, (-)-α-curcumene.
Herout *et al., Collect. Czech. Chem. Commun.,* **18,** 297 (1953)

O. brachybotra.
APORPHINE ALKALOID: predicentrine.
Chen *et al., Phytochemistry,* **15,** 1161 (1976)

O. bucherii.
APORPHINE ALKALOID: 3-hydroxyglaucine.
Roensch *et al., Justus Liebigs Ann. Chem.,* 744 (1983)

O. caparrapi.
SESQUITERPENOIDS: caparrapidiol, caparrapidoxide, caparrapitriol.
Borges-del-Castillo *et al., Tetrahedron Lett.,* 3731 (1966)
Brooks, Campbell, *Phytochemistry,* **8,** 215 (1969)

O. cymbarum: dehydroeugenol B.
DITERPENOID: dehydroeugenol B.
de Diaz *et al., Phytochemistry,* **19,** 681 (1980)

O. glaziovii.
APORPHINE ALKALOID: (R)-(-)-3,5-dihydroxy-6-methoxyaporphine.
Gilbert *et al., J. Am. Chem. Soc.,* **86,** 694 (1964)

O. leucoxylon.
APORPHINE ALKALOID: ocoxylonine.
Ahmad, Cava, *Heterocycles,* **7,** 927 (1977)

O. macropoda (H.B.K.) Mez.
APORPHINE ALKALOIDS: dehydrodicentrine, dehydroocopodine, dicentrinone, ocopodine, predicentrine.
Cava, Venkateswarlu, *Tetrahedron,* **27,** 2639 (1971)

O. miniarum.
APORPHINE ALKALOIDS: norleucoxylonine, ocominarine, ocominarone, ocotominarine.
Vecchietti *et al., Farmaco Ed. Sci.,* **34,** 829 (1979)

O. porosa.
BENZOPYRAN: porosin.
Aiba *et al., Phytochemistry,* **12,** 413 (1973)

O. puberula Nees.

APORPHINE ALKALOIDS: dehydroocoteine, ocoteine, thalmine, thalminine.
Baralle *et al.*, *Phytochemistry*, **12**, 948 (1973)

O. rodiaei.

BISBENZYLISOQUINOLINE ALKALOIDS: demerarine, norrodiasine, 2-nortetrandrine, ocodemerine, ocotine, ocotosine, otocamine, rodiasine, sepeerine.
Hearst, *J. Org. Chem.*, **29**, 466 (1965)
Chen *et al.*, *J. Chem. Soc., C*, 2479 (1967)

O. variabilis.

APORPHINE ALKALOID: variabiline.
UNCHARACTERIZED ALKALOID: apoglaziovine.
Cava *et al.*, *Tetrahedron Lett.*, 4647 (1972)

ODONTEA

O. bicolor.

MONOTERPENOID: 4,5-decadiene-7,9-diyn-1-ol.
Bew *et al.*, *J. Chem. Soc., C*, 129 (1966)

ODONTITES (Schophulariaceae)

O. serotina (Lam.) Dum.

Two uncharacterized alkaloids.
Platonova *et al.*, *Med. Prom.*, **3**, 21 (1963)

O. verna (Bellardi) Dum.

IRIDOIDS: aucubigenin-1-O-β-cellobioside, 8-epi-loganin.
Bianco *et al.*, *Gazz. Chim. Ital.*, **111**, 91 (1981)

O. verna subsp. serotina.

IRIDOIDS: aucubigenin-1-O-β-serotinoside, 2-O-benzoylaucubin.
Bianco *et al.*, *J. Nat. Prod.*, **44**, 448 (1981)

OENANTHE (Umbelliferae)

O. aquatica (L.) Poir.

SESQUITERPENOID: trans-pentadeca-7,13-diene-9,11-diyn-4-one.
Bohlmann *et al.*, *Chem. Ber.*, **104**, 1322 (1971)

O. crocata L.

ACETYLENIC: oenanthotoxin.
SESQUITERPENOIDS: trans-pentadeca-2,8-diene-4,6-diyn-1-ol, trans-pentadeca-2,8,10-triene-4,6-diyn, trans-pentadeca-2,8,10-triene-4,6-diyn-12-ol.
Bohlmann, Rode, *Chem. Ber.*, **101**, 1163 (1968)

OENOTHERA (Onagraceae)

O. biennis L.

ACID: cis-6,9,12-octadecatrienoic acid.
STEROIDS: campesterol, β-sitosterol.
Fedeli *et al.*, *Riv. Ital. Sostanze Grasse*, **53**, 25 (1976); *Chem. Abstr.*, **85** 59636v

OIDIODENDRON

O. truncatum.

TETRATERPENOID: clerocidin.
Andersen *et al.*, *Tetrahedron Lett.*, 469 (1984)

OKOUBAKA

O. aubrevillei.

PHENOLICS: (-)-epi-allocatechin gallate, (+)-catechin, (+)-epi-catechin, (-)-epi-catechin gallate, (+)-gallocatehcin.
Wagner *et al.*, *Planta Med.*, 404 (1985)

OLAX

O. andronensis.

SAPONIN: olaxoside.
Forgacs *et al.*, *Phytochemistry*, **20**, 1689 (1981)

O. glabriflora.

SAPONIN: olaxoside.
Forgacs *et al.*, *Phytochemistry*, **20**, 1689 (1981)

O. psitticorum.

SAPONIN: olaxoside.
Forgacs *et al.*, *Phytochemistry*, **20**, 1689 (1981)

OLDENLANDIA (Rubiaceae)

O. biflora.

UNCHARACTERIZED ALKALOID: biflorine.
Chaudan, Tiwari, *J. Indian Chem. Soc.*, **29**, 386 (1952)

OLEA (Oleaceae)

O. europaea.

MONOTERPENOID: oleuropeic acid.
TERPENE GLYCOSIDE: oleuropein.
TRITERPENOID: erythrodiol.
Huneck, Lehn, *Bull. Soc. Chim.*, 321 (1963)
Panizzi *et al.*, *Gazz. Chim. Ital.*, **95**, 1279 (1965)

OLEANDRA (Cryptogrammataceae)

O. nerifolia.

TRITERPENOIDS: 23-ethoxyhopane, nerifoliol.
Goswami *et al.*, *Tetrahedron Lett.*, 287 (1979)

O. wallichi.

TRITERPENE: wallichienene.
Tsuda, Isobe, *Tetrahedron Lett.*, 3337 (1965)

OLEARIA (Compositae)

O. heterocarpa.

DITERPENOID: olearin.
Pinkey, Simpson, *J. Chem. Soc., Chem. Commun.*, 9 (1967)

O. paniculata.

DITERPENOID: 13-epi-manoyl oxide.
TRITERPENOID: isoursenol.
Cross *et al.*, *J. Chem. Soc.*, 2937 (1963)

OLEUM

O. cinae.

MONOTERPENE: (±)-dipentene.
Semmler, *Ber. Dtsch. Chem. Ges.*, **33**, 1457 (1900)

OLIVIA

O. miniata.

CAROTENOID: lutein.
Zechmeister, Tuzson, *Ber. Dtsch. Chem. Ges.*, **67**, 170 (1934)

ONCINOTIS

O. inandensis.

MACROCYCLIC ALKALOIDS: inandenine A, inandenine B.
Veith *et al.*, *Helv. Chim. Acta*, **53**, 1355 (1970)

O. nitida Benth.

PYRIDYL ALKALOID: oncinotine.
Badawi *et al.*, *Helv. Chim. Acta*, **51**, 1813 (1968)

ONOBRYCHIS (Papilionaceae)

O. amoena.
FLAVONOIDS: kaempferol 3-O-glucoside, quercetin
3-O-galactopyranoside, quercetin 3-O-glucopyranoside,
quercetin 3-O-rhamnopyranoside, quercetin 3-O-rutinoside.
Luk'yanchikov et al., Khim. Prir. Soedin., 564 (1985)

O. angustifolia.
FLAVONOIDS: hyperin, isorhamnetin, isorhamnetin
3-O-β-D-galactofuranoside, rutin, vitexin.
Moniava et al., Izv. Akad. Nauk Gruz. SSR, 2, 119 (1976)

O. bobrovi.
FLAVONOIDS: arbutin, astragalin, hyperoside, quercetin, rutin.
Nurmukhamedova et al., Khim. Prir. Soedin., 261 (1982)

O. chlorassanica.
FLAVONOIDS: kaempferol 3-O-glucoside, quercetin
3-O-galactopyranoside, quercetin 3-O-glucopyranoside,
quercetin 3-O-rhamnopyranoside, quercetin 3-O-rutinoside.
Luk'yanchikov et al., Khim. Prir. Soedin., 564 (1985)

O. cyri.
FLAVONOID: rutin.
Moniava et al., Izv. Akad. Nauk Gruz. SSR, 2, 119 (1976)

O. echidna.
FLAVONOIDS: kaempferol 3-O-glucoside, quercetin
3-O-galactopyranoside, quercetin 3-O-glucopyranoside,
quercetin 3-O-rhamnopyranoside, quercetin 3-O-rutinoside.
Luk'yanchikov et al., Khim. Prir. Soedin., 564 (1985)

O. ferganica.
FLAVONOIDS: kaempferol 3-O-glucoside, quercetin
3-O-galactopyranoside, quercetin 3-O-glucopyranoside,
quercetin 3-O-rhamnopyranoside, quercetin 3-O-rutinoside.
Luk'yanchikov et al., Khim. Prir. Soedin., 564 (1985)

O. grandis.
FLAVONOIDS: kaempferol 3-O-glucoside, quercetin
3-O-galactopyranoside, quercetin 3-O-glucopyranoside,
quercetin 3-O-rhamnopyranoside, quercetin 3-O-rutinoside.
Luk'yanchikov et al., Khim. Prir. Soedin., 564 (1985)

O. iberica.
FLAVONOIDS: astragalin, isoastragalin, isoquercitrin,
kaempferol, kaempferol 3-O-rutinoside, rutin.
Moniava et al., Izv. Akad. Nauk Gruz. SSR, 2, 119 (1976)

O. inermis.
FLAVONOIDS: astragalin, isoquercetrin, rutin.
Moniava et al., Izv. Akad. Nauk Gruz. SSR, 2, 119 (1976)

O. kacketica.
FLAVONOIDS: astragalin, isoquercitrin, kaempferol, kaempferol
3-O-β-D-glucofuranoside, kaempferol 3-O-rutinoside,
quercetin 3-β-D-glucopyranoside-7-O-β-L-rhamnofructoside.
Moniava et al., Izv. Akad. Nauk Gruz. SSR, 2, 119 (1976)

O. kemularia.
COUMARINS: scopoletin, umbelliferone.
Moniava, Khim. Prir. Soedin., 513 (1975)

O. seravschanica.
FLAVONOIDS: kaempferol 3-O-glucoside, quercetin
3-O-galactopyranoside, quercetin 3-O-glucopyranoside,
quercetin 3-O-rhamnopyranoside, quercetin 3-O-rutinoside.
Luk'yanchikov et al., Khim. Prir. Soedin., 564 (1985)

O. sosnowski.
FLAVONOIDS: apigenin 6,7-diglucoside, isoquercitrin, quercetin,
quercetin 3-O-β-L-arabinofuranoside
7-β-D-glucopyranoside.
Moniava et al., Izv. Akad. Nauk Gruz. SSR, 2, 119 (1976)

O. vassiltschenkoi.
FLAVONOIDS: austragalin, hyperoside, kaempferol, quercetin.
Bol et al., Khim. Prir. Soedin., 100 (1976)

O. viciifolia Scop. (O. sativa Lam.).
FLAVONOIDS: afrormosin, formononetin, garbanzol,
isoliquiritigenin, liquiritigenin, medicarpin, vestitol,
vestitone.
Ingham, Z. Naturforsch., 33C, 146 (1978)

ONOCERIS

O. albicans.
SESQUITERPENOID: onoseriolide.
Bohlmann et al., Phytochemistry, 19, 689 (1980)

O. hyssipifolia.
COUMARIN: 4-desmethoxy-4-methylmercaptopereflorin.
Bohlmann, Zdero, Phytochemistry, 16, 239 (1977)

ONOCLEA (Athyriaceae)

O. sensibilis L.
STEROID: pterosterone.
Rimpler, Arch. Pharm., 305, 746 (1972)

ONONIS (Papilionaceae)

O. arvensis.
ISOFLAVONE: onogenin.
PHENOLIC:
7-O-β-D-glucopyranoside-3',4'-methylenedioxypterocarpan.
(I) Kovalev et al., Khim. Prir. Soedin., 354 (1975)
(P) Kovalev et al., ibid, 104 (1976)

O. leiosperma.
FLAVONOIDS: astragalin, hispidulin, hyperoside, kaempferol,
populin, quercetin.
Kovalev et al., Khim. Prir. Soedin., 253 (1982)

O. spinosa L.
TRITERPENOID: onocerol.
Schultz, Z. Physiol. Chem., 238, 35 (1936)

ONOPORDON (Compositae)

O. acanthum L.
ACIDS: caffeic acid, chlorogenic acid, succinic acid.
ANTHOCYANIN: cyanidin 3,5-diglucoside.
FLAVONOIDS: apigenin, apigenin 7-rutinoside, apigenin
7-O-glucuronide, chrysoeriol, eriodictyol, luteolin,
quercetin, quercetin-3-glucoside.
AMINO ACIDS: choline, proline, valine.
SESQUITERPENOID: onopordopicrin.
Drozdz et al., Collect. Czech. Chem. Commun., 33, 1730 (1968)
Karl et al., Dtsch. ApothZtg., 116, 57 (1976); Chem. Abstr., 84 147639h

ONOSMA (Boraginaceae)

O. echoides.
BENZOQUINONE: (S)-shikonin.
Shukla et al., Phytochemistry, 10, 1909 (1971)

ONYCHIUM (Cryptogrammataceae)

O. aurantum.
SESQUITERPENOIDS: onitin, onitisin.
Banerji, Tamakrishnan, Tetrahedron Lett., 1369 (1974)

O. siliculosum.
CHALCONES: 2',6'-dihydroxy-4'-methoxychalcone, pashanone.
Wollenweber, Phytochemistry, 21, 1462 (1982)

OPERCULINA (Convolvulaceae)

O. aurea.
DITERPENOID GLYCOSIDES: aureoside, isoaureoside.
Canonica et al., Gazz. Chim. Ital., 107, 223 (1977)

OPHIOGLOSSUM

O. costatum.
AMINO ACIDS: alanine, arginine, diaminobutyric acid, glutamate, lysine, serine, threonine.
Goswami, *Isr. J. Chem.*, **25**, 211 (1976)

O. gramineum.
AMINO ACIDS: alanine, arginine, diaminobutyric acid, glutamate, lysine, threonine.
Goswami, *Isr. J. Chem.*, **25**, 211 (1976)

O. lusitanicum.
AMINO ACIDS: alanine, arginine, diaminobutyric acid, glutamate, lysine, threonine.
Goswami, *Isr. J. Chem.*, **25**, 211 (1976)

O. nudicaule.
AMINO ACIDS: alanine, arginine, diaminobutyric acid, glutamate, lysine, threonine.
Goswami, *Isr. J. Chem.*, **25**, 211 (1976)

O. petiolatum.
AMINO ACIDS: alanine, arginine, diaminobutyric acid, glutamate, lysine, threonine.
Goswami, *Isr. J. Chem.*, **25**, 211 (1976)

O. polyphyllum.
AMINO ACIDS: alanine, arginine, diaminobutyric acid, glutamate, lysine, threonine.
Goswami, *Isr. J. Chem.*, **25**, 211 (1976)

O. vulgatum.
AMINO ACIDS: alanine, arginine, diaminobutyric acid, glutamate, lysine, threonine.
Goswami, *Isr. J. Chem.*, **25**, 211 (1976)

OPHIOPOGON (Liliaceae)

O. japonicus Ker-Gawles.
ALDEHYDES: dotriacontanal, octacosanal, triacontanal.
Isono et al., *J. Pharm. Soc., Japan*, **96**, 86 (1976)

O. japonicus Ker.-Gawles var. genuinus Maxim.
SAPONINS: ophiopogonins A, B, B′, C, C′, D and D′.
STEROIDS: ophiopogenin A, ophiopogenin B, ophiopogenin B′, ophiopogenin C, ophiopogenin C′, ophiopogenin D, ophiopogenin D′, ophiopogenin V.
Watanabe et al., *Chem. Pharm. Bull.*, **25**, 3049 (1977)

OPHIORRHIZA (Rubiaceae)

O. japonica Bl.
INDOLE ALKALOIDS: harman, ophiorine A, ophiorine B.
Aimi et al., *Tetrahedron Lett.*, 5299 (1985)

O. kuroiwai Mak.
INDOLE ALKALOIDS: harman, ophiorine A, ophiorine B.
Aimi et al., *Tetrahedron Lett.*, 5299 (1985)

OPHOSORIA

O. quadripinnata.
ALIPHATIC: 10-nonacosanone.
Soeder, Hodgin, *Phytochemistry*, **14**, 2508 (1975)

OPHYROSPORUS

O. chilca.
DITERPENOID: 2-oxolabda-8(17),13-dien-15-ol.
Bohlmann et al., *Phytochemistry*, **23**, 1513 (1984)

OPLOPANAX (Araliaceae)

O. japonicus.
SESQUITERPENOIDS: oplodiol, oplopanone.
Minato, Ishikawa, *J. Chem. Soc., C*, 423 (1967)

OPUNTIA (Cactaceae)

O. ficus-indica (L.) Mill.
PHENOLICS: neobetanin, piscidic acid, piscidic acid methyl ester.
Nordal et al., *Acta Chem. Scand.*, **20**, 1431 (1966)
Strack et al., *Phytochemistry*, **26**, 2399 (1987)

ORCHIS (Orchidaceae)

O. militaris.
GLUCOSIDES: loroglossin, miliarin.
Aasen et al., *Acta Chem. Scand.*, B **29**, 1002 (1975)

O. papilionacea L.
FLAVONOIDS: quercetin 3,7-diglucoside, quercetin 7-O-glucoside, quercetin 3-O-glucoside-7-O-rhamnoside, quercetin 3-O-glucoside-7-O-rutinoside.
Pagani et al., *Boll. Chim. Farm.*, **121**, 174 (1982); *Chem. Abstr.*, **97** 141709c

ORICIA

O. gabonensis.
ACRIDONE ALKALOID: evoxanthine.
QUINOLONE ALKALOID: 7-methoxy-N-methylflindersine.
TRITERPENOID: lupeol.
Khalid, Waterman, *Phytochemistry*, **20**, 2761 (1981)

O. renieri.
QUINOLONE ALKALOID: 7-methoxy-N-methylflindersine.
DIMERIC QUINOLONE ALKALOIDS: vepridimerines A, B, C and D.
Khalid, Waterman, *Phytochemistry*, **20**, 2761 (1981)
Njadju et al., *Tetrahedron*, **23**, 2041 (1982)

O. suaveolens.
QUINOLONE ALKALOIDS: 6,7-dimethoxy-N-methylflindersine, orcine.
Abe, Taylor, *Phytochemistry*, **10**, 1167 (1971)

ORICIOPSIS

O. glaberrima.
TERPENOID: oriciopsin.
Ayafor et al., *Phytochemistry*, **21**, 2602 (1982)

ORIGANUM (Labiatae)

O. heracleoticum.
MONOTERPENOID: 4,5-epoxy-p-menth-1-ene.
Lawrence et al., *Phytochemistry*, **13**, 1012 (1974)

ORIXA (Rutaceae)

O. japonica Thunb.
FURANOQUINOLINE ALKALOIDS: kokusagine, kokusaginine, kokusaginoline, skimmianine.
QUINOLONE ALKALOIDS: japonine, nororixine, orixin orixinone.
UNCHARACTERIZED ALKALOID: orixidine.
Terasaka et al., *J. Pharm. Soc., Japan*, **51**, 99 (1931)
Terasaka et al., *Chem. Pharm. Bull.*, **8**, 1142 (1960)

ORLANDO

O. tangelo x clementina.
TRITERPENOID: citraurinene.
Leuenberger, Stewart, *Phytochemistry*, **15**, 227 (1976)

ORMENSIS

O. multicaulis.
MONOTERPENOID: santolina alcohol.
Cretien-Bessiere et al., *Bull. Soc. Chim. Fr.*, 2018 (1969)

ORMOSIA

O. coutinhai.
SPARTEINE ALKALOID: 11-oxotetrahydrorhombifoline.
McLean et al., Can. J. Chem., **45**, 751 (1967)

O. dasycarpa Jacks.
UNCHARACTERIZED ALKALOID: dasycarpine.
Clarke, Grundon, J. Chem. Soc., 41 (1960)

O. jamaicensis (Urb.).
LUPANINE ALKALOIDS: jamaidine, jamine, ormojanine, ormojine, ormosajine, ormosanine, ormosinine.
Hassall, Wilson, J. Chem. Soc., 2657 (1964)

O. panamensis.
LUPANINE ALKALOID: jamine.
Naegali et al., Tetrahedron Lett., 2069, (1964)
Naegali et al., ibid., 2075 (1964)

O. semicastata.
LUPANINE ALKALOIDS: ormocastrine, 18-epi-ormosanine.
Konovalova et al., J. Gen. Chem., USSR, **21**, 773 (1951)
McLean et al., Can. J. Chem., **49**, 1976 (1971)

ORNITHOGLOSSUM (Liliaceae/Iphigenieae)

O. glaucum Salisb. var. grandiflora.
COLCHICINE ALKALOID: 3-demethylcolchicine.
Reichstein et al., Planta Med., **16**, 357 (1968)

O. viride Ait.
COLCHICINE ALKALOIDS: colchicine, N-formyl-N-deacetylcolchicine.
Reichstein et al., Planta Med., **16**, 357 (1968)

OROBANCHE (Orobanchaceae)

O. aegyptiaca Pers.
TROPANE ALKALOIDS: apoatropine, atropine, belladonnine, meteloidine.
Balbaa et al., Chem. Abstr., 87 164210r

O. arenaria.
PHENOLIC ESTERS: arenarioside, pheliposide.
Andary et al., J. Nat. Prod., **48**, 778 (1985)

O. lutea Baumg.
UNCHARACTERIZED ALKALOID: orobanhamine.
Rubinshtein et al., J. Gen. Chem., USSR, **23**, 157 (1953)

O. owerinii.
CAROTENOIDS: auroxanthin, violaxanthin.
Dzhumyrko et al., Khim. Prir. Soedin., 712 (1985)

O. rapum Thuill.
ESTERS: orobanchoside, verbascoside.
Andary et al., Chem. Abstr., 86 40165q

O. rapum-genistae Thuill.
ESTERS: orobanchoside, verbascoside.
Andary et al., Phytochemistry, **21**, 1123 (1982)

OROBUS (Papilionaceae)

O. tuberosus L. (Lathyrus montana Bernh.)
ISOFLAVONE: oroboside.
Charaux et al., Bull. Soc. Chim. Biol., **21**, 1330 (1939)

OROXYLUM

O. indicum.
FLAVONES: chrysin, oroxylin A.
Asakawa et al., Bull. Chem. Soc., Japan, **44**, 2761 (1971)

ORTHENTHERA

O. viminea.
SAPONIN: ornine.
STEROIDS: ornogenin, orgogenin, orthenine.
(Sap, S) Kaur et al., J. Nat. Prod., **48**, 938 (1985)
(S) Tiwari et al., Phytochemistry, **24**, 2391 (1985)

ORTHOPRETYGIUM

O. huanauy.
TRITERPENOID: 3-oxo-6-hydroxyurs-18-en-28-oic acid methyl ester.
Gonzalez et al., Phytochemistry, **22**, 1828 (1983)

ORTHOSIPHON

O. spicatus.
COUMARINS: eupatorin, scutellarein 5,6,7,4'-tetramethyl ether, sinensetin.
Wollenweber, Mann, Planta Med., 459 (1985)

ORTHOSPHENIA (Celastraceae)

O. mexicana Standley.
TRITERPENOID: orthosphenic acid.
Gonzalez et al., J. Org. Chem., **48**, 3759 (1983)

ORYCTES (Solanaceae)

O. nevadensis.
FLAVONOIDS: quercetin 3-O-rutinoside, quercetin 3-O-rutinoside-7-O-glucoside.
Averett et al., Phytochemistry, **22**, 2325 (1983)

ORYZA (Poaceae)

O. sativa.
STEROIDS: cellopentaosylsitosterol, cellotetraosylsitosterol.
DITERPENOID: oryzalexin A.
(S) Oinishi et al., Agr. Biol. Chem., **44**, 333 (1980)
(T) Akatsuka et al., ibid., **47**, 445 (1983)

O. sativa var. koshikikeri.
DITERPENOIDS: ineketone, momilactones A, B and C.
NORSESQUITERPENOID: (S)(+)-dehydrovomifoliol.
Kato et al., Phytochemistry, **16**, 45 (1977)

OSCILLATORIA (Cyanophyceae/Oscillatoriaceae)

O. limosa.
CAROTENOID: 4-oxomyxol-2'-glucoside.
Liaaen-Jensen et al., Phytochemistry, **10**, 3132 (1971)

O. rubescens.
CAROTENOIDS: oscillaxanthin, myxoxanthophyll.
Hertzberg, Liaaen-Jensen, Phytochemistry, **5**, 557 (1966)

OSMANTHUS (Oleaceae)

O. absolute.
DINORSESQUITERPENOID:
(8E)-4,7-epoxymegastigma-5(11),8-diene.
SESQUITERPENOIDS:
(7R,10S,5E)-trimethyl-7,10-epoxy-2,5,11-dodecatriene,
(7S,10S,5E)-trimethyl-7,10-epoxy-2,5,11-dodecatriene.
Kaiser, Lamparsky, Helv. Chim. Acta, **62**, 1878 (1979)

O. fortunei Carr.
LIGNANS: echinacoside,
guaiacylglycerol-4-O-β-D-glucopyranoside,

O. fragrans (Thunb.) Lour.
ALCOHOL: benzyl alcohol.
ALDEHYDE: pelargonaldehyde.
KETONE: β-ionone.
MONOTERPENOIDS: linalool, linalool-cis-furanoid oxide, linalool-trans-furanoid oxide.
SESQUITERPENOID: farnesol.
IRIDOIDS: 10-acetoxyoleuropein, 10-acetoxyligustroside.
Gogiya *et al.*, *Rastit. Resur.*, **22**, 243 (1986)

OSMITOPSIS (Compositae/Inuleae)
O. asteriscoides (L.) Cass.
SESQUITERPENOIDS: 1,8-epoxyosmitopsin, 4,5-epoxyosmitopsin, osmitopsin.
Bohlmann, Zdero, *Chem. Ber.*, **107**, 1409 (1974)

OSMUNDA (Osmundaceae)
O. cinnamomea.
STEROIDS: 24-ethyllathosterol, β-sitosterol, spinasterol.
Nes *et al.*, *Lipids*, **12**, 511 (1977)

OSTEOSPERMUM (Calendulaceae)
O. clandestinum (Less.) T.Norl.
One acetylenic compound.
Bohlmann, Zdero, *Chem. Ber.*, **108**, 362 (1975)

O. corymbosum L.
DITERPENOIDS:
11α-acetoxy-12β-hydroxysandaracopimar-15-en-8β-ol,
11α-hydroxy-12β-acetoxysandaracopimar-15-en-8β-ol,
sandaracopimar-15-en-8β-ol.
Bohlmann, Zdero, *Chem. Ber.*, **108**, 362 (1975)

O. eklonis.
Two polyacetylenes.
Bohlmann, Zdero, *Chem. Ber.*, **100**, 362 (1975)

O. fruticosum (L.) Norb.
DITERPENOIDS: 11α-acetoxysandaracopimar-15-ene-8β,12-diol.
Bohlmann *et al.*, *Chem. Ber.*, **106**, 826 (1973)

O. junceum Beig.
DITERPENOID: 12β-acetoxysandaracopimar-15-ene-8β,11α-diol.
Bohlmann *et al.*, *Chem. Ber.*, **106**, 826 (1973)

O. oppositifolium (Ait.) Norl.
DITERPENOIDS: 11α-acetoxysandaracopimar-15-en-8β,12-diol,
12-acetoxysandaracopimar-15-en-8β,11α-diol.
Bohlmann *et al.*, *Chem. Ber.*, **106**, 826 (1973)

O. polygalioides L.
DITERPENOID:
11α-hydroxy-12β-acetoxysandaracopimar-15-en-8β-ol.
Bohlmann, Zdero, *Chem. Ber*, **108**, 362 (1975)

O. rotundifolium (DC.) Norl.
DITERPENOIDS:
11α-acetoxy-12β-hydroxysandaracopimar-15-en-8β-ol,
11α-hydroxy-12β-acetoxysandaracopimar-15-en-8β-ol.
Bohlmann, Zdero, *Chem. Ber.*, **108**, 362 (1975)

O. subulatum DC.
DITERPENOID: 11α,12β-diacetoxysandaracopimar-15-en-8β-ol.
Bohlmann, Zdero, *Chem. Ber.*, **108**, 362 (1975)

OSTODES
O. paniculata.
PHORBOL TERPENOIDS: ostodin,
12-O-undecadienoylphorbol-13-acetate.
Handa *et al.*, *J. Nat. Prod.*, **46**, 123 (1983)

OSTRYODERIA
O. chevalieri Dunn.
One uncharacterized alkaloid.
Blaansard, Martin, *Bull. Sci. Pharmacol.*, **46**, 268 (1939)

OSYRIS
O. tenuifolia.
SESQUITERPENOID: lanceol.
Bradfield *et al.*, *J. Chem. Soc.*, 1619 (1936)

OTANTHUS (Compositae)
O. maritimus (L.) Hoffm. et Link.
Three sesquiterpenoids.
Price, Pinder, *J. Org. Chem.*, **35**, 2568 (1970)

OTHONNA
O. amplexicaulis Thunb.
SESQUITERPENOIDS:
6β-angeloyloxy-3β-isobutyryloxyfuranoeremophilan-15-oic acid,
3β-angeloyloxy-15β-isovaleryloxyfuranoeremophilan-15-oic acid.
Bohlmann *et al.*, *Chem. Ber*, **107**, 3928 (1974)

O. arborescens.
SESQUITERPENOID:
3β,6β-diangeloyloxyfuranoeremophilan-15-oic acid.
Bohlmann *et al.*, *Chem. Ber.*, **107**, 3928 (1974)

O. barkerae.
SESQUITERPENOIDS:
3β,6α-diangeloyloxyfuranoeremophilan-15-oic acid,
3β,6β-diangeloyloxyfuranoeremophilan-15-oic acid.
Bohlmann *et al.*, *Chem. Ber.*, **107**, 3928 (1974)

O. bulbosa.
SESQUITERPENOID:
3β-acetoxy-2β-angeloyloxy-10αH-furanoeremophilane.
Bohlmann *et al.*, *Chem. Ber.*, **107**, 3928 (1974)

O. complexicaulis.
SESQUITERPENOID:
10β-hydroxy-6β-isovaleryloxyfuranoeremophilane.
Bohlmann *et al.*, *Chem. Ber.*, **107**, 3928 (1974)

O. coronopifolia.
SESQUITERPENOIDS: 3β-angeloyloxyfuranoeremophilan-15-oic acid, 3β,6β-diangeloyloxyfuranoeremophilan-15-oic acid isoeremophilene, isohumulene.
Bohlmann *et al.*, *Chem. Ber.*, **107**, 3928 (1974)

O. dentata.
SESQUITERPENOIDS: 3β-angeloyloxyfuranoeremophilan-15-oic acid, 3β,6β-diangeloyloxyfuranoeremophilan-15-oic acid.
Bohlmann *et al.*, *Chem. Ber.*, **107**, 3928 (1974)

O. filicaulis.
SESQUITERPENOIDS:
3β-angeloyloxy-6β-acetoxyfuranoeremophilane,
3β-angeloyloxyfuranoeremophilane,
3β-angeloyloxy-6β-hydroxyfuranoeremophilane.
Bohlmann *et al.*, *Chem. Ber.*, **107**, 3928 (1974)

O. quercifolia.
SESQUITERPENOIDS:
6β-angeloyloxy-3β-senecioylfuranoeremophilan-15-oic acid,
3β,6β-diangeloyloxyfuranoeremophilan-15-oic acid.
Bohlmann *et al.*, *Chem. Ber.*, **107**, 3928 (1974)

OUROUPARIA

O. formosana Mats.
UNCHARACTERIZED ALKALOID: formosanine.

INDOLONE ALKALOID: uncarine A.
> Raymond-Hamet, *C. R. Acad. Sci.,* **203**, 1383 (1936)
> Hendrickson *et al., J. Am. Chem. Soc.,* **84**, 650 (1962)

O. gambir Baill. (Uncaria gambier Roxb.).
CARBOLINE ALKALOID: (-)-dihydrogambertannine.
> Merlin *et al., Tetrahedron,* **23**, 3129 (1967)

O. guianensis.
INDOLONE ALKALOID: rhyncophylline.
> Seaton, Marion, *Can. J. Chem.,* **35**, 1102 (1957)

O. kawakami Hayata.
UNCHARACTERIZED ALKALOID: hanadamine.
> Kondo, Oshima, *J. Pharm. Soc., Japan,* **52**, 63 (1957)

O. rhyncophylla Mats.
INDOLONE ALKALOIDS: isorhyncophylline, isorhyncophylline
N-oxide, rhyncophylline.
> Raymond-Hamet *et al., C. R. Acad. Sci.,* **199**, 587 (1934)

OXALIS (Oxalidaceae)

O. acetosella L.
FLAVANOID: 2″-glucoisovitexin.
> Tschesche *et al., Chem. Ber.,* **109**, 2901 (1976)

O. cernua.
GLUCOSIDE: cernuoside.
> Geissman *et al., J. Am. Chem. Soc.,* **77**, 4622 (1955)

OXYCOCCUS (Ericaceae)

O. quadripetalis.
FLAVANOIDS: hyperoside, myricetin, quercetin.
> Shelyuto *et al., Khim. Prir. Soedin.,* 515 (1975)

OXYSTIGMA

O. oxyphyllum.
DITERPENOIDS: enantio-labda-8(20),13-dien-15-oic acid,
eperua-7,13-dien-15-oic acid, eperu-7-en-15-oic acid,
labda-8(20),13-dien-15-oic acid.
> Ekong, Okogun, *J. Chem. Soc., Chem. Commun.,* 72 (1967)

OXYTROPIS (Leguminosae)

O. almaatensis.
Two unidentified phenolcarboxylic acids and three
unidentified flavonoids.
> Baimukhambetov, Chem. Abstr., *87 2356r*

O. komarovii.
FLAVONOIDS: astragalin, kaempferol, kaempferol
7-O-β-L-glucorhamnoside, neoisorutin, nicotiflorin,
oxytroside, quercetin.

STEROID: β-sitosterol.
> Baimukhambetov, Chem. Abstr., 87 2356r

O. muricata (Pall.) DC.
AMINE ALKALOID: N-benzoylphenylaminoethylcarbinol.
> Duboshina, Proskurnina, *J. Gen. Chem. USSR,* **33**, 2071 (1963)

O. myriophylla (Pall.) DC.
FLAVONOIDS: acetylmyrioside, coumaroylisooxymyrioside,
oxymyrioside, oxytroside.
> Blinova *et al., Rastit. Resur.,* **13**, 466 (1977)

O. pseudoglandulosa.
AMINES: cinnamoyl-β-phenethylamine,
(+)-(R)-N-benzoyl-2-phenyl-2-hydroxyethylamine,
(+)-(S)-benzoyl-2-phenyl-2-hydroxyethylamine.
FLAVONOIDS: chrysin, isoliquiritigenin.
> Huneck *et al., Fitoterapia,* **57**, 423 (1987)

P

PACHYCARPUS

P. lineolatus.
STEROID: utendin.

Abisch *et al.*, *Helv. Chim. Acta*, **42**, 1014 (1959)

PACHYCEREUS (Cactaceae)

P. pectenaboriginum.
ISOQUINOLINE ALKALOID: heliamine.

Strombon, Bruhn, *Acta Pharm. Suec.*, **15**, 127 (1978)

P. tehauntepecanus.
ISOQUINOLINE ALKALOID: tehaunine.

Kapadia *et al.*, *J. Chem. Soc., Chem. Commun.*, 856 (1970)

P. weberi.
ISOQUINOLINE ALKALOIDS: lemaireocereine, nortehaunine, weberidine, weberine.

Mata, McLaughlin, *Phytochemistry*, **19**, 633 (1980)
Mata, McLaughlin, *ibid.*, **19**, 673 (1980)

PACHYDICTYON (Phaeophyceae)

P. coriaceum.
DITERPENOIDS: acetoxycrenulatin, acetoxypachydiol, acetylcoriacenone, 19-epi-acetylcoriacenone, acetyldictyol C, acetyldictyolal, 6-acetylsanadiol, dilophol acetate, 3-hydroxyacetyldilophol, neodictyolactone, pachyaldehyde, pachydictyol A, pachylactone, sanadaol.

Ishitsuja *et al.*, *J. Chem. Soc., Chem. Commun.*, 906 (1982)
Ishitsuka *et al.*, *Tetrahedron Lett.*, 2639 (1986)

PACHYGONE

P. ovata.
APORPHINE ALKALOID: reticuline N-oxide.
BISBENZYLISOQUINOLINE ALKALOIDS: N-methylpachygonamine, pachygonamine, pachyovatamine, tiliamosine.

Dasgupta *et al.*, *Lloydia*, **42**, 399 (1979)
Uvais *et al.*, *Heterocycles*, **20**, 1927 (1983)
Sultanbawa *et al.*, *Phytochemistry*, **24**, 589 (1985)

PACHYPODANTHIUM (Annonaceae)

P. confine Engl. et Diels.
STYRENE: 2,4,5-trimethoxystyrene.

Bevalot *et al.*, *Plant Med. Phytother.*, **10**, 179 (1976)

P. staudtii Engl. et Diels.
APORPHINE ALKALOID: pachypodanthine.
QUATERNARY ALKALOID: staudine.

Cave *et al.*, *J. Nat. Prod.*, **43**, 103 (1980)

PACHYRRHIZUS (Leguminosae/Papilionaceae)

P. erosus.
ISOFLAVANS: dehydroneotenone, erosenone, neotenone, pachyrrhizine.
ROTENOIDS: erosnin, erosone, 12α-hydroxyerosone, 12α-hydroxymunduserone, 12α-hydroxyrotenon, pachyrrhizone, rotenone.

(I) Crombie *et al.*, *J. Chem. Soc.*, 1569 (1963)
(R) Kalra *et al.*, *Indian J. Chem.*, **15B**, 1084 (1977)

PACHYSANDRA (Buxaceae)

P. terminalis Sieb. et Zucc.
STEROIDAL ALKALOIDS: O-deacetylpachysandrine A, O-deacetylpachysandrine B, epipachysamines A, B, C, D, E and F, pachysamine A, pachysamine B, pachysandrines A, B, C, D and E, pachysantermine A, pachysantermine B, pachystermine A, pachystermine B, spiropachysine, terminaline.
TRITERPENOIDS: pachysana-16,21-diene-3β,28-diol, pachysandienol A, pachysandienol B, pachysandiol A, pachysandiol B, pachysan-16-en-3β,28-diol, pachysantriol, pachysonol, 2α,3β,16β-trihydroxyfriedelane.

(A) Tomita *et al.*, *Tetrahedron Lett.*, 1053 (1964)
(A) Kikuchi *et al.*, *ibid.*, 2077 (1968)
(T) Yokoi *et al.*, *Chem. Pharm. Bull.*, **33**, 4223 (1985)

PAEDERIA (Rubiaceae)

P. foetida.
ALIPHATICS: ceryl alcohol, hentriacontane, hentriacontanol.
IRIDOIDS: asperuloside, paederoside, scandoside.
STEROIDS: campesterol, β-sitosterol, stigmasterol.

Shukla *et al.*, *Phytochemistry*, **15**, 1989 (1976)

P. scandens.
IRIDOID: scandoside.
THIOIRIDOIDS: paederoside, paederosidic acid.

Inouye *et al.*, *Phytochemistry*, **13**, 2219 (1974)

P. scandens (Lour.) Merrill var. mairei (Lev.)
HARA IRIDOID: scandoside.

Inouye *et al.*, *Tetrahedron Lett.*, 683 (1968)

PAEONIA (Ranunculaceae/Paeoniaceae)

P. albiflora.
TERPENOIDS: albiflorin, paeoniflorigenone, paeoniflorin, paeonifloriquinone.

Shimizu *et al.*, *Tetrahedron Lett.*, 3069 (1981)

P. chinensis.
TERPENOID: paeoniflorin.

Shibata *et al.*, *Tetrahedron Lett.*, 1991 (1964)

P. lactiflora Pall.
PINENES: lactifloric acid, lactiflorin, (Z)-1S,5R-α-pinen-10-yl-α-vicianoside.

Lang *et al.*, *Yaoxue Xuebao*, **18**, 551 (1983); *Chem. Abstr.*, 100 82714f

PAGIANTHA (Apocynaceae)

P. cerifera (Pasncher et Sebert) Mkgf.
INDOLE ALKALOIDS: (-)-ibogaine, pagicerine, (-)-voacangine, (-)-voacangine hydroxyindolenine.
THIOINDANE: pagisulfine.

(A) Bert *et al.*, *Heterocycles*, **23**, 2505 (1985)
(Th) Bert *et al.*, *Heterocycles*, **24**, 1567 (1986)

PAJANELIA

P. multijuga P.DC.
FLAVONOIDS: chrysin, oroxylin. A

Joshi *et al.*, *Proc. Indian Acad. Sci.*, **86A**, 41 (1977)

PALEFOXIA

P. arida.
DITERPENOID: darutigenol.
Herz et al., Phytochemistry, **17**, 1060 (1978)

P. rosea.
DITERPENOIDS: 18-benzylcardenasol, jesromotetrol, norjulslimdiolone acetonide, 7-pimarene-3β,15,16,18-tetraol, 7-pimarene-3,15,16-triol.
Dominguez et al., Rev. Latinoam. Quim., **9**, 99 (1978)
Bohlmann, Czerson, Phytochemistry, **18**, 115 (1979)

PALAQUIM

P. trembi.
TRITERPENOID: palireubin.
Jungfleisch, Leroux, C. R. Acad. Sci., **142**, 1218 (1906)

PALICOURIA (Rubiaceae)

P. alpina (Sw.) DC.
DIAZAPHENANTHRENE ALKALOID: calycanthine.
CARBOLINE ALKALOID.: palinine.
Stuart, Woo-Ming, Tetrahedron Lett., 3853 (1974)

P. rigida H.B.K.
UNCHARACTERIZED ALKALOID: douradine.
Santesson, Arch. Pharm., **235**, 143 (1897)

PALYTHOA

P. tuberculosa.
AZINE ALKALOIDS: isopalythazine, palythazine.
Uemura et al., Chem. Lett., 1481 (1979)

PANACIS

P. japonicus.
SAPONIN: glycoside F-1.
Lui et al., Chem. Pharm. Bull., **24**, 253 (1976)

PANAX (Araliaceae)

P. ginseng C.A.Meyer.
FLAVONOIDS: kaempferol, panasenoside, trifolin.
SAPONINS: ginsenolide Lb2, ginsenolide Lc, ginsenolide Lc2, ginsenolide Ld, ginsenolide Le, ginsenolide Lg1, ginsenolide Lg2, ginsenoside Rb1, ginsenoside Rd, ginsenoside Re, ginsenoside Rf, ginsenoside Rg, 3-O-β-glucopyranosyl-20S-protopanaxadiol, glycoside P-1, notoginsenoside R-1, notoginsenoside R-2, panaxoside A.
STEROIDS: (20R)-protopanaxadiol, (20S0-protopanaxadiol.
SESQUITERPENOIDS: neoclovene, α-panasinsene, β-panasinsene.
TRITERPENOIDS: panaxadiol, panaxatriol.
(F) Wang et al., Chem. Abstr., 104 85413a
(Sap, S) Lin, Lin, J. Chin. Chem. Soc., **23**, 107 (1981)
(Sap) Cai et al., Yaoxue Tongbao, **17**, 500 (1982); Chem. Abstr., 97 212635g
(T) Zhou et al., Chem. Pharm. Bull., **29**, 2844 (1981)

P. ginseng radix rubra.
SAPONIN: ginsenoside Rh-2.
Kitagawa et al., J. Pharm. Soc., Japan, **103**, 612 (1983)

P. japonica Major.
SAPONINS: chikusetsusaponins I, Ia, II, III, IV, IVa, L-5, L-9a, L-10, LT-5 and LT-8, majonoside R-1, majonoside R-2.
Lin, Shoji, J. Chin. Chem. Soc., **26**, 29 (1979)
Morita et al., Chem. Pharm. Bull., **30**, 4341 (1982)

P. notoginseng.
SAPONINS: notoginsenoside Fa, notoginsenoside Fc, notoginsenoside Fe, notoginsenoside R-1, notoginsenoside R-2.
Zhou et al., Chem. Pharm. Bull., **29**, 2844 (1981)
Yang et al., Phytochemistry, **22**, 1473 (1983)

P. pseudoginseng subsp. himalaicus.
SAPONINS: pseusoginsenoside F-8, pseudoginsenoside F-11.
Tomaka, Yohara, Phytochemistry, **17**, 1353 (1978)

P. pseudoginseng subsp. himalaicus var. angustifolius.
ALIPHATICS: 24-hydroxyhexatetracontanoic acid, 2-methylhexatetracont-1-en-3,21-diol, tritriacontenyl octacosanoate.
SAPONINS: chikusetsusaponin IV, chikusetsusaponin V, ginsenoside R-0, ginsenoside R-b1.
(Al) Shukla et al., Phytochemistry, **25**, 2201 (1986)
(Sap) Kondo et al., Chem. Pharm. Bull., **21**, 2705 (1973)

P. quinquefolium L.
SAPONINS: ginsenoside R-0, ginsenoside Rb-1, ginsenoside Re, ginsenoside Rg-1, pseudoginsenoside F-11, quinquenoside R-1.
Basso et al., Chem. Pharm. Bull., **30**, 4534 (1980)
Wei et al., Zhongcaoyao, **14**, 389 (1983); Chem. Abstr., 100 3501h

P. zingiberensis Wu et Feng.
SAPONINS: chikusetsusaponin IV, ginsenoside R-0, ginsenoside Rh-1, zingibroside R-1.
Yang et al., Yaoxue Xuebao, **19**, 232 (1984); Chem. Abstr., 103 3686j

PANCRATIUM (Amaryllidaceae)

P. arabicum.
UNCHARACTERIZED ALKALOID: sickenbergine.
Ahmad et al., Lloydia, **27**, 115 (1964)

P. biflorum.
BENZOPYRAN: biflorin.
Ghosal et al., Phytochemistry, **22**, 2591 (1983)

P. longiflorum Roxb.
INDOLE ALKALOID: norneronine.
Rangaswami, Krishna, Tetrahedron Lett., 4481 (1966)

P. maritimum L.
ISOQUINOLINE ALKALOID: lycorine.
LUPINANE ALKALOID: pancracine.
UNCHARACTERIZED ALKALOID: sickenbergine.
Sandberg, Michel, Lloydia, **26**, 78 (1963)
Ahmad et al., ibid., **27**, 115 (1964)

P. sickenbergeri.
UNCHARACTERIZED ALKALOID: sickenbergine.
Ahmad et al., Lloydia, **27**, 115 (1964)

P. tortuosum.
UNCHARACTERIZED ALKALOID: sickenbergine.
Ahmad et al., Lloydia, **27**, 115 (1964)

PANDA

P. oleosa Pierre.
PEPTIDE ALKALOID: pandamine.
Pais et al., Bull. Soc. Chim. Fr., 817 (1964)

PANDACA (Apocynaceae)

P. boiteau.
INDOLE ALKALOID: ervitsine.
Adrianisiferana et al., Tetrahedron Lett., 2587 (1977)

P. caducifolia.
CARBAZOLE ALKALOIDS: pandine, pandoline.
LeMen et al., Tetrahedron Lett., **483**, 3119 (1974)

P. ensepala.

IMIDAZOLIDINOCARBAZOLE ALKALOID:
(+)-20S-pseudoaspidospermidine.
Quirin et al., Phytochemistry, **14**, 812 (1975)

P. mocquerysii var. pendula.

INDOLE ALKALOIDS: coronaridine, (-)-heyneanine,
(-)-19-epi-heyneanine, 19-hydroxycoronaridine,
(-)-voacangarine, (-)-19-epi-voacangarine, (-)-voacangine.
Beilefon et al., Phytochemistry, **14**, 1649 (1975)

P. multiflora.

One uncharacterized indole alkaloid.
Petitfrere et al., Phytochemistry, **14**, 1648 (1975)

P. ochrascens.

CARBAZOLE ALKALOID: 19,20-dihydrocondylocarpine.
HOMOCARBOLINE ALKALOIDS,: 19epi-iboxygaine,
19-epi-iboxygaline.
Panas et al., Phytochemistry, **13**, 1369 (1974)

P. retusa.

UNCHARACTERIZED ALKALOID: (-)-pachysiphine.
LeMen et al., Phytochemistry, **13**, 280 (1974)

PANELLUS (Agaricales)

P. serotinus.

PIGMENT: 3-methylriboflavin.
Steglich et al., Z. Naturforsch., **32C**, 520 (1977)

PANGIUM (Flacourtiaceae)

P. edule Reinw.

CARBONITRILE: gynocardin, hydrocyanic acid.
Kim et al., J. Chem. Soc., Chem. Commun., 381 (1970)

PANICUM (Gramineae)

P. miliaceum.

TRITERPENOID: miliacin.
Ito, J. Chem. Soc., Japan, **55**, 910 (1934)

PAPAVER (Papaveraceae)

P. alborosum Hulten.

QUINAZOLONE ALKALOID: alborine.
PROTOBERBERINE ALKALOID: mecambridine.
Pfeifer, Thomas, Naturwissenschaften, **21**, 701 (1966)

P. alpinum L.

HOMOISOQUINOLINE ALKALOIDS: alpinigenine, alpinine.
ISOQUINOLINE ALKALOID: amurensine, amurensinine.
Maturova et al., Collect. Czech. Chem. Commun., **32**, 419 (1967)

P. alpinum L. subsp. alpinum.

HOMOISOQUINOLINE ALKALOID: papaverrubine G.
Pfeifer, Dohnert, Pharmazie, **22**, 343 (1967)

P. anomalum.

ESTER ALKALOID: pavanoline.
Pfeifer, Thomas, Pharmazie, **22**, 454 (1967)

P. arenarium.

ISOQUINOLINE ALKALOID: arenine.
Israilov et al., Khim. Prir. Soedin., 417 (1978)

P. armenaicum N.Busch.

BENZYLISOQUINOLINE ALKALOID: armepavine.
Yunusov et al., J. Gen. Chem., USSR, **10**, 641 (1940)

P. atlanticum.

OXOQUINOLIZIDINE ALKALOID: oxoprotopine.
Preininger, Santavy, Pharmazie, **25**, 356 (1970)

P. bracteatum Lindl.

APORPHINE ALKALOIDS: bracetoline, isothebaine.
BENZYLISOQUINOLINE ALKALOID.: papaverine Homoisoquinoline
alkaloid alpinigenine.
ISOQUINOLINE ALKALOID: N-methylcorydaldine.
PHENANTHRENE ALKALOIDS: 14β-hydroxycodeine,
14β-hydroxycodeinone.
SPIROISOQUINOLINE ALKALOIDS: bracteine, (-)-orientalinone.
UNCHARACTERIZED ALKALOIDS: bractamine, oripavine.
Heydenreich, Pfeifer, Pharmazie, **22**, 124 (1967)
Theuns et al., Phytochemistry, **16**, 753 (1977)

P. californicum.

Uncharacterized glycoalkaloid.
Nemeckova et al., Collect. Czech. Chem. Commun., **35**, 1733 (1970)

P. caucasicum Marsch.-Bieb.

APORPHINE ALKALOIDS: L-(+)-nuciferoline, (-)-nuciferoline.
ISOQUINOLINE ALKALOID: papaverrubine C.
SPIROISOQUINOLINE ALKALOIDS: fungipavine, mecambridine,
pronuciferine.
Slavik, Slavikova, Collect. Czech. Chem. Commun., **28**, 1720 (1963)
Pfeifer et al., Pharmazie, **23**, 267 (1968)

P. commutatum Fisch et Mey.

Three uncharacterized non-phenolic alkaloids.
Slavik et al., Collect. Czech. Chem. Commun., **30**, 3961 (1965)

P. dubium L.

HOMOISOQUINOLINE ALKALOID: dubirheine.
UNCHARACTERIZED ALKALOIDS: aporheidine, aporheine.
Paves, J. Chem. Soc. Abstr., **1**, 368 (1905)

P. dubium var. glabrum.

SPIROISOQUINOLINE ALKALOID: oxyhydrastinine.
Hussain et al., Phytochemistry, **22**, 319 (1983)

P. floribundum Desf.

UNCHARACTERIZED ALKALOIDS: floripavidine, floripavine.
Yunusov et al., J. Gen. Chem., USSR, **10**, 641 (1940)

P. fugax Poir.

BENZYLISOQUINOLINE ALKALOID: armepavine.
APORPHINE ALKALOID: (-)-isoroemerine.
SPIROISOQUINOLINE ALKALOIDS: fungipavine, mecambridine.
UNCHARACTERIZED ALKALOID: floripavine.
Bick, Experientia, **20**, 362 (1964)

P. glaucum Boiss. et Hauskn.

HOMOISOQUINOLINE ALKALOIDS: glaucamine, epi-glaucamine,
glaudine, papaverrubine C.
Slavik et al., Collect. Czech. Chem. Commun., **30**, 2864 (1965)
Slavik et al., ibid., **30**, 3687 (1965)

P. hybridum L.

UNCHARACTERIZED ALKALOID: pahybrine.
Platonova et al., J. Gen. Chem., USSR, **26**, 181 (1956)

P. lisae.

BENZYLISOQUINOLINE ALKALOID: macrantaline.
Melik-Guseinov et al., Khim. Prir. Soedin., 239 (1979)

P. macrostomum Boiss.

BENZYLISOQUINOLINE ALKALOID: sevanine.
BENZYLNAPHTHALENE ALKALOID: macrostomine.
Mnatsakanyan et al., Tetrahedron Lett., 851 (1974)

P. nudicaule L.

PROTOBERBERINE ALKALOID: mecambridine.
OXOPROTOBERBERINE ALKALOID: muramine.
Maturova et al., Planta Med., **14**, 22 (1966)

P. nudicaule L. var. amurense.

LUPINANE ALKALOIDS: amurensine, amurensinine.
PHENANTHRENE ALKALOIDS: amurine, nudaurine.
SPIROISOQUINOLINE ALKALOIDS: amuroline, amuronine.
Maturova et al., Planta Med., **14**, 22 (1966)

P. nudicaule L. var. auranticum.
UNCHARACTERIZED ALKALOID: nudaurine.
Boit, Flentje, *Naturwissenschaften*, **47**, 180 (1960)

P. oreophilum Rupr.
HOMOISOQUINOLINE ALKALOIDS: oreodine, oreogenine, papaverrubine F.
PROTOBERBERINE ALKALOID: mecambridine.
OXOPROTOBERBERINE ALKALOID: oreophiline.
SPIROISOQUINOLINE ALKALOIDS: oreoline, oridine, isooridine, N-methyloreoline.
Maturova et al., *Planta Med.*, **14**, 22 (1966)
Mann, Pfeifer, *Pharmazie*, **22**, 124 (1967)

P. orientale L.
APORPHINE ALKALOID: orientinine.
PHENANTHRENE ALKALOIDS: isothebaine, oripavine, thebaine.
PROTOBERBERINE ALKALOID: orientalidine.
SPIROISOQUINOLINE ALKALOID: dihydroorientalinone.
OXOQUINOLIZIDINE ALKALOID: protopine.
Israilov et al., *Khim. Prir. Soedin.*, 258 (1984)

P. pavonium Schrenk.
OXOQUINOLIZIDINE ALKALOID: protopine.
OXOPROTOBERBERINE ALKALOID: α-allocryptopine.
UNCHARACTERIZED ALKALOID: roemeridine.
Platonova et al., *J. Gen. Chem., USSR*, **26**, 173 (1956)

P. persicum Lindl.
BENZYLISOQUINOLINE ALKALOID: armepavine.
APORPHINE ALKALOIDS: O-demethylnuciferine, funpavine, L-(+)-nuciferolin, roemerine.
OXOQUINOLIZIDINE ALKALOID: protopine.
Preininger et al., *Collect. Czech. Chem. Commun.*, **32**, 2682 (1967)
Israilov et al., *Khim. Prir. Soedin.*, 258 (1984)

P. polychaetum.
APORPHINE ALKALOID: L-(+)-nuciferoline.
Pfeifer et al., *Pharmazie*, **23**, 267 (1968)

P. pseudocanescens.
QUATERNARY ALKALOID: mecambridine methohydroxide.
Novac, Slavik, *Collect. Czech. Chem. Commun.*, **39**, 883 (1974)

P. pseudoorientale.
BENZYLISOQUINOLINE ALKALOID: macrantaline.
Melik-Guseinov et al., *Khim. Prir. Soedin.*, 239 (1979)

P. radicatum.
HOMOISOQUINOLINE ALKALOID: papaverrubine.
ISOQUINOLINE ALKALOIDS: amurensinine, O-methylthalisopavine.
NAPHTHOISOQUINOLINE ALKALOID: sanguinarine.
OXOQUINOLIZIDINE ALKALOID: protopine.
OXOPROTOBERBERINE ALKALOIDS: β-allocryptopine, cryptopine.
PROTOBERBERINE ALKALOID: berberine.
Boehm et al., *Planta Med.*, **28**, 210 (1975)

P. rhoeas L.
AMINE ALKALOID: adlumiceine.
HOMOISOQUINOLINE ALKALOIDS: isorhoeadine, isorhoeagenin, papaverrubines A, B, D and E, rhoeadine.
GLYCOALKALOID: isorhoeagenine glucoside.
Preininger et al., *Phytochemistry*, **12**, 2513 (1973)

P. somniferum L.
AMINE ALKALOID: narceine.
BENZYLISOQUINOLINE ALKALOIDS: (+)-codamine, (+)-reticuline.
ISOQUINOLINE ALKALOIDS: hydrocotarnine, narcotaline, narcotine, narcotoline, palaudine, papaverine, (±)-reticuline, thebaine.
HOMOISOQUINOLINE ALKALOIDS: N-methyl-14-deacetylisoporphyroxine, papaverrubines A, B and C, porphyroxine.

NAPHTHOISOQUINOLINE ALKALOIDS: 6-acetonyldihydrosanguinarine, norsanguinarine.
PHENANTHRENE ALKALOIDS: codeine, 16-hydroxythebaine, morphine.
PROTOBERBERINE ALKALOIDS: (-)-canadine, (-)-scoulerine.
OXOPROTOBERBERINE ALKALOID: 13-oxocryptopine.
UNCHARACTERIZED ALKALOIDS: lanthopine, papaveramine, pseudomorphine.
Manske, *Can. J. Res.*, **18B**, 75 (1940)
Manske, *ibid.*, **18B**, 288 (1940)
Brockmann-Hanssen et al., *Planta Med.*, **18**, 366 (1970)
Furuya et al., *Phytochemistry*, **11**, 3041 (1972)

P. triniifolium.
APORPHINE ALKALOIDS: floripavidine, L-(+)-nuciferoline.
BENZYLISOQUINOLINE ALKALOID: papaverine.
MORPHINANDIENONE ALKALOIDS: amurine, salutaridine.
PROAPORPHINE ALKALOIDS: mecambridine, sinactine.
PROTOBERBERINE ALKALOIDS: cheilanthifoline, scoulerine.
Pfeifer et al., *Pharmazie*, **23**, 276 (1968)
Sariyar, *Planta Med.*, **49**, 43 (1983)

P. urbanianum.
ISOQUINOLINE ALKALOID: N-methyl-6,7-dimethyltetrahydroisoquinoline.
PHENANTHRENE ALKALOID: O-methylsalutaridine.
UNCHARACTERIZED ALKALOID: 6,7-didehydroaporheine.
Preininger, Tosnavova, *Planta Med.*, **23**, 233 (1973)
Manustakyan et al., *Khim. Prir. Soedin.*, 849 (1980)

PAPUACEDRUS (Cupressaceae)
P. torricellenis (Schlechter) Li.
TROPOLONE: 3-isopropyl-7-methoxytropone.
Zavarin, *J. Org. Chem.*, **27**, 3368 (1962)

PARABENZOIN
P. praecox.
SESQUITERPENOID: 8α-acetoxyelemol.
Ohara et al., *Bull. Chem. Soc., Japan*, **46**, 641 (1973)

P. trilobum Nakai.
SESQUITERPENOIDS: shiromodiol 6-acetate, shiromodiol diacetate, shiromool.
Wada et al., *Tetrahedron Lett.*, 4673 (1968)

PARACARYUM (Boraginaceae)
P. himalayense (Klotsch.) C.B.Clarke.
PYRROLIZIDINE ALKALOIDS: viridiflorine, viridiflorine N-oxide.
Men'shikov et al., *J. Gen. Chem., USSR*, **18**, 1736 (1948)

PARACOFFEA
P. bengalensis.
No caffeine present.
Charrier, Berthand, *Chem. Abstr.*, 84 118445m

P. erbracteolata.
No caffeine present.
Charrier, Berthand, *Chem. Abstr.*, 84 118445m

P. humbertii.
No caffeine present.
Charrier, Berthand, *Chem. Abstr.*, 84 118445m

PARACYNOGLOSSUM (Boraginaceae)
P. imeretinum (Kusn.) M.Pop.
PYRROLIZIDINE ALKALOIDS: echinatine, echinatine N-oxide, heliosupine, heliosupine N-oxide.
Man'ko, Marchenko, *Khim. Prir. Soedin.*, 537 (1971)

PARAGEUM

P. montanum.

FLAVONOIDS: hyperin, quercetin, rutin.

Vainagii *et al., Rastit. Resur.,* 12, 199 (1976); *Chem. Abstr.,* 85 59886d

PARASTREPHIA

P. lepidophylla.

BENZOFURAN: tremetone.

Bohlmann *et al., Phytochemistry,* 16, 1973 (1977)

PARAVALLARIS

P. maingayi.

STEROID ALKALOID: maingayine.

Davis *et al., Chem. Ind.,* 627 (1970)

P. microphylla Pitard.

STEROIDAL ALKALOIDS: 7β-hydroxyparavallardine, 7α-hydroxyparavallarine, 11α-hydroxyparavallarine, N-methylparavallaridine, N-methylparavallarine, paravallaridine, paravallarine.

STEROIDS:
10β,20S-dihydroxy-3-oxopregna-1,4-diene-18-carboxylic acid lactone, 16α,20S-dihydroxy-3-oxopregn-4-ene-18-carboxylic acid lactone, 20S-hydroxy-3,6-dioxopregnane-18-carboxylic acid-18,20-lactone, 20S-hydroxy-3,6-dioxopregn-4-ene-18-carboxylic acid-18,20-lactone, 20S-hydroxy-3-oxo-4,6-pregnadiene-18-carboxylic acid-18,20-lactone, 20S-hydroxy-3-oxo-4-pregnene-18-carboxylic acid-18,20-lactone.

(A) Husson *et al., Bull. Soc. Chim. Fr.,* 3162 (1969)

(S) Husson *et al., ibid.,* 1686 (1971)

PARENTUCELLA (Schophulariaceae)

P. flaviflora (Boiss.) Nevski.

One uncharacterized alkaloid.

Platonova *et al., Med. Prom.,* 3, 21 (1963)

PARIS (Liliaceae)

P. axialis.

SAPONIN:
24α-hydroxypennogenin-3-O-α-L-rhamnopyranosyl-α-arabinofuranosyl-β-D-glucopyranoside.

Chen *et al., Yunnan Zhiwu Yanjiu,* 9, 239 (1987); *Chem. Abstr.,* 107 194916z

P. polyphylla L.

SAPONINS: 3-hydroxypregna-5,16-dien-20-one 3-O-chacotrioside, pariphyllin A, pariphyllin B.

Khanna *et al., Indian J. Chem.,* 13, 781 (1975)

PARMELIA (Parmeliaceae)

P. caperata (L.) Arch.

STEROID: ergosterol.

TERPENOID: pinastric acid.

Serra *et al., J. Pharm. Sci.,* 65, 737 (1976)

P. caraccensis.

DIOXAPIN: galbinic acid.

TERPENOIDS: acetylsalazinic acid, salazinic acid, usnic acid.

Huneck, Schreiber, *Phytochemistry,* 15, 437 (1976)

P. consperma (Ehrh. ex Ach.) Ach.

TERPENOID: salazinic acid.

Malik *et al., Indian J. Chem.,* 10, 1040 (1972)

P. entotheichroa.

TRITERPENOIDS: 6-O-acetylleucotylin, 6-deoxy-16-acetylleucotylin.

Yosioka *et al., Chem. Pharm. Bull.,* 14, 804 (1966)

P. formosana.

XANTHONE: lichexanthone.

Ruben *et al., Phytochemistry,* 15, 1093 (1976)

P. horrescens Tayl.

DEPSIDES: chloroatranorin, 4,5-di-O-methylhiascic acid, gyrophoric acid, lecanoric acid.

Elix, Engkaninan, *Aust. J. Chem.,* 29, 2701 (1976)

P. leucotyliza.

TRITERPENOIDS: leucotylic acid, leucotylin, zeorin.

Huneck, *Chem. Ber.,* 94, 614 (1961)

P. livida.

DIOXEPINS: hydroxycolensoic acid, methoxycolensoic acid.

Culberson *et al., Aust. J. Chem.,* 30, 599 (1977)

P. pseudofatiscens.

DEPSIDES: atranorin, chloroatranorin, 4,5-di-O-methylhiascic acid, gyrophoric acid, lecanoric acid.

Elix, Engkaninan, *Aust. J. Chem.,* 29, 2701 (1976)

P. reptans.

ACID: succinoprotocetraric acid.

Baker *et al., Aust. J. Chem.,* 21, 137 (1973)

P. tinctorium Despr.

TRITERPENOIDS: 3β-acetoxy-12β,22-dihydroxyhopane 3β,22-dihydroxyhopane, 6α,22-dihydroxyhopane.

Goto *et al., Chem. Abstr.,* 107 194875k

PARMOTREME

P. conformatum.

DEPSIDONE: malonprotocetraric acid.

Keogh, *Phytochemistry,* 16, 1102 (1977)

PARNARA

P. guttata Bremer.

STEROIDS: cholestan-3-one, cholesterol, β-sitosterol.

Endo, Mitsuhashi, *Chem. Abstr.,* 86 2381z

PARQUETINA

P. nigrescens (Afzel.) Bullock.

STEROIDS: convalloside, nigrescigenin.

Schenker *et al., Helv. Chim. Acta,* 37, 1004 (1954)

PARSONIA (Apocynaceae)

P. heterophylla A.Cunn.

PYRROLIZIDINE ALKALOIDS: heterophylline, parsonine.

SESQUITERPENOID: heterophyllol.

Edgar *et al., Tetrahedron Lett.,* 3053 (1979)

P. spiralis Wall.

PYRROLIZIDINE ALKALOIDS: heterophylline, spiracine, spiraline, spiranine.

Edgar *et al., Tetrahedron Lett.,* 2657 (1980)

PARTHENIUM (Compositae)

P. alpinum var. tetraneuris.

SESQUITERPENOIDS: tetraneurins A, B and C.

Ruesch, Mabry, *Tetrahedron,* 25, 805 (1969)

P. argentatum.

SESQUITERPENOID: parthenol.

TRITERPENOIDS: argentatines A, B and C.

Rodriguez-Hahn *et al., Rev. Latinoam. Quim.,* 1, 24 (1970)

P. bipinnatifidum.
FLAVONOID: quercetagetin-3,7,3'-trimethyl ether.
Shen et al., Phytochemistry, **15**, 1045 (1976)

P. confertum.
SESQUITERPENOIDS: conchosin A, conchosin B.
Romo de Vivar et al., Phytochemistry, **17**, 279 (1978)

P. confertum var. lyratum.
SESQUITERPENOID: tetraneurin F.
Yoshioka et al., J. Org. Chem., **35**, 2888 (1970)

P. fruticosum.
SESQUITERPENOIDS: fruticin A, fruticin B, incaulin, tetraneurin B.
Matsubara, Romo de Vivar, Phytochemistry, **24**, 613 (1985)

P. glomeratum.
FLAVONOID: quercetagetin-3,7,3'-trimethyl ether.
Shen et al., Phytochemistry, **15**, 1045 (1976)

P. hysterophorus.
FLAVONOID: quercetagetin-3,7-dimethyl ether.
TERPENOIDS: dihydroisoparthenin, 11-hydroxyguaien, isoquaiene, parthenin.
(F) Shen et al., Phytochemistry, **15**, 1045 (1976)
(T) Bohlmann et al., Phytochemistry, **16**, 575 (1977)
(T) Picman et al., ibid., **21**, 1801 (1982)

P. lozanianum.
SESQUITERPENOIDS: tetraneurins B, C and D.
Yoshioka et al., Tetrahedron, **26**, 2167 (1970)

P. rollinsianum.
FLAVONOIDS: artemetin, crysoplenol, penduletin, polycladin, quercetagin-3,6,7-trimethyl ether, quercetin-3,3'-dimethyl ether.
Maurice et al., Phytochemistry, **15**, 517 (1976)
Shen et al., ibid, **15**, 1045 (1976)

P. tomentosum.
SESQUITERPENOID: stramonin B.
Grieco et al., J. Org. Chem., **43**, 4552 (1978)

P. tomentosum var. stramonium.
SESQUITERPENOIDS: stramonin A stramonin B.
Grieco et al., J. Org. Chem., **43**, 4552 (1978)

PASANIA
P. edulis.
HYDROCARBONS: ditriacontane, heptacosane, hexacosane, monotriacontane, nonacosane, octacosane, pentacosane, triacontane, tritriacontane.
STEROID: β-sitosterol.
TRITERPENOIDS: friedelanol, epi-friedelanol, friedelin, glutinol, lupeol.
Tachi et al., J. Pharm. Soc., Japan, **98**, 349 (1978)

PASPALIUM
P. deltatum.
TRITERPENOID: lupeol methyl ether.
Ohmoto et al., J. Chem. Soc., D, 601 (1969)

PASSIFLORA (Passifloraceae)
P. capsularis.
GLYCOSIDE: passicapsin.
Fischer et al., Planta Med., **45**, 42 (1982)

P. coccinea.
GLYCOSIDE: passicoccin.
Spencer et al., Phytochemistry, **24**, 2615 (1985)

P. edulis Sims.
BENZOPYRANS: 8aα-edulan, 8aβ-edulan, edulan I, edulan II, 6,7-epoxyedulan.
TERPENOID: passiflorin.
(B) Itoh et al., Agr. Biol. Chem., **44**, 2871 (1980)
(T) Murray et al., Aust. J. Chem., **25**, 1921 (1972)

P. foetida.
FLAVONOIDS: 7,4'-dimethoxyapigenin, ermanin, pachypodol.
Echeverri et al., Chem. Abstr., 105 187580r

P. incarnata.
HARMAN ALKALOID: passiflorine.
Neu, Arzneim. Forsch., **4**, 601 (1954)

P. platyloba.
COUMARIN: esculentin.
FLAVONOIDS: isomollupentin, isovitexin 7-rhamnoglucoside.
Ajanoglu et al., Phytochemistry, **21**, 799 (1982)

P. serratodigitata.
COUMARINS: serratin, serratin 7-O-β-glucoside.
FLAVONOIDS: chrysin, isovitexin, orientin, vitexin, 2''-xylosylvitexin.
Ulubelen et al., Phytochemistry, **21**, 1145 (1982)

P. warmingii.
GLYCOSIDE: linamarin.
Fischer et al., Planta Med., **45**, 42 (1982)

PASTINACA (Pastinaceae)
P. sativa.
COUMARINS: bergapten, imperatorin, isobergapten, isopimpinellin, pimpinellin, sphondin, umbelliferone.
Gusak et al., Chem. Abstr., 86 68378k

PASTINACOPSIS (Pastinaceae)
P. glacialis.
COUMARINS: bergapten, imperatorin, isobergapten, isopimpinellin, pimpinellin, sphondin, umbelliferone.
Gusak et al., Chem. Abstr., 86 68378k

PATAGONULA (Boraginaceae)
P. americana L.
SESQUITERPENOIDS: cordiachrome G, cordiachrome H, peuchdiachrome H.
Moir, Thomson, J. Chem. Soc., Perkin I, 1556 (1973)

PATENA
P. lucida.
TRITERPENOID: bassic acid.
King et al., J. Chem. Soc., 1338 (1955)

PATRINIA
P. intermedia.
SAPONINS: patrinosides C, C-1 and D.
Bucharov et al., Khim. Prir. Soedin., 17 (1967)

P. scabiosaefolia.
IRIDOID: patrinoside.
SAPONINS: scabiosides A, B, C, D, E, F and G.
Taguchi, Endo, Chem. Pharm. Bull., **22**, 1935 (1974)

P. sibirica.
SAPONINS: sibirosides A, B and C.
Bukharov, Karlin, Mater. Nauch. Konf. Inst. Org. Fiz. Khim. Akad. Nauk, SSSR, 55 (1969)

P. villosa.
IRIDOIDS: villoside, villosol, villosolside.
Taguchi et al., J. Pharm. Soc., Japan, **93**, 607 (1973)
Xu et al., Yaoxue Xuebao, **20**, 652 (1985); Chem. Abstr., 104 65906k

PAULLINIA (Sapindaceae)

P. cupana.
PURINE ALKALOIDS: theobromine, theophylline.
Blout *et al.*, *J. Am. Chem. Soc.*, **72**, 479 (1950)

PAULOWNIA (Scrophulariaceae)

P. tomentosa (Thunb.) Steud.
IRIDOID: paulownioside.
Adrian *et al.*, *J. Nat. Prod.*, **44**, 739 (1981)

PAURIDIANTHA (Rubiaceae)

P. callicarpoides.
CARBOLINE ALKALOIDS: pauridianthine, pauridianthinine.
Pousset *et al.*, *C. R. Acad. Sci.*, **272C**, 665 (1971)

P. lyalli Brem.
CARBOLINE ALKALOIDS: 6-trans-feruloyllyaloside, hydroxylyalidine, lyalidine, lyaloside.
Levesque *et al.*, *Tetrahedron*, **38**, 1417 (1982)

PAUSINYSTALIA

P. angolensis.
One carboline alkaloid.
Van der Meulen, Van der Kerk, *Recl. Trav. Chim. Pays-Bas*, **83**, 141, 148 (1964)

P. macroceras.
CARBOLINE ALKALOID: yohimbine.
Witkop, *Justus Liebigs Ann. Chem.*, **554**, 83 (1943)
Witkop, *ibid.*, **554**, 127 (1943)

P. mayumbensis.
CARBOLINE ALKALOID: mayumbine.
Raymond-Hamer, *C. R. Acad. Sci.*, **232**, 2354 (1951)

P. paniculata Welw.
CARBOLINE ALKALOIDS: paniculatine, yohimbine.
Clemo, Swan, *J. Chem. Soc.*, 617 (1946)

P. trillesi.
CARBOLINE ALKALOID: yohimbine.
Schomer, *Chem. Zentralbl., II*, 2309 (1927)

P. yomhinbe (Corynanthe yohimbe).
CARBOLINE ALKALOID: dihydrositsirikine, yohimbine, α-yohimbine.
Witkop, *Justus Liebigs Ann. Chem.*, **554**, 83 (1943)
Witkop, *ibid.*, **554**, 127 (1943)
Le Hir, Goutarel, *Bull. Soc. Chim. Fr.*, 1023 (1953)

PAVETTA

P. lanceolata.
CARBOLINE ALKALOID: pavettine.
Jordaan *et al.*, *J. S. Afr. Chem. Inst.*, **21**, 22 (1968)

PAVONIA (Malvaceae)

P. zeylanica.
SAPONIN: paraphylline.
Tiwari, Minocha, *Phytochemistry*, **19**, 701 (1980)

PAXILLUS

P. atrotomentosus.
PIGMENT: xerocomic acid.
Besl *et al.*, *Chem. Abstr.*, **88** 117773n

PECHUEL-LOESCHEA

P. leibnitziae.
SESQUITERPENOIDS: methylpechueloate, norpechuelol, norpechuelone, pechuelol, pechuelone.
Bohlmann *et al.*, *Phytochemistry*, **21**, 1160 (1982)

PECTUMCULUS

P. glycymeris.
CAROTENOID: glycymerin.
Fabre, Lederer, *Bull. Soc. Chim. Biol.*, **36**, 105 (1934)

PEDICULARIS (Schophulariaceae)

P. dolichorrhiza Schrenk.
QUINOLIZIDINE ALKALOID: N-methylcytisine.
UNCHARACTERIZED ALKALOIDS: indicamine, plantagonine.
Danilova, Konovalova, *J. Gen. Chem. USSR*, **22**, 2237 (1952)

P. lapponica.
IRIDOIDS: aucubin, euphroside, mussaenoside.
Berg *et al.*, *Phytochemistry*, **24**, 491 (1985)

P. ludwigii Rgl.
UNCHACTERIZED ALKALOIDS: indicamine, plantagonine.
Danilova, Konovalova. *J. Gen. Chem. USSR*, **22**, 2237 (1952)

P. macrochloa Vved.
PYRIDINE ALKALOIDS: gentianamine, noractinidine.
UNCHARACTERIZED ALKALOID: plantagonine.
Abdusamatov, *Avtoref. Dokl. Diss. Tashkent*, (1972)

P. olgae Rgl.
QUATERNARY ALKALOID: indicainine.
QUINOLIZIDINE ALKALOID: N-methylcytisine.
PYRIDINE ALKALOIDS: pediculonaridine, pedicularine, pedicularinine, pediculidine, pediculine, pediculinine, plantagonidine.
UNCHARACTERIZED ALKALOIDS: indicamine, plantagonine.
Abdusamatov, Yunusov, *Khim. Prir. Soedin.*, 334 (1969)

P. palustris.
IRIDOIDS: aucubin, boschnaloside, 3-hydroxyboschnaloside, scandoside methyl ester.
Berg *et al.*, *Phytochemistry*, **24**, 491 (1985)

P. rhinanthoides Schrenk.
PYRIDINE ALKALOID: tecostidine.
UNCHARACTERIZED ALKALOID: plantagonine.
Danilova, Konovalova, *J. Gen. Chem., USSR*, **22**, 2237 (1952)

P. silvatica.
IRIDOIDS: euphroside, epi-loganin, plantarenaloside.
Berg *et al.*, *Phytochemistry*, **24**, 491 (1985)

P. violescens Schrenk.
PYRIDINE ALKALOID: incanine.
UNCHARACTERIZED ALKALOID: plantagonine.
Danilova, Konovalova, *J. Gen. Chem., SSSR*, **22**, 2237 (1952)

PEGANUM (Zygophyllaceae)

P. harmala L.
ACID: 9,14-dihydroxyoctadecanoic acid.
CARBOLINE ALKALOIDS: harmaline, harmalol, ruine.
QUINAZOLINE ALKALOIDS: deoxypeganidine, deoxypeganine, dipegnine, deoxyvascinone, peganine, (±)-peganine, (-)-peganine, pegamine, (+)-peganine, peganol, (+)-vascinone, (±)-vascinone.
ANTHRAQUINONES: peganone 1, peganone 2.
(Ac) Ahmad *et al.*, *Phytochemistry*, **16**, 1761 (1977)
(A) Khashmirov *et al.*, *Khim. Prir. Soedin.*, 453 (1970)
(A) Telezhenetskaya *et al.*, *ibid.*, 849 (1971)
(An) Pitre *et al.*, *Planta Med.*, **53**, 106 (1987)

PEGOLETTIA

P. senegalensis.
SESQUITERPENOIDS: 8-hydroxypegolettiolide, pegolettiolide.
Bohlmann et al., Phytochemistry, 22, 1637 (1983)

PELARGONIUM (Geraniaceae)

P. graveolens L'Herit. ex Ait.
MONOTERPENOID: citronellyl diethylamine.
Rojohn, Klein, Dragoco Rep. (Engl. Ed.), 24, 150 (1977)

P. odoratissimum (L.) L'Herit. ex Ait.
MONOTERPENOIDS: (-)-menthone, geraniol.
Samophralov, Dokl. Akad. Nauk SSSR, 84, 1179 (1952)

P. roseum.
SESQUITERPENOIDS: furopelargones A, B, C and D.
Gianotti et al., Tetrahedron, 24, 2055 (1968)

PELLIA

P. barbigera.
FURANOQUINOLINE ALKALOID: isoplatydesmine.
Higa, Scheuer, Phytochemistry, 13, 1269 (1974)

P. endiviaefolia.
DITERPENOID: succulatal.
Asakawa et al., Planta Med., 32, 362 (1977)

PELTIGERA (Peltigeraceae)

P. aphthosa (L.) Willd.
PHENOLICS: aphtosin, tenuiorin.
TRITERPENOIDS: phlebic acid A, phlebic acid B.
(P) Bryan et al., Aust. J. Chem., 29, 1079 (1976)
(P) Bryan et al., ibid., 29, 1147 (1976)
(T) Takahashi et al., Phytochemistry, 9, 2037 (1970)

PELTOBOYKINIA

P. watanabii.
TERPENOIDS: α-peltoboykinolic acid, β-peltoboykinolic acid.
Izawa et al., Phytochemistry, 12, 1508 (1973)

PELTOPHORUM

P. africanum.
FLAVONOIDS: kaempferol, myricetin, myricetin 3-rutinoside, quercetin, quercetin 3-glucoside, quercetin 3-rutinoside.
PHENOLICS: chlorogenic acid, gallic acid, trans-4-hydroxypipecolic acid 4-sulphate.
El-Sherbeiny et al., Planta Med., 32, 165 (1977)
Evans et al., Phytochemistry, 24, 2593 (1985)

P. ferrugineum.
ANTHOCYANIN: (+)-leucocyanidin.
CATECHIN: (-)-epi-catechin.
FLAVONOIDS: bergenin, quercetin.
Sastry et al., Chem. Abstr. 87 98803h

PELTOPHYLLUM (Saxifragaceae)

P. peltatum.
FLAVONOIDS: kaempferol, myricetin 3-O-arabinoside, myricetin 3-O-glucoside, myricetin 3-O-rhamnoside, myricetin 3-O-xyloside, quercetin 3-O-arabinoside, quercetin 3-O-glucoside, quercetin 3-O-rhamnoside.
Bohm et al., Phytochemistry, 15, 2012 (1976)

PELVETIA (Phaeophyceae)

P. canalicaulata Dane et Thur.
STEROID: 24-oxocholesterol.
Motzfeldz, Acta Chem. Scand., 24, 1846 (1970)

PENIOCEREUS (Cactaceae)

P. fosterianus Cut.
STEROIDS: peniocerol, peniosterol.
Djerassi et al., Proc. Chem. Soc., 450 (1961)
Djerassi et al., J. Chem. Soc., 1160 (1965)

P. macdougalli.
STEROID: macdougallin.
Djerassi et al., J. Am. Chem. Soc., 85, 835 (1963)

PENIOPHORA

P. aurantiaca.
CAROTENOID: astaxanthin.
Tischer, Z. Physiol. Chem., 240, 191 (1941)

PENSTEMON (Scrophulariaceae)

P. barbatus (Cav.) Roth.
IRIDOIDS: barbatoside, penstemonoside, penstemoside.
Junior, Planta Med., 45, 127 (1962)

P. deutus.
SPIROIRIDOID: penstemide.
Jolad et al., Tetrahedron Lett., 4119 (1976)

P. serrulatus.
IRIDOIDS: serrulatoside, serruloside, 8-epi-valerosidate.
Junior, Planta Med., 50, 417 (1984)

PENTACERAS

P. australis Hook f.
CARBOLINE ALKALOIDS: canthin-6-one, 5-methoxycanthin-6-one.
Haynes et al., Aust. J. Sci. Res., 5A, 387 (1952)

PENTACLETHRA

P. eetveldeana Wild et Th. Dur.
Uncharacterized saponins.
Welter et al., Bull. Soc. R. Sci. Liege, 44, 687 (1975)

P. macrophylla Benth.
AMIDE ALKALOID: paucine.
Merck, Merck's Jahresbl., 11 (1894)

PENTADESMA (Guttiferae/Clusiaceae)

P. butyracea Sabine.
XANTHONES: jacarubein, osajaxanthone, pentadesmaxanthone.
Gunasekera et al., J. Chem. Soc., Perkin I, 11 (1977)

PENTADIPLANDRA

P. brazzeana.
THIOUREA: N,N'-bis(4-methoxybenzyl)-thiourea.
Migirab et al., Phytochemistry, 16, 1719 (1977)

PENTAPHALANGIUM (Guttiferae/Clusiaceae)

P. solomonse Warb.
BIPHENYL:
3,5-dihydroxy-4-(3-hydroxy-3-methylbutyryl)-biphenyl.
STEROIDS: campesterol, α-sitosterol, stigmasterol.
XANTHONES: 6-deoxyisojacareubin,
1-hydroxy-3,5-dimethoxy-4(3-methylbut-2-enyl)-xanthen-9-one, 1-hydroxy-3,6,7-trimethoxyxanthen-9-one,
1,3,5-trihydroxy-2-(3-methylbut-2-enyl)-xanthen-9-one,
1,3,7-trihydroxy-2-(3-methylbut-2-enyl)-xanthone.
(B) Cotterill et al., J. Chem. Soc., Perkin 1, 2423 (1974)
(X) Owen, Scheinmann, ibid., 1018 (1974)

PENTZIA

P. elegans.
SESQUITERPENOIDS: dehydroleucodin, isoepoxyestafietin.
Bohlmann, Zdero, *Tetrahedron Lett.*, 621 (1972)

PEREIRA

P. madagascariensis.
CARBOLINE ALKALOID: 4,7-dimethoxy-1-vinyl-β-carboline.
Boutgaigynon-Zylber, Polonsky, *Chim. Ther.*, 5, 396 (1970)

PERESKIA (Cactaceae)

P. grandiflora.
SAPONIN: saponin P-0.
Sahu et al., *Phytochemistry*, 13, 529 (1974)

PEREZIA (Compositae/Mutisieae)

P. alamani var. oolepis.
COUMARIN: 8-hydroxypereflorin.
SESQUITERPENOIDS: diperezone, isoparvifolinone.
Joseph-Nathan et al., *Phytochemistry*, 21, 1129 (1982)

P. carpholepis.
SESQUITERPENOIDS: curcuquinol isovalerate, cyperene,
cyperenone, parvifoline, perezone, α-pipitzol, β-pipitzol.
Joseph-Nathan et al., *Phytochemistry*, 21, 669 (1982)

P. coerulescens.
COUMARIN: 5-formyl-3,4,8-trimethoxycoumarin.
Angeles et al., *Phytochemistry*, 23, 2094 (1984)

P. cuernavacana.
SESQUITERPENOIDS: α-pipitzol, β-pipitzol.
Walls et al., *Tetrahedron Lett.*, 1577 (1965)

P. hebeclada.
SESQUITERPENOIDS: α-perezol, β-perezol, τ-perezol.
Gonzalez et al., *Phytochemistry*, 11, 1803 (1972)

P. multiflora (H. et B.) Less.
COUMARINS: 3,8-dimethoxypereflorin, 3-methoxypereflorin,
8-methoxypereflorin, pereflorin.
SESQUITERPENOID: 8-angeloyloxyproustianol acetate.
(C) Bohlmann et al., *Phytochemistry*, 16, 239 (1977)
(T) Bohlmann, Zdero, *Chem. Ber.*, 112, 427 (1979)

P. runcinata.
QUINONES: hydroxyperezone isovalerate, O-isovalerylperezone.
Joseph-Nathan et al., *Phytochemistry*, 16, 1086 (1977)

PERGULARIA

P. extensa.
TRITERPENOIDS: 3-hydroxyfriedelan-7-one, roxburghalone.
Garg et al., *Phytochemistry*, 7, 2035 (1968)

P. pallida.
PHENANTHROINDOLIZIDINE ALKALOIDS: deoxypergularinine,
pergularine, pergularinine.
Mulchandam, Venkatachalan, *Phytochemistry*, 15, 1561 (1976)

PERIANDRA

P. dulcis.
SAPONINS: periandrins I, II and III.
TRITERPENOIDS: 3-dehydroperiandric acid, periandric acid.
(S) Hashimoto et al., *Phytochemistry*, 21, 2335 (1982)
(T) Yamasa Shoyu Co., *Japanese Patent, 82 144, 240*

PERICOPSIS

P. angolensis Baker.
BENZOPYRAN: 3,4,9-trimethoxypterocarpan.
ISOFLAVANOID: 7,4'-dihydroxyisoflavanone.
STILBENE: astringenin.
(B) Harper et al., *J. Chem. Soc., Chem. Commun.*, 309 (1965)
(F) Fitzgerald et al., *J. Chem. Soc., Perkin I*, 186 (1976)
(St.) Grassman et al., *Chem. Ber.*, 89, 2523 (1956)

P. elata Harms.
ISOFLAVANOID: 7,4'-dihydroxyisoflavanone.
PHENOLIC: 2-O-methylangolensin.
Fitzgerald et al., *J. Chem. Soc., Perkin I*, 186 (1976)

P. laxiflora Benth.
ISOFLAVANOID: 7,4'-dihydroxyflavanone.
Fitzgerald et al., *J. Chem. Soc., Perkin I*, 186 (1976)

P. mooniana Thw.
ISOFLAVANOID: 7,4'-dihydroxyisoflavanone.
Fitzgerald et al., *J. Chem. Soc., Perkin I*, 186 (1976)

P. schliebenii Harms.
QUINOLIZIDINE ALKALOID: N-methylcytisine.
Fitzgerald et al., *J. Chem. Soc., Perkin I*, 186 (1976)

PERIDINIUM (Pyrrhophyceae/Peridiniaceae)

P. balticum UTEX 1563.
CAROTENOID: peridinin.
STEROID: peridinosterol.
Swenson et al., *Tetrahedron Lett.*, 4663 (1980)

P. foliaceum UTEX 1688.
STEROIDS: 4α-methyl-5α-gorgostanol, peridinosterol.
Swenson et al., *Tetrahedron Lett.*, 4663 (1980)

PERILLA

P. frutescens.
MONOTERPENOIDS: limonene, linalool, α-pinene.
SESQUITERPENOIDS: (2E,4E,6E,)-allofarnesene,
(2Z,4E,6E)-allofarnesene.
Sakai et al., *Bull. Chem. Soc., Japan*, 42, 3615 (1969)
Sugisawa et al., *Agr. Biol. Chem.*, 40, 231 (1976)

P. frutescens var. acuta.
COUMARIN: scutellerein.
GLYCOSIDES:
(R)-2-(2-O-β-D-glucopyranosyl-β-D-glucopyranosyloxy)-
phenylacetonitrile, prunasin.
PHENOLICS: caffeic acid, rosmarinic acid.
Aritomi et al., *Phytochemistry*, 24, 2438 (1985)

P. frutescens f. viridis.
SESQUITERPENOIDS: 2-cis-4-trans-allofarnesene,
2-trans-4-trans-allofarnesene.
Naves, *Helv. Chim. Acta*, 49, 1029 (1966)

P. nankinensis.
GLYCOSIDE: shisonin.
MONOTERPENOID: (-)-perillaldehyde.
(G) Takeda et al., *Proc. Japan. Acad.*, 40, 510 (1964)
(T) Schimmel, *Chem. Zentralbl., II*, 1758 (1910)

P. ocimoides.
GLYCOSIDE: shisonin.
SESQUITERPENOIDS: β-caryophyllene, germacrene D, perilla
ketone, perillene, rosefuran.
(Gly) Takeda et al., *Proc. Japan. Acad.*, 40, 510 (1964)
(T) Misra, Husain, *Planta Med.*, 53, 379 (1987)

PERIPENTADENIA

P. mearsii.
ALKALOID: peripentadenine.
TROPANE ALKALOID: 3-acetyl-6-hydroxytropane.
Lamberton et al., J. Nat. Prod., **46**, 235 (1983)

PERIPLOCA (Periplocaceae/Asclepiadaceae)

P. calophylla.
GLYCOSIDES: calocin, plocinin.
Srivastava et al., J. Nat. Prod., **45**, 211 (1982)
Deepak et al., Phytochemistry, **24**, 3015 (1985)

P. sepium.
STEROIDS: glycoside E, glycoside F,
21-O-methylpregn-5-ene-3,14,17,21-tetrahydroxy-20-one.
Ishizone et al., Chem. Pharm. Bull., **20**, 2402 (1972)

PERIPTERYGIA (Celastraceae)

P. marginata Loes.
MACROCYCLIC ALKALOIDS: dihydroperiphylline isoperiphylline, neoperiphylline, periphylline.
Hocquemiller et al., Tetrahedron, **33**, 645 (1977)

PERITASSA (Celastraceae)

P. campestris.
MAYTENSANOID: maytensin.
De Lima et al., Rev. Inst. Antibiotics, **9**, 17 (1969)

PERNETTYA (Ericaceae)

P. furens.
SESQUITERPENOIDS: pernetal, pernetic acids A, B, C, D and E, pernetol.
Hosozawa et al., Phytochemistry, **24**, 2317 (1985)

PEROVSKIA (Labiatae)

P. abrotanoides.
SESQUITERPENOID: abrotanol.
Serkabaeva, Bazalitskaya, Izv. Akad. Nauk Kaz. SSR, Ser. Tekh. Nauk, 3 (1963)

P. scrophulariaefolia.
SESQUITERPENOID: alloaromadendrene.
Dolejs et al., Collect. Czech. Chem. Commun., **25**, 1483 (1960)
Dolejs et al., ibid., **25**, 1877 (1960)

PERSICA

P. vulgaris.
ALIPHATICS: dotricontane, hentriacontane, heptacosane, hexacosane, hexacosanol, nonacosane, octacosane, octacosanol, pentacosane, tetracosanol, triacontane, triacontanol, tritriacontane.
STEROID: β-sitosterol.
Sadykov, Khim. Prir. Soedin., 133 (1977)

PERSOONIA

P. elliptica.
ALCOHOL: persoonol.
Cannon et al., Aust. J. Chem., **24**, 1925 (1971)

PERTUSARIA (Pertusariaceae)

P. coraxillina.
PHENOLIC: thamnolic acid.
Wachtmeister et al., Acta Chem. Scand., **24**, 1395 (1955)

P. globularis.
ACID: confluentin acid.
Culberson et al., Bryologist, **75**, 362 (1972)

P. rhodesiaca.
DEPSIDE: haemathamnolic acid.
Harper et al., J. Chem. Soc., C, 1603 (1967)

P. sulphurata.
XANTHONE: lichexanthone.
Ruben et al., Phytochemistry, **15**, 1093 (1976)

P. tuberculifera.
ACID: confluentin acid.
Culberson et al., Bryologist, **75**, 362 (1972)

PERTYA (Compositae)

P. glabrescens.
SESQUITERPENOID: pertilide.
Nagumo et al., Chem. Pharm. Bull., **30**, 586 (1982)

P. robusta (Maxim.) Beauv.
SESQUITERPENOID: glucozaluzanin C.
DITERPENOID: pertoyl methyl ether.
Nagumo et al., J. Pharm. Soc., Japan, **100**, 427 (1980)

PERYMENIUM

P. ecuadoricum.
DITERPENOIDS: perymenic acid, perymenuimic acid.
Bohlmann, Zdero, Phytochemistry, **16**, 986 (1977)

PESCHIERA (Apocynaceae)

P. affinis (Muell. Arg.) Miers.
CARBOLINE ALKALOIDS: affinine, affinisine.
Cava et al., Chem. Ind. (London), 1193 (1964)

P. lacta.
CARBOLINE ALKALOIDS: affinine, akuammidine, conodurine, geissochizol, tombozine, voacamine, vobasine.
Voticky et al., Collect. Czech. Chem. Commun., **42**, 1403 (1977)

P. lundii (DC.) Miers.
HOMOCARBOLINE ALKALOID: iboxygaine hydroxyindolenine.
Whang et al., J. Org. Chem., **34**, 412 (1969)

PETALOSTYLIS

P. labicheoides.
CARBOLINE ALKALOID: tetrahydroharman.
Paris et al., Bull. Soc. Chim. Fr., 780 (1957)

PETASITES (Compositae)

P. albus (L.) Gaertner.
SESQUITERPENOID: eremophilene.
DITERPENOIDS: petasitolide A, petasitolide B.
THIODITERPENOIDS: S-petasitolide A, S-petasitolide B.
Hochmannova et al., Collect. Czech. Chem. Commun., **27**, 1870 (1962)

P. hybridus (L.) Gaertner, Meyer et Scherb.
SESQUITERPENOIDS:
2-angeloyloxy-9-oxo-10αH-furanoeremophilane, isobutyrylneopetasyl ester, isopetasane, isopetasoside, isopetasyl ester isobutyrate, methylacrylylneopetasyl ester, methylacrylylpetasyl ester, cis-2-methyl-2-butenyl-3-deoxy-13-petasyl ester, cis-2-methyl-2-butenylneopetasyl ester, cis-2-methyl-2-butenylpetasyl ester, (Z)-3-methylthioacryloylisopetasyl ester, neopetasan, senecioylneopetasyl ester, senecioylpetasyl ester.
Novotny et al., Phytochemistry, **11**, 2795 (1972)
Neuenschwander et al., Helv. Chim. Acta, **62**, 605 (1979)

P. japonicus Maxim.

PYRROLIZIDINE ALKALOIDS: fukinotoxin, petasinine, petasinoside.

SESQUITERPENOIDS: 9-acetoxyfukinanolide, 6-acetylfuranofukinol, bakkenolides A, B, C, D and E, eremofukinone, isopetasoside, S-japonin, petasitin, petasitalone, petasoside.

(A) Yamada *et al., Tetrahedron Lett.*, 4543 (1978)
(T) Naya *et al., ibid.*, 2961 (1971)

P. kablikianus.

SESQUITERPENOIDS: eremophilene, kablicin.

Hochmannova *et al., Collect. Czech. Chem. Commun.*, 24, 360 (1976)

P. officinalis L.

SESQUITERPENOIDS: eremophilene, eremophilenolide, petasene, petasitolides A and B.

THIODITERPENOIDS: S-petasitolide A, S-petasitolide B.

Hochmannova *et al., Collect. Czech. Chem. Commun.*, 24, 360 (1976)

P. paradoxus.

SESQUITERPENOID: kablicin.

Novotny *et al., Tetrahedron Lett.*, 1401 (1968)
Novotny *et al., ibid.*, 2281 (1968)

PETCHIA (Apocynaceae)

P. ceylanica Wight.

DIMERIC ALKALOIDS: peceylanine, peceyline, pelankine.

Kunesch *et al., Tetrahedron Lett.*, 1727 (1980)

PETERAVANIA

P. schultzii.

Three sesquiterpenoids.

Bohlmann, Suwita, *Phytochemistry*, 17, 567 (1978)

PETILIUM (Liliaceae)

P. eduardi (Rgl.) Vved.

STEROIDAL ALKALOIDS: edpetilidine, edpetiline, edpetilinine, edpetine, edpetisidinine, eduardine, imperialine, imperialone, peisimine.

Nuriddinov *et al., Khim. Prir. Soedin.*, 333 (1968)
Shakirov *et al., ibid.*, 584 (1979)

P. raddeana (Rgl.) Vved.

STEROIDAL ALKALOIDS: alvanidine, alvanine, edpetiline, imperialine, petilidine, petilidinine, petilimine, petiline, petilinine, raddeanine.

Babaev, Shakirov, *Khim. Prir. Soedin.*, 682 (1972)
Nakhatov *et al., ibid.*, 747 (1983)

PETROSELINUM (Umbelliferae)

P. crispum.

FLAVONOID: myristicin.

MONOTERPENOIDS: apiole, 4-isopropenyl-1-methylbenzene, p-mentha-1,3,8-triene, β-phellandrene, 2-(p-tolyl)propan-2-ol.

MacLeod *et al., Phytochemistry*, 24, 2623 (1985)

PETTERIA (Leguminosae)

P. ramentacea.

QUINOLIZIDINE ALKALOIDS: anagyrine, cytisine, 5,6-dehydrolupanine, lupanine, N-methylcytisine, rhombifoline.

Wink, Withe, *Phytochemistry*, 24, 2567 (1985)

PETUNIA

P. hybrida.

CHALCONE: 4,2',4',6'-tetrahydroxychalcone.

De Vleming, Kho, *Phytochemistry*, 15, 348 (1976)

PEUCEDANUM (Umbelliferae)

P. arenarium.

COUMARINS: peuarin, peuchloridin, peuchlorin, peuchlorinin.

Zheleva *et al., Phytochemistry*, 15, 209 (1976)

P. baicalense (Redow.) Koch.

COUMARINS: bergapten, deltoin, isoimperatorin, 5-methoxy-8-hydroxypsoralen, phellopterin.

Avramenko *et al., Khim. Prir. Soedin.*, 421 (1975)

P. bourgaei Lange in Wilk. et Lange.

COUMARINS: athamantin, 3'-hydroxy-4'-isovaleryloxy-2',3'-dihydrooroselol, 3'-hydroxy-4'-senecioyloxy-2',3'-dihydrooroselol, oroselol, oroselone, umbelliferone.

Gonzalez *et al., An. Quim.*, 72, 568 (1976)

P. dhana.

COUMARINS: 5-geranoxypsoralenn, imperatorin, isoimperatorin.

Agarwal *et al., Indian J. Chem.*, 13, 1108 (1975)

P. grande.

COUMARIN: columbianadin.

Willette *et al., J. Pharm. Sci.*, 53, 275 (1964)

P. mogoltavicum.

COUMARIN TERPENOIDS: mogoltavinin, tavimolidin.

Khasanov *et al., Khim. Prir. Soedin.*, 480 (1979)

P. oreoselinum.

BENZOPYRAN: 9-acetoxy-O-isovaleryldihydrooreoselol.

COUMARIN: vaginidin.

DITERPENOID: peucelinendiol.

(B) Bohlmann *et al., Chem. Ber.*, 102, 1963 (1969)
(C) Lemmich *et al., Acta Chem. Scand.*, 24, 2893 (1970)
(T) Lemmich, *Phytochemistry*, 18, 1195 (1979)

P. ostruthium (L.) Koch.

BENZOPYRAN: alatol.

COUMARINS: isooxypeucedanin, tert-O-methoxypeucedanin hydrate, oxypeucedanin, oxypeucedanin hydrate, pabulenone.

(B) Abyshev *et al., Khim. Prir. Soedin.*, 722 (1973)
(C) Reisch *et al., Phytochemistry*, 14, 1889 (1975)

P. palustre (L.) Moench.

COUMARINS: columbianadin, columbianadinoxide.

Nielsen *et al., Acta Chem. Scand.*, 18, 1379 (1964)
Nielsen *et al., ibid.*, 18, 2111 (1964)

P. pauciradiatum.

COUMARINS: crocalone, radiatinal, radiatinin.

Bagirov, Belyi, *Khim. Prir. Soedin.*, 250 (1982)

P. praeruptorum Dunn.

COUMARIN: praeruptin E.

Ye *et al., Yaoxue Xuebao*, 17, 431 (1982); *Chem. Abstr.*, 97 178696x

P. stenocarpum.

COUMARINS: byakangelicin, 6-carboxy-7-isobutyryloxycoumarin, gosferol, heraclenol, 5-hydroxy-8-(1,1-dimethylallyl)-psoralen, stenocarpin, umbelliprenin.

Gonzalez *et al., An. Quim.*, 73, 858 (1977)

P. terebinthaceum Fisher et Turcz.

COUMARINS: decursin, umbelliferone.

STEROIDS: α-sitosterol, stigmasterol.

Yook *et al., Yakhak Hoechi*, 30, 73 (1986); *Chem. Abstr.*, 105 102407s

PEUCEDANUM (Umbelliferae)

P. turcomanicum.
NORDITERPENOID: oxypeucedanin hydrate.
Abyshev *et al.*, *Khim. Prir. Soedin.*, 847 (1979)

P. turgeniifolium Wolff.
COUMARINS: (±)-diisovalerylkhellactone,
(±)-7-hydroxy-8-(2',3'-dihydroxy-3'-methylbutyl)coumarin,
isopeucedanin, (±)-cis-khellactone, (±)-transkhellactone,
(±)-peuformosin, turgeniifolin A, turgeniifolin B.
Sun *et al.*, *Chem. Abstr.*, 100 82755v

PEUCEPHYLLUM

P. schottii Gray.
SESQUITERPENOID: peucephyllin.
Begley *et al.*, *Tetrahedron Lett.*, 1105 (1975)

PEUMUS (Monimiaceae)

P. boldus Molina.
APORPHINE ALKALOID: norisocorydine.
Ruegger, *Helv. Chim. Acta*, 42, 754 (1959)

PFAFFIA

P. paniculata.
NORTRITERPENOIDS: 3-acetoxypfaffic acid, 3-acetoxypfaffic
acid methyl ester, pfaffic acid, pfaffic acid methyl ester.
SAPONINS: pfaffosides A, B and C.
(Sap) Nishimoto *et al.*, *Phytochemistry*, 23, 139 (1984)
(T) Takemoto *et al.*, *Tetrahedron Lett.*, 1057 (1983)

PHAEANTHUS

P. ebracteolatus.
BISBENZYLISOQUINOLINE ALKALOID: phaeanthine.
Santes, *Ber. Dtsch. Chem. Ges.*, 65, 472 (1932)

PHALAENOPSIS (Orchidaceae)

P. amabilis (L.) Blume.
One uncharacterized pyrrolizidine alkaloid.
Brandage, Luning, *Acta Chem. Scand.*, 23, 1151 (1969)

P. cernu-cervi Reichb.
PYRROLIZIDINE ALKALOID: cornucervine.
Brandage *et al.*, *Acta Chem. Scand.*, 25, 349 (1971)

P. mannii.
One uncharacterized pyrrolizidine alkaloid.
Brandage, Luning, *Acta Chem. Scand.*, 230, 1151 (1969)

PHALARIS (Gramineae)

P. aquatica.
INDOLE ALKALOID: 7-methoxygramine.
Mulvena *et al.*, *Phytochemistry*, 22, 2885 (1983)

P. arundinacea L.
CARBOLINE ALKALOIDS:
1,2,3,4-tetrahydro-6-methoxy-2,9-dimethyl-β-carboline,
1,2,3,4-tetrahydro-6-methoxy-2-methyl-β-carboline
INDOLE ALKALOID,: dimethyltryptamine.
Culvenor *et al.*, *Aust. J. Chem.*, 17, 1301 (1964)
Vijayanagar *et al.*, *Lloydia*, 38, 442 (1975)

P. tuberosa.
CARBOLINE ALKALOID:
1,2,3,4-tetrahydro-6-methoxy-2-methyl-β-carboline.
INDOLE ALKALOIDS: dimethyltryptamine,
5-methoxy-N,N-dimethyltryptamine,
5-methoxy-N-methyltryptamine.
Culvenor *et al.*, *Aust. J. Chem.*, 17, 1301 (1964)
Filho *et al.*, *Phytochemistry*, 14, 821 (1975)

PHALLINUS

P. gilvas.
STEROIDS: ergosta-7,22-dien-3β-ol, ergostadienol acetate,
ergosterol.
TRITERPENOID: trametenolic acid.
Ahmad *et al.*, *Phytochemistry*, 15, 2000 (1976)

P. pomaceus.
COUMARINS: 3,14'-bihispidinyl, hispidin.
Klaar *et al.*, *Chem. Ber.*, 103, 1058 (1977)

PHARBITIS

P. nil.
GIBBERELLINS: gibberellin A-26, gibberellin A-27, gibberellin
A-29, pharbitic acid.
Yokota *et al.*, *Planta*, 87, 180 (1969)

PHASCOLION

P. strombi.
GUANIDINES: (-)-phascoline, phascolosomine.
Guillou *et al.*, *J. Biol. Chem.*, 248, 5668 (1973)

PHASEOLUS (Papilionaceae)

P. aureus Roxb.
AMINO ACID: τ-glutamylcystenyl-β-alanine.
COUMESTANS: aureol, phaseol.
(Am) Carnegie, *Biochem. J.*, 89, 459 (1963)
(Am) Carnegie, *ibid.*, 89, 471 (1963)
(C) O'Neill, *Z. Naturforsch.*, 38C, 698 (1983)

P. coccineus.
ISOFLAVONOIDS: cyclokievitone, daidzein, demethylvestitol,
7,4'-dihydroxy-5,2'-dimethoxyisoflavone, genistein, glycuol,
2'-hydroxydihydrodaidzein, 2'-hydroxygenistein,
2'-hydroxyisoprunetin, isoferreirin, isoprunetin, kievitone,
phaseollidin, phaseollin, phaseollinisoflavone,
phaseoluteone.
GIBBERELLIN: 3-O-β-D-glucopyranosylgibberellin A-8.
DITERPENOIDS: abscisic acid, 7β,13-dihydroxykaurenolide,
16,17-mono-O-propylidene-6α,7α,16β,17-
tetrahydroxykauranic acid, ent-6α,7α,13-trihydroxykaurenoic
acid, ent-6α,7α,17-trihydroxykauranoic acid.
(Gib) Schreiber *et al.*, *Tetrahedron Lett.*, 4285 (1967)
(I) Adesanya *et al.*, *Phytochemistry*, 24, 2699 (1985)
(T) Gaskin, McMillan, *ibid.*, 14, 1575 (1975)

P. multiflorus.
GIBBERELLINS: bamboo gibberellin, gibberellin A-18,
gibberellin A-20.
SESQUITERPENOID: phaseic acid.
Pryce *et al.*, *Tetrahedron Lett.*, 5009 (1967)
MacMillan *et al.*, *J. Chem. Soc., Chem. Commun.*, 124 (1968)

P. radiatus.
SAPONINS: adzukisapogenin, adzukisaponin.
Miyamichi, Onishi, *Chem. Zentralbl.*, I, 3195 (1932)

P. vulgaris.
ACID: phaselic acid.
BENZOPYRANS: phaseollidin, phaseollin.
FLAVAN: phaseollinisoflavan.
GIBBERELLINS: gibberellin A-37, gibberellin A-38.
SAPONINS: phaseolides A, B, C, D and E.
(Acid) Scarpati *et al.*, *Gazz. Chim. Ital.*, 90, 212 (1960)
(B, F) Burden *et al.*, *Tetrahedron Lett.*, 4175 (1972)
(G, Sap) Hiraga *et al.*, *Agr. Biol. Chem.*, 36, 354 (1972)

PHEBALIUM (Rutaceae)

P. drummondi.
COUMARIN: phebalosin.
Chow *et al.*, *Aust. J. Chem.*, **19**, 483 (1966)

P. nudum.
COUMARIN: phebalin.
ESTER: phebalarin.
Briggs *et al.*, *Tetrahedron*, **2**, 256 (1958)

P. rude.
DITERPENOIDS: 15α-hydroxykaur-16-en-19-oic acid,
kaur-16-en-19-oic acid.
Carman *et al.*, *Aust. J. Chem.*, **19**, 861 (1966)

P. squamulosum Vent.
SESQUITERPENOID: squamulosone.
Batey *et al.*, *Aust. J. Chem.*, **24**, 2173 (1971)

P. tuberculosum.
COUMARIN: phebalosin.
Chow *et al.*, *Aust. J. Chem.*, **19**, 483 (1966)

PHEGOPTERIS

P. polypodioides.
FLAVONE: phegopolin.
Dawson *et al.*, *Aust. J. Chem.*, **18**, 1871 (1965)

PHELLINE

P. billardieri.
HOMOERYTHRINA ALKALOID: phellibidine.
Langlois *et al.*, *Phytochemistry*, **19**, 1279 (1980)

P. brachyphylla.
HOMOERYTHRINA ALKALOIDS: homoerythratine,
O-methylphellinine, phellinine.
Debourges, Langlois, *J. Nat. Prod.*, **45**, 163 (1982)

P. comosa.
HOMOERYTHRINA ALKALOIDS: schelhammericine,
3-epi-schelhammericine, schelhammerine,
3-epi-schelhammerine.
Langlois *et al.*, *Bull. Soc. Chim. Fr.*, **10**, 3535 (1970)

P. lucida.
HOMOERYTHRINA ALKALOID: 1,2-dihydrocomosidine,
halidinine.
Razofimbelo *et al.*, *C. R. Acad. Sci.*, **301**, 519 (1985)

PHELLINUS (Poriales)

P. pomaceus (Pers.) Maire.
TRITERPENOIDS: javeroic acid, phellinic acid.
Gonzalez *et al.*, *J. Chem. Soc., Perkin I*, 551 (1986)

PHELLODENDRON (Rutaceae)

P. amurense Rupr.
BENZOPYRAN: amurensin.
FLAVONOIDS: phellamurin, phellatin.
PROTOBERBERINE ALKALOIDS: berberine, palmatine.
TRITERPENOID: limonin.
(A) Gupta, Spenser, *Can. J. Chem.*, **43**, 133 (1965)
(B) Hasegawa *et al.*, *J. Am. Chem. Soc.*, **75**, 5507 (1953)
(F) Glyzin *et al.*, *Khim. Prir. Soedin.*, 762 (1970)
(F) Otryahenkova *et al.*, *ibid.*, 662 (1976)
(T) Enina, *Nauka-Prakt. Farm.*, 90 (1974); *Chem. Abstr.*, 86 40157p

P. chinense.
FLAVONOID: phellarin.
Ostryashenkova *et al.*, *Khim. Prir. Soedin.*, 662 (1976)

P. japonicum.
FLAVONOIDS: phellarin, phellodendroside.
Ostryachenkova *et al.*, *Khim. Prir. Soedin.*, 662 (1976)

P. lavallei.
FLAVONE: phellatin.
Glyzin *et al.*, *Khim. Prir. Soedin.*, 762 (1970)

P. piriforme.
FLAVONOID: phellarin.
Ostryachenkova *et al.*, *Khim. Prir. Soedin.*, 662 (1976)

P. sacchalense.
FLAVONOID: phellarin.
Ostryachenkova *et al.*, *Khim. Prir. Soedin.*, 662 (1976)

PHELLOPTERUS

P. littoralis.
BENZOPYRAN: phellopterin.
Chatterjee *et al.*, *Arch. Pharm.*, **295**, 248 (1962)

PHILLIPSIA

P. carminia.
CAROTENOID: phillipsiaxanthin.
Arpin, Liaaen-Jensen, *Bull. Soc. Chim. Biol.*, **49**, 527 (1967)

PHILLYREA (Oleaceae)

P. latifolia L.
TRITERPENOIDS: betulin, lupeol, ursolic acid.
Stout, Stevens, *J. Org. Chem.*, **28**, 1259 (1963)

P. media.
GLUCOSIDE: oleuropeine.
IRIDOID: ligustroside.
Popov *et al.*, *Dokl. Bolg. Akad. Nauk*, **28**, 1509 (1975); *Chem. Abstr.*, **84** 102359a

PHLOJODICARPUS

P. sibiricus.
PYRANOCOUMARINS: isoimperatorin,
cis-kellactone-3'-O-acetyl-4'-(2-methylbutanoate),
saxdorphine, umbelliferone.
Gantimur *et al.*, *Khim. Prir. Soedin.*, 108 (1986)

PHLOMIS (Labiatae)

P. fruticosa L.
IRIDOIDS: lamiide, lamiidioside, phlomiol.
Bianco *et al.*, *Gazz. Chim. Ital.*, **107**, 67 (1977)

P. umbrosa Turcz.
IRIDOID: umbroside.
Chung *et al.*, *Soengyak Hakhoe Chi*, **12**, 82 (1981)

PHLOX (Polemonidaceae)

P. decussata.
AMINO ACID: oxypinnatanine.
Pollard, Steward, *Plant Physiol.*, *Suppl.31 (XI)* (1956)

PHOEBE (Lauraceae)

P. cinnamomifolia.
PHENOLIC: (-)-catechin.
SESQUITERPENOID: α-farnesene.
Martinez-Alvares *et al.*, *Rev. Latinoam. Quim.*, **14**, 76 (1983)

P. clemensii Allan.
APORPHINE ALKALOID:
2,11-dihydroxy-1,10-dimethoxyaporphine.
Johns, Lamberton, *Aust. J. Chem.*, **20**, 1277 (1967)

P. formosana.
SPIROISOQUINOLINE ALKALOID: lauformine.
Lu, Tsei, *Heterocycles*, **22**, 1031 (1984)

P. molicella.
APORPHINE ALKALOID: norpreocoteine.
Stermitz, Castro, *J. Nat. Prod.*, **46**, 913 (1983)

PHOENIX (Palmae)
P. dactylifera.
STEROIDS: 5α-campesterane-3,6-dione,
5α-stigmast-22-ene-3,6-dione.
DITERPENOID: 3-O-caffeoylshikimic acid.
(S) Fernandez *et al.*, *Phytochemistry*, **22**, 2087 (1983)
(T) Harbourne *et al.*, *Phytochemistry*, **13**, 1557 (1974)

P. dactylifera var. zahdi.
PHENOLICS: chlorogenic acid, isochlorogenic acid.
Abdul-Wahab *et al.*, *Bull. Coll. Sci. Univ. Baghdad*, **14**, 239 (1973);
Chem. Abstr., 84 176722k

PHOLIDOTA
P. chinensis.
TRITERPENOIDS: cyclopholidone, cyclopholidonol.
Lin *et al.*, *Planta Med.*, 4 (1986)

PHOLIOTA (Agericaceae)
P. articulata.
PHENANTHROPYRANS: isoflavidinin, isooxoflavidinin.
Majumder *et al.*, *Phytochemistry*, **21**, 2713 (1982)

P. imbricata.
PHENANTHROPYRAN: imbricatin.
Majumder *et al.*, *Phytochemistry*, **21**, 2713 (1982)

PHOMA
P. exigua var. non-oxydabilis.
SESQUITERPENOID: phomerone.
Riche *et al.*, *Tetrahedron Lett.*, 2765 (1974)

P. lingam.
ALKALOID: phomamide.
Ferezou *et al.*, *J. Chem. Soc., Perkin I*, 113 (1980)

PHORMIUM (Agavaceae/Liliaceae)
P. tenax J.R. et G. Forst.
TRITERPENOIDS: isocucurbitacin D, 3-epi-cucurbitacine D.
Kupchan *et al.*, *Phytochemistry*, **17**, 767 (1978)

PHOTINIA (Rosaceae)
P. glabra.
TRITERPENOID: epi-friedelenol.
Brownlie *et al.*, *Chem. Ind.*, **686**, 1156 (1955)

PHRYMA
P. leptostachya.
DIOXOBICYCLOOCTANES: phrymarolin I, phrymarolin II.
Taniguchi *et al.*, *Agr. Biol. Chem.*, **36**, 1489 (1972)

PHYLICA
P. fengeriana.
APORPHINE ALKALOID: lauroscholtzine.
Baarschers, Arndt, *Tetrahedron*, **21**, 2155 (1965)

P. rogersii Pillans.
APORPHINE ALKALOIDS: lauroscholtzine, N-methyllaurotetanine,
(+)-reticuline, rogersine.
Baarschers, Arndt, *Tetrahedron*, **21**, 2155 (1965)
Baarschers, Arndt, *ibid.*, **21**, 2159 (1965)

PHYLLANTHUS (Euphorbiaceae)
P. brasiliensis.
SESQUITERPENE ESTER: phyllanthoside
Kupchan *et al.*, *J. Am. Chem. Soc.*, **99**, 3199 (1977)

P. discoides Muell. Arg.
ALKALOIDS: phyllalbine, phyllanthine, phyllantidine,
allo-securinine.
Horii *et al.*, *Tetrahedron Lett.*, 1877 (1972)

P. niruri.
ALKALOIDS: 4-methoxynorsecurinine, nirurine.
FLAVONOIDS: astragalin, isoquercitrin, quercetin, quercitrin,
rutin.
LIGNANS: hypophyllanthin, lintetralin, niranthin, nirtetralin,
phyllanthin, phyltetralin.
PHTHALIC ESTER: phyllester. Steroid β-sitosterol.
(A) Mulchandani, Hassarajani, *Planta Med.*, **50**, 104 (1984)
(A) Petchnaree *et al.*, *J. Chem. Soc., Perkin 1*, 1551 (1986)
(F) Nara *et al.*, *Plant Med. Phytother.*, **11**, 82 (1977)
(L, P, S) Singh *et al.*, *Indian J. Chem.*, **25B**, 600 (1986)

P. orbiculatus L.C.Rich.
FLAVONOIDS: astragalin, isoquercitrin, quercetin, quercitrin,
rutin.
Nara *et al.*, *Plant Med. Phytother.*, **11**, 82 (1977)

P. reticulatus.
STEROID: sitosterol.
TRITERPENOIDS: betulinic acid, friedelin,
21α-hydroxyfriedel-4(23)-en-3-one.
Hui *et al.*, *Phytochemistry*, **15**, 797 (1976)

P. urinaria L.
FLAVONOIDS: astragalin, isoquercitrin, quercetin, quercitrin,
rutin.
Nara *et al.*, *Plant Med. Phytother.*, **11**, 82 (1977)

PHYLLANTHRON
P. comorense.
NAPHTHOPYRANS: dehydrotectol, tectol.
Joshi *et al.*, *Phytochemistry*, **13**, 663 (1944)

PHYLLOCLADUS (Podocarpaceae)
P. alpinus.
DITERPENE: phyllocladene.
Briggs, *J. Soc. Chem. Ind.*, **56**, 137 (1937)

P. rhomboideus.
DITERPENE: phyllocladene.
Briggs, *J. Soc. Chem. Ind.*, **56**, 137 (1937)

PHYLLODOCE (Ericaceae)
P. nipponica.
STEROID: β-sitosterol.
TRITERPENOIDS: α-amyrin, betulinic acid, lupeol, oleanolic acid,
taraxerol, ursolic acid.
Kawai *et al.*, *Chem. Abstr.*, 88 34499b

PHYLLOGEITON
P. zeyheri Sond.
FLAVANONES: aromadendrin, kaempferol, naesopsin, zeyherin.
Volstaedt, Roux, *Tetrahedron Lett.*, 1647 (1971)

PHYLLOSPADIX
P. iwatensis.
ALKALOID: phyllospadine.
Takagi *et al.*, *Agr. Biol. Chem.*, **44**, 3019 (1980)

PHYLLOSTACHYS (Gramineae)

P. edulis.

GIBBERELLIN: gibberellin A-19.

Tamura *et al.*, *Tetrahedron Lett.*, 2465 (1966)

PHYMASPERMUM (Compositae/Anthemideae)

P. parvifolium.

SESQUITERPENOID: phymaspermone.

Bohlmann, Zdero, *Tetrahedron Lett.*, 851 (1972)

PHYSALIS (Solanaceae)

P. alkekengi L.

TROPANE ALKALOIDS: hygrine, 3-tigloyltropane.
CAROTENOID: physalien.
STEROID: physalactone.

(A) Zito, Leary, *J. Pharm. Sci.*, **55**, 1150 (1966)
(Car.) Kuhn, Wiegund, *Helv. Chim. Acta*, **12**, 499 (1929)
(S) Masslenikova *et al.*, *Khim. Prir. Soedin.*, 531 (1977)

P. alkekengi var. francheti.

BITTER PRINCIPLES: physalins A and B.

Matsuura *et al.*, *Phytochemistry*, **19**, 1175 (1980)

P. angulata.

BITTER PRINCIPLES: physalins E, F, G, H, I, J and K.

Row *et al.*, *Phytochemistry*, **17**, 1641 (1978)

P. francheti.

BITTER PRINCIPLES: physalins A, B and C.
CAROTENOID: physalien.
STEROIDS: physanol A, physanol B.

Sharma *et al.*, *Phytochemistry*, **13**, 2239 (1974)

P. ixocarpa Brot.

TERPENOIDS: ixocarpalactone A, ixocarpalactone B.

Kirson *et al.*, *J. Chem. Res.*, 1178 (1979)

P. lancifolia.

BITTER PRINCIPLES: physalins E, F, G, H, I and J.

Row *et al.*, *Phytochemistry*, **17**, 1641 (1978)
Row *et al.*, *ibid.*, **17**, 1647 (1978)

P. minima.

BITTER PRINCIPLES: physalins A, B, B-1 and D.
STEROIDS: withaphysalins A, B and C.

Glotter *et al.*, *J. Chem. Soc., Perkin I*, 1370 (1975)

P. peruviana.

STEROIDS: 4-deoxyphysalolactone, 2,3-dihydrowithanolide E,
3-O-β-D-glucopyranoside, 4β-hydroxywithanolide E.
4β-HYDROXYWITHANOLIDE F,: physalactone B, physalin A,
physalolactone, physalolactone B, withanolide E,
withaperuvin A.

Kirson *et al.*, *Phytochemistry*, **15**, 340 (1976)
Salai *et al.*, *J. Chem. Res. Synop.*, 152 (1983)

P. peruviana L. var. 'Varansai'.

STEROIDS: 4-deoxyphysalolactone, perulactone.

Frolow *et al.*, *J. Chem. Soc., Perkin I*, 1029 (1981)

P. pubescens L.

STEROIDS: physapubenolide, physapubescin, pubescenin,
pubescenol.

Row *et al.*, *Phytochemistry*, **23**, 427 (1984)
Glotter *et al.*, *J. Chem. Soc., Perkin I*, 2241 (1985)

P. viscosa.

STEROIDS: viscosalactone A, viscosalactone B,
withphysanolide.

Pelletier *et al.*, *Heterocycles*, **15**, 317 (1981)

PHYSCIA (Hepaticae)

P. aipolia (Ehrh. ex Humb.) Furnrohr.

TRITERPENOIDS: 6,22-hopanediol 20-acetate, leucotylin
6-acetate, leucotylin 6,16-diacetate.

Elix *et al.*, *Aust. J. Chem.*, **35**, 641 (1982)

P. caesia (Hoffm.) Furnrohr.

TRITERPENOID: zeorin.

Huneck, Lehn, *Bull. Soc. Chim.*, 1702 (1963)

PHYSOCHLAINA (Solanaceae)

P. alaica E.Korst.

TROPANE ALKALOIDS: apoatropine, apohyoscine, α-belladonine,
β-belladonine, 6-hydroxyhyoscyamine N-oxide,
hyoscyamine, 6-oxohyoscyamine, 6-oxyatropine,
physochlaine, scopolamine.

Mirzamatov *et al.*, *Khim. Prir. Soedin.*, 415 (1974)

P. dubia Pasch.

TROPANE ALKALOIDS: 6-hydroxyhyoscyamine,
6-hydroxyhyoscyamine diacetate.

Mirzamatov *et al.*, *Khim. Prir. Soedin.*, 493 (1972)

P. orientalis (MB.) G.Don.

TROPANE ALKALOIDS: apoatropine, hyoscyamine, scopolamine,
tropine.

Mirzamatov *et al.*, *Khim. Prir. Soedin.*, 493 (1972)

PHYSOSTIGMA (Leguminosae)

P. venenosum Balf.

PHYSOSTIGMINE ALKALOIDS: eseramine, eseridine,
physostigmine, physovenine.
UNCHARACTERIZED ALKALOIDS: calabacine, calabatine.

Dopke, *Naturwissenschaften*, **50**, 713 (1963)

PHYTOLACCA (Phytolaccaceae)

P. acinosa.

TRITERPENOIDS: epi-acetylaleuritiolic acid, acinosolic acid,
jaligonic acid, phytolaccanol, spergulagenic acid.

Razdan *et al.*, *Phytochemistry*, **21**, 2339 (1982)

P. americana.

BENZODIOXIN: americanin.
SAPONINS: phytolaccasaponins A, B, C, D, E and G,
phytolaccoside A, phytolaccoside B, phytolaccatoxin.
TRITERPENOIDS: 3-acetylaleuritolic acid,
3-oxo-30-carbomethoxy-24-norolean-12-en-29-oic acid,
phytolaccinic acid, phytolaccogenic acid, pokeberrygenin.

(B) Woo *et al.*, *Tetrahedron Lett.*, 3239 (1978)
(Sap, T) Kang, Woo, *J. Nat. Prod.*, **43**, 516 (1980)

P. dodecandra L.

SAPONINS: lemnatoxin, lemnatoxin C.

Parkhurst *et al.*, *Can. J. Chem.*, **52**, 702 (1974)

P. esculenta.

TRITERPENOIDS: esculentosides A, B, C, D, E and F, jaligonic
acid.

Woo, *Lloydia*, **36**, 326 (1973)

P. octandra.

SAPONINS: yiamoloside A, yiamoloside B.

Moreno, Rodriguez, *Phytochemistry*, **20**, 1446 (1981)

P. rivinoides.

TRITERPENOID: 20-carbomethoxyoleanic acid.

Gonzalez *et al.*, *An. Quim.*, **68**, 1057 (1972)

PICEA (Pinaceae)

P. abies (L.) Karsten.
DITERPENOID: (3R)-(-)-geranyllinalool.
TRITERPENOIDS: 4α-methoxyserrat-14-en-21β-ol,
21α-methoxyserrat-14-en-3-one.
Kimland, Norin, *Acta Chem. Scand.*, **21**, 825 (1967)
Norin, Winell, *ibid.*, **26**, 2289 (1972)

P. ajanensis.
ACIDS: p-coumaric acid, ferulic acid, p-hydroxybenzoic acid,
protocatechuic acid, vanillic acid.
DITERPENOIDS: abietic acid, dehydroabietic acid,
isopimara-8,15-dien-18-oic acid, isopimaric acid,
laevopimaric acid, neoabietic acid, palustric acid, pimaric
acid, sandaracopimaric acid.
GLUCOSIDES: p-coumaric acid β-glucoside, ferulic acid
β-glucoside, p-hydroxybenzoic acid β-glucoside,
protocatechuic acid β-glucoside, vanillic acid β-glucoside.
MONOTERPENOIDS: camphene, 3-carene, p-cymol, limonene,
myrcene, β-phellandrene, α-pinene, β-pinene, τ-terpinene.
SESQUITERPENOID: ajanol.
(A, G) Ivanova *et al.*, *Khim. Prir. Soedin*, 415 (1975)
(T) Babkin *et al.*, *ibid.*, 736 (1978)
(T) Schmidt *et al.*, *Izv. Sib. Otd. Akad. Nauk SSSR*, 133 (1978); *Chem. Abstr.*, 88 166736m

P. glehnii L.
MONOTERPENOIDS: camphene, 3-carene, p-cymol, limonene,
myrcene, β-phellandrene, α-pinene, β-pinene, τ-terpinene,
terpinolene.
SESQUITERPENOID: gleenol.
DITERPENOIDS: abietic acid, dehydroabietic acid,
isopimara-8,15-dien-18-oic acid, isopimaric acid,
laevopimaric acid, neoabietic acid, palustric acid, pimaric
acid, sandaracopimaric acid.
Kurvyakov *et al.*, *Khim. Prir. Soedin.*, 164 (1979)

P. koraiensis.
MONOTERPENOIDS: camphene, 3-carene, p-cymol, limonene,
myrcene, β-phellandrene, α-pinene, β-pinene, τ-terpinene,
terpineol.
DITERPENOID: abietic acid, dehydroabietic acid,
isopimara-8(15)-en-18-oic acid, isopimaric acid,
laevopimaric acid, neoabietic acid, neocembrene, palustric
acid, pimaric acid, sandaracopimaric acid.
Schmidt *et al.*, *Khim. Prir. Soedin.*, 694 (1970)
Schmidt *et al.*, *Izv. Sib. Otd. Akad. Nauk SSSR*, 133 (1978); *Chem. Abstr.*, 88 166736m

P. obovata.
DITERPENOIDS: R(-)-cembrene A, neocembrene.
FLAVANOIDS: dihydroquercetin, kaempferol, quercetin.
GLUCOSIDES: p-coumaric acid β-glucoside,
3,4-dihydroxyacetophenone glucoside, ferulic acid
β-glucoside, p-hydroxyacetophenone glucoside,
p-hydroxybenzoic acid β-glucoside,
3-methoxy-4-hydroxyacetophenone glucoside, vanillic acid
β-glucoside.
MONOTERPENOIDS: 3-carene, β-phellandrene, α-pinene,
β-pinene.
PHENOLICS: p-coumaric acid, ferulic acid, p-hydroxybenzoic
acid, protocatechuic acid, vanillic acid.
STILBENES: astringenin, astringin, isorhapontigenin,
isorhapontin, resveratrol.
(G, P) Ivanova *et al.*, *Khim. Prir. Soedin.*, 107 (1976)
(F, P) Shibanova *et al.*, *Chem. Abstr.*, 87 197315g
(T) Schmidt *et al.*, *ibid.*, 694 (1970)

P. schrenkiana.
MONOTERPENOIDS: (+)-cis-α-bisabolene, camphene, β-carene,
p-cymol, dipentene, geraniol, linalool, β-phellandrene,
α-pinene, β-pinene, terpinolene.
Pentegova *et al.*, *Khim. Prir. Soedin.*, 459 (1987)

P. sitchensis L.
TRITERPENOIDS: 3β,21α-dimethoxyserrat-14-ene,
3α-methoxy-21β-hydroxyserratene,
3β-methoxy-21α-hydroxyserratene,
3β-methoxy-21β-hydroxyserratene,
3α-methoxyserrat-13-en-21β-ol,
3α-methoxyserrat-14-en-21β-ol,
3β-methoxyserrat-14-en-21β-ol,
3β-methoxyserrat-14-en-21-one.
Rogers, Rozon, *Can. J. Chem.*, **48**, 1021 (1970)

PICRADENIOPSIS

P. woodhousei.
SESQUITERPENOID:
3,10-dihydroxy-2-oxo-8-tigloyloxy-3,11(13)-guaiadien-12,6-
olide.
Herz *et al.*, *J. Org. Chem.*, **45**, 3163 (1980)

PICRALIMA

P. klaineanas Pierre.
CARBOLINE ALKALOIDS: akuammenine, (±)-akuammicine,
akuammidine, akuammigine, akuammigine hydrate,
akuammiline, akuammine, picraline.
Clinquart, *J. Pharm. Belg.*, **9**, 187 (1927)
Henry, Sharp, *J. Chem. Soc.*, 1950 (1927)

P. nitida Stapf.
CARBOLINE ALKALOIDS: burnamine, picraphylline.
Taylor *et al.*, *Bull. Soc. Chim. Fr.*, 392 (1964)

PICRAMNIA

P. pentandra.
TRITERPENOID: 3-epi-betulinic acid.
Robinson *et al.*, *Phytochemistry*, **9**, 907 (1970)

PICRASMA (Simaroubaceae)

P.
AILANTHOIDES: Carboline alkaloid
1-hydroxymethyl-β-carboline.
TERPENOIDS: nigakialcotol, nigakiheniacetals A, B, C, D, E and
F, nigakilactones A, B, C, D, E, F, G, H, I, J, K, L, M and
N, picrasinisides A, B, C, D, E, F and G.
(A) Koike *et al.*, *Chem. Pharm. Bull.*, **21**, 837 (1973)
(T) Okano *et al.*, *Tennen Yuki Kagobutsu Toronkai Koen Yoshishu*, **26**,
54 (1983)

P. crenata.
UNCHARACTERIZED ALKALOID: sigmine.
Pereira, *Ann. Fac. Med., Univ. San Paulo*, **14**, 269 (1938)

P. excelsa.
CARBOLINE ALKALOID: N-methoxy-1-vinyl-β-carboline.
Wagner *et al.*, *Planta Med.*, **36**, 113 (1979)

P. javanica Bl.
CARBOLINE ALKALOID: 4-methoxy-1-vinyl-β-carboline.
Johns *et al.*, *Aust. J. Chem.*, **23**, 629 (1970)

P. quassioides D.Don. Benn.
CARBOLINE ALKALOIDS: 1-carboxy-β-carboline, picrasidines A,
B, L, M and P.
QUASSINOIDS: kumijians A, B, C and G, picrasins A, B, C, D,
E, F, G, N, O and Q, kusulactone.
(A) Koike *et al.*, *Chem. Pharm. Bull.*, **34**, 2090 (1986)
(Q) Ohmoto *et al.*, *ibid.*, **33**, 4901 (1985)

PICRIS (Compositae)

P. echioides L. (Helmintia echioides (L.) Gaertn.).
Two sesquiterpenoids.
Bohlmann et al., Phytochemistry, **20**, 2029 (1981)

P. fel-terrae.
TRITERPENOIDS: picfeltarragenins I, II, III, IV, V and VI.
Gan et al., Yuaxue Xuebao, **40**, 812 (1982)

P. hieracioides L. var. japonica Regel.
SESQUITERPENOIDS: picrisides A, B and C.
PHENOLICS: crepidiaside A, 1β,13-dihydrolactucin, ixerin F, lactucin.
Nishimura et al., Chem. Pharm. Bull., **34**, 2518 (1986)

PICRIDIUM

P. crystallinum.
SESQUITERPENOID: picridine.
Gonzalez et al., An. Quim., **66**, 419 (1970)

PICROLEMMA

P. pseudocoffea.
TERPENOIDS: 15-deacetylsergeolide, sergeolide.
Polonsky et al., J. Nat. Prod., **47**, 994 (1984)

PICRORHIZA

P. kurrooa.
IRIDOID: picroside I and II.
SAPONIN:
25-acetoxy-2β-glucosyloxy-3,16,20-trihydroxy-9-methyl-18-norlanosta-5,23-dien-22-one.
TERPENOID: kutkiol.
Laurie et al., Phytochemistry, **24**, 2659 (1985)

PIERIS (Ericaceae)

P. japonica D.Don.
STEROID: β-sitosterol.
TOXINS: asebotoxins I, II, III, IV, V, VI, VII, VIII, IX and X, picroside A, pieristoxins A, B, C, D, E, F, G, H, I, J and K.
TRITERPENOIDS: α-amyrin, betulinic acid, lupeol, oleanolic acid, taraxerol, ursolic acid.
(S, T) Kawai et al., Chem. Abstr., 88 34499b
(Tox) Hikino et al., Chem. Pharm. Bull., **28**, 3124 (1980)

P. taiwanensis.
FLAVONOIDS: asebotin, hyperin, phloridzin, quercetin 3-arabinoside.
STEROIDS: campesterol, β-sitosterol.
TRITERPENOIDS: β-amyrin, lupeol, oleanolic acid.
Chou et al., Chem. Abstr., 100 82782b

PIERREODENDRON (Simaroubaceae)

P. kerstingii Little.
AMIDE: aurantiamide acetate.
CARBAZOLE ALKALOID: 8-hydroxyxanthin-6-one.
QUASSINOIDS: 2'-acetylglaucarubin, ailanthinone, dehydroailanthinone, excelsin, glaucarubin, glaucarubinone.
Kupchan et al., J. Org. Chem., **40**, 654 (1975)
Ampofo et al., J. Nat. Prod., **48**, 863 (1985)

PILGERODENDRON

P. uniferus.
SESQUITERPENOID: (-)-δ-cadinol.
Motl et al., Collect. Czech. Chem. Commun., **23**, 1297 (1958)

PILOCARPUS (Rutaceae)

P. jaborandi Holmes.
FURANE ALKALOIDS: isopilocarpidine, isopilocarpine, pilocarpidine, pilocarpine.
Wagenaar, Pharm. Weekbl., **67**, 285 (1930)

P. microphyllus Stapf.
FURANONE ALKALOIDS: iso-epi-pilosine, isopilosine, pilocarpine, pilosine.
Roche, Lynch, Analyst, **73**, 311 (1948)

P. pennatifolius Lem.
FURANE ALKALOIDS: isopilocarpine, pilocarpine.
Jowett, J. Chem. Soc., **77**, 483 (1900)

P. racemosus Wahl.
FURANE ALKALOID: pilocarpine.
Preobrashenski et al., Ber. Dtsch. Chem. Ges., **63**, 460 (1930)

P. selloanus Engl. (P. pennatifolius Lem.)
FURANE ALKALOID: pilocarpine.
Roche, Lynch, Analyst, **73**, 311 (1948)

P. spicatus.
FURANE ALKALOID: pseudopilocarpine.
Petit, Polonsky, J. Pharm., **5**, 369 (1897)

P. trachyophus.
FURANE ALKALOID: pilocarpine.
Gerrard, Pharm. J., **5**, 865 (1875)
Gerrard, ibid., 965 (1875)

PILOCEREUS (Cactaceae)

P. chrysacanthus.
AMINE ALKALOID: N-methyl-3,4-dimethoxyphenethylamine.
Bruhn et al., Phytochemistry, **16**, 622 (1977)

P. guerroronis.
ISOQUINOLINE ALKALOID: O-methylcorypalline.
Lindgren et al., Lloydia, **39**, 464 (1976)

PILOPLEURA

P. kozo-poljanskii.
BENZOPYRAN: piopleurine.
Avvamenko et al., Khim. Prir. Soedin., 18 (1973)

PIMELEA (Thymelaeaceae)

P. prostrata (J.R. et G.Forst.) Willd.
DITERPENOID: prostratin.
McCormick et al., Tetrahedron Lett., 1735 (1976)

P. simplex.
TRITERPENOID: simplexin.
Roberts et al., Aust. J. Chem., **51**, 325 (1975)

PIMPINELLA (Umbelliferae)

P. diversifolia.
ALKALOID: arnottianamide.
COUMARIN: isoangenomelin.
(A) Ishii et al., Tetrahedron Lett., 1203 (1976)
(C) Bohlmann et al., Chem. Ber., **108**, 433 (1975)

P. thellungiana Wolff.
PHENOLIC: 3-methoxy-5-(1-ethoxy-2'-hydroxyphenyl)phenol.
STEROIDS: β-sitosterol, τ-sitosterol.
Wang et al., Yaoxue Xuebao, **18**, 522 (1983); Chem. Abstr., 100 64982n

PINELLIA

P. pedatisecta.

ACID: palmitic acid.

AMINE ALKALOIDS: pedatisectine A, pedatisectine B, pedisectine A, pedisectine B.

CARBOLINE ALKALOIDS: 1-acetyl-β-carboline, β-carboline.

PYRIDINE ALKALOID: 2-methyl-3-hydroxypyridine.

AMIDE: nicotinamide.

STEROID: β-sitosterol.

URACILS: 5-methyluracil, uracil.

Qin et al., Zhongcaoyao, 17, 197 (1986); Chem. Abstr., 105 94501a

PINGUICULA (Lentibulariaceae)

P. vulgaris L.

FLAVONE: scutellarin

IRIDOIDS: globularin, globularicisin, scutellarioside II.

(F) Arisawa et al., Chem. Pharm. Bull., 18, 916 (1970)

(I) Marco, J. Nat. Prod., 48, 338 (1985)

PINUS (Pinaceae)

P. albicans.

SESQUITERPENOID: albicaulene.

Haagen-Smith et al., J. Am. Pharm. Assoc., 40, 557 (1951)

P. banksiana.

DITERPENOID: agathadiol.

NORDITERPENOIDS: 18-norabieta-8,11,13-trien-4-ol, 19-norabieta-8,11,13-trien-4-ol.

TRITERPENOIDS: serrat-14-ene-3α,21β-diol, serrat-14-ene-3β,21α-diol.

Enzell, Acta Chem. Scand., 15, 1303 (1961)

Tsuda et al., Tetrahedron Lett., 1279 (1964)

Rowe, ibid., 2347 (1964)

P. caribea.

DITERPENOID: isodextropimarinal.

Harris, Sanderson, J. Am. Chem. Soc., 70, 3870 (1948)

P. clausena.

FLAVANONE: pinobanksin 7-methyl ether.

Lindstedt et al., Acta Chem. Scand., 4, 1042 (1950)

P. contorta.

DITERPENOIDS: agathadiol, agatholal, (+)-13-epi-manool, 18-norlabda-8(17),13-diene-4α,15-diol, epi-torulosol.

GLUCOSIDES: benzoic acid β-D-glucopyranoside, chavicol β-D-glucopyranoside, rheosmin β-D-glucopyranoside, zingerone β-D-glucopyranoside.

(Gly) iguchi et al., Phytochemistry, 16, 1587 (1977)

(T) Rowe, Scroggins, J. Org. Chem., 29, 1554 (1964)

(T) Manning, Aust. J. Chem., 26, 2735 (1973)

P. densiflora.

SESQUITERPENE: longifolene.

Corey et al., J. Am. Chem. Soc., 86, 438 (1964)

P. edulis.

DITERPENOID: 8,9-isopimaric acid.

Joye et al., J. Org. Chem., 31, 320 (1964)

P. eldarica.

DITERPENOIDS: δ-cadinene, τ-cadinene, caryophyllene, chamazulene, humulene, longifolene, α-muurolene.

MONOTERPENOIDS: bornyl acetate, camphor, citral, terpineal.

Kolesnikova et al., Rastit. Resur., 13, 351 (1977); Chem. Abstr., 87 35863g

P. elliottii.

DITERPENOID: trans-communic acid.

Joye et al., J. Org. Chem., 30, 429 (1965)

P. jeffreyi.

PIPERIDINE ALKALOID: pinidine.

Tallent, Horning, J. Am. Chem. Soc., 78, 4467 (1956)

P. kochiana.

MONOTERPENOIDS: bornyl acetate, citral, terpineal.

DITERPENOIDS: δ-cadinene, τ-cadinene, caryophyllene, chamazulene, humulene, longifolene, α-muurolene.

Kolesnikova et al., Rastit. Resur., 13, 351 (1977); Chem. Abstr., 87 35863g

P. koraiensis.

DITERPENOIDS: (-)-cembrene A, α-pinacene, β-pinacene, τ-pinacene, pinusolide.

Birch et al., J. Chem. Soc., Perkin 1, 2653 1972)

P. lambertiana.

DITERPENOID: lambertianic acid.

TRITERPENOIDS: serrat-14-ene-3β,21β-diol, serrat-14-ene-3β,21α-diol 3-methyl ether, 21-epi-serratenol 21 methyl ether.

Dauben, Gorman, Tetrahedron, 22, 679 (1966)

Rowe et al., Phytochemistry, 11, 365 (1972)

P. longifolia L.

SESQUITERPENE: longicyclene.

Nayak, Dev, Tetrahedron Lett., 243 (1963)

P. luchuensis.

TRITERPENOIDS: 3β,21α-dimethoxyserrat-14-ene, 3β-hydroxyserrat-14-en-21-one, 3β-methoxyserrat-14-en-21α-ol, 3β-methoxyserrat-14-en-21-one, serrat-14-en-3.21-dione.

Cheng et al., J. Chin. Chem. Soc., 22, 341 (1975); Chem. Abstr., 84 132658q

P. merkusi.

DITERPENOID: mercusic acid.

Weissman, Holtzforschung, 28, 186 (1974)

P. monticola L.

DITERPENOIDS: 8,11,13-abietatrien-7-ol, 8,11,13-abietatrien-15-ol, anticopalic acid, 9,10-epoxy-9,10-seco-8,11,13-abietatriene.

NORTRITERPENOID: 30-nor-3β-methoxyserrat-14-en-21-one.

Conner et al., J. Org. Chem., 46, 2987 (1981)

P. pallasiana L.

DITERPENOIDS: 8,11,13-abietatriene, δ-cadinene, τ-cadinene, caryophyllene, chamazulene, dehydroabietane, humulene, longifolene, α-muurolene.

MONOTERPENOIDS: bornyl acetate, camphor, citral, terpineal.

Vlad et al., Khim. Prir. Soedin., 20 (1971)

Kolesnikova et al., Rastit. Resur., 13, 351 (1977); Chem. Abstr., 87 35863g

P. palustris L.

DITERPENOIDS: isodextropimarinal, palustric acid.

Wenkert et al., J. Am. Chem. Soc., 86, 2038 (1935)

P. ponderosa.

MONOTERPENOIDS: 3-carene, limonene, myrcene, α-pinene, β-pinene.

Smith, U.S. Dept. Agric. Tech. Bull., 1532 (1977); Chem. Abstr., 87 98798k

P. pumila.

NORDITERPENOID: fichtelite.

Ruzicka, Waldman, Helv. Chim. Acta, 18, 611 (1935)

P. radiata.

51 terpenoids.

Simpson et al., Phytochemistry, 15, 328 (1976)

P. roxburghii.

PHENOLIC: hexacosylferulate.

Chatterjee et al., Phytochemistry, 16, 397 (1977)

P. sabiniana Dougl.

PIPERIDINE ALKALOIDS: pinidine, (-)-α-pipecoline.
Tallant et al., J. Am. Chem. Soc., 77, 6361 (1955)

P. sibirica L.

DITERPENOIDS: agathenediol, isocembrene, isocembrol,
methylisocupressate, methyl trans-sciadopate, α-pinacene,
β-pinacene, τ-pinacene, pinusolide.
CATECHINS: (+)-catechin, (-)-epi-catechin.
MONOTERPENOIDS: 3-carene, β-phellandrene, α-pinene,
β-pinene.
SESQUITERPENOID: sibirene.
(C) Polyakova et al., Aktual. Vopr. Bot. Resur. Sib., 190 (1976); Chem.
Abstr., 85 189201k
(T) Kashianova et al., Khim. Prir. Soedin., 52 (1968)
(T) Lisina et al., ibid., 300 (1972)

P. sosnowskyi.

DITERPENOIDS: δ-cadinene, τ-cadinene, caryophyllene,
chamazulene, humulene, longifolene, α-muurolene.
MONOTERPENOIDS: bornyl acetate, camphor, citral, terpineal.
Kolesnikova et al., Rastit. Resur., 13, 351 (1977); Chem. Abstr., 87
35863g

P. strobus L.

DITERPENOIDS: strobal, strobic acid, strobol.
Zinkel, Evans, Phytochemistry, 11, 3387 (1972)

P. sylvestris L.

CATECHINS: (+)-catechin, (-)-epi-catechin.
DITERPENOIDS: abieta-8,11,13-triene-15,18-diol,
dehydropinifolic acid, isopimara-7,15-dien-18-al,
isopimara-7,15-diene, isopimaric acid, 8,9-isopimaric acid,
α-longipinene, 18-nordehydroabietan-4α-ol,
18-nordehydroabietan-4(19)-ene,
19-norisopimara-8(14),15-dien-3-one,
18-norpimara-8,15-dien-4α-ol,
19-norpimara-8(14),15-dien-3-one, (+)-pinifolic acid,
sapietic acid.
MONOTERPENOIDS: 3-carene, β-phellandrene, α-pinene,
β-pinene.
(C) Polyakova et al., Aktual. Vorp. Bot. Resur. Sib., 190 (1976); Chem.
Abstr., 85 189201k
(T) Conner et al., Phytochemistry, 16, 1777 (1977)
(T) Norin et al., Acta Chem. Scand., B 34, 301 (1980)

P. sylvestris L. var. sylvestris.

DITERPENOIDS: δ-cadinene, τ-cadinene, caryophyllene,
chamazulene, humulene, longifolene, α-muurolene.
MONOTERPENOIDS: bornyl acetate, camphor, citral, terpineal.
Kolesnikobva et al., Rastit. Resur., 13, 351 (1977); Chem. Abstr., 87
35863g

P. thunbergi L.

STEROID: 2-deoxycastasterone.
SESQUITERPENOIDS: longifolene, longifol-7(15)-en-5-ol.
(S) Yokota et al., Agr. Biol. Chem., 47, 2419 (1983)
(T) Simonsen, J. Chem. Soc., 123, 2642 (1923)

P. torreyana.

PIPERIDINE ALKALOID: pinidine.
Corey et al., J. Am. Chem. Soc., 83, 1251 (1961)

PIPER (Piperaceae)

P. auranticum.

ALKALOIDS: aurantiamide, aurantiamide acetate, piperettine.
Banerji, Das, Ind. J. Chem., 13, 1234 (1975)

P. auritum.

APORPHINE ALKALOIDS: cepharadione A, cepharadione B.
Haensel et al., Lloydia, 38, 529 (1975)

P. beetle L.

PYRIDINE ALKALOIDS: arecaidine. arecoline Diterpenoid
chavibetol.
Bertrum, Gildmeister, J. Prakt. Chem., 39, 349 (1889)

P. callosum.

AMIDE ALKALOIDS: pipercallosidine, pipercallosine,
piperovatine.
Pring et al., J. Chem. Soc., Perkin I, 1493 (1982)

P. clusii.

PIPERIDINE ALKALOID: piperine.
Peinmann, Arch. Pharm., 234, 245 (1896)

P. communis L.

TRITERPENOID: 19(20)-dehydroursolic acid.
Brieskorn, Suss, Tetrahedron Lett., 515 (1972)

P. cubeba.

SESQUITERPENOID: bicyclosesquiphellandrene.
Terhune et al., Phytochemistry, 13, 1183 (1974)

P. fadyenii.

PYRONE: fadyenolide.
Pelter et al., Tetrahedron Lett., 1545 (1981)

P. farnechoni.

PIPERIDINE ALKALOID: piperine.
Rugheuner, Ber. Dtsch. Chem. Ges., 15, 1391 (1882)

P. futokadsura Sieb. et Zucc.

DITERPENOIDS: futoenone, piperenone.
NEOLIGNANS: kadsurenone, kadsurin A, kadsurin B.
QUINOL: futoquinol.
(N) Chang et al., Phytochemistry, 24, 2079 (1985)
(Q) Takahashi et al., Chem. Pharm. Bull., 18, 100 (1970)
(T) Matsui, Munakata, Tetrahedron Lett., 1905 (1975)

P. guineense Schum. et Thom.

PIPERIDINE ALKALOIDS: dihydrowisanidine, dihydrowisanine,
2'-methoxypiperine, olalasine, wisanidine, wisanine.
AMIDES: dihydropiperine,
N-isobutyryloctadeca-trans-2-trans-4-dienamide,
dihydropiperlonguminine, piperine, sylvatine, trichostachine.
Dwuma-Badu et al., Lloydia, 39, 60 (1976)
Sondengam et al., Phytochemistry, 16, 1121 (1977)
Addae-Mensa et al., Planta Med., 41, 200 (1981)

P. jaborandi.

UNCHARACTERIZED ALKALOID: jaborandine.
Wiesner, Can. J. Chem., 29, 353 (1951)

P. longum L.

PIPERIDINE ALKALOIDS: chavicine, piperine, piperlongumine,
piperonaline, piperundecalidine.
AMIDE: piperlonguminine.
LIGNANS: diaeudesmin, sesamin, sylvatin.
(A, Am) Ott et al., Ber. Dtsch. Chem. Ges., 55, 2653 (1922)
(A, Am) Tabuneng et al., Chem. Pharm. Bull., 31, 3562 (1983)
(L) Dutta et al., Phytochemistry, 14, 2090 (1975)

P. marginatum.

PIPERIDINE ALKALOID: piperine.
Peinemann, Arch. Pharm., 234, 245 (1896)

P. methysticum Forst.

PYRIDINE ALKALOID: pipermethystine.
CHALCONE: flavokawain C.
NORSESQUITERPENOID: dihydrokawain-5-ol.
PYRAN: 5,6,7,8-tetrahydroyangonin.
(A) Smith, Tetrahedron, 35, 437 (1979)
(C) Dutta et al., Indian J. Chem., 11, 509 (1973)
(P) Achenbach et al., Chem. Ber., 104, 2688 (1971)
(T) Achenbach, Wittmann, Tetrahedron Lett., 3259 (1970)

P. nigrum L.

PIPERIDINE ALKALOIDS: chavicine, piperine, piperoleins A and B.

PYRROLIDINE ALKALOID: piperyline.

MONOTERPENOIDS: 1(7),2-p-menthadien-6-ol, 3,8(9)-menthadien-1-ol, methyl cyclohepta-2,4-dien-6-one, β-pinone.

SESQUITERPENOID: sesquisabinene.

Ott *et al., Ber. Dtsch. Chem. Ges.,* **57**, 214 (1924)

Terhune *et al., Can. J. Chem.,* **53**, 3285 (1975)

P. novae-hollandiae.

PIPERIDINE ALKALOID: dihydropiperine.

Loder *et al., Aust. J. Chem.,* **22**, 1531 (1969)

P. officinarum L.

PIPERIDINE ALKALOIDS: chavicine, piperine.

AMIDES: eicosa-2,4,8-trienoic acid isobutylamide, N-isobutyltrideca-13-(3,4-methylenedioxyphenyl)-trienamide.

Gupta *et al., Phytochemistry,* **15**, 425 (1976)

P. ovatum.

PIPERIDINE ALKALOID: piperovatine.

Dunstan, Garnett, *J. Chem. Soc.,* **67**, 94 (1895)

P. peepuloides Roxb.

PIPERIDINE ALKALOID: peepuloidine.

Atal *et al., Tetrahedron Lett.,* 1397 (1968)

P. retrofractum.

AMIDE ALKALOIDS: retrofractamide A, retrofractamide C.

Banerji *et al., Phytochemistry,* **24**, 279 (1985)

P. sanctum.

FURAN: epoxypiperolide.

Pelter, Haansal, *Z. Naturforsch.,* **27B**, 1186 (1972)

P. sylvaticum Roxb.

AMIDES: N-isobutyldeca-2,4-dienamide, sylvatine.

FLAVONOIDS: 5-hydroxy-7-methoxyflavone, 5-hydroxy-7,3′,4′-trimethoxyflavone.

(Am) Banerji *et al., Experientia,* **30**, 223 (1974)

(F) Banerji *et al., Indian J. Chem.,* **15B**, 495 (1977)

P. trichostachya DC.

PYRROLIDINE ALKALOIDS: cyclostachines A and B, piperstachine, trichonine, trichostachine.

Joshi *et al., Helv. Chim. Acta,* **58**, 2295 (1975)

P. tuberculatum.

DIMERIC ALKALOID: piplartine dimer A.

Filho *et al., Phytochemistry,* **20**, 345 (1981)

PIPTADENIA (Leguminosae)

P. macrocarpa.

FLAVAN: 3,3′,4′,7,8-pentahydroxyflavan.

PHENOLICS: dalbergin, 3,54-dimethoxydalbergione, kuhlmannin.

STEROIDS: β-sitosterol, β-sitosterol glucoside, β-sitosterol palmitate.

TRITERPENOIDS: lupenone, lupeol.

Miyauchi *et al., Mokuzai Gakkaishi,* **22**, 47 (1976); *Chem. Abstr.,* **84** 147704a

P. peregrina Benth.

INDOLE ALKALOID: bufotenine.

Stromberg *et al., J. Am. Chem. Soc.,* **76**, 1707 (1954)

PIPTANTHUS (Leguminosae)

P. nanus M.Pop.

ALKALOIDS: isopiptanthine, piptamine, piptanthine.

UNCHARACTERIZED ALKALOID: nanine.

Yakovleva, Proskurnina, *J. Gen. Chem., USSR,* **34**, 3841 (1964)

P. nepalensis (Hook.) D.Don.

QUINOLIZIDINE ALKALOIDS: anagyrine, cytisine, lupanine, N-methylcytisine, (+)-sparteine.

ISOFLAVONES: daidzein, daidzein β-diglucoside, daidzein β-monoglucoside, formononetin, formononetin β-diglucoside, formononetin β-monoglucoside, genistein, genistein β-diglucoside, genistein β-monoglucoside, ononin, ononin β-diglucoside, ononin β-monoglucoside.

(A) Faugeras *et al., Plant Med. Phytother.,* **9**, 273 (1975)

(I) Markham *et al., Phytochemistry,* **7**, 791 (1968)

(I) Paris *et al., Planta Med.,* **29**, 32 (1976)

PIPTOCALYX (Trimeniaceae)

P. moorei.

GLUCOSIDE: piptoside.

LIGNAN: calopiptin.

STEROID: α-sitosterol.

McAlpine *et al., Aust. J. Chem.,* **21**, 2095 (1968)

PIPTOCARPHA (Asteraceae)

P. chontalensis Pall.

SESQUITERPENOIDS: piptocarphins A, B, C, D, E, F, G and H.

Cowall *et al., J. Org. Chem.,* **46**, 1108 (1981)

P. paradoxa.

TRITERPENOIDS: bazzanene, bazzanenol, lupenyl acetate, lupeol.

Mathur *et al., J. Nat. Prod.,* **45**, 495 (1982)

PIPTOLEPIS

P. ericoides.

SESQUITERPENOIDS: 8-angeloyloxyzexbrevanolide, 5-hydroxy-6-methacryloxy-14,15-isogoyazansanolide.

DITERPENOID: piptolepolide.

Bohlmann *et al., Phytochemistry,* **20**, 731 (1981)

P. leptospermoides.

SESQUITERPENOIDS: 15-hydroxyeremantholide C, piptospermolide.

Bohlmann *et al., Phytochemistry,* **21**, 1439 (1982)

PIPTOPORUS (Basidiomycetae)

P. australiensis (Wakefield) Cunn.

PIGMENTS: piptoporic acid, piptoporic acid acetate, piptoporic acid methyl ester.

POLYENE: 3-acetoxy-18-methyl-5,7,9,11,13,15,17-icosoheptanoic acid.

Gill *et al., J. Chem. Soc., Perkin I,* 1449 (1982)

PIPTOTRIX

P. aureolare.

MONOTERPENOID: aureolal.

Hernandez *et al., Phytochemistry,* **25**, 1743 (1986)

PIQUIERA

P. trinerva Cav.

DITERPENOIDS: α-santolal, trinervinol.

Jiminez *et al., Rev. Latinoam. Quim.,* **14**, 20 (1983)

PISCIDIA (Papilionaceae)

P. erythrina L. One uncharacterized alkaloid.

BENZOPYRAN: piscidone.

TERPENOIDS: 3-O-methylfukiic acid, piscidic acid methyl ester.

    (A) Dankworth, Schutte, *Arch. Pharm.*, **272**, 701 (1934)

    (B) Falshaw *et al.*, *Tetrahedron, Suppl. No.* **7**, 333 (1966)

    (T) Heller, Tamm, *Helv. Chim. Acta*, **58**, 974 (1975)

PISOLITHUS (Basidiomycetae)

P. tinctorius.

STEROID: ergosterol peroxide.

TRITERPENOID: pisolactone.

    (S) Voigt *et al.*, *Z. Chem.*, **25**, 405 (1985)

    (T) Lobo *et al.*, *Tetrahedron Lett.*, 2205 (1983)

PISTACIA (Anacardiaceae)

P. terebinthus L.

TRITERPENOIDS: dihydromasticadiendiol, dihydro-3-epi-masticadienoic acid methyl ester, dihydromasticadienolic acid methyl ester, isomasticadiendiol, isomasticadienolic acid, 3-epi-isomasticadienolic acid, masticadienolic acid, 3-epi-masticadienolic acid, oleanolic acid, oleanolic aldehyde, oleanonic aldehyde.

    Caputo, Mangoni, *Gazz. Chim. Ital.*, **100**, 317 (1970)

P. vera.

ANTHOCYANIN: cyanidin 3-O-β-galactoside.

MONOTERPENOIDS: 9,10-cyclo-2,4-menthanediol, 9,10-cyclo-2,4-menthen-4-ol. Triterpenoid 3,11,13-trihydroxyoleanane.

    (An) Miniati, *Fitoterapia*, **52**, 267 (1981)

    (T) Mangoni *et al.*, *Phytochemistry*, **21**, 811 (1982)

    (T) Monaco *et al.*, *ibid.*, **21**, 2408 (1982)

PISUM (Leguminosae)

P. sativum L.

BENZOPYRANS: 3-hydroxy-2,9-dimethoxypterocarpan, 4-hydroxy-2,3,9-trimethoxypterocarpan, 2,3,9-trimethoxypterocarpan.

    Pueppke *et al.*, *J. Chem. Soc., Perkin I*, 946 (1975)

PITARIA

P. punctata.

COUMARINS: braylin, neobraylin.

    Hlubucek *et al.*, *Aust. J. Chem.*, **24**, 2347 (1971)

PITHECILLOBIUM

P. arboreum.

TRITERPENE ALKALOID: O(3)-acetylamino-2-deoxy-β-D-glucopyranosyloleanolic acid.

    Ripperger *et al.*, *Phytochemistry*, **20**, 2434 (1981)

P. dulce.

STEROID: 7,22-stigmastadien-3-ol-β-D-glucopyranoside.

    Nigam *et al.*, *Indian J. Chem.*, **5**, 395 (1967)

P. saman.

SAPONIN: samanin C.

    Varshney *et al.*, *Indian J. Pharm.*, **39**, 80 (1977)

PITHECOLOBIUM

P. saman Benth.

MACROCYCLIC ALKALOID: pithecolobine.

SAPONIN: samanin B.

    (A) Greshoff, *Ber. Dtsch. Chem. Ges.*, **23**, 3541 (1890)

    (Sap) Varshney, Vyas, *J. Indian Chem. Soc.*, 992 (1978)

PITTOSPORUM (Pittosporaceae)

P. brevicalyx.

TRITERPENOIDS: barrigenol R-1, pittobrevigenin, theasapogenol D.

    Yang *et al.*, *Yaoxue Tongbao*, **21**, 12 (1986); *Chem. Abstr.*, 105 94488b

P. buchananii.

ACETYLENICS: trans-pentadeca-1,8,10-triene-4,6-diyn-3-ol, trans-pentadeca-1,8,10-triene-4,6-diyn-3-one.

P. nilghrense Wr. et Arn.

SAPONINS: pittoside A, pittoside B.

    Jain *et al.*, *Indian J. Pharm.*, **42**, 12 (1980)

P. phillyraeoides.

TRITERPENOID: phillyrigenin.

    Beckwith *et al.*, *Aust. J. Chem.*, **9**, 428 (1956)

P. tobira.

TRITERPENOID: barrigenol R-1.

    Ito *et al.*, *Tetrahedron Lett.*, 2289 (1967)

P. undulatum Vent.

TRITERPENOIDS: barrigenol A-1, pittosapogenin.

    Cornforth, Earl, *J. Proc. Roy. Soc., N. S. W.*, **72**, 249 (1938)

    Cornforth, Earl, *ibid.*, **72**, 252 (1938)

PITURANTHUS

P. tortuosus (Desf.) Benth. et Hook.

ACETYLENES: falcarinone, trans-heptadeca-1,8-diene-4,6-diyne-3-ol-10-one.

    Schulte *et al.*, *Arch. Pharm.*, **310**, 945 (1977)

PITYRODIA (Verbenaceae)

P. lepidota (F.Muell.) Pritzel.

DITERPENOID: 6,18-dihydroxy-3,13-clerodien-15-oic acid methyl ester.

    Ghisalberti *et al.*, *Aust. J. Chem.*, **34**, 1009 (1981)

PITYROGRAMMA

P. austroamericana.

CHALCONE: 2',6'-dihydroxy-4,4'-dimethoxychalcone.

    Wollenweber, *Z. Pflanzenphysiol*, **78**, 344 (1976); *Chem. Abstr.*, **85** 17060x

P. calomelanus.

SESQUITERPENOID: calomelanolactone.

    Bardouille *et al.*, *Phytochemistry*, **17**, 275 (1978)

P. triangularis.

CHALCONE: triangularin.

FLAVONES: ceroptin, 3,5-dihydroxy-7,4'-dimethoxyflavone, 5,7-dihydroxy-3-methoxy-6,8-dimethylflavone, 4'-methoxy-3,5,7-trihydroxyflavone, 6-methyl-8-methoxy-3,5,7-trihydroxyflavone, pityrogrammin.

    (C) Star *et al.*, *Phytochemistry*, **17**, 586 (1978)

    (F) Dietz *et al.*, *ibid.*, **20**, 1181 (1981)

P. triangularis var. triangularis.

FLAVONE: 5,7-dihydroxy-6,8-dimethoxyflavone.

    Dietz *et al.*, *Phytochemistry*, **20**, 1181 (1981)

P. trifoliata.
COUMARINS:
8-cinnamoyl-3,4-dihydro-5,6-dihydroxy-4-phenylcoumarin,
3',4',5,7-tetrahydroxy-8-cinnamoyl-4-phenylcoumarin,
4',5,7'-trihydroxy-8-cinnamoyl-4-phenylcoumarin.
Dietz et al., Z. Naturforsch., 35, 36 (1980)

PLACODIUM
P. saxicolum.
TRITERPENOID: zeorin.
Huneck, Lehn, Bull. Soc. Chim., 1702 (1963)

PLAGIOCHILA (Hepaticae)
P. acanthophylla.
SESQUITERPENOIDS: bicyclohumulenone, (-)-maalioxide.
Matsuo et al., J. Chem. Soc., Chem. Commun., 174 (1979)
P. acanthophylla subsp. japonica.
SESQUITERPENOIDS: (+)-bicyclohumulenone, fuscoplagin A, fuscoplagin B.
Hashimoto et al., Tetrahedron Lett., 6473 (1985)
P. asplenioides (L.) Dumort.
SESQUITERPENOIDS: plagiochilins C, D, E and F.
Asakawa et al., Phytochemistry, 18, 1355 (1979)
P. hattoriana.
SESQUITERPNOIDS: hanegokedial, plagiochilin B.
Asakawa et al., Phytochemistry, 17, 1794 (1978)
P. ovalifolia.
SESQUITERPENOIDS: (+)-acetoxyovalifoliene, (+)-hanegoketrial, (-)-maalian-5-ol, (+)-ovalifolienal, (+)-ovalifolienalone, plagiochilin G.
Matsuo et al., J. Chem. Soc., Perkin I, 2816 (1981)
P. semidecurrens.
SESQUITERPENOIDS: (+)-acetoxyovalifoliene, (+)-ovalifolienal, (+)-ovalifolienalone, (+)-ovalifoliene, ovalimethoxy I, ovalimethoxy II, (+)-oxaplagiochene A, (-)-plagiochendial.
Atsumi et al., Koen Yoshishu - Koryo Terupen Oyobi, Seiyu Kagakuin, Kansuru Toronkai, 23rd, 267 (1979)
P. yokogurensis.
SESQUITERPENOIDS: furanoplagiochilal, plagiochilide, plagiochilin A, plagiochilin H, plagiochilin I.
Asakawa et al., Phytochemistry, 19, 2147 (1980)

PLANALTOA
P. lychnophoroides.
DITERPENOID:
ent-15,16-epoxy-3-hydroxy-8(17),13(16)-labdadien-2,12-dione.
Bohlmann et al., Phytochemistry, 21, 465 (1982)

PLANCHONELLA (Sapotaceae)
P. anteridifera (White et Francis) Lam.
PYRROLIZIDINE ALKALOIDS: 1-benzoylmethyl-8H-pyrrolizidine, planchonelline, 1β-tigloylmethyl-8Hβ-pyrrolizidine.
Hart, Lamberton, Aust. J. Chem., 19, 1259 (1966)
P. chartacea Lam.
No alkaloids present.
Hart, Lamberton, Aust. J. Chem., 19, 1259 (1966)
P. cotinifolia (A.DC.) Dubard.
No alkaloids present.
Hart, Lamberton, Aust. J. Chem., 19, 1259 (1966)
P. myrsinoides (A.Cunn. ex Benth.) S.T.Blake.
No alkaloids present.
Hart, Lamberton, Aust. J. Chem., 19, 1259 (1966)

P. thyrsoides White.
PYRROLIZIDINE ALKALOID: planchonelline.
Hart, Lamberton, Aust. J. Chem., 19, 1259 (1966)
P. torricellensis.
No alkaloids present.
Hart, Lamberton, Aust. J. Chem., 19, 1259 (1966)

PLANCHONIA (Barringtoniaceae)
P. caraya F.Muell.
TRITERPENOID: 16,21,22,28-tetrahydroxyolean-12-en-3-one.
Khong, Lewis, Aust. J. Chem., 30, 1311 (1977)

PLANCKIA
P. populnea.
TERPENOID: populnonic acid.
Delle Monache et al., Gazz. Chim. Ital., 102, 317 (1972)

PLANTAGO (Plantaginaceae)
P. arenaria Waldst. et Kit.
ALKALOID: arenaine.
IRIDOID: plantarenaloside.
(A) Rabaron et al., J. Am. Chem. Soc., 93, 6270 (1971)
(T) Popov et al., Izv. Khim., 14, 175 (1981)
P. indica L.
PYRIDINE ALKALOIDS: indicaine, indicamine, plantagonine.
Danilova, Konovalova, J. Gen. Chem., USSR, 22, 2237 (1952)
P. lanceolata.
FLAVONOID: apigenin 7-O-glucoside.
INDANONE: loliolide.
IRIDOID: aucuboside.
PHENOLICS: caffeic acid, β-coumaric acid, ferulic acid, genistic acid, p-hydroxybenzoic acid, p-hydroxyphenylacetic acid, protocatechuic acid, syringic acid, vanillic acid.
(F) Haznagy et al., Pharmazie, 31, 482 (1976)
(In) Toth et al., Pharmazie, 31, 51 (1976)
(Ir, P) Swiatek, Herba Pol., 23, 20 (1977); Chem. Abstr., 88 148981h
P. major.
FLAVONOIDS: apigenin, luteolin 7-O-β-D-glucoside, luteolin 7-O-β-D-glucuronide.
Lebedev-Kosov, Khim. Prir. Soedin., 812 (1976)
P. media.
IRIDOID: aucuboside.
PHENOLICS: caffeic acid, p-coumaric acid, ferulic acid, genistic acid, p-hydroxybenzoic acid, p-hydroxyphenylacetic acid, protocatechuic acid, synringic acid, vanillic acid.
Swiatek, Herba Pol., 23, 201 (1977); Chem. Abstr., 88 148981h
P. olgae.
PYRIDINE ALKALOID: plantagonine.
Abdusamatov, Yunusov, Khim. Prir. Soedin., 334 (1969)
P. psyllum L.
IRIDOID: plantarenaloside.
Popov et al., Izv. Khim., 14, 175 (1981)
P. ramosa Aschers.
PYRIDINE ALKALOID: indicamine.
Danilova, Konovalova, J. Gen. Chem., USSR, 22, 2237 (1952)

PLATANUS (Platanaceae)
P. acerifolia.
FLAVONE: platanetin.
Egger et al., Z. Pflanzenphysiol., 68, 92 (1972)
P. hybrida.
TRITERPENOID: 3-oxoplatanic acid.
Aplin et al., J. Chem. Soc., 3269 (1963)

P. occidentalis.
   TRITERPENOID: platanic acid.
      Ruzicka, Lamberton, *Helv. Chim. Acta,* **23,** 1338 (1940)
P. orientalis.
   ALIPHATICS: n-hebtriacontanol, 16-hentriacontanone,
      12-triacosanol.
   BENZOPYRAN: amurensin.
   GLUCOSIDE: amurensin.
   STEROID: β-sitosterol stearide.
      Dhar, Munjal, *Planta Med.,* **29,** 91 (1976)

PLATHYNEMIA
P. reticulata.
   DITERPENOIDS: plathyterpol, vinhaticoic acid methyl ester.
      King, Rodrigo, *J. Chem. Soc., Chem. Commun.,* 575 (1967)

PLATISMATIA (Hepaticae)
P. glauca (L.) Culb. et Culb.
   NORTRITERPENOID: 30-nor-21α-hopen-22-one.
      Hveding-Bergseth *et al., Phytochemistry,* **22,** 1826 (1983)

PLATITENIA
P. absinthifolia.
   COUMARIN: suberosin.
      Sood *et al., J. Indian Chem. Soc.,* **55,** 850 (1978)

PLATONIA (Guttiferae/Clusiaceae)
P. insignis.
   XANTHONE: 1,7-dihydroxyxanthone.
      Locksley *et al., J. Chem. Soc.,* C, 430 (1966)

PLATYCARYA (Juglandaceae)
P. glomerata.
   SESQUITERPENOIDS: 8-desoxysalonitenolide,
      11,13-dihydro-8-desoxysalonitenolide.
      Bohlmann, Zdero, *Phytochemistry,* **16,** 1832 (1977)

PLATYCODON
P. grandiflorum.
   SAPONINS: deaploplatycodin, deaploplatycodin D-3,
      2-methoxy-methylplatyconate A, platycodin, platycodins A,
      B, C, D and D-2, platycodosides A, B and C.
   TRITERPENOIDS: platycodigenin, platycogenic acids A, B and C.
      Ishii *et al., Chem. Lett.,* 719 (1978)

PLATYDESMA (Rutaceae)
P. campanulata Mann.
   FURANOQUINOLINE ALKALOIDS: pilokeanine, platydesmine.
      Werner, Scheuer, *Tetrahedron,* **19,** 1293 (1963)
P. spathulatum (A.Gray) DC.
   FURANOQUINOLINE ALKALOID: 6-methoxydictamnine.
      Scheuer, Werner, *Symp. Phytochem., Hong Kong* (1964)

PLATYMISCIUM
P. praecox.
   FLAVANONE: (±)-liquiritigenin.
      Shimoda *et al., Ber. Dtsch. Chem. Ges.,* **67,** 434 (1934)

PLATYTAENIA
P. dasycarpa.
   COUMARINS: bergapten, scopoletin, suberosin,
      O-tigloyl-4′,5′-dihydrooroselol, umbelliferone, zosimin.
      Zhukov *et al., Khim. Prir. Soedin,* 574 (1977)

PLECTANIA
P. coccinea.
   CAROTENOIDS: 2′-dehydroplectaniaxanthin, plectaniaxanthin.
      Arpin, Liaaen-Jensen, *Phytochemistry,* **6,** 995 (1967)

PLECTOCOMIOPSIS
P. geminiiflorus.
   CARBOLINE ALKALOID: plectocomine.
      Kiang *et al., Lloydia,* **30,** 189 (1967)

PLECTRANTHUS (Labiatae)
P. caninus.
   DITERPENOIDS: coleons M, N, O, P, Q, R, S and T.
      Arihara *et al., Helv. Chim. Acta,* **58,** 343 (1975)
P. ecklonii Benth.
   DITERPENOID: parviflorone.
   FLAVONES: cirsimaritin,
      2(S)-5,4′-dihydroxy-6,7-dimethoxyflavone.
   QUINONE: ecklonoquinone.
      Uchida *et al., Helv. Chim. Acta,* **63,** 225 (1980)
P. edulis.
   DITERPENOID: edulone A.
      Buchbauer *et al., Helv. Chim. Acta,* **61,** 1969 (1978)
P. grandidentatus.
   ANTHRAQUINONES: grandidones A, B, C and D,
      2-epi-grandidone B.
      Ruedi *et al., Helv. Chim. Acta,* **64,** 2227 (1981)
      Ruedi et el., *ibid,* **64,** 2251 (1981)
P. lanuginosus.
   DITERPENOIDS: 6,7-dehydroroyleanone, lanugones A, B, C, D,
      E, F, G, H, I, J, K, K′, L, M, N, O, P, Q, R and S.
      Schmid *et al., Helv. Chim. Acta,* **62,** 2136 (1982)
P. madagascariensis.
   DITERPENOID: coleon I.
      Ruedi *et al., Helv. Chim. Acta,* **58,** 1899 (1975)
P. myrianthus Briq.
   DITERPENOIDS: coleons V and W.
      Miyase *et al., Helv. Chim. Acta,* **60,** 2770 (1977)
P. nilgherricus.
   DITERPENOIDS: 3-acetoxyfuerstone, nilgherrons A and B.
      Miyase *et al., Helv. Chim. Acta,* **60,** 2789 (1977)
P. parviflorus Willd.
   DITERPENOIDS: parviflorones B, D, E and F.
      Ruedi, Eugster, *Helv. Chim. Acta,* **61,** 709 (1978)
P. purpuratus Harv.
   DITERPENOID:
      16,17,19-trihydroxy-3-phyllocladanone-17,19-diacetate.
      Katti *et al., Helv. Chim. Acta,* **65,** 2189 (1982)
P. rugosus.
   ALIPHATICS: ceryl alcohol, cetyl alcohol, triacontanol.
   FLAVONOIDS: 5,7-dihydroxy-6,3′-dimethoxyflavone,
      5,4′-dihydroxy-3,6,7-trimethoxyflavone.
   PHENOLIC: E-3,4-dihydroxycinnamic acid.
   STEROID: β-sitosterol.
   TRITERPENOIDS: 3-acetylplectranthic acid,
      3β-hydroxyurs-12-en-28-oic acid, plectranthic acid,
      plectranthoic acid A, plectranthoic acid B,
      12-ursane-3,29-diol.
      (Al, F, P, S) Kumar *et al., Chem. Abstr.,* 100 64974m
      (T) Razdan *et al., Phytochemistry,* **21,** 408 (1982)
P. strigosus Benth.
   DITERPENOIDS: parviflorone G, parviflorone H.
      Alder *et al., Helv. Chim. Acta,* **67,** 1534 (1984)

PLEIOCARPA (Apocynaceae)

P. mutica Benth.

CARBOLINE ALKALOIDS: isotuboflavine, N(a)-methylcarpagine, norisotuboflavine.

DIMERIC ALKALOID: pleiomutine.

IMIDAZOLIDINOCARBAZOLE ALKALOIDS:
N(a)-carbomethoxy-10,22-isoxokopsane, huntabrine, kopsinilam, kopsinoline, N(a)-methyl-10,22-dioxokopsane, pleiocarpamine, pleiocarpine, pleiocarpinilam, pleiocarpinine, pleiocarpoline, pleiocarpolinine.

Khan et al., Helv. Chim. Acta, 48, 1957 (1965)
Kump et al., ibid., 48, 1007 (1965)

P. pycnantha (K.Schum.) Stapf.

DIMERIC IMIDAZOLIDINOCARBAZOLE ALKALOID: pycnanthinine.

orman, Schmid, Monatsh. Chem., 98, 1554 (1967)

P. pycnantha (K.Schum.) Stapf. var. pycnantha M.Pichon.

IMIDAZOLIDINOCARBAZOLE ALKALOIDS: (+)-pycnanthine, pycnanthinol.

Gorman et al., Helv. Chim. Acta, 52, 33 (1969)

P. pycnantha (K.Schum.) Stapf. var. tubicina (Stapf.) M.Pichon.

CARBAZOLE ALKALOIDS: 19,20-dihydroakuammicine, tubifolidine, tubifoline.

Weismann et al., Helv. Chim. Acta, 44, 1977 (1961)

P. talbotii Wernham.

CARBOLINE ALKALOIDS: deformyltalbotinic acid, deformyltalbotinic acid methyl ester, didehydrotalbotine, 16-epi-affinine, talbotine, talcarpine, talpinine, tetrahydrotalbotine.

Naranjo et al., Helv. Chim. Acta, 55, 752 (1972)
Pinar et al., ibid., 56, 2719 (1973)

P. tubicina Stapf.

IMIDAZOLIDINOCARBAZOLE ALKALOIDS: aspidofractine, (-)-1,2-dehydroaspidospermatine, kopsinoline, pleiocarpoline, pleiocarpolinine, (+)-quebrachamine, tubiflavine, tuboxenine.

CARBAZOLE ALKALOID: tubotaiwine.

Bycroft et al., Helv. Chim. Acta, 47, 1147 (1964)
Kump et al., ibid., 48, 1002 (1965)

PLEIOSPERMUM (Rutaceae)

P. alatum (Wight et Arn.) Swingle.

ACRIDONES:
1,5-dihydroxy-2,3-dimethoxy-10-methyl-9-acridone, 1,6-dihydroxy-2,3,5-trimethoxy-10-methyl-9-acridone.

ALKALOID: alatamide.

(Ac) Bowen et al., Phytochemistry, 25, 429 (1986)
(A) Chatterjee et al., Aust. J. Chem., 28, 457 (1975)

PLENASIUM

P. banksiaefolium.

STEROIDS: ecdysterone, ponasterone A.

Hikino et al., Planta Med., 31, 71 (1977)

PLEOCARPUS

P. revolutens.

SESQUITERPENOID: hydroxyisopatchoulenone.

de Silva et al., Phytochemistry, 16, 379 (1977)

PLEOPELTIS

P. farinosa.

TRITERPENOID: 24-methyldammara-12,25-diene.

Wij, Rangaswami, Indian J. Chem., 13, 748 (1975)

PLEUROCORONIS

P. pluriseta.

DITERPENOID: dihydrodendroidinic acid.

Bohlmann et al., Phytochemistry, 20, 2433 (1981)

PLEUROSPERMUM (Umbelliferae)

P. angelicoides.

MONOTERPENOID: angelicoidenol.

Mahmood et al., Phytochemistry, 22, 774 (1983)

P. lindleyanum.

FLAVONOIDS: bergapten, galuteolin, glucoluteolin, isoimperatorin, isopimpinellin, (±)-oxypeucedanin.

Chen et al., Zhongcaoyao, 18, 290 (1987); Chem. Abstr., 107 172506p

PLEUROSTYLA (Celastraceae)

P. africana.

MACROCYCLIC ALKALOID: pleurostyline.

Wagner et al., Tetrahedron Lett., 781 (1978)

P. opposita (Wall.) Alston.

Four tetraterpenoids.

Dantanarayana et al., J. Chem. Soc., Perkin I, 2717 (1981)

PLEUROTUS

P. sajor-caju (Fr.) Singer.

AMINO ACIDS: alanine, arginine, aspartic acid, glutamic acid, histidine, leucine, lysine, phenylalanine, threonine, valine.

Jendaik, Kapoor, Mushroom Sci., 41, 154 (1976); Chem. Abstr., 87 98800e

PLOCAMIUM (Rhodophyceae/Plocamiaceae)

P. cartilagineum (L.) Dixon.

HALOGENATED MONOTERPENOIDS:
7-(bromochloromethyl)-3,4,8-trichloro-3-methyl-1,5,7-octatriene,
4-bromo-1,5-dichloro-2-chlorovinyl-1,5-dimethylcyclohexane,
2-chloro-4-bromo-1-chlorovinyl-1-methyl-5-methylenecyclohexane,
2-chloro-trans-1-chlorovinyl-4-cyclohexane, plocamene A, plocamene B, plocamenone.

Stierle, Sims, Tetrahedron, 35, 1261 (1979)
Crews et al., J. Org. Chem., 49, 1371 (1984)

P. coccineum.

MONOTERPENOID: coccinene.

Castedo et al., J. Nat. Prod., 47, 724 (1984)

P. costata.

HALOGENATED MONOTERPENOIDS: costatolide, costatone.

Stierle et al., Tetrahedron Lett., 4455 (1976)

P. mertensii.

HALOGENATED MONOTERPENOIDS:
2-bromo-1-chloro-4-(2-chlorovinyl)-1-methyl-5-methylenecyclohexane,
2-bromo-4,5-dichloro-1-(2-chlorovinyl)-1,5-dimethylcyclohexane, mertensene.

Capon et al., Aust. J. Chem., 37, 537 (1984)

P. oregonum.

HALOGENATED MONOTERPENOID: oregonene A.

Crews, J. Org. Chem., 42, 2674 (1977)

P. violaceum.

HALOGENATED MONOTERPENOIDS: plocamenes D and E, violacene.

Crews et al., J. Org. Chem., 43, 116 (1978)

PLUCHEA (Compositae/Inuleae)

P. chingoyo.
SESQUITERPENOIDS: pluchein, plucheinol, 3-epi-plucheinol.
Chiang *et al.*, *Rev. Latinoam. Quim.*, **10**, 95 (1979)
Chiang *et al.*, *Phytochemistry*, **18**, 2033 (1979)

P. dioscordia.
One eudesmolide.
Omar *et al.*, *Phytochemistry*, **22**, 779 (1983)

P. foetida.
SESQUITERPENOID:
4-acetoxy-3-(2-methyl-2,3-epoxybutyryloxy)-11-hydroxy-6,7-dehydroeudesman-8-one.
Bohlmann, Mahanta, *Phytochemistry*, **17**, 1189 (1978)

P. indica.
SESQUITERPENOID:
3-(2′,3′-diacetoxy-2′-methylbutyryl)-cuauhtenone.
Mukhoparhyay *et al.*, *J. Nat. Prod.*, **46**, 671 (1983)

P. odorata.
SESQUITERPENOID: cuauhtenone.
Nakanishi *et al.*, *J. Am. Chem. Soc.*, **96**, 609 (1974)

P. rosea.
SESQUITERPENOIDS: ixtlixochilin I, ixtlixochilin II, ixtlixochilin III.
Dominguez *et al.*, *Phytochemistry*, **20**, 2297 (1981)

P. sagittalis.
FLAVONOIDS: centaureidin, chrysosplenol, 3′,4′,5,7-tetrahydroxy-3,6,8-trimethoxyflavone.
Martino *et al.*, *Phytochemistry*, **15**, 1086 (1976)
Martino *et al.*, *An. Asoc. Quim. Argent.*, **73**, 401 (1985); *Chem. Abstr.*, 104 31793f

P. sericea.
FLAVONOID: quercetin 3,3′-dimethyl ether.
SESQUITERPENOIDS: (11S)-11,13-dihydrotessaric acid, tessaric acid.
TRITERPENOID: taraxeryl acetate.
Romo de Vivar *et al.*, *Chem. Lett.*, 957 (1982)

PLUMBAGO (Plumbaginaceae)

P. zeylanica.
AMINO ACIDS: alanine, glycine, histidine, hydroxyproline, methionine, threonine, tryptophan.
BENZOQUINONES: 3,3′-biplumbagin, chitranone, isoshinanolone, 1,2(3)-tetrahydro-3,3′-biplumbagin.
STEROID: β-sitosterol.
(Am) Dinda *et al.*, *J. Indian Chem. Soc.*, **64**, 261 (1987)
(B) Sankaram *et al.*, *Phytochemistry*, **15**, 237 (1976)
(B, S) Gunaherath *et al.*, *ibid.*, **22**, 1245 (1983)

PLUMIERA (Apocynaceae)

P. acutifolia.
ANTIBIOTIC: fulvoplumerin.
IRIDOID: plumeride.
(Antib) Schmid, Bencze, *Helv. Chim. Acta*, **36**, 205 (1953)
(Antib) Schmid, Benze, *ibid.*, **36**, 1468 (1953)

P. lancifolia.
IRIDOID: plumeride.
Boorsma, *Meded. Lds. Pttuin.*, **13**, 27 (1894)

P. rubra L. var. alba.
IRIDOID: plumeride.
Franchimont, *Recl. Trav. Chim. Pays-Bas*, **18**, 334 (1899)
Franchimont, *ibid.*, **18**, 477 (1899)

P. sericifolia.
ALKALOID: vincubine.
Cueller *et al.*, *Rev. Cubana Farm.*, **10**, 31 (1976); *Chem. Abstr.*, 86 1285878j

PODALYRIA

P. buxifolia.
LUPANE ALKALOIDS: (+)-lupanine, (±)-lupanine.
Clemo *et al.*, *J. Chem. Soc.*, 432 (1931)

P. calyptrapa.
LUPANE ALKALOID: (-)-lupanine.
Davis, *Arch. Pharm.*, **235**, 199 (1897)
Davis, *ibid.*, **235**, 218 (1897)
Davis, *ibid.*, **235**, 229 (1897)

P. sericea.
LUPANE ALKALOIDS: (+)-lupanine, (±)-lupanine.
Clemo *et al.*, *J. Chem. Soc.*, 432 (1931)

PODANTHUS

P. mitiqui.
SESQUITERPENOIDS: arturin, deacetylovalifolin, erioflorin methacrylate.
Hoeneisen *et al.*, *Phytochemistry*, **19**, 2765 (1980)

P. ovalifolium.
SESQUITERPENOIDS: erioflorin, erioflorin acetate, erioflorin methacrylate, ovalifolin.
Gneccio *et al.*, *Phytochemistry*, **12**, 2469 (1973)

PODOCARPUS (Podocarpaceae)

P. amarus.
COUMARINS: biochanin A, daidzein, genistein.
Carman *et al.*, *Aust. J. Chem.*, **38**, 485 (1985)

P. blumei.
TERPENOIDS: blumenols A, B and C.
Galbraith *et al.*, *J. Chem. Soc., Chem. Commun.*, 113 (1972)

P. cupressina var. imbricata.
TERPENOID: podocarpic acid.
Haworth, Moore, *J. Chem. Soc.*, 633 (1946)

P. dacrydioides.
DITERPENOID: pododacric acid.
SESQUITERPENOID: kongol.
Corbett *et al.*, *Tetrahedron Lett.*, 1009 (1967)

P. dactyloides.
TERPENOIDS: podocarpic acid, pododacric acid, selin-11-en-4β-ol.
Corbett, Smith, *Tetrahedron Lett.*, 1009 (1967)

P. elatus R.Br.
DITERPENOID: ar-abietatriene.
STEROIDS: crustecdysone, podocdysones A, B and C.
(S) Galbraith *et al.*, *Experientia*, **29**, 782 (1973)
(T) Kitadani *et al.*, *Chem. Pharm. Bull.*, **18**, 402 (1975)

P. elongatus.
FLAVONOIDS: armentoflavone, bilobetin, isoginkgetin, podocarpusflavone.
Prasad, Krishnamurty, *Indian J. Chem.*, **14B**, 727 (1976)

P. ferrugineus.
DITERPENOIDS: 8,11,13-abietatriene-12,19-diol, cryptojaponol, isomiropinic acid, (+)-kaurene, mirene, 2-oxoferruginol.
Cambie *et al.*, *Phytochemistry*, **23**, 333 (1984)

P. gracilior.
FLAVONES: 4,7-dimethylamentoflavone, podocarpusflavone, podocarpusflavone A.
NORDITERPENOID: podolide.
    (F) Prasad et al., Indian J. Chem., 16B, 727 (1976)
    (F) Kubo et al., Rev. Latinoam. Quim., 14, 59 (1983)
    (T) Kupchan et al., Experientia, 31, 137 (1975)

P. hallii.
NORDITERPENOIDS: hallactone A, hallactone B, halliol, sellowin A.
    Cambie et al., J. Org. Chem., 40, 3789 (1975)

P. intermedium.
STEROID: ponasterone C 2-cinnamate.
    Russell et al., Aust. J. Chem., 25, 1935 (1972)

P. lambertianus.
STEROID: 3,5-dihydroxy-5-stigmastan-6-one.
DITERPENOIDS: 5α-carboxytotarol, 17-isophyllocladenol, lambertic acid, macrophyllic acid.
    Campello et al., Phytochemistry, 14, 243 (1975)

P. macrophyllus.
FLAVONES: podocarpusflavone A, podocarpusflavone B, sciadopitysin.
STEROIDS: 24-ethyl-5-cholestene-3β,26-diol, makisterone A, makisterone B.
TERPENOIDS: macrophyllic acid, numakilactones A, B, C, D and E, numakilactone A glycoside.
    (F) Chexal et al., Chem. Ind. (London), 28 (1970)
    (S) Ohmoto et al., Chem. Pharm. Bull., 28, 1894 (1980)
    (T) Hayashi et al., Tetrahedron Lett., 3385 (1972)

P. mannii.
DITERPENOID: 16-hydroxytotarol.
    Brandt, Thomas, Nature, 170, 1018 (1952)

P. milanjianus.
NORDITERPENOIDS: milanjilactone A, milanjilactone B, nagilactone A, nagilactone G, podolactones A, B and C.
    Hembree et al., Phytochemistry, 18, 1691 (1979)
    Cassady et al., J. Org. Chem., 49, 942 (1984)

P. nagi.
FLAVONE: podocarpusflavone A.
NORDITERPENOIDS: 1-deoxy-2-hydroxynagilactone B, 16-hydroxypodolide, 15-methoxycarbonylnagilactone D, nagilactones A, B and C, ponalactone A, ponalactone C, ponalactone A-β-D-glucopyranoside.
    (F) Chexal et al., Chem. Ind. (London) 28 (1970)
    (T) Hembree et al., Phytochemistry, 18, 1691 (1979)

P. nakaii Hay.
STEROIDS: ponasterones A, B and C.
    Nakanishi et al., Tetrahedron Lett., 1105 (1968)
    Nakanishi et al., ibid., 1111 (1968)

P. neriifolius.
FLAVONE: podocarpusflavone.
NORDITERPENOIDS: podolactone B, podolactone C, podolactone D, podolactone E.
    (F) Chexal et al., Chem. Ind. (London), 28 (1970)
    (T) Galbraith et al., J. Chem. Soc., Chem. Commun., 1362 (1971)

P. nubigenus.
DITERPENOID: nubilactone A.
    de Silva et al., Phytochemistry, 12, 883 (1973)

P. odocarpus.
NORDITERPENOID: nagilactone C.
    Hembree et al., Phytochemistry, 18, 1691 (1979)

P. salignus D.Don.
TERPENOIDS: salignone, salignones A, B, C, D, H and J.
    Martin et al., Phytochemistry, 23, 2867 (1984)

P. sellowii.
NORDITERPENOID: nagilactone G.
    Hembree et al., Phytochemistry, 18, 1691 (1979)

P. spicatus.
ISOFLAVONE: podospicatin.
PHENOLIC: isolariciresinol.
    (I) Briggs et al., Tetrahedron, 6, 143 (1959)
    (I) Briggs et al., ibid., 6, 145 (1959)
    (P) Freudenberg, Weinges, Tetrahedron Lett., 17, 19 (1959)

P. totara L.
DITERPENOIDS: 17-hydroxytotarol, pododacric acid, podototarin, totarol.
    Cambie et al., Tetrahedron, 19, 209 (1963)

P. urbanii.
NORDITERPENOIDS: 2,3-dihydroxypodolide, urbalactone.
    Dasgupta et al., Phytochemistry, 20, 153 (1981)

P. vulgare.
STEROID: crustecdysone.
    Hoffmeister, Grutzmacher, Tetrahedron Lett., 407 (1966)

PODOCHAENIUM
P. eminens.
SESQUITERPENOIDS:
13-acetoxy-11,13-dihydro-7,11-dehydro-3-deoxyzaluzanin C, 11,13-dihydro-7,11-dehydro-3-deoxyzaluzanin C, 7-hydroxy-3-deoxyzaluzanin C, 7-hydroxy-3-deoxy-11,13-dihydrozaluzanin C.
    Bohlmann, Ngo Le Van, Phytochemistry, 16, 1304 (1977)

PODOPETALUM (Leguminosae)
P. ormondii F.Muell.
QUINOLIZIDINE ALKALOID: podopetaline.
    Hart et al., Tetrahedron Lett., 5333 (1972)

PODOPHYLLUM (Podophyllaceae/Berberidaceae)
P. emodi Wall. ex Royle. (P. hexandrum Royle).
TOXINS:
4'-demethyldeoxypodophyllotoxin-4'-(O-β-D-glucopyranoside), podophyllotoxin-O-β-D-glucopyranoside.
    Kupchan et al., J. Pharm. Sci., 54, 459 (1965)

P. peltatum.
TOXINS:
4'-demethyldeoxypodophyllotoxin-4'-(O-β-D-glucopyranoside), podophyllotoxin-O-β-D-glucopyranoside.
    Kupchan et al., J. Pharm. Sci., 54, 459 (1965)

P. pleianthum.
TOXINS: desoxyphyllotoxin, desoxyphyllotoxin 4'-methyl ether, isopicropodophyllone, isopicropodophyllone 4'-methyl ether, podophyllotoxin, podophyllotoxin 4'-methyl ether, podophyllotoxone, podophyllotoxone 4'-methyl ether.
    Jackson, Dewick, Phytochemistry, 24, 2407 (1985)

POGONOPUS
P. tubuklosus (DC.) Schumann.
UNCHARACTERIZED ALKALOIDS: pogonopamine, pogonopeine, pogonopidine, pogonopine.
CARBOLINE ALKALOID: tubulosine.
    Brauchi et al., J. Am. Chem. Soc., 86, 1895 (1964)

POGOSTEMON (Labiatae)

P. cablin.
SESQUITERPENOIDS: cycloseychellene, pogostol.
Terhune *et al.*, *Tetrahedron Lett.*, 4705 (1973)

P. formosanus.
SESQUITERPENOID: germacrone.
Ognyanov *et al.*, *Collect. Czech. Chem. Commun.*, **23**, 2033 (1958)

P. parviflorus Benth.
SESQUITERPENOID: parvinolide.
Nanda *et al.*, *J. Chem. Res. Synop.*, *(12)*, 394 (1984)

P. patchouli Pellet.
PYRIDINE ALKALOIDS: guaipyridine, epi-guaipyridine, patchoulipyridine.
PYRANONE: dhelwangin.
TERPENOIDS: norpatchoulenol, patchoulene, patchouli alcohol, pogostol, seychellene.
(A) Buchi *et al.*, *J. Am. Chem. Soc.*, **88**, 3109 (1966)
(P) Klein *et al.*, *Tetrahedron Lett.*, 2279 (1969)
(T) Tesseire *et al.*, *Recherches*, **19**, 8 (1974)

P. plectranthoides Desf.
SESQUITERPENOID: stemonolone.
Phadnis *et al.*, *J. Chem. Soc., Perkin I*, 937 (1984)

POINSETTIA

P. pulcherrima.
ALIPHATIC: octacosanol.
STEROID: β-sitosterol.
TRITERPENOIDS: germanicol, germanicyl acetate, epi-germanicyl acetate.
Gupta *et al.*, *J. Nat. Prod.*, **46**, 937 (1983)

POLEMONIUM (Polemonaceae)

P. coeruleum L.
TRITERPENOID: polemorniogenin.
Tandon *et al.*, *Indian J. Chem.*, **20B**, 46 (1981)

POLIANTHES

P. tuberosa L.
FURANONES: 5-(1-butenyl)-dihydro-2(3H)-furanone, 5-(2-hexenyl)-dihydro-2(3H)-furanone.
MONOTERPENOIDS: (+)-jasmine lactone, tuberolide.
(F) Maurer *et al.*, *Helv. Chim. Acta*, **65**, 462 (1982)
(T) Kaiser, Lamparsky, *Tetrahedron Lett.*, 1659 (1976)

P. tuberosa var. azucena.
STEROID: 9(11)-dehydrohecogenin.
Wagner *et al.*, *J. Am. Chem. Soc.*, **73**, 2494 (1951)

POLYALTHIA (Annonaceae)

P. fragrans.
DITERPENOID: polyalthic acid.
Gopinath *et al.*, *Helv. Chim. Acta*, **44**, 1040 (1961)

P. nitidissima.
BISBENZYLISOQUINOLINE ALKALOIDS: N,'-dimethyllindoldhamine, 7-O-methyllinddoldhamine.
Jossang *et al.*, *Planta Med.*, **49**, 20 (1983)

P. oligosperma.
APORPHINE ALKALOIDS: noroconovine, oconovine, polygospermine.
Guinedeau *et al.*, *Plant Med. Phytother.*, **12**, 166 (1978)

P. olivieri Engl.
APORPHINE ALKALOIDS: oliveridine, oliverine, oliveroline.
TRITERPENOIDS: polyanthenol, polycarpol.
(A) Hamonniere *et al.*, *Phytochemistry*, **16**, 1029 (1977)
(T) Leboeuf *et al.*, *Tetrahedron Lett.*, 3559 (1976)

P. suaveolens.
APORPHINE ALKALOIDS: noroliverine, pachypodanthine, polyalthine, polyveoline.
INDOLE ALKALOIDS: isopolyalthenol, neopolyalthenol, polyavolensine, polyavolensinol, polyavolensinone, polyavolinamide.
Okorie *et al.*, *Phytochemistry*, **20**, 2575 (1981)
Kunesch *et al.*, *Tetrahedron Lett.*, 4937 (1985)

POLYANTHUS

P. tuberosa var. azucena.
STEROID: 9(11)-dehydrohecogenin.
Padilla *et al.*, *Rev. Cubana Farm.*, **9**, 251 (1975)

POLYGALA (Polygalaceae)

P. arillata.
XANTHONE: 1,2,3-trimethoxyxanthone.
Ghosal *et al.*, *J. Chem. Soc., Perkin I*, 740 (1977)

P. caudata Rehd. et Wils.
XANTHONES: wubangziside A, wubangziside, B wubangziside C.
Pan *et al.*, *Yaoxue Xuebao*, **20**, 662 (1985); *Chem. Abstr.*, **104** 31792e

P. chinensis.
NAPHTHALENES: chinensin, chinensinaphthol, 4-methoxychinensin.
SAPONIN: saponin PC.
(N) Ghosal *et al.*, *Phytochemistry*, **13**, 2281 (1974)
(Sap) Brieskorn, Kilbinger, *Arch. Pharm.*, **306**, 824 (1975)

P. macradenia.
XANTHONES: 1,2,3,4,6,7-hexamethoxyxanthone, 1-methoxy-2,3:6,7-bis(methylenedioxy)xanthone, polygalaxanthone A, 1,2,3-trihydroxy-6,7-methylenedioxyxanthone, 1,3,4-trihydroxy-6,7-methylenedioxyxanthone.
Dreyer *et al.*, *Tetrahedron*, **25**, 4415 (1969)

P. paenea.
TOXIN: 7-dehydroxy-4-demethylpodophyllotoxin.
TRITERPENOID: polygalacic acid.
XANTHONES: polygalaxanthone A, polygalaxanthone B, 1,2,3-trihydroxy-6,7-methylenedioxixanthone.
(Tox) Polonsky *et al.*, *Bull. Soc. Chim. Fr.*, 1722 (1962)
(T) Rondest, Polonsky, *ibid.*, 1253 (1963)
(X) Dreyer *et al.*, *Tetrahedron*, **25**, 4415 (1969)

P. senega.
TRITERPENOIDS: cyclosenegin, polygalacic acid, presenegin, senegenin, senegin, tenuifolin.
Yosioka *et al.*, *Tetrahedron Lett.*, 6303 (1966)

P. senega var. latifolia.
SAPONINS: senegines I, II, III and IV.
TRITERPENOID: senegenic acid.
Pelletier *et al.*, *Tetrahedron Lett.*, 3065 (1964)

P. spectabilis DC.
XANTHONES: 2-hydroxy-1,3-dimethoxy-7,8-methylenedioxyxanthone, 1,2,3,7,8-pentamethoxyxanthone, 1,2,3-trihydroxy-7,8-methylenedioxyxanthone.
Andreade *et al.*, *Lloydia*, **40**, 344 (1977)

P. tenuifolia Willd.
PHENOLIC: 3,4,5-trimethoxycinnamic acid.
TRITERPENOID: tenuifolin.
XANTHONES: 6-hydroxy-1,2,3,7-tetramethoxyxanthone, 1,2,3,6,7-pentamethoxyxanthone, 1,2,3,7-tetramethoxyxanthone.
(P, X) Ito *et al.*, *Phytochemistry*, **16**, 1614 (1977)
(T) Pelletier *et al.*, *Tetrahedron*, **27**, 4417 (1971)

POLYGONATUM (Liliaceae)

P. multiflorum (L.) All.
FLAVONOIDS: 8-C-galactosylapigenin,
6-C-galactosyl-8-C-arabinosylapigenin.
STEROIDS: saponoside A, saponoside B.
(F) Chopin *et al.*, *Phytochemistry*, **16**, 1999 (1977)
(S) Janeczko, *Acta Pol. Pharm.*, **37**, 559 (1980)

P. odoratum (Mill.) Druce var. pluriflorum (Mig.) Ohwi.
COUMARINS: cosmosiin, saponarin, vitexin, vitexin
2''-O-glucoside, vitexin 2''-O-sophoroside.
Morita *et al.*, *J. Pharm. Soc., Japan*, **96**, 1180 (1976)

P. officinale All.
FLAVONOIDS: kaempferol glucoside, kaempferol
glucorhamnoside, quercetin glucorhamnoside, quercetin
rhamnoside.
Nakov *et al.*, *Farmatsiya (Sofia)*, **26**, 43 (1976); *Chem. Abstr.*, **87**
65311t

P. sewerzowii.
GLUCOFRUCTAN: sewerin.
Rakhmanberdyeva *et al.*, *Khim. Prir. Soedin.*, 146 (1986)

P. stenophyllum.
SAPONINS: polygonatosides C-1 and C-2.
Strigina *et al.*, *Khim. Prir. Soedin.*, 848 (1980)

POLYGONUM (Polygonaceae)

P. affine D.Don.
FLAVONOIDS: isoaffinetin, isoorientin, isoorientin
X''-O-arabinoside, isovitexin, isovitexin X''-O-arabinoside.
Krause *et al.*, *Z. Pflanzenphysiol.*, **79**, 372 (1976); *Chem. Abstr.*, **85**
156479d

P. hydropiper L.
SESQUITERPENOIDS: isodrimenol, isodrimeninol, polygodial,
polygonal.
Asakawa, Takemoto, *Experientia*, **35**, 1420 (1979)

P. multiflorum.
GLUCOSIDE: polygoacetophenoside.
Yoshizaki *et al.*, *Planta Med.*, **53**, 273 (1987)

P. nepalense.
FLAVONES: 5,4'-dimethoxy-6,7-methylenedioxyflavanone,
5,6,7,2',3',4',5'-heptamethoxyflavone,
5,6,7,4'-tetramethoxyflavanone.
STEROID: α-sitosterol.
TRITERPENOID: taraxerone.
Rathore *et al.*, *Phytochemistry*, **25**, 2223 (1986)

P. sachalinense.
PHENOLICS: emodin, emodin 8-O-β-D-glucoside, physcion,
physcion 8-O-β-D-glucoside.
Kang *et al.*, *Chem. Abstr.*, **97** 141734g

P. weyrichii.
PHENOLICS: (-)-epi-allocatechol, (-)-epi-catechol,
(-)-epi-catechol gallate, gallic acid, (+)-gallocatechol,
(-)-epi-gallocatechol gallate.
Varnaite, Cekelyte, *Chem. Abstr.*, **87** 197347u

POLYMNIA (Compositae)

P. canadensis.
SESQUITERPENOID: polymnolide.
Bohlmann *et al.*, *Phytochemistry*, **19**, 115 (1980)

P. fructicosa.
FLAVONOIDS: naringenin, sakuranetin.
PHENOLIC: cinnamic acid 2-phenethyl ester.
DITERPENOIDS: kaurenoic acid, kaurenoic aldehyde.
Bohlmann, Zdero, *Phytochemistry*, **16**, 492 (1977)

P. maculata Cav.
DITERPENOID: maculatin.
Herz, Bhat, *Phytochemistry*, **12**, 1737 (1973)

P. maculata Cav. var. maculata.
SESQUITERPENOIDS: polymatins A, B and C.
Le Ven Ngo, Fischer, *Phytochemistry*, **18**, 851 (1979)

P. uvedalia (L.) L.
SESQUITERPENOIDS: polydalin, uvedalin.
Herz, Bhat, *J. Org. Chem.*, **35**, 2605 (1970)

POLYPODIUM (Polypodiaceae)

P. aureum.
STEROID: polypodaurein.
Jizba *et al.*, *Phytochemistry*, **13**, 1915 (1974)

P. faurei.
TRITERPENOIDS: α-polypodatetraene, β-polypodatetraene.
Shiojima *et al.*, *Tetrahedron Lett.*, 3733 (1983)

P. formosanum.
TRITERPENOIDS: cyclomargenol, cyclomargenols A, B and C,
cyclomargenone.
Ageta *et al.*, *Chem. Lett.*, 881 (1982)

P. intermedium.
STEROIDS: polypodine B 2-cinnamate, polypodine D
2-cinnamate.
Russell *et al.*, *Aust. J. Chem.*, **25**, 1935 (1972)

P. juglandifolium.
TRITERPENOIDS:
4-desmethyl-24,24-dimethyllanost-20(21)-en-3-ol,
9(11)-fernen-6-ol, 9(11)-fernen-20-ol, polypodinols A, B
and C.
Sunder, Rangaswami, *Indian J. Chem.*, **15B**, 541 (1977)

P. someyeh.
TRITERPENOID: eupha-7,24-diene.
Arai *et al.*, *Chem. Pharm. Bull.*, **30**, 2419 (1980)

P. subpetiolatus.
TRITERPENOID: farnesol palmitate.
Anderson *et al.*, *J. Nat. Prod.*, **42**, 168 (1979)

P. virginianum.
STEROIDS: ecdysone, ecdysterone.
Hikino *et al.*, *Lloydia*, **39**, 246 (1976)

P. vulgare L.
NORTRITERPENOID: 31-norcycloartenol.
SAPONINS: osladin, polypodosaponin.
STEROIDS: crustecdysone, 5β,20-dihydroxyecdysone,
5α-hydroxycrustecdysone.
TRITERPENOID: serratene.
(Sap) Jizba *et al.*, *Collect. Czech. Chem. Commun.*, **32**, 2867 (1971)
(S) Russell *et al.*, *J. Chem. Soc., Chem. Commun.*, 71 (1971)
(T) Inubushi *et al.*, *Tetrahedron Lett.*, 1303 (1964)

POLYPORUS (Polyporaceae)

P. anthracophilus.
MONOTERPENOIDS: 2,8-decadiene-4,6-diynedioic acid,
(E,E)-decadiene-4,6-diynedioic acid.
TRITERPENOIDS: acetyleburicoic acid, eburicoic acid, eburicoic
acid methyl ester.
Yokoyama *et al.*, *Chem. Pharm. Bull.*, **22**, 877 (1974)

P. australiensis.
TRITERPENOID: tumulosic acid.
Cort *et al.*, *J. Chem. Soc.*, 3713 (1954)

P. betulinus.
TRITERPENOIDS: polyporenic acids, A, B and C, tumulosic acid.
Cort *et al.*, *J. Chem. Soc.*, 3713 (1954)

P. guttulatus.
SESQUITERPENOID: cis and trans-matricarianol.
Bohlmann et al., Chem. Ber., **105**, 1919 (1979)

P. officinalis L.
TRITERPENOID: agaricolic acid.
Valentin, Knulte, Pharm. Zeit., **96**, 478 (1957)

P. pinicola.
TRITERPENOID: lanosta-7,9(11),24-triene-3β,21-diol.
Halsall, Sayer, J. Chem. Soc., 2031 (1959)

P. sulphureus.
TRITERPENOID: eburicoic acid.
Holker et al., J. Chem. Soc, 2422 (1953)

P. tumulosus.
TRITERPENOID: tumulosic acid.
Cort et al., J. Chem. Soc., 3713 (1954)

P. versicolor.
STEROIDS: cerivisterol, 5α,6-dihydroergosterol, ergosterol, 3β,5α,6β,9α-tetrahydroxyergosta-7,22-diene.
Valisolalao et al., Tetrahedron, **39**, 2779 (1983)

POLYPTERIS

P. texana.
SESQUITERPENOID:
9-angeloyloxy-2,3-dihydro-7-hydroxy-8-isovaleryloxy-1-oxo-α-longipinene.
Bohlmann, Zdero, Chem. Ber., **108**, 3543 (1975)

POLYSIPHONIA (Rhodophyceae)

P. brodiaei.
CAROTENOIDS: α-carotene, β-carotene, α-cryptoxanthin, β-cryptoxanthin, lutein, zeaxanthin.
Bjoernland et al., Phytochemistry, **15**, 291 (1976)

P. nigrescens.
HALOGENATED PHENOLICS: 2,3-dibromo-4,5-dihydroxybenzyl alcohol, 3,5-dibromo-p-hydroxybenzyl alcohol, 2,3,2′,3′-tetrabromo-4,5,4′,5′-tetrahydroxyphenylmethane, 2,3,6-tribromo-4,5-dihydroxybenzyl alcohol.
Pedersen, Phytochemistry, **17**, 291 (1978)

P. urceolata.
CAROTENOIDS: α-carotene, β-carotene, α-cryptoxanthin, β-cryptoxanthin, lutein, zeaxanthin.
Bjoernland et al., Phytochemistry, **15**, 291 (1976)

POLYSTHICUM (Dryopteridaceae)

P. aculeatum (L.) Roth.
TRITERPENOID: polysthicol.
Laonigro et al., Tetrahedron Lett., 3109 (1980)

POLYSTICHUM

P. acrostichoides.
STEROIDS: 2-ethyllathosterol, β-sitosterol, spinasterol.
Nes et al., Lipids, **12**, 511 (1977)

POLYSTICTUS

P. cinnabarinus.
PIGMENT: cinnabarine.
Lemberg et al., Austral. J. Expt. Biol. Med. Sci., **30**, 271 (1952)

POLYSTIGMA

P. rubricum.
CAROTENOID: lycoxanthin.
Karrer et al., Helv. Chim. Acta, **13**, 268 (1930)
Karrer et al., ibid., **13**, 1084 (1930)

PONCIRUS (Rutaceae)

P. trifoliata.
COUMARINS: bergapten, 7-geranyloxycoumarin, imperatorin, isoponcimarin, 6-methoxy-7-geranyloxycoumarin, poncimarin, poncitrin, ponfolin.
Guiotto et al., Phytochemistry, **16**, 1257 (1977)
Furukawa et al., Chem. Pharm. Bull., **34**, 3922 (1986)

PONDRANEA

P. ricasoliana.
IRIDOID: pondraneoside.
Guiso et al., J. Nat. Prod., **45**, 462 (1982)

PONGAMIA

P. glabra.
PIPERIDINE ALKALOID: glabrin.
BENZOFURANS: pongamol, pongapin.
CHALCONES: pongachalcones I and II.
CHROMENE: glabrochromene.
FLAVANOIDS: gematin, kanjone, karanjin, 2′-methoxyfurano[2″,3″:7,8]flavone, 7-methoxyfurano-(4″,5″,6,5-)-flavone, 8-methoxyfurano-(4″,5″,6,7)-flavone, 3′-methoxypongapin pinnatin, pongaglabrone.
(A) Aneja et al., J. Chem. Soc., 163 (1963)
(B, F) Subrahmanyan et al., Indian J. Chem., **15B**, 12 (1977)
(C, Chr) Saini et al., J. Nat. Prod., **46**, 936 (1983)

P. pinnata.
BENZOPYRAN: gematin.
FLAVONES: pinnatin, pongapin.
(B) Row, Aust. J. Sci. Res., 754 (1952); Chem. Abstr., 47 12372
(F) Mahey et al., Indian J. Chem., **10**, 585 (1972)

POPOWIA (Annonaceae)

P. cauliflora.
FLAVANONE: 5,7,8-trimethoxyflavanone.
Ponichpol et al., Phytochemistry, **17**, 1363 (1978)

POPULUS (Salicaceae)

P. balsamifera.
SESQUITERPENOIDS: α-bisabolol, (-)-curcumene, (-)-α-curcumene.
Kergomard et al., Tetrahedron, **33**, 2215 (1977)

P. grandidentata.
GLUCOSIDE: populoside.
Erickson et al., Phytochemistry, **9**, 857 (1970)

P. lasiocarpa.
COUMARIN: 1,3-p-coumaryl-2-acetylglycerol.
PHENOLICS: lasiocarpin A, lasiocarpin B, lasiocarpin C.
(C) Asakawa et al., Phytochemistry, **15**, 811 (1976)
(P) Asakawa et al., ibid., **16**, 1791 (1977)

P. pseudosimonii.
PHENOLICS: azelaic acid, caffeic acid, o-dihydroxybenzene, 2,6-dimethoxy-p-benzoquinone, ferulic acid, p-hydroxybenzoic acid, p-hydroxycinnamic acid, salicylic acid, saligenin, salireposide, vanillic acid.
STEROID: β-sitosterol.
Pei et al., Zhongcaoyao, **14**, 433 (1983); Chem. Abstr., 100 32203t

P. tremula L.
STEROIDS: citrostadienol, (Z)-α-sitosterol.
TRITERPENOIDS: α-amyrin, β-amyrin, butyrospermol, cycloartenol, lupeol.
Roshcin et al., Khim. Prir. Soedin., 516 (1986)

P. tremuloides L.

STEROIDS: citrostadienol, stigmasta-3,5-dien-7-one, tremulone.
Abramovitch, Micetich, *Can. J. Chem.*, **40**, 2017 (1962)

PORANTHERA

P. corymbosa.

QUINOLIZIDINE ALKALOIDS: acetylporanthericine, poranthericine, porantheridine, porantheriline, porantherine.
Johns *et al.*, *Aust. J. Chem.*, **27**, 2025 (1974)

PORELLA (Musci)

P. arboris-vitae (With.) Grolle. (P. laevigata (Schrad) Lindb.)

SESQUITERPENOID: isodriminol.
Asakawa *et al.*, *J. Hattori Bot. Lab.*, **46**, 163 (1979)

P. fauriei.

SESQUITERPENOID: tadeonal.
Asakawa *et al.*, *Chem. Abstr.*, 88 166708d

P. gracillima.

SESQUITERPENOIDS: ent-gurjunene, β-pinguisene, tadeonal.
Asakawa *et al.*, *Phytochemistry*, **17**, 457 (1978)

P. japonica.

SESQUITERPENOID: porelladiolide.
Asakawa *et al.*, *Phytochemistry*, **20**, 257 (1981)

P. perrottetiana.

DITERPENOIDS: perrottianals A and B.
Asakawa *et al.*, *Phytochemistry*, **18**, 1681 (1979)

P. platyphylla (L.) Pfeiff.

SESQUITERPENOIDS: pinguisanin, β-pinguisanediol, pinguisanolide.
Asakawa *et al.*, *Phytochem* **18**, 1348 (1979)

P. vernicosa.

SESQUITERPENOIDS: ent-bicyclogermacrene, ent-gurjunene, β-pinguisene.
Asakawa *et al.*, *Phytochemistry*, **17**, 457 (1978)

PORIA (Polyporaceae)

P. carbonica.

TRITERPENOID: eburicoic acid.
Kariyone, Kurono, *J. Pharm. Soc., Japan*, **60**, 318 (1940)

P. sinuosa Fr.

ACETYLENES: trans-dec-2-ene-4,6,8-triynedioic acid,
trans-dec-2-ene-4,6,8-triyne-1,10-diol,
trans-dec-2-ene-4,6,8-triyn-1-ol,
2,10-dihydroxydec-4-ene-6,8-diynoic acid,
trans-non-2-ene-4,6-diyne-1,9-diol,
trans-non-2-ene-4,6,8-triynoic acid,
trans-non-2-ene-4,6,8-triyn-1-ol,
(N)-trans-non-2-ene-4,6,8-triynolyl valine,
cis-tetradec-9-ene-2,4,6-triynedioic acid.
Cambie *et al.*, *J. Chem. Soc.*, 2056 (1963)

PORONIA

P. punctata.

ANTIBIOTICS: punctatins A, B and C.
Anderson *et al.*, *J. Chem. Soc., Chem. Commun.*, 917 (1984)

PORPHYRIDIUM (Rhodophyceae/Porphyridiaceae)

P. cruentum.

STEROIDS: 4,24-dimethyl-5α-cholesta-8,22-dien-3β-ol,
4-methyl-5α-cholesta-5,22-dien-3β-ol.
Beastall *et al.*, *Eur. J. Biochem.*, **41**, 301 (1973)

PORTULACA (Portulacaceae)

P. grandiflora Hook.

DITERPENOIDS: portulanol, portulene, portulenol, portulenone.
Ohsaki *et al.*, *Chem. Lett.*, 1585 (1986)

P. oleracea.

PIGMENTS: betanidin 5-O-β-cellobioside, isobetanidin 5-O-β-cellobioside.
Imperato, *Phytochemistry*, **14**, 2091 (1975)

POSOQUERIA

P. latifolia.

IRIDOIDS: latifonin, posoquenin.
Chao, Svoboda, *J. Nat. Prod* **43**, 571 (1980)

POTAMOGETON (Potamogetonaceae)

P. ferrugineus.

DITERPENOID: potamogetonin.
Smith *et al.*, *J. Org. Chem.*, **41**, 593 (1976)

POTENTILLA (Rosaceae)

P. asiatica.

ACIDS: linoleic acid, linolenic acid, oleic acid.
Grigienova *et al.*, *Khim. Prir. Soedin.*, 491 (1977)

P. desertorum.

ACIDS: linoleic acid, linolenic acid, oleic acid.
Grigienova *et al.*, *Khim. Prir. Soedin.*, 491 (1977)

P. erecta (L.) Raenschel.

PROCYANIDINS: procyanidin B-3, cis-procyanidin B-3, procyanidin B-6.
Schleep *et al.*, *J. Chem. Soc., Chem. Commun.*, 392 (1986)

P. orientalis.

ACIDS: linoleic acid, linolenic acid, oleic acid.
Grigienova *et al.*, *Khim. Prir. Soedin.*, 491 (1977)

P. tormentilla Stokes (P. erecta (L.) Rausch.).

TRITERPENOID: tormentic acid.
Potier *et al.*, *Bull. Soc. Chim. Fr.*, 3458 (1966)

POTERIUM (Rosaceae)

P. lasiocarpum.

SAPONIN: poterioside.
Shukyurov *et al.*, *Khim. Prir. Soedin.*, 531 (1974)

P. polygamum Waldst. et Kit.

ACIDS: linoleic acid, linolenic acid, oleic acid.
SAPONIN: poterioside.
(Ac) Grigienova *et al.*, *Khim. Prir. Soedin.*, 491 (1977)
(Sap) Shukyurov *et al.*, *Khim. Prir. Soedin.*, 531 (1974)

POUTERIA

P. caimito.

TRITERPENOID: dammarenediol.
Velliccari *et al.*, *Planta Med.*, **22**, 196 (1972)

PRANGOS (Umbelliferae)

P. alata.

BENZOPYRAN: alatol.
Abyshev *et al.*, *Khim. Prir. Soedin.*, 722 (1973)

P. biebersteinii.

COUMARIN: prandiol.
Abyshev *et al.*, *Khim. Prir. Soedin.*, 574 (1974)

P. bucharica.

COUMARINS: deltoin, imperatorin, isoimperatorin, marmesin, osthole, oxypeucedanin hydrate, pranchingin, prangenin, prangenin hydrate, scopoletin, suberosin.

Danchul et al., Khim. Prir. Soedin., 575 (1977)

P. ferulacea.

COUMARIN: pranferin.

NORSESQUITERPENOID: prangoferol.

(C) Nikonov et al., Khim. Prir. Soedin., 255 (1971)

(T) Abyshev et al., ibid., 114 (1972)

P. latiloba.

COUMARINS: isoimperatorin, isooxypeucedanin, isopeucedanin, latilobinol, marmesin, oxypeucedanin, oxypeucedanin hydrate, pranchimgin.

Abyshev et al., Khim. Prir. Soedin., 90 (1979)

P. lophotera.

BENZOPYRAN: isogosphorol.

Abyshev et al., Khim. Prir. Soedin., 83 (1974)

P. pabularia Lindl.

COUMARINS: prangosin, suberosin.

COUMARIN ALKALOID: prangosine.

(A) Akramov et al., Khim. Prir. Soedin., 287 (1967)

(C) Unaev et al., ibid., 248 (1974)

PREMNA (Verbenaceae)

P. integrifolia L.

UNCHARACTERIZED ALKALOIDS: ganiarine, premnine.

FLAVONE:

6-C-β-D-glucopyranosyl-8-C-β-D-xylopyranosylapigenin.

(A) Basu, Dandiya, J. Am. Pharm. Assoc., 36, 389 (1947)

(F) Ramana et al., Indian J. Pharm. Sci., 47, 165 (1985); Chem. Abstr., 105 94492y

P. latifolia Roxb.

SESQUITERPENOIDS:

2-isopropenyl-6,10-dimethylspiro-[4,5]-dec-6-ene, 2-isopropenyl-6-formyl-10-methylspiro-[4,5]-dec-6-ene.

DITERPENOIDS: 5,6-dehydronellionol, nellionol, premnaspiradiene, premnaspiral, premnolal.

Rao et al., Indian J. Chem., 21B, 267 (1982)

PRIMULA (Primulaceae)

P. denticulata.

TRITERPENOIDS: pridentagenins A, B, C and D.

Ahmad et al., Sci. Pharm., 50, 26 (1982)

P. elatior (L.) Hill.

SAPONINS: elatioric acid, primula sapogenins Sc, Sd and Sf, primula saponin A, saponin PS4.

TRITERPENOID: elatigenin.

(Sap) Tschesche, Ziegler, Justus Liebigs Ann. Chem., 674, 185 (1964)

(Sap) Kitagawa et al., Tetrahedron Lett., 5377 (1968)

(T) Rhykopf, Mohs, Ber. Dtsch. Chem. Ges., 69, 1522 (1936)

P. farinosa.

FLAVONOIDS: 5-hydroxy-2'-methoxyflavone, 2'-methoxyflavone.

Wollenweber, Mann, Biochem. Physiol. Pflanz., 181, 667 (1986); Chem. Abstr., 105 222754z

P. florindae.

FLAVONE: 2',5-dihydroxyflavone.

Bouillant et al., C. R. Acad. Sci., 273C, 1629 (1971)

P. hirsuta All.

ANTHOCYANIN: hirsutin.

Karrer et al., Helv. Chim. Acta, 10, 758 (1927)

P. officinalis L.

GLYCOSIDES: primeverin, primulaverin.

TRITERPENOIDS: primulagenins A, B, F and G.

(G) Jones et al., J. Chem. Soc., 1618 (1933)

(T) Tschesche, Ziegler, Justus Liebigs Ann. Chem., 674, 185 (1964)

P. sieboldii E.Morren.

SAPONIN: sakuraso saponin 1.

TRITERPENOID: protoprimulagenin A.

Kitagawa et al., Tetrahedron Lett., 5377 (1968)

P. veris L.

SAPONIN: primulaverin.

TRITERPENOIDS: priverogenin A, priverogenin A acetate, priverogenin B, priverogenin B acetate, protoprimulagenin A.

Tschesche et al., Justus Liebigs Ann. Chem., 696, 160 (1966)

PRINTZIA

P. laxa.

DITERPENOIDS: isoprintzianic acid, printzianic acid.

Bohlmann, Zdero, Phytochemistry, 17, 487 (1978)

PRIONOSTEMMA

P. aspera.

TRITERPENOIDS: prionostemmadione, pristimerinine.

Delle Monache et al., J. Chem. Soc., Perkin I, 2649 (1979)

PRIORIA

P. copaifera.

DITERPENOID: cativic acid.

Kalman, J. Am. Chem. Soc., 60, 1423 (1938)

PROCLEMA (Rubiaceae)

P. pendula Ait.

ANTHRAQUINONES: alizarin 1-methyl ether, damnecanthol ethyl ether, 1,5-dihydroxy-2-methylanthraquinone, 1-hydroxy-2-methylanthraquinone, morindone 5-methyl ether, rubiadin.

Gonzalez et al., An. Quim., 73, 869 (1977)

PRONEPHRIUM

P. triphyllum.

FLAVANOIDS: triphyllins A, B and C.

Tanaka et al., Chem. Pharm. Bull., 33, 5231 (1985)

PROSOPIS

P. africanas Taub.

PIPERIDINE ALKALOIDS: isoprosopinines A and B, prosafrine, prosafrinine, prosopine, prosopinine.

Khuong-Huu et al., Bull. Soc. Chim. Belg., 81, 425 (1972)

Khuong-Huo et al., ibid., 81, 443, (1972)

P. glandulosa.

AMINE ALKALOID: N-methyltyramine.

STEROID: β-sitosterol.

TRITERPENOID: oleanolic acid.

(A) Keller et al., J. Pharm. Sci., 62, 408 (1973)

S, T) Ahmed et al., Fitoterapia, 57, 457 (1986)

P. juliflora.

PIPERIDINE ALKALOIDS: julifloricine, julifloridine, juliflorine, juliprosopine.

Ott-Longoni et al., Helv. Chim. Acta, 63, 2119 (1980)

P. kuntzei.

PHENOLIC: 3,4-dimethoxydalbergiobe.

Yoshimoto et al., Mokkuzai Gakkaishi, 21, 686 (1975); Chem. Abstr., 84 147705b

P. nigra.
AMINE ALKALOID: N-methyltryptamine.
> Moro et al., Phytochemistry, 14, 827 (1975)

P. spicigera.
PIPERIDINE ALKALOID: spicigerine.
FLAVONE: prosogerin E.
STEROIDS: campesterol, cholesterol, β-sitosterol, stigmasterol.
> (A, S) Jewers et al., Phytochemistry, 15, 238 (1976)
> (F) Bhardwaj et al., Indian J. Chem., 20B, 446 (1981)

PROSOPIDASTRUM

P. globosum.
FLAVANOIDS: corniculatusin 3β-O-robinobioside, limocitrin 3β-O-galactoside.
> Agnese et al., J. Nat. Prod., 49, 528 (1986)

PROSTANTHERA (Labiatae)

P. prunellioides.
SESQUITERPENOID: maaliol.
> Hellyer, Aust. J. Chem., 15, 157 (1962)

PROTEA (Proteaceae)

P. rubropilosa.
SUGAR: D-allose.
> Christensen et al., Carbohyd. Res., 7, 510 (1968)

PROTUSNEA

P. malacea.
PHENOLICS: divaric acid, divaricatic acid, ethyl divaricatinate, usnic acid.
> Chamy et al., J. Nat. Prod., 48, 307 (1985)

PROUSTIA (Compositae/Mutisieae)

P. pyrifolia Hosas.
SESQUITERPENOID: proustianol angelate.
> Bohlmann, Zdero, Chem. Ber., 112, 427 (1979)

PRUNUS (Rosaceae)

P. avium.
ACIDS: caffeic acid, chlorogenic acid, o-coumaric acid, p-coumaric acid, p-coumarylquinic acid.
FLAVONE: jaceidin.
> (Ac) Yu et al., J. Am. Soc. Hortic. Sci., 100, 536 (1975); Chem. Abstr., 84 56510t
> (F) Wollenweber et al., Z. Naturforsch., 27B, 567 (1972)

P. cerasus.
FLAVANONES: aromadendrol, cerasidin, cerasin, cerasinone.
> Nagarajan et al., Phytochemistry, 16, 1317 (1977)

P. domestica.
FLAVONE: prudomestin.
PIGMENT: persicaxanthin.
> (F) Nagajaran et al., Phytochemistry, 3, 477 (1964)
> (P) Gross et al., ibid., 20, 2267 (1981)

P. lusitanica.
TRITERPENOID: 2α,3α-dihydroxyurs-12-en-28-oic acid.
> Biessels et al., Phytochemistry, 13, 203 (1974)

P. mahaleb.
ACID: o-coumaric acid.
COUMARINS: herniarin, kaempferol.
> Yu et al., J. Am. Soc. Hortic. Sci., 100, 536 (1975); Chem. Abstr., 84 56510t

P. mume Sieb. et Zucc.
ACIDS: caproic acid, caprylic acid, isovaleric acid, linoleic acid, oleic acid, valeric acid.
MONOTERPENOIDS: o-cresol, p-cresol, p-cymene, eugenol, guaiacol, cis-3-hexen-1-ol, linalool, cis-linalool oxide, trans-linalool oxide, α-terpineol.
> Kameoka et al., Nippon Kogei Kagaku Kaishi, 50, 389 (1976); Chem. Abstr., 86 2358x

P. persica (L.) Batsch.
FLAVANONES: 3',5-dihydroxy-4',7-dimethoxyflavanone, kaempferol 3-rhamnoside, multiflorin A, multiflorin B, multinoside, quercetin.
GIBBERELLINS: gibberellin A-5, gibberellin A-32, gibberellin A-32 acetonide.
SESQUITERPENOID: (+)-abscisic acid.
> (F) Takagi et al., J. Pharm. Soc., Japan, 97, 109 (1977)
> (G) Yamaguchi et al., Agr. Biol. Chem., 34, 1439 (1970)
> (T) Yamaguchi et al., ibid., 39, 2399 (1975)

P. puddum.
ISOFLAVONE: prunetin.
> Farkas et al., Tetrahedron, 25, 1013 (1969)

P. serotina.
TRITERPENOID: 2α,3α-dihydroxyurs-12-en-28-oic acid.
> Biessels et al., Phytochemistry, 13, 203 (1974)

P. verecunda.
FLAVONE: glucogenkwanin.
> Hasegawa et al., J. Am. Chem. Soc., 76, 5559 (1954)

P. yedoensis.
FLAVANONES: naringenin 7-methyl ether, sakuranin.
> Asahina et al., Chem. Zentralbl., I, 1672 (1928)

PSAMMOSILENE

P. unicoides.
NORDITERPENOID: 3β-hydroxy-12,17-diene-28-norolean-23-al.
> Pu et al., Yunnan Zhiwu Yanjiu, 6, 463 (1984)

PSEUDEVERNIA (Lichenes)

P. furfuracea (L.) Ach.
DEPSIDONE: furfuric acid.
> Gunzinger et al., Helv. Chim. Acta, 68, 1936 (1985)

PSEUDOCEDRELA

P. kotschyii.
SAPONINS: pseudrelone A1, pseudrelone A2.
> Taylor, Phytochemistry, 18, 1574 (1979)

PSEUDOCINCHONA

P. africana.
YOHIMBINE ALKALOIDS: corynantheine, corynanthidine, (-)-corynanthidine, corynanthine, corynoxeine, α-yohimbine.
> Janot, Goutarel, C. R. Acad. Sci., 218, 852 (1944)

PSEUDOCYPHELLARIA (Stictaceae)

P. amphisticta.
STEROID: amphistictinic acid.
> Ronaldson et al., Aust. J. Chem., 31, 215 (1978)

P. degelii.
Three terpenoids.
> Goh et al., J. Chem. Soc., Perkin I, 1560 (1978)

P. durville.
TRITERPENOIDS: durvilldiol, durvillonol.
> Huneck, Williams, Z. Naturforsch., 22B, 1182 (1967)

P. impressa.
TRITERPENOID: stictic acid.
Fox *et al.*, *Phytochemistry*, **9**, 2057 (1970)

P. intricata var. thoursii (Delise) Vainio.
TRITERPENOID: 7β-acetoxy-22-hydroxyhopane.
Huneck, Follinam, *Z. Naturforsch.*, **22B**, 1182 (1967)

P. physciospora.
DEPSIDONE: methyl 5-chlorovirensate.
DIOXEPIN: physciosporin.
Maass *et al.*, *Can. J. Chem.*, **55**, 2839 (1977)

P. quercifolia.
ACID: O-methylgyrophoric acid.
Maass, *Bryologist*, **78**, 178 (1975)

PSEUDOEROTUM

P. ovalis.
STEROIDS: pseurotins A, B, C, D and E.
SESQUITERPENOID: ovalicin.
Sigg, Weber, *Helv. Chim. Acta*, **51**, 1395 (1968)

PSEUDOLARIX (Pinaceae)

P. kaempferi.
DITERPENOIDS: pseudolaric acids A, B, C, C-1 and C-2.
Shou *et al.*, *Planta Med.*, **47**, 35 (1983)

PSEUDOLOBIVIA

P. kermesina.
AMINE ALKALOIDS: hordenine, N-methyltyramine, tyramine.
Follas *et al.*, *Phytochemistry*, **16**, 1459 (1977)

PSEUDOPANAX (Araliaceae)

P. arboreum.
TRITERPENOIDS: 2,3-seco-24-nor-12-ursen-2,3,28-trioic acid, 4-oxo-3,24-dinor-2,3-secours-12-ene-2,28-dioic acid, 3-oxo-24-norurs-12-en-28-oic acid, 2α,3α,23-trihydroxy-12-oleanen-28-oic acid.
Bowden *et al.*, *Aust. J. Chem.*, **28**, 91 (1975)

PSEUDOPARMELIA

P. texana.
TRITERPENOID: 3-acetoxyhopan-1,22-siol.
Huneck *et al.*, *Phytochemistry*, **22**, 2027 (1983)

PSEUDOPLEXAURA

P. porosa.
DITERPENOID: crassin.
STEROIDS: 24-ethyl-22-dehydro-19-norcholesterol, 24-methyl-19-norcholesterol, 19-norcholesterol.
(S) Popov *et al.*, *Tetrahedron Lett.*, 3421 (1976)
(T) Hossain, vander Helm, *Recl. Trav. Chim. Pays-Bas*, **88**, 1413 (1969)

PSEUDOPOTAMILLA

P. occelata.
STEROID: occelasterol.
Kobayashi, Mitsuhashi, *Steroids*, **24**, 399 (1974)

PSEUDOTSUGA (Rambusiodaceae)

P. japonicum.
SESQUITERPENOID: germacrene D.
Yoshihara *et al.*, *Tetrahedron Lett.*, 2263 (1969)

P. menziesii (Mirb.) Franco.
BENZOPYRAN: methylnarigenin.
FLAVANONE: 3',4',5,7-tetrahydroxy-6-methylflavanone.
GLUCOSIDE: poriolin.
TERPENOIDS: pseudotsugonal, ar-pseudotsugonal, pseudotsugonoxide, ar-pseudotsugonoxide, thunbergol, ar-todomatuic acid.
(B, Gl) Barton *et al.*, *Can. J. Chem.*, **45**, 1020 (1967)
(F) Barton *et al.*, *ibid.*, **47**, 869 (1969)
(T) Sakai, Hirose, *Chem. Lett.*, 491 (1973)

PSEUDOWINTERA (Winteraceae)

P. colorata.
SESQUITERPENOID: colorata-4(13),8-dienolide.
Corbett, Chee, *J. Chem. Soc., Perkin I*, 850 (1976)

PSEUDOXANDRA

P. aff. lucida.
BENZYLISOQUINOLINE ALKALOIDS: oxandrine, oxandrinine, pseudoxandrine, pseudoxandrinine.
Cave *et al.*, *Can. J. Chem.*, **64**, 1390 (1986)

PSEUDOXINYSSA

P. pitys.
Four sesquiterpenoids.
Wratten, Faulkner, *Tetrahedron Lett.*, 1395 (1978)

PSEUDUVARIA

P. cf. dolichonema J.Sinclair.
NORAPORPHINE ALKALOID: 1,2,9,10-tetramethoxynoraporphine.
Johns *et al.*, *Aust. J. Chem.*, **23**, 423 (1970)

PSIADIA

P. altissima.
DITERPENOIDS: 6-deoxypsiadiol, isopsiadiol, psiadiol.
Canonica *et al.*, *Gazz. Chim. Ital.*, **99**, 260 (1969)
Canonica *et al.*, *ibid.*, **99**, 276 (1969)

PSIDIUM (Myrtaceae)

P. guaijava.
TRITERPENOID: guaijavolic acid.
UNCHARACTERIZED TERPENOID: sesquiguavene.
Arthur, Hui, *J. Chem. Soc.*, 1403 (1954)

PSILOCYBE

P. pelliculosa.
TOXIN: psilocybin.
Repke *et al.*, *J. Pharm. Sci.*, **66**, 113 (1976)

P. semilanceata.
TOXINS: baeocystin, psilocybin.
Repke *et al.*, *J. Pharm. Sci.*, **66**, 113 (1976)

PSILOTHONNA

P. tagetes.
ACETYLENIC: 1,3-undecadiene-5,7,9-triyne.
Bohlmann *et al.*, *Chem. Ber.*, **104**, 954 (1971)

PSILOTROPHE

P. cookeri.
SESQUITERPENOID: psilotropin.
de Silva *et al.*, *Phytochemistry*, **9**, 59 (1970)

P. villosa.
SESQUITERPENOID: 4-hydroxypsilotropine.
Bohlmann *et al.*, *Phytochemistry*, **19**, 1491 (1980)

PSILOTUM
P. nudum.
PHENOLIC: psilotic acid.

Shamsuddin *et al.*, *Phytochemistry*, **24**, 2458 (1985)

PSORALEA (Leguminosae)
P. corylifolia.
CHALCONE: neobavachalcone.

CHROMENES: psoralen, psoralidin.

DITERPENOIDS: (S)-(E)-bakuchiol, bavachalone, bavachin, (-)-bavachinin, corylidin, isobavachalcone, isobavachin.

ISOFLAVONE: corylinal.

(Chal) Seshadri *et al.*, *Phytochemistry*, **16**, 1995 (1977)

(Chr) Austin *et al.*, *ibid.*, **12**, 1657 (1973)

(I) Gupta *et al.*, *ibid.*, **17**, 164 (1978)

(T) Gupta *et al.*, *ibid.*, **16**, 403 (1977)

(T) Jungk *et al.*, *Experientia*, **34**, 1121 (1978)

PSOROMA (Pannariaceae)
P. contorta.
BIPHENYL: contortin.

Elix *et al.*, *Aust. J. Chem.*, **37**, 1531 (1984)

P. dimorphum.
DEPSIDONES: diploicin, pannarin, vicanicin.

Piovano *et al.*, *J. Nat. Prod.*, **48**, 854 (1985)

P. pallidum.
DEPSIDONES: dechloropannarin, pannarin, vicanicin.

Piovano *et al.*, *J. Nat. Prod.*, **48**, 854 (1985)

P. pulchrum.
DEPSIDONES: pannarin, vicanicin.

Piovano *et al.*, *J. Nat. Prod.*, **48**, 854 (1985)

P. reticulatum.
DEPSIDONES: pannarin, vicanicin.

Piovano *et al.*, *J. Nat. Prod.*, **48**, 854 (1985)

PSYCHOTRIA
P. adenophylla.
STEROID: β-sitosterol.

TRITERPENOIDS: α-amyrin, bauerenol, betulinic acid.

Dan *et al.*, *Fitoterapia*, **57**, 445 (1986)

P. beccarioides.
PENTAMERIC ALKALOID: psychotridine.

Hart *et al.*, *Aust. J. Chem.*, **27**, 639 (1974)

P. ipecacuanha.
ISOQUINOLINE ALKALOID: emetamine.

Oyman, *J. Chem. Soc.*, **111**, 428 (1917)

PTAEROXYLON
P. obliquum.
BENZOPYRANS: prenyletin, umtatin.

SESQUITERPENOIDS: nieshoutin, obliquetin, obliquetol, obliquin.

(B) Dean *et al.*, *J. Chem. Soc., C*, 114 (1966)

(T) McCabe *et al.*, *ibid.*, 145 (1967)

PTELEA (Rutaceae)
P. trifoliata L.
FURANOQUINOLINE ALKALOIDS: (+)-hydroxyruine, maculosidine, preleine, ptelefolidine, ptelefoline, ptelefructine, pteleoline, skimmianine.

QUATERNARY ALKALOID: O-methylptelefolium.

QUINOLONE ALKALOIDS: N-methylflindersine, ptelecortine, pteledimeridine, 1,2,3,4-tetrahydro-2-isopropenyl-5,7-dimethoxy-4-methylfuro-[2,3b]-quinolin-9-one.

COUMARINS: aurapten, imperatorin, phellopterin.

(A) Brown *et al.*, *Aust. J. Chem.*, **7**, 181 (1954)

(A) Mester *et al.*, *Justus Liebigs Ann. Chem.*, 1785 (1979)

(C) Reisch *et al.*, *Phytochemistry*, **14**, 1678 (1975)

PTERIDIUM
P. aquilinum (L.) Kuhn.
OXOINDANE: 2,4,6-trimethyl-3-oxo-5-indanacetic acid.

TERPENOIDS: aquilide A, pterosides A, B, C, D, M and Z.

(O) Yoshihara *et al.*, *Chem. Pharm. Bull.*, **19**, 1491 (1971)

(T) Van der Howven *et al.*, *Carcinogenesis (London)*, **4**, 1587 (1983)

P. aquilinum Kuhn. var. latisculum Underw.
STEROID: ponasteroside A.

SESQUITERPENOIDS: ptaquiloside, pterosins A, B, C, D, E, G, I, J, K, L, N, O, P and Z, pteroside P.

(S) Takemoto *et al.*, *Tetrahedron Lett.*, 4199 (1968)

(T) Kuroyama *et al.*, *Chem. Pharm. Bull.*, **22**, 723 (1974)

(T) Niwa *et al.*, *Tetrahedron Lett.*, 4117 (1983)

PTERIDOPHYLLUM
P. racemosum.
SPIROISOQUINOLINE ALKALOID: corydalispirone.

Itokawa *et al.*, *Phytochemistry*, **15**, 577 (1976)

PTERIS (Pteridaceae)
P. aquilinum.
GLYCOSIDE: (3R)-pterosin 3-O-β-D-glucopyranoside.

Tanaka *et al.*, *Chem. Pharm. Bull.*, **30**, 3640 (1982)

P. bella.
SESQUITERPENOIDS: 2-hydroxypterosin C, 11-hydroxypterosin C, (3R)-hydroxypterosin H, 11-hydroxypterosin T.

Tanaka *et al.*, *Chem. Pharm. Bull.*, **30**, 3640 (1982)

P. cretica.
DITERPENOIDS: creticosides A, B, C and E.

Murakami *et al.*, *Chem. Pharm. Bull.*, **22**, 1686 (1974)

P. dispar.
DITERPENOID: ent-7β,9-dihydroxy-15-oxokaur-16-en-19,6β-olide.

Murakami *et al.*, *Chem. Pharm. Bull.*, **24**, 549 (1976)

P. fauriei.
SESQUITERPENOIDS: pterosin W, pterosin X.

Murakami *et al.*, *Chem. Pharm. Bull.*, **24**, 1961 (1976)

P. kiushinensis.
SESQUITERPENOIDS: pterosins S, T and U.

Murakami, Satake, *Chem. Pharm. Bull.*, **23**, 936 (1975)

P. livida.
DITERPENOID: ent-6β,9-dihydroxy-15-oxokaur-16-en-19-oic acid 19-O-β-D-glucopyranoside.

Tanaka *et al.*, *Chem. Pharm. Bull.*, **29**, 3455 (1981)

P. longipes.
DITERPENOIDS: pretokaurenes L-2 and L-4.

Murakami *et al.*, *Chem. Pharm. Bull.*, **29**, 657 (1981)

P. longipinna.
CHROMENES: pterochromenes L-1, L-2, L-3 and L-4.

Tanaka *et al.*, *Chem. Pharm. Bull.*, **26**, 1339 (1978)

P. oshinensis.
SESQUITERPENOIDS: pterosin C 3-α-L-arabinopyranoside, pterosin Q 3-α-L-arabinopyranoside, pteroside Q, pterosin Q.

Mutakami, Tanaka, *Chem. Pharm. Bull.*, **23**, 1890 (1975)

P. plumbaea.
DITERPENOIDS: pterokaurane P-1, pterokaurane P-1 2-O-β-D-glucopyranoside, pterokauranes -2, P-3 and P-4.

Tanaka *et al.*, *Chem. Pharm. Bull.*, **26**, 3260 (1978)

PTERIS (Pteridaceae)

P. ryukyensis.
DITERPENOID: pterokaurane R.
Tanaka et al., Chem. Pharm. Bull., **26**, 1339 (1978)

P. wallichiana.
SESQUITERPENOIDS: isopteroside C, pterosin B, pterosin C, pterosin C 3-O-β-D-glucopyranoside, wallichioside.
STEROIDS: β-sitosterol, β-sitosterol palmitate.
(S) Sengupta et al., Indian J. Chem., **14B**, 817 (1976)
(T) Murakami et al., Chem. Pharm. Bull., **24**, 173 (1976)
(T) Sengupta et al., Phytochemistry, **15**, 995 (1976)

PTEROCARPUS (Juglandaceae)

P. erinaceus.
SAPOGENIN: pseudobaptigenin.
Schmidt et al., Monatsh. Chem., **53**, 454 (1929)

P. indicus.
BENZOFURAN: pterofuran.
Cooke et al., Aust. J. Chem., **17**, 379 (1964)

P. macrocarpus.
ISOFLAVANONE: macrocarposide.
TRITERPENOID: eudes-4(15)-ene-2,11-diol.
(I) Verma et al., Planta Med., 315 (1986)
(T) Bahl et al., Tetrahedron, **24**, 6231 (1968)

P. marsupium.
FLAVONOIDS: 7-hydroxy-5,4'-dimethoxy-8-methylisoflavone 7-rhamnoside, irisolidol 7-O-β-D-glucopyranoside, retusin 7-O-β-D-glucopyranoside.
Mitra, Joshi, Phytochemistry, **22**, 2326 (1983)

P. osun.
ISOFLAVANONE: santal.
Navasimhachari et al., Proc. Indian Acad. Sci., **37A**, 531 (1953)

P. santalinus.
STEROIDS: β-sitosterol, stigmasterol.
SESQUITERPENOID: isopterocarpolone.
TRITERPENOIDS: β-amyrin, β-amyrone, betulin, erythrodiol, eudes-4(15)-ene-2,11-diol, lup-20(29)-ene-2α,3β-diol, lupenone, lupeol, epi-lupeol, pterocarpdiolone, pterocarpol, pterocarptriol.
Kumar, Seshadri, Phytochemistry, **15**, 1417 (1976)

PTEROCARYA

P. rhoifolia.
NAPHTHOQUINONES: 3,3'-bijuglone, cyclotrijuglone, juglone.
Hirakawa et al., Phytochemistry, **25**, 1494 (1986)

PTEROCAULON

P. balansae.
COUMARIN: 2',3'-dihydroxypuberulin.
Magalhaes et al., Phytochemistry, **20**, 1369 (1981)

P. lanatum. Coumarins,
2',3'-dihydroxypuberulin. 2',3'-epoxypuberulin.
Magalhaes et al., Phytochemistry, **20**, 1369 (1981)

PTEROCEPHALUS (Dipsacaceae)

P. plumosus.
PHENOLIC: swertiajaponin.
Zemtzova et al., Khim. Prir. Soedin., 705 (1977)

PTEROCEREUS (Cactaceae)

P. gaumeri.
ISOQUINOLINE ALKALOID: pterocereine.
Mohamed et al., Lloydia, **42**, 197 (1979)

PTERODON (Leguminosae)

P. apraricioi.
ISOFLAVONES: 3',7-dihydroxy-4',6-dimethoxyisoflavone, 4',7-dihydroxyisoflavone, 4',7-dimethoxyisoflavone, 7-hydroxy-2',4',5',6-tetramethoxyisoflavone, 2',3',4',6,7-pentamethoxyisoflavone, 3',4',5',6,7-pentamethoxyisoflavone, 3',4',6,7-tetramethoxyisoflavone, 2',7,8-trimethoxy-4',5'-methylenedioxyisoflavone.
STEROIDS: β-sitosterol, stigmasterol.
Galina et al., Phytochemistry, **13**, 2593 (1974)
de Almeida et al., Phytochemistry, **14**, 2716 (1975)

P. emarginatus.
DITERPENOIDS: 6-acetoxy-6,7-dihydroxyvouacapan-17-oic acid, 7-acetoxyvouacapane, 6-acetoxyvouacapano-17,7-lactone, methyl 7β-acetoxy-6α-hydroxycouacapan-17β-oate, methyl-6α,7β-dihydroxyvouacapan-17-oate.
Mahajan, Montiero, J. Chem. Soc., Perkin I, 520 (1973)

P. obliquum.
CHROMONES: ptaerochromenol, ptaerocyclin, ptaeroglycol, ptaeroxylin, ptaeroxylone.
Dean et al., Tetrahedron Lett., 2737 (1967)

P. poligaiaflorus.
DITERPENOID: methyl 7-hydroxy-6-acetoxyvouacapan-17-oate.
Mahajan, Montiero, J. Chem. Soc., Perkin I, 520 (1973)

P. pubescens.
ISOFLAVONE: 3',4',6,7-tetramethoxyisoflavone.
Campbell et al., J. Chem. Soc., C, 1787 (1969)

PTEROGYNE

P. nitens.
GUANIDINE ALKALOIDS: pterogynidine, pterogynine.
Corral et al., Rev. Latinoam. Quim., **2**, 179 (1972)

PTERONIA (Umbelliferae)

P. ciliata Thunb.
COUMARIN: isofraxidin-(3,3-dimethylallyl ether).
Bohlmann et al., Chem. Ber., **108**, 2955 (1975)

P. glabra L.
COUMARIN: -(3-methyl-2,3-epoxybutyloxy)-5,6-methylenedioxycoumarin.
Bohlmann et al., Chem. Ber., **108**, 2955 (1975)

PTEROXYLON

P. loblquum.
SESQUITERPENOIDS: heteropeucenin, heteropeucenin dimethyl ether.
Dean, Taylor, J. Chem. Soc., C, 114 (1966)

PTERYXIA

P. terebinthina.
BENZOPYRANS: pterybinthinone, pteryxin.
Thompson et al., J. Nat. Prod., **42**, 120 (1979)

PTILIDIUM (Musci)

P. ciliare (Hedw.) Hampe.
SESQUITERPENOID: deoxopinguisone.
Krutov et al., Phytochemistry, **12**, 1405 (1973)

PTILOSARCUS

P. gurneyi.
DITERPENOID: ptilosarcone.
Wratten et al., Tetrahedron Lett., 1559 (1977)

PTILOSTEMON

P. chamepuce.
SESQUITERPENOIDS: ptilostemonal, ptilostemonol.
Takeda *et al., Bull. Chem. Soc., Japan,* **56,** 1125 (1983)

PUCCINIA

P. dispersa.
CAROTENOIDS: β-carotene, τ-carotene, lycopene.
Serova *et al., Chem. Abstr.,* 84 102303c

PUERARIA

P. chinensis.
ISOFLAVANONE: kakkatin.
Kubo *et al., Chem. Pharm. Bull.,* **23,** 2449 (1975)

P. lobata.
ISOFLAVONES: daidzin, 4′,7-dihydroxyisoflavone.
Markham *et al., Phytochemistry,* **7,** 791 (1968)

P. mirifica.
ISOFLAVANOIDS: coumestrol, daidzein, daidzin, genisteine, mirificin, puerarin.
Ingham *et al., Z. Naturforsch.,* **41C,** 403 (1986)

P. radix.
TERPENOID: kadsusapogenol C.
Kinjo *et al., Chem. Pharm. Bull.,* **33,** 1293 (1985)

P. thunbergiana.
FLAVONOIDS: biochanin A, formononetin, genistin, kakkalide, ononin, puerarin, sissotrin.
STEROIDS: β-asitosterol, β-sitosterol 3-O-β-D-glucopyranoside.
(F) Kurihara *et al., J. Pharm. Soc., Japan,* **95,** 1283 (1975)
(F, S) Kurihara *et al., ibid.,* **96,** 1486 (1976)

P. tuberosa.
BENZOPYRAN: tuberoside.
Joshi *et al., Indian J. Chem.,* **10,** 1112 (1972)

PULICARIA (Compositae)

P. crispa.
SESQUITERPENOIDS: pulicariolide, scocrispolide.
Bohlmann, Knoll, *Phytochemistry,* **18,** 1112 (1972)

P. dysenterica (L.) Bernh.
SESQUITERPENOIDS: 1α-acetoxyisocomene, 12-deoxy-14-hydroxy-9-caryophyllenone 14-acetate, (E)-12,14-dihydroxy-9-caryophyllenone, (E)-12,14-dihydroxycaryophyllenone 1-acetate, (E)-12,14-dihydroxycaryophyllenone 14-acetate, (E)-12,14-dihydroxycaryophyllenone 12,14-diacetate, (Z)-12,14-dihydroxy-9-caryophyllenone, (Z)-12,14-dihydroxycaryophyllenone 12-acetate, (Z)-12,14-dihydroxycaryophyllenone 14-acetate, (Z)12,14-dihydroxycaryophyllenone 12,14-diacetate.
TRITERPENOID: pulidysenterin.
Bohlmann *et al., Phytochemistry,* **20,** 2529 (1981)

P. gnaphaloides.
DITERPENOID:
ent-15,16-epoxy-3,13(16),14-clerodatriene-18,6-olide.
Ruataiyan *et al., Phytochemistry,* **20,** 2772 (1981)

P. salviifolia.
FLAVONOID: rutin.
DITERPENOID: salvicin.
Nurmukhamedova *et al., Khim. Prir. Soedin.,* 299 (1986)

P. scabra.
DIMERIC SESQUITERPENOID: puliscabrin.
Bohlmann *et al., Phytochemistry,* **21,** 1659 (1982)

P. undulata.
ACETOPHENONE: 2-hydroxy-4,6-dimethoxyacetophenone.
Ayoub *et al., Fitoterapia,* **52,** 247 (1981)

PULSATILLA (Ranunculaceae)

P. cernua.
SAPONIN: pulsatilla saponin I.
Shimizu *et al., Chem. Pharm. Bull.,* **26,** 1666 (1978)

P. rubra D.
FLAVONOIDS: kaempferol 3-p-coumaroylglcuoside, kaempferol 3-glucoside, kaempferol 7-glucoside.
Reynaud *et al., Plant Med. Phytother.,* **11,** 87 (1977)

PULTNAEA (Leguminosae/Podalyriae)

P. altissima F.Muell. ex Benth.
INDOLE ALKALOID: (S)(+)-N(b)-dimethyltryptophan methyl ester.
Johns *et al., Aust. J. Chem.,* **24,** 439 (1971)

PUNICA (Punicaceae)

P. granatum L.
PIPERIDINE ALKALOIDS: isopelletierine, methylisopelletierine, N-methylpelletierine, (±)-pelletierine, (-)-pelletierine, pseudopelletierine.
TANNINS: 2-O-galloylpunicalin, punicafolin, punicalagin, punicalin.
Tanaka *et al., Chem. Pharm. Bull.,* **34,** 650 (1986)

PUTRANJIVA

P. roxburghii.
FLAVONE: podocarpusflavone.
SAPONINS: putranjiva saponins A, B, C and D, putranosides A, B, C and D.
TRITERPENOIDS: friedelane-3,7-dione, putranjic acid, iputranjivadione, putranjivic acid, putrankivanonol, putrone, roxburghonic acid.
(F) Chexal *et al., Chem. Ind. (London),* 28 (1970)
(Sap) Seshadri *et al., Indian J. Chem.,* **13,** 444 (1975)
(T) Sengupta *et al., Tetrahedron,* **24,** 1205 (1968)

PUTTERLICKIA

P. verrucosa.
MACROCYCLIC ALKALOIDS: maytanacine, maytansine.
Kupchan *et al., J. Org. Chem.,* **42,** 2349 (1977)

PYCNANTHUS

P. kombo.
POLYPRENOID: kombic acid.
Groenawegan *et al., Phytochemistry,* **22,** 1973 (1983)

PYCNARRHENA

P. australiana.
BISBENZYLISOQUINOLINE ALKALOID: 2-N-norberbamine.
Sioumis, Vashist, *Aust. J. Chem.,* **25,** 2251 (1972)

P. longifolia.
ISOQUINOLINE ALKALOID: pycnarrhine.
BISBENZYLISOQUINOLINE ALKALOIDS: berbacolorflammine, colorflammine, krukovine.
Siwon *et al., Phytochemistry,* **20,** 323 (1981)
Van Beek, Verpoorte, *J. Org. Chem.,* **47,** 898 (1982)

P. manillensis Vidal.
UNCHARACTERIZED ALKALOIDS: ambaline, ambalinine.
Santos *et al., Univ. Philipp. Nat. Appl. Sci. Bull.,* **3,** 353 (1933)

P. ozantha Diels.
BISBENZYLISOQUINOLINE ALKALOIDS: N,N-bisnoraromoline,
2'-N-noromabegine.
Loder, Nearn, *Aust. J. Chem.*, **25**, 2193 (1972)

PYGEUM
P. acuminatum.
TRITERPENOIDS: acuminatic acid,
2,3,19-trihydroxy-12-ursen-28-oic acid.
Chandel, Rastogi, *Indian J. Chem.*, **15B**, 914 (1977)

PYGMACOPREMNA
P. herbacea.
TERPENOID: bharangin.
Sankaram *et al.*, *IUPAC Int. Symp. Chem. Nat. Prod., 11th*, **2**, 97 (1978)

PYRETHRUM (Compositae)
P. parthenifolium.
SESQUITERPENOID: pyrethrin.
Yunusov *et al.*, *Khim. Prir. Soedin.*, 532 (1983)

P. pyrethroides.
SESQUITERPENOIDS: isopyrethroidinin, pyrethroidinin.
Abduazimov *et al.*, *Khim. Prir. Soedin.*, 574 (1985)

PYRACANTHA
P. angustifolia.
CAROTENOID: phytofluene.
Goodwin, *Biochem. J.*, **53**, 538 (1953)

PYROLA (Pyrolaceae)
P. media Swartz.
SESQUITERPENOIDS: 5-geranyl-2-methyl-1,4-benzoquinone,
2-methy-5-(3-methylbut-2-enyl)-1,4-benzoquinone. Burnett,
Thomson, J. Chem. Soc., 847 (1968)

P. renifolia.
GLUCOSIDE: renifolin.
Inouye *et al.*, *Chem. Pharm. Bull.*, **12**, 533 (1964)

PYRUS (Rosaceae)
P. communis L.
GIBBERELLIN: gibberellin A-45.
GLUCOSIDE: 6-O-acetylarbutin.
PHYTOALEXINS: α-pyrufuran, β-pyrufuran.
(Gib) Bearder *et al.*, *Tetrahedron Lett.*, 669 (1975)
(Glu) Durkea *et al.*, *J. Food Sci.*, **33**, 461 (1968)
(Ph) Kemp *et al.*, *J. Chem. Soc., Perkin I*, 2267 (1983)

P. malus L. (Malus sylvestris Mill.).
TRITERPENOIDS: 19-hydroxyursolic acid, 20-hydroxyursolic
acid, 19-hydroxyursonic acid, pomolic acid, pomonic acid.
Brieskorn, Wunderer, *Z. Naturforsch.*, **21**, 1005 (1966)
Lawrie *et al.*, *J. Chem. Soc., C*, 851 (1967)

PYXINE
P. coccifera.
TRITERPENOID: (20S,24R)-epoxy-25-acetoxydammaran-3-one.
Huneck, *Phytochemistry*, **15**, 799 (1976)

P. endochrysina.
TRITERPENOIDS: pyxinic acid, pyxinic acid methyl ester,
pyxinic acid 3-acetate, pyxinic acid methyl ester 3-acetate,
pyxinol, pyxinol 3-acetate, pyxinol 3,25-diacetate.
Yosioka *et al.*, *Chem. Pharm. Bull.*, **20**, 502 (1972)

Q

QUALEA
Q. labouriauanas.
BENZOPHENONE:
bis-(2-hydroxy-4,6-dimethoxy-3-methylphenyl)-methanone.
Correa *et al., Phytochemistry*, **20**, 305 (1981)

QUARARIBEA (Bombacaceae)
Q. funebris (Lieve.) Vischer.
ALKALOID: funebrin.
Raffauf *et al., J. Org. Chem.*, **49**, 2714 (1984)

QUASSIA (Simarubaceae)
Q. africana.
QUASSINOIDS: simalikilatones A, B, C and D.
Tresca *et al., C. R. Acad. Sci.*, **273C**, 601 (1971)

Q. amara.
CARBOLINE ALKALOIDS: 1-methoxycarbonyl-β-carboline,
3-methylcanthin-2,6-dione,
1-vinyl-4,8-dimethoxy-β-carboline.
QUASSINOIDS: quassimarin, simalikilactones A, B, C and D.
STEROIDS: α-sitostenone, stigmast-4-en-3-one.
(A) Barbetti *et al., Planta Med.*, **53**, 289 (1987)
(S) Lavie, Kaye, *J. Chem. Soc.*, 5001 (1963)
(T) Kupchan, Streelman, *J. Org. Chem.*, **41**, 3481 (1978)

QUERCUS (Fagaceae)
Q. acuta.
ALIPHATICS: hentriacontane, heptacosane, nonacosane,
tritriacontane.
STEROID: β-sitosterol.
TRITERPENOIDS: β-amyrin, β-amyrin acetate, friedelanol acetate,
epi-friedelanol, friedelin, lupeol, taraxanol.
Kamano *et al., J. Pharm. Soc., Japan*, **96**, 1202 (1976)

Q. acutissima.
HYDROCARBONS: ditriacontane, heptacosane, hexacosane,
monotriacontane, nonacosane, octacosane, pentacosane,
triacontane, tritriacontane.
STEROID: β-sitosterol.
TRITERPENOIDS: friedelanol, epi-friedelanol, friedelin, glutinol,
lupeol.
Tachi *et al., J. Pharm. Soc., Japan*, **98**, 349 (1978)

Q. aegilops.
TANNIN: valolaginic acid.
Mayer *et al., Ann. Chem.*, 987 (1976)

Q. aliena.
HYDROCARBONS: ditriacontane, ,i.heptacosane, hexacosane,
monotriacontane, nonacosane, octacosane, pentacosane,
triacontane, tritriacontane.
STEROID: β-sitosterol.
TRITERPENOIDS: friedelanol, epi-friedelanol, friedelin, glutinol,
lupeol.
Tachi *et al., J. Pharm. Soc., Japan*, **98**, 349 (1978)

Q. bambusaefolia.
TRITERPENOIDS: taraxer-1,4-dien-3-one,
14-taraxerene-3,16-diol.
Hui, Li, *J. Chem. Soc., Perkin I*, 897 (1977)

Q. championi.
TRITERPENOIDS: 21-(H)-hopane-3β,22-diol,
hop-17(21)-en-3β-ol, hop-17(21)-en-3β-ol 3-acetate.
Hui, Li, *J. Chem. Soc., Perkin I*, 897 (1977)

Q. dentata.
HYDROCARBONS: ditriacontane, heptacosane, hexacosane,
monotriacontane, nonacosane, octacosane, pentacosane,
triacontane, tritriacontane.
STEROID: β-sitosterol.
TRITERPENOIDS: friedelinol, epi-friedelinol, friedelin, glutinol,
lupeol.
Tachi *et al., J. Pharm. Soc., Japan*, **98**, 349 (1978)

Q. gilva Blume.
ALIPHATICS: hentriacontane, heptacosane, nonacosane
tritriacontane.
STEROID: β-sitosterol.
TRITERPENOIDS: β-amyrin, β-amyrin acetate,
24,24-dimethyllanosta-9(11),25-dien-3-acetate, friedelanol
acetate, epi-friedelanol, friedelin, gilvanol, lupeol, taraxanol.
(Al, S ,T) Kamano *et al., J. Pharm. Soc., Japan.*, **96**, 1202 (1976)
(T) Tackie *et al., J. Pharm. Soc., Japan*, **96**, 1213 (1976)

Q. glauca Thunb.
ALIPHATICS: hentriacontane, heptacosane, nonacosane,
tritriacontane.
STEROID: β-sitosterol.
TRITERPENOIDS: β-amyrin, β-amyrin acetate, cyclobalanone,
friedelanol acetate, epi-friedelanol, friedelin, taraxanol.
(Al, S, T) Kamano *et al., J. Pharm. Soc., Japan*, **96**, 1202 (1976)
(T) Tackie *et al., Chem. Pharm. Bull.*, **19**, 2193 (1971)

Q. ilex.
TRITERPENOID: 2,3,19,23-tetrahydroxy-12-ursen-28-oic
acid-β-D-glucopyranosyl ester.
Ramussi *et al., Ann. Chim.*, 1448 (1983)

Q. lusitanica.
GLUCOSIDE: tannin.
Armitage *et al., J. Chem. Soc.*, 1842 (1961)

Q. macrolepis.
PHENOLIC: valoneaic acid.
Schmidt *et al., Justus Liebigs Ann. Chem.*, **591**, 156 (1955)

Q. microlepis.
TANNIN: valolaginic acid.
Mayer *et al., Ann. Chim.*, 987 (1976)

Q. mongolica.
ALIPHATICS: ditriacontane, heptocosane, hexacosane,
monotriacontane, nonacosane, octacosane, pentacosane,
triacontane, tritriacontane.
STEROID: β-sitosterol.
TRITERPENOIDS: friedelanol, epi-friedelanol, friedelin, glutinol,
lupeol.
Tachi *et al., J. Pharm. Soc., Japan*, **98**, 349 (1978)

Q. mongolica var. grosseserrata.
ALIPHATICS: ditriacontane, heptacosane, hexacosane,
monotriacontane, nonacosane, octacosane, pentacosane,
triacontane, tritriacontane.
STEROID: β-sitosterol.
TRITERPENOIDS: friedelanol, epi-friedelanol, friedelin, glutinol,
lupeol.
Tachi *et al., J. Pharm. Soc., Japan*, **98**, 349 (1978)

Q. myrsenaefolia.

ALIPHATICS: hentriacontane, heptacosane, nonacosane, tritriacosane.

STEROID: β-sitosterol.

TRITERPENOIDS:
3-acetoxy-24,25-dimethyllanosta-9(11),23-diene, β-amyrin, β-amyrin acetate, friedelanol acetate, epi-friedelanol, friedelin, lupeol, taraxanol, taraxera-1,14-dien-3-ol.

(P, S, T) Kamano *et al., J. Pharm. Soc., Japan.,* **96,** 1202 (1976)

(T) Hui, Li, *J. Chem. Soc., Perkin I,* 897 (1977)

Q. phyllyracoides.

ALIPHATICS: hentriacontane, heptacosane, nonacosane, tritriacontane.

STEROID: β-sitosterol.

TRITERPENOIDS: β-amyrin, β-amyrin acetate, friedelanol acetate, epi-friedelanol, friedelin, lupeol, taraxanol.

Kamano *et al., J. Pharm. Soc., Japan.,* **96,** 1202 (1976)

Q. robur.

TRITERPENOID: 2α-acetoxyfriedel-3-one.

Talapatra *et al., Indian J. Chem.,* **16B,** 361 (1978)

Q. sessilifolia Salisb. (Q. petraea (Mattuschka) Liebl.).

ALIPHATICS: hentriacontane, nonacosane, tritriacontane.

BENZOFURANS: vescalegin, vescalin.

STEROID: β-sitosterol.

TRITERPENOIDS: β-amyrin, β-amyrin acetate, friedelanol acetate, epi-friedelanol, friedelin, lupeol, taraxanol.

(A, S, T) Kamano *et al., J. Pharm. Soc., Japan,* **96,** 1202 (1976)

(T) Mayer *et al., Justus Liebigs Ann. Chem.,* **707,** 177 (1967)

Q. serrata.

ALIPHATICS: ditriacontane, heptacosane hexacosane, monotriacontane, nonacosane, octacosane, pentacosane, triacontane, tritriacontane.

STEROID: β-sitosterol.

TRITERPENOIDS: friedelanol, epi-friedelanol, friedelin, glutinol, lupeol.

Tachi *et al., J. Pharm. Soc., Japan,* **98,** 349 (1978)

Q. stenophylla Makino.

ALIPHATICS: hentriacontane, heptacosane, nonacosane, tritriacontane.

STEROID: β-sitosterol.

TANNINS: stenophynin A, stenophynin B.

TRITERPENOIDS: β-amyrin, β-amyrin acetate, friedelanol acetate, epi-friedelanol, friedelin, lupeol, taraxanol.

(A, S, T) Kamano *et al., J. Pharm. Soc., Japan,* **96,** 1202 (1976)

(B) Nishimura *et al., Chem. Pharm. Bull.,* **34,** 3223 (1986)

Q. valonea.

PHENOLIC: valoneaic acid.

TANNIN: valolaginic acid.

(P) Schmidt *et al., Justus Liebigs Ann. Chem.,* **591,** 156 (1955)

(Tan) Mayer *et al., Ann. Chem.,* 987 (1976)

Q. variabilis.

ALIPHATICS: ditriacontane, heptacosane, hexacosane, monotriacontane, nonacosane, octacosane, pentacosane, triacontane, tritriacontane.

STEROID: β-sitosterol.

TRITERPENOIDS: friedelanol, epi-friedelanol, friedelin, glutinol, lupeol.

Tachi *et al., J. Pharm. Soc., Japan,* **98,** 349 (1978)

QUISQUALIS (Combretaceae)

Q. fructus.

OXODIAZOLIDINE: quisqualic acid.

Takemoto, *J. Pharm. Soc., Japan,* **95,** 326 (1975)

Q. indica.

OXODIAZOLIDINE: quisqualic acid.

Takemoto, *J. Pharm. Soc., Japan,* **95,** 326 (1975)

R

RABDOSIA (Labiatae)

R. adenantha.
DITERPENOID: adenanthin.
Xu et al., Chem. Nat. Prod. Sin-Am. Symp., 275 (1980)

R. amthystoides.
DITERPENOID: amethystoidin A.
Cheng et al., Chem. Abstr., 96 139670

R. effusa.
DITERPENOIDS: effusanins A, B, C, D and E, effusin.
Fujita et al., Chem. Lett., 1635 (1980)

R. eriocalyx.
DITERPENOIDS: eriocalyxin A, eriocalyxin B.
Wang et al., Yunnan Zhiwu Yanjiu, 4, 407 (1982)

R. excisa.
DITERPENOIDS: excisanin A, excisanin B.
Taiho Pharmaceuticals Co., Japanese Patent, 82 167, 938

R. henryi.
DITERPENOID: henryin.
Li et al., Yunnan Zhiwu Yanjiu, 6, 453 (1984)

R. inflexa.
DITERPENOIDS: inflexin, inflexinol.
Fujita et al., Phytochemistry, 21, 903 (1982)

R. japonica.
DITERPENOIDS: rabdosins A, B and C.
Liu et al., Yaoxue Xuebao, 17, 682, (1982)
Liu et al., ibid., 17, 750 (1982)

R. japonica var. glaucocalyx.
DITERPENOIDS: glaucocalyxin A, glaucocalyxin B.
Zu et al., Yunnan Zhiwu Yanjiu, 3, 283 (1981)

R. lasiocarpa.
DITERPENOIDS: carpalasionin, rabdolasional.
Takeda et al., Chem. Lett., 833 (1982)

R. latifolia var. reniformis.
DITERPENOIDS: reniformins A, B and C.
Wang et al., Zhiwu Xuebao, 28, 292 (1986); Chem. Abstr., 105 222711h

R. longituba.
DITERPENOIDS: longikaurins A, B, C, D, E and F.
Fujita et al., J. Chem. Soc., Chem. Commun., 205 (1980)

R. macrocalyx.
DITERPENOIDS: macrocalin A, macrocalin B, macrocalyxins A, B, C, D and E.
Wang et al., Zhiwu Xuebao, 28, 415 (1986); Chem. Abstr., 106 47185m

R. macrocalyx var. jiuhua Z.W.Xue et X.W.Wang.
DITERPENOIDS: excisanin A, excisanin B, jiuhuanin A, macrocalyxin A.
Wang et al., Zhiwu Xuebao, 28, 185 (1986); Chem. Abstr., 105 75888r

R. macrophylla.
DITERPENOID: rabdophyllin G.
Cheng et al., Yaoxue Tongbao, 17, 174 (1982)

R. microphylla.
DITERPENOID: rabdophyllin G.
Cheng et al., Yaoxue Xuebao, 17, 917 (1982)

R. rosthornii.
DITERPENOID: rosthorin.
Li, Wang, Yaoxue Xuebao, 19, 590 (1984)

R. rubescens.
DITERPENOIDS: ludongin, rubescensins A, B and C.
Zheng et al., Yunnan Zhiwu Yanjiu, 6, 316 (1984)

R. sculponeata.
DITERPENOIDS: sculponeata A, B and C, sculponeatins A, B and C.
Wang et al., Zhongcaoyao, 13, 491 (1982)

R. shikokiana (Makino) Hara.
DITERPENOIDS: rabdo-epi-gibberellolide, rabdosians A, B, C and D, shikodomedin, shikodiamedin, shikokianoic acid, shikokiaside, shikokoanal acetate.
Ochi et al., J. Chem. Soc. Chem. Commun., 810 (1982)

R. shikokiana var. occidentalis.
DITERPENOIDS: O-methylshikoccin, shikoccin, shikoccidin.
Node et al., Chem. Pharm. Bull., 33, 1029 (1985)

R. ternifolia.
DITERPENOIDS: isodonic acid, sodoponin, ternifolin.
Wang et al., Zhiwu Xuebao, 27, 520 (1985); Chem. Abstr., 104 31723h

R. trichocarpa.
DITERPENOIDS: trichorabdials A, B and C.
Node et al., Chem. Lett., 2023 (1982)

R. umbrosa.
DITERPENOIDS: leukannenins A, B, C, D, E and F, rabdohakusin.
Kubo et al., Chem. Lett., 1613 (1984)

R. umbrosa var. latifolia.
DITERPENOID: rabdolatifolin.
Takeda et al., Phytochemistry, 22, 2531 (1983)

R. umbrosa var. leucantha f. kameba.
DITERPENOIDS: leukamenins A, B, C, D, E and F.
Takeda et al., Chem. Lett., 1229 (1981)

RADERMACHERA (Bignoniaceae)

R. sinica Hemsl.
SESQUITERPENOIDS: (2R,3R)-3-hydroxydehydroiso-α-lapachone, (2S,3R)-3-hydroxydehydroiso-α-lapachone.
Inouye et al., J. Chem. Soc., Perkin I, 2764 (1981)

RADIX (Cucurbitaceae)

R. bryonica alba.
One uncharacterized alkaloid.
Gulubov, Venkov, Nauch. Tr. Vissh. Pedagog. Inst. Resvdv. Mat. Fis. Khim. Biol., 8, 137 (1970)

R. hellebore niger.
STEROID: hellebrigenin.
Karrer, Helv. Chim. Acta, 26, 1353 (1943)

R. pareira brava.
BISBENZYLISOQUINOLINE ALKALOIDS: (-)-curine, isococlaurine.
King, J. Chem. Soc., 744 (1940)

R. sarsaparillae.
STEROIDS: parillin, sarsasapogenin, sarsasaponin.
Marker, Lopez, J. Am. Chem. Soc., 69, 2389 (1947)

RADULA (Hepaticae)

R. perrottetii.
SESQUITERPENOID: 2,3-cuparenediol.
Asakawa et al., Phytochemistry, 21, 2481 (1982)

RAMALINA (Usneaceae)

R. ceruchis.
TRITERPENOID: ceruchdiol.
Bendz et al., Acta Chem. Scand., **19**, 1185 (1965)

R. stenospora.
PHENOLIC: tenosporic acid.
Culberson et al., Phytochemistry, **9**, 841 (1970)

R. tingitana.
ACID: usnic acid.
Dean et al., J. Chem. Soc., 1250 (1953)

RANDIA (Rubiaceae)

R. canthioides.
IRIDOID: randioside.
Uesato et al., Phytochemistry, **21**, 353 (1982)

R. dumetorum.
SAPONINS: dumetoronins A, B, C, D, E and F.
TRITERPENOIDS: randialic acid A, randialic acid B.
(Sap) Varshney et al., J. Indian Chem. Soc., **55**, 397 (1978)
(T) Tandon et al., Indian J. Chem., **4**, 483 (1966)

R. formosa.
IRIDOID: 10-caffeoyldeacetyldaphylloside.
Sainty et al., J. Nat. Prod., **45**, 676 (1982)

R. spinosa.
TRITERPENOID: spinosic acid, spinosic acid A, spinosic acid B.
Aplin et al., J. Chem. Soc., C, 1067 (1971)

RANUNCULUS (Ranunculaceae)

R. acer.
CAROTENOID: taraxanthin.
Kuhn, Lederer, J. Physiol. Chem., **200**, 108 (1931)

R. baudotii.
FLAVONOID: quercetin.
PHENOLICS: caffeic acid, p-coumaric acid, ferulic acid, p-hydroxybenzoic acid, vanillic acid.
Barandiaran et al., Fitoterapia, **58**, 59 (1987)

RAPANEA (Myrsinaceae)

R. laetevirens.
BENZOQUINONES: embelin, rapanone.
PHENOLIC: tridecenylresorcinol.
Murthy et al., Tetrahedron, **21**, 1445 (1965)
Madrigal et al., Lipids, **12**, 402 (1977)

RAUWOLFIA (Apocynaceae)

R. caffra Sonder.
CARBOLINE ALKALOIDS: raucaffricine, raucaffridine, raucaffriline, raucaffrinoline, rauwolfine.
Khan, Siddiqui, Experientia, **28**, 127 (1972)

R. canescens L.
CARBOLINE ALKALOIDS: ajmalicine, ajmaline, aricine, carpagine, deserpidine, isoreserpinine, raujemidine, reserpidine, reserpiline, rauwolscine, α-yohimbine.
Khan et al., Chem. Ind. (London), 1264 (1954)
Ulshafer et al., J. Org. Chem., **21**, 923 (1956)

R. confertiflora.
CARBOLINE ALKALOIDS: raufloricine, raufloridine, rauflorine.
Daniel et al., Chem. Ind. (Milan), **51**, 1042 (1971)

R. cummingsii.
CARBOLINE ALKALOIDS: N-demethyldihydropurpeline, purpeline, raupurpeline.
Iwu, Court, Experientia, **33**, 1268 (1977)

R. discolor M.Pichon.
CARBOLINE ALKALOID: tetraphyllinine.
Combes et al., Phytochemistry, **5**, 1065 (1966)

R. heterophylla Roem. et Sch.
UNCHARACTERIZED ALKALOIDS: chalchupine A, chalchupine B, heterophylline.
Hochstein et al., J. Am. Chem. Soc., **77**, 3351 (1955)

R. ligustrina.
CARBOLINE ALKALOID: isoreserpiline pseudoindozyl.
Finch et al., Experientia, **19**, 296 (1963)

R. mauiensis Sherff.
CARBOLINE ALKALOID: sandwicine.
INDOLE ALKALOID: mauiensine.
Borman et al., Tetrahedron, **1**, 328 (1957)

R. media.
CARBOLINE ALKALOIDS: cabucine, 12-hydroxymauiensine, mauiensine, reserpiline.
Kan et al., Phytochemistry, **25**, 1783 (1986)

R. mombasiana.
CARBOLINE ALKALOID: endolobine.
Iwu, Court, Planta Med., **33**, 232 (1978)

R. nitida Jacq.
CARBOLINE ALKALOIDS: acetylalloyohimbine, 17-O-acetylnortetraphyllicine, deserpidine, N-methylvellosimine, rauniticine, raunitidine.
Smith et al., Lloydia, **27**, 440 (1964)
Ayer, Court, Phytochemistry, **20**, 2569 (1981)

R. obscura.
CARBOLINE ALKALOID: 10-methoxygeissoschizol.
Timmins, Court, Planta Med., **27**, 105 (1975)

R. oreogiton.
Six carboline alkaloids.
Akinloye, Court, Phytochemistry, **19**, 2741 (1980)

R. parakensis.
CARBOLINE ALKALOIDS: pelirine, perakine, peraksine.
Kiang et al., J. Chem. Soc., 1394 (1960)

R. reflexa.
CARBOLINE ALKALOID: flexicorine.
Chatterjee et al., J. Org. Chem., **47**, 1732 (1982)

R. salicifolia Griseb.
CARBOLINE ALKALOID: raucubaine.
Kutney et al., Heterocycles, **14**, 1309 (1980)

R. sandwicensis A.DC.
CARBOLINE ALKALOID: sandwicine.
Gorman et al., Tetrahedron, **1**, 328 (1957)

R. sellowii Muell.
CARBOLINE ALKALOID: ajmalicine.
Pakrashi et al., J. Am. Chem. Soc., **77**, 6687 (1955)

R. serpentina Benth ex Kurz.
CARBOLINE ALKALOIDS: ajmalimine, ajmaline, ajmalinine, chandrine, isoraunitidine, raubasine, raugalline, rauhimbine, raupine, rauwolfinine, rescidine, reserpiline, reserpine, reserpinine, sandwicolidine, sarpagine, serpentine, serpentinine, serpine, serpinine.
Stoll et al., Helv. Chim. Acta, **38**, 270 (1955)
Shamma et al., J. Am. Chem. Soc., **85**, 2507 (1963)
Siddiqui et al., Planta Med., **53**, 288 (1987)

R. suaveolens S.Moore.
CARBOLINE ALKALOID: suaveoline.
Majumder et al., Tetrahedron Lett., 1563 (1972)

R. sumatrana Jack.
CARBOLINE ALKALOID: N-demethylseradamine.
Hanaoka et al., Helv. Chim. Acta, **53**, 1723 (1970)

R. tetraphylla.
CARBOLINE ALKALOIDS: tetraphyllicine, tetraphylline.
Djerassi, Fishman, *Chem. Ind.,* 627 (1955)

R. volkensii.
CARBOLINE ALKALOIDS: O-acetylpreperakine, volkensine.
Akinloye, Court, *Planta Med.,* **37**, 361 (1979)
Akinloye *et al., Phytochemistry,* **19**, 307 (1980)

R. vomitora Afz.
CARBOLINE ALKALOIDS: 17-acetylajmaline, ajmalinine, deacetyldeformylakuammiline, deacetyldeformylpicraline, isocarapanaubine, isosandwicine, mitoridine, neonorreserpine, isoreserpiline pseudoindoxyl, perakine, raumitorine, raunitidine, rauvanine, rauvomiline, rauvomitine, rauvoxine, rauvoxinine, renoxydine, rescidine, rescinamine, seredine, serpenticine.
Voelter, *Z. Naturforsch.,* **35B**, 920 (1980)
Malik *et al., Heterocycles,* **16**, 1727 (1981)

RAVENIA (Rutaceae)
R. spectabilis (Lemonia spectabilis).
QUINOLONE ALKALOIDS: atanine, ravenine, ravenoline, spectabiline.
Paul *et al., Indian J. Chem.,* **7**, 678 (1969)
Talapatra *et al., Tetrahedron Lett.,* 4789 (1969)

READEA (Rubiaceae)
R. membranacea Gillespie.
LUPANE ALKALOIDS:
2,3-dehydro-O-(2-pyrrolylcarbonyl)-virgiline,
O-(2-pyrrolylcarbonyl)-virgiline.
Manchand *et al., J. Chem. Soc., C,* 615 (1968)

REAUMURIA
R. mucronata.
FLAVONOID: kaempferol 3,7-disulphate.
Nawwar *et al., Phytochemistry,* **16**, 1319 (1977)

REHMANNIA (Scrophulariaceae)
R. glutinosa (Gaertn.) Libosch. ex DC.
IRIDOIDS: rehmaglutins A, B, C and D, rehmanniosides A, B, C and D.
Kitagawa *et al., Chem. Pharm. Bull.,* **34**, 1399, (1986)
Kitagawa *et al., ibid.,* **34**, 1403 (1986)

REICHARDIA
R. tingitana L.Roth. var. orientalis (L.) Asch. et Schweinf.
SESQUITERPENOID: 14-deoxylactucin.
El-Masry *et al., Acta Pharm. Suec.,* **17**, 137 (1980)

REINECKIA
R. carnes Kunth.
STEROIDS: carneagenin, convallamarogenin, isocarnesgenin, isoreineckiagenin, kitagenin, pentologenin, reineckiagenin.
Tschesche *et al., Ber. Dtsch. Chem. Ges.,* **94**, 1699 (1961)
Takeda *et al., Tetrahedron Lett.,* 1107 (1962)

REINKELLA (Lichenes)
R. parishii.
BENZOFURAN: schizopeltic acid.
Santesson *et al., Acta Chem. Scand.,* **21**, 1111 (1967)

REJOUA (Apocynaceae)
R. aurantiaca Gaud.
INDOLONE ALKALOID: iboluteine.
Guise *et al., Aust. J. Chem.,* **18**, 927 (1965)

RELHANIA
R. acerosa.
DITERPENOID: relhanic acid.
Bohlmann *et al., Phytochemistry,* **18**, 631 (1979)

REMERIA
R. maritima.
NORSESQUITERPENOIDS: isoevodionol, preremirol, remirol.
Allan *et al., Tetrahedron Lett.,* 4669, (1969)
Allan *et al., ibid.,* 4673 (1969)

REMIJA
R. pedunculata.
ISOQUINOLINE ALKALOIDS: cupreine, quinidine, quinine.
Woodward *et al., J. Am. Chem. Soc.,* **67**, 860 (1945)

R. purdieana.
INDOLE ALKALOID: cinchonamidine.
UNCHARACTERIZED ALKALOIDS: chairamidine, chairamine.
Hesse, *Justus Liebigs Ann. Chem.,* **185**, 296, (1877)
Hesse, *ibid.,* **185**, 323 (1877)

REMIREA (Cyperaceae)
R. maritima.
BENZOFURANS: remiridiol, remirol.
BENZOPYRAN: remirin.
Allan *et al., Tetrahedron Lett.,* 4673 (1969)

RESEDA (Resedaceae)
R. lutea.
NAPHTHYLAMINE: phenyl-β-naphthylamine.
Sultankhodzhaev *et al., Khim. Prir. Soedin.,* 406 (1976)

R. luteola L.
ALKALOIDS: resedine, residinine.
AMINE: β-hydroxyphenylethylamine.
NAPHTHYLAMINE: phenyl-β-naphthylamine.
(A) Tadshibaev *et al., Khim. Prir. Soedin.,* 270 (1976)
(Am) Lutfullin *et al., ibid.,* 625 (1976)
(N) Sultankhodzhaev *et al., ibid.,* 406 (1976)

RETAMA
R. raetam.
FLAVONOIDS: apigenin 6,8-di-C-glucoside, apigenin 7-glucoside, chrysoeriol 7-glucoside, daidzein, daidzein 7,4'-dimethyl ether, luteolin 7-glucoside, orientin, orientin 4'-glucoside.
Abdalla, Saleh, *J. Nat. Prod.,* **46**, 755 (1983)

R. sphaerocarpa Boiss.
SPARTEINE ALKALOIDS: (-)-epi-baptifoline, retamine.
UNCHARACTERIZED ALKALOID: sphaerocarrine.
Ribas, Vega, *Ion (London),* **13**, 148 (1953)

RETANILLA
R. ephedra.
UNCHARACTERIZED ALKALOID: integeressine.
Bhakuni *et al., Rev. Latinoam. Quim.,* **5**, 158 (1974)

RHAMNUS (Rhamnaceae)
R. alaternus.
PIGMENTS: alaternin, aloe emodin.
> Briggs et al., J. Chem. Soc., 3069 (1953)

R. catharticus L.
ANTHRACENE: emodinanthrone.
> Perkin, J. Chem. Soc., 65, 937 (1894)

R. formosana.
ANTHRAQUINONE: emodin.
FLAVANOLS: kaempferol, rhamnetin, rhamnocitrin.
> Lin et al., Chem. Abstr., 105 39398x

R. frangula L.
PEPTIDE ALKALOIDS: franganine, frangufoline, frangulanine B.
GLUCOSIDE: emodin-8-glucoside.
NAPHTHALENES: glucofrangulin A, glucofrangulin.
> (A) Tschesche et al., Tetrahedron Lett., 2993 (1968)
> (G, N) Rosca, Cucu, Planta Med., 28, 178, (1975)
> (G, N) Rosca, Cucu, ibid., 28, 343 (1975)

R. japonica.
FURANONES: α-sorigenin, β-sorigenin.
> Haber et al., Helv. Chim. Acta, 39, 1654 (1956)

R. nakaharai.
GLYCOSIDE: rhamnustrioside.
> Lin et al., Phytochemistry, 21, 1466 (1982)

R. oleoides L. subsp. graecus (Boiss. et Reut.) Holmboe.
ANTHRACENES: chrysophanol, chrysophanol glucoside, emodin, emodin glucoside, physcion, physcion glucoside.
> Baytop et al., Chem. Abstr., 88 148948c

R. virgata.
PHENOLICS: maesopin, 5-methoxy-7-hydroxyphthalide, 6-O-methylalaterin.
> Kalidhar, Sharma, J. Indian Chem. Soc., 62, 411 (1985)

RHAPONTICUM
R. carthamoides.
SAPONINS: chaponticosides A, B, C, D, E, F, G and H.
> Vereskovskii et al., Khim. Prir. Soedin., 578 (1977)

R. integrifolium.
STEROIDS: 24(25)-dehydromalsisterone, 24(28)-dehydromalsisterone, integristerone A, integristerone B.
> Baltaev et al., Khim. Prir. Soedin., 463 (1978)

RHAZYA
R. orientalis A.DC.
CARBOLINE ALKALOIDS: 5α-carboxystrictosidine, strictosidine.
INDOLE ALKALOIDS: secamine, tetrahydropresecamine, tetrahydrosecamine, 16,17:15,20-tetrahydrosecodine, 16,17:15,20-tetrahydrosecodin-17-ol.
> Atta-ur-Rahman, Zaman, Heterocycles, 22, 2023 (1984)

R. stricta Decaisne.
CARBAZOLE ALKALOID: sewarine.
CARBOLINE ALKALOIDS: akuammidine, 5α-carboxystrictosidine, (+)-1,2-dehydroaspidospermidine, N(b)-methylstrictamine, rhazimal, rhazimol, strictalamine, strictamine, stricticine, strictosidine.
INDOLE ALKALOIDS: presecamine, secamine, tetrahydropresecamine, tetrahydrosecamine, 16,17:15,20-tetrahydrosecodine, 16,17:15,20-tetrahydrosecodin-17-ol.
IMIDAZOLIDINOCARBAZOLE ALKALOIDS: (+)-vincadifformine, (±)-vincadifformine.

UNCHARACTERIZED ALKALOIDS: rhazidine, rhazinaline, rhazinilam, rhazinine.
> Ahmad et al., J. Chem. Soc., Pakistan, 1, 69 (1979)
> Atta-ur-Rahman, Zaman, Heterocycles, 22, 2023 (1984)
> Atta-ur-Rahman et al., Z. Naturforsch, 42, 91 (1987)

RHEEDIA
R. gardneriana Planch. et Triana.
XANTHONES: 4',5'-dihydro-1,5,6-trihydroxy-4',5',5'-trimethylfurano-(2',3':3,2)-4-(1,1-dihydro-2-enyl)xanthone, 1,5-dihydroxy-6',6'-dimethylpyrano-(2',3':6,7)xanthone, 1,5-dihydroxyxanthone, 1,7-dihydroxyxanthone.
> Delle Monache et al., J. Nat. Prod., 46, 655 (1983)

RHEI
R. rhizoma.
PHENOLIC: (+)-catechin.
STILBENES: 3,5,4'-trihydroxystilbene 4'-O-β-D-(6''-O-galloyl)-glucopyranoside, 3,5,4'-trihydroxystilbene 4'-O-β-D-glucopyranoside.
> Nonaka et al., Chem. Pharm. Bull., 25, 2300 (1977)

RHEUM (Polygonaceae)
R. palmatum.
PIGMENTS: aloe emodin, aloe emodin 8-β-D-glucoside.
> Tutin et al., J. Chem. Soc., 99, 946 (1911)

R. undulatum.
PIGMENT: aloe emodin
> Tutin et al., J. Chem. Soc., 99, 946 (1911)

RHIGIOCARYA
R. racemifera.
UNCHARACTERIZED ALKALOID: O-methylflavinantine.
> Tackie et al., Phytochemistry, 13, 2884 (1974)

RHINOPETALUM
R. bucharicum.
STEROIDAL ALKALOIDS: rhinoline, rhinolinine.
> Samikov et al., Khim. Prir. Soedin., 350 (1979)

R. stenantherum.
STEROIDAL ALKALOIDS: stenantidine, stenantine.
> Samikov et al., Khim. Prir. Soedin., 349 (1981)

RHIPOCEPHALUS (Chlorophyceae)
R. phoenix.
SESQUITERPENOIDS: rhipocephalin, rhipocephenal.
> Sun, Fenical, Tetrahedron Lett., 685 (1979)

RHIZOBIUM
R. japonicum.
AMINO ACIDS: dihydrorhizobitoxine, rhizobitoxine.
> Owens et al., Plant Physiol., 40, 927 (1965)

RHIZOCTONIA
R. leguminicola.
INDOLIZIDINE ALKALOID: slaframine.
PYRIDINE ALKALOID: 3,4,5-trihydroxyoctahydro-1-pyridine.
> Rainet et al., Nature, 205, 203 (1965)

RHIZOPUS
R. nigricans.
PTERIDINE: rhizopterine.
> Rickes et al., J. Am. Chem. Soc., 69, 2751 (1947)

RHODEA
R. japonica Roth.
GLYCOSIDE: rhodexoside.
STEROIDS: digitoxigenin, periplogenin, rhodeasapogenin, rhodexins A, B, C and D, steroid Br-1, 25(27)-spirostene-1,2,3,4,5,6,7-heptaol.
(Gly) Kuchukhidze *et al., Izv. Akad. Nauk Gruz. SSR,* **8**, 157 (1982); *Chem. Abstr.,* 97 141773u
(S) Miyahara *et al., Tetrahedron Lett.,* **83** (1980)

RHODIOLA
R. algida.
LACTONE: alginoside.
Pangarova *et al., Khim. Prir. Soedin.,* 334 (1975)

RHODNIUS
R. prelixus.
PIGMENT: rhodnitin.
Visconti *et al., Helv. Chim. Acta,* **46**, 2509 (1963)

RHODODENDRON (Ericaceae/Rhododendraceae)
R. catanbiensis.
DITERPENOIDS: acetylandromed-10(18)-enol, acetylandromedol, andromedol.
von Kuerten *et al., Arch. Pharm.,* **304**, 753 (1971)
R. caucasicum.
STEROIDS: cholesterol, β-sitosterol, stigmasterol.
Ushishvili *et al., Soobshch. Akad. Nauk Gruz. SSR,* **80**, 180 (1975); *Chem. Abstr.,* 84 86734m
R. dauricum L.
COUMARINS: andromedotoxin, scopoletin, umbelliferone.
FLAVONOIDS: 8-demethylfarrerol, farrerol, hyperoside, isohyperoside, kaempferol, myricetin, quercetin.
MONOTERPENOIDS: α-eudesmol, β-eudesmol, τ-eudesmol, germacrone, juniper camphor, menthol.
QUINONE: hydroquinone.
(C, Q) Fu *et al., Chem. Abstr.,* 88 60107s
(F) Liu *et al., Hua Hsueh Hsueh Pao,* **34**, 211 (1976); *Chem. Abstr.,* 88 85995t
(T) Hsu *et al., Chem. Abstr.,* 88 60139d
R. degronianum Carr.
TRITERPENOIDS: α-amyrin, β-amyrin, α-amyrin acetate, β-amyrin acetate, dendropanoxide, friedelin, epi-friedelinol, ursolic acid, uvaol.
Kurihara *et al., J. Pharm. Soc., Japan,* **96**, 1407 (1976)
R. dilatatum.
STEROID: β-sitosterol.
TRITERPENOIDS: α-amyrin, betulinic acid, lupeol, oleanolic acid, taraxerol, ursolic acid.
Kawai *et al., Chem. Abstr.,* 88 34499b
R. falconeri.
TRITERPENOID: uvaol 3-acetate.
Sengupta *et al., J. Indian Chem. Soc.,* **46**, 775 (1969)
R. farrerae.
FLAVANONE: farrerol.
Arthur, *J. Chem. Soc.,* 3740 (1955)
R. hymenanthus.
TRITERPENOID: ursolic acid.
Stout, Stevens, *J. Org. Chem.,* **28**, 1259 (1963)
R. japonicum.
STEROID: β-sitosterol.
DITERPENOIDS: rhodojaponins I, II, III, IV, V, VI and VII.
TRITERPENOIDS: α-amyrin, betulinic acid, lupeol, oleanolic acid, taraxerol, ursolic acid.
Kawai *et al., Chem. Abstr.,* 88 34499b

R. linearifolium.
TRITERPENOIDS: adianene-2β,3β-diol, 7-fernene-2,3-diol, 9(11)-fernene-2,3-diol, 7-fernen-3-ol, motiol, motodiol, neomotiol.
Nakamura *et al., Tetrahedron Lett.,* 3461 (1966)
R. macrocephalum.
TRITERPENOID: 7-fernene-2,3-diol.
Nakamura *et al., Tetrahedron Lett.,* 2017 (1965)
R. macrophyllum.
TRITERPENOID: campanulin.
Kimura *et al., Chem. Pharm. Bull.,* **8**, 1145 (1960)
R. metternichii.
STEROID: β-sitosterol.
TRITERPENOIDS: α-amyrin, betulinic acid, lupeol, oleanolic acid, taraxerol, ursolic acid.
Kawai *et al., Chem. Abstr.,* 88 34499b
R. molle G.Don.
DITERPENOID: rhomotoxin.
Pu, *Zhongcaoyao,* **14**, 293 (1983)
R. simiarum.
TRITERPENOID: simiarenol.
Aplin *et al., J. Chem. Soc., C,* 1251 (1966)
R. simsii.
BENZOPYRANS: matteucinin, matteucinol.
Arthur *et al., J. Chem. Soc.,* 3197 (1960)
R. westlandii.
TRITERPENOID: epoxyglutinane.
Arthur, Hui, *J. Chem. Soc.,* 55 (1961)

RHODOMELA (Rhodophyceae)
R. confervoides.
HALOGENATED PHENOLICS: 2,3-dibromo-4,5-dihydroxybenzyl alcohol, 3,5-dibromo-p-hydroxybenzyl alcohol, 2,3,2',3'-tetrabromo-4,5,4',5'-tetrahydroxydiphenylmethane, 2,3,6-tribromo-4,5-dihydroxybenzyl alcohol.
Pedersen. Phytochemistry, **17**, 291 (1978)
R. larix.
HALOGENATED PHENOLICS: 2,2',3,3'-tetrabromo-4,4',5,5'-tetrahydroxydiphenylmethane, 2,2',3-tribromo-3',4,4',5-tetrahydroxy-6'-methoxymethyldiphenylmethane.
Kurata *et al., Chem. Lett.,* 1435 (1977)

RHODOMYRTUS
R. tomentosa.
TRITERPENOIDS: 3β-acetoxy-11α,12α-epoxyolean-28,13β-olide, 3β-acetoxy-12α-hydroxyolean-28,13-olide.
Hui, Li, *Phytochemistry,* **15**, 1741 (1976)

RHODOPAXILLUS
R. nudus (Bull. ex Fr.) Maire.
STEROID: portensterol.
Regerat *et al., Ann. Pharm. Fr.,* **34**, 323 (1976)

RHODOPHIALA
R. bifida.
ISOQUINOLINE ALKALOIDS: O-acetylmontanine, pancracine.
Wildman *et al., Pharmazie,* **22**, 725 (1967)

RHODOPHYLLUS
R. crassipes.
AMINO ACID: 2-amino-3-butenoic acid.
Matsumoto, *Chem. Abstr.,* 103 3724v

RHODYMENIA (Rhodophyceae/Rhodymeniaceae))

R. palmata.
CAROTENOIDS: α-carotene, β-carotene, α-cryptoxanthin, β-cryptoxanthin, lutein, zeaxanthin.
STEROIDS: cholest-5,25-dien-3β,14-diol, desmosterol.
(Car) Bjoernland et al., Phytochemistry, 15, 291 (1976)
(S) Morisaki et al., Chem. Pharm. Bull., 24, 3214 (1976)

RHUS (Anacardiaceae)

R. parviflora.
FLAVONOIDS: afzelin, kaempferol, myricetin, myricitrin, quercetin, quercitrin.
Nair et al., Curr. Sci., 46, 448 (1977)

R. succedanea.
FLAVANONES: rhusflavanone, succedaneaflavanone.
GLYCOSIDE: rhoifolin.
(F) Chen et al., J. Chem. Soc., Perkin 1, 98 (1976)
(G) Coussio et al., Experientia, 20, 562 (1964)

R. typhina.
GLUCOSIDE: tannin.
Armitage et al., J. Chem. Soc., 1842 (1961)

R. wallichi.
ACID: gallic acid.
FLAVANONES: kaempferol, myricetin, quercetin, quercetin 3-O-arabinopyranoside, quercetin 3-O-galactoside, quercetin 3-O-xyloside.
STEROIDS: β-sitosterol, β-sitosterol glucoside.
Sinha et al., J. Nat. Prod., 49, 546 (1986)

RHYNCHOSIA

R. minima.
FLAVONOIDS: 8-C-arabinosylapigenin, 6,8-di-C-glucoside, 6-C-glucosylapigenin, 6-C-glucosyl-8-C-xylosylapigenin, di-C-hexosylapigenin, 6-C-hexosyl-8-C-pentosylapigenin, C-pentosyl-C-hexosylapigenin.
Besson et al., Phytochemistry, 16, 498 (1977)

RICCARDIA (Hepaticae)

R. incurvata Lindb.
INDOLE ALKALOIDS: 6-3-methyl-2-butenyl)-indole, 7-(3-methyl-2-butenyl)-indole.
Benesova et al., Collect. Czech. Chem. Commun., 34, 1807 (1969)

R. sinuata (Hook.) Trev.
INDOLE ALKALOIDS: 6-(3-methyl-2-butenyl)-indole, 7-(3-methyl-2-butenyl)-indole.
Benesova et al., Collect. Czech. Chem. Commun., 34, 1807 (1969)

RICCIA

R. fluitans L.
ACETYLENIC: 9-octadecen-6-ynoic acid.
FLAVONOIDS: apigenin 7-O-glucuronide, lucenin, luteolin, luteolin 7-O-glucuronide, luteolin 7-O-glucuronide-3'-O-rhamnoside.
(A) Kohn et al., Phytochemistry, 26, 2101 (1987)
(F) Vandekerkhove, Z. Pflanzenphysiol, 86, 217 (1978); Chem. Abstr., 88 86025p

RICINOCARPUS (Euphorbiaceae)

R. muricatus.
DITERPENOIDS: eperuane-8,15-diol, eperuane-8,15,18-triol, 8(17)-eperuene-15,18-dioic acid, enantio-13-epi-labdane-8,15-diol.
FLAVONE: 3',4',5-trihydroxy-3,7,8-trimethoxyflavone.
Henrick, Jefferies, Tetrahedron, 21, 1175, (1965)
Henrick, Jefferies, ibid., 21, 3219 (1965)

R. stylosus.
DITERPENOIDS: (-)-1α,19-dihydroxy-16α-kauran-17-oic acid, (-)-16α,17-dihydroxy-16α-kauran-19-oic acid, 16α-kaurane-17,19-dioic acid, 16α-kaurane-16α,17,19-triol, kaur-16-en-19-oic acid.
FLAVONES: 3,3',4',5,5',7-hexamethoxyflavone, 3',5,5'-trihydroxy-3,4',7-trimethoxyflavone.
(F) Henrick et al., Aust. J. Chem., 17, 934 (1964)
(T) Henrick, Jefferies, Tetrahedron Lett., 1507 (1964)

RICINUS (Euphorbiaceae).

R. communis L.
PYRIDONE ALKALOID: ricinine.
DITERPENE: casbene.
NORTRITERPENOID: 30-norlupan-3β-ol-20-one.
TOXIN: ricin.
Robinson, West, Biochem. J., 9, 70 (1970)

RIELLA

R. affinis.
FLAVONOIDS: apigenin 7-O-glucoside, chrysoeriol, luteolin 7-O-glucoside.
Markham et al., Phytochemistry, 15, 151 (1976)

R. americana.
FLAVONOIDS: apigenin 7-O-glucoside, chrysoeriol, luteolin 3'-O-glucoside, luteolin 7-O-glucoside.
Markham et al., Phytochemistry, 15, 151 (1976)

RINDERA (Boraginaceae)

R. austroechinata M.Pop.
PYRROLIZIDINE ALKALOIDS: echinatine, echinatine N-oxide.
Tsuda, Marion, Can. J. Chem., 41, 1919 (1968)

R. baldchuanica Kuzn.
PYRROLIZIDINE ALKALOIDS: echinatine, rinderine, trachelanthamine, turkestanine.
Akramov et al., Dokl. Akad. Nauk Uzb. SSR, 6, 28 (1964)

R. cyclodonta Bge.
PYRROLIZIDINE ALKALOIDS: echinatine, lindelofine, lindelofine N-oxide.
Akramov et al., Dokl. Akad. Nauk Uzb. SSR, 6, 28 (1964)

R. echinata Rgl.
PYRROLIZIDINE ALKALOIDS: echinatine, trachelanthamine, trachelanthine N-oxide.
Akramov et al., Dokl. Akad. Nauk Uzb. SSR, 6, 28 (1964)

R. oblongifolia M.Pop.
PYRROLIZIDINE ALKALOIDS: echinatine, turkestanine.
UNCHARACTERIZED ALKALOID: caratagine.
Akramov et al., Dokl. Akad. Nauk Uzb. SSR, 6, 28 (1964)

ROBINIA (Leguminosae)

R. pseudacacia L.
BENZOPYRAN: acacin.
CHALCONE: robtein.
FLAVANONE: 3',4',5',7-tetrahydroxyflavanone.
DITERPENOID: hydroxymethylglutamylhydroxyabscicic acid.
(B) Plouvier et al., C. R. Acad. Sci., 269, 646 (1969)
(Ch, F) Roux et al., Biochem. J., 82, 324 (1962)
(T) Hirai et al., Phytochemistry, 17, 1625 (1978)

ROCCELLA (Roccellaceae)

R. canariensis.
PEPTIDE: roccanin.
> Bohlmann et al., Tetrahedron Lett., 3065 (1970)

R. fuciformis (L.) DC.
DIAZINE ALKALOID: picrocelline.
CHROMENE:
6-ethoxymethyl-5-hydroxy-7-methoxy-2-methylchromene.
> (A) Forster et al., J. Chem. Soc., **121**, 816 (1922)
> (C) Aberhart et al., J. Chem. Soc., C, 704 (1969)

R. portentosa.
CHROMENE: portenol.
> Aberhart et al., J. Chem. Soc., C, 1612 (1970)

ROEMERIA (Papaveraceae)

R. carica.
BENZYLISOQUINOLINE ALKALOID: (+)-roemecasrine.
> Gozler et al., Heterocycles, **24**, 1227 (1986)

R. hybrida (L.) DC. (R. orientalis Boiss.).
OXOQUINOLIZIDINE ALKALOID: protopine.
UNCHARACTERIZED ALKALOID: roemeridine.
> Platonova et al., J. Gen. Chem., USSR, **26**, 173 (1956)

R. refracta (Stev.) DC.
AMINE ALKALOIDS: (-)-ephedrine, (+)-pseudoephedrine.
APORPHINE ALKALOIDS: anonaine, isoroemerine,
(-)-mecambridine, remrefidine, roemerine, roemerinine.
ISOQUINOLINE ALKALOIDS: reframidine, reframine, reframoline.
SPIROISOQUINOLINE ALKALOID: roemeramine.
> Cooke, Haynes, Aust. J. Chem., **7**, 99 (1954)
> Slavik, Slavikova, Collect. Czech. Chem. Commun., **33**, 4066 (1968)

ROLANDRA (Compositae/Vernonieae)

R. fruticosa (L.) Kuntze.
SESQUITERPENOIDS: 13-acetoxyrolandrolide,
13-ethoxyrolandrolide, isorolandrolide, rolandrolide.
> Herz et al., J. Org. Chem., **46**, 761 (1981)

R. heterogama.
Three sesquiterpenoids.
> Bohlmann, Zdero, Phytochemistry, **17**, 565 (1978)

ROLLINIA (Annonaceae)

R. emarginata.
APORPHINE ALKALOIDS: (-)-anonaine, (-)-asimilobine,
(+)-reticuline.
> Nieto, J. Nat. Prod., **49**, 717 (1986)

R. mucosa.
TETRAHYDROFURAN: rolliniastatin 1.
> Pettit et al., Can. J. Chem., **65**, 1433 (1987)

ROMNEYA (Papaveraceae)

R. coulteri var. trichocalyx (Eastwood) Jepson.
APORPHINE ALKALOIDS: (-)-reticuline, romneine.
OXOPROTOBERBERINE ALKALOID: coulteropine.
> Stermitz et al., Tetrahedron, **22**, 1167 (1980)

RONDELETIA (Rubiaceae)

R. panamensis.
DITERPENOIDS: oxidopanamensin, panamensin, rondeletin.
> Koike et al., Tetrahedron, **36**, 1167 (1980)

ROSA (Rosaceae)

R. acicularis.
CATECHINS: (-)-1-gallocatechin, (-)-1-epi-gallocatechin.
> Pankov, Safronova, Rastit. Resur., **11**, 520 (1975)

R. amblyotis.
CATECHINS: (+)-catechin, (-)-epi-catechin gallate,
(-)-gallocatechin, (-)-epi-gallocatechin.
> Pankov, Safronova, Rastit. Resur., **11**, 520 (1975)

R. canina L.
CAROTENOID: rubixanthin.
> Lederer, Bull. Soc. Chim. Belg., **20**, 611 (1938)

R. damascena Mill.
AMINE: β-phenethylamine.
CAROTENOID: rubixanthin.
FLAVONES: kaempferol, quercetin.
FLAVONE GLUCOSIDES: kaempferol 3-O-β-galactopyranoside,
kaempferol 3-O-β-D-glucopyranoside.
GENIN: pectolinarigenin.
MONOTERPENOIDS: citronellol, geraniol, nerol, methyleugenol.
NORSESQUITERPENOID: damascenone.
STEROID: β-sitosterol.
> Papanov et al., Chem. Abstr., 105 94467u

R. davurica.
CATECHINS: (-)-gallocatechin, (-)-1-epi-gallocatechin.
> Pankov, Safronova, Rastit. Resur., **11**, 520 (1975)

R. foetida.
CAROTENOIDS: antheroxanthin, auroxanthin, β-carotene, lutein,
luteoxanthin, mutatoxanthin, violaxanthin, zeaxanthin.
> Buchecker, Eugster, Helv. Chim. Acta, **60**, 1754 (1977)

R. gallica.
GLUCOSIDE: β-phenethyl-β-D-glucopyranoside.
GLYCOSIDES: galloside A, galloside B, galloside C.
> (Glu) Nel'nikov et al., Khim. Prir. Soedin., 807 (1975)
> (Gly) Bugorskii et al., ibid., 420 (1977)

R. gracilipes.
CATECHINS: (-)-gallocatechin, (-)-epi-gallocatechin.
> Pankov, Safronova, Rastit. Resur., **11**, 520 (1975)

R. koreana.
CATECHINS: (-)-gallocatechin, (-)-epi-gallocatehcin.
> Pankov, Safronova, Rastit. Resur., **11**, 520 (1975)

R. maximowicziana.
CATECHINS: (+)-catechin, (-)-epi-catechin gallate,
(-)-gallocatechin, (-)-epi-gallocatechin.
> Pankov, Safronova, Rastit. Resur., **11**, 520 (1975)

R. multiflora Thunb.
FLAVONOIDS: glycoside f-1, glycoside f-3, multiflorin A,
multiflorin B, quercitrin.
STEROID: β-sitosterol.
TRITERPENE GLUCOSIDES: 2α,19α-dihydroxyursolic acid
β-D-glucopyranoside, rosamultin.
TRITERPENOID: 2α,19α-dihydroxyursolic acid.
> (F) Yamaki et al., J. Pharm. Soc., Japan, **96**, 284 (1976)
> (T) Du et al., Yaoxue Xuebao, **18**, 314 (1983)

R. roxburgii.
TRITERPENOID: roxburgic acid.
> Liang, Yaoxue Xuebao, **22**, 121 (1987); Chem. Abstr., 107 4276v

R. rubinosa L.
CAROTENOID: rubixanthin.
> de Ville et al., J. Chem. Soc., Chem. Commun., 1311 (1969)

R. rugosa.

CATECHINS: (-)-gallocatechin, (-)-epi-gallocatechin.

GLYCOSIDE: β-phenylethyl-β-D-glucopyranoside.

(C) Pankov, Safronova, *Rastit. Resur.*, **11**, 520 (1975)

(Gl) Sikharulidze *et al.*, *Prikl. Biokhim. Mikrobiol.*, **12**, 759 (1976); *Chem. Abstr.*, **86** 13813t

ROSMARINUS (Labiatae)

R. officinalis L.

DITERPENE ALKALOIDS: isorosmaricine, N-methylrosmaricine, rosmaricine.

FLAVONE: repitrin.

DITERPENOIDS: picrosalvin, rosmadiol, rosmanol, rosmarinol, rosmariquinone.

(F) Imre *et al.*, *Phytochemistry*, **16**, 799 (1977)

(F) Nakatani *et al.*, *Agr. Biol. Chem.*, **47**, 353 (1983)

(T) Houlihan *et al.*, *J. Am. Oil Chem. Soc.*, **62**, 98 (1985)

ROTHMANNIA

R. globosa.

IRIDOIDS: gardenoside, α-gardiol, β-gardiol, genipin, geniposide, mussaenoside, scandoside methyl ester.

Jensen *et al.*, *Phytochemistry*, **22**, 1761 (1983)

ROTTLERA

R. tinctoria.

BENZOPYRAN: rottlerin.

Reio, *J. Chromatogr.*, **68**, 183 (1972)

ROUREA (Connaraceae)

R. santaloides.

ALIPHATIC: hentricontane.

AMINO ACID: methionine sulfoximine.

ANTHOCYANIN: leucopelargonidin.

BENZOQUINONE: rapanone.

STEROID: β-sitosterol.

(Al, An, S) Ramiah *et al.*, *J. Inst. Chem. (India)*, **48**, 196 (1976); *Chem. Abstr.*, **86** 68377j

(Am) Jeannoda *et al.*, *J. Ethnopharmacol.*, **14**, 11 (1985); *Chem. Abstr.*, **104** 31778e

(B) Murthy *et al.*, *Tetrahedron*, **21**, 1445 (1965)

ROYLEA (Labiatae)

R. calycina (Roxb.) Briq.

DITERPENOIDS: calyenone, calyone, epi-calyone, precalyone.

Prakesh *et al.*, *J. Chem. Soc., Perkin I*, 1305 (1979)

ROZITES

R. caperta.

AMINO ACID: S-2-aminoethyl-4-cysteine.

Matsumoto, Chem. Abstr., *103 3724u*

RUBIA (Rubiaceae)

R. cordifolia L.

ANTHRAQUINONES:

1-acetoxy-6-hydroxy-2-methylanthraquinone 3-O-α-rhamnosyl-α-glucoside,

1,4-dihydroxy-2-carboethoxyanthraquinone,

1-hydroxy-2-carboxy-3-methoxyanthraquinone,

1-hydroxy-2-methyl-6-methoxyanthraquinone,

1-hydroxy-2-methyl-7-methoxyanthraquinone.

COUMARIN: scopoletol.

PEPTIDES: hexapeptides I, II, III and IV.

STEROID: β-sitosterol.

TRITERPENOIDS: oleanolic acid acetate, rubiconmaric acid, rubifolic acid.

(An, C, S) Vidal-Tessier *et al.*, *Ann. Pharm. Fr.*, **44**, 117 (1986)

(P) Itokawa *et al.*, *Chem. Pharm. Bull.*, **34**, 3762 (1986)

(T) Talapatra *et al.*, *Phytochemistry*, **20**, 1923 (1981)

R. iberica.

PIGMENT: nordamnacanthal.

Brew *et al.*, *J. Chem. Soc., C*, 2001 (1971)

R. peregrina L.

IRIDOIDS: asperulosidic acid, 10-deacetylasperulosidic acid.

Bianco *et al.*, *Gazz. Chim. Ital.*, **108**, 13 (1978)

R. tinctorum L.

IRIDOIDS: asperulosidic acid, 10-deacetylasperulosidic acid.

PIGMENTS: digiferrol, galiosin.

(I) Bianco *et al.*, *Gazz. Chim. Ital.*, **108**, 13 (1978)

(P) Imre *et al.*, *Tetrahedron Lett.*, 4681 (1971)

RUBUS (Rosaceae)

R. arctica.

FURANONE: 2,5-dimethyl-4-methoxy-2,3-dihydro-3-furanone.

Honkanen *et al.*, *Kem. Kemi*, **3**, 180 (1976); *Chem. Abstr.*, **85** 59551p

R. chingii.

DITERPENOID: rubusoside.

Tanaka *et al.*, *Agr. Biol. Chem.*, **45**, 2165 (1981)

R. crataegifolius.

STEROIDS: campesterol, cholestanol, β-sitosterol, stigmasterol.

TRITERPENOID: ursolic acid.

Moon *et al.*, *Yakhak Hoe Chi*, **20**, 37 (1976); *Chem. Abstr.*, **85** 119629g

R. ellipticus.

ALIPHATICS: octacosanol, octacosanoic acid.

STEROIDS: β-sitosterol, β-sitosterol β-D-glucoside.

TRITERPENOIDS: rubitic acid, ursolic acid.

Bhakuni *et al.*, *Indian Drugs*, **24**, 272 (1987); *Chem. Abstr.*, **107** 151197y

R. fruticosus agg.

TRITERPENOIDS: rubinic acid, rubitic acid.

Mukherjee *et al.*, *Phytochemistry*, **23**, 2581 (1984)

R. moluccanus.

TRITERPENOID: rubisic acid.

Bhattacharyya, Dutta, *J. Indian Chem. Soc.*, **46**, 381 (1969)

R. rosiofolius.

SESQUITERPENOIDS: pregeigerene, rosifoliol.

Jones, Sutherland, *Aust. J. Chem.*, **21**, 2255 (1968)

R. suavissimus.

TRITERPENOID: suavissimoside R-1.

Gao *et al.*, *Chem. Pharm. Bull.*, **33**, 37 (1985)

RUDBECKIA (Compositae)

R. bicolor Nutt.

FLAVONE: tambulin.

Ahuja *et al.*, *Indian J. Chem.*, **13**, 1134 (1975)

R. laciniata L.

SESQUITERPENOIDS: bisabolen-1,4-endo-peroxide, rudbeckianone, rudbeckiolide.

Bohlmann *et al.*, *Phytochemistry*, **17**, 2034 (1978)

Jakupovic *et al.*, *Justus Liebigs Ann. Chem.*, 1474 (1986)

R. mollis.

FLAVONOID: eupatolitin 3-O-rhamnoside.

SESQUITERPENOID: rudmollin.

(F) Iyengar *et al.*, *Indian J. Chem.*, **14B**, 906 (1976)

(T) Herz *et al.*, *J. Org. Chem.*, **46**, 1356 (1981)

RUELLIA (Acanthaceae)

R. tuberosa L.
ALIPHATICS: hentriacontane, nonacosane.
STEROIDS: campesterol, β-sitosterol, stigmasterol.
Andiwal et al., Indian Drugs, 23, 48 (1985); Chem. Abstr., 104 3417t

RUILOPEZIA

R. marginata.
Four diterpenoids.
Usubillaga et al., Planta Med., 35, 331 (1979)

RUMEX (Polygonaceae)

R. acetosa L.
FLAVONOIDS: chrysophanol, chrysophanol anthrone, emodin, physcion,
Tamamo et al., Agr. Biol. Chem., 46, 1913 (1982)

R. alpinus L.
NAPHTHALENES: chrysophanol, 3-methylnaphthalene-1,8-diol, nepodin.
Bauch et al., J. Chem. Soc., Perkin I, 689 (1975)

R. denticulata.
PIGMENT: denticulatol.
Chi et al., J. Chin. Chem. Soc., 15, 21 (1947); Chem. Abstr., 42 552g

R. nepalensis.
ANTHRAQUINONES: emodin, physcion.
Suri et al., J. Indian Chem. Soc., 53, 1158 (1976)

R. obtusifolius L.
DITERPENOID: lapathinic acid.
Tschirch, Weil, Arch. Pharm., 250, 30 (1912)

R. orientalis.
NAPHTHOQUINONE: orientalone.
Sharma et al., Indian J. Chem., 15B, 544 (1977)

R. patientia L.
NAPHTHALENES: chrysaphanol, 3-methylnaphthalene-1,8-diol, nepodin.
Bauch et al., J. Chem. Soc., Perkin I, 689 (1975)

R. pulcher.
ANTHRAQUINONES: emodin dianthrone diglucoside sulphate, emodin 8-glucoside sulphate.
Harborne et al., Phytochemistry, 16, 1314 (1977)

R. vesicarius.
AMINO ACIDS: cystine, glutamic acid, histidine, phenylalanine, proline.
Tiwari et al., Chem. Abstr., 88 19010s

RUPUS

R. rosiofolius.
SESQUITERPENOID: bicyclogermacra-1(10),4(5)-diene.
Southwell, Tetrahedron Lett., 873 (1977)

RUSCUS (Liliaceae/Ruscaceae)

R. aculeatus L.
BENZOFURAN: ruscodibenzofuran.
STEROIDS: deglycoruscoside, ruscoside, neoruscogenin, ruscogenin, rusconin.
(B) El-Sohly et al., Tetrahedron, 33, 1711 (1977)
(S) Bombardelli et al., Fitoterapia, 43, 3 (1972)

R. hypoglossum L.
FLAVANOLS: isorhamnetin, isovitexin C-glucoside, isovitexin O-glucoside, nicotiflorin.
El-alfy et al., Plant Med. Phytother., 9, 308 (1975)

R. ponticus.
SAPONINS: ruscoponticosides C, D and E.
Pkheidze et al., Soobshch. Akad. Nauk Gruz SSR, 120, 561 (1985);
Chem. Abstr., 105 3499s

RUSPOLIA (Acanthaceae)

R. hypercrateriiformis M.R.
PYRROLIZIDINE ALKALOIDS: norruspoline, norruspolinone, ruspoline, ruspolinone.
Roessler et al., Helv. Chim. Acta, 61, 1200 (1978)

RUSSULA (Russulaceae)

R. ochroleuca.
AMINO ACID: uncharacterised acid.
Welter et al., Phytochemistry, 15, 1984 (1976)

R. sardonia.
SESQUITERPENOIDS: furanether A, furosardonin A, sardonialactone A.
Andina et al., Phytochemistry, 19, 93 (1980)

RUTA (Rutaceae)

R. angustifolia.
AROMATICS: moskachans A, B, C and D.
COUMARIN: angustifolin.
(Ar) Borges del Castillo et al., Phytochemistry, 25, 2209 (1986)
(C) Borges del Castillo et al., ibid., 23, 2095 (1984)

R. chalepensis L.
ACRIDONE ALKALOIDS: arborinine, ribinalidine, rutacridone.
FURANOQUINOLINE ALKALOIDS: dictamnine, kokusaginine, skimmianine.
QUINOLINE ALKALOID: graveolinine.
COUMARINS: bergapten, byakangelicin, chalepensin, chalepin, heliettin, isopimpinellin, psoralen, rutamarin, xanthotoxin, xanthyletin.
SESQUITERPENOID: isoimperiatorin.
(A, C) Gonzalez et al., An. Quim., 73, 430 (1977)
(C) Reisch et al., Tetrahedron Lett., 4395 (1968)
(T) El-Tawil et al., Pharmazie, 35, 503 (1980)

R. graveolens L.
ACRIDONE ALKALOIDS: furacridone, gravacridonechlorine, gravacridonediol, gravacridonediol O-β-D-glucoside, gravacridonediol O-methyl ether, gravacridonetriol, rutacridone.
QUINOLINE ALKALOIDS: graveoline, graveolinine, methylenedioxyphenyl-butyl-4-quinolone.
COUMARINS: gravelliferone, isorutarin, 8-methoxygravelliferone, rutacultin, rutaretin, suberenone.
(A) Reisch et al., Phytochemistry, 11, 2359 (1972)
(C) Garg et al., ibid., 17, 2135 (1978)

R. oreojasme Weeb.
COUMARINS: fatagarin, oreojasmin, sabandinin.
Gonzalez et al., An. Quim., 71, 842 (1975); Chem. Abstr., 84 161766

R. pinnata.
COUMARINS:
6-(3'-ethoxy-2'-hydroxy-3'-methylbutyryl)-7-methoxycoumarin, furopinnarin, sabandinin, sabandinone, tamarin.
Gonzalez et al., Herba Hung., 10, 95 (1971) Chem Abstr. 79 15854m
Gonzalez et al., Phytochemistry, 16, 2033 (1977)

RYANIA

R. speciosa Vahl.
DITERPENE ALKALOIDS: dehydroryanodine ryania diterpene esters A, C, C-1, C-2 and D, ryanodine.
Ruest et al., Can. J. Chem., 63, 2840 (1985)

RYTIPHLEA (Rhodophyceae)

R. tinctoria.

HALOGENATED PHENOLICS: 5,6-dibromo-3,4-dimethoxybenzyl alcohol, 5,6-dibromo-3,4-dimethoxybenzyl alcohol ethyl ether, 2,4-dibromo-1,3,5-trimethoxybenzene, 3′,5,5′,6-tetrabromo-2′,3,4,4′,6′-pentamethoxydiphenylmethane.

Chevrolet-Magueur *et al.*, *Phytochemistry*, **15**, 767 (1976)

SABADILLA

S. minor.

UNCHARACTERIZED ALKALOID: hydroalkamine S.

Auterhoff, *Arch. Pharm.*, **288**, 549 (1955)

SABINA

S. officinalis.

MONOTERPENOID: sabinol.

Short, Read, *J. Chem. Soc.*, 1040 (1939)

SACCHARUM

S. officinarum L.

FLAVONOID: 5,7-dimethylapigenin 4'-O-β-D-glucopyranoside.

STEROIDS: ikshusterol, 7-epi-ikshusterol, 24-methylenelophenol, stigmastane, 5α-stigmastane-3,5,6-triol.

TRITERPENOID: isosawamilletin.

(F) Misra *et al.*, *Curr. Sci.*, **47**, 152 (1978)

(S) Deshmane, Dev, *Tetrahedron*, **27**, 1109 (1971)

(T) Eglinton *et al.*, *Tetrahedron Lett.*, 2323 (1964)

SAELANIA

S. glaucescens.

DITERPENOID: 16α-kauran-1-ol.

Nilsson, Mattensson, *Acta Chem. Scand.*, **25**, 1486 (1971)

SAFRAN

S. aquila.

MONOTERPENE GLUCOSIDE: picrocrocin.

Kuhn, Winterstein, *Ber. Dtsch. Chem. Ges.*, **67**, 344 (1934)

S. naturalis.

CAROTENOIDS: crocetin-(β-D-gentiobiosyl)-(β-D-glucosyl) ester, crocetin di-(β-D-glucosyl) ester, crocin.

Pfander *et al.*, *Helv. Chim. Acta*, **58**, 1608 (1975)

SAGINA

S. japonica.

FLAVONOIDS: apigenin 6-C-arabinosyl-8-C-glucopyranoside, apigenin 6,8-di-glucoside, apigenin 8-glucoside, apigenin 6-O-rhamnosyl-C-glucoside.

Zhuang, *Zhongcaoyao*, **14**, 295 (1983) *Chem. Abstr.*, **100** 3466a

SAGOTIA

S. racemosa.

PHENANTHRENES: microndrol E, microndrol F.

De Alvarenga *et al.*, *Phytochemistry*, **15**, 844 (1976)

SALACIA

S. macrosperma.

STEROIDS: salacia quinonemethide, salaspermic acid.

Reddy *et al.*, *Indian J. Chem.*, **20B**, 197 (1981)

S. madagascariensis.

STEROID: isoguesterin.

Sneden *et al.*, *J. Nat. Prod.*, **44**, 503 (1981)

S. prenoides.

TRITERPENOIDS: 1,3-dioxofriedelen-24-al, 25,26-epoxyfriedelene-1,3-dione, 7,24-oxidafriedelane-1,3-dione, 25,26-oxidafriedelane-1,3-dione.

Tewari *et al.*, *J. Chem. Soc., Perkin I*, 146 (1974)

SALICORNIA (Chenopodiaceae)

S. europaea L.

CHROMONES: 6,7-dimethoxychromone, 6,7-methylenedioxychromone.

FLAVONOIDS: (malonyl-6''-β-D-glucoside)-3-quercetol, quercetol, +i.quercetol 3-β-D-glucoside, rutoside.

(Ch) Chiji *et al.*, *Agr. Biol. Chem.*, **42**, 159 (1978)

(F) Geslin *et al.*, *J. Nat. Prod.*, **48**, 111 (1985)

S. herbacea.

UNCHARACTERIZED ALKALOIDS: salicornine, saliherbine.

Brokowski, Drost, *Pharmazie*, **20**, 390 (1965)

SALIX (Salicaceae)

S. songorica.

FLAVONOIDS: quercetin, rutin, salicin, triandrin.

Kompanysev *et al.*, *Khim. Prir. Soedin.*, 813 (1976)

S. triandra L.

GLUCOSIDE: salidroside.

Bridel *et al.*, *C. R. Acad. Sci.*, **183**, 231 (1926)

S. viminalis.

FLAVONOIDS: apigenin 7-O-glucoside, isoquercitrin, isorhamnetin 3-O-(6-acetylglucoside), isorhamnetin 3-O-glucoside.

Karl *et al.*, *Phytochemistry*, **16**, 1117 (1977)

SALMALIA (Bombacaceae)

S. malabarica (DC.) Schott et Endl. (Bombax ceiba).

SESQUITERPENOID: 1,6-dihydroxy-3-methyl-5-(methylethyl)-7-methoxy-8-naphthalene carboxylic acid.

Sood *et al.*, *Phytochemistry*, **21**, 2125 (1982)

SALMEA

S. scandens.

AMIDES: deca-2Z,8Z-dien-6-yn-1-oic acid phenylethylamide, 6-oxohexa-2E,4E-dien-1-oic acid N-phenylethylamide, nona-6,8-diyn-2,3-epoxy-1-oic acid phenylethylamide.

Bohlmann *et al.*, *Phytochemistry*, **24**, 595 (1985)

SALSOLA (Chenopodiaceae)

S. kali L.

FLAVONES: isorhamnetin 3-O-glucoside, isorhamnetin 3-O-rutinoside.

Tomas *et al.*, *Fitoterapia*, **56**, 365 (1985)

S. laricifolia.

COUMARINS: calycanthoside, fraxetin, fraxidin, fraxidin 8-O-β-D-glucopyranoside, isofraxidin, lariside, scopoletin 7-O-β-D-glucopyranoside.

Narantuyan *et al.*, *Khim. Prir. Soedin.*, 243, 288 (1986)

S. micranthan.
SAPONINS: salsolides A, B, C, D and E.
Savona et al., Phytochemistry, **21**, 2563 (1982)

S. richteri Karel.
ISOQUINOLINE ALKALOIDS: (-)-salsolidine, (±)-salsolidine, (-)-salsoline, (±)-salsoline.
UNCHARACTERIZED ALKALOID: salsamine.
Orekhov et al., Bull. Soc. Chim. Fr., **4**, 1265 (1937)

S. subaphylla SAM. (Aellenia subaphylla SAM Aellen.)
AMINE: ferulylputrescine.
AMINE ALKALOID: subaphylline.
Ryabinin et al., Dokl. Akad. Nauk SSSR, **67**, 513 (1949)

SALVADORA

S. persica.
UREA ALKALOID: N,N'-bis(3-methoxybenzyl)urea.
Ezmirly et al., Planta Med., **35**, 191 (1979)

SALVIA (Labiatae)

S. aethiopis.
ALIPHATIC: eicosane.
DITERPENOIDS: aethiopinone, salvipinone.
FLAVONES: 5-hydroxy-3',4',7-trimethoxyflavone, salvigenin.
STEROIDS: β-sitosterol, β-sitosterol 3-β-D-glucoside.
(Al, F, S) Ulubelen et al., Planta Med., **29**, 318 (1976)
(T) Boya et al., Phytochemistry, **20**, 1367 (1981)
(T) Rodriguez et al., ibid., **23**, 1803 (1984)

S. argentea.
DITERPENOIDS: aethiopinone, arucadiol, ferruginol, 1R-hydroxymultirone, 1-ketoaethiopinone, salvipisone.
Michavila et al., Phytochemistry, **25**, 1935 (1986)

S. ballotaeflora.
DITERPENOIDS: conactyone, icetexone, romulogarzone.
Dominguez et al., Planta Med., **30**, 237 (1976)

S. breviflora.
DITERPENOIDS: brevifloralactone, brevifloralactone acetate.
TRITERPENOIDS: oleanolic acid, ursolic acid.
Cuevas et al., Phytochemistry, **26**, 2019 (1987)

S. broussonetii.
TRITERPENOID: anagadiol.
Gonzalez et al., J. Chem. Soc., Chem. Commun., 567 (1971)

S. canariensis L.
STEROIDS: β-sitosterol, β-sitosterol β-D-glucoside.
DITERPENOIDS: arucatriol, caryophyllene, clovanediol, galdosol.
TETRATERPENOID: squalene.
TRITERPENOIDS: oleanolic acid, ursolic acid.
Fraga et al., An. Quim., **70**, 701 (1975)
Gonzalez et al., ibid., **71**, 705 (1976)

S. carnosa.
DITERPENOID: carnosol.
Brieskorn et al., J. Org. Chem., **29**, 2293 (1964)

S. coccinea.
DITERPENOID: salviacoccin.
Savona et al., Phytochemistry, **21**, 2563 (1982)

S. digitaloides.
DITERPENE GLYCOSIDE: baiyunoside.
Tanaka et al., Chem. Pharm. Bull., **31**, 780 (1983)

S. divinorum.
DITERPENOID: salvinorin.
Ortega et al., J. Chem. Soc., Perkin I, 2505 (1982)

S. drobovii.
QUINONES: cryptotanshinone, miltirone.
Romanova et al., Khim. Prir. Soedin., 414 (1977)

S. farinacea.
NORDITERPENOIDS: salvifaricin, salvifarin.
Savona et al., Phytochemistry, **22**, 784 (1983)

S. fulgens.
DITERPENOID: salvigenolide.
Esquivel et al., Tetrahedron, **41**, 3213 (1985)

S. gesneraefolia.
TERPENOIDS: gesnerifolin A, gesnerifolin B.
Jiminez et al., Rev. Latinoam. Quim., **10**, 165 (1979)

S. glutinosa.
TRITERPENOID: olean-12-ene-3β,11α-diol.
Beynon et al., J. Chem. Soc., 1233 (1938)

S. horminium.
TRITERPENOID: olean-13,18-ene-2α,3β-diol, olean-13(18)-ene-2β,3β-diol.
Ulubelen et al., Phytochemistry, **16**, 790 (1977)

S. hypoleuca.
SESTERTERPENOIDS: salvialeucolide, salvialeucolide-2,3-lactone, salvialeucolide methyl ester.
Rustaiyan et al., Phytochemistry, **21**, 1812 (1982)

S. karabachensis.
QUINONES: cryptotanshinone, tanshinone I.
Romanova et al., Khim. Prir. Soedin., 414 (1977)

S. lanata.
TRITERPENOID: 3-epi-ursolic acid.
Huneck et al., Phytochemistry, **20**, 3279 (1981)

S. lanigera.
ESTER: methylsarnosate.
Al-Hazimi et al., Phytochemistry, **25**, 1238 (1986)

S. lavandulaefolia.
DITERPENOIDS: 7-ethoxy-12-O-methyl royleanone, 7-ethoxyroyleanone.
Michavila et al., Phytochemistry, **25**, 266 (1986)

S. leucantha.
TRITERPENOID: 3-epi-erythrodiol.
Urbanek et al., Curr. Sci., **48**, 107 (1979)

S. melissodora.
DITERPENOID: melissodoric acid.
Rodriguez-Hahn et al., Rev. Latinoam. Quim., **4**, 93 (1973)

S. moorcroftiana.
DITERPENOID: deoxyfuerstione.
Bakshi et al., Plants Med., 408 (1986)
Simoes et al., Phytochemistry, **25**, 755 (1986)

S. multiorrhiza Bunge.
PIGMENTS: cryptotanshinone, 6-hydroxytanshinone II, isocryptotanshinone, isotanshinone I, isotanshinone II, isotanshinone III, tanshinolactone, tanshinone I, tanshinone II, tanshinone IIIB, tanshinonic acid, tanshinlactone, Diterpenoid salviol.
TRINORDITERPENOID: danshenspiroketallactone.
(P) Kakisawa et al., J. Chem. Soc., Perkin **1**, 1716 (1976)
(P) Luo et al., Chem. Pharm. Bull., **34**, 3166 (1986)
(T) Hayashi et al., J. Chem. Soc., Chem. Commun., 541 (1971)
(T) Kong et al., Yaoxue Xuebao, **20**, 747 (1985); Chem. Abstr., **104** 85377s

S. nemorosa.
DITERPENOID: nemorone.
Romanova et al., Khim. Prir. Soedin., 199 (1971)

S. nicolsoniana.
STEROIDS: β-sitosterol, β-sitosterol 3-glucoside.
TRITERPENOIDS: betulinic acid, 3,24-dihydroxyolean-12-en-28,30-dioic acid, 3,24-dihydroxyolean-12-en-28-oic acid, oleanolic acid,

ursolic acid, 3-epi-ursolic acid.
Pereda-Miranda *et al.*, *J. Nat. Prod.*, **49**, 225 (1986)

S. officinalis L.

DITERPENOIDS: picrosalvin, salvin.
FLAVONOIDS: apigenin 7-O-glucuronide, luteolin
7-O-glucuronide, salvigenin 7-O-glucuronide.
PHENOLICS: caffeic acid, chlorogenic acid.
(F, P) Prokopenko *et al.*, *Farm. Zh. (Kiev)*, 75 (1982); *Chem. Abstr.*, 97 141716c
(T) Linde, *Helv. Chim. Acta*, **47**, 1234 (1964)
(T) Brieskorn *et al.*, *J. Org. Chem.*, **29**, 2293 (1964)

S. oxyodon.

Three diterpenoids.
Escudero *et al.*, *Phytochemistry*, **22**, 585 (1983)

S. phlomoides.

DITERPENOIDS: demethylcryptojaponol, salviphlomone.
Rodriguez-Hahn *et al.*, *Phytochemistry*, **22**, 2011 (1983)

S. plebeia.

FLAVONE: salvitin.
DITERPENOID: epoxysalviacoccin.
(F) Gupta *et al.*, *Indian J. Chem.*, **13**, 215 (1975)
(T) Garcia-Alvarez *et al.*, *Phytochemistry*, **25**, 272 (1986)

S. przewalskii Maxim. var. mandarinorum (Diels) Stib.

DITERPENOIDS: przewaquinone C, przewaquinone D,
przewaquinone E, przewaquinone F.
Yang *et al.*, *Yaoxue Xuebao*, **19**, 274 (1984); *Chem. Abstr.*, 103 19811g

S. sapinae.

DITERPENOID: 7β,15-dihydroxyabietatriene.
Pereda-Miranda *et al.*, *Phytochemistry*, **25**, 1931 (1986)

S. sclarea L.

DITERPENOID: sclareol.
SESQUITERPENOIDS: (2R,5E)-epoxycaryophyll-5-en-12-al,
(2S,5E)-epoxycaryophyll-5-en-12-al,
(2R,5E-epoxycaryophyll-5-ene,
(1R,5R)-epoxysalvial-4(14)-en-1-one, isospathulenol,
salvial-4(14)-en-1-one.
Maurer, Hauser, *Helv. Chim. Acta*, **66**, 2223 (1983)

S. splendens.

DITERPENOIDS: salviarin, splendidin.
Savona *et al.*, *J. Chem. Soc., Perkin I*, 533 (1979)

S. tomentosa.

DITERPENOID: 3-hydroxy-8,11,13,15-abietatetraen-18-oin acid.
Ulubelen *et al.*, *J. Nat. Prod.*, **44**, 119 (1981)

S. trautvetteri.

QUINONE: cryptotanshinone.
Romanova *et al.*, *Khim. Prir. Soedin.*, 414 (1977)

S. trijuga Diels.

QUINONE: hydroxymethylenetanshinquinone.
Yang *et al.*, *Yaoxue Xuebao*, **17**, 517 (1982); *Chem. Abstr.*, 97 195769r

S. triloba.

FLAVONOIDS: apigenin 7-glucuronide, apigenin 7-glucoside,
chrysoeriol 7-glucuronide, 6,8-di-C-glucosylapigenin,
6-hydroxyluteolin 6,3'-dimethyl ether, luteolin 7-glucoside,
luteolin 7-glucuronide, luteoline diglucoside, luteolin
7-glucuronide-3'-glucoside, 6-methoxyapigenin 7-glucoside,
6-methoxyapigenin 7-glucuronide, 6-methoxyluteolin
7-glucoside, 6-methoxyluteolin 7-glucuronide.
DITERPENOID: picrosalvin.
(F) Abdalla *et al.*, *Phytochemistry*, **22**, 2057 (1983)
(T) Brieskorn *et al.*, *J. Org. Chem.*, **29**, 2293 (1964)

S. virgata.

FLAVONES: 5-hydroxy-3',4',7-trimethoxyflavone, salvigenin,
salvigenin 5-glucoside.
TRITERPENOID: virgatic acid.
(F) Ulubelen *et al.*, *Lloydia*, **38**, 446 (1975)
(T) Ulubelen, Agamoglu, *Phytochemistry*, **15**, 309 (1976)

SAMADERA

S. indica.

DITERPENOIDS: samaderins A, B, C, D and E, samaderoside B.
Wani *et al.*, *J. Chem. Soc., Chem. Commun.*, 295 (1977)

S. madagascariensis.

TRITERPENOIDS: melianodiol, melianodiol diacetate.
Ferrier, Polonsky, *J. Chem. Soc., Chem. Commun.*, 261 (1971)

SAMBUCUS (Caprifoliaceae)

S. canadensis.

STEROIDS: campesterol, β-sitosterol, stigmasterol.
TRITERPENOIDS: α-amyrin palmitate, β-amyrin palmitate.
Inoue, Sato, *Phytochemistry*, **14**, 1871 (1975)

S. ebulus L.

IRIDOID: ebuloside.
STEROID: stigmast-4-ene-3,6-dione.
(I) Gross *et al.*, *Helv. Chim. Acta*, **69**, 156 (1986)
(S) Tunmann, Grimm., *Arch. Pharm.*, **307**, 891 (1974)

S. nigra.

STEROIDS: campesterol, β-sitosterol, stigmasterol.
Inoue, Sato, *Phytochemistry*, **14**, 1871 (1975)

SAMUELA (Cactaceae)

S. carnoserana Trel.

STEROIDS: kammogenin, manogenin, mexogenin, samogenin.
Marker *et al.*, *J. Am. Chem. Soc.*, **69**, 2167 (1947)
Marker *et al.*, *ibid.*, **69**, 2401 (1947)

SANBARIA

S. stellipila.

FLAVONES: sorbarin, sorbifolin.
Arisawa *et al.*, *Chem. Pharm. Bull.*, **18**, 916 (1970)

SANDORICUM

S. indicum.

TRITERPENOIDS: bryonolic acid, bryononic acid.
Tunmann, Dadry, *Naturwissenschaften*, **26B**, 620 (1971)

SANGUINARIA (Papaveraceae).

S. canadensis L.

NAPHTHOISOQUINOLINE ALKALOIDS: dihydrosanguilutine,
oxysanguinarine. sanguidimerine, sanguilutine,
sanguinarine, sanguirubine.
OXOPROTOBERBERINE ALKALOID: α-allocryptopine.
Stermitz *et al.*, *Phytochemistry*, **14**, 834 (1975)

SANGUIS

S. draconis.

PIGMENT: deacorubin.
Brockmann *et al.*, *Ber. Dtsch. Chem. Ges.*, **77**, 279 (1944)

SANGUISORBA (Rosaceae)

S. officinalis L.

TANNIN: sanguiin H-6.
TRITERPENOID: tomentosolic acid.
(Tan) Nonaka *et al.*, *Chem. Pharm. Bull.*, **30**, 2255 (1982)
(T) Barton *et al.*, *J. Chem. Soc.*, 5163 (1962)

SANSEVIERA (Liliaceae)

S. trifasciata.
STEROIDS: (25S)-ruscogenin, sansevierigenin.
Tada et al., Chem. Pharm. Bull., 21, 308 (1973)

SANTALUM

S. album.
MONOTERPENOIDS: 2-methyl-3-methylenenorbornan-2-ol, exo-norbicycloekasantalal.
SESQUITERPENOIDS: α-santalol, β-santalol, santalone.
Demole et al., Helv. Chim. Acta, 59, 737 (1976)

S. lanceolatum.
SESQUITERPENOID: lanceol.
Birch, Murray., J. Chem. Soc., 1888 (1951)

S. spicatum.
SESQUITERPENOID: 2,6,10-trimethyl-2,6,10-dodecatriene.
Birch et al., Aust. J. Chem., 23, 2337 (1970)

SANTOLINA (Compositae)

S. chamaecyperissus.
FLAVONOIDS: apigenin, apigenin 7-glucoside, apigenin 7-glucuronide, apigenin 7-rhamnoglucoside, chrysoeriol 7-glucoside, luteolin, luteolin 7-glucuronide, luteolin 7-rhamnoglucoside.
MONOTERPENOID: atemisia ketone.
(F) Becchi et al., Plant Med. Phytother., 10, 139 (1976)
(T) Zelkow et al., J. Org. Chem., 29, 2786 (1964)

S. oblongifolia.
SESQUITERPENOIDS: cis-bejarol, α-trans-bejarol, β-trans-bejarol, oplopenone.
de Pascual Teresa et al., Phytochemistry, 22, 2235 (1983)

SAPINDUS (Sapindaceae)

S. mukorossi.
SAPONINS: mukuroziosides Ia, Ib, IIa, IIb, sapinosides A, B, C and D.
Chirva et al., Khim. Prir. Soedin., 316, 374 (1970)
Kasai et al., Phytochemistry, 25, 871 (1986)

S. saponaria.
FLAVONOIDS: luteolin, 4'-methoxyflavone, rutin.
STEROID: β-sitosterol. Triterpenoids α-amyrin, β-amyrin.
Abdel Wahab, Selim, Fitoterapia, 56, 167 (1985)

SAPIUM

S. baccatum.
TRITERPENOIDS: acetoxyaleuritolic acid, baccatin.
Saha et al., Tetrahedron Lett., 3095 (1977)

S. indicum.
TOXINS: sapintoxins A, B, C and D.
Taylor et al., Experientia, 37, 681 (1981)

S. klotzschianum.
UNCHARACTERIZED ALKALOID: sapinine.
Ribero, Mochado, Bol. Inst. Quim. Agr., 27, 13 (1952)

S. sebiferum.
TRITERPENOIDS: moretenol, 3-epi-moretenol, moretenone, sebiferenic acid.
Khastgir et al., J. Chem. Soc., Chem. Commun., 1217 (1967)
Pradhan et al., Phytochemistry, 23, 2593 (1984)

SAPONARIA (Caryophyllaceae)

S. officinalis L.
PYRAN GLUCOSIDE: sapopyroside.
SAPONINS: saponaria saponin, saponaroside, saponoside, saponosides A, B, C and D.
(G) Plouvier et al., Phytochemistry, 25, 546 (1986)
(Sap) Bukharov, Shcherbak, Khim. Prir. Soedin., 389 (1969)
(Sap) Chirva et al., ibid., 188 (1969)

SAPRANTHUS (Annonaceae)

S. palanga.
ALKALOIDS: liriodenine, 9S-sebiferine.
LACTONE: sapranthin.
Etse, Waterman, Phytochemistry, 25, 1903 (1986)

SAPROSPIRA

S. grandis.
CAROTENOIDS: carotenoid S-434, carotenoid S-460, carotenoid S-500.
Aasen, Liaaen-Jensen, Acta Chem. Scand., 20, 811 (1966)

SARCOCAPNOS

S. crassifolia.
ISOQUINOLINE ALKALOID: oxocompostelline.
SECOCULARINE ALKALOIDS: norsecocularidine, norsecosarcocapnidine, norsecosarcocapnine, secosarcocapnidine, secosarcocapnine.
Castedo et al., Heterocycles, 26, 591 (1987)

S. enneaphylla.
HOMOISOQUINOLINE ALKALOID: ribasidine.
Boente et al., Tetrahedron Lett., 4481 (1983)

SARACOCOCCA

S. pruniformis Lindl.
STEROIDAL ALKALOIDS:
3α-(τ,τ-dimethylacrylyl)amino-20α-dimethylamino-5α-pregnane, 3 β-(τ,τ-dimethylacrylyl)amino-20α-dimethylamino-5α-pregnane, kurchessine, saracodine, saracosine, saracosinine.
UNCHARACTERIZED ALKALOID: saracosidinine.
Chatterjee, Mukherjee, Chem. Ind., 769 (1966)

S. saligna.
STEROIDAL ALKALOID: salignine.
Kiamuddin, Haque, Pakistan J. Sci. Ind. Res., 9, 103 (1966)

SARCOLAEANA

S. multiflora.
ACIDS: caffeic acid, p-coumaric acid, gallic acid, gentisic acid, p-hydroxybenzoic acid, vanillic acid.
FLAVONOIDS: kaempferol 3-rhamnoside, myricetol 3-rhamnoside, myricitrin, quercetin 3-rhamnoside.
Paris et al., Plant Med. Phytother., 9, 230 (1975)

SARCOMELICOPE (Rutaceae)

S. dogniensis Hartley.
ACRIDONE ALKALOIDS:
(-)-cis-1,2-dihydroxy-1,2-dihydro-N-demethylacronycine, (+)-1-hydroxy-1,2-dihydro-N-demethylacronycine, 1-methoxy-3-(2-methylpropanal-2-oxy)-acridin-9-one-4-carbaldehyde, 1-oxo-1,2-dihydro-N-demethylacronycine.
Mitaku et al., Heterocycles, 26, 2057 (1976)

SARGASSUM (Phaeophyceae/Sargassaceae)

S. crassifolium J.Asgardh.
ACID: palmitic acid.
MONOTERPENOIDS: (-)-loliolide, (-)-loliolide acetate, (+)-epi-loliolide.
Kuniyoshi *et al., Bull. Coll. Sci. Univ. Ryukyus*, **40**, 41 (1975); *Chem. Abstr.*, 105 75867h

S. kjellamanianum.
ALKALOID: sargassum lactam.
Nozaki *et al., Koen Yoshishu - Terupen Oyabi Seiyu Kagakuni Kansuru Toronkai, 23rd*, 19 (1979)

S. ringgoldianum.
STEROIDS: sargasterol, saringosterol.
Tsuda *et al., J. Am. Chem. Soc.*, **80**, 921 (1958)

S. sagamianum var. yezoense.
PLASTOQUINONES: sargahydroquinoic acid, yezoquinolide.
Segawa *et al., Chem. Lett.*, 1365 (1987)

S. serratifolium.
CHROMENE: sargachromenol.
QUINONES: sargaquinal, sargaquinoic acid.
Nozaki *et al., Koen Yoshishu - Terupen Oyabi Seiyu Kagakuni Kansuru Toronkai, 23rd*, 19 (1979)

S. thunbergii.
STEROID: 24-vinyloxycholesta-5,23-dien-3β-ol.
Kobayashi *et al., Chem. Pharm. Bull.*, **33**, 4012 (1985)

S. tortile.
SESQUITERPENE COUMARINS: sargaquinol, sargatetraol, sargatriol.
Ishitsuka *et al., Chem. Lett.*, 1269 (1979)

SARGENTIA

S. greggii.
FLAVONE: cerrosillin B.
Dominguez *et al., Rev. Latinoam. Quim.*, **7**, 41 (1976)

SARGENTODOXA

S. cuneata.
PHENOLICS: emodin, physcion.
STEROIDS: daucasterol, β-sitosterol.
Wang *et al., Zhongcaoyao*, **13**, 7 (1982); *Chem. Abstr.*, 97 107021n

SAROTHAMNUS (Leguminosae)

S. catalaunicus Webb.
LUPANE ALKALOIDS: catalaudesmine latalauverine, sarodesmine.
Faugeras, Paris, *Plant Med. Phytother.*, **5**, 134 (1971)

S. patens. (Cytisus striatus).
LUPANE ALKALOIDS: isocinevanine, 13-isovanilloyloxylupanine.
FLAVONE: chrysin 7-β-glucoside.
(A) Faugeras *et al., Ann. Pharm. Fr.*, **30**, 527 (1972)
(F) Brum-Bousquet *et al., Planta Med.*, **30**, 375 (1976)

S. scoparius.
SPARTEINE ALKALOID: sparteine.
FLAVONOIDS: 6-O-acetylscoparoside, genitoside, scoparoside.
(A) D'Aragona, *Boll. Chim. Farm.*, **87**, 240 (1948)
(F) Brum-Bousquet *et al., Lloydia*, **40**, 591 (1977)

SARRACENIA (Sarraceniaceae)

S. flava.
TRITERPENOID: betulinaldehyde.
Bhattacharyya *et al., Phytochemistry*, **15**, 431 (1976)

SASSAFRAS (Lauraceae)

S. albidum.
TERPENOID: apiol.
Sethi *et al., Phytochemistry*, **15**, 1773 (1976)

S. officinalis L.
MONOTERPENOID: 4-allyl-1,2-(methylenedioxy)-benzene.
Perkin *et al., J. Chem. Soc.*, 1663 (1927)

SATUREJA (Labiatae)

S. acinos.
FLAVONOID: naringenin 7-(6″-O-α-L-rhamnopyranosyl-β-D-glucopyranoside).
STEROID: β-sitosterol β-D-glucopyranoside.
TRITERPENOID: ursolic acid.
Escudero *et al., J. Nat. Prod.*, **48**, 128 (1985)

S. calamintha.
TRITERPENOIDS: calaminthadiol, isocalaminthadiol, ursolic acid, 3-epi-ursolic acid.
Giannetto *et al., Phytochemistry*, **18**, 1203 (1979)

S. douglasii.
MONOTERPENOIDS: camphene, camphor, carvone, isomenthone, menthone, pulegone.
Lincoln *et al., Biochem. Syst. Ecol.*, **4**, 237 (1976); *Chem. Abstr.*, 86 117614v

S. graeca.
TRITERPENOIDS: calaminthadiol, isocalaminthadiol.
Giannetto *et al., Phytochemistry*, **18**, 1203 (1979)

S. montana.
TRITERPENOIDS: crataegolic acid, oleanolic acid.
Escudero *et al., J. Nat. Prod.*, **48**, 128 (1985)

S. obovata.
TRITERPENOID: oleanolic acid.
Escudero *et al., J. Nat. Prod.*, **48**, 128 (1985)

SAURURUS

S. cernuus.
LIGNOIDS: manassantin A, manassantin B, saucerneol, saucernetin.
Rao, Alvarez, *Tetrahedron Lett.*, 4947 (1983)

SAUSSUREA (Compositae)

S. affinis.
SESQUITERPENOID: saussureolide.
Kalsi *et al., Phytochemistry*, **22**, 1993 (1983)

S. elegans (Ldb.) M.B.Spreng.
SESQUITERPENE ALKALOID: elegantine.
SESQUITERPENOID: saelin.
HALOGENATED SESQUITERPENOIDS: elegin, salegin.
(A) Sham'yanov *et al., Khim. Prir. Soedin.*, 819 (1976)
(A) Sham'yanov et el., *ibid.*, 865 (1976)
(T) Sham'yanov *et al., ibid.*, 258 (1980)

S. frolowii.
UNCHARACTERIZED TERPENOID: sosyurol.
Troshchenko, Kobrin, *Khim. Prir. Soedin.*, 256 (1965)

S. involucrata ar. et Kin.
FLAVONOIDS: quercetin, rutin, 5,7,4'-trihydroxy-6,3'-dimethoxyflavone, 5,7,4'-trihydroxy-6-methoxyflavone.
Jia *et al., Gaodeng Xuexiao Huaxue Xuebao*, **4**, 581 (1983); *Chem. Abstr.*, 100 32225b

S. laniceps.
ACETOPHENONE: p-hydroxyacetophenone.
COUMARINS: physcion, scopoletin, umbelliferone.
STEROID: β-sitosterol.
Zhao et al., Zhongcaoyao, **17**, 296 (1986); Chem. Abstr., 105 168932r

S. lappa Clarke.
NORSESQUITERPENOID:
(E)-9-isopropyl-6-methyldeca-5,9-dien-2-one.
SESQUITERPENOIDS: costic acid, α-costol, β-costol, (+)-costol, costunolide, isozaluzanin, isozaluzanin C, 12-methoxydihydrocostunolide, sausurrea lactone.
TRITERPENOIDS: α-amyrin stearate, β-amyrin stearate, lupeol.
Maurer et al., J. Chem. Soc., Chem. Commun., 353 (1977)
Kalsi et al., Phytochemistry, **22**, 1993 (1983)

S. neopulchella.
SESQUITERPENOID: saupirine.
Chugunov et al., Khim. Prir. Soedin., 727 (1971)

S. pulchella.
SESQUITERPENOIDS: saupirine, saurine.
Chugunov et al., Khim. Prir. Soedin., 727 (1971)

SAUTELLARIA
S. orientalis.
FLAVONE: 5,7-dihydroxyflavone-7-β-D-glucoside.
Asakawa et al., Bull. Chem. Soc., Japan, **44**, 2761 (1971)

SAXIFRAGA (Saxifragaceae)
S. aizoon Jacq.
ANTHOCYANINS: cyanidin 3-rhamnoside, cyanidin 3-xyloside.
FLAVONOIDS: kaempferol, quercetin, quercetin 3-rhamnoglucoside, quercetin 3-rhamnoside.
PHENOLICS: chlorogenic acid, ellagic acid.
Pawlowska, Acta Soc. Bot. Pol., **45**, 383 (1976); Chem. Abstr., 87 2369x

S. stolonifera Meerb.
ACIDS: capric acid, caprylic acid, isovaleric acid, linoleic acid, myristic acid, oleic acid, palmitic acid, stearic acid.
ALCOHOL: benzyl alcohol.
BENZOPYRAN: norbergenin.
MONOTERPENOIDS: citronellol, p-cresol, cumene, p-cymene, geraniol, geranyl acetate, linalool, α-pinene, α-terpineol, terpinolene.
(Ac, Al, T) Kameoka et al., Yukagaku, **25**, 490 (1976); Chem. Abstr., 85 119616a
(B) Tanayama et al., Phytochemistry, **22**, 1053 (1983)

SCABIOSA (Dipsacaceae)
S. atropurpurea.
GLYCOSIDE: swertisin.
Zemtsova et al., Khim. Prir. Soedin., 705 (1977)

S. comosa.
COUMARINS: bergapten, coumarin, umbelliprenin.
Dargaeva et al., Khim. Prir. Soedin., 387 (1976)

S. japonica Miq.
IRIDOIDS: cantleyoside, loganin, sweroside.
Endo et al., J. Pharm. Soc., Japan, **96**, 246 (1976)

S. ochroleuca.
ACIDS: caffeic acid, chlorogenic acid.
Zemtsova, Aktual. Vopr. Farm., **2**, 82 (1974); Chem. Abstr., 84 147618a

S. soongorica.
COUMARIN: angenomalin.
SAPONINS: songoroside M, songoroside O.
STEROID: β-sitosterol.

TRITERPENOID: oleanolic acid.
(C, Sap) Akimaliev et al., Khim. Prir. Soedin., 476 (1976)
(S, T) Akimaliev, ibid., 711 (1977)

SCAPANIA (Musci)
S. aequiloba.
SESQUITERPENE: annastreptene.
Andersen et al., Tetrahedron, **33**, 617 (1977)

S. aspera.
SESQUITERPENENE: anastreptene.
Andersen et al., Tetrahedron, **33**, 617 (1977)

S. nemorosa (L.) Dum.
SESQUITERPENE: anastreptene.
Andersen et al., Tetrahedron, **33**, 617 (1977)

S. paludosa.
SESQUITERPENE: anastreptene.
Andersen et al., Tetrahedron, **33**, 617 (1977)

S. undulata (L.) Dum.
SESQUITERPENOIDS: anastreptene, (+)-ent-epi-cubenol, (-)-longiborneol, (-)-longifolene, (-)-longipinanol, scapanin A, scapanin B.
Andersen et al., Tetrahedron, **33**, 617 (1977)
Huneck et al., J. Hattori Bot. Lab., **54**, 125 (1983); Chem. Abstr., 100 20398x

SCAVEOLA (Goodeniaceae)
S. lobelia (Th.) Murr.
COUMARINS: angenomalin, isoangenomalin.
SESQUITERPENOID: 19H,13(18)-dehydrogermanicol acetate.
(C) Bhan et al., Phytochemistry, **12**, 3010 (1973)
(C, T) Bohlmann et al., Chem. Ber., **108**, 433 (1975)

SCELETIUM
S. joubertii L.Bol.
SCELETIUM ALKALOIDS: dehydrojoubertiamine, dihydrojoubertiamine, joubertiamine, mesembrane, 3'-methoxy-4'-O-methyljoubertiamine.
Arndt, Kruger, Tetrahedron Lett., 3727 (1970)

S. namaquense.
SCELETIUM ALKALOIDS: N-acetyltortuosamine, mesembrine, 3'-methoxy-4'-O-methyljoubertiamine, sceletium alkaloid A-4.
Jeffs et al., J. Org. Chem., **47**, 3611 (1982)

S. strictum.
INDOLE ALKALOIDS: O-acetylmesembrenol, channaine.
Abou-Doma et al., J. Chem. Soc., Chem. Commun., 1078 (1978)

S. subvelutinum. Bol.
SCELETIUM ALKALOID: O-methyldihydrojoubertiamine.
Neuwenhuis et al., J. Chem. Soc., Perkin I, 284 (1981)

S. tortuosum.
QUINOLINE ALKALOID: tortuosamine.
Snyckers et al., J. Chem. Soc., Chem. Commun., 1467 (1971)

SCENEDESMUS (Chlorophyceae/Coelastraceae)
S. obliquus (Turpin) Kuetz.
CAROTENOID: loroxanthin.
STEROID: chondrillasterol.
(S) Bergmann, Feeny, J. Org. Chem., **15**, 812 (1950)
(T) Aitzetmuller et al., Phytochemistry, **8**, 1761 (1969)

SCEPTRIDIUM
S. ternatum var. ternatum.
BIBENZYL: ternatin.
Tanaka et al., Chem. Pharm. Bull., **34**, 3727 (1986)

SCHEFFLERA

S. capitata Harms.
SAPONIN: schemeroside.
STEROID: β-sitosterol β-D-glucoside.
TRITERPENOIDS: capitogenic acid, hederagenin
    3-O-α-L-arabinopyranoside, oleanolic acid
    3-O-α-L-arabinopyranoside.
    Jain *et al.*, *Indian J. Chem.*, **21B**, 622 (1982)

S. octophylla.
TRITERPENOID: 3,11-dihydroxy-20(29)-lupene-23,28-dioic acid.
    Lischewski *et al.*, *Phytochemistry*, **23**, 1695 (1984)

SCHELHAMMERA

S. multiflora.
HOMOERYTHRINA ALKALOID:
    3-epi-20,21-seco-schelhammericine.
    Sioumis, *Aust. J. Chem.*, **24**, 2737 (1971)

S. pedunculata. F.Muell.
HOMOERYTHRINA ALKALOIDS: schelhammericine,
    schelhammeridine, schelhammerine,
    3-epi-20,21-secoschelhammericine.
    Johns *et al.*, *Aust. J. Chem.*, **22**, 2219 (1969)
    Sioumis, *ibid.*, **24**, 2737 (1971)

S. undulata.
HOMOERYTHRINA ALKALOID:
    3-epi-20,21-secoschelhammericine.
    Sioumis, *Aust. J. Chem.*, **24**, 2737 (1971)

SCHINUS (Anacardiaceae)

S. kanjaoensis.
TRITERPENOID: barrigenol A-1.
    Cole *et al.*, *Chem. Ind. (London)*, 254 (1955)

S. mertensiana.
SAPONIN: desacetylboninsaponin A.
    Kitagawa *et al.*, *Chem. Pharm. Bull.*, **24**, 1260 (1976)

S. molle.
SESQUITERPENOID: β-spathulene.
TRITERPENOIDS: isomasticadienonic acid, isomasticadienonalic
    acid, 3-epi-masticadienolalic acid.
    Piozzo-Balbi *et al.*, *Phytochemistry*, **17**, 2107 (1978)

S. terebinthifolius.
TRITERPENOIDS: baurenone, masticadienonic acid, schinol,
    terebinthifolic acid.
    De Paivo Campello, Marsaiol, *Phytochemistry*, **14**, 2300 (1975)

S. terebinthus.
MONOTERPENOIDS: 3-carene, p-cymene, limonene,
    α-phellandrene, β-phellandrene, α-pinene, β-pinene,
    sabinene, terpinolene.
SESQUITERPENOIDS: carvotanacetone, caryophyllene,
    cis-sabinene.
TRITERPENOIDS: α-amyrin, α-amyrinone, simiarenol,
    simiarenone.
    Lloyd *et al.*, *Phytochemistry*, **16**, 1301 (1977)

SCHISANDRA (Schizandra) (Schisandraceae)

S. chinensis Baill.
CYCLOOCTANES: wuweizichu B, wuweizisu A, wuweizisu B,
    wuweizisu C.
LIGNANS: gomisin, gomisin A, gomisin B, gomisin C,
    pregomisin.
SESQUITERPENOIDS: chamigrenal, (-)-chamigrene.

DITERPENOIDS: gomisin R, pregomisin, schisandrin,
    sesquicarene.
    (Cy) Kochetkov *et al.*, *J. Gen. Chem., USSR.*, **31**, 3454 (1961)
    (L) Ikeya *et al.*, *Chem. Pharm. Bull.*, **26**, 682 (1978)
    (T) Ikeya *et al.*, *Chem. Pharm. Bull.*, **30**, 3207 (1982)

S. grandiflora.
TRITERPENOID: schizandroflorin.
    Talapatra *et al.*, *Indian J. Chem.*, **21B**, 76 (1982)

S. henryi.
LIGNAN: enshizhisu.
DITERPENOID: enschicine.
    Liu *et al.*, *Phytochemistry*, **23**, 1143 (1984)

S. nigra.
SESQUITERPENOID: schizandronol.
TRITERPENOIDS: schizandronic acid, schizanlactones A and B.
    Takahashi, Takana, *Chem. Pharm. Bull.*, **24**, 2000 (1976)

S. rubriflora Rehd. et Wils.
LIGNANS: deoxyschizandrin, meso-dihydroguaiaretic acid,
    gomisin O, (-)-rubschizandrin, rubschizantherin, pregomisin,
    schizanhenol, schizanhenol acetate, schizanhenol B,
    wuweizisu C.
    Wang *et al.*, *Yaoxue Xuebao*, **20**, 832 (1985); *Chem. Abstr.*, 105 3521t

SCHISTOSTEPHIUM

S. heptalobum.
SESQUITERPENOID: 1,3-dihydroxyarbusculin B.
    Bohlmann *et al.*, *Phytochemistry*, **22**, 1623 (1983)

SCHIZANTHUS

S. hookeri.
TROPANE ALKALOIDS: (-)-6β-angeloyloxytropan-3α-ol,
    (-)-hygroline, (+)-pseudohygrine,
    9-0-3α-senecioyloxytropan-6β-ol,
    (-)-6β-tigloyloxytropan-3α-ol, (3R,6R)-tropan-3α,6β-diol.
    Gambaro *et al.*, *Phytochemistry*, **22**, 1838 (1983)

SCHIZONEPETA

S. tenuifolia Briq.
FLAVONOIDS: apigenin 7-O-D-glucoside, luteolin
    7-O-β-D-glucoside.
Monoterpenoids schizanepetosides A, B and C.
    Kubo *et al.*, *Chem. Pharm. Bull.*, **34**, 3097 (1986)

SCHIZOPELTE (Lichenes)

S. caffaeoides (Boj.) Baill.
CARBOLINE ALKALOIDS: caffaeoschizin, isoschizogaline,
    isoschizogamine, schizogamine, schizolutine,
    schizophylline, schizozygine, α-schizozygol, β-schizozygol,
    tabernoschizine.
    Tenner, Karnweisz, *Experientia*, **19**, 244 (1963)

SCHIZOZYGIA

S. caffaeoidea.
CARBOLINE ALKALOIDS: schigamine, schizogaline,
    schizophylline, schizozygine.
    Renner, Kernweisz, *Experientia*, **19**, 244 (1963)

SCHKUHRIA (Compositae)

S. multiflora.
DITERPENOID: schkuhrianol.
    Bohlmann *et al.*, *Phytochemistry*, **19**, 881 (1980)

S. pinnata.
SESQUITERPENOIDS: schkuhrin I, schkuhrin II.
    Boltmann *et al.*, *Phytochemistry*, **19**, 881 (1980)

S. schkuhrioides.
SESQUITERPENOIDS: 11,13-dehydroeriolin, elemanschkuhriolide, frutescin, schkuhrioidin, schkuhriolide.
Delgado *et al., J. Org. Chem.,* **49,** 2994 (1984)

SCHLEICHERA
S. oleosa.
FLAVONOID: scopoletin.
STEROID: β-sitosterol, Triterpenoids betulin, betulinic acid, lupeol, lupeol acetate.
Dan *et al., Fitoterapia,* **57,** 445 (1986)
S. trijuga Willd.
STEROIDS: brassicasterol, campesterol, cholesterol, β-sitosterol, stigmasterol.
Sharma *et al., Collect. Czech. Chem. Commun.,* **42,** 3487 (1977)

SCHOENOCAULON
S. officinale A.Gray.
STEROIDAL ALKALOIDS: cevadine, sabadinine.
UNCHARACTERIZED ALKALOID: sabadine.
Hennig *et al., J. Am. Pharm. Assoc.,* **40,** 168 (1951)

SCHOTIA
S. brachypetala.
PHENOLICS:
(E)-1-(3,5-dihydroxyphenyl)-2-(3,4,5-trihydroxyphenyl)-ethylene,
(Z)-1-(3,5-dihydroxyphenyl)-2-(3,4,5-trihydroxyphenyl)-ethylene.
STILBENES: (E)-3,3′,4,5,5′-pentahydroxystilbene, (Z)-3,3′,4,5,5′-pentahydroxystilbene.
King *et al., J. Chem. Soc.,* 4477 (1956)

SCHUMMANIOPHYTON
S. problematicum.
ALKALOID: schummaniophytine.
Schittler, Spiteller, *Tetrahedron Lett.,* 2911 (1978)

SCIADOPITYS
S. verticillata.
FLAVONE: sciadopitysin.
SESQUITERPENE: 2,5-diepi-cedrene.
DITERPENOIDS:
15,16-epoxy-20-oxo-8(17),13(16),14-labdatrien-19-oic acid methyl ester, (-)-isophyllocladene, methyl trans-sciadpate, sciadin, sciadinone, verticillol.
(F) Beckmann *et al., Phytochemistry,* **10,** 2465 (1971)
(T) Hasegawa *et al., Chem. Lett.,* 1 (1983)

SCIADOTENIA
S. toxifera.
BISBENZYLISOQUINOLINE ALKALOIDS: sciadoferine, sciadoline.
Takahashi *et al., Heterocycles,* **5,** 367 (1976)

SCIRPUS (Cyperaceae)
S. fluviatilis.
TERPENOIDS: scirpusins A and B.
Nakajima *et al., Chem. Pharm. Bull.,* **26,** 3050 (1978)

SCLERODERMA (Basidiomycetae)
S. aurantia.
TRITERPENOID: lanost-8,24-diene-3,23-diol.
Entwhistle, Pratt, *Tetrahedron,* **24,** 3949 (1968)

SCLEROTINA
S. fruticola.
SESQUITERPENOID: sclerosporal.
TRITERPENOID: 10,11-epoxysqualene.
Katayama *et al., Tetrahedron Lett.,* 1773 (1979)

SCOPARIA (Scrophulariaceae)
S. dulcis L.
ACID: dulcicoic acid.
DITERPENOIDS: scopadulcic acid A, scopadulcic acid B, scoparic acid A.
(Ac) Mahato *et al., Phytochemistry,* **20,** 171 (1981)
(T) Hayashi *et al., Tetrahedron Lett.,* 3693 (1987)

SCOPOLIA (Solanaceae)
S. carnoilica Jacq.
TROPANE ALKALOIDS: atropine, (+)-hyoscyamine, hyoscine, scopolamine, 3-tigloyltropane, tropane.
Marion, Thomas, *Can. J. Chem.,* **33,** 1853 (1955)
Nikolic *et al., Acta Pharm. Jugosl.,* **26,** 257 (1976); *Chem. Abstr.,* 86 13807u
S. japonica.
TROPANE ALKALOIDS: (-)-hyoscyamine, norhyoscyamine.
Marion, Spenser, *Can. J. Chem.,* **32,** 1116 (1954)
S. livida.
TROPANE ALKALOID: scopine.
Willstatter *et al., Ber. Dtsch. Chem. Ges.,* **56,** 1079 (1923)
S. lurida.
TROPANE ALKALOID: (-)-hyoscyamine.
Marion, Thomas, *Can. J. Chem.,* **22,** 1853 (1955)
S. tangutica Maxim.
TROPANE ALKALOIDS: atropine, cuskhygrine, (-)-hyoscyamine, scopolamine.
Gadamer, *Arch. Pharm.,* **239,** 294 (1901)

SCROPHULARIA (Scrophulariaceae)
S. lateriflora.
IRIDOID: laterioside.
Siatek *et al., Pharm. Acta Helv.,* **56,** 37 (1981)
S. smithii.
One triterpenoid.
Breton, Gonzalez, *J. Chem. Soc.,* 1401 (1963)
S. vernalis.
IRIDOIDS: harpagide 8-acetate, harpagoside, 6-methylcatalpol.
PHENOLICS: caffeic acid, p-coumaric acid, ferulic acid, p-hydroxybenzoic acid, 1-hydroxyphenylacetic acid, neoferulic acid, vanillic acid.
Swiatek *et al., Acta Pol. Pharm.,* **33,** 658 (1977); *Chem. Abstr.,* 87 180763d

SCUTELLARIA (Labiatae)
S. altissima.
FLAVONOIDS: baicalein, altisin, oroxylin, oroxyloside, wogonin, wogonoside.
TERPENOID GLUCOSIDES: scutellarioside I, scutellarioside II.
(F) Beshko *et al., Khim. Prir. Soedin.,* 514 (1975)
(T) Weinges *et al., Justus Liebigs Ann. Chem.,* 2190 (1975)
S. araxensis Grossh.
AMINO ACIDS: alanine, arginine, asparagine, aspartic acid, glutamic acid, glycine, histidine, hydroxyproline, isoleucine, leucine, lysine, methionine, proline, serine, threonine.
Akhmed-Zade *et al., Chem. Abstr.,* 86 103116x

S. baicalensis.
FLAVONES: carthamidin, isocarthamidin, isowogonin, oroxylin A, wogonin.
Takido *et al., J. Pharm. Soc., Japan,* **96**, 381 (1976)

S. granulosa.
FLAVONOIDS: 2'-methoxychrysin, 2'-methoxychrysin 7-β-D-glucuronide.
Litvenenko *et al., Khim. Prir. Soedin.,* 730 (1976)

S. irzewalskii.
FLAVONE: hispidulin.
Doherty *et al., J. Chem. Soc.,* 5577 (1963)

S. karjagini.
FLAVONOIDS: apigenin, luteolin 7-β-glucuronide, scutellarin 7-β-glucuronide.
Glyzin *et al., Khim. Prir. Soedin.,* 98 (1975)

S. oreophila.
FLAVONOIDS: baicalein, baicalin, chrysin, cynaroside, luteolin.
Nasundan, *Khim. Prir. Soedin.,* 805 (1975)

S. orientalis.
FLAVONOIDS: apigenin, chrysin, chrysin 7-β-D-glucoside, chrysin 7-β-glucuronide, scutellarin 7-β-glucuronide.
Glyzin *et al., Khim. Prir. Soedin.,* 98 (1975)

S. prilipkoana Grossh.
AMINO ACIDS: alanine, arginine, aspartic acid, glutamic acid, glycine, hydroxyproline, isoleucine, leucine, lysine, methionine, proline, serine, threonine.
Akhmed-Zade *et al., Chem. Abstr.,* 86 103116x

S. scordiifolia Fisch.
FLAVONOIDS: apigenin, apigenin glucuronide, baicalein, baicalein glucuronide, chrysin, chrysin 7-β-D-glucuronide, luteolin, luteolin glucuronide, oroxylin, oroxylin glucuronide, scutellarin, scutellarin glucuronide, wogonin, wogonin glucuronide.
Popova *et al., Farm. Zh. (Kiev),* **31**, 89 (1976); *Chem. Abstr.,* 85 90145h

S. smithii.
TRITERPENOIDS: oleana-11,13(18)-diene-3,22,28-triol, smithiandienol.
Breton, Gonzalez, *J. Chem. Soc.,* 1401 (1963)

SCUTIA
S. buxifolia Reiss.
PEPTIDE ALKALOIDS: aralionine, hysodricanine A, mauritine H, scutianine, scutianines A, B, C, D, E, F, G and H.
Tschesche *et al., Phytochemistry,* **16**, 1025 (1977)
Tschesche *et al., ibid.,* **16**, 1817 (1977)
Morel *et al., ibid.,* **18**, 473 (1979)

SEBASTIANA
S. pavonia.
TRITERPENOID: epi-β-amyrin.
Reyes *et al., Rev. Latinoam. Quim.,* **14**, 67 (1983)

SECHIUM
S. edule.
AMINO ACIDS: α-amino-τ-ureidobutyric acid, m-carboxyphenylalanine, citrulline.
Inatomi, Adachi, *Meiji Daigaku Nogakubu Kenkyu Hokoku,* **34**, 1 (1975); *Chem. Abstr.,* 84 71478g

SECURIDACA
S. diversifolia.
Twelve kaempferol and quercetin glycosides.
Hamburger *et al., Phytochemistry,* **24**, 2689 (1985)

SECURINEGA (Euphorbiaceae)
S. suffruticosa (Pall.) Rehd.
ALKALOIDS: securinegine, (-)-securinine, securinols A, B and C, securitinine, suffruticodine, suffruticonine.
Shabana *et al., Arch. Pharm. Chem. Sci. Ed.,* **7**, 158 (1979)

S. virosa Pax. et Hoffm.
ALKALOIDS: virosecurianine, virosecurinine.
Nakano *et al., Chem. Ind.,* 1651 (1962)

SEDUM (Crassulaceae)
S. acre L.
PIPERIDINE ALKALOIDS: sedacrine, sedacryptine, sedamine, sedinone, sedridine.
PYRIDINE ALKALOID: nicotine.
Colau, Hootele, *Can. J. Chem.,* **61**, 470 (1983)

S. caucasicum.
Twenty flavonoids and phenolics.
Zaitsev *et al., Khim. Prir. Soedin.,* 527 (1983)

S. ewersii.
COUMARINS: 6,7-dioxycoumarin, umbelliferone.
FLAVONOIDS: kaempferol, kaempferol 4-O-β-D-glucopyranoside, kaempferol 7-O-α-L-rhamnopyranoside, quercetin, rutin.
Krasnov *et al., Khim. Prir. Soedin.,* 389 (1976)

S. sarmentosum.
PIPERIDINE ALKALOID: (+)-methylallosedridine.
Bayerman *et al., Recl. Trav. Chim. Pays-Bas* **91**, 1441 (1972)

SELINUM (Umbelliferae)
S. carvifolium (L.) L.
MONOTERPENOIDS:
5-(2-acetoxymethyl)-2-butenoyloxy-3,6,6-trimethyl-1,3-cyclohexadiene-1-carboxylic acid,
2-formyl-1,1,5-trimethylcyclohexa-2,4-dien-6-ol,
2-formyl-1,1,5-trimethylcyclohexa-2,5-dien-4-ol,
2-formyl-1,1,5-trimethylcyclohexa-2,4-dien-6-one,
2-formyl-1,1,5-trimethylcyclohexa-2,5-dien-4-one.
Bohlmann, Zdero, *Chem. Ber.,* **102**, 2211 (1969)
Lemmich *et al., Acta Chem. Scand.,* **25**, 995 (1971)

S. vaginatum.
BENZOPYRANS: salinone, selinidin.
COUMARINS: vaginidin, vaginidiol, vaginol.
DITERPENOID: vaginatin.
Rajendran *et al., Indian J. Chem.,* **16B**, 4 (1978)

SEMECARPUS (Anacardiaceae)
S. anacardium L.
FLAVONOIDS: galluflavanone, tetrahydroamentoflavone, tetrahydrorobustaflavone.
Murthy *et al., Phytochemistry,* **22**, 2636 (1983)

S. kurzii.
ACID: isoricinoleic acid.
FLAVONE: semecarpetin.
Alam *et al., Chem. Ind. (London)*

S. vitiensis.
FLAVONOIDS: kaempferide, kaempferide 7-O-β-diglucoside, kaempferide 7-O-β-glucoside, kaempferol, kaempferol 7-O-β-diglucoside, kaempferol 7-O-β-glucoside, myricetin, myricetin 7-0-β-diglucoside, myricetin 7-O-β-glucoside, quercetin, quercetin 7-O-β-diglucoside, quercetin 7-O-β-glucoside.
PHENOLICS: gallic acid, 3-(n-pentadecenyl)phenol, 3-(n-pentadecenyl)-pyrocatechol, pyrocatechol.
Pramono *et al., Plant Med. Phytother.,* **19**, 159 (1985)

SEMELE

S. androgyna L.
STEROIDS: androgenin A, androgenin B,
25S-dihydrodracogenin, 27-hydroxyruscogenin,
isoandrogenin A, isoandrogenin B.
Gonzalez et al., Phytochemistry, 14, 2257 (1975)

SEMENOVIA (Pastinaceae)

S. rubtazovii.
COUMARINS: bergapten, imperatorin, isobergapten,
isopimpinellin, pimpinellin, sphondin, umbelliferone.
Gusak et al., Chem. Abstr., 86 68378k

S. transiliensis.
COUMARINS: bergapten, imperatorin, isobergapten,
isopimpinellin, pimpinellin, sphondin, umbelliferone.
Gusak et al., Chem. Abstr., 86 68378k

SEMPERVIVUM (Crassulaceae)

S. ruthenicum.
FLAVONOID: scutellarein 7-rutinoside.
Gumenyuk, Khim. Prir. Soedin., 428 (1975)

SENECIO (Compositae)

S. abrotanifolius.
SESQUITERPENOIDS: abrotanifolone, 5,8-bis-(angeloyloxy)-2,3
10,11-diepoxy-2,3,10,11-tetrahydro-α-bisabol-4-one.
Bohlmann et al., Chem. Ber., 109, 2014 (1976)

S. adnatus DC.
PYRROLIZIDINE ALKALOID: platyphylline.
Orekhov et al., Ber. Dtsch. Chem. Ges., 68, 653 (1935)

S. alpinus.
PYRROLIZIDINE ALKALOID: seneciphylline.
Konovalova et al., Bull. Soc. Chim. Fr., 4, 2037 (1937)

S. amphibolus C.Koch.
PYRROLIZIDINE ALKALOID: macrophylline.
Danilov et al., J. Gen. Chem., USSR., 23, 1417 (1953)
Danilov et al., ibid., 23, 1597 (1953)

S. angulatus.
PYRROLIZIDINE ALKALOID: angulatine.
Porter, Geissman, J. Org. Chem., 27, 4132 (1962)

S. anteuphorbium.
SESQUITERPENOID:
5-acetoxy-2-angeloyloxy-8-hydroxypresilphiperfolane.
Bohlmann et al., Phytochemistry, 21, 1331 (1982)

S. aquaticus Hill.
PYRROLIZIDINE ALKALOID: aquaticine.
Evans, Evans, Nature, 164, 30 (1949)

S. argentino.
PYRROLIZIDINE ALKALOIDS: retrorsine, senecionine.
Pestchanker, Giordano, J. Nat. Prod., 49, 722 (1986)

S. aureus.
PYRROLIZIDINE ALKALOIDS: floridanine, florosenine, otosenine,
senecionine.
Roeder et al., Planta Med., 49, 57 (1983)

S. auricula.
PYRROLIZIDINE ALKALOID: neosenkirkine.
Martin Panizo, Rodriguez, An. Quim., 70, 1043 (1974)

S. borysthenicus Andr. (S. praealthus Bertol.).
PYRROLIZIDINE ALKALOID: seneciphylline.
Konovalova et al., Bull. Soc. Chim. Fr., 4, 2037 (1937)

S. brasiliensis.
PYRROLIZIDINE ALKALOID: brasilinecine.
De Camargo Fonseca, An. Fac. Farm. Odontol. Univ. San Paulo, 9, 85 (1951)

S. campestris var. maritima.
PYRROLIZIDINE ALKALOID: campestrine.
Blackie, Pharm. J., 138, 102 (1937)

S. cannabifolius.
PYRROLIZIDINE ALKALOID: senecicannabine.
Asada et al., Tetrahedron Lett., 189 (1982)

S. carthamoides.
PYRROLIZIDINE ALKALOID: carthamoidine.
Adams et al., J. Am. Chem. Soc., 64, 571 (1942)

S. cathcariensis.
SESQUITERPENOIDS: seneremophilondiol
3-angelate-9-senecionate, seneremophilondiol diangelate.
Bohlmann, Zdero, Phytochemistry, 17, 1337 (1978)

S. cineraria DC.
PYRROLIZIDINE ALKALOIDS: jacobine, seneciphylline.
Konovalova, Orekhov, Bull. Soc. Chim. Fr., 4, 2037 (1937)

S. cissampelinum.
PYRROLIZIDINE ALKALOIDS: senampelines A, B, C and D.
Bohlmann et al., Chem. Ber., 110, 474 (1977)

S. clevelandii.
FURANONE: cleviolide.
Bohlmann et al., Phytochemistry, 21, 2425 (1982)

S. crassissimum.
SESQUITERPENOID: senecrassidiol.
Bohlmann et al., Phytochemistry, 20, 468 (1981)

S. crispus.
SESQUITERPENOID: 3-acetoxy-14-(angeloyloxy)-cacalohastin.
Bohlmann, Zdero, Chem. Ber., 111, 3140 (1978)

S. crucifolia.
PYRROLIZIDINE ALKALOID: jacobine.
Geissman et al., Aust. J. Chem., 12, 1247 (1959)

S. cymosus.
FLAVONOIDS: kaempferol 3-O-robinoside-7-O-rhamnoside,
quercetin 3-O-galactoside.
Reyes et al., Rev. Latinoam. Quim., 8, 134 (1977)

S. digitalifolius.
NORSESQUITERPENOIDS: senedigitalene, ar-senedigitalene
Bohlmann et al., Phytochemistry, 17, 759 (1978)

S. doronicum.
PYRROLIZIDINE ALKALOID: doronine.
Roder et al., Phytochemistry, 19, 1275 (1980)

S. elegans.
SESQUITERPENOIDS: 9-acetoxy-1,10-didehydrofukinone,
6β-isovaleryloxy-9-oxo-10αH-furanoeremophilane,
6β-senecioyloxy-4,5-didehydro-5,6-secoeuryopsin.
Bohlmann, Zdero, Chem. Ber., 107, 2912 (1974)

S. erraticus Berthol.
PYRROLIZIDINE ALKALOIDS: erucifoline, floridanine,
(+)-otosenine, senecionine.
Sedmera et al., Collect. Czech. Chem. Commun., 37, 4112 (1972)

S. filaginoides.
PYRROLIZIDINE ALKALOID: retrorsine, senecionine.
SESQUITERPENOID: 6-acetoxy-10αH-furanoeremophil-1-one.
(A) Pestchanker, Giordano, J. Nat. Prod., 49, 722 (1986)
(T) de Salmeron et al., Planta Med., 47, 221 (1983)

S. fistulosus.
FLAVONOID: kaempferol 3-O-robinoside-7-O-rhamnoside.
SESQUITERPENOID:
4α-hydroxy-6β-angeloyloxy-10β-acetoxy-9-
oxofuranoeremophilane.
(F) Reyes et al., Rev. Latinoam. Quim., 8, 134 (1977)
(T) Villarro, Torres, J. Nat. Prod., 48, 841 (1985)

S. franchetii C.Winkl.
PYRROLIZIDINE ALKALOIDS: franchetine, sarracine N-oxide.
Akramov et al., Khim. Prir. Soedin., 351 (1967)

S. fuchsii.
PYRROLIZIDINE ALKALOID: fuchsisenecionine.
Muller, Chem. Zentralbl., II, 1049 (1925)

S. fulgens (Hook.f.) Nicholson. (Kleinia fulgens Hook.f.)
MONOTERPENOID:
3,4-epoxy-7(14),10-bisaboladiene-1,2-diangelate 8-acetate.
Bohlmann et al., Phytochemistry, 20, 469 (1981)

S. gerardii.
SESQUITERPENOIDS: seneremophilondiol
3-angelate-9-senecioate, seneremophilondiol diangelate,
seneremophilondiol disenecionate, seneremophilondiol
3-tiglate-9-senecioate.
Bohlmann, Zdero, Phytochemistry, 17, 1337 (1978)

S. gillesiano.
PYRROLIZIDINE ALKALOIDS: retrorsine, senecionine.
NORSESQUITERPENOIDS: 1-hydroxyplatyphyllide, platyphyllide.
(A) Pestchanker, Giordano, J. Nat. Prod., 49, 722 (1986)
(T) Guidugli et al., Phytochemistry, 25, 1923 (1986)

S. glabellum.
PYRROLIZIDINE ALKALOIDS: integerrimine, integerrimine
N-oxide, senecionine, senecionine N-oxide.
Ray et al., Phytochemistry, 26, 2431 (1987)

S. glanduloso-pisolus.
SESQUITERPENOID:
1-hydroperoxy-4(15),5,10(14)-germacratriene.
Bohlmann et al., Phytochemistry, 21, 2595 (1982)

S. glandulosus.
PYRROLIZIDINE ALKALOIDS: retrorsine, senecionine.
Pestchanker, Giordano, J. Nat. Prod., 49, 722 (1986)

S. glastifolius.
PYRROLIZIDINE ALKALOID: graminifoline.
SESQUITERPENOIDS: 1,10-epoxyfuranoeremophilane,
9-oxo-6β-senecioyloxy-1β,10-epoxyfuranoeremophilane.
(A) de Waal, Onderstepoort J., 16, 149 (1947)
(T) Bohlmann, Zdero, Chem. Ber., 107, 2912 (1974)

S. grandiflorus.
SESQUITERPENOID:
6α-angeloyloxy-9,10-dehydrofuranoeremophil-1-one.
Bohlmann, Zdero, Phytochemistry, 17, 1161 (1978)

S. halminifolius.
SESQUITERPENOID:
6α-angeloyloxy-9,10-dehydrofuranoeremophil-1-one.
Bohlmann, Zdero, Phytochemistry, 17, 1161 (1978)

S. hieracioides.
SESQUITERPENOIDS: 7(11),9-eremophiladien-12,8-olide,
8α-hydroxy-7(11),9-eremophiladien-12,8-olide.
Bohlmann et al., J. Nat. Prod., 47, 718 (1984)

S. hygrophylus Dyer et Sm.
PYRROLIZIDINE ALKALOIDS: hygrophylline, platyphylline.
Orekhov et al., Ber. Dtsch. Chem. Ges., 68, 1886 (1935)

S. ilicifolius Thunb.
PYRROLIZIDINE ALKALOIDS: pterophine, senecionine.
de Waal, Onderstepoort J., 16, 158 (1941)

S. illinitus.
PYRROLIZIDINE ALKALOIDS: acetylsenkirkine, senecionine,
senkirkine.
Gonzalez et al., Planta Med., 160 (1986)

S. inaequidens.
SESQUITERPENOIDS:
4-hydroxy-3-methoxy-1-oxomethyl-2,3-dehydrocacalol,
14-hydroxy-3-methoxy-1-oxomethyl-2,3-dehydrocacalol,
14-isovaleryloxy-6-methyl-1,2-dehydrocacalol.
Bohlmann, Zdero, Phytochemistry, 17, 1161 (1978)

S. incanus.
PYRROLIZIDINE ALKALOID: seneciphylline.
Konovalova et al., Bull.Soc. Chim. Fr., 4, 2037 (1937)

S. inortus.
SESQUITERPENOID: 14-acetoxycacalol propionate.
Bohlmann, Zdero, Phytochemistry, 17, 1161 (1978)

S. integerrimus Nutt.
PYRROLIZIDINE ALKALOIDS: integerrimine, senecionine,
Manske, Can. J. Res., 17B, 1, 8 (1939)

S. isatideus DC.
PYRROLIZIDINE ALKALOIDS: isatidine, retrorsine.
Blackie, Pharm. J., 138, 102 (1937)

S. jacobaea L.
PYRROLIZIDINE ALKALOIDS: jacobine, jacoline, jaconine,
jacozine, otosenine, renardine.
SESQUITERPENOID:
6-angeloyloxy-9,10-dehydrofuranoeremophil-1-one.
(A) Konovalova et al., Bull. Soc. Chim. Fr., 4, 2037 (1937)
(T) Bohlmann, Zdero, Phytochemistry, 17, 1161 (1978)

S. jacquinianus.
SESQUITERPENOID:
6-angeloyloxy-9,10-dehydrofuranoeremophil-1-one.
Bohlmann, Zdero, Phytochemistry, 17, 1161 (1978)

S. kirkii Hook.
PYRROLIZIDINE ALKALOIDS: acetylsenkirkine, senkirkine.
Briggs et al., J. Chem. Soc., 1891 (1978)

S. kubensis Grossh.
PYRROLIZIDINE ALKALOID: seneciphylline.
Konovalova, Orekhov, Bull. Soc. Chim. Fr., 4, 2037 (1937)

S. lanceus.
SESQUITERPENOID:
6α-senecioyloxy-1,10-dehydrofuranoeremophil-9-one.
Bohlmann et al., Chem. Ber., 110, 474 (1977)

S. latifolius.
PYRROLIZIDINE ALKALOIDS: senecifolidine, senecifoline.
Watt, J. Chem. Soc., 95, 466 (1909)
Watt, ibid., 95, 469 (1909)

S. leucostachys.
PYRROLIZIDINE ALKALOIDS: retrorsine, senecionine.
Pestchanker, Giordano, J. Nat. Prod., 49, 722 (1986)

S. longibolus.
PYRROLIZIDINE ALKALOIDS: integerrimine, integerrimine
N-oxide, longiboline, retronecanol, retrorsine, retrorsine
N-oxide, riddelline, senecionine, senecionine N-oxide,
seneciphylline, seneciphylline N-oxide.
Manske, Can. J. Res., 17B, 8 (1939)
Ray et al., Phytochemistry, 26, 2431 (1987)

S. lydenburgensis.
SESQUITERPENOIDS:
14-acetoxy-9-propanoyl-2α,14-dihydrocacalol,
3-acetoxy-9-propanoyl-14-angelyl-3α,14-dihydrocacalol,
2,14-diacetoxy-9-propanoyl-2α,14-dihydrocacalol,
2,14-diacetoxy-9-propanoyl-2β,14-dihydrocacalol,
2-(3-methylbutanoyl)-14-acetoxy-9-propanoyl-2α,14-
dihydrocacalol,
14-(2-methylpropanoyl)-9-propanoyl-2α,14-dihydrocacalol.
Bohlmann et al., Phytochemistry, 21, 681 (1982)

S. macrophyllus MB.
PYRROLIZIDINE ALKALOID: macrophylline.
Danilova et al., J. Gen. Chem., USSR., **23**, 1417 (1953)
Danilove et al., ibid., **23**, 1597 (1953)

S. macrospermus.
SESQUITERPENOIDS: 14-angeloyloxy-cacahastin-3β-ol,
14-angeloyloxydehydrocacalohastin.
Bohlmann, Zdero, Chem. Ber., **111**, 3140 (1978)

S. macrotis.
Six sesquiterpenoids.
Bohlmann et al., Phytochemistry, **20**, 1155 (1981)

S. mauriei.
SESQUITERPENOIDS: 2,3-desoxohillardinol
isobutyrate-12-methacrylate, hillardinol isobutyrate,
hillardinol 2′-methacrylate, senmauricinol isobutyrate,
senmauricinol 2′-methacrylate.
Bohlmann, Zdero, Phytochemistry, **17**, 1333 (1978)

S. medley-woodii.
SESQUITERPENOIDS:
3β-acetoxy-6β-(2-methylbutyryloxy)-10βH-
furanoeremophilane,
3β-acetoxy-6-(2-methylbutyryloxy)-10βH-furanoeremophil-
9-one.
Bohlmann, Zdero, Phytochemistry, **17**, 1161 (1978)

S. microglossus.
SESQUITERPENOID: 4(15)-eudesmene-1,6-diol.
Bohlmann et al., Phytochemistry, **22**, 1675 (1983)

S. mikanoides (Walp.) Otto.
PYRROLIZIDINE ALKALOID: mikanoidine.
Manske, Can. J. Res., **14**, 8 (1936)

S. nemorensis.
ACIDS: chlorogenic acid, fumaric acid.
FLAVONOIDS: quercetin, rutin.
PHENOLIC: pyrogallol.
SESQUITERPENOIDS: nemosenins A, B, C and D, senemorin.
(A, F, P) Nguyen Thi Nhia, Farmatsiya (Sofia), **25**, 39 (1975); Chem.
Abstr., 84 86738r
(T) Novotny et al., Collect. Czech. Chem. Commun., **38**, 739 (1973)

S. nemorensis var. bulgaricus.
SESQUITERPENOIDS: 3-oxo-8α-eremophila-1,7-dien-8,12-olide,
3-oxo-8α-ethoxyeremophila-1,7-dien-8,12-olide,
3-oxo-8α-hydroxyeremophila-1,7-dien-8,12-olide,
3-oxo-8α-methoxyeremophila-1,7-dien-8,12-olide.
Jizba et al., Collect. Czech. Chem. Commun., **43**, 1113 (1978)

S. nemorensis L. var. subdecurrens Griseb.
PYRROLIZIDINE ALKALOIDS: nemorensine, oxynemorensine.
Jizba et al., Collect. Czech. Chem. Commun., **46**, 1048 (1981)

S. othonnae MB.
PYRROLIZIDINE ALKALOIDS: floridanine, onetine, otosenine,
seneciphylline.
Danilova et al., J. Gen. Chem., USSR, **32**, 647 (1962)

S. othonniformis.
PYRROLIZIDINE ALKALOIDS: bisline, isoline.
Concourakis et al., J. Chem. Soc., C., 2312 (1970)

S. otiles.
FLAVONOIDS: kaempferol 7-O-glucoside, kaempferol
3-O-robinoside-7-O-rhamnoside, quercetin 3-O-galactoside.
Reyes et al., Rev. Latinoam. Quim., **8**, 134 (1977)

S. oxyodontus.
SESQUITERPENOIDS: 3-O-angeloyl-2,5-dehydrosenecioodontol,
8-angeloyloxy-3,4-dihydro-3,4-epoxy-bisabolen-5-one,
3-O-angeloyloxysenecioodontol,
3-O-methyl-2,5-senecioodontol, senoxyden.
Bohlmann, Zdero, Phytochemistry, **17**, 1591 (1978)

S. oxyriifolius.
One uncharacterized sesquiterpenoid.
Bohlmann, Zdero, Phytochemistry, **17**, 1669 (1978)

S. palmatus Pall.
PYRROLIZIDINE ALKALOID: seneciphylline.
Konovalova, Orekhov, Bull. Soc. Chim. Fr., **4**, 2037 (1937)

S. paludaffinis.
SESQUITERPENOIDS: senedigitalene,
senedigitalen-1,4-endo-peroxide.
Bohlmann et al., Phytochemistry, **17**, 2134 (1978)

S. panduriformis.
SESQUITERPENOIDS: 3-acetoxycacalohastin, 13-dehydromaturin,
4β-isobutyryloxy-1α-senecioyloxy-10αH-furanoeremophil-9-
one, isomaturinin, maturin acetate,
1-senecioyloxy-4-isobutyryloxy-10βH-furanoeremophil-9-
one.
Bohlmann et al., Chem. Ber., **110**, 474 (1977)

S. paucicalyculatus.
PYRROLIZIDINE ALKALOID: paucicaline.
Pretorius, Onderstepoort J., **22**, 297 (1949)

S. petasis.
PYRROLIZIDINE ALKALOID: bisline.
Coucourakis et al., J. Chem. Soc., C., 2312 (1970)

S. phillipicus.
PYRROLIZIDINE ALKALOIDS: retrorsine, seneciphylline.
Gonzalez et al., Planta Med., 160 (1986)

S. pinnatus.
SESQUITERPENOID:
1α-hydroxy-6β-angeloyloxy-10αH-furanoeremophil-9-one.
de Salmeron et al., Planta Med., **47**, 221 (1983)

S. platyphylloides Somm. et Lev.
PYRROLIZIDINE ALKALOIDS: neoplatyphylline, platyphylline,
platyphylline N-oxide, sarracine, seneciphylline,
seneciphylline N-oxide.
Orekhov et al., Ber. Dtsch. Chem. Ges., **68**, 653 (1935)
Orekhov et al., ibid., **68**, 1886 (1935)
Konovalova et al., Bull. Soc. Chim. Fr., **4**, 2037 (1937)

S. platyphyllus DC.
PYRROLIZIDINE ALKALOIDS: platyphylline, platyphylline
N-oxide, seneciphylline.
Keokemoer, Warren, J. Chem. Soc., **63**, (1955)

S. pojarkovae.
PYRROLIZIDINE ALKALOIDS: sarracine, seneciphylline.
Chernova et al., Aktual. Vopr. Farm., **2**, 36 (1974); Chem. Abstr., 84
102348w

S. polyanthemoides.
SESQUITERPENOID:
6α-angeloyloxy-9,10-dehydrofuranoeremophil-1-one.
Bohlmann, Zdero, Phytochemistry, **17**, 1161 (1978)

S. praecox.
SESQUITERPENOID: precoxilin A.
Ortega et al., Rev. Latinoam. Quim., **6**, 136 (1975)

S. procerus var. procerus.
PYRROLIZIDINE ALKALOID: procerine.
Jovcera, Collect. Czech. Chem. Commun., **43**, 2312 (1978)

S. propinquus Schischk.
PYRROLIZIDINE ALKALOID: seneciphylline.
Konovalova, Orekhov, Bull. Soc. Chim. Fr., **4**, 2037 (1937)

S. pseudoarnica.
PYRROLIZIDINE ALKALOID: senecionine.
Blackie, Pharm. J., **138**, 102 (1937)

S. pterophorus DC.
PYRROLIZIDINE ALKALOID: pterophine.
SESQUITERPENOID: 10αH-furanoeremophila-6,9-dione.
  (A) de Waal, *Onderstepoort J.*, **16**, 158 (1941)
  (T) Bohlmann et al., *Chem. Ber.*, **110**, 474 (1977)

S. ragonesi.
PYRROLIZIDINE ALKALOIDS: retrorsine, senecionine, uspallatine.
  Pestchanker, Giordanos, *J. Nat. Prod.*, **49**, 722 (1986)

S. renardi C.Winkl.
PYRROLIZIDINE ALKALOIDS: renardine, seneciphylline, senkirkine.
  Danilova, Konovalova, *J. Gen. Chem.*, *USSR*, **20**, 1921 (1950)
  Briggs et al., *J. Chem. Soc.*, 2492 (1965)

S. retrorsus DC.
PYRROLIZIDINE ALKALOIDS: isatidine, retrorsine.
  de Waal, *Onderstepoort J.*, **146**, 777 (1940)

S. rhombifolius (Willd.) Sch. Bip.
PYRROLIZIDINE ALKALOIDS: neoplatyphylline, platyphylline, sarracine, sarracine N-oxide, seneciphylline.
  Valilov et al., *Khim. Prir. Soedin.*, 128 (1973)

S. rhomboideus.
SESQUITERPENOIDS: rhomboidol, rhomboidol acetate.
  Bohlmann, Zdero, *Phytochemistry*, **17**, 1337 (1978)

S. Riddellii.
PYRROLIZIDINE ALKALOID: riddelline.
  Manske, *Can. J. Res.*, **17B**, 1 (1939)

S. rigidus.
SESQUITERPENOID:
  1α-acetoxy-6β-angeloyloxy-9-oxo-10αH-furanoeremophilane,
  6β-(2-methylacryloyloxy)-1β,10-epoxyfuranoeremophilane,
  9-oxo-6β-tigloyloxy-1β,10-epoxyfuranoeremophilane,
  6β-senecioyloxy-1β,10-epoxyfuranoeremophilane.
  Bohlmann, Zdero, *Chem. Ber.*, **107**, 2912 (1975)

S. rivularis.
PYRROLIZIDINE ALKALOID: 3-angeloyloxyheliotrine.
  Klasek et al., *Collect. Czech. Chem. Commun.*, **32**, 2512 (1967)

S. rosmarinifolius L.
PYRROLIZIDINE ALKALOID: rosmarinine.
  Richardson, Warren, *J. Chem. Soc.*, 452 (1943)

S. ruwenzoriensis.
PYRROLIZIDINE ALKALOIDS: ruwenine, ruzorine.
  Sapiro, *J. Chem. Soc.*, 1942 (1953)

S. salignus.
SESQUITERPENOID:
  1,10-epoxy-4α-hydroxy-6β-isobutyryloxyfuranoeremophil-9-one.
  Bohlmann et al., *Chem. Ber.*, **109**, 819 (1976)

S. sarracenicus L.
PYRROLIZIDINE ALKALOIDS: sarracine, sarracine N-oxide.
  Danilova et al., *Dokl. Akad. Nauk SSSR*, **89**, 865 (1953)

S. scaposus.
SESQUITERPENOIDS: isoscaposone, isosenescaposone, senescaposone.
  Bohlmann, Zdero, *Phytochemistry*, **17**, 1337 (1978)

S. scleratus.
PYRROLIZIDINE ALKALOIDS: isatidine, scleratine.
  de Waal, *Nature*, **146**, 777 (1940)

S. seratophiloides.
PYRROLIZIDINE ALKALOID: senecivernine.
  Pestchanker, Giordano, *J. Nat. Prod.*, **49**, 722 (1986)

S. silvaticus.
SESQUITERPENOIDS: 8,8α-epoxyfuranoligularan, silvasenecine.
  Bohlmann et al., *Chem. Ber.*, **107**, 2912 (1974)

S. smithii.
SESQUITERPENOIDS: bisabolal, bisabol-1-ene 12-acetate, 1-oxo-12-bisabolal, senecioyleuryopsonol.
  Bohlmann et al., *Phytochemistry*, **20**, 2839 (1981)

S. spartioides.
PYRROLIZIDINE ALKALOIDS: seneciphylline, spartoidine.
  Manske, *Can. J. Res.*, **17**, 1 (1939)

S. squalidus L.
PYRROLIZIDINE ALKALOIDS: senecionine, squalidine.
  Blackie, *Pharm. J.*, **138**, 102 (1937)

S. stenocephalus Maxim.
PYRROLIZIDINE ALKALOID: seneciphylline.
  Konovalova et al., *Bull. Soc. Chim. Fr.*, **4**, 2037 (1937)

S. subalpinus.
PYRROLIZIDINE ALKALOID: seneciphylline.
  Konovalova et al., *Bull. Soc. Chim. Fr.*, **4**, 2037 (1937)

S. subrubiflorus.
SESQUITERPENOID: 8-hydroxy-15-isopimaren-11-one.
  Bohlmann et al., *Phytochemistry*, **21**, 1697 (1982)

S. subulatus.
PYRROLIZIDINE ALKALOIDS: retrorsine, senecionine.
  Pestchanker, Giordano, *J. Nat. Prod.*, **49**, 722 (1986)

S. swaziensis Compton.
PYRROLIZIDINE ALKALOID: swazine.
  Gordon-Gray et al., *Tetrahedron Lett.*, 707 (1972)

S. triangularis.
PYRROLIZIDINE ALKALOID: triangularine.
  Roitman, *Aust. J. Chem.*, **36**, 1203 (1983)

S. umbellatus.
SESQUITERPENOID:
  1α-acetoxy-6-isobutyryloxy-9-oxo-10αH-furanoeremophilane,
  6β-isobutyryloxy-9-oxo-10αH-furanoeremophilane,
  6β-isovaleryloxy-9-oxo-10αH-furanoeremophilane.
  Bohlmann, Zdero, *Chem. Ber.*, **107**, 2912 (1974)

S. uspallatensis.
PYRROLIZIDINE ALKALOIDS: retrorsine, senecionine, uspallatine.
  Pestchanker, Giordano, *J. Nat. Prod.*, **49**, 722 (1986)

S. vernalis.
PYRROLIZIDINE ALKALOID: senecivernine.
  Roeder et al., *Planta Med.*, **37**, 131 (1979)

S. vira-vira.
PYRROLIZIDINE ALKALOIDS: anacrotine, neoplatyphylline, uspallatine.
FLAVONOIDS: quercetin, rutin.
STEROIDS: campesterol, β-sitosterol, stigmasterol, stigmasta-3,5-dien-7-one, stigmasta-4,6-dien-3-one.
TRITERPENOIDS: α-amyrin, β-amyrin.
  Jares et al., *J. Nat. Prod.*, **50**, 514 (1987)

S. viscosus L.
PYRROLIZIDINE ALKALOID: senecionine.
  Blackie, *Pharm. J.*, **138**, 102 (1937)

S. vulgaris L.
PYRROLIZIDINE ALKALOID: senecionine.
  Blackie, *Pharm. J.*, **138**, 102 (1937)

S. yegua.
FLAVONOIDS: kaempferol 7-O-glucoside, kaempferol 3-O-robinoside-7-O-rhamnoside, quercetin 3-O-galactoside, quercetin 3-O-rutinoside.
  Reyes et al., *Rev. Latinoam. Quim.*, **8**, 134 (1977)

SEPTORIA
S. nodorum.
PYRIDINE ALKALOID: septorine.
Devys et al., C. R. Acad. Sci., **286C**, 457 (1978)

SEQUOIA
S. sempervirens.
TERPENOIDS: sequierins A, B, C and D, polyalthic acid.
Ohto et al., Agr. Biol. Chem., **42**, 1957 (1978)

SEQUOIADENDRON
S. gigantea.
NORPIGNANS: sequrins A, B, C, D, E, F and G.
PHENOLIC: agatharesinol.
Henley-Smith et al., Phytochemistry, **15**, 1285 (1976)

SERRATULA
S. wolfi Andrae.
ESTERS: methyl linoleate, methyl linolenate, methyl palmitate.
FLAVONOID: kaempferol 3-methyl ether.
SESQUITERPENOIDS: caryophyllene, caryophyllene epoxide, germacrene D.
Czerson et al., Chem. Ber., **109**, 2291 (1976)

SESAMUM (Pedaliaceae)
S. angolense Welw.
ROTENONES: sesamolin, sesangolin.
Jones et al., J. Org. Chem., **27**, 3232 (1962)
S. indicum.
FLAVONE: pedalin.
LIGNANS: sesaminol, sesanol.
(F) Morita et al., Chem. Pharm. Bull., **8**, 59 (1960)
(L) Osawa et al., Agr. Biol. Chem., **49**, 3351 (1985)

SESBANIA (Fagaceae)
S. drummondii.
DIAZANAPHTHALENE ALKALOID: sesbanine.
DINORSESQUITERPENOIDS: drummondone A, drummondone B.
(A) Powell et al., J. Am. Chem. Soc., **101**, 2785 (1979)
(T) Powell et al., J. Org. Chem., **51**, 1074 (1986)

SESELI (Umbelliferae)
S. bocconi subsp. bocconi.
COUMARIN: bocconin.
Bellino et al., Phytochemistry, **25**, 1195 (1986)
S. dichotomum.
COUMARINS: anomalin, bergapten,
Dikhovlinova et al., Khim. Prir. Soedin., 811 (1976)
S. eriocephalum.
COUMARINS: anomalin, isoimperatorin, pterixin, sukadorfin.
Sokolova et al., Khim. Farm. Zh., **11**, 53 (1977)
S. grandivittatum.
COUMARINS: grandivitin, grandivitinol.
Abyshev et al., Khim. Prir. Soedin., 640 (1977)
S. gummiferum.
COUMARINS:
4'-angeloyloxy-3'-(3,3-dimethylacryloyloxy)-3',4'-dihydroseselin,
(+)-2'',3-dihydro-4'-angeloyloxy-3',4'-dihydro-3'-isovaleryloxyseselin.
SESQUITERPENOID: isosamidin.
(C) Nielsen et al., Acta Chem. Scand., **25**, 529 (1971)
(T) Lemmich et al., ibid., **20**, 2497 (1966)

S. indicum.
STEROIDS: indosterol, 9(11),22-stigmastadien-3-ol.
MONOTERPENOIDS: β-cyclolavandulal, β-cyclolavandulic acid.
(S) Gupta et al., Tetrahedron Lett., 1221 (1974)
(T) Dixit et al., Chem. Ind. (London), 1256 (1967)
S. libanotis (L.) Koch.
SESQUITERPENOID: isosamidin.
Nielsen et al., Acta Chem. Scand., **25**, 529 (1971)
S. montanum.
BENZOPYRAN: apterin.
Lemmich et al., Phytochemistry, **17**, 139 (1978)
S. petraeum.
COUMARIN: anomalin.
Dikhovlinova et al., Khim. Prir. Soedin., 811 (1976)
S. ponticum.
COUMARINS: anomalin, bergapten, xanthogalin.
Dikhovlimova et al., Khim. Prir. Soedin., 811 (1976)
S. seseliflorum.
BENZOPYRAN: sechulin.
Aminov et al., Khim. Prir. Soedin., 152 (1974)
S. sibiricum.
BENZOPYRANS: sesebrin, sesebrinol, seseburicin.
COUMARINS: osthol, sibiricin, siribicol.
(B) Atal et al., Phytochemistry, **17**, 2111 (1978)
(C) Austin et al., Tetrahedron, **24**, 3247 (1968)
S. tortuosum LBS. Eur.
COUMARINS: agasyllin, bergapten, columbianetin, decursinol, dihydroxanthyletin, imperatorin, oroselol, pranferin, xanthotoxin.
PHENOLIC: 4-hydroxy-3,5-dimethoxybenzaldehyde.
STEROID: β-sitosterol.
Gonzalez et al., An. Quim., **78C**, 184 (1982)
S. webbii Cosson.
COUMARINS: bergapten, byakangelicin, imperatorin, isoimperatorin, isoooxypeucedanin, oxypeucedanin hydrate, umbelliferone, xanthotoxin.
Gonzalez et al., An. Quim., **78C**, 274 (1982)

SEVERINA
S. buxifolia (Poir.) Ten.
AMIDE ALKALOIDS: hexadecanoylseverine, severine.
Dreyer et al., Tetrahedron, **26**, 5745 (1970)

SHEPHERDIA
S. argentea.
CARBOLINE ALKALOID: tetrahydroharmol.
Ayer, Browne, Can. J. Chem., **48**, 1980 (1970)
S. canadensis.
CARBOLINE ALKALOID: shepherdine.
Ayer, Browne, Can. J. Chem., **48**, 1980 (1970)

SHEPNOPAPPUS
S. pickelii.
SESQUITERPENOID:
8-O-(2-methylpropenoyl)-stilpnotomentolide.
Bohlmann et al., Phytochemistry, **21**, 1045 (1982)
S. tomentosus.
SESQUITERPENOID:
8-O-(2-methylpropenoyl)-stilpnotomentolide.
Bohlmann et al., Phytochemistry, **21**, 1045 (1982)

SHOREA

S. acuminata.
TRITERPENOIDS: 2α,3α-dihydroxyolean-12-en-28-oic acid, mangiferonic acid.
Cheung, Yan, *Aust. J. Chem.*, **25**, 2003 (1972)

S. meristopteryx.
TRITERPENOID: shoreic acid.
Lantz, Wolff, *Bull. Soc. Chim. Fr.*, 2131 (1968)

S. tellura.
ANTHRAQUINONE: deoxyerythrolaccin.
Mehandale *et al.*, *Tetrahedron Lett.*, 2231 (1968)

SIDERITIS (Labiatae)

S. angustifolia.
DITERPENOIDS: isosideritol, jativatriol, lagascatriol, sideritol.
von Carstenn-Lichterfeld *et al.*, *Aust. J. Chem.*, **27**, 517 (1974)
Ayer *et al.*, *Can. J. Chem.*, **52**, 2792 (1974)

S. arborescens subsp. paulii.
DITERPENOIDS: ent-6α,8α-dihydroxylabda-13(16),14-diene, ent-13-epi-16,18-dihydroxymanool oxide, ent-16,18-dihydroxymanool oxide, ent-18-hydroxy-15(16)-peroxylabd-13-ene, ent-6α,16,18-trihydroxymanool oxide.
Garcia-Granados *et al.*, *Phytochemistry*, **24**, 517 (1985)

S. biflora.
DITERPENOIDS: 3β-acetoxy-7α,11-dihydroxy-15β,16-epoxykaurene, epoxyisosidol.
Garcia-Alvarez, Rodriguez, *Phytochemistry*, **12**, 1113 (1973)

S. canariensis.
DITERPENOIDS: 7α-acetoxy-18-hydroxytrachylobanol, ribenol, trachinodiol, trachinol, vierol.
NORDITERPENOID: tiganone.
Gonzalez *et al.*, *Phytochemistry*, **12**, 1113 (1973)

S. candicans.
DITERPENOIDS: candidiol, candol A, candol B.
Gonzalez *et al.*, *Phytochemistry*, **12**, 2721 (1973)

S. candicans var. eriocephala.
DITERPENOID: candicandol.
Breton *et al.*, *Tetrahedron Lett.*, 599 (1969)

S. catillaris.
PHENOLICS: caffeic acid, chlorogenic acid, neochlorogenic acid.
Fefer *et al.*, *Farm. Zh. (Kiev)*, **92** (1977); *Chem. Abstr.*, **87** 197310b

S. chamaedrifolia.
DITERPENOIDS: ent-kaur-15-ene-7,18-diol, ent-kaur-16-ene-3,7,18-triol.
NORDITERPENOIDS: villenatriol, villenatriolone, villenol, villenol 19-acetate, villenolone, villenolone 19-acetate.
Rodriguez *et al.*, *Phytochemistry*, **17**, 281 (1978)

S. euboea.
DITERPENOIDS: eubol, eubotriol.
Venturella, Belling, *Experientia*, **33**, 1270 (1977)

S. foetens.
DITERPENOIDS: 6-acetylandalusol, 6-acetylisoandalusol, 18-acetylisoandalusol.
Rodriguez *et al.*, *Phytochemistry*, **19**, 2405 (1980)

S. funkiana Willk.
DITERPENOIDS: funkiol, isofunkiol, sidofunkiol.
Granados-Garcia, *An. Quim.*, **76**, 178 (1980)

S. gomerae.
FLAVONE: eupatorin 3-methyl ether.
DITERPENOIDS: gomeraldehyde, 13-epi-gomeraldehyde, gomeric acid, 13-epi-gomeric acid.
(F) Kupchan *et al.*, *J. Pharm. Sci.*, **54**, 929 (1965)
(T) Gonzalez *et al.*, *Phytochemistry*, **14**, 2655 (1975)

S. grandiflora.
DITERPENOID: tartessol.
Rabanal *et al.*, *Experientia*, **30**, 977 (1974)

S. leucantha.
DITERPENOIDS: isofoliol, isoleucanthol, isolinearol, isosidol, leucanthol, linearol, sidol.
de Quesada *et al.*, *Tetrahedron Lett.*, 2187 (1972)

S. linearifolia.
DITERPENOIDS: isofoliol, isolinearol, isosidol, sidol.
de Quesada *et al.*, *Tetrahedron Lett.*, 2187 (1972)

S. linearis.
DITERPENOIDS: isolinearol, linearol.
de Quesada *et al.*, *Tetrahedron Lett.*, 2187 (1972)

S. luteola.
DITERPENOIDS: ent-18-acetoxy-16-kaurene-3,7-diol, ent-15-kaurene-7,18-diol, ent-16-kaurene-3,7,18-triol.
Rodriguez *et al.*, *Phytochemistry*, **14**, 1670 (1975)

S. montana L.
QUATERNARY ALKALOID: stachydrine.
Schultz, Trier, *Z. Physiol. Chem.*, **67**, 59 (1910)

S. mugronensis.
DITERPENOID: borjatriol.
Rodriguez, Valverde, *Tetrahedron*, **29**, 2837 (1973)

S. paulii.
DITERPENOIDS: epoxyisofoliol, ent-3β,7α,18-trihydroxy-15β,16-epoxykaurane.
Rodriguez, Valverde, *An. Quim.*, **72**, 189 (1976)

S. pusilla (Lag.) Pau.
DITERPENOIDS: isopusillatriol, isosideritol.
Granados-Garcia *et al.*, *Tetrahedron Lett.*, 3611 (1980)

S. reverchoni.
DITERPENOIDS: 12-acetyljativatriol, benuol, conchitriol, jativatriol, lagascatriol, lagascol, serradiol, sideritol, tobarrol.
Marquez *et al.*, *Phytochemistry*, **14**, 2713 (1975)

S. serrata.
DITERPENOIDS: 12-acetyljativatriol, 1,12-diacetyljativatriol, 1,17-diacetyljativatriol, lagascol, serradiol.
Escamilla, Rodriguez, *Phytochemistry*, **19**, 463 (1980)

S. sicula.
DITERPENOIDS: epoxysideritriol, epoxysiderol, sideridiol, sideripol, siderol, sideroxal.
Venturella *et al.*, *Phytochemistry*, **17**, 811 (1978)

S. syriaca.
DITERPENOIDS: ent-15,16-epoxy-7,18-kauranediol, siderone, ucriol.
Venturella *et al.*, *Phytochemistry*, **22**, 600 (1983)

S. theazans.
DITERPENOID: epoxyisolinearol.
Venturella *et al.*, *Phytochemistry*, **14**, 1451 (1975)

S. tragoriganum.
Five ent-kauranes.
Carrascal *et al.*, *An. Quim.*, **74**, 1547 (1978)

S. varoi subsp. oriensis.
SESQUITERPENOIDS: andalusol, 6-deoxyandalusal,
6-deoxyandalusoic acid, 6-deoxyandalusol,
1β,4α-dihydroxy-6β-acetoxyeudesmane,
1β,4β-dihydroxy-6β-acetoxyeudesmane,
1β-hydroxy-6β-acetoxyeudesm-3-ene,
1β-hydroxy-6β-acetoxyeudesm-4-ene,
1β-hydroxy-6β-acetoxyeudesm-4(15)-ene, varodiol.
   Granadas-Garcia et al., Phytochemistry, **25**, 2171 (1986)

SIEGESBECKIA
S. orientalis.
SESQUITERPENOID: orientalide.
DITERPENOID: darutin.
   Baruah et al., Phytochemistry, **18**, 991 (1979)

S. pubescens.
DITERPENOIDS: kirenol, pimar-8(14)-ene-6β,15,16,18-tetraol.
   Murakami et al., Tetrahedron Lett., 4991 (1973)

SIGMOSCEPTRELLA
S. laevis.
SESQUITERPENOID: sigmosceptrellin A methyl ester.
   Albericci et al., Tetrahedron Lett., 2687 (1979)

SILAUM (Umbelliferae/Apiaceae)
S. silaus (L.) Schinz et Thell. (S. flavescens Bernh.).
PHTHALIDES: Z-ligustilide, neocnidilide.
   Gijbels et al., Sci. Pharm., **50**, 158 (1982)

SILENE (Caryophyllaceae/Dianthaceae)
S. armeria.
FLAVONOIDS: vicenin, vicenin diglycoside, vicenin glycoside.
   Darmograi et al., Khim. Prir. Soedin., 114 (1977)

S. artemisetorum.
FLAVONOIDS: vicenin, vicenin diglycoside, vicenin glycoside.
   Darmograi et al., Khim. Prir. Soedin., 114 (1977)

S. boissieri.
FLAVONOIDS: vicenin, vicenin diglycoside, vicenin glycoside.
   Darmograi et al., Khim. Prir. Soedin., 114 (1977)

S. brahuica.
FLAVONOIDS: saparonetin, saparonetin diglycoside, saparonetin
glycoside, vitexin, vitexin diglycoside, vitexin glycoside.
STEROIDS: sileneosides A, B and C.
   (F) Darmograi et al., Khim. Prir. Soedin., 114 (1977)
   (S) Saatov et al., ibid., 211 (1982)

S. bupleuroides.
FLAVONOIDS: adonivernite, homoadonivernite, isosaponarin.
   Darmograi et al., Khim. Prir. Soedin., 114 (1977)

S. chersonensis.
FLAVONOIDS: vicenin, vicenin diglycoside, vicenin glycoside.
   Darmograi et al., Khim. Prir. Soedin., 114 (1977)

S. chlorantha.
FLAVONOIDS: vicenin, vicenin diglycoside, vicenin glycoside.
   Darmograi et al., Khim. Prir. Soedin., 114 (1977)

S. chlorifolia.
FLAVONOIDS: adonivernite, homoadonivernite, isosaponarin.
   Darmograi et al., Khim. Prir. Soedin., 114 (1977)

S. compacta.
FLAVONOIDS: adonivernite, homoadonivernite, isosaponarin.
   Darmograi et al., Khim. Prir. Soedin., 114 (1977)

S. comutata.
FLAVONOIDS: vicenin, vicenin diglycoside, vicenin glycoside.
   Darmograi et al., Khim. Prir. Soedin., 114 (1977)

S. cretica.
FLAVONOIDS: adonivernite, homoadonivernite, isosaponarin.
   Darmograi et al., Khim. Prir. Soedin., 114 (1977)

S. cubanensis.
FLAVONOIDS: adonivernite, homoadonivernite, isosaponarin.
   Darmograi et al., Khim. Prir. Soedin., 114 (1977)

S. cyri.
FLAVONOIDS: vicenin, vicenin diglycoside, vicenin glycoside.
   Darmograi et al., Khim. Prir. Soedin., 114 (1977)

S. dolichocarpa.
FLAVONOIDS: vicenin, vicenin diglycoside, vicenin glycoside.
   Darmograi et al., Khim. Prir. Soedin., 114 (1977)

S. foliosa.
FLAVONOIDS: vicenin, vicenin diglycoside, vicenin glycoside.
   Darmograi et al., Khim. Prir. Soedin., 114 (1977)

S. graminifolia.
FLAVONOIDS: vicenin, vicenin diglycoside, vicenin glycoside.
   Darmograi et al., Khim. Prir. Soedin., 114 (1977)

S. italica.
FLAVONOIDS: vicenin, vicenin diglycoside, vicenin glycoside.
   Darmograi et al., Khim. Prir. Soedin., 114 (1977)

S. jenisseensis.
FLAVONOIDS: vicenin, vicenin diglycoside, vicenin glycoside.
   Darmograi et al., Khim. Prir. Soedin., 114 (1977)

S. macrostyla.
FLAVONOIDS: vicenin, vicenin diglycoside, vicenin glycoside.
   Darmograi et al., Khim. Prir. Soedin., 114 (1977)

S. multijida.
FLAVONOIDS: saparonetin, saparonetin diglycoside, saparonetin
glycoside, vitexin, vitexin diglycoside, vitexin glycoside.
   Darmograi et al., Khim. Prir. Soedin., 114 (1977)

S. nutans.
FLAVONOIDS: vicenin, vicenin diglycoside, vicenin glycoside.
   Darmograi et al., Khim. Prir. Soedin., 114 (1977)

S. polaris.
FLAVONOIDS: adonivernite, homoadonivernite, isosaponarin.
   Darmograi et al., Khim. Prir. Soedin., 114 (1977)

S. praemixta.
STEROID: silenosterone.
   Saatov et al., Khim. Prir. Soedin., 793 (1979)

S. repens.
FLAVONOIDS: saparonetin, saparonetin diglycoside, saparonetin
glycoside, vitexin, vitexin diglycoside, vitexin glycoside.
   Darmograi et al., Khim. Prir Soedin., 114 (1977)

S. scabrifolia.
STEROIDS: 2-deoxy-α-ecdysone, 2-deoxy-α-ecdysone-3-acetate,
ecdysterone, ecdysterone-22-O-benzoate.
   Saatov et al., Khim. Prir. Soedin., **77**, 439 (1986)

S. schafta.
BENZOPYRAN: schaftoside.
   Chopin et al., Phytochemistry, **13**, 2583 (1974)

S. supina.
FLAVONOIDS: saparonetin, saparonetin diglycoside, saparonetin
glycoside, vitexin, vitexin diglycoside, vitexin glycoside.
   Darmograi et al., Khim. Prir. Soedin., 114 (1977)

S. turgida.
FLAVONOIDS: saparonetin, saparonetin diglycoside, saparonetin
glycoside, vitexin, vitexin diglycoside, vitexin glycoside.
   Darmograi et al., Khim. Prir. Soedin., 114 (1977)

S. wolygensis.
FLAVONOIDS: vicenin, vicenin diglycoside, vicenin glycoside.
   Darmograi et al., Khim. Prir. Soedin., 114 (1977)

SILPHIUM

S. perfoliatum.
SAPONINS: silphiosides A, B, C, D and E.
SESQUITERPENOIDS: precarabrene, precarabrenone, 7αH-silphinene, 7βH-silphinene, 6-silphiperfolene.
DITERPENOIDS: 16-acetoxycarterochaetol, 16-acetoxycarterochaetol-15-acetate, 16-acetoxy-8,12-epoxy-14-oxo-12-labden-16-ol, 15,16-dioxocarterochaetol, 13S,14S-epoxycarterochaetol, 6-guaiene-4,10-diol, 6-guaien-4-ol, 16-hydroxycarterochaetol, 16-oxocarterochaetol.
(Sap) Davidyants et al., Khim. Prir. Soedin., 63 (1986)
(T) Bohlmann et al., Phytochemistry, 19, 259 (1980)

SILYBUM (Compositae)

S. marianum (L.) Gaertn.
ACIDS: arachidic acid, behenic acid, linoleic acid, linolenic acid, palmitic acid, stearic acid.
LIGNANS: 2,3-dehydrosilychristin, 2,3-dehydrosilymarin, silandrin, silybin, silychristin, silydianin, silymonin.
(A) Kaczmarek et al., Herba Pol., 21, 213 (1975); Chem. Abstr., 84 14685y
(L) Pelter et al., Chem. Ber., 108, 790 (1975)
(L) Szilagyi et al., Chem. Abstr., 97 212666z

SIMABA

S. cuspidata.
CARBOLINE ALKALOIDS: 3-methoxycanthin-2,6-dione, 8-methoxycanthin-6-one.
Geibrecht et al., Phytochemistry, 19, 313 (1980)

S. multiflora.
CARBOLINE ALKALOIDS: 10-hydroxycanthin-6-one, 10-methoxycanthin-6-one.
QUASSINOIDS: 12-dehydro-6α-senecioyloxychaparrin, 13,18-dehydro-6α-senecoiyloxychaparrin, karinolide.
(A) Arisawa et al., J. Nat. Prod., 46, 222 (1983)
(Q) Moretti et al., ibid., 49, 440 (1986)

S. cf. orinocensis.
STEROIDS: guanepolide, simarinolide.
Polonsky et al., Tetrahedron Lett., 2306 (1983)

SIMAROUBA (Simaroubaceae)

S. amara.
BITTER PRINCIPLE: simarolide.
QUASSINOID: 2'-acetylglaucarubin.
TRITERPENOIDS: tirucalla-7,24-diene-3,21-dione, tirucalla-7,24-dien-3-one.
(B) Polonsky et al., Bull. Soc. Chim. Fr., 1546 (1959)
(Q) Polonsky et al., Experientia, 34, 1122 (1978)
(T) Polonsky et al., Phytochemistry, 15, 337 (1976)

S. glauca.
QUASSINOIDS: 15-O-β-D-glucopyranosylglaucarubol, 15-O-β-D-glucopyranosylglaucorubolone.
Bhatnager et al., Tetrahedron Lett., 299 (1984)

S. versicolor.
QUASSINOIDS: amarolide 11-acetate, amarolide-11,12-diacetate.
Ghosh et al., Lloydia, 40, 364 (1977)

SIMMONDSIA

S. californica.
GLUCOSIDE: simmondsin.
Elliger et al., J. Chem. Soc., Perkin I, 2209 (1973)

SIMSIA

S. dombeyana.
SESQUITERPENOID: simsiolide.
Bohlmann, Zdero, Phytochemistry, 16, 776 (1977)

S. foetida.
ISOFLAVONOID: 7,4'-dimethyl-3'-hydroxygenistein.
Romo de Vivar et al., Rev. Latinoam. Quim., 13, 18 (1982)

SINAPIS (Cruciferae)

S. alba L. (Brassica alba (L.) Rabenh.
AMINO ACID: 4-hydroxybenzoylcholine.
THIOCYANATE: sinalbin.
(Am) Clausen et al., Phytochemistry, 21, 917 (1982)
(Th) Ettlinger et al., J. Am. Chem. Soc., 78, 4172 (1956)

SINDORA

S. wallichii.
SESQUITERPENE: copaene.
Buchi et al., Proc. Chem. Soc., 214 (1963)

SINOMENIUM

S. acutum Rehd. et Wils.
HALOGENATED ALKALOIDS: acutumidine, acutumine.
PHENANTHRENE ALKALOIDS: dehydrosinomenine, isosinomenine, sinactine, sinoacutine, sinomenine, tuduranine.
Pkamoto et al., Tetrahedron Lett., 1933 (1969)

SINTON

S. citrangequat (Citrus sinensis x Poncirus trifoliata x Fortunella marginata).
CAROTENOIDS: citraxanthin, 8',9'-dihydro-8'-hydroxycitraxanthin, 3-hydroxycitraxanthin.
Yokoyama, White, J. Org. Chem., 31, 3452 (1966)

SIPHONODON

S. australe.
TRITERPENOIDS: friedelane-3,x-dione, friedelane-3,y-dione, friedelane-3,x,y-trione.
Courtney, Gascoigne, J. Chem. Soc., 2115 (1956)

SIPHULA (Usneaceae)

S. ceratites (Wahlenb.) Fr.
BENZOPYRONE: siphulin.
Bruun et al., Acta Chem. Scand., 19, 1677 (1965)

SISYMBRIUM (Brassicaceae/Cruciferae)

S. altissimum L. (S. pannonicum Jacq.).
17 FATTY ACIDS:
Dolya et al., Rastit. Resur., 22, 249 (1986); Chem. Abstr., 105 39401t

SIUM (Umbelliferae/Apiaceae)

S. latifolium L.
SESQUITERPENOIDS: siolactone, siolacetate.
Casinovi et al., Collect. Czech. Chem. Commun., 48, 2411 (1983)

SKIMMIA (Rutaceae)

S. japonica.
FURANOQUINOLINE ALKALOID: skimmianine.
QUATERNARY ALKALOID: (+)-platydesminium.
BENZOPYRAN: alatol.

TRITERPENOID: taraxerol.

(A) Boyd, Grunson, *J. Chem. Soc., C,* 556 (1970)
(B) Abyshev *et al., Khim. Prir. Soedin.,* 722 (1973)
(T) Takemoto *et al., J. Pharm. Soc., Japan,* **74,** 1271 (1954)

S. laureola.
FURANOQUINOLINE ALKALOID: skimmianine.
Asahina, Inubuse, *Ber. Dtsch. Chem. Ges.,* **63,** 2052 (1930)

S. repens.
FURANOQUINOLINE ALKALOID: dictamnine.
Asahina *et al., Ber. Dtsch. Chem. Ges.,* **63,** 2045 (1930)

S. wallachii.
STEROIDS: β-sitosterol, skimmiwallichin, taraxerol, 3-epi-taraxerol, taraxerone.
Ray *et al., Indian J. Chem.,* **14,** 1876 (1975)

SKYTANTHUS

S. acutus Meyen.
PIPERIDINE ALKALOIDS: hydroxyskytanthine I, hydroxyskytanthine II, α-skytanthine, β-skytanthine, δ-skytanthine, τ-skytanthine, β-skytanthine N-oxide.
Adolphen *et al., Tetrahedron,* **23,** 3147 (1967)
Streeter *et al., Chem. Ind. (London),* **45,** 1631 (1969)

SMALLANTHUS

S. uvedalia.
DITERPENOIDS: 16,17-epoxy-3-hydroxy-19-kauranal, ent-12-hydroxy-18-(3-methylbutanoyl)-16-kauren-19-oic acid, ent-12-hydroxy-18-(3-methyl-2-butenoyl)-16-kauren-19-oic acid.
Bohlmann *et al., Phytochemistry,* **19,** 107 (1980)

SMILACENA (Liliaceae)

S. stellata.
STEROID: diosgenin.
Marker *et al., J. Am. Chem. Soc.,* **65,** 1199 (1943)

SMILAX (Smilacaceae)

S. aristolochaefolia Mill.
SAPONINS: parillin, sarsaparilloside.
Tschesche *et al., Tetrahedron Lett.,* 2785 (1967)

S. aspera L.
SAPONIN: pseudosarsasapogenin.
TRITERPENOID: 31-norcycloartanol.
(Sap) Ayengar *et al., Curr. Sci.,* **36,** 653 (1967)
(T) Berti *et al., Tetrahedron Lett.,* 125 (1967)

S. excelsa.
STEROIDAL GLYCOSIDE: diosgenin-3-O-α-L-rhamnopyranoside.
Iskendarov *et al., Khim. Prir. Soedin.,* 805 (1975)

S. lanceolata.
STEROID: sarsasapogenin.
Marker *et al., J. Am. Chem. Soc.,* **65,** 1199 (1943)

S. ornata Hook.
STEROID: smilagenin.
Kintya *et al., Khim. Prir. Soedin.,* 615 (1972)

S. rotundifolia.
STEROID: sarsasapogenin.
Marker *et al., J. Am. Chem. Soc.,* **65,** 1199 (1943)

S. sieboldi Miq.
STEROID: laxogenin.
Akahari *et al., Chem. Pharm. Bull.,* **13,** 545 (1965)

SMIRNIOPSIS

S. aucheri.
BENZOPYRANONES: smirnioridin, smirniorin.
Nikonov *et al., Khim. Prir. Soedin.,* 592 (1969)

SMIRNOVIA (Leguminosae)

S. turkestana Bge.
ALKALOIDAL AMINES: smirnovidine, smirnovine, sphaerophysine.
Ryabinin *et al., Dokl. Akad. Nauk SSSR,* **76,** 851 (1951)

SMYRNIUM (Umbelliferae)

S. connatum.
SESQUITERPENOID: istanbulin C.
Ulubelen *et al., Phytochemistry,* **18,** 338 (1979)

S. cordifolium.
SESQUITERPENOID: 2-acetylfurodien.
Ulubelen *et al., Phytochemistry,* **23,** 1793 (1984)

S. creticum.
SESQUITERPENOIDS: istanbulins D, E and F.
Ulubelen *et al., Phytochemistry,* **21,** 2128 (1982)

S. olastrum L.
SESQUITERPENOIDS: 1,10-epoxygermacrone, furanoeremophil-1-one, istanbulins A and B.
Ulubelen *et al., Tetrahedron Lett.,* 4455 (1971)

SOLANDRA (Solanaceae)

S. longiflora.
TROPANE ALKALOID: norhyoscyamine.
Petrie, *Proc. Linn. Soc., N.S.W,* **41,** 815 (1917)

S. nitida.
PYRANONE: dehydracetic acid.
Rivera *et al., Experientia,* **32,** 1490 (1976)

SOLANOSTOMA

S. triste.
DITERPENOID: ent-kaur-16-ene-11α-acetoxy-15α-ol.
Connolly, Thornton, *J. Chem. Soc., Perkin I,* 736 (1973)

SOLANUM (Solanaceae)

S. abutiloides.
STEROID: isonuatigenin.
Evans *et al., Planta Med.,* **41,** 166 (1981)

S. acaule.
STEROIDAL ALKALOID: solacauline.
Schreiber, *Chem. Ber.,* **87,** 1007 (1954)

S. acaule x S. ajanhuiri (clones).
GLYCOALKALOIDS: commersonine, sisunine.
Osman *et al., Phytochemistry,* **25,** 967 (1986)

S. acculeastrissimum.
STEROIDAL ALKALOID: solasodine.
Kadkade *et al., Lloydia,* **40,** 224 (1977)

S. aculeatissimum.
SAPONINS: aculeatoside A, aculeatoside B.
Saijo *et al., Phytochemistry,* **22,** 733 (1983)

S. aculeatum.
STEROIDAL ALKALOID: 3-desamino-3-hydroxysolacapsine.
Coll *et al., Phytochemistry,* **22,** 2099 (1983)

S. adoense Hochst.
STEROIDAL ALKALOID: solasodine.
Ntahomvukiya *et al., Plant Med. Phytother.,* **17,** 18 (1983)

S. affin-ecuadorensis.

STEROIDAL ALKALOIDS: deacetylsolafilidine, solafilidine.

Martinod *et al., Politecnica,* **3**, 46 (1977)

S. angustifolium.

STEROIDAL ALKALOID: solangustine.

Tutin, Cleaver, *J. Chem. Soc.,* **105**, 164 (1914)

S. aurantiacobaccatum de Wild.

STEROID: diosgenin.

Carle *et al., Planta Med.,* **32**, 195 (1977)

S. auriculatum.

STEROIDAL ALKALOID: solauricine.

Anderson, Briggs, *J. Chem. Soc.,* 1036 (1937)

S. aviculare.

STEROIDAL ALKALOIDS: purapurine, solanaviol, solasodine.

Kaneko *et al., Phytochemistry,* **19**, 299 (1980)

S. bahamense.

STEROID: behamgenin.

Coll *et al., Phytochemistry,* **22**, 787 (1983)

S. callium.

STEROIDAL ALKALOIDS: isosolafloridine, solacallinidine.

Bird *et al., Tetrahedron Lett.,* 3653 (1976)

S. carolinense.

AMINE ALKALOID: solaurethine.

Evans, Somanabandhu, *Phytochemistry,* **16**, 1852 (1977)

S. chacoense Bitt.

STEROIDAL ALKALOIDS: dehydrocommersonine, leptine I.

STEROIDS: campesterol, β-sitosterol, stigmasterol.

TRITERPENOIDS: cycloartanol, 24-methylenecycloartanol.

(A) Zacharias, Osman, *Plant Sci. Lett.,* **10**, 283 (1977)

(S, T) Mierzwa *et al., Rocz. Chem.,* **51**, 391 (1977)

S. congestiflorum Dun. Dc.

STEROIDAL ALKALOIDS: 23-oxosolacongestidine, 24-oxosolacongestidine, solacongestidine, solafloridine.

Sato *et al., J. Org. Chem.,* **34**, 1577 (1969)

S. crispum.

UNCHARACTERIZED ALKALOID: natrine.

Bianchi *et al., Ann. Chim. (Rome),* **43**, 208 (1953)

S. dasyphyllum.

One glycosteroidal alkaloid.

Couve, Demole, *Planta Med.,* **28**, 168 (1975)

S. demissum.

STEROIDAL ALKALOIDS: demissidine, demissine.

SESQUITERPENOIDS: rishitin, rishitinol.

(A) Kuhn, Low, *Ber. Dtsch. Chem. Ges.,* **80**, 406 (1947)

(T) Tamiyama *et al., Phytopathology,* **58**, 115 (1968)

S. dulcamara L.

STEROIDAL ALKALOIDS: 15α-hydroxysoladulcidine, 15α-hydroxytomatidine, solaceine, soladulcidine, α-soladulcine, β-soladulcine, α-solamarine, β-solamarine, δ-solamarine, τ(1)-solamarine, τ(2)-solamarine, tomatid-5-en-3β-ol.

Tukalo *et al., Khim. Prir. Soedin.,* 207 (1971)

S. dunalianum.

STEROIDAL ALKALOIDS: soladunalidine, soladunalinidine.

Bird *et al., Tetrahedron Lett.,* 159 (1978)

S. ecuadorense.

STEROIDAL ALKALOID: deacetylsolaphyllidine.

Usubillaga *et al., Politecnica,* **3**, 107 (1973)

S. erianthum G.Don.

STEROIDAL ALKALOIDS: solamargine, solanidine, solanine, solasodine, solasonine.

Moreira *et al., Cienc. Cult. (Sao Paulo),* 114 (1981); *Chem. Abstr.,* **97** 195831e

S. giganteum Jacq.

STEROIDAL ALKALOIDS: solanogantamine, solanogantine.

Pakrashi *et al., J. Indian Chem. Soc.,* **55**, 1109 (1978)

S. glaucophyllum.

STEROID: 1,25-dihydroxyvitamin D3.

Napoli *et al., Chem. Abstr.,* 88 3055z

S. globiferum.

STEROIDAL ALKALOIDS: solasodiene, solasodine.

STEROIDS: 3-anhydronuatigenin, diosgenin, 3-isoanhydronuatigenin, isonuatigenin, nuatigenin.

Doepke *et al., Pharmazie,* **31**, 133 (1976)

S. grandiflorum var. pulverulentum.

STEROIDAL ALKALOID: grandiflorine.

Friere, *C. R. Acad. Sci.,* **105**, 1074 (1887)

S. havanense.

STEROIDAL ALKALOIDS: etioline, tomatidenol.

Ripperger, *Pharmazie,* **32**, 537 (1977)

S. hispidum.

STEROIDAL ALKALOID: juripidine.

STEROIDS: hispidogenin, hispidin, hispigenin, neosolaspigenin, solaspigenin.

DITERPENOID: solanolide.

(A) Chakravarti *et al., Phytochemistry,* **22**, 2843 (1983)

(S, T) Chakravarti *et al., ibid.,* **19**, 1249 (1980)

S. hypomalacophyllum.

STEROIDAL ALKALOID: solaphyllidine.

STEROID: andesgenin.

(A) Usabillaga *et al., J. Am. Chem. Soc.,* **92**, 700 (1970)

(S) Gonzalez *et al., Phytochemistry,* **14**, 2483 (1975)

S. jamaicense.

STEROIDS: caelogenin, isocaelogenin.

Dopke *et al., Z. Chem.,* **16**, 104 (1976)

S. khasianum var. chatterjeeum sengupta.

STEROIDAL ALKALOID: solamargine.

Sath, *J. Inst. Chem., (Calcutta),* **43**, 116 (1971)

S. kieseritzkii SAM.

STEROIDAL ALKALOIDS: solamargine, tomatidine.

Briggs, Vining, *J. Chem. Soc.,* 2809 (1953)

S. laciniatum Ait.

STEROIDAL ALKALOIDS: solasodine, solaradixine.

STEROID: diosgenin.

(A) Schreiber *et al., Tetrahedron, 20, 1939* (1964)

(S) Irismetov *et al., Khim. Prir. Soedin.,* 668 (1976)

S. lainanense.

STEROIDAL ALKALOID: solasodenone.

Adam *et al., Phytochemistry,* **17**, 1070 (1978)

S. lycopersicum L.

STEROIDAL ALKALOID: solanocapsine.

Breyer-Brandwyk, *Bull. Sci. Pharmacol.,* **36**, 541 (1929)

S. lyratum.

GLYCOSIDES: aspidistrin, methylprotoaspidistrin, glycoside SL-0.

Yahara *et al., Phytochemistry,* **24**, 2748 (1985)

S. mammosum.

Two uncharacterized alkaloids.

Saelkoff, *Arch. Pharm.,* **301**, 111 (1968)

S. marginatum.

STEROIDAL ALKALOID: solamargine.

Briggs *et al., J. Chem. Soc.,* 3587 (1952)

S. megacarpum Koidz.

STEROIDAL ALKALOIDS: megacarpidine, megacarpine.

Briggs *et al., J. Chem. Soc.,* 3 (1942)

S. melongena.
SESQUITERPENOIDS: aubergenone, lubimin, 9-oxonerolidol.
  Stoessl et al., Can. J. Chem., **53**, 3361 (1975)

S. melongena var. esculentum.
STEROIDAL ALKALOIDS: solamargine, solasodine, solasonine.
STEROID: β-sitosterol.
  El-Khrisy et al., Fitoterapia, **57**, 440 (1986)

S. nemorensis var. bulgarica.
SESQUITERPENOIDS:
  3-oxo-8-epoxyeremophila-1,7-dien-8,12-olide,
  3-oxo-8-eremophila-1,7-dien-8,12-olide.
  Jizba et al., Collect. Czech. Chem. Commun., **43**, 1113 (1978)

S. nigrum L.
STEROIDAL ALKALOIDS: solamargine, α-solamargine, solasonine.
SAPONIN: uttronin A.
STEROIDS: uttroside A, uttroside B.
  (A) Aslanov et al., Khim. Prir. Soedin., 674 (1971)
  (Sap, S) Sharma et al., Phytochemistry, **22**, 1241 (1983)

S. oleraceum.
STEROIDAL ALKALOIDS: solamargine, solasonine.
  Bite, Shabana, Egypt. J. Pharm. Sci., **16**, 85 (1975); Chem. Abstr., 87 35893s

S. panduraeforme.
One uncharacterized steroidal alkaloid.
  Velsman, Louw, J. S. Afr. Chem. Inst., **2**, 119 (1949)

S. paniculatum L.
STEROIDAL ALKALOIDS: isojuribidine, isojuripidine, isopaniculidine, juribine, paniculidine.
STEROIDS: neochlorogenin, peniculogenin, tigogenin, epi-tigogenin.
  (A) Schreiber, Ripperger, Tetrahedron Lett., 5997 (1966)
  (S) Ripperger et al., Chem. Ber., **100**, 1741 (1967)

S. persicum.
STEROIDAL ALKALOID: solapersine.
  Novruzov et al., Khim. Prir. Soedin., 434 (1975)

S. platanifolium.
STEROIDAL ALKALOIDS: salatifoline, solamargine, solasonine.
  Puri, Bhatnagar, Phytochemistry, **14**, 2096 (1975)

S. polyadenium Greenm.
STEROIDAL GLYCOALKALOID: polyanine.
  Schreiber, Annalen, **674**, 168 (1964)

S. pseudocapsicum L.
STEROIDAL ALKALOIDS: isosolanocapine, solanocapine, epi-solanocapine, solanocapsidine, solanocapsine.
  Chakravarti et al., J. Chem. Soc., Perkin I, 467 (1984)

S. pseudoquina.
STEROIDAL ALKALOID: solaquidine.
  Usubillaga et al., Phytochemistry, **16**, 1961 (1976)

S. pubescens.
STEROIDAL ALKALOIDS: solanopubamides A and B.
  Kumari et al., Phytochemistry, **25**, 2003 (1986)

S. rostratum Dun.
STEROIDAL ALKALOIDS: solamargine, solasonine.
  Nabiev, Shakirov, Khim. Prir. Soedin., 116 (1974)

S. seaforthianum.
STEROIDAL ALKALOIDS: isosolaseaforthine, solanoforthine, solaseaforthine.
  Ali et al., Tetrahedron Lett., 3871 (1978)

S. sisymbriifolium.
STEROIDS: isonuatigenin, nuatigenin.
  Tschesche et al., Tetrahedron, **20**, 287 (1954)

S. sodameum.
STEROIDAL ALKALOID: solasonine.
  Briggs, Vining, J. Chem. Soc., 2809 (1953)

S. surattense.
STEROIDAL ALKALOID: solasurine.
  Seth, Chatterjee, J. Inst. Chem. (Calcutta), **41**, 194 (1968)

S. tomatillo.
STEROIDAL ALKALOIDS: dihydrotomatillidine, tomatillidine.
  Kusano et al., Chem. Pharm. Bull., **24**, 661 (1976)

S. tomentosa Decne.
STEROIDS: diosgenone, yamagenone.
  Morales-Mendez et al., Rev. Fac. Farm. Univ. Los Andes, **15**, 133 (1974)

S. torvum.
STEROIDAL ALKALOIDS: solasodeine, solasodine.
GENINS: chlorogenin, neochlorogenin, torvogenin.
SAPONIN: torvonin A.
  (A, G) Doepke et al., Pharmazie, **30**, 755 (1975)
  (Sap) Mahmood et al., Phytochemistry, **24**, 2456 (1985)

S. tovarense.
STEROIDS: diosgenone, yamagenone.
  Morales-Mendez et al., Rev. Fac. Farm. Univ. Los Andes, **15**, 133 (1974)

S. transcaucasicum Pojark.
STEROIDAL ALKALOIDS: solamargine, α-solamargine, solasonine.
  Aslanov, Khim. Prir. Soedin., 132 (1972)

S. tripartitum Dunal.
SOLAMINE ALKALOIDS: solapalmitenine, solapalmitine,
  Kupchan et al., J. Am. Chem. Soc., **89**, 5718 (1967)

S. tuberosum L.
STEROIDAL ALKALOIDS: solanidol, allo-solanidol, solanidine, solanidine-T, solanine, tolatid-5-en-3β-ol.
FLAVONOIDS: cyanoside, isoquercitrin, rutin.
PHENOLIC: chlorogenic acid.
STEROIDS: barogenin, citrostadienol, 24-methylenelophenol, β-sitosterol.
SESQUITERPENOIDS: lubimin, epi-lubimin, rishitin, rishitinol, 1(10),3,11-spirovetivatrien-2-one.
  (A) Rosen, Rosen, Chem. Ind., 1581 (1954)
  (F, P) Bondarenko et al., Khim. Prir. Soedin., 647 (1983)
  (S) Kaneko et al., Phytochemistry, **16**, 791 (1977)
  (T) Coxon et al., Tetrahedron Lett., 2929 (1974)

S. verbascifolium.
STEROIDAL ALKALOIDS: solasonine, solaverbascine.
  Adam et al., Phytochemistry, **19**, 1002 (1980)

S. vespertilio.
STEROIDS: anosmagenin, bejamarin, 15-dehydro-14β-anosmagenin, 20S-hydroxyvespertilin, vespertilin.
  Gonzalez et al., An. Quim., **70**, 250 (1974)

S. villosum.
STEROIDAL ALKALOID: solavilline.
  Bogner et al., Dtsch. Akad. Landsirtschatswiss, Berlin, **27**, 87 (1961)

S. xanthocarpum.
STEROIDAL ALKALOIDS: solasodine, solasonine.
STEROIDS: carpesterol, 4α-methyl-24-methylcholest-7-ene-3β,22-diol, 4α-methyl-24-methylcholest-7-ene-3α,25-diol, norcarpesterol.
  (A) Briggs, Vining, J. Chem. Soc., 2809 (1953)
  (S) Kusano et al., Phytochemistry, **14**, 1679 (1975)

S. xylocarpum.
STEROIDAL ALKALOID: solasonine.
STEROID: 4α-methyl-24-methylcholest-7-ene-3β,22-diol.
  Briggs, Vining, J. Chem. Soc., 2809 (1953)

SOLENANTHUS (Boraginaceae)

S. circinnatus Lbd.
PYRROLIZIDINE ALKALOID: echinatine.
Akramov *et al.*, *Dokl. Akad. Nauk SSSR*, **6**, 28 (1964)

S. coronatus Rgl.
PYRROLIZIDINE ALKALOID: echinatine.
Kiyamitdinova *et al.*, *Khim. Prir. Soedin.*, 911 (1967)

S. hirsutus Rgl.
PYRROLIZIDINE ALKALOID: echinatine.
Akramov *et al.*, *Dokl. Akad. Nauk SSSR*, **6**, 28 (1964)

S. karateginus Lipsky.
PYRROLIZIDINE ALKALOIDS: echinatine, turkestanine.
Akramov *et al.*, *Dokl. Akad. Nauk SSSR*, **6**, 28 (1964)

S. turkestanicus (Rgl. et Smirn.) Kuzn.
PYRROLIZIDINE ALKALOIDS: rindlerine, turkestanine.
Akramov *et al.*, *Dokl. Akad. Nauk SSSR*, **10**, 29 (1970)

SOLENOSTEMON

S. monostachys.
DITERPENOIDS: Mon-A, Mon-B, Mon-C.
Miyase *et al.*, *J. Chem. Soc., Chem. Commun.*, 859 (1977)

S. sylvaticus.
DITERPENOIDS: 19-acetoxycoleon O, 17-acetoxycoleon P, 14-acetoxycoleon Q, 16-acetoxycoleon W, 16-acetoxycoleon X, coleon X, coleon Z, 12-deacetyl-7,14-diacetoxycoleon Q, 12-deacetyl-3,7,14-triacetoxycoleon Q, 6,14-diacetoxycoleon Q, syl-A, syl-B, syl-C, syl-D.
Miyase *et al.*, *Helv. Chim. Acta*, **62**, 2374 (1979)

SOLENOSTOMA (Musci)

S. triste (Nees) K.Mull.
DITERPENOIDS: (16R)-ent-11α-hydroxykauran-15-one, ent-kaur-16-ene-11α,15-diol.
Connolly, Thornton, *J. Chem. Soc., Perkin I*, 736 (1973)

SOLIDAGO (Compositae)

S. altissima L.
ACETYLENICS: dehydromatricarin ester, dehydromatricarin lactone, methyl-10-(2-methyl-2-butenoyloxy)-cis-2-cis-8-decadiene-4,6-diynoate.
DITERPENOIDS: 6-angeloyloxykolavenic acid, solidagolactones I, II, III, IV, V, VI, VII and VIII, solidagonic acid, tricyclosolidagolactone.
Nishima *et al.*, *Tetrahedron Lett.*, 2809 (1984)

S. arguta.
DITERPENOIDS: 15,16-epoxy-cis-cleroda-3,13(16),14-triene, 15,16-epoxy-cis-cleroda-3,13(16),14-triene-19,6-olide.
McCrindle *et al.*, *J. Chem. Soc., Perkin I*, 1590 (1976)

S. canadensis.
DITERPENOIDS: solidagenone, solidago diterpene A.
Anthonsen *et al.*, *Tetrahedron*, **25**, 2233 (1969)

S. elongata.
DITERPENOIDS: elongatolides A, B, C, D, E and F.
Anthonsen *et al.*, *Acta Chem. Scand.*, **23**, 1068 (1969)

S. gigantea.
DITERPENOID: solidago diterpene A.
Gerlach, *Pharmazie*, **20**, 523 (1965)

S. gigantea var. serotina.
DITERPENOIDS: solidagoic acids A and B.
Anthonsen *et al.*, *Acta Chem. Scand.*, **22**, 351 (1968)

S. juncea.
DITERPENOIDS: epoxyjunceic acid, junceic acid, junceanols W, X and Y.
Henderson *et al.*, *Can. J. Chem.*, **51**, 1322 (1973)

S. missouriensis.
DITERPENOIDS: ent-7,13-abietadiene, ent-7,13-abietadien-2-ol acetate, ent-7,13-abietadien-3-ol, ent-7,13-abietadien-3-one, ent-8,13S-epoxy-14-labden-3-one, ent-13-epi-manoyl oxide, missourienols A, B and C.
Anthonsen, Bergland, *Acta Chem. Scand.*, **27**, 1073 (1973)

S. nemoralis.
DITERPENOIDS: ent-7,13-abietadien-18-oic acid, ent-7,13-abietadien-5-ol, rugosolide.
Bohlmann *et al.*, *Phytochemistry*, **19**, 2655 (1980)

S. sempervirens.
DITERPENOIDS: sempervirenic acid, solidagenone.
Purushothaman *et al.*, *Phytochemistry*, **22**, 1042 (1983)

S. serotina.
DITERPENOIDS: solidagoic acid A, solidagoic acid B.
Anthonsen *et al.*, *Can. J. Chem.*, **51**, 1332 (1973)

S. virgaurea L.
DITERPENOIDS: virgaureagenins A, B, C, D, E, F and G.
Hiller *et al.*, *Herba Hung.*, **19**, 91 (1980)

SOLORINA

S. crocea.
ANTHRAQUINONE PIGMENTS: averantin, averantin 6-methyl ether, 4,4'-disolarinic acid, norsolorinic acid, solorinic acid.
Steglich, Jedtke, *Z. Naturforsch.*, **31C**, 97 (1976)

SONCHUS (Compositae)

S. arvensis L.
FLAVONOIDS: linarin, luteolin 7-glucoside.
Bondarenko *et al.*, *Chem. Abstr.*, 88 186099j

S. jacquini.
SESQUITERPENOID: jacquinelin.
Barrera *et al.*, *J. Chem. Soc., C*, 1298 (1966)

S. macrocarpus.
SESQUITERPENOID: sonchucarpolide.
Mahmood *et al.*, *Phytochemistry*, **22**, 1290 (1983)

S. pinnatus.
SESQUITERPENOID: jacquinelin.
Barrera *et al.*, *J. Chem. Soc., C*, 1298 (1966)

S. radicatus.
SESQUITERPENOID: jacquinelin.
Barrera *et al.*, *J. Chem. Soc., C*, 1298 (1966)

S. tubifer.
SESQUITERPENOID: tubiferine.
Barrera *et al.*, *Tetrahedron Lett.*, 3475 (1967)

SONNERATIA

S. apetala.
GIBBERELLIN: tetrahydrogibberellin A-3.
Gaskin *et al.*, *Chem. Ind.*, 424 (1972)

SOPHORA (Leguminosae)

S. alopecurioides L. (Goebelia alopeculioides (L.) Bge.).
MATRINE ALKALOIDS: 13,14-dehydrosophoridine, 3-hydroxysophoridine, matrine, neosophoramine, sophocarpine, sophoramine, (-)-sophoridine, Sparteine alkaloid baptifoline.
Orekhov *et al.*, *Ber. Dtsch. Chem. Ges.*, **68**, 431 (1935)
Kusmuradov *et al.*, *Khim. Prir. Soedin.*, 231 (1978)

S. angustifolia var. flavescens Sieb. et Zucc.
MATRINE ALKALOIDS: matrine, oxymatrine,
Kondo et al., Arch. Pharm., 275, 493 (1937)

S. chrysophylla.
QUINOLIZIDINE ALKALOIDS: cytisine, mamanine, mamanine N-oxide.
Kadooka et al., Tetrahedron, 32, 919 (1976)

S. flavescens Ait.
MATRINE ALKALOIDS: 7-dehydrosophoramine, (+)-kuraramine, matrine, sophocarpine, sophoranol.
FLAVONOIDS: kushenol E, kushenol F.
SAPONIN: sophoraflavoside I.
(A) Kondo et al., Chem. Pharm. Bull., 26, 1832 (1978)
(F) Wu et al., J. Pharm. Soc., Japan, 105, 736 (1985)
(S) Yoshikawa et al., Chem. Pharm. Bull., 33, 4267 (1985)

S. franchetiana.
QUINOLIZIDINE ALKALOIDS: tsukushinamines A, B and C.
Ohmiya et al., Phytochemistry, 20, 1997 (1981)

S. griffithii Stocks.
MATRINE ALKALOIDS: argentine, matrine, pachycarpine, sophoramine.
QUINOLIZIDINE ALKALOIDS: cytisine, N-methylcytisine.
Aslanov et al., Khim. Prir. Soedin., 398 (1972)

S. japonica L.
MATRINE ALKALOIDS: matrine, sophocarpine.
QUINOLIZIDINE ALKALOIDS: cytisine N-methylcytisine.
FLAVONOIDS: anhydropisatin, biochanin A 7-β-D-gentiobioside, biochanin A 7-β-D-xylosylglucoside, irisolidone, irisolidone 7-O-glucoside, rutin, sissotrin.
ISOFLAVAN: sophoraflavanoloside.
ISOFLAVONES: biochanin A, 5,7-dihydroxy-3′,4′-methylenedioxyisoflavone, sopharol.
PHYTOALEXINS: (+)-maackiain, (+)-medicarpin.
STEROID: β-sitosterol.
(A) Abdusalamov et al., Khim. Prir. Soedin., 658 (1972)
(F) Takeda et al., Phytochemistry, 16, 619 (1977)
(I, S) Komatsu et al., J. Pharm. Soc., Japan, 96, 254 (1976)
(P) Van Etten et al., Phytochemistry, 22, 2291 (1983)

S. microphylla Ait.
MATRINE ALKALOID: matrine.
QUINOLIZIDINE ALKALOIDS: methylcytisine, cytisine.
Govindachari et al., J. Chem. Soc., 3839 (1957)

S. moorcroftiana Benth.
MATRINE ALKALOIDS: α-matrine, oxymatrine, sophocarpine.
Faugeras et al., Plant Med. Phytother., 10, 85 (1976)

S. pachycarpa Schrenk. (Goebelia pachycarpa (Schrenk) Bge.).
MATRINE ALKALOIDS: goebeline, isosophocarpine, matrine, matrine N-oxide, sophoramine, sophocarpine N-oxide, sophoramine.
QUINOLIZIDINE ALKALOIDS: cytisine, (+)-sparteine.
FLAVONOID: genistein 7-O-xyloglucoside.
(A) Pakanaev, Sadykov, J. Gen. Chem., USSR, 33, 1374 (1963)
(F) Sattikulov et al., Uzb. Khim. Zh., 11 (1983); Chem. Abstr., 100 82724j

S. secundiflora.
QUINOLIZIDINE ALKALOIDS: 11-allylcytisine, cytisine, N-methylcytisine, sparteine.
ISOFLAVONE: unanisoflavone.
(A) Keller et al., Phytochemistry, 14, 2305 (1975)
(I) Minhaj et al., Tetrahedron Lett., 2391 (1976)

S. speciosa.
QUINOLIZIDINE ALKALOID: cytisine.
Govindachari et al., J. Chem. Soc., 3839 (1957)

S. subprostrata.
PHENOLICS: sophoradin, sophoradochromene, sophoranachromene, sophoranone.
Komatsu et al., Chem. Pharm. Bull., 18, 602 (1970)

S. tetraptera J.Miller.
MATRINE ALKALOID: matrine.
QUINOLIZIDINE ALKALOIDS: cytisine, methylcytisine.
Kondo et al., J. Pharm. Soc., Japan, 266, 1 (1928)

S. tomentosa.
QUINOLIZIDINE ALKALOIDS: cytisine, (-)-epi-lamprolobine, (+)-lamprolobine N-oxide, 5-(3′-methoxycarbonylbutyryl)-aminoethyl-trans-quinolizidine.
ISOFLAVONES: isosophoranone, isosophoronol.
PTEROCARPANS: sophoracarpan A, sophoracarpan B.
(A) Murakoshi et al., Phytochemistry, 20, 1725 (1981)
(I) Delle Monache et al., Gazz. Chim. Ital., 107, 189 (1977)
(P) Kinoshita et al., Chem. Pharm. Bull., 34, 3067 (1986)

SORBARIA (Rosaceae)
S. arborea.
GLYCOSIDE: cardiospermin-p-hydroxybenzoate.
Nahrstedt, Z. Naturforsch., 31C, 397 (1976)

SORBUS (Rosaceae)
S. aucaparia L.
UNCHARACTERIZED TERPENOIDS: sorbicortol I, sorbicortol II.
Danoff, Zellner, Monatsh. Chem., 59, 307 (1932)

S. commixta.
FLAVONOIDS: luteolin 7-glucoside, quercetin, quercetin 3-glucoside, quercetin 3-glucoxyloside.
PHENOLICS: amygdalin, prunasin.
(F) Bylka et al., Herba Pol., 23, 105 (1977); Chem. Abstr., 88 186090z
(P) Takaishi, Kuwajima, Phytochemistry, 15, 1984 (1976)

SOULAMEA
S. amara Lam.
QUASSINOID: 15-O-benzoylbrucein D.
Bhatnager et al., Tetrahedron Lett., 1225 (1985)

S. fraxinifolia.
CARBOLINE ALKALOID: 1-(2′-hydroxyethyl)-β-carboline.
QUASSINOID: 6-acetoxypicrasin.
Charles et al., J. Nat. Prod., 49, 303 (1986)

S. muelleri.
DITERPENOIDS: soulameanone, soulameanone-1,12-diacetate.
Polonsky et al., Tetrahedron, 36, 2983 (1980)

S. pancheri.
QUASSINOID: picrasin B.
Hikino et al., Phytochemistry, 14, 2473 (1975)

S. tomentosa.
QUASSINOIDS: isobrucein A, solarubinone, souleameolide.
Polonsky et al., J. Chem. Soc., Chem. Commun., 641 (1979)

SOYMIDA
S. febrifuga.
TRITERPENOIDS: febrifugin, febrinins A and B, febrinolide.
Mallavarapu et al., Phytochemistry, 24, 305 (1985)

SPARTIUM (Leguminosae/Papilionaceae)

S. junceum L.
SPARTEINE ALKALOID: sparteine.
COUMARINS: chrysin, chrysin 7-gentiobioside, chrysin 7-β-D-glucopyranoside.
(A) Bellavita et al., Farm. Sci. e Tec., 3, 424 (1948)
(C) De Rosa et al., Phytochemistry, 22, 2323 (1983)

S. scoparium.
SPARTEINE ALKALOID: sparteine.
Bellavita et al., Farm. Sci. Tec., 3, 424 (1948)

SPATHELIA

S. glabrescens.
CHROMENE: sorbifolin.
Chan et al., J. Chem. Soc., 2548 (1967)

S. sorbifolia.
CHROMENES: allopteroxylin, sorbifolin.
NORTRITERPENOID: spathelin.
(C) Taylor et al., J. Chem. Soc., Perkin I, 397 (1977)
(T) Burke et al., Tetrahedron Lett., 5299 (1968)

S. wrightii.
DIPYRAN:
5-methoxy-2,2,8-trimethyl-10-senecioyl-2H,6H-benzo[1,2-b 5-4-b']dipyran-6-one.
Diaz et al., Phytochemistry, 22, 2090 (1983)

SPHAERANTHUS (Compositae)

S. indicus L.
UNCHARACTERIZED ALKALOID: sphaeranthine.
Basu, Lamsal, J. Am. Pharm. Assoc., 35, 174 (1946)

SPHAEROCARPOS

S. texanus.
FLAVONOIDS: luteolin 7,4'-di-O-glucuronide, luteolin 7-O-glucuronide.
Markham et al., Phytochemistry, 15, 151 (1976)

SPHAEROCOCCUS (Rhodophyceae)

S. coronipifolius L. (Goodenough et Woodward) C.A. Asgardh.
DITERPENOID: presphaerol.
HALOGENATED DITERPENOIDS: bromosphaerodiol, bromosphaerol, sphaerococcenol A.
Cafieri et al., Tetrahedron Lett., 963 (1979)

SPHAEROPHORUS

S. scrobiculatus.
DEPSIDONE: 4-O-methylhypoprotocetraric acid.
TRITERPENOIDS: 7β-acetoxyhopan-22-ol, 15,22-dihydroxyhopane.
Huneck, Tibell, J. Hattori Bot. Lab., 58, 203 (1985); Chem. Abstr., 104 17660d

SPHAEROPHYSA (Leguminosae)

S. salsula (Pall.) DC.
ALKALOIDAL AMINE: sphaerophysine.
Rubinshtein, Men'shikov, J. Gen. Chem., USSR., 14, 161 (1944)
Rubinshtein, Men'shikov, ibid., 14, 172 (1944)

SPHAGNUM (Musci)

S. magellanicum Brid.
LIGNINS: none present. Phenanthrofuran sphagnorubin.
(L) Nimz, Tutschek, Holzforschung, 31, 101 (1977)
(Ph) Rudolph et al., Z. Naturforsch., 24B, 1211 (1969)

SPILANTHES

S. oleracea Jacq.
FLAVONOIDS: apigenin 7-glucoside, apigenin 7-neohesperidoside, quercetin 3-glucoside, rutin.
Verykokidou-Vitsaropoulos et al., Arch. Pharm., 316, 815 (1983)

SPINACEA

S. oleracea.
SAPONINS: spinasaponin A, spinasaponin B.
Tschesche et al., Justus Liebigs Ann. Chem., 726, 125 (1969)

SPIRAEA (Rosaceae)

S. cantoneniensis.
TRITERPENOIDS: betulinic acid, 3-epi-betulinic acid.
Tanabe et al., J. Pharm. Soc., Japan, 96, 248 (1976)

S. hypericifolia.
CATECHINS: (+)-catechol 7-β-L-arabinofuranoside, (+)-catechol 7-α-L-rhamnofuranoside, catechol 7-xyloside.
FLAVONOIDS: apigenin, apigenin 5-β-D-glucopyranoside, luteolin, luteolin 5-β-D-glucopyranoside.
Chumbalov et al., Khim. Prir. Soedin., 103 (1976)

S. japonica L.
DITERPENE ALKALOIDS: spiradines A, B, C, D, E, F and G, spiredine, spireine.
Gorbunov et al., Khim. Prir. Soedin., 124 (1976)

S. koreana.
DITERPENE ALKALOIDS: spirajine, spireine.
Jin, Chem. Abstr., 70 26539

S. thunbergii.
TRITERPENOIDS: glutinol, taraxerol.
Tanabe et al., J. Pharm. Soc., Japan, 96, 248 (1976)

S. tosaensis.
TRITERPENOIDS: glutinone, taraxerol.
Tanabe et al., J. Pharm. Soc., Japan, 96, 248 (1976)

SPIROSTACHYS

S. africana.
DITERPENOID: stachenone.
Baarschers et al., J. Chem. Soc., 4046 (1962)

SPONDIAS

S. mangifera.
AMINO ACIDS: alanine, cystine, glycine, leucine, serine.
Saxena et al., J. Inst. Chem., (India), 49, 107 (1977); Chem. Abstr., 87 65354j

SPORORMIA

S. affinis.
TERPENOID: terpenoid LL-N313τ.
McGahren et al., J. Am. Chem. Soc., 96, 1616 (1974)

SPREKELIA (Amaryllidaceae)

S. formosissima Herb.
ERYTHRINA ALKALOIDS: 3-epi-macronine, pretazettine.
ISOQUINOLINE ALKALOID: lycorine.
Wildman, Bailey, J. Org. Chem 33, 3749 (1968)

STACHYS (Labiatae)

S. annua (L.) L.
DITERPENOIDS: annuanone, stachylone, stachysolone.
Popa et al., Khim. Prir. Soedin., 324 (1974)

S. atherocalyx.
FLAVONOIDS: acetylspectabiflaside, acetylstachyflaside, spectabiflaside, stachyflaside.
IRIDOIDS: harpagide, harpagide acetate.
    (F) Kostyuchenko *et al., Khim. Prir. Soedin.,* **98**, 106 (1976)
    (I) Komissarenko *et al., ibid.,* 187 (1982)

S. balansae Boiss. et Kotschy.
QUATERNARY ALKALOID: stachydrine.
    Pulatova, *Khim. Prir. Soedin.,* 62 (1969)

S. betoniciflora Rupr.
QUATERNARY ALKALOID: stachydrine.
    Pulatova, *Khim. Prir. Soedin.,* 62 (1969)

S. grossheimii.
IRIDOIDS: harpagide, harpagide acetate.
    Komissarenko *et al., Khim. Prir. Soedin.,* 109 (1976)

S. hissarica Rgl.
QUATERNARY ALKALOID: stachydrine.
    Pulatova, *Khim. Prir. Soedin.,* 62 (1969)

S. iberica.
IRIDOIDS: harpagide, harpagide acetate.
    Komissarenko *et al., Khim. Prir. Soedin.,* 109 (1976)

S. inflata.
FLAVONOID: stachyflaside.
    Komissarenko *et al., Khim. Prir. Soedin.,* 98 (1976)

S. lanata Jacq.
QUATERNARY ALKALOID: stachydrine.
    Pulatova, *Khim. Prir. Soedin.,* 62 (1969)

S. lavandullillia.
IRIDOIDS: harpagide, harpagide acetate.
    Komissarenko *et al., Khim. Prir. Soedin.,* 109 (1976)

S. palustris.
FLAVAN GLYCOSIDE:
acetyl-4',5,8-trihydroxyflavan-7-α-D-glucopyranoside-β-D-glucopyranoside.
    Bishara *et al., Ukr. Khim. Zh.,* **42**, 284 (1976); *Chem. Abstr.,* 85 2514z

S. spectabilis.
IRIDOIDS: harpagide, harpagide acetate.
    Komissarenko *et al., Khim. Prir. Soedin.,* 109 (1976)

S. sylvatica.
QUATERNARY ALKALOIDS: betonicine, turicine.
IRIDOIDS: harpagide, harpagide acetate.
DITERPENOID: stachysic acid.
    (A) Paudler, Wagner, *Chem. Ind.,* 1693 (1962)
    (I) Komissarenko *et al., Khim. Prir. Soedin.,* 109 (1976)
    (T) Popa *et al., Khim. Prir. Soedin.,* 447 (1974)

S. tubifera Naudin.
QUATERNARY ALKALOID: stachydrine.
    Henry, King, *J. Chem. Soc.,* 2866 (1950)

STACHYTARPHETA (Verbenaceae)

S. jamaicensis (L.) Vahl. (S. indica Vahl.).
FLAVONOIDS: apigenol 7-glucuronide, 6-hydroxyluteolol 7-glucuronide, luteolol 7-glucuronide.
IRIDOID: tarphetalin.
PHENOLIC: chlorogenic acid.
    (F, P) Duret *et al., Plant Med. Phytother.,* **10**, 96 (1976)
    (I) Jawad *et al., Egypt. J. Pharm. Sci.,* **18**, 511 (1977)

S. mutabilis.
IRIDOIDS: 6-hydroxyipolamiide, ipolamiide.
    de Luca *et al., Phytochemistry,* **22**, 1185 (1983)

STACHYURUS (Stachyuraceae)

S. praecox Sieb. et Zucc.
ELLAGITANNINS: casuarictin, casuarinin, casuriin, pedunculagin, stachyurin, strictinin, tellimagrandin I.
STEROIDS: stachysterones A, B, C and D.
    (E) Okuda *et al., J. Chem. Soc., Perkin I,* 1265 (1983)
    (S) Hikino *et al., Chem. Pharm. Bull.,* **23**, 1458 (1975)

STAPHYLEA (Staphyleaceae)

S. pinnata L.
AMINO ACIDS: isoleucine, oxypinnatane, pinnatenine.
    Grove *et al., Tetrahedron,* **29**, 2715 (1973)

STAUNTONIA

S. hexaphylla.
ANTHOCYANINS: chrysanthemin, cyanidin 3-xylosylglucoside.
GLYCOSIDES: mubenins A, B and C.
    (An) Ishikura *et al., Phytochemistry,* **15**, 442 (1976)
    (Gl) Takemoto, Kometani, *Ann. Chim.,* **685**, 237 (1965)

STEGANOTAENIA

S. araliacea Hochst.
LIGNANS: araliangin, steganacin, steganangin, steganol steganone.
    Kupchan *et al., J. Am. Chem. Soc.,* **95**, 1335 (1973)

STEIRACTINIA

S. mollis.
SESQUITERPENOIDS: steiractinolides A, B, C, D, E and F, 6-tiglylivangustin, 6-tiglylivangustin 1-acetate.
    Bohlmann *et al., Ann. Chim.,* 962 (1983)

STELLERA

S. chamaejasme L.
COUMARINS: daphnetin, daphnoretin, umbelliferone.
FLAVANONES: chamaejasmine, isochamaejasmine.
    (C) Modonova *et al., Khim. Prir. Soedin.,* 709 (1985)
    (F) Niwa *et al., Chem. Pharm. Bull.,* **34**, 3249 (1986)

STEMMADENIA

S. belle.
TRITERPENOID: taraxasterol acetate.
    Talapatra *et al., Indian J. Chem.,* **11**, 977 (1973)

S. donnell-smithii R.E.Woodson.
INDOLE ALKALOIDS: (+)-quebrachamine, tabernanthine.
    Bartlett *et al., J. Am. Chem. Soc.,* **80**, 126 (1958)

STEMODIA (Scrophulariaceae)

S. maritima.
DITERPENOIDS: maritimol, stemarin, stemodin, stemodinol, stemodinone, stemolide.
    Hufford *et al., J. Pharm. Sci.,* **65**, 778 (1976)

STEMONA

S. collinsae.
BENZOPYRONES: stemonal, stemonacetal.
    Shiengthong *et al., Tetrahedron Lett.,* 2015 (1974)

S. japonica Miq.
ALKALOIDS: isostemonamine, protostemonine, protostephanine, stemmonamine, stemofoline, stemonidine.
    Iuzuka *et al., J. Chem. Soc., Chem. Commun.* 125 (1973)

S. ovata Nakai.
ALKALOIDS: stemonidine, stemonine.
    Suzuki, *J. Pharm. Soc., Japan,* **49**, 78 (1929)

S. sessilifolia Miq.
UNCHARACTERIZED ALKALOIDS: stenine, tuberostemonine, tuberostemonine A.
Furuya, *Chem. Zentralbl.*, 1823 (1913)
Edwards *et al.*, *Can. J. Chem.*, **40**, 455 (1962)

S. tuberosa.
ALKALOIDS: oxotuberostemonine, stemonine, stenine.
Uyeo *et al.*, *Chem. Pharm. Bull.*, **15**, 768 (1967)

STENACHAENIUM
S. macrocephalum.
ACID: 9,10-epoxy-3,12-octadecadienoic acid.
Kleinman *et al.*, *Lipids*, **6**, 617 (1971)

STENOCALYX
S. michelii.
NORSESQUITERPENOIDS: τ-elemene, furanoelemene, isofuranodiene.
SESQUITERPENOID: selina-4(14),7(11)-diene.
Ruecker *et al.*, *Planta Med.*, **31**, 322 (1977)

STENOCARPUS
S. salignus.
LAPACHONES: stenocarpinoquinone A, stenocarpinoquinone B, stenocarpinoquinone C.
Mock *et al.*, *Aust. J. Chem.*, **26**, 1121 (1973)

STENOSOLON
S. heterophylla (Vahl.) Mgf.
IMIDAZOLIDINOCARBAZOLE ALKALOIDS: ervafolene, ervafolidine, 3-epi-ervafolidine, ervafoline, 19'-hydroxyervafolene, 19'-hydroxyervafoline.
Henriques *et al.*, *Acta Chem. Scand.*, B **34**, 509 (1980)

STEPHANIA (Menispermaceae)
S. abyssinica (Dill. et Rich.) Walp.
OXOAPORPHINE ALKALOID: oxoxylopine.
PHENANTHRENE ALKALOIDS:
6-acetyl-6-dihydro-epi-stephamiersine, metaphanine, protostephabyssine, staphaboline, stephabyssine, stephavanine.
Tomita *et al.*, *Tetrahedron Lett.*, 3605 (1964)
Van Wyk *et al.*, *J. S. Afr. Chem. Inst.*, **28**, 284 (1975)

S. capilin Spreng.
APORPHINE ALKALOID: stephanine.
Tomita, Shirai, *J. Pharm. Soc., Japan*, **62**, 381 (1942)

S. capitata.
APORPHINE ALKALOID: crebanine.
Tomita, Shirai, *J. Pharm. Soc., Japan*, **62**, 381 (1942)

S. cepharantha Hayata.
APORPHINE ALKALOIDS: cepharadiones A and B, dehydrostephanine.
BISBENZYLISOQUINOLINE ALKALOIDS: cepharanthine, cepharoline, cycleanine, isotetrandrine, phaeanthine.
NORAPORPHINE ALKALOIDS: cepharanone A, cepharanone B.
PHENANTHRENE ALKALOID: cepharamine.
Akasu *et al.*, *Tetrahedron Lett.*, 3609 (1974)
Kunimoto *et al.*, *J. Pharm. Soc., Japan*, **101**, 1951 (1981)

S. delavayi Diels.
PHENANTHRENE ALKALOIDS: delavanine, 16-oxodelavanine.
Fedeeva *et al.*, *Khim. Prir. Soedin.*, 140 (1970)

S. dinklagei.
APORPHINE ALKALOID: stephalegine.
Tackie *et al.*, *Lloydia*, **37**, 6 (1974)

S. elegans.
PHENANTHRENE ALKALOID: isosinoacutine.
Khosa *et al.*, *Chem. Ind. (London)*, 662 (1980)

S. epigeae H.S.Lo.
APORPHINE ALKALOID: (-)-cassythicine.
PHENANTHRENE ALKALOIDS: sinoacutine, sinomenine.
Ding *et al.*, *Chem. Abstr.*, 105 168903g

S. glabra (Roxb.) Miers.
BISBENZYLISOQUINOLINE ALKALOIDS: cycleanine, N-demethylcycleanine.
PROTOBERBERINE ALKALOIDS: stepharamine, (-)-stepholidine.
QUATERNARY ALKALOIDS: dehydrocorydalmine, stepharanine, (-)-tetrahydropalmatine.
SPIROAPORPHINE ALKALOIDS: pronuciferine, stepharine.
UNCHARACTERIZED ALKALOIDS: gindarine, gindaricine, gindarinine.
Tomita, Kunimoto, *J. Pharm. Soc., Japan*, **58**, 4613 (1962)
Bhakuni, Gupta, *J. Nat. Prod.*, **45**, 407 (1982)

S. hernandifolia (Walp.) Willd. (S. discolor Spreng.).
BISBENZYLISOQUINOLINE ALKALOID: fangchinoline.
PHENANTHRENE ALKALOIDS: aknadine, aknadinine, 4-demethylhasubonine, 3-O-demethylhernandifoline, hernandifoline, hernandine, hernandoline, hernandolinol, methylhernandine, stephisoferuline.
Fadeeva *et al.*, *Farmatsiya*, **19**, 28 (1970)
Fesenko *et al.*, *Khim. Prir. Soedin.*, 158 (1971)

S. japonica Miers.
APORPHINE ALKALOIDS: oxostephanine, (-)-stephanine.
BISBENZYLISOQUINOLINE ALKALOIDS: homostephanoline, stebisimine, (+)-epi-stephanine, stephanoline, stepinonine.
PHENANTHRENE ALKALOIDS: hasubonine, hypo-epi-stephanine, metaphanine, 16-oxohasubonine, 16-oxoprometaphanine, prometaphanine, protometaphanine, protostephanaberrine, stephabenine, stephamiersine, epi-stephamiersine, stephanaberrine, stephasunoline.
Kondo *et al.*, *Chem. Pharm. Bull.*, **31**, 2574 (1983)
Matsui *et al.*, *J. Nat. Prod.*, **49**, 588 (1986)

S. longa L.
PHENANTHRENE ALKALOID: longanone.
Lao *et al.*, *Yaoxue Xuebao*, **16**, 940 (1981)

S. rotunda Lour.
ISOQUINOLINE ALKALOID: rotundine.
PROTOBERBERINE ALKALOID: stepharotine.
Cava *et al.*, *J. Org. Chem.*, **33**, 2785 (1968)

S. sasaki Hayata.
APORPHINE ALKALOIDS: crebanine, oxocrebanine, phanostenine, (R)-roermeroline, steporphine, stesakine.
PHENANTHRENE ALKALOID: aknadilactam.
QUATERNARY ALKALOID: N-methylpapaveraldinium.
SPIROAPORPHINE ALKALOID: N-acetylstephamarine.
Kunimoto *et al.*, *J. Pharm. Soc., Japan*, **101**, 431 (1981)

S. tetrandra S.Moore.
BISBENZYLISOQUINOLINE ALKALOIDS: menisidine, menisine, tetrandrine.
UNCHARACTERIZED ALKALOIDS: hanfanchines A, B and C.
Hsu *et al.*, *J. Chin. Chem. Soc.*, **7**, 123 (1940)

S. venosa.
APORPHINE ALKALOIDS: ayuthianine, (+)-corydine, sukhodianine, thailandine, uthongine.
Guinaudeau *et al.*, *J. Nat. Prod.*, **45**, 355 (1982)

STEPHANOTIS

S. japonica Makino.
STEROIDAL ALKALOIDS: stephanthranilines A, B and C.
STEROID: dihydrogagamin.
DITERPENOID: stephanol.
(A, S) Terada *et al.*, *Chem. Pharm. Bull.*, **27**, 2304 (1979)
(T) Fukoka, Mitsuhashi, *ibid.*, **16**, 553 (1968)

STERCULIA (Sterculiaceae)

S. alata.
ACID: sterculynic acid.
Jevans *et al.*, *Tetrahedron Lett.*, 2167 (1968)

S. colorata.
FLAVONOIDS: 6-glucuronosyloxykluteolin,
5,7,3',4'-tetrahydroxy-6-O-β-D-flucuronylflavone.
Nair *et al.*, *Phytochemistry*, **15**, 839 (1976)

S. foetida.
ANTHOCYANIN: cyanidin 3-O-glucoside.
FLAVONOIDS: isoscutellarin, luteolin 6-O-β-D-glucopyranoside,
procyanidin β-D-glucopyranoside.
Nair *et al.*, *Curr. Sci.*, **46**, 14 (1976)

S. pallens.
ACIDS: malvalic acid, sterculic acid.
Mustafa *et al.*, *J. Am. Oil Chem. Soc.*, **63**, 1191 (1986); *Chem. Abstr.*,
105 206282p

STEREOCAULON (Hepaticae)

S. colensoi.
DIOXEPIN: colensoic acid.
Fox *et al.*, *Phytochemistry*, **9**, 2567 (1970)

STEREOSPERMUM

S. kunthianum.
LIGNANS: (+)-cycloolivil, (-)-olivil.
Ghogomu-Tih *et al.*, *Planta Med.*, 464 (1985)

S. personatum.
IRIDOID: specioside.
De Silva *et al.*, *Indian J. Chem.*, **21B**, 703 (1982)

S. suaveolens DC.
STEROID: β-sitosterol.
SESQUITERPENOIDS: dehydro-α-lapachone, dehydrotectol,
lapachol.
Purushothaman *et al.*, *J. Res. Indian Med.*, **9**, 107 (1974); *Chem. Abstr.*,
85 2558s

STEREUM (Stereaceae/Thelephoraceae)

S. hirsutum (Willd. ex Fr.) Fr.
AMINE: mycosporine.
Favre-Bonvin *et al.*, *Can. J. Chem.*, **54**, 1105 (1976)

S. purpureum.
SESQUITERPENOIDS: methylhydroxysterpurate,
methylhydroxysterpurate ethylidene acetate, stereopolide,
sterpuric acid.
Ayer *et al.*, *Tetrahedron Suppl.*, 379 (1981)

STERNBERGIA (Amaryllidaceae)

S. fischeriana (Herb.) Roem.
ISOQUINOLINE ALKALOIDS: lycorine, sternine.
Proskurnina, Ismailov, *J. Gen. Chem., USSR.*, **23**, 2056 (1953)

S. lutea (L.) Ker.-Gawl.
ISOQUINOLINE ALKALOIDS: luteine, lutessine, lycorine, tazettine.
Evidente, *J. Nat. Prod.*, **49**, 90 (1986)

STEVIA

S. achalensis.
SESQUITERPENOIDS: achalensolide, achalensolone.
Oberti *et al.*, *J. Org. Chem.*, **48**, 4038 (1983)

S. amambayensis.
TWO BISABOLONES: Two melampolides Three sesquiterpenoids.
Zdero, Bohlmann, *Phytochemistry*, **25**, 1755 (1986)

S. baudiana.
SAPONIN: stevioside.
DITERPENOID: steviol.
Bridel, Lavialle, *J. Pharm. Chem.*, **14**, 99 (1931)
Bridel, Lavialle, *ibid.*, **14**, 154 (1931)
Bridel, Lavialle, *ibid.*, **14**, 321 (1931)
Bridel, Lavialle, *ibid.*, **14**, 369 (1931)

S. breviaristata.
DITERPENOID: breviarolide.
Oberti *et al.*, *Phytochemistry*, **25**, 1479 (1986)

S. jaliscensis.
SESQUITERPENOIDS:
7α-angeloyloxy-9β-tigloyloxy-1-oxo-longipinene,
9β-angeloyloxy-7α-tigloyloxo-1-oxo-longipinene.
Bohlmann *et al.*, *Chem. Ber.*, **109**, 3366 (1976)

S. japonica.
ALKALOID: stemospironine.
Sakata *et al.*, *Agr. Biol. Chem.*, **42**, 457 (1978)

S. lucida.
DITERPENOID: ent-15,16-epoxy-7,13(16),14-labdatrien-18-oic
acid.
Salmon *et al.*, *Phytochemistry*, **22**, 1412 (1983)

S. monardaefolia.
SESQUITERPENOIDS: 6-angeloyloxynidorellol,
11,13-dihydroeucannabinolide.
Gomez *et al.*, *Phytochemistry*, **22**, 197 (1983)

S. myriadenia.
SESQUITERPENOID:
8-hydroxy-1(10),3,11(13)-guaiatrien-12,6-olide.
Bohlmann *et al.*, *Phytochemistry*, **21**, 2021 (1982)

S. paniculata.
DITERPENE GLUCOSIDES: paniculosides I, II, III and IV.
Yamasaki *et al.*, *Chem. Pharm. Bull.*, **25**, 2895 (1977)

S. polycephala.
SESQUITERPENOID: stephalic acid.
Angeles *et al.*, *Phytochemistry*, **21**, 1804 (1982)

S. purpurea.
SESQUITERPENOIDS: bisabol-1-one,
2,3-epoxy-2,3-dihydrobisabol-1-one.
Bohlmann *et al.*, *Chem. Ber.*, **109**, 3366 (1976)

S. rebaudiana.
SAPONINS: dulcosides A and B, rebaudiosides A, B, C, D, E
and F.
DITERPENOID: austroinulin.
(Sap) Kobayashi *et al.*, *Phytochemistry*, **16**, 1405 (1977)
(T) Sakamoto *et al.*, *Chem. Pharm. Bull.*, **25**, 3437 (1977)

S. rhombifolia.
SESQUITERPENOID: stevin.
Rios *et al.*, *Tetrahedron*, **23**, 4265 (1967)

S. salicifolia. Cav.
DITERPENOIDS: salicifoliol, stevinsol.
Ortega *et al.*, *Rev. Latinoam. Quim.*, **11**, 45 (1980)

S. serrata.

ALKALOID: carmeline.

SESQUITERPENOIDS: christianins I, II and III, raistevione.

(A) Salmon *et al.*, *Rev. Latinoam. Quim.*, **6**, 45 (1975)

(T) Salmon *et al.*, *ibid.*, **8**, 172 (1978)

(T) Roman *et al.*, *Tetrahedron*, **37**, 2769 (1981)

STICTA (Hepaticae)

S. aurata.

PIGMENT: calycin.

Letcher *et al.*, *Tetrahedron Lett.*, 3541 (1967)

S. billardieri.

TRITERPENOIDS: hopane-7β,22-diol-7-acetate, hopane-15,22-diol.

Corbett, Young, *J. Chem. Soc., C*, 1564 (1966)

S. colensoi.

TRITERPENOIDS: stictane-3β,22α-diol, stictane-3β,22α-diol diacetate, stictane-2α,3β,22α-triol, stictane-2α,3β,22α-triol 2-acetate, stictane-2α,3β,22α-triol 3-acetate, stictane-2α,3β,2-triol 2,3-diacetate, stictane-2α,3β,22α-triol triacetate.

Chin *et al.*, *J. Chem. Soc., Perkin I*, 1437 (1973)

S. coronata.

TRITERPENOIDS: stictane-3β,22α-diol, stictane-3β,22α-diol 3-acetate, stictane-3β,22α-diol diacetate, stictane-2α,3β,22α-triol, stictane-2α,3β,22α-triol 2-acetate, stictane-2α,3β,2-triol 3-acetate, stictane-2α,3β,2-triol 2,3-diacetate, stictane-2α,3β,22α-triol triacetate, stictan-22-ol 3β-acetate.

Chin *et al.*, *J. Chem. Soc., Perkin I*, 1437 (1973)

S. flavicans.

TRITERPENOIDS: stictane-3β,22α-diol, stictane-3β,22α-diol diacetate, stictane-2α,3β,22α-triol, stictane-2α,3β,22α-triol 2-acetate, stictane-2α,3β,22α-triol 3-acetate, stictane-2α,3β,22α-triol 2,3-diacetate.

Chin *et al.*, *J. Chem. Soc., Perkin I*, 1437 (1973)

S. mougeotiana.

TRITERPENOIDS: 6α-acetoxy-7α,22α-hopane, 7α-acetoxy-6α,22α-hopane, hopane-11α,22-diol, hopane-6α,7α,22-triol.

Corbett, Cumming, *J. Chem. Soc., C*, 955 (1971)

STILBE

S. ericoides.

IRIDOID: stilbericoside.

Rimpler, Pistor, *Z. Naturforsch.*, **29C**, 368 (1974)

STILLINGIA

S. sylvatica.

PHORBOL TERPENOIDS: stillingia factors S-1, S-2, S-3, S-4, S-5, S-6, S-7 and S-8.

Adolf, Hecker, *Tetrahedron Lett.*, 2887 (1980)

STILPNOPAPPUS

S. pickelii.

SESQUITERPENOIDS:
8-O-(2-methylpeopenoyl)-stilpnotomentolide, 8-O-(2-tiglyl)-stilpnotomentolide.

Bohlmann *et al.*, *Phytochemistry*, **21**, 1045 (1982)

S. tomentosus.

SESQUITERPENOIDS:
8-O-(2-methylpropenoyl)-stilpnotomentolide, 8-O-(2-tiglyl)-stilpnotomentolide.

Bohlmann *et al.*, *Phytochemistry*, **21**, 1045 (1982)

STILPNOPHYTUM (Compositae/Anthemideae)

S. linifolium (Thunb.) Less.

SESQUITERPENOID: 12-acetoxy-10,11-dehydrongaione.

Bohlmann, Zdero, *Chem. Ber.*, **107**, 1071 (1974)

STIZOLOBIUM

S. hassjoo.

PYRAZINE ALKALOID: stizolanine.

AMINO ACID: stizolobic acid.

Yoshida *et al.*, *Phytochemistry*, **15**, 1723 (1976)

STIZOLOPHUS

S. balsamita (Lam.) Cass. (Centaurea balsamita Lam.).

UNCHARACTERIZED ALKALOID: stizolophine.

SESQUITERPENOID: balsamin.

(A) Kuzuvkov *et al.*, *J. Gen. Chem. USSR*, **23**, 157 (1953)

(T) Rybalko *et al.*, *Khim. Prir. Soedin.*, 467 (1976)

S. coronopifolius.

SESQUITERPENOID: stizolicin.

Mukhametzhanov *et al.*, *Khim. Prir. Soedin.*, 505 (1970)

STOEBE

S. plumosa.

PHENOLIC: stoebenone.

Bohlmann *et al.*, *Chem. Ber.*, **105**, 2604 (1972)

STOECHOSPERMUM

S. marginatum.

DITERPENOIDS:
19-acetoxy-5R,15,16-trihydroxyspata-13,17Z-diene, 19-acetoxy-5R,15,18R-trihydroxyspata-13,16E-diene, 5R,16-dihydroxyspata-13,17-diene, 5R,17R-dihydroxyspata-13,16E-diene, 5R-hydroxyspata-13,17-diene, 5-oxo-15,18R,19-trihydroxyspata-13,16E-diene, 5-oxo-15,18S,19-trihydroxyspata-13,16E-diene, 5R,15S,18R,19-tetrahydroxyspata-13,16E-diene, stoechospermol.

Gerwick *et al.*, *J. Org. Chem.*, **46**, 2233 (1981)

STOKESIA (Compositae)

S. laevis (Hill) Greene.

SESQUITERPENOIDS: 8,10-diacetoxy-1-hydroxyhirsutinolide, 8,10-diacetoxy-1-hydroxyhirsutinolide 13-acetate, 8,10-diacetoxy-1-hydroxyhirsutinolide 13-acetate-1-mthyl ether.

Bohlmann *et al.*, *Phytochemistry*, **18**, 987 (1979)

STOMANTHUS

S. africanus.

SESQUITERPENOID: 9-angeloyl-10,11-epoxydihydrofarnesol.

Bohlmann *et al.*, *Phytochemistry*, **21**, 1155 (1982)

STREBLUS

S. asper Lour.

GLYCOSIDES: asperoside, vijaloside.

STEROIDS: glucokamaloside, glucostrebloside, kamaloside, strebloside, strophalloside, strophanalloside.

(Gly) Saxena *et al.*, *Planta Med.*, 343 (1985)

(S) Manzetti, Reichstein, *Helv. Chim. Acta*, **47**, 2303 (1964)

(S) Manzetti, Reichstein, *ibid.*, **47**, 2320 (1964)

STREMPELIOPSIS (Apocynaceae)

S. strempelioides K.Schum.

IMIDAZOLIDINOCARBAZOLE ALKALOID: fannine.

BISINDOLE ALKALOID: (+)-strempeliopsidine.

INDOLE ALKALOIDS: (-)-aspidospermine, (+)-eburnamonine, (+)-haplocidine, (-)-haplocidine, pleiocarpamine, (-)-strempeliopine, (+)-tubotaiwine, (+)-tubotaiwine N-oxide.

Laguna et al., Planta Med., 50, 285 (1984)

STROPHANTHUS (Apocynaceae)

S. africanus.

SESQUITERPENOID: 9-angeloyloxy-10,11-epoxydihydrofarnesol.

Bohlmann et al., Phytochemistry, 21, 1155 (1982)

S. amboensis (Schinz) Engl. et Pax.

STEROIDS: amboside, sarmentocymarin.

Rochter et al., Helv. Chim. Acta, 37, 76 (1954)

S. barteri.

STEROID: sarmentocymarin.

Jacobs, Heidelberger, J. Biol. Chem., 81, 365 (1928)

S. courmontii.

STEROIDS: courmontosides A, B and C sarmentocymarin.

Schindler, Reichstein, Helv. Chim. Acta, 34, 1732 (1951)

S. divaricatus (Lour.) Hook et Arn.

SAPONINS: caudoside, divaricoside.

STEROIDS: caudogenin, divostroside, musaroside, panstroside, pseudocaudoside, sarmutoside.

Renkonen et al., Helv. Chim. Acta, 42, 182 (1959)

S. emeni.

STEROIDS: alloperiplogenin, emicymarin, isoemicymarin, isoperiplogenin.

Spenser et al., Experientia, 3, 323 (1947)

S. gerrardii.

STEROID: sarmutocymarin.

Reichstein et al., Helv. Chim. Acta, 33, 365 (1950)

S. grandiflorus.

STEROID: sarmentocymarin.

Reichstein et al., Helv. Chim. Acta, 33, 365 (1950)

S. intermedius Pax.

STEROID: inertoside.

TRITERPENOIDS: arriagoside, leptoside.

(S) Hegedus et al., Helv. Chim. Acta, 36, 357 (1953)

(T) Hegedus, Reichstein, ibid., 38, 1133 (1955)

S. kombe.

STEROIDS: erysimoside, erysimosol, glucocymarol, isoemicymarin, isoperiplocymarin, musaroside, panstroside, strophanthidin.

Nagata et al., Helv. Chim. Acta, 40, 41 (1957)

Kaiser et al., Justus Liebigs Ann. Chem., 643, 192 (1961)

S. nicholsonii.

STEROIDS: ericymarin, periplocin, periplocymarin.

Katz, Reichstein, Helv. Chim. Acta, 28, 478 (1945)

S. petersianus.

STEROID: sarmentocymarin.

Reichstein et al., Helv. Chim. Acta, 33, 365 (1950)

S. preussii Engler et Pax.

STEROID: sarmentocymarin.

Reichstein et al., Helv. Chim. Acta, 33, 365 (1950)

S. sarmentosus.

STEROIDS: sarmentocymarin, sarmentogenin, sarmentosides A, B, C, D and E, sarmentoside A 11-acetate, sarnovide, sarverogenin, sarveroside.

Richter et al., Helv. Chim. Acta, 37, 76 (1954)

STRYCHNOS (Loganiaceae)

S. aculeata Solered.

STRYCHNOS ALKALOID: brucine.

Robinson, Prog. Org. Chem., 1, 1 (1952)

S. alviminiana.

CARBAZOLE ALKALOIDS: alvimine, alviminine.

Marini-Bettolo et al., An. Assoc. Quim. Argent., 70, 263 (1982)

S. amazonica.

CARBAZOLE ALKALOID: nordihydrotoxiferine.

CARBOLINE ALKALOID: macusine B.

Marini-Bettolo et al., Gazz. Chim. Ital., 103, 523 (1973)

S. angolensis.

CARBOLINE ALKALOID: 11-methoxymacusine A.

Verpoor et al., J. Nat. Prod., 46, 578 (1983)

S. angustiflora Benth.

CARBOLINE ALKALOIDS: angusitidine, angustine, angustoline.

Au et al., J. Chem. Soc., Perkin I, 13 (1973)

S. brachiata.

CARBAZOLE ALKALOID: 11-methoxydiaboline.

Galeffi et al., Ann. Chim. (Rome), 63, 849 (1973)

S. brasiliensis Mart.

CARBAZOLE ALKALOIDS:

12-hydroxy-11-methoxy-spermostrychnine, 12-hydroxy-11-methoxytrychnobrasiline, strychnobrasiline, strychnosiline, strychnosolidine.

Iwataki et al., Tetrahedron, 27, 2541 (1971)

S. camptoneura.

CARBOLINE ALKALOID: camptoneurine.

Koch, Ann. Pharm. Fr., 30, 299 (1972)

S. castelenaeana.

CARBAZOLE ALKALOID: 3-hydroxydiaboline.

Galeffi et al., Phytochemistry, 21, 2393 (1982)

S. cinnamomifolia Thw.

CARBAZOLE ALKALOID: strychnine.

Chakravarti, Robinson, Nature, 160, 18 (1947)

S. columbrina L.

CARBAZOLE ALKALOID: strychnine.

Chakravarti, Robinson, Nature, 160, 18 (1947)

S. dale.

BISCARBOLINE ALKALOIDS:

10,10'-dimethoxy-3S,17S-Z-N(4)-methyltetrahydrousambarensine, 10,10'-dimethoxy-3S,17S-Z-tetrahydrousambarensine.

Verpoort et al., Tetrahedron Lett., 239 (1986)

S. decussata.

CARBOLINE ALKALOIDS: decussine, 3,14-dihydrodecussine, 10-hydroxy-3,14-dihydrodecussine, 10-hydroxy-17-O-methylakagerine, 10-hydroxy-21-O-methylkribine, 10-hydroxy-21-epi-O-methylkribine, malindine.

Rolfsen et al., J. Nat. Prod., 44, 415 (1981)

S. diaboli Sandwith.

CARBAZOLE ALKALOID: diaboline.

King, J. Chem. Soc., 955 (1949)

S. dinklagei Gilg.

CARBAZOLE ALKALOID: 17-oxoellipticine.

MONOTERPENE ALKALOIDS: dinklageine, strychnovoline.

Skaltsounig et al., Tetrahedron Lett., 2783 (1984)

Michel et al., J. Nat. Prod., 48, 86 (1985)

S. divaricans.

BISCARBAZOLE ALKALOID: calebassine.

Bernauer et al., Helv. Chim. Acta, 41, 673 (1958)

S. dolichothyrsa.
STEROIDS: campesterol, β-sitosterol, stigmasterol.
TRITERPENOID: filican-3-one.
　(S) Siwon et al., *Planta Med.*, **31**, 57 (1977)
　(T) Verpoorte et al., *Phytochemistry*, **17**, 817 (1978)

S. erichsonii.
INDOLE ALKALOID: erichsonine.
　Forgacs et al., *Phytochemistry*, **25**, 969 (1986)

S. fendleri.
CARBAZOLE ALKALOIDS: N-acetylstrychnosplendine,
　11-methoxystrychnofendlerine, strychnofendlerine.
　Galeffi et al., *Gazz. Chim. Ital.*, **106**, 773 (1976)
　Galeffi, Marini-Bettolo, *ibid.*, **110**, 81 (1980)

S. froesii.
CARBAZOLE ALKALOID: nordihydrotoxiferine.
　Marini-Bettolo, Delle Monache, *Gazz. Chim. Ital.*, **103**, 543 (1973)

S. gossweileri Excell.
UNCHARACTERIZED ALKALOIDS: strychnochromine,
　strychnoxanthine,
　Coune, *Plant Med. Phytother.*, **12**, 106 (1978)

S. helotii var. reticulata.
UNCHARACTERIZED ALKALOID: holstiine.
　Janot, Goutarel, *C. R. Acad. Sci.*, **232**, 852 (1951)

S. henningsii Gilg.
CARBAZOLE ALKALOIDS: 17-acetylhenningsoline,
　acetyl-11-methoxystrychnosplendine, O-acetylretuline,
　henningsamine, henninsoline, 11-methoxyhenningsamine,
　10-methoxytsilanine, rindline, tsilanine.
　Grossert et al., *J. Chem. Soc.*, 2812 (1965)
　Chapya et al., *Gazz. Chim. Ital.*, **113**, 773 (1983)

S. hirsuta Spruce et Benth.
CARBOLINE ALKALOIDS: strychnohirsutine,
　tetrahydrostrychnohirsutine.
　Galeffi, Marini-Bettolo, *Tetrahedron*, **37**, 3167 (1981)

S. holstii.
CARBAZOLE ALKALOID: condensamine.
　Denoel et al., *Arch. Int. Physiol.*, **59**, 341 (1951)

S. holstii var. reticulata.
UNCHARACTERIZED ALKALOID: holstiine.
　Janot, Goutarel, *C. R. Acad. Sci.*, **232**, 852 (1951)

S. holstii var. reticulata f. condensata.
CARBAZOLE ALKALOIDS: reticuline, retuline.
　Bosley, *J. Pharm. Belg.*, **6**, 150 (1951)
　Bosley, *J. ibid.*, **6**, 243 (1951)

S. icaja Baill.
CARBAZOLE ALKALOIDS:
　21,23-epoxy-N-methyl-sec-pseudobrucine,
　4-hydroxystrychnine, icajine, sungucine.
　Jaminet, *J. Pharm. Belg.*, **8**, 449 (1953)
　Lamotte et al., *Tetrahedron Lett.*, 4227 (1979)

S. ignatii Berg.
CARBAZOLE ALKALOID: strychnine.
　Woodward, *Nature*, **162**, 155 (1948)

S. jobertiana.
CARBAZOLE ALKALOID: jobertine.
　Delle Monache et al., *Ann. Inst. Super-Sanita*, **3**, 564 (1967)

S. kasengaensis.
BISCARBAZOLE ALKALOID: matopensine.
　Massiot et al., *Heterocycles*, **20**, 2339 (1983)

S. lethalis Barb.
UNCHARACTERIZED ALKALOID: strychnolethaline.
　Carneiro, *C. R. Acad. Sci.*, **206**, 1202 (1938)

S. ligustrina Blume.
CARBAZOLE ALKALOIDS: brucine, strychnine.
　Govindachari, Robinson, *Nature*, **160**, 18 (1947)

S. lucida.
CARBAZOLE ALKALOID: strychnine.
　Woodward, *Nature*, **162**, 155 (1948)

S. matopensis.
BISCARBAZOLE ALKALOID: matopensine.
　Massiot et al., *Heterocycles*, **20**, 2339 (1983)

S. melioniana Baill.
CARBOLINE ALKALOIDS: flavopereirine, C-mavacurine,
　meliononines A, B, E, F, H and L.
　Bickel et al., *Helv. Chim. Acta*, **38**, 649 (1955)

S. nedeola.
CARBOLINE ALKALOID: normacusine B.
　Patel et al., *Phytochemistry*, **12**, 451 (1973)

S. ngouniense.
INDOLE ALKALOIDS: ngouniensine, epi-ngouniensine.
　Massiot et al., *J. Chem. Soc., Chem. Commun.*, 368 (1982)

S. nigritana.
BISCARBOLINE ALKALOIDS: 18-dehydro-10-hydroxynigritanine,
　dehydronigritanine, nigritanine.
　Okuakwa et al., *Gazz. Chim. Ital.*, **108**, 615 (1978)

S. nux-vomica L.
CARBAZOLE ALKALOIDS: brucine, α-colubrine, β-colubrine,
　16-hydroxy-α-colubrine, 16-hydroxy-β-colubrine,
　3-methoxyicajine, novacine, pseudostrychnine, strychnine,
　vomicine.
UNCHARACTERIZED ALKALOIDS: struxine, strychnicine.
IRIDOID: 4-deoxyloganin.
　(A) Warnat, *Helv. Chim. Acta*, **14**, 997 (1931)
　(A) Woodward, *J. Am. Chem. Soc.*, **70**, 2107 (1948)
　(I) Bisset et al., *Phytochemistry*, **13**, 265 (1974)

S. nux-vomica L. var. javanica.
CARBAZOLE ALKALOID: strychnine.
　van Boorsma, Bull. Inst. Bot. Buitenz., *XIV*, 3 (1902)

S. potatorum.
STEROIDS: β-sitosterol, stigmasterol.
TRITERPENOIDS: 3β-acetoxyoleanolic acid, α-amyrin, β-amyrin,
　isomotiol, lupeol, oleanolic acid.
　Singh et al., *Phytochemistry*, **17**, 154 (1978)

S. psilosperma.
CARBAZOLE ALKALOIDS: deacetylstrychnospermine,
　spermostrychnine, strychnospermine.
　Anet et al., *Aust. J. Chem.*, **6**, 58 (1953)

S. rheedei Clarke.
CARBAZOLE ALKALOID: brucine.
　Leuchs, *Ber. Dtsch. Chem. Ges.*, **65**, 1230 (1932)

S. romeubelenii.
CARBAZOLE ALKALOID: 11-methoxydiaboline.
　Marini-Bettolo et al., *Gazz. Chim. Ital.*, **101**, 971 (1971)

S. rubiginosa.
CARBOLINE ALKALOID: strychnorubigine.
　Marini-Bettolo et al., *Atti. Acad. Naz. Lincei. Cl. Sci. Fis. Mat. Nat. Rend.*, **65**, 293 (1978)

S. spinosa Kam.
CARBOLINE ALKALOID: 10-hydroxyakagerine.
MONOTERPENOID: stryspinolactone.
　(A) Oguakwa et al., *Gazz. Chim. Ital.*, **110**, 37 (1980)
　(T) Adesogan et al., *Phytochemistry*, **20**, 2585 (1981)

S. splendens Gilg.

CARBAZOLE ALKALOIDS: N(a)-acetylstrychnosplendine, isosplendine, isosplendoline, isostrychnosplendine, splendoline, strychnosplendine.

Koch *et al.*, *Tetrahedron Lett.*, 2353 (1966)

Koch *et al.*, *Bull. Soc. Chim. Fr.*, 3250 (1968)

S. tabascana.

CARBOLINE ALKALOIDS: acetyltabascanine, tabascanine.

Galeffi *et al.*, *Farmaco Ed. Sci.*, **26**, 1100 (1971)

S. tchibangensis.

BISCARBOLINE ALKALOID: tchibangensine.

Richard *et al.*, *Phytochemistry*, **17**, 539 (1978)

S. tieute Lesch.

CARBAZOLE ALKALOID: strychnine.

Woodward, *Nature*, **162**, 155 (1948)

S. toxifera Schomb.

BISBENZYLISOQUINOLINE ALKALOID: (+)-curine.

BISCARBAZOLE ALKALOIDS: C-alkaloids A, B, C, D, E, F and G, alloferine, calebassine, calebassines A, F and I, calebassinine, caracurines II, V, VI and VIII, C-isotoxiferine I, toxiferine I.

CARBAZOLE ALKALOIDS: caracurine VII, C-curarine I, C-curarine II, C-curarine III, hemitoxiferine I, strychnolethaline, C-xanthocurine.

CARBOLINE ALKALOIDS: C-alkaloid Y, lochnerine, macusines A, B and C.

UNCHARACTERIZED ALKALOIDS: C-alkaloids I, J, L, M, O, P, Q, R, S, X and UB, cryptocurine, fedamazine, C-guaianine, C-isodihydrotoxiferine I, lochneram, toxiferines II and IIB.

Meyer *et al.*, *Helv. Chim. Acta*, **39**, 1214 (1956)

Karrer *et al.*, *Angew. Chem.*, **70**, 644 (1958)

Battersby, Yeowell, *J. Chem. Soc.*, 4419 (1964)

S. usambarensis.

BISCARBOLINE ALKALOIDS: 6,7-dihydroflavopereirine, dihydrousambarensine, N-methylusambarensine, strychnobaridine, strychnopentamine, strychnophylline, usambarenine, usambaridine Br, usambaridine Vi, usambarine.

CARBOLINE ALKALOIDS: isomalindine, malindine, O-methylmacusine B, strychnofoline.

Dupont *et al.*, *Acta Cryst.*, **33B**, 3801 (1977)

Angenot *et al.*, *J. Pharm. Belg.*, **33**, 11 (1978)

Caprasse *et al.*, *Planta Med.*, **50**, 27 (1984)

S. vacaoua Baill.

GLUCOALKALOID: bacancosine.

Bourquelot, Herissey, *C. R. Acad. Sci.*, **144**, 575 (1907)

S. variabilis.

BISCARBAZOLE ALKALOID: 12'-hydroxyisostrychnobiline.

CARBAZOLE ALKALOIDS: rosibiline, strychnopivotine.

Tits *et al.*, *Phytochemistry*, **19**, 1531 (1980)

S. wallichiana Stend.

CARBAZOLE ALKALOIDS: brucine N-oxide, 14-hydroxy-N-methyl-sec-pseudostrychnine, 14-hydroxynovacine, icajine N-oxide.

UNCHARACTERIZED ALKALOID: pseudobrucine.

Bisset, Chaudhury, *Phytochemistry*, **13**, 259 (1974)

STUARTIA (Theaceae)

S. monadelpha.

TRITERPENOID: camelliagenin A 16-cinnamate.

Ogino *et al.*, *Chem. Pharm. Bull.*, **16**, 1866 (1968)

STYLOPHORUM

S. diphyllum Nutt.

NAPHTHOISOQUINOLINE ALKALOID: chelidonine.

Chen, MacLean, *Can. J. Chem.*, **45**, 3001 (1967)

STYLOTRICHUM

S. rotundifolium.

Six sesquiterpenoids.

Bohlmann *et al.*, *Phytochemistry*, **20**, 1887 (1981)

STYPANDRA (Liliaceae/Stypandraceae)

S. grandis C.T.White.

CHROMONE: 5,7-dihydroxy-6-methyl-2-nonacosylchromone.

NAPHTHOQUINONE: stypandrone.

(C) Cooke *et al.*, *Tetrahedron Lett.*, 1039 (1970)

(N) Cooke *et al.*, *Aust. J. Chem.*, **18**, 218 (1965)

S. imbricata.

BINAPHTHALENE: stypandrine.

Colegate *et al.*, *Aust. J. Chem.*, **38**, 1233 (1985)

STYPHNODENDRON

S. coriaceum.

TRITERPENOIDS: 2-hydroxymachaerinic acid lactone, sapogenin K.

Tursch *et al.*, *Bull. Soc. Chim. Belg.*, **75**, 127 (1966)

STYPOPODIUM (Phaeophyceae)

S. zonale.

ICHTHYOTOXINS: stypoldione, stypotriol.

Gerwick, Fenical, *J. Org. Chem.*, **46**, 22 (1981)

STYRAX (Styracaceae)

S. japonica.

TRITERPENOIDS: desacetyljegosaponin, jegosapogenol A, jegosapogenol B, jegosaponin.

Sugiyama *et al.*, *Nippon Kagaku Zasshi*, **88**, 794 (1967)

Sugitama *et al.*, *ibid.*, **88**, 1316 (1967)

SUAEDA

S. nudiflora Willd.

AMINO ACIDS: alanine, aspartic acid, glutamic acid, glycine, leucine, proline.

COUMARIN: 6,7-dimethoxycoumarin.

Kapadia *et al.*, *Chem. Abstr.*, 107 151208c

SUBTRIBUS

S. espeletiinae.

DITERPENOID: ent-8,13-epoxy-14-labden-3β-ol.

Bohlmann *et al.*, *Phytochemistry*, **19**, 267 (1980)

SUILLUS (Agaricaceae)

S. bovinus.

PIGMENT: amitenone.

Minami *et al.*, *Tetrahedron Lett.*, 5067 (1968)

S. collinus.

PIGMENTS: variegatic acid, variegatorubin, xerocomic acid.

Edwards *et al.*, *J. Chem. Soc.*, 2582 (1971)

S. grevillei.

BISBENZOFURAN: theleporic acid.

FLAVANONE: 3',4,4'-trihydroxypulvinone.

PIGMENT:
3-(3,4-dihydroxyphenyl)-2,7,8-trihydroxy-1,4-dibenzofurandione.
(B) Sullivan et al., J. Pharm. Sci., **60**, 1727 (1971)
(F) Edwards et al., J. Chem. Soc., Perkin I, 1921 (1973)
(P) Edwards et al., ibid., 351 (1975)

S. leptoopus.
PIGMENT: grevillin D.
Besl et al., Chem. Abstr., 88 117773n

S. tridentinus.
PIGMENT: tridentoquinone.
Besl et al., Chem. Ber., **108**, 3675 (1975)

SURIANA

S. maritima L.
STEROID: surianol.
Mitchell, Geissman, Phytochemistry, **10**, 1559 (1971)

SWAINSONIA (Leguminosae)

S. canescens.
ALKALOID: swainsonine.
Colegate et al., Aust. J. Chem., **32**, 2257 (1979)

SWARTZIA

S. madagascariensis.
BENZOPYRANS: demethylpterocarpin, 3,4,9-trimethoxypterocarpin.
Harper et al., J. Chem. Soc., Chem. Commun., 309 (1965)

SWEETIA

S. elegans.
LUPANE ALKALOID: sweetinine.
Balandrin et al., J. Nat. Prod., **44**, 619 (1984)

SWERTIA (Gentianaceae)

S. carolinensis.
IRIDOID: loganic acid.
Coscia, Huarneccia, J. Chem. Soc., Chem. Commun., 138 (1968)

S. graciliflora Gontsch.
PYRIDINE ALKALOIDS: gentianamine, gentianine, gentioflavine.
Rulko, Nadler, Diss. Pharm. Pharmacol., **22**, 329 (1972)

S. japonica.
BENZOPYRANS: swertiajaponin, swertisin.
GLUCOSIDE: swertiamarin.
MONOTERPENE GLYCOSIDES: amarogentin, amaroswerin, swertiaside.
SAPONIN: swericinactoside.
(B) Komatsu et al., Tetrahedron Lett., 1611 (1966)
(G) Inouye et al., Chem. Pharm. Bull., **18**, 1856 (1970)
(Sap) Shang, Mao, Yaoxue Xuebao, **19**, 819 (1984)
(T) Ikeshiro et al., Planta Med., **50**, 485 (1984)

S. marginata Schrenk.
PYRIDINE ALKALOIDS: gentiananamine, gentianine, gentioflavine.
Rakhmatullaev et al., Khim. Prir. Soedin., 64 (1969)

S. perennis.
XANTHONE: 1,3,7,8-tetrahydroxyxanthone-1,3-diglucoside.
Chaudhuri et al., Phytochemistry, **10**, 2425 (1971)

S. petiolata Royle.
STEROID: β-sitosterol.
TRITERPENOID: ursolic acid.
XANTHONES: 1,7-dihydroxy-3,8-dimethoxyxanthone, 1,7-dihydroxy-3-methoxyxanthone-8-O-glucoside.
Bhan et al., Chem. Abstr., 100 64973k

S. purpurascens.
BENZOPYRAN: swertisin.
Kamatsu et al., Tetrahedron Lett., 1611 (1966)

S. randaiensis.
PYRIDINE ALKALOID: gentianine.
TRITERPENOID: oleanolic acid.
Wu et al., J. Chin. Chem. Soc., **23**, 53 (1976); Chem. Abstr., 85 2572s

SWIETENIA

S. macrophylla.
TERPENOIDS: 8,30-epoxyswietenine acetate, swietenine, swietenolide, swietenolide diacetate.
Taylor et al., Phytochemistry, **22**, 2870 (1983)

S. mahogani.
TRITERPENOIDS: cycloartenol, cyclomahogenol, cycloswietenol, hederagenin, lupeol benzoate, mahogenol, 31-norcycloswietenol.
Anjaneyulu et al., Indian J. Chem., **17B**, 423 (1979)

SWINGLEA

S. glutinosa Merr.
AMINE ALKALOID:
4-(4-acetoxy-3,7-dimethylocta-2,6-dienoyloxy)-N-benzoylphenethylamine.
Dreyer, Tetrahedron, **26**, 5745 (1970)

SYMPHONIA (Guttiferae/Clusiaceae)

S. globifera L.
XANTHONES: 1,7-dihydroxyxanthone, mesuxanthones A and B, symphoxanthone, 1,3,5,6-tetrahydroxyxanthone, 1,3,6,7-tetrahydroxyxanthone, 1,3,7-trihydroxy-4-(3-methylbut-2-enyl)xanthone, ugaxanthone.
Jackson et al., Tetrahedron, **25**, 1603 (1969)
Bandaranayake et al., Phytochemistry, **14**, 265, 1878 (1975)

SYMPHOPAPPUS (Compositae)

S. compressus.
ONE GUAIANOLIDE:
Bohlmann et al., Planta Med., **50**, 276 (1984)

S. itatiayensis.
DITERPENOID: ent-cis-clerodene-15,16,19,6-diolide.
Vichnewski et al., Phytochemistry, **18**, 129 (1979)

S. polystachyus.
FLAVANOID: genkwanin.
Mesquita et al., Phytochemistry, **25**, 1255 (1986)

SYMPHORICARPUS

S. racemosus (Michaux.)
FLAVANOIDS: luteolin 7-β-glucoside, luteolin 7-rhamnoglucoside, quercetin 3-glucoside.
Chavant et al., Plant Med. Phytother., **9**, 267 (1975)

SYMPHYTUM (Boraginaceae)

S. asperum Lepech.
PYRROLIZIDINE ALKALOIDS: asperumine, echinatine, heliosupine N-oxide.
Man'ko, Kotovskii, J. Gen. Chem. USSR, **40**, 2519 (1970)

S. caucasicum M.Bieb.
PYRROLIZIDINE ALKALOIDS: echinatine, heliosupine N-oxide, lasiocarpine, symphytine, viridiflorine.
Man'ko et al., Rastit. Resur., **5**, 508 (1969)

S. officinalis L.

PYRROLIZIDINE ALKALOIDS: echinatine, heliosupine N-oxide, lasiocarpine, symphytine, viridiflorine.

Man'ko *et al.*, *Rastit. Resur.*, **6**, 409 (1970)

S. orientale L.

PYRROLIZIDINE ALKALOID: anadoline.

Ulubelen, Doganca, *Tetrahedron Lett.*, 2583 (1970)

S. tuberosum.

ALIPHATICS: heneicosane, tricosane.

PYRROLIZIDINE ALKALOIDS: anadoline, echimidine.

STEROID: β-sitosterol.

Ulubelen *et al.*, *Phytochemistry*, **16**, 499 (1977)

S. x uplandicum Nyman. (S. asperum x S. officinale).

PYRROLIZIDINE ALKALOIDS: 7-acetylintermedine, 7-acetyllycopsamine, uplandicine.

Culvenor *et al.*, *Aust. J. Chem.*, **33**, 1105 (1980)

SYMPLOCOS (Symplocaceae)

S. celastrinea Mart.

APORPHINE ALKALOID: caaverine.

Tschesche *et al.*, *Tetrahedron*, **20**, 1435 (1954)

S. confusa.

CHALCONE: confusoside.

Tanoka *et al.*, *Chem. Pharm. Bull.*, **30**, 2421 (1982)

S. glauca.

PHENOLIC: verbenalin.

Tanoka *et al.*, *Chem. Pharm. Bull.*, **30**, 2421 (1982)

S. spicata.

FLAVONOID: rhamnetin 3-O-β-D-galactosyl-4-O-β-D-galactopyranoside. Triterpenoids 19α-hydroxyarjunolic acid 3,28-bis-β-D-glucopyranoside, 19α-hydroxyasiatic acid 3,28-bis-β-D-glucopyranoside.

(F) Tiwari *et al.*, *Phytochemistry*, **15**, 833 (1976)

(T) Higuchi *et al.*, *ibid.*, **21**, 907 (1982)

SYNCARPIA

S. laurifolia.

ACID: syncarpic acid.

Hodgson *et al.*, *Aust. J. Chem.*, **13**, 385 (1960)

SYNEILENSIS

S. aconitifolia.

MONOTERPENE GLUCOSIDE: D-linalool β-D-glucoside-3,4-diangelate.

Bohlmann, Zdero, *Phytochemistry*, **16**, 1057 (1977)

S. palmata.

PYRROLIZIDINE ALKALOID: syneilesine.

SESQUITERPENOID: 3-angeloyloxy-6-(3-methylbutanoyloxy)-furanoeremophilan-14-oic acid.

(A) Kikuchi, Furuya, *Tetrahedron Lett.*, 3657 (1974)

(T) Kuroda *et al.*, *Chem. Lett.*, 1313 (1978)

SYNTHERISMA

S. sanguinalis.

TRITERPENOID: miliacin.

Suguyama, Abe, *Nippon Kagaku Zasshi*, **83**, 1051 (1961)

SYRENIA

S. sessiliflora.

FLAVONOID: isorhamnetin 3-β-glucopyranosyl-7-α-L-rhamnopyranoside.

STEROIDS: corchoroside A, erysimotoxin, strophanthidin.

Komissarenko *et al.*, *Khim. Prir. Soedin.*, 671 (1976)

SYRINGA (Oleaceae)

S. vulgaris L.

PYRIDINE ALKALOIDS: jasmidine, jasminidine, jasminine.

FLAVONOID: rutin.

IRIDOIDS: deoxysyringoxide, hydroxynuezhiside, ligustroside, nuezhiside, oleuropein, syringopicroside, syringoxide.

PYRAN: syringenone.

(A) Ripperger *et al.*, *Phytochemistry*, **17**, 1069 (1978)

(F, P) Marekov *et al.*, *Khim. Ind. (Sofia)*, **58**, 132 (1986); *Chem. Abstr.*, **105** 168854s

(I) Asaka *et al.*, *Tetrahedron*, **26**, 2365 (1970)

SYZYGIUM

S. aromaticum.

STEROIDS: campesterol, β-sitosterol glucoside, stigmasterol glucoside.

TRITERPENOID: 2α-hydroxyoleanolic acid methyl ester.

Brieskorn *et al.*, *Phytochemistry*, **14**, 2308 (1975)

S. cumina.

FLAVONOIDS: 1-galloylglucose, 3-galloylglucose, 3,6-hydroxyphenoylglucose, 4,6-hydroxyphenoylglucose.

Bhatia *et al.*, *Planta Med.*, **28**, 346 (1975)

T

TABEBUIA (Bignoniaceae)

T. avellanedae Lorentz ex Griseb.
ANTHRACENE: tabebuin.
SESQUITERPENOID: lapachenole.
Burnett, Thomson, *J. Chem. Soc., C,* 2100 (1967)

T. chrysantha.
SESQUITERPENOIDS: dehydrotectol, dihydrolapachenole,
nordihydrolapachenole.
Burnett, Thomson, *J. Chem. Soc., C,* 850 (1968)

T. guayacan Hemsl.
BENXANTHONES: guayacanin, guayin.
NAPHTHAQUINONE:
2,3-bis(3-methyl-2-butenoyl)-1,4-naphthoquinone.
SESQUITERPENOIDS: dehydro-α-lapachone, lapachol,
α-lapachone, β-lapachone.
Manners *et al., Tetrahedron,* 32, 543 (1976)
Manners *et al., Phytochemistry,* 15, 225 (1976)

T. rosea.
SESQUITERPENOID: dehydrotectol.
Joshi *et al., Phytochemistry,* 12, 952 (1973)

TABERNAEMONTANA (Apocynaceae)

T. accedens.
BISCARBOLINE ALKALOIDS: accedinine, accedinisine.
CARBOLINE ALKALOIDS: accedine, N-demethyl-epi-accedine,
N-demethyl-epi-affinine.
HOMOCARBOLINE ALKALOID: N-demethylvoacamine.
INDOLE ALKALOID: N-methyl-16-epi-affinine.
Achenbach, Schaller, *Chem. Ber.,* 109, 3627 (1976)

T. albiflora (Miq.) Pull.
IMIDAZOLIDINOCARBAZOLE ALKALOIDS:
18-hydroxy-20-epi-ibophyllidine, 19-hydroxyibophyllidine,
19R-hydroxy-20-epi-ibophyllidine,
19S-hydroxy-20-epi-ibophyllidine.
Christiane *et al., Tetrahedron Lett.,* 3363 (1980)

T. amblyocarpa Urb.
INDOLE ALKALOIDS: (-)-ibogamine, (-)-iboxygaine,
(-)-kreyneanine, (-)-19-oxovoacangine, (-)-tabersonine.
Perez *et al., Rev. Latinoam. Quim.,* 16, 73 (1985)

T. amygdalifolia.
IMIDAZOLIDINOCARBAZOLE ALKALOIDS:
demethoxycylindrocarpidine, O-demethylpalosine,
10-oxocylindrocarpidine.
Achenbach, *Tetrahedron Lett.,* 1793 (1967)

T. apoda Wr. ex Sauv.
IMIDAZOLIDINOCARBAZOLE ALKALOID: apodinine.
Inglesias, *Rev. CENIC Cienc. Fis.,* 10, 357 (1979)

T. armenica.
IMIDAZOLIDINOCARBAZOLE ALKALOID: apodine.
Inglesias, Diatta, *Rev. CENIC Cienc. Fis.,* 6, 141 (1975)

T. brachyacantha.
CARBOLINE ALKALOIDS: anhydrovobasinediol, normacusine B.
Patel *et al., Phytochemistry,* 12, 451 (1973)

T. brachyphylla.
CARBOLINE ALKALOID: normacusine B.
Patel *et al., Phytochemistry,* 12, 451 (1973)

T. chippii.
BISINDOLE ALKALOID: vobparicine.
Van Beek *et al., Tetrahedron Lett.,* 2057 (1984)

T. citrifolia.
HOMOCARBOLINE ALKALOID: 14-dehydrotetrastachyne.
Abaul *et al., C.R. Acad. Sci.,* 29B, 627 (1984)

T. coronaria Br. (Ervatamia coronaria.).
HOMOCARBOLINE ALKALOID: coronaridine.
INDOLE ALKALOID: tabernaemontanine.
UNCHARACTERIZED ALKALOID: coronarine.
Gorman *et al., J. Am. Chem. Soc.,* 82, 1142 (1960)
Cava *et al., Tetrahedron Lett.,* 53 (1963)

T. crassa.
INDOLONE ALKALOID: crassanine.
Cava *et al., J. Org. Chem.,* 33, 3350 (1968)

T. cummingsii.
INDOLE ALKALOID: 2-ethyl-3-(2-ethylpiperidinoethyl)-indole.
Crooks *et al., J. Chem. Soc., Chem. Commun.,* 1210 (1968)

T. dichotoma.
IMIDAZOLIDINOCARBAZOLE ALKALOID: dichomine.
INDOLE ALKALOIDS: (-)-apparicine, 19-epi-iboxygaine,
isomethuenine, 12-methoxyvoaphylline, perivine,
19-epi-voacristine, vobasine.
Perera *et al., Planta Med.,* 49, 148, 232 (1983)

T. divaricata.
Three indole alkaloids.
Rastogi *et al., Phytochemistry,* 19, 1209 (1980)

T. eglandulosa.
IMIDAZOLIDINOCARBAZOLE ALKALOIDS: dichomine, tecamine.
Perera *et al., Planta Med.,* 49, 232 (1983)

T. elegans Stapf.
INDOLE ALKALOIDS: apparicine, dregamine, dregaminol,
dregaminol methyl ether, (3R)-hydroxyconodurine,
(3S)-hydroxyconodurine,
16S-hydroxy-16,22-dihydroapparicine,
3-hydroxytabernaelegantine B, isovoacangine,
3-methoxytabernaelegantine B, tabernaelegantines A, B, C
and D, tabernaelegantinidine, tabernaemontanine,
tabernaemontaninol, tubotaiwine, vobasine,
tabernaelegantinines A and B.
Van der Heijden *et al., Planta Med.,* 144 (1986)

T. fuchsiaefolia.
CARBOLINE ALKALOID: affinisine.
Cava *et al., Chem. Ind.,* 1193 (1964)

T. genenho.
HOMOCARBOLINE ALKALOIDS: 19-hydroxycoronaridine,
20-hydroxycoronaridine.
Agwada *et al., Helv. Chim. Acta,* 58, 1001 (1975)

T. glandulosa.
CARBOLINE ALKALOIDS: 12-demethoxytabernulosine,
tabernulosine.
HOMOCARBOLINE ALKALOID: 19-ethoxycoronaridine.
Achenbach *et al., Justus Liebigs Ann. Chem.,* 830 (1982)

T. heyneana Wall.
INDOLONE ALKALOID: tabernoxidine.
Joshi *et al., Indian J. Chem.,* 23B, 101 (1984)

T. johnstonii.

BISINDOLE ALKALOIDS: gubunamine, tabernamine.

Kingston *et al.*, *J. Pharm. Sci.*, **67**, 249 (1978)

T. laurifolia.

HOMOCARBOLINE ALKALOID: isovoacristine.

Cava *et al.*, *Chem. Ind.*, 2064 (1965)

T. oliveracea.

CARBAZOLE ALKALOID: condylocarpine N-oxide.

Achenbach, Rafflesberger, *Z. Naturforsch.*, **35B**, 885 (1980)

T. oppositifolia.

HOMOCARBOLINE ALKALOID: coronaridine.

Gorman *et al.*, *J. Am. Chem. Soc.*, **82**, 1142 (1960)

T. pachysiphon Stapf.

INDOLONE ALKALOID: conopharyngine pseudoindoxyl.

Crooks, Robinson, *J. Pharm. Pharmacol.*, 25, 820 *(193)* (193)

T. penduliflora.

INDOLE ALKALOID: coronaridine.

Anbujam *et al.*, *Planta Med.*, 463 (1985)

T. psychotrifolia H.B.K.

INDOLE ALKALOIDS: olivacine, taberpsychine.
TERPENOID: 16-epi-vobasinic acid.

Burnell, Medina, *Can. J. Chem.*, **49**, 307 (1971)

T. quadraggulensis.

HOMOCARBOLINE ALKALOID: (20R)-20-hydroxyibogamine.

Achenbach, Rafflesberger, *Z. Naturforsch.*, **35B**, 219 (1980)

T. recurva Rox.

INDOLE ALKALOID: tabernaemontanine.

Khan *et al.*, *J. Bangladesh Acad. Sci.*, **9**, 143 (1985); *Chem. Abstr.*, 105 3519y

T. riedelii.

CARBOLINE ALKALOID: (+)-apovincamine.

Cava *et al.*, *J. Org. Chem.*, **33**, 1055 (1968)

T. rigida.

CARBOLINE ALKALOIDS: (+)-apovincamine, (±)-vincamine.

Cava *et al.*, *J. Org. Chem.*, **33**, 1055 (1968)

T. rupicola Benth.

INDOLONE ALKALOIDS: montanine, rupicoline.

Niemann, Kessel, *J. Org. Chem.*, **31**, 2265 (1966)

T. undulata.

CARBOLINE ALKALOID: quebrachidine.

Gorman *et al.*, *Tetrahedron Lett.*, 39 (1963)

TABERNANTHE

T. iboga Baill.

DIMERIC ALKALOID: gabonine.
HOMOCARBOLINE ALKALOIDS: epi-ibogamine, ibogaine, iboxygaine, tabernanthine.
IMIDAZOLIDINOCARBAZOLE ALKALOIDS: ibophyllidine, iboxyphylline.
INDOLONE ALKALOID: iboluteine.

Khuong-Huu *et al.*, *Tetrahedron*, **32**, 2539 (1976)

T. pubescens.

HOMOCARBOLINE ALKALOIDS: 3,6-oxidoibogaine, 3,6-oxidoiboxygaine.

Mulamba *et al.*, *Lloydia*, **44**, 184 (1981)

T. subsessilis.

IMIDAZOLIDINOCARBAZOLE ALKALOIDS: ibophyllidine, iboxyphylline.

Khuong-Huu *et al.*, *Tetrahedron*, **32**, 2539 (1976)

TACCA

T. plantaginea Hance.

BITTER PRINCIPLES: taccalin, taccalonolide A, taccanolide B.

Chen *et al.*, *Tetrahedron Lett.*, 1673 (1987)

TAGETES

T. florida.

COUMARINS: 7-hydroxycoumarin dimethylallyl ether, 7-methoxycoumarin, 6,7,8-trimethoxycoumarin.

Rios *et al.*, *Rev. Latinoam. Quim.*, **7**, 33 (1976)

T. glandulifera.

MONOTERPENOIDS: 5-isobutyl-3-methyl-2-furancarbaldehyde, cis-tagetone, trans-tagetone.

Jones, Smith, *J. Chem. Soc.*, **127**, 2530 (1925)

T. gracilis.

DITERPENOID: bis-trans-ocimenone.

Bohlmann *et al.*, *Phytochemistry*, **18**, 341 (1979)

T. minuta L.

THIOPHEN: 5-(but-3-en-1-ynyl)-2,2'-bithienyl.

Atkinson *et al.*, *J. Chem. Soc.*, 7109 (1965)

TAIWANIA

T. cryptomerioides.

FLAVONE: taiwaniaflavone.
FURANONES: taiwanins A, B and D.
LACTONES: taiwanin C, taiwanin E.
DITERPENOID: 6β-acetoxy-7α-hydroxyroyleanone.
SESQUITERPENOIDS: angustifolene, cadinane-3β,9α-diol, cadinane-3β,9β-diol, cadinan-3-en-9-ol-2-one, cadin-2-ene-4α,9α-diol, T-cadinol, muurol-3-ene-2α,9β-diol, muurol-3-ene-2β,9β-diol, muurol-3-en-9β-ol-2-one, τ-muurolol.

(F, Fu, L) Kamil *et al.*, *Chem. Ind. (London)*, 160 (1977)
(T) Kuo *et al.*, *J. Chin. Chem. Soc.*, **26**, 71 (1979)

TALASSA

T. transilensis.

SESQUITERPENOIDS: talasin A, talasin B.
UNCHARACTERIZED TERPENOID: talassin.

Konovalova *et al.*, *Khim. Prir. Soedin.*, 690 (1975)

TALAUMA

T. hodgsoni.

LIGNAN: (+)-sesamin.

Krajniak *et al.*, *Aust. J. Chem.*, **26**, 687 (1973)

T. mexicana Don.

BISBENZYLISOQUINOLINE ALKALOID: aztequine.
OXONORAPORPHINE ALKALOID: liriodenine.

Pallares, Garza, *Arch. Biochem.*, **16**, 275 (1948)
Kametani *et al.*, *Phytochemistry*, **14**, 1884 (1975)

TAMARIX (Tamaricaceae)

T. aphylla.

FLAVONOIDS: kaempferol 4',7-dimethyl ether, kaempferol 4',7-dimethyl ether-3-sulphate, quercetin, quercetin 3-O-isoferulyl-β-glucuronate, rhamnocitrin, rhamnocitrin 3-glucoside, rhamnocitrin 3'-glucuronide-3,5,4'-trisulphate, rhamnocitrin 3-rhamnoside.

Nawwar *et al.*, *Experientia*, **31**, 1118 (1975)
El Ansari *et al.*, *Phytochemistry*, **15**, 231 (1976)

TAMUS (Dioscoreaceae)

T. communis L.
CAROTENOID: lycoxanthin.
ISOFLAVONE: 2,3,4-trimethoxy-7,8-methylenedioxyisoflavone.
PHENANTHRENES:
4,7-dihydroxy-2,6-dimethoxy-9,10-dihydrophenanthrene,
2-hydroxy-1,7-dimethoxy-5,6-methylenedioxyphenanthrene,
4-hydroxy-2,6,7-trimethoxy-9,10-dihydrophenanthrene.
STEROIDS: campesterol, β-sitosterol, stigmasterol.
(C) Winterstein *et al., Angew. Chem.,* **32,** 902 (1960)
(I, P) Reisch *et al., Tetrahedron Lett.,* 67 (1969)
(P) Aquino *et al., J. Nat. Prod.,* **48,** 811 (1985)
(S) Capasso *et al., J. Ethnopharmacol.,* **8,** 327 (1983); *Chem. Abstr.,* 100 48585e

T. edulis Moew.
STEROIDS: afurigenin, eduligenin, 25S-hydroxytamusgenin, lowegenin, 7-oxodiosgenin, 7-oxotamusgenin, tamusgenin.
Barreira *et al., Phytochemistry,* **9,** 1641 (1970)
Gonzalez *et al., ibid.,* **10,** 1339 (1971)

TANACETOPSIS

T. mucronata.
SESQUITERPENOID: micurin.
Abduazimov *et al., Khim. Prir. Soedin.,* 398 (1981)

TANACETUM (Compositae)

T. balsamita subsp. balsamitoides.
SESQUITERPENOID: balsamitone.
Bohlmann *et al., Chem. Ber.,* **108,** 1369 (1975)

T. macrophyllum Willd.
SESQUITERPENOID: chrysartemin A.
Ristic *et al., Glas. Hem. Drus. Beograd.,* **47,** 319 (1982); *Chem. Abstr.,* 97 178746p

T. odessanum.
Two homosesquiterpenoids.
Bohlmann, Knoll, *Phytochemistry,* **17,** 319 (1978)

T. parthenium L.
SESQUITERPENOIDS: chrysanthemolide, chrysanthemonin, 10-epi-canin, 1β-hydroxyarbusculin, 8β-hydroxyreynosin, mangoliolide, partholide, reynosin, santamarin, tanaparthin, tanaparthin-1α,4α-epoxide, tanaparthin-1β,4β-epoxide.
Johnson *et al., Eur. Pat. Appl.,* **98** 041 (1984)
Stefanovic *et al., J. Serb. Chem. Soc.,* **50,** 435 (1985); *Chem. Abstr.,* 105 75914w

T. pseudoachillea.
SESQUITERPENOIDS: tanacin, tanadin, tanakhin, tanaspin.
Yunusov *et al., Khim. Prir. Soedin.,* **170,** 309 (1976)
Yunusov, Sidyakin, *ibid.,* 411 (1979)

T. santalinoides (DC.) Feinbr. et Fertig.
SESQUITERPENOIDS: dihydroridentin, 4-Z-1-epi-dihydroridentin, erivanin.
El-Sebakhy *et al., Pharmazie,* **41,** 298, (1986)
El-Sebakhy *et al., ibid.,* **41,** 525 (1986)

T. serotinum.
SESQUITERPENOIDS: beogradolides A and B.
Stefanovic *et al., Glas. Ham. Drus. Beograd.,* **47,** 13 (1982)

T. tanacetioides.
SESQUITERPENOID: 1-oxo-α-longipinene.
Bohlmann *et al., Chem. Ber.,* **110,** 3572 (1977)

T. vulgare L.
MONOTERPENOIDS: chrysanthemyl acetate, thuj-4-en-2-yl acetate.
SESQUITERPENOIDS: longipinane-2,7-dione, reynosin, tabulin, tanacetals A and B.
Dembitskii, Sulaeva, *Khim. Prir. Soedin.,* 527 (1984)

T. vulgare L. var. crispum.
SESQUITERPENOID: crispolide.
Appendino *et al., Phytochemistry,* **21,** 1099 (1982)

TANGHINIA

T. madagascariensis.
STEROID: tanghinin.
Sigg *et al., Helv. Chim. Acta,* **38,** 166 (1955)

T. venenifera.
STEROIDS: tanghiferin, tanghin, tanghinigenin, tanghinigenin 3-O-(2-O-acetyl-6-deoxy-3-O-methyl-α-L-glycopyranoside, veneniferin.
Abe *et al., Chem. Pharm. Bull.,* **25,** 2744 (1977)

TAONIA (Phaeophyceae)

T. atomaria J.G.Asgardh.
PYRAN: tondiol.
TERPENOIDS: atomaric acid, taondiol, taondiol dimer.
(P) Gonzalez *et al., Tetrahedron Lett.,* 2729 (1971)
(T) Gonzalez *et al., Tetrahedron Lett.,* 3951 (1974)

T. australasica.
DITERPENOID: taonianone.
Murphy *et al., Tetrahedron Lett.,* 1555 (1981)

TARAXACUM (Compositae)

T. dens leonis.
STEROID: 31-nordihydrolanosterol.
Atallah, Nicholas, *Steroids,* **17,** 611 (1971)

T. japonicum.
TRITERPENOIDS: taralupenol, taralupenol acetate.
Ageta *et al., Tetrahedron Lett.,* 2289 (1981)

T. kok-saghyz.
TRITERPENOID: taraxasterol.
Bauer, Brunner, *Arch. Pharm.,* **276,** 605 (1938)

T. officinale (L.) Weber.
CAROTENOID: flavoxanthin.
SESQUITERPENOIDS: 11,13-dihydrotaraxin acid-β-D-glucopyranosyl ester, taraxin acid, taraxin acid-β-D-glucopyranosyl ester.
TRITERPENOIDS: pseudotaraxasterol, taraxasterol, 14-taraxeren-3β-ol, taraxerol, taraxol.
(C) Kuhn, Brockmann, *Z. Physiol. Chem.,* **213,** 191 (1932)
(T) Hansel *et al., Phytochemistry,* **19,** 857 (1980)

TARENNA

T. bipindensis.
CARBOLINE ALKALOID: dihydroelaecarpidine.
Boissier *et al., Experientia,* **27,** 677 (1971)

T. kotoensis var. gyokushenko.
IRIDOID: tarrenoside.
Takeda *et al., Chem. Pharm. Bull.,* **24,** 1216 (1976)

TAXILLUS

T. kaempferi.
FLAVONOID: isoglucodistylin.
Sakurai *et al., Bull. Chem. Soc., Japan,* **55,** 3051 (1982)

TAXODIUM (Taxodiaceae)

T. distichum.
QUINONE: taxoquinone.
DITERPENOIDS: 5-dehydrosugiol, taxodione, taxodone.
(Q) Kupchan *et al., J. Am. Chem. Soc.,* **90,** 6923 (1968)
(T) Kupchan *et al., J. Org. Chem.,* **34,** 3912 (1969)

T. lanceolata.
STEROID: β-sitosterol.
TRITERPENOIDS: ellagic acid, oleanolic acid.
　　Ahmad et al., J. Indian Chem. Soc., 53, 1165 (1976)

T. mucronatum.
FLAVONOIDS: cryptomerin A, cryptomerin B, hinokiflavone, isocryptomerin, podocarpusflavone, sciadopitysin.
DITERPENOID: 8-hydroxypimar-15-en-19-oic acid.
　　(F) Pelter et al., Tetrahedron, 27, 1627 (1971)
　　(F) Ahmad et al., J. Indian Chem. Soc., 53, 1165 (1976)
　　(T) Ramos et al., Phytochemistry, 23, 1329 (1984)

TAXUS (Taxaceae)
T. baccata L.
TAXINE ALKALOIDS: taxine I, taxine B.
ESTER:: myo-inositol-p-coumaric ester. Fu
　　(A) de Marcano, Halsall, J. Chem. Soc., Chem. Commun., 365 (1975)
　　(E) Dittrich et al., Plant Physiol., 59, 279 (1977); Chem. Abstr., 86 136344y
　　(F) Majumdar et al., Indian J. Chem., 10, 677 (1972)
　　(P) Khan et al., Planta Med., 30, 82 (1976)
　　(T) Chauviere et al., C. R. Acad. Sci., 293, 501 (1981)

T. baccata subsp. cuspidata.
DITERPENOID: taxinin.
　　Kurono et al., Tetrahedron Lett., 2153 (1963)

T. brevifolia.
TAXINE ALKALOID: taxol.
　　Wani et al., J. Am. Chem. Soc., 93, 2325 (1971)

T. canadensis.
TAXINE ALKALOID: taxinine.
　　Bourbeau, Laval Med., 19, 511, (1954)
　　Bourbeau, ibid., 19, 637 (1954)

T. chinensis.
ALKALOIDS: taxinine, taxinine A.
STEROID: β-sitosterol.
　　Chiang, Chem. Abstr., 84 118448q

T. cuspidata.
STEROID: taxisterone.
DITERPENOIDS: 4(20),11-taxadiene-2α,5α,7β,9α,10β,13α-hexaol, 4(20),11-taxadiene-2α,5α,7β,9α,10β,13α-hexaol-5α-cinnamoyl-2α,7β,9α,10β,13α-pentaacetate, 4(20),11-taxadiene-2α,5α,7β,9α,10β,13α-hexaol-5-cinnamoyl-2α,7β,9α,10β-tetraacetate-13-one, taxinine, taxinines A, B, C, D, E, F, G, H, K and L, taxusin.
　　(S) Nakano et al., Phytochemistry, 21, 2749 (1982)
　　(T) de Marcano et al., J. Chem. Soc., Chem. Commun., 1282 (1969)

TECLEA (Rutaceae)
T. boviniana.
ACRIDONE ALKALOIDS: 2-methoxytecleanthine, 1,3,4-trimethoxy-N-methylacridone.
　　Vaquette et al., Plant Med. Phytother., 8, 57 (1974)

T. grandifolia Engl.
SESTERTERPENOIDS: 7-deacetylazadirone, 7-deacetylproceranone, tecleanin, tecleanone.
　　Ayafor et al., J. Chem. Soc., Perkin I, 1750 (1981)

T. natalensis (Sond.) Engl.
ACRIDONE ALKALOIDS: arborinine, tecleanthine.
UNCHARACTERIZED ALKALOID: tecleanine.
　　Pegel, Weight, J. Chem. Soc., C, 2327 (1969)

T. simplicifolia.
AMINE ALKALOID: O-methylhordenine.
　　Kirkwood, Marion, J. Am. Chem. Soc., 72, 2522 (1950)

T. sudanica.
FURANOQUINOLINE ALKALOIDS: flindersiamine, maculine, skimmianin.
QUATERNARY ALKALOID: candicine.
TRITERPENOID: lupeol.
　　Fish et al., Fitoterapia, 48, 170 (1977)

T. verdoorniana.
ACRIDONE ALKALOID: tecleanone.
FURANOQUINOLINE ALKALOIDS: tecleaverdine, tecleaverdoornine, tecleine.
　　Ayafor, Okogun, J. Chem. Soc., Perkin I, 909 (1982)

TECOMA (Bignoniaceae)
T. capensis.
IRIDOIDS: 7-O-(p-hydroxybenzoyl)-tecomoside, 7-O-(p-hydroxy)-trans-cinnamoyltecomoside, 7-O-trans-cinnamoyltecomoside, 7-O-(p-methoxy)-trans-cinnamoyltecomoside, tecomoside.
　　Bianco et al., J. Nat. Prod., 46, 314 (1983)

T. chrysantha.
IRIDOID: arameloside.
　　Bianco et al., Planta Med., 46, 33 (1982)

T. fulva.
IRIDOIDS: plantarenaloside, stansioside.
　　Gambarino et al., J. Nat. Prod., 48, 992 (1985)

T. heptaphylla.
IRIDOIDS: 6-O-(p-hydroxybenzoyl)-6-epi-monomelittoside, 6-O-(p-methoxybenzoyl)-6-epi-monomelittoside, 6-epi-monomelittoside.
　　Bianco et al., Gazz. Chim. Ital., 113, 465 (1983)

T. stans (Stenolobium stans).
PIPERIDINE ALKALOIDS: 5-dehydroskytanthine, 4-hydroxyskytanthine, 7(a)-hydroxyskytanthine, tecomanine, tecostanine, tecostidine.
IRIDOIDS: amareloside, plantarenaloside, stansioside.
　　(A) Gross et al., Phytochemistry, 12, 201 (1973)
　　(T) Bianco et al., Gazz. Chim. Ital., 112, 199 (1982)

TECOMELLA
T. undulata (G.Don.) Seem.
ESTERS: octacosonylacetylferulate, octacosonyl ferulate.
IRIDOIDS: tecoside, 6-O-veratrylcatalposide P.
PHENOLIC: veratric acid.
TERPENOIDS: dehydro-α-lapachone, dehydrotectol, lapachol, tectol, tectoquinone.
　　(I) Joshi et al., Phytochemistry, 14, 1441 (1975)
　　(EP) Joshi et al., Curr. Sci., 46, 145 (1977)
　　(T) Joshi et al., Planta Med., 71 ((1986)

TECTONA (Rubiaceae)
T. grandis L.
PIGMENT: tectoleafquinone.
DITERPENOID: tectograndinol.
　　(P) Agarwal et al., Tetrahedron Lett., 2623 (1965)
　　(T) Rimpler, Christiansen, Z. Naturforsch., 32C, 724 (1977)

TEDANIA
T. digitata.
CAROTENOID: tedanin.
　　Okukado et al., Bull. Chem. Soc., Japan., 48, 1061 (1975)

TELEKIA (Compositae)
T. speciosa (Schreb.) Baumg.
UNCHARACTERIZED ALKALOID: telekine.
SESQUITERPENOIDS: 2,3-dihydroaromaticin, isotelekin, telekin.
  (A) Brutko, Massagetov, *Khim. Prir. Soedin.*, 57 (1968)
  (T) Bohlmann *et al.*, *Phytochemistry*, **18**, 887 (1979)

TELETOXICUM
T. peruvianum.
ISOQUINOLINE ALKALOIDS: peruvianine, teletoxine.
NORAPORPHINE ALKALOID: telezoline.
  Menachery, Cava, *J. Nat. Prod.*, **44**, 320 (1981)

TELLIMA (Saxifragaceae)
T. grandiflora.
PHENOLICS:
  2,3-digallyl-4,6-hexahydrophenoyl-β-D-glucopyranose,
  1,2,3-trigallyl-4,6-hexahydrophenoyl-β-D-glucopyranose.
  Wilkins *et al.*, *Phytochemistry*, **15**, 211 (1976)

TELOSCHISTES (Teloschistaceae)
T. flavicans (Swartz) Norman.
ANTHROQUINONE: fallacinal.
DIOXEPIN: vicanicin.
  (An) Nakano *et al.*, *Phytochemistry*, **11**, 3505 (1972)
  (D) Neelakabntan *et al.*, *Tetrahedron*, **18**, 597 (1962)

TEMPLETONIA (Leguminosae)
T. retusa (Vent.) R.Br.
QUINOLIZIDINE ALKALOID: templetine.
  Cannon *et al.*, *Tetrahedron Lett.*, 1683 (1974)

TEPHROSIA (Papilionaceae)
T. apollinea.
FLAVONES: pseudosemiglabrinol, semiglabrinol.
  Ahmed, *Phytochemistry*, **25**, 955 (1986)
T. candida.
FLAVANONES: candidone, dehydrorotenone,
  6-hydroxykaempferol 4'-methylether, ovalichalcone.
ISOCOUMARIN: candidin.
  (F) Roy *et al.*, *Phytochemistry*, **25**, 961 (1986)
  (I) Chibber *et al.*, *ibid.*, **20**, 1460 (1981)
T. elongata.
BENZOPYRAN: elongatin.
  Smalberger *et al.*, *Tetrahedron*, **31**, 2297 (1975)
T. fulvinervis.
FLAVONOIDS: fulvinervin A, fulvinervin B.
  Rao *et al.*, *Phytochemistry*, **24**, 2427 (1985)
T. hamiltonii Drumm.
FLAVONOID: rutin.
PHENOLICS: caffeic acid, p-hydroxybenzoic acid, protocatechuic
  acid.
  Basu *et al.*, *Chem. Abstr.*, 88 71492q
T. hildebrandtii Vatke.
FLAVONOIDS: hildgardtene, hildgardtol A, hildgardtol B,
  methylhildgardtol A, methylhildgardtol B.
PTEROCARPANS: hildecarpidin, hildecarpin.
  (F) Delle Monache *et al.*, *Phytochemistry*, **25**, 1711 (1986)
  (P) Lwande *et al.*, *Insect Sci. Appl.*, **7**, 501 (1986); *Chem. Abstr.*, 106
    15746d
  Lwande *et al.*, *Phytochemistry*, **26**, 2425 (1987)
T. lupinifolia.
BENZOPYRANS: lupinifolin, lupinifolinol.
  Smalberger *et al.*, *Tetrahedron*, **30**, 3927 (1974)

T. madrensis.
FLAVAN: 5,7-dimethoxy-8-prenylflavan.
  Gomez *et al.*, *Phytochemistry*, **22**, 1305 (1983)
T. maxima.
ISOFLAVONES: maximaisoflavone B, maximaisoflavone J.
T. multijuga.
TERPENOIDS: multijugin, multijunginol.
  Vleggaar *et al.*, *Tetrahedron*, **31**, 2571 (1975)
  Murthy *et al.*, *J. Nat. Prod.*, **48**, 967 (1985)
T. nubica.
FLAVONOIDS: apollinine, lanceolatin, pseudosemiglabrin,
  semiglabrin.
  Ammar, Jarvis, *J. Nat. Prod.*, **49**, 719 (1986)
T. polystachyoides.
DIPYRAN: tephrodin.
FLAVONES: tachrosin, tephrostachin.
  Vleggaar *et al.*, *Tetrahedron Lett.*, 703 (1972)
T. purpurea. One uncharacterized alkaloid.
STEROID: β-sitosterol.
TRITERPENOID: lupeol.
  Basu *et al.*, *Indian J. Chem.*, **15B**, 971 (1977)
T. semiglabrata.
BENZOPYRANS: semiglabrin, semiglabrinol.
  Smalberger *et al.*, *Tetrahedron*, **29**, 3099 (1973)
T. villosa (Linn.).
ROTENOIDS: villinol, villosin, villosinol, villosol, villosone.
  Sarma *et al.*, *Indian J. Chem.*, **14B**, 152 (1976)
T. woodii.
CHALCONE: oaxacacin.
FLAVONOID: mixtecacin.
  Dominguez *et al.*, *Phytochemistry*, **22**, 2047 (1983)

TERMINALIA (Combretaceae)
T. arjuna.
TRITERPENOIDS: arjungenin, arjunic acid, arjunolic acid,
  terminoic acid.
TRITERPENE GLYCOSIDES: arjunetin, arjunglucosides I, II and
  III.
  Tsuyuki *et al.*, *Bull. Chem. Soc., Japan*, **52**, 3127 (1979)
T. chebula.
TANNINS: chebulagic acid, chebulic acid, chebulinic acid,
  neochebulagic acid, neochebulic acid.
  Haslam *et al.*, *J. Chem. Soc., C*, 2381 (1967)
T. conferta.
TRITERPENOID: arjunolic acid.
  King *et al.*, *J. Chem. Soc.*, 3995 (1955)
T. ivorensis.
TRITERPENOID: terminolic acid.
  King *et al.*, *J. Chem. Soc.*, 1333 (1955)
T. officinalis L.
TRITERPENE GLUCOSIDE: sericoside.
  Bombardelli *et al.*, *Phytochemistry*, **13**, 2559 (1974)
T. paniculata.
BENZOPYRAN: 3,3',4-tri-O-methylflavellagic acid.
  Ramachandra *et al.*, *Tetrahedron*, **23**, 879 (1967)
T. sericea.
TRITERPENOIDS: sericic acid, sericoside.
  Bombardelli *et al.*, *Phytochemistry*, **13**, 2559 (1974)
T. tomentosa.
TRITERPENOID: tomentosic acid.
  Rao, Rao, *Tetrahedron Lett.*, **27**, 12 (1960)

TERNSTROEMIA (Ternstroemiaceae)

T. cherryi.
TRITERPENOIDS: betulic acid, betulinic acid.
Courtney *et al.*, *Aust. J. Chem.*, **18**, 591 (1965)

T. japonica.
CAROTENOID: ternstroemiaxanthin.
Kikuchi, Yamaguchi, *Bull. Chem. Soc., Japan*, **47**, 885 (1974)

T. merrilliana.
TRITERPENOIDS: betulic acid, betulinic acid.
Courtney *et al.*, *Aust. J. Chem.*, **18**, 591 (1965)

TESSARIA

T. absinthioides.
SESQUITERPENOIDS: 12-acetoxyeremophilene,
1(10)-eremophiladien-13-oic acid, eremophilen-12-oic acid,
methyleremophilen-12-oic acid, methyleremophilen-12-oic
acid methyl ester, tessaric acid.
Bohlmann *et al.*, *Phytochemistry*, **16**, 1302 (1977)

T. dodoneifolia (Hook et Arn.) Cabr.
FLAVONOIDS: dihydroquercetin 3-acetate, eriodictyol, luteolin,
naringenin 7,4'-dimethyl ether, quercetin 3,6-diethyl ether,
quercetin 3-ethyl ether.
SESQUITERPENOID: tedonodiol.
(F) Kavak *et al.*, *An. Quim.*, **73**, 305 (1977)
(T) Guerreiro *et al.*, *An. Asoc. Quim. Argent.*, **67**, 119 (1979)

TESTULEA

T. gabonensis Pellegr.
AMINE ALKALOID: N-methyltyramine.
CARBOLINE ALKALOID: N-methyltetrahydro-β-carboline.
TRITERPENOID: epi-friedelinol.
Leboeuf *et al.*, *Plant Med. Phytother.*, **11**, 230 (1977)

TETRACHYRON

T. orizabaensis.
DITERPENOID: tetrachysin.
Ohno *et al.*, *Phytochemistry*, **18**, 1687 (1979)

TETRACLINUS

T. articulata. (Totara articulata).
DITERPENOIDS: 12-acetylsandaracopimaric acid, hinokiol,
hinokione, sandaracopimaric acid, torulosic acid, torulosol,
torulosol 18-acetate, totarolenone, totarolone.
Edwards *et al.*, *Can. J. Chem.*, **38**, 663 (1960)
Gough *et al.*, *Chem. Ind. (London)*, 2059 (1964)

TETRADENIA

T. riparia.
DITERPENOIDS: 8(14),15-sandaracopimaradiene-2α,18-diol,
8(14),15-sandaracopimaradiene-7,18-diol.
van Puyvelde *et al.*, *J. Ethnopharmacol.*, **17**, 269 (1986); *Chem. Abstr.*,
106 29950c
van Puyvelde *et al.*, *Phytochemistry*, **26**, 493 (1987)

TETRADYMIA

T. glabrata.
SESQUITERPENOIDS: 6β,10β-dihydroxyfuranoeremophilane,
10β-hydroxy-6β-isobutyrylfuranoeremophilane,
6-isobutyryl-10-hydroxyfuranoeremophilane, tetradymol.
Jennings *et al.*, *J. Org. Chem.*, **41**, 4078 (1976)

TETRAGONIA

T. tetragonoides.
NORDITERPENOIDS: (2E)-3,7,11,15-tetramethylhexadecanal,
6,10,14-trimethyl-2-methylenepentadecanal.
Aoki *et al.*, *Phytochemistry*, **21**, 1361 (1982)

TETRAGONOTHAEA

T. ludoviciana.
SESQUITERPENOIDS: tetraludins A, B, C, D, E, F, G, H, I, J, K,
L and M.
Quijano *et al.*, *Phytochemistry*, **19**, 1485 (1980)

T. repanda.
SESQUITERPENOIDS: repandins A, B, C and D.
Saman *et al.*, *J. Org. Chem.*, **44**, 3400 (1979)

TETRAPANAX

T. papyriferum.
SAPONINS: papyriosides L-IIa, L-IIb, L-IIc and L-IId.
TRITERPENOIDS: papyriogenins A, B, C, D, E, F and G.
Asada *et al.*, *J. Chem. Soc., Perkin 1*, 325 (1980)

TETRAPATHAEA (Passifloraceae)

T. tetrandra.
CYCLOPENTANOIDS: deidaclin, tetraphyllin A, tetraphyllin B,
epi-tetraphyllin B.
Spencer *et al.*, *Phytochemistry*, **22**, 1815 (1983)

TETRATAENIUM (Pastinaceae)

T. olgae.
COUMARINS: bergapten, imperatorin, isobergapten,
isopimpinellin, pimpinellin, sphondin, umbeliiferone.
Gusak *et al.*, *Chem. Abstr.*, 86 68378k

TEUCRIUM (Labiatae)

T. africanum.
HALOGENATED DITERPENOIDS: tafricanin A, tafricanin B.
Hanson *et al.*, *J. Chem. Soc., Perkin I*, 1005 (1982)

T. amarum.
MONOTERPENOID: teucrein.
Bellebia *et al.*, *J. Chem. Res. Synop.*, 328 (1983)

T. arduini.
IRIDOIDS: teucardoside, teuhircoside.
Ruhdorfer *et al.*, *Z. Naturforsch.*, **36**, 697 (1981)

T. boreale.
MONOTERPENOID: chrysanthemyl acetate.
Dembitskii, Suleeva, *Khim. Prir. Soedin.*, 527 (1984).

T. botrys L.
DITERPENOIDS: 19-deacetylteuscorodol, 6-hydroxy- teuscordin,
montanin D, teubotrin, teuchamaedrin, teucvidin.
De La Torre *et al.*, *Phytochemistry*, **25**, 2385 (1986)

T. capitatum.
DITERPENOID: lolin.
Marquez *et al.*, *Tetrahedron Lett.*, 2823 (1981)

T. carolipaui.
SESQUITERPENOID: 11-hydroxyvalenc-1(10)-en-2-one.
Savona *et al.*, *Phytochemistry*, **26**, 571 (1987)

T. chamaedrys L.
DITERPENOIDS: teuchamaedryins A, B and C, 6-epi-teucrin A,
teucroxide, teugin.
Malakov *et al.*, *Phytochemistry*, **24**, 301 (1985)

T. eriocephalum.
DITERPENOID: eriocephalin.
Fayos *et al.*, *J. Org. Chem.*, **44**, 4992 (1979)

T. flavum.
DITERPENOID: teuflin.
Savona *et al., J. Chem. Soc., Perkin I,* 1915 (1979)

T. fragile.
DITERPENOID: teugin.
Bruno *et al., Phytochemistry,* **20,** 2259 (1981)

T. fruticans L.
DITERPENOIDS: fruticolone, isofruticolone.
Savona *et al., J. Chem. Soc., Perkin I,* 356 (1978)

T. gnaphalodes.
DITERPENOIDS: 19-acetylgnaphalin, gnaphilidin, gnaphalin.
Savona *et al., Tetrahedron Lett.,* 379 (1979)

T. hyrcanicum.
DITERPENOIDS: teucrins H-1,H-2, H-3 and H-4.
IRIDOID: teuhircoside.
Ruhdorfer *et al., Z. Naturforsch.,* **36C,** 697 (1981)

T. japonicum.
DITERPENOIDS: teucjaponins A and B.
Miyase *et al., Chem. Pharm. Bull.,* **29,** 3561 (1981)

T. marum.
DITERPENOIDS: allodolicholactone, dolicholactone.
Pagnoni *et al., Aust. J. Chem.,* **29,** 1375 (1976)

T. montanum.
DITERPENOIDS: montanins A, B, C and D.
Malakov *et al., Z. Naturforsch.,* **33,** 1142 (1978)

T. montanum subsp. skorpilii.
DITERPENOID: montanin D.
Papanov, Malakov, *Phytochemistry,* **22,** 2787 (1983)

T. polium.
DITERPENOIDS: 6-acetylpicropoline, teupolins I, II, III, IV and V.
Malakov, Papanov, *Phytochemistry,* **22,** 2791 (1983)

T. polium var. alba.
FLAVANOIDS: cirsidiol, salvigenin.
IRIDOID: teucardoside.
Rizk *et al., Planta Med.,* 87 (1986)

T. polium subsp. aureum (Schreber) Archangeli.
CLERODANES: 19-acetylgnaphalin, aurofolin, gnaphilidin, teucrin P-1.
Aguren *et al., J. Chem. Soc.,* 3364 (1981)

T. polium subsp. pilosum.
FLAVONOIDS: cirsidiol, salvigenin.
IRIDOID: teucardoside.
DITERPENOID: 19-acetylteupolin U.
Rizk *et al., Planta Med.,* 87 (1986)
De La Torre *et al., Phytochemistry,* **25,** 2239 (1986)

T. pyrenaicum.
DITERPENOIDS: teupyreinidin, teupyrein, teupyrenone, teupyrin A, teupyrin B.
Fernandes *et al., Phytochemistry,* **25,** 1405 (1986)

T. ramosissimum Desf.
FLAVONOID: apigenin 6,8-diglucoside.
Reynaud *et al., Plant Med. Phytother.,* **10,** 199 (1976)

T. scordium L.
DITERPENOIDS: 2-keto-19-hydroxyteuscordin, teuscordinone.
Papanov *et al., Phytochemistry,* **24,** 297 (1985)

T. scorodonia L.
DITERPENOIDS: teuscoridin, teuscorodal, teuscorodol, teuscorodonin, teuscorolide.
Marco *et al., Phytochemistry,* **22,** 727 (1983)

T. spinosum.
DITERPENOIDS: 19-acetylteuspinin, teuspinin.
Savona *et al., Heterocycles,* **14,** 193 (1980)

T. subspinosum.
DITERPENOIDS: 6α-hydroxyteuscordin, teucrin H-2, teucvin, teuflin.
Fernandez *et al., Phytochemistry,* **25,** 1405 (1986)

T. viscidum var. Miquelianum.
DITERPENOIDS: teuflin, teucvin.
Node *et al., J. Chem. Res. Synop.,* (2), 32 (1981)

TEVENOTICA
T. persica DC.
TRITERPENOIDS: α-amyrin, lupeol.
Rustaiyan *et al., Fitoterapia,* **48,** 173 (1977)

THALICTRUM (Ranunculaceae)
T. alpinum L.
BISBENZYLISOQUINOLINE ALKALOIDS: N-desmethylthalirugosidine, N-desmethylthalirugisine, hernandezine, thalpindione.
Wu *et al., Lloydia,* **43,** 372 (1980)

T. aquilegifolium.
GLUCOSIDE: thalictoside.
TRITERPENOID: aquilegifolin.
Ina *et al., Chem. Pharm. Bull.,* **34,** 726 (1986)

T. baicalense.
BISBENZYLISOQUINOLINE ALKALOID: N-demethylthalistyline.
Lu, *Zhongcaoyao,* **15,** 195 (1984)

T. cultratum.
APORPHINE-BENZYLISOQUINOLINE ALKALOIDS: (+)-thalibulamine, (+)-thalifaramine, (+)-thalifaretine, (+)-thalifaricine, (+)-thalifaroline, (+)-thalifaronine.
BISBENZYLISOQUINOLINE ALKALOIDS: (-)-5-hydroxythalidasine, (-)-5-hydroxythalmine, (+)-thalmiculatimine, (-)-thalmiculimine, (-)-thalmiculine.
Hussain *et al., J. Nat. Prod.,* **49,** 488, (1986)
Hussain *et al., ibid.,* **49,** 494 (1986)

T. dasycarpum Fisch. et Lall.
BISBENZYLISOQUINOLINE ALKALOIDS: dehydrothalicarpine, thalicarpine, thalidasine.
Kupchan *et al., J. Am. Chem. Soc.,* **89,** 3075 (1967)

T. dioicum.
BISBENZYLISOQUINOLINE ALKALOIDS: thalidicne, thalidine, thalidoxine.
Shamma *et al., Lloydia,* **39,** 395 (1976)

T. faberi.
BISBENZYLISOQUINOLINE ALKALOIDS: huangshanine, (+)-thalifaberine, (+)-thalifabine.
Linet *et al., Planta Med.,* **49,** 55 (1983)

T. fauriei.
BISBENZYLISOQUINOLINE ALKALOID: thalifaurine.
Chen *et al., J. Pharm. Sci.,* **69,** 1061 (1980)

T. fendleri Engelm.
APORPHINE ALKALOIDS: preocoteine, veronamine.
BISBENZYLISOQUINOLINE ALKALOIDS: tetrahydrothalifendine, thalidaldine, thalidastine, thalidezine, thalifendine, thalifendlerine.
ISOQUINOLINE ALKALOIDS: N-methylcorydaline, N-methylthalidaldine.
Shamma, Podczasy, *Tetrahedron,* **27,** 727 (1971)

T. flavum L.
AMINE ALKALOIDS: thalicsine, thalflavidine.
APORPHINE ALKALOID: magnoflorine.
BISBENZYLISOQUINOLINE ALKALOIDS: thalicarpine, thaliflavine.

ISOQUINOLONE ALKALOID: thalflavine.
PROTOBERBERINE ALKALOIDS: berberine, cryptopine.
Umarov *et al.*, *Khim. Prir. Soedin.*, 683 (1973)

T. foetidum L.
APORPHINE ALKALOID: magnoflorine.
BISBENZYLISOQUINOLINE ALKALOIDS: fetidine, thalfine, thalfinine, thalfoetidine.
SAPONINS: cyclofoetoside A, foetosides A, B and C.
(A) Mollov, Georgiev, *Chem. Ind.*, 1178 (1966)
(Sap) Ganenko *et al.*, *Khim. Prir. Soedin.*, 66 (1986)

T. foliosum.
APORPHINE ALKALOID: N,O,O-trimethylsparsiflorine.
PROTOBERBERINE ALKALOIDS: berberine, thalictrine.
Bhakuni, Singh, *J. Nat. Prod.*, **45**, 252 (1982)

T. hernandezi.
BISBENZYLISOQUINOLINE ALKALOID: hernandezine.
Padilla, Herran, *Tetrahedron*, **18**, 427 (1962)

T. isopyroides SAM.
APORPHINE ALKALOIDS: cabudine, dehydrothalicamine, magnoflorine, thalicamine, thalicaminine.
BISBENZYLISOQUINOLINE ALKALOIDS: thalisopidine, thalisopine, thalisopinine.
PROTOBERBERINE ALKALOID: protopine.
Pulatova *et al.*, *Khim. Prir. Soedin.*, 609 (1969)
Kurbanov *et al.*, *Dokl. Akad. Nauk Tadzh. SSR*, **18**, 20 (1980)

T. longipedunculatum E.Nik.
AMINE ALKALOID: thalicsine.
APORPHINE ALKALOID: magnoflorine.
BISBENZYLISOQUINOLINE ALKALOIDS: thalidezine, thalifetidine.
PROTOBERBERINE ALKALOID: berberine.
Khodzhaev *et al.*, *Khim. Prir. Soedin.*, 441 (1973)

T. longistylum DC.
BISBENZYLISOQUINOLINE ALKALOIDS: N-desmethylthalistyline, thalistyline, thalistyline methioidide.
Wu *et al.*, *Tetrahedron Lett.*, 3687 (1976)

T. minus L.
AMINE ALKALOID: thaliglucinone.
APORPHINE ALKALOIDS: (+)-glaucine, magnoflorine, preocoteine, thalicmidine, thalicmidine N-oxide, thalicmine, thalicminine.
APORPHINE-BENZYLISOQUINOLINE ALKALOIDS: (+)-iznikine, thalmelatidine.
BISBENZYLISOQUINOLINE ALKALOIDS: istambulamine, O-methylthalicberine, takatonine, thaisivasin, thaliadanine, thalmelatine, thalmetine, thalmine, thalmineline.
UNCHARACTERIZED ALKALOIDS: talminelline, thaliglucinone, thalirabine, thaliracebine, thalmidine, thalrugosamine.
Gromova *et al.*, *Khim. Prir. Soedin.*, 670 (1985)
Baser, Kirmer, *Planta Med.*, 448 (1985)

T. minus Race B.
APORPHINE-BENZYLISOQUINOLINE ALKALOID: (+)-bursanine.
BISBENZYLISOQUINOLINE ALKALOIDS: O-methylthalibrunine, thalistine, thalmirabine.
Guinaudeau *et al.*, *Tetrahedron Lett.*, 2523 (1982)

T. minus L. var. adiantifolium.
APORPHINE ALKALOID: adiantifoline.
ISOQUINOLINE ALKALOID: noroxyhydrastinine.
Doskotch *et al.*, *Tetrahedron Lett.*, 4999 (1968)
Doskotch *et al.*, *Tetrahedron*, **25**, 469 (1969)

T. minus L. var. elatum Koch.
APORPHINE-BENZYLISOQUINOLINE ALKALOID: thalmelatine.
BISBENZYLISOQUINOLINE ALKALOIDS: dehydrothalicarpine, elatrine, thaliblastine, thalmineline.
Kupchan *et al.*, *J. Org. Chem.*, **33**, 1052 (1968)

T. minus L. var. microphyllum.
BISBENZYLISOQUINOLINE ALKALOID: uskudaramine.
Guinaudeau *et al.*, *J. Org. Chem.*, **47**, 5406 (1982)

T. minus L. var. minus.
BISBENZYLISOQUINOLINE ALKALOIDS: thaivaisine, thalvarmine.
Baser, Kirmer, *Planta Med.*, 448 (1985)

T. podocarpum Hunb.
BISBENZYLISOQUINOLINE ALKALOIDS: N-desmethylthalistyline, thalistyline, thalistyline methioidide.
Wu *et al.*, *Tetrahedron Lett.*, 3687 (1976)

T. polygamum.
APORPHINE ALKALOIDS: N-methylpalaudinium chloride, thalphenine.
APORPHINE-BENZYLISOQUINOLINE ALKALOIDS: pennsylvanamine, pennsylvanine, thalictrogamine, thalictropine.
BISBENZYLISOQUINOLINE ALKALOIDS: thaligmine, thalpine.
Shamma, Moniot, *Tetrahedron Lett.*, 775 (1973)
Shamma *et al.*, *Heterocycles*, **43**, 270 (1980)

T. revolutum DC.
BISBENZYLISOQUINOLINE ALKALOIDS: neothalibrine, revolutopine, thalilutidine, thalpine.
ISOQUINOLINE ALKALOID: revolutinone.
Wu *et al.*, *Lloydia*, **43**, 270 (1980)

T. rochebrunianum.
BISBENZYLISOQUINOLINE ALKALOIDS: dihydrothalictrinine, N'-norhernandezine, N'-northalibrunine, oxothalibrunimine, oxothalibrunine, thalibrunine, thalictrinine.
Wu *et al.*, *J. Org. Chem.*, **45**, 213 (1980)

T. rugosum Ait.
AMINE ALKALOIDS: thaliglucine, thaliglucinone.
BISBENZYLISOQUINOLINE ALKALOIDS: thaligosidine, thaligosine, thaligosinine, thalilutine, thalirugidin, thalirugine, thaliruginine, thalirugosamine, thalirugosidine, thalirugosine, thalirugosinone.
Wu *et al.*, *J. Org. Chem.*, **43**, 580 (1978)
Wu *et al.*, *Lloydia*, **43**, 143 (1980)

T. simplex L.
AMINE ALKALOID: thalicsine.
APORPHINE ALKALOIDS: magnoflorine, thalicsimidine.
BISBENZYLISOQUINOLINE ALKALOIDS: hernandezine, thalfetidine, thalsimidine, thalsimine, thalsinine.
OXOPROTOBERBERINE ALKALOID: thalictrisine.
PROTOBERBERINE ALKALOIDS: α-allocryptopine, berberine.
UNCHARACTERIZED ALKALOIDS: thalictricine, thalictrinine.
Umarov *et al.*, *Khim. Prir. Soedin.*, 329 (1968)

T. strictum.
PAVINANE ALKALOID:
2,3,7-trimethoxy-N-methyl-8,9-methylenedioxypavinane.
Yunusov *et al.*, *Khim. Prir. Soedin.*, 116 (1976)

T. sultanbadense.
BISBENZYLISOQUINOLINE ALKALOIDS: hernandezine N-oxide, thalbadenzine.
Abizabbarova *et al.*, *Khim. Prir. Soedin.*, 139 (1978)

T. thunbergii DC.
APORPHINE ALKALOIDS: thalicmine, thalictruberine.
BISBENZYLISOQUINOLINE ALKALOIDS: methylthalicberine, thalicberberine, thalicberine, thalictrine.
Tomita, Kitamura, *J. Pharm. Soc., Japan*, **79**, 1092 (1959)

T. urbaini Hayata.
APORPHINE ALKALOIDS: -(+)-isocorydine, S-(+)-oconovine.
Wu *et al.*, *Chem. Abstr.*, 88 34563t

THAMNOLIA (Usneaceae)

T. subrosmicularis.
TERPENOID: baeomycesic acid.
Koller et al., Monatsh. Chem., **66**, 57 (1935)

T. subuniformis.
STEROID: ergosterol peroxide.
Strigina et al., Khim. Prir. Soedin., 551 (1976)

T. vermicularis Ach. ex Schaer.
PHENOLIC: thamnolic acid.
Wachmeister et al., Acta Chem. Scand., **9**, 1395 (1955)

THAMNOSMA

T. montana.
QUINOLINE ALKALOID: robustine.
COUMARINS: thamnosin, thamnosmin, thamnosmonin, thamontanin, umbelliprenin.
Chang et al., Lloydia, **39**, 134 (1976)

THAPSIA (Umbelliferae)

T. garganica L.
COUMARIN: 6-methoxy-7-geranyloxycoumarin.
SESQUITERPENOIDS: thapsigargicin, thapsigargin.
(C) Carsen, Sandberg, Acta Chem. Scand., **24**, 1113 (1978)
(T) Christensen et al., Tetrahedron Lett., 3829 (1980)

T. villosa var. minor.
SESQUITERPENOIDS: 15-acetoxythapsan-14-al, (1S,6R)-1-senecioyloxy-6,14-epoxythapsan-15-ol, (1S,3R)-1-senecioyloxy-6,14-epoxythapsan-15-ol 5-acetate, (1S)-1-senecioyloxy-6(14)-thapsen-15-ol.
de Pascual Teresa et al., Phytochemistry, **25**, 1171 (1986)

THEA (Theaceae)

T. sinensis.
TRITERPENOIDS: barringtogenin C, camelliagenin D, theasapogenol A, theasapogenol E, 5-tirucalla-7,24-dien-3β-ol.
Nakano et al., Tetrahedron Lett., 1675 (1967)
Itoh et al., Lipids, **11**, 434 (1976)

THELEPHORA (Lichenes)

T. palmata.
BISBENZOFURAN: thelephoric acid.
Gripenberg, Tetrahedron, **10**, 135 (1960)

THELEPOGON (Gramineae/Poaceae)

T. elegans.
PYRROLIZIDINE ALKALOID: thelepogine.
UNCHARACTERIZED ALKALOID: thelepogidine.
Crow, Aust. J. Chem., **15**, 159 (1962)

THELEPUS

T. setosus.
HALOGENATED NORSESQUITERPENOID: thelepin.
Higa, Scheuer, J. Am. Chem. Soc., **96**, 1146 (1974)

THELESPERMA

T. megapotamicum.
FLAVONOIDS: luteolin, luteolin 7-O-β-D-glucoside, marein.
Ateya et al., Planta Med., **45**, 247 (1982)

THELONOTA

T. ananas.
TRITERPENOIDS: 23-acetoxyholosta-8,25-dien-3β-ol, 23-acetoxyholosta-9(11),25-dien-3β-ol, 23-acetoxyholost-9(11)-ene-3β,25-diol, 23-acetoxyholost-8-en-3β-ol, 23-acetoxyholost-9(11)-en-3β-ol, holosta-9(11),25-diene-3β,23-diol, thelothurins A and B.
Kelecom et al., Tetrahedron, **32**, **2313**, 2353 (1976)

THEOBROMA (Sterculiaceae)

T. cacao L.
PURINE ALKALOIDS: theobromine, theophylline.
FLAVONOIDS: apigenin 7-O-glucoside, chrysoeriol 7-O-glucoside, isovitexin, luteolin, luteolin 7-O-glucoside, quercetin 3-O-galactoside, quercetin 3-O-glucoside, vitexin.
(A) Biltz, Max., Justus Liebigs Ann. Chem., **423**, 320 (1921)
(F) Jalal et al., Phytochemistry, **16**, 1377 (1977)

THEONELLA

T. swinhoei.
STEROID: theonellasterol.
Kho et al., J. Org. Chem., **46**, 1836 (1981)

THERMOPSIS (Leguminosae)

T. alpina (Pall.) Ldb.
SPARTEINE ALKALOIDS: alpine, argettine, cytisine, N-methylcytisine, pachycarpine, thermopsine.
Iskandarov et al., Avtoref. Dokl. Diss. Tashkent, (1973)

T. alterniflora Rgl. et Schmalh.
SPARTEINE ALKALOIDS: alteramine, anagyrine, argentamine, argentine, cytisine, dimetamine, N-methylcytisine, pachycarpine, thermopsine.
Iskandarov et al., Khim. Prir. Soedin., 218 (1972)

T. dolichocarpa V.Nikit.
SPARTEINE ALKALOIDS: cytisine, pachycarpine, thermopsine.
Normatov et al., J. Gen. Chem., USSR, **3**, 45 (1962)

T. fabacea (Pall.) DC.
SPARTEINE ALKALOIDS: cytisine, N-methylcytisine, pachycarpine, thermopsine.
Ryabinin et al., J. Prakt. Chem., **28**, 663 (1955)

T. lanceolata R.Br. (Sophora lupinoides Link.).
SPARTEINE ALKALOIDS: anagyrine, argentine, cytisine, dithermanine, homothermopsine, N-methylcytisine, pachycarpine, rhombifoline, thermopsamine, thermopsine.
Cockburn, Marion, Can. J. Chem. **30**, 92 (1952)
Vinogradova et al., Khim. Prir. Soedin., 87 (1972)

T. rhombifolia (Watt.) Richards.
SPARTEINE ALKALOIDS: cytisine, 5,6-dehydrolupanine, methylcytisine, rhombifoline, rhombinine.
Cho, Martin, Can. J. Chem., **49**, 265 (1971)

THESIUM (Santalaceae)

T. minkwitzianum B.Fedtsch.
DIPYRROLIZIDINE ALKALOID: thesine.
Arendaruk et al., J. Gen. Chem. USSR., **30**, 670 (1960)

THESPESIA (Malvaceae)

T. populnea (L.) Soland. ex Correa.
FLAVONOIDS: calycopterin, quercetin.
NAPHTHALENE: (+)-gossypol.
(F) Kasim et al., Curr. Sci., **44**, 888 (1975)
(N) Bhakuni et al., Experientia, **24**, 109 (1968)

THEVETIA

T. ahouia A.DC.

IRIDOID: 3'-O-methylevomonoside.

Jolad et al., J. Org. Chem., 46, 1946 (1981)

T. neriifolia Juss.

STEROID: thevetin A.

Bloch et al., Helv. Chim. Acta, 43, 652 (1960)

T. peruviana (Pers.) K.Schum.

FLAVONOIDS: hesperitin 7-glucoside, kaempferol, epi-peruviol
acetate, quercetin.

IRIDOIDS: theveside, theviridoside.

STEROIDS: peruvoside, β-sitosterol, thevefoline.

TRITERPENOIDS: α-amyrin, β-amyrin, betulin, lupeol, lupenyl
acetate.

(F, T) Rao et al., Indian J. Pharm., 37, 124 (1975)

(I) Sticher, Tetrahedron Lett., 3195 (1970)

(S) Frerejacque, Durgeat, C. R. Acad. Sci., 272B, 2620 (1971)

THLASPI (Brassicaceae/Cruciferae)

T. arvense L.

20 fatty acids.

Dolya et al., Rastit. Resur., 22, 249 (1986); Chem. Abstr., 105 39401

THRYPTOMENE

T. kochii.

SESQUITERPENOID: globulol.

Dolejs et al., Collect. Czech. Chem. Commun., 25, 1487 (1960)

THUJA (Cupressaceae)

T. orientalis.

DITERPENOIDS: neothujic acid III.

SESQUITERPENOIDS: α-cuparenone, β-cuparenone, occidenol,
occidentalol, occidol.

Parker et al., J. Chem. Soc., 1558 (1962)

Balansard et al., Trav. Soc. Pharm., Montpellier, 36, 307 (1976)

T. plicata.

NAPHTHALENES: plicatic acid plicatinanaphthalene,
plicatananaphthol.

PHENOLIC: τ-thujaplicatene.

DITERPENOIDS: isopimarinol,
19-norisopimara-8(14),15-dien-4α-ol,
19-norisopimara-8(14),15-dien-4β-ol,
19-norisopimara-8(14),15-dien-3-one, (+)-sugiol.

MONOTERPENOIDS: α-thujaplicin, β-thujaplicin, τ-thujaplicin,
thujic acid, α-thujone, β-thujone.

TOXIN: α-apoplicatitoxin.

(N) Gardner et al., Can. J. Chem., 37, 1703 (1959)

(P) MacDonald, Barton, ibid., 48, 3144 (1970)

(T) Senter et al., Phytochemistry, 14, 2233 (1975)

(Tox) MacDonald et al., ibid., 51, 482 (1973)

T. standishii.

DITERPENOID: nezukol.

MONOTERPENOID: nezukone.

Kitadani, J. Pharm. Soc., Japan, 91, 664 (1970)

THUJOPSIS

T. dolabrata.

DITERPENOIDS: ar-abietatriene, 8,11,13-abietatrien-12,16-oxide,
costal, α-costal, deoxypodophyllic acid,
16-hydroxyferruginol.

SESQUITERPENOIDS: α-pseudowiddrene, thujopsene, thujyl
alcohol, (+)-δ-thujyl alcohol.

Hasegawa et al., Phytochemistry, 22, 643 (1983)

THYMBRA

T. spicata.

FLAVONOIDS: 6-hydroxyluteolin 7,3'-dimethyl ether,
6-hydroxyluteolin 7,3',4'-trimethyl ether.

Miski et al., Phytochemistry, 22, 2093 (1983)

THYMELAEA

T. hirsuta.

FLAVONOID: vicenin-2.

Nawwar et al., Phytochemistry, 16, 1319 (1977)

THYMUS (Labiatae)

T. loscosii Willk.

FLAVONOIDS: 6-hydroxyluteolin 7-glucoside, luteolin
7-glucoside.

Adzet et al., Trav. Soc. Pharm. Montpellier, 41, 193 (1981); Chem.
Abstr., 97 141778z

T. satureioides.

FLAVONE: 5,6,4'-trihydroxy-7,3'-dimethoxyflavone.

Voirin et al., Planta Med., 523 (1985)

THYRSANOSPERMUM

T. diffusum Champ.

STEROID: β-sitosterol.

TRITERPENOID: thyrsanolactone.

(S) Chen et al., Chem. Abstr., 84 102341p

(T) Aimi et al., Tetrahedron, 37, 983 (1981)

T. diffusum Champ. var. longitubum Ohwi.

TRITERPENOID: thyrsanolactone.

Aimi et al., Tetrahedron, 37, 983 (1981)

TILIA (Sapindaceae/Tiliaceae)

T. argentea.

BENZOPYRAN: tiliroside.

Lin, J. Chin. Chem. Soc., 22, 383 (1975); Chem. Abstr., 84 147653

T. cordata Mill.

ACIDS: eicosanoic acid, eicosenoic acid, linoleic acid, linolenic
acid, myristic acid, oleic acid, palmitic acid, palmitoleic
acid.

TRITERPENOID: taraxerol.

(Ac) Mruk-Luczkiewicz, Chem. Abstr., 86 103052y

(T) Beaton et al., J. Chem. Soc., 2131 (1955)

T. x europaea L. (T. vulgaris Hayne).

SESQUITERPENOID: (-)-7-hydroxycalamene.

Burden et al., Phytochemistry, 22, 1039 (1983)

TILIACORA (Menispermaceae)

T. dinklagei.

APORPHINE ALKALOID: oblongine.

BISBENZYLISOQUINOLINE ALKALOIDS: dinklacorine,
nortiliacorine A, nortiliacorinine A.

Dwuma-Badu et al., J. Nat. Prod., 46, 742 (1983)

T. funifera.

BISBENZYLISOQUINOLINE ALKALOID: funiferine.

Tackie, Thomas, Ghana J. Sci., 5, 11 (1965)

T. racemosa.

BISBENZYLISOQUINOLINE ALKALOIDS: N-acetyltiliacorine A,
misine, nortiliacorine A, nortiliacorinine A, tiliacoridine,
tiliacorine, tiliarine.

Guha et al., Tetrahedron Lett., 4241 (1976)

T. triandra Diels.

BISBENZYLISOQUINOLINE ALKALOIDS: nortiliacorinine A, tiliacorine, tiliacorine N-oxide, tiliacorinine, tiliandrine, yanangcorinine, yanangine.

Pachaly, Tan, *Arch. Pharm.*, **319**, 126 (1986)

Pachaly, Tan, *ibid.*, **319**, 841 (1986)

TILLANDSIA (Bromeliaceae)

T. usneoides L.

FLAVONE: 4',5,7-trihydroxy-3,3',5',6-tetramethoxyflavone.

TRITERPENOIDS: cycloart-23-en-3β,25-diol, cycloart-23-en-3,25-diol methyl ether, cycloart-25-ene-3,24-diol.

(F) Lewis *et al.*, *Phytochemistry*, **16**, 1114 (1977)

(T) McCrindle, Djerassi, *J. Chem. Soc.*, 4034 (1962)

T. utriculata L.

FLAVONOIDS: cirsilineol, 6-hydroxykaempferol 3,6,7,4'-tetramethyl ether, jaceosidin.

Ulubelen *et al.*, *Rev. Latinoam. Quim.*, **13**, 35 (1982)

TIMONIUS

T. kaniensis.

QUINOLINE ALKALOID: dihydrocupreine.

Johns, Lamberton, *Aust. J. Chem.*, **23**, 211 (1970)

TINOMISCIUM (Menispermaceae)

T. petiolare.

PROTOBERBERINE ALKALOID: (-)-isocorypalmine.

Tomita, Furukawa, *J. Pharm. Soc., Japan*, **87**, 881 (1967)

T. philippinense.

DITERPENOID: tinophyllone.

Gombolvej *et al.*, *J. Philipp. Pharm. Assoc.*, **45**, 303 (1958)

TINOSPORA

T. dentata.

BERBERINE ALKALOIDS: jetrorrhizine, palmatine.

DITERPENOID: columbin.

Chen, *J. Chin. Chem. Soc.*, **22**, 271 (1975); *Chem. Abstr.*, 84 28054d

T. tuberculata Beumee.

DITERPENOIDS: boropetol B, boropetoside B.

Fukuda *et al.*, *Chem. Pharm. Bull.*, **34**, 2868 (1986)

TITHONIA (Compositae)

T. diversifolia.

SESQUITERPENOIDS: diversifolin, tagitins A, B, C, D, E and F.

Pal *et al.*, *Indian J. Chem.*, **14B**, 259 (1976)

Ciccio *et al.*, *Rev. Latinoam. Quim.*, **10**, 134 (1979)

T. fruticosa.

SESQUITERPENOIDS: deoxyfruticin, tifruticin.

Herz, Shamma, *J. Org. Chem.*, **40**, 3118 (1975)

T. rotundifolia (Mill.) Blake.

SESQUITERPENOIDS: 8-angelyl-14-hydroxytithifolin, 8-angelyl-14-hydroxytithifolin 14-acetate, 19-methylspaerocephalin, rotundin, tirotundin, tithifolin.

Bohlmann *et al.*, *Phytochemistry*, **20**, 267 (1981)

Romo de Vivar *et al.*, *ibid.*, **21**, 375 (1982)

T. tagitiflora.

SESQUITERPENOIDS: tagitinins A, B, C, D, E and F.

Rastogi *et al.*, *J. Pharm. Sci.*, **65**, 918 (1976)

TODDALIA (Rutaceae)

T. aculeata Pers.

PROTOBERBERINE ALKALOIDS: berberine, chelerythrine.

UNCHARACTERIZED ALKALOIDS: toddaline, toddalinine, Coumarins aculeatin, norbraylin, toddaculin, toddalolactone, 5,7,8-trimethoxycoumarin.

(A) Awe, *Arch. Pharm.*, **284**, 152 (1951)

(A) Silva *et al.*, *Phytochemistry*, **10**, 3255 (1971)

(C) Desmukh *et al.*, *ibid.*, **15**, 1419 (1976)

T. asiatica.

ALKALOIDS: arnuttinamide, 8-hydroxydihydrochelerythrine, toddalidimerine.

COUMARIN: toddalenone.

(A) Sharma *et al.*, *Phytochemistry*, **21**, 252 (1982)

(C) Ishii *et al.*, *Chem. Pharm. Bull.*, **31**, 3330 (1983)

TONDUZA

T. longifolia.

CARBOLINE ALKALOID: deserpidine.

Neuss *et al.*, *J. Am. Chem. Soc.*, **77**, 4087 (1955)

TOONA

T. alata var. australis.

TERPENOID: toonafolin.

Kraus *et al.*, *Ann. Chim.*, 1838 (1981)

T. ciliata.

TERPENOIDS: 6-acetoxytoonacilin, toonacilin.

Kraus *et al.*, *Angew. Chem.*, **90**, 476 (1978)

T. sureni.

TERPENOID: surenolactone.

TRITERPENOID: surenin.

Kraus *et al.*, *Ann. Chem.*, 87 (1982)

TORILIS (Umbelliferae)

T. japonica (Houtt.) DC.

SESQUITERPENOIDS: 6,8-cyclo-4(14)-eudesmen-1-ol, 6-methoxy-4(14)-eudesmen-1-ol, torilin.

Itokawa *et al.*, *Chem. Lett.*, 1253 (1983)

T. scabra DC.

SESQUITERPENOIDS: 1-acetoxy-6,7-epoxy-2Z-humulene, 6,7-epoxy-2Z-humulen-1-ol.

Itokawa *et al.*, *Chem. Lett.*, 1581 (1983)

TORREYA (Taxaceae)

T. nucifera.

DITERPENOID: kayadiol.

FURAN: dendrolasin.

SESQUITERPENOIDS: nuciferal, nuciferol, torreyal.

(F) Bernardi *et al.*, *Tetrahedron Lett.*, 3893 (1967)

(T) Sayama *et al.*, *Agr. Biol. Chem.*, **35**, 1069 (1971)

TOURNEFORTIA (Boraginaceae)

T. sibirica L.

UNCHARACTERIZED ALKALOIDS: tourneforcidine, tourneforcine.

Men'shikov *et al.*, *J. Gen. Chem., USSR*, **22**, 1465 (1952)

TOVARIA

T. pendula Ruiz. et Pav.

GLUCOSIDES: N-acetyl-3-indolylmethylglucosinolate, 4-hydroxyglucobrassicin, glucobrassicin, 4-methoxyglucobrassicin, neoglucobrassicin.

Schraudolf *et al.*, *Z. Naturforsch.*, **41C**, 526 (1986)

TOVOMITA (Guttiferae/Clusiaceae)

T. choisyana Planch et Triana.
XANTHONES: tovopyrifolin A, tovoxanthone.
Gabriel et al., Phytochemistry, 11, 3035 (1972)

T. macrophylla (Planch et Triana) Walp.
XANTHONES: tovophyllin A, tovophyllin B.
Goncales de Oliviera et al., Phytochemistry, 11, 3323 (1972)

T. pyrifolium Planch et Triana.
XANTHONES: stovophyllin A, tovophyllin B, tovopyrifolins A, B and C.
Mesquita et al., Phytochemistry, 14, 803 (1975)

TOXICODENDRON

T. capense.
DITERPENOIDS: capencin, pretoxin.
SESQUITERPENOID: isodihydroheynanchin.
Corbella et al., Tetrahedron, 25, 4835 (1969)

T. diversilobum.
CATECHOL: urushiol.
Corbett, J. Pharm. Sci., 64, 1715 (1975)

TOXYLON (Moraceae)

T. pomiferum Rafin.
FLAVONOID: 5,7,2',4'-tetraacetoxyflavone.
PHENOLIC: oxyresveratrol.
XANTHONES: 8-deoxygartanin, 6-deoxyjacareubin, toxylxanthones A, B, C and D.
(F, P) Gerber, Phytochemistry, 25, 1697 (1986)
(X) Deshpande et al., Indian J. Chem., 11, 518 (1973)

TRACHELANTHUS (Boraginaceae)

T. hissoricus Lipsky.
PYRROLIZIDINE ALKALOIDS: trachelantamine, trachelantamine N-oxide, viridiflorine, viridiflorine N-oxide.
Akramov et al., Dokl. Akad. Nauk Uzb. SSR, 6, 35 (1961)

T. korolkovii. (Lipsky). Fedtsch.
PYRROLIZIDINE ALKALOIDS: trachelantamine, trachelantamine N-oxide, trachelantine.
Akramov et al., Khim. Prir. Soedin., 351 (1967)

TRACHELIUM (Campanulaceae)

T. caeruleum L.
PYRANS:
trans-8,9-epoxy-9-(tetrahydropyran-2-yl)-non-2-ene-4,6-diynl,
trans-8,9-epoxy-9-(tetrahydropyran-2-yl)-non-trans-2-ene-4,6-diyn-1-ol,
9-(tetrahydropyran-2-yl)-non-2-ene-4,6-diyn-1,8,9-triol.
Bentley et al., J. Chem. Soc., Perkin I, 1987 (1974)

TRACHELOSPERMUM

T. asiaticum.
STEROIDS: steikaside A, teikaside B
Abe, Yamauchi, Chem. Pharm. Bull., 29, 416 (1981)

T. asiaticum var. intermedium.
GLYCOSIDE: arctigenic acid-4'-β-gentiobioside.
Nishibe et al., Experientia, 29, 17 (1973)

TRACHYCALYMNA

T. fimbriatum Hook.
STEROID: 3,8,14-trihydroxy-5,8,17-pregnan-20-one.
Elber et al., Helv. Chim. Acta, 52, 2583 (1969)

TRACHYLOBIUM

T. verrucosum (Gaertn.) Oliv.
BENZOFURANS: 2-benzyl-2-hydroxybenzo-[b]-furan-3(2H)-one, 2-benzyl-2,3',4',6'-tetrahydroxybenzo-[b]-furan-3(2H)-one.
CHALCONES: (+)-mopanol, (+)-mopanol B, (+)-peltogynol, (+)-peltogynol B, 2',3,4,4'-pentahydroxychalcone.
FLAVANONES: (±)-3-O-methyl-2,3-cis-fustin, ±)-3-O-methyl-2,3-trans-fustin.
DITERPENOIDS: enantio-18-acetoxy-8(20),13-labdadien-15-oic acid, enantio-labd-8(20)-ene-15,18-dioic acid, enantio-labd-8(20)-ene-15,18-diol, enantio-labd-8(20)-en-15-oic acid, enantio-pinifolic acid, trachylobanic acid, kaurenic acid.
(B, C, F) Ferreira et al., J. Chem. Soc., Perkin 1, 1492 (1974)
(T) Hugel et al., Tetrahedron, 8th Suppl., 203 (1966)

TRACHYSPERMUM

T. roxburghianum.
MONOTERPENOIDS: carvacrol, linalool, α-phellandrene, β-phellandrene, α-pinene, β-pinene, sabinene, terpinene, α-terpineol.
Saito, Planta Med., 30, 349 (1976)

TRAMETES

T. cinnabarina.
PIGMENTS: cinnabarin, tramesanguin.
Gripenberg et al., Acta Chem. Scand., 17, 703 (1963)

T. dickensii.
TRITERPENOID: carboxyacetylquercinic acid.
Adam et al., Tetrahedron Lett., 1461 (1967)

T. feeii.
TRITERPENOID: 6-hydroxypolyporenic acid.
Simes et al., J. Chem. Soc., Chem. Commun., 1150 (1969)

T. lilacino.
TRITERPENOID: dehydrotumulosic acid.
Pinhey et al., Aust. J. Chem., 23, 2141 (1970)

T. odorata.
TRITERPENOID: 3β-hydroxylanosta-8,24-dien-21-oic acid.
Halsall, Rogers, J. Chem. Soc., 2036 (1959)

T. stowardi.
TRITERPENOIDS: carboxyacetylstowardolic acid, stowardolic acid.
Cheung et al., Aust. J. Chem, 26, 609 (1973)

TRAVERSIA (Asteraceae/Senecionae)

T. baccharoides Hook f.
ESTERS: butyl caffeate, dibutyl traversate.
FLAVANONES: quercetin 3-isobutyrate, quercetin 3'-isobutyrate, quercetin 4-isobutyrate, 5,3',4'-trihydroxy-7-methoxyflavanone.
DITERPENOIDS: labd-13-ene-5,18-diol, 13-epi-ent-manoyl oxide.
Kulanthaival, Benn, Can. J. Chem., 64, 514 (1986)

TREMIA

T. micrantha.
UNCHARACTERIZED ALKALOIDS: tremidine, tremine.
Ribiero, Machado, Bol. Inst. Quim. Agr., 27, 7 (1952)

T. orientalis.
TRITERPENOID: trematol.
Ogunkota et al., Phytochemistry, 16, 1606 (1977)

TRENTEPOHLIA
(Chlorophyceae/Trentepohliaceae)
T. iolithus.
CAROTENOIDS: β,ε-carotene, β,β-carotene-2,2'-diol,
β,β-caroten-2-ol, β,ε-caroten-2-ol, β,β-caroten-2-ol epoxide.
Kjosen *et al.*, *Acta Chem. Scand.*, **26**, 3053 (1972)
Buchecker *et al.*, *ibid.* **28B**, 449 (1974)

TREVOA
T. trinervis.
TRITERPENOIDS: trevogenins A, B and C.
Betancor *et al.*, *Tetrahedron Lett.*, 1125 (1982)

TREWIA (Euphorbiaceae)
T. nudiflora.
MAYTANSOID ALKALOIDS: dehydrotrewiasine,
N-methyltrenudone, treflorine, trenudine, trewiasine.
PYRIDONE ALKALOID: nudiflorine.
Powell *et al.*, *J. Am. Chem. Soc.*, **104**, 4829 (1982)

TRIANTHEMA
T. monogyna L.
One uncharacterized alkaloid.
Basu, Sharma, *Quart. J. Pharm.*, **20**, 39 (1947)

TRIBONEMA (Xanthophyceae/Tribonemataceae)
T. aequale.
CAROTENOID: heteroxanthin.
Kleinig, Egger, *Z. Naturforsch.*, **22B**, 868 (1967)

TRIBULUS (Zygophyllaceae)
T. terrestris L.
SAPONIN: saponoside C.
STEROIDS: hecogenin, neohecogenin glucoside, tribulosin.
(Sap) Tomova, Gyulematova, *Planta Med.*, **34**, 188 (1978)
(S) Mahato *et al.*, *J. Chem. Soc., Perkin I*, 2405 (1981)

TRICACHNA
T. vestita.
UNCHARACTERIZED ALKALOID: tricachnine.
Ribiero, Machado, *Eigenharia Quim.*, **4**, 26 (1952)

TRICHADENIA
T. zeylandica.
TRITERPENOIDS: acetoxytrichadenic acid, trichadenal,
trichadenic acid, trichadenic acid A, trichadenic acid B,
trichadonic acid.
Gunasekera, Sultanbawa, *Tetrahedron Lett.*, 2837 (1973)
Gunasekera, Sultanbawa, *J. Chem. Soc., Perkin I*, 483 (1977)

TRICHILIA (Meliaceae)
T. dregeana.
SESTERTERPENOIDS: dregeana-1, dregeana-2, dregeana-3,
dregeana-4.
Mulholland *et al.*, *Phytochemistry*, **19**, 2421 (1980)
T. havanensis.
LIMONOIDS: havanensin, neohavanensin, trichilenone acetate.
Chan *et al.*, *J. Chem. Soc., Chem. Commun.*, 720 (1967)
T. heudelottii.
LIMONOIDS: heudebolin, heudelottins A, B, C, D, E and F.
Okorie, Taylor, *J. Chem. Soc., C*, 1828 (1968)
T. hirta.
LIMONOIDS: deacetylhirtin, hirtin.
Chan, Taylor, *J. Chem. Soc., Chem. Commun.*, 206 (1966)

T. hispida Penning.
TRITERPENOIDS: bourjotinolone A, hispidols A and B,
hispidone.
Jolad *et al.*, *J. Org. Chem.*, **46**, 4085 (1981)
T. prieruiana.
LIMONOID: prieruianine.
Bevan *et al.*, *Nature*, **206**, 1323 (1965)
T. roka.
LIMONOIDS: 7-acetyltrichilin, sendanin, trichilin.
Nakatani *et al.*, *Phytochemistry*, **24**, 195 (1985)

TRICHOCEREUS (Cactaceae)
T. candicans.
AMINE ALKALOIDS: N-methyltyramine, tyramine.
QUATERNARY ALKALOID: candicine.
Luduena, *C. R. Soc. Biol.*, **121**, 368 (1936)
Mata *et al.*, *Lloydia*, **39**, 461 (1976)
T. spachianus.
AMINE ALKALOIDS: N-methyltyramine, tyramine.
Mata *et al.*, *Lloydia*, **39**, 461 (1976)
T. terscheki.
AMINE ALKALOID: trichocereine.
Luduena, *C. R. Soc. Biol.*, **121**, 368 (1974)

THICHOCLINE
T. incana.
COUMARIN: trichoclin.
Miyakado *et al.*, *Phytochemistry*, **17**, 143 (1978)

TRICHOCOLEA (Musci)
T. tomentella (Ehrh.) Dum.
ESTERS: isotomentellin, tomentellin, trichocolein.
Asakawa *et al.*, *Experientia*, **34**, 155 (1978)

TRICHOCOLEOPSIS
T. sacculata.
DITERPENOIDS: isosacculatal, sacculatal.
Asakawa *et al.*, *Tetrahedron Lett.*, 1407 (1977)

TRICHODERMA
T. viride.
PYRONE: 6-(pent-1-enyl)-α-pyrone.
Moss *et al.*, *Phytochemistry*, **14**, 2706 (1975)

TRICHODESMA (Boraginaceae)
T. incanum (Bge.) DC.
PYRROLIZIDINE ALKALOIDS: incanine, incanine N-oxide,
trichodesmine, trichodesmine N-oxide.
Men'shikov, Rubinshtein, *Ber. Dtsch. Chem. Ges.*, **68**, 2039 (1935)

TRICHOGONIA
T. gardneri.
SESQUITERPENOID:
(5R,6R,7S,8R,9R)-14-acetoxy-3-chloro-9-hydroxy-2-oxo-8-
tigloyloxyguaia-1(10),3-dien-6,12-olide.
Vichnewski *et al.*, *Phytochemistry*, **24**, 291 (1985)
T. grazielae.
SESQUITERPENOID: α-muurolen-15-oic acid.
Bohlmann *et al.*, *Phytochemistry*, **20**, 1323 (1981)

TRICHOLOMA
T. portentosum (Fr.) Qu.
STEROID: portensterol.
TRITERPENOID: tricholidic acid.
> Nozoe et al., Chem. Lett., 1679 (1982)

TRICHOLOMOPSIS
T. rutilans.
ACETYLENICS: L-2-aminohex-4-ynoic acid,
L-2-amino-3-hydroxyhex-4-ynoic acid.
AMINO ACIDS: L-3-(3-carboxy-4-furyl)-alanine,
τ-glutamyl-L-2-aminohex-4-ynoic acid,
τ-L-glutamyl-L-erythro-2-amino-3-hydroxyhex-4-ynoic acid.
> Niimura et al., Phytochemistry, 16, 1435 (1977)

TRICHOSANTHES (Cucurbitoideae)
T. kirilowii.
STEROIDS: campesterol, 7-campesterol,
24-ethylcholesta-5,25-dien-3β-ol,
24-ethylcholesta-7,24(25)-dien-3β-ol,
24-ethylcholesta-7,25-dien-3β-ol, β-sitosterol,
stigmasta-7,22-dien-3β-ol, stigmasterol, 7-stigmasterol.
> Homberg et al., Phytochemistry, 16, 288 (1977)

T. palmata.
SAPONIN: trichonin.
TRITERPENOID: cycloeucalenol.
> Bhander, Rastogi, Indian J. Chem., 22B, 252 (1983)

TRICHOSTEMA
T. lanceolatum.
MONOTERPENOIDS: caryophyllene, p-cymene, β-phellandrene,
α-pinene, α-terpinene, τ-terpinene, α-terpineol, terpin-4-ol,
terpinolene.
> Schultz et al., J. Agric. Food Chem., 24, 862 (1976); Chem. Abstr., 85
> 43684y

TRICLISIA
T. dictyophylla.
PHENANTHRENE ALKALOID: tridictyophylline.
> Spiff et al., Lloydia, 44, 160 (1981)

T. gilletii.
BISBENZYLISOQUINOLINE ALKALOIDS: efirine, tricordatine,
trigelletine.
OXOAPORPHINE ALKALOID: O-methylmoschatoline.
UNCHARACTERIZED ALKALOIDS: tricliseine, triclisine.
> Huls, Detry, Bull. Soc. R. Sci. Liege, 42, 73 (1973)

T. patens.
BISBENZYLISOQUINOLINE ALKALOIDS: aromoline,
homoschatoline, tetrandrine.
> Dwuma-Badu et al., Phytochemistry, 14, 2423 (1975)

T. subcordata.
BISBENZYLISOQUINOLINE ALKALOIDS: aromoline,
homoschatoline, tetrandrine.
> Dwuma-Badu et al., Phytochemistry, 14, 2423 (1975)

TRIDACHIA
T. crispata.
PYRONES: crispatene, crispolone.
> Ireland et al., Tetrahedron, Suppl. No. 1, 233 (1980)

TRIFOLIUM (Papilionaceae)
T. alpestre.
PHENOLICS: chlorogenic acid, p-coumaric acid,
3-p-coumaroylquinine.
> Kazakov et al., Khim. Prir. Soedin., 258 (1976)

T. arvense.
AMINO ACIDS: α-alanine, L-asparagine, L-cysteine, L-glutamic
acid, L-histidine, leucine, L-lysine, methionine, serine,
threonine, tryptophan.
> Peterson et al., Nauka Prakt. Farm., 106 (1976); Chem. Abstr., 86
> 40158q

T. medium.
PHENOLICS: chlorogenic acid, p-coumaric acid,
3-p-coumaroylquinine.
> Kazakov et al., Khim. Prir. Soedin, 258 (1976)

T. montanum.
FLAVONOIDS: hyperoside, ononin, populnin, quercetin.
> Kazakov et al., Khim. Prir. Soedin., 415 (1977)

T. pratense L.
AMINO ACIDS: α-alanine, L-asparagine, cis-clovamide,
trans-clovamide, L-cysteine, L-glutamic acid, L-histidine,
leucine, L-lysine, methionine, serine, tryptophan.
FLAVONES: isorhamnetin, pratensein.
PHENOLICS: chlorogenic acid, p-coumaric acid,
3-p-coumaroylquinine.
PTEROCARPIN: demethylpterocarpin.
> (Ac) Yoshihara et al., Agr. Biol. Chem., 41, 1679 (1977)
> (F) Wong, Chem. Ind. (London), 1963 (1961)
> (P) Harper et al., J. Chem. Soc., Chem. Commun., 310 (1965)
> (Ph) Kazakov et al., Khim. Prir. Soedin., 258 (1976)

T. repens L.
AMINO ACIDS: α-alanine, L-asparagine, L-cysteine, L-glutamic
acid, L-histidine, leucine, L-kysine, methionine, serine,
tryptophan.
MONOTERPENOIDS: camphene, β-caryophyllene, p-cymene,
limonene, linalool, cis-linalool oxide, trans-linalool oxide,
linalyl acetate, 1-p-menthene, α-pinene, β-pinene,
β-sesquiphellandrene, α-terpinene, α-terpineol, α-terpinyl
acetate.
TRITERPENOIDS: soyasapogenols A, B, C, D and E,
soyasaponins I, II and III.
> (Ac) Peterson et al., Nauka Prakt. Farm., 106 (1976); Chem. Abstr., 86
> 40158q
> (T) Kitagawa et al., Chem. Pharm. Bull., 24, 121 (1976)
> (T) Kameoka et al., Agr. Biol. Chem., 41, 1785 (1977)

F. repens L. cv Louisiana Nolin.
LECTIN: trifolin.
> Truchet et al., Physiol. Plant, 66, 575 (1986); Chem. Abstr., 105
> 39367m

TRIGONELLA (Leguminosae)
T. coerulea.
SAPONIN: protodioscin 22-methyl ether.
STEROIDS: diosgenin, gitogenin, neogitogenin, spirostadiene.
> (Sap) Bogacheva et al., Khim. Prir. Soedin., 421 (1977)
> (S) Bogacheva et al., Khim.-Farm. Zh., 10, 78 (1976); Chem. Abstr., 86
> 52662m

T. foenum-graecum.
PYRIDINE ALKALOID: trigonelline.
FLAVONOIDS: kaempferol, quercetin.
SAPONINS: trigofoenosides A, B, C and D, trigonellosides A, B
and C, yamogenin-3,26-bisglycoside, yamogenin tetrosides
A, B and C.

STEROIDS: fenugreekine, β-sitosterol.
  (A) Johns *et al., Ber.,* **18,** 2518 (1885)
  (F) Sood *et al., Indian J. Pharm.,* **37,** 100 (1975)
  (Sap) Gupta *et al., Phytochemistry,* **24,** 2399 (1985)
  (S) Ghosal *et al., ibid.,* **13,** 2247 (1974)

T. occulata.
FLAVONOID: quercetin.
STEROID: β-sitosterol.
  Jain, *Indian J. Pharm.,* **38,** 25 (1976)

TRILLIUM (Liliaceae)
T. erectum L.
SAPONINS: nolonin, trillarin, trillin.
STEROIDS: bethogenin, fesogenin, kryptogenin, nologenin, nolonin, trillenogenin, trillogenin.
  (Sap) Grove *et al., J. Am. Pharm. Assoc.,* **27,** 457 (1938)
  (S) Marker *et al., J. Am. Chem. Soc.,* **69,** 2167, (1947)
  (S) Marker *et al., ibid.,* **69,** 2386 (1947)

T. grandiflorum (Michx) Salisb.
SAPONIN: trilloside.
  Mockle, *Can. J. Pharm.,* **90,** 162 (1957)

T. kamtschaticum.
SAPONINS: 18-norspirostanol oligoglycoside, trillenoside A.
STEROIDS: bethogenin, 26-chloro-26-deoxycryptogenin, pennogenin, trillenogenin.
  (Sap) Nohara *et al., Tetrahedron Lett.,* 4381 (1975)
  (S) Nohara *et al., Chem. Pharm. Bull.,* 1772 (1974)

T. sessile subsp. californicum.
STEROID: diosgenin.
  Marker *et al., J. Am. Chem. Soc.,* **65,** 1199 (1943)

T. stylorum.
STEROID: diosgenin.
  Marker *et al., J. Am. Chem. Soc.,* **65,** 1199 (1943)

T. tschonoskii.
Two uncharacterized saponins.
  Nakano *et al., Phytochemistry,* **22,** 1047 (1983)

TRIPETALIA (Ericaceae)
T. paniculata.
STEROID: β-sitosterol.
TRITERPENOIDS: α-amyrin, betulinic acid, lupeol, oleanolic acid, taraxerol, ursolic acid.
  Kawai *et al., Chem. Abstr.,* 88 24499b

TRIPHASIA
T. trifolia.
CAROTENOID: triphasiaxanthin.
  Yokoyama *et al., J. Org. Chem.,* **35,** 2080 (1970)

TRIPHYOPHYLLUM
T. peltatum.
ISOQUINOLINE ALKALOIDS: isotriphyophylline, 8-methyltetrahydrotriphyophylline, N-methyltriphyophylline, triphyopeltine, triphyophylline.
  Lavault, Bruneton, *C. R. Acad. Sci.,* **287C,** 129 (1978)

TRIPTEROSPERMUM
T. taiwanense.
TRITERPENOIDS: norathyriol, oleanolic acid.
XANTHONE: tripteroside.
  Lin *et al., Phytochemistry,* **21,** 948 (1982)

TRIPTERYGIUM
T. hypoglucum.
DITERPENOID: hypolide.
  Wu *et al., Yunnan Chih We Yen Chiu,* **1,** 29 (1979)

T. regelii.
DITERPENOID: tripterolide.
  Wu *et al., Yunnan Chih We Yen Chiu,* **1,** 29 (1979)

T. wilfordii Hook.
MACROCYCLIC ALKALOIDS: celabenzine, celafaurine, wilfordine, wilforgine, wilforine, wilfortrine, wilforzine.
DITERPENOIDS: celastrol, neotriptophenolide, tripdiolide, triptolide, triptolidenol, triptonide, triptonolide, triptonoterpene, triptonoterpene methyl ether, triptophenolide, triptophenolide methyl ether, wilforlides A and B.
LIGNAN: (-)-lirioresinol.
  (A) Kupchan *et al., J. Org. Chem.,* **42,** 3660 (1977)
  (L) Briggs *et al., J. Chem. Soc.,* C, 3042 (1968)
  (T) Deng *et al., Zhiwu Xuebao,* **27,** 516 (1985); *Chem. Abstr.,* **104** 31722g

TRISTANIA
T. conferta.
TRITERPENOIDS: 24-methylenecycloartenol, 2,3,23-trihydroxy-12-oleanen-28-oic acid.
  Bowden *et al., Aust. J. Chem.,* **28,** 91 (1975)

TRITICUM (Gramineae)
T. aestivum.
GIBBERELLINS: gibberellin A-60, gibberellin A-61, gibberellin A-62.
  Kirkwood *et al., J. Chem. Soc., Perkin I,* 689 (1982)

TRIUMFETTA
T. rhomboidea.
FLAVONOIDS: scutellarein 7-O-α-L-rhamnoside, triumboidin.
  Srinavasan *et al., Fitoterapia,* **52,** 285 (1982)

TRIXIS
T. inula.
SESQUITERPENOIDS: 9-acetoxytrixikingolide 2'-methylbutyrate, 9α-isovaleryloxytrixikingolide-2'-methylbutyrate, 9α-senecioloxy-12-desoxotrixikingolide-2'-methylbutyrate, 9-valeryloxytrixikingolide 2'-methylbutyrate.
  Bohlmann *et al., Phytochemistry,* **20,** 1649 (1981)

T. paradoxa Cass.
SESQUITERPENOID: trixol.
  Gonzalez *et al., An. Quim.,* **80C,** 319 (1984).

T. vantheri.
SESQUITERPENOID: methylmethoxytrixate.
  Bohlmann *et al., Phytochemistry,* **20,** 1649 (1981)

T. wrightii.
SESQUITERPENOIDS: 9α-senecioyloxytrixikingolide-2'-methylbutyrate, trixikingolide 2'-isovalerate, trixikingolide 2'-methylbutyrate.
  Bohlmann, Zdero, *Chem. Ber.,* **112,** 435 (1979)

TROCHODENDRON
T. aralioides.
TRITERPENOIDS: betulin palmitate, trochoic acid, trochol.
  Davy *et al., J. Chem. Soc.,* 2702 (1951)

TROCHOLEJEUNEA

T. sandwicensis.

SESQUITERPENOIDS: dehydropinguisanin, dehydropinguisenol, dehydropinsuisenal, pinguisenal, pinsuisenol.

Asakawa et al., Phytochemistry, 19, 2651 (1980)

TRYPTODENDRON

T. lineatum.

MONOTERPENOID: lineatin.

Mori et al., Tetrahedron, 36, 2197 (1980)

TSUGA

T. canadensis.

SESQUITERPENE: canadene.

Kritchevsky, Anderson, J. Am. Pharm. Assoc., 44, 535 (1955)

T. chinensis Pritz var. formosana Hay.

ACID: p-hydroxybenzoic acid.

ESTER: (3,5-dihydroxy-2-methyl)-phenyl-4'-hydroxybenzoate.

FLAVONES: epi-afzelechin, epi-catechin, kaempferol.

STEROIDS: campesterol, β-sitosterol.

MONOTERPENOID: dihydroconiferyl alcohol.

DITERPENOIDS: larixol, 13-epi-manool.

Fang et al., J. Chin. Chem. Soc., 32, 477 (1985); Chem. Abstr., 105 75891m

T. sieboldii.

NAPHTHOFURAN: tsugaresinol.

Haworth et al., J. Chem. Soc., 636 (1935)

TULIPA (Liliaceae)

T. clusiana cv. 'Cynthia'.

SAPONINS: tuliposide A, tuliposide B.

Slob et al., Chem. Abstr., 87 98805k

T. gesneriana.

ACID: 3,4-dihydroxy-2-methylenebutanoic acid.

GLUCOSIDES: 1-tuliposide A, 1-tuliposide B, 6-tuliposide A.

(Ac) Tschesche et al., Tetrahedron Lett., 701 (1968)

(G) Tschesche et al., Chem. Ber., 102, 2057 (1969)

TUPISTRA

T. aurantiaca.

STEROIDS: 3-epi-neoruscogenin, 3-epi-ruscogenin, 25,27-pentrogenin, rammogenin A, rammogenin B, rammogenin C, rammogenin D,

Yang et al., Yunnan Zhiwu Yanjiu, 9, 217 (1987); Chem. Abstr., 107 194913w

TURBINA

T. corymbosa.

DITERPENOIDS: corymbol, corymbosin, turbicoryn.

Garcia-Jiminez et al., Tetrahedron, 23, 2557 (1967)

TURNEFORCIS

T. sibirica.

UNCHARACTERIZED ALKALOIDS: turneforcidine, turneforcine.

Men'shikov et al., J. Gen. Chem. USSR, 22, 1465 (1952)

TURNERA

T. diffusa.

FLAVONE: 5-hydroxy-7,3',4'-trimethoxyflavone.

Dominguez et al., Planta Med., 30, 68 (1976)

TURPINIA

T. arguta.

TRITERPENOIDS: 2,3-dihydroxyurs-12-en-18-oic acid, 2,19-dihydroxyursolic acid.

Fang et al., Zhongcaoyao, 16, 53 (1985)

TURREANTHUS

T. africanus.

TRITERPENOID: turreanthin.

Bevan et al., J. Chem. Soc., Chem. Commun., 636 (1965)

TUSSILAGO (Compositae)

T. farfara L.

TRITERPENOIDS: arnidenediol, faradiol.

Santer, Stevenson, J. Org. Chem., 27, 3204 (1962)

TYDEMANIA

T. expeditionitis.

Three triterpenoids.

Paul et al., Tetrahedron Lett., 3459 (1982)

TYLECODON (Crassulaceae)

T. grandiflorus (Burm.F.) Toelken.

TOXINS: tyledosides A, B, C, D, E, F and G..

Steyn et al., J. Chem. Soc., Perkin I, 429 (1986)

T. wallichii.

SAPONIN: cotyledoside.

Steyn et al., J. Chem. Soc., Perkin I, 965 (1984)

TYLOPHORA

T. asthmatica Wight et Arn.

PHENANTHROINDOLIZIDINE ALKALOIDS: (+)-isotylocrebine, tylopherine, tylophorinicine, tylophorinidine, tylophorinine.

Mulchandani et al., Phytochemistry, 23, 1206 (1984)

T. crebiflora S.T.Blake.

PHENANTHROINDOLIZIDINE ALKALOID: (-)-tylocrebine.

Gellert et al., J. Chem. Soc., 1008 (1962)

T. dalzellii.

One uncharacterized alkaloid.

Rao et al., J. Pharm. Sci., 60, 1725 (1971)

T. indica.

Two uncharacterized alkaloids.

Rao et al., J. Pharm. Sci., 60, 1725 (1962)

T. kerrii Craib.

TRITERPENOIDS: tylolupenol A, tylolupenol B.

Xu et al., Zhongcaoyao, 14, 49 (1983)

TYPHA (Typhaceae)

T. angustata L.

ACIDS: trans-p-hydroxycinnamic acid, trans-propenoic acid 3-hydroxyphenyl-2,3-dihydroxypropyl ester, protocatechuic acid, succinic acid, vanillic acid.

Xu et al., Zhiwu Xuebao, 28, 523 (1986); Chem. Abstr., 106 30035q

T. latifolia L.

STEROID: typhasterol.

Schneider et al., Tetrahedron Lett., 3859 (1983)

U

UDOTEA (Chlorophyceae/Udotaceae)

U. argentea.
LACTONE: udoteal epoxylactone.
Paul, Fenical, *Phytochemistry*, **24**, 2239 (1985)

U. petiolata (Turr.) Bergesen.
DITERPENOID: petiodial.
Fattorusso *et al.*, *Experientia*, **30**, 1275 (1983)

ULEX (Leguminosae)

U. europaeus L.
Two uncharacterized alkaloids.
Ribas, Besanta, *Anales Real. Soc. Esp. Fis. Quim.*, **48B**, 161 (1952)

ULMUS (Ulmaceae)

U. campestris.
TRITERPENOID: friedelin.
Walther *et al.*, *Pharmazie*, **31**, 578 (1976)

U. glabra Huds.
SESQUITERPENOIDS: 7-hydroxycalamenal,
7-hydroxycalamenene.
Rowe, Toda, *Chem. Ind. (London)*, 922 (1969)

U. laciniata.
COUMARINS: lacinilenes A, B, C and D.
Nishikawa *et al.*, *Mokuzai Gakkaishi*, **18**, 623 (1972)

U. rubra Muhl.
STEROID: α-sitosterol.
SESQUITERPENOIDS: 7-hydroxycalamenal,
7-hydroxycalamenene.
(S) Bates *et al.*, *Tetrahedron Lett.*, 6163 (1968)
(T) Rowe, Toda, *Chem. Ind., (London)*, 922 (1969)

U. sieboldiana.
FLAVONE: chrysin.
Marsh, *Biochem. J.*, **59**, 58 (1955)

U. thomasii.
SESQUITERPENOID: 7-hydroxycalamenene.
Rowe, Toda, *Chem. Ind. (London)*, 922 (1969)

ULUGBEKIA

U. tschimganica.
PYRROLIZIDINE ALKALOID: uluganine.
PHENOLICS: ulugbekic acid, ulugbinic acid.
(A) Khasanova *et al.*, *Khim. Prir. Soedin.*, 809 (1974)
(P) Khasanova *et al.*, *ibid.*, 601 (1976)

ULVA (Chlorophyceae/Ulvaceae)

U. rigida.
CAROTENOID: loroxanthin.
Aitzetmuller *et al.*, *Phytochemistry*, **8**, 1761 (1969)

ULVARIA (Chlorophyceae/Monostromataceae)

U. chamae L.
COUMARIN: chamaernen.
FLAVANONES: isouvaretin, uvaretin.
(C) Lasswell *et al.*, *Phytochemistry*, **16**, 1439 (1977)
(F) Hufford *et al.*, *J. Org. Chem.*, **41**, 1297 (1976)

UMBELLULARIA (Lauraceae)

U. californica.
MONOTERPENOID: umbellulone.
Power, Lees, *J. Chem. Soc.*, **85**, 636 (1904)

UMBILICARIA (Umbilicariaceae)

U. pustulata.
POLYSACCHARIDES: pustulan, umbilicin.
Lindberg *et al.*, *Acta Chem. Scand.*, **8**, 985 (1954)

UNCARIA (Rubiaceae)

U. attenuata.
PROTOBERBERINE ALKALOID: 3-iso-19-epi-ajmalicine.
Phillipson, Hemingway, *Phytochemistry*, **14**, 1855 (1975)

U. bernaysii.
SPIROINDOLE ALKALOIDS: uncarines A, B, C, D, E and F.
uncarine C N-oxide, uncarine D N-oxide, uncarine E
N-oxide, uncarine F N-oxide.
Hemingway, Phillipson, *J. Pharm. Pharmacol.*, **24**, 169P (1972)

U. calophylla.
CARBOLINE ALKALOIDS: dihydrocorynantheine, gambirine,
pseudoyohimbine.
Goh *et al.*, *Phytochemistry*, **24**, 880 (1985)

U. elliptica.
ALKALOID: roxburghine X.
Herath *et al.*, *Phytochemistry*, **18**, 1385 (1979)

U. ferrea DC.
SPIROINDOLE ALKALOIDS: uncarine C, uncarine D, uncarine E.
Hart *et al.*, *J. Chem. Soc., Chem. Commun.*, 87 (1967)

U. gambir (Hunter) Roxb.
ALKALOIDS: roxburghines A, B, C, D and E.
QUATERNARY ALKALOID: ouroupanine.
SPIROINDOLE ALKALOIDS: gambirdine, gambirine,
gambirtannine, isogambirdine, oxogambirtannine.
LIMONOID: 7-acetoxy-dihydronomilin.
(A) Chan *et al.*, *Tetrahedron Lett.*, 3403 (1968)
(T) Ahmed *et al.*, *Can. J. Chem.*, **56**, 1020 (1978)

U. guianensis.
SPIROINDOLONE ALKALOIDS: isorhyncophylline N-oxide,
rhyncophylline N-oxide.
Hemingway, Phillipson, *J. Pharm. Pharmacol.*, **26**, 113P (1974)

U. kawakamii.
SPIROINDOLE ALKALOIDS: uncarine A, uncarine B.
Raymond-Hamet, *C. R. Acad. Sci.*, **203**, 1383 (1936)

U. orientalis Guill.
OXINDOLE ALKALOIDS: isomitraphylline, isorhyncophylline,
(-)-mitraphylline, rhyncophylline, uncarine.
Croquelois *et al.*, *Ann. Pharm. Fr.*, **35**, 417 (1977)

U. pteropoda.
SPIROINDOLE ALKALOIDS: isopteropodine, pteropodine.
Yeoh *et al.*, *Tetrahedron Lett.*, 931 (1966)

U. rhyncophylla.
CARBOLINE ALKALOID: dihydroxycorynantheine.
Haginawa *et al.*, *J. Pharm. Soc., Japan*, **93**, 448 (1973)

U. thwaitesii.

TRITERPENOIDS: acetyluncaric acid, ketouncaric acid, uncaric acid.

Herath *et al.*, *Phytochemistry*, **17**, 1979 (1978)

U. tomentosa.

CARBOLINE ALKALOID: hirsutine N-oxide.

SPIROINDOLE ALKALOIDS: isomitraphylline, isomitraphylline N-oxide, rhyncophylline N-oxide.

Hemingway, Phillipson, *J. Pharm. Pharmacol.*, **26**, 113P (1974)

UNGERNIA (Amaryllidaceae)

U. ferganica Vved.

ALKALOIDS: galantamine, hippeastrine, hordenine, lycorine, pancratine, tazettine.

Israilov *et al.*, *J. Gen. Chem.*, *USSR*, **3**, 18 (1965)

U. minor Vved.

ALKALOIDS: lycorine, tazettine, ungeramine, ungminoridine, ungminorine.

Abduasimov *et al.*, *J. Gen. Chem.*, *USSR*, **5**, 31 (1967)

U. sewerzowii (Rgl.) B.Fedtsch.

ALKALOIDS: galantamine, hippeastrine, lycorine, (+)-narvedine, (±)-narvedine, (-)-narvedine, tazettine, ungeridine, ungerine, unsevine.

Sadykov, Shakirov, *Khim. Prir. Soedin.*, 134 (1972)

U. spiralis.

ALKALOIDS: hippeastrine, galantamine, tazettine, ungeramine.

Allayarov *et al.*, *Khim. Prir. Soedin.*, 143 (1970)

U. tadshikorum Vved.

ALKALOIDS: galantamine, hippeastrine, lycorine, pancratine, tazettine.

Abdusamatov *et al.*, *J. Gen. Chem.*, *USSR*, **1**, 18 (1963)

U. trisphaera Bge.

ALKALOIDS: galantamine, hippeastrine, lycorine, pancratine, tazettine, trisphaeridine, trisphaerine, ungerine.

MANNAN: ungeromannan.

(A) Yagudaev, Yunusov, *Khim. Prir. Soedin.*, 505 (1972)

(M) Malikov, Rakhimov, *ibid.*, 21 (1986)

U. victoris Vved.

ALKALOIDS: galantamine, hippeastrine, hordenine, lycorine, (±)-narvedine, tazettine.

Abdusamatov *et al.*, *J. Gen. Chem.*, *USSR.*, **2**, 45 (1962)

U. vvedensky.

ALKALOID: ungvedine.

Kadirov *et al.*, *Khim. Prir. Soedin.*, 585 (1979)

UNONA

U. lawii.

FLAVONES: unonal, unonal 7-methyl ether.

Joshi *et al.*, *Indian J. Chem.*, **14B**, 9 (1976)

URCEOLARIA

U. cretacea.

TRITERPENOID: zeorin.

Barton *et al.*, *J. Chem. Soc.*, 2239 (1958)

URECHTITES (Apocynaceae)

U. karwinskii Muell.

NECINE ALKALOID: loroquine.

Del Castillo *et al.*, *Tetrahedron Lett.*, 1219 (1970)

URGINEA (Amaryllidaceae)

U. altissima.

ISOQUINOLINE ALKALOIDS: acetylcaranine, caranine, lycorine.

Miyakado *et al.*, *Phytochemistry*, **14**, 2717 (1975)

U. indica (Roxb.) Kunth.

BUFADIENOLIDES: scillarenin, scillarenin 3-O-α-L-rhamnoside, scillarenin 3-O-α-L-rhamnoside-β-D-glucoside, scilliglaucoside, scilliglaucoside 3-O-β-D-glucoside, scilliphaosidin, scilliphaosidin 3-O-α-L-rhamnoside.

Kopp, Danner, *Sci. Pharm.*, **51**, 227 (1983); *Chem. Abstr.*, 100 32202s

U. maritima.

FLAVONOIDS: kaempferol 7-glucoside-3-diglucoside, kaempferol 7-glucoside-3-rhamnoglucoside, kaempferol 7-glucoside-3-triglucoside, kaempferol 7-hamnoside-3-rhamnoglucoside, Steroid scillirosidin.

(F) Fernandez *et al.*, *Sci. Pharm.*, **44**, 307 (1976); *Chem. Abstr.*, 86 103073f

(S) Wartburg *et al.*, *Helv. Chim. Acta*, **42**, 1620 (1959)

UROSPERMUM (Compositae)

U. dolechampi.

SESQUITERPENOID: urospermal.

Bentley *et al.*, *J. Chem. Soc.*, *Chem. Commun.*, 475 (1970)

URSINIA (Compositae)

U. alpina.

SESQUITERPENOID: (2E,6E)-12-hydroxynerolidol.

Bohlmann *et al.*, *Phytochemistry*, **19**, 587 (1980)

U. anthemoides.

DITERPENOIDS: ursialpinolide, ursinolides A, B and C.

SESQUITERPENOIDS: 1-angeloyloxy-12-hydroxynerolidol, (2E,6E)-12-hydroxynerolidol.

Bohlmann *et al.*, *Phytochemistry*, **19**, 587 (1980)

U. anthemoides (L.) Poiret.

DITERPENOIDS: ursinolides A, B and C.

Grabarczyl *et al.*, *Pol. J. Pharmacol. Pharm.*, **25**, 469 (1973)

U. nana.

SESQUITERPENOIDS:

2-angeloyloxy-8-hydroxy-5-presilphiperfolone, athamanolide, 9-desacetoxyanhydroathamantholide, 9-desacetoxy-5-hydroxyathamantholide, ursinanolide.

Bohlmann *et al.*, *Phytochemistry*, **21**, 1309, (1982)

Bohlmann *et al.*, *ibid.*, **21**, 1331 (1982)

U. saxatilis.

SESQUITERPENOIDS: desacetoxymatricin, 5-hydroxyachillin.

Bohlmann *et al.*, *Phytochemistry*, **21**, 1357 (1982)

USNEA (Usneaceae)

U. aciculifera.

ACID: constictic acid.

Yoshioka *et al.*, *Chem. Pharm. Bull.*, **18**, 2364 (1970)

U. aliphatica.

ALIPHATIC: α-(15-hydroxyhexadecyl)-itaconic acid.

Keogh *et al.*, *Phytochemistry*, **16**, 134 (1977)

U. annulata.

STEROIDS: 5α,8α-epidioxy-5α-ergosta-6,22-dien-3β-ol, usnic acid.

Strigina, Sviridov, *Phytochemistry*, **17**, 327 (1978)

U. barbata.

NAPHTHOPYRAN: barbalolin.

Duntze *et al.*, *Ann. Chem.*, **491**, 220 (1931)

U. canariensis.

NAPHTHOPYRAN: canarione.

Huneck *et al.*, *Phytochemistry*, **16**, 121 (1977)

U. dentrictica.
BENZODIOXEPIN: salazinic acid.
Malik *et al.*, *Indian J. Chem.*, **10**, 1040 (1972)

U. undulata.
DIOXEPIN: galbinic acid.
Mendiondo *et al.*, *Phytochemistry*, **11**, 424 (1972)

USTILAGO
U. maydis.
UNCHARACTERIZED ALKALOIDS: ustilaginine, ustilagotoxine.
Mas, *Bol. Soc. Quim.*, *Peru*, **4**, 3 (1938)

UVARIA (Annoniaceae)
U. afzelii Scot Elliot.
BENZOPYRANS: 2-hydroxy-7,8-dehydrograndiflorone, uvafzelic acid, uvafzelin, vafzelin.
FLAVONE: emorgdone.
ISOCOUMARINS: demethoxymatteucinol, 2'-hydroxy-2-methoxymatteucinol.
MONOTERPENOID: syncarpic acid.
Hufford, Oguntimein, *J. Org. Chem.*, **46**, 3073 (1981)

U. angolensis.
FLAVONOIDS: anguvetin, (±)-chamanetin 5-methyl ether, (+)-6,8-dimethylpinocembrin 5-methyl ether, flavokawin B.
Hufford *et al.*, *J. Nat. Prod.*, **45**, 337 (1982)

U. catocarpa.
ESTERS: seneol, senepoxide.
Hollands *et al.*, *Tetrahedron*, **24**, 1633 (1968)

U. chamae.
CHALCONES: chamuvarin, chamuvaritin, diuvaretin, isouvaretin, uvaretin.
ESTERS: chamauvarin, chamauvaritin.
FLAVONOIDS: chamanetin, dichamanetin, isochamanetin, pinocembrin, pinostrobin.
MONOTERPENOID: chamanen.
(C, F) Lasswell *et al.*, *J. Org. Chem.*, **42**, 1285 (1977)
(E) Okorie *et al.*, *Phytochemistry*, **16**, 1591 (1977)

U. elliotiana.
INDOLE ALKALOID: 3,6-bis(3-methyl-2-propenyl) indole.
Achenbach *et al.*, *Tetrahedron Lett.*, 2571 (1979)

U. ferrugina.
ESTERS: senepoxide, tingitanoxide.
Kodpinid *et al.*, *Tetrahedron Lett.*, 2019 (1983)

U. kirkii.
XANTHONES: 1,2,3,4,6,7-hexamethoxyxanthone, uvaretin.
Tammami *et al.*, *Phytochemistry*, **16**, 2040 (1977)

U. ovata.
BISBENZYLISOQUINOLINE ALKALOID: chondrofoline.
PHENOLIC: 2,3,6-trimethoxybenzoate.
(A) Panichpol *et al.*, *Phytochemistry*, **16**, 621 (1977)
(P) Panichpol *et al.*, *J. Pharm. Pharmacol.*, **28**, 71P (1976)

U. pandensis Verde.
SESQUITERPENOIDS: farnesol, 3-farnesylindole.
TRITERPENOID: α-tocopherol.
Nkunya *et al.*, *Phytochemistry*, **26**, 2402 (1987)

U. rufas.
PHENOLIC: 3,7-bisbenzoyloxyhept-4-ene-1,2,6-triol.
Chayunkiat *et al.*, *J. Sci. Soc. Thailand*, **10**, 239 (1984); *Chem. Abstr.*, 103 3676f

UVARIODENDRON
U. connivens.
PHENOLICS: elemicin, 3',4',5'-trimethoxycinnamaldehyde, 3',4',5-trimethoxycinnamyl alcohol.
Mohammed *et al.*, *J. Nat. Prod.*, **48**, 328 (1985)

UVARIOPSIS
U. guineense Keay.
PHENANTHRENE ALKALOIDS: 8-methoxyuvariopsine, noruvariopsamine, uvariopsamine, uvariopsamine N-oxide.
Leboeuf, Cave, *Phytochemistry*, **11**, 2833 (1972)

U. solheidii Robyns et Ghesq.
PHENANTHRENE ALKALOID: uvariopsine.
Bouquet *et al.*, *C. R. Acad. Sci.*, **271C**, 1100 (1970)

UVULARIA (Liliaceae)
U. cirrhosa Thunb.
UNCHARACTERIZED ALKALOID: peimunine.
Li, *J. Chin. Pharm. Assoc.*, **2**, 235 (1940)

V

VACCARIA (Caryophyllaceae)
V. segetalis.
SAPONINS: vaccagosides A, B and C.

Baeva et al., Khim. Prir. Soedin., 658 (1975)

VACCINIUM (Ericaceae)
V. arboreum Marsh.
ANTHOCYANINS: cyanidin 3-O-arabinoside, cyanidin
3-O-galactoside, delphinidin 3-O-arabinoside, delphinidin
3-O-galactoside, delphinidin 3-O-glucoside, malvidin
3-O-arabinoside, malvidin 3-O-galactoside, paeonidin
3-O-arabinoside, paeonidin 3-O-galactoside, paeonidin
3-O-glucoside, petunidin 3-O-arabinoside, petunidin
3-O-galactoside, petunidin 3-O-glucoside.

Ballinger et al., Can. J. Plant Sci., 62, 683 (1982); Chem. Abstr., 97
195820a

V. bracteatum.
IRIDOID: vaccinoside.

Sakakibara et al., Chem. Pharm. Bull., 19, 1979 (1971)

V. myrtillus L.
QUINOLIZIDONE ALKALOIDS: myrtine, epi-myrtine.

Slosse, Hootele, Tetrahedron, 37, 4287 (1982)

V. oxycoccus L. (Oxycoccus palustris Pers.).
CARBOLINE ALKALOID: cannagunine.

Jankowski et al., Experientia, 27, 1141 (1971)

V. vitisi-daea L.
COUMARIN: 2-O-caffeoylarbutin.

Haslam et al., J. Chem. Soc., 5649 (1964)

VALERIANA (Valerianaceae)
V. alliariifolia.
VALEPOTRIATE: 1-acetylvaltratum.

Hoelze, Kock, Planta Med., 50, 458 (1984)

V. chionophila.
FLAVONOIDS: acacetin, diosnetin, diosnetin 7-O-β-D-glucoside,
luteolin, luteolin 7-O-β-D-glucoside, quercetin.

Trzhetsinskii et al., Khim. Prir. Soedin., 255 (1982)

V. condamoana.
VALEPOTRIATES: didrovaltrate, homodidrovaltrate,
homovaltrate, isovaleryloxyhydroxydidrovaltrate,
isovaltrate, valtrate.

Chavadej et al., Pharm. Weekbl., 7, 167 (1985); Chem. Abstr., 104
17640x

V. fauriei.
SESQUITERPENOID: cyclokessyl acetate.

Oshima et al., Tetrahedron Lett., 1839 (1986)

V. fedtschenkoi.
FLAVONOIDS: acacetin, diosnetin, diosnetin 7-O-β-D-glucoside,
luteolin, luteolin 7-O-β-D-glucoside, quercetin.

Trzhetsinskii et al., Khim. Prir. Soedin., 255 (1982)

V. ficariifolia.
FLAVONOID: luteolin 7-O-β-D-glucoside.

Chavadej et al., Pharm. Weekbl., 7, 167 (1985); Chem. Abstr., 104
17640x

V. japonica.
SESQUITERPENOID: cryptofauranol.

Hikino et al., Chem. Pharm. Bull., 13, 631 (1965)

V. micropterina.
VALEPOTRIATES: didrovaltrate, homodidrovaltrate,
homovaltrate, isovaleroxyhydroxydidovaltrate, isovaltrate,
valtrate.

Chavadej et al., Pharm. Weekbl., 7, 167 (1985); Chem. Abstr., 104
17640x

V. officinalis L.
PYRIDINE ALKALOID: valerianine.
PYRROLE ALKALOIDS: 2-acetylpyrrole, chatinine.
UNCHARACTERIZED ALKALOID: valerine.
IRIDOIDS: kanokosides A, B, C and D.
SESQUITERPENOIDS: faurinone, kessane, kessene,
8-epi-kessanol, α-kessyl alcohol, kessyl glycol, kongol,
maaliol, (+)-maalioxide, valeranone, valerenal, valerenic
acid, valerenol, valerenolic acid, valerianol.
VALEPOTRIATES: valepotriate valtrate, valerosidatum.

(A) Franck et al., Angew. Chem., 82, 875 (1970)
(T) Holzl et al., Tetrahedron Lett., 1171 (1976)
(T) Endo et al., Chem. Pharm. Bull., 25, 2140 (1977)

V. priophylla.
VALEPOTRIATES: didrovaltrate, homodidrovaltrate,
homovaltrate, isovaleryloxyhydroxydidrovaltrate,
isovaltrate, valtrate.

Chavadej et al., Pharm. Weekbl., 7, 167 (1985); Chem. Abstr., 104
17640x

V. pulchella.
VALEPOTRIATES: didrovaltrate, homodidrovaltrate,
homovaltrate, isovaleroxyhydroxydidrovaltrate, isovaltrate,
valtrate.

Chavadej et al., Pharm. Weekbl., 7, 167 (1985); Chem. Abstr., 104
17640x

V. tilaefolia.
PYRONE ESTERS: valtrathydrines B1, B2 and B3.

Holzl et al., Tetrahedron Lett., 1171 (1976)

V. vaginata.
VALEPOTRIATE: isodihydrovaltrate.

Kucaba et al., Phytochemistry, 19, 575 (1980)

V. wallichi.
SESQUITERPENOIDS: acetylkanokonol, trans-bergamotene,
kanokonol, sesquifenchene.
VALEPOTRIATES: acevaltrate, dihydrovaltrate, valtrate, IV
HD-valtrate.

(T) Nozoe et al., Tetrahedron Lett., 4625 (1976)
(V) Funke et al., Planta Med., 28, 215 (1978)

VALLARIS
V. solanacea.
STEROIDS: solanoside, valleroside.

Kaufmann, Helv. Chim. Acta, 48, 83 (1965)

VALLESIA
V. antillana.
INDOLE ALKALOID: (-)-apparicine.

Cuellar et al., Rev. Cubana Farm., 10, 217 (1976); Chem. Abstr., 88
19009y

V. dichotoma Ruiz. et Pav.
CARBOLINE ALKALOID: vallesamidine.
IMIDAZOLIDINOCARBAZOLE ALKALOIDS: 1-acetylaspidoalbine, 1-acetyl-17-hydroxyaslidoalbine, aspidospermine, deformylvallesine, dichotamine, dichotine, 11-methoxydichotamine, vallesine. Brown et al., Tetrahedron Lett., 1731 (1963)
Lins, Djerassi, *J. Am. Chem. Soc.*, **82**, 6019 (1970)

V. glabra.
CARBOLINE ALKALOID: vallesamidine.
IMIDAZOLIDINOCARBAZOLE ALKALOIDS: aspidospermine, vallesine. Conroy et al., J. Am. Chem. Soc., 79, 1763 (1957)

VANDA (Orchidaceae)
V. cristata.
PYRROLIZIDINE ALKALOID: acetyllaburnine.
Lindstrom, Luning, *Acta Chem. Scand.*, **23**, 3352 (1969)

VANDOPSIS
V. longicaulis.
PIPERIDINE ALKALOIDS: N-methylpiperidine N-oxide, N-methylpiperidinium iodide.
Brandage, Luning, *Acta Chem. Scand*, **24**, 353 (1970)

VANGUERIA
V. tomentosa.
TRITERPENOIDS: tomentosolic acid, vanguerin, vanguerolic acid.
Barton et al., *J. Chem. Soc.*, 5163 (1962)

VANILLOSMOPSIS
V. erythropappa.
SESQUITERPENOIDS: 15-deoxygoyazansolide, eremanthin, vanillosmin.
Vichnewski et al., *Phytochemistry*, **15**, 1775 (1976)

VASCANELLA
V. hastata.
MACROCYCLIC ALKALOID: carpaine.
Govindachari, Narasimhan, *J. Chem. Soc.*, 2635 (1953)

VELLOZIA (Velloziaceae)
V. bicolor.
DITERPENOID: 13S-hydroxy-15,16-bisnorpimaran-20,8-olide.
Pinto et al., *J. Chem. Soc., Chem. Commun.*, 464 (1983)
V. candida.
DITERPENOIDS: epoxycorcovadin, epoxyvellozin.
Pinto et al., *Tetrahedron Lett.*, 5043 (1983)
V. caput-ardaea.
DITERPENOIDS: 11,12-dehydrovelloziolone, velloziolone.
Pinto et al., *Quim. Nova.*, **6**, 71 (1983)
V. compacta.
DITERPENOIDS: compactone, seven diterpenoids.
Rodriguez et al., *Phytochemistry*, **22**, 2005 (1983)
V. flavicans.
DITERPENOIDS: veadeiroic acid, veadeirol.
Pinchin et al., *Phytochemistry*, **17**, 1671 (1978)
V. glabra.
TRITERPENOID: (E)-masticadienonic acid.
Pinto et al., *An. Acad. Brasil Cienc.*, **53**, 69 (1981)
V. piresiana.
DITERPENOID: 1,8,11,13-cleistanthatetraene-3,7-dione.
Pinto et al., *Phytochemistry*, **23**, 1293 (1984)

V. pusilla.
One diterpenoid.
Pinto et al., *J. Chem. Res. Synop.*, 104 (1983)
V. stipitata.
DITERPENOID: 7,16-epoxy-20-nor-5(10),6,8,11,13-cleistanthapentaene-3-one.
Pinto et al., *Tetrahedron Lett.*, 405 (1979)

VENTILAGO
V. calyculata.
ANTHRAQUINONE: calyculatone.
GLYCOSIDES: emodin-1-O-α-L-rhamnoside, emotin-8-O-β-D-glucoside.
ISOCHROMANES: ventilein A, ventilein B.
ISOCHROMAQUINONES: ventilactone A, ventilactone B.
(A) Misra et al., *Indian J. Chem.*, **20B**, 721 (1981)
(G) Rao et al., *J. Nat. Prod.*, **49**, 343 (1986)
(I) Rao et al., *Phytochemistry*, **24**, 2669 (1985)
V. madraspatana.
ANTHRACENE: physcihydrone.
CHROMAQUINONES: ventiloquinones A, B, C, D, E, F, G and H.
(An) Perkin, *J. Chem. Soc.*, **65**, 937 (1984)
(C) Marshall et al., *Phytochemistry*, **24**, 2373 (1985)

VEPRIS
V. ampody H.Perr.
QUINOLONE ALKALOIDS: 2-(hydroxynonyl)-4-quinolone, 2-(nona-3,6-dienyl)-4-quinolone, 2-(oxoundecyl)-4-quinolone.
Kan-Fan et al., *Phytochemistry*, **9**, 1283 (1970)
V. bilocularis.
FURANOQUINOLINE ALKALOID: flindersamine.
TERPENOID: veprisone.
(A) Brown et al., *Aust. J. Chem.*, **7**, 181 (1954)
(T) Govindachari et al., *Tetrahedron*, **20**, 2985 (1964)
V. louisii.
BISQUINOLINE ALKALOIDS: vepridimerine A, vepridimerine B.
QUINOLONE ALKALOID: N-methylpreskimmianine.
QUINOLONE: veprisilone.
(A) Ayafor et al., *Tetrahedron Lett.*, 3293 (1980)
(A) Ngadju et al., *ibid.*, 2041 (1982)
(Q) Ayafor et al., *Phytochemistry*, **21**, 955 (1982)

VERATRUM (Liliaceae)
V. album L.
STEROIDAL ALKALOIDS: angeloylzygadenine, deacetylgermitetrine, deacetylneoprotoveratrine, deacetylprotoveratrine, germerine, germine, germitetrine A, germitetrine B, isogermine, protoveratrine A, protoveratrine B, pseudojervine, rubijervine, veracintine, veralbidine, veratrobasine, veratroylzygadenine.
Kutney et al., *Intersci. Chem. Rep.*, **4**, 265 (1970)
V. album L. var. grandiflorum Maxim.
STEROIDAL ALKALOID: 11-deoxyjervine.
Masamune et al., *Tetrahedron Lett.*, 913 (1964)
V. album L. subsp. lobelianum (Bernh.) Suessenguth.
GLYCOALKALOID: rhamnoveracintine.
STEROIDAL ALKALOIDS: glucoveracintine, 20-(2-methyl-1-pyrrolin-5-oyl)-4-pregnen-3-one, veralinine, veralobine, veralosidine, veralosine, veralosinine, veramarine, veramerine, veramine, verazine.

STEROID: 20-(2-methyl-1-pyrrolin-5-yl)-4-pregnen-3-one.
  (A) Bondarenko et al., Khim. Prir. Soedin., 132 (1973)
  (A) Grancai et al., Chem. Pap., 40, 835 (1986); Chem. Abstr., 107
    194870e
  (S) Vassova et al., Collect. Czech. Chem. Commun., 40, 895 (1975)

V. californicum (Durand).
STEROIDAL ALKALOIDS: cyclopamine, cycloposine, muldamine.
  Keeler, Steroids, 18, 741 (1974)

V. dahuricum.
STEROIDAL ALKALOID: verdine.
  Nakhatov et al., Khim. Prir. Soedin., 131 (1980)

V. escholtzii.
STEROIDAL ALKALOIDS: escholerine, isorubijervine,
  isorubijervosine.
  Klohs et al., J. Am. Chem. Soc., 75, 2133 (1953)

V. fimbriatum.
STEROIDAL ALKALOIDS: germanitrine, germinitrine.
  Klohs et al., J. Am. Chem. Soc., 74, 4473 (1952)

V. grandiflorum.
STEROIDAL ALKALOIDS: baikeideine, baikeine, etioline,
  hosukinidine, epi-rubijervine, shinonomenine, teinemine,
  veraflorizine, veratramine.
STEROIDS: cholesterol, dormantinol.
  (A) Ito et al., Tetrahedron Lett., 2961 (1972)
  (A) Kaneko et al., Chem. Pharm. Bull., 27, 2534 (1979)
  (S) Tanaka et al., Phytochemistry, 16, 1247 (1977)

V. lobelianum Bernh.
STEROIDAL ALKALOIDS: deacetylveratrine A,
  dideacetylprotoveratrine A, germerine, germidine,
  germinaline, isorubijervine, loveraine, jervine, protoverine,
  rubijervine, veralinine, veralkamine, veralodine,
  veralodinine, veralodisine, veraloidine, veralosidinine,
  veramarine, veratroylzygadenine, verazine.
UNCHARACTERIZED ALKALOID: veralazine.
  Bondarenko et al., Khim. Prir. Soedin., 854 (1971)
  Bondarenko, ibid., 132 (1973)

V. maackii.
STEROIDAL ALKALOIDS: angeloylzygadenine, verazine.
  Ikawa et al., J. Biol. Chem., 159, 517 (1945)

V. sabadilla.
STEROIDAL ALKALOID: cerine.
  Ikawa et al., J. Biol. Chem., 159, 517 (1945)

V. stamineum.
STEROIDAL ALKALOID: 3-(1-2-methylbutyryl)-zygadenine.
  Yagi, Kawasaki, J. Pharm. Soc., Japan, 82, 210 (1968)

V. viride Ait.
STEROIDAL ALKALOIDS: cevadine, cevagenine,
  deacetylprotoveratrine, germbudine, germidine, germine,
  germitrine, neogermbudine, pseudojervine, rubijervine,
  veratramine, veratridine, veratrine, veratrosine.
  Vejdelek et al., Chem. Listy, 49, 1538 (1955)
  Kupchan et al., J. Am. Chem. Soc., 81, 1921 (1959)

V. viriole.
STEROIDAL ALKALOID: isogermbudine.
  Myers et al., J. Am. Chem. Soc., 74, 3198 (1952)

VERBASCUM (Scrophulariaceae)
V. georgicum.
IRIDOIDS: verbascoside A, verbascoside B.
  Agababyan et al., Khim. Prir. Soedin., 446 (1982)

V. lychnitis.
CAROTENOID: crocetin.
FLAVONOIDS: apigenin, luteolin 5-glucoside,
  3',4',5-trihydroxy-7-methoxyflavone.
IRIDOID: aucuboside.
  Serdyuk et al., Khim. Prir. Soedin, 545 (1976)

V. nigrum L.
IRIDOIDS: nigroside 1, nigroside 2.
  Siefert et al., Helv. Chim. Acta, 65, 1678 (1982)

V. nobile Vel.
PYRIDINE ALKALOID: pediculinine.
UNCHARACTERIZED ALKALOIDS: verbascine, verbasine.
  Ninova et al., Khim. Prir. Soedin., 540 (1971)

V. phlomoides.
SAPONIN: verbascosaponin.
  Tschesche et al., Chem. Ber., 113, 1754 (1980)

V. pseudonobile.
MACROCYCLIC ALKALOID: verbaskine.
  Koblikova et al., Tetrahedron Lett., 4831 (1983)

V. sinuatum.
IRIDOID: sinuatol.
  Bianco et al., Planta Med., 41, 75 (1981)

V. songoricum Schrenk.
PYRIDINE ALKALOIDS: anabasine, plantagonine.
  Abdusamatov, Yunusov, Khim. Prir. Soedin., 334 (1969)

V. thapsiforme.
POLYSACCHARIDE: ajugose.
  Herissey et al., C. R. Acad. Sci., 259, 824 (1954)

VERBENA (Verbenaceae)
V. hastata.
IRIDOID: hastatoside.
  Rimpler, Scheuer, Z. Naturforsch., 34C, 311 (1979)

V. officinalis L.
FLAVONE: 5-hydroxy-3,6,7,3',4'-pentamethoxyflavone.
IRIDOIDS: aucubin, hastatoside, verbenalin, (+)-verbenone.
STEROID: α-sitosterol.
TRITERPENOIDS: lupeol, ursolic acid.
  Makboul, Fitoterapia, 57, 50 (1986)

V. pulchella.
IRIDOIDS: pulchellosides I and II.
  Milz, Rimpler, Tetrahedron Lett., 895 (1978)

V. rupestris.
SESQUITERPENOIDS: 10-epi-campherenol, rupestrinol, rupestrol.
  Box, Chen, Phytochemistry, 16, 987 (1977)

VERBESINA
V. aff. coahiulensis.
SESQUITERPENOIDS: verafinin A, verafinin B.
  Guererro, Diaz, Rev. Latinoam. Quim., 14, 70 (1983)

V. coahiulensis.
SESQUITERPENOIDS: verafinin A, verafinin B, veratinin.
  Guerrero et al., Rev. Latinoam. Quim., 6, 53, 119 (1975)

V. eggersii.
SESQUITERPENOIDS:
  6β-cinnamoyloxy-1α-acetoxyeudesman-14-oic acid,
  6β-cinnamoyloxy-3,4-didehydroeudesman-14-al,
  6β-cinnamoyloxy-3,4-didehydroeudesmane,
  6β-cinnamoyloxyeudesman-14-al,
  6β-cinnamoyloxy-1β,4α-dihydroxyeudesmane,
  6β-cinnamoyloxy-3α-hydroxyeudesman-14-al,
  6β-cinnamoyloxy-3β-hydroxyeudesman-14β-al,
  6β-cinnamoyloxy-14α-hydroxyeudesmane,
  6β-cinnamoyloxy-1α-hydroxyeudesman-14-oic acid,

6β-cinnamoyloxy-hydroxyeudesman-14-oic acid.
Bohlmann, Lonitz, *Chem. Ber.*, **111**, 254 (1978)

V. occidentalis.
SESQUITERPENOID: 8β-acetoxy-9β-hydroxyverboccidentene.
Bohlmann, Lonitz, *Phytochemistry*, **17**, 453 (1978)

V. rupestris.
SESQUITERPENOIDS: chaenocephalol, iso-epi-camphenol,
rupestrinol, rupestrol, rupestrol cinnamate, rupestrol
orthocinnamate.
Box *et al.*, *Phytochemistry*, **16**, 987 (1977)

V. virgata.
Two sesquiterpenoids.
Martinez *et al.*, *Phytochemistry*, **22**, 979 (1983)

V. virginica.
SESQUITERPENOIDS: verbasindiol, α-verbesinol coumarate,
β-verbesinol coumarate.
Herz, Kumar, *Phytochemistry*, **20**, 247 (1981)

VERNONIA (Compositae)
V. adoensis.
Two glaucolide sesquiterpenoids.
Bohlmann *et al.*, *Phytochemistry*, **23**, 1795 (1984)

V. amygdalina.
SESQUITERPENOID: 11,13-dihydrovernodalin, vernodalin,
vernomygin.
Ganjian *et al.*, *Phytochemistry*, **22**, 2525 (1983)

V. angusticeps.
SESQUITERPENOIDS: 3-hydroxygrandolide, 3-oxograndolide.
Perez Souto, *Rev. Cubana Farm.*, **17**, 17 (1983); *Chem. Abstr.*, **100**
20448p

V. anthelmintica.
ACID: vernolic acid.
STEROID: vernosterol.
SESQUITERPENOID: vernodalol.
(A) Hopkins *et al.*, *J. Amer. Oil Chem. Soc.*, **37**, 682 (1960)
(S) Fioriti *et al.*, *Tetrahedron Lett.*, 2971 (1970)
(T) Asaka *et al.*, *Phytochemistry*, **16**, 1838 (1977)

V. arkansana.
SESQUITERPENOIDS:
4,7-epoxy-1,5-dihydro-11(13)-bourbonen-12,6-olide,
4,7-epoxy-1,5-dihydro-8-methylpropenyl-11(13)-bourbonen-
12,6-olide,
4,7-epoxy-1,5-dihydro-8-tiglyl-11(13)-bourbonen-12,6-olide.
Bohlmann, *Phytochemistry*, **20**, 473 (1981)

V. baldwinii.
SESQUITERPENOID: glaucolide B.
Padolina, *Ph.D. Thesis Univ. Texas & Austin*, (1973)

V. colorata.
SESQUITERPENOID: vernolide.
Toubiana, Gadamer, *Tetrahedron Lett.*, 1333 (1967)

V. compactiflora.
SESQUITERPENOIDS: compactifloride,
8-hexanoylcompactifloride.
Bohlmann *et al.*, *Phytochemistry*, **21**, 695 (1982)

V. conerea.
TRITERPENOID: 24-hydroxytaraxer-14-ene.
Misra *et al.*, *J. Nat. Prod.*, **47**, 865 (1984)

V. conferta.
SESQUITERPENOID: confertolide.
Toubiana *et al.*, *C. R. Acad. Sci.*, **270C**, 1033 (1970)

V. cotoneaster.
SESQUITERPENOIDS:
2,8,13-triacetoxy-1(10),4,7(11)-germacratrien-12,6-olide,
vernonallenoid.
Bohlmann *et al.*, *Phytochemistry*, **21**, 695 (1982)

V. echitifolia.
Four sesquiterpenoids.
Bohlmann *et al.*, *Phytochemistry*, **20**, 2233 (1981)

V. erdverbengii.
SESQUITERPENOIDS: 8-desacetylglaucolide A,
8-desacetylglaucolide A tiglate, glaucolide A, three
hirsutinolides. Triterpenoids lupeol, lupeyl acetate.
Dominguez *et al.*, *J. Nat. Prod.*, **49**, 704 (1986)

V. fasciculata.
FLAVONE: fasciculatin.
Narain *et al.*, *J. Chem. Soc., Perkin I*, 1018 (1977)

V. flexuosa.
SESQUITERPENOIDS: vernoflexin, vernoflexuoside.
Kiesel, *Pol. J. Pharmacol. Pharm.*, **27**, 461 (1975)

V. glauca.
SESQUITERPENOID: glaucolide A.
Padolina *et al.*, *Ph.D. Thesis, Univ. Texas & Austin*, (1973)

V. guineensis.
SESQUITERPENOIDS: vernodalin, vernolepin.
Kupchan *et al.*, *J. Org. Chem.*, **34**, 3908 (1969)

V. hymenolepis.
SESQUITERPENOIDS: vernolepin, vernomerin.
Kupchan *et al.*, *J. Am. Chem. Soc.*, **90**, 3596 (1968)

V. jaciculata.
SESQUITERPENOID: marginatin.
Padolina *et al.*, *Phytochemistry*, **13**, 225 (1974)

V. marginata.
SESQUITERPENOID: marginatin, four glaucolides, two
jalcaguaianolides, one pseudoelephantopide and four
vernomargalides.
Jakupovic *et al.*, *Phytochemistry*, **25**, 1179 (1986)

V. molissima.
SESQUITERPENOIDS: 8-acetyl-13-ethoxypiptocarphol,
diacetylpiptocarphol, loliolide, piptocarphin A.
Catalan *et al.*, *J. Nat. Prod.*, **49**, 351 (1986)

V. monocephala.
SESQUITERPENOID: 8-hexanoylcompactifloride.
Bohlmann *et al.*, *Phytochemistry*, **21**, 695 (1982)

V. nudiflora.
SESQUITERPENOIDS: 8α-senecioyloxydehydrocostus lactone,
vernudiflorin.
Bohlmann, Zdero, *Phytochemistry*, **16**, 778 (1977)

V. officinalis L.
IRIDOIDS: vadroside, veproside.
Afifi-Yazar *et al.*, *Helv. Chim. Acta*, **64**, 16 (1981)

V. oligocephala.
Two glaucolide sesquiterpenoids.
Bohlmann *et al.*, *Phytochemistry*, **23**, 1795 (1984)

V. pectoralis.
SESQUITERPENOIDS: pectorolide, vernopectolide A,
vernopectolide B.
Monpon, Toubiana, *Tetrahedron*, **32**, 2545 (1976)

V. polyanthes.
SESQUITERPENOIDS: vernopolyanthofuran, vernopolyanthone.
Bohlmann *et al.*, *Phytochemistry*, **20**, 473 (1981)

V. saltensis.
SESQUITERPENOIDS: hirsutinolide-8,13-diacetate, hirsutinolide-1,8,13-triacetate.
Bohlmann et al., Phytochemistry, **18**, 289 (1979)

V. scorpioides.
SESQUITERPENOIDS: hirsutinolide-1,13-diacetate, hirsutinolide-8-propionate, scorpiolide.
Bohlmann et al., Phytochemistry, **18**, 289 (1979)
Warning et al., Ann. Chim., 467 (1987)

V. squamulosa.
SESQUITERPENOIDS: glaucolides A, B, C, D, E, F and G, piptocarphin E.
Cayalan et al., J. Nat. Prod., **49**, 351 (1986)

V. stachenoides.
SESQUITERPENOIDS: 3-hydroxystilpnotomentolide 8-O-4-acetoxy-3-methyl-2-butenoate, 3-hydroxystilpnotomentolide 8-O-5-acetoxysenecionate.
Bohlmann et al., Phytochemistry, **21**, 1445 (1982)

V. sublutea Scott Elliot.
SESQUITERPENOID: subluteolide.
Mompon, Toubiana, Tetrahedron, **23**, 2199 (1977)

V. sutherlandii.
Two glaucolide sesquiterpenoids.
Bohlmann et al., Phytochemistry, **23**, 1795 (1984)

V. uniflora.
SESQUITERPENOIDS: glaucolide D, glaucolide E.
Betkouski et al., Rev. Latinoam. Quim., **6**, 191 (1975)

VERONICA (Scrophulariaceae)
V. anthelmintica.
STEROIDS: 7,24(28)-stigmastadien-3-ol, 8,14,24(28)-stigmastatrien-3-ol.
Lin et al., Phytochemistry, **11**, 2319 (1972)

V. bellidioides.
PHENYLPROPANOID: ehrenoside.
Lahloub et al., Planta Med., 325 (1986)

V. officinalis L.
FLAVONOID: 6-hydroxyluteolin-7-monoglycoside.
IRIDOIDS: minecoside, verminoside, veronicoside, verproside.
(F) Kojcik, Acta Pol. Pharm., **34**, 295 (1977); Chem. Abstr., 88 86009m
(I) Sticher et al., Helv. Chim. Acta, **62**, 530 (1979)
(I) Afifi-Yazar et al., ibid., **63**, 905 (1980)

V. peregrina L.
FLAVONOID: luteolin.
PHENOLICS: protocatechuic acid, vanillic acid.
Jin et al., Zhongcaoyao, **13**, 10 (1982); Chem. Abstr., 97 17871g

V. spicata.
FLAVONOIDS: apigenin, apigenin 7-β-D-glucuronide, cymaroside, luteolin.
Kusev et al., Khim. Prir. Soedin., 704 (1977)

VERTICILLIUM
V. agaricum.
CAROTENOID: apo-4,4'-carotenic acid methyl ester.
Valadon, Mummery, Phytochemistry, **16**, 613 (1977)

V. dahliae.
NAPHTHALENONE: 4-hydroxyscytalone.
SESQUITERPENOID: dehydrohemigossypol.
Iwasaki et al., Tetrahedron Lett., 13 (1972)

V. lecani.
STEROID:
2β-hydroxy-4,4,14-trimethyl-5α-pregna-2,9(11)-diene-20S-carboxylic acid.
TRITERPENOID: 23,24,25,26,27-pentanorlanost-8-en-3,22-diol.
Gore, Phytochemistry, **23**, 1721 (1984)

VESTIA
V. lycioides.
CARBOLINE ALKALOIDS: 1-acetyl-3-carbomethoxy-β-carboline, 1-acetyl-8-carbomethoxy-β-carboline.
Fain et al., Phytochemistry, **17**, 338 (1978)

VETIVERIA
V. zizanioides.
NORSESQUITERPENOID: norkhusinol oxide.
SESQUITERPENOIDS: allokhusinol, bicyclodecenone, (-)-cadinane, cyclocopacamphene, cyclocopacamphorenic acid, epi-cyclocopacamphorenic acid, isobisabolene, isokhusenic acid, isokhusinol, isokhusinol oxide, isovalencenic acid, isovalencenol, isovetiselinenol, khusenol, khusilal, khusimene, khusimone, khusinol, epi-khusinol, khusinoloxide, khusiol, khusol, (+)-khusitene, khusitone, khusol, selina-4(14),7(11)-diene, tricyclovetivenol, veticadinol, vetiselinene, vetiselineno, α-vetispirene, β-vetispirene, α-vetivenine, β-vetivenine, α-vetivone, β-vetivone, zizanal, 7-epi-zizanal, zizanoic acid, 5-epi-zizanoic acid, zizanol.
Jain et al., Tetrahedron Lett., 4639 (1982)
Kalsi et al., Tetrahedron, **41**, 3387 (1985)

VIBURNUM (Caprifoliaceae)
V. awabuki K.Koch.
COUMARINS: scopolitin, scopolitin glucoside.
Kuroyanagi et al., Chem. Pharm. Bull., **34**, 4012 (1986)

V. betulifolium.
IRIDOIDS: decapetaloside, viburnalloside.
Jensen et al., Phytochemistry, **24**, 487 (1985)

V. davidii.
CHALCONES: davidigenin, davidioside, davidioside-2'-O-β-glucoside, 4'-O-methyldavidigenin.
Jensen et al., Phytochemistry, **16**, 2536 (1977)

V. dilatatum.
GLUCOSIDE: dilaspirolactone.
Iwagawa et al., Phytochemistry, **23**, 2299 (1984)

V. japonicum.
IRIDOIDS: acetylalloside, adenoxide, 2',3'-O-diacetylfurcatoside, furcatoside A.
Iwagawa et al., Phytochemistry, **25**, 1227 (1986)

V. lantana L.
ANTHOCYANIN: cyanidin 3-glucoside.
FLAVONOID: kaempferol 3-diglucoside.
TRITERPENOID: ursolic acid.
(An, F) Guichard et al., Plant Med. Phytother., **10**, 105 (1976)
(T) Cucu et al., Pharmazie, **32**, 542 (1977)

V. lantanoides.
CHALCONES: davidioside, davidioside 4'-O-methyl ether.
Jensen et al., Phytochemistry, **16**, 2036 (1977)

V. odoratissimum.
DITERPENOIDS: vibsanthins A, B, C, D, E and F.
Kawazu et al., Agr. Biol. Chem., **44**, 1367 (1980)

V. opulus L.
IRIDOIDS: opulus iridoids I, II, III and IV.
Bock et al., Phytochemistry, **17**, 753 (1978)

V. phlebotrichum.
MONOTERPENE GLYCOSIDE: phlebotricoside.
Hase *et al.*, *Phytochemistry*, **21**, 1435 (1982)

V. suspensum.
IRIDOID: suspensolide A.
Hase *et al.*, *Chem. Lett.*, 13 (1982)

V. urceolatum.
IRIDOIDS: iridoids VU-A, VU-B, VU-C and VU-D, urceolide.
Hase *et al.*, *Phytochemistry*, **22**, 255, 1977 (1983)
Hase *et al.*, *ibid.*, **22**, 1977 (1983)

VICIA (Leguminosae/Fumariaceae)
V. angustifolia.
STEROIDS: β-sitosterol, β-sitosterol β-D-glucoside, stigmasterol.
Kameoka *et al.*, *Chem. Abstr.*, 104 17659k

V. faba L.
ACIDS: (+)-7-isojasmonic acid, (-)-jasmonic acid.
GIBBERELLIN: gibberellin A-53.
PHYTOALEXIN: medicarpin.
SESQUITERPENOID: wyerone epoxide.
URACIL: convicine.
(Acid) Miersch *et al.*, *J. Plant Growth Regul.*, **5**, 91 (1986); *Chem. Abstr.*, 105 222748a
(G, T) Hargreaves *et al.*, *Phytochemistry*, **15**, 1119 (1976)
(Ph) Hargreaves *et al.*, *Nature*, **262**, 318 (1976)
(U) Brown *et al.*, *ibid.*, **11**, 3203 (1972)

V. pseudoorobus.
AMINO ACID: α-(hydroxymethyl)serine.
Saito *et al.*, *Phytochemistry*, **24**, 853 (1985)

V. sativa.
AMINO ACIDS: τ-glutamyl-τ-aminobutyric acid, τ-glutamyl-β-cyanoalanine.
URACIL: convicine.
(Am) Larsen *et al.*, *Acta Chem. Scand.*, **19**, 1071 (1965)
(U) Brown *et al.*, *Phytochemistry*, **11**, 3203 (1972)

VICOA
V. indica DC.
FLAVONE: 3′,5,6-trihydroxy-4′,7-dimethoxyflavone.
SESQUITERPENOIDS: viscolides A, B, C and D.
(F) Purushothaman *et al.*, *Indian Drugs*, **23**, 480 (1986); *Chem. Abstr.*, 105 75921w
(T) Purushothaman *et al.*, *Indian J. Chem.*, **25B**, 417 (1986)

V. vestita.
SESQUITERPENOID: vestolide.
Sachdev, Kulshreshtha, *Phytochemistry*, **23**, 2379 (1984)

VIGNA (Papilionaceae)
V. angularis.
SAPOGENIN: azukisapogenol.
Kitagawa *et al.*, *Chem. Pharm. Bull.*, **31**, 664 (1983)

V. marginata.
FLAVONOID: robinin.
Morita *et al.*, *Chem. Abstr.*, 88 14894b

V. unguiculata (L.) Walp.
AMINO ACIDS: arginine, glutamic acid, glycine, methionine.
BENZOFURAN: vignafuran.
(Am) Otoul, *Bull. Rech. Agron. Gembloux*, **9**, 247 (1974); *Chem. Abstr.*, 84 176694c
(B) Preston *et al.*, *Phytochemistry*, **14**, 1843 (1975)

VIGUIERA
V. angustifolia.
SESQUITERPENOIDS: budlein A, budlein B.
Romo de Vivar *et al.*, *Phytochemistry*, **15**, 525 (1976)

V. bishopii.
DITERPENOIDS: ent-9(11)-beyeradien-19-oic acid, ent-9(11)-trachyloben-19-oic acid.
Bohlmann *et al.*, *Phytochemistry*, **20**, 113 (1981)

V. buddleiaeformis.
SESQUITERPENOIDS: budlein A, budlein B.
Romo de Vivar *et al.*, *Phytochemistry*, **15**, 525 (1976)

V. dentata.
Three cycloartenone triterpenoids.
Gao *et al.*, *Phytochemistry*, **25**, 1489 (1986)

V. eriophora.
SESQUITERPENOID: 17,18-dehydrovinguiepinin.
Delgado *et al.*, *Phytochemistry*, **21**, 1305 (1982)

V. insignis.
DITERPENOID: ent-15-beyerene-3,12-diol.
Delgado *et al.*, *Phytochemistry*, **22**, 1227 (1983)

V. linearis.
SESQUITERPENOIDS: 15-hydroxyacetylerioflorin, viguilenin.
Delgado *et al.*, *Phytochemistry*, **24**, 2736 (1985)

V. macrophylla. One
germacranolide and two heliangolides. Gershenzon et al., Phytochemistry, 23, 1281 (1984)

V. maculata.
DITERPENOID: 15-oxozoapatlin.
Delgado *et al.*, *Phytochemistry*, **23**, 2674 (1984)

V. oblongifolia.
SESQUITERPENOID: 4,11-amorphadiene.
Bohlmann *et al.*, *Phytochemistry*, **23**, 1183 (1984)

V. pinnatilobata.
SESQUITERPENOID: vinguipinin.
Romo de Vivar *et al.*, *Rev. Latinoam. Quim.*, **9**, 171 (1978)

V. sphaerocephala.
SESQUITERPENOID: sphaerocephalin.
Ortega *et al.*, *Phytochemistry*, **19**, 1545 (1980)

V. stenoloba.
DITERPENOID: stenolobin.
SESQUITERPENOIDS: deacetylvinguistenin, viguiestenin.
Romo de Vivar *et al.*, *Rev. Latinoam. Quim.*, **9**, 171 (1978)

VILEX
V. rotundifolia.
DITERPENOID: dehydroabietane.
MONOTERPENOIDS: bornyl acetate, camphene, 1,8-cineole, limonene, linalool, β-myrcene, β-pinene, terpinen-4-ol, α-terpinyl acetate, τ-terpinyl acetate.
SESQUITERPENOIDS: δ-cadinene, α-cadinol, α-copaene, δ-elemene, β-guaiene, α-muurolene.
Iwamura *et al.*, *Nippon Nogei Kogaku Kaishi*, **52**, 45 (1978); *Chem. Abstr.*, 88 186084a

VILLANOVA
V. titicaensis.
DITERPENOID: 1,10-epoxycannabinolide.
Gomez *et al.*, *Phytochemistry*, **22**, 197 (1983)

VINCA (Apocynaceae)

V. difformis.
IMIDAZOLIDINOCARBAZOLE ALKALOIDS: difformine, normacusine B, (+)-vincadifformine, (±)-vincadifformine, vincamedine.
Garnier *et al.*, *Planta Med.*, 66 (1986)

V. elegantissima Hort.
SPIROINDOLONE ALKALOIDS: elegantine, elegantissine, isoelegantissine.
Bhattacharyya, Pakrashi, *Tetrahedron Lett.*, 159 (1972)

V. erecta Rgl. et Schmalh.
CARBAZOLE ALKALOIDS: N-acetylvinervine, akuammicine, akuammine, 10-methoxyvinorine, O-methylakuammine, norfluorocurarine, vincandidine, vincanicine, vincanidine, vincanine, vinervidine, vinervidine N-oxide, vinervine, vinervinine.
CARBOLINE ALKALOIDS: apovincamine, eburnamine, eburnamonine, ervincine, ervine, 11-hydroxypleiocarpamine, isoreserpiline, picrinine, tombozine, vincamine, vincaricine, vincaridine, vincarine, vincarinine, vincarpine, vincine.
FURANOQUINOLINE ALKALOID: skimmianine.
IMIDAZOLIDINOCARBAZOLE ALKALOIDS: epoxykopsinine, epoxykopsinone, ervamycine, ervinadine, ervinadinine, ervinceine, ervincinine, kopsanone, kopsanone lactam, kopsinilam, kopsinilamine, kopsinine, pseudokopsinine.
SPIROINDOLONE ALKALOIDS: ervinidine, erycinine, vineridine, vineridine N-oxide, vinerine.
Kuchenkova *et al.*, *Izv. Akad. Nauk SSSR*, 12, 2152 (1965)
Rakhimov *et al.*, *Khim. Prir. Soedin.*, 461, 521 (1969)
Il'yasova *et al.*, *ibid.*, 164 (1971)

V. herbacea Waldst. et Kit.
CARBAZOLE ALKALOIDS: akuammicine, akuammine, vincanine.
CARBOLINE ALKALOIDS: herbaceine, herbadine, herbamine, norflurocurarine, reserpinine, vincamine, vincarine.
FURANOQUINOLINE ALKALOID: skimmianine.
IMIDAZOLIDINOCARBAZOLE ALKALOIDS: 16-methoxy-(-)-tabersonine, N-methyl-6,7-dehydroaspidospermidine, tabersonine.
SPIROINDOLONE ALKALOIDS: herbaline, herbavine, herbozine.
SPIROQUINOLONE ALKALOID: isomajdine.
UNCHARACTERIZED ALKALOIDS: herbavinine, vincaherbine, vincaherbinine, vincaline I, vincaline II.
Zabolstnaya *et al.*, *J. Gen. Chem., USSR*, 33, 3780 (1963)
Aynilian *et al.*, *J. Pharm. Sci.*, 64, 341 (1975)
Chkhivadze *et al.*, *Khim. Prir. Soedin.*, 227 (1976)

V. libanotica.
CARBOLINE ALKALOID: herbadine.
Aynilian *et al.*, *J. Pharm. Sci.*, 64, 341 (1975)

V. major L.
CARBAZOLE ALKALOID: akuammine.
CARBOLINE ALKALOIDS: lochniverine, majoridine, majorinine, majvinine, 10-methoxyvellosimine, reserpinine, vincamajine, vincamajinine.
SPIROINDOLONE ALKALOIDS: isomajdine, majdine, majdinine.
UNCHARACTERIZED ALKALOIDS: vincamajoreine, vinine.
Janot, LeMen, *Ann. Pharm. Fr.*, 13, 325 (1953)
Yagudaev *et al.*, *Khim. Prir. Soedin.*, 197 (1968)
Zhukovich, Vachnadze, *ibid.*, 720 (1985)

V. major L. var. elegantissima Hort.
CARBOLINE ALKALOIDS: dehydrovincarpine, reserpinine, vincarpine.
Ali *et al.*, *Tetrahedron Lett.*, 4887 (1976)

V. major L. var. major.
CARBOLINE ALKALOID: majvinine.
Banerji, Chakravarti, *Phytochemistry*, 16, 1124 (1977)

V. minor L.
CARBOLINE ALKALOIDS: 10-methoxydeacetylakuammiline, isovincamine, 10-methoxyeburnamonine, epi-pleiocarpamine N(4)-oxide, strictamine, vincamedine, 14-epi-vincamine, vincaminine, vincanorine, vincinine.
DIMERIC ALKALOIDS: vincamidine, vincristine.
INDOLE ALKALOIDS: (+)-vincadine, vincaminoreine, vincaminoridine, vincaminorine, vincatine.
IMIDAZOLIDINOCARBAZOLE ALKALOIDS: 16-methoxyminovincine, 16-methoxyminovincinine, 16-methoxyvincadifformine, 11-methoxyvincamine, (+)-N-methylaspidospermidine, N-methylquebrachamine, minovincine, (-)-minovincinine, minovine, vincatine.
QUATERNARY ALKALOID: N(1)-methyl-2β,16β-dihydroakuammicine N(4)-methochloride.
UNCHARACTERIZED ALKALOIDS: minovinceine, vincareine, vincoridine, vincorine, vinomine, vinoxine.
Schlittler, Furlmeier, *Helv. Chim. Acta*, 36, 2017 (1953)
Mokry *et al.*, *Tetrahedron Lett.*, 1185 (1962)
Dopke *et al.*, *Tetrahedron Lett.*, 6065 (1968)

V. pubescens Urv.
CARBOLINE ALKALOID: reserpinine.
UNCHARACTERIZED ALKALOIDS: pubescine, vinine.
Orekhov *et al.*, *Arch. Pharm.*, 272, 70 (1934)

V. pusilla.
CARBOLINE ALKALOID: vincapusine.
Mitra *et al.*, *Phytochemistry*, 20, 865 (1981)

V. rosea L.
CARBAZOLE ALKALOIDS: lochneridine, preakuammicine.
CARBOLINE ALKALOIDS: isositsirikine, lochnerine, lochnervine, pericyclivine,
(A) Svoboda *et al.*, *J. Pharm. Sci.*, 50, 409 (1961)
(A) Abdurakhimova *et al.*, *Khim. Prir. Soedin.*, 224 (1965)
(A) Neuss *et al.*, *Helv. Chim. Acta*, 56, 2660 (1973)
(A) Tafue *et al.*, *J. Pharm. Sci.*, 64, 1953 (1975)
(Cy) Chen *et al.*, *Phytochemistry*, 15, 1565 (1976)
(G) Bhakuni *et al.*, *Phytochemistry*, 13, 2541 (1974)
(I) Guarneccia *et al.*, *J. Am. Chem. Soc.*, 96, 7079 (1974)

VINCETOXICUM (Asclepiadaceae)

V. funebre (Boiss. et Kotschy) Pobed.
PHENANTHROINDOLIZIDINE ALKALOID: antofine.
Platonova *et al.*, *J. Gen. Chem., USSR*, 28, 3131 (1958)

V. hirundinaria.
STEROID: hirundoside A.
Stockel *et al.*, *Helv. Chim. Acta*, 52, 338 (1969)

V. officinale Monsch.
STEROIDS: anhydrohirundigenin, hirundigenin.
Kennard *et al.*, *Tetrahedron Lett.*, 3799 (1968)

VIOLA (Violaceae)

V. cornuta L.
SACCHARIDE: violutin.
Robertson *et al.*, *J. Chem. Soc.*, 2770 (1932)

V. tricolor L.
CAROTENOIDS: auroxanthin, violaxanthin.
GLYCOSIDE: violanthin.
Horhammer *et al.*, *Tetrahedron Lett.*, 1707 (1965)

VIRGILIA

V. capensis.
LUPANE ALKALOIDS: virgilidine, virgiline.

White, *N. Z. J. Sci. Tech.*, **27B**, 474 (1946)

V. divaricata.
LUPANE ALKALOID: virgiboldine.

Van Eijk, Radema, *Planta Med.*, **44**, 224 (1982)

V. oroboides (Berg.) Salter.
LUPANE ALKALOIDS: O-(2-pyrrolcarbonyl)-virgiline, virgiboldine, virgiline.

Van Eijk, Radema, *Planta Med.*, **44**, 224 (1982)

VIROLA (Myristicaceae)

V. caducifolia.
ISOFLAVONES: biochanin A, dehydroferreirin, 5,7-dihydroxy-2',4'-dimethoxyisoflavone, 5,7,2'-trihydroxy-4'-methoxyisoflavone.

Braz et al., *Phytochemistry*, **15**, 1029 (1976)

V. carinata.
FLAVONE: 7,4'-dimethoxyflavone.

NAPHTHALENE: galcatin.

NEOLIGNANS: carinatidin, carinatidiol, carinatin, (-)-carinatone, dehydroeugenol B, dihydrocarinatidin, (-)-galcatin, (+)-guaiacin, (-)-isootobaphenol.

(Na) Kawanishi et al., *Phytochemistry*, **22**, 2277 (1983)

(F, N) Gottlieb et al., *Phytochemistry*, **15**, 773 (1976)

V. cuspidata.
CARBOLINE ALKALOIDS: 6-methoxyharmalan, 6-methoxyharman.

Cassady et al., *Lloydia*, **34**, 161 (1971)

V. elongata.
PHENOLICS: virolanols A, B and C.

Kijjoa et al., *Phytochemistry*, **20**, 1385 (1981)

V. multinervia.
PROPANOID: 1-(4'-hydroxy-2'-methoxyphenyl)-3-(3''-hydroxy-4''-methoxyphenyl)-propane.

Braz Filho et al., *Phytochemistry*, **15**, 567 (1976)

V. rufula.
CARBOLINE ALKALOID: 1,2,3,4-tetrahydro-6-methoxy-2-methyl-β-carboline.

Ghosal, Mukherjee, *J. Org. Chem.*, **31**, 2284 (1986)

V. surinamensis.
LIGNINS: galbacin, veraguensin.

Barata, Baker, *An. Acad. Bras. Cienc.*, **49**, 387 (1977); *Chem. Abstr.*, **88** 148966g

V. theiodora Warb.
CARBOLINE ALKALOID: 1,2,3,4-tetrahydro-6-methoxy-2-methyl-β-carboline.

Ghosal, Mukherjee, *J. Org. Chem.*, **31**, 2284 (1986)

VISCARIA

V. viscosa.
FLAVONOIDS: homoorientin, isosaponarin, orientin, saponaretin, saponaretin 6''-O-galactoside, vitexin.

Darmograi et al., *Sb. Nauch. Tr. Ryazan. Med. Inst.*, **50**, 22 (1975); *Chem. Abstr.*, **84** 56522y

VISCUM (Loranthacaea)

V. album L. One uncharacterized alkaloid.
LIGNANS: eleutheroside E, syringaresinol mono-O-glucoside, syringenin-4'-apiosyl-glucoside, syringenin-4'-O-glucoside, syringin.

(A) LePrince, *C. R. Acad. Sci.*, **145**, 940 (1907)

(L) Rao et al., *Lloydia*, **41**, 56 (1978)

(L) Wagner et al., *Planta Med.*, 102 (1986)

V. liquidambaricola.
STEROID: β-sitosterol.

TRITERPENOID: β-amyrin.

Chen et al., *Chem. Abstr.*, 84 102342q

V. multinerve Hayata.
COUMARINS: homoeriodictyol, naringenin, rhamnazin 3-glucoside.

STEROIDS: campesterol, β-sitosterol, stigmasterol.

TRITERPENOIDS: β-amyrenone, β-amyrin, β-amyrin acetate, betulic acid, lupenone, lupeol, lupenyl acetate, oleanolic acid.

Lin, *Chem. Abstr.*, 100 82780z

VISMIA

V. guarimirangae.
LIGNAN: 5,5'-dimethoxysesamin.

Camele et al., *Phytochemistry*, **21**, 417 (1982)

V. guineensis.
XANTHONES: bianthrone A-1, geranyloxyemodin, geranyloxyemodin anthrone, madagascin anthrone, vismione H.

Botta et al., *Phytochemistry*, **25**, 1217 (1986)

VITEX (Verbenaceae)

V. cannabinifolia.
DITERPENOID: vitexilactone.

Taguchi, *Chem. Pharm. Bull.*, **24**, 1602 (1976)

V. littoralis.
FLAVANONE: vitexin.

Evans et al., *J. Chem. Soc.*, 3510 (1957)

V. megapotamica.
STEROIDS: 5β,20R-dihydroxyecdysone, inokosterone, pterosterone, viticosterone E.

Rimpler, *Arch. Pharm.*, **305**, 746 (1972)

V. negundo.
UNCHARACTERIZED ALKALOID: nishindine.

IRIDOIDS: 6'-p-hydroxybenzoylmussaenosidic acid, negundoside, nishindaside.

MONOTERPENOIDS: camphene, citral, α-pinene.

SESQUITERPENOID: caryophyllene.

(A) Basu, Singh, *Quart. J. Pharm. Pharmacol.*, **20**, 136 (1947)

(I) Dutta et al., *Tetrahedron*, **39**, 3067 (1985)

(T) Manalo, *Philipp. J. Sci.*, **111**, 79 (1982); *Chem. Abstr.*, 100 48569c

V. rotundifolia.
DITERPENOIDS: prerotundifuran, rotundifuran.

Asaka et al., *Chem. Lett.*, 937 (1973)

V. trifolia.
STEROIDS: β-sitosterol, β-sitosterol-β-D-glucoside.

TRITERPENOID: friedelin.

Vedantham et al., *Indian J. Pharm.*, **38**, 13 (1976)

V. vinifera.
PHENOLICS: α-viniferin, β-viniferin.

Langcake et al., *Experientia*, **33**, 151 (1977)

VITIS (Vitaceae)

V. vinifera.
ACIDS: linoleic acid, linolenic acid, oleic acid, palmitic acid, stearic acid.

Cherrad et al., *Connaiss. Vigne Vin.*, **8**, 233 (1974); *Chem. Abstr.*, 86 722f

V. vinifera var. 'Trebbiano'.

ANTHOCYANINS: procyanidin B-1, procyanidin B-2, procyanidin B-3, procyanidin B-4.

FLAVONOIDS: kaempferol 3-glucoside, quercetin 3-glucoside.

PHENOLICS: (+)-catechin, (-)-epi-catehcin, (+)-gallocatehin.

Piretti *et al., Ann. Chim. (Rome),* **66,** 429 (1976)

VOACANGA (Apocynaceae)

V. africana Stapf.

DIMERIC ALKALOIDS: deoxyvobtusine, deoxyvobtusine lactone, folicangine, isovoafoline, voacamidine, voacamine, voacamine N-oxide, voacorine, voafolidine, voafoline, vobtusine, vobtusine lactone.

HOMOCARBOLINE ALKALOIDS: voacangine, voacangine hydroxyindolenine, voacangine pseudoindoxyl, voacristine, voacristine hydroxyindolenine, voacryptine.

INDOLE ALKALOIDS: voacamine, voacamine N-oxide, voaphylline, iboaphylline hydroxyindolenine, vobasine.

UNCHARACTERIZED ALKALOIDS: voacafricine, voacafrine, voacamidine, voacamine.

Kunesch, Poisson, *Tetrahedron Lett.,* 1745 (1968)

Stock *et al., Helv. Chim. Acta,* **66,** 2525 (1983)

V. chalotiana Pierre et Stapf.

CARBOLINE ALKALOIDS:
17-O-acetyl-19,20-dihydrovoachalotine, dehydrovoachalotine, voacarpine, voachalotine, voacoline.

UNCHARACTERIZED ALKALOID: voamonine.

Bombardelli *et al., Phytochemistry,* **15,** 2021 (1976)

Daniel *et al., Heterocycles,* **14,** 201 (1980)

V. dichotoma.

HOMOCARBOLINE ALKALOID: voacryptine.

Renner, Prins, *Experientia,* **17,** 106 (1961)

V. dregei E.M.

HOMOCARBOLINE ALKALOID: voacangine.

INDOLE ALKALOID: dregamine.

Cava *et al., Tetrahedron Lett.,* 53 (1963)

V. howarsii var. obtusa.

INDOLE ALKALOIDS: 18-decarbomethoxyvoacamine, voaluteine.

Goldblatt *et al., Phytochemistry,* **9,** 1293 (1970)

V. obtusa.

DIMERIC ALKALOID: vobtusine.

Stauffacher, Seebeck, *Helv. Chim. Acta,* **41,** 169 (1958)

V. papuana (F.Muell.) K.Schum.

INDOLE ALKALOIDS: voacamidine, voacamine.

Stock *et al., Helv. Chim. Acta,* **66,** 2625 (1983)

VOCHYSIA

V. guaianensis.

FLAVANOID: vochysine.

Baudouin *et al., J. Nat. Prod.,* **46,** 681 (1983)

VOUACAPOUA

V. americana.

DITERPENOID: vouacapenic acid.

King, Goodson, *J. Chem. Soc.,* 1117 (1955)

V. macropetala.

PHENOLIC:
(E)-1-(3,5-dihydroxyphenyl)-2-(3,4,5-trihydroxyphenyl)-ethylene.

STILBENES: astringenin, (E)-3,3',4,5,5'-pentahydroxystilbene.

TRITERPENOIDS: acetylvouacaprenol, vouacapenol.

(P) King *et al., J. Chem. Soc.,* 4477 (1956)

(St) Grassman *et al., Chem. Ber.,* **89,** 2523 (1956)

(T) King *et al., J. Chem. Soc.,* 1117 (1955)

W

WALPURGIA
W. stuhlmannii.
SESQUITERPENOID: mukaadial.

Kubo et al., Chem. Lett., 979 (1983)

WALSURA
W. piscidia.
TRITERPENOIDS: piscidinols A, B, C, D and E, piscidofuran.

Purushothaman et al., Phytochemistry, 24, 2349 (1985)

W. tabulata.
TRITERPENOID: walsurenol.

Chatterjee et al., J. Chem. Soc., Chem. Commun., 418 (1968)

WALTHERIA (Sterculiaceae)
W. americana.
PEPTIDE ALKALOIDS: adouetin X, adouetin Y, adouetin Y',
adouetin Z.

Pais et al., Bull. Soc. Chim. Fr., 1145 (1968)

WARBURGHIA
W. ugandensis.
SESQUITERPENOID: muzigadial, ugandensidial, ugandensolide.

Nakanishi, Kubo, Isr. J. Chem., 16, 28 (1977)

WEDELIA (Compositae)
W. asperrima.
DITERPENOID: wedeloside.

Oelrichs et al., J. Nat. Prod., 43, 414 (1980)

W. calendulaceae.
BENZOFURAN: norwedelic acid.

Govindachari et al., Phytochem., 24, 3068 (1985)

W. grandiflora.
SESQUITERPENOIDS: 6-methylpropenoylwedelifloride,
6-methylpropenoylwedelefloride 4-acetate.

Bohlmann et al., Phytochemistry, 19, 2047 (1980)

W. regis.
DITERPENOIDS: wederegiolides A, B and C.

Bohlmann et al., Phytochemistry, 23, 1673 (1984)

W. rugosa Greenm.
TRITERPENOIDS: β-amyrin, β-amyrin acetate, β-amyrin oleate,
β-amyrin palmitate.

Alemane et al., Rev. Cubana Farm., 11, 47 (1977); Chem. Abstr., 88
3076g

W. scaberrima.
TRITERPENE GLYCOSIDE: wedelin.

Maros et al., J. Nat. Prod., 46, 836 (1983)

W. trilobata.
DITERPENOIDS: 3a-angeloyloxykaur-16-en-19-oic acid methyl
ester, 15-angeloyloxykaur-16-en-19-oic acid methyl ester,
3a-cinnamoyloxykaur-16-en-19-oic acid methyl ester,
3a-tigloyloxykaur-16-en-19-oic acid methyl ester,
simsiolide, wedeliasecokaurenolide.
SESQUITERPENOIDS: 6-angeloyloxytrilobolide, 6-methyl
propanoyltrilobolide.

Bohlmann et al., Phytochemistry, 20, 751 (1981)

WELWITSCHIA
W. mirabilis.
PHENOLICS: gnetin F, gnetin G, gnetin H, gnetin I.

Lins et al., Bull. Soc. Chim. Belg., 95, 737 (1986)

WIDDRINGTONIA (Cupressaceae)
W. dracomontana.
SESQUITERPENOIDS: cryptomeridiol, cryptomerione,
cryptonetone.

Nagahama et al., Bull. Chem. Soc., Japan, 37, 1029 (1964)
Ito et al., Tetrahedron Lett., 3185 (1969)

W. juniperoides.
SESQUITERPENOIDS: cuparene, cuparenic acid.

Parker et al., J. Chem. Soc., 1558 (1962)

W. schwartzii.
SESQUITERPENOIDS: cuparene, cuparenic acid.

Parker et al., J. Chem. Soc., 1558 (1962)

W. whytei.
SESQUITERPENOIDS: cuparene, cuparenic acid.

Parker et al., J. Chem. Soc., 1558 (1962)

WIESNERELLA
W. denudata.
SESQUITERPENOID: dehydrotulipinolide.

Asakawa et al., Phytochemistry, 19, 567 (1980)

WIGANDIA
W. kunthii.
HOMOSESQUITERPENOID: wigandol.
QUINONE: farnesylhydroquinone.

Gomez et al., Phytochemistry, 19, 2202 (1980)

WIKSTROEMIA
W. elliptica.
COUMARINS: daphnoretin, umbelliferone.
LIGNANS: (±)-lariciresinol, (±)-5'-methoxylariciresinol,
syringaresinol.

Duh et al., J. Nat. Prod., 49, 706 (1986)

W. monticola.
TOXINS: wikstrotoxins A, B, C and D.

Jolad et al., J. Nat. Prod., 46, 675 (1983)

W. shikokiana.
BIFLAVANONES: shikokianins A and B.

Niwa et al., Chem. Pharm. Bull., 34, 3631 (1986)

W. viridiflora.
COUMARINS: daphnoretin, wikstrosin.
FURANONE: arctigenin.
LIGNAN: wikstromol.

(C, F) Inagaki et al., Chem. Pharm. Bull., 20, 2710 (1972)
(F) Tandon et al., Phytochemistry, 16, 1991 (1977)
(L) Tandon et al., Phytochemistry, 15, 1789 (1976)

WILCOXIA
W. viperiana.
STEROIDS: 3β-hydroxycholest-7-en-6-one, viperidinone,
viperidone.

Djerassi et al., Ber. Dtsch. Chem. Ges., 97, 3118 (1964)

WITHANIA (Solanaceae)

W. coagulans.

STEROIDS:
6,7-epoxy-5,20-dihydroxy-1-oxo-2,24-withadienolide,
6,7-epoxy-5,27-dihydroxy-1-oxo-2,24-withadienolide,
5,25-ergostadiene-3,24-diol, withaferin A.
Velde et al., Phytochemistry, 22, 2253 (1983)

W. frutescens.

STEROIDS:
2,3-dihydro-4,17,27-trihydroxy-1-oxo-22R-witha-2,5,24-trienolide,
4,5,27-trihydroxy-6-chloro-22R-witha-2,24-dienolide,
4,5,27-trihydroxy-1-oxo-6-chloro-22R-witha-2,5,24-trienolide,
4,17,27-trihydroxy-1-oxo-22R-witha-2,5,24-trienolide.
Gonzalez et al., An. Quim., 70, 69 (1974)

W. somnifera L.

UNCHARACTERIZED ALKALOIDS: pseudowithanine,
somniferianine, somniferine, somniferinine, somnine,
withananine, withanine, withananinine.
STEROIDS:
5,6-epoxy-4,17-dihydroxo-1-oxo-2,24-withadienolide,
6,7-epoxy-5,27-dihydroxy-1-oxo-2,24-withadienolide,
5,6-epoxy-4,20-dihydroxy-1-oxo-2-withenolide,
5,6-epoxy-4,20-dihydroxy-1-oxo-24-withenolide,
5,6-epoxy-20-hydroxy-1,4-dioxo-2,24-withadienolide,
5,6-epoxy-20-hydroxy-1,4-dioxo-2-withenolide,
14-hydroxywithanolide D, 17-hydroxywithanolide D,
27-hydroxywithanolide D,
24-methylcholesta-5,24-dien-3β-ol, 5,24-stigmastadien-3-ol,
14,20R,27-trihydroxy-1-oxo-3,5,24-withatrienolide,
1,3,20-trihydroxy-5,24-withadienolide, withaferin A,
withananine, α-withananine, withanine, withanolides A, B,
C, D, E, F, G, H, I, J, K, L, M, N, O, P, Q, R, S, T, U and
WS-1.
(A) Majumder et al., Indian J. Pharm., 17, 158 (1955)
(S) Glotter et al., J. Chem. Soc., Perkin I, 341 (1979)
(S) Velde et al., Phytochemistry, 20, 1359 (1981)

W. somnifera Dunal.

PIPERIDINE ALKALOIDS: anaferine, anahygrine.
Leary et al., Chem. Ind. (London), 283 (1964)

WITHERINGIA

W. coccoloboides.

STEROID: 25,26-dihydrophysalin C.
Antoun et al., Lloydia, 44, 579 (1981)

WOODFORDIA (Lythraceae)

W. floribunda Salisb.

GLUCOSIDE: bergenin.
St. Pyrek et al., Rocz. Chem., 51, 1679 (1977)

W. fruticosa.

BENZOPYRAN: norbergenin.
Tanayama et al., Phytochemistry, 22, 1053 (1983)

WOODWARDIA

W. orientalis.

TRITERPENOID: woodwardinic acid.
Murakami, Chen, Chem. Pharm. Bull., 19, 25 (1971)

WRIGHTIA

W. tomentosa.

STEROIDAL ALKALOIDS: conessidine, conessine, conkurchine,
holarrhimine, kurchine.
Jayaswal, Indian J. Pharm., 39, 37 (1977)

WUNDERLICHIA

W. mirabilis.

SESQUITERPENOID: wunderolide.
Bohlmann et al., Phytochemistry, 20, 1631 (1981)

WURMBEA (Liliaceae)

W. capensis Thunb.

COLCHICINE ALKALOID: 3-demethylcolchicine.
Pijewska et al., Collect. Czech. Chem. Commun., 32, 158 (1967)

W. dioica F. Muell.

COLCHICINE ALKALOIDS: colchicine, 2-demethylcolchicine,
3-demethylcolchicine, N-formyl-N-deacetylcolchicine,
β-lumicolchicine, τ-lumicolchicine.
FLAVONOID: luteolin.
Pijewska et al., Collect. Czech. Chem. Commun., 32, 158 (1967)

W. purpurea Banks.

COLCHICINE ALKALOIDS: colchicine, 2-demethylcolchicine.
Pijewska et al., Collect. Czech. Chem. Commun., 32, 158 (1967)

W. spicata.

Uncharacterized alkaloid.
Pijewska et al., Collect. Czech. Chem. Commun., 32, 158 (1967)

WYETHIA

W. arizonica.

One sesquiterpenoid.
Bohlmann et al., Planta Med., 50, 195 (1985)

W. mollis.

TRITERPENOID: 22,25-epoxylanosta-7,9(11)-dien-3-one.
Waddell et al., Phytochemistry, 21, 1631 (1982)

X

XANTHIUM (Compositae)

X. canadense Mill.
SESQUITERPENOIDS: isoalantolactone, xanthanene, xanthanodiene, xantholide A, xantholide B.
TRITERPENOIDS: taraxerol, taraxerol acetate.
- Tanaka *et al., Chem. Pharm. Bull.,* **24,** 1419 (1976)
- Tahora *et al., Tetrahedron Lett.,* 1861 (1980)

X. indicum.
SESQUITERPENOIDS: xanthumin, 2-epi-xanthumin, 8-epi-xanthumin-5-epoxide.
- Bohlmann *et al., Phytochemistry,* **21,** 1441 (1982)

X. mongolicum.
SAPONIN: atractyloside.
- Wang *et al., Zhongcaoyao,* **14,** 529 (1983)

X. orientale.
SESQUITERPENOIDS: xanthanol, 2-epi-xanthanol.
- Bohlmann *et al., Phytochemistry,* **20,** 2429 (1981)

X. strumarium.
SAPONIN: carboxyatractyloside.
SESQUITERPENOIDS: isoxanthanol, xanthanol.
STEROID: campesterol, β-sitosterol, stigmasterol.
- (Sap, T) Cole *et al., J. Agric. Food Chem.,* **28,** 1330 (1980)
- (S) Chawla *et al., Chem. Abstr.,* **87** 197348v

XANTHOCEPHALUM

X. linearifolium.
DITERPENOID: daniellol.
- Bohlmann *et al., Phytochemistry, b* **18,** 2040 (1979)

XANTHOCERAS

X. sorbifolium.
CATECHINS: 2β,3β-epi-catechin, 2β,3β-epi-gallocatechin.
FLAVONOIDS: 2α,3β-dihydromyricetin, 2α,3β-dihydroquercetin.
- Huang, Feng, *Zhongcaoyao,* **18,** 199 (1987); *Chem. Abstr.,* **107** 194890m

X. zambesiaca.
ISOFLAVONES: 7,3'-dihydroxy-8,4'-dimethoxyisoflavone,
3',4'-dimethoxy-6,7-methylenedioxyisoflavone,
7-hydroxy-8,4'-dimethoxyisoflavone,
7-hydroxy-8,3',4'-trimethoxyisoflavone,
3',4',8-trimethoxy-6,7-methylenedioxyisoflavone.
- Harper *et al., Phytochemistry,* **15,** 1019 (1976)

XANTHOGALUM

X. purpurascens.
COUMARIN: tomazin.
- Sokolova *et al., Khim. Prir. Soedin.,* 359 (1969)

XANTHORIA (Teloschistaceae)

X. elegans.
CAROTENOID: mutatoxanthin.
- Czeczuga, *Biochem. Syst. Ecol.,* **11,** 329 (1983)

X. fallax.
ANTHRAQUINONE: fallacinal.
- Murakami *et al., Chem. Pharm. Bull.,* **4,** 298 (1956)

X. parietina (L.) Beltram.
ACID: parietinic acid.
ANTHRAQUINONE: parietin.
STEROIDS: ergosta-5,7,22-trien-3β-ol, ergosta-5,8,22-trien-3β-ol, lichesterol.
- (Ac) Sargent *et al., J. Chem. Soc., C,* 307 (1970)
- (S) Lenton *et al., Phytochemistry,* **12,** 1135 (1973)

X. resendei.
ANTHRAQUINONE: fallacinal.
TRITERPENOIDS: 12α-acetoxy-3β-hydroxyfern-9(11)-ene, 9(11)-fernene-3,12-diol.
- (An) Murakami, *Chem. Pharm. Bull.,* **4,** 298 (1956)
- (T) Gonzalez *et al., Phytochemistry,* **13,** 1547 (1974)

XANTHORRHOEA (Xanthorrhoeaceae/Liliaceae)

X. hastile.
PHENOLIC: cinnamyl p-coumarate.
- Birch *et al., Tetrahedron Lett.,* 2211 (1964)

X. preissi.
FLAVAN: 4',5,7-trimethoxyflavan.
NAPHTHALENE: xanthorrhoein.
- (F) Birch *et al., Tetrahedron Lett.,* 2211 (1964)
- (N) Birch *et al., ibid.,* 1623 (1964)

X. reflexa.
NAPHTHALENE: xanthorrhoein.
- Birch *et al., Tetrahedron Lett.,* 1623 (1964)

X. resinosa.
NAPHTHALENES: xanthorrhoeol, xanthorrhone.
- Birch *et al., Tetrahedron Lett.,* 481 (1967)

XANTHOXYLUM

X. arnottianum Maxim.
ALKALOIDS: arnottianamide, chelerythrine chloride, decarine, N-desmethylchelerythrine, iwamide, oxychelerythrine, skimmianine.
COUMARINS: columbianetin, marmesin, xanthoarnol.
LIGNANS: (-)-asarinin, (-)-sesamin.
TRITERPENOID: β-amyrin.
- Ishii *et al., J. Pharm. Soc., Japan,* **97,** 890 (1977)

X. belizense.
QUINOLINE ALKALOID: platydesmine.
- Rideau *et al., Phytochemistry,* **18,** 155 (1979)

X. brachyacanthum.
APORPHINE ALKALOID: isocorydine.
- Kuck, Frydman, *J. Org. Chem.,* **26,** 5253 (1961)

X. cuspidatum Champ. (Fagara cuspidata Engl.)
FURANOQUINOLINE ALKALOIDS: dictamine, τ-fagarine, haplopine, robustine, skimmianine.
NAPHTHALENE ALKALOIDS: arnottianamide, isoarnottianamide.
QUATERNARY ALKALOIDS: de-N-methylavicine, nitidine chloride.
PHENYLPROPANOID: cuspidiol.
STEROID: β-sitosterol.
- Ishii *et al., J. Pharm. Soc., Japan,* **96,** 1458 (1976)

X. integrifoliolum (Merr.) Merr.
QUINOLONE ALKALOID: integriquinolone.
- Ishii *et al., Chem. Pharm. Bull.,* **30,** 1992 (1982)

X. macrophyllum.
AMIDE ALKALOID: fagaramide.
Thoms *et al., Ber. Dtsch. Chem. Ges.,* **44,** 3717 (1911)

X. piperitum.
FURAN: sanshodiol.
LIGNAN: xanthoxylol.
(F) Abe *et al., Chem. Pharm. Bull.,* **22,** 2650 (1974)
(L) Abe *et al., ibid.,* **21,** 1617 (1973)

X. senegalense.
LIGNAN: fagarol.
Abe *et al., Chem. Pharm. Bull.,* **22,** 2650 (1974)

X. tingoassuiba.
LIGNAN: (-)-sesamin.
Abe *et al., Chem. Pharm. Bull.,* **22,** 2650 (1974)

XERANTHEMUM (Compositae/Cynareae)

X. cylindraceum Sibth. et Smith.
SESQUITERPENOID: xerantholide.
Samek *et al., Collect. Czech. Chem. Commun.,* **42,** 2441 (1977)

XEROCOMUS

X. chrysetenteron.
PHENOLIC: xercomic acid.
Edwards *et al., J. Chem. Soc.,* 2582 (1971)

XEROMPHIS

X. spinosa.
TRITERPENOID:
3-O-(β-D-xylopyranosyloxy)-olean-12-en-28-oic acid.
Salieja *et al., Planta Med.,* 72 (1986)

XYLIA

X. dolabriformis.
PROANTHOCYANIDIN: dolabriproanthocyanidin.
DITERPENOIDS: 3-oxomanoyl oxide,
8(14),15-sandaracopimaradiene,
8(14),15-sandaracopimaradiene-3,18-diol,
8(14),15-sandaracopimaradien-3-ol,
8(14),15-sandaracopimaradien-3-one. (P) Kumar et al.,
Indian J. Chem., 14B, 654 (1976)
(T) Laidlaw, Morgan, *J. Chem. Soc.,* 644 (1963)

XYLOCARPUS (Meliaceae)

X. granatus Koenig. (Carapa granata).
LIMONOIDS: 7-acetoxydihydronomilin,
3-acetoxy-1-oxomeliac-8,30-enoate, gedunin, methyl
3β-acetoxy-1-oxomeliac-8,30-enoate, xylocarpin.
Okorie, Taylor, *J. Chem. Soc., C,* 211 (1970)
Ahmed *et al., Can. J. Chem.,* **56,** 1020 (1978)

X. molluscensis.
LIMONOIDS: xyloccensins A, B, C, D, E, F, G and H.
MONOTERPENOID: xylomollin.
Connolly *et al., J. Chem. Soc., Perkin* **1,** 1993 (1976)

XYLOPIA

X. aethiopica.
DITERPENOIDS: ent-7-oxo-16-kauren-19-oic acid, xylopic acid.
Hasan *et al., Phytochemistry, b* **21,** 1365 (1982)

X. brasiliensis.
NORAPORPHINE ALKALOID: 9-methoxyliriodenine.
Casagrande, Marotte, *Farmaco Ed. Sci.,* **25,** 799 (1970)

X. discreta.
APORPHINE ALKALOID: xylopine.
Schmutz *et al., Helv. Chim. Acta,* **42,** 335 (1959)

X. papuana Diels.
ISOQUINOLINE ALKALOID: coclaurine.
Kidd *et al., Chem. Ind.,* 748 (1955)

X. quintasii.
DITERPENOID: ent-7-acetoxy-18-trachylobanic acid.
Hasan *et al., Phytochemistry,* **21,** 177 (1982)

XYLOSMA (Flacourtiaceae)

X. velutina.
PHENOLIC: xylosmacin.
TRITERPENOID: velutinic acid.
(P) Cordell *et al., Lloydia,* **40,** 340 (1977)
(T) Chang *et al., Phytochemistry,* **16,** 1443 (1977)

XYRIS (Xyridaceae)

X. semifuscata.
ANTHRAQUINONES: 3-methoxychrysazin,
1,3,8-trihydroxyanthraquinone.
Bercht *et al., Phytochemistry,* **14,** 2099 (1975)

XYSMALOBIUM

X. undulatum R.Br.
SAPONIN: uzarin.
STEROID: ascleposide.
(Sap) Tschesche, Brathge, *Chem. Ber.,* **85,** 1042 (1952)
(S) Huber *et al., Helv. Chim. Acta,* **34,** 46 (1951)

Y

YOUNGIA

Y. denticulata (Houtt.) Kitan.
SESQUITERPENOID GLYCOSIDES: youngiasides A, B, C and D.
Adegawa et al., Chem. Pharm. Bull., **34**, 3769 (1986)

YUCCA (Agavaceae)

Y. aloifolia Naudin.
STEROID: smilagenin.
Marker et al., J. Am. Chem. Soc., **65** 1199 (1943)

Y. angustissinma.
STEROID: sarsasapogenin.
Marker et al., J. Am. Chem. Soc., **65**, 1199 (1943)

Y. arizonica McKel.
STEROID: sarsasapogenin.
Marker et al., J. Am. Chem. Soc., **65**, 1199 (1943)

Y. arkansana Trel.
STEROID: smilagenin.
Marker et al., J. Am. Chem. Soc., **65**, 1199 (1943)

Y. australis Engelm.
STEROID: smilagenin.
Marker et al., J. Am. Chem. Soc., **65**, 1199 (1943)

Y. baccata Torr.
STEROID: sarsasapogenin.
Marker et al., J. Am. Chem. Soc., **65**, 1199 (1943)

Y. baleyi Wood-Standl.
STEROID: sarsasapogenin.
Marker et al., J. Am. Chem. Soc., **65**, 1199 (1943)

Y. brevifolia Engelm.
STEROIDS: kammogenin, sarsasapogenin, smilagenin.
Marker et al., J. Am. Chem. Soc., **65**, 1199 (1943)

Y. confinis McKel.
STEROID: sarsasapogenin.
Marker et al., J. Am. Chem. Soc., **65**, 1199 (1943)

Y. decipiens Trel.
STEROID: sarsasapogenin.
Marker et al., J. Am. Chem. Soc., **65**, 1199 (1943)

Y. elata Engelm.
STEROID: sarsasapogenin.
Marker et al., J. Am. Chem. Soc., **65**, 1199 (1943)

Y. elephantipes Regel.
STEROID: smilagenin.
Marker et al., J. Am. Chem. Soc., **65**, 1199 (1943)

Y. filifera Chaband.
STEROIDS: filiferin A, filiferin B, sarsasapogenin, smilagenin, tigogenin, willagenin.
Marker et al., J. Am. Chem. Soc., **65**, 1199 (1943)
Lemieuz et al., Carbohyd. Res., **55**, 113 (1977)
Rivera et al., Q. J. Crude Drug Res., **15**, 139 (1977); Chem. Abstr., 88 148964e

Y. flaccida Haw.
STEROIDS: furcogenin, smilagenin, yuccogenin.
Marker et al., J. Am. Chem. Soc., **65**, 1199 (1943)

Y. glauca Nutt.
STEROID: sarsasapogenin.
Marker et al., J. Am. Chem. Soc., **65**, 1199 (1943)

Y. gloriosa L.
STEROIDS: gloriogenin, 12-hydroxysmilagenin, smilagenin.
Marker et al., J. Am. Chem. Soc., **65**, 1199 (1943)
Gonzalez et al., An. Quim., **68**, 309 (1972)

Y. harrimannii Trel.
STEROIDS: kammogenin, sarsasapogenin.
Marker et al., J. Am. Chem. Soc., **69**, 2167 (1947)
Marker et al., ibid., **69**, 2401 (1947)

Y. jalicensis Trel.
STEROIDS: sarsasapogenin, smilagenin.
Marker et al., J. Am. Chem. Soc., **65**, 1199 (1943)

Y. louisianensis Trel.
STEROID: smilagenin.
Marker et al., J. Am. Chem. Soc., **65**, 1199 (1943)

Y. recurvifolia Salisb.
STEROID: smilagenin.
Marker et al., J. Am. Chem. Soc., **65**, 1199 (1943)

Y. reverchoni Trel.
STEROID: smilagenin.
Marker et al., J. Am. Chem. Soc., **65**, 1199 (1943)

Y. rigida.
STEROID: sarsasapogenin.
Marker et al., J. Am. Chem. Soc., **65**, 1199 (1943)

Y. schidigera.
STEROID: sarsasapogenin.
Marker et al., J. Am. Chem. Soc., **65**, 1199 (1943)

Y. schottii Engelm.
STEROIDS: mexogenin, neomanogenin, samogenin, sarsasapogenin, texogenin.
Marker et al., J. Am. Chem. Soc., **69**, 2167 (1947)

Y. thornberi McKel.
STEROID: sarsasapogenin.
Marker et al., J. Am. Chem. Soc., **65**, 1199 (1943)

Y. torreyi Schafer.
STEROID: sarsasapogenin.
Marker et al., J. Am. Chem. Soc., **65**, 1199 (1943)

Y. treculeana Hook.
STEROIDS: sarsasapogenin, smilagenin, willagenin.
Rivera et al., Q. J. Crude Drug Res., **15**, 139 (1977); Chem. Abstr., 88 148964e

Y. valida.
STEROID: sarsasapogenin.
Marker et al., J. Am. Chem. Soc., **65**, 1199 (1943)

Y. whipplei subsp. intermedia.
STEROIDS: gitogenin, tigogenin.
Marker et al., J. Am. Chem. Soc., **65**, 1199 (1943)

Y. whipplei subsp. parishii.
STEROID: tigogenin.
Marker et al., J. Am. Chem. Soc., **65**, 1199 (1943)

Y. whipplei subsp. precursa.
STEROID: gitogenin.
Marker et al., J. Am. Chem. Soc., **65**, 1199 (1943)

Y. whipplei subsp. typiva.
STEROID: tigogenin.
Marker et al., J. Am. Chem. Soc., **65**, 1199 (1943)

Z

ZALUZANIA

Z. angusta.

SESQUITERPENOIDS: zaluzanin A, zaluzanin B, zaluzanin C, zaluzanin D.

Romo de Vivar et al., Tetrahedron, 23, 3093 (1967)

ZANTHOXYLUM (Rutaceae)

Z. acanthopodium DC.

FLAVONOIDS: tambulin, tambulol.

TOXIN: acanthotoxin.

TRITERPENOID: β-amyrenone.

(F) Ahuja et al., Indian J. Chem., 13, 1174 (1975)

(T) Pal et al., Curr. Sci., 45, 739 (1976)

(Tox) Roy et al., Chem. Ind., (London), 231 (1977)

Z. arborescens Rose.

FURANOQUINOLINE ALKALOIDS:
8-hydroxy-4,7-dimethoxyfuranoquinoline,
8-(2-isopentenoyloxy)-4,7-dimethoxyfuranoquinoline.

Grina et al., J. Org. Chem., 47, 2648 (1982)

Z. avicennae (Lam.) DC.

NAPHTHOPHENANTHRIDINE ALKALOIDS: avicine.

PROTOBERBERINE ALKALOIDS: α-allocryptopine, (-)-canadine.

Fish et al., Phytochemistry, 14, 841 (1975)

Z. belizense.

CARBOLINE ALKALOID: canthin-6-one.

FURANOQUINOLINE ALKALOIDS: dictamnine, platydesmine, skimmianine.

COUMARINS: citropten, isopimpinellin, marmesin.

Najjar et al., Phytochemistry, 14, 2309 (1975)

Z. brachyacanthum.

PROTOBERBERINE ALKALOIDS: α-allocryptopine, (-)-canadine.

MONOTERPENOID: menthane-1S,2S,4S-triol.

(A) Manske, Can. J. Res., 20B, 53 (1942)

(T) Thappa et al., Phytochemistry, 15, 1568 (1976)

Z. budrunga.

UNCHARACTERIZED ALKALOIDS: budrungaine, budrungainine.

MONOTERPENOID: menthane-1S,2S,4S-triol.

(A) Khastgir et al., Curr. Sci., 16, 185 (1947)

(T) Thappa et al., Phytochemistry, 15, 1568 (1976)

Z. caribaeum Lam.

PROTOBERBERINE ALKALOID: berberine.

Awe, Arch. Pharm., 284, 362 (1951)

Z. conspersipunctatum Merr. et Perry.

OXOPROTOBERBERINE ALKALOIDS: isoprotopine, pseudoprotopine.

DIOXEPIN: psoromic acid.

(A) Johns et al., Aust. J. Chem., 22, 2233 (1969)

(D) Huneck et al., ibid., 29, 1059 (1976)

Z. culantrillo.

ALKALOID: culantraramine.

Swineheart et al., Phytochemistry, 19, 1219 (1980)

Z. decaryi.

NAPHTHOPHENANTHRIDINE ALKALOID: decarine.

/MONOTERPENOIDS: eucalyptol, 4-hydroxyterpinene, linalool, β-pinene, α-terpineol.

SESQUITERPENOID: caryophyllene epoxide.

(A) Vaquette et al., Phytochemistry, 13, 1257 (1974)

(T) Vaquette et al., Plant Med. Phytother., 9, 315 (1975)

Z. dipetalum.

CARBOLINE ALKALOID: canthin-6-one.

QUATERNARY ALKALOIDS: chelerythrine, nitidine, tembetarine.

COUMARINS: avicennol, xanthoxyletin.

FLAVONOID: hesperidin.

TRITERPENOID: lupeol.

Fish et al., Phytochemistry, 14, 2073 (1975)

Z. elephantiasis.

COUMARIN: cis-avicennol.

Gray et al., Phytochemistry, 16, 1017 (1977)

Z. fagara Sarg.

COUMARIN: castanaguyone.

Snyder et al., Tetrahedron Lett., 5015 (1981)

Z. flavum.

ALKALOID: dihydrorutaecarpine.

Waterman, Phytochemistry, 15, 578 (1976)

Z. leprieurii.

ACRIDONE ALKALOIDS: arborinine, xanthoxoline.

MONOTERPENOID: α-pinene.

SESQUITERPENOIDS: α-cadinane, β-farnesene, humulene.

Reisch et al., Pharmazie, 40, 811 (1985)

Z. microcarpum.

NAPHTHOISOQUINOLINE ALKALOID: nornitidine.

Boulware, Stermitz, Lloydia, 44, 200 (1981)

Z. myriacanthum.

QUATERNARY ALKALOIDS: dihydronitidine, magnoflorine, nitidine nitrate, tembetarine.

STEROID: β-sitosterol.

Waterman, Phytochemistry, 14, 2530 (1975)

Z. nitidum (Lam.) DC.

NAPHTHOISOQUINOLINE ALKALOIDS: nitidine, oxynitidine.

Arthur et al., J. Chem. Soc., 1840 (1959)

Z. ochroxylon DC.

UNCHARACTERIZED ALKALOIDS: α-xantherine, β-xantherine.

LePrince et al., Bull. Soc. Pharmacol., 18, 343 (1911)

Z. oxyphyllum.

QUATERNARY ALKALOIDS: zanthoxyline, zanthoxyphylline.

Tiwari, Masood, Phytochemistry, 16, 1068 (1977)

Z. pluviatile.

SESQUITERPNOIDS: pluviatilol, pluviatolide.

Corrie et al., Aust. J. Chem., 23, 133 (1970)

Z. podocarpum Hemsl.

AMIDE: neoherculin.

STEROID: β-sitosterol.

TRITERPENOID: β-amyrin.

Renei et al., Yaoxue Xuebao, 19, 268 (1984); Chem. Abstr., 103 19810f

Z. punctatum Vahl.

ALKALOID: alfileramine.

Caolo et al., Tetrahedron, 35, 1487 (1979)

Z. rhetsa.

One monoterpenoid.

Mathur et al., Tetrahedron, 35, 2495 (1979)

Z. senegalense DC. (Fagara xanthoxyloides).

UNCHARACTERIZED ALKALOID: artarine.

Giacosa, Monari, Gazzetta, 17, 362 (1887)

Z. thomense.

AMIDE: zanthomanide.

Simeray et al., Phytochemistry, 24, 2720 (1985)

Z. tsihanimposa.

NAPHTHOISOQUINOLINE ALKALOID: dihydrochelerythrine 11-acetone.

Decaudin et al., Phytochemistry, 13, 505 (1974)

Z. veneficum F.M.Bail.

PROTOBERBERINE ALKALOIDS: (-)-canadine, N-methylcanadine.

Manske, Can. J. Res., 20B, 53 (1942)

ZEA (Gramineae/Poaceae)

Z. mays.

CYTOKININ: zeatin.

TRITERPENOID: cyclosadol acetate.

(C) Shaw et al., J. Chem. Soc., C, 921 (1966)

(T) Pinhas, Bull. Soc. Chim. Fr., 2037 (1969)

ZEDOARIA

Z. rhizome.

SESQUITERPENOIDS: furanogermenone, germacrone-4,5-epoxide.

Shibuya et al., Heterocycles, 215 (1982)

ZELKOVA (Ulmaceae)

Z. serrata Makino.

NAPHTHALENIC: 7-hydroxy-3,3'-dimethoxy-8,7'-bicadalenyl ether.

SESQUITERPENOIDS: homozelkoserratone, zelkoserratone.

Hayashi, Takahashi, Mokuzai Gakkaishi, 26, 54 (1980)

ZEPHYRANTHES (Amaryllidaceae)

Z. candida.

INDOLE ALKALOIDS: nerinine, zephyranthine.

Jeffs, Hawksworth, Tetrahedron Lett., 217 (1963)

Ozaki, Chem. Pharm. Bull., 12, 253 (1964)

Z. carinata.

ISOQUINOLINE ALKALOID: carinatine.

Kobayashi et al., Chem. Pharm. Bull., 25, 2244 1977) 1977)

Z. rosea Lindl.

ISOQUINOLINE ALKALOID: lycorine.

Greathouse, Rigler, Am. J. Bot., 28, 702 (1941)

Z. texana.

ISOQUINOLINE ALKALOID: lycorine.

Greathouse, Rigler, Am. J. Bot., 28, 702 (1941)

ZEXMENIA (Compositae/Heliantheae)

Z. brevifolia.

SESQUITERPENOID: orizabin.

Ortega et al., Rev. Latinoam. Quim., 2, 38 (1971)

Z. gnaphaloides Gray.

SESQUITERPENOIDS: (7E)-10-epi-valerenol, (7Z)-10-epi-valerenol, 10-epi-α-valerenol.

Bohlmann et al., Chem. Ber., 113, 2410 (1980)

Z. phyllocephala.

DITERPENOIDS:

7α-acetoxy-1β,11α-dihydroxysandaracopimara-8(14),15-diene,

1β-acetoxy-7α-hydroxysandaracopimara-8(14),15-diene,

7α-acetoxy-1β-hydroxysandaracopimara-8(14),15-diene,

sandaracopimara-8(14),15-diene-1,7-diol,

sandaracopimara-8(14),15-diene-1β,7α,11α-triol,

sandaracopimara-8(14),15-dien-7-ol,

sandaracopimara-8(14),15-1,11-diol.

Bohlmann, Lonitz, Chem. Ber., 111, 843 (1978)

ZEYHERA

Z. digitalis.

PHENOLIC: zeyherol.

de Silveira et al., Phytochemistry, 14, 1829 (1975)

Z. tuberculosa.

PHENOLIC: zeyherol.

SESQUITERPENOID: dehydro-α-lapachone.

(P) de Silveira et al., Phytochemistry, 14, 1829 (1975)

(T) Manners et al., ibid., 15, 225 (1976)

ZEYSIA

Z. matrella.

TRITERPENOID: fern-9(11)-en-3-one.

Nishimoto et al., Tetrahedron, 24, 735 (1968)

ZIERIA (Rutaceae)

Z. aspalathoides A.Cunn. ex Benth.

MONOTERPENOID: car-3-en-2-one.

Lassak et al., Aust. J. Chem., 27, 2061 (1974)

Z. macrophylla.

SESQUITERPENOID: zierone.

Bradfield et al., Proc. R. Soc. N.S.W., 67, 200 (1933)

Z. smithii.

MONOTERPENOIDS: (-)-linalool, O-methyleugenol, safrole.

Fletcher et al., J. Econ. Entomol., 68, 815 (1975); Chem. Abstr., 84 56529f

ZINGIBER (Zingiberaceae)

Z. americana.

SESQUITERPENOID: zerumbone.

van Veer, Recl. Trav. Chim. Pays Bas, 58, 691 (1939)

Z. officinale.

PHENOLIC: (R)-gingerol.

Connell et al., Aust. J. Chem., 22, 1033 (1969)

Z. zerumbet Smith.

SESQUITERPENOIDS: humenol, humulene epoxide I, humulene epoxide II, humulenol II, zerumbone.

Desmodaran, Dev, Tetrahedron, 24, 4123 (1968)

ZINNIA

Z. acerosa.

SESQUITERPENOIDS: zinarosin, zinniadilactone.

Romo et al., Rev. Latinoam. Quim., 2, 24 (1971)

Z. angustifolia.

SESQUITERPENOID: zinangustolide.

Bohlmann et al., Phytochemistry, 20, 1623 (1981)

Z. elegans.

FLAVONOIDS: apigenin 7-glucoside, apigenin 4'-glucoside, cosnosiin, kaempferol 3-glucoside, kaempferol 3-xyloside-7-glucoside, luteolin 7-glucoside, quercetin 3-glucoside.

Dembinska-Migas et al., Herba Pol., 29, 197 (1983); Chem. Abstr., 103 3694k

Z. flavicoma.

SESQUITERPENOIDS: zinaflorin A-1, zinaflorin A-2, zinaflorin A-3, zinaflorin A-4, zinaflorin A-5.

Ortega et al., Phytochemistry, 24, 2635 (1985)

Z. haageana.

SESQUITERPENOID: haageanolide.

Kessel, Phytochemistry, 17, 1059 (1978)

Z. peruviana.

SESQUITERPENOIDS: zinaflorin I, zinaflorin II, zinaflorin III, zinaflorin IV.

Ortega et al., Chem. Lett., 1607 (1983)

Z. tenuifolia.

DITERPENOID: 12,18-dihydroxygeranylgeraniol.

Bohlmann *et al.*, *Phytochemistry*, **20**, 1623 (1981)

ZIZIA

Z. aptera.

BENZOPYRAN: apterin.
SESQUITERPENOID: isosamidin.

(B) Steck *et al.*, *Phytochemistry*, **13**, 1925 (1974)
(T) Hata *et al.*, *Chem. Pharm. Bull.*, **22**, 957 (1974)

ZIZIFLORA

Z. bungeana.

TRITERPENOID: bungeolic acid.

Bimurzaev, Nikonov, *Khim. Prir. Soedin.*, 530 (1980)

ZIZYPHUS (Rhamnaceae)

Z. abyssinica.

MACROCYCLIC ALKALOIDS: abyssenine A, abyssenine B,
abyssenine C.

Tschesche *et al.*, *Phytochemistry*, **13**, 2328 (1974)

Z. aenophila Mill.

PEPTIDE ALKALOIDS: zizyphine, zizyphinine.

Menard *et al.*, *Helv. Chim. Acta*, **46**, 1801 (1963)

Z. amphibia A.Cheval.

PEPTIDE ALKALOIDS: amphibins A, B, C, D, E, F, G and H.

Tschesche *et al.*, *Chem. Ber.*, **107**, 686 (1974)

Z. hysodrica.

PEPTIDE ALKALOID: hysodricanine A.

Tschesche *et al.*, *Phytochemistry*, **16** 1025 (1977)

Z. joazeiro.

SAPONIN: jujubogenin
3-O-α-L-arabinofuranosyl-β-D-glucopyranosyl-α-L-
arabinopyranoside.
TRITERPENOID: betulinic acid.

(Sap) Higuchi *et al.*, *Phytochemistry*, **23**, 2597 (1984)
(T) Trigueiro *et al.*, *Cienc. Cult. (Sao Paulo)*, 44 (1981); *Chem. Abstr.*,
97 195830d

Z. jujuba.

SAPONINS: jujuboside A, jujuboside B, jujuboside B1,
jujuboside C.

Inouye *et al.*, *J. Chem. Res.*, 144 (1978)

Z. mauritiana Lasm.

PEPTIDE ALKALOIDS: mauritines A, B, C, D, E and F.
ANTHRAQUINONE:
3,6,8-trihydroxy-1-methylanthraquinone-2-carboxylic acid.
STEROID: zizogenin.

(A) Tschesche *et al.*, *Justus Liebigs Ann. Chem.*, 1915 (1974)
(An) Gadgil *et al.*, *Tetrahedron Lett.*, 2223 (1968)
(S) Srivastava, Srivastava, *Phytochemistry*, **18**, 1758 (1979)

Z. mucronata Willd.

PEPTIDE ALKALOIDS: mucronines A, B, C and D.

Tschesche *et al.*, *Chem. Ber.*, **105**, 3106 (1972)

Z. nummularia.

PEPTIDE ALKALOIDS: N-desmethyljubanine A, jubanine A,
jubanine B, mauritine C, nummularines A, B, C, D, E, F, G,
H and K.
STEROID: zizynummin.

(A) Miana *et al.*, *Fitoterapia*, **56**, 363 (1985)
(S) Sharma, Kumar, *Phytochemistry*, **22**, 1469 (1983)

Z. oenoplia Mill.

PEPTIDE ALKALOIDS: zizyphines A, B, C, D, E, F and G.

Cassells *et al.*, *Tetrahedron*, **30**, 2461 (1974)
Tschesche *et al.*, *Tetrahedron Lett.*, 2941 (1974)

Z. sativa.

PEPTIDE ALKALOIDS: sativanines A, B, C, D and F.
PHENOLICS: terephthalic acid, terephthalic acid dimethyl ester,
terephthalic acid methyl ester.

(A) Shah *et al.*, *Phytochemistry*, **22**, 931 (1983)
(P) Thakur *et al.*, *Planta Med.*, **28**, 172 (1975)

Z. spina-christi.

TRITERPENOIDS: betulic acid, ceanothic acid.

Ikram *et al.*, *Planta Med.*, **29**, 289 (1976)

Z. vulgaris.

TRITERPENOID: betulinic acid.

Krajniak *et al.*, *Aust. J. Chem.*, **22**, 1331 (1969)

Z. vulgaris var. spinosus.

TRITERPENOID: betulin.

Zimmermann, Helv. Chim. Acta, 27, *332*++ (1944)
Zimmermann, *ibid.*, 27, 334 (1944)

ZOLLERNIA

Z. poraensis.

CHALCONE: 2,2',4,4'-tetrahydroxydihydrochalcone.
FLAVONOIDS: formononetin,
(±)-7-hydroxy-4'-methoxyisoflavone, isoliquiritigenin,
(+)-liquiritigenin, (+)-medicarpin, (+)-vestitol.

Ferrari *et al.*, *Phytochemistry*, **22**, 1663 (1983)

ZOYSIA

Z. matrella.

TRITERPENOIDS: fern-9(11)-en-3-one, 12-ketoarundin.

Obafemi *et al.*, *Phytochemistry*, **18**, 496 (1979)

ZYGADENUS (Liliaceae)

Z. elegans Pursh.

STEROIDAL ALKALOIDS: isogermine, zygadenine.

Feofilaklov, Alekseeva, *J. Gen. Chem. SSSR*, **28**, 989 (1955)

Z. intermedius.

STEROIDAL ALKALOIDS: isogermine, zygadenine.

Heyl *et al.*, *J. Am. Chem. Soc.*, **35**, 258 (1913)

Z. venenosum.

STEROIDAL ALKALOIDS: isogermine, zygacine.

Kupchan *et al.*, *J. Am. Chem. Soc.*, **77**, 689 (1953)

INDEX